Oxford Textbook of

Clinical Neurophysiology

Oxford Textbooks in Clinical Neurology

PUBLISHED

Oxford Textbook of Epilepsy and Epileptic Seizures
Edited by Simon Shorvon, Renzo Guerrini, Mark Cook, and Samden Lhatoo

Oxford Textbook of Vertigo and Imbalance
Edited by Adolfo Bronstein

Oxford Textbook of Movement Disorders
Edited by David Burn

Oxford Textbook of Stroke and Cerebrovascular Disease
Edited by Bo Norrving

Oxford Textbook of Neuromuscular Disorders
Edited by David Hilton-Jones and Martin R. Turner

Oxford Textbook of Neurorehabilitation
Edited by Volker Dietz and Nick Ward

Oxford Textbook of Neuroimaging
Edited by Massimo Filippi

Oxford Textbook of Cognitive Neurology and Dementia
Edited by Masud Husain and Jonathan M. Schott

FORTHCOMING

Oxford Textbook of Sleep Disorders
Edited by Sudhansu Chokroverty and Luigi Ferini-Strambi

Oxford Textbook of Headache Syndromes
Edited by Michel Ferrari, Joost Haan, Andrew Charles, David Dodick, and Fumihiko Sakai

Oxford Textbook of Neuro-ophthalmology
Edited by Fion Bremner

Oxford Textbook of Neuro-oncology
Edited by Tracy Batchelor, Ryo Nishikawa, Nancy Tarbell, and Michael Weller

Oxford Textbook of Clinical Neuropathology
Edited by Sebastian Brandner and Tamas Revesz

Oxford Textbook of Clinical Neurophysiology

Edited by

Kerry R. Mills
Department of Basic and Clinical Neuroscience, King's College London, UK

Series Editor

Christopher Kennard

OXFORD
UNIVERSITY PRESS

Great Clarendon Street, Oxford, OX2 6DP,
United Kingdom

Oxford University Press is a department of the University of Oxford.
It furthers the University's objective of excellence in research, scholarship, and education by publishing worldwide. Oxford is a registered trade mark of Oxford University Press in the UK and in certain other countries

Impression: 1

Published in the United States of America by Oxford University Press
198 Madison Avenue, New York, NY 10016, United States of America

British Library Cataloguing in Publication Data

Data available

Library of Congress Control Number: 2016937972

ISBN 978–0–19–968839–5

Printed in Great Britain by Bell & Bain Ltd., Glasgow

Preface

Clinical Neurophysiology provides a wide range of techniques for investigating patients with neurological conditions and is crucial in some, such as entrapment neuropathies and epilepsy. The discipline has advanced in recent years and can now provide information in the clinic on such diverse aspects as the function of single ion channels in peripheral nerve disease to the degree of degeneration of the corticospinal tract in patients with, say, motor neurone disease. We also have techniques to investigate brain function on a millisecond timescale with EEG and MEG and are able to co-register this activity with brain structure using MRI and fMRI. Clinical Neurophysiologists, therefore, need not only to have an in-depth knowledge of the neurological conditions that they will encounter, but also to understand the technical aspects of their techniques and to appreciate how signals are recorded and dealt with in the digital domain by our modern computer systems. They should be particularly aware of the pitfalls that can befall them. One might say that the modern clinical neurophysiologist is a cross between a clinician and a digital signal processor.

One must never lose sight, however, of the fact that the patient is the prime concern for diagnosis and prognosis. It is a much quoted aphorism that clinical neurophysiology is an extension of the neurological examination and this is still true. Despite the welcome development of test protocols, algorithms, flow charts, minimal standards, and so on, they come to naught unless they provide an answer for the patient in front of you. Only the well-trained clinical neurophysiologist will be able to assess both the clinical information and then put the neurophysiological data in context.

Successful practice of neurophysiology also demands a knowledge of the basic science underlying the investigations performed, in order to reach a conclusion with regard to the pathology likely to be involved. This volume provides sections giving a summary of the basic science underlying the techniques, a description of the techniques themselves and a section describing the use of the techniques in clinical situations. The subject has traditionally been subdivided into EEG, the recording of brain electrical activity through electrodes placed on the scalp, nerve conduction studies, and EMG, the electrical activity deriving from nerve and muscle, and evoked potentials where brain activity locked in time to a peripheral nerve, visual or auditory stimulus is averaged to give information on afferent pathways. However, many basic principles are common to all techniques and these are emphasized in the first section.

I hope this will prove a useful textbook and reference source for trainees in neurophysiology, neurology, and neurosurgery but also that it will appeal to general neurologists, paediatricians and technologists undertaking neurophysiology. This venture would not have been possible without the willingness of authors to provide their excellent contributions for which I thank them. I would also like to thank Oxford University Press for continual encouragement.

Kerry R. Mills

Preface

Contents

Abbreviations

3,4-DAP	3,4-diaminopyridine
AASM	American Academy of Sleep Medicine
AC	alpha coma
Ach	acetylcholine
AChR	acetylcholine receptor
ADC	analogue-to-digital converter
ADHD	attention deficit hyperactivity disorder
ADM	abductor digiti minimi
ADP	adenosine diphosphate
AED	anti-epileptic drug
AEP	auditory evoked potential
AFP	acute flaccid paralysis
AHI	apnoea–hypopnoea index
AHP	after-hyperpolarization
AIDP	acute inflammatory demyelinating polyneuropathy
ALS	amyotrophic lateral sclerosis
AMAN	acute motor axonal neuropathy
AMSAN	acute motor and sensory axonal neuropathy
APB	abductor pollicis brevis
ASFAP	apparent single fibre action potential
ASPS	advanced sleep phase syndrome
ATC	alpha-theta coma
ATP	adenosine triphosphate
BAEP	brain stem auditory evoked potential
BCR	bulbocavernosus reflex
BECTS	benign epilepsy with centrotemporal spikes
BF	body floating
BFNC	benign familial neonatal seizures
BiPED	bilateral periodic epileptiform discharges
BIS	bispectral index
BMI	body mass index
BOLD	blood oxygenation level dependent
BP	blood pressure
BRE	benign Rolandic epilepsy
CAE	childhood absence epilepsy
CAP	cyclical alternating pattern
CB	conduction block
CBF	cerebral blood flow
CCEP	corticocortical-evoked responses
cEEG	continuous EEG monitoring
CF	cardiac floating
CGA	corrected gestational age
CHEP	cold and hot evoked potential
CIDP	chronic inflammatory demyelinating polyneuropathy
CIPN	chemotherapy-induced peripheral neuropathy
CMAP	compound muscle action potential
CMCT	central motor conduction time
CMRR	common mode rejection ratio
CMS	congenital (genetic) myasthenic syndromes
CMT	Charcot–Marie–Tooth neuropathy
CNE	concentric needle electrode
CNS	central nervous system
CPAP	continuous positive airways pressure
CPSE	complex partial status epilepticus
CPT	carnitine palmityl transferase
CRD	complex repetitive discharges
CSA	central sleep apnoea
CSAP	compound sensory action potentials
CSR	common system reference
CTS	carpal tunnel syndrome
CV	conduction velocity
cVEMP	cervical vestibular evoked myogenic potential
DHPR	dihydropyridine receptor
DICS	dynamic imaging of coherent sources
DM	dermatomyositis
DML	distal motor latency
DNET	dysembryoplastic neuroepithelial tumour
dSPM	dynamic statistical parametric mapping
DSPS	delayed sleep phase syndrome
EAS	external anal sphincter
ECD	equivalent-current dipole
ECG	electrocardiograph
ECoG	electrocorticography
ED	emergency departments
EDS	excessive daytime sleepiness
EEG	electroencephalograph
EEG-fMRI	EEG and functional MRI
EGTCSA	epilepsy with generalized tonic–clonic seizures on awakening
EIEE	early infantile epileptic encephalopathy
EIM	electrical impedance myography
ELMA	eyelid myoclonia with absences
EM	epileptic myoclonus
EMA	epilepsy with myoclonic absences
EME	early (neonatal) myoclonic encephalopathy
EMG	electromyograph
ENM	epileptic negative myoclonus
EOG	electro-oculogram
EP	evoked potential

EPP endplate potential
EPSP excitatory post-synaptic potential
ER event-related
ESES electrical status epilepticus in slow wave sleep
ESS Epworth Sleepiness Scale
EUS external urethral sphincter
FARR Friedreich's ataxia with retained reflexes
FCD focal cortical dysplasia
FCR flexor carpi radialis
FD fibre density
FDI first dorsal interosseous
FFT Fast Fourier Transform
FIRDA frontal intermittent delta activity
FLE frontal lobe epilepsy
fMRI functional magnetic resonance imaging
FO foramen ovale
FOG fibres containing oxidative and glycolytic enzymes
FOLD female, occipital, low amplitude, drowsy
FOS fixation-off sensitivity
FP Fasciculation Potential
FR fatigue-resistant muscle fibres
FS febrile seizures
FS+ febrile seizures plus
FWCV F wave conduction velocity
GA gestational age
GABA gamma amino butyric acid
GBS Guillain–Barré syndrome
GEFS genetic epilepsy with febrile seizures plus
GGE genetic generalised epilepsy
GN geniculate nucleus
GPED generalized periodic epileptiform discharge
GPPR generalized photoparoxysmal response
GSWD generalized spike-wave discharge
GTCS generalized tonic-clonic seizures
HCN hyperpolarisation-activated cyclic nucleotide-gated
HCP half-cell potentials
HFF high frequency filter
HFO high frequency oscillations
HIBI hypoxic–ischaemic brain injury
HIE hypoxic ischaemic encephalopathy
HLA human leucocyte antigen
HNA hereditary neuralgic amyotrophy
HNPP hereditary neuropathy with pressure palsies
HR heart rate
HSE herpes simplex encephalitis
HUT head-up tilting
IBM inclusion body myositis
IC intracranial
ICC intra-class correlation coefficiency
ICEEG intracranial EEG
ICSD International Classification of Sleep Disorders
IED inter-electrode distance
IEE infantile epileptic encephalopathies
IGE idiopathic generalized epilepsy
IIED interictal epileptiform discharges
ILAE International League Against Epilepsy
IoC impairment of cognition
IONM intraoperative neurophysiological monitoring
IOZ ictal onset zone
IP interference pattern
IPA interference pattern analysis
IPS intermittent photic stimulation
IPSP inhibitory post-synaptic potentials
IS infantile spasms
iv intravenous
IVIg intravenous immunoglobulin
JAE juvenile absence epilepsy
JME juvenile myoclonic epilepsy
JMG juvenile MG
KCC2 K+–Cl– cotransporter
LE limbic encephalitis
LEMS Lambert Eaton myasthenic syndrome
LFF low frequency filters
LICI long interval intracortical inhibition
LMN lower motor neuron
LOC loss of consciousness
LSB least-significant-bit
M1 primary motor cortex
MADSAM multifocal acquired demyelinating sensory and motor neuropathy
MAE myoclonic astatic epilepsy
MAG myelin associated glycoprotein
MCS minimally conscious state
MCV motor conduction velocity
ME myoclonic encephalopathy
MEG magnetoencephalography
MEP motor evoked potential
MEPP miniature endplate potentials
MG myasthenia gravis
MLAEPs middle latency auditory evoked potentials
MMN multifocal motor neuropathy
MMNCB multifocal motor neuropathy with conduction block
MN motoneuron
MND motor neurone disease
MNE minimum norm estimate
MRI magnetic resonance imaging
MS multiple sclerosis
MSA multiple system atrophy
MSLT multiple sleep latency test
MTLE mesial temporal lobe epilepsy
MTS mesial temporal sclerosis
MU motor unit
MUAP motor unit action potential
MUNE motor unit number estimation
MUNIX motor unit number index
MUP motor unit potential
MuSK muscle specific kinase
MVC maximum voluntary contraction
MWT maintenance of wakefulness test
NCS nerve conduction study
NCSE non-convulsive status epilepticus
NCSz non-convulsive seizures
NFLE nocturnal frontal lobe epilepsy
NMJ Neuromuscular Junction
NM negative myoclonus
NMDA N-methyl D-aspartate
NREM non-rapid eye movement sleep
NRGC nucleus reticularis gigantocellularis

NTLE	neocortical temporal lobe epilepsy
OBPP	obstetric brachial plexus palsy
ODI	Oxygen Desaturation Index
OM	ocular myasthenia
ORIDA	occipital intermittent rhythmic delta activity
OS	Ohtahara's syndrome
OSA	obstructive sleep apnoea
PAD	primary afferent depolarization
PC	periodic complex
PCA	principle component analysis
pCO_2	partial pressure of CO_2
PED	periodic epileptiform discharge
PET	positron emission tomography
Pi	inorganic phosphate
PIC	persistent inward current
PLED	periodic lateralized epileptiform discharge
PLM	periodic limb movement
PLMD	periodic limb movement disorder
PLMS	periodic leg movements in sleep
PLMW	periodic limb movements during resting wakefulness
PM	polymyositis
PME	progressive myoclonic encephalopathies
PMEI	partial migrating epilepsy in infancy
PMP22	peripheral myelin protein 22
PNTML	pudendal nerve terminal motor latency
PoTS	postural orthostatic tachycardia syndrome
PPR	photo paroxysmal response
PROMM	proximal myotonic myopathy
PSG	polysomnography
PSP	progressive supranuclear palsy
PSTH	peri-stimulus time histogram
ipsw	positive sharp waves
PTT	pulse transit time
PVS	persistent vegetative state
R&K	Rechtschaffen and Kales
RDI	respiratory disturbance index
REM	rapid eye movement
RERA	respiratory effort related arousal
RIP	respiratory inductance plethysmography
RIV	relative intertrial variation
RL	residual latency
RLS	restless leg syndrome
RMTD	rhythmic mid-temporal theta bursts of drowsiness
RNS	repetitive nerve stimulation
RSBD	REM sleep behaviour disorder
RSN	resting-state networks
rTMS	repetitive TMS
RyR	ryanodine receptor
SAE	sepsis-associated encephalopathy
SAM	synthetic-aperture magnetometry
SAP	sensory action potential
SBS	secondary bilateral synchrony
sCJD	Creutzfeldt–Jacob disease
SCLC	small cell lung cancer
SCV	sensory conduction velocity
SDB	sleep disordered breathing
SDTC	strength duration time-constant
SE	status epilepticus
sEMG	surface EMG
SEP	somatosensory evoked potential
SEPT9	septin gene
SFAP	single fibre action potentials
SFEMG	single fibre electromyography
SFN	small fibre neuropathy
SICF	short interval intracortical facilitation
SICI	short-interval intracortical inhibition
SIRPID	stimulus-induced rhythmic, periodic, or ictal discharge
SMA	supplementary motor areas
SMEI	severe myoclonic epilepsy of infancy
SNAP	sensory nerve action potential
SNR	signal-to-noise ratio
SNRI	serotonin–norepinephrine reuptake inhibitors
SOL	sleep onset latency
SOREMP	sleep onset REM period
SOZ	seizure onset zone
SP	silent period
SPECT	single photon emission computed tomography
SPES	single pulse electrical stimulation
SQUIDS	superconducting quantum interference devices
SREDA	subclinical rhythmical electrographic discharges in adults
SSEPs	short-latency somatosensory evoked potentials
SSMA	supplementary sensorimotor area
SSR	sympathetic skin response
SSRI	serotonin-specific reuptake inhibitors
TA	tibialis anterior
TC	theta coma
tcACS	transcranial alternating current stimulation
tcCO2	transcutaneous carbon dioxide monitoring
tcDCS	transcranial direct current stimulation
tce	transcranial electrical
TD	temporal dispersion
TES	transcranial electrical stimulation
TH	therapeutic hypothermia
TIA	transient ischaemic attack
TILDA	the Irish Longitudinal Study on Aging
TIVA	total intravenous anaesthesia
TLE	temporal lobe epilepsy
TLI	terminal latency index
TLOC	transient loss of consciousness
TMS	transcranial magnetic stimulation
TNMG	transient neonatal myasthenia gravis
TOF	train of four
TPE	therapeutic plasma exchange
TRP	transient receptor potential
UARS	upper airway resistance syndrome
UMN	upper motor neuron
UW	unresponsive wakefulness
VC	vital capacity
vEEG	video EEG
VEP	visual evoked potential
VGCC	voltage-gated calcium channels
VGKC	voltage gated potassium channel
WHAM	wakefulness, high amplitude, anterior, male

Contributors

Gonzalo Alarcón, Comprehensive Epilepsy Center Neuroscience Institute, Academic Health Systems, Hamad Medical Corporation, Doha, Qatar; Department of Basic and Clinical Neuroscience, Institute of Psychiatry, Psychology and Neuroscience, King's College London, UK; Department of Clinical Neurophysiology, King's College Hospital NHS Trust, London, UK; Departamento de Fisiología, Facultad de Medicina, Universidad Complutense, Madrid, Spain

Sidra Aurangzeb, Specialty Registrar in Neurology and Clinical Neurophysiology, Oxford University Hospitals NHS Trust, UK

Jeremy D. P. Bland, Consultant in Clinical Neurophysiology, East Kent Hospitals University NHS Foundation Trust, UK

Helmut Buchner, Department of Neurology and Clinical Neurophysiology, Klinikum Vest Hospital, Recklinghausen, Germany

David Burke, Department of Neurology, Royal Prince Alfred Hospital and University of Sydney, Australia

Mamede de Carvalho, Professor of Physiology, Institute of Physiology, Faculty of Medicine, Instituto de Medicina Molecular, University of Lisbon, Portugal; Department of Neurosciences, Hospital de Santa Maria-CHLN, Lisbon, Portugal

Santiago Catania, Consultant Clinical Neurophysiologist, Department of Clinical Neurophysiology, The National Hospital for Neurology and Neurosurgery, London, UK

J. Helen Cross, Clinical Neurosciences Section, UCL Institute of Child Health, Great Ormond Street Hospital for Children, London, UK

R. Shane Delamont, Consultant Neurologist, Departments of Neurology and Neurophysiology, King's College Hospital, London, UK

Beate Diehl, Department of Clinical and Experimental Epilepsy, Institute of Neurology, University College London, UK; Department of Clinical Neurophysiology, National Hospital for Neurology and Neurosurgery, Queen Square, London, UK

Robert Elwes, Consultant Neurologist and Clinical Neurophysiologist, King's College Hospital, London, UK

Elaine Foley, Aston Brain Centre, School of Life and Health Sciences, Aston University, Birmingham, UK

Adrian J. Fowle, West Surrey Clinical Neurophysiology, Ashford and St Peter's Hospitals NHS Foundation Trust, Chertsey, UK

Hessel Franssen, Associate Professor of Neurology and Clinical Neurophysiology, Department of Neuromuscular Disorders, Brain Center Rudolf Magnus, University Medical Center Utrecht, the Netherlands

Anders Fuglsang-Frederiksen, Department of Clinical Neurophysiology, Aarhus University Hospital, Denmark

Paul L. Furlong, Aston Brain Centre, School of Life and Health Sciences, Aston University, Birmingham, UK

Sushma Goyal, Consultant Paediatric Clinical Neurophysiologist, Evelina London Children's Hospital, Puffin EEG Department, St Thomas' Hospital, London, UK

Robin Howard, Department of Neurology, The National Hospital for Neurology, Neurosurgery and Psychiatry, London, UK

James Howells, Bill Gole MND Fellow, Brain and Mind Centre, and Sydney Medical School, University of Sydney, Australia

John G. R. Jefferys, Professor of Neuroscience and Senior Research Fellow, Department of Pharmacology, University of Oxford, UK

Robin P. Kennett, Consultant in Clinical Neurophysiology, Oxford University Hospitals NHS Trust, UK

Matthew C. Kiernan, Brain and Mind Centre, Sydney Medical School, University of Sydney, Australia

Roo Killick, Sleep Physician & Postdoctoral Research Fellow, Woolcock Institute of Medical Research and Sydney Medical School, University of Sydney, NSW, Australia

Jun Kimura, Department of Neurology, University of Iowa Health Care, Iowa City, USA

Michalis Koutroumanidis, Department of Neurology and Department of Clinical Neurophysiology and Epilepsy, Guy's and St Thomas' NHS Foundation Trust, London, UK

Marian Lazaro, Nerve and Brain Studies Department, Guy's and St Thomas NHS Foundation Trust, London, UK; Neurophysiology Department, Evelina London Children's Hospital, St Thomas' Hospital, London, UK

Cindy S-Y Lin, Translational Neuroscience Facility, School of Medical Sciences, University of New South Wales, Australia

Kerry R. Mills, Emeritus Professor of Clinical Neurophysiology, Department of Basic and Clinical Neuroscience, King's College London, UK

V. Peter Misra, Consultant Clinical Neurophysiologist, Imperial College Heathcare NHS Trust, Hammersmith Hospital, London, UK

Friederike Moeller, Department of Neurophysiology, Great Ormond Street Hospital for Children, London, UK

Marc R. Nuwer, Professor and Vice Chair, Department of Neurology, David Geffen School of Medicine, Department Head, Clinical Neurophysiology, Ronald Reagan UCLA Medical Center, University of California, Los Angeles, USA

Susanna B. Park, Brain and Mind Centre, Sydney Medical School, University of Sydney, Australia

Matthew Pitt, Consultant Clinical Neurophysiologist, Department of Clinical Neurophysiology, Great Ormond Street Hospital for Children, London, UK

Ronit M. Pressler, Clinical Neurosciences Section, UCL Institute of Child Health, Great Ormond Street Hospital for Children, London, UK

Kirsten Pugdahl, Department of Clinical Neurophysiology, Aarhus University Hospital, Denmark

Michel J. A. M van Putten, Neurologist/Clinical neurophysiologist, Professor of Clinical Neurophysiology, Hospital Medisch Spectrum Twente, Department of Clinical Neurophysiology and Clinical Neurophysiology (CNPH), MIRA-Institute for Biomedical Technology and Technical Medicine, Faculty of Science and Technology, University of Twente, Enschede, the Netherlands

Dimitrios Sakellariou, Department of Clinical Neurophysiology and Epilepsy, Guy's and St Thomas' NHS Foundation Trust, London, UK; Department of Academic Neurosciences, King's College London, UK; Neurophysiology Unit, Department of Physiology, School of Medicine, University of Patras, Greece

Donald B. Sanders, Professor of Neurology, Duke University Medical Center, Durham, NC, USA

Stefano Seri, Aston Brain Centre, School of Life and Health Sciences, Aston University, Birmingham, UK

Erik Stålberg, Department of Clinical Neurophysiology, Section of Neuroscience, Uppsala University, Sweden

Dick F. Stegeman, Medical Physicist, Radboud University Medical Center Nijmegen, Department of Neurology/Clinical Neurophysiology, Donders Institute for Brain, Cognition and Behaviour, Nijmegen, the Netherlands

Michael Swash, Emeritus Professor of Neurology, Barts & the London School of Medicine, Queen Mary University of London, UK; Consultant Neurologist, Royal London Hospital, UK; Professor of Neurology, University of Lisbon, Portugal

Hatice Tankisi, Department of Clinical Neurophysiology, Aarhus University Hospital, Denmark

Vasiliki Tsirka, Department of Clinical Neurophysiology and Epilepsy, Guy's and St Thomas' NHS Foundation Trust, London, UK; Department of Academic Neurosciences, King's College London, UK

Kanjana Unnwongse, Section of Adult Epilepsy, Department of Neurology, Prasat Neurological Institute, Bangkok, Thailand

Antonio Valentín, Department of Basic and Clinical Neuroscience, Institute of Psychiatry, Psychology and Neuroscience, King's College London, UK; Department of Clinical Neurophysiology, King's College Hospital NHS Trust, London, UK; Departamento de Fisiología, Facultad de Medicina, Universidad Complutense, Madrid, Spain

Josep Valls-Solé, EMG Unit, Neurology Department, University of Barcelona, Hospital Clinic, Barcelona, Spain

Matthew C. Walker, Professor of Neurology, UCL Institute of Neurology and National Hospital for Neurology and Neurosurgery, Queen Square, London, UK

Tim Wehner, Department of Clinical and Experimental Epilepsy, Institute of Neurology, University College London, UK; Department of Clinical Neurophysiology, National Hospital for Neurology and Neurosurgery, Queen Square, London, UK

Caroline Witton, Aston Brain Centre, School of Life and Health Sciences, Aston University, Birmingham, UK

Zenobia Zaiwalla, Consultant in Clinical Neurophysiology and Non Respiratory Paediatric and Adult, Sleep Disorders Lead, John Radcliffe Hospital, Oxford, UK

Machiel J. Zwarts, Kempenhaeghe, Academic Centre for Epilepsy, Heeze, The Netherlands

SECTION 1

Scientific basis of clinical neurophysiology

CHAPTER 1

Nerve, muscle, and neuromuscular junction

Machiel J. Zwarts

Introduction

A thorough understanding of anatomy and physiology is essential for every clinical neurophysiologist. In this chapter, the basics of the physiology of nerve and muscle function with respect to signal transduction are presented. Although this information is also essential for the workings of the brain, the measurements of the electromyographer have a more direct relevance with these basics. For example, with single fibre electromyograph (EMG) the transmission at a single neuromuscular junction can be studied in detail. The same applies to pathology. Many peripheral nerve disorders and abnormalities, for example, conduction block, can be directly understood from the physiology of nerve action potential generation and conduction. Therefore, some emphasis is made upon the peripheral nervous system.

The membrane potential

Every cell is surrounded by a membrane, typically a lipid bilayer. In this membrane numerous special proteins and receptors reside. They all have different functions related to transport, either active or passive. The inner part of the cell is usually electro-negative with respect to the outer part. Using a micropipette one can measure this potential difference of about 60–100 mV (negative inside). This is known as the resting membrane potential or transmembrane potential. A large part of this potential difference is explained by the different concentrations of ions between extracellular and intracellular compartments (1–3).

The classic example to explain this is taken from a large container filled with water and divided by a semipermeable membrane (Fig. 1.1). On one side of the membrane, a salt is added to the water, which dissolves into positively and negatively charged ions. Consequently, depending on the permeability of the membrane for the ion, some of the ions move through the membrane because of the concentration gradient. However, this is immediately accompanied by an opposing force due to the developing electrical gradient (if the ions are positively charged, the diffusion of ions will result in a negatively charge in the compartment from which they leave). A new equilibrium will arise between the forces due to concentration gradient and the electrochemical gradient. This equilibrium is described by the Nernst equation:

$$E_I = RT \ln [I]_O / ZF[I]_i$$

where E is the transmembrane potential, R the gas constant, F the Faraday constant, T the absolute temperature, Z the charge of the ions, and $[I]_o$ and $[I]_i$ the outer and inner concentrations of the ion, respectively.

Using physiological concentrations of potassium, the calculated transmembrane potential inside the cell is −84 mV with respect to the outside, which is close the normal resting value of the membrane potential.

Indeed, the resting membrane potential of nerve cells, including axons is usually governed mainly by the concentration gradient of potassium, with low extracellular concentrations and high intracellular concentrations. For muscles, the situation is slightly different (vide infra).

The concentrations gradients for different ions are usually rather high. Another example is sodium, being high extracellular (~140 mM) and low intracellular (~12 mM). These gradients

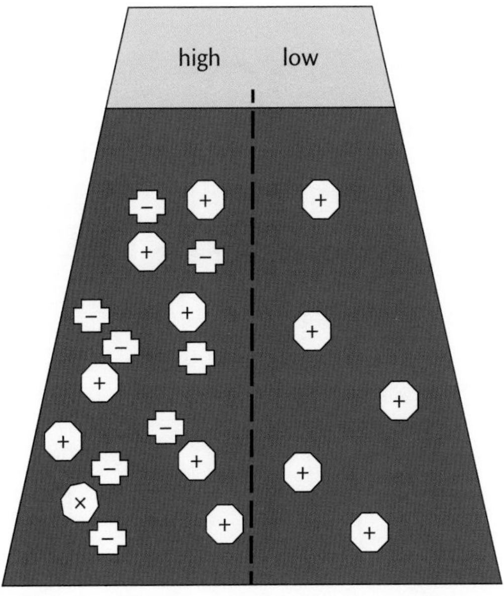

Fig. 1.1 Schematic presentation of a container with fluid divided by a semipermeable membrane. The membrane does not permit diffusion of the anions. The positively charged cations follow the concentration gradient, moving to the right, but thereby create an electrochemical gradient. The left part of the container becomes negative with respect to the right side. The two opposing forces reach an equilibrium described by the Nernst equation.

are maintained by active and thus energy dependent-pumping processes. The most important is the Na–K pump that transports 3 Na^+ ions from inside to outside, for 2 K^+ ions from outside to inside the cell. This is an energy-dependent process consuming adenosine triphosphate (ATP).

It should be realized that we speak here of bio-electricity. In fact, the currents and voltage differences in the body are based on the concentration and diffusion of charged ions. This is completely different from the electricity as we know it in our world of toasters and television sets. Here, the electricity is based on the flow of electrons. The transition of bio-electricity to the electricity, as we know it, occurs at the electrode-skin interface.

Ion channels

Specialized proteins in the cell membrane make it possible to use the transmembrane potential as the driving force to generate action potentials that conduct along the neurons, dendrites and axons, as well as along the muscle membrane (4–8). These proteins span the axon, neuron, or muscle membrane, and consist of different subunits that allow conformational changes—ion channels. Nowadays, for every ion, whole families of channels are known (9). The channel usually consists of subunits, and one or more smaller subunits. Each subunit is formed of multiple (often six) transmembrane-spanning elements (see Fig. 1.2). The central part of the channel (the pore) is water filled and allows passage of ions. Usually, the channel is ion specific to a greater or lesser degree. If the inner pore is open, the ion can enter or leave the cell without hindrance (following its electrochemical gradient) and, if closed, the permeability for the ion is very low. The opening and closing of the channel is called gating. Depending on the channel, the state can be closed as default, e.g. Na^+ channels, these open on a change in membrane potential, so-called voltage-dependent channels. Some K^+ channels are not voltage dependent, but they open and close in a random matter. Other channels are ligand specific and respond to opening when in contact with a certain neurotransmitter.

The voltage-dependent Na^+ channel consists of one principal subunit, the alpha subunit with four homologous repeats. In addition, one beta subunit is present. The fourth segment of the alpha subunit is the voltage sensing part (Fig. 1.2). This voltage sensing part consists of charged amino acids that are displaced by the transmembrane voltage field. As a result, the pore of the channel undergoes a conformational change and the net result is a large increase in Na^+ conductance. Following this activation, the Na^+ channel inactivates, probably by another part of the alpha subunit (repeats III–IV). Both a fast and slow inactivation of the channel has been described.

The action potential

The Na^+ and K^+ channels play key roles in the generation of action potentials (2,3,6).

Following a local depolarization of the membrane of around –20 mV to a threshold value of –70 mV, the Na^+ channel undergoes a conformational change accompanied by a considerable increase in Na^+ permeability. Consequently, the membrane potential approaches the resting state potential of Na^+ (about +40 mV; Fig. 1.3). Within 1 ms following the above described events, the Na^+ channels are inactivated; at the same time, K^+ channels start to open allowing K^+ to leave the cell. These processes permit the membrane potential to be restored. In the period of inactivation of the Na channels, the membrane is completely refractory to a new depolarization (absolute refractory period) and afterwards

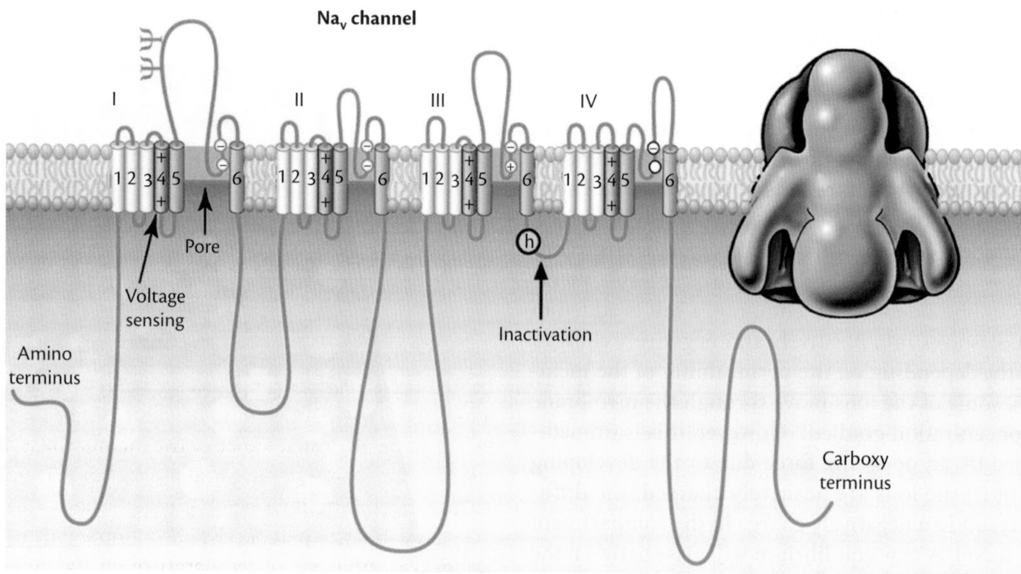

Fig. 1.2 Schematic presentation of the voltage gated Na^+ channel with subunits. Voltage sensing is accomplished by subunit I, the fourth segment, indicated on the left.

Adapted from *Nature*, **409**, Sato C, Ueno Y, Asai K, Takahashi K, Sato M, Engel A, and Fujiyoshi Y, The voltage-sensitive sodium channel is a bell-shaped molecule with several cavities, pp. 1047–51, copyright (2001), with permission from Nature Publishing Group; *Science* STKE, 2004(253), Yu FH and Catterall WA, The VGL chanome: a protein superfamily specialized for electrical signaling and ionic homeostasis, pp. re15, Copyright (2004), with permission from American Association for the Advancement of Science.

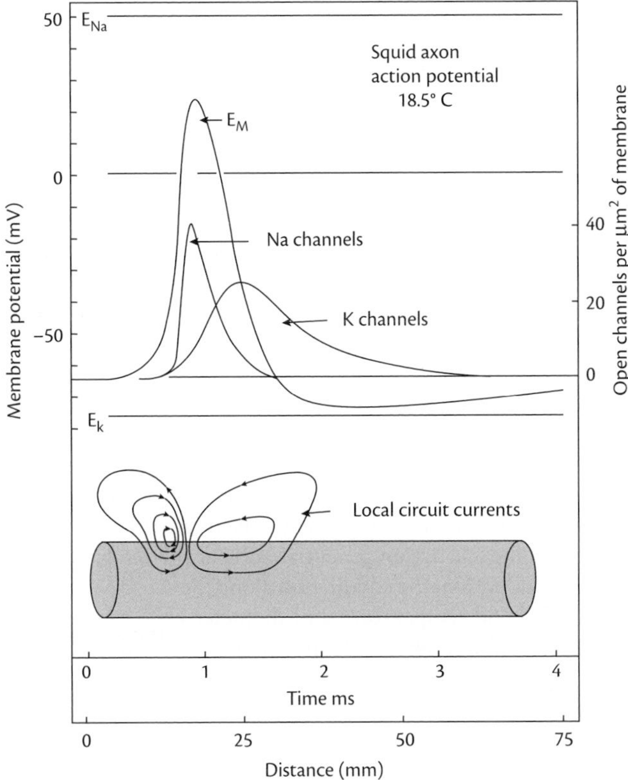

Fig. 1.3 Action potential of a giant squid axon. Note the increase of open Na channels, producing the rise of the membrane potential and the subsequent rise in K channels responsible for the repolarization and overshoot of the membrane potential.

Adapted from Hille B, Introduction to physiology of excitable cells. In: Patton HD, Fuchs AF, Hille B, et al. (Eds), *Textbook of physiology*, 21st edn. copyright (1989), with permission from Elsevier.

is partially refractory. Following an overshoot of the membrane potential (after-hyperpolarization), the initial equilibrium emerges again (Fig. 1.4). Intracellularly, the action potential is a monophasic, positive potential. Extracellularly, however, the situation is more complicated. The current is spread in a large volume conductor and in addition travels along the nerve. The result is basically a triphasic waveform. The leading part of the action potential approaching a recording electrode first generates

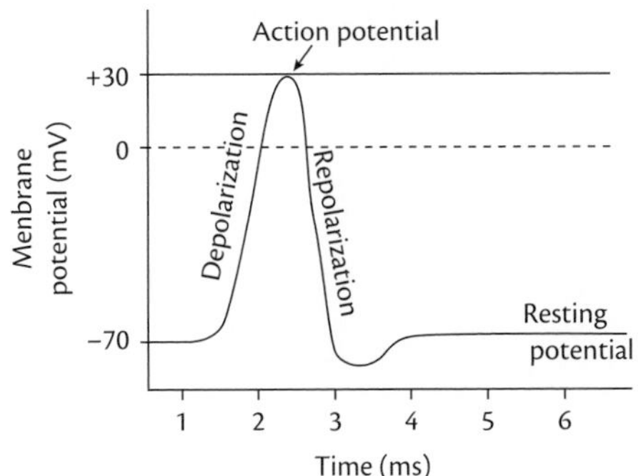

Fig. 1.4 Change in membrane potential during an action potential. The duration of the action potential is roughly 1–1.5 ms.

https://psychlopedia.wikispaces.com/action+potential

a positive potential, at the place of the depolarization itself this reverses to negative and in the trailing phase of the potential, a positive deflection occurs again (see also Chapter 6 and ref. (2)). Essentially, this triphasic form is the basic form of all extracellular measurements near an axon or muscle fibre. Another important point is that the display of action potentials and the resulting measurements are usually provided as voltage changes against time (see Fig. 1.3), based upon a point measurement at a certain position. In *space*, the action potential has a considerable physical spread. For example, an action potential of 1.5 ms duration and a conduction velocity of 50 m/s has a physical spread along the nerve of 75 mm (50 m/s = 50 mm/ms, in 1.5 ms a distance of 1.5 × 50 mm = 75 mm is travelled). Since extracellular action potentials are always measured as the difference between two electrodes, this large spatial spread can easily make both recording electrodes active with respect to the action potential and thereby influence the recorded amplitude.

The conduction of the action potential is generated by the passive spread of current that induces a depolarization further ahead in the membrane. This process generates the same changes in nearby voltage-dependent Na channels, thereby producing a regenerative potential travelling along the axon.

Saltatory conduction and conduction velocity

The above described processes of conduction hold for the unmyelinated axon. However, many axons are myelinated. Schwann cells (in the peripheral nervous system) lying near the axon are wrapped around the axon, thereby creating an electrical insulation of the axon. In the brain, the oligodendrocyte fulfils the same purpose (5,7). The resistance of the membrane at the insulated parts is up to 300 times higher. Between the covered parts of the axon, small bare parts of axon exist at regular intervals that are called the nodes of Ranvier. Due to this configuration, the action potential leaps from node to node, the insulated parts of the axon cannot be depolarized. This way of conduction is less energy consuming. Secondly, the conduction velocity rises enormously, typically from 2–4 ms to 40–60 m/s in the human peripheral nervous system. The axon membrane at the nodes of Ranvier is highly specialized to make this possible, with a very high concentration of voltage-gated Na channels, which facilitates uninterrupted conduction.

The conduction velocity depends on a large number of factors, an important factor being the temperature. Since all these processes are dependent on metabolism, a low temperature will slow the conduction. Furthermore, the resistance of the axon is important; this depends largely on the internal resistance, which is related in a square root manner to the diameter. In myelinated axons, the conduction velocity is linearly related to the diameter of the axon. The capacitance of the membrane is also important—a large capacitance implies that more charge is needed to depolarize the membrane at a certain point.

From a clinical perspective, it is important to understand the important influence of temperature on the conduction velocity. Temperature at the feet can easily reach 20–25°C in countries with temperate or more harsh climates. This not only influences the conduction, but also the amplitude of the recorded potentials. In pathology, demyelination is an important factor causing reduced velocities or even conduction block.

In addition specialized areas of the nodes of Ranvier and on the axon cell body have membrane domains with specialized function. One of them is the initial axon segment. It contains clusters of ion channels, such that the segment has the lowest threshold for firing, thus serving to integrate synaptic input to the nerve cell.

Neuromuscular junction and other synapses

Another highly specialized region of the motor axon is its terminal part where it connects to the muscle (10,11). Here, the action potential is transferred to the muscle by means of chemical neurotransmission. The motor nerve branches into many small twigs, each innervating a single muscle fibre. The end of the nerve twig is called a bouton. It contains many mitochondria and microtubules on one side, and vesicles clustered about membrane thickenings at the other (Fig. 1.5). The muscle membrane underlying this bouton is also specialized, it contains many folds, thereby increasing its surface area to about eight times the axon terminal side. At the summit of the junctional folds, a very high number of acetylcholine receptors (Ach) receptors are present. The synaptic cleft between muscle and axon is about 20–30 nm wide.

The action potentials arriving at the neuromuscular junction give rise to a strong influx of calcium through voltage-gated Ca channels. This, in turn, triggers the release of ACh through a cascade of intracellular signalling events terminating in the induction of exocytosis of small vesicles which contain about 5000–10,000 ACh molecules each. About 100 vesicles are released by every action potential. Every neuromuscular junction (NMJ) contains about 200,000–400,000 vesicles of which 1000–30,000 lie in the immediate surroundings of the so-called active zones. These active zones contain the vesicles that are released by an action potential. The much larger pool of vesicles lying further away from the membrane, serve as a reserve pool. Following the release of ACh containing vesicles, ACh crosses the synaptic cleft in a fraction of a millisecond and binds to ligand gated ACh receptors in the muscle membrane resulting in a depolarization of about 90 mV. This is called the endplate potential (EPP). Usually, a depolarization of about 10–15 mV is enough to generate an action potential in the extrasynaptic muscle membrane. This 90 mV depolarization is more than enough to trigger voltage gated Na channels in the extrasynaptic muscle membrane leading to a propagated action potential along the muscle fibre in both directions. In the depth of the folds of the NMJ a very high concentration of Na channels is present to facilitate this process. The propagated muscle action potential halts at the tendon. Similar to nerve conduction, the conduction along the muscle membrane also depends on temperature and diameter of the muscle fibre (12,13). It is comparable with the conduction of unmyelinated nerve axons (~4 m/s).

It should be noted that the magnitude of the EPP is not fixed. It is dependent on factors related to presynaptic (number of vesicles released and vesicle content), post-synaptic (number of Ach receptors), and synaptic cleft factors, such as the acetylcholinesterase (AChase) activity. Furthermore, the EPP amplitude is also

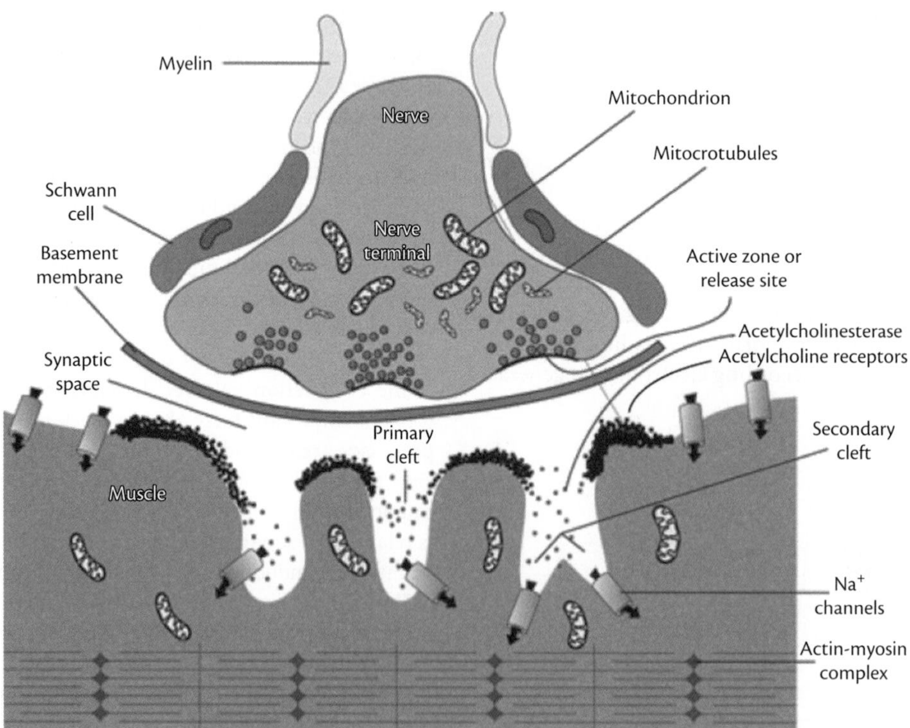

Fig. 1.5 The neuromuscular junction with above the ending of the motor axon in a specialized region, where Ca^{2+} influx following an action potential is the trigger for Ach-containing vesicles to be released. The released ACh crosses the synaptic cleft and reaches the specialized area of the muscle with many folds containing ACh receptors responsible for a local depolarization of the muscle membrane, which is the trigger of a propagated action potential along the muscle fibre.

Reproduced from *Anaesthesia*, **64**(Suppl. 1), Martyn JAJ, Jonsson Fagerlund M, Eriksson LI, Basic principles of neuromuscular transmission, pp. 1–9, copyright (2009), with permission from John Wiley and Sons.

dependent on the firing history of the axon. Following repetitive stimulation several phases of potentiation are recognized with different time courses.

After its release, the ACh is hydrolysed by AChase in a rapid and effective way. The resulting choline is used by the nerve terminal again to form ACh after its uptake. Except for the release of a large number of ACh containing vesicles following an action potential, single vesicles are also released spontaneously in a random manner. This results in small depolarizations of the muscle membrane of about 1 mV; these are called miniature endplate potentials (MEPP). They result in a small depolarization of the muscle membrane, insufficient to produce a self-sustaining action potential, but only giving electrotonic spread of the current. However, with needle EMG these depolarizations can be recorded as endplate noise.

The NMJ is just an example of a specialized synapse. In the brain, the connection between neurons consists of a multitude of different synapses. However, the basic layout is the same as for the NMJ. Many different neurotransmitters are known, either excitatory or inhibitory. Examples are glutamate, serotonin, acetylcholine, etc. Glutamate is by far the most important excitatory neurotransmitter in the brain. Every neuron in the brain receives many thousands of excitatory and inhibitory inputs via these synapses on the dendritic tree, cell body, but also the axon hillock. The net result of all these influences determines the membrane potential and thus the firing behaviour of the cell.

Muscle membrane, t-tubules, and excitation-contraction coupling

As stated above, the action potential originating at the NMJ travels in both directions along the muscle membrane with a conduction velocity of about 4 m/s (12,13). Since most NMJs are placed more or less at the centre of the muscle fibre, an even contraction of the muscle is assured. Since the motor axons divide into many twigs each innervating one muscle fibre, both during voluntary activation and as a result of stimulating the motor nerve, a summation of hundreds of more or less synchronously activated muscle fibre action potentials can be recorded by the clinical neurophysiologist. This amplification effect is very helpful, since in this way it is easy to record (for example, with needle EMG), the firings of a single motor neuron.

The action potential of the sarcolemma is generated in a similar manner as the nerve action potential. An important difference, however, is the conductance of the chloride channels, which is high in muscle (as opposed to axons) and important in determining the resting membrane potential and in the repolarization phase of the muscle membrane (14–16). Therefore, the afterdepolarization of the muscle action potential is slow. The propagated action potential along the sarcolemma also enters the t-tubule system, specialized invaginations of the sarcolemma (Fig. 1.6A). Here, we encounter a special situation where the extracellular space is confined to a small tubule. As a consequence, the concentration of extracellular ions that accumulate after action potentials can rise to high levels, especially during rapid firing of the motor unit. The high conductance of the muscle membrane for chloride assures that this will not happen, as it passively follows the Na^+, thereby restoring the membrane depolarization quickly (14).

Depolarization of the transverse tubules is detected by voltage-sensitive Ca^{2+} channels (Cav), also known as dihydropyridine receptors (DHPRs), and these communicate directly with Ca^{2+} channels, also known as ryanodine receptors (RyRs), which can deliver far more calcium than can enter through the Ca^{2+} channels themselves (17,18) (Fig. 1.6B).

The consequent increased concentration of intracellular Ca^{2+} is bound by troponin C (TnC), which triggers contraction. Consequently, a segment of actin becomes available to myosin as

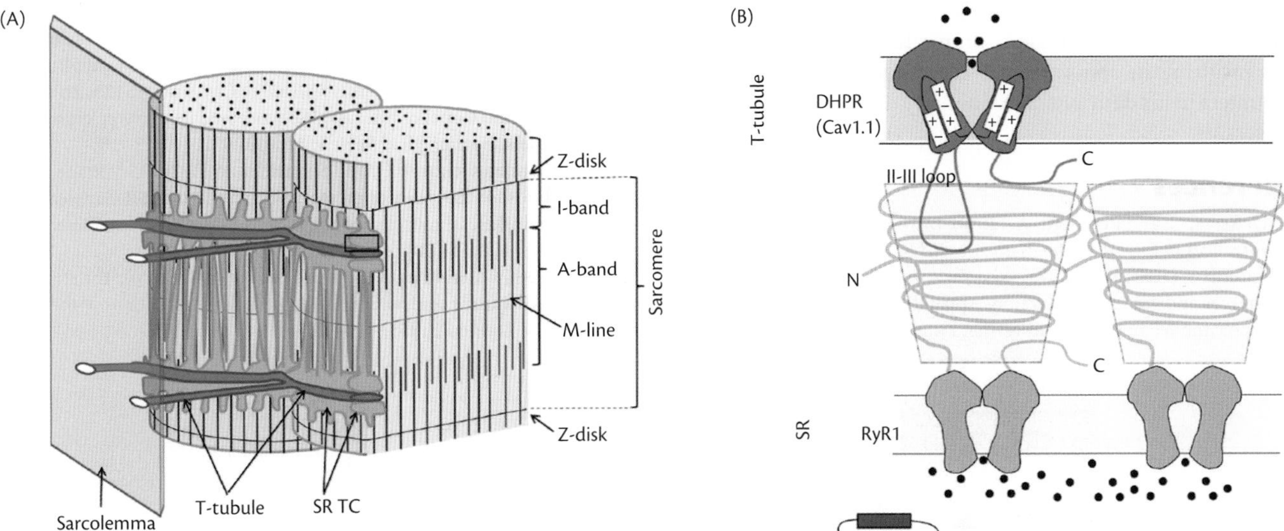

Fig. 1.6 (A) The propagated action potential along the sarcolemma (on the left) is conducted into the T tubule, which leads to extrusion of Ca^{2+} from the sarcoplasmatic reticulum and triggers the actin myosine proteins to contract by sliding of the filaments and shortening the sarcomere. (B) The DHPR receptors are shown, which are voltage sensitive and activate the release of Ca^{2+} via the RyR1 receptors.

Adapted from *Prog Biophys Mol Biol*, **108**(3), Hernandez-Ochoa EO and Schneider MF, Voltage clamp methods for the study of membrane currents and SR Ca release in adult skeletal muscle fibers, pp. 98–118, copyright (2012), with permission from Elsevier.

one tropomyosin is moved away from the myosin-binding site on actin. The subsequent binding of myosin to actin generates force and shortening of sarcomeres. At low intracellular Ca^{2+} concentrations, tropomyosin blocks the myosin-binding site on actin.

The contraction of the muscle itself is realized by supramolecular structures, the sarcomeres (19,20). These consist of thick and thin filaments composed of myosin and actin, respectively. These are positioned intertwined with each other, partly overlapping. The molecular motor is myosin, which is attached at the thin filaments and is an elongated molecule with a tail and a head. The head is the moving part of the molecule and responsible for force generation. It is attached at actin (cross-bridge). After binding to ATP actin and myosin dissociate, ATP hydrolysis occurs (producing inorganic phosphate (Pi) and adenosine diphosphate (ADP)), which is followed by a conformational change of the head and binding of actin to myosin again. The release of Pi is followed by another conformational change in the head of the myosin molecule and transmitted to the actin molecule resulting in a power stroke. As long as Ca is freely bound within the thin filaments and ATP is available this cross-bridge cycling continues. However, Ca^{2+} is taken up again swiftly following an action potential by the sarcoplasmatic reticulum and the force production ceases.

As described above, a single action potential along a motor axon gives rise to the simultaneous contractions of a few hundred muscle fibres, which are all governed by this motor neuron (the motor unit) resulting in a twitch. The peak force of a single twitch is about 10–20% of the maximal force of the motor unit. At high frequencies a fused tetanic contraction can be obtained resulting in maximal force, this occurs at frequencies around 40 Hz.

In conclusion, a survey is provided of the essential workings of the excitable cell, the way membrane potentials originates, and the depolarization of the membrane through voltage-dependent ion channels. The consequent transmission of action potentials and conveying of information to other neurons is described. Finally, neuromuscular transmission and the activation of the muscle membrane are discussed, resulting in contraction. At all levels, neurological diseases can affect proper functioning of these systems resulting in abnormalities the clinical neurophysiologist encounters in his daily practice. These are discussed in subsequent chapters.

References

1. Di Resta, C. and Becchetti, A. (2010). Introduction to ion channels. *Advances in Experimental Medicine and Biology*, 674, 9–21.
2. Dumitru, D., Stegeman, D.F., and Zwarts, M.J. (2002). Electric sources and volume conduction; Appendix: the leading/trailing dipole model and near-field/far-field waveforms. In: *Electrodiagnostic medicine* (Eds D. Dumitru, A. M. Amato, and M. J. Zwarts), 2nd edn, pp. 27–67. Philadelphia, PA: Hanley and Belfus, Inc.
3. Putten, van M.J.A.M. (2009). *Essentials of neurophysiology*. Berlin: Springer.
4. Ashcroft, F.M. (2000). How channels work. In: *Ion channels and disease* (Eds F. M. Ashcroft), pp. 21–158. London: Academic Press.
5. Rasband, M.N. (2011). Composition, assembly, and maintenance of excitable membrane domains in myelinated axons. *Seminars in Cell & Developmental Biology*, **22**(2), 178–84.
6. Debanne, D., Campanac, E., Bialowas, A., Carlier, E., and Alcaraz, G. (2011). Axon physiology. *Physiological Review*, **91**, 555–602.
7. Rasband, M.N. (2011). Composition, assembly, and maintenance of excitable membrane domains in myelinated axons. *Seminars in Cell & Developmental Biology*, **22**(2), 178–84.
8. Hille, B. (1989). Introduction to physiology of excitable cells. In: *Textbook of physiology* (Eds) H. D. Patton, A. F. Fuchs, B. Hille, et al., 21st edn, pp. 1–80. Philadelphia, PA: WB Saunders.
9. Yu, F.H., Yarov-Yarovoy, V., Gutman, G.A., and Catterall, W.A. (2005). Overview of molecular relationships in the voltage-gated ion channel superfamily. *Pharmacological Reviews*, **57**, 387–9510
10. Martyn, J.A.J., Jonsson Fagerlund, M., and Eriksson, L.I. (2009). Basic principles of neuromuscular transmission. *Anaesthesia*, **64**(Suppl. 1), 1–9.
11. Nishimune, H. (2012). Active zones of mammalian neuromuscular junctions: formation, density, and aging. *Annals of the New York Academy of Sciences*, **1274**(1), 24–32.
12. Blijham, P.J., ter Laak, H.J., Schelhaas, H.J., van Engelen, B.G.M., Stegeman, D.F., and Zwarts, M.J. (2006). Relation between muscle fiber conduction velocity and fiber size in neuromuscular disorders. *Journal of Applied Physiology*, **100**, 1837–41.
13. Arendt-Nielsen, L. and Zwarts, M.J. (1989). Measurement of muscle fiber conduction velocity in human—techniques and applications. *Journal of Clinical Neurophysiology*, **6**, 173–90.
14. Bretag, A.H. (1987). Muscle chloride channels. *Physiological Reviews*, **67**, 618–724.
15. Jurkat-Rott, K., Fauler, M., and Lehmann-Horn, F. (2006). Ion channels and ion transporters of the transverse tubular system of skeletal muscle. *Journal of Muscle Research in Cell Motility*, **27**, 275–90.
16. Dulhunty, A.F. (1978). The dependence of membrane potential on extracellular chloride concentration in mammalian skeletal muscle fibres. *Journal of Physiology*, **276**, 67–82.
17. MacIntosh, B.R., Holash, R.J., and Renaud, J-M. (2012). Skeletal muscle fatigue—regulation of excitation–contraction coupling to avoid metabolic catastrophe. *Journal of Cell Science*, **125**, 2105–14.
18. Fill, M. and Copello, J.A. (2002). Ryanodine receptor calcium release channels. *Physiological Review*, **82**, 893–922.
19. Hernandez-Ochoa, E.O. and Schneider, M.F. (2012). Voltage clamp methods for the study of membrane currents and SR Ca release in adult skeletal muscle fibers. *Progress in Biophysics & Molecular Biology*, **108**, 98–118.
20. Kenneth, C., Holmes, K.C., and Geeves, M.A. (2000). The structural basis of muscle contraction. *Philosophical Transactions of the Royal Society, London B*, **355**, 419–31.

CHAPTER 2

The motor unit

David Burke and James Howells

What is a motor unit?

The term 'motor unit' refers to the lower motor neuron (or 'motoneuron'), its axon, and the few hundred muscle fibres that it innervates. The term is generally reserved for α motoneurons, which innervate the force-producing muscle fibres of skeletal muscle (termed 'extrafusal' muscle fibres, as distinct from the 'intrafusal' muscle fibres of muscle spindles, which are innervated by γ motoneurons). So-called β motoneurons branch to innervate both extrafusal and intrafusal muscle. The properties of the motoneuron, its axon, and the muscle fibres that it innervates are matched. This chapter will consider the motor units of skeletal muscle and will focus on those innervating the limbs.

The α motoneurons recruited first in a voluntary contraction, by transcranial stimulation of the motor cortex or by reflex action are of small size, in accordance with the so-called 'size principle' (1). They have axons of relatively small diameter (and consequently slow conduction velocity) and innervate a relatively small number of 'red' muscle fibres, which produce small twitch contractions that are resistant to fatigue (Fig. 2.1). On the other hand, the first axons activated by weak electrical stimulation of motor axons in the peripheral nerve tend to be of large diameter and conduction velocity. The parent motoneurons are larger, and the twitch contractions produced by their 'white' muscle fibres produce greater force, with a greater tendency to fatigue.

The term 'motor unit' is often used to refer to the motor unit action potential (MUAP), which is the electromyograph (EMG) signature of the motor unit, due to the summed activity of the muscle fibres (or single fibre action potentials (SFAPs); see Chapter 7) innervated by the single motor axon. The size and shape of individual MUAPs vary with the distribution of muscle fibres within the muscle, their proximity to the recording electrodes, and their depth within the muscle, whether one uses surface electrodes or intramuscular needles or wires. High-density recording techniques have been developed that allow the activity of multiple single units to be identified in surface recordings, using multiple closely spaced electrodes fixed to the skin over the contracting muscle (2–4). In routine diagnostic studies using intramuscular needle electrodes, commercial EMG machines now allow the activity of different motor units to be identified during voluntary contractions using spike identification and interference pattern decomposition.

EMG activity can be used to measure the activity of α motoneurons, termed the 'final common path' by Sherrington (5), thus allowing insight into the descending and reflex influences that play on the motoneuron pool. EMG potentials are the usual output measured in motor control studies, whether the EMG activity is part of a voluntary contraction, due to a reflex process or due to transcranial stimulation. As mentioned above, the activity of a single motor unit can often be recorded, and this provides information about the behaviour of a single α motoneuron. Unless selective recording techniques are used, this will be a low-threshold motor unit because it is difficult to distinguish individual MUAPs when many are active. Studying the behaviour of single units can be important because conclusions based on the behaviour of the whole motoneuron pool rely on the assumption that it behaves in a homogeneous way to inputs and that it does so with a linear input/output relationship (6). This does not always occur, the best example being that cutaneous inputs can alter motor unit recruitment threshold by raising the threshold for voluntary activation of small motor units and lowering it for larger motor units (7).

Motoneuron

Motoneurons have been considered passive responders to the net sum of the excitatory and inhibitory inputs acting on them, but it is clear that they can respond to neuromodulators and intracellular messengers by changing the intrinsic properties of the cell membrane and their responsiveness to inputs. The motoneurons considered in this Chapter are located in the grey matter of the anterior horn of the spinal cord, predominantly in Rexed's lamina IX (8), in columns extending over a number of segments overlapping with those for adjacent muscles. In early development, there are more motoneurons than in adulthood, the excess being lost by apoptosis, in part because they do not receive neurotrophic factor support from the innervated muscle. As mentioned above, motoneurons fall into two groups, correlated with their size. α motoneurons are larger and innervate skeletal muscle (the so-called 'extrafusal' muscle). Small γ motoneurons innervate the intrafusal muscle of the muscle spindle and will not be considered further in this chapter. [The so-called β motoneurons, which innervate both extra- and intrafusal muscle, are of size similar to that of α motoneurons.]

Morphology

The motoneuron has an extensive pattern of dendrites, which increase the extent and surface area over which excitatory and inhibitory inputs to the motoneuron form synapses that regulate its excitability and discharge. Each motoneuron has one axon, which arises from the cell body at a specialized region, the axon hillock, and which extends from the anterior horn through the anterior root into the peripheral nerve to innervate the relevant muscle—a length of as much as 1 m or more (Fig. 2.1). Acetylcholine is the

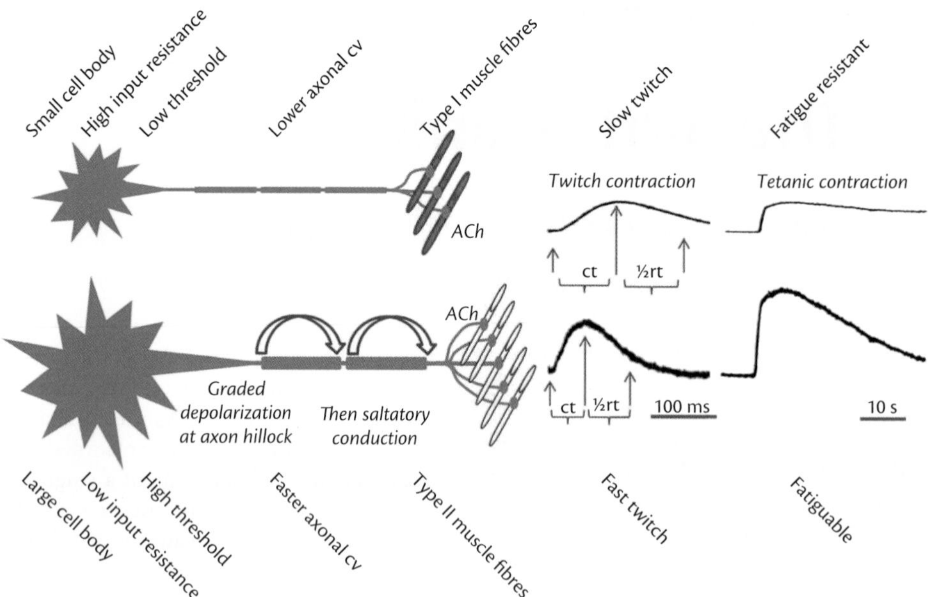

Fig. 2.1 The motor unit. Cell body, axon, innervated muscle fibres, twitch contraction, and force of tetanic contraction for a small and for a large motoneuron. The muscle fibres are 'red' (type I) for the small motoneuron and 'white' (type II) for the large motoneuron. 'Threshold' refers to the threshold for recruitment of the motoneuron in voluntary and reflex contractions, not the threshold of the axon to electrical stimulation (which tends to be lower the larger the axon).
ct = contraction time; ½rt = half relaxation time; cv = conduction velocity; *ACh* = acetylcholine, the transmitter released at the neuromuscular junction. Force profiles are illustrative and uncalibrated.

excitatory transmitter at its terminals at the neuromuscular junction and is also the transmitter on the recurrent axon collaterals that project to Renshaw cells.

Synaptic actions

Excitatory post-synaptic potentials (EPSPs) and inhibitory post-synaptic potentials (IPSPs) are summed at the axon hillock. The axon hillock terminates with the first myelin segment, at a 'heminode' (Fig. 2.1). An action potential is triggered if the membrane potential threshold is exceeded. It then propagates in an 'all-or-none' manner along the motor axon 'jumping' from node of Ranvier to node of Ranvier (saltatory conduction), ultimately invading the motor nerve terminals innervating the muscle fibres of the motor unit. EPSPs are brief depolarizing potentials, resulting from the opening of ligand-gated channels by the excitatory transmitter, glutamate, which is released from the presynaptic terminals. The reaction of glutamate with its receptors on the post-synaptic membrane opens the channel and allows an inward flow of positive ions (Na^+, K^+, Ca^{2+}) into the post-synaptic cell. EPSPs may be graded in size, dependent on the number of active boutons that the presynaptic axon has on the cell. They have a slow decay, allowing temporal summation of the excitatory effects of closely spaced inputs in different axons. Not all synaptic boutons are necessarily invaded every time there is an impulse in the parent axon and its branches. The phenomenon of post-tetanic potentiation is probably due to the activation of a greater percentage of the synapses from an afferent on the motoneuron, i.e. to the activation of previously 'silent' synapses.

Similarly, IPSPs are brief hyperpolarizing potentials, resulting from the opening of ligand-gated channels by the inhibitory transmitter, glycine, which is released from the presynaptic terminals and triggers a Cl^- current that causes transient hyperpolarization. The action of glycine is antagonized by strychnine. A major contribution to the IPSP is the conductance change in the motoneuron, independent of the change in membrane potential that it produces (9).

Gamma amino butyric acid (GABA) is the neurotransmitter responsible for 'presynaptic inhibition' of group Ia terminals on motoneurons. The anatomical substrate is an axo-axonal synapse in which the axons from a presynaptic inhibitory interneuron synapse with the presynaptic terminals of the incoming afferent fibres. Activity at this synapse produces a depolarization of the afferent terminals (so-called 'primary afferent depolarization'). This reduces the amount of excitatory transmitter released at the synapse and thereby decreases transmission at that synapse, and does so without altering the excitability of the motoneuron. At the Ia-motoneuron synapse, presynaptic inhibition is mediated primarily through the activation of $GABA_A$ receptors, and can be blocked by bicuculline (10). Baclofen blocks $GABA_B$ receptors, and its effects on spinal cord function in patients with spasticity are not effects on presynaptic inhibition (as was once thought). It is of interest that the synapses of corticospinal axons on the motoneuron are not affected by presynaptic inhibition (6).

Motoneuron recruitment

The recruitment of α motoneurons by voluntary effort, corticospinal volleys, and reflex action proceeds in an orderly sequence from small to large as the excitatory input grows. This is largely because the smaller the motoneuron the greater the input resistance. In accordance with Ohm's law, the potential produced by a given current is greater in small motoneurons because of their higher input resistance (Fig. 2.1). In addition to a slower conduction velocity and a higher input resistance, small low-threshold motoneurons have a longer membrane time constant, lower rheobase, and longer after-hyperpolarization (AHP). These properties can change after exercise and in diseases, such as stroke or spinal cord injury. Inhibitory actions are also generally matched to the size of the motoneuron, unless the distribution of synapses is non-uniform (see 'What is a motor unit?' and (7)).

When the force of a voluntary contraction is increased, new previously silent motoneurons are recruited and the discharge rate of active motoneurons increases. The extent to which these strategies are used varies for different muscles. The discharge rates of motoneurons are also related to size, with higher rates achievable by large motoneurons (which produce large twitch contractions with a greater tendency to fatigue; Fig. 2.1). This is so whether the contraction is a steady submaximal effort or a brief intense effort, and is not surprising because the shorter contraction times for these motor units would require a higher rate to produce 'fusion' (see 'Muscle twitches, tetanic contractions, and voluntary force').

Input–output relationship

The gain of the input–output relationship for the motoneuron pool can be altered, and this has major implications for motor control studies because the same volley would then produce a different output, not necessarily related to the experimental condition or manipulation. The best documented mechanism for changing the gain of a motoneuron is the development of persistent inward currents (PICs) in motoneurons (Fig. 2.2; for reviews, see (11,12). PICs are due to dendritic persistent Na^+ currents and L-type Ca^{2+} currents, and can result in 'plateau potentials', which long outlast the stimulus that triggered them. They can result in an apparently self-sustained motoneuron discharge, and have been implicated in the spasms of spinal spasticity in rats (13) and humans (14). Intact descending monoaminergic drives are necessary for PICs to develop in motoneurons (11), but these pathways are transected in complete spinal lesions. The PICs then occur because of plastic changes in spinal circuitry. Over time there is greater 'expression of 5-HT_{2C} receptor isoforms that are spontaneously active (constitutively active) without 5-HT. Such constitutive receptor activity restores large persistent calcium currents in motoneurons in the absence of 5-HT' (15).

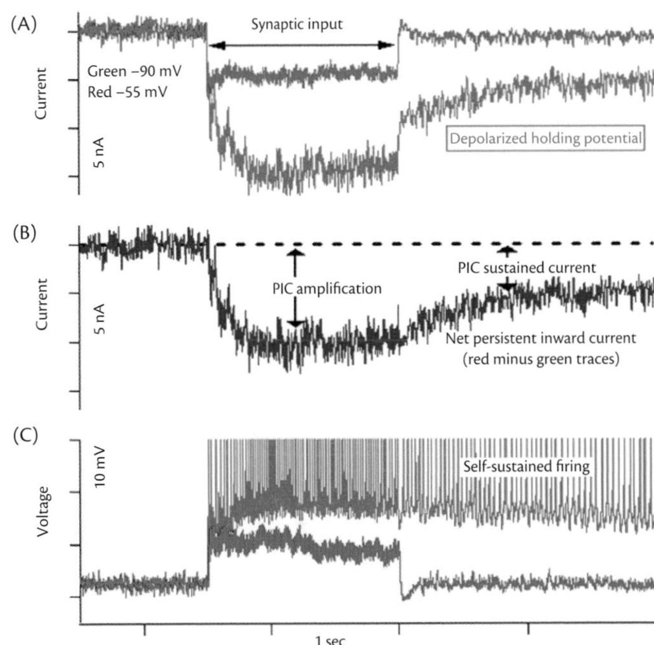

Fig. 2.2 Persistent inward currents, plateau potentials, and self-sustained firing. The dendritic persistent inward current (PIC) amplified and prolonged synaptic input in a low-threshold, type S motoneuron. (A) At a hyperpolarized holding potential (−90 mV; green trace), synaptic input produced a steady current with a sharp onset and offset. At a depolarized holding potential (~−55 mV; red trace), the PIC is activated, and amplifies and prolongs the same input. Baseline holding currents are removed to allow the traces to be superimposed. (B) The difference between the currents in A reflects the net PIC contribution. (C) Under current clamp conditions, the same input produces a steady excitatory post-synaptic potential (EPSP) when the cell is hyperpolarized (~−90 mV; green trace). At a more depolarized level (−70 mV; red trace, offset removed), the input evokes repetitive firing and then slower self-sustained firing when the input is removed.

Repolarization and the after-hyperpolarization

Repolarization of the motoneuron following an action potential involves fast K^+ currents, but is followed by an AHP, the duration of which is a major determinant of the neuron's firing rate. In different neurons, the AHP can last tens to some hundreds of milliseconds, and has two components, both dependent on the Ca^{2+}-activated K^+ channel family, which has a slow action. The AHP is longest in small motoneurons, and shortens with increasing discharge rate, the speed of recovery dependent on the decay of the currents driving the hyperpolarization, and particularly the speed of buffering of internal calcium. Recovery from the AHP may also be facilitated by the hyperpolarization-activated current, I_h, although its role in motoneurons is largely related to their resonant properties (12). I_h is more active in larger motoneurons and, not surprisingly, is more active on the axons of low threshold to electrical stimulation (see 'Biophysical properties and their implications"). As discussed further under 'Motor axon', I_h is a depolarizing current that is activated by hyperpolarization and operates to oppose that hyperpolarization. The **h**yperpolarisation-activated **c**yclic **n**ucleotide-gated (HCN) channels underlying this current are among the most sensitive to metabolic demands, and contribute to the plasticity of neural connections. In an analogous way, the HCN isoform present in the heart, HCN4, is highly expressed in the pacemaker region of the heart and is responsible for normal cardiac rhythmicity.

Motor axon

In human studies with skin and subcutaneous tissue between the surface stimulation and the nerve trunk, the activation of motor axons in descending order of size is not as reliable as in experiments in animals where the nerve can be exposed. The sequence of activation with electrical stimulation is still broadly in order of descending conduction velocity (16), and presumably therefore axonal diameter. The stimulus–response curve is sigmoidal or 'S-shaped', and relatively steep when compared with the stimulus-response curve for cutaneous afferent axons [(17,18); see Fig. 2.3]. The stimulus–response curve is not routinely measured in routine diagnostic studies, but in chronic demyelinating diseases, an increase in threshold and a decrease in the slope of the stimulus–response curve can be valuable diagnostic findings (19).

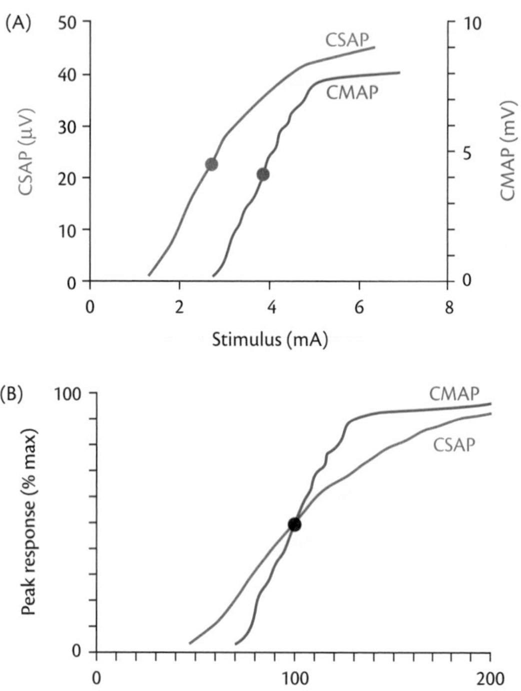

Fig. 2.3 Activation of motor units and cutaneous afferents by graded electrical stimulation. (A) Amplitudes of compound sensory action potentials (CSAP; shown in red) and compound muscle action potentials (CMAP; shown in green) in response to graded stimuli. The filled circles mark the half-maximal compound response. Note the earlier, but more gradual CSAP response to increasing stimulus intensity than motor. (B) Normalized stimulus–response relationships for the data in (A) with the peak response normalized to the maximal compound response. The stimulus is normalized as a percentage of the mean threshold (marked as filled circles in (A)).

The function, growth, and survival of an axon depends on axoplasmic transport to and from the cell body of nutrients, proteins, lipids, and other essential factors. Orthograde (anterograde) transport from the cell body for vesicles is relatively fast, 200–400 mm/day, but that of proteins, such as channels and cytoskeletal neurofilaments is relatively slow, at an overall rate of <10 mm/day, because of frequent pauses in the transport. Retrograde axoplasmic transport returns debris, such as used vesicles to the cell body, and this pathway provides a route for certain neurotropic viruses, such as herpes simplex, rabies, and poliomyelitis, to invade the cell body.

Conduction velocity of the fastest axons

In human nerves, the fastest motor axons innervating upper-limb muscles have conduction velocities of 50–60 m/s in the distal upper limb and 40–50 m/s in the distal lower limb. These velocities are 5–10 m/s slower over the same nerve segment than the fastest cutaneous and muscle afferents (which have similar maximal velocities in humans (20)). However, the threshold for afferents is much lower than that of motor axons—for example, human group Ia afferents can be activated using stimuli 0.5–0.6 times that required for the lowest-threshold motor axons (6). While differences in size contribute, the axons have different biophysical properties (see 'Biophysical properties and their implications'), and these are major reasons for the higher threshold for motor axons.

Conduction velocity distribution

Routine motor conduction studies rely on measuring the latencies for the onset of the compound muscle action potential (CMAP), and the conduction velocities are therefore those for the fastest axons in the nerve. The range of conduction velocities for motor axons innervating a muscle has been determined using collision techniques, often referred to as the Hopf technique (21), but preceded by the report of Thomas et al. (22). In the latter study, velocities for the slowest motor axons were 30–50% less than those of the fastest. This range has not been confirmed in later studies, e.g. a 12% spread of conduction velocities, from the fastest 5% to the slowest 5% was reported using a computer-based collision technique (23). The range of conduction velocities for a motor nerve has also been estimated using decomposition techniques, the spread of velocities being 12–24 m/s for the thenar muscles (24), with the fastest motor axons slower than the fastest cutaneous afferents and the slowest motor axons faster than the slowest cutaneous afferents. This is consistent with the steeper stimulus-response curve of motor axons in Fig. 2.3.

Biophysical properties and their implications

Axons have only one function, the secure transmission of impulses for minimal expenditure of energy and, to do this, their biophysical properties are adapted to the pattern and rate of discharge that they are normally required to maintain. Motor axons have properties that differ from those of cutaneous afferent axons (25,26) and group I muscle afferents (27). Specifically, motor axons are less excitable (i.e. they require stronger stimuli), and this is largely because they have less activity of two depolarizing conductances: I_h (the hyperpolarization-activated current, due to activity of hyperpolarization-activated cyclic-nucleotide gated channels) and I_{NaP} (the persistent Na^+ current). The latter is more active on sensory axons because they are ~4 mV more depolarized than motor axons (26). The greater I_{NaP} underlies the longer strength-duration time constant and lower rheobase of sensory axons (28). These differences dictate that sensory axons can be activated at lower threshold than motor axons, and this reason is arguably more important than any size difference. One implication of the lesser excitability of motor axons is that diseased or damaged motor axons would be more likely to undergo conduction block than sensory axons for the same impulse load (25,29).

There are differences in the properties of motor axons innervating different muscles (30), presumably because they discharge at different rates and in different patterns. In addition, as discussed above, high- and low-threshold motoneurons innervating a single muscle maintain different discharge rates and patterns and, not surprisingly, the biophysical properties of their axons are graded with the threshold for stimulation (31,32). In particular, there is greater activity of HCN channels on the motor axons of lowest threshold to electrical stimulation (i.e. the larger, faster-conducting motor axons). If these differences extend throughout the axon and its terminals, fasciculation would be more likely to arise in the more excitable larger motor units. The graded differences in biophysical properties also have implications for some motor unit number estimation (MUNE) techniques, particularly those that rely on the assumption that the properties of a select group of motor units are representative of the pool. Not only does motor unit size (and CMAP amplitude) vary systematically with

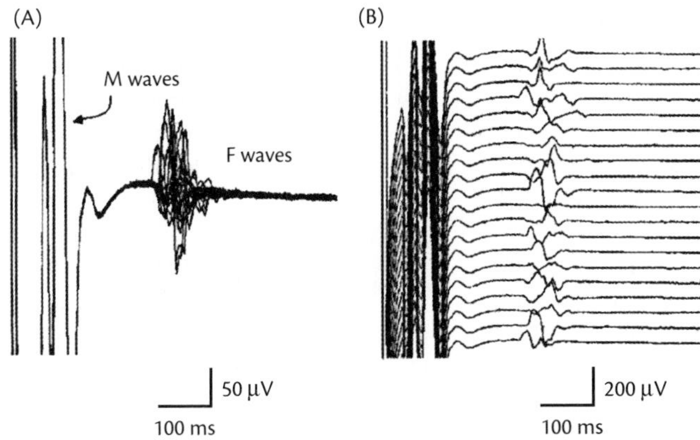

Fig. 2.4 F-waves. F waves of the thenar muscles. (A) Superimposition of 20 consecutive responses. (B) Raster display of the same 20 responses. Note the variability of waveform and latency, and in particular the potential for phase cancellation if the raw traces were averaged. High-pass filter 100 Hz.
Reproduced from Pierrot-Deseilligny E and Burke D, *The Circuitry of the Human Spinal Cord. Its Role in Motor Control and Movement Disorders*, copyright (2005), with permission from Cambridge University Press.

motoneuron size and axonal conduction velocity, but the properties of the axons differ.

F waves

When an axon is stimulated electrically, impulses are conducted in both directions from the stimulation site, orthodromically to the muscle, producing the MUAP (also known as the M wave), and antidromically back to the motoneuron. On 2–5% of occasions, the antidromic impulse produces a 'backfiring' of the motoneuron and this results in an orthodromic impulse being conducted in that axon back to the muscle. This produces a late discharge, at ~30 ms for the intrinsic muscle of the hand, due to antidromic conduction to the motoneuron, the 'turn-around' time, and then orthodromic conduction from motoneuron to muscle.

F waves produced by a sequence of 10–20 stimuli vary in latency, onset polarity, and morphology for most muscles (Fig. 2.4), and this indicates that there is no simple summation of motor unit action potentials into the CMAP. This is relevant for some MUNE techniques. It also implies that overall F-wave activity cannot be measured by averaging raw EMG traces to a sequence of stimuli—the traces need to be full-wave rectified before averaging and, before rectification, it is essential that the EMG trace has returned to baseline after the M wave (33).

While many different measures of F wave activity have been advocated, the best validated is the shortest latency of a sequence of F waves. Given a representative sample, this measure reflects the conduction time for the fastest (largest) motor axons in the pool, the motor axons that are responsible for conventional conduction velocity measurements. The spread of F wave latencies ('chronodispersion') is also often reported (as are mean and maximal F wave latencies), justified because all motoneurons can generate F waves (34). However, motoneurons will do so only if they do not discharge reflexly in response to the intense afferent volley and, in practice, F waves are generated mainly by motoneurons with rapidly conducting axons (33,35). This is presumably because reflex discharges in low-threshold motoneurons will prevent the antidromic motor volleys from generating the F wave discharges (see Fig. 3.7, Chapter 3). Either way, F wave studies are an important component of diagnostic testing because they provide information about the whole length of fast motor axons, from motoneuron to muscle. Such studies complement routine motor nerve conduction studies, which are commonly performed only on the more accessible distal segments.

F waves are often used in motor control studies to control for the excitability of the motoneuron pool. This practice is flawed (36), as discussed in Chapter 3.

Motor unit number estimates

Most techniques used to estimate the number of surviving motor units in a given muscle rely on EMG recordings and dividing the maximal CMAP by the 'mean' MUAP (Fig. 2.5). It is rare for a distal nerve to innervate only one muscle and, for example, for the median-innervated thenar muscles, some median-innervated motor units may be relatively deep or relatively remote from recording electrodes over abductor pollicis brevis. There will be significant phase cancellation when single MUAPs are summed (Fig. 2.3) and, given the issues raised in the previous section, it can be concluded that incremental stimulation methods of MUNE are intrinsically flawed. It is also possible to perform similar measurements using the force produced by the twitch contraction of the muscle, but again the twitch forces for the different motor units in a pool vary greatly from small to large, and the transducer needs to be able to resolve the forces produced in different axes by the various muscles innervated by the stimulated nerve.

A number of MUNE techniques have been advocated: the incremental stimulation method in Fig. 2.5, multiple point stimulation, spike-triggered averaging, and versions of the statistical method introduced by Daube (37). These techniques are reviewed elsewhere (38,39). Other techniques with various advantages have been reported or are under development (4, 40–42). In myopathic diseases, the decrease in size of the single motor unit means that small MUAPs may be close to the noise level for detection, and incremental techniques cannot be recommended. However, in neuropathic diseases, such as motor neuron disease/amyotrophic lateral sclerosis (ALS), the loss of motor units is accompanied by enlargement of the surviving MUAPs, improving the signal-to-noise ratio. However, this is partially offset by the greater instability of individual MUAPs in ALS.

Clinical utility does not require that a measure be accurate in absolute terms—merely that changes in the measure mirror changes in clinical status, and that the technical limitations are recognized by clinicians. In ALS, changes in MUNE can provide good evidence of progression of disease (40,41), even if the absolute values are debatable. 'MUNE values were the most sensitive index for documenting changes in disease progression over time', more sensitive than other measures, including clinical assessment and handgrip dynamometry (43).

Muscle

In mammals, each muscle fibre is innervated by a single motor axon, any polyneuronal innervation being eliminated during the neonatal period (44). The number of muscle fibres innervated by a motor axon (the 'innervation ratio') varies with the muscle: <20 for extraocular muscles, 100–200 for the intrinsic muscles of the hand and >1000 for medial gastrocnemius. The number of motor

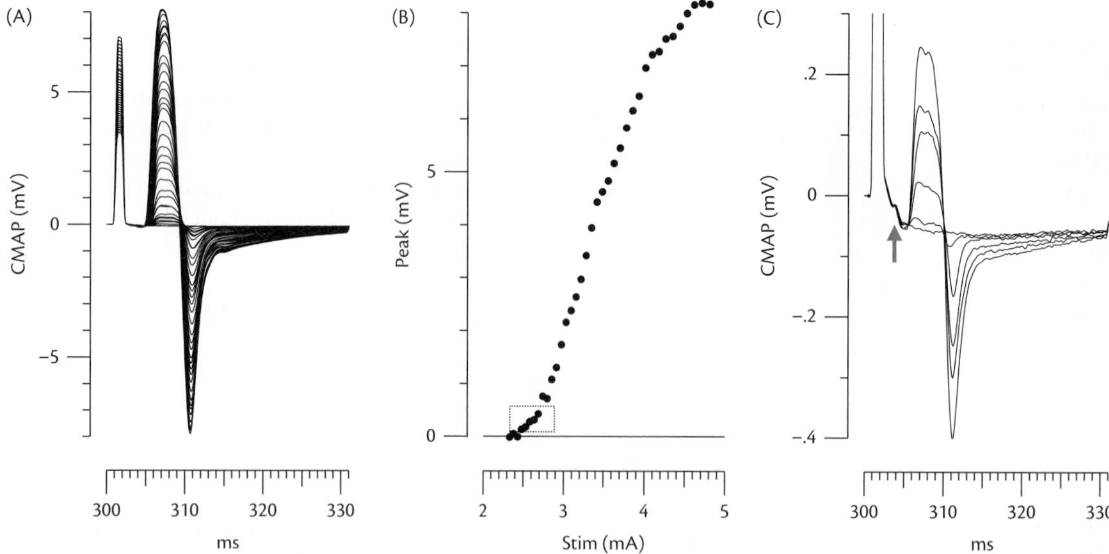

Fig. 2.5 CMAP to incremental stimulation. (A) Compound muscle action potentials recorded over the thenar eminence in response to graded stimuli to the median nerve at the wrist. (B) Baseline-to-peak measurements of the CMAP shown in (A). (C) Fine detail of the recruitment of low-electrical threshold motor units into the compound response. Note that with this scaling the antidromic volley in sensory axons in the digital nerve can be seen in recordings over abductor pollicis brevis (APB), and is indicated by the red arrow.

axons innervating any one muscle also varies, from a hundred or more for individual intrinsic muscles of the hand to 500–600 for medial gastrocnemius.

Fibre types

Muscle fibres are classified as type I or II by their histochemical profile in muscle biopsy tissue, but in diagnostic biopsies it is not possible to distinguish between fibres belonging to different motor units of the same type. In rat muscle, tetanisation of single motor axons allowed Edström and Kugelberg (45) to demonstrate glycogen depletion in muscle fibres innervated by a single motor axon, and thereby to identify the cross-sectional area of muscle occupied by the glycogen-depleted fibres. These pioneering studies demonstrated that the muscle fibres belonging to a single motor unit were scattered over a substantial cross-section of the muscle, 15–20%, with fibres from different motor units intermixed, so that a gradually rising 'ramp' contraction of the muscle would have a smooth force profile. It is unusual for more than two muscle fibres from the same motor unit to be immediately adjacent one another. These findings in animals have been confirmed for human muscle using single fibre EMG and scanning EMG techniques by Stålberg and colleagues (46).

Physiological properties

Motor units can also be classified by the physiological properties of the innervated muscle fibres. Physiological and histochemical properties are causally matched, but the two approaches involve different terminologies, dependent on the approach used to classify the muscle fibres. For example, large motoneurons have large, fast-conducting motor axons and tend to innervate 'white' muscle fibres, which produce twitch contractions that are fast and of high amplitude (Fig. 2.1, lower unit). However, these twitch contractions are more susceptible to fatigue when activity is sustained (hence type FF [fast-twitch fatiguable] motor units.) The fibres in the motor unit are rich in glycolytic enzymes (and the unit is therefore also termed FG [fast-twitch glycolytic], with histochemically type II muscle fibres). Small motoneurons have smaller, more slowly conducting motor axons that innervate relatively few 'red' muscle fibres (Fig. 2.1, upper unit). Together, they produce twitch contractions that are slow and of low amplitude, but are resistant to fatigue (hence type S [slow] motor units). These muscle fibres are rich in oxidative enzymes (and are therefore also termed SO [slow oxidative] motor units, and the muscle fibres are histochemically referred to as type I). In between these two extremes there are large motoneurons that innervate fast-twitch, but fatigue-resistant muscle fibres (FR), containing oxidative and glycolytic enzymes (FOG units). In most studies on human subjects, the physiological properties are continuously graded, rather than falling into two or three distinct entities.

The slowly contracting 'red' muscle fibres of the first recruited type S motor units underlie steady sustained contractions, such as those involved in maintaining a posture. They are red because of their myoglobin content and blood supply, and they have more mitochondria than 'white' muscle fibres. The fast 'white' muscle fibres of later recruited type F motor units have more sarcoplasmic reticulum and this allows them to take up and release Ca^{2+} more rapidly. Accordingly, these fibres are designed more for rapid movement and brief bursts of strength. However, lactic acid builds up more in these fibres and they are more prone to fatigue.

Muscle twitches, tetanic contractions, and voluntary force

The twitch contractions of the first voluntarily recruited motor units have a slow contraction time (~100 ms in human muscles, measured from the EMG potential to the peak of the evoked force) and a slow half-relaxation time (~150 ms, measured from the force peak to the time when force has decreased by 50% from its peak) and produce relatively small forces. Later recruited motor units have contraction times around 50 ms and faster half-relaxation times and produce greater force. When repetitively stimulated the force produced by a single twitch increases greatly in the so-called 'staircase phenomenon', perhaps by 50% in response to stimuli at low rate (~2/s) for 1–2 min (47). With higher stimulation rates, the individual twitch contractions summate, ultimately 'fusing' to give a smooth force profile (a 'tetanic' contraction). The tetanic fusion frequency is lower for the low-threshold slow-twitch units than for higher-threshold fast twitch units, as expected given their slower twitch profile. When motor units are activated asynchronously, the frequency of stimulation required to produce a smooth force profile is lower.

With continued activation the force produced by the active motor units begins to decline, and does so more rapidly and more extensively for fast-twitch motor units. Whether fatigue is due to a failure of the drive to the muscle or a failure of the muscle to maintain its strength can be assessed in patients using the 'twitch interpolation technique', originally developed by Merton (48). In this technique, a supramaximal stimulus is delivered to the motor nerve innervating the target muscle, while the subject maintains a contraction of that muscle using maximal effort. If the interpolated stimulus produces a detectable increase in force, the contraction was not the maximum that could be produced by the muscle. If there is no detectable twitch, the fatigue resides in the machinery of the contracting muscle fibres. This procedure has been refined by Gandevia and colleagues (for review, see (49)), who have extended it to study fatigue occurring at different levels of the neuraxis, up to motor cortex (e.g. 50).

While the complaint of 'fatigue' commonly raises concerns about failure of transmission at the neuromuscular junction due to myasthenia gravis or a myasthenic syndrome, these conditions are relatively rare. The term 'fatigue' is non-specific and can result from many conditions—ranging from psychological disinclination to exert maximal effort at one extreme to weakness due to a failure of the force-producing machinery at the other. Any cause of weakness can be associated with the perception by the patient that they fatigue more easily than they should. In part, this is because they require greater effort to perform routine tasks and, presumably, also because they then use more fatiguable motor units. Objective methods have been developed, not only for testing transmission across the neuromuscular junction (see Chapter 27), but also for other causes of the failure of a contracting muscle to maintain a steady force level (see 49–51), but these are outside the scope of this chapter.

Muscle fibre conduction velocity and the velocity recovery function

Depolarization of a muscle fibre produces an action potential that propagates slowly along the muscle fibre, the term 'propagation velocity' being generally used, rather than 'conduction velocity'. In biceps brachii of healthy subjects at 36.5°C, Buchthal and colleagues (52) reported values of 4.02 ± 0.13 m/s. Normally, the action potential propagates in both directions from the neuromuscular junction and, as a result, the potential recorded by electrodes at or over the motor point will begin with a negative deflection. If the recording electrode is not over the motor point, the initial deflection will be positive, as the action potential propagates towards the recording electrode, and then negative as it moves away (Fig. 2.6).

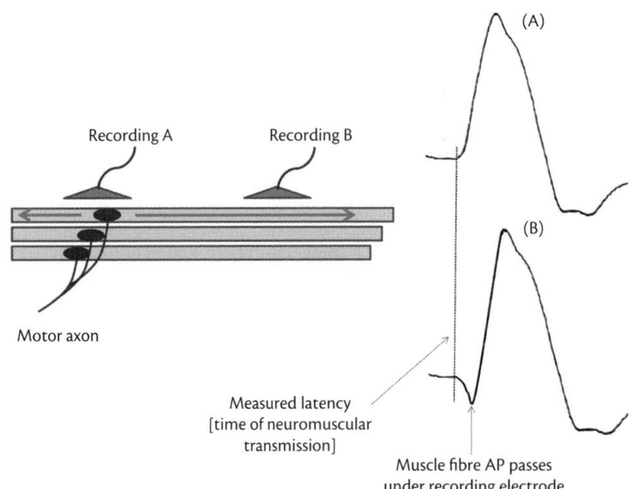

Fig. 2.6 CMAP latency. (A) When the active recording electrode is over the motor point, the muscle fibre action potential propagates along the fibre in both directions away from the electrode. The CMAP has a negative (upwards) onset. (B) When the recording electrode is not over the motor point propagation of the action potential from the motor point to the recording electrode produces an initial positivity (downwards) followed by negativity as it propagates along the muscle fibre away from the electrode.

The important measure in motor conduction studies is the time of neuromuscular transmission, not the time that the action potential passes under the recording electrode, and the correct latency to measure is the latency of the first deflection from baseline. There are two exceptions to this rule. First, sometimes the initial positive deflection is a far-field potential from another muscle innervated earlier by the stimulated nerve, not the target muscle. Secondly, stimulation can produce an antidromic volley in sensory axons that may be seen, particularly in recordings over the thenar eminence when the amplification is high (see the red arrow in Fig. 2.5).

The propagation velocity is dependent on the diameter of the muscle fibre, and it decreases in amplitude and duration in diseases that lead to muscle fibre atrophy, even in disuse atrophy. As a result, the MUAP may become smaller and have a less smooth profile because of less complete summation of the individual muscle fibre potentials that make up the MUAP. This also occurs in endocrine and metabolic causes of muscle disease, where the changes in MUAP morphology are often suggestive of myopathy, but not diagnostic.

There are changes in propagation velocity during normal activity (i.e. when muscle fibre dimensions do not change). The excitability of the muscle fibre membrane changes after an action potential, with phases of refractoriness and supernormality reminiscent of those seen with axons (46,53). More recently, techniques for studying the recovery of excitability of the muscle fibre membrane after impulses have been extended using brief trains of impulses, the underlying principles validated and the reproducibility of recordings confirmed (54,55). These studies provide novel insights *in vivo* into membrane properties and its abnormalities in patients with primary muscle disease, whether genetic or acquired. With fatigue, there is a slowing of the propagation velocity and this largely accounts for the shift to lower frequencies of the power spectrum of the EMG activity of the fatiguing muscle (56).

References

1. Henneman, E., Somjen, G., and Carpenter, D.O. (1965). Functional significance of cell size in spinal motoneurons. *Journal of Neurophysiology*, **28**, 560–80.
2. Masuda, T. and Sadoyama, T. (1987). Skeletal muscles from which the propagation of motor unit action potentials is detectable with a surface electrode array. *Electroencephalography and Clinical Neurophysiology*, **67**, 421–7.
3. Kleine, B.U., Schumann, N.P., Stegeman, D.F., and Scholle, H.C. (2000). Surface EMG mapping of the human trapezius muscle: the topography of monopolar and bipolar surface EMG amplitude and spectrum parameters at varied forces and in fatigue. *Clinical Neurophysiology*, **111**, 686–93.
4. van Dijk, J.P., Blok, J.H., Lapatki, B.G., van Schaik, I.N., Zwarts, M.J., and Stegeman, D.F. (2008). Motor unit number estimation using high-density surface electromyography. *Clinical Neurophysiology*, **119**, 33–42.
5. Sherrington, C.S. (1906). *The integrative action of the nervous system*. New Haven, CT: Yale University Press.
6. Pierrot-Deseilligny, E. and Burke, D. (2012). *The circuitry of the human spinal cord: spinal and corticospinal mechanisms of movement*. New York, NY: Cambridge University Press.
7. Garnett, R. and Stephens, J.A. (1981). Changes in the recruitment threshold of motor units produced by cutaneous stimulation in man. *Journal of Physiology*, **311**, 463–73.
8. Rexed, B. (1952).The cytoarchitectonic organization of the spinal cord in the cat. *Journal of Comparative Neurology*, **96**, 414–95.
9. Coombs, J.S., Eccles, J.C., and Fatt, P. (1955). The specific ionic conductances and the ionic movements across the motoneuronal membrane that produce the inhibitory post-synaptic potential. *Journal of Physiology*, **130**, 326–73.
10. Stuart, G.J. and Redman, S.J. (1992). The role of $GABA_A$ and $GABA_B$ receptors in presynaptic inhibition of Ia EPSPs in cat spinal motoneurones. *Journal of Physiology*, **447**, 675–92.
11. ElBasiouny, S.M., Schuster, J.E., and Heckman, C.J. (2010). Persistent inward currents in spinal motoneurons: important for normal function but potentially harmful after spinal cord injury and in amyotrophic lateral sclerosis. *Clinical Neurophysiology*, **121**, 1669–79.
12. Manuel, M. and Zytnicki, D. (2011). Alpha, beta and gamma motoneurons: Functional diversity in the motor system's final pathway. *Journal of Integrative Neuroscience*, **10**, 243–76.
13. Li, Y., Gorassini, M.A., and Bennett, D.J. (2004). Role of persistent sodium and calcium currents in motoneuron firing and spasticity in chronic spinal rats. *Journal of Neurophysiology*, **91**, 767–83.
14. Nickolls, P., Collins, D.F., Gorman, R.B., Burke, D., and Gandevia, S.C. (2004). Forces consistent with plateau potentials evoked in patients with chronic spinal cord injury. *Brain*, **127**, 660–70.
15. Murray, K.C., Nakae, A., Stephens, M.J., et al. (2010). Recovery of motoneuron and locomotor function after spinal cord injury depends on constitutive activity in $5\text{-}HT_{2C}$ receptors. *Nature Medicine*, **16**, 694–700.
16. Hennings, K., Kamavuako, E.N., and Farina, D. (2007). The recruitment order of electrically activated motor neurons investigated with a novel collision technique. *Clinical Neurophysiology*, **118**, 283–91.
17. Kiernan, M.C., Burke, D., Andersen, K.V., and Bostock, H. (2000). Multiple measures of axonal excitability: a new approach in clinical testing. *Muscle & Nerve*, **23**, 399–409.
18. Kiernan, M.C., Lin, C.S-Y., Andersen, K.V., Murray, N.M.F., and Bostock, H. (2001). Clinical evaluation of excitability measures in sensory nerve. *Muscle & Nerve*, **24**, 883–92.
19. Cappelen-Smith, C., Kuwabara, S., Lin, C.S-Y., Mogyoros, I., and Burke, D. (2001). Membrane properties in chronic inflammatory demyelinating polyneuropathy. *Brain*, **124**, 2439–47.
20. Macefield, G., Gandevia, S.C., Burke, D. (1989). Conduction velocities of muscle and cutaneous afferents in the upper and lower limbs of human subjects. *Brain*, **112**, 1519–32.
21. Hopf, H.C. (1962). Untersuchungen uber die Unterschiede in der Leitgeschwindigkeit motorischer Nervenfasern beim Menschen. *Deutsche Zeitschrift für Nervenheilkunde*, **183**, 579–89.
22. Thomas, P.K., Sears, T.A., and Gilliatt, R.W. (1959). The range of conduction velocity in normal motor nerve fibres in the small muscles of the hand and foot. *Journal of Neurology, Neurosurgery & Psychiatry*, **22**, 175–81.
23. Ingram, D.A., Davis, G.R., and Swash, M. (1987). Motor nerve conduction velocity distributions in man: results of a new computer-based collision technique. *Electroencephalography and Clinical Neurophysiology*, **66**, 235–43.
24. Dorfman, L.J. (1984). The distribution of conduction velocities (DCV) in peripheral nerves: a review. *Muscle & Nerve*, **7**, 2–11.
25. Bostock, H., Cikurel, K., and Burke, D. (1998). Threshold tracking techniques in the study of human peripheral nerve. *Muscle & Nerve*, **21**, 137–58.
26. Howells, J., Trevillion, L., Bostock, H., and Burke, D. (2012). The voltage dependence of I_h in human myelinated axons. *Journal of Physiology*, **590**, 1625–40.
27. Lin, C.S-Y., Chan, J.H.L., Pierrot-Deseilligny, E., and Burke, D. (2002). Excitability of human muscle afferents studied using threshold tracking of the H reflex. *Journal of Physiology*, **545**, 661–9.
28. Mogyoros, I., Kiernan, M.C., and Burke, D. (1996). Strength-duration properties of human peripheral nerve. *Brain*, **119**, 439–47.
29. Kiernan, M.C., Lin, C.S-Y., and Burke, D. (2004). Differences in activity-dependent hyperpolarization in human sensory and motor axons. *Journal of Physiology*, **558**, 341–9.
30. Kuwabara, S., Cappelen-Smith, C., Lin, C.S-Y., Mogyoros, I., Bostock, H., and Burke, D. (2000). Excitability properties of median and peroneal motor axons. *Muscle & Nerve*, **23**, 1365–73.
31. Shibuta, Y., Nodera, H., Mori, A., Okita, T., and Kaji, R. (2010). Peripheral nerve excitability measures at different target levels: the effects of aging and diabetic neuropathy. *Journal of Clinical Neurophysiology*, **27**, 350–7.
32. Trevillion, L., Howells, J., Bostock, H., and Burke, D. (2010). Properties of low-threshold motor axons in the human median nerve. *Journal of Physiology*, **588**, 2503–15.
33. Espiritu, M.G., Lin, C.S-Y., and Burke, D. (2003). Motoneuron excitability and the F wave. *Muscle & Nerve*, **27**, 720–7.
34. Kimura, J., Yanagisawa, H., Yamada, T., Mitsudome, A., Sasaki, H., and Kimura, A. (1984). Is the F wave elicited in a select group of motoneurons? *Muscle & Nerve*, **7**, 392–9.
35. Guiloff, R. and Modarres-Sadeghi, H. (1991). Preferential generation of recurrent responses by groups of motor neurons in man: conventional and single unit F wave studies. *Brain*, **114**, 1771–801.
36. Burke, D. (2014). Inability of F waves to control for changes in the excitability of the motoneurone pool in motor control studies. *Clinical Neurophysiology*, **125**, 221–2.
37. Daube, J.R. (1995). Estimating the number of motor units in a muscle. *Journal of Clinical Neurophysiology*, **12**, 585–94.
38. Doherty, T., Simmons, Z., O'Connell, B., et al. (1995). Methods for estimating the numbers of motor units in human muscles. *Journal of Clinical Neurophysiology*, **12**, 565–84.
39. Bromberg, M.B. (Ed.). (20031). Motor Unit Number Estimation (MUNE): Proceedings of the First International Symposium on MUNE. *Clinical Neurophysiology*, **55**(Suppl.), 1–340.
40. Baumann, F., Henderson, R.D., Gareth Ridall, P., Pettitt, A.N., and McCombe, P.A. (2012). Quantitative studies of lower motor neuron degeneration in amyotrophic lateral sclerosis: evidence for exponential decay of motor unit numbers and greatest rate of loss at the site of onset. *Clinical Neurophysiology*, **123**, 2092–8.

41. Baumann, F., Henderson, R.D., Ridall, P.G., Pettitt, A.N., and McCombe, P.A. (2012) Use of Bayesian MUNE to show differing rate of loss of motor units in subgroups of ALS. *Clinical Neurophysiology*, **123**, 2446–53.
42. Ives, C.T. and Doherty, T.J. (2014). Intra-rater reliability of motor unit number estimation and quantitative motor unit analysis in subjects with amyotrophic lateral sclerosis. *Clinical Neurophysiology*, **125**, 170–8.
43. Felice, K.J. (1997). A longitudinal study comparing thenar motor unit number estimates to other quantitative tests in patients with amyotrophic lateral sclerosis. *Muscle & Nerve*, **20**, 179–85.
44. Brown, M.C., Jansen, J.K.S., and Van Essen, D. (1976). Polyneuronal innervation of skeletal muscle in new-born rats and its elimination during maturation. *Journal of Physiology*, **261**, 387–422.
45. Edström, L. and Kugelberg, E. (1968). Histochemical composition, distribution of fibres and fatiguability of single motor units. Anterior tibial muscle of the rat. *Journal of Neurology, Neurosurgery, & Psychiatry*, **31**, 424–43.
46. Stålberg, E., Trontelj, J.V., and Sanders, D.B. (2010). *Single fiber EMG*, 3rd edn. Fiskebäckskil: Edshagen Publishing Company.
47. Slomić, A., Rosenfalck, A., and Buchthal, F. (1968). Electrical and mechanical responses of normal and myasthenic muscle—with particular reference to the staircase phenomenon. *Brain Research*, **10**, 1–78.
48. Merton, P.A. (1954). Voluntary strength and fatigue. *Journal of Physiology*, **123**, 553–64.
49. Gandevia, S.C. (2001). Spinal and supraspinal factors in human muscle fatigue. *Physiological Reviews*, **81**, 1725–89.
50. Taylor, J.L. and Gandevia, S.C. (2001). Transcranial magnetic stimulation and human muscle fatigue. *Muscle & Nerve*, **24**, 18–29.
51. Zwarts, M.J., Bleijenberg, G., and van Engelen, B.G. (2008). Clinical neurophysiology of fatigue. *Clinical Neurophysiology*, **119**, 2–10.
52. Buchthal, F., Guld, C., and Rosenfalck, P. (1955). Propagation velocity in electrically activated muscle fibres in man. *Acta Physiologica Scandinavica*, **34**, 75–89.
53. Stålberg, E. (1966). Propagation velocity in human muscle fibers in situ. *Acta Physiologica Scandinavica*, **287**(Suppl.), 1–112.
54. Z'Graggen, W.J., Troller, R., Ackermann, K.A., Humm, A.M., Bostock, H. (2011). Velocity recovery cycles of human muscle action potentials: repeatability and variability. *Clinical Neurophysiology*, **122**, 2294–9.
55. Bostock, H., Tan, S.V., Boërio, D., and Z'Graggen, W.J. (2012). Validity of multi-fiber muscle velocity recovery cycles recorded at a single site using submaximal stimuli. *Clinical Neurophysiology*, **123**, 2296–305.
56. Arendt-Nielsen, L. and Mills, K.R. (1985). The relationship between mean power frequency of the EMG spectrum and muscle fibre conduction velocity. *Electroencephalography and Clinical Neurophysiology*, **60**, 130–4.

CHAPTER 3

Motor control: spinal and cortical mechanisms

David Burke

Introduction

Movement can be initiated without conscious action and can be modified at all levels of the neuraxis. In lower animal species, there is a remarkable capacity of the spinal cord to maintain movement and even to play a role in its initiation, knowledge that dates back to Sherrington (1,2). For example, in a study entitled 'Walking spinal carnivores', animals subjected to thoracic or lumbar transections of the spinal cord followed by post-operative exercise of the hind limbs showed remarkable functional recovery: 'Some of the animals showed a recovered ability to walk, run or jump in a coordinated manner. Recovery appeared to be directly related to amount of stimulated exercise and inversely to the age at transection. In two animals a second transection was performed, with no detriment to walking' (3). These findings have been confirmed in subsequent studies, and Rossignol and colleagues summarized their findings on the recovery of locomotion following complete spinal transaction in the cat as follows: 'The results from the spinal cats emphasize the capacity of the spinal cord, in isolation from all descending influences to generate a well-organized pattern of locomotion' (4).

With the evolution of manual skill, it is popularly believed that there has been increasing 'encephalization' of motor control, such that many of the functions that occur at spinal level in lower animals are either redundant or directed from cerebral level in humans. This belief is accentuated by the ability of transcranial stimulation to reveal only the monosynaptic corticospinal projection to spinal motoneuron pools unless specific experiments are undertaken to reveal disynaptic projections (5). However, using Sherrington's terminology from 1906 (6), the motoneurons in the spinal cord are 'the final common path' for movements of the limbs, and movements will occur only if the motoneuron pool can respond to the descending command. 'The final movement is only that part of the supraspinally derived programme that the spinal cord circuitry deems appropriate. While the capacity of the spinal cord to generate or sustain even simple movements, particularly in human subjects, is limited, the influence that it plays in shaping the final motor output should not be underestimated' (5). Conclusions that changes in motor behaviour occur at a cortical level require the demonstration that spinal mechanisms are not altered appropriately to explain the motor behaviour. This chapter will focus on motor control mechanisms in human subjects. The techniques that can be used to assess different mechanisms in humans are detailed elsewhere (5), as have their advantages and their limitations.

Cortical mechanisms

Where voluntary movements are conceived can provoke philosophical as well as scientific arguments. The primary motor cortex (M1) is located in the precentral gyrus, Brodmann's area 4 and, together with the premotor cortex and supplementary motor areas (SMA), it forms the 'motor cortex' (Fig. 3.1A). The premotor cortex is in area 6, immediately anterior to area 4, while the latter consists of the superiomedial part of area 6, extending onto the medial side of hemisphere. M1 is an important site of integration of cortical mechanisms controlling voluntary motor commands and the corticospinal system is ultimately the executive pathway for all voluntary movement. However, 'volition' does not reside in motor cortex and even simple movements are not initiated there. The premotor cortex and supplementary motor areas are higher-level areas, capable of encoding complex movement patterns and of selecting appropriate motor plans for the task at hand. The premotor area is involved in movements triggered by external cues, such as vision. It has extensive interconnections with primary sensory areas, and its outputs are channelled through M1. Lesions of the premotor cortex result in apraxia—the inability to perform a movement in the absence of paralysis. The supplementary motor area is involved in willed (internally generated) movement. Unilateral movements are associated with activity in SMA bilaterally, and this occurs even when just thinking about movement.

M1 is somatotopically organized (Fig. 3.1B), much as are the somatosensory functions of the postcentral gyrus, and there is a less well-defined somatotopic organization of both the premotor and the supplementary motor areas. Within M1, the cortical areas projecting to different motoneuron pools make up the so-called 'homunculus'. There are relatively large representations for lips, tongue and face, fingers and hand, and feet, and relatively sparse representations for more proximal limb muscles and the trunk (Fig. 3.1B). The body representation is upside down, with the hip near the vertex, the leg area extending down the interhemispheric fissure, and the face and hands extending laterally along the precentral gyrus. Accordingly, there is a distorted scale model of the human body, with greater representation for those body parts most involved in skilled movement and densely innervated by the corticospinal system.

The cells of origin of ~30% corticospinal axons are giant pyramidal neurons (Betz cells) in layer V of the primary motor cortex

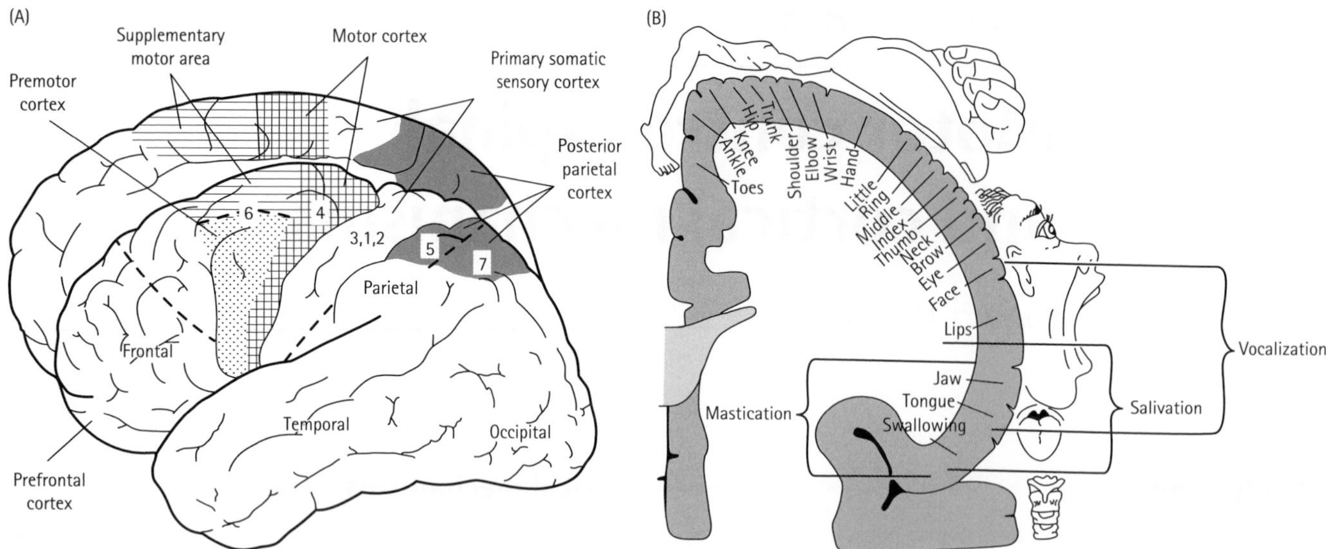

Fig. 3.1 Sketch of (A) MI, SMA, and PMA, and (B) homunculus. (A) Cerebral cortex viewed from the left with the interhemispheric extension of motor and sensory areas shown for the right hemisphere. (B) The motor homunculus. This differs from the sensory homunculus in fine details.
Reproduced from http://www.acbrown.com/neuro/Lectures/Motr/NrMotrPrmr.htm courtesy of Professor Arthur C. Brown, Oregon Health & Science University, USA.

(Fig. 3.2). Betz cells have one apical dendrite and many basal dendrites, the latter extending laterally into all cortical areas, particularly layers V and VI. Other corticospinal axons arise from the supplementary motor area, the premotor area and even the post-central 'sensory' cortical areas. The majority of corticospinal axons decussate in the medullary pyramids and then traverse the dorsolateral fasciculus as the 'lateral corticospinal tract' to innervate the motoneuron pools of limb and trunk muscles on the opposite side of the body to the originating motor cortex. It is worth noting that corticospinal axons have functions other than the activation of motoneurons, e.g. they modulate spinal reflex function (see 'Muscle mechanoreceptors and the γ efferent ("fusimotor") system' and 'Spinal mechanisms that can be tested reliably in the human subjects') and are involved in the centrifugal control of ascending sensory pathways (7), largely through separate axons, rather than axons that branch to fulfil the different roles. Relatively few axons in the corticospinal pathways (<5%) are large and fast conducting. These are the axons responsible for motor evoked

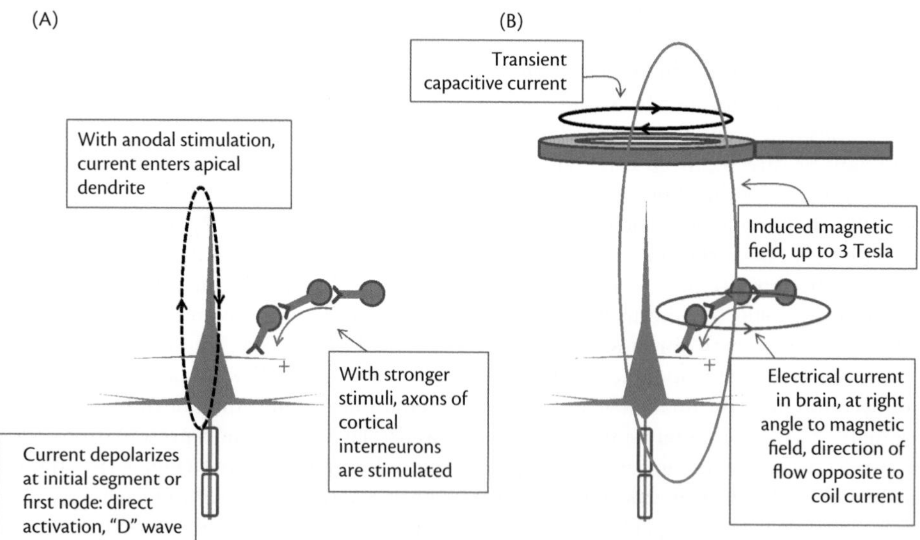

Fig. 3.2 Betz cell, and its activation by TES and TMS. The giant pyramidal cells of Betz are shown diagrammatically with a long apical dendrite and a profusion of basal dendrites directed horizontally into all cortical areas, particularly layers V and VI. (A) With TES, anodal current enters the apical dendrite and depolarizes at initial segment or first node, directly activating the corticospinal axon, producing a 'D' wave. With stronger stimuli, axons of cortical interneurons are stimulated to activate corticospinal neurons trans-synaptically, an indirect form of activation, hence 'I' waves. This produces later volleys in the same population of axons. (B) With TMS, a transient capacitive current pulse flows through a coil and induces a rapidly changing magnetic field at right angles to the coil. This produces electrical currents in cortical tissue with a current flow opposite to that in the coil, stimulating the axons of cortical interneurons at lowest intensity. Electrical fields produced by TMS are parallel to the head surface. TMS is insensitive to the skull conductivity.

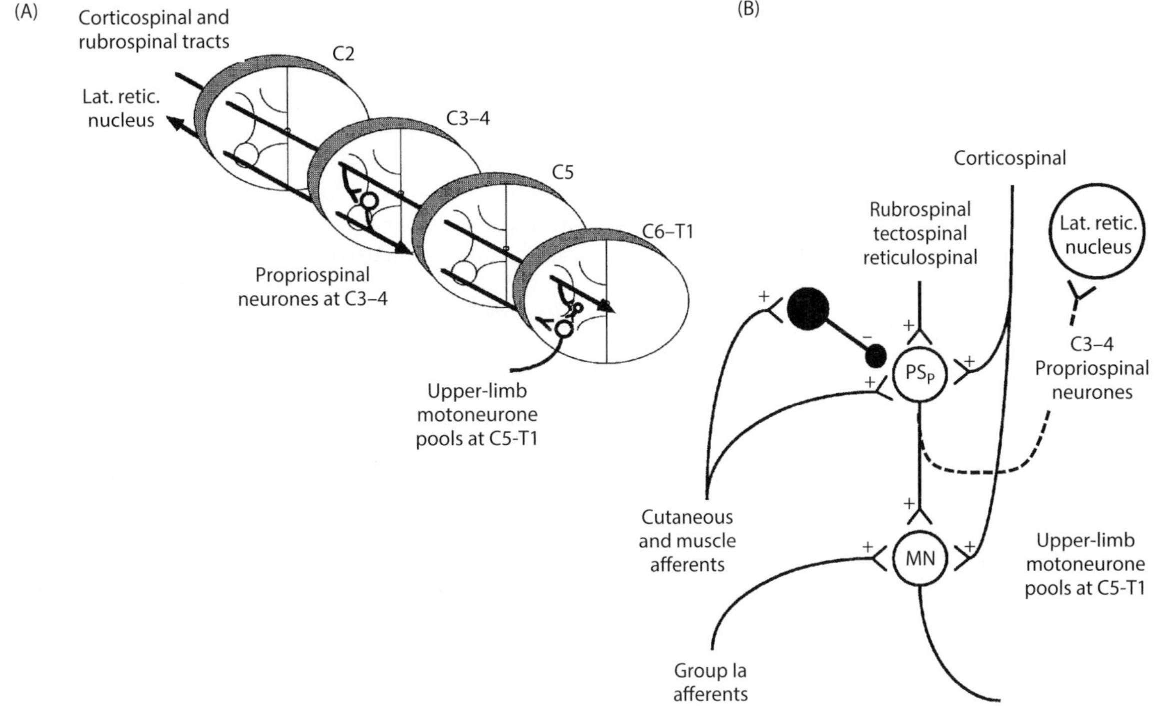

Fig. 3.3 C3–C4 propriospinal system. (A) Propriospinal neurons located at C3–4 receive monosynaptic activation from the corticospinal (and other) tracts. The axons of the neurons feed back to the lateral reticular nucleus and pass down the anterior half of the lateral column to the motoneuron pools innervating the upper limb. (B) Some synaptic connections of C3–4 propriospinal neurons: excitation from various descending systems, and excitation and inhibition from cutaneous and muscle afferents in the limb. The neurons produce monosynaptic excitation of motoneurons in the C5–T1 segments and provide feedback to the lateral reticular nucleus.

potentials. Accordingly, studies in humans using motor evoked potentials to transcranial stimulation of the motor cortex provide insight into the function of only a minority of corticospinal axons. Nevertheless, it is likely that these axons are those involved in the control of movement (or at least are representative of them).

The motor cortex receives input from the basal ganglia and cerebellum via the ventrolateral nucleus of the thalamus. It also receives direct cortico-cortical inputs from other cortical areas, including the post-central 'sensory' cortex, transcallosal, and more deeply relayed inputs from the contralateral motor cortex, and indirect inputs via the thalamus.

Monosynaptic and disynaptic corticospinal projections

In macaque monkeys (and presumably humans), the corticospinal system is the last major neural system to develop: conduction velocity matures to adult levels only during the postnatal period: 'full myelination of corticospinal axons in the spinal cord would not occur until ~36 months' (8), and this is presumably the reason why the Babinski response is a normal finding in neonates. The corticospinal projection in higher primates has a major monosynaptic projection to motoneurons, a projection that is absent in the cat and is less well developed in lower primates (7). In the cat corticospinal actions are transmitted at segmental level through one or two segmental interneurons to the target motoneurons.

In higher primates (macaque monkey) and humans, the segmental connection is monosynaptic. However there are disynaptic propriospinal projections to all tested motoneuron pools in the human upper limb (with the exception of the intrinsic muscles of the hand) through a system of neurons located at the C3-C4 levels (5,9). These C3–4 propriospinal neurons form an integration centre that transmits some of the corticospinal command for upper limb movement (Fig. 3.3), and is discussed in greater detail later ('Propriospinal transmission of the cortical command'). The propriospinally transmitted disynaptic excitatory post-synaptic potentials (EPSPs) appear in motoneurons ~1 ms after the monosynaptic corticospinal EPSP, and cannot be revealed by conventional techniques using transcranial stimulation of the motor cortex unless special conditioning-test protocols are used (5). This is largely because the corticospinal volley produced by transcranial stimulation consists of multiple components (so-called D and I waves, dependent on stimulus type, orientation, and intensity—Figs 3.4 and 3.5), and the motoneuron discharge usually involves temporal summation.

Other descending motor pathways

The contralateral projection of the corticospinal system is the major, but not the sole pathway responsible for voluntary movement. In some humans, up to 30% of corticospinal fibres may be uncrossed, descending in the ventral funiculus on the same side as the cortex of origin. At least some of these axons cross in the spinal cord, but their final destinations are not clear. Other pathways have been implicated in the recovery from, e.g. stroke (10). There are ipsilateral projections of the corticospinal tracts to motoneurons via reticulospinal neurons, projecting through segmental interneurons to spinal motoneurons. Cortico-reticulo-spinal pathways may receive projections from both right and left motor cortices (11).

There are a number of other descending pathways that could also be valuable in the recovery of movement after corticospinal lesions. *Rubrospinal pathways* are important pathways involved in voluntary movement in animals, but are rudimentary in humans. The rubrospinal tract arises from the red nucleus, crosses to the opposite side in the brainstem and projects to the cervical spinal cord together with the corticospinal system. It relays information from the cerebellum and striatum predominantly to upper limb flexor motoneurons. *Tectospinal pathways* are involved in movements of the head and eyes to visual stimuli. They arise from the superior colliculus, and decussate to coordinate contralateral head and eye movements. The *vestibular system* and its connections are designed to maintain head and eye coordination, and support balance and the upright posture. The vestibular nuclei have projections cranially via the medial longitudinal fasciculus to the oculomotor nuclei and caudally to the motoneuron pools of cervical, truncal, and lower-limb segments. Stabilization of the head in space and head/eye coordination is facilitated by the *medial vestibulospinal tract*, which projects from the medial vestibular nucleus down the anterior funiculus of the spinal cord to upper cervical motoneurons. The projection is bilateral and cervical only. It underlies the cervical vestibular evoked myogenic potential (cVEMP; (12)). The *lateral vestibulospinal tract* is unilateral and uncrossed, projecting to interneurons and motoneurons controlling antigravity muscles. It is probably involved in the increased muscle tone responsible for the 'hemiplegic posture' (13).

Corticospinal volleys and motor evoked potentials produced by transcranial stimulation

Transcranial stimulation of M1 can be produced by high-voltage electrical stimuli ('transcranial electrical stimulation' (TES)) or using a 'magnetic' coil ('transcranial magnetic stimulation' (TMS); see also Chapter 14). The latter method is relatively free of discomfort and has now superseded TES for clinical purposes. A number of different coils is available for transcranial magnetic stimulation, developed to focus the stimulation (figure-of-eight or butterfly coil) or to penetrate deeper, thereby accessing lower-limb cortical areas better (double-cone coil; H-shaped coil). The interested reader is referred to reviews for further detail (14–19). Safety issues are covered by Rossi et al. (20), and current diagnostic usage by Groppa et al. (21). The corticospinal volleys evoked in awake unanaesthetized patients with epidural leads for the control of pain are reviewed by Di Lazzaro (18). A checklist has been developed to help readers assess the quality of published articles using TMS (22). The following text will focus on the innervation of the upper limb.

Only at threshold does a single transcranial stimulus produce a single volley in corticospinal axons. With *transcranial electrical stimulation*, anodal stimulation is used (Fig. 3.2A), and at threshold corticospinal axons are stimulated directly at the axon hillock or the first nodes of Ranvier, producing a 'D wave'. With increases in stimulus intensity later waves appear at longer latency (Fig. 3.4), due to activation of cortical interneurons which project to the corticospinal neuron (23). These late higher-threshold waves are termed 'I waves' because they involve indirect transynaptic activation of the corticospinal axon. Where identifiable at two levels, the D and I wave components have much the same conduction velocity in the spinal cord (Fig. 3.5), indicating that they arise at

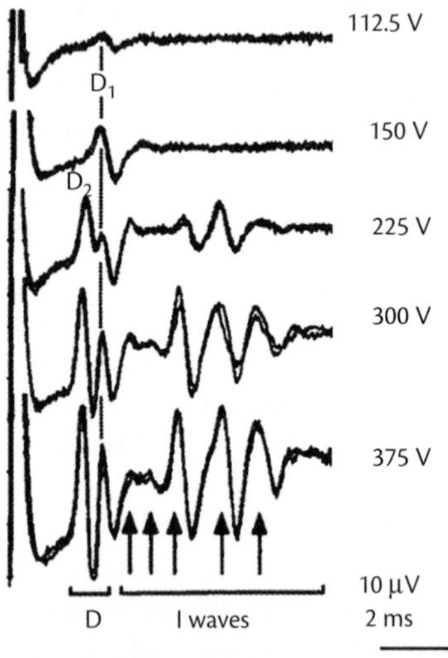

Fig. 3.4 D and I waves with increasing stimulus level. Corticospinal volleys produced by transcranial electrical stimulation of increasing strength, recorded at the low-cervical level using epidural leads. The lowest threshold component (112.5 V) represents the D wave. With stronger stimuli components of longer latency appear and grow with stimulus strength. These are I waves. In addition, the D wave becomes bifid as some corticospinal axons are activated at a subcortical site.

Reproduced from *J Physiol*, **456**, Hicks R, Burke D, Stephen J, Woodforth I, Crawford M, Corticospinal volleys evoked by electrical stimulation of human motor cortex after withdrawal of volatile anaesthetics, pp. 393–404, copyright (1992), with permission from John Wiley and Sons.

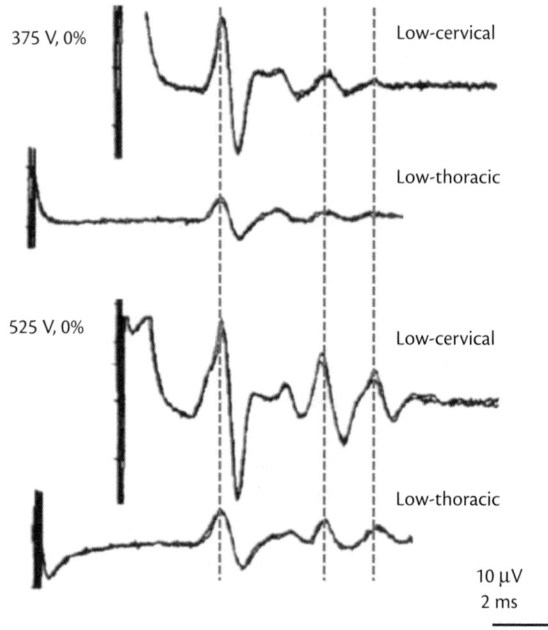

Fig. 3.5 D and I waves have same conduction velocity. Corticofugal volleys evoked by vertex anodal stimulation at two stimulus intensities, 375 and 525 V (female, age 15). Recordings of the descending volley were made at a low-cervical level (upper traces) and at a low-thoracic level (lower traces), the latter offset to align the peaks of the D waves. The I waves have the same latencies after the D waves at the two levels despite the longer conduction time to the low-thoracic level. This indicates that the conduction velocity of the axons in the I waves was the same as for those in the D wave.

Adapted from *J Physiol*, **456**, Hicks R, Burke D, Stephen J, Woodforth I, Crawford M, Corticospinal volleys evoked by electrical stimulation of human motor cortex after withdrawal of volatile anaesthetics, pp. 393–404, copyright (1992), with permission from John Wiley and Sons.

different, progressively 'higher' levels within the corticospinal system as latency increases.

Dependent on coil orientation transcranial magnetic stimulation at threshold can excite I waves in isolation (18). As a result, if there is a change in the response to threshold magnetic stimulation, but not to threshold anodal stimulation, it is likely that the change occurred in the motor cortex, a conclusion that is secure only with threshold stimuli. Over recent years, this use of transcranial electrical stimulation has decreased as researchers have opted to exclude a spinal locus of action, rather than prove that the change in excitability occurred at a cortical site. This is regrettable because tests that exclude a phenomenon are less convincing than tests that establish a phenomenon (and also because the available tests are often not conclusive; see 'F waves'). Transcranial electrical stimulation is now used mainly for intraoperative monitoring of the integrity of corticospinal pathways during neurosurgical and orthopaedic operations. This involves anaesthetized patients for whom discomfort due to the stimulus is not an issue.

With TMS, a transient large capacitive pulse of current flows through a coil held over the subject's head (Fig. 3.2B). This induces a rapidly changing magnetic field of up to 3 Tesla at right angles to the coil. In turn, this produces electrical currents in cortical tissue with a current flow opposite to that in the coil, stimulating the axons of cortical interneurons at lowest intensity (18). Activating these axons results in transynaptic activation of the corticospinal neurons, an indirect activation that produces I waves. As with TES, I waves are conducted in the same population of axons and have the same conduction velocity as the D wave, but because they have a longer latency, the latency of the motor evoked potential is ~2 ms longer with threshold TMS than with threshold TES.

With TMS, the effective stimulus site is anywhere under the annulus of the coil. This renders focal stimulation virtually impossible with the 11-cm circular coil, but the figure-of-eight coil effectively doubles the strength of the magnetic field under the crossing so that, at threshold, cortical regions can be mapped. A clockwise direction of current flow in the coil is optimal for stimulation of the right hemisphere and counter-clockwise for the left hemisphere. Using the figure-of-eight coil, 'The largest responses were obtained with the coil at about 50° to the parasagittal plane with a backward flowing inducing current', producing an opposite current flow in the cortex, i.e. posterior-anterior (24). Epidural recordings of the corticospinal volleys suggest that different cortical elements are activated by posterior-anterior and anterior-posterior current flows (18).

The motoneuron discharge in response to transcranial stimulation usually requires temporal summation of successive components of the descending volley. As a result, the first motoneurons in the evoked discharge may be activated some milliseconds after the onset of corticospinal excitation, and the discharge of different motoneurons will be dispersed. This has led to the development of a *triple-stimulation technique* (25), which uses collision to minimize the desynchronization of the compound electromyograph (EMG) potential.

Based on the motor evoked potential (MEP) to TMS and given a measure of conduction in the peripheral pathway, 'central motor conduction time' (CMCT) can be estimated. Two measures of peripheral conduction have been advocated. Both provide approximate values only. High-voltage (cathodal) electrical or magnetic stimulation over appropriate spinous processes in the cervical region activates axons at the intervertebral foramen (if not so strong that the point of stimulation moves more distally). Subtraction of the latency of the compound muscle action potential (CMAP) so evoked from the latency of the MEP to TMS provides the estimate of CMCT. However, this measure overestimates CMCT because it includes conduction in the spinal and proximal portions of the motor root. These sites could be sites of pathology. The F wave technique involves measuring the fastest F wave for the test muscle (usually intrinsic muscles of the hand or foot), adding the M-wave latency, subtracting a nominal 1 ms (for the turnaround time at the motoneuron, see Chapter 2), and dividing the residuum by two to give a 'pure' measure of motor conduction from the motoneuron to the muscle. In practice, this technique cannot be used on proximal muscles (such as biceps brachii) because of the difficulty in identifying F waves for such short segments. Both methods, but particularly the F wave technique, measure the conduction time for the fastest axons contributing to the CMAP, but their motoneurons are not those preferentially activated by corticospinal volleys. This issue will be of little importance if the MEP is a relatively large percentage of the maximal CMAP.

Cortical excitability and interneuronal mechanisms

TMS can be used to estimate the excitability of motor cortex, commonly used measures being the threshold for the MEP and the stimulus-response curve for the MEP. A number of different stimulating paradigms has been developed to provide insight into cortical interneurons and their influence on the corticospinal neuron (19), these measures have been applied to study the effects of drugs on cortical mechanisms (17), and the corticospinal volleys underlying each protocol have been documented in awake human subjects (18). The best known measures are arguably the *cortical silent period* and *short-interval intracortical inhibition* (SICI). The latter requires double-pulse stimulation, the conditioning stimulus subthreshold for a motor evoked potential and the test stimulus set to produce a MEP of predetermined amplitude (26). Different paradigms varying the conditioning-test interval and/or the intensity of the conditioning stimulus have been introduced, and allow the measurement of other facilitatory and inhibitory processes, presumably mediated by different populations of cortical interneurons.

The 'cortical silent period' is assessed during a steady contraction of the target muscle, and is manifested as EMG silence lasting 300–400 ms following the MEP produced by suprathreshold TMS. The depth and duration of the silence depends on the strength of the conditioning stimulus and the orientation of the coil. The initial part of the EMG early silence, up to ~50 ms, is largely due to spinal mechanisms, and the later suppression, after 100 ms, is due to reduced excitability of motor cortex. Presumably, the same inhibitory phenomenon underlies 'long-interval intracortical inhibition', for which a test MEP is conditioned by a suprathreshold MEP. Here, the conditioning-test paradigm allows studies in subjects at rest.

SICI consists of two superimposed waves of inhibition with peaks at conditioning-test intervals of 1 and 2.5 ms, occurring on a background of refractoriness of the stimulated axons of cortical interneurons. Synaptic mechanisms appear to be involved in both phases of inhibition, the second mediated by $GABA_A$ receptors (17). SICI has been reported to be decreased prior to the onset of movement and in stroke (decreased resting SICI and decreased modulation prior to voluntary movement, the latter correlated with prognosis), dystonia (both blepharospasm and dystonia of the affected upper limb, the latter with decreased movement-related modulation), cortical myoclonus, and amyotrophic lateral sclerosis (ALS). SICI may also be decreased in traumatic brain injury and incomplete spinal cord injury. With the exception of sleep disturbances, SICI seems always to be decreased, when affected, and the changes are clearly not disease specific. However such studies are clinically beneficial where the loss of SICI or its modulation is correlated with prognosis. In ALS abnormal SICI represents a cortical abnormality, and this can be valuable evidence of upper motor neuron dysfunction in a patient who presents with predominantly lower motor neuron signs.

In a refinement adapted from studies of axonal excitability (the technique of threshold tracking, see (28)), the test stimulus is controlled by a computer to produce a MEP of constant amplitude. Here inhibition then requires a stronger stimulus and facilitation a weaker stimulus to produce the target MEP. This technique has been used extensively by Vucic, Kiernan, and colleagues to probe SICI in ALS (decreased), presymptomatic familial ALS (may be decreased before disease onset is apparent clinically), Kennedy's disease (not decreased) and other lower motoneuron conditions mimicking ALS (not decreased; for reviews, see (29,30).

Conditioning stimulation of M1 results in robust suppression lasting up to 50 ms of the MEP produced by stimulation of opposite side, probably consisting of two separate phenomena with maxima at ~10 ms and ~40 ms. Both are probably gamma amino butyric acid B ($GABA_B$)-mediated within the M1 opposite the conditioning stimuli. However under specific conditions interhemispheric facilitation can be demonstrated at relatively short intervals (19). These interactions appear to be transcallosal. It is not known whether activating transcallosal pathways has rehabilitation benefits.

Movements of distal muscles operating on the hand and fingers are controlled from the contralateral hemisphere, but midline muscles receive bilateral innervation. For clinicians the best known example is frontalis. Projections from M1 to most ipsilateral motoneuron pools innervating the upper limb can be demonstrated during a voluntary contraction using TMS in healthy children up to the age of ~10 years, more often for proximal than distal muscles. Ipsilateral MEPs have been recorded in healthy adults, but then only during strong background contraction of the target muscle. An ipsilateral projection has been demonstrated in patients with cerebral palsy, congenital mirror movements and after hemispherectomy. The ipsilateral projections to proximal muscles are direct, through the uncrossed corticospinal tract. In hand muscles, the ipsilateral MEP is commonly associated with an ipsilateral silent period and appears at longer latency than the MEP in homologous contralateral muscles, consistent with the view that the pathway is indirect, possibly via the reticular formation. Ipsilateral MEPs in hand muscles are weaker, and may have a role in complex tasks such those involving bimanual coordination. Techniques to mobilize these pathways could be a potential rehabilitation strategy.

Plasticity of cortical mechanisms

Plasticity can occur at all levels of the nervous system, as a result of long-term changes in synaptic function. Plasticity of spinal reflex mechanisms has been extensively documented by Wolpaw's group (for a review see (31)), in mice, in healthy human subjects (32) and in patients with incomplete spinal cord injury (33). The induced changes in reflex behaviour are associated with changes in motoneuron properties, perhaps analogous to the changes in motor axon excitability that occur after stroke (see Chapter 2).

Different TMS protocols can produce changes in motor cortex excitability that significantly outlast the conditioning stimuli, and these have provoked great interest in plasticity at cerebral level and the potential for using the research protocols in rehabilitation. When adequate experiments have been undertaken, virtually all of these protocols have been shown to produce changes at other levels of the neuraxis, particularly the spinal cord. The TMS protocols have been involved repetitive afferent inputs, pairing afferent inputs and TMS ('paired associative stimulation'), repetitive TMS, both continuous trains (rTMS) and stimuli delivered in repetitive bursts ('theta burst stimulation'), transcranial direct current stimulation (tcDCS), and transcranial alternating current stimulation (tcACS). Such protocols are outside the scope of this chapter, and interested readers are referred to other works (19,27,34–38).

Muscle mechanoreceptors and the γ efferent ('fusimotor') system

The muscle spindle and the Golgi tendon organ are often grouped together as 'stretch receptors', but they transduce different properties (5). Spindle endings lie *in parallel* with the extrafusal muscle fibres and are exquisitely sensitive to stretch that produces changes in length of those fibres or the derivatives of length. A tendon organ is *in series* with the 5–10 motor units that have muscle fibres that insert into the tendon organ capsule. It is relatively insensitive to any stimulus except contractile force, unless the muscle is contracting when stretched. Golgi tendon organs are located at musculo-fascial and musculo-tendinious junctions, not within the tendon, and receive no efferent innervation. Each tendon organ gives rise to a single afferent fibre, the group Ib afferent. This axon has the same dimensions and much the same excitability to electrical stimulation as group Ia afferents from the muscle spindle.

The muscle spindle is an unusual receptor because it is innervated by motoneurons what can alter the responsiveness of the receptors on the modified muscle fibres of the muscle spindle ('intrafusal' muscle fibres). Most of these motoneurons are termed γ motoneurons because they are smaller than α motoneurons, which innervate the force-generating 'extrafusal' muscle. However, some spindles also receive innervation from larger motoneurons (so-called β motoneurons), which branch to innervate both extrafusal muscle and nearby muscle spindles.

There are two receptors on the muscle spindle, the primary spindle ending, which is sensitive to changes in the length of that region of muscle and to the derivatives of length. It gives rise to a single group Ia afferent axon. There may be a number of secondary endings in a single spindle, and they respond mainly to changes in muscle length. Each spindle gives rise to a number of group II afferents. Group Ia and Ib afferents are traditionally thought to be faster than any other in the peripheral nerve, based on findings for the hind limb of the cat. However, in humans, the fastest muscle afferents and the fastest cutaneous afferents have similar conduction velocities in both the upper limb and the lower limb (39), perhaps 50–65 m/s in the distal upper limb and 40–55 m/s in the distal lower limb. Group II afferents have a conduction velocity of roughly two-thirds that of group I afferents, ~35–40 m/s (5). The reflex effects of group II afferents are excitatory and can be particularly powerful, but as a result of the difference in conduction velocities, the group II excitation will occur later than group Ia excitation. There will inevitably be some temporal summation of group II excitation with the earlier Ia effects, whether the afferents are activated electrically or by muscle stretch. In the lower limb, medium-latency responses to muscle stretch are largely group II-mediated. In the upper limb, responses at comparable medium latencies may involve long-loop transcortical responses (40) and/or a spinal reflex of group II origin (41), although this varies with the target muscle group.

The efferent innervation of the spindle is referred to as 'fusimotor' and, whether γ or β, there are basically two types: 'dynamic' and 'static'. The former enhances the dynamic response of primary endings to stretch; the latter enhances the overall discharge of both primary and secondary endings. Dynamic γ motoneurons have been implicated in a number of physiological processes and pathologies, e.g. reinforcement of the tendon jerk by performing the Jendrassik manoeuvre, and the hyperrreflexia of spasticity, but the balance of evidence is against a role in both of these phenomena (5). Static γ motoneurons innervating a contracting muscle (and only that muscle) are activated by voluntary effort, more so the greater the effort, and this is usually sufficient to enhance the discharge of spindle endings in the contracting muscle (5,42,43). During unrestrained movements, the increase in static fusimotor drive will normally be insufficient to maintain the discharge of spindle endings when the contracting muscle shortens, unless the movement is slow or is performed against resistance. Despite considerable research (and even more speculation), the role of the fusimotor system in human movement is not yet resolved. One hypothesis is that it is more important when learning a motor skill, particularly during development, when slow precise movements may involve bracing the slowly moving joint by co-contracting antagonists (43). These are the circumstances when the fusimotor drive that accompanies voluntary contractions can provide significant muscle spindle feedback to the central nervous system about the state of the muscle.

Motoneurons, 'monosynaptic' reflexes, and F waves

Motoneuron recruitment and reflex gain

Motor unit properties and the basis of EPSPs and inhibitory post-synaptic potentials (IPSPs) are discussed in Chapter 2. In brief, motoneurons are normally recruited in order of size (44). Accordingly graded reflex inputs recruit motoneurons in the test pool in an orderly sequence, inversely related to motoneuron size, and the EPSPs and IPSPs produced by group Ia afferents are larger the smaller the motoneuron (45). This same 'size principle' applies for most inputs to the motoneuron pool in human subjects—voluntary drives, corticospinal volleys, and segmental afferent inputs are distributed homogeneously across the pool. However, recruitment according to size is not always the case, e.g. Garnett and Stephens (46) have demonstrated that cutaneous afferent volleys from the index finger can increase the threshold for voluntary recruitment of low-threshold small motor units and decrease the threshold for high-threshold large motor units. As a result cutaneous volleys can alter the sequence of motoneuron recruitment.

The motoneuron is not just a passive relay of descending and segmental commands—its properties can change. At a single motor unit level, the best documented change in responsiveness is that due to the development of 'plateau potentials' following the activation of persistent inward currents through TTX-sensitive Na^+ and L-type Ca^{2+} channels located on the dendrites (see Fig. 2.2, Chapter 2). Plateau potentials develop during normal motor behaviour, are modulated by descending monoaminergic pathways, can amplify inputs to the affected motoneuron, and can cause self-sustained firing in them. They have been implicated in the spasms and spasticity of spinal cord injury (47–49).

The H reflex and tendon jerk

These spinal reflexes are commonly considered to be virtually identical, subject to the same controlling influences, except that:

- the tendon jerk is more variable from trial to trial because it depends on tendon percussion, the efficacy of which varies with the stiffness of the muscle and tendon;
- the tendon jerk is dependent on the level of fusimotor drive.

It is debatable whether changes in fusimotor drive can change the muscle spindle afferent volley sufficiently to alter the tendon jerk (5) but, either way, the H reflex is not immune to changes in fusimotor drive. Any background fusimotor activity would increase the background activity of spindle afferents, and this would condition the motoneuron pool, raising its excitability. On the other hand, the greater background spindle discharge would result in post-activation depression of transmission at the synapse between the active Ia afferents and the motoneuron (see later), thereby decreasing the excitation of the motoneuron pool produced by the afferent volley for the H reflex. As a result, the reflex response could actually be smaller despite the increased excitability of the motoneuron pool.

There are many other differences between the tendon jerk and H reflex:

- the nature of afferents (Ia and Ib equally for H reflex; mainly Ia for tendon jerk);
- the number of discharges/afferent (one with the H reflex; multiple discharges in each activated afferent, at high frequency, for the tendon jerk);
- the dispersion of the afferent volley (much greater for the tendon jerk);
- the source of afferents (mainly muscle afferents innervating triceps surae and the intrinsic muscles of the foot for the soleus H reflex—mechanosensitive receptors in skin and muscles of the calf, pretibial flexors and foot for the Achilles tendon jerk);
- the extent of homosynaptic Ia depression (greater for the H reflex than for 'stretch reflexes' (50));
- the duration of the rising phase of the resulting EPSPs in motoneurons (2–3 ms for the H reflex; >10 ms for the tendon jerk);

- the susceptibility to di- and oligo-synaptic inputs to motoneurons (present if EPSP lasts >1 ms, as they do for both reflexes, but greater the longer the EPSP);
- the susceptibility to Renshaw feedback (present if EPSP lasts >1 ms, as they do for both reflexes, but greater the longer the EPSP).

Few of these differences are controllable, and it is unlikely that comparisons of the H reflex and the tendon jerk can ever be used as a reliable measure of fusimotor drive. Comparing two α motoneuron discharges to make conclusions about γ motoneurons is very indirect, and the reader is cautioned against this practice.

Disynaptic inhibition and recruitment of motoneurons into the H reflex

The H reflex and the tendon jerk are commonly referred to as 'monosynaptic reflexes'. There is little argument that the excitation that produces these reflexes is predominantly, if not exclusively monosynaptic excitation from group Ia muscle spindle afferents. However, as discussed below, the H reflex is susceptible to disynaptic inhibition, whereas the term 'monosynaptic reflex' carries the implication that it is immune to changes in interneuron excitability.

When the H reflex is elicited, the stimulus activates a minority of the group I afferents. A liminal EPSP can be produced in motoneurons with stimuli as low as 0.5–0.6 times the threshold stimulus for the M wave (5), but a maximal Ia volley requires a stimulus of 4 times threshold for the M wave, at least for the femoral nerve and the H reflex of quadriceps femoris (51). Fig. 3.6 shows the growth of the stimulus-response curves for the M wave and the H reflex. The curve for the M wave has a sigmoidal shape, but that for the H reflex reaches a peak and then declines as the M wave grows. The decline is largely due to the collision between the reflex discharge and the antidromic volley in motor axons, illustrated for the two smaller motoneurons on the left in Fig. 3.7.

A stimulus to the tibial nerve does not activate only group Ia afferents. Ib afferents will be recruited by any stimulus sufficient to produce detectable Ia effects, and their inhibitory input will reach the motoneuron pool ~1 ms after the group Ia input. If all Ia afferents are activated by a stimulus for the H reflex (an unlikely event), the slowest would reach the motoneuron pool ~7–8 ms after the fastest. The rising phase of the group I EPSP in soleus motoneurons lasts longer than 1 ms (<3 ms), but is still relatively brief, as if the monosynaptic EPSP were curtailed by a later-arriving inhibitory input. There is precedence for this: in intracellular recordings in the cat (figs 2 and 3 in Araki et al., (52)) and in human reflex studies (fig. 1A in (53)). Recently, it has been shown that inhibitory inputs in the group I volley for the test H reflex can curtail the excitation due to that volley (54).

When a reflex response is facilitated or inhibited, the changes involve either inhibiting motoneurons that were last recruited into the control reflex, or recruiting motoneurons that just failed to be activated in the control reflex. In other words, the motoneurons most likely to be affected by disynaptic inputs will be those likely to be recruited near the peak of the EPSP, i.e. those motoneurons more dependent on the balance between monosynaptic excitation and disynaptic inhibition. As shown by Marchand-Pauvert and colleagues (54), the 'monosynaptic' reflex can be altered by varying the amount

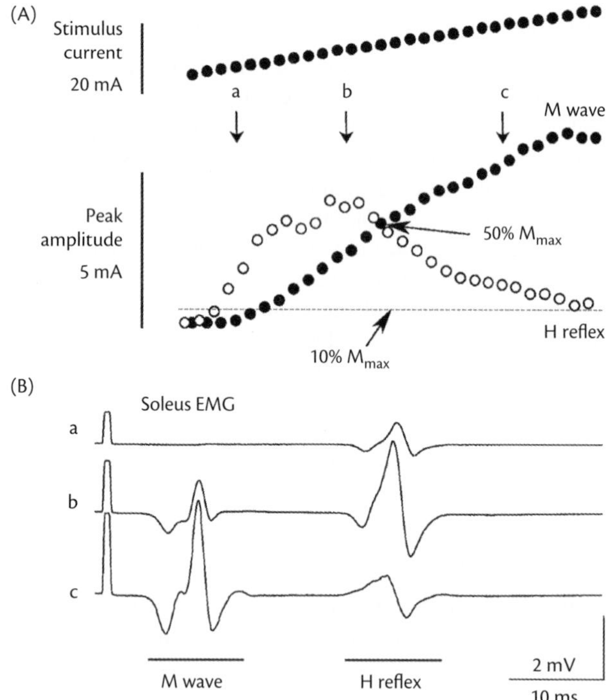

Fig. 3.6 The H reflex of soleus. (A) the top trace shows the steady increase in the stimulus current from below motor threshold to the maximal M wave (M_{max}), producing the H reflexes and M waves shown in the bottom traces in A. Note that the current needed to produce an H reflex that was 10% M_{max} was less than needed for an M wave of the same size, i.e. that the H reflex had a lower threshold than the M wave. (B) Three traces of raw data obtained at the letters indicated in (A) *a*, an H reflex, but no M wave; *b*, a larger H reflex and an M wave; *c*, a larger M wave and smaller H reflex.

Reproduced from *J Neurophysiol*, **100**(6), McNulty PA, Jankelowitz SK, Wiendels TM, Burke D, Post-activation depression of the soleus H reflex measured using threshold tracking, pp. 3275–84, copyright (2008), with permission from The American Physiological Society.

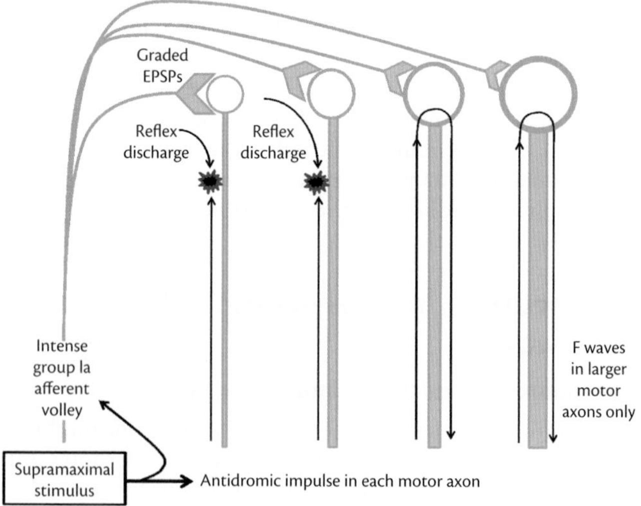

Fig. 3.7 Reflexly activated motoneurons cannot produce F waves. In human subjects, the supramaximal stimulus necessary for F wave studies will produce an intense afferent volley. If motoneurons are activated by this afferent input, the reflex discharge will collide with the antidromic volley in motor axons, thus preventing the antidromic invasion of those motoneurons (the two motoneurons on the left). As a result, F waves will occur only for higher-threshold motoneurons that have a smaller compound EPSP and do not discharge reflexly in response to the afferent volley.

of disynaptic inhibition due to Ib afferents in the test volley, i.e. the reflex may depend on monosynaptic excitation but it is not immune to interneuronally transmitted effects. Suppression of transmission across Ib inhibitory interneurons (now called non-reciprocal group I inhibitory interneurons) is one factor responsible for the potentiation of reflexes by voluntary contraction of the test motoneuron pool.

F waves

The F wave is a poor measure of motoneuron excitability, as argued elsewhere (5,55–58). Despite this, F waves are frequently used to demonstrate the absence of changes in the excitability of the motoneuron pool. It is therefore worth repeating the reasons why this otherwise useful clinical test should not be used for that purpose in motor control studies.

F waves for the thenar muscles are illustrated in Fig. 2.4 of Chapter 2. As indicated above, volition, reflex inputs, and transcranial stimulation recruit motoneurons in a pool in order of increasing size. However, as illustrated in Fig. 3.7, F waves cannot appear in motoneurons that have discharged reflexly in response to the supramaximal stimulus required for F wave studies (56). Even though it cannot be appreciated in surface recordings, low-threshold motoneurons innervating the thenar muscles can discharge in an H reflex when the muscle is at rest (59), and do so overtly when the thenar muscles are contracting (60). F waves cannot appear in the very motoneurons most likely to be affected by an experimental manoeuvre and, as a result, the frequency of F wave discharges is greater in larger motoneurons with faster-conducting axons (61). Hultborn and Nielsen (55) have elaborated on the reasons why the F wave is less sensitive to changes in motoneuron excitability than the H reflex and, in a recent modelling study (62), it was concluded '... both an increase and a decrease of excitability can promote a recurrent discharge when absent or can inhibit it when present'.

Spinal mechanisms that can be tested reliably in the human subjects

The activity of all reflex circuits within the spinal cord is subject to supraspinal control. This is exerted on interneurons that, with the exception of the presynaptic inhibitory interneuron, are intercalated in the reflex pathway. However, the extent of presynaptic inhibition can be modulated by descending commands by altering the excitability of presynaptic inhibitory interneurons, termed 'PAD INs' in Fig. 3.8. A number of spinal reflex phenomena can

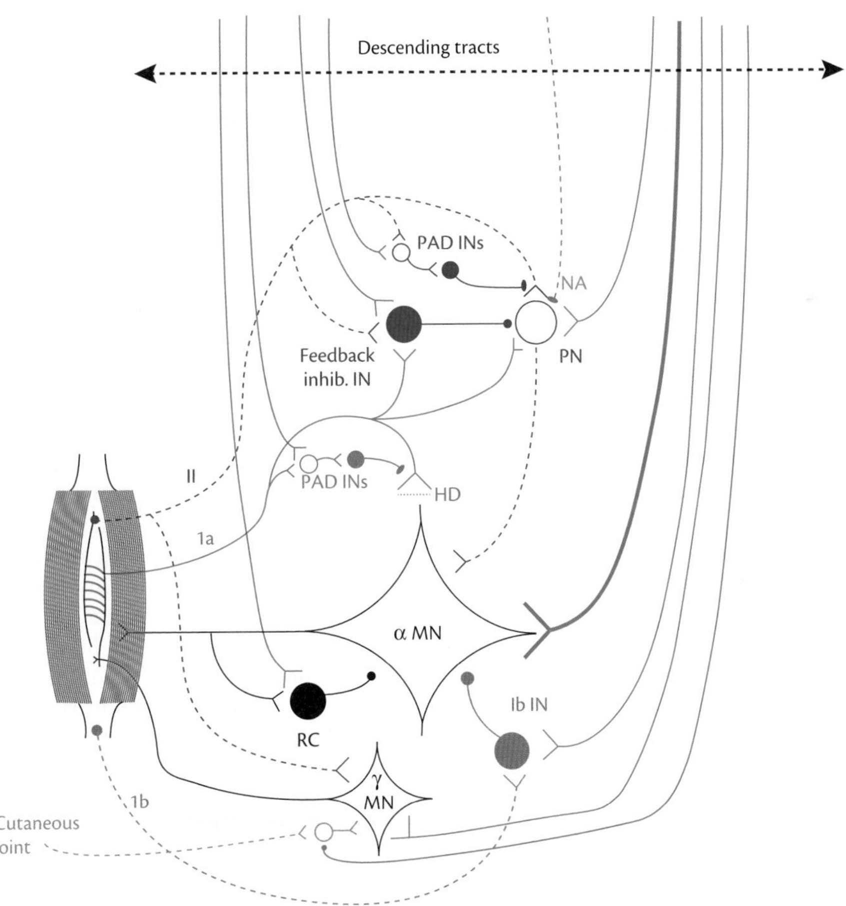

Fig. 3.8 Spinal circuits that can be tested reliably in human subjects. In this Figure, 'PAD Ins' are the interneurons that make axo-axonic synapses on Ia afferent terminals and are responsible for presynaptic inhibition. 'HD' refers to homosynaptic depression, the depression of transmitter release at the Ia-motoneuron synapse responsible for 'post-activation depression'.

IN = interneuron; MN = motoneuron; PN = propriospinal neuron.

Reproduced from Pierrot-Deseilligny E, Burke D, *The Circuitry of the Human Spinal Cord. Its Role in Motor Control and Movement Disorders*, copyright (2005), with permission from Cambridge University Press.

now be tested reliably in human subjects (Fig. 3.8). The background physiology and the details of testing protocols are provided elsewhere (5). In addition to fusimotor function (5,42,43), validated procedures have been developed to assess monosynaptic group Ia excitation (homonymous and heteronymous), post-activation depression of Ia excitation, presynaptic inhibition of the Ia excitation, recurrent inhibition (and heteronymous recurrent inhibition), reciprocal Ia inhibition, non-reciprocal group I inhibition (formerly termed Ib inhibition), group II excitation, cutaneo-muscular reflexes, and transmission through the C3–4 propriospinal system.

There are alterations in these spinal circuits in patients with spinal cord lesions and in many circuits in patients with stroke. In both spinal-injured and hemiplegic patients there is a loss of the normal control of reflex circuitry during voluntary movement, and this contributes to the loss of dexterity. Rather than 'spasticity', weakness and loss of dexterity are the disabling symptoms of stroke (13). In spinal cord-injured patients weakness, loss of bladder and bowel function, and spasms are the major causes of disability.

Post-activation depression of transmission at the group Ia synapse on the motoneuron

This phenomenon is also referred to as 'homosynaptic depression' because activity in an afferent depresses transmission at the synapses made by that afferent and only at those synapses. The depressed reflex response is believed to result from transmitter depletion in the presynaptic terminals of the Ia afferents. The reflex depression is marked by its long time course, decaying over some 10 s, much longer than occurs with inhibitory synaptic events: 5–10 ms for disynaptic Ia or Ib inhibition, perhaps 35 ms for recurrent inhibition, 200–300 ms for presynaptic inhibition. The reflex depression can be produced by a single conditioning stimulus (and tested in a conditioning test paradigm in which the interstimulus interval is increased) or by using trains of stimuli at different frequencies (so-called 'low-frequency depression' (63)). This phenomenon underlies the habituation of spinal reflexes such as the H reflex and tendon jerk when they are repetitively elicited, and is notable because the reflex depression can be attenuated and even abolished by a background contraction of the test muscle. In addition, loss of post-activation depression is the sole phenomenon that is directly correlated with the degree of spasticity following stroke or spinal cord injury or in multiple sclerosis. How voluntary contraction and these central nervous system disorders can alter the phenomena responsible for post-activation depression is not clear.

Post-synaptic inhibition

Recurrent inhibition is due to an axon collateral from the α motoneuron which excites an inhibitory interneuron (the Renshaw cell) at a cholinergic synapse. The Renshaw cell provides a negative feedback onto motoneurons of the contracting and synergistic muscles and onto Ia inhibitory interneurons projecting to the antagonist. As a result, it down-regulates the contraction of the agonist motoneurons and removes inhibition on the activity of the antagonist, a situation appropriate for alternating movements at a joint. The duration of the resultant inhibition is relatively long compared with other postsynaptic inhibitory phenomena because, when excited, Renshaw cells respond with more than a single impulse. Recurrent inhibition can be facilitated by the administration of L-acetyl carnitine. *Reciprocal Ia inhibition* refers to the inhibitory effects of group Ia afferents from an agonist on the antagonist motoneuron pool, and this was the first reflex circuit to be documented experimentally in human subjects (64). In addition to descending controls, the responsible interneurons can be inhibited by Renshaw cells coupled to the agonist. Interestingly, the disynaptic inhibition between the motoneurons of flexor and extensor carpi radialis is not true reciprocal Ia inhibition (as was commonly thought). The inhibition cannot be altered by activating recurrent inhibition, and is, instead, 'non-reciprocal group I inhibition'. This is intuitively reasonable because these muscles are antagonists only in flexion/extension movements at the wrist: they are synergists when the wrist is moved into radial deviation or ulnar deviation. *'Ib' inhibition* is now commonly referred to a *non-reciprocal group I inhibition* in recognition that group Ia afferents also have excitatory projections to these interneurons. As mentioned earlier, during voluntary contractions there is a depression of this inhibition of the contracting muscle for the duration of the contraction, and this is greater the stronger the contraction. This differs from all other inhibitory phenomena (leaving post-activation depression aside), and is largely because the afferent feedback coming from the contracting muscle depresses transmission across the inhibitory interneuron.

Presynaptic inhibition of group Ia terminals on motoneurons

The axon of the last-order presynaptic inhibitory interneuron makes 'axo-axonic' synapses on the presynaptic terminals of the Ia afferent axon. Depolarization of the presynaptic terminals (so-called 'primary afferent depolarization') decreases the release of the excitatory transmitter, glutamate, through an action that involves $GABA_A$ receptors. This depresses transmission in the reflex pathway without altering the excitability of the motoneuron or its ability to respond to other inputs, such as corticospinal volleys. As mentioned above, the inhibition lasts 200–300 ms (although in some tests used in humans the true duration can be obscured by a transient facilitation, giving the appearance of two separate inhibitory phenomena). Voluntary contractions switch off presynaptic inhibition of the Ia excitation directed to the contracting muscle, thus allowing stronger Ia excitation to reinforce the contraction. The effect is in proportion to the strength of the contraction, but occurs only at the onset of the contraction: it gradually returns over some hundreds of milliseconds back to the pre-contraction level, even if the contraction is maintained. A loss of presynaptic inhibition occurs in patients with spinal cord injury, but probably not in hemiplegic patients, once allowance is made for age.

Group II excitation

Group II muscle afferents can have inhibitory effects on extensor motoneurons, associated with excitation of the flexors, in at least some animal preparations (e.g. after spinal cord transaction). This pattern is commonly referred to as a 'flexor reflex afferent' pattern. In humans, however, group II volleys produce excitation of pure extensors, whether the afferents are activated by electrical stimulation or muscle stretch. The group II volleys project to a population of interneurons which, for the lower limb, are located above the segments innervating limb muscles, but project down the cord to the relevant motoneuron pools (i.e. they are, by definition, 'propriospinal'). These interneurons also receive excitation from group Ia afferents, such that there is convergent excitation from afferents with widely differing conduction velocities. Group II reflex effects will have a longer latency, but they will occur on a background of conditioning by Ia effects. Oral administration of tizanidine suppresses transmission across group II interneuron, depressing the resultant reflex responses. Group II excitation is increased in patients at rest with stroke and spinal cord injury and, in Parkinson's disease, there is a loss of the normal modulation of group II reflex function with changes in task.

Propriospinal transmission of the cortical command for upper-limb movement

A component of the corticospinal command for upper-limb movement is transmitted in humans to the motoneuron pool through a population of interneurons that is located above the C5 segment (Fig. 3.3). This system receives excitatory and particularly powerful inhibitory inputs from cutaneous and muscle afferents in the limb and may also be activated by descending tracts other than corticospinal (rubro-, tecto-, reticulospinal, but not vestibulospinal). These propriospinal neurons have recurrent collaterals to the lateral reticular nucleus (although probably fewer than in the cat), and they project down the cord to segmental motoneuron pools through the anterior part of the lateral funiculus (Fig. 3.3A). The propriospinal system projects to different muscle groups in the upper limb in order to coordinate the non-willed contractions of muscles throughout the limb necessary when one puts out the hand to grasp an object. No propriospinal input has yet been defined for motoneurons of the intrinsic muscles of the hand, and it is notable that the MEPs produced by TMS are largest there. It has been shown that, in patients recovering from stroke, more of the voluntary command passes through the propriospinal system, presumably both corticospinal and non-corticospinal, and this is probably an adaptive change to compensate for the paresis. However here as elsewhere, not all of the plastic change is beneficial: the improvement in the ability to move comes at the expense of unwanted synergistic contractions. This is a maladaptive response due to the fact that the commands transmitted through propriospinal circuitry cannot be focused and must involve other motoneuron pools.

References

1. Sherrington, C.S. (1910). Remarks on the reflex mechanism of the step. *Brain*, **33**, 1–25.
2. Creed, R.S., Denny-Brown, D., Eccles, J.C., Liddell, E.G.T., and Sherrington, C.S. (1938). *Reflex activity of the spinal cord.* London: Oxford University Press.
3. Shurrager, P.S. and Dykman, R.A. (1951). Walking spinal carnivores. *Journal of Comparative Physiology and Psychology*, **44**, 252–62.
4. Rossignol, S., Drew, T., Brustein, E., and Jiang, W. (1999). Locomotor performance and adaptation after partial or complete spinal cord lesions in the cat. *Progress in Brain Research*, **123**, 349–65.
5. Pierrot-Deseilligny, E. and Burke, D. (2012). *The circuitry of the human spinal cord: spinal and corticospinal mechanisms of movement.* New York, NY: Cambridge University Press.
6. Sherrington, C.S. (1906). *The integrative action of the nervous system.* New Haven, CT: Yale University Press.
7. Lemon, R.N., and Griffiths, J. (2005). Comparing the function of the corticospinal system in different species: organizational differences for motor specialization? *Muscle & Nerve*, **32**, 261–79.
8. Olivier, E., Edgley, S.A., Armand, J., and Lemon, R.N. (1997). An electrophysiological study of the postnatal development of the corticospinal system in the macaque monkey. *Journal of Neuroscience*, **17**, 267–76.
9. Alstermark, B., Isa, T., Pettersson, L-G., and Sasaki, S. (2007). The C3–C4 propriospinal system in the cat and monkey: a spinal premotoneuronal centre for voluntary motor control. *Acta Physiologica*, **189**, 123–40.
10. Jankowska, E. and Edgley, S.A. (2006). How can corticospinal tract neurons contribute to ipsilateral movements? A question with implications for recovery of motor functions. *Neuroscientist*, **12**, 67–79.
11. Edgley, S.A., Jankowska, E., and Hammar, I. (2004). Ipsilateral actions of feline corticospinal tract neurons on limb motoneurons. *Journal of Neuroscience*, **24**, 7804–13.
12. Rosengren, S.M., Welgampola, M.S., and Colebatch, J.G. (2010). Vestibular evoked myogenic potentials: past, present and future. *Clinical Neurophysiology*, **121**, 636–51.
13. Burke, D., Wissel, J., and Donnan, G.A. (2013). Pathophysiology of spasticity in stroke. *Neurology*, **80**(Suppl. 2), S20–6.
14. Mills, K.R. (1991). Magnetic brain stimulation: a tool to explore the action of the motor cortex on single human spinal motoneurones. *Trends in Neuroscience*, **14**, 401–5.
15. Rothwell, J.C. (1997). Techniques and mechanisms of action of transcranial stimulation of the human motor cortex. *Journal of Neuroscience Methods*, **74**, 113–22.
16. Kobayashi, M. and Pascual-Leone, A. (2003). Transcranial magnetic stimulation in neurology. *Lancet Neurology*, **2**, 145–56.
17. Ziemann, U. (2004). TMS and drugs. *Clinical Neurophysiology*, **115**, 1717–29.
18. Di Lazzaro, V. (2008). Transcranial stimulation measures explored by epidural spinal cord recordings. In: E. Wassermann, C. Epstein, U. Ziemann, V. Walsh, T. Paus, and S. H. Lisanby (Eds) *Oxford handbook of transcranial stimulation*, pp. 153–69. Oxford: Oxford University Press.
19. Reis, J., Swayne, O.B., Vandermeeren, Y., et al. (2008). Contribution of transcranial magnetic stimulation to the understanding of cortical mechanisms involved in motor control. *Journal of Physiology*, **586**, 325–51.
20. Rossi, S., Hallett, M., Rossini, P.M., Pascual-Leone, A., and the Safety of TMS Consensus Group. (2009). Safety, ethical considerations, and application guidelines for the use of transcranial magnetic stimulation in clinical practice and research. *Clinical Neurophysiology*, **120**, 2008–39.
21. Groppa, S., Oliviero, A., Eisen, A., et al. (2012). A practical guide to diagnostic transcranial magnetic stimulation: report of an IFCN committee. *Clinical Neurophysiology*, **123**, 858–82.
22. Chipchase, L., Schabrun, S., Cohen, L., et al. (2012). A checklist for assessing the methodological quality of studies using transcranial magnetic stimulation to study the motor system: an international consensus study. *Clinical Neurophysiology*, **123**, 1698–704.
23. Hicks, R., Burke, D., Stephen, J., Woodforth, I., and Crawford, M. (1992). Corticospinal volleys evoked by electrical stimulation of human motor cortex after withdrawal of volatile anaesthetics. *Journal of Physiology*, **456**, 393–404.
24. Mills, K.R., Boniface, S.J., and Schubert, M. (1992). Magnetic brain stimulation with a double coil: the importance of coil orientation. *Electroencephalography and Clinical Neurophysiology*, **85**, 17–21.
25. Magistris, M.R., Rösler, K.M., Truffert, A., Landis, T., and Hess, C.W. (1999). A clinical study of motor evoked potentials using a triple stimulation technique. *Brain*, **122**, 265–79.
26. Kujirai, T., Caramia, M.D., Rothwell, J.C., et al. (1993). Corticocortical inhibition in human motor cortex. *Journal of Physiology*, **471**, 501–19.

27. Ziemann, U., Paulus, W., Nitsche, M.A., et al. (2008). Consensus: motor cortex plasticity protocols. *Brain stimulation*, **1**, 164–82.
28. Bostock, H., Cikurel, K., and Burke, D. (1998). Threshold tracking techniques in the study of human peripheral nerve. *Muscle & Nerve*, **21**, 137–58.
29. Vucic, S., Ziemann, U., Eisen, A., Hallett, M., and Kiernan, M.C. (2013). Transcranial magnetic stimulation and amyotrophic lateral sclerosis: pathophysiological insights. *Journal of Neurology, Neurosurgery, and Psychiatry*, **84**, 1161–70.
30. Vucic, S. and Kiernan, M.C. (2013). Unility of transcranial magnetic stimulation in delineating amyotrophic lateral sclerosis pathophysiology. In: A. M. Lozano and M. Hallett (Eds) *Handbook of Clinical Neurology*, Vol. 116, *Brain Stimulation*, pp. 561–75. Amsterdam: Elsevier.
31. Wolpaw, J.R. and Tennissen, A.M. (2001). Activity-dependent spinal cord plasticity in health and disease. *Annual Review of Neuroscience*, **24**, 807–43.
32. Thompson, A.K., Chen, X.Y., and Wolpaw, J.R. (2009). Acquisition of a simple motor skill: task-dependent adaptation plus long-term change in the human soleus H-reflex. *Journal of Neuroscience*, **29**, 5784–92.
33. Thompson, A.K., Pomerantz, F.R., and Wolpaw, J.R. (2013). Operant conditioning of a spinal reflex can improve locomotion after spinal cord injury in humans. *Journal of Neuroscience*, **33**, 2365–75.
34. Hallett, M. (2001). Plasticity of the human motor cortex and recovery from stroke. *Brain Research Review*, **36**, 169–74.
35. Huang, Y.Z., Edwards, M.J., Rounis, E., Bhatia, K.P., and Rothwell, J.C. (2005). Theta burst stimulation of the human motor cortex. *Neuron*, **45**, 201–6.
36. Chipchase, L.S., Schabrun, S.M., and Hodges, P.W. (2011). Peripheral electrical stimulation to induce cortical plasticity: a systematic review of stimulus parameters. *Clinical Neurophysiology*, **122**, 456–63.
37. Fricke, K., Seeber, A.A, Thirugnanasambandam, N., Paulus, W., Nitsche, M.A., and Rothwell, J.C. (2011). Time course of the induction of homeostatic plasticity generated by repeated transcranial direct current stimulation of the human motor cortex. *Journal of Neurophysiology*, **105**, 1141–9.
38. Nitsche, M.A., Müller-Dahlhaus, F., Paulus, W., and Ziemann, U. (2012). The pharmacology of neuroplasticity induced by non-invasive brain stimulation: building models for the clinical use of CNS active drugs. *Journal of Physiology*, **590**, 4641–62.
39. Macefield, G., Gandevia, S.C., and Burke, D. (1989). Conduction velocities of muscle and cutaneous afferents in the upper and lower limbs of human subjects. *Brain*, **112**, 1519–32.
40. Matthews, P.B.C. (1989). Long-latency stretch reflexes of two intrinsic muscles of the human hand analysed by cooling the arm. *Journal of Physiology*, **419**, 519–38.
41. Lourenço, G., Iglesias, C., Cavallari, P., Pierrot-Deseilligny, E., and Marchand-Pauvert, V. (2006). Mediation of late excitation from human hand muscles via parallel group II spinal and group I transcortical pathways. *Journal of Physiology*, **572**, 585–603.
42. Vallbo, Å.B., Hagbarth, K.E., Torebjörk, H.E., and Wallin, B.G. (1979). Somatosensory, proprioceptive, and sympathetic activity in human peripheral nerves. *Physiological Review*, **59**, 919–57.
43. Burke, D. (1981). The activity of human muscle spindle endings in normal motor behavior. *International Reviews in Physiology*, **25**, 91–126.
44. Henneman, E., Somjen, G., and Carpenter, D.O. (1965). Functional significance of cell size in spinal motoneurons. *Journal of Neurophysiology*, **28**, 560–80.
45. Burke, R.E. and Rymer, W.Z. (1976). Relative strength of synaptic input from short-latency pathways to motor units of defined type in cat medial gastrocnemius. *Journal of Neurophysiology*, **39**, 447–58.
46. Garnett, R. and Stephens, J.A. (1981). Changes in the recruitment threshold of motor units produced by cutaneous stimulation in man. *Journal of Physiology*, **311**, 463–73.
47. Li, Y., Gorassini, M.A., and Bennett, D.J. (2004). Role of persistent sodium and calcium currents in motoneuron firing and spasticity in chronic spinal rats. *Journal of Neurophysiology*, **91**, 767–83.
48. Nickolls, P., Collins, D.F., Gorman, R.B., Burke, D., and Gandevia, S.C. (2004). Forces consistent with plateau potentials evoked in patients with chronic spinal cord injury. *Brain*, **127**, 660–70.
49. Murray, K.C., Nakae, A., Stephens, M.J., et al. (2010). Recovery of motoneuron and locomotor function after spinal cord injury depends on constitutive activity in 5-HT2C receptors. *Nature Medicine*, **16**, 694–700.
50. Morita, H., Petersen, N., Christensen, L.O., Sinkjaer, T., and Nielsen, J. (1998). Sensitivity of H-reflexes and stretch reflexes to presynaptic inhibition in humans. *Journal of Neurophysiology*, **80**, 610–20.
51. Gracies, J.M., Pierrot-Deseilligny, E., and Robain, G. (1994). Evidence for further recruitment of group I fibres with high stimulus intensities when using surface electrodes in man. *Electroencephalography and Clinical Neurophysiology*, **93**, 353–7.
52. Araki, T., Eccles, J.C., and Ito, M. (1960). Correlation of the inhibitory post-synaptic potential of motoneurones with the latency and time course of inhibition of monosynaptic reflexes. *Journal of Physiology*, **154**, 354–77.
53. Pierrot-Deseilligny, E., Morin, C., Bergego, C., and Tankov, N. (1981). Pattern of group I fibre projections from ankle flexor and extensor muscles in man. *Experimental Brain Research*, **42**, 337–50.
54. Marchand-Pauvert, V., Nicolas, G., Burke, D., and Pierrot-Deseilligny, E. (2002). Suppression of the H reflex in humans by disynaptic autogenetic inhibitory pathways activated by the test volley. *Journal of Physiology*, **542**, 963–76.
55. Hultborn, H. and Nielsen, J.B. (1995). H-reflex and F-responses are not equally sensitive to changes in motoneuronal excitability. *Muscle & Nerve*, **18**, 1471–4.
56. Espiritu, M.G., Lin, C.S-Y., and Burke, D. (2003). Motoneuron excitability and the F wave. *Muscle & Nerve*, **27**, 720–7.
57. Lin, J.Z. and Floeter, M.K. Do F-wave measurements detect changes in motor neuron excitability? *Muscle & Nerve*, **30**, 289–94.
58. Burke, D. (2014). Inability of F waves to control for changes in the excitability of the motoneurone pool in motor control studies. *Clinical Neurophysiology*, **125**, 221–2.
59. Trontelj, J.V. (1973). A study of the F response by single fibre electromyography. In: D. E. Desmedt (Ed.) *New developments in electromyography and clinical neurophysiology*, vol. 3, pp. 318–22. Basel: Karger.
60. Burke, D., Adams, R.W., and Skuse, N.F. (1989). The effects of voluntary contraction on the H reflex of human limb muscles. *Brain*, **112**, 417–33.
61. Guiloff, R.J. and Modarres-Sadeghi, H. (1991). Preferential generation of recurrent responses by groups of motor neurons in man. Conventional and single unit F wave studies. *Brain*, **114**, 1771–801.
62. Balbi, P., Martinoia, S., Colombo, R., and Massobrio, P. (2014). Modelling recurrent discharge in the spinal α-motoneuron: reappraisal of the F wave. *Clinical Neurophysiology*, **125**, 427–9.
63. McNulty, P.A., Jankelowitz, S.K., Wiendels, T.M., and Burke, D. (2008). Post-activation depression of the soleus H reflex measured using threshold tracking. *Journal of Neurophysiology*, **100**, 3275–84.
64. Mizuno, Y., Tanaka, R., and Yanagisawa, N. (1971). Reciprocal group I inhibition on triceps surae motoneurons in man. *Journal of Neurophysiology*, **34**, 1010–7.

CHAPTER 4

Cortical activity: single cell, cell assemblages, and networks

John G. R. Jefferys

Introduction

Neuronal signalling depends crucially on the movement of ions across plasma (external) membranes through various kinds of specialized ion channel. The first part of this chapter will outline cellular mechanisms responsible for cortical (and other neural) activity. An important consequence of these movements of ions is that the displacement of charge results in electric currents, which are the basis of many of the methods described in this volume, and are introduced in the second part of this chapter, from 'Recording groups of neurons'.

Electrical (and, less commonly, magnetic) recordings provide the fastest temporal resolution for detecting neuronal activity in the brain. The classical means of recording cortical activity is electroencephalography or electroencephalogram (EEG). The distance between the recording sites on the scalp and the cortical tissue, which generates the signal, means that spatial resolution is limited, with each electrode recording from an area of several square centimetres. EEGs are particularly useful when electrical activity is synchronized over large areas of the cortex, as occurs during seizures and certain stages of sleep. EEGs are recorded from the scalp and detect activity in superficial cortex, but are insensitive to activity in the sulci, in which case magnetoencephalography (see Chapter 13) may be more useful due to the orthogonal vectors of magnetic and electric fields. Higher spatial resolution can be achieved by placing electrodes onto the cortical surface (e.g. sub-dural arrays), or into the brain tissue (e.g. depth electrodes). Intracortical recordings with microwires and other microscopic recording electrodes can detect the activity of single neurons or groups of neurons, although in general these are used more for research than for clinical investigation.

How does neural tissue generate electrical signals?

Single cells

Neurons, like all living cells, are bounded by a phospholipid membrane that prevents the free movement of chemicals, including charged ions. Electrophysiological signalling relies on the existence of gradients of ions across the cell membrane and on the controlled flow of those ions through specialized ion channel and transporter molecules.

The extracellular fluid contains relatively low concentrations (~3 mM) of potassium ions (K^+) and high concentrations (~135 mM) of sodium ions (Na^+), mainly controlled by the blood–brain barrier. Neurons, in contrast, contain higher concentrations of K^+ and lower concentrations of Na^+ than the extracellular fluid. Establishing these ion gradients needs work, mainly by a 'sodium pump' or sodium-potassium ATPase, a transmembrane ATPase which uses metabolic energy to pump Na^+ outwards and K^+ inwards. These ion gradients are crucial for signalling by neurons (Fig. 4.1).

The neuronal membrane contains many classes of ion channel. By definition, ion channels are selective for what they allow through the otherwise impermeable membrane. They may open in response to changes in membrane potential and/or the binding of chemicals to receptor sites within the ion channel protein, or they may be constitutively open. At rest neuronal membranes are primarily permeable to K^+ ions due to the presence of what are called leak channels (or more specifically, tandem pore domain K^+ channels). Given the greater concentration of K^+ ions inside

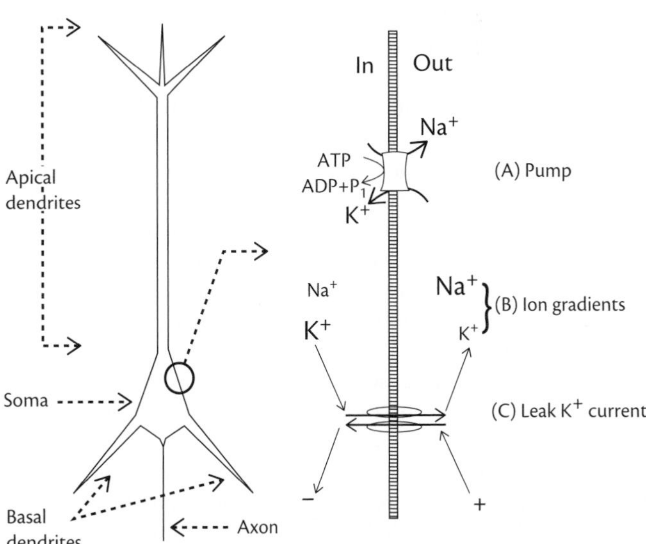

Fig. 4.1 Schematic neuron (left) with major components labelled, and expanded membrane (right). (A) Na^+-K^+ ATPase or sodium *pump* uses metabolic energy to remove Na^+ from, and load K^+ into, the neuron, resulting in *ion gradients* across the cell membrane (B). K^+ diffuses through the *leak channel* under its chemical gradient (C, top), making the inside of the neuron more negative than the outside. The resulting electrical gradient drives the positively charged K^+ ions into the cell (C, bottom). When the influx and efflux are equal the Nernst equilibrium applies.

the neuron, they diffuse down their concentration gradient to the extracellular space. The transfer of positive charge outwards results in the inside of the neuron becoming more negative than the outside. This continues until the voltage difference builds up sufficiently to cause movement of positively charged K^+ ions back into the neuron at a rate that balances the movement due to the outward chemical gradient. This is known as the reversal potential because, in this case, reductions in potential (depolarization) cause a net efflux of K^+, while increases in potential (hyperpolarization) cause a net influx, In the case of a single charge carrier the reversal potential is described by the Nernst equation:

$$E_m = \frac{RT}{zF}\ln\left(\frac{[K^+]_o}{[K^+]_i}\right) \quad \text{[eqn 4.1]}$$

where E_m is the membrane potential at equilibrium, R is the ideal gas constant, T is the temperature in degrees Kelvin, z is the valency of the ion (here 1), F is Faraday's constant linking charge to amount of ion, ln is the natural logarithm, $(K^+)_o$ is the concentration of potassium ions outside the cell, and $(K^+)_i$ is the concentration inside.

The real resting potential of neurons (approximately −65 mV) is more depolarized than the Nernst equilibrium potential for the known difference between intracellular and extracellular K^+ concentrations (of the order of −90 mV). In practice other ion channels contribute to the resting potential measured in central neurons, with notable contributions from Na^+ and Cl^- ions. In this case, the reversal potential is described by the Goldman–Hodgkin–Katz equation:

$$E_m = \frac{RT}{F}\ln\left(\frac{P_{Na}[Na^+]_o + P_K[K^+]_o + P_{Cl}[Cl^-]_i}{P_{Na}[Na^+]_i + P_K[K^+]_i + P_{Cl}[Cl^-]_o}\right) \quad \text{[eqn 4.2]}$$

where P_{ion} is the permeability of that ion through the membrane, and $z = 1$ for each of the three ions and can be removed.

During normal neuronal function the permeability of membranes to Na^+, K^+, Cl, and other ions varies considerably due to the effects of a large variety ion channels. The action potential is a critical aspect of neuronal function, being responsible for carrying information quickly and (usually) reliably along axons. The details of the ion channels responsible for action potentials in the squid giant axon were revealed in considerable detail by Hodgkin and Huxley in a seminal series of papers (1), which continue to dominate our thinking on the biophysics of excitable membranes over 60 years later (2).

The key players in the action potential are voltage-gated Na^+ and K^+ channels. The pore forming α subunits of voltage-gated Na^+ channels are encoded by nine genes, four of which are expressed in the CNS and three in the PNS. The molecular biology of K^+ channels is even more complex and beyond the scope of this chapter. In very brief summary (Fig. 4.2), at resting membrane potential voltage-gated Na^+ channels are mostly in a closed, impermeable state. On depolarization from rest, voltage-gated Na^+ channels switch to an active or open state, allowing Na^+ ions to enter the neuron, which results in further depolarization. This

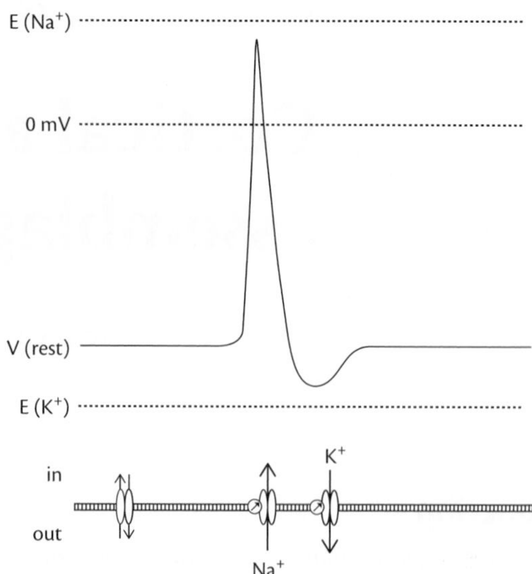

Fig. 4.2 Cartoon of action potential generation. Schematic action potential (top) and main components on membrane (below). From left to right: other leak channels maintain resting potential at about −65 mV, if membrane depolarizes towards −55 mV, voltage-gated sodium channels open (the 'voltmeter' dial indicates voltage sensors within the channel protein), Na^+ ions enter and depolarize the cell towards the equilibrium potential for Na^+ (E(Na^+). Sodium channels close after a short (~ms) period, while a voltage-gated potassium channel opens resulting in an influx of K^+ ions, which hyperpolarize the neuron towards K^+ equilibrium potential (E(K^+)).

sets up a positive feedback loop rapidly opening almost all Na^+ channels and causing a rapid shift towards a positive reversal potential dominated by the transmembrane Na^+ gradient (reversal for Na^+ is around +60 mV). Most voltage-gated Na^+ channels have a voltage- and time-dependent switch from the active to an inactive state, which limits the duration of the action potential, and is responsible for the absolute refractory period, which prevents neurons from firing again for a period of a few milliseconds. Repolarization of the neuron back to, or more usually beyond, resting potential is mediated by voltage-gated delayed rectifier K^+ channels. The additional K^+ permeability brings the membrane potential closer to the reversal potential for K^+. A variety of other channels mediate longer kinds of after-hyperpolarization, including those due to Ca^{2+}-activated K^+ currents.

Many other ion channels influence the excitability properties of neurons, modulating their excitability, defining their firing characteristics, mediating oscillatory activity and many other aspects, but these are beyond the scope of this chapter (3,4).

Synapses

Communication between neurons is crucial for the functioning of the cortex, and for the brain in general. The most important element in this communication is the synapse (Fig. 4.3). The concept of the synapse is close to that of the neuromuscular junction (see Chapter 1). A chemical neurotransmitter is released from a specialized part of one neuron (the presynaptic terminal), diffuses across a short distance (the synaptic cleft), and binds to receptors on the target neuron (the post-synaptic membrane). Receptors also occur on the presynaptic terminal or at sites outside the synapse, where they have specialized functions beyond the scope of

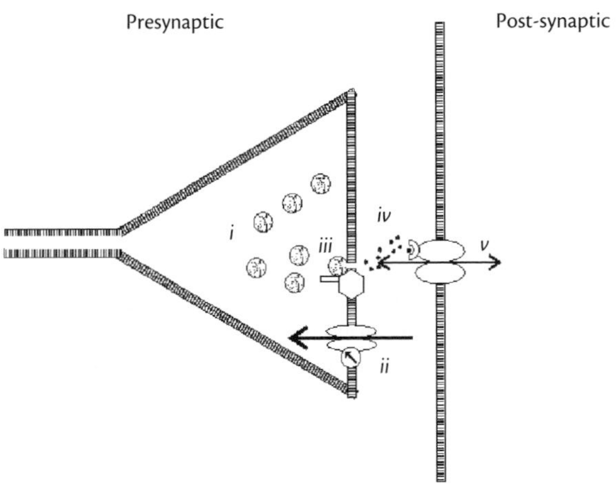

Fig. 4.3 Cartoon of synapse with ligand gated channel. Synaptic terminals contain neurotransmitter within vesicles ((*i*) circles containing dots). Action potentials propagating along the axon (left) depolarize the synaptic terminal, activating voltage-gated Ca^{2+} channels (*ii*). The increase in intracellular Ca^{2+} triggers a rapid chemical interactions between synaptic proteins which make a vesicle fuse with the presynaptic membrane and release its contents (*iii*). Transmitter diffuses across the synaptic cleft and binds to the receptor site of the ligand-gated channel in the post-synaptic membrane (*iv*). The channel opens and allows ions to diffuse down their electrochemical gradients (*v*).

this chapter. The transmitter is contained in synaptic vesicles and released by molecular mechanisms triggered by entry of Ca^{2+}, through voltage-gated Ca^{2+} channels, into the synaptic terminal as a result of depolarization by incoming action potentials. In the cortex, the major excitatory neurotransmitter is glutamate, and the major inhibitory neurotransmitter is GABA. Over 80% of neocortical neurons are excitatory, in that they depolarize their post-synaptic targets and increase the probability that they will reach threshold. Most of the remainder are inhibitory, in that they reduce the probability that their post-synaptic targets will fire, and either hyperpolarize or clamp usually membrane potential to resting potential.

The post-synaptic receptors are either ligand-gated ion channels or G protein-coupled receptors. The pharmacological and molecular classification of both classes of receptor is extensive and complex. Ligand-gated ion channels consist of assemblies of several receptor subunits, which link together into a barrel-like structure with a central ion channel, with transmitter binding sites in their extracellular domains (Fig. 4.3). In general, excitatory ligand-gated channels are permeable to cations (Na^+, K^+ and in some cases Ca^{2+}) when the transmitter is bound to its receptor site—they have reversal potentials close to zero. Ligand-gated inhibitory gamma amino butyric acid A ($GABA_A$) receptors are permeable to Cl^- ions, and to some extent to bicarbonate ions. In adult mammalian brain the reversal potential for Cl^- ions is hyperpolarized to resting potential due to the presence of a K^+–Cl^- cotransporter known as KCC2, which uses the concentration gradient of K^+ to maintain intracellular Cl^- concentration below its equilibrium potential. Ligand-gated channels provide the rapid signalling that mediates information processing in the cortex, for instance, in perception or in motor control.

G protein-couple receptors differ fundamentally from ligand-gated channels. They are a single protein with seven transmembrane domains that rely on other signalling molecules to alter neuronal function. These chemical interactions make the effects of G protein-couple receptors much slower than those of ligand-gated channels. The transmitter binds to a binding site in an extracellular part of the molecule, inducing a conformational change that activates a G protein associated with an intracellular site on the receptor protein by splitting it into α and βυ subunits. Gα acts through one of several second messenger systems (cyclic adenosine monophosphate, phospholipase C, rho-GTPases), impacting on intracellular signalling pathways, metabolism, gene transcription and so on. Gβυ also affects ion channels including certain classes of K^+ and Ca^{2+} channel. Inhibitory $GABA_B$ receptors work by opening K^+ channels, and cause considerable hyperpolarization—approaching the reversal potential for K^+. Excitatory metabotropic glutamate receptors act through transient receptor potential channels to produce slow depolarizations. The involvement of second messenger systems and intracellular signalling pathways means that the resulting synaptic potentials are slow.

Glutamate and GABA are the major neurotransmitters in the cortex, but the story is inevitably more complex. Cortical neurons may release more than one neuroactive chemical, for instance, many GABA-ergic inhibitory neurons also release neuropeptides (e.g. cholecystokinin, neuropeptide Y, somatostatin). Specific inputs to the cortex from other parts of the CNS usually use glutamate as their neurotransmitter, such as in sensory inputs. However, many subcortical structures use amine rather than amino acid neurotransmitters, and they act predominantly through G protein-couple receptors, e.g. serotonin from the raphe nucleus, noradrenaline (norepinephrine in the USA) from the locus coeruleus, acetylcholine from the septum and elsewhere, and dopamine from the substantia nigra. In electrophysiological terms these inputs to the cortex usually modulate firing patterns and excitability.

Neurotransmitters need to be removed from the extracellular space to stop their actions accumulating and, in some cases, to provide the precise timing required for aspects of neural processing. Removing glutamate from the extracellular space is particularly important because in high enough amounts it can kill neurons through a process known as excitotoxicity. Each transmitter has its own mechanisms for inactivation. Acetylcholine and neuropeptides are degraded by enzymes in the extracellular space, but many neurotransmitters are taken into neurons and/or glia. Perhaps the architype reuptake system is for serotonin, which is important in the use of serotonin specific reuptake inhibitors in depression. Similar transporter systems are important for glutamate and other excitatory transmitters, and for GABA. Much of this uptake is into glia, which then can recycle the metabolites of transmitters. Glia are important in other ways, notably in controlling extracellular ions (especially K^+, which is important for neuronal excitability) and through the release of 'gliotransmitters', such as glutamate and adenosine triphosphate.

Neuronal structure and function

Neurons are complex structures. The widest part is the cell body, or soma, which contains the nucleus (Fig. 4.1). Several thin processes normally extend from the soma—the dendrites, which receive much of the incoming synaptic input, and the (usually longer) axon/s, which transmit action potentials to synaptic terminals. Neurons in the cortex can be classified by their microscopic morphology, physiological properties, content of neurotransmitter and other materials (e.g. calcium binding proteins), and their post-synaptic targets. In general, cortical neurons have extensive dendrites that receive large numbers of excitatory synaptic inputs. These can amount to tens of thousands in the case of large pyramidal neurons, the largest class of excitatory neuron of neocortex, hippocampus and similar cortices. Inhibitory inputs occur on specific regions of pyramidal and other neurons, depending on the class of inhibitory neuron. For instance basket cells target pyramidal cells somata, and axo-axonic cells target the axon initial segment, which is a specialized region where the axon connects to the soma. Most commonly, action potentials are initiated at or near the axon initial segment. This arrangement makes basket and axo-axonic cells especially effective at preventing their post-synaptic neurons from firing in response to excitatory synaptic input. Other inhibitory neurons terminate on other parts of pyramidal neurons, or in some cases exclusively on other classes of (non-pyramidal) cortical neuron. The spatial organization of the synaptic and voltage-gated channels is important for cortical electrophysiology.

Recording groups of neurons

Recording the collective activity of large numbers of cortical neurons is fundamental to clinical neurophysiology. The most common and least invasive recording of cortical activity is the electroencephalogram, where potential differences are recorded from relatively large electrodes on the scalp. This method was developed early in the twentieth century, with the first recordings from the human scalp made by Berger (5) following similar recordings in experimental animals. There is a fundamental challenge in recording neuronal activity through many millimetres of non-neural tissues. Each neuron generates a tiny electrical signal, but fortunately the anatomical organization of the cortex can summate the signals generated by many neurons acting in concert provides relatively large electrical signals.

Laminar organization of the cortex

The various parts of the cerebral cortex share a feature that is crucial for much of electrophysiology. Its neurons are organized into layers, with the cell bodies of major classes of neurons located at specific depths beneath the cortical surface. Pyramidal neurons are particularly important here because of the organization of their dendrites broadly perpendicular to the plane of the cortex (Fig. 4.4). The details of this 'laminar' organization is important because it determines how electrical currents flow during activity. Differences exist between neocortex and phylogenetically older cortices, such as the hippocampus, for instance in the number of distinct cellular layers (conventionally described as 6 and 3, respectively), and in the trajectories of incoming 'afferent' axons (perpendicularly from deep white matter and parallel to the cell layers, respectively). Despite these differences, cerebral cortices share selective localization of ion channels in neuronal membranes, both the ligand-caged ion channels at the various kinds of synapse made onto neurons and the voltage-gated ion channels responsible for action potentials and other 'intrinsic' neuronal properties. The laminar organization of the cortex means that selective localization on neurons translates to selective localization at particular depths in the cortex.

Electrical field potentials generated by laminar structures

The subcellular localization of synapses and voltage-gated ion channels together with the laminar organization of neurons within cortex are important for the generation of macroscopic potentials such as seen in the EEG and evoked potentials. To explain how this works let us first consider a pyramidal neuron, with an excitatory pathway synapsing part way along its long apical dendrites (Fig. 4.4, left-hand schematic). Excitatory synapses in the brain mostly use glutamate as their transmitter, which binds to receptors, changing the conformation of the receptor proteins to allow positively charged ions (Na^+, K^+ and in some cases Ca^{2+}) to pass through the post-synaptic membrane. The net result of the movements of these ions is an influx of positive charge, which leads to depolarization from the negative resting potential. This influx leads to a current along the inside of the dendrite towards the rest of the neuron due to the voltage difference: following Ohm's law, voltage (V) is proportion to current (I) with the constant relationship defined as resistance (R) so that $V = IR$. This intracellular current is responsible for transmitting the synaptic excitation to the spike initiation zone (the region where voltage-gated Na^+ channels first reach threshold for an action potential). As mentioned above, the spike initiation zone usually is at or near the axon initial segment where the axon joins the cell body or soma of the neuron. Even at rest neuronal membranes are permeable to some ions. This causes attenuation of synaptic signals with distance due to the proportion of intracellular current leaking across the neuronal membrane into the extracellular space (Fig. 4.4A, grey arrows). In some neurons biology corrects for

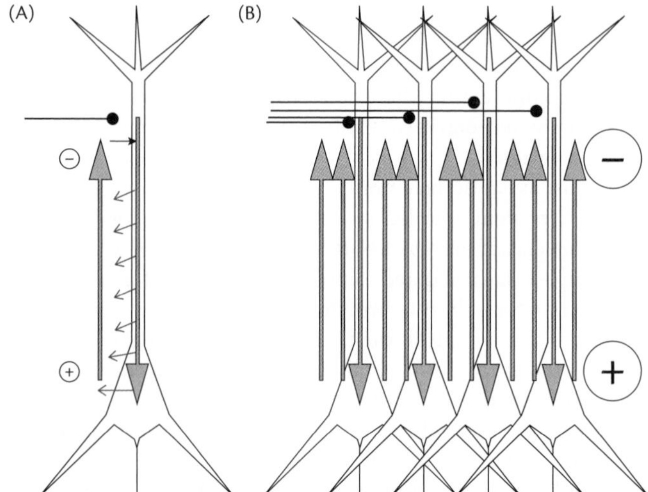

Fig. 4.4 Generation of field potentials. (A) Cartoon of a pyramidal cell receiving an excitatory synaptic input onto its apical dendrites. The net movement of cations causes a current into the post-synaptic region. Current mainly flows intracellularly some distance along the dendrites because the resistance of the intracellular fluid is lower than that of the cell membrane. If the excitatory input is large enough it triggers an action potential, usually at or near the axon initial segment. The current exits the neuron and completes the circuit to replace the charge that entered through the post-synaptic membrane. The extracellular current produces a voltage gradient. (B) When many neurons are organized in layers receiving similar inputs, the resulting current flow summates, leading to large 'field potentials', which can be large enough to be recorded through the skull and scalp as the EEG.

this attenuation by increasing the numbers of receptors at more remote synapses (6). As outlined above, because neuronal membranes are impermeable lipid bilayers they have the property of capacitance, as well as the properties of resistors conferred by the ion channels, but we will ignore time-dependent changes apart from noting that transients (e.g. synaptic potentials) slow with distance along dendrites, an effect that can be modified by some classes of voltage-gated ion channels.

From the point of view of recording EEGs or evoked potentials the organization of inward current in one part of a neuron with the resulting outward current in other parts is crucial because the loop is closed by an extracellular current, which creates an extracellular voltage gradient that can be recorded extracellularly. Single neurons produce highly localized and minute voltage gradients (a fraction of a millivolt, extending over a few hundred microns (7), but if large populations of cortical neurons work in parallel then the extracellular currents reinforce each other leading to substantial voltage gradients that can be recorded some distance away (Fig. 4.4B). These are generally called field potentials. What is meant by 'working in parallel', is that large numbers of neurons are doing similar things at roughly the same time, so that they generate currents within restricted layers, which can add up to a greater current density, and therefore larger voltage gradients. This can be described as neuronal synchrony, although variability in timing of the activity of individual neurons and the mechanisms by which neurons communicate mean that synchrony usually is imprecise. Cortical neurons have sufficient synchrony to produce measurable activity during several physiological conditions and pathological states such as epileptic seizures (see Chapter 11). They also can be synchronized by sensory, electrical, or magnetic stimulation to produce evoked responses (see Chapter 15).

In the case of the classes of inhibitory neuron mentioned above, basket cells and axo-axonic cells, synapse within the pyramidal layers (III and V in neocortex). When GABA binds to $GABA_A$ receptors, negative Cl^- ions flow into the neurons. The net current flows in the opposite direction to that produced by excitatory synapses so that the extracellular potential at the level of the inhibitory synapses is more positive than elsewhere round the neuron. The net effect for a scalp recording is that both correspond to surface negativity.

Action potentials in large cortical neurons usually are initiated at or near the axon initial segment, a specialized structure at the junction between the cell body, or soma, and the axon. They then propagate down the axon to the neuron's presynaptic terminals. In some cases, they also propagate up the dendrites as back-propagating action potentials. As we have outlined, action potentials are caused by the transient opening of voltage-gated Na^+ channels, which represent an influx of positive charge and similar intracellular, distributed outward membrane currents, and a return extracellular current, which results in extracellular voltage gradients.

In practice, signals such as the EEG are predominantly due to synaptic currents because much more charge moves through post-synaptic synaptic channels, partly because they remain open longer than the voltage-gated channels of the action potential. The longer durations of post-synaptic currents also increases the likelihood of summation of synaptic currents generated by large numbers of neurons receiving similar inputs. This is particularly important for the EEG where individual electrodes detect activity in several square centimetres of cortex. Action potentials over this area will be so dispersed in time that they will have minimal impact on the overall signal.

Relationship of neuronal firing with ECoG recordings

More invasive recordings with smaller electrodes can detect action potentials. 'Single unit' recordings from microwire electrodes go back to the 1980s (8) which, amongst other discoveries, provided the first evidence for the counterintuitive idea that most neurons decreased their firing during epileptic seizures. Microwires are particularly good for recording action potentials because their frequency response characteristics make them most effective at detecting the rapidly changing voltages found during brief action potentials. Clusters of microwires can be inserted down clinical depth electrodes, and have led to exciting work on the correlations of the firing of individual neurons in the temporal lobe with concepts, for instance, where a person or an object triggers a response irrespective of whether the stimulus is an image, a written word or even a spoken word (9).

During the late twentieth and early twenty-first centuries, technology developed to allow the manufacture of multi-electrode arrays comprising large numbers of penetrating microelectrodes (e.g. 96 in Utah arrays). They have been inserted through or under subdural mat electrodes during presurgical evaluation of people with medically intractable epilepsy. Early reports confirmed the earlier work that neurons in general reduced their firing rates during electrographic seizures (10). Subsequently, it turned out that there could be a sampling problem. Placing multi-electrode arrays in areas identified by conventional sub-dural recording as potential seizure onset zones identified acceleration of multi-unit activity in very focal areas of the cortex, which were much smaller than the extensive areas generating the classic EEG patterns of seizures (11). Experimental evidence from rodent brain slices in vitro suggested that this was due to synaptic inhibition constraining synaptic excitation in tissue that received inputs from the more focal region of neuronal hyperactivity (12,13). If the cause of the epilepsy is the pathologically accelerated firing of very local areas of cortex then it would appear to follow that removal of the tissue that starts the pathological firing should prevent seizures starting.

Penetrating arrays can damage the cortex, which is why they are best placed in tissue that is likely to be resected. However, less invasive recording methods may detect consequences of high-frequency neuronal firing. Perhaps the most obvious is high-frequency oscillations, which are defined by being faster than frequencies of 100–300 Hz depending on recording conditions and locations. In experimental contexts high-frequency oscillations can be generated by action potentials in the absence of synaptic activity (14), where their spatial extent of co-firing clusters of neurons was a fraction of a millimetre. This can explain why microelectrodes detect high frequency oscillations than classic clinical depth or sub-dural contacts (15). The distinction between synaptic and action potentials in clinical signals is more than an intellectual distraction. There already is evidence that successful outcomes for surgical resection of epileptic foci increase with the amount of tissue generating high-frequency oscillations removed (16).

Slow potentials

Conventional metal electrodes are not well-suited to recording very slow signals. For most clinical applications this does not matter, but the brain does generate signals that are stable for many seconds. Perhaps the clearest example comes from slow potentials recorded with intracranial electrodes during seizures (17–19). Together with high-frequency oscillations, slow potentials may provide more precise localization of epileptic foci. Slow potentials are readily recorded with glass micropipettes under experimental conditions, where they are associated with increased extracellular K^+ concentrations that can be measured directly with appropriate sensors (20). The most straightforward idea is that the negative potential shift is related to the removal of K^+ ions from the extracellular fluid.

Conclusions

Movement of ions across neuronal surface membranes are crucial to the signalling and computational processing of cortical neurons. Inputs to neurons are provided by synapses, using a variety of neurotransmitters, mainly onto the dendrites, somata, and axon initial segments. The many inputs each neuron receives are integrated to determine whether or not the neuron will trigger an action potential, which transmits the neurons output to its target neurons. All these membrane currents produce extracellular currents and associated voltage gradients.

The laminar structure of the cortex leads to the summation of the small currents produced by individual neurons, as long as large numbers of neurons receive similar synaptic inputs, as occurs during the rhythmic and/or synchronous activities found during epileptic seizures, sleep, and evoked potentials. These kind of summations result in high current densities within the tissue leading to substantial voltage gradients that can be recorded through the scalp. More invasive recordings can detect the firing of individual neurons or small clusters of neurons. The distinction between the field potentials/EEGs generated by synaptic currents from those generated by action potential currents can be critical for, e.g. identification of epileptic foci.

References

1. Hodgkin, A.L. and Huxley, A.F. (1952). A quantitative description of membrane current and its application to conduction and excitation in nerve. *Journal of Physiology*, **117**(4), 500–44.
2. Nelson, M.E. (2011). Electrophysiological models of neural processing. *Wiley Interdisciplinary Reviews: Systems Biology and Medicine*, 3(1), 74–92.
3. Contreras, D. (2004). Electrophysiological classes of neocortical neurons. *Neural Networks*, **17**(5–6), 633–46.
4. He, C., Chen, F., Li, B., and Hu, Z. (1929). Neurophysiology of HCN channels: from cellular functions to multiple regulations. *Progress in Neurobiology*, **112**, 1–23.
5. Berger, H. (1929). Uber das elektroenkephalogramm des menschen. *Archives of Psychiatry*, **87**, 527–70.
6. Magee, J.C. and Cook, E.P. (2003). Somatic EPSP amplitude is independent of synapse location in hippocampal pyramidal neurons. *Nature Neuroscience*, 3(9), 895–903.
7. Fatt P. (1957). Electric potentials occurring around a neurone during its antidromic activation. *Journal of Neurophysiology*, **20**, 27–60.
8. Babb, T.L., Wilson, C.L., and Isokawa-Akesson, M. (1987). Firing patterns of human limbic neurons during stereoencephalography (SEEG) and clinical temporal lobe seizures. *Electroencephalography & Clinical Neurophysiology*, **66**, 467–82.
9. Quiroga, R.Q. (2012). Concept cells: the building blocks of declarative memory functions. *Nature Reviews: Neuroscience*, **13**(8), 587–97.
10. Truccolo, W., Donoghue, J.A., Hochberg, L.R., et al. (2011). Single-neuron dynamics in human focal epilepsy. *Nature: Neuroscience*, **14**(5), 635–U130.
11. Schevon, C.A., Weiss, S.A., McKhann, G., et al. (2012). Evidence of an inhibitory restraint of seizure activity in humans. *Nature Communications*, **3**, 1060.
12. Trevelyan, A.J., Sussillo, D., and Yuste, R. (2007). Feedforward inhibition contributes to the control of epileptiform propagation speed. *Journal of Neuroscience*, **27**(13), 3383–7.
13. Mormann, F. and Jefferys, J.G.R. (2013). Neuronal firing in human epileptic cortex: the ins and outs of synchrony during seizures. *Epilepsy Currents*, **13**(2), 100–2.
14. Jiruska, P., Csicsvari, J., Powell, A.D., et al. (2010). High-frequency network activity, global increase in neuronal activity and synchrony expansion precede epileptic seizures in vitro. *Journal of Neuroscience*, **30**(16), 5690–701.
15. Worrell, G.A., Gardner, A.B., Stead, S.M., et al. (2008). High-frequency oscillations in human temporal lobe: simultaneous microwire and clinical macroelectrode recordings. *Brain*, **131**, 928–37.
16. Jacobs, J., Zijlmans, M., Zelmann, R., et al. (2010). High-frequency electroencephalographic oscillations correlate with outcome of epilepsy surgery. *Annals of Neurology*, **67**(2), 209–20.
17. Wu, S.S., Veedu, H.P.K., Lhatoo, S.D., Koubeissi, M.Z., Miller, J.P., and Luders, H.O. (2014). Role of ictal baseline shifts and ictal high-frequency oscillations in stereo-electroencephalography analysis of mesial temporal lobe seizures. *Epilepsia*, **55**(5), 690–8.
18. Kanazawa, K., Matsumoto, R., Imamura, H., et al. (2015). Intracranially recorded ictal direct current shifts may precede high frequency oscillations in human epilepsy. *Clinical Neurophysiology*, **126**(1), 47–59.
19. Ikeda, A., Terada, K., Mikuni, N., et al. (1996). Subdural recording of ictal DC shifts in neocortical seizures in humans. *Epilepsia*, **37**(7), 662–74.
20. Heinemann, U., Konnerth, A., Pumain, R., and Wadman, W.J. (1986). Extracellular calcium and potassium changes in chronic epileptic brain tissue. *Advances in Neurology*, **44**, 641–61.

CHAPTER 5

Recording of neural signals, neural activation, and signal processing

Dick F. Stegeman and Michel J. A. M. Van Putten

Introduction

In this chapter we will discuss the recording of electrical signals in the context of clinical neurophysiology. We present various basic concepts, including volume conduction, recording of electrical signals, electrical stimulation and safety, and aspects that are relevant in the digitization and post-processing of data.

Measuring in the extracellular space, mass activity, and volume conduction in clinical neurophysiology

Neurophysiological signals are traditionally recorded in an area around the actual bioelectrical sources. The voltages, therefore, result from the extracellular currents around the sources, since there is no clinical technique whereby the data can be recorded from inside a neuron, axon, or muscle fibre.

For the central nervous system this has primarily to do with the relative inaccessibility of the relevant structures, i.e. neurons and glial cells, as this would require opening of the skull or (at least) necessitates skin and dura penetration for access to the spinal cord. Also, the risk of damage during the acquisition of neural signals should be minimal. Clearly, this eliminates single cell recordings for routine clinical practice.

The peripheral neuromuscular system can physically be approached by a recording electrode with relative ease, but one should realize that when an axon or a muscle fibre is penetrated by the relatively large needle electrode, the cell membrane is damaged, and the action potential propagation is disrupted.

Although the absence of access to intracellular recordings may at first sight seem disadvantageous, extracellular recordings provide useful information at a global level as it results from synchronized activity of neural assemblies or axon bundles, which has been proven to be very useful for diagnostic purposes. Of course, recording individual cell activity provides diagnostic information, as well, e.g. to characterize particular channelopathies or the detailed role of a neuron in a network.

Indeed, almost all measurements in clinical neurophysiology record 'mass activity', the more or less simultaneous and similar activity of large groups of similarly orientated excited dendrites, axons, or muscle fibres. Also, analysis and explanatory models are mostly based on the assumption of the mass character of the sources underlying the recorded signals (e.g. (1)). Only a few exceptions exist, like the measurement of jitter with single fibre electromyograph (EMG) in the evaluation of the neuromuscular junction, where the signals result from one or a few muscle fibres only (ref. Chapter 7).

A disadvantage of extracellular recording is the spread of extracellular currents that depend, in a complex way, on the passive electrical properties (e.g. the different electrical conductivities of muscle, subcutaneous fat and bone) of the extracellular structures (2). This phenomenon is known as 'volume conduction'.

The recorded signals, therefore, represent a deformed 'image' of what is happening inside the cell population. The clinical neurophysiologist should always be aware of these aspects to prevent unjustified conclusions. A good example of mistakes that can be made is erroneous interpretation of signals that result from 'cross-talk'—the current from a certain neuronal population or muscle has no strict electric barrier to an electrode that was presumed to measure activity from another group of neurons or another muscle. Indeed, to define where a signal has been generated and whether it is coming from the expected or another source can be quite challenging (3). Several methods have been developed to find the sources of electrical activity. An example is the source localization of spikes or rhythms from electroencephalography (EEG) or magnetoencephalography (MEG) recordings (4,5). In the area of EMG, it concerns the discussions whether activity of synergists or both agonist and antagonist muscles are mixed in the same EMG recording during gait analysis.

Moving and non-moving sources

The basic source of electrical activity in the neurophysiology of the peripheral nervous system is the action potential, an almost monophasic transient change in membrane voltage, with a duration of 1–2 ms (see Chapters 1 and 4). This is different from the actual sources of electrical activity in the EEG that mainly result from synchronous activity of excitatory and inhibitory postsynaptic currents of cortical pyramidal cells (6).

Outside the cell, this source (i.e. the change in membrane voltage) is seen as a transmembrane current profile entering and leaving the neuron or the muscle fibre or nerve fibre. The character

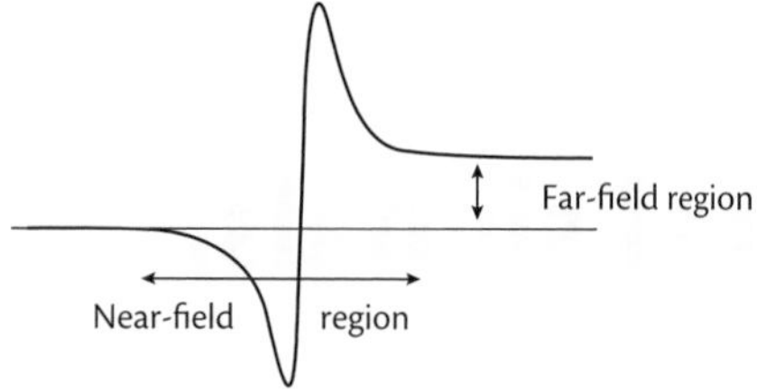

Fig. 5.1 Near- and far-field regions around a dipolar current source in a finite volume conductor like the human arm.

Reproduced from Dumitru D, Amato, AA, Zwarts, MJ. (Eds), *Electrodiagnostic medicine*, 2nd edn, copyright (2002), with permission from Elsevier.

of the transmembrane source essentially depends on whether the sources are in a fixed position, as in the cortex[1], or are propagating, as in the white matter of the brain and in peripheral nerves and muscles. In the first case the source can be described physically as a so-called current dipole. In the latter case the source is a linear quadrupole, also indicated as a tripole (8,9).

These two types of source differ significantly in their capacity to generate potentials at a distance in the above mentioned volume conduction. For the idealized case in which a dipole source is embedded in an infinite and also homogenous volume conductor, which is biologically impossible, of course, basic physics states that the potential drops as $1/R^2$ where R is the distance between the source and measuring electrode. For a quadrupolar source, the potential drops as $1/R^3$. So, also in an ideal volume conductor the dipolar source has a 'longer arm' than a quadrupolar source. In a finite volume conductor, such as (part of) the human body, there is an even stronger difference between the two source types. A dipolar source can generate so-called 'far-field 'potentials (Fig. 5.1). Such a potential component can be recorded over very long distances, even over the whole body. The best example of a dipole as a far-field generator is the heart as a bioelectric source. The electrocardiogram (ECG) can be measured all over the body. This is not so much related to the fact that the heart is a strong source, but it essentially results from the notion that the heart is a dipolar current source. One of the reasons why the EEG signal mainly reflects the dipolar, mainly excitatory post-synaptic potentials and inhibitory post-synaptic potentials, activity of the cortical layers and not the white matter, is due to the fact that the white matter consists of axonal structures for which the propagating sources are of the short-distance quadrupolar type and not as synchronous as in, e.g. muscle fibre activity. Extracellular recordings of propagating sources are essentially triphasic because of the +/–/+ character of such sources (Fig. 5.2, from figure 1.1 in Dumitru et al. (10)).

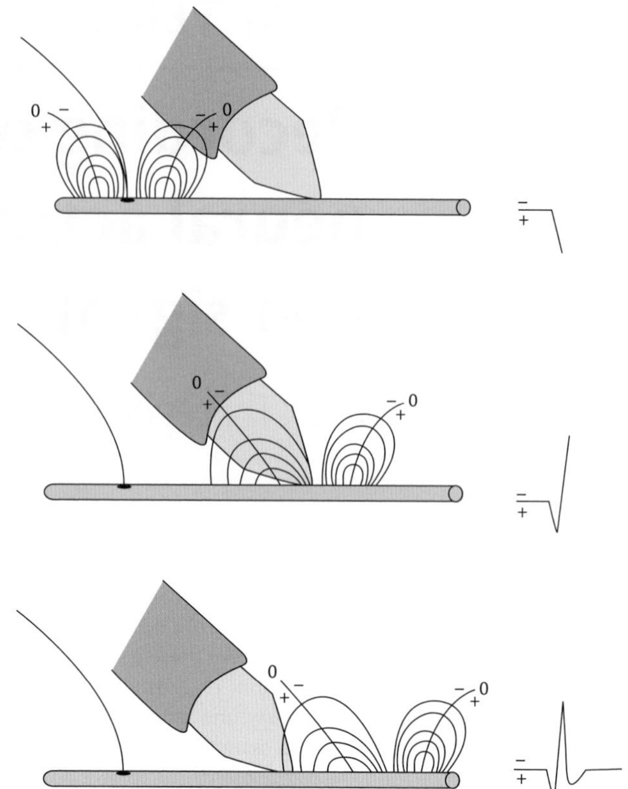

Fig. 5.2 The development of a triphasic muscle fibre extracellular action potential from a passing quadrupolar (source–sink–source) transmembrane current profile.

Reproduced from Dumitru D, Amato, AA, Zwarts, MJ. (Eds), *Electrodiagnostic medicine*, 2nd edn, copyright (2002), with permission from Elsevier.

Electrodes, electrode-(skin)-tissue interface, and amplifiers

We have already mentioned that electrodes in clinical neurophysiology, including the relatively small EMG-needles, are large compared to cellular dimensions, being of the order of 1–20 μm in man. That also contributes to the fact that most measured activity comes from large cell populations.

Electrodes can be made of different metals. The general principle of measuring electric potentials caused by the flow of ions in a fluid is to transfer the ionic current and resulting potential to the movement of electrons in the metal electrode and wire connected to the amplifier. This transfer uses the interface between skin, ions in the tissue, and electrode material (Fig. 5.3, (6)). It is essential to understand that both resistive elements (R) and capacitive elements (C) are part of the circuit. Together, these elements make up the electrode impedance. The electrode, therefore, also acts as a filter, with a frequency response depending on the relative contribution of the resistive and capacitative elements.

Dependent on the material used, the capacitative (C) elements may dominate the transfer of the electric charge as occurs in the so-called 'polarizable 'electrodes. Noble metal electrodes made of gold, silver, platinum are examples of this category, and thus actual charge transfer is limited: the electrical current is mainly a capacitative current. When the resistive (R) component dominates, the electrode approaches a 'non-polarizable 'electrode, and actual charge transfer does take place as an exchange of electrons in a so-called redox reaction. The well-known silver/silver chloride electrode approaches the idealized non-polarizable type.

An important difference between the two categories is how low frequency signals are transferred. Because of the dominance of the capacitive elements in the polarizable electrodes (Fig. 5.3), DC, and low frequency changes in the tissue meet a very high

[1] Although neurons are fixed, the dynamic interplay of cortical neurons may also result in traveling waves with velocities of 1–10 m/s (7).

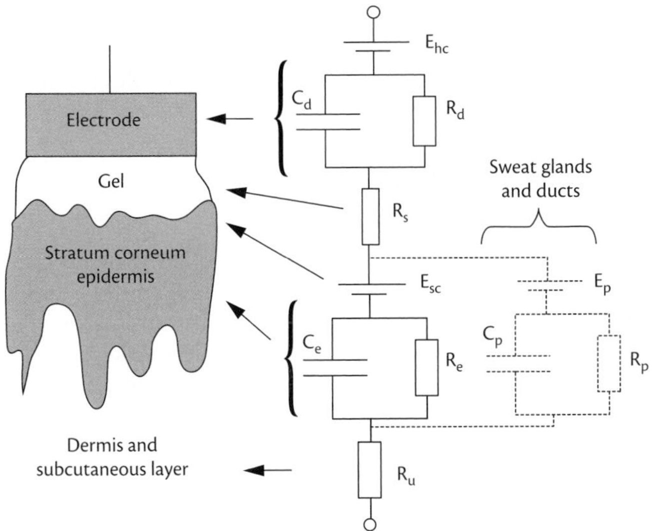

Fig. 5.3 The equivalent circuit of a biopotential electrode in terms of the electrode–skin–tissue interface. It can be expressed in an electronic circuit as in the right-hand part of that figure. Note the resistive (R) and the capacitive (C) elements, parallel and in series.

Reproduced from Webster JG (ed.), Medical Instrumentation, application and design, 4th edition, copyright (2010), with permission from John Wiley and Sons.

tissue-electrode impedance with detrimental consequences like a decreased signal amplitude and an increased noise level in the low frequency range.

In particular for EEG measurements, this may be critical (as signals of interest include 0–4 Hz), which explains the use of non-polarisable electrodes, like the Ag/AgCl electrode in EEG.

Another property of electrodes is that they, just like a standard battery, generate a potential difference at the electrode skin-interface, i.e. between the ion solutions in the tissue and the metal of the electrode. This potential difference can be in the order of several 100s of mV, certainly large compared with the signals the amplifiers are expected to measure (10s of μVs to 10–20 mV) (Chapter 6 (6)) Since these so-called half-cell potentials (HCPs) are at both sides of the amplifier inputs (see later), they are expected to cancel each other. Because the HCP magnitudes depend on the electrode material, it is advised always to use the same material at both inputs. A problem, existing even when all precautions are taken, is that the HCP can cause movement artefacts. A movement of the electrode over the skin disturbs the electrode interface and then the HCP changes abruptly after which it builds up again. This reflects as a high amplitude low frequency artefact in the signal. For further reading see Neuman (11).

The quality of the signal depends not only on the impedance of the electrode–tissue interface, but also on the quality of the amplifier system. In recent decades, (very) high-input impedance amplifiers have become available. Together with the electrode impedances mentioned above, the input impedance determines the quality of the data to a large extent. The combination of as small as possible electrode-skin impedance, preferably equal for the + and the – side of the input, with as high as possible amplifier input impedance is therefore the most ideal for reliable recordings. The high amplifier impedance allows only a minimal current into the amplifier circuit. On top of that, the minimal current limits additional voltage changes at the electrode–skin interface resulting from the finite electrode resistance. Both factors make sure that the amplifier measures the potential at the skin surface as faithfully as possible.

The general principle is that the input impedance of the amplifier should at least be several orders of magnitude higher than the electrode–skin impedances. The ratio between both impedances is usually in the order of e.g. 5–10 kOhm for the electrodes versus 50–100 MOhms for the input of the amplifier (a range of 10^4). Most amplifier systems are capable of measuring electrode–skin impedances, often at a single relevant frequency (e.g. 20 Hz for EEG). The latter is relevant since the capacitive elements in the electrode–skin interface cause the impedance to be frequency dependent, as discussed previously.

Physics dictates that to measure biopotentials requires measuring the difference in electric potential between two sites. The development of the differential amplifier with a well-balanced pair of inputs (+ and –), an electronic principle derived in the 1930s, was a major contribution to reliably measure biopotentials (12–14). In the design of such amplifiers it is especially important that non-desired potential differences between the body and the amplifier system, such as the 50 or 60 Hz power-line interference, are suppressed as much as possible. The quality of that suppression is expressed in the common mode rejection ratio (CMRR). This is the ratio between the amplification of the desired difference between the + and – electrodes and the unavoidable presence of a small remainder of the often large common mode signal, equally present at both inputs. The CMRR is expressed in the logarithmic dB scale (20 dB per factor 10 suppression). Today's amplifiers have CMRR values over 100–120 dB, i.e. at least a factor of 10^5–10^6. For more detailed information we refer to Webster (15), Van Putten (6), and Van Putten (14).

Electrode montages

Measuring 'a potential' always imples the measurement of a potential difference, which asks for the choice of the position of two electrodes to record an EEG or EMG. Together these two electrodes define a montage. When we measure the voltage difference between pairs (2) of electrodes, typically positioned close to the source of the activity of interest, we call it a bipolar montage (Fig. 5.4). In case the counter electrode is chosen to be the same for all other electrodes over the skull or over a muscle, it is called the (single) 'reference' electrode and the set-up a 'monopolar' recording montage (Fig. 5.4).

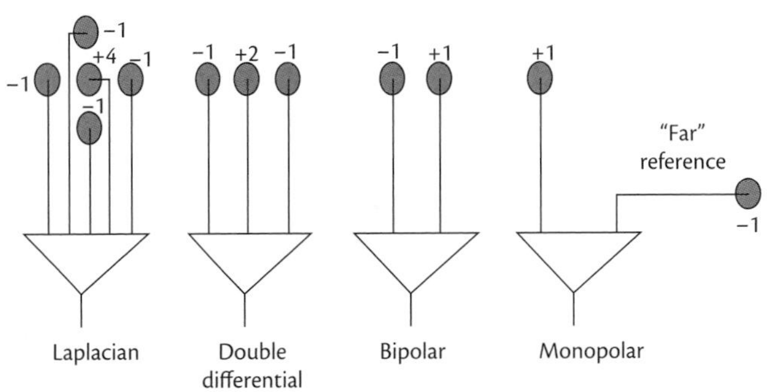

Fig. 5.4 Four often used different electrode montages and their names: from left to right a decreasing order, i.e. a more global view on the activity. The double differential montage is sometimes used in EMG interpretation.

The spatial distribution of neurophysiological sources and the effect of volume conduction together, cause a potential in tissue or at the skin surface not only to evolve in the time domain as a time sequence, but also changes as a function of the location. Indeed, we deal with spatio-temporal signals, in fact a three-dimensional signal (time–position–amplitude). This remark is made in the section on electrode montages because the spatial aspect is the most essential property in the choice of the position of each of the two (+ and –) inputs to the amplifier. As a rule, when one is interested in a local phenomenon, a short distance (inter-electrode distance (IED)) over that location between the two electrodes should be chosen. The advantage of a short distance is not only the local view of the recording, but also the avoidance of the influence of non-desired potential fluctuations from distant sources through volume conduction (e.g. 'cross talk', as mentioned previously).

In fact, the bipolar montage with a short IED records an estimate of the local spatial derivative of the spatial potential profile below both electrodes. To increase the focality of the view on the spatial profile of the potential, an even shorter IED can be chosen, but also or even better a 'higher order' montage than a bipolar one as, for instance, the double differential and the Laplacian montage (Fig. 5.4). The latter was first described for EEG signals (16), but higher order montages were also introduced in EMG after the development of so-called high-density EMG (17,18).

In most multichannel EEG and EMG systems, either the monopolar or average reference montage is used in the recording and storage of the data. A monopolar montage has the serious disadvantage that there is a large influence of cross-talk and uncertainty on the location of the sources of activity. The obvious advantage of both recording montages is that after having recorded the data digitally, all other montages can be realized.

In reviewing the EEG, most electroencephalographers use several montages, as a single montage may not always suffice to detect abnormal waveforms, as a particular montage represents a single projection of the 'three-dimensional' extracellular currents. Furthermore, by combining montages, it is often possible to define the direction of the net dipole associated with epileptiform discharges, which may assist in the classification. An example is Rolandic epilepsy, where the orientation of the dipole associated with the centrotemporal spikes is tangential to the skull. This results from the fact that the epileptic focus in Rolandic epilepsy is in the wall of a sulcus, and, together with the horizontal (with respect to the skull) orientation of the pyramidal cells, the dipole field is horizontal, as well.

Analogue-to-digital conversion and anti-aliasing filtering

Electrophysiological signals are essentially of an analogue character, which means that they can be measured at any time instant and have an amplitude in a certain range with any possible value within this range. The use of digital computers makes it necessary that these signals are converted to the digital domain, which is done by an analogue-to-digital converter (ADC). Two characteristics have to be defined in this process: the sampling interval (ms) or sampling frequency (Hz), and the AD-resolution (number of bits).

An important criterion with respect to the sampling frequency is to limit signal frequencies to the so-called 'Nyquist frequency' (19). It is half the sampling frequency and thus determines the highest frequency that is allowed to be present in the measured signals. Signals with higher frequencies than the Nyquist frequency cause 'aliasing', a spurious signal, resulting from erroneous sampling of another signal as illustrated in Fig. 5.5. To avoid aliasing, it should be ensured that the signal is low-pass filtered before it is captured by the ADC. Of course, this first anti-aliasing filtering stage cannot be implemented digitally.

The amplitude resolution is determined by the number of bits of the ADC. N bits stands for 2^N levels, e.g. $2^{12} = 4096$ in case of 12 bits (Fig. 5.6). The actual precision of the stored data is determined by the 'least-significant-bit' (LSB). Assuming a noise level of the amplifier and the electrodes of some μVs, this LSB should represent about 1.0 μV for an EEG or voluntary EMG recording. For a 12-bit ADC that means that the largest amplitude that can be measured is 4096*1.0 = 4096 μV or about 4 mV. Since that is less than, for instance, a normal compound muscle action potential (CMAP) amplitude, the amplification should be turned down before a CMAP signal can enter the ADC without clipping. Nowadays, ADCs use between 16 and 24 bits. For a 16 bits ADC, the range would be 66 mV, a large enough range for all routine

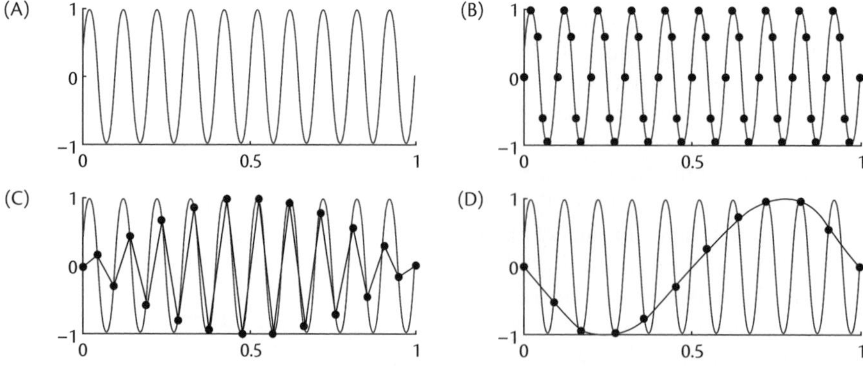

Fig. 5.5 Illustration of aliasing. A 1-s continuous signal segment containing a 10-Hz sinusoid (A) is sampled with 50 Hz (B), 21 Hz (C). In case of undersampling (D) 11 Hz < 20 Hz (Nyquist frequency for 10 Hz), a 'phantom' sinusoid appears in this case of 1 Hz, the difference between the original sinusoid of 10 and 11 Hz sampling rate.
Reproduced from Van Putten MJAM, *Essentials of neurophysiology: basic concepts and clinical applications for scientists and engineers*, copyright (2009), with permission from Springer.

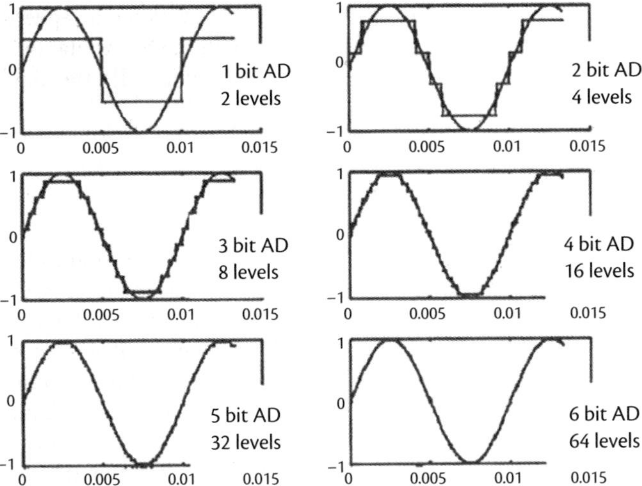

Fig. 5.6 Example of different AD conversion resolutions (N = 1, 2, 3, 4, 5, and 6 bits) and the corresponding number of levels (2^N) in converting a sinusoid. Typical sampling resolutions nowadays lie between 12 and 16 up to 24 bits.
Reproduced from Van Putten MJAM, *Essentials of neurophysiology: basic concepts and clinical applications for scientists and engineers*, copyright (2009), with permission from Springer.

recordings. With a 22–24 bits ADC, it is even possible to avoid clipping from drift of DC recorded signals with large fluctuations, for instance from the above mentioned half-cell potential asymmetries at the electrodes.

Currently, therefore amplifiers with a variable gain before the ADC conversion are no longer needed, nor is it necessary to be concerned about the Nyquist frequency. Most commercial electrodiagnostic systems take care of the requirements of digitization. Nevertheless, it is useful to be aware of the underlying principles.

Aliasing in the *spatial domain* might be of more relevance in today's practice. For instance the complete spatial profile of the EEG distribution over the scalp is under-sampled by a 32-channel EEG recording. That means that the Nyquist criterion is not obeyed in the spatial sense. Interpolation of the data into a colour map may have distortions as exemplified in the image processing example in Fig. 5.7.

Fig. 5.7 Example of spatial aliasing. The left-hand picture is spatially down-sampled straightforwardly to get the central picture. That means that high spatial frequencies become under-sampled and introduce spatial aliasing. The right-hand figure has been made after low pass filtering before the down-sampling. This results in a less sharp, but otherwise undistorted figure.

Signal processing

As soon as data are available in digital form, a virtually unlimited world of processing tools can be used. We will discuss a few of the more common techniques. The calculation of electrode montages is a signal processing step already discussed. Averaging and digital filtering form other classes of processing tools that can be applied to isolate certain signal components in the time or in the frequency domain. As mentioned earlier, electrophysiological signals are often measured in three dimensions (two spatial dimensions and time). Such data structures are well known in fields like physics, mathematics, and genetic engineering, and many algorithms have been developed. Well known in the EEG literature is so-called principle component analysis (PCA) to disentangle a signal in main determinants and Fast Fourier Transform (FFT)-based techniques to estimate the power spectrum or hemispheric asymmetries (20,21). Other examples are beamforming techniques for source localization or graphical theoretical approaches to study brain network behaviour (22,23). The ongoing changes in the frequency content of signals as a function of ongoing time have resulted in development of time-frequency processing tools. Well known in the research domain on EEG and MEG are the software packages EEGlab (24) and Fieldtrip (25). Many methods to remove unwanted artefacts from the target signals (e.g. power line interference, remove EMG from EEG, remove ECG from EMG) are available at different levels of complexity. Recent developments also include techniques to assist in the interpretation of the EEG background patterns, the detection of interictal discharges (26–28) and real-time processing of EEG signals in the ICU (29).

Averaging

Essential in the context of small, but relevant, potential components is averaging of repeated time locked occurrences of such components to 'lift' them from the background activity. Most well-known is the isolation of evoked potentials from the ongoing EEG, but also the quantification of sensory nerve potentials may need an averaged signal to improve the signal quality. Averaging does its work as expected when a number of conditions are met. There should be evidence that a certain stimulus does evoke a response in time that has a relevant deterministic component. That means that there should be a fixed latency with respect to the onset of the stimulus and a stable waveform from the brain structures or from a peripheral nerve. This response should be accompanied by activity that is not related to the stimulus onset (thus randomly occurring in time) with an equal chance of adding a positive and a negative signal contribution at any moment. When this is the case, the increase of the signal-to-noise ratio (SNR, the goal of the process) goes up with the square root of the number N of stimulus repetitions (Fig. 5.8).

An important condition to make averaging useful, is that the SNR at the start is not so low that the necessary number of stimuli can never be realized in practice. The use of additional stimuli will not assist, as these as these will only marginally increase the SNR due to the square root relationship between the two (e.g. a factor of only 3.16 ($=\sqrt{10}$) improvement when increasing N from 100 to 1000).

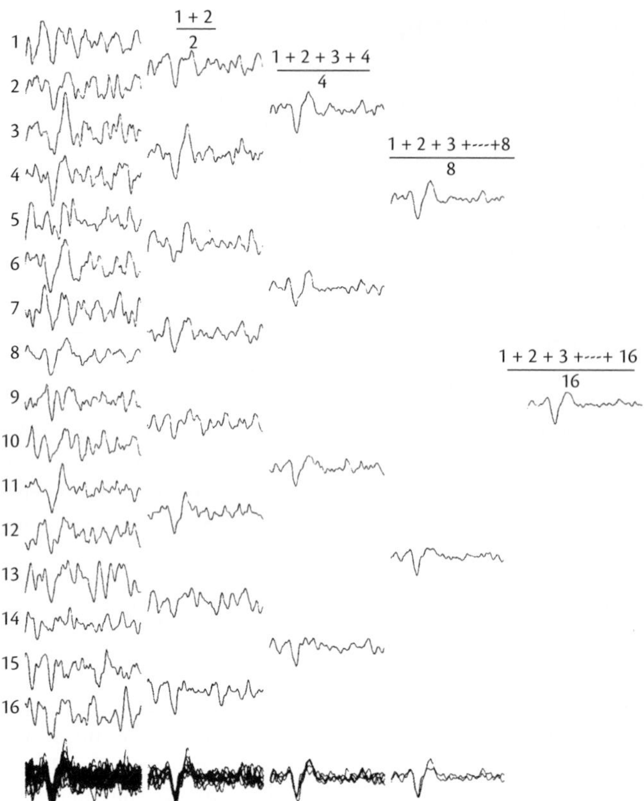

Fig. 5.8 Example of the working of an averaging procedure. Sixteen raw recorded evoked potential responses are averaged two by two in the second column and so on. After having averaged all 16 responses, the main component of the response is clearly visible, whereas it was not or hardly in the original raw data.

Rectification

Since electrophysiological signals are mostly multiphasic with positive and negative components, a phenomenon called 'phase cancellation' often arises. Negative potential components from a certain source might interfere with positive components from another source. These components then destructively interfere with each other. A way to avoid that problem, for instance, when averaging evoked components that are unstable in time or have high frequency components, is to rectify the incoming wave shape. All negative potential values are turned positive and added as positive components to those from other wave shapes. Rectification is also a standard operation to estimate the ongoing amplitude variation of surface EMG signals. Despite the obvious advantages of rectification, it should be realized that rectification is a strongly non-linear operation with major effects on properties of the signal such as frequency content, amplitude, and the relation with other simultaneous signals (30,31).

Filtering and spectral analysis

Another way to remove unwanted aspects of the signal is to filter frequency components that for physiological reasons cannot be part of the desired signal. Filtering of digital signals is a very fast operation with modern computer hardware because spectral analysis of a signal with Fourier-analysis algorithms can be done very efficiently. The development of computer power is in general the reason that signals should be captured in the computer without any previous filtering or other manipulation, apart from the previously mentioned anti-aliasing filter. Almost all manipulations to improve signal quality or to identify specific aspects can be done on-line or very fast post-hoc. Artefact reduction is often done in the sense of more or less advanced filtering. For instance, the frequency content of most relevant EEG signal components in relation to the EMG signal frequency band allows the suppression of EMG artefact by simply low-pass filtering at for instance 30 Hz. When the surface EMG of trunk muscles is the signal and the ECG is the unwanted artefact, it is often appropriate to use the same cut-off frequency of 30 Hz, but now defined as a high-pass filter.

Filtering is based on the manipulations of the frequency content of signals after spectral analysis. This is one way to use spectral analysis. It can also be used to study the electrophysiological signals in different frequency bands (e.g. 32). Next to the avoidable phenomenon of aliasing, spectral analysis obeys some other basic rules worthwhile knowing. The result of spectral analysis via the Fourier transform, just as the digitized time signal, also has a digitized representation. When a time segment of T s is analysed, the resolution in the frequency domain $\Delta f = 1/T$. That means the longer the segment that is analysed, the more precise the presentation in the frequency domain. Let us call the time step Δt, then $N = T/\Delta t$ data points are available in time. Taking into account the Nyquist criterion the frequency transformation returns with $N/2$, instead of N relevant frequency data points, up to a frequency of $fN = N/2^{*}\Delta f$, the Nyquist frequency. From rewriting the above follows $fN = 1/(2\Delta t)$. Thus, similar to how T relates to Δf, the smaller the time step Δt, the higher the frequencies (up to fN) can be derived from the signal. This is not very surprising, but it is important to realize this equivalence between resolutions in the time- and frequency-domains.

As a final point, it should be mentioned that a reliable spectral resolution needs the averaging over more frequency points, which makes the effective spectral resolution less than 1/T (6). An equivalent alternative is, however, when there is a longer duration of stationary signal available, to average the spectrum over several non-overlapping epochs.

Correlation, coherence, and signal decomposition

There are several situations where it is relevant to decide whether two signals look alike. Examples include the estimation of conduction velocities in motor conduction studies or in fatigue studies using surface EMG. Also, particular EEG phenomena exist where similarity of wave shapes is relevant (33). Techniques to estimate similarity can be summarized as correlation techniques that quantifies in how far two signals are linearly related (24,25). Correlation can be done in the time domain, but also has variants in the frequency domain. In that case, the term coherence is used. Coherence indicates how far a specific frequency component in a signal is followed, especially with a stable phase shift, by the same frequency component in another signal. Recently, especially the studies of coherence between brain rhythms from EEG or MEG, on the one hand, and frequency components in EMG signals, on the other, are often quantified to gain insight into the 'communication' between cortical structures and muscles as effectors of movement (34). Alternatively, complex non-linear relations are tested that makes sense, since it is likely that there are many non-linear processes involved. In fact, already at the level of the action potential, non-linear equations are needed to describe its generation, mainly resulting from the non-linear average dynamics of the various voltage-gated channels (see also (35)).

In the field of EMG, the identification of a known signal template (a motor unit potential) in a time series (the ongoing EMG) is the basis of a vast amount of research. These techniques are called 'template matching' techniques. The goal is to identify reoccurring firings of as many as possible motor units (motor unit decomposition). These techniques have been developed for both needle EMG recording and for high density EMG signals (36,37).

Electrical stimulation

The measurement of electrophysiological variables often uses electrical stimulation because excitable nerve and muscle tissue can easily be activated by short (0.1–1.0 ms) electrical pulses. Since these electric current pulses are large compared with the electrophysiological current sources, and are entering the volume conducting tissue, they cause significant artefact signals, which can even block the amplifiers for a while. Usually, these short pulses do not interfere with electrophysiological responses arriving several milliseconds later. However, the large input shock to the amplifier can lead to a longer-lasting amplifier response (ringing), which may interfere with the physiological responses. Properly designed amplifiers can reduce this effect to an acceptable level.

In recent years, there has been interest in investigating the properties of nervous structures by the use of small subthreshold ongoing electrical currents, mostly of a constant (DC) current type. In the diagnosis of peripheral nerve pathology, the technique of 'threshold tracking' using a polarizing subthreshold current to study excitability changes of peripheral nerves is the best known example (38). More recently, the technique of subthreshold transcranial DC stimulation (tcDCS) has become popular. With a small (1–2 mA) current for 5–20 min, a change in the excitability of cortical regions can be produced with the potential for therapeutic applications for instance in stroke rehabilitation (39). Commercial equipment has become available for these subthreshold adaptive stimulation modalities in recent years, but the actual contribution to rehabilitation still needs to be defined.

Electrical safety

Electrical currents may cause harm in the body, depending on the current density and the structures involved. The actual current through the body is a function of the voltage applied and the impedance, where the latter depends on the frequency of the applied voltage. Up to 1 kHz, the major contribution to the impedance is the intact skin. At larger frequencies, in particular >10 kHz, the blood, and the intra- and extracellular fluid are the main contributors to the impedance.

The perception threshold for currents of 50 Hz is about 1–5 mA; for a direct current it is about 2–10 mA. When currents are larger, muscle and nerve stimulation occurs, which may result in muscle contraction and pain. When the current reaches a particular threshold, it is no longer possible to make voluntary movements. This current strength is called the '(cannot) let-go-current'. At 50 Hz, this is about 6 mA. At larger currents, 18–22 mA, a continuous contraction of the diaphragm and other respiratory muscles occurs, resulting in asphyxia. Even larger currents may cause ventricular fibrillation, where the critical current strength depends on the current density through the heart. If current enters through the hands, the critical current to cause ventricular fibrillation is about 75–400 mA. Continuous cardiac contractions occur with currents as large as 1–6 A. With currents larger than 10 A, overheating of tissue occurs due to the dissipation caused by these large currents, in particular at those places where the current enters the body, as current densities are largest at these positions. An overview of these effects is presented in Table 5.1.

For completeness, we mention that all these situations are macro-shock conditions—current enters and leaves the body. In situations using implantable devices, such as a pacemaker, leak currents from the electrode may enter the heart muscle, as well. These conditions are known as microshock situations, and effects on the heart may already occur with currents larger than 10 mA. For more details about electrical safety, we recommend National Institute for Occupational Safety and Health (40), and Fish and Geddes (41).

Medical instrumentation needs to be safe, and must avoid undesired macro-shock conditions—undesired leak currents must be small, therefore. Leak currents include earth leak-currents and patient leak-currents. Finally, there is a functional small current, too, that flows in the patient from the amplifier in order to realize the measurement. Although this 'leaks' out of the instrument, it is typically not referred to as a leakage current.

The class of a medical instrument defines how the isolation is realized to prevent undesired leak currents. The type of an instrument defines how the patient and the operator are protected from leak currents. As an example, a class II instrument has a double isolation, whereas class I instruments is equipped with a ground wire. Type B (Body) instrumentation can be connected to the patient, but not with the heart or blood vessels. Body floating (BF) instrumentation is allowed to do so, whereas cardiac floating (CF) is designed for direct measurement in the heart or large vessels.

Table 5.1 Effects of 50 Hz currents applied for 1–3 s with two copper wires, one held in each hand. Part of these data is extrapolated from experiments in dogs

Current (mA)	Effect	Remark
1–5	Tingling sensation	Detection threshold
5–8	Pain	
6–20	Continuous muscle contraction	(Cannot) let-go-current
18–22	Respiratory arrest/asphyxia	
75–400	Ventricular fibrillation	
1000–6000	Continuous heart contraction	
>10,000	Burning of tissue	

EMG, EEG, ECG amplifiers for instance, are typically Type B. Catheters used in the ICU to measure cardiac output (Schwan–Ganz) are CF. Finally, H instrumentations are not designed for patients, and have electrical safety similar to household instruments (like a dish washer or a PC).

Conclusions

We have discussed the recording of electrical signals in the context of clinical neurophysiology. Aspects that are relevant in the digitization and post-processing of data, although not a problem for daily routine, have been briefly discussed. We took some steps towards the interpretation of signals and differences between signals in terms of their underlying (electro)physiology and the biophysics of volume conduction. Some basics of advanced procedures to analyse data are discussed. Aspects of electrical stimulation were included, with also some recent developments in diagnostic and therapeutic constant current stimulation. The background of hazardous electric currents and the safety of bioelectric equipment have been given.

References

1. David, O. and Friston, K.J. (2003). A neural mass model for MEG/EEG: coupling and neuronal dynamics. *NeuroImage*, **20**(3), 1743–55.
2. Dumitru, D. and DeLisa, J.A. (1991). AAEM Minimonograph #10: volume conduction. *Muscle & Nerve*, **14**(7), 605–24.
3. Dimitrova, N.A., Dimitrov, G.V., and Nikitin, O.A. (2002). Neither high-pass filtering nor mathematical differentiation of the EMG signals can considerably reduce cross-talk. *Journal of Electromyography & Kinesiology*, **12**(4), 235–46.
4. Lopes da Silva, F. (2004). Functional localization of brain sources using EEG and/or MEG data: volume conductor and source models. *Magnetic Resonance Imaging*, **22**(10), 1533–8.
5. Cheyne, D.O. (2013). MEG studies of sensorimotor rhythms: a review. *Experimental Neurology*, **245**, 27–39.
6. Van Putten, M.J.A.M. (2009). *Essentials of neurophysiology. Basic concepts and clinical applications for scientists and engineers*, Series in Biomedical Engineering. Berlin: Springer.
7. Hindriks, R., Van Putten, M.J.A.M., and Deco, G. (2014). Intra-cortical propagation of EEG alpha oscillations. *NeuroImage*, **103**, 444–53.
8. Stegeman, D.F., Dumitru, D., King, J.C., and Roeleveld, K. (1997). Near- and far-fields: source characteristics and the conducting medium in neurophysiology. *Journal of Clinical Neurophysiology*, **14**(5), 429–42.
9. Stegeman, D.F., Blok, J.H., Hermens, H.J., and Roeleveld, K. (2000). Surface EMG models: properties and applications. *Journal of Electromyography and Kinesio*logy, Oct;**10**(5):313–26.
10. Dumitru, D., Stegeman, D.F., and Zwarts, M.J. (2002). Electric sources and volume conduction. In: D. Dumitru, A. A. Amato, M. J. Zwarts (Eds) *Electrodiagnostic medicine*, 2nd edn, Chapter 2, pp. 27–53. Philadelphia, PA: Hanley & Belfus.
11. Neuman, M.R. (2000). Biopotential electrodes. In: Joseph D. Bronzino (Ed.) *The biomedical engineering handbook*, 2nd edn. Boca Raton, Chapter 48, pp. 1–13, FL: CRC Press LLC.
12. Geddes, L.A. (1996). Who invented the differential amplifier? *IEEE Engineering Medicine and Biology Magazine*, **15**, 116–17.
13. Pallas-Areny, R. and Casas, O. (2007). A hands-on approach to differential circuit measurements. *Measurement*, **40**, 8–14.
14. Van Putten, A.F.P. (1988). *Electronic measurement systems*. New York, NY: Prentice-Hall.
15. Webster, J.G. (Ed.) (2010). *Medical instrumentation, application and design*, 4th edn. New York, NY: John Wiley & Sons.
16. Hjorth, B. (1975). An on-line transformation of EEG scalp potentials into orthogonal source derivations. *Electroencephalography and Clinical Neurophysiology*, **39**(5), 526–30.
17. Zwarts, M.J. and Stegeman, D.F. (2003). Multichannel surface EMG: basic aspects and clinical utility. *Muscle & Nerve*, **28**(1), 1–17.
18. Merletti, R. and Parker, P.J. (Eds) (2004). *Electromyography: physiology, engineering, and non-invasive applications*. Hoboken, NJ: Wiley-IEEE Press.
19. Nyquist, H. (1928). Certain topics in telegraph transmission theory. *Transactions of the American Institute of Electrical Engineers*, **47**, 617–44.
20. Pourtois, G., Delplanque, S., Michel, C., and Vuilleumier, P. (2008). Beyond conventional event-related brain potential (ERP): exploring the time-course of visual emotion processing using topographic and principal component analyses. *Brain Topography*, **20**(4), 265–77.
21. Finnigan, S. and Van Putten, M.J. (2013). EEG in ischaemic stroke: quantitative EEG can uniquely inform (sub-)acute prognoses and clinical management. *Clinical Neurophysiology*, **124**(1), 10–19.
22. Litvak, V., Eusebio, A., Jha, A., et al. (2010). Optimized beamforming for simultaneous MEG and intracranial local field potential recordings in deep brain stimulation patients. *NeuroImage*, **50**(4), 1578–88.
23. Stam, C.J. (2014). Modern network science of neurological disorders. *Nature Review Neuroscience*, **15**(10), 683–95.
24. Brunner, C., Delorme, A., and Makeig, S. (2013). EEGlab—an open source Matlab Toolbox for electrophysiological research. *Biomedizinische Technik (Berlin)*, **58**, (Suppl. 1).
25. Oostenveld, R., Fries, P., Maris, E., and Schoffelen, J.M. (2011). FieldTrip: open source software for advanced analysis of MEG, EEG, and invasive electrophysiological data. *Computational Intelligence and Neuroscience*, **2011**, 156869 (open access).
26. Lodder, S.S. and Van Putten, M.J.A.M. (2013). Quantification of the adult EEG background pattern. *Clinical Neurophysiology*, **124**(2), 228–37.
27. Lodder, S.S., Askamp, J., and Van Putten, M.J.A.M. (2014). Computer-assisted interpretation of the EEG background pattern: a clinical evaluation. *PlosOne*, **9**(1), e85966.
28. Lodder, S.S. and Van Putten, M.J.A.M. (2014). A self-adapting system for the automated detection of inter-ictal epileptiform discharges. *PlosOne*, **9**(1), e85180.
29. Tjepkema-Cloostermans, M.C., Van Meulen, F., Meinsma, G., and Van Putten, M.J.A.M. (2013). A Cerebral Recovery Index (CRI) for early prognosis in patients after cardiac arrest. *Critical Care*, **17**(5), R252.
30. Alaid, S. and Kornhuber, M.E. (2013). The impact of rectification on the electrically evoked long-latency reflex of the biceps brachii muscle. *Neuroscience Letters*, **556**, 84–8.
31. McClelland, V.M., Cvetkovic, Z., and Mills, K.R. (2014). Inconsistent effects of EMG rectification on coherence analysis. *Journal of Physiology*, **592**(1), 249–50.
32. Schomer, D.L. and Lopes da Silva, F.H. (2011). *Electroencephalography: basic principles, clinical applications, and related fields*, 6th edn. Philadelphia, PA: Lippincott Williams & Wilkins.
33. Hofmeijer, J., Tjepkema-Cloostermans, M.C., and Van Putten, M.J.A.M. (2014). Burst-suppression with identical bursts: a distinct EEG pattern with poor outcome in postanoxic coma. *Clinical Neurophysiology*, **125**(5), 947–54.
34. Van Elswijk, G., Maij, F., Schoffelen, J.M., Overeem, S., Stegeman, D.F., and Fries, P. (2010). Corticospinal beta-band synchronization entails rhythmic gain modulation. *Journal of Neuroscience*, **30**, 4481–8.
35. Klonowski, W. (2009). Everything you wanted to ask about EEG but were afraid to get the right answer. *Nonlinear Biomedical Physics*, **3**(1), 2.

36. McGill, K.C., Lateva, Z.C., and Marateb, H.R. (2005). EMGLAB: an interactive EMG decomposition program. *Journal of Neuroscience Methods*, **149**(2), 121–33.
37. Holobar, A. and Farina, D. (2014). Blind source identification from the multichannel surface electromyogram. *Physiological Measurement*, **35**, R143–65.
38. Bostock, H., Cikurel, K., and Burke, D. (1998). Threshold tracking techniques in the study of human peripheral nerve. *Muscle & Nerve*, **21**(2), 137–58.
39. Hummel, F.C. and Cohen, L.G. (2006). Non-invasive brain stimulation: a new strategy to improve neurorehabilitation after stroke? *Lancet: Neurology*, **5**, 708–12.
40. NIOSH (1998). Available at: http://www.cdc.gov/niosh/docs/98-131/pdfs/98-131.pdf (accessed April 2016).
41. Fish, R.M. and Geddes, L.A. (2009). Conduction of electrical current to and through the human body: a review. *Eplasty*, **9**, e44 (open access).

SECTION 2

Techniques of clinical neurophysiology

CHAPTER 6

Nerve conduction studies

Jun Kimura

Fundamentals and techniques

Methods for stimulation and recording

Stimulus intensity and possible risk

A single shock of 0.1 ms duration, if delivered over the skin surface, can activate a healthy nerve fully with intensities ranging from 100 to 300 V or, assuming a tissue resistance of 10 KΩ, 10–30 mA. It may require 400–500 V or 40–50 mA in a diseased nerve with decreased excitability (1). Electrical stimulation of this magnitude causes no particular risks, unless the current directly reaches the cardiac tissue through an indwelling catheter or central venous pressure line inserted directly into the heart (2). Patients with cardioverters and defibrillators deserve special care to avoid electric stimulation near the heart (3).

Surface and needle electrodes

Surface electrodes are better suited to record a compound muscle action potential (CMAP). Here, the onset latency reflects the fastest fibres, and the amplitude, the number of available motor axons. In contrast, the use of needle electrodes helps identify a small response from an atrophic muscle, differentiating confounding discharges from outside the target muscle after proximal stimulation. Single fibre recording allows measurement of single motor axon conduction characteristics (4).

Surface electrodes also serve better for recording sensory and mixed nerve action potentials. The use of ring electrodes placed over the proximal and distal interphalangeal joints work well for antidromic sensory potentials, which are large enough to require no averaging. The use of needle electrodes enhances the response by decreasing the noise from the electrode tissue surface (5–7). In effect, near nerve recording increases the signal-to-noise ratio by about five times and, during averaging, reduces the time required to reach the same resolution.

Averaging technique

Small signals within the expected noise level of the system necessitate an averaging technique to improve the signal-to-noise ratio. Here, digitized signals time-locked to the stimulus sum after repetitive stimulation, whereas the random noise with no temporal relationship is cancelled during successive trials.

Motor nerve conduction studies

Waveform, amplitude, and duration

A submaximal stimulus may not necessarily cause orderly activation of motor axons (8); a supramaximal shock is required to elicit a maximal CMAP. The nerve impulses reach the muscle in a slightly dispersed fashion, rather than synchronously because the individual nerve axons vary in length and velocity. The size of the CMAP reflects the number of motor units within the recording radius of the active electrode, i.e. about 20 mm from the skin surface. The location of the recording electrodes determines the contribution of the constituent motor units to the acquired potential (9).

An initially negative biphasic waveform results (Figs 6.1 and 6.2) from an active lead (E1) placed on the belly of the muscle and an indifferent lead (E2) on the tendon. Usual measurements include:

- latency, from the stimulus to the onset;
- amplitude, from the baseline to the negative peak;
- duration, from the onset to the baseline crossing or the final return to the baseline;
- area under the waveform measured by electronic integration.

The onset latency reflects the fast conducting motor fibres, although the shortest, but not necessarily fastest, axons may give rise to the initial potential. Amplifier gain settings alter the identifiable point of deviation from the baseline and consequently the measured value (10,11).

Latency and conduction velocity

The latency difference between the two CMAPs elicited at proximal and distal stimulation points, in effect, excludes nerve activation time, neuromuscular transmission, and time required for generation of the muscle action potential, yielding the conduction time from one point of stimulation to the next (Fig. 6.2). Dividing the corresponding distance by the latency difference results in the conduction velocity (CV) in metres per second as follows:

$$\mathrm{CV(m/s)} = \frac{\mathrm{D(mm)}}{Lp - Ld\mathrm{(ms)}} \qquad \text{[eqn 6.1]}$$

where D represents the distance between the two stimulus points in millimetres, and *Lp* and *Ld*, the proximal and distal latencies in milliseconds.

Separation of the successive points by several centimetres improves the accuracy of surface measurement (12). The inclusion of longer unaffected segments, however, lowers the sensitivity of detecting a focal lesion seen in a compressive neuropathy (13). Here, 'inching' studies with shorter incremental stimulation steps helps isolate a localized abnormality by an abrupt change in latency and waveform (14,15).

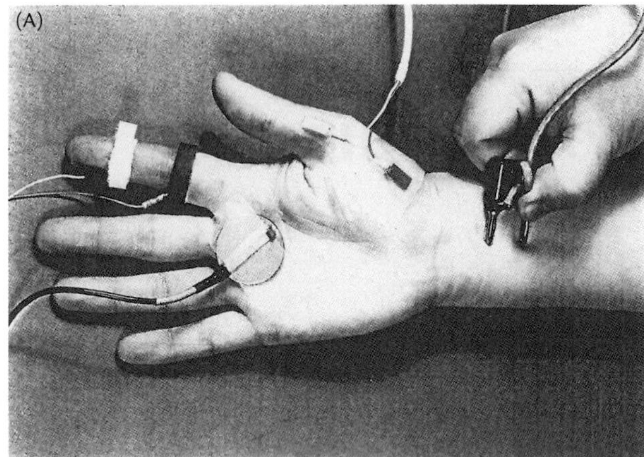

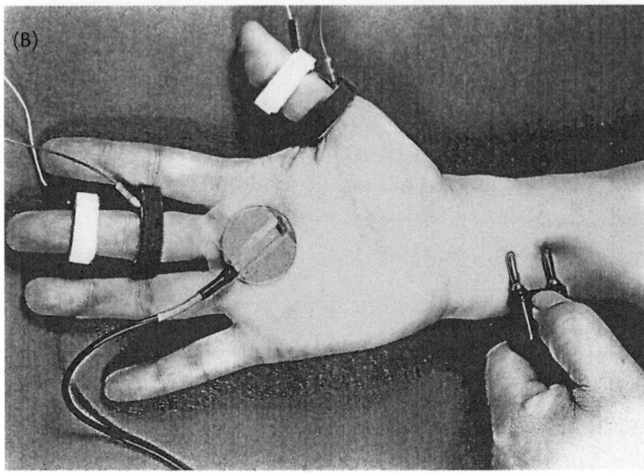

Fig. 6.1 (A) Motor and sensory conduction studies of the median nerve. The photo shows stimulation at the wrist, 3 cm proximal to the distal crease and recording over the belly (E1) and tendon (E2) of the abductor pollicis brevis for motor conduction, and around the proximal (E1) and distal (E2) interphalangeal joints of the second digit for antidromic sensory conduction. The ground electrode is located in the palm. (B) Alternative recording sites for sensory conduction study of the median nerve with the ring electrodes placed around the proximal (E1) and distal (E2) interphalangeal joints of the third digit, or the base (E1) and the interphalangeal joint (E2) of the first digit.

Reproduced from Kimura, J, *Electrodiagnosis in diseases of nerve and muscle: principles and practice*, 4th edn, copyright (2013), with permission from Oxford University Press.

The neuromuscular transmission time precludes calculation of conduction velocity for the most distal segment. The use of a fixed distance for electrode placement reduces individual differences of the distal latency for improved accuracy. The distal latency (L_d) actually measured slightly exceeds the estimated value (L_d') derived by dividing the terminal distance (D) by proximal CV. The terminal latency index, calculated as the ratio between the estimated and measured latency (L_d'/L_d), reflects the distal conduction delay (16,17). In a study of the median nerve with L_d of 4.0 ms, D of 8 cm, and CV of 50 m/s, the calculated values would include L_d' of 1.6 ms (8 cm/50 m/s), and terminal index ratio of 0.4 (1.6 ms/4.0 ms).

Types of abnormalities

A single motor fibre may show either axonal damage or demyelination (4). The nerve as a whole may reveal combinations of coexisting pathology. Three basic types of conduction abnormality comprise:

- reduced amplitude with a relatively normal latency;
- increased latency with relatively normal amplitude;
- absent response.

With axonotmesis, a shock applied proximal to the site of lesion evokes a small response in proportion to the number of surviving axons. In contrast, shocks applied distal to the lesion elicit a normal response initially. During the first 5–10 days, however, distal stimulation also progressively fails to elicit a full response as Wallerian degeneration causes the failure of neuromuscular transmission and loss of nerve excitability. Therefore, stimulation above and below the lesion eventually elicits an equally small response during the second week of injury. In neurapraxia, which usually suggests demyelination, distal stimulation gives rise to a normal CMAP, but proximal stimulation reveals slowed conduction.

An absent response to proximal stimulation indicates either a complete conduction block or axonal degeneration involving most motor axons. Again, stimulation below the point of the lesion can differentiate the two during the second week, when Wallerian degeneration has evolved—a normal or near normal potential in neurapraxia versus absence of response in axonotmesis. Serial electrophysiological studies help delineate progressive recovery in amplitude of the evoked potentials proportional to the degree of nerve regeneration after axonal loss.

A loss of the fast conducting fibres causes a mild prolongation of latency and slowing of conduction velocity in the absence of demyelination. Solely based on an axonal loss, amplitude reduction to 80% of the control value may cause a decrease in conduction velocity to 80% of the lower limit of normal. If the amplitude falls further, to less than half the mean normal value, the conduction velocity may fall to 70% of the lower limits. For the same reason, slowing of motor conduction may result from a loss of the anterior horn cells in a patient with myelopathy or amyotrophic lateral sclerosis.

F waves as a measure of motor conduction

F wave, which results from the backfiring of antidromically activated anterior horn cells (15) serves as a measure of the entire motor pathway of the peripheral axons. A few limiting factors allow backfiring to materialize in only 1–5% of the motor neuron population invaded by antidromic impulses. The presumed impedance mismatch tends to prevent the entry of antidromic impulses into the cell body. The recurrent discharge cannot propagate distally during the refractory period of 1.0 ms at the axon hillock after the passage of an antidromic impulse. Somatodendrite spikes would abate when the Renshaw cells, activated by antidromic impulses, inhibit the anterior horn cells with a synaptic delay of 2.0 ms.

Recording procedures

Clinical studies include stimulation of the median and ulnar nerves at the wrist and of the tibial and peroneal nerves at the ankle. With stimulation at the elbow or knee, M latency increases, whereas F latency decreases by the same amount. Thus, Fd + Md = Fp + Mp, where Fd and Fp, and Md and Mp represent the latencies of F wave and M response elicited by distal and proximal stimulation. This equation allows calculation of proximal F latency by Fp = Fd + Md – Mp.

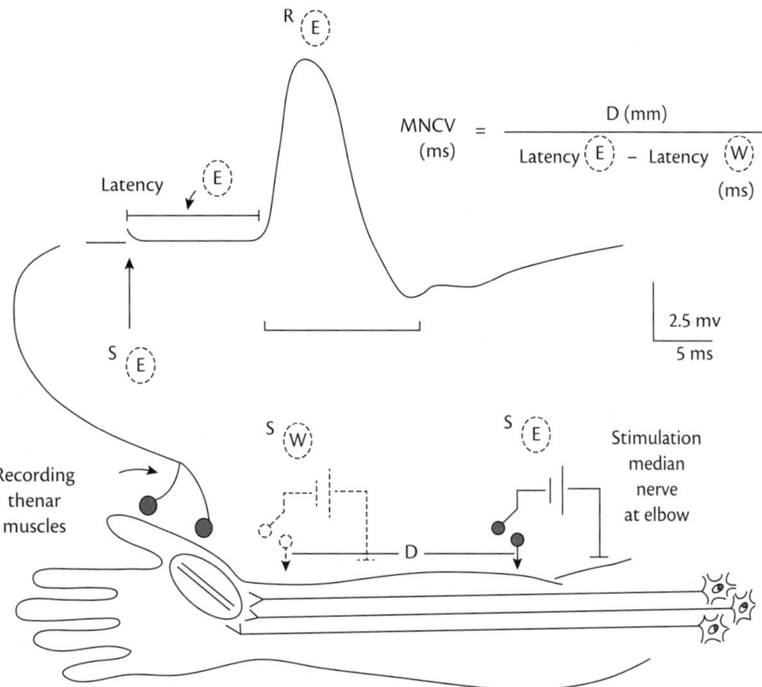

Fig. 6.2 Compound muscle action potential recorded from the thenar eminence following the stimulation of the median nerve at the elbow. The nerve conduction time from the elbow to the wrist equals the latency difference between the two responses elicited by the distal and proximal sites of stimulation. The motor nerve conduction velocity (MNCV) calculated by dividing the surface distance between the stimulus points by the subtracted times concerns the fastest fibres.

Proximal latency

The latency difference between the F wave and the M response (F – M), equals the proximal latency from the stimulus point to and from the spinal cord. Subtracting an estimated delay of 1.0 ms for the turnaround time at the cell and dividing by two, (F – M – 1)/2, yields the conduction time along the proximal segment from the stimulus site to the spinal cord. The use of a very proximal M response elicited by magnetic stimulation applied over the spine allows calculation of root conduction time (18).

F wave conduction velocity

F wave latencies suffice for sequential changes in the same subject or for unilateral lesions affecting one nerve by comparison with the normal side or to another nerve in the same limb as the baseline. Otherwise, surface determination of the limb length or the patient's height helps adjust F wave latencies among different individuals. We estimate the nerve length corresponding to the proximal latency by measuring from the stimulus point to the C7 spinous process via the axilla and midclavicular point in the upper limbs and from the stimulus site to the T12 spinous process by way of the knee and greater trochanter of the femur in the lower limb. The estimated nerve length (D) divided by the calculated proximal conduction time from the stimulus point to the spinal cord (F – M – 1)/2 yields the F wave conduction velocity (FWCV) as follows:

$$\text{FWCV(m/s)} = \frac{2\text{D(mm)}}{\text{F} - \text{M} - 1\text{(ms)}} \quad \text{[eqn 6.2]}$$

A waves

Another late response is the A wave. A decrease in A-wave latency with more proximal stimulation indicates antidromic passage of the impulse up to the point of division before turning around to proceed distally along another branch not excited by the stimulus. Pathophysiological mechanisms include, in addition to collateral sprouting, ephaptic, and ectopic discharges generated at a hyperexcitable site along the proximal portion of the nerve (19).

Shocks of higher intensity, activating both branches distally, can eliminate a collateral A wave by collision. An ephaptic A wave, however, may persist if the antidromic impulse of the fast conducting axon has already passed the site of ephaptic transmission induced by a neighbouring slow conducting axon. An ectopic A wave generated by antidromic passage of an impulse across a hyperexcitable segment is abolished by paired shocks because of the collision between the ectopically generated orthodromic impulse and the second antidromic impulse.

Sensory nerve conduction studies

Waveform, amplitude, and duration

Sensory nerve conduction studies in the upper limbs comprise stimulation of the nerve trunk and recording of an antidromic digital potential distally (Fig. 6.1) or mixed nerve potential proximally. Alternatively, stimulation of the digital nerves elicits an orthodromic sensory potential in the nerve trunk proximally (20,21). Although surface recording provides reliable results (22,23), the use of needle electrodes improves signal-to-noise ratios, especially if combined with signal averaging (5,24). Small late components thus recorded originate from the affected fibres of about 4 μm in diameter, and 15 m/s in velocity. These abnormalities may constitute the only findings in some neuropathies which otherwise would escape detection.

Antidromic potentials from the digital nerves, which lie superficially, yield a higher amplitude than the orthodromic response from the nerve trunk. The position of the recording electrodes relative to the nerve of interest dictates the waveform of sensory nerve action potentials. Antidromic potentials generally have an initially negative biphasic waveform, whereas orthodromic potentials show an initially positive triphasic waveform. Both surface and needle recordings vary substantially among different subjects and to a lesser extent between the two sides in the same individual. A gender difference, favouring women, probably reflects a more superficial location of the nerve (25).

Latency and conduction velocity

A sensory latency consists only of nerve propagation time to the recording point, allowing calculation of sensory conduction velocity with stimulation at a single site. In studying a focal pathology, 'inching' of the stimulus in short increments can isolate the lesion more precisely based on an abrupt change in latency and waveform of the antidromic sensory potential (15). An orthodromic approach using multiple recording electrodes placed over a short segment can also document latency abnormalities (26). Unlike the antidromic recording, however, orthodromic recording from different sites poses difficulty in assessing the changes in waveform or amplitude, which primarily reflects the depth of the nerve from the skin surface.

Type of abnormality

The same discussion, in principle, applies to both motor and sensory conduction abnormalities. Studies of the sural nerve serves as one of the most sensitive measures to diagnose a length-dependent distal axonal polyneuropathy (27). Patients with neuropathy often have a reduction of the sural to radial nerve ratio of sensory potentials to 0.40 or less compared with the normal mean of 0.71 (28). Unlike plexopathy, root avulsion spares the sensory fibres still attached to the ganglion. Intraforaminal radicular lesions, however, may occasionally involve the ganglion or post-ganglionic portion of the roots affecting the digital nerve potential. In this case, radiculopathy shows selective sensory change of specific digits; the first digit by C6, the second and third digits by C7, and the fourth and fifth digits by C8 lesions, in contrast to plexopathy, which tends to affect multiple digits.

Clinical value and limitations

Physiological variation among different nerve segments

The nerves in the leg with longer axons conduct 7–10 m/s slower than those in the arms with shorter axons. Studies of conduction velocity show no statistical difference between median and ulnar nerves, or between tibial and peroneal nerves, and a high degree of symmetry between the two sides for each nerve.

The F-wave conduction velocity between cord and axilla exceeds the motor nerve conduction velocity between elbow and wrist. The two proximal segments, cord-to-axilla and axilla-to-elbow, however, show no significant difference. These findings suggest a faster conduction in the proximal than in the distal nerve segments with a non-linear slowing over the most distal segment.

Table 6.1 Normal motor nerve conduction velocities (m/s)

Age	Ulnar nerve	Median nerve	Peroneal nerve
0–1 week	32 (21–39)	29 (21–38)	39 (19–31)
1 week to 4 months	42 (27–53)	34 (22–42)	36 (23–53)
4 months to 1 year	49 (40–63)	40 (26–58)	48 (31–61)
1–3 years	59 (47–73)	50 (41–62)	54 (44–74)
3–8 years	66 (51–76)	58 (47–72)	57 (46–70)
8–16 years	68 (58–78)	64 (54–72)	57 (45–74)
Adults	63 (52–75)	63 (51–75)	56 (47–63)

Adapted from *Acta Paediatrica*, **52**(Suppl. S146), Gamstorp I, Normal conduction velocity of ulnar, median and peroneal nerves in infancy, childhood and adolescence, pp. 68–76, copyright (1963), with permission from John Wiley and Sons.

Effects of temperature and myelination

A lower body temperature causes slowing of conduction velocity and augmentation in amplitude of nerve and muscle (29–31). Distal latencies increase by 0.3 ms per degree for both median and ulnar nerves upon cooling the hand. Cold-induced delay in Na^+ channel opening probably causes slowing of conduction, whereas its slow inactivation accounts for an increase in amplitude. In demyelinating axons, a temperature rise quickens the activation of Na^+ channels, facilitating impulse propagation over the length of a fibre. Fast inactivation of Na^+ channels, however, reduces the action potential to below the critical level leading to a conduction block (32). Thus, change in temperature induces two completely separate effects in latency and amplitude.

Maintaining ambient temperature between 31°C (70°F) and 33°C (74°F) reduces this type of variability. The use of an infrared heat lamp, prior immersion of the limb in warm water for 30 min or standardized exercise (33) helps maintain the skin temperature above 32°C as measured with a thermistor. If the body temperature remains low, the addition of 4% of the calculated conduction velocity for each degree below 32°C normalizes the result. This formula, established in a normal population, however, may not necessarily apply in the diseased peripheral nerve.

The process of maturational myelination accompanies a rapid increase in nerve conduction velocities from roughly half the adult value in full-term infants to the adult range at age 3–5 years. One series (Table 6.1) showed a steep increase in conduction of the peroneal nerve in infancy as compared to a slower maturation of the median nerve during early childhood (34). Conduction velocities in premature infants have an even slower range. The values reported at 23–24 weeks of foetal life averaged roughly one-third those of newborns of normal gestational age (35). Studies based on the expected date of birth of premature infants showed a different time course of maturation for motor and proprioceptive conduction (36).

In later childhood and adolescents from age 3 to 19 years, both motor and sensory conduction velocities change as a function of age and growth in length, showing a slight increase in the upper

Table 6.2 Normal latencies (ms) and conduction velocities (m/s) in different age groups (mean ± SD)

Nerve	Age 10–35 years (30 cases)		Age 36–50 years (16 cases)		Age 51–80 years (18 cases)	
	Sensory	Motor	Sensory	Motor	Sensory	Motor
Median nerve						
Digit–wrist	67.5 ± 4.7		65.8 ± 5.7		59.4 ± 4.9	
Wrist–muscle		3.2 ± 0.3*		3.7 ± 0.3*		3.5 ± 0.2*
Wrist–elbow	67.7 ± 4.4	59.3 ± 3.5	65.8 ± 3.1	55.9 ± 2.6	62.8 ± 5.4	54.5 ± 4.0
Elbow–axilla	70.4 ± 4.8	65.9 ± 5.0	70.4 ± 3.4	65.1 ± 4.2	66.2 ± 3.6	63.6 ± 4.4
Ulnar nerve						
Digit–wrist	64.7 ± 3.9			66.5 ± 3.4		57.5 ± 6.6
Wrist–muscle		2.7 ± 0.3*		2.7 ± 0.3*		3.0 ± 0.35*
Wrist–elbow	64.8 ± 3.8	58.9 ± 2.2	67.1 ± 4.7	57.8 ± 2.1	56.7 ± 3.7	53.3 ± 3.2
Elbow–axilla	69.1 ± 4.3	64.4 ± 2.6	70.6 ± 2.4	63.3 ± 2.0	64.4 ± 3.0	59.9 ± 0.7
Common peroneal nerve						
Ankle–muscle		4.3 ± 0.9*		4.8 ± 0.5*		4.6 ± 0.6*
Ankle–knee	53.0 ± 5.9	49.5 ± 5.6	50.4 ± 1.0	43.6 ± 5.1	46.1 ± 4.0	43.9 ± 4.3
Posterior tibial nerve						
Ankle–muscle		5.9 ± 1.3*		7.3 ± 1.7*		6.0 ± 1.2*
Ankle–knee	56.9 ± 4.4	45.5 ± 3.8	49.0 ± 3.8	42.9 ± 4.9	48.9 ± 2.6	41.8 ± 5.1
H reflex, popliteal		71.0 ± 4.0		64.0 ± 2.1		60.4 ± 5.0
Fossa						
		27.9 ± 2.2*		28.2 ± 1.5*		32.0 ± 2.1*

*Latency in milliseconds.

Values are means ± 1 SD.

Adapted from *Neurology*, **13**(12), Mayer, RF, Nerve conduction studies in man, pp. 1021–30, copyright (1963), with permission from Wolters Kluwer Health, Inc.

limb, and slight decrease in the lower limb. Conduction velocities begin to decline after 30–40 years of age, but not more than 10 m/s by 60 years or even by 80 years of age. In one study (37), a reduction in the mean conduction rate averaged 10% at 60 years of age (Table 6.2). Preferential loss of the largest and fastest conducting motor units probably results in a gradual increase in latencies of the F wave and somatosensory evoked potentials with advancing age.

Height and other factors

Nerve conduction measures also reflect various physical characteristics. Studies of sural, peroneal, and tibial nerves show an inverse relationship between height and conduction velocity (38).

Clinical characteristics

Nerve conduction studies may provide certain characteristic patterns of abnormality, which can indicate the relatively specific nature of clinical disorders (39,40). Patients with amyotrophic lateral sclerosis typically show reduced amplitude of compound muscle, but not sensory, action potential with increased F wave latencies and slowing of distal motor conduction velocities (41). Hereditary neuropathies show little difference from one nerve to another in the same patient and among different members in the same family. Similar involvement of different nerve fibres causes a small temporal dispersion, despite a very prolonged latency. In contrast, acquired demyelination with disproportionate involvement of various nerve segments results in more asymmetrical abnormalities with pathological temporal dispersion. The reverse, however, does not always hold, as acquired disorders may show diffuse abnormalities reminiscent of hereditary pathology. A disproportionate reduction in sensory amplitude of the median as compared with sural nerve supports the diagnosis of a primary demyelination (42), whereas a greater reduction of sural over radial nerve response suggests an axonal polyneuropathy (43,44).

Studies of individual nerves

Cranial nerves

Facial nerve

After stimulation of the facial nerve anterior to the mastoid process, the evoked response rarely reveals a clear delay in latency, even with substantial axonal degeneration because the remaining axons conduct normally. Amplitude comparison between the two

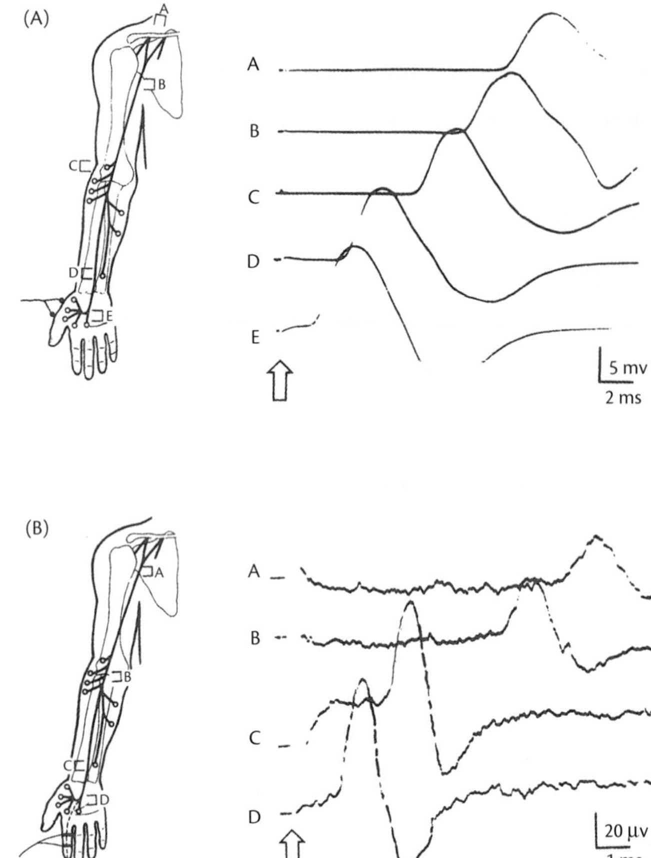

Fig. 6.3 (A) Motor nerve conduction study of the median nerve. The sites of stimulation include Erb's point (A), axilla (B), elbow (C), wrist (D), and palm (E). Compound muscle action potentials are recorded with surface electrodes placed on the thenar eminence. (B) Sensory nerve conduction study of the median nerve. The sites of stimulation include axilla (A), elbow (B), wrist (C), and palm (D). Antidromic sensory potentials are recorded with a pair of ring electrodes placed around the second digit.

sides, however, serves well in determining the degree of axonal loss, which in turn predicts the prognosis.

Accessory nerve

Stimulation of the accessory nerve along the posterior border of the sternocleidomastoid elicits a motor response of the trapezius recordable with a pair of electrodes over the belly of the muscles and the tendon.

Other nerves

Studies of hypoglossal nerve may help characterize obstructive sleep apnoea (45).

Upper limb nerves

Median nerve

The stimulation sites include Erb's point, axilla, elbow, wrist and palm (46,47,48) along the relatively superficial course of the nerve (Fig. 6.3). Table 6.3 (upper half) summarizes normal values of responses recorded from the thenar eminence. Recording from the second lumbrical and the first palmar interosseous using the same electrodes after stimulation of the median and ulnar nerve at the wrist allows latency comparison between the two (49–52). Recording from the abductor pollicis brevis and adductor pollicis by the same electrode placed on the thenar eminence renders an equally sensitive comparison between these two nerves. A difference exceeding 0.5 ms suggests an abnormal delay over the distal segment as might be seen in the carpal tunnel syndrome.

Table 6.3 (lower half) summarizes our normal values for the antidromic digital potentials recorded with ring electrodes placed 2 cm apart around the proximal (E1) and distal (E2) interphalangeal joints of the index finger. Median and ulnar nerve sensory potentials of the ring finger have nearly the identical latencies when elicited with stimulation at the wrist for the same conduction distance. In our study, the latency difference between the two nerves greater than 0.4 ms (mean + 2SD) indicates an abnormality. Spread of the radial sensory potential may mask the abnormality in the orthodromic sensory conduction (53).

Ulnar nerve

Routine motor conduction studies (Table 6.4 upper half) use the motor response (Fig. 6.4) from the hypothenar muscle (54,55). An additional recording from the first dorsal interosseous or adductor pollicis helps evaluate the deep palmar nerve (11,56,57). The muscle potential elicited by stimulation of this branch serves as a good measure of surviving motor axons. With a lesion at the wrist, the lumbrical–interosseous comparison described earlier often shows a latency difference greater than 0.5 ms, documenting abnormal slowing of the ulnar nerve compared to the median nerve (58).

Table 6.4 (lower half) summarizes our normal values for an antidromic sensory potential recorded from the little finger after stimulation of the nerve above and below the elbow, 3 cm proximal to the distal crease at the wrist, and 5 cm distal to the crease in the palm. Alternatively, stimulation of the little finger with ring electrodes placed around the interphalangeal joints gives rise to an orthodromic sensory potential recordable at various sites along the course of the nerve. Radial nerve cutaneous branch may show anomalous innervation to the ulnar dorsum of the hand (59,60).

Martin–Gruber anastomosis

This motor anastomosis, reported in 15–50% of an unselected population, originates from the anterior interosseous nerve a few centimetre distal to the medial humeral epicondyle and, usually, though not always, supplies ordinarily ulnar-innervated intrinsic hand muscles (61). Other rare anomalies in the forearm include communication crossing from the ulnar to the median nerve (62), which occasionally involves only the sensory axons.

In the presence of Martin–Gruber anastomosis, stimulation of the median nerve at the elbow coactivates the communicating ulnar nerve fibres, producing a larger amplitude compared with the distally elicited response, which lacks the ulnar component. Stimulation of the ulnar nerve at the elbow spares the communicating branch still attached to the median nerve, evoking only a partial response. Stimulation at the wrist, activating the additional anomalous fibres, gives rise to a full response, mimicking a conduction block at the elbow.

In a rare condition, called all-median hand, all the intrinsic hand muscles usually supplied by the ulnar nerve receive

Table 6.3 Median nerve*

Site of stimulation	Amplitude†: motor (mv) sensory (μv)	Latency‡ to recording site (ms)	Difference between right and left (ms)	Conduction time between two points (ms)	Conduction velocity (m/s)
Motor fibres					
Palm	6.9 ± 3.2(3.5)§	1.86 ± 0.28 (2.4)¶	0.19 ± 0.17 (0.5)¶		
				1.65 ± 0.25 (2.2)¶	48.8 ± 5.3 (38)**
Wrist	7.0 ± 3.0 (3.5)	3.49 ± 0.34 (4.2)	0.24 ± 0.22 (0.7)		
				3.92 ± 0.49 (4.9)	57.7 ± 4.9 (48)
Elbow	7.0 ± 2.7 (3.5)	7.39 ± 0.69 (8.8)	0.31 ± 0.24 (0.8)		
				2.42 ± 0.39 (3.2)	63.5 ± 6.2 (51)
Axilla	7.2 ± 2.9 (3.5)	9.81 ± 0.89 (11.6)	0.42 ± 0.33 (1.1)		
Sensory fibres					
Digit					
				1.37 ± 0.24 (1.9)	58.8 ± 5.8 (47)
Palm	39.0 ± 16.8 (20)	1.37 ± 0.24 (1.9)	0.15 ± 0.11 (0.4)		
				1.48 ± 0.18 (1.8)	56.2 ± 5.8 (44)
Wrist	38.5 ± 15.6 (19)	2.84 ± 0.34 (3.5)	0.18 ± 0.14 (0.5)		
				3.61 ± 0.48 (4.6)	61.9 ± 4.2 (53)
Elbow	32.0 ± 15.5 (16)	6.46 ± 0.71 (7.9)	0.29 ± 0.21 (0.7)		

*Mean ± standard deviation (SD) in 122 nerves from 61 healthy subjects, 11 to 74 years of age (average. 40), with no apparent disease of the peripheral nerves.

†Amplitude of the evoked response, measured from the baseline to the negative peak.

‡Latency, measured to the onset of the evoked response.

§Lower limits of normal, based on the distribution of the normative data.

¶Upper limits of normal, calculated as the mean + 2 SD.

**Lower limits of normal, calculated as the mean − 2 SD.

innervation via the communicating fibres, thus escaping a severe damage to the ulnar groove at the elbow. An injury to the median nerve at the elbow may cause spontaneous discharges in the ulnar-innervated intrinsic hand muscles.

Radial nerve

Stimulation at the supraclavicular fossa, spinal groove near the axilla, above the elbow and forearm (Fig. 6.5) elicits evoked potentials recordable from the extensor digitorum communis or the extensor indicis. The sensory branches, after giving off the posterior antebrachial cutaneous nerve in the forearm, emerge near the surface about 10 cm above the lateral styloid process. Surface stimulation here evokes an antidromic sensory potential recordable by a pair of ring electrodes placed around the thumb, or by the disc electrode over the first web space or slightly more proximally in the snuffbox (63,64).

Other nerves

Other nerves occasionally tested include palmar cutaneous (65), medial (55), lateral and posterior (66), antebrachial cutaneous, median palmar cutaneous (67), and anterior interosseous nerves (68).

Nerves of the shoulder girdle

Phrenic nerve

Stimulation of the phrenic nerve along the posterior border of the sternocleidomastoid muscle induces a diaphragmatic action potential, which is seen as a negativity over the sternum and a positivity over the 8th rib along the anterior axillary line. A pair of surface electrodes placed over these recording sites yields the largest amplitude being the difference of the two potentials of opposite polarity (69). Diaphragmatic contraction causes a hiccup or interruption of voluntarily sustained vocalization. Excessive stimulation may coactivate the brachial plexus behind the anterior scalene muscle. Normal values (Table 6.5) decrease from 6 to 8 ms in infancy to about 5 ms at 1 year of age, and increase again to 6–7 ms at 10 and 18 years of (70–72). In addition to conduction studies, the technique may have applications for diaphragmatic pacing in patients with spinal cord injuries (73).

Brachial plexus

Stimulation at Erb's point activates the anterior rami of the spinal nerves derived from the C5 to C8 and T1 roots, which innervate proximal muscles of the shoulder girdle. Commonly tested

Table 6.4 Ulnar nerve*

Site of stimulation	Amplitude†: motor (mv) sensory (μv)	Latency‡ to recording site (ms)	Difference between right and left (ms)	Conduction time between two points (ms)	Conduction velocity (m/s)
Motor fibres					
Wrist	5.7 ± 2.0 (2.8)§	2.59 ± 0.39 (3.4)¶	0.28 ± 0.27 (0.8)¶		
				3.51 ± 0.51 (4.5)¶	58.7 ± 5.1 (49)**
Below elbow	5.5 ± 2.0 (2.7)	6.10 ± 0.69 (7.5)	0.29 ± 0.27 (0.8)		
				1.94 ± 0.37 (2.7)	61.0 ± 5.5 (50)
Above elbow	5.5 ± 1.9 (2.7)	8.04 ± 0.76 (9.6)	0.34 ± 0.28 (0.9)		
				1.88 ± 0.35 (2.6)	66.5 ± 6.3 (54)
Axilla	5.6 ± 2.1 (2.7)	9.90 ± 0.91 (11.7)	0.45 ± 0.39 (1.2)		
Sensory fibres					
Digit					
				2.54 ± 0.29 (3.1)	54.8 ± 5.3 (44)
Wrist	35.0 ± 14.7 (18)	2.54 ± 0.29 (3.1)	0.18 ± 0.13 (0.4)		
				3.22 ± 0.42 (4.1)	64.7 ± 5.4 (53)
Below elbow	28.8 ± 12.2 (15)	5.67 ± 0.59 (6.9)	0.26 ± 0.21 (0.5)		
				1.79 ± 0.30 (2.4)	66.7 ± 6.4 (54)
Above elbow	28.3 ± 11.8 (14)	7.46 ± 0.64 (8.7)	0.28 ± 0.27 (0.8)		

*Mean ± standard deviation (SD) in 130 nerves from 65 healthy subjects, 13 to 74 years of age (average, 39), with no apparent disease of the peripheral nerves.

†Amplitude of the evoked response, measured from the baseline to the negative peak.

‡Latency, measured to the onset of the evoked response, with the cathode 3 cm above the distal crease in the wrist.

§Lower limits of normal, based on the distribution of the normative data.

¶Upper limits of normal, calculated as the mean + 2 SD.

**Lower limits of normal, calculated as the mean − 2 SD.

Reproduced from Kimura, J, *Electrodiagnosis in diseases of nerve and muscle: principles and practice*, 4th edn, copyright (2013), with permission from Oxford University Press.

nerves by this means include the axillary nerves with recording from the deltoid and the long thoracic nerve with recording from the serratus anterior (74). Although the remaining axons tend to show a relatively normal latency, the amplitude of the recorded response decreases in proportion to the loss of axons. Normal values vary considerably among different subjects and between the two sides in the same individual. The amplitude preservation above one half of the normal side usually suggests a good prognosis.

Musculocutaneous nerve

Motor conduction studies consist of stimulation of the nerve above the clavicle just behind the sternocleidomastoid muscle and at the axilla medial to the axillary artery, and recording the muscle action potentials from the biceps brachii. Stimulation of the terminal sensory branch, lateral antebrachial cutaneous nerve above the elbow elicits an antidromic sensory potential recordable over the lateral aspect of the forearm as a test of the C6 dermatome.

Lower limb nerves

Tibial nerve

Stimulation of the tibial nerve at the popliteal fossa and medial to the medial malleolus elicits the muscle response recordable from the abductor hallucis medially and abductor digiti quinti laterally (Fig. 6.6). Table 6.6 summarizes the normal values in our laboratory.

Common and deep peroneal nerve

Routine studies consist of stimulating the nerve above and below the head of the fibula, and half way between the medial and lateral malleolus and recording the muscle action potential from the extensor digitorum brevis (Fig. 6.7). Separating the two stimulus sites across the knee improves the accuracy in calculating the conduction velocity. A shorter incremental stimulation, however, delineates a focal conduction abnormality much better, documenting an abrupt change in latency and waveform. The extensor digitorum longus or tibialis anterior serve as useful substitutes in patients with an atrophic extensor digitorum brevis. Table 6.7 summarizes the normal values in our laboratory.

Accessory deep peroneal nerve

In 20–30% of the general population, the extensor digitorum brevis receives partial innervation from the superficial peroneal nerve via a communicating branch called the accessory deep peroneal nerve. The anomalous fibres descend on the lateral aspect of the leg before passing behind the lateral malleolus to supply the lateral portion of the muscle. The anomaly, inherited as dominant trait, may innervate the extensor digitorum brevis exclusively without contribution from the deep peroneal nerve.

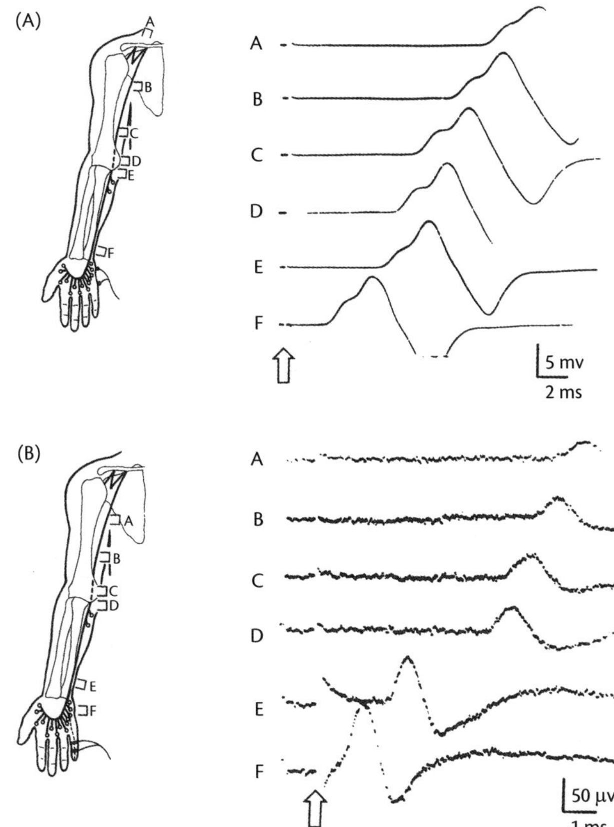

Fig. 6.4 (A) Motor nerve conduction study of the ulnar nerve. The sites of stimulation include Erb's point (A), axilla (B), above the elbow (C), elbow (D), below the elbow (E), and wrist (F). Compound muscle action potentials are recorded with surface electrodes placed on the hypothenar eminence. (B) Sensory nerve conduction study of the ulnar nerve. The sites of stimulation include axilla (A), above the elbow (B), elbow (C), below the elbow (D), wrist (E), and palm (F). The tracings show antidromic sensory potentials recorded with the ring electrodes placed around the fifth digit.

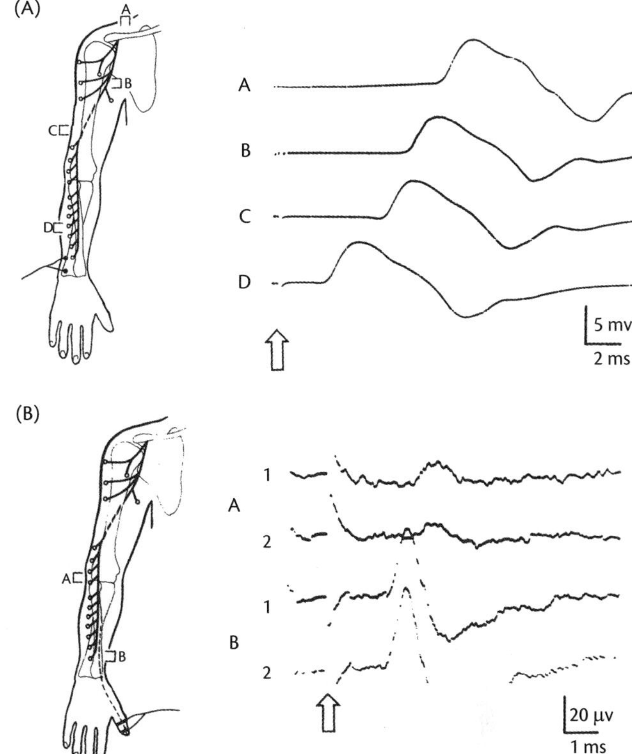

Fig. 6.5 (A) Motor nerve conduction study of the radial nerve. The sites of stimulation include Erb's point (A), axilla (B), above the elbow (C), and mid-forearm (D). Compound muscle action potentials are recorded from the extensor indicis with a pair of surface electrodes. (B) Sensory nerve conduction study of the radial nerve. The sites of stimulation include elbow (A) and distal forearm (B). Antidromic sensory potentials are recorded using the ring electrodes placed around the first digit.

Injury to the deep peroneal nerve, causing weakness of the dorsiflexors of the foot, may spare the lateral part of the extensor digitorum brevis supplied by the anastomosis. Stimulation of the common peroneal nerve at the knee excites a full response, which equals to the sum of responses from the deep peroneal nerve stimulated at the ankle and from the accessory deep peroneal nerve stimulated behind the lateral malleolus. The collision technique may help identify isolated abnormalities by selective blocking of unwanted impulses via the communicating branch (75).

Table 6.5 Phrenic nerve latency of different age groups

Age range	N	Mean ± SD
0–6 months	45	6.0 ±1.6
6 months to 1 year	34	5.0 ± 1.2
1–2 years	34	4.8 ± 0.8
2–5 years	34	4.9 ± 0.8
5–10 years	34	5.5 ± 0.8
10–18 years	20	6.3 ± 1.2

Superficial peroneal nerve

This mixed nerve, when stimulated with the cathode placed against the anterior edge of the fibula, elicits the antidromic sensory potential at the ankle just medial to the lateral malleolus. The study helps distinguish a distal lesion from an L5 radiculopathy, which spares sensory potentials (76–78).

Sural nerve

This sensory nerve originates from the tibial nerve in the popliteal fossa, receiving a communicating branch from the common peroneal nerve. Stimulation of the nerve in the lower third of the leg posteriorly lateral to the midline elicits an antidromic sensory potentials along the posterior edge of the lateral malleolus (Fig. 6.8). The study of this nerve allows comparison between electrophysiological and histological findings (79).

Table 6.6 Tibial nerve*

Site of stimulation	Amplitude[†] (mv)	Latency[‡] to recording site (ms)	Difference between two sides (ms)	Conduction time between two points (ms)	Conduction velocity (m/s)
Ankle	5.8 ± 1.9 (2.9)[§]	3.96 ± 1.00 (6.0)[¶]	0.66 ± 0.57 (1.8)[¶]		
				8.09 ± 1.09 (10.3)[¶]	48.5 ± 3.6 (41)[**]
Knee	5.1 ± 2.2 (2.5)	12.05 ± 1.53 (15.1)	0.79 ± 0.61 (2.0)		

*Mean ± standard deviation (SD) in 118 nerves from 59 healthy subjects, 11 to 78 years of age (average, 39), with no apparent disease of the peripheral nerves.

†Amplitude of the evoked response, measured from the baseline to the negative peak.

‡Latency, measured to the onset of the evoked response, with a standard distance of 10 cm between the cathode and the recording electrode.

§Lower limits of normal, based on the distribution of the normative data.

¶Upper limits of normal, calculated as the mean + 2 SD.

**Lower limits of normal, calculated as the mean − 2 SD.

Sural nerve studies serve as one of the most sensitive measures for detecting various types of neurogenic (80,81) abnormality and response to therapy (82). The sural to radial amplitude ratio may help document abnormalities not apparent based on the absolute values. Preganglionic lesions spare the sensory action potentials despite a clinical sensory loss in an S1 or S2 radiculopathy or with cauda equina lesion.

Nerves of the pelvic girdle

Lumbosacral plexus

Needle or high voltage surface stimulation (83,84) of the L4, L5, or S1 spinal nerves helps evaluate the lumbar plexus derived from the L2, L3, and L4 roots, and the sacral plexus arising from the L5, S1, and S2 roots. This, combined with distal stimulation of the plexus, allows calculation of the latency difference, which equals the conduction time through the plexus. The F wave and H reflex serve as alternative, indirect measures of nerve conduction across this region.

Femoral nerve

Stimulation of the femoral nerve above or below the inguinal ligament elicits a muscle potential in the rectus femoris. When recorded at various distances from the stimulus site, the latency increases progressively reflecting the vertically orientated end-plate region.

Other nerves

Other nerves of interest for conduction studies include (85) medial and lateral plantar (86–89) lateral femoral cutaneous (90–92) and digital nerves of the foot (93).

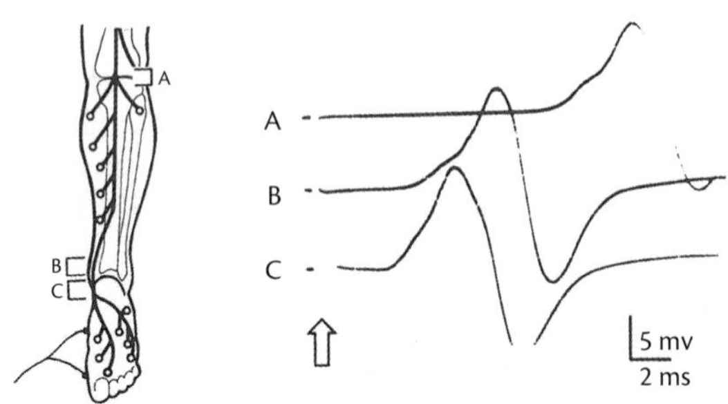

Fig. 6.6 Motor nerve conduction study of the tibial nerve. The sites of stimulation include knee (A), above the medial malleolus (B), and below the medial malleolus (C). Compound muscle action potentials are recorded with surface electrodes placed over the abductor halluces.

Waveform analysis and other aspects

Technical errors

Often overlooked sources of error include technical problems, which account for most unexpected findings. These include intermittent power failure, excessive spread of stimulation current, anomalous innervation, temporal dispersion, inaccuracy of surface measurement and inadvertent anodal stimulation (94). Isolated abnormalities may lead to an erroneous conclusion unless interpreted with caution and in the clinical context. Composite scores (95,96), rather than individual attributes and the use of percentiles and normal deviates (97) may prove more useful for overall assessment of dysfunction. An automated hand-held nerve conduction device usually provides inadequate information, lacking waveform analysis (98,99).

Current spread and collision technique

The current intended for the median nerve may spread to the neighbouring ulnar nerve at the axilla. The measured onset latency will then reflect the normal ulnar response, which precedes the slow median response in the carpal tunnel syndrome. A second stimulus delivered to the ulnar nerve at the wrist induces an antidromic volley, which collides with the orthodromic ulnar volley from the axilla, leaving only the median impulses to reach the muscle. Similarly, the use of a distal stimulus can block the median nerve in selective recording of the ulnar response after coactivation of both nerves at the axilla in the study of a tardy ulnar palsy.

In either case, distal stimulation achieves a physiological nerve block through collision, allowing selective recording of the median or ulnar component despite coactivation of both nerves proximally (15). Delaying the proximal stimulus by a few milliseconds optimally separates the incidental response induced by the

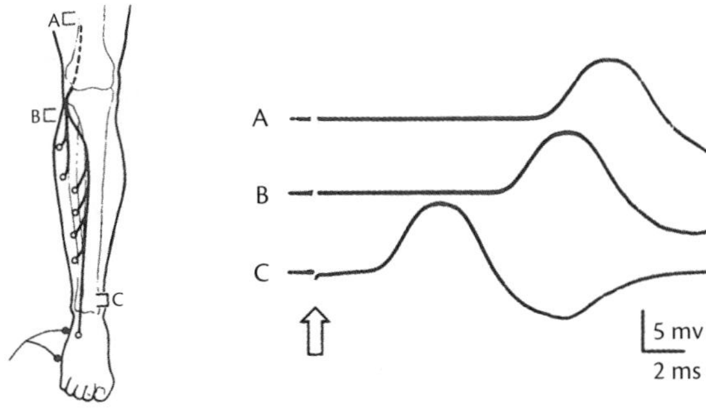

Fig. 6.7 Motor nerve conduction study of the common peroneal nerve. The sites of stimulation include above the knee (A), below the knee (B), and ankle (C). Compound muscle action potentials are recorded with surface electrodes over the extensor digitorum brevis.

distal stimulus from the intended response elicited by the proximal stimulus. To achieve a collision, this delay should not exceed the conduction time between the distal and proximal points of stimulation plus 1 ms or the absolute refractory period.

Temporal dispersion and phase cancellation

Physiological temporal dispersion

The physiologically slower impulses lag behind the fast ones in proportion to the distance of the conduction path. Thus, the longer the distance, the greater the desynchronization among different fibres, leading to a response with smaller amplitude and longer duration. Contrary to common belief, reduction in area of a diphasic action potential also results from phase cancellation between the opposite peaks. The physiological phase cancellation affects muscle responses relatively little because, with the same slight shift in latency, longer duration motor unit potentials still superimpose nearly in phase rather than out of phase (15). This rule does not apply to short duration muscle action potentials such as those recorded from the intrinsic foot muscles.

Pathological temporal dispersion

In evaluating peripheral neuropathies with segmental block, the size of the recorded response serves as a measure of the number of excitable nerve axons. A difference between proximal and distal responses, however, does not necessarily imply a conduction block.

In a demyelinating neuropathy, a long duration motor response also diminishes dramatically simply due to phase cancellation between normally conducting and pathologically slow fibres (15). This type of phase cancellation reduces the amplitude of the muscle response well beyond the usual physiological limits, giving rise to a false impression of motor conduction block. A maximal phase cancellation results from a latency shift in the order of one-half the total duration of unit discharge. With further separation, excessive desynchronization may now counter the physiological phase cancellation, sometimes paradoxically increasing the size of the response.

The commonly used criteria based on percentage reduction to distinguish pathological from physiological temporal dispersion actually holds only in entirely standardized studies because variables such as the interelectrode distance between the two recording electrodes influences the outcome substantially (100). As an alternative means, segmental stimulation at more than two sites allow testing the linear relationship between the latency and the size of the recorded responses in physiological phase cancellation (101). A non-linear change in amplitude or waveform indicates either a conduction block associated with clinical weakness or a pathological temporal dispersion, which by itself, should not affect muscle strength. Composite scores as opposed to individual values may improve sensitivity and reproducibility of nerve conduction abnormalities (95).

Clinical assessment of conduction block

Criteria for conduction block

A disproportionately small compound muscle action potential elicited by proximal as compared with distal stimulation serves as a measure of conduction block. This finding usually suggests a demyelinating lesion, although it may result from other reversible

Table 6.7 Common peroneal nerve*

Site of stimulation	Amplitude† (mv)	Latency‡ to recording site (ms)	Difference between right and left (ms)	Conduction time between two points (ms)	Conduction velocity (m/s)
Ankle	5.1 ± 2.3(2.5)§	3.77 ± 0.86 (5.5)¶	0.62 ± 0.61 (1.8)¶		
				7.01 ± 0.89 (8.8)¶	48.3 ± 3.9 (40)**
Below knee	5.1 ± 2.0 (2.5)	10.79 ± 1.06 (12.9)	0.65 ± 0.65 (2.0)		
				1.72 ± 0.40 (2.5)	52.0 ± 6.2 (40)
Above knee	5.1 ± 1.9 (2.5)	12.51 ± 1.17 (14.9)	0.65 ± 0.60 (1.9)		

*Mean ± standard deviation (SD) in 120 nerves from 60 healthy subjects, 16 to 86 years of age (average, 41), with no apparent disease of the peripheral nerves.

†Amplitude of the evoked response, measured from the baseline to the negative peak.

‡Latency, measured to the onset of the evoked response, with a standard distance of 7 cm between the cathode and the recording electrode.

§Lower limits of normal, based on the distribution of the normative data.

¶ Upper limits of normal, calculated as the mean + 2 SD.

**Lower limits of normal, calculated as the mean − 2 SD.

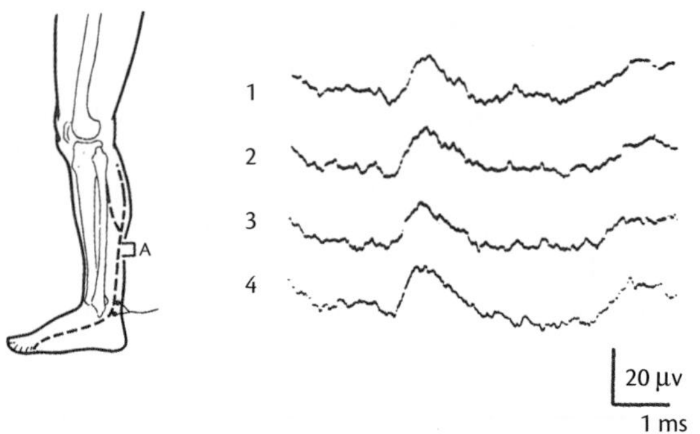

Fig. 6.8 Antidromic sensory nerve conduction study of the sural nerve. The diagram shows stimulation on the calf slightly lateral to the midline in the lower third of the leg, and recording with surface electrodes placed behind the lateral malleolus.

Reproduced from Kimura, J, *Electrodiagnosis in diseases of nerve and muscle: principles and practice*, 4th edn, copyright (2013), with permission from Oxford University Press.

conditions, such as ischaemia and channelopathies (102–104). Electromyography reveals little or no evidence of denervation despite poor recruitment of motor unit potentials, which fire rapidly to compensate for the blocked fibres. Selective damage of the myelin sheath can also cause pathological temporal dispersion and repetitive discharges, broadening the evoked action potential.

The usual criterion for motor conduction block consists of a reduction in amplitude ratio greater than 50% with less than 15% increase in duration. This approach, however, may lead to a false conclusion because phase cancellation alone can induce a similar waveform changes in the absence of conduction block. The combination of clinical and electrophysiological findings delineates motor conduction block more conclusively (105). A vigorous twitch and a large amplitude response elicited by distal stimulation help document conduction block, if associated with clinical weakness and a paucity of voluntarily activated motor unit potentials. The absence of F waves usually implies proximal conduction block (15), although a sustained period of immobility can also results in reversible inexcitability of the anterior horn cells.

Dissimilar responses elicited by distal and proximal stimuli may result from the use of a submaximal stimulus. Structural abnormalities may render the nerve segment inexcitable despite the use of ordinarily adequate stimulation as reported in some cases of multifocal motor neuropathy. Such a failure to excite the involved segment maximally may erroneously suggest a conduction block. In these cases, however, stimulation of a more proximal, unaffected nerve segment will evoke a relatively normal, albeit pathologically dispersed response. The presence of anomalous branches such as the Martin–Gruber anastomosis may also lead to a confusing pattern of responses, sometimes mimicking a conduction block.

Studies of short versus long segments

Segmental stimulation in short increments

'Inching' the stimulus in short increments (15) can isolate a focal lesion more effectively than standard conduction studies, which by including the unaffected segments lowers the sensitivity of

Table 6.8 F waves in normal subject*

No. of nerves tested	Site of stimulation	F-wave latency to recording site (ms)	Difference between right and left (ms)	Central latency† to and from the spinal cord (ms)	Difference between right and left (ms)	Conduction velocity‡ to and from the spinal cord (m/s)
122 median nerves from 61 subjects	Wrist	26.6 ± 2.2 (31)**	0.95 ± 0.67 (2.3)**	23.0 ± 2.1 (27)**	0.93 ± 0.62 (2.2)**	65.3 ± 4.7 (56)††
	Elbow	22.8 ± 1.9 (27)	0.76 ± 0.56 (1.9)	15.4 ± 1.4 (18)	0.71 ± 0.52 (1.8)	67.8 ± 5.8 (56)
	Axilla¶	20.4 ± 1.9 (24)	0.85 ± 0.61 (2.1)	10.6 ± 1.5 (14)	0.85 ± 0.58 (2.0)	
130 ulnar nerves from 65 subjects	Wrist	27.6 ± 2.2 (32)	1.00 ± 0.83 (2.7)	25.0 ± 2.1 (29)	0.84 ± 0.59 (2.0)	65.3 ± 4.8 (55)
	Above elbow	23.1 ± 1.7 (27)	0.68 ± 0.48 (1.6)	16.0 ± 1.2 (18)	0.73 ± 0.52 (1.8)	65.7 ± 5.3 (55)
	Axilla¶	20.3 ± 1.6 (24)	0.73 ± 0.54 (1.8)	10.4 ± 1.1 (13)	0.76 ± 0.52 (1.8)	
120 peroneal nerves from 60 subjects	Ankle	48.4 ± 4.0 (56)	1.42 ± 1.03 (3.5)	44.7 ± 3.8 (52)	1.28 ± 0.90 (3.1)	49.8 ± 3.6 (43)
	Above knee	39.9 ± 3.2 (46)	1.28 ± 0.91 (3.1)	27.3 ± 2.4 (32)	1.18 ± 0.89 (3.0)	55.1 ± 4.6 (46)
118 tibial nerves from 59 subjects	Ankle	47.7 ± 5.0 (58)	1.40 ± 1.04 (3.5)	43.8 ± 4.5 (53)	1.52 ± 1.02 (3.6)	52.6 ± 4.3 (44)
	Knee	39.6 ± 4.4 (48)	1.25 ± 0.92 (3.1)	27.6 ± 3.2 (34)	1.23 ± 0.88 (3.0)	53.7 ± 4.8 (44)

*Mean ± standard deviation (SD) in the same patients shown in Tables 6.1, 6.4, 6.11, and 6.13.

†Central latency = F – M, where F and M represent latencies of the F wave and M response.

‡Conduction velocity = 2D/ (F – M –1), where D indicates the distance from the stimulus point to C7 or T12 spinous process.

¶F (A) = F (E) + M(E) – M(A), where F(A) and F(E) represent latencies of the F wave with stimulation at the axilla and elbow, and M(A) and M(E), latencies of the corresponding M response.

**Upper limits of normal calculated as mean + 2 SD.

††Lower limits of normal calculated as mean – 2 SD.

Reproduced from Kimura, J, *Electrodiagnosis in diseases of nerve and muscle: principles and practice*, 4th edn, copyright (2013), with permission from Oxford University Press.

the study. If the nerve impulse conducting at a rate of 0.2 ms/cm (50 m/s) drops to 0.4 ms/cm in the involved 1-cm segment, the conduction time over a 10-cm segment increases only 10% from 2.0 to 2.2 ms, whereas the same 0.2-ms increase causes a 100% change from 0.2 to 0.4 ms if measured over a 1-cm segment. The large per unit increase in latency more than compensates for the inherent measurement error that might have resulted from inaccurate step changes of the stimulating electrodes (106–108). Even if technical difficulties prevent sequential stimulation near the site of the lesion, a series of responses from above and below the affected zone can characterize a local change by forming two, rather than one, parallel lines accompanied by an abrupt waveform change between the two series of responses.

Late responses for evaluation of long pathways

Table 6.8 summarizes the normal ranges and the upper and lower limits of F wave latency and other aspects of the F-wave values. Fig. 6.9 shows F-wave latencies plotted against the subject's height, indicating the upper limit of normal as mean +2 SD.

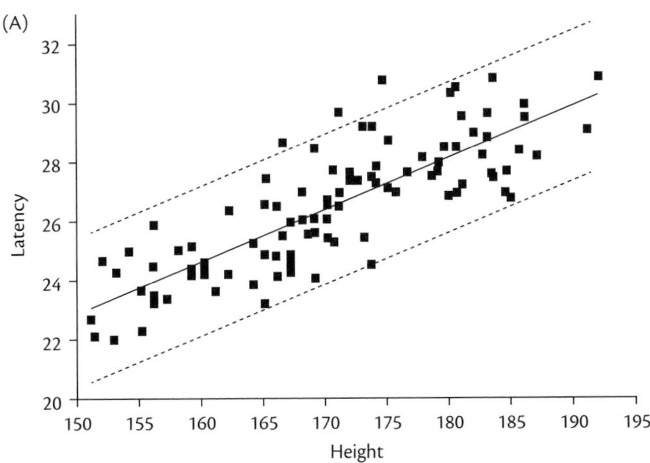

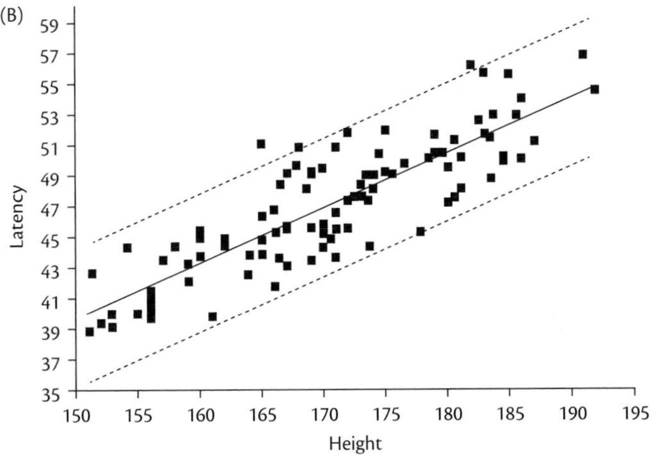

Fig. 6.9 Latency (abscissa) vs height (ordinate) with the oblique line indicating upper limit of normal (mean + 2 SD). (A) Minimum latency of the ulnar nerve. (B) Minimum latency of the tibial nerve.

Reproduced from Kimura, J, *Electrodiagnosis in diseases of nerve and muscle: principles and practice*, 4th edn, copyright (2013), with permission from Oxford University Press.

In a diffuse process, the conduction delay becomes more evident since for a longer path the delays in segmental abnormalities accumulate. With a nerve impulse propagating at a rate of 0.2 ms/cm (50 m/s), a 10% delay causes only 0.2 ms increase in latency for a 10-cm segment but as much as a 2.0-ms change for a 100-cm segment. Study of a longer path also improves the accuracy of latency and distance determination because the same absolute error constitutes a smaller percentage of the total measurement. A 1-cm error for a 10-cm segment amounts to 10% of the actual value, making the calculated conduction velocity vary between 50 and 55 m/s. The same 1-cm error for a 100-cm segment represents only 1% change, with conduction velocity varying between 50 and 50.5 m/s. Thus, the blink reflex after trigeminal nerve stimulation serves better than the direct response elicited by facial nerve stimulation in detecting a delay caused by polyneuropathies (Table 6.9). Similarly, the study of F waves and the H reflexes, covering a longer path, offers a better sensitivity and accuracy for a diffuse process than the conventional conduction study (Table 6.10).

The presence of A waves, if detected in F-wave studies, also points to an acute or chronic neuropathy with demyelination or axonal degeneration or regeneration (109). These include entrapment syndromes, brachial plexus lesions, diabetic neuropathy, hereditary motor sensory neuropathy, facial neuropathy, amyotrophic lateral sclerosis, Guillain–Barré syndrome, and cervical root lesions.

Reproducibility of various measures

Relative intertrial variation (RIV) serves as a measure of reproducibility, expressed in percentages of the difference between the two measures over the mean value of the two,

$$\text{RIV}(\%) = 100(V2 - V1) / 0.5(V1 + V2) \qquad \text{[eqn 6.3]}$$

where $V1$ and $V2$ represent the values of the first and the second measurements of the pair. The ranges of RIV between −10% to +10% indicate an acceptably low variability.

In another independent index, intra-class correlation coefficiency (ICC), a large among-subject difference offsets a great intra-individual variability as follows:

$$\text{ICC} = \sigma s^2 / (\sigma s^2 + \sigma \varepsilon^2) \qquad \text{[eqn 6.4]}$$

where σs^2 and $\sigma\varepsilon^2$ represent intra-subject variance and experimental error. Values exceeding 0.9, usually considered as a reliable measure, may result from a large intra-subject variance, rather than a small experimental error.

In our study of the inter-trial variability (105,110) we repeated all measurements twice at a time interval of 1–4 weeks. In all, 32 centres participated in the study of 132 healthy subjects (63 men) and 65 centres in the evaluation of 172 patients (99 men) with a mild diabetic polyneuropathy.

Fig. 6.10 shows ICC and the 5th–95th percentiles of RIVs in both groups. Both the controls and the patient group showed the most variability in amplitude followed by terminal latency, and motor and sensory conduction velocity. The measures meeting the RIV criteria of less than ±10% included F-wave latency and F-wave conduction velocity of median and tibial nerves, and sensory

Table 6.9 Blink reflex elicited by electrical stimulation of the supraorbital nerve in normal subjects and patients with bilateral neurological disease (mean ± SD)

Category	No. of patients	Direct response right and left combined			R_1 right and left combined			Direct response (ms)	R_1 (ms)	R/d ratio	Ipsi-lateral r_2 (ms)	Contra-lateral r_2 (ms)
		ABS	*DELAY*	*NI*	*ABS*	*DELAY*	*NI*					
Normal	83 (glabellar tap 21)*	0	0	166	0	0	166	2.9 ± 0.4	10.5 ± 0.8 (12.5 ± 1.4)	3.6 ± 0.5	30.5 ± 3.4	30.5 ± 4.4
Guillain–Barré syndrome	90	12	63	105	20	78	82	4.2 ± 2.1	15.1 ± 5.9	3.9 ± 1.3	37.4 ± 8.9	37.7 ± 8.4
Chronic inflammatory polyneuropathy	14	4	13	11	7	13	8	5.8 ± 2.6	16.4 ± 6.4	3.1 ± 0.5	39.5 ± 9.4	42.0 ± 10.3
Fisher syndrome	4	0	0	8	0	1	7	2.7 ± 0.2	10.7 ± 0.8	3.9 ± 0.4	31.8 ± 1.3	31.4 ± 1.9
Hereditary motor sensory neuropathy type 1	62	9	88	27	0	105	19	6.7 ± 2.7	17.0 ± 3.7	2.8 ± 0.9	39.5 ± 5.7	39.3 ± 6.4
Hereditary motor sensory neuropathy type II	17	0	0	34	1	0	33	2.9 ± 0.4	10.1 ± 0.6	3.6 ± 0.6	30.1 ± 3.8	30.1 ± 3.7
Diabetic polyneuropathy	86	2	20	150	1	17	154	3.4 ± 0.6	11.4 ± 1.2	3.4 ± 0.5	33.7 ± 4.6	34.8 ± 5.3
Multiple sclerosis	62	0	0	124	1	44	79	2.9 ± 0.5	12.3 ± 2.7	4.3 ± 0.9	35.8 ± 8.4	37.7 ± 8.0

Abs, absent response; NI, normal.

*R_1 elicited by a midline glabellar tap in another group of 21 healthy subjects.

Reproduced from Kimura, J, *Electrodiagnosis in diseases of nerve and muscle: principles and practice*, 4th edn, Copyright (2013), with permission from Oxford University Press.

Table 6.10 Percentage of abnormalities of conduction studies versus F-wave latencies

EDX results of 132 patients with diabetes mellitus	64 patients without clinical signs or symptoms of polyneuropathy	68 patients with clinical signs and symptoms of polyneuropathy
Abnormal NCS and F-wave latency	2/64 (3%)	43/68 (63%)
Prolonged F-wave latency only	21/64 (33%)	14/68 (21%)
Total abnormality	23/64 (36%)	57/68 (84%)

NCS: motor nerve conduction studies of the median, ulnar, peroneal, tibial nerves and sensory nerve conduction studies of the median, ulnar, superficial peroneal, sural and plantar nerves F-wave latency: minimal F-wave latency of the median, ulnar, peroneal and tibial nerves assessed using height-latency nomogram.

Adapted from *Muscle Nerve*, **49**(6), Pan L, Jian F, Lin J, et al, F-wave latencies in patients with diabetes mellitus, pp. 804–8, copyright (2014), with permission from John Wiley and Sons.

conduction velocity of the median nerve in both healthy subjects and the patients. Similarly, measures meeting the ICC criteria of over 0.9 included F-wave latency of the median and tibial nerves in both groups. In some amplitude measurements, a large intra-subject variance concealed a large experimental error, leading to a high ICC despite a large RIV.

This finding suggests that a high ICC indicating a statistical correlation between two measurements does not necessarily imply a good reproducibility. Our data indicates that F-wave latency of the median and tibial nerves meet the criteria for a reliable measure showing a large ICC (>0.9) with a small RIV (±10%). When evaluating single patients against a normal range established in a group of subjects, F-wave conduction velocity, or the latency nomogram against the height are more appropriate to minimize the effect of the limb length.

To summarize, short distances magnify focal conduction abnormalities despite increased measurement error, and long distances, though insensitive to focal lesions, accumulate diffuse or multisegmental abnormalities for better sensitivity, accuracy, and reliability (111).

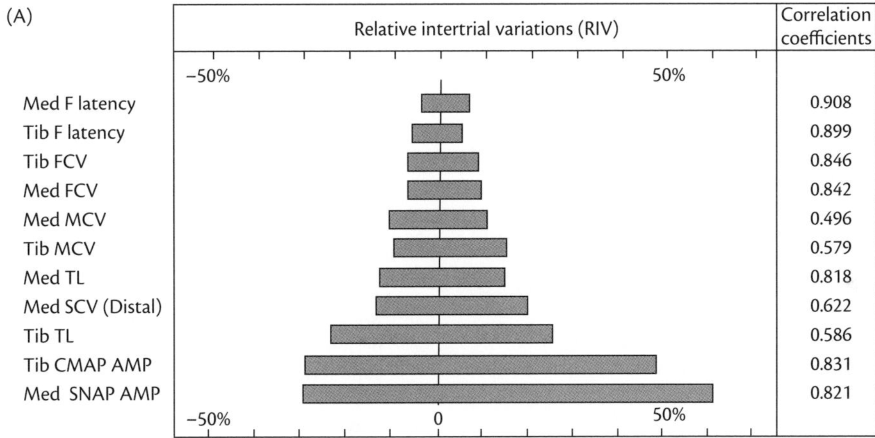

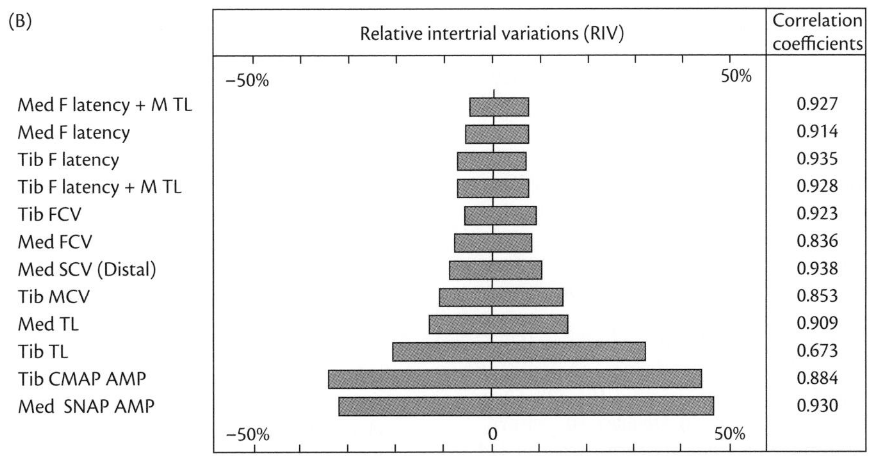

Fig. 6.10 Reproducibility of various measures in (A) healthy volunteers and (B) patients with diabetic neuropathy. All studies were repeated twice at a time interval of 1 to 4 weeks to calculate relative intertrial variations as an index of comparison.

Reproduced from *Muscle Nerve*, **20**, Kimura J, Facts, fallacies, and fancies of nerve conduction studies: twenty-first annual Edward H. Lambert lecture, pp. 777–87, copyright (1997), with permission from John Wiley and Sons.

References

1. Al-Shekhlee, A., Shapiro, B.E., and Presto, D.C. (2003). Iatrogenic complications and risks of nerve conduction studies and needle electromyography. *Muscle & Nerve*, **27**, 517–26.
2. Mellion, M.L., Buxton, A.E., Iyer, V., et al. (2010). Safety of nerve conduction studies in patients with peripheral intravenous lines. *Muscle & Nerve*, **42**, 189–91.
3. Schoeck, A.P., Mellion, M.L., Gilchrist, J.M., et al. (2007) Safety of nerve conduction studies in patients with implanted cardiac devices. *Muscle & Nerve*, **35**, 521–4.
4. Padua, L., Caliandro, P., and Stalberg, E. (2007). A novel approach to the measurement of motor conduction velocity using a Single Fibre EMG electrode. *Clinical Neurophysiology*, **118**, 1985–90.
5. Park, K.S., Lee, S.H., Lee, K.W., et al. (2003). Interdigital nerve conduction study of the foot for an early detection of diabetic sensory polyneuropathy. *Clinical Neurophysiology*, **114**, 894–7.
6. Seo, J.H. and Oh, J.H. (2002). Near-nerve needle sensory conduction study of the medial calcaneal nerve: new method and report of four ceases of medial calcaneal neuropathy. *Muscle & Nerve*, **26**, 654–8.
7. Uluc, K., Temucin, C.M., Ozdamar, S.E., et al. (2008). Near-nerve needle sensory and medial plantar nerve conduction studies in patients with small-fiber sensory neuropathy. *European Journal of Neurology*, **15**, 928–32.
8. Hennings, K., Arendt-Nielsen, L., and Andersen, O.K. (2005). Orderly activation of human motor neurons using electrical ramp prepulses. *Clinical Neurophysiology*, **116**, 597–604.
9. Nandedkar, S.D. and Barkhaus, P.E. (2007). Contribution of reference electrode to the compound muscle action potential. *Muscle & Nerve*, **36**, 87–92.
10. Goldfarb, A.R., Saadeh, P.B., and Sander, H.W. (2005). Effect of amplifier gain setting on distal motor latency in normal subjects and CTS patients. *Clinical Neurophysiology*, **116**, 1581–4.
11. Takahashi, N. and Robinson, L.R. (2010). Does display sensitivity influence motor latency determination? *Muscle & Nerve*, **41**, 309–12.
12. Landau, M.E., Diaz, M.I., Barner, K.C., et al. (2003). Optimal distance for segmental nerve conduction studies revisited. *Muscle & Nerve*, **27**, 367–9.
13. Azriel, Y., Weimer, L., Lovelace R., et al. (2003). The utility of segmental nerve conduction studies in ulnar mononeuropathy at the elbow. *Muscle & Nerve*, **27**, 46–50.
14. Herman, D.N., Preston, D.C., McIntosh, K.A., et al. (2001). Localization of ulnar neuropathy with conduction block across the elbow. *Muscle & Nerve*, **24**, 698–700.
15. Kimura, J. (2013). *Electrodiagnosis in diseases of nerve and muscle, principles and practice*, 4th edn. New York, NY: Oxford University Press.
16. Shahani, B.T., Young, R.R., Potts, F., et al. (1979). Terminal latency index and late response studies in motor neuron disease, peripheral neuropathies and entrapment syndromes. *Acta Neurologica Scandinavica*, **60**(Suppl. 73), 118.
17. Lee, K.Y., Lee, Y.J., and Koh, S.H. (2009). Usefulness of the median terminal latency ratio in the diagnosis of carpal tunnel syndrome. *Clinical Neurophysiology*, **120**, 765–9.
18. Inaba, A., Yokota, T., Otagiri, A., et al. (2002). Electrophysiological evaluation of conduction in the most proximal motor root segment. *Muscle & Nerve*, **25**, 608–11.
19. Bischoff, C., Stålberg, E., Falck, B., et al. (1996). Significance of A-waves recorded in routine motor nerve conduction studies. *Electroencephalography & Clinical Neurophysiology*, **101**, 528–53.
20. Aprile, I., Stalberg, E., Tonali, P., et al. (2003). Double peak sensory responses at submaximal stimulation. *Clinical Neurophysiology*, **114**, 256–62.
21. Metso, A.J., Palmu, K., and Partanen, J.V. (2008). Compound nerve conduction velocity – A reflection of proprioceptive afferents? *Clinical Neurophysiology*, **119**, 29–32.
22. Albers, J.W. (1997). Principles of sensory nerve conduction studies. In: The 49th meeting of the American Academy of Neurology, held in Boston, MA. April 12–19, 1997.
23. Ven, A.A., Van Hees, J.G., and Stappaerts, K.H. (2004). Effect of size and pressure of surface recording electrodes on amplitude of sensory nerve action potentials. *Muscle & Nerve*, **30**, 234–8.
24. Krarup, C. (2004). Compound sensory action potential in normal and pathological human nerves. *Muscle & Nerve*, **29**, 465–83.
25. Hasanzadeh, P., Oveisgharan, S., Sedighi, N., et al. (2008). Effect of skin thickness on sensory nerve action potential amplitude. *Clinical Neurophysiology*, **119**, 1824–8.
26. Seror, P. (2001). Simplified orthodromic inching test in mild carpal tunnel syndrome. *Muscle & Nerve*, **24**, 1595–600.
27. Albers, J.W. (1993). Clinical neurophysiology of generalized polyneuropathy. J *Clinical Neurophysiology*, **10**, 149–66.
28. Rutkove, S.B., Kothari, M.J., Raynor, E.M., et al. (1997). Sural/radial amplitude ratio in the diagnosis of mild axonal polyneuropathy. *Muscle & Nerve*, **20**, 1236–41.
29. Drenthen, J., Blok, J.H., Dudok van Heel, E.B.M., et al. (2008). Limb temperature and nerve conduction velocity during warming with hot water blankets. *Journal of Clinical Neurophysiology*, **25**, 104–10.
30. Landau, M.E., Barner, K.C., Murray, E.D., et al. (2005). Cold elbow syndrome: spurious slowing of ulnar nerve conduction velocity. *Muscle & Nerve* 2005, **32**, 815–17.
31. Rutkove, S.B. (2001). Effects of temperature on neuromuscular electrophysiology. *Muscle & Nerve*, **24**, 867–82.
32. Rutkove, S.B. (2001). Focal cooling improves neuronal conduction in peroneal neuropathy at the fibular neck. *Muscle & Nerve*, **24**, 1622–6.
33. Sanden, H., Edblom, M., Hagberg, M., and Wallin, B.G. (2005). Bicycle ergometer test to obtain adequate skin temperature when measuring nerve conduction velocity. *Clinical Neurophysiology*, **116**, 25–7.
34. Gamstorp, I. (1963). Normal conduction velocity of ulnar, median and peroneal nerves in infancy, childhood and adolescence. *Acta Paediatrica*, Suppl. 146, 68–76.
35. Smit, B.J., Kok, J.H., De Vries, L.S., et al. (1999). Motor nerve conduction velocity in very preterm infants. *Muscle & Nerve*, **22**, 372–7.
36. Bougle, D., Denise, P., Yaseen, H., et al. (1990). Maturation of peripheral nerves in preterm infants. Motor and proprioceptive nerve conduction. *Electroencephalography & Clinical Neurophysiology*, **75**, 118–21.
37. Mayer, R.F. (1963). Nerve conduction studies in man. *Neurology (Minneapolis)*, **13**, 1021–30.
38. Rivner, M.H., Swift, T.R., and Malik, K. (2000). Influence of age and height on nerve conduction. *Muscle & Nerve*, **24**, 1134–41.
39. Rajabally, Y.A., Beri, S., and Bankart, J. (2009). Electrophysiological markers of large fibre sensory neuropathy: a study of sensory and motor conduction parameters. *European Journal of Neurology*, **16**, 1053–9.
40. Rigler, I. and Podnar, S. (2007). Impact of electromyographic findings on choice of treatment and outcome. *European Journal of Neurology*, **14**, 783–7.
41. De Carvalho, M. and Swash, M. (2000). Nerve conduction studies in amyotrophic lateral sclerosis. *Muscle & Nerve*, **23**, 344–52.
42. Bromberg, M.B. and Albers, J.W. (1993). Patterns of sensory nerve conduction abnormalities in demyelinating and axonal peripheral nerve disorders. *Muscle & Nerve*, **16**, 262–6.
43. Esper, G.J., Nardin, R.A., Benatar, M., et al. (2005). Sural and radial sensory responses in healthy adults: diagnostic implications for polyneuropathy. *Muscle & Nerve*, **31**, 628–32.
44. Overbeek, B.U.H., van Alfen, N., Bo, J.A., et al. (2005). Sural/radial nerve amplitude ratio: reference values in healthy subjects. *Muscle & Nerve*, **32**, 613–18.
45. Ramchandren, S., Gruis, K.L., Chervin, R.D., et al. (2010). Hypoglossal nerve conduction findings in obstructive sleep apnea. *Muscle & Nerve*, **42**, 257–61.

46. Chang, M.H., Liu, L.H., Lee, Y.C., et al. (2006). Comparison of sensitivity of transcarpal median motor conduction velocity and conventional conduction techniques in electrodiagnosis of carpal tunnel syndrome. *Clinical Neurophysiology*, **117**, 984–91.
47. Ali, E., Delamont, R.S., Jenkins, D., Bland, J.D. and Mills, K.R. (2013). Bilateral recurrent motor branch of median nerve neuropathy following long-distance cycling. *Clinical Neurophysiology*, **124**, 1258–60.
48. Smith, E.P.V.W., Chan, Y.H., and Kannan, T.A. (2007). Medial thenar recording in normal subjects and carpal tunnel syndrome. *Clinical Neurophysiology*, **118**, 757–61.
49. Chang, M.H., Wei, S.J., Chang, H.L., et al. (2002). Comparison of motor conductions technique in the diagnosis of carpal tunnel syndrome. *Neurology*, **58**, 1603–7.
50. Al-Shekhlee, A., Fernandes Filho, J.A., Sukul, D., et al. (2006). Optimal recording electrode placement in the lumbrical-interossei comparison study. *Muscle & Nerve*, **33**, 289–93.
51. Takahashi, N., Takahash, O., Ogawa, S., et al. (2006). What is the origin of the premotor potential recorded from the second lumbrical? *Muscle & Nerve*, **34**, 779–81.
52. Therimadasmy, A.K., Li, E., and Wilder-Smith, E.P. (2007). Can studies of the second lumbrical interossei and its premotor potential reduce the number of tests for carpal tunnel syndrome? *Muscle & Nerve*, **36**, 491–6.
53. Sonno, M., Tsaiweichao-Shozawa, Y., Oshimi-Sekiguchi, M., et al. (2006). Spread of the radial SNAP: a pitfall in the diagnosis of carpal tunnel syndrome using standard orthodromic sensory conduction study. *Clinical Neurophysiology*, **117**, 604–9.
54. Sonoo, M., Kurokawa, K., Higashihara, M., et al. (2001). Origin of far-field potentials in the ulnar compound muscle action potential. *Muscle & Nerve*, **43**, 671–6.
55. Higashihara, M., Sonoo, M., Imafuku, I., et al. (2010). Origin of ulnar compound muscle action potential investigated in patients with ulnar neuropathy at the wrist. *Muscle & Nerve*, **41**, 704–6.
56. Wang, F.C. (2011). Can we accurately measure the onset latency to the first dorsal interosseous? *Muscle & Nerve*, **43**, 769–70.
57. Shakir, A., Micklesen, P.J., and Robinson, L.R. (2004). Which motor nerve conduction study is best in ulnar neuropathy at the elbow? *Muscle & Nerve*, **29**, 585–90.
58. Kothari, M.J., Preston, D.C., and Logigian, E.L. (1996). Lumbrical-interossei motor studies localize ulnar neuropathy at the wrist. *Muscle & Nerve*, **19**, 170–4.
59. Leis, A.A. and Wells, J.K. (2008) Radial nerve cutaneous innervation to the ulnar dorsum of the hand. *Clinical Neurophysiology* 2008, **119**, 662–6.
60. McCluskey, L.F. (1996). Anomalous superficial radial sensory innervation of the ulnar dorsum of the hand: a cause of 'paradoxical' preservation of ulnar sensory nerve function. *Muscle & Nerve*, **19**, 923–5.
61. Amoirid, G., and Vlachonikolis, I.G. (2003). Verification of the median-to-ulnar and ulnar-to-median nerve motor fiber anastomosis in the forearm: an electrophysiological study. *Clinical Neurophysiology*, **114**, 94–8.
62. Oh, S.J., Claussen, G.C., and Ahmad, B.K. (1995). Double anastomosis of median-ulnar and ulnar-median nerves: report of an electrophysiologically proven case. *Muscle & Nerve*, **18**, 1332–4.
63. Tamura, N., Kuwabara, S., Misawa, S., et al. (2005). Superficial radial sensory nerve potentials in immune-mediated and diabetic neuropathies. *Clinical Neurophysiology*, **116**, 2330–3.
64. Park, B.K., Bun, H.R., Hwang, M., et al. (2010). Medial and lateral branches of the superficial radial nerve: cadaver and nerve conduction studies. *Clinical Neurophysiology*, **121**, 228–32.
65. Imai, T., Wada, T., and Matsumoto, H. (2004). Entrapment neuropathy of the palmar cutaneous branch of the median nerve in carpal tunnel syndrome. *Clinical Neurophysiology*, **115**, 2514–17.
66. Prakash, K.M., Leoh, T.H., Dan, Y.F., et al. (2004). Posterior antebrachial cutaneous nerve conduction studies in normal subjects. *Clinical Neurophysiology* **115**, 752–4.
67. Rathakrishnan, R., Therimadasamy, A.K, Chan, Y.H., et al. (2007). The median palmar cutaneous nerve in normal subjects and CTS. *Clinical Neurophysiology*, **118**, 776–80.
68. Vucic, S. and Yiannikas, C. (2007). Anterior interosseous nervous conduction study: normative data. *Muscle & Nerve*, **35**, 119–21.
69. Resman-Gaspersc, A. and Podnar, S. (25008). Phrenic nerve conduction studies: technical aspects and normative data. *Muscle & Nerve*, **37**, 36–41.
70. Imai, T., Shizukawa, H., Imaizumi, H., et al. (2000). Phrenic nerve conduction in infancy and early childhood. *Muscle & Nerve*, **23**, 915–18.
71. Russell, R.I., Helps, B.A., Helm, P.J. (2001). Normal values for phrenic nerve latency in children. *Muscle & Nerve*, **24**, 1548–50.
72. Imai, T., Yuasa, H., Kat, Y., and Matsumoto, H. (2005). Aging of phrenic nerve conduction in the elderly. *Clinical Neurophysiology*, **116**, 2560–4.
73. Al-Shekhlee, A., Onders, R.P., Syed, T.U., et al. (2008). Phrenic nerve conduction studies in the spinal cord injury: applications for diaphragmatic pacing. *Muscle & Nerve*, **38**, 1546–52.
74. Seror, P. (2006). The long thoracic conduction study revisited in 2006. *Clinical Neurophysiology*, **117**, 2446–50.
75. Sander, H.W., Quinto, C., and Chokroverty, S. (1998). Accessory deep peroneal neuropathy: collision technique diagnosis. *Muscle & Nerve*, **21**, 121–3.
76. Oh, S.J., Demirci, M.D., Dajani, B., et al. (2001). Distal sensory nerve conduction of the superficial peroneal nerve: new method and its clinical application. *Muscle & Nerve*, **24**, 689–94.
77. Kushnir, M., Klein, C., Kimiagar, Y., et al. (2005). Medial dorsal superficial peroneal nerve studies in patients with polyneuropathy and normal sural responses. *Muscle & Nerve*, **31**, 386–9.
78. Park, G.Y., Im, S., Lee, J.I., et al. (2010). Effect of superficial peroneal nerve fascial penetration site of nerve conduction studies. *Muscle & Nerve*, **41**, 227–33.
79. Campbell, C.A., Turza, K.C., and Morgan, R.F. (2009). Postoperative outcomes and reliability of 'sensation-sparing' sural nerve biopsy. *Muscle & Nerve*, **40**, 603–9.
80. Killian, J.M. and Foreman, P.J. (2001). Clinical utility of dorsal sural nerve conduction studies. *Muscle & Nerve*, **24**, 817–20.
81. Kokotis, P., Mandellos, D., Papagianni, A., et al. (2010). Nomogram for determining lower limit for the sural response. *Clinical Neurophysiology*, **121**, 561–3.
82. Vinik, A.I., Bril, V., Litchy, W.J., et al. (2005). Sural sensory action potential identifies diabetic peripheral neuropathy responders to therapy. *Muscle & Nerve*, **32**, 619–25.
83. Alfonsi, E., Merlo, I.M., Clerici, A.M., et al. (1999). Proximal nerve conduction by high-voltage electrical stimulation. *Clinical Neurophysiology*, **114**, 239–47.
84. Inaba, A., Yokota, T., Komori, T., et al. (1996). Proximal and segmental motor nerve conduction in the sciatic nerve produced by percutaneous high voltage electrical stimulation. *Electroencephalography & Clinical Neurophysiology*, **101**, 100–4.
85. Bademkiran, F., Obay, B., Aydogdu, I., et al. (2007). Sensory conduction study of the infrapatellar branch of the saphenous nerve. *Muscle & Nerve*, **35**, 224–7.
86. Hemmi, S., Inoue, K., Murakami, T., et al. (2007). Simple and novel method to measure distal sensory nerve conduction of the medial plantar nerve. *Muscle & Nerve*, **36**, 307–12.
87. Im, S., Park, J.H., Kim, H.W., et al. (2010). New method to perform medial plantar proper digital nerve conduction studies. *Clinical Neurophysiology*, **121**, 1059–65.
88. Loseth, S., Nebuchennykh, M., Stalberg, E., et al. (2007). Medial plantar nerve conduction studies in healthy controls and diabetics. *Clinical Neurophysiology*, **118**, 1155–61.
89. Sylantiev, C., Schwartz, R., Chapman, J., et al. (2008). Medial plantar nerve testing facilitates identification of polyneuropathy. *Muscle & Nerve*, **38**, 1595–8.
90. Kushnir, M., Klein, C., Kimiagar, Y., et al. (2008). Distal lesion of the lateral femoral cutaneous nerve. *Muscle & Nerve*, **37**, 101–3.

91. Seror, P. and Seror, R. (2006). Meralgia paresthetica: clinical and electrophysiological diagnosis in 120 cases. *Muscle & Nerve*, **33**, 650–4.
92. Shin, Y.B., Park, J.H., Kwon, D.R., et al. (2006). Variability in conduction of the lateral femoral cutaneous nerve. *Muscle & Nerve*, **33**, 645–9.
93. Oh, S.J. (2007). Neuropathies of the foot. *Clinical Neurophysiology*, **118**, 954–80.
94. Yasunami, T., Miyawaki, Y., Kitano, K., et al. (2005). Shortening of distal motor latency in anode distal stimulation. *Clinical Neurophysiology*, **116**, 1355–61.
95. Dyck, P.J., Litchy, W.J., Daube, J.R., et al. (2003). Individual attributes versus composite scores of nerve conduction abnormality: sensitivity, reproducibility, and concordance with impairment. *Muscle & Nerve*, **27**, 202–10.
96. Lew, H.L., Wang, L., and Robinson, L.R. (2000). Test-retest reliability of combined sensory index: implications for diagnosing carpal tunnel syndrome. *Muscle & Nerve*, **23**, 1261–4.
97. Dyck, P.J., O'Brien, P.C., Litchy, W.J., et al. (2001). Use of percentiles and normal deviates to express nerve conduction and other test abnormalities. *Muscle & Nerve*, **24**, 307–10.
98. Schmidt, K., Chinea, N.M., Sorenson, E.J., et al. (2011). Accuracy of diagnoses delivered by an automated hand-held nerve conduction device in comparison to standard electrophysiological testing in patients with unilateral leg symptoms. *Muscle & Nerve*, **43**, 9–13.
99. Tan, S.V., Sandford, F., Stevenson, M., et al. (2012). Hand-held nerve conduction device in carpal tunnel syndrome: a prospective study. *Muscle & Nerve*, **45**, 275–92.
100. Van Auken, S.F. and van Dijk, J.G. (2003). Two approaches to measure amplitude changes of the sensory nerve action potential over a length of nerve. *Muscle & Nerve*, **27**, 297–301.
101. Kimura, J. (1998). Kugelberg lecture: principles and pitfalls of nerve conduction studies. *Electroencephalography & Clinical Neurophysiology*, **106**, 470–6.
102. Cappelen-Smith, C., Lin, C.S., and Burke, D. (2003). Activity-dependent hyperpolarization and impulse conduction in motor axons in patients with carpal tunnel syndrome. *Brain*, **126**, 1001–8.
103. Kaji, R. (2003). Physiology of conduction block in multifocal motor neuropathy and other demyelinating neuropathies. *Muscle & Nerve*, **27**, 285–96.
104. Kaji, R., Bostock, H., Kohara, N., et al. (2000). Activity-dependent conduction block in multifocal motor neuropathy. *Brain*, **123**, 1602–11.
105. Kimura, J. (1997). Facts, fallacies, and fancies of nerve conduction studies: twenty-first annual Edward H. Lambert lecture. *Muscle & Nerve*, **20**, 777–87.
106. Kim, D.H., Kang, Y.K., Hwang, M., et al. (2005). Reference values of fractioned neurography of the ulnar nerve at the wrist in healthy subjects. *Clinical Neurophysiology*, **116**, 2853–7.
107. Kim, B.J., Koh, S.B., Park, K.W., et al. (2008). Pearls & oysters: false positives in short-segment nerve conduction studies due to ulnar nerve dislocation. *Neurology*, **70**, 9–13.
108. Lo, Y.L., Leoh, T.H., Xu, L.Q., et al. (2005). Short-segment nerve conduction studies in the localization of ulnar neuropathy of the elbow: use of flexor carpi ulnaris recordings. *Muscle & Nerve*, **31**, 633–6.
109. Bischoff, C. (2002). Neurography: late responses. *Muscle & Nerve*, **11**(Suppl.), S59–65.
110. Kohara, N., Kimura, J., Kaji, R., et al. (1996). Multicenter analysis on intertrial variability of nerve conduction studies: healthy subjects and patients with diabetic polyneuropathy. In: J. Kimura and H. Shibasaki (Eds) *Recent advances in clinical neurophysiology*, pp. 809–15. Amsterdam: Elsevier Science BV.
111. Kimura, J. (2001). Long and short of nerve conduction measures: reproducibility for sequential assessments. *Journal of Neurology, Neurosurgery & Psychiatry*, **71**, 427–30.
112. Pan, L., Jian, F., Lin, J., et al. (2014). F-wave latencies in patients with diabetes mellitus. *Muscle & Nerve*, **49**(6), 804–8.

CHAPTER 7

Electromyography

Erik Stålberg

The motor unit

Normal conditions

The functional unit in the muscle is the so-called motor unit (MU), i.e. one motor neuron, its axon and all fibres innervated by this axon (1). The total number of MUs is not exactly known, but has been estimated with the techniques called motor unit number estimation (MUNE). The first such method was developed by McComas et al. (2). Different MUNE methods have estimated the number in limb muscle to be between 100 and 500. This value differs between muscles and is difficult to verify with morphological data. There are also few anatomical studies on the number of muscle fibres in a MU. One morphological study (3) reported a number of 9 for eye muscles and up to 2000 in large muscles. This latter high number could not be verified with electrophysiological, but indirect techniques show 200–300 muscles fibre in the tibialis anterior muscle (4).

All muscle fibres in a given MU have the same histochemical, metabolic and mechanical characteristics. They are randomly scattered within an area of 5–15 mm in diameter (5). In limb muscles the fibres in one MU are separated by about 250 μm. Histochemical classification is based the concentration of oxidative enzymes and myofibrillar ATPase. The classes are called type I, type IIA, type IIB, type C. Mechanical parameters are also used to separate slow twitch fibres (type I fibres, oxidative) and fast twitch fibres (type II fibres glycolytic with subgroups). These groups have been most extensively studied in animals, which differ from human muscles. The size of MUs vary regarding number of fibres, size of fibres, size of axonal diameter (6). During increasing force, the small MUs are the first to be recruited, followed by the others in regular order, the so-called size principle (7). The size of the MU cannot be assessed by routine electromyograph (EMG) methods, but is well reflected in the macro EMG recording (8).

Motor unit in pathology

The MU has a restricted number of ways to respond to pathological situations.

Reinnervation

Two different types of reinnervation take place, dependent on the initial cause. In the rare cases of complete nerve damage, axonotomy, the only way to achieve reinnervation is via outgrowth of axons that hopefully find the old track in the Schwann sheath. The time for the first sign of reinnervation depends on the site of lesion, proximal or distal. When the first axon reaches the muscle, it will innervate the most adjacent muscle fibres, and produce very small motor unit potentials (MUPs). These have been called nascent MUPs. It is a misunderstanding that these are typical for all situations of reinnervation. In the case of partial nerve lesion, so-called collateral sprouting takes place. Here, distal nerve twigs grow to reach neighbouring denervated muscle fibres. This starts quickly and relatively independent on the site of lesion, since it is triggered by the Wallerian retrograde degeneration. In this case, the surviving motor unit innervates denervated muscle fibres in competition with other MUs within the area. Both electrophysiological (5) and morphological methods (9) indicate that a reinnervating MU does not extend outside its original borders, i.e. outside its territory. The reason is unknown. Electrically, one gets the impression that the territory has increased since the reinnervation MUPs have higher amplitude and long duration, and therefore heard by the electrode over a longer distance. This is true, but does not mean that the morphological territory has increased. In this type of reinnervation there are never initially abnormally small MUPs, as above, since the reinnervation takes place from a normal existing motor unit. This will influence the shape of the MUP in reinnervation.

Myopathy

The change in morphology among muscle fibres depends on type of myopathy. In dystrophies, there are fibre diameter variation, loss of muscle fibres, focal necrosis, longitudinal muscle fibre splitting, and increased interstitial tissue. All these factors may change the electrical signals around the muscle fibre, which is reflected in the MUP shape. In metabolic myopathies, there may be very little of membrane destruction or fibre degeneration and, therefore, the MUPs are less affected, even normal. In channelopathies different types of spontaneous activity may be seen, but the MUPs may be normal.

Neuromuscular junction disorders

Unless there is a secondary denervation, the MUPs are normal, except that they show a variation in shape since the individual muscle fibre signals summate with an inherent jitter.

Central disorder

Here, the MUPs are normal, but are activated with an abnormal pattern.

Methods to study the neuromuscular transmission

Repetitive nerve stimulation

As the first test of choice in cases of suspected disturbances of neuromuscular transmission is the repetitive nerve stimulation (RNS). A surface electrode is used to stimulate the nerve and the

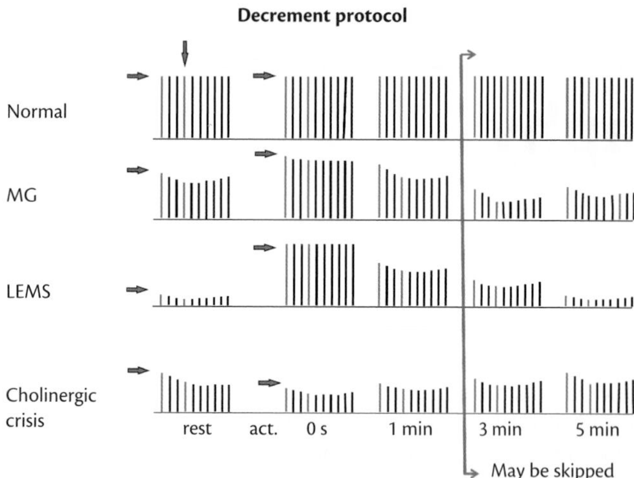

Fig. 7.1 Schematic representation of CMAP amplitudes with RNS in different conditions. A train of 10 stimuli of 3 Hz is given. Maximal voluntary activation for 10 (or 20–30s) s is immediately (within 10 s) followed by a new train of stimuli with the muscle at rest. This is repeated after 1 min. We no longer suggest tests after 3 and 5 min. This protocol needs modifications according to the situation, particularly the severity of symptoms and the ability of patient collaboration. In non-cooperative patients, high frequency (20 Hz) is used instead of voluntary activation.
In normal conditions, no decrement or increment is seen. In myasthenia gravis, resting amplitude may be reduced, a decrement (saddle shaped) is seen and facilitation after activation is seen, both as increase in amplitude (if this was diminished at rest) and repair of the decrement. In LEMS the initial amplitude is low, a decrement is seen. After activation (activation in LEMS is preferably 10 s), the amplitude increases by more than 60% and decrement is less pronounced. In severe MG exhaustion may be seen after 1–5 min. In cholinergic crisis, the initial amplitude may be reduced and a decrement is seen. After activation the amplitude decreases and more decrement is seen. A slow return to baseline condition is seen after 5 min.
Stålberg ©, 2013.

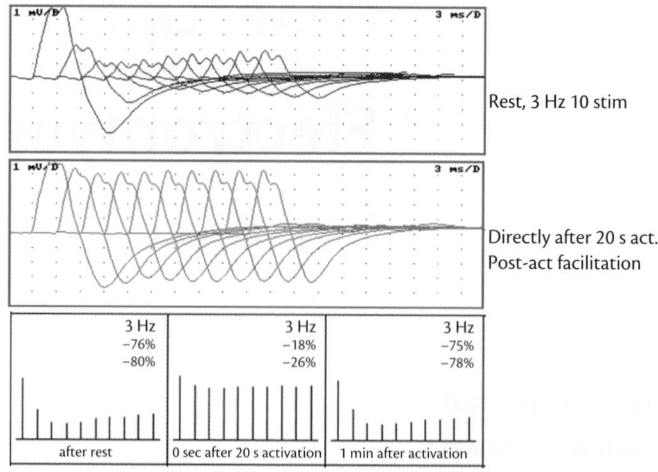

Fig. 7.2 Repetitive nerve stimulation and recording from the ADM muscle in a patient with severe myasthenia. Note the relative repair in decrement after 20 s of activity, and return to pre-activation values after 1 min. The exhaustion is not shown here.
Stålberg ©, 2013.

evoked muscle response is recorded with surface electrodes over the target muscle. The nerve is stimulated 5–10 times with a frequency of 2–5 Hz. The relative change in amplitude of the compound muscle action potential (CMAP) between the 1st and 4th stimulus is calculated, and denoted as the decrement. In healthy subjects and with rigorous technical precautions (temperature, muscle fixation, supramaximal stimulation, absolute relaxation of the muscle), the decrement is typically less than 5%. After this, the patient is asked to perform a maximal muscle contraction for 10–30 s (this is as effective as high frequency stimulation, which is painful, but used when patient cooperation is poor). Immediate relaxation is then required, and the RNS is repeated to detect possible repair of a previously increased decrement, so-called post-exercise facilitation. In healthy subjects, there may be a slight increase in CMAP amplitude, but no decrement or increment occurs. The test may then be repeated after 1 min. Earlier we recommended additional test after 3 and 5 mins, but this is not routine any longer for the detection of a possible so-called post-exercise exhaustion. The protocol is shown in Fig 7.1.

In myasthenia gravis (MG), a variable degree of decrement is seen, predominantly in proximal and facial muscles (anconeus, deltoid, trapezius, nasalis, frontalis). The resting compound muscle action potential (CMAP) amplitude may be diminished due to transmission block in some motor endplates, even at rest. Immediately after activity, the amplitude may increase and the decrement becomes less than before (Fig. 7.2). The absolute amplitude carries important information. A low resting amplitude indicates inactivity in a number of endplates. The amplitude after activation gives the information whether all motor endplates can be activated, or whether the condition is so severe that even this facilitation did not activate all motor endplates. In this situation one cannot expect full effect of anti-Ach-esterase. Immunosuppression may allow the formation of new endplates, and anti-Ach-esterase may then give full effect. After 3 min, the exhaustion may be seen as a decrease of the CMAP amplitude and a more pronounced decrement. To see exhaustion in MG an activation time of 30 s or even more is required and is best seen in relatively pronounced neuromuscular defects.

In Lambert Eaton myasthenic syndrome (LEMS), a presynaptic disorder, the CMAP amplitude at rest is low (less than 2 mV in hand muscles) and a decrement is seen. After activation for 10 s (better than 20 or 30 s of activation (10)) a pronounced amplitude increment is seen, typically more than 60% (11) with repair of the decrement. Exhaustion may then follow (12).

The changes in amplitude as described, reflect the summated effect of exhaustion and facilitation of individual motor-endplates during the study (13). The physiological events will be briefly described.

The arrival of an action potential at the nerve terminal will open the voltage gated Ca^{2+} ion channels initiating the influx of Ca^{2+} ions into the terminal. This triggers the release of ACh quanta from specific sites in the preterminal membrane. The ACh diffuses across the synaptic gap of 500 Å and reaches the post-synaptic ACh receptors on the muscle fibre. Na^{+} channels are opened and an endplate potential (EPP) is generated. When this has produced sufficient membrane depolarization the threshold for impulse generation is reached and an action potential travels down the muscle fibre in both directions from the endplate.

The ACh is immediately hydrolysed and the choline is reabsorbed via endocytosis to the nerve terminal.

The EPP increases in amplitude during activation as a sign of intratetanic facilitation. A counteracting process is the run-down of ACh stores, reducing the EPP. Normally, the EPP is many times greater than is needed to generate an action potential and these dynamic changes are not observed clinically or with RNS. However, when there is a reduced safety margin of neuromuscular transmission, these effects may be unmasked. This may happen due to presynaptic defects, such as nerve disorders, reduced ACh production during early reinnervation, in some forms of congenital myasthenia and with impaired release, as in LEMS. It also happens in post-synaptic abnormalities, such as MG with reduction of responding receptors, high levels of magnesium, and in congenital myasthenic disorders with paucity of ACh receptors, lack of ACh esterase and abnormal K^+ or Na^+ channels.

Details of these physiological events cannot be studied with RNS, but require intracellular recordings or patch clamp experiments, which are not applicable in the clinical routine. Even so, RNS gives a useful summation of the physiological events.

The physiology of individual motor endplates can, however, be studied with single fibre EMG, described next.

Single fibre EMG

Single fibre EMG is a method developed to study the microphysiology of the motor unit. (14,15). It has been described in great detail in a monograph (16). The method is based on the use of a small recording surface, 25 μm in diameter (compare with mean muscle fibre diameter of 50 μm) and a cut-off for low frequencies in the signal, which have a relative dominance for distant fibres. In this way, a practical uptake radius is reduced to about 300 μm. With this method it is possible to study the propagation velocity of individual muscle fibres and see the temporary facilitation of velocity after the passage of one impulse (called velocity recovery function) and the fatigue with long-term activity. Another parameter is the local concentration of muscle fibres in the MU, the so-called fibre density (FD). One muscle fibre is approached with the electrode and its action potentials trigger the sweep. The number of time locked fibres is noticed in 20 different recording sites. The mean values of time-locked signals (triggering signal is included), gives the FD. The third parameter is the neuromuscular jitter. This is seen as time variability in arriving time of two signals from fibres in the same MU. This variability is found to reflect the variation in transmission time in the two motor endplates at consecutive discharges (Fig. 7.3). It is of the order of 15–50 μsec, which is the combined jitter in the two involved motor endplates. If electrical stimulation is used (see later), the jitter represents events in one motor-endplate. The normal values are therefore lower than with voluntary activation.

Jitter in neuromuscular disorders, voluntary activation

This parameter has been the most useful in routine work in question of disturbed neuromuscular transmission, i.e. mainly in myasthenia gravis. In this diagnosis, the jitter is increased, and with increasing degree of disturbance, also temporary impulse blockings occur in one of the other of the recorded fibres. Blocking is related to muscle weakness and decrement on repetitive nerve stimulation. Increased jitter alone does not give symptoms or decrement. This is the reason for the high sensitivity in this method since it detects even subclinical changes. A number of articles have been published showing the use of jitter analysis, both for diagnosis and for monitoring over time (16). The sensitivity is high—97% in ocular myasthenia and 99% in generalized when two muscles are studied, usually orbicularis oculi and frontalis, or extensor digitorum (16). The specificity is not as high, since all conditions that cause disturbed neuromuscular transmission such as Lambert–Eaton syndrome, early reinnervation, and electrolyte disturbance will give increased jitter.

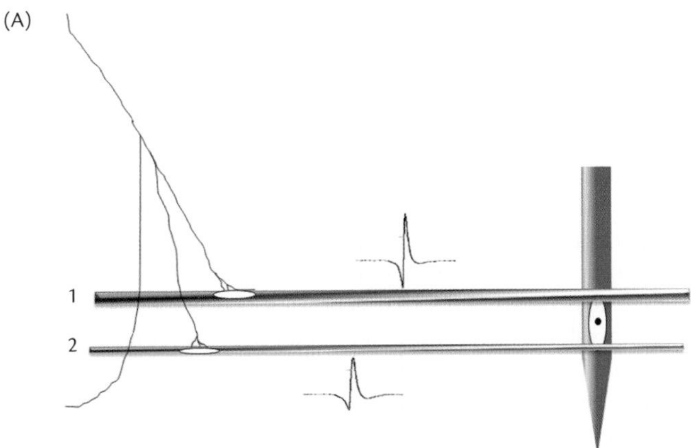

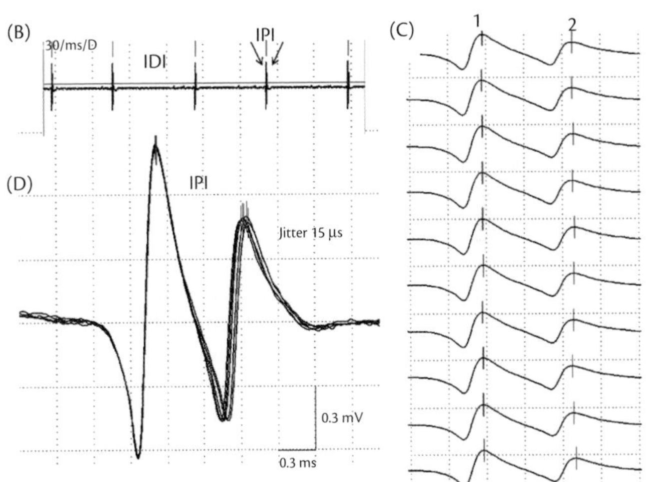

Fig. 7.3 (A) Schematic presentation of the recording conditions for jitter measurement in voluntarily activated muscle. The SFEMG needle electrode is positioned to record action potentials (APs) from a pair of muscle fibres (1 and 2) innervated by the same motor axon. (B) AP pair firing at low degree of voluntary effort with interdischarge (IDI) and interpotential (IPI) interval indicated. (C) The two potentials at a higher sweep speed with the sweep triggered by the first potential and moving down after each discharge, and (D) shows several discharges superimposed to demonstrate variability in the interpotential intervals (the neuromuscular jitter).
Stålberg ©, 2013.

Stimulation single fibre electromyograph

The muscle fibres can be activated also by electrical stimulation and the jitter between stimulus and response is calculated (17). This jitter includes only one motor endplate and the resulting stimulation jitter is equal to the jitter during voluntary contraction divided by square root of 2. In practice, the intramuscular axon is stimulated by a monopolar electrode as cathode, and a surface electrode close by as anode. With increasing stimulation strength first one axon is activated. If the stimulation strength is just liminal, there is stimulation site jitter and also blocking due to insufficient axonal depolarization. With a slight increase in stimulus, say from 3 to 3.5 mV, this axon is now fully stimulated. At the same time, an adjacent axon may be activated, again initially in a subliminal fashion. So, just watching signals on the screen, some of which show jitter and impulse blocking, does not tell that the patient has a neuromuscular problem if the displayed jitter is increased. The operator must make a decision for each accepted signal, that he stimulus that generated this particular signal is suprathreshold. This is the major pitfall with the stimulation jitter analysis. The advantage is that patient cooperation is unnecessary, i.e. can be used on unconscious patients, in patients with movement disorders, in children and in animal experiments. Another advantage is that the discharge frequency is well controlled, thus on one hand removing the effect of VRF and on the other to test jitter at high and low frequency activation. In myasthenia gravis, the jitter increase with increasing stimulation frequency, in LEMS it decreases (18). Properly done, the result from stimulation jitter analysis is the same as for voluntary activation when it comes to sensitivity to detect disturbed transmission defect.

Single fibre electromyograph in other disorders

Single fibre electromyograph (SFEMG) is also applied in other disorders to study the microphysiology of the motor unit (16). In ongoing reinnervation, the FD is increasing, and there is increased jitter and blocking. With time, 3–6 months after a focal nerve damage, the jitter returns towards normal, although is often remains slightly increased. FD may revert towards normal if the time to successful reinnervation is short (e.g. after botulinum denervation, which is very distal) or remains high as in sequel after polio. In myopathies with dystrophy, the FD is increased—possibly astonishingly. The explanation may be collateral sprouting to muscle segments that have lost their innervation after focal necrosis. It may also be longitudinal fibre splitting, innervation of myoblasts as a regenerative phenomenon, or secondary neurogenic changes in severe myopathies. This also means that reinnervation is ongoing, which gives a certain degree of jitter. Often this is of moderate degree, except in very fast progressive cases, e.g. in myositis.

Jitter with concentric needle electrode

Over the last years, the use of reusable material has been banned in many countries. This includes EMG needle electrodes. For SFEMG this has become a major problem, since the specifics of smallness has not been possible to copy in disposable electrodes yet. Therefore a surrogate has been found, namely the smallest facial needle electrode (19). This electrode often records from more than one fibre, but with careful repositioning the electrode so-called 'apparent single fibre action potentials' (ASFAP) can be obtained. There are certain criteria to accept these signal, they should be constant in shape on consecutive discharges, and have no notches or shoulder on their rising slope. These criteria are often difficult to meet, particularly in large limb muscles, but are easier in facial muscles. A more careful editing must be undertaken with this technique than with SFEMG and is therefore somewhat more time consuming, certainly in limb muscles. The method of electrical stimulation can also be applied, with even more risk of interference between muscle fibres. It is not only the response from a given axon that may be complex, but it is also usually more than one axon that is activated. Activity from different motor units cannot be separated by triggering as in voluntary activation, but occur together after the stimulus. Some reference material has been collected to date (20–23). A multicentre study [23] is a more complete reference data both for voluntary and stimulated activation. The conclusion is that with fulfilling the quality criteria, the jitter values from the concentric needle electrode (CNE) can be used to study neuromuscular transmission. It seems to be equally useful as SFEMG to diagnose myasthenia gravis. FD cannot be measured with CNE.

Methods to study the motor unit

Macro-EMG

Another method, opposite to SFEMG is macro-EMG. This method is very non-selective and developed to give an overall view of the MU comprising number of muscle fibres and size of muscle fibres (8). The recording is performed from the cannula, which has a restricted exposed surface of 15 mm in order to get a standardized electrode size. A remote reference is used. This signal gives a summation pattern of all activated MUs and contributions from just one MU cannot be appreciated. Therefore, the macro-electrode has a SFEMG surface 7.5 mm behind the tip. This is now recording activity from one muscle fibre and this signal is used to trigger an averager to which the cannula signal is fed. After a number of discharges, the contribution from one MU is obtained, the macro-MUP. The recording is performed from 20 MUs and an average is given as result. It should be noted that this value represents the low threshold MUs in the muscle since the muscle is activated with only slight contraction strength. This has been used to study the size of MU in normal condition and in various neurogenic condition, both as instant values and the dynamic change with time, e.g. in patients with late polio (24). This method has not been used in clinical routine, but has been useful in questionable cases and in follow up studies.

Over the last years, disposable material is required in some countries. A modification of the electrode has already earlier been suggested (25). Here, a macro-needle electrode is modified by using a concentric needle electrode, with the tip recording surface used for triggering, the Conmac method. A theoretical drawback is that the main part of the cannula may not be inside the MU under study. This probably influences the absolute values.

Concentric or monopolar electrodes in routine EMG

Relationship between the EMG signal and generating muscle fibres

Each muscle fibre produced an action potential, mainly biphasic, which travels in both directions along the muscle fibre from the endplate to the tendon with a speed of 1.5–6 m/s (26). The electrical field around the muscle fibre can be recorded extracellularly by means of an EMG electrode. The examples in this Chapter will be taken from the concentric needle electrode, the most commonly used in Europe and elsewhere except the USA, where monopolar electrodes probably are more commonly used. The principles are the same, but the quantitative MUP analysis may show some differences. Fibres far away from the electrode contribute will less amplitude, compared to those in close vicinity of the recording tip. As indicated in Fig. 7.4, the electrical field is heard before the depolarization reached the electrode as slow positive going waves. Notice that when the fields from the action potentials still are far

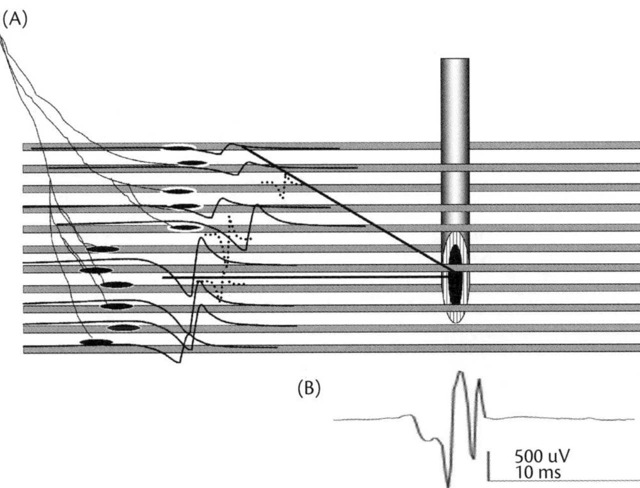

Fig. 7.4 Schematic presentation of the generation of a MUP. (A) The electrical field around individual muscle fibres is heard as slow wave components by the electrode already before the depolarization has reached the electrode. The distance to 'remote' and 'close' fibre (lines) is relatively similar, compared with the distance when the depolarization is passing the electrode in which case the closest fibres have a dominating contribution. (B) The summated result (real recording), the so-called motor unit potential (MUP).
Stålberg ©, 2013.

away from the electrode, perhaps 20 mm away, the relative distance to the 'remote fibres' is nearly the same as that of the fibres that pass the tip. Therefore, the slow wave contribution from both distant and close fibres is similar. When the field is just passing the electrode, the closest fibres are relatively much closer to the electrode than the distant ones. Therefore, the duration of the MUP, defined by onset of slow wave component and their disappearance, reflects fibres over a larger area than the sharp components in the MUP.

The MUP is the temporal and spatial summation of these fibre action potentials. If they are synchronous, a high amplitude homogenous signal is recorded. With temporal dispersion, as effect of different propagation velocity in the individual muscle fibres, the MUP is more irregular. This happens if the recording is performed away from the endplate and in pathology if there is an increased fibre diameter variation (the fibre diameter is directly related to propagation velocity) (Fig. 7.5).

With more fibres in the uptake are, the amplitude increases. The relationship between number of fibres and amplitude is linear but with a slope indicating that full summation is not present. This is due to so-called phase cancellation which makes the sum smaller than expected from the amplitude of individual spikes. This is more pronounced with temporal dispersion.

The MUP is the signal difference between the electrode tip and the cannula which is used as reference (Fig. 7.6). With increasing force, it may well happen that next recruited MU has such a position in the muscle that it is not recorded at all by the tip, but only by the cannula. In this case, a MUP with low amplitude and inversed polarity will be seen, not an uncommon experience. These signals should not be used for quantitation, since the recording area is different which gives a different signal shape.

Practical hints for recording

EMG is a medical consultation where we use other tools than at the neurological investigation. History, clinical exam, and study of earlier files and EMG reports should precede the exam. All tools should be prepared in the EMG room for a smooth flow of the exam.

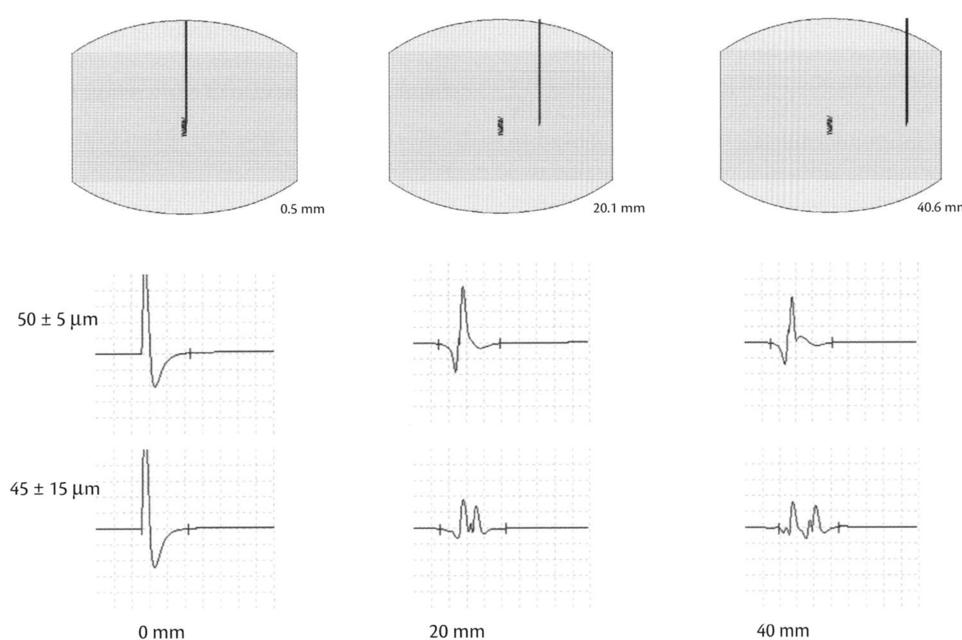

Fig. 7.5 Simulation to show the effect of end-plate to recording distance (0, 20, and 40 mm) and effect of fibre diameters (indicated to the left). In the lower row of recordings, the fibre diameter variation is increased. Note that the MUPs are the same when recording is performed at the endplate zone. With increasing distance, the amplitude decreases, particularly when the fibre diameter variation is increased (myopathy).
Stålberg ©, 2013.

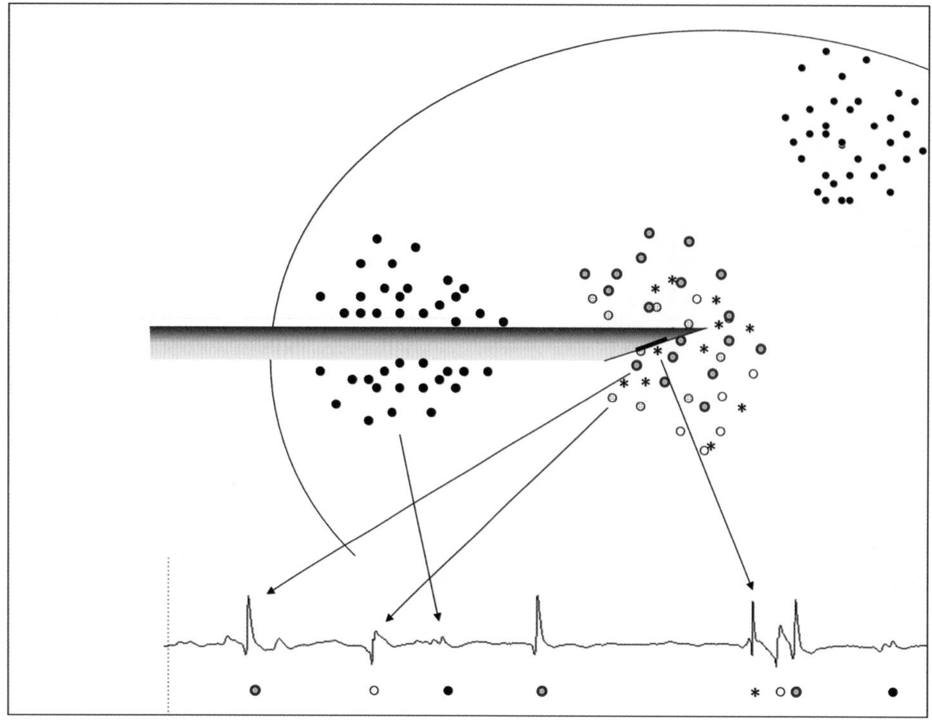

Fig. 7.6 Recording during moderate contraction with five different MUs active. One remote, is not contributing to the MUP and one MU is positioned over the cannula. In this case, the fibres from the MU with lowest threshold (three repetitions in the figure) have higher amplitude than the others, simply due to the position of the muscle fibres in the MUs.
Stålberg ©, 2013.

The electrode should be held like a pen and inserted with a small but brisk movement, with the examiners hand supported on the limb of the patient. Insertion should not be made in the endplate zone (if this is known), but 10–20 mm distal to that. End-plate area is recognized by the presence of endplate noise, sharply rising MUPs in negative direction and often painful spots since the endplates are close to nociceptors. The electrode should initially be inserted just of few mm under the fascia, and then moved in small steps, like following the sides of a pyramid. Usually, two skin insertions should be performed, if possible spaced by at least 2 cm, to avoid that the same MU is investigated again.

Before spontaneous activity is recorded, make sure that the electrode is in muscle tissue, by asking the patient for a short activation. Before recording at strong contraction, remove the electrode outside the fascia to avoid mechanical bend of the electrode. Insert the needle during contraction and remove it slowly during ongoing contraction, to give minimal pain.

Parameters to measure

The quantitation of EMG should focus on such parameters that have a known counterpart morphologically (27) and that change with pathology. This will be discussed in detail in another Chapter, but will briefly mentioned here:

- Amplitude, measured peak to peak. It is related to number of fibres within 0.5 mm from the electrode tip.
- Duration, measured from first deviation from baseline to return of the slow wave. It is related to number of fibres within 2 cm of the MU.
- Number of phases and turns. Both reflect temporal dispersion. More than four phases is called polyphasic MUP and more than five turns is called serrated or complex MUPs.
- Shape instability at consecutive discharges, jiggle. This reflects the neuromuscular transmission, abnormal not only in myasthenia gravis, but also in early reinnervation.
- Satellites, late components that appear after a MUP, separated by a baseline of 2 ms. It usually indicate a slowly conducting muscle fibre, i.e. atrophy in myopathy or in neurogenic conditions.
- Thickness and size index which have similar meaning (28), and complexity (29) are some other parameters that sometimes are used to classify MUPs.

Routine electromyograph with muscle at rest

Spontaneous activity, recorded in the muscle, may be generated in the muscle fibre, in the axon or motor neuron. In the first case, the signals represent single fibre action potentials, in the latter two, MUPs are recorded.

Normal condition

Insertional activity

When the needle electrode is inserted into the muscle, a short burst of activity occurs, sometimes with some positive waves at the end. This activity is generated by the tip of the electrode and has no pathological significance. Attempts have been made to quantify

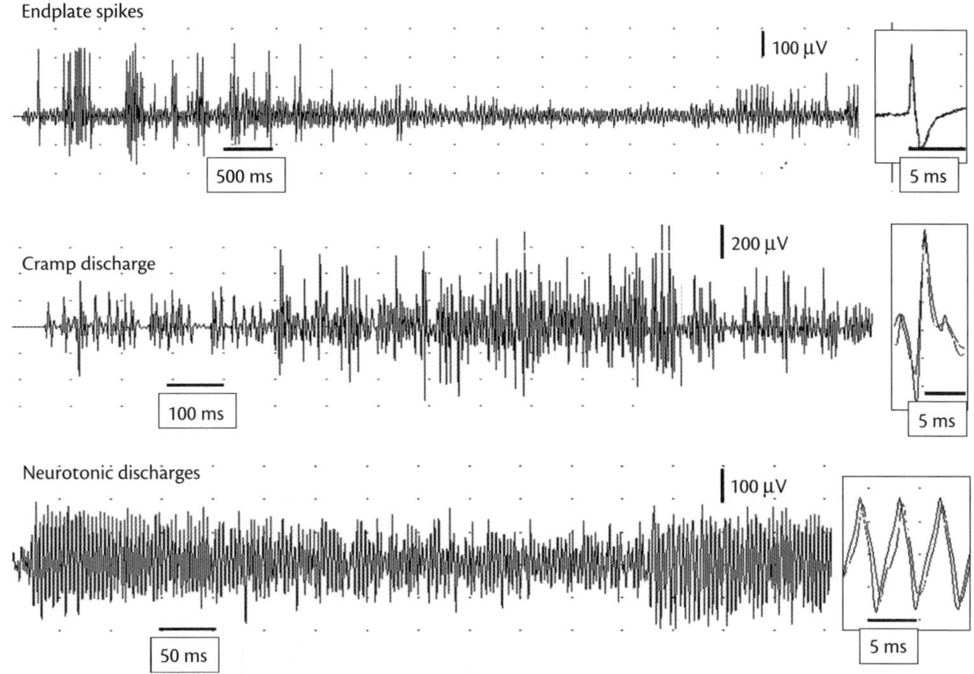

Fig. 7.7 Activity with muscle at rest. Trace one shows both endplate noise, causing a blurring of the baseline, and irregularly firing negative going end-plate spikes. Trace two shows signals with muscle in cramp. Trace three shows high frequency discharges in neurotonia.

Reproduced from Stålberg E, Daube JR: Electromyographic methods; in Stålberg E, (ed.) *Handbook of Clinical Neurophysiology*, Volume 2, pp. 147–185, copyright (2003), with permission from Elsevier.

the duration of this activity to see an assumedly longer duration in a denervated muscle. This has not turned out to be a useful parameter and is therefore not recommended. During attack of paralysis in hyperkalaemic periodic paresis and in fibrotic muscles there is no insertional activity.

Endplate noise and endplate spikes

When a recording is performed just under the motor endplate, an irregular low amplitude noise is heard (see Fig. 7.7, trace 1). This corresponds to the ongoing release of acetylcholine generating miniature endplate potentials. This activity has not been used to assess the condition of the neuromuscular junction. A concomitant finding is the so-called endplate spikes (Fig. 7.7, trace 1). They are irregular biphasic spikes with an initial negative take-off, probably generated by the irritation of the terminal nerve twig by the electrode and have no pathological significance. These should not be mistaken for fibrillation potentials. Their shape and firing pattern discriminates the two.

Findings in pathology

Fibrillation potentials and positive waves

In conditions with membrane instability such as after denervation and after focal necrosis of muscle fibres in myopathy where one segment then is without endplate, fibrillation potentials are seen (Fig. 7.8). The activity starts 7–21 days after denervation and may remain for very long time until the muscle fibre is reinnervated. Positive sharp waves (PSW; Fig. 7.8) have the same significance as fibrillation potentials although they may occur somewhat earlier. The generation is not absolutely clear, but the shape suggests that conduction is decreased or even blocked at the tip of the electrode. Transition from fibrillation potentials to positive waves can be seen (30). The spontaneous activity is more pronounced in warm muscles than in cold. There is also an impression that steroids and beta-blockers suppress the spontaneous activity.

Myotonic discharges

These discharges are often positive waves, but sometimes biphasic spikes that appear with waxing and waning amplitude and frequency between 20 and 100 Hz (Fig. 7.8). They are seen in primary myotonic disorders and can be provoked by muscle percussion or needle movements. Depending on type of myotonia they may disappear after a few minutes of muscle activation (warm up effect) or increase (paramyotonia). It is often important to study the MUP in patients with myotonic disorders, to assess whether they are normal or myopathic. This is difficult in the noise caused by the myotonic signal. A trick is to activate the muscle strongly for a minute. After this there is a silence for a couple of minutes corresponding to the warm up effect and now the MUPs can be studied, since voluntary activity can be elicited undisturbed.

Complex repetitive discharges

The complex repetitive discharges (CRD) are striking in shape and sound. They start and stop abruptly and consist typically of many single fibre components closely time locked to each other, without jitter (Fig. 7.8, trace 5). They appear with a discharge frequency of 1-60/s and may last for many seconds. These discharges are generated by so-called ephaptic transmission between hyperexcitable muscle fibres, e.g. after denervation (Fig. 7.9) (31). They are seen both in chronic neurogenic and myopathic conditions (e.g. myositis, Duchenne dystrophy, distal hereditary myopathy, Pompe's disease) and do not help in classifying a given condition.

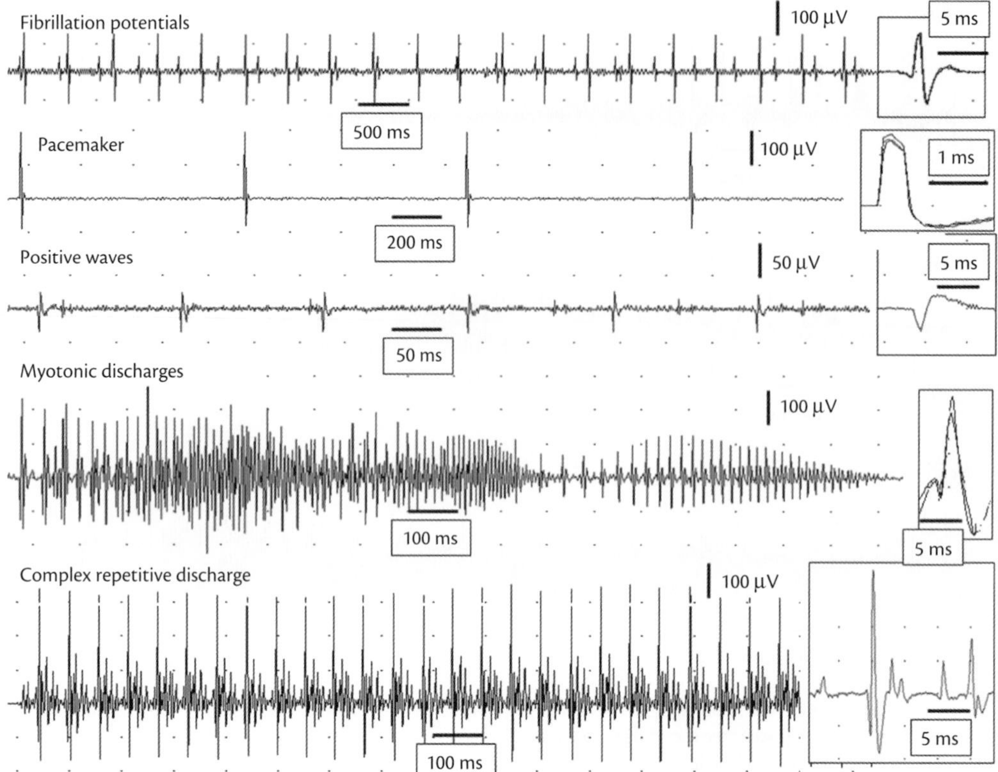

Fig. 7.8 Activity recorded from muscle at rest. Artefacts from a pacemaker (trace 2) must be differentiated from other activity. Other examples are generated in the muscle fibres.

Reproduced from Stålberg E, Daube JR: Electromyographic methods; in Stålberg E, (ed.) *Handbook of Clinical Neurophysiology*, Volume 2, pp. 147–185, copyright (2003), with permission from Elsevier.

Fasciculation potentials

Fasciculation potentials are seen as MUPs occurring at irregular and low frequency of 0.1–10 Hz in healthy subjects and with some higher frequency in pathological conditions, up to 30 Hz (Fig. 7.10). This difference cannot usually be used to separate malignant from benign fasciculations. Most of them are generated in the peripheral nerve tree, although some in the motor neuron. In normal subjects they are seen, particular in triceps surae muscles after exercise. In pathological conditions they are seen in motor neuron disease, many chronic neurogenic processes, thyrotoxicosis, tetany and anti-Ach-esterase therapy. Since they start in the axon, they sometimes give rise to F-waves, seen as a double discharge of the fasciculation, with an interval corresponding to the F-latency for the studied muscle (seen in Fig. 7.10, beginning of trace 3).

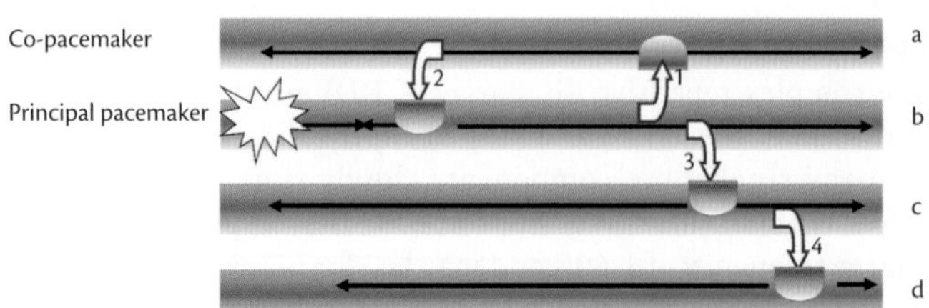

Fig. 7.9 Schematic explanation of complex repetitive discharge (CRD) in a denervated muscle. The drawing shows a primary pace-maker which is a loop that has been triggered by a fibrillation potential. Hot spots react to the passing electrical field and activate neighbouring muscle fibres.

Stålberg ©, 2013.

Since they sometimes occur with very low frequency, it is advisable to wait with the electrode in constant position for up to 1 min. A few sites must be studied if the study aims for exclusion of fasciculation in the tested muscle. In MND, the shape of fasciculations may vary mainly due to the jiggle in the MUP, but more important are the other characteristics such as signs of denervation and the unstable MUPs indicating ongoing reinnervation.

Double and multiple discharges

Double discharges are defined as a MUP firing twice within the relative refractory period (3–20 ms) from the previous discharge. They occur in radiculopathies, late stage of Guillain–Barré syndrome (GBS), inflammatory myopathies, but are not specific for a given diagnosis. They are often related to a tendency for muscle cramps. The double discharge is distinguished from late components of a MUP in its latency, always more than 3 ms, and that it usually produces a block of next regular firing of the MUP. Its shape is very similar to the first, although not identical since it occurs in relative refractoriness, during which time the amplitude is reduced compared with the first discharge. The amplitude drop is well correlated to the interval between the two discharges. Finally, when the electrode is slightly moved, the shape changes in parallel between the two discharges. If they had represented different MUPs, their amplitude should change independent of each other.

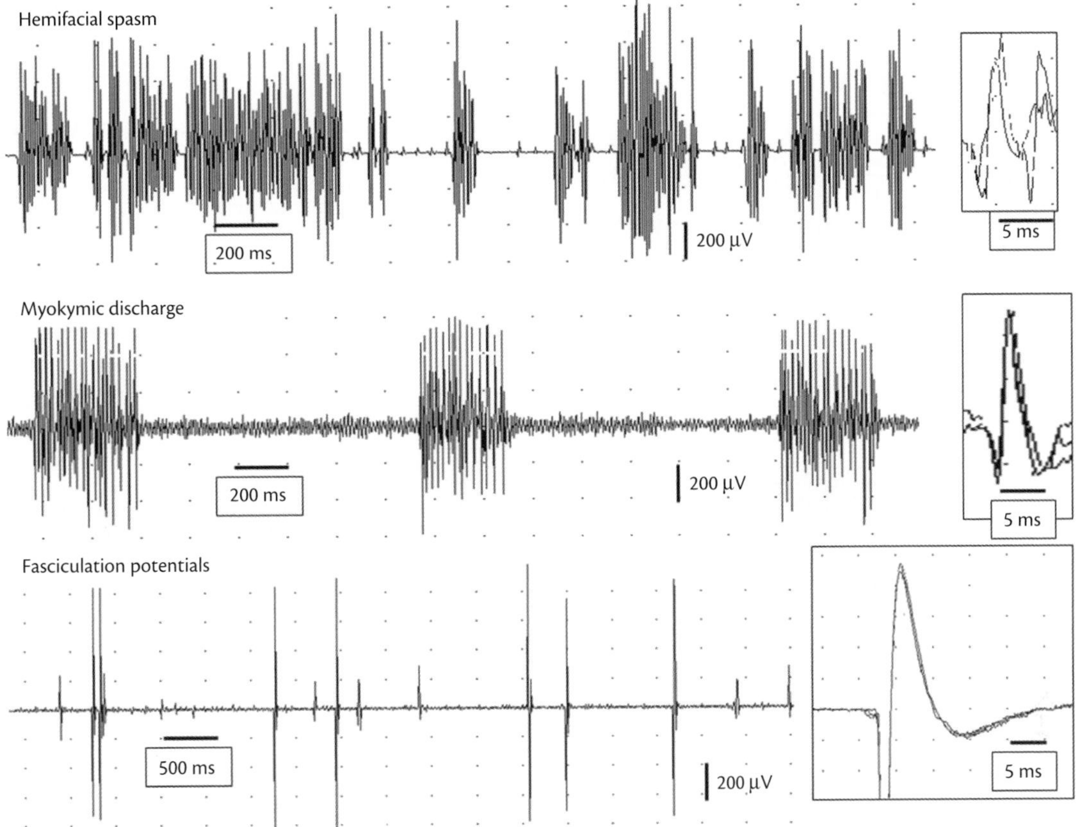

Fig. 7.10 Spontaneous activity starting in the axon or motor neuron. Note in trace 3, that the fasciculation potentials fires twice in the beginning of the sweep. The second discharge is probably an F-response generated by the initial fasciculation.

Reproduced from Stålberg E, Daube JR: Electromyographic methods; in Stålberg E, (ed.) *Handbook of Clinical Neurophysiology*, Volume 2, pp. 147–185, copyright (2003), with permission from Elsevier.

Multiple discharge may be seen as triplets, e.g. in hyperventilation and as multiplets in ischaemia.

A special double discharge is seen in normal muscle at the initiation of activity after more than 10 s of rest. The first discharge is often double, so-called initial doublet, and may be explained as an initial escape from the normal Renshaw inhibition.

Myokymic discharges

Myokymic discharges usually start in the axon, occur in burst of 2–20 with an inherent frequency of 20–80 Hz. They occur at intervals of 0.5–10 s (Fig. 7.10, trace 2). These discharges are seen in demyelinating disorders and in radiation plexopathy, and are usually seen on the surface as flickering or worm-like movements.

Neurotonic discharges

Neurotonic discharges (earlier neuromyotonic) are seen as doublets, triplets or bursts of activity with a frequency of 150–300 Hz (Fig. 7.11). They can be seen as extra discharges after voluntary activation, but more often spontaneously. They start in the axon (32,33), probably distally and indicate a potassium channel block. They occur in neurotonia, but also as a paramalignant phenomenon.

Synkinesis

These are defined as activity in a muscle activated at contraction of another muscle. It may be due to aberrant innervation or to ephatic transmission between axons. This activity is seen after Bell palsy and plexus lesions.

Muscle cramps

Muscle cramps are often painful. The activity starts peripherally in the nerve and causes an EMG pattern similar to that of strong voluntary contraction (Fig 7.7, trace 2). This occurs both in healthy subjects (during the night) and in chronic neurogenic conditions.

Findings during moderate contraction

Normal condition

With increasing voluntary contraction one MU after the other is recruited, first with low firing rate (6–7 Hz) and then with increasing rate as the force increases (Fig. 7.12). At a given moment a few MUs are active, and the one with the highest firing rate, is most likely the one that was first recruited, i.e. with the lowest threshold. There is an impression that higher amplitude MUPs are recruited later with increasing force, which should support the so-called size principle (7). This interpretation is probably not true. The small electrode does not record from the entire MU and does not reflect its total size (34). Instead, the explanation may be that there is an increasing statistical change that more MUs just close to the recording tip are recorded with increasing effort, seen with higher amplitudes. In the normal muscle the amplitude is usually not due

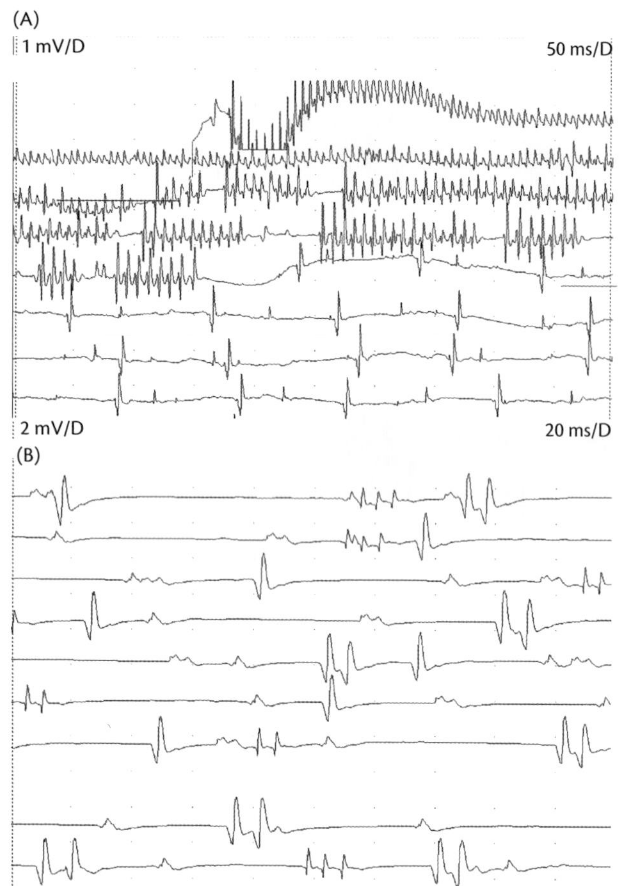

Fig. 7.11 Recording in a patient with neurotonia. (A) Muscle at rest. High frequency discharges and myokymic discharges are seen after initial electrode movement. (B) Voluntary activity produces double discharge or triple discharges.
Stålberg ©, 2013.

to summation but more related to recording distance for a few fibres, since the short MUPs do not summate very well. About 10% of the highest individual spikes may be due to summation (35).

A given MUP has the same shape during continuous activity and shows only minor variability. Other simultaneously recorded MUPs are most likely of different shape. When the needle electrode is moved to a new recording position, the earlier studied MUs may again be recorded, although with a completely different shape. The shape cannot classify the MU. The only hint may be that a given MU is recorded with the same firing rate at different position if the degree of activity is constant. Therefore, the technique of so-called multi-MUP analysis is a great advantage. Here, up to six MUPs of different shapes are recorded from one site with a constant electrode position. These cannot be generated by the same MU. Exceptions occur in pathological situation, where a given MUP may have a large jiggle and, therefore, be erroneously classified as two or more MUP in the same recording site. So, does it cause a major problem that 20 recorded MUP only represent 10–15 different MUs? Probably not, the pathology even with in a given MU may vary; a result with occasional recordings from two areas in a given MU does probably not differ statistically from a result obtained from definitely different MUs. The other question is whether the study of mainly low threshold MUs, which is the case in EMG, gives a correct picture of pathology. There are only

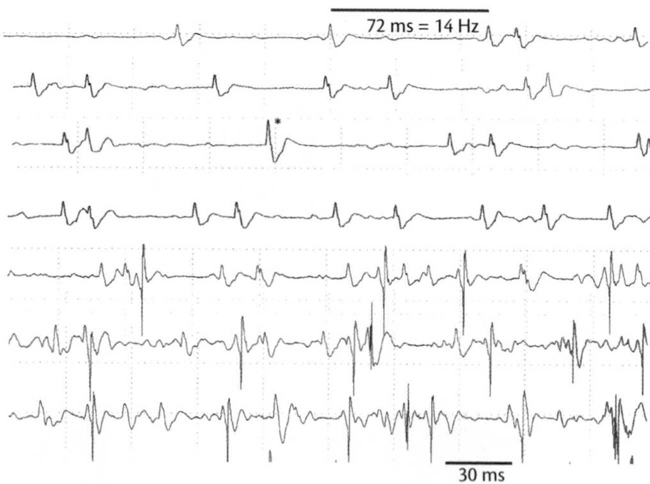

Fig. 7.12 Recording during increasing voluntary activation. More MUs come into action. The newly recruited MUPs may have low amplitude, and sometimes higher than the first due to the fact that their fibres are closer to the recording tip that earlier recruited MUs. In the fig we see one occasion of summation (*) of the two already active MUPs, a signal that never reoccurs. The so-called recruitment rate is defined as the firing rate of the first MU when the second is activated. This is seen on top trace, 14 Hz.
Stålberg ©, 2013.

exceptional situations where only high threshold motor units are abnormal, and this bias therefore does not influence the interpretation significantly. In disuse atrophy (and steroid myopathy) there is type II atrophy. However, fibre atrophy and disuse atrophy does not give any distinct EMG changes, and the inability to record from small type II fibres, is an insignificant problem.

Findings in pathology

Neuropathy

EMG in a neurogenic condition depends on the stage of the condition. After a partial nerve lesion the MUPs are normal during the first week (but the interference pattern at strong contraction is reduced). After 1–3-week fibrillation potentials and PSW appear. After one month or more, collateral sprouting gives the first signs of reinnervation, first with late components with large jitter, and then with increasing number of initially late components that then merge and form an increasing irregular MUP. During the first 6 month this shows a large jiggle, but in the chronic stage, the signals are increasingly stable with high amplitude and long duration. The multi-MUP analysis shows typically long duration MUP with increased amplitude and area (Fig. 7.13).

The situation of complete nerve cut show signs of denervation and loss of voluntary activity. The first signs of reinnervation as small, nascent MUPs appear after a time that depends on the site of lesion. These follow then the dynamics described above.

Myopathy

Depending on the type of myopathy the EMG is more or less abnormal. In muscular dystrophies, it is common to see some type of spontaneous activity, such as fibrillation potentials, PSW, CRD. This should not be misinterpreted to mean that the condition is primarily neurogenic. An early sign is polyphasicity of the MUP, which is due to muscle fibre diameter variation, a common

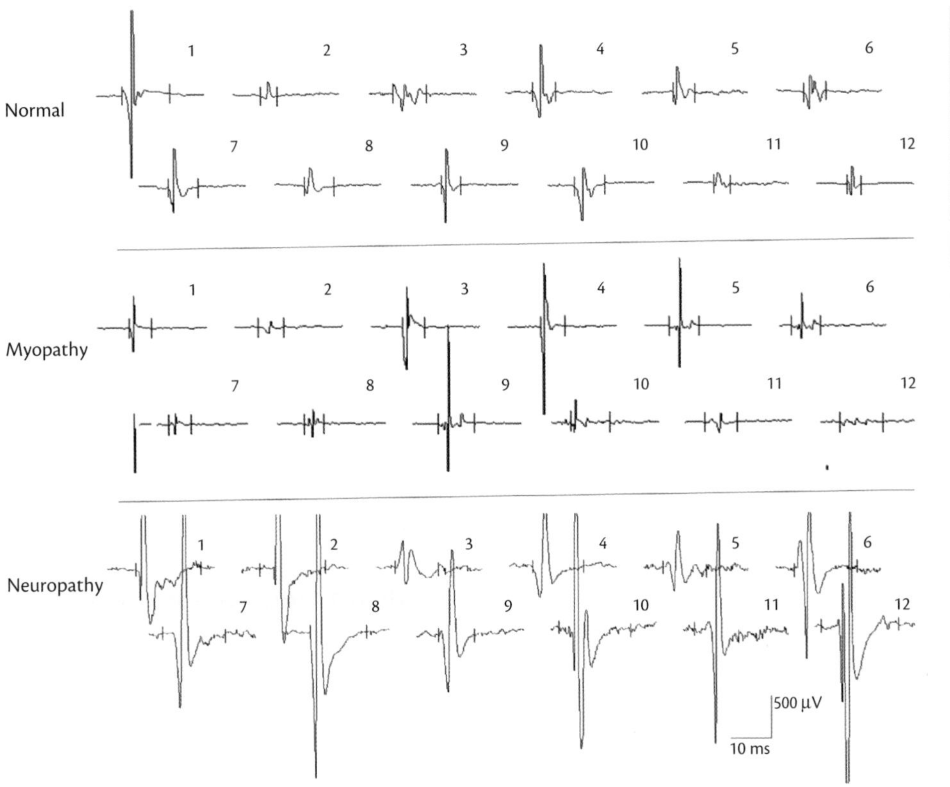

Fig. 7.13 Multi-MUP recordings from normal, myopathic and neurogenic conditions.
Stålberg ©, 2013.

morphological finding in myopathy. The amplitude is decreased due to the fibre diameter variation with less summation of individual spike components. Depending on the myopathy, abnormalities may dominate in proximal or distal muscles. With increasing pathology, loss of muscle fibres theoretical produced short and low amplitude MUPs. The MUP is also influenced by other factors, such as regeneration of muscle fibres, fibre hypertrophy, secondary neurogenic changes followed by real reinnervation and fibrosis that changes the volume conduction characteristics of the muscle. In this context it is important to mention that a relatively common finding in myopathy is high amplitude, but short MUPs (Fig. 7.13). This can interpreted as a recording from a large diameter muscle fibre. It is probably not a generally large fibre along its entire course, since these signals should then occur early in the MUP, which is not the case. It is possible that there is a focal swelling of the muscle fibre, seen in biopsies, but considered as artefact. Another possibility is that the recording is performed from at segment along a muscle fibre where splitting starts. The fibre is enlarged here generating two or more absolute synchronous action potentials from each of the branches, which will give high amplitude MUP (Fig. 7.14). In any case, one should not misinterpret this finding as neurogenic, in which case the duration is increased.

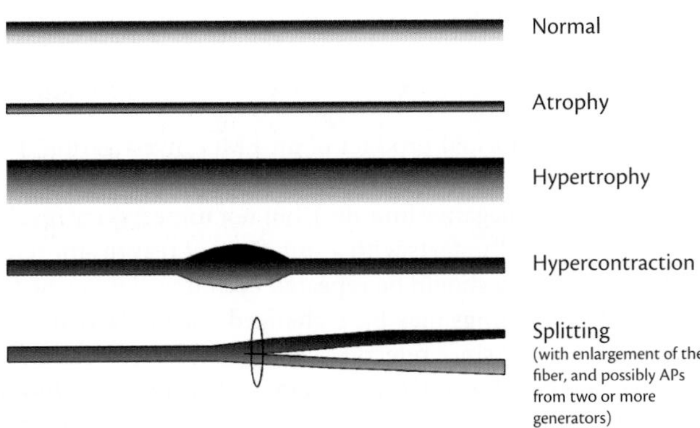

Fig. 7.14 Schematic alternatives that may give high amplitude spikes in myopathy.
Stålberg ©, 2013.

Neuromuscular junction dysfunction

In myasthenia gravis and other disorder the neuromuscular transmission, the MUPs may be normal, except for a variation in shape from one discharge to the other. Due to variable summation of individual spike, the amplitude will vary. This was reported as the first sign of myasthenia gravis in EMG (36).

Findings during strong contraction

Normal condition

During increasing force, new MUs are recruited in an orderly manner. Each MU starts with a firing rate of 6–8 Hz and increases to 30–50 Hz. The successive MUs are introduced into activity when the previous MU has reached a certain frequency. This can be quantitated, see other chapters in this book. At full effort, many MUs contribute to a mixed signal, where the detailed shape of individual MUP is obliterated. Still the shape of individual MUP remains since summation is not very frequent due to the short duration of each signal. Therefore, amplitude and number of turning points

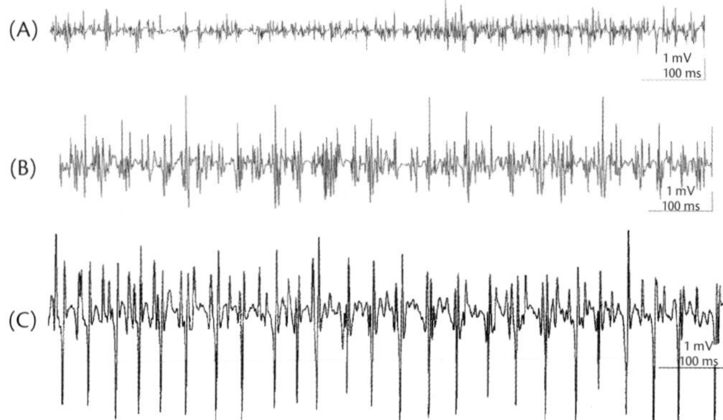

Fig. 7.15 Interference pattern in myopathy (A), in healthy control (B) and in a patient with status post-polio (C).
Reproduced from Stålberg E, Daube JR: Electromyographic methods; in Stålberg E, (ed.) *Handbook of Clinical Neurophysiology*, Volume 2, pp. 147–185, copyright (2003), with permission from Elsevier.

relatively well describes the characteristics of individual MUPs. This is used in the quantitation of the interference pattern.

As a visual assessment, we can estimate reasonably the maximal amplitude, but omitting the few amplitude outliers on negative and positive signals. Remaining amplitude is called the Envelope amplitude (Fig. 7.15). The other parameter is the fullness of the signal. Is the baseline well filled with signals, or are there gaps, the latter indicating loss of active MUs. The sound is also helpful, but not directly quantified, other than in spectral analysis.

Findings in pathology

Neuropathy

In cases of neurogenic condition, that also means loss of active exons, it should be possible to detect a reduced pattern. The investigation often starts with slight and increasing activation of the muscle. It may be easily seen, that the second and following MUPs occur when the first already has a higher firing rate. This is called late recruitment. The advantage with this analysis approach is that the loss of MUs is seen already before the patient is asked for a full contraction. This is true even day one in Guillain–Barré syndrome or an acute radiculopathy. The next step in many laboratories is to analyse the pattern at full contraction. As soon as there is weakness due to neurogenic reasons, the interference patterns should be reduced. At later stage, when reinnervation has occurred, high amplitude MUPs are seen and the interference may show a large variation in amplitude, a general increase in envelope amplitude and a reduced interference. Note that the degree of neurogenic involvement in a given muscle may vary considerable. This is partly dependent on the fact that reinnervation of neighbouring denervated MUs may only occupy a part the surviving MU. The other factor is that all muscles (except rhomboideus) have dual innervation. If one of the roots to a muscle is involved, the other may be normal. The roots have different distribution within the muscle. Therefore it is important the EMG analysis is performed from at least two well-separated sites.

Myopathy

Each MUP is polyphasic, and generally lower in amplitude and with short duration, unless there is an excessive polyphasicity. The number of MUs is unchanged, unless the muscle is in an end-stage. With increasing force, new MUs are recruited earlier than in normal subjects. Speculatively, this may be due to weakness of the muscle in which more MUs need to cooperate to give a certain force. With strong contraction, the interference is full and the envelope amplitude is reduced. The sound is higher pitched than in the normal and certainly the neurogenic conditions. In cases when the electromyographer is uncertain on neuropathy or myopathy, the interference pattern is crucial and very helpful.

Central disorders

In a muscle with weakness due to insufficient of central drive the MUP is normal, but the firing pattern is irregular. This is appreciated already during increasing force, but also as inability to keep a steady and full contraction. In general there are no signs of denervation but a slight degree of fibrillations have been reported that may come from secondary nerve compression or some form of transsynaptic disturbance. Other methods than EMG must be used to further study central mechanisms.

Pitfalls in electromyography

Omissions

- Inadequate patient evaluation.
- Testing incorrect muscles.
- Testing insufficient number of muscles.
- Inadequate sampling in a muscle.
- Incorrect parameter assessment.

Commissions

- Performance of unnecessary steps.
- Wrong electrode.
- Inappropriate needling procedure.
- Inadequate muscle contraction.

Wrong conclusions

- Suggesting diagnosis without adequate information.
- Identifying normal variants as abnormality.
- Misinterpreting unusual waveforms.
- Fail to detect pathology.
- Misinterprete findings.

Reporting

The Report is the formal product of an EMG investigation. It is important that it is comprehensible, strict, contains relevant facts (both positive and negative findings), but not unnecessary results.

The report typically starts with a summary of reasons for referral. Then the history should be repeated even if well described in the referral since things may have changed, particularly if some months have passed since referral. A focused neurological examination should help formulate a strategy for the neurography and EMG exam. Often it is wise to start with neurography and sometimes this study is even enough. Next step is usually the EMG and perhaps other methods necessary to get a complete picture, such as autonomic tests, studies of evoked potentials (TMS, SEP,

VEP, BAEP). The report of the study then contains subsections. First a summary of neurography, perhaps data (with indication of deviation from reference values) and then an interpretation of the neurography. Then a summary of EMG findings, in tables or charts followed by a narrative interpretation. Finally, a general summary of the entire study, where the most important conclusion should be noticed in the first sentence, as eye catching information. Supportive data may then be elaborated.

Depending on the general skill of the electromyographer, a comment related to the clinical relevance may be added, clearly stating that this last comment is an overall summary, also including clinical data.

References

1. Sherrington, C.S. (1929). Ferrier Lecture—some functional problems attaching to convergence. *Proceedings of the Royal Society, London*, **105**, 332–62.
2. McComas, A.J., Fawcett, P.R.W., Campbell, M.J., and Sica, R.E.P. (1971). Electrophysiological estimation of the number of motor units within a human muscle. *Journal of Neurology, Neurosurgery, & Psychiatry*, **34**, 121–31.
3. Feinstein, B., Lindegård, B., Nyman, E., and Wohlfart, G. (1955). Morphologic studies of motor units in normal human muscles. *Acta Anatomica*, **23**, 127–42.
4. Gath, I. and Stålberg, E. (1981). In situ measurements of the innervation ratio of motor units in human muscles. *Experimental Brain Research*, **43**, 377–82.
5. Stålberg, E. and Dioszeghy, P. (1991). Scanning EMG in normal muscle and in neuromuscular disorders. *Electroencephalography and Clinical Neurophysiology*, **81**, 403–16.
6. Burke, R.E. (1980). Motor units in mammalian muscle. In: A. J. Sumner (Ed.) *The physiology of peripheral nerve disease*, pp. 133–94. Philadelphia, PA: WB Saunders.
7. Henneman, E., Somjen, G., and Carpenter, D.O. (1965). Functional significance of cell size in spinal motor neurones. *Journal of Neurophysiology*, **28**, 560–89.
8. Stålberg, E. (2011). Macro electromyography, an update. *Muscle & Nerve*, **44**, 292–302.
9. Kugelberg, E., Edström, L., and Abbruzzese, M. (1970). Mapping of motor units in experimentally reinnervated rat muscle. Interpretation of histochemical and atrophic fibre patterns in neurogenic lesions. *Journal of Neurology, Neurosurgery, & Psychiatry*, **33**, 319–29.
10. Hatanaka, Y. and Oh, S.J. (2008). Ten-second exercise is superior to 30 second exercise for post-exercise facilitation in diagnosing Lambert–Eaton myasthic syndrom. *Muscle & Nerve*, **37**, 572–5.
11. Tim, R.W. and Sanders, D.B. (1994). Repetitive nerve stimulation studies in the Lambert-Eaton myasthenic syndrome. *Muscle & Nerve*, **17**, 995–1001.
12. Oh, S.J., Hatanaka, Y., Ito, E., and Nagai, T. (2014). Post-exercise exhaustion in Lambert-Eaton myasthenic syndrome. *Clinical Neurophysiology*, **125**, 411–14.
13. Trontelj, J.V., Sanders, D.B., and Stålberg, E. (2001). Electrophysiological methods for assessing neuromuscular transmission. In: W. F. Brown, C. F. Bolton, and M. J. Aminoff (Eds) *Neuromuscular function and disease*, pp. 414–32. Philadelphia, PA: Saunders.
14. Stålberg, E. and Ekstedt, J. (1973). Single fibre EMG and microphysiology of the motor unit in normal and diseased human muscle. In: J. Desmedt (Ed.) *New developments in EMG and clinical neurophysiology*, pp. 113–29. Basel: S. Karger.
15. Ekstedt, J. and Stålberg, E. (1973). Single-fibre electromyography for the study of the microphysiology of the human muscle. In: J. Desmedt (Ed.) *New developments in electromyography and clinical neurophysiology*, pp. 89–112. Basel: Karger.
16. Stålberg, E., Trontelj, J.V., and Sanders, D.B. (2010). *Single fiber EMG*, 3rd edn. Uppsala: Edshagen Publishing House.
17. Trontelj, J.V. and Stålberg, E. (1992). Jitter measurement by axonal stimulation. Guidelines and technical notes. *Electroencephalography and Clinical Neurophysiology*, **85**, 30–7.
18. Trontelj, J.V. and Stålberg, E. (1990). Single motor end-plates in myasthenia gravis and LEMS at different firing rates. *Muscle & Nerve*, **14**, 226–32.
19. Stålberg, E. and Sanders, D.B. (2009). Jitter recordings with concentric needle electrodes. *Muscle & Nerve*, **40**, 331–9.
20. Kouyoumdjian, J.A. and Stålberg, E. (2008). Reference jitter values for concentric needle electrodes in voluntarily activated extensor digitorum communis and orbicularis oculi muscles. *Muscle & Nerve*, **37**, 694–9.
21. Kouyoumdjian, J.A. and Stålberg, E. (2011). Concentric needle jitter on stimulated orbicularis oculi in 50 healthy subjects. *Clinical Neurophysiology*, **122**, 617–22.
22. Kokubun, N., Sonoo, M., Imai, T., et al. (2012). Reference values for voluntary and stimulated single-fibre EMG using concentric needle electrodes: a multicentre prospective study. *Clinical Neurophysiology*, **123**, 613–20.
23. Stålberg, E., Sanders, D.B., Ali, S., et al. (2016). Reference values for jitter recorded by concentric needle electrodes in healthy controls: A multicenter study. *Muscle & Nerve*, **53**(3), 351–62.
24. Grimby, G., Stålberg, E., Sandberg, A., and Stibrant-Sunnerhagen, K. (1998). An 8-year longitudinal study of muscle strength, muscle fibre size, and dynamic electromyogram in individuals with late polio. *Muscle & Nerve*, **21**, 1428–37.
25. Jabre, J.F. (1991). Concentric macro electromyography. *Muscle & Nerve*, **14**, 820–5.
26. Stålberg, E. (1966). Propagation velocity in single human muscle fibres. *Acta Physiologica Scandinavica*, (Suppl. 287), 1–112.
27. Nandedkar, S.D., Sanders, D.B., Stålberg, E., and Andreassen, S. (1988). Simulation of concentric needle EMG motor unit action potentials. *Muscle & Nerve*, **2**, 151–9.
28. Sonoo, M. and Stålberg, E. (1993). The ability of MUP parameters to discriminate between normal and neurogenic MUPs in concentric EMG: analysis of the MUP 'thickness' and the proposal of 'size index'. *Electroencephalography & Clinical Neurophysiology*, **89**, 291–303.
29. Zalewska, E. and Hausmanowa-Petrusewicz, I. (1995). Evaluation of MUAP shape irregularity—a new concept of quantification. *IEEE*, **42**, 616–20.
30. Nandedkar, S.D., Barkhaus, P.E., Sanders, D.B., and Stålberg, E. (2000). Some observations of Fibrillations & Positive sharp waves. *Muscle & Nerve*, **23**, 888–94.
31. Trontelj, J.V. and Stålberg, E. (1983). Bizarre repetitive discharges recorded with single fibre EMG. *Journal of Neurology, Neurosurgery & Psychiatry*, **46**, 310–16.
32. Torbergsen, T., Stålberg, E., and Brautaset, N.J. (1996). Generator sites for spontaneous activity in neuromyotonia. An EMG study. *Electroencephalography & Clinical Neurophysiology*, **101**, 69–78.
33. Newsom-Davis, J. (1993). Immunological association of acquired neuromyotonia (Isaac's syndrome). Report of five cases and literature review. *Brain*, **116**, 453–69.
34. Ertas, M., Stålberg, E., and Falck, B. (1995). Can the size principle be detected in conventional EMG recordings? *Muscle & Nerve*, **18**, 435–9.
35. Nandedkar, S.D., Sanders, D.B., and Stålberg, E. (1986). Simulation and analysis of the electromyographic interference pattern in normal muscle. Part II: activity, upper centile amplitude, and number of small segments. *Muscle & Nerve*, **9**, 486–90.
36. Harvey, A.M. and Masland, R.L. (1941). The electromyogram in myasthenia gravis. *Bulletin of the Johns Hopkins Hospital*, **48**, 1–13.

CHAPTER 8

Quantitative electromyography

Anders Fuglsang-Frederiksen, Kirsten Pugdahl, and Hatice Tankisi

Introduction to quantitative electromyography

For more than 70 years, electromyography (EMG) has been used as a diagnostic tool in patients with myopathy and neurogenic disorders. Following the introduction of manual motor unit action potential (MUP) analysis at weak effort, several other methods have been introduced, including automatic analysis of MUPs and analysis of the interference pattern (IP) at stronger effort.

Qualitative EMG analysis with visual and auditory evaluation of the properties of a number of MUPs sampled at weak effort, the motor unit firing rate and the IP at maximal effort, seems of value in evaluating myopathic and neurogenic changes in the muscle if the changes are clear-cut or the examiner is experienced. The drawbacks using qualitative visual and auditory EMG analysis are:

- The judgment of the properties of the MUP may be biased by the examiner's expectations of the diagnosis.
- The method requires years of training and experience with respect to delineating abnormal changes against normal findings.

The advantage is that it is quickly done, in a few minutes, and can be performed on all muscles, which are possible to reach with a needle.

With quantitative MUP analysis, the findings are reproducible and evidence based if a representative normal material is available (1,2) and therefore comparable between laboratories and over time. Quantitative EMG (QEMG) can additionally be useful in borderline cases, in the guidance of botulinum toxin treatment (3), in monitoring of treatment (4), and even in children a high diagnostic accuracy can be obtained by an experienced examiner (5). The method requires normal values, which are best obtained in the examiner's laboratory, and that the examiner is trained in the method. This limits the number of muscles, which can be examined to those where normal values exist.

Quantitative techniques are also more time-consuming and studies have shown that physicians using quantitative techniques generally study fewer muscles using a goal-directed strategy guided by a detailed clinical investigation and the on-going EMG results, while the physicians using qualitative methods tend to use a screening approach with many muscles examined (6). The advantages and limitations of quantitative EMG are summarized in Box 8.1.

With an increase in force from zero up to maximum voluntary contraction (MVC) the EMG shows an increase in number of spikes and amplitude. At weak effort one can identify individual MUPs, while at higher efforts there is interference with summation and cancellation of action potential components. MUP analysis may therefore only reflect pathophysiological and physiological changes in motor units recruited at weak effort (7), whereas IP analysis (IPA) may reflect changes in motor units recruited in the whole force range, as it describes the electrical activity of the muscle from a weak effort up to MVC. In order to quantify the IP, automatic methods are needed (8).

One should realize that while EMG is useful in detecting myopathic or neurogenic changes in general, it is most often of limited help in the differentiation of different forms of myopathy and neurogenic disorders, although it can usually discriminate between acute and chronic processes.

Quantitative motor unit potential analysis

Quantitative analysis of MUPs was introduced in the diagnosis of neuromuscular disorders in the 1940s (9–11). The mean duration and the mean amplitude of the MUPs and the incidence of polyphasic potentials from a representative number of motor units sampled at weak effort were measured manually. This method was further developed and applied on muscles from patients and controls in the 1950s and 60s (1,12–15). During the 1970s and 80s

Box 8.1 Advantages and limitations of quantitative EMG

Advantages

- Less subjectivity in the analysis compared with qualitative analysis.
- Increased sensitivity and specificity.
- More reliable in borderline cases.
- Reproducible and evidence-based.
- Comparable data over time and between laboratories.

Limitations

- Time-consuming.
- Requires a trained examiner.
- Requires good patient cooperation.
- Available muscles limited to those with reference material.

Box 8.2 Definitions and procedure for motor unit potential (MUP) analysis

Quantitative MUP analysis

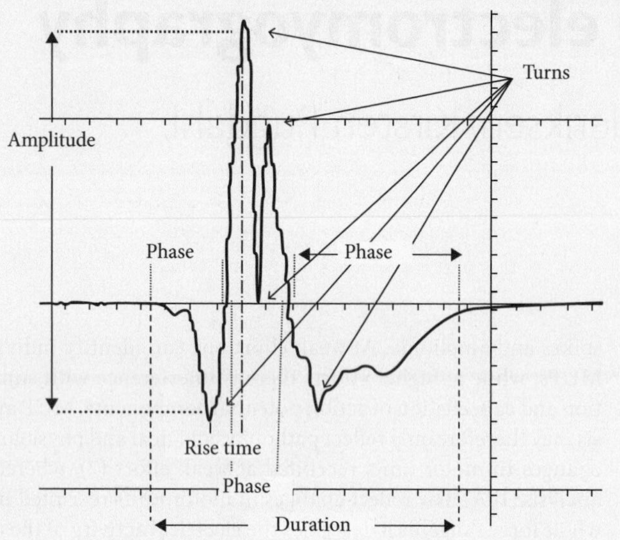

Fig. B8.1 Simple triphasic MUP. Definitions for duration, peak-to-peak amplitude, phases, turns, and rise time are indicated. Sweep speed: 10 ms/div, gain: 100 μV/div.

Definitions and parameters (204)

- *Amplitude (peak-to-peak)*: distance between the maximum positive to maximum negative peak
- *Area*: the space under the curve of the waveform, usually measuring only the negative components)
- *Baseline crossing*: changes in voltage from negative to positive or vice versa
- *Duration*: the time from the first deflection from the baseline to the final return to the baseline
- *Number of phases*: number of baseline crossings plus one (≥ 5 in polyphasic potentials)
- *Number of turns*: number of changes in direction above noise level (≥ 20–50 μV)
- *Rise time*: the time between the maximal negative peak and the preceding maximal positive peak
- *Size index*: 2 × $\log_{10}(\text{amp}_{(mV)})$ + area/$\text{amp}_{(ms)}$ (50)
- *Thickness*: area/amplitude ratio of the MUP (49). Corresponds to the duration

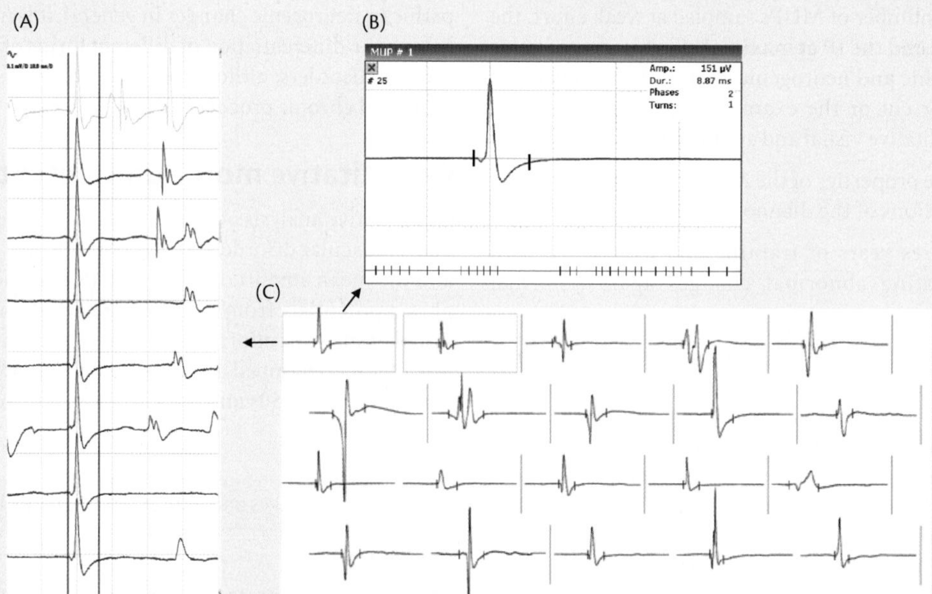

Fig. B8.2 Collection of 20 MUPs from a normal muscle (C). The cursors are placed either (A) manually using the individual MUPs (I-MUP) or (B) automatically on the averaged potential (Multi-MUP method). The MUP shown in (B) is averaged ×25.

Procedure

- Concentric needle electrodes are generally used in Europe, while monopolar electrodes are more commonly used in North America.
- Filter settings: 2–20 Hz to 10 kHz.

(continued)

Box 8.2 Continued

Individual MUP (12)

- The electrode is placed within the muscle at random without optimizing the amplitude.
- Recordings are made during slight voluntary contraction of the muscle with only one or a few motor units active.
- MUPs are extracted using a template matching algorithm and the cursors are placed by the examiner using the individual MUP (Fig. B8.2A).
- 1–4 motor units are randomly sampled at each of several needle insertion sites by movement of the needle at least 5 mm between the individual sites.
- At least 20 different motor units are sampled by 2–5 new insertions perpendicular and transversally to the course of the muscle fibres. MUPs with a peak-to-peak amplitude >50 μV are accepted.
- The averaged duration and amplitude of the MUPs, and the percentage of polyphasic potentials are compared with the mean of age-matched controls.
- The duration has to be measured at the same amplification and same time window to ensure reproducibility (in the original Buchthal method 0.1 mV and 10 ms/division).
- One muscle may be analysed within 15–20 min.

Multi-MUP (33)

- The electrode is inserted randomly in the muscle, but positioned so that the EMG signal is crisp, i.e. with a rise time <500 μs.
- Recordings are made at slight to moderate contraction (5–30% of maximum force) for at least 5 s.
- At each recording 1–6 MUPs are extracted based on template matching
- Signals with a negative peak amplitude >30 μV are recognized (approx. corresponding to peak-to-peak amplitude >50 μV). The slope of the rising phase at the triggering point >30 μV/0.1 ms and time between successive MUPs >2.5 ms.
- The cursors are placed automatically using the averaged MUP (Fig. B8.2B). Manual correction is possible.
- Usually two or three skin insertions and a few different sites at each insertion are sampled to collect 20–30 MUPs.
- Data are analysed by comparing the mean of parameters to age-matched controls or by outlier analysis.
- Total recording, analysis and editing time of 20–30 MUPs about 4–8 min.

several analyses were developed using automatic measurements, averaging of several action potentials from the same motor unit, template matching, and decomposition (16–26). An artificial neural network technique, including wavelet analysis (27,28), and different multivariate methods, e.g. factor analysis (29) and discriminant analysis (30–32), have been suggested in order to increase the specificity and the sensitivity of the MUP analysis. A quantitative automatic MUP analysis with measurements on averaged MUPs selected by a simple form of decomposition and using template matching, i.e. the multi-MUP analysis was introduced in a useable form in the 1990s (33–35). Recently, a new decomposition technique and a decomposition-enhanced spike-triggered averaging technique have been developed (36–39).

For a description of individual MUP (I-MUP) and multi-MUP analysis procedure and diagnostic criteria for the measured parameters see Box 8.2 and Table 8.1.

MUP parameters and their significance

In order to obtain a representative sample of a muscle, action potentials from at least 20 different motor units should be sampled (1,40,41). MUPs are sampled at weak effort, on average corresponding to about 4% of MVC (42). The conventional parameters of the MUP are the duration, which reflects the size of the motor unit, the amplitude that is dependent on the two or three muscle fibres nearest the actual recording site of the electrode, and the percentage of polyphasic potentials (12). MUP duration is the most robust feature to differentiate between myogenic and neurogenic conditions (25). In control subjects the mean duration of MUPs increases continuously with age (12,14,43). Polyphasic potentials or satellite components result from loss of muscle fibres (7), regeneration of muscle fibres (44), or variation in muscle fibre diameter (45,46), but may also be seen in disuse atrophy, where they are associated with duration and amplitude within normal limits (47,48).

In order to increase the diagnostic yield, several other parameters of the MUP have been suggested:

- the number of turns, the area and the thickness, i.e. area divided by amplitude (49);
- the size index (50,51);
- the instability of the MUP (jiggle) (51,52);
- shape irregularity (53);
- energy (54).

Table 8.1 Diagnostic criteria for the most used quantitative EMG parameters

Technique	Specific criteria		Non-specific criteria
	Myopathic	*Neurogenic*	
Motor unit potential analysis			
◆ Mean duration	↓	↑	
◆ Mean amplitude	↓	↑ (>200%)	
◆ % polyphasic pot.			↑
◆ Satellite potentials			↑
◆ Thickness	↓	↑	
◆ Size index	↓	↑	
Macro-EMG			
◆ Median amplitude	↓	↑	
◆ Median area	↓	↑	
◆ Fibre density			↑
Power spectrum analysis			
◆ Mean power frequency	↑	↓	
◆ Median power frequency	↑	↓	
◆ Relative power at 1400 Hz		↓	↑
Turns-amplitude analysis			
Peak-ratio			
◆ Turns-amplitude ratio	↑	↓	
◆ Number of small time intervals	↑	↓	
◆ Turns/s	↑	↓	
◆ Mean amplitude	↓	↑	
Cloud			
◆ Mean amplitude or turns/s	↓*	↑*	
◆ Number of small segments or activity	↑*	↓*	
◆ Upper centile amplitude/activity	↓*	↑*	

*≥2 points (10%) outside the cloud of healthy controls.

MUP wavelet analysis seems a potential diagnostic tool to differentiate neurogenic myopathic and normal muscles (55,56).

Quantitative analysis of concentric MUPs in myopathy

The random and diffuse degeneration as well as regeneration of muscle fibres in myopathic muscles are reflected by short duration, low-amplitude, and a polyphasic shape of individual MUPs (7) (Fig. 8.1). When only one muscle in each patient was studied, 59% of patients with progressive muscular dystrophy had a decreased mean duration of individual MUPs (57), while examination of several muscles increased the sensitivity to 72–85% with a high specificity (44,58). Of 80 muscles from a group of patients with various types of myopathy, 46% showed a decreased duration of individual MUPs and the specificity was low (59). The diagnostic sensitivity of the duration measurement may be increased if the mean duration is obtained from simple potentials excluding polyphasic potentials (60).

When applying automatic analysis of averaged MUPs in the brachial biceps muscle, 67% of 30 patients with inflammatory myopathy had a decreased mean duration or decreased area:amplitude ratio (61). In patients with inflammatory myopathy or inclusion body myositis, multi-MUP analysis showed a high diagnostic yield of 94–100% (34,62). Similarly, reasonably predictive values for a diagnosis of myopathy was found in 31 patients with facioscapulohumeral muscular dystrophy (63,64). Averaged MUP analysis may be useful in monitoring changes, e.g. dengue myopathy (65).

Quantitative analysis of concentric MUPs in neurogenic lesions

In patients with neurogenic disorders the individual MUP analysis shows increased mean duration indicating chronic partial denervation with secondary peripheral sprouting, i.e. reinnervation, with a high sensitivity (70–80%) and specificity (44,66–69) (Fig. 8.1). Averaged MUP analysis showed a poor sensitivity of MUP mean duration (15–30%) in 11 patients with neurogenic disease, while a combination of mean amplitude and thickness increased the sensitivity to 70% (50). In 220 patients with suspected amyotrophic lateral sclerosis (ALS), individual MUP analysis showed signs of reinnervation in 87–91% of weak, and in 44% of non-weak muscles in the absence of denervation activity (69). Mean duration, mean amplitude, and mean area all showed a linear increase over time in 15 patients with ALS (70). Combining decomposition based MUP analysis with Bayes' rule into a Bayesian muscle characterization in 14 patients with neurogenic disorders and 16 healthy control subjects, a categorization accuracy of 80–95% was obtained (71).

Quantitative analysis of concentric MUPs in neuromuscular transmission failure

Quantitative MUP analysis showed myopathic changes (decreased duration) in 20–50% patients with myasthenia gravis, highest in facial muscles (72–74) and myopathic changes were seen both in patients with MuSK antibody positive, as well as acetylcholine receptor antibody positive patients. Similar findings may be seen in patients with Lambert–Eaton syndrome (75). The decreased MUP duration seen in myasthenia gravis may be due to an associated myopathy (44) or random blocking of muscle fibres giving a picture of myopathy (73).

Macro-EMG MUP analysis

To obtain a relative indication of the motor unit size the macro-EMG was introduced (76,77). The signal on a single fibre electrode is used to trigger the EMG signal from the cannula for averaging of the macro-EMG MUP. The information obtained by the macro-EMG MUP may already be included in the information obtained by the concentric MUP. The macro-EMG amplitude and area are linearly related to the duration of the concentric MUPs, both in controls and in patients with neuromuscular disorders (78,79).

The reported sensitivity of the macro-EMG analysis with a decrease in area or amplitude of the MUP is low compared with concentric or monopolar MUP analysis (30,61,79), i.e. 10–44% in patients with myopathy (61,79,80). In patients with mitochondrial cytopathy both myopathic and neurogenic changes may be seen (81,82). In a few patients with myopathy, the amplitude or the area of the macro-EMG MUPs may even be increased (79,83).

In neurogenic muscle, high amplitude and area reflect peripheral motor unit sprouting (84,85). The diagnostic sensitivity of

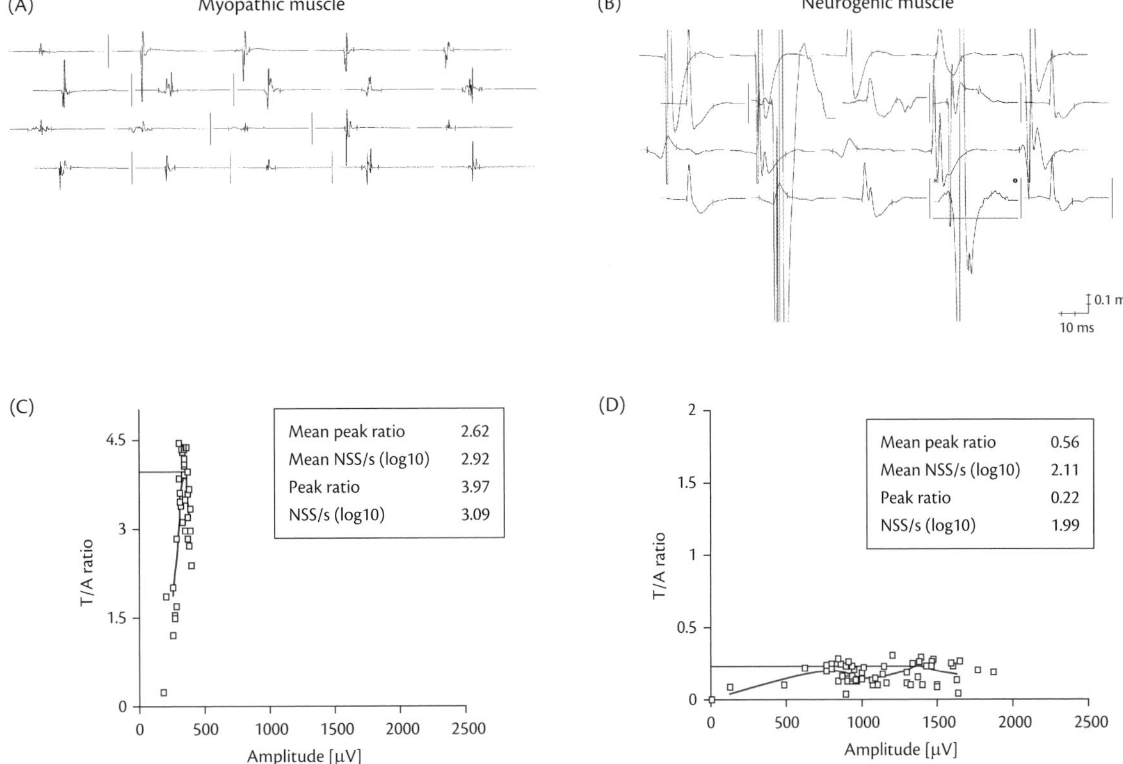

Fig. 8.1 MUP (A,B) and peak-ratio (C,D) analysis in myopathic and neurogenic muscle.

amplitude or area of macro MUP is high (62–92%) in neurogenic muscle from patients with an L4 root lesion (86), ALS (87), prior polio (88), and Guillain–Barré syndrome (89). Macro-EMG is easy to perform and is recommended for following disease progression (90,91).

For a description of macro-EMG analysis procedure and diagnostic criteria of the measured parameters (see Box 8.3 and Table 8.1 and Chapter 8).

Analysis of motor unit firing rate

Visual qualitative analysis of the firing rate of motor units has been related to the degree of muscle contraction (92,93). In myopathy, early recruitment may be seen, i.e. more MUPs are present for the level of muscle contraction compared with controls. Owing to the weakness of the muscle, the patient has to use a relatively higher force associated with a higher firing frequency or early recruitment (7,94). The literature about the diagnostic value of quantitative analysis of motor unit firing rate is conflicting. At a force of 10% of MVC, a similar firing rate was found in patients with myopathy, neurogenic disorders, and controls (95), while at 30% of MVC or higher force levels, the motor unit firing rate was increased in patients with myopathy (17,96). Using the decomposition technique, there was no difference in firing rate between controls and patients with myopathy or neurogenic disorders (34,97,98).

Quantitative interference pattern analysis

In patients with myopathy, the IP at MVC may be full with decreased amplitude, even in a muscle with decreased force, while the pattern is reduced or discrete with increased amplitude in neurogenic conditions. In clear-cut myopathic or neurogenic cases visual analysis of the interference pattern may be diagnostic, while in less clear-cut cases the use of a quantitative method is recommended. Quantitative methods may also be useful in muscles from patients with myopathy at a very advanced stage, where the IP at MVC may show reduced recruitment due to pronounced loss of muscle fibres resulting in loss of whole motor units (99).

Power spectrum analysis

The findings of an increase in high frequencies in muscles of patients with myopathy (100,101) and an increase in low frequencies in muscles of patients with neuropathy (102) suggested that spectrum analysis with analogue octave band filters could be used as a diagnostic tool. It is now possible to use fast Fourier transformation with a higher resolution. The best diagnostic parameters of spectrum analysis were the relative power at 1400 Hz and the mean power frequency (see Table 8.1) (103). The sensitivity depends on the force, being higher at 30% of maximal force than at 10%, with sensitivities reported as 42–55% in myopathy (98,103,104) and 64–73% in neurogenic disorders (103). Neural network analysis of the frequency analysis of the IP at MVC in patients with neuromuscular disorders gave diagnostic yields of 60–80%, similar to that of the turns-amplitude analysis (105) and in hypertrophic cardiomyopathy the power spectrum analysis together with turns-amplitude analysis and MUP analysis detected subclinical myopathy in 15% of the patients (106). Methods using the wavelet transform technique to analyse the IP seem promising (107,108).

Box 8.3 Procedure for macro-EMG MUP analysis

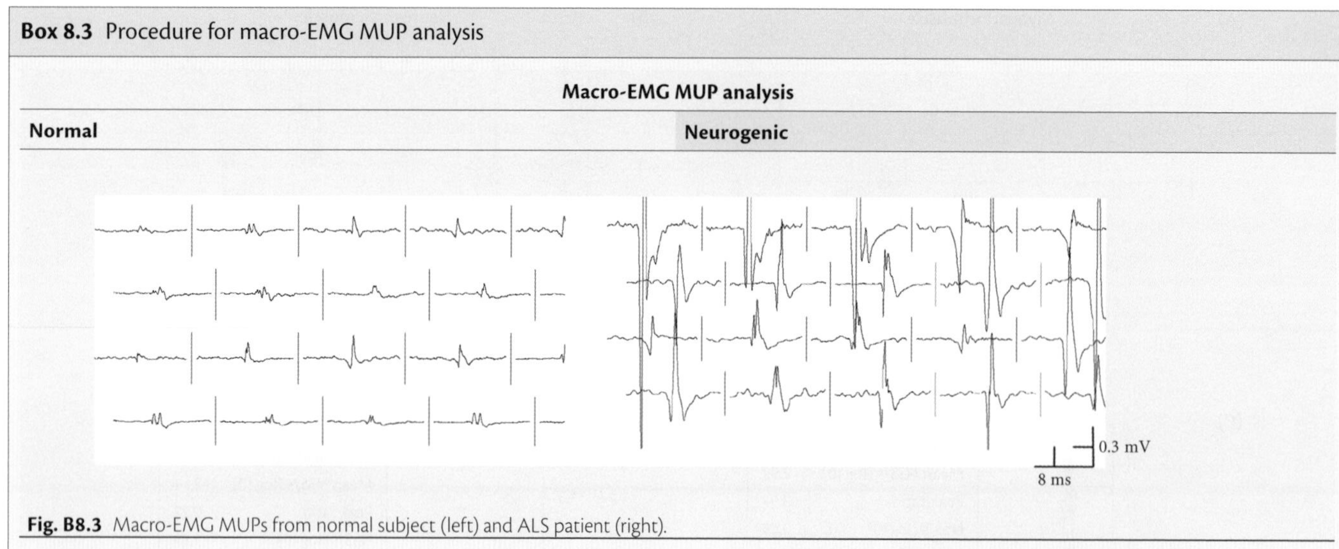

Fig. B8.3 Macro-EMG MUPs from normal subject (left) and ALS patient (right).

Procedure (76,77)

- A modified single-fibre EMG electrode with a platinum wire with a diameter of 25 μm exposed in a sideport 7.5 mm proximal to the tip is used. The 0.55 mm diameter steel cannula is insulated to within 15 mm of the tip; the non-insulated part is used as reference. The insertion length is at least 15 mm.
- Recordings are made on two channels: One channel displays a single fibre EMG (SFEMG) signal recorded between the platinum wire and the shaft of the same electrode; filter settings: 500–10 kHz. The other channel records the signal between the electrode shaft and a surface or monopolar needle electrode placed remotely from the investigated muscle; filter settings: 5–10 kHz.
- The SFEMG recording is used to trigger the sweep. In this way, the action potential of the synchronously firing motor unit from the cannula is extracted and the resultant averaged potential is called the macro MUP.
- The average amplitude and area of 20 different macro-MUPs recorded at different depths of the muscle using 2–5 insertions are calculated.
- All recordings are made with slight muscle activation—less than 30% of maximal force.

Turns-amplitude analysis

A technique for manual turns-amplitude analysis was introduced in the 1960s (109) and a few years later this was automated (110). The number of motor units and their firing rate influence the number of turns/s (95,111–113). The degree of summation due to overlap of MUPs, or due to synchronization, influences both the number of turns and the amplitude (114,115). The more summation, then the more is the cancellation of small spikes and the higher the amplitude of the IP. The duration of individual MUPs influences the number of turns secondary to the influence on the degree of summation. In muscles from patients with neuromuscular disorders, a decrease in duration results in less summation and cancellation of small spikes, and may give rise to a high number of turns/s and vice versa (68,104,116). The amplitude of the IP is also dependent on the amplitude of the individual MUPs (113,117). The shape of individual MUPs, i.e. the incidence of polyphasic potentials, has an influence on the IP—the higher the incidence of polyphasic potentials, the higher the number of turns (68,116). The turns-amplitude parameters are little affected by age, sex, and temperature (111,115,118–120).

Turns-amplitude analysis at a force of 2 kg showed increased turns in myopathies (109). Analysis of turns-amplitude at a force fixed relative to maximum (e.g. 30% of MVC) indicated that this method was better at discriminating between myopathic, neurogenic, and normal muscle than analysis at a force of, e.g. 2 kg, and gave a similar diagnostic yield as the individual MUP analysis (57,68,121,122).

Analysis of turns-amplitude at a given force level of the muscle requires cooperation of the patient. Maintaining a given force is time-consuming and is not possible to obtain for all muscles. Therefore, several modifications of turns-amplitude analysis have been suggested, eliminating the need to estimate muscle force (120,123–126). For a description of the practice procedure of the two most used methods—cloud and peak-ratio—and the diagnostic criteria for the measured parameters please see Box 8.4 and Table 8.1.

Cloud, activity, and number of small segments

A plot of mean amplitude as a function of turns can be obtained without measurement of force. The distribution of 20 values from control subjects measured at a range of muscle forces, has been called 'the cloud' (120). If a subject has two points outside this cloud, it is considered pathological. In patients with myopathy, mean amplitude/number of turns values were found to be low, whereas patients with neurogenic lesions had high values. The cloud method may give rise to false-negative findings in patients with myopathy. In 8 of 17 patients with myopathy,

Box 8.4 Definitions and procedure for turns-amplitude analysis

Turns-amplitude analysis

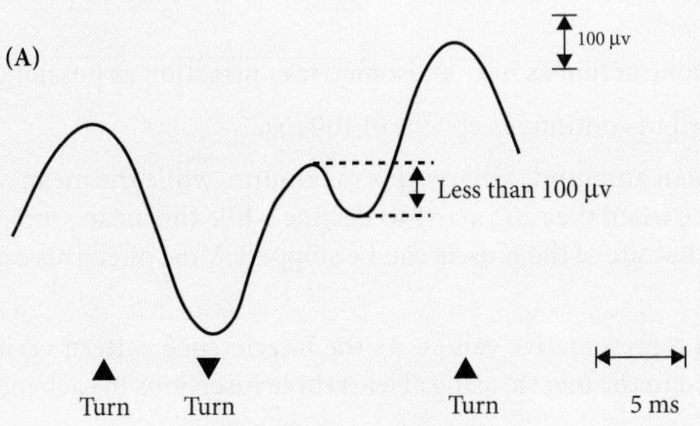

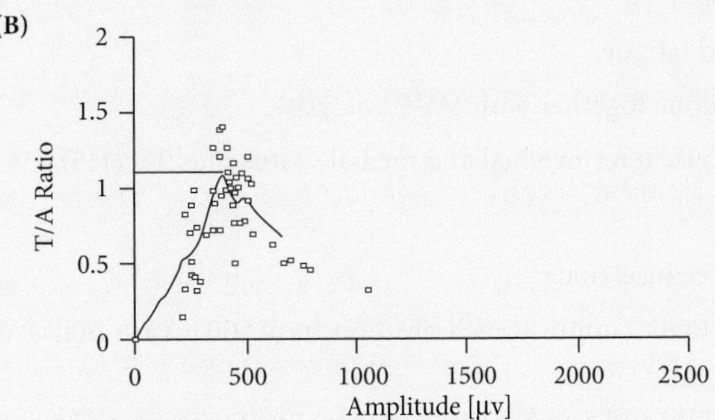

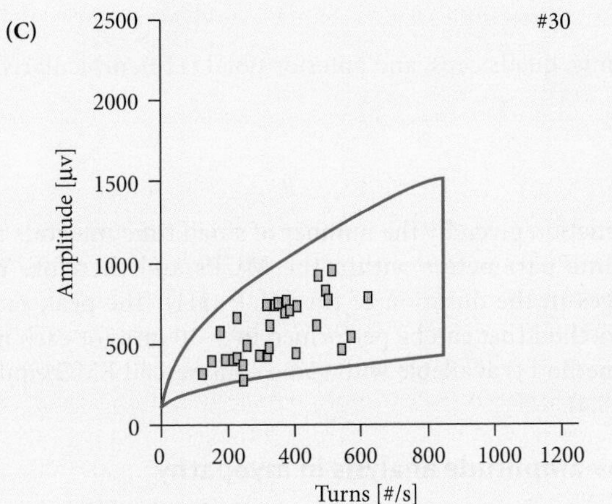

Fig. B8.4 (A) Definition of turns. (B) Peak-ratio analysis in normal muscle. (C) Cloud analysis in a normal muscle.

Definitions and parameters

(A) Turns-amplitude analysis

- *Turns/s*: number of potential reversals >100 μV counted over a certain time unit, often converted to number of turns/s (109).
- *Cumulative amplitude*: the summation of amplitude between turns per time unit (109).
- *Mean amplitude*: the average amplitude between two turns (109).
- *Time intervals between turns*: the duration of a segment between two turns (109).

(B) Peak-ratio method (123)

- *Turns-amplitude ratio*: number of turns divided by the mean amplitude per time unit (57).
- *Mean peak ratio*: the average of the maximum value of the turns-amplitude ratio obtained from 10 sites (123).
- *Mean small time intervals (NSS)*: number of time intervals <1.5 ms/sec. Calculated from the ten 100-ms epochs selected for the peak ratio. Logarithmic transformation is performed (57,115).

(C) Cloud method (120)

Activity: time within an epoch where MUPs are present, i.e. quantification of the fullness of the IP (129).

Upper centile amplitude: the amplitude exceeded by only 1% of segments. Quantifies the largest spikes of the IP (129).

Number of small segments (NSS): the sum of segments below a certain duration, i.e. <1.5 ms for segments with amplitude <0.5 mV and <3 ms for segments between 0.5 and 2 mV, all segments >2 mV being excluded (129).

Procedure

- Subject in a supine position.
- Concentric needle electrode with a recording area of 0.07 mm is commonly used.
- Filter settings: 20–10 kHz; Sampling frequency: 50 kHz or at least 25 kHz.

(continued)

Box 8.4 Continued

Peak-ratio method (B)

- The peak ratio is obtained from the IP during a gradual increase in force from zero over 10–20 s, obtained by careful instruction of the subject.
- The force is exerted against the hand of the examiner with the muscle contraction as near an isometric contraction as possible.
- During the gradual increase in force, the interference pattern is analysed in continuous epochs of 100 ms.
- At moderate force level the ratio of turns to mean amplitude against mean amplitude will reach a maximum, while the mean amplitude continues to increase. At that moment of gradual increase in force when the ratio starts to decline while the mean amplitude continues to increase, the peak ratio has been obtained and analysis at that site of the muscle can be stopped. Most often this occurs at about 10–30% of maximum force.
- The peak ratio is measured in at least 10 sites in each muscle to obtain representative values. As the interference pattern varies in different parts of the muscle, these 10 sites should be carefully distributed in the muscle using at least three insertions in each muscle (111,114).
- The depth of the electrode should be changed in steps of at least 5 mm.
- The pause between contractions has to be at least 30 s in order to avoid fatigue.
- The peak ratio analysis takes less than 10 min per muscle and can be done together with MUP analysis.
- Reference material exists for the brachial biceps, abductor pollicis brevis, anterior tibial and medial vastus muscles (115).

Cloud method (C)

- Epochs of the IP sampled at 6–9 sites in each muscle, using two or three insertions.
- The IP is analysed at 3–5 different force levels ranging from minimum to maximum at each site. Epochs of 500 ms are analysed from at least 20 recordings.
- Plots of turns against amplitude, upper centile amplitude against activity and number of small segments against activity are compared to 'the cloud', i.e. the distribution of values from control subjects measured at low and high forces of the muscle.
- Two or more points (10%) outside the cloud is considered pathological.
- Reference material for the brachial biceps, extensor digitorum communis, quadriceps, and anterior tibial (120), orbicularis oculi and oris (205), and frontalis and mentalis muscles (139).

the force level had an influence on the sensitivity of the cloud method. In these patients, values obtained between 10 and 30% of the MVC were abnormal indicating myopathy, but at force levels above 50% of MVC values fell into the normal range and the pathology was obscured (127). Another study showed that mean amplitude/turns values obtained at near MVC in control subjects were actually above the limits of the original cloud, and thus falsely indicating neurogenic lesion (128). A modified version of the cloud analysis was later introduced (113,126,129,130). The cloud method is available as part of commercial EMG equipment (Box 8.4).

Ratio of turns to amplitude

The peak ratio is the maximum value of the ratio of turns/s to mean amplitude obtained by using the mean amplitude per turn as an indicator of force (123). Mean amplitude can be substituted for force, as it increases linearly with force (114,115,131). The peak ratio probably reflects the point in the increase of force where summation has obliterated the silent periods between MUP discharges, while at the same time there is still too little summation and cancellation to obscure abnormality (132). This point may be reached at different force levels at different sites in the muscle, depending on the density of the motor units (133). The information given by the number of small time intervals reflects the time parameters within the MUPs and therefore reflects changes in the duration of the MUP (111). The peak ratio is a fast method that can be performed in 5–10 min for each muscle. The method is available with some commercial EMG equipment (Box 8.4).

Turns-amplitude analysis in myopathy

In the muscles of patients with myopathy, the random and diffuse degeneration of muscle fibres as well as regeneration are reflected in the short duration, low-amplitude, and the polyphasic shape of individual MUPs (7). As indicated above, these changes in individual motor unit potentials influence the IP, resulting in an increase in number of turns/s, a decrease in mean amplitude per turn, an increase in ratio of turns to mean amplitude, and an increase in number of small time intervals between turns (Fig. 8.1 and Table 8.1).

Cloud analysis in muscles from patients with myopathy showed increased number of small segments, although the data points were usually within the normal cloud of number of small segments against activity. In all patients, the interpretation of the plots agreed with qualitative assessments of the IP made by a clinical neurophysiologist (126). In muscles from patients with myopathy, the sensitivity of the IPA was high (75–100%) with a specificity of 85% and cloud analysis was more sensitive than the MUP analysis (59,61,62,134,135). In young patients with myopathy, the cloud analysis supplemented the qualitative EMG analysis in four of 33 patients (136).

Peak-ratio analysis in the brachial biceps muscle of 25 patients with different kinds of myopathy showed an increased peak ratio in 92%, and 84% had an increased number of small time intervals less than 1.5 ms, while the individual MUPs sampled at weak effort showed that 72% had decreased mean duration of simple potentials (137). Finsterer et al. also found the peak-ratio method to be better than the MUP analysis, although the sensitivity for both methods was lower than in the former study, probably due to inclusion of more borderline cases (67). The analysis of peak values of turns/s, mean amplitude and ratio was shown to be more sensitive than the cloud method (95% versus 80%) (138). In 19 patients with myopathy, analysis of the ratio of turns to mean amplitude in the anterior tibial muscle showed myopathic features in 68% (98). In facial muscles the slope of mean amplitude versus turns per second (i.e. inverted ratio) in facial muscles had the same diagnostic yield as the cloud analysis (50–60%). However, the specificity of the first method was higher (95% versus 80%) (139). In the frontalis muscle of patients with myopathy, the ratio of turns to mean amplitude had a sensitivity of 62% and a specificity of 97% when compared with controls and patients with neurogenic conditions (140).

Turns-amplitude analysis in neurogenic lesions

In muscles from patients with neurogenic disorders, peripheral sprouting is reflected as increased duration and high amplitude of individual motor unit potentials. The large motor unit potentials influence the degree of summation in the IP in the opposite direction to that seen in myopathy. The large motor unit potentials are associated with a decreased number of turns/s, increased mean amplitude per turn, decreased ratio of turns to mean amplitude, and decreased number of small time intervals (68). This is reflected in a proportional increase in amplitude divided by turns/s with increasing fibre density (141). In neurogenic muscle, the loss of whole motor units results in a further decrease in the number of turns/s (Fig. 8.1 and Table 8.1) (68).

Cloud analysis of the brachial biceps in 8 patients with motor neuron disorders showed that the upper centile amplitude was increased compared with controls in all patients. All patients had data points above the cloud of upper centile amplitude against activity, while the number of small segments was decreased or normal. Fullness of the IP, i.e., activity, was incomplete in two patients, but full in the others (126). In patients with recurrent laryngeal mononeuropathy only a few patients showed data outside the normal cloud boundaries for laryngeal EMG (142).

Peak ratio analysis in 21 patients with neurogenic disorders showed decreased peak ratio in 86% and decreased number of small time intervals less than 1.5 ms in 43% with one or both of these abnormalities seen in all patients (137). MUP analysis in the same muscle showed increased mean duration in 95% of the patients. In another study of the anterior tibial muscle of 21 patients with neurogenic disorders, 81% had a neurogenic peak ratio analysis (143). In patients with bulbar onset amyotrophic lateral sclerosis, peak ratio analysis in the tongue had a higher sensitivity than MUP analysis, while in patients with limb onset the motor unit potential analysis had the higher sensitivity (144).

Turns-amplitude analysis in neuromuscular transmission failure

Turns amplitude analysis of the IP indicated abnormality in 40–60% of the patients with myasthenia gravis, most often with similar changes as in myopathic muscle and less often as in neurogenic muscle (72–74,145). This pattern suggests either random block of muscle fibres giving a myopathic picture, or functional block of the whole or a major part of the motor unit or changed recruitment pattern giving a neurogenic picture (73,74) or an associated myopathy (44).

Disuse atrophy

The disused quadriceps muscle examined at a force of 30% of mvc in patients having the leg immobilized for 4 weeks by a plaster cast showed decreased turns/s and mean amplitude consistent with functional loss of active motor units (146). Eight days later, when most of the force had been regained, turns/s and mean amplitude were again normal. Such reversible motor dysfunction may be due to lack of sensory feedback affecting recruitment. The rapid reactivation of motor units is an example of the plasticity of the nervous system in the regulation of voluntary effort (146).

Dystonic muscles

Quantitative EMG seems to be helpful in the localization and evaluation of dystonic muscles in patients with torticollis (147,148). In a comparison of the effect of botulinum toxin treatment of dystonic muscles with and without EMG assistance in patients with torticollis, a significant greater magnitude of improvement was present in the group with EMG assistance (147). Similarly, patients treated with botulinum guided by EMG had a better clinical outcome than those treated without EMG guidance (3). An increase in ratio of turns to amplitude 6 weeks or more after botulinum toxin treatment suggest a reversible, random loss of muscle fibres (149–151). This finding accords with the decreased mean duration of individual motor unit potentials that occurs after treatment (152). Turns-amplitude analysis is also useful in botulinum toxin treatment of patients with laryngeal dystonia (153,154) and oromandibular dystonia (155–157), and has been used to characterize the EMG of patients with diaphragmatic myoclonus (158,159).

An argument in favour of using IPA during treatment of dystonic muscle is that the amount of botulinum toxin may be minimized by the selection of muscles with proven involvement and to some extent by the quantification of abnormal activity (3).

Surface electromyography

There have been many attempts to use surface EMG in the diagnosis of neuromuscular disorders. The advantage of surface EMG, especially in children, is that it is painless. The surface electrode has, however, an inherent problem especially with respect to diagnosis of patients with myopathy. The distance and the tissue between the muscle fibres and the skin will decrease the amplitude of the EMG signal and acts as a filter decreasing the high-frequency components of the EMG signal (160,161). The development from analysis of a single channel to analysis of multiple electrodes has probably increased the possibility of providing a diagnostic surface EMG method.

Single-channel surface EMG

In a study of 16 children with muscle disease, Barry et al. (162) claimed a high sensitivity of 82% measuring the ratio between acoustic myography (an estimate of the mechanical output of the muscle from muscle sounds) and amplitude of surface EMG at different fixed forces. It is unknown if this method is specific for myopathy. A quantitative surface EMG method 'clustering index (CI)' uses CI to quantify the degree of uneven distribution of the S EMG signal (163,164). The EMG is analysed at various forces of the muscle. The technique is an attempt to quantify the visual analysis of the IP. The diagnostic yield of the 'clustering index' method seems moderate to high with a sensitivity of 61% in myopathy and 92–100% in neurogenic patients (163,164). The method seems promising, but has not been fully evaluated.

A new method uses the application of high frequency, low intensity alternating current to a limb and measures the resulting surface voltages over a muscle given as impedance changes, i.e. electrical impedance myography (EIM) (165,166). The EIM may give information on the anisotropy of the muscles and seems to give structural information on membranes in the muscle (167,168).

EIM has shown promise as a sensitive measure of disease severity and progression especially in children (166,169,170). With respect to differentiation between myopathic and neurogenic muscle changes it seems that more studies are needed (167).

Multi-channel surface EMG

Surface electrode arrays or high-density surface EMG has been tested sporadically in patients with neuromuscular disorders (171–175). One study obtaining high spatial resolution EMG via a multi-electrode array and using advanced parameters including the autocorrelation function as well as muscle fibre conduction velocity found a high sensitivity of 85% in 41 patients with myopathy (176). A technique analysing the MUPs using a 16-channel surface electrode array had a sensitivity of 92% and a specificity of 80% in 12 patients with motor neuron disorders and 10 normal controls (177).

In a high-density surface EMG study of nine patients with post-poliomyelitis with 126 channels the area under the curve of the extracted single MUPs was higher in all patients compared with that of control subjects (172). Spatial filtering using multi-channel EMG actually reduced the detection of fasciculations in muscles from patients with ALS (178). High density surface EMG can be used for measuring changes in motor unit potential size over time, i.e. in polio patients (179).

Literature reviews of the clinical utility of surface EMG cast doubt on surface EMG as a diagnostic tool in the diagnosis of neuromuscular disorders (180–182).

Muscle fibre conduction velocity

The conduction velocity of propagation along the muscle fibre has been determined during voluntary effort (15,183) and by stimulation (184).

Surface electrodes and voluntary contraction

In patients with muscular dystrophy and patients with hypokalaemic periodic paralysis the muscle fibre conduction velocity (MFCV) obtained by a multi-channel electrode was decreased (173,175,185). In contrast, one study using analysis of EMG power spectrum found increased MFCVs in patients with myopathy (186).

Needle electrodes and electrical stimulation

Of 14 patients with Duchenne muscular dystrophy, 33% had decreased MFCV, while the mean MUP duration was decreased in 50% (187). Using intramuscular stimulation and needle recording in patients with different types of myopathy, the MFCV was decreased in most patients and complemented quantitative MUP analysis (188) and turns amplitude analysis (189). The high sensitivity of MFCV in myopathy was later confirmed (190). In patients with myopathy or hypokalaemic periodic paralysis, the MFCV obtained by intramuscular needle electrodes was reduced more often than the MFCV obtained by surface electrodes (185,191,192).

The method cannot be used to differentiate between patients with myopathy and patients with neurogenic muscle lesions, as MFCV in the latter group may also be reduced (55,56,187,191,193–195).

Direct muscle stimulation

In critical illness patients with paralysis, it may be difficult to evaluate the EMG as the patient may not be able to cooperate. Fibrillation potentials may be seen both in critical illness neuropathy and in critical illness myopathy. In order to differentiate between neuropathy and myopathy in these cases, a method using direct muscle stimulation was suggested (196,197). In muscles with myopathy, the electrical muscle excitability may be decreased while it is normal in neurogenic muscles. Thus, in myopathic muscles the ratio between compound muscle action potential amplitude obtained by nerve stimulation and the action potential obtained by direct muscle stimulation is high, up to 1, while the ratio in neurogenic muscles is small and may even be zero. In 13 patients with critical illness associated weakness, a ratio of more than 0.5 indicating myopathy was found in all patients (198). With direct muscle stimulation supramaximal stimulation may be hard to achieve and it is difficult to be sure that intramuscular nerve twigs have not also been activated; this may have an impact on the reproducibility of the method (199). The duration of the CMAP induced by nerve stimulation may also be useful being prolonged in critical illness myopathy (200).

In patients with critical illness myopathy in whom the voluntary contraction is possible, analysis of the EMG is valuable and e.g. MUP analysis will often reveal myopathic changes (198). The diagnostic challenge is especially in tetraparalytic patients in whom voluntary activity cannot be analysed. In these patients the direct muscle stimulation may support the diagnosis.

Advantages, limitations, and pitfalls of different quantitative electromyography methods

The MUP analysis methods using templates and averaging techniques like the multi-MUP analysis are quickly performed, in

5–10 min, but some information may be lost in the averaging. Twenty-five per cent or more of the MUPs obtained by multi-MUP analysis may require manual correction (2,33,201,202). Some automatic MUP analyses, such as the multi-MUP analysis, require a rise-time of 500 μs or less from peak to peak of the potential, or a corresponding slope, for sampling of MUPs. This criterion results in the loss of some MUPs for analysis, which in turn may give rise to a systematic error in the selection of MUPs. The criterion is probably too restrictive (203). The only criterion for selection of MUPs in the original Buchthal method is a random sampling of MUPs with peak-to-peak amplitude of more than 50 μV. The individual MUP analysis is probably the most precise MUP method, but it is time-consuming, takes 15–20 min, and may require at least a year's training. For the quantitative MUP analyses the common drawbacks are:

- One cannot be sure that selection bias is avoided, i.e. the examiner unconsciously selects short or long duration MUPs.
- Only motor units recruited at weak effort are included and on average only 4% of maximum (range 0–20%) can be analysed, i.e. motor units recruited at higher forces are not analysed.

The turns-amplitude analysis requires reference material best obtained in the examiner's laboratory. Both the cloud analysis and the peak ratio methods are rapid (5–10 min), as they do not need force measurements. Moreover, motor units recruited both at low effort and at higher effort are included in the analysis. These IPA methods are easy to learn. The pitfalls are:

- Development of fatigue if too high an effort is performed for too long, and the rests between contractions are too short. For example, a contraction of up to 50% of maximum requires a 30-s pause in order to avoid fatigue.
- If the needle is outside the muscle, the analysis is performed on volume-conducted EMG signals losing high-frequency signals.

Both fatigue and sampling outside the muscle may mimic changes seen in a neurogenic lesion. IPA, which has electrodiagnostic possibilities in the future such as the power spectrum analysis, motor unit firing rate analysis, the cloud activity method, and the peak ratio method are only available in some commercial EMG machines and control values are, at best, available for only a few muscles in the upper and lower limbs, and the cranial muscles.

Some EMG machines offer the possibility of performing both MUP analysis and turns-amplitude analysis simultaneously in the same muscle. In this way the MUPs can be sampled during the rest periods between turns-amplitude recordings, thus saving time. New methods, such as some of the surface EMG methods, impedance measurement, MUP energy measurement, firing rate methods, and MFCV measurement require further studies.

Conclusions

It is a fact that some physicians use QEMG and others use qualitative EMG for analysis of MUP properties and the IP, while all physicians analyse the spontaneous activity qualitatively (6). What is preferable depends on the disorder and the chosen strategy of the examination. Qualitative visual analysis of MUPs and the interference pattern may be diagnostic in patients with clear-cut changes, but may be biased and misleading in patients with borderline changes in the muscle. In patients with, e.g. nerve entrapments or polyneuropathy, qualitative EMG is fast and helpful, and may give additional information to the nerve conduction velocity studies. In other disorders, such as motor neuron disorders and myopathies, there may not be methods available for diagnostic support other than the EMG. The authors therefore recommend using QEMG in at least two or three selected muscles in these patients, perhaps supplemented with qualitative EMG in other muscles. In order to obtain supplementary information, the authors' laboratory usually applies both the quantitative individual MUP analysis and the peak ratio method to a few muscles in patients with myopathy or motor neuron disorders, as these methods diagnostically supplement each other.

In electrodiagnosis of neuromuscular disorders it is recommended that experience is obtained with a MUP analysis applied at weak effort, and an IP analysis applied at stronger effort.

References

1. Buchthal, F., Guld, C., and Rosenfalck, P. (1954). Action potential parameters in normal human muscle and their dependence on physical variables. *Acta Physiologica Scandinavica*, **32**, 200–18.
2. Takehara, I., Chu, J., Li, T.C., and Schwartz, I. (2004). Reliability of quantitative motor unit action potential parameters. *Muscle & Nerve*, **30**(1), 111–13.
3. Werdelin, L., Dalager, T., Fuglsang-Frederiksen, A., et al. (2011). The utility of EMG interference pattern analysis in botulinum toxin treatment of torticollis: a randomised, controlled and blinded study. *Clinical Neurophysiology*, **122**(11), 2305–9.
4. Sikjaer, T., Rolighed, L., Hess, A., Fuglsang-Frederiksen, A., Mosekilde, L., and Rejnmark, L. (2014). Effects of PTH (1–84) therapy on muscle function and quality of life in hypoparathyroidism: results from a randomized controlled trial. *Osteoporosis International*, **25**(6), 1717–26.
5. Hellmann, M., Kleist-Retzow, J.C., Haupt, W.F., Herkenrath, P., and Schauseil-Zipf, U. (2005). Diagnostic value of electromyography in children and adolescents. *J Clinical Neurophysiology*, **22**(1), 43–8.
6. Fuglsang-Frederiksen, A., Johnsen, B., Vingtoft, S., et al. (1995). Variation in performance of the EMG examination at six European laboratories. *Electroencephalography & Clinical Neurophysiology*, **97**, 444–50.
7. Buchthal, F. (1991). Electromyography in the evaluation of muscle diseases. *Methods in Clinical Neurophysiology*, **2**, 25–45.
8. Fuglsang-Frederiksen, A. (2000). The utility of interference pattern analysis. *Muscle & Nerve*, **23**(1), 18–36.
9. Buchthal, F. and Clemmensen, S. (1941). On the differentiation of muscle atrophy by electromyography. *Acta Psychiatrica Neurologica*, **16**, 143–81.
10. Kugelberg, E. (1947). Electromyograms in muscular disorders. *Journal of Neurology, Neurosurgery & Psychiatry*, **10**, 122–33.
11. Kugelberg, E. (1949). Electromyography in muscular dystrophies. Differentiation between dystrophies and chronic lower motor neurone lesions. *Journal of Neurology, Neurosurgery & Psychiatry*, **12**, 129–36.
12. Buchthal, F., Pinelli, P., and Rosenfalck, P. (1954). Action potential parameters in normal human muscle and their physiological determinants. *Acta Physiologica Scandinavica*, **32**, 219–29.
13. Pinelli, P. and Buchthal, F. (1953). Muscle action potentials in myopathies with special regard to progressive muscular dystrophy. *Neurology*, **3**, 347–59.
14. Sacco, G., Buchthal, F., and Rosenfalck, P. (1962). Motor unit potentials at different ages. *Archives of Neurology*, **6**, 366–73.
15. Stålberg, E. (1966). Propagation velocity in human muscle fibers in situ. *Acta Physiologica Scandinavica*, **70**(Suppl. 287), 1–112.
16. Bergmans, J. (1971). Computer assisted on line measurement of motor unit potential parameters in human electromyography. *Electromyography*, **11**(2), 161–81.

17. Dorfman, L.J., Howard, J.E., and McGill, K.C. (1989). Motor unit firing rates and firing rate variability in the detection of neuromuscular disorders. *Electroencephalography & Clinical Neurophysiology*, **73**, 215–24.
18. Falck, B. (1983). Automatic analysis of individual motor unit potentials with a special two channel electrode [Dissertation]. University of Turku, Turku.
19. Guiheneuc, P. (1989). Results and reliability controls in computer analysis of needle EMG recordings. In: J. E. Desmedt (Ed.) *Computer-aided electromyography and expert systems*, pp. 55–66. Amsterdam: Elsevier.
20. Kopec, J. and Hausmanowa-Petrusewicz, I. (1976). On-line computer application in clinical quantitative electromyography. *Electromyography Clinical Neurophysiology*, **16**, 49–64.
21. Lang, A.H. and Falck, B. (1980). A two-channel method for sampling, averaging and quantifying motor unit potentials. *Journal of Neurology*, **223**, 199–206.
22. LeFever RS, De Luca CJ. A procedure for decomposing the myoelectric signal into its constituent action potentials—Part I: Technique, theory, and implementation. *IEEE Transactions on Biomedical Engineering*, **29**(3), 149–57.
23. LeFever, R., Xenakis, A., and DeLuca, C.J. (1982). A procedure for decomposing the myoelectric signal into its constituent action potentials-Part II: Execution and Test for Accuracy. *IEEE Transactions on Biomedical Engineering*, **29**, 158–64.
24. McGill, K.C., Cummins, K.L., and Dorfman, L.J. (1985). Automatic decomposition of the clinical electromyogram. *IEEE Transactions on Biomedical Engineering*, **32**(7), 470–7.
25. Stewart, C.R., Nandedkar, S.D., Massey, J.M., Gilchrist, J.M., Barkhaus, P.E., and Sanders, D.B. (1989). Evaluation of an automatic method of measuring features of motor unit action potentials. *Muscle & Nerve*, **12**, 141–8.
26. Stålberg, E. and Antoni, L. (1983). Computer aided EMG analysis. In: J. E. Desmedt (Ed.) *Progress in clinical neurophysiology. Computer aided analysis of EMG*, 10th edn, pp. 186–234. Basel: Karger.
27. Christodoulou, C.I. and Pattichis, C.S. (1999). Unsupervided pattern recognition for the classification of EMG signals. *IEEE Transactions on Biomedical Engineering*, **46**(2), 169–78.
28. Pattichis, C.S. and Pattichis, M.S. (1999). Time-scale analysis of motor unit action potentials. *IEEE Transactions on Biomedical Engineering*, **46**(11), 1320–9.
29. Nandedkar, S.D. and Sanders, D.B. (1989). Principal component analysis of the features of concentric needle EMG motor unit action potentials. *Muscle & Nerve*, **12**, 288–93.
30. Cengiz, B., Ozdag, F., Ulas, U.H., Odabasi, Z., and Vural, O. (2002). Discriminant analysis of various concentric needle EMG and macro-EMG parameters in detecting myopathic abnormality. *Clinical Neurophysiology*, **113**(9), 1423–8.
31. Pfeiffer, G. and Kunze, K. (1995). Discriminant classification of motor unit potentials (MUPs) successfully separates neurogenic and myopathic conditions. A comparison of multi- and univariate diagnostical algorithms for MUP analysis. *Electroencephalography & Clinical Neurophysiology*, **97**(5), 191–207.
32. Pfeiffer, G. (1999). The diagnostic power of motor unit potential analysis: an objective Bayesian approach. *Muscle & Nerve*, **22**(5), 584–91.
33. Bischoff, C., Stålberg, E., Falck, B., and Eeg-Olofsson, K.E. (1994). Reference values of motor unit action potentials obtained with multi-MUAP analysis. *Muscle & Nerve*, **17**, 842–51.
34. Nandedkar, S., Barkhaus, P.E., and Charles, A. (1995). Multi-motor unit action potential analysis (MMA). *Muscle & Nerve*, **18**, 1155–66.
35. Stålberg, E., Falck, B., Sonoo, M., Stalberg, S., and Astrom, M. (1995). Multi-MUP EMG analysis—a two year experience in daily clinical work. *Electroencephalography & Clinical Neurophysiology*, **97**(3), 145–54.
36. Boe, S.G., Stashuk, D.W., Brown, W.F., and Doherty, T.J. (2005). Decomposition-based quantitative electromyography: effect of force on motor unit potentials and motor unit number estimates. *Muscle & Nerve*, **31**(3), 365–73.
37. Ives, C. and Doherty, T. (2014). Intra-rater reliability of motor unit number estimation and quantitative motor unit analysis in subjects with amyotrophic lateral sclerosis. *Clinical Neurophysiology*, **125**(1), 170–8.
38. Lawson, V.H., Bromberg, M.B., and Stashuk, D. (2004). Comparison of conventional and decomposition-enhanced spike triggered averaging techniques. *Clinical Neurophysiology*, **115**(3), 564–8.
39. Nikolic, M. and Krarup, C. (2011). EMGTools, an adaptive and versatile tool for detailed EMG analysis. *IEEE Transactions on Biomedical Engineering*, **58**(10), 2707–18.
40. Podnar, S. and Mrkaic, M. (2003). Size of motor unit potential sample. *Muscle & Nerve*, **27**(2), 196–201.
41. Rosenfalck, A. and Rosenfalck, P. (1975). *Electromyography-sensory and motor conduction. Findings in normal subjects.* Copenhagen: Rigshospitalet.
42. Fuglsang-Frederiksen, A. (1989). Interference EMG analysis. In: J. E. Desmedt (Ed.) *Computer-aided electromyography and expert systems*, pp. 161–79. Amsterdam: Elsevier.
43. Petersén, I. and Kugelberg, E. (1949). Duration and form of action potentials in the normal human muscle. *Journal of Neurology, Neurosurgery, & Psychiatry*, **12**, 124–8.
44. Buchthal, F. and Kamieniecka, Z. (1982). The diagnostic yield of quantified electromyography and quantified muscle biopsy in neuromuscular disorders. *Muscle & Nerve*, **5**, 265–80.
45. Stålberg, E. and Karlsson, L. (2001). Simulation of EMG in pathological situations. *Clinical Neurophysiology*, **112**(5), 869–78.
46. Zalewska, E., Rowinska-Marcinska, K., Gawel, M., and Hausmanowa-Petrusewicz, I. (2012). Simulation studies on the motor unit potentials with satellite components in amyotrophic lateral sclerosis and spinal muscle atrophy. *Muscle & Nerve*, **45**(4), 514–21.
47. Baumann F. (1968). Contribution on the electromyographic diagnosis of the so-called disuse atrophy of skeletal muscles. *Electromyography*, **8**(4), 383–95.
48. Thage, O. (1974). Quadriceps weakness and wasting: a neurological, electrophysiological and histological study [Dissertation]. University of Michigan Ann Arbor, MI.
49. Nandedkar, S.D., Barkhaus, P.E., Sanders, D.B., and Stålberg, E.V. (1988). Analysis of amplitude and area of concentric needle EMG motor unit action potentials. *Electroencephalography & Clinical Neurophysiology*, **69**, 561–7.
50. Sonoo, M. and Stålberg, E. (1993). The ability of MUP parameters to discriminate between normal and neurogenic MUPs in concentric EMG: analysis of the MUP 'thickness' and the proposal of 'size index'. *Electroencephalography & Clinical Neurophysiology*, **89**, 291–303.
51. Sonoo, M. (2002). New attempts to quantify concentric needle electromyography. *Muscle & Nerve*, **11**(Suppl.), S98–102.
52. Stålberg, E.V. and Sonoo, M. (1994). Assessment of variability in the shape of the motor unit action potential, the 'jiggle,' at consecutive discharges. *Muscle & Nerve*, **17**(10), 1135–44.
53. Zalewska, E. and Hausmanowa-Petrusewicz, I. (2005). The SIIR index—a non-linear combination of waveform size and irregularity parameters for classification of motor unit potentials. *Clinical Neurophysiology*, **116**(4), 957–64.
54. Sheean, G.L. (2012). Quantification of motor unit action potential energy. *Clinical Neurophysiology*, **123**(3), 621–5.
55. Rodriguez-Carreno, I., Gila-Useros, L., Malanda-Trigueros, A., Gurtubay, I.G., Navallas-Irujo, J., and Rodriguez-Falces, J. (2010). Application of a novel automatic duration method measurement based on the wavelet transform on pathological motor unit action potentials. *Clinical Neurophysiology*, **121**(9), 1574–83.
56. Tomczykiewicz, K., Dobrowolski, A.P., and Wierzbowski, M. (2012). Evaluation of motor unit potential wavelet analysis in the electrodiagnosis of neuromuscular disorders. *Muscle & Nerve*, **46**(1), 63–9.

57. Fuglsang-Frederiksen, A., Scheel, U., and Buchthal, F. (1976). Diagnostic yield of analysis of the pattern of electrical activity and of individual motor unit potentials in myopathy. *Journal of Neurology, Neurosurgery & Psychiatry*, **39**, 742–50.
58. Buchthal, F., Kamieniecka, Z., and Schmalbruch, H. (1974). *Fibre types in normal and diseased human muscles and their physiological correlates*. Amsterdam: Excerpta Medica.
59. Nirkko, A.C., Rosler, K.M., and Hess, C.W. (1995). Sensitivity and specificity of needle electromyography: a prospective study comparing automated interference pattern analysis with single motor unit potential analysis. *Electroencephalography & Clinical Neurophysiology*, **97**, 1–10.
60. Buchthal, F. (1977). Electrophysiological signs of myopathy as related with muscle biopsy. *Acta Neurologica Napoli*, **32**, 1–29.
61. Barkhaus, P.E., Nandedkar, S.D., and Sanders, D.B. (1990). Quantitative EMG in inflammatory myopathy. *Muscle & Nerve*, **13**, 247–53.
62. Barkhaus, P.E., Periquet, M.I., and Nandedkar, S.D. (1999). Quantitative electrophysiologic studies in sporadic inclusion body myositis. *Muscle & Nerve*, **22**, 480–7.
63. Podnar, S. and Zidar, J. (2006). Sensitivity of motor unit potential analysis in facioscapulohumeral muscular dystrophy. *Muscle & Nerve*, **34**(4), 451–6.
64. Podnar, S. (2009). Predictive values of motor unit potential analysis in limb muscles. *Clinical Neurophysiology*, **120**(5), 937–40.
65. Kalita, J., Misra, U.K., Maurya, P.K., Shankar, S.K., and Mahadevan, A. (2012). Quantitative electromyography in dengue-associated muscle dysfunction. *Journal of Clinical Neurophysiology*, **29**(5), 468–71.
66. Buchthal, F. and Clemmensen, S. (1941). Electromyographical observations in congential myotonia. *Acta Psychiatrica et Neurologica*, **16**, 389–403.
67. Finsterer, J., Mamoli, B., and Fuglsang-Frederiksen, A. (1997). Peak-ratio interference pattern analysis in the detection of neuromuscular disorders. *Electroencephalography & Clinical Neurophysiology*, **105**(5), 379–84.
68. Fuglsang-Frederiksen, A., Scheel, U., and Buchthal, F. (1977). Diagnostic yield of the analysis of the pattern of electrical activity of muscle and of individual motor unit potentials in neurogenic involvement. *Journal of Neurology, Neurosurgery, & Psychiatry*, **40**, 544–54.
69. Krarup, C. (2011). Lower motor neuron involvement examined by quantitative electromyography in amyotrophic lateral sclerosis. *Clinical Neurophysiology*, **122**(2), 414–22.
70. de Carvalho, M., Turkman, A., and Swash, M. (2014). Sensitivity of MUP parameters in detecting change in early ALS. *Clinical Neurophysiology*, **125**(1), 166–9.
71. Pino, L.J., Stashuk, D.W, Boe, S.G., and Doherty, T.J. (2010). Probabilistic muscle characterization using QEMG: application to neuropathic muscle. *Muscle & Nerve*, **41**(1), 18–31.
72. Farrugia, M.E., Kennett, R.P., Hilton-Jones, D., Newsom-Davis, J., and Vincent, A. (2007). Quantitative EMG of facial muscles in myasthenia patients with MuSK antibodies. *Clinical Neurophysiology*, **118**(2), 269–77.
73. Lo Monaco, M., Christensen, H., and Fuglsang-Frederiksen, A. (1993). Quantitative EMG findings at different force levels in patients with myasthenia gravis. *Neurophysiology Clinic*, **23**, 353–61.
74. Odabasi, Z., Kuruoglu, R., and Oh, S.J. (2000). Turns-amplitude analysis and motor unit potential analysis in myasthenia gravis. *Acta Neurologica Scandinavica*, **101**(5), 315–20.
75. Crone, C., Christiansen, I., and Vissing, J. (2013). Myopathic EMG findings and type II muscle fiber atrophy in patients with Lambert–Eaton myasthenic syndrome. *Clinical Neurophysiology*, **124**(9), 1889–92.
76. Stålberg, E. (1980). Macro EMG, a new recording technique. *Journal of Neurology, Neurosurgery, & Psychiatry*, **43**(6), 475–82.
77. Stålberg, E. and Fawcett, P.R. (1982). Macro EMG in healthy subjects of different ages. *Journal of Neurology, Neurosurgery, & Psychiatry*, **45**(10), 870–8.
78. Finsterer, J. and Fuglsang-Frederiksen, A. (2000). Concentric needle EMG versus macro EMG I. Relation in healthy subjects. *Clinical Neurophysiology*, **111**(7), 1211–15.
79. Finsterer, J. and Fuglsang-Frederiksen, A. (2001). Concentric-needle versus macro EMG. II. Detection of neuromuscular disorders. *Clinical Neurophysiology*, **112**(5), 853–60.
80. Hilton-Brown, P. and Stålberg, E. (1983). Motor unit size in muscular dystrophy, a macro EMG and scanning EMG study. *Journal of Neurology, Neurosurgery, & Psychiatry*, **46**(11), 996–1005.
81. Finsterer, J. and Fuglsang-Frederiksen, A. (1999). Macro-EMG in mitochondriopathy. *Clinical Neurophysiology*, **110**(8), 1466–70.
82. Torbergsen, T., Stålberg, E., and Bless, J.K. (1991). Nerve-muscle involvement in a large family with mitochondrial cytopathy: electrophysiological studies. *Muscle & Nerve*, **14**, 35–41.
83. Fawcett, P.R.W., Kennett, R.P., and Johnson, M.A. (1990). A neurophysiological investigation of the facio-sacpulo-humeral (FSH) syndrome. *Journal of the Neurological Sciences*, **98**(Suppl.), 195.
84. Dengler, R., Konstanzer, A., Kuther, G., Hesse, S., Wolf, W., and Struppler, A. (1990). Amyotrophic lateral sclerosis: macro-EMG and twitch forces of single motor units. *Muscle & Nerve*, **13**(6), 545–50.
85. Sandberg, A., Hansson, B., and Stalberg, E. (1999). Comparison between concentric needle EMG and macro EMG in patients with a history of polio. *Clinical Neurophysiology*, **110**(11), 1900–8.
86. Ulas, U.H., Cengiz, B., Alanoglu, E., Ozdag, M.F., Odabasi, Z., and Vural, O. (2003). Comparison of sensitivities of macro EMG and concentric needle EMG in L4 radiculopathy. *Neurological Sciences*, **24**(4), 258–60.
87. Tackmann, W. and Vogel, P. (1988). Fibre density, amplitudes of macro-EMG motor unit potentials and conventional EMG recordings from the anterior tibial muscle in patients with amyotrophic lateral sclerosis. A study on 51 cases. *Journal of Neurology*, **235**(3), 149–54.
88. Sandberg, A., Nandedkar, S.D., and Stalberg, E. (2011). Macro electromyography and motor unit number index in the tibialis anterior muscle: differences and similarities in characterizing motor unit properties in prior polio. *Muscle & Nerve*, **43**(3), 335–41.
89. Dornonville de la Cour, C., Andersen, H., Stalberg, E., Fuglsang-Frederiksen. A., and Jakobsen, J. (2005). Electrophysiological signs of permanent axonal loss in a follow-up study of patients with Guillain-Barre syndrome. *Muscle & Nerve*, **31**(1), 70–7.
90. Dornonville de la Cour, C. (2004). Permanent neuropathy and reinnervation in the Guillain–Barré Syndrome [PhD thesis]. Aarhus University, Aarhus.
91. Sandberg, A. and Stalberg, E. (2004). Changes in macro electromyography over time in patients with a history of polio: a comparison of 2 muscles. *Archives of Physical Medicine and Rehabilitation*, **85**(7), 1174–82.
92. Petajan, J. (1974). Clinical electromyographic studies of diseases of the motor unit. *Electroencephalography & Clinical Neurophysiology*, **36**, 395.
93. Petajan, J.H. (1991). AAEM minimonograph #3: motor unit recruitment. *Muscle & Nerve*, **14**, 489–502.
94. Fuglsang-Frederiksen, A. and Lo Monaco, M. (1981). Pattern of electrical activity in normal and pathological muscle during increasing force. *Electroencephalography & Clinical Neurophysiology*, **52**, 83.
95. Fuglsang-Frederiksen, A., Smith, T., and Hogenhaven, H. (1987). Motor unit firing intervals and other parameters of electrical activity in normal and pathological muscle. *Journal of Neurological Sciences*, **78**, 51–62.
96. Piotrkiewicz, M., Hausmanowa-Petrusewicz, I., and Mierzejewska, J. (1999). Are motoneurons involved in muscular dystrophy? *Clinical Neurophysiology*, **110**, 1111–22.
97. Boe, S.G., Stashuk, D.W., and Doherty, T.J. (2007). Motor unit number estimates and quantitative motor unit analysis in healthy subjects and patients with amyotrophic lateral sclerosis. *Muscle & Nerve*, **36**(1), 62–70.

98. Kurca, E. and Drobny, M. (2000). Four quantitative EMG methods and theirs individual parameter diagnostic value. *Electromyography & Clinical Neurophysiology*, **40**(8), 451–8.
99. Buchthal, F. (1976). *Diagnostic significance of the myopathic EMG.* Amsterdam: Excerpta Medica.
100. Walton, J. (1952). Analysis with the audio-frequency spectrometer. *Journal of Neurology, Neurosurgery, & Psychiatry*, **15**, 219–26.
101. Richardson, A.T. (1951). Newer concepts of electrodiagnosis. *St Thomas Hosp Rep*, 164–174.
102. Larsson, L.E. (1975). On the relation between the EMG frequency spectrum and the duration of symptoms in lesions of the peripheral motor neuron. *Electroencephalography & Clinical Neurophysiology*, **38**, 69–78.
103. Ronager, J., Christensen, H., and Fuglsang-Frederiksen, A. (1989). Power spectrum analysis of the EMG pattern in normal and diseased muscles. *Journal of Neurological Sciences*, **94**(1–3), 283–94.
104. Fuglsang-Frederiksen, A. and Rønager, J. (1990). EMG power spectrum, turns-amplitude analysis and motor unit potential duration in neuromuscular disorders. *Journal of Neurological Sciences*, **97**, 81–91.
105. Abel, E.W., Zacharia, P.C., Forster, A., and Farrow, T.L. (1996). Neural network analysis of the EMG interference pattern. *Medical Engineering & Physics*, **18**, 12–17.
106. Karandreas, N., Stathis, P., Anastasakis, A., et al. (2000). Electromyographic evidence of subclinical myopathy in hypertrophic cardiomyopathy. *Muscle & Nerve*, **23**(12), 1856–61.
107. von Tscharner, V. (2002). Time-frequency and principal-component methods for the analysis of EMGs recorded during a mildly fatiguing exercise on a cycle ergometer. *Journal of Electromyography and Kinesiology*, **12**(6), 479–92.
108. Abel, E.W., Meng, H., Forster, A., and Holder, D. (2006). Singularity characteristics of needle EMG IP signals. *IEEE Transactions on Biomedical Engineering*, **53**(2), 219–25.
109. Willison, R.G. (1964). Analysis of electrical activity in healthy and dystrophic muscle in man. *Journal of Neurology, Neurosurgery, & Psychiatry*, **27**, 386–94.
110. Fitch, P. (1967). An analyser for use in human electromyography. *Electronic Engineering*, **39**, 240–3.
111. Fuglsang-Frederiksen, A. and Månsson, A. (1975). Analysis of electrical activity of normal muscle in man at different degrees of voluntary effort. *Journal of Neurology, Neurosurgery, & Psychiatry*, **38**, 683–94.
112. McGill, K.C., Lau, K., and Dorfman, L.J. (1991). A comparison of turns analysis and motor unit analysis in electromyography. *Electroencephalography & Clinical Neurophysiology*, **81**, 8–17.
113. Nandedkar, S.D., Sanders, D.B., and Stålberg, E.V. (1986). Simulation and analysis of the electromyographic interference pattern in normal muscle. Part I: turns and amplitude measurements. *Muscle & Nerve*, **9**, 423–30.
114. Christensen, H., Lo Monaco, M., Dahl, K., and Fuglsang-Frederiksen, A. (1984). Processing of electrical activity in human muscle during a gradual increase in force. *Electroencephalography & Clinical Neurophysiology*, **58**, 230–9.
115. Liguori, R., Dahl, K., and Fuglsang-Frederiksen, A. (1992). Turns-amplitude analysis of the electromyographic recruitment pattern disregarding force measurement. I. Method and reference values in healthy subjects. *Muscle & Nerve*, **15**, 1314–18.
116. Fuglsang-Frederiksen, A. (1981). Electrical activity and force during voluntary contraction of normal and diseased muscle. *Acta Neurologica Scandinavica*, **83**(Suppl.), 1–60.
117. Hayward, M. and Willison, R.G. (1977). Automatic analysis of the electromyogram in patients with chronic partial denervation. *Journal of Neurological Sciences*, **33**, 415–23.
118. Finsterer, J. and Mamoli, B. (1994). T/A-analysis with or without measuring force: problems in the evaluation of normal limits. *Electroencephalography & Clinical Neurophysiology*, **34**, 215–24.
119. Finsterer, J. and Mamoli, B. (1996). Turn/amplitude parameter changes during sustained effort. *Electroencephalography & Clinical Neurophysiology*, **101**, 438–45.
120. Stålberg, E., Chu, J., Bril, V., Nandedkar, S., Stålberg, S., and Ericsson, M. (1983). Automatic analysis of the EMG interference pattern. *Electromyography & Clinical Neurophysiology*, **56**, 672–81.
121. Bril, V. and Fuglsang-Frederiksen, A. (1984). Number of potential reversals (turns) and amplitude of the pattern of electrical activity of the abductor pollicis brevis muscle in patients with neurogenic diseases. *Acta Neurologica Scandinavica*, **70**, 169–75.
122. Haridasan, G., Sanghvi, S.H., Jindal, G.D., Joshi, V.M., and Desai, A.D. (1979). Quantitative electromyography using automatic analysis. A comparative study with a fixed fraction of a subject's maximum effort and two levels of thresholds for analysis. *Journal of Neurological Sciences*, **42**, 53–64.
123. Fuglsang-Frederiksen, A., Lo Monaco, M., and Dahl, K. (1985). Turns analysis (peak ratio) in EMG using the mean amplitude as a substitute of force measurement. *Electroencephalography & Clinical Neurophysiology*, **60**, 225–7.
124. Fuglsang-Frederiksen, A., Christensen, H., Lo Monaco, M., and Dahl, K. (1985). Pattern of electrical activity and force in normal and pathological muscle: S-index of turns and amplitude. *Electroencephalography & Clinical Neurophysiology*, **60**, 30–2.
125. Gilai, A., Hadas-Halperin, I., and Roth, V.G. (1985). Continuous EMG interference pattern analysis and computerised tomography of muscles. *Electroencephalography & Clinical Neurophysiology*, **61**, S57.
126. Nandedkar, S.D., Sanders, D.B., and Stålberg, E.V. (1986). Automatic analysis of the electromyographic interference pattern. Part II: findings in control subjects and in some neuromuscular diseases. *Muscle & Nerve*, **9**, 491–500.
127. Fuglsang-Frederiksen, A., Dahl, K., and Lo Monaco, M. (1984). Electrical muscle activity during a gradual increase in force in patients with neuromuscular diseases. *Electroencephalography & Clinical Neurophysiology*, **57**, 320–9.
128. Nandedkar, S.D., Sanders, D.B., and Stålberg, E.V. (1991). On the shape of the normal turns-amplitude cloud. *Muscle & Nerve*, **14**, 8–13.
129. Nandedkar, S.D., Sanders, D.B., and Stålberg, E.V. (1986). Simulation and analysis of the electromyographic interference pattern in normal muscle. Part II: activity, upper centile amplitude, and number of small segments. *Muscle & Nerve*, **9**, 486–90.
130. Nandedkar, S.D., Sanders, D.B., and Stålberg, E.V. (1986). Automatic analysis of the electromyographic interference pattern. Part I: development of quantitative features. *Muscle & Nerve*, **9**, 431–9.
131. Liguori, R., Dahl, K., Vingtoft, S., and Fuglsang-Frederiksen, A. (1990). Determination of peak-ratio by digital turns-amplitude analysis on line. *Electromyography & Clinical Neurophysiology*, **30**, 371–8.
132. Fuglsang-Frederiksen, A. (1993). Turns-amplitude analysis of the EMG interference pattern. *Methods in Clinical Neurophysiology*, **4**, 81–100.
133. Fuglsang-Frederiksen, A. (2006). The role of different EMG methods in evaluating myopathy. *Clinical Neurophysiology*, **117**(6), 1173–89.
134. Dioszeghy, P., Egerhazi, A., Molnar, M., and Mechler, F. (1996). Turn-amplitude analysis in neuromuscular diseases. *Electromyography & Clinical Neurophysiology*, **36**(8), 463–8.
135. Hokkoku, K., Sonoo, M., Higashihara, M., Stalberg, E., and Shimizu, T. (2012). Electromyographs of the flexor digitorum profundus muscle are useful for the diagnosis of inclusion body myositis. *Muscle & Nerve*, **46**(2), 181–6.
136. Chang, J., Park, Y.G., Choi, Y.C., Choi, J.H., and Moon, J.H. (2011). Correlation of electromyogram and muscle biopsy in myopathy of young age. *Archives of Physical and Medical Rehabilitation*, **92**(5), 780–4.
137. Liguori, R., Dahl, K., Fuglsang-Frederiksen, A., and Trojaborg, W. (1992). Turns-amplitude analysis of the electromyographic

recruitment pattern disregarding force measurement. II. Findings in patients with neuromuscular disorders. *Muscle & Nerve*, **15**, 1319–24.

138. Lefaucheur, J.P., Verroust, J., and Gherardi, R.K. (1996). Turns-amplitude analysis assessment of myopathies in HIV-infected patients. *Journal of Neurological Sciences*, **136**, 148–53.
139. Karandreas, N., Kararizou, E., Papagianni, A., Zambelis, T., and Kokotis, P. (2011). Turns-amplitude analysis in normal and myopathic facial muscles. *Muscle & Nerve*, **43**(3), 342–7.
140. Matur, Z., Baslo, M.B., and Oge, A.E. (2014). Quantitative electromyography of the frontalis muscle. *Journal of Clinical Neurophysiology*, **31**(1), 48–54.
141. Fisher, M.A. (1997). Root mean square voltage/turns in chronic neuropathies is related to increase in fiber density. *Muscle & Nerve*, **20**(2), 241–3.
142. Statham, M.M., Rosen, C.A., Nandedkar, S.D., and Munin, M.C. (2010). Quantitative laryngeal electromyography: turns and amplitude analysis. *Laryngoscope*, **120**(10), 2036–41.
143. Finsterer, J. (1998). Influence of spontaneous activity on peak-ratio analysis. *Electroencephalography & Clinical Neurophysiology*, **107**(4), 254–7.
144. Finsterer, J., Fuglsang-Frederiksen, A., and Mamoli, B. (1997). Needle EMG of the tongue: motor unit action potential versus peak ratio analysis in limb and bulbar onset amyotrophic lateral sclerosis. *Journal of Neurology, Neurosurgery, & Psychiatry*, **63**(2), 175–80.
145. Fuglsang-Frederiksen, A. and Krarup, C. (1982). The electrical activity during sustained voluntary contraction in myasthenia. *Proceedings of the Vth International Congress in Neuromuscular Diseases*, Marseille, 39.2.
146. Fuglsang-Frederiksen, A. and Scheel, U. (1978). Transient decrease in number of motor units after immobilisation in man. *Journal of Neurology, Neurosurgery, & Psychiatry*, **41**, 924–9.
147. Comella, C.L., Buchman, A.S., Tanner, C.M., Brown-Toms, N.C., and Goetz, C.G. (1992). Botulinum toxin injection for spasmodic torticollis: increased magnitude of benefit with electromyographic assistance. *Neurology*, **42**, 878–82.
148. Ostergaard, L., Fuglsang-Frederiksen, A., Sjo, O., Werdelin, L., and Winkel, H. (1996). Quantitative EMG in cervical dystonia. *Electromyography & Clinical Neurophysiology*, **36**(3), 179–85.
149. Fuglsang-Frederiksen, A., Sjö, O., Östergaard, L., and Winkel, H. (1992). Quantitative EMG changes in torticollis after treatment with botulinum toxin. IX *International Congress of Electromyography and Clinical Neurophysiology, Jerusalem*, 217.
150. Fuglsang-Frederiksen, A., Ostergaard, L., Sjo, O., Werdelin, L., and Winkel, H. (1998). Quantitative electromyographical changes in cervical dystonia after treatment with botulinum toxin. *Electromyography & Clinical Neurophysiology*, **38**(2), 75–9.
151. Ostergaard, L., Fuglsang-Frederiksen, A., Werdelin, L., Sjo, O., and Winkel, H. (1994). Quantitative EMG in botulinum toxin treatment of cervical dystonia. A double-blind, placebo-controlled study. *Electroencephalography & Clinical Neurophysiology*, **93**(6), 434–9.
152. Odergren, T., Tollback, A., and Borg, J. (1994). Electromyographic single motor unit potentials after repeated botulinum toxin treatments in cervical dystonia. *Electroencephalography & Clinical Neurophysiology*, **93**(5), 325–9.
153. Fuglsang-Frederiksen, A., Östergaard, L., Sjö, O., et al. (1993). Turns-amplitude analysis in botulinum toxin treatment of focal dystonia. *Canadian Journal of Neurological Sciences*, **20**, 350.
154. Lovelace, R.E., Blitzer, A., and Brin, M.F. (1993). Quantitations of the effect of Botox on laryngeal/spasmodic dysphonia with the turns analysis method. *Canadian Journal of Neurological Sciences*, **20**, 365.
155. Bakke, M., Werdelin, L.M., Dalager, T., Fuglsang-Frederiksen, A., Prytz, S., and Moller, E. (2003). Reduced jaw opening from paradoxical activity of mandibular elevator muscles treated with botulinum toxin. *European Journal of Neurology*, **10**, 695–9.
156. Davidson, B.J. and Ludlow, C.L. (1996). Long-term effects of botulinum toxin injections in spasmodic dysphonia. *Annals of Otology, Rhinology & Laryngology*, **105**(1), 33–42.
157. Erdal, J., Werdelin, L., and Fuglsang-Frederiksen, A. (1996). Experience with long-term botulinum toxin treatment of oromandibular dystonia, guided by quantitative EMG. *Acta Neurologica Scandinavica*, **94**, 424.
158. Chen, R., Remtulla, H., and Bolton, C.F. (1995). Electrophysiological study of diaphragmatic myoclonus. *Journal of Neurology, Neurosurgery & Psychiatry*, **58**(4), 480–3.
159. Collins, S.J., Chen, R.E., Remtulla, H., Parkes, A., and Bolton, C.F. (1997). Novel measurement for automated interference pattern analysis of the diaphragm. *Muscle & Nerve*, **20**(8), 1038–40.
160. Christensen, H. and Fuglsang-Frederiksen, A. (1986). Power spectrum and turns analysis of EMG at different voluntary efforts in normal subjects. *Electroencephalography & Clinical Neurophysiology*, **64**, 528–35.
161. Rosenfalck, P. (1969). *Intra- and extracellular potential fields of active nerve and muscle fibres. A physico-mathematical analysis of different models*. København: Akademisk Forlag.
162. Barry, D.T., Gordon, K.E., and Hinton, G.G. (1990). Acoustic and surface EMG diagnosis of pediatric muscle disease. *Muscle & Nerve*, **13**(4), 286–90.
163. Uesugi, H., Sonoo, M., Stalberg, E., et al. (2011). 'Clustering Index method': a new technique for differentiation between neurogenic and myopathic changes using surface EMG. *Clinical Neurophysiology*, **122**(5), 1032–41.
164. Higashihara, M., Sonoo, M., Yamamoto, T., et al. (2011). Evaluation of spinal and bulbar muscular atrophy by the clustering index method. *Muscle & Nerve*, **44**(4), 539–46.
165. Rutkove, S.B. (2009). Electrical impedance myography: Background, current state, and future directions. *Muscle & Nerve*, **40**(6), 936–46.
166. Rutkove, S.B., Caress, J.B., Cartwright, M.S., et al. (2014). Electrical impedance myography correlates with standard measures of Als severity. *Muscle & Nerve*, **49**(3), 441–3.
167. Garmirian, L.P., Chin, A.B., and Rutkove, S.B. (2009). Discriminating neurogenic from myopathic disease via measurement of muscle anisotropy. *Muscle & Nerve*, **39**(1), 16–24.
168. Stegeman, D.F., Pillen, S., Kleine, B.U., and Zwarts, M.J. (2006). Bridging function and structure of the neuromuscular system. *Clinical Neurophysiology*, **117**(6), 1169–72.
169. Rutkove, S.B., Gregas, M.C., and Darras, B.T. (2012). Electrical impedance myography in spinal muscular atrophy: a longitudinal study. *Muscle & Nerve*, **45**(5), 642–7.
170. Rutkove, S.B., Caress, J.B., Cartwright, M.S., et al. (2013). Electrical impedance myography correlates with standard measures of Als severity. *Muscle & Nerve*, **49**(3), 441–3.
171. Drost, G., Blok, J.H., Stegeman, D.F., van Dijk, J.P., van Engelen, B.G., and Zwarts, M.J. (2001). Propagation disturbance of motor unit action potentials during transient paresis in generalized myotonia: a high-density surface EMG study. *Brain*, **124**(Pt 2), 352–60.
172. Drost, G., Stegeman, D.F., Schillings, M.L., et al. (2004). Motor unit characteristics in healthy subjects and those with postpoliomyelitis syndrome: a high-density surface EMG study. *Muscle & Nerve*, **30**(3), 269–76.
173. Hilfiker, P. and Meyer, M. (1984). Normal and myopathic propagation of surface motor unit action potentials. *Electroencephalography & Clinical Neurophysiology*, **57**(1), 21–31.
174. Sun, T.Y., Lin, T.S., and Chen, J.J. (1999). Multielectrode surface EMG for noninvasive estimation of motor unit size. *Muscle & Nerve*, **22**(8), 1063–70.
175. Yamada, M., Kumagai, K., and Uchiyama, A. (1987). The distribution and propagation pattern of motor unit action potentials studied by multi-channel surface EMG. *Electroencephalography & Clinical Neurophysiology*, **67**(5), 395–401.

176. Huppertz, H.J., Disselhorst-Klug, C., Silny, J., Rau, G., and Heimann, G. (1997). Diagnostic yield of noninvasive high spatial resolution electromyography in neuromuscular diseases. *Muscle & Nerve*, **20**(11), 1360–70.
177. Wood, S.M., Jarratt, J.A., Barker, A.T., and Brown, B.H. (2001). Surface electromyography using electrode arrays: a study of motor neuron disease. *Muscle & Nerve*, **24**(2), 223–30.
178. Jahanmiri-Nezhad, F., Barkhaus, P.E., Rymer, W.Z., and Zhou, P. (2014). Sensitivity of fasciculation potential detection is dramatically reduced by spatial filtering of surface electromyography. *Clinical Neurophysiology*, **125**(7), 1498–500.
179. Bickerstaffe, A., van Dijk, J.P., Beelen, A., Zwarts, M.J., and Nollet, F. (2014) Loss of motor unit size and quadriceps strength over 10 years in post-polio syndrome. *Clinical Neurophysiology*, **125**(6), 1255–60.
180. Haig, A.J., Gelblum, J.B., Rechtien, J.J., and Gitter, A.J. (1996). Technology assessment: the use of surface EMG in the diagnosis and treatment of nerve and muscle disorders. *Muscle & Nerve*, **19**, 392–5.
181. Pullman, S.L., Goodin, D.S., Marquinez, A.I., Tabbal, S., Rubin, M. (2000). Clinical utility of surface EMG: report of the therapeutics and technology assessment subcommittee of the American Academy of Neurology. *Neurology*, **55**(2), 171–7.
182. Meekins, G.D., So, Y., and Quan, D. (2008). American Association of Neuromuscular & Electrodiagnostic Medicine evidenced-based review: use of surface electromyography in the diagnosis and study of neuromuscular disorders. *Muscle & Nerve*, **38**(4), 1219–24.
183. Denslow, J.S. and Hassett, C.C. (1943). The polyphasic action currents of the motor unit complex. *American Journal of Physiology*, **139**, 652–60.
184. Buchthal, F., Guld, C., and Rosenfalck, P. (1955). Propagation velocity in electrically activated muscle fibres in man. *Acta Physiologica Scandinavica*, **34**(1), 75–89.
185. Zwarts, M.J., van Weerden, T.W., Links, T.P., Haenen, H.T., and Oosterhuis, H.J. The muscle fiber conduction velocity and power spectra in familial hypokalemic periodic paralysis. *Muscle & Nerve*, **11**(2), 166–73.
186. Ebersteinm A. and Goodgold, J. (1985). Muscle fiber conduction velocity calculated from EMG power spectra. *Electromyography & Clinical Neurophysiology*, **25**(7–8), 533–8.
187. Cruz Martinez, A. and Lopez Terradas, J.M. (1990). Conduction velocity along muscle fibers in situ in Duchenne muscular dystrophy. Archives of Physical and Medical Rehabilitation, **71**(8), 558–61.
188. Naumann, M. and Reiners, K. (1996). Diagnostic value of in situ muscle fiber conduction velocity measurements in myopathies. *Acta Neurologica Scandinavica*, **93**(2–3), 193–7.
189. Blijham, P.J., Hengstman, G.J., Ter Laak, H.J., van Engelen, B.G., and Zwarts, M.J. (2004). Muscle-fiber conduction velocity and electromyography as diagnostic tools in patients with suspected inflammatory myopathy: a prospective study. *Muscle & Nerve*, **29**(1), 46–50.
190. Blijham, P.J., van Engelen, B.G., Drost, G., Stegeman, D.F., Schelhaas, H.J., and Zwarts, M.J. (2011). Diagnostic yield of muscle fibre conduction velocity in myopathies. *Journal of Neurological Sciences*, **309**(1–2), 40–4.
191. Van der Hoeven, J.H., Zwarts, M.J., and van Weerden, T.W. (1993). Muscle fiber conduction velocity in amyotrophic lateral sclerosis and traumatic lesions of the plexus brachialis. *Electroencephalography & Clinical Neurophysiology*, **89**(5), 304–10.
192. Zwarts, M.J. (1989). Evaluation of the estimation of muscle fiber conduction velocity. Surface versus needle method. *Electroencephalography & Clinical Neurophysiology*, **73**(6), 544–8.
193. Cruz-Martinez, A. and Arpa, J. (1999). Muscle fiber conduction velocity in situ (MFCV) in denervation, reinnervation and disuse atrophy. *Acta Neurologica Scandinavica*, **100**(5), 337–40.
194. Gruener, R., Stern, L.Z., and Weisz, R.R. (1979). Conduction velocities in single fibers of diseased human muscle. *Neurology*, **29**(9 Pt 1), 1293–7.
195. Blijham, P.J., Schelhaas, H.J., Ter Laak, H.J., van Engelen, B.G., and Zwarts, M.J. (2007). Early diagnosis of ALS: the search for signs of denervation in clinically normal muscles. *Journal of Neurological Sciences*, **263**(1–2), 154–7.
196. Rich, M.M., Teener, J.W., Raps, E.C., Schotland, D.L., and Bird, S.J. (1996). Muscle is electrically inexcitable in acute quadriplegic myopathy. *Neurology*, **46**(3), 731–6.
197. Rich, M.M., Bird, S.J., Raps, E.C., McCluskey, L.F., and Teener, J.W. (1997). Direct muscle stimulation in acute quadriplegic myopathy. *Muscle & Nerve*, **20**(6), 665–73.
198. Trojaborg, W., Weimer, L.H., and Hays, A.P. (2001). Electrophysiologic studies in critical illness associated weakness: myopathy or neuropathy—a reappraisal. *Clinical Neurophysiology*, **112**(9), 1586–93.
199. Duez, L., Qerama, E., Fuglsang-Frederiksen, A., Bangsbo, J., Jensen, T.S. (2010). Electrophysiological characteristics of motor units and muscle fibers in trained and untrained young male subjects. *Muscle & Nerve*, **42**(2), 177–83.
200. Allen, D.C., Arunachalam, R., and Mills, K.R. (2008). Critical illness myopathy: Further evidence from muscle-fiber excitability studies of an acquired channelopathy. *Muscle and Nerve*, **37**(1), 14–22.
201. Bromberg, M.B., Smith, A.G., and Bauerle, J. (1999). A comparison of two commercial quantitative electromyographic algorithms with manual analysis. *Muscle & Nerve*, **22**(9), 1244–8.
202. Papagianni, A.E., Kokotis, P., Zambelis, T., and Karandreas, N. (2012). MUAP values of two facial muscles in normal subjects and comparison of two methods for data analysis. *Muscle & Nerve*, **46**(3), 346–50.
203. Barkhaus, P.E. and Nandedkar, S.D. (1996). On the selection of concentric needle electromyogram motor unit action potentials: is the rise time criterion too restrictive? *Muscle & Nerve*, **19**(12), 1554–60.
204. Stålberg, E., Andreassen, S., Falck, B., Lang, H., Rosenfalck, A., and Trojaborg, W. Quantitative analysis of individual motor unit potentials: a proposition for standardized terminology and criteria for measurement. J *Clinical Neurophysiology* 1986;3:313–48.
205. Farrugia, M.E. and Kennett, R.P. (2005). Turns amplitude analysis of the orbicularis oculi and oris muscles. *Clinical Neurophysiology*, **116**(11), 2550–9.

CHAPTER 9

Axonal excitability: molecular basis and assessment in the clinic

Susanna B. Park, Cindy S-Y. Lin, and Matthew C. Kiernan

Introduction

Nerve conduction studies remain the 'gold standard' technique for the assessment of peripheral nerve pathology, providing information about compound action potential amplitude, latency, and conduction velocity. Nerve conduction studies utilize supramaximal stimuli to provide information regarding the number of conducting fibres and the speed of transmission. However, this only provides limited and non-specific characterization of the underlying basis for nerve pathology and does not provide information about the function of the conducting fibres or their excitability properties. While slowed conduction velocity may suggest demyelination, it can also be produced by temperature effects, membrane potential changes, Na^+ channel blockade, or conversely remyelination. In addition, such findings may not correlate with the clinical picture. As such, the development of a complimentary technique to provide for assessment of axonal resting membrane potential and excitability would provide greater molecular understanding of the activity of voltage-gated ion channels and ion pumps present on the axonal membrane. While the study of peripheral axonal excitability has a long history, it has only been during the last 20 years that this technique has been more broadly available for the clinical neurophysiologist. This chapter will briefly cover the background and history of axonal excitability with a focus on technique description and practical considerations. An overview of excitability measures and the key ion channels contributing to membrane potential will be outlined and several examples of the utility of axonal excitability studies in clinical practice will be discussed.

Description of axonal excitability techniques

Background and history

The role of electricity in the physiology and function of the nervous system was an exciting topic of scientific inquiry in the 18th and 19th centuries, extending from Galvani's experiments with animal electricity in the 1780s to Helmholtz's examination of nerve conduction velocity in the 1850s (1,2). The assessment of the excitability properties of axons was instigated in the early 1900s, and Georges Weiss coined the fundamental law of electrostimulation regarding the relationship between current strength and current duration (3). In 1909, Louis Lapicque defined the classical term 'rheobase' in nerve tissue as the threshold current required for a stimulus of infinite duration, and 'chronaxie' as the pulse duration for a current of twice rheobasic strength (4). While in the 1930s, excitability was assessed through measurement of chronaxie and rheobase, by 1952, Hodgkin and Huxley had developed a complete model of axonal excitability using experimental data from voltage-clamp recordings and modelling of the giant squid axon (5). The properties of the unmyelinated giant squid axon are remarkably similar to those of myelinated mammalian axons (6), and the Hodgkin–Huxley model remains the predominant explanation of membrane excitability.

While these techniques were developed using in vitro preparations, electrodiagnostic techniques for clinical assessment of peripheral nerve function were developed during the 1940s and nerve conduction techniques were implemented as part of clinical peripheral nerve assessment from the 1950s onwards. The excitability changes in single human motor axons were assessed in situ by Joseph Bergmans in the 1970s, utilizing surface electrodes to assess threshold and the effects of experimental manoeuvres to alter membrane potential (7). The technique was further developed by Hugh Bostock to enable use of threshold measurements of compound potentials (8,9). Furthermore, Bostock and colleagues developed specialized software and semi-automated protocols, improving the speed of testing. In 1999, at the Nordic course in clinical axonal electrophysiology, the TROND protocol was developed to optimize the technique for clinical use (10). Current axonal excitability protocols derived from this initial TROND protocol assess multiple measures of axonal excitability in motor and sensory axons.

Axonal excitability techniques in current practice

Axonal excitability studies provide complementary information about axonal membrane potential and ion channel function, using submaximal stimuli to examine the properties underlying the excitability of the axon. Similar to nerve conduction studies, excitability studies assess large myelinated axons. However, axonal excitability is assessed at the site of stimulation and does not provide information about properties along the length of the nerve (8).

Despite these differences, axonal excitability studies are undertaken in a similar fashion to nerve conduction studies, with surface electrodes for stimulating and recording (Fig. 9.1). Required equipment includes a bipolar constant current stimulator, preamplifier and specialized software and recordings are made on a computer with a data acquisition board. These items are all now

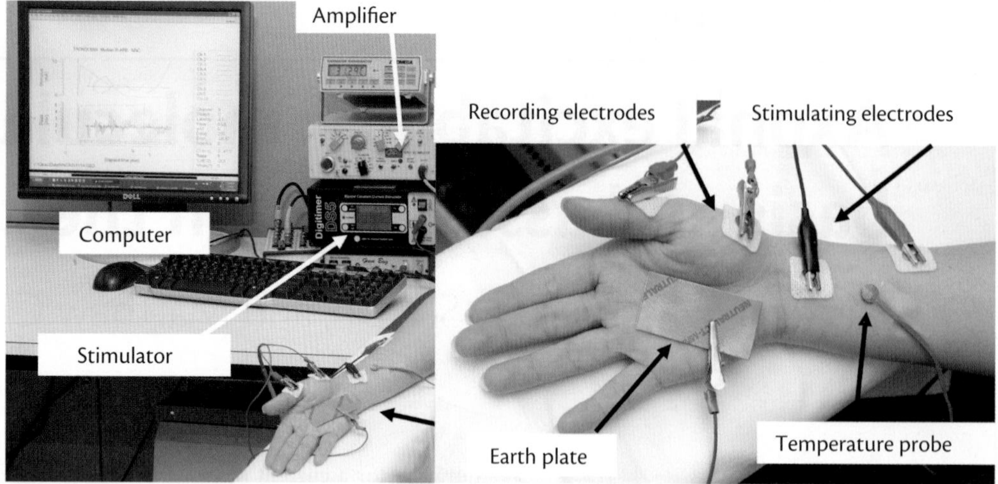

Fig. 9.1 Axonal excitability system and equipment (Left), with depiction of recording undertaken in the median nerve (Right).

broadly commercially available, improving the establishment of the technique for clinical use.

Electrodes placements are similar to nerve conduction studies; however, the anode is placed ~10 cm proximally to the cathode and diagonally off the path of the nerve to allow for polarization. The majority of axonal excitability studies have been undertaken in accessible upper limb nerves including median and ulnar nerves, although studies in the sural (11), peroneal, tibial (12), and facial nerves (13) have all been published.

The main principle of current axonal excitability protocols is threshold tracking. Threshold is defined, in this setting, as the stimulus required to produce a compound potential of a prespecified amplitude (8). Typically, the target selected corresponds to 40% of the maximum compound amplitude, which matches to the steepest segment of the stimulus response curve and is thereby most responsive to change (Fig. 9.2; (9)). Proportional tracking allows the size of the tracking step to be determined continuously online, with the change in stimulus current proportional to the error between the target and the previous response (8). Threshold tracking, as opposed to tracking changes in response amplitude to a constant stimulus, allow for greater responsiveness to change and for the same population of axons to be monitored (8). Because these techniques examine the relative difference between resting threshold and threshold following a manoeuvre designed to alter excitability, most excitability parameters are presented as percentage threshold change, to improve comparability across groups.

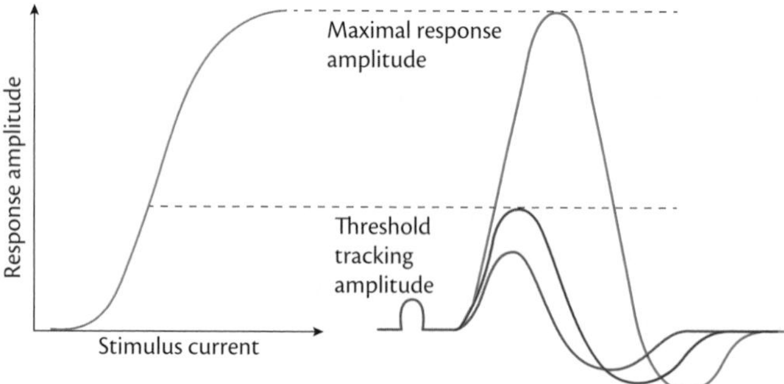

Fig. 9.2 Stimulus response properties and threshold tracking. (Right) Stimulus–response curve demonstrating the pattern of increase in response amplitude with increasing stimulus current. (Left) Dashed lines represent maximal response amplitude to a supramaximal stimulus and threshold tracking amplitude corresponding to ~40% maximal response amplitude and the steepest part of the stimulus–response curve. During the threshold tracking procedure, the stimulus current is adjusted online to achieve the preset threshold tracking amplitude.

Axonal excitability properties may be altered by changes in membrane potential, or ion channel dysfunction, degeneration, or demyelination. Coherent changes across a suite of excitability parameters may identify potential alterations in membrane potential. Axonal excitability studies can also identify changes in the function of a specific ion channel, for instance with the Na^+ channel blocker tetrodotoxin (14) or genetic mutations in K_v1.1 potassium channels associated with episodic ataxia type 1 (15). Accordingly, axonal excitability studies have developed a role both as a research technique to examine disease pathophysiology and as a clinical investigation technique.

A number of experimental manoeuvres including transient limb ischemia, hyperventilation and maximal voluntary contraction have been utilized to explore further determinants of axonal excitability in disease states and healthy axons (15–18).

A complete motor axonal excitability assessment protocol can be undertaken in less than 10 min, with sensory protocols taking up to 15 min. Because axonal excitability studies rely on tracking the stimulus current over several minutes, it is important that the recording is stable over this time period, without major changes in threshold or amplitude of the unconditioned response. In addition, care should be taken during electrode placement to find the site of lowest threshold and appropriate skin preparation to reduce skin/electrode impedance. As axonal excitability measures are sensitive to temperature, it is important to ensure a stable and warm temperature at the testing site (19).

Semi-automated protocols begin with recording a stimulus–response curve, whereby the stimulus current is increased gradually in a stepwise manner until the response amplitude fails to increase with additional current. As the stimulus–response curve is utilized to determine the level of threshold tracking, the peak response must remain truly maximal and stable.

Towards a molecular understanding

Voltage-gated ion channels and ion pumps contribute to resting membrane potential and determine the level of axonal excitability. There are several key conductances.

Voltage-gated sodium (Na^+) channels

Voltage-gated Na^+ channels are critical for action potential propagation in human myelinated axons (6,9,20). Voltage-gated Na^+ channels are clustered at high density at the nodes of Ranvier in myelinated axons (21,22), with the predominant isoform $Na_V1.6$ (14,23,24). The majority of the nodal Na^+ current is transient, demonstrating fast activation and inactivation kinetics (9,25). The nodal regions of the axonal membrane are highly enriched with Na^+ channels, with the density at the node ensuring saltatory conduction, while the internode demonstrates a lower density of sodium channels (26). A persistent Na^+ current which represents only a small fraction of the total current also influences membrane excitability (27). $Na_V1.6$ can also produce a persistent, non-inactivating current (28), which is active near threshold (25,29).

Voltage-gated potassium channels

Voltage-gated potassium channels are key determinants of axonal excitability. Fast K^+ currents display fast activation and inactivation kinetics. In the PNS, the voltage-gated potassium channel isoform $K_V1.1$ is located in high density at the juxtaparanode and provides a fast K^+ current, which acts to dampen excitability at the node following impulse conduction (30–34). In contrast, slow K^+ currents are located preferentially at the nodes of Ranvier, corresponding to KCNQ2 channels labelled strongly at the nodes of Ranvier in large fibers in rat sciatic nerve (35). Slow K^+ currents are active at resting membrane potential and act to limit repetitive firing (30,36,37).

Hyperpolarization-activated cation conductance (I_h)

The hyperepolarization-activated cation conductance consists of a mixed cation conductance of Na^+ and K^+ ions, activated by membrane hyperpolarization (38). The I_h current is mediated by hyperpolarization-activated cyclic nucleotide-gated channels (HCN) which are activated at membrane potentials more negative than −50mV and can be active at resting membrane potential (38,39,40). HCN channels act to stabilize and maintain membrane potential and input resistance (40).

Sodium/potassium ATPase

Sodium/potassium ATPase (Na^+/K^+ pump) functions as an energy dependent ion pump (41). The ratio of pump exchange is fixed and voltage-independent: three Na^+ ions are exported to every two K^+ ions imported into the cell, yielding a net positive electrogenic charge (42) and maintaining the ratio of sodium to potassium ions.

Clinical measures of axonal excitability

Strength-duration time-constant

The strength–duration time constant (SDTC) reflects the relationship between the strength and duration of a stimulus (Fig. 9.3; (43)). As stimulus duration is shortened, greater stimulus intensity is required to produce a response. Rheobase is defined as the lowest current strength that can produce an impulse, even for a pulse of infinite duration. Chronaxie is termed the stimulus duration required to produce twice the rheobasic current. SDTC is equivalent to chronaxie as determined by Weiss' law (3,10,45) and reflects the rate of decrease of threshold current as stimulus duration is lengthened. The relationship between stimulus duration and stimulus charge is linear and can be estimated with only two stimulus widths, although in current protocols more widths are used (10,14). SDTC reflects persistent Na^+ conductances which are active near threshold and is voltage-dependent (45), as well as passive membrane properties indicating the capacitance of the nodal membrane (46).

Recovery cycle of excitability

The recovery cycle utilizes a paired pulse paradigm to assess the recovery of excitability following a single impulse, as the interstimulus interval is varied between 2 and 200ms (Fig. 9.4; (47)). At short interstimulus intervals (~2–4 ms), threshold reflects the inactivation and gradual recovery of transient Na^+ channels (17). Action potential conduction produces a period when the axon is absolutely refractory (0.5–1ms) due to inactivation of transient Na^+ channels. When the channel is inactivated, ions cannot pass

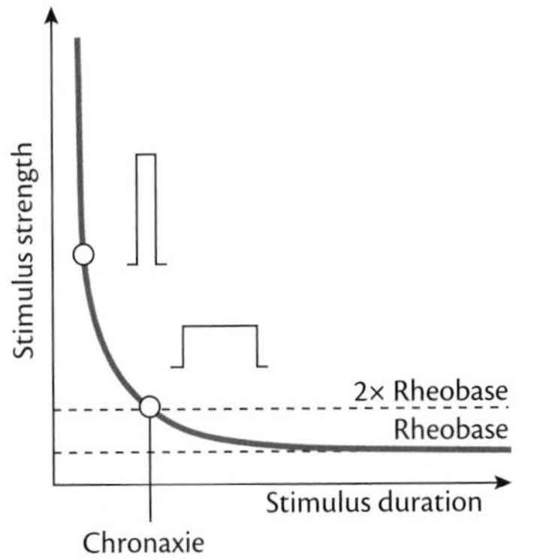

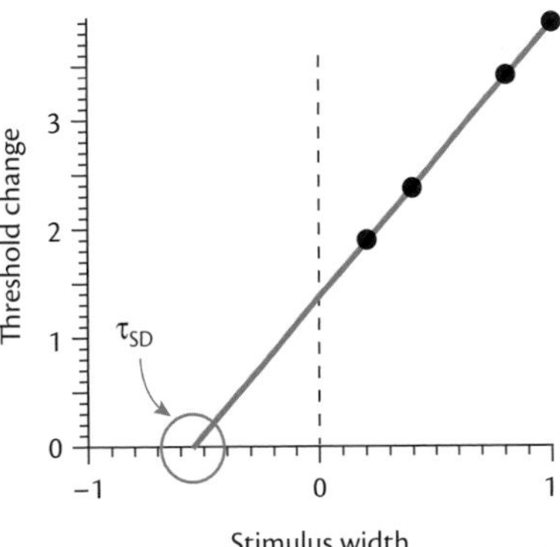

Fig. 9.3 Strength-duration properties. (Left) Strength–duration curve of stimulus strength versus stimulus duration, demonstrating rheobase as the minimum stimulus strength required to produce a response and chronaxie as the stimulus duration corresponding to twice rheobasic current. (Right) Strength–duration time constant (τ_{SD}; SDTC) determined as the negative intercept on the x-axis of the line produced by stimulus width and threshold charge.

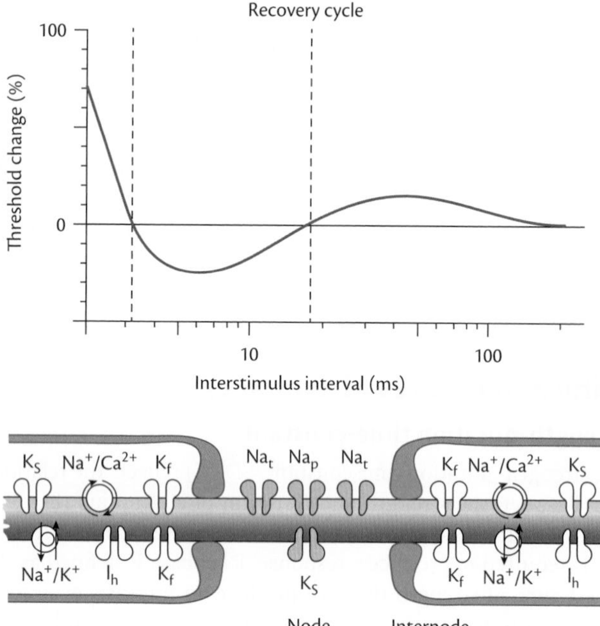

Fig. 9.4 Molecular characterization of the recovery cycle of excitability (Top) Recovery cycle of excitability, demonstrating the changes in threshold in response to differing interstimulus intervals (2–200 ms). To the left of the first dotted line indicates refractoriness, where the threshold current is increased, followed by superexcitability, with reduced threshold current between the dotted lines. Subexcitability is to the right of the second dotted line, with greater threshold current. (Bottom) Depiction of the ion channels, pumps and exchangers on the axonal membrane. Ion channel important for the recovery cycle of excitability are highlighted in purple. At the node of Ranvier, these include Na^+ channels responsible for inactivation and refractoriness, and K_s^+ which slowly activate to produce subexcitability.

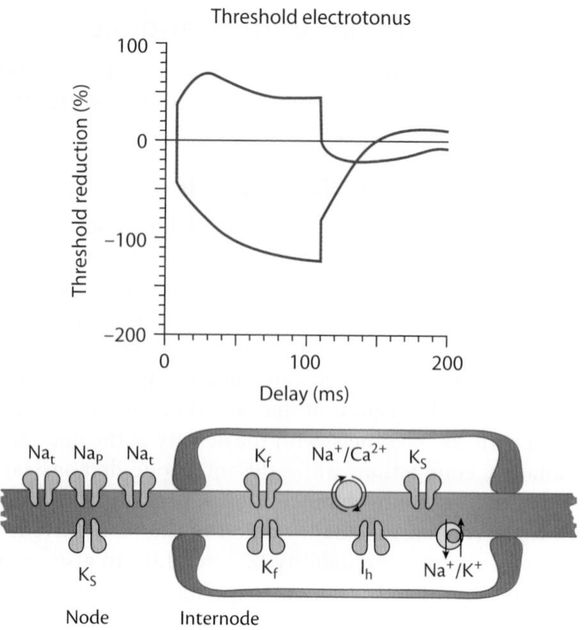

Fig. 9.5 Molecular characterization of threshold electrotonus (Top) Threshold electrotonus waveform, with depolarizing threshold electrotonus depicted at the top and hyperpolarizing threshold electrotonus below, in response to ±40% of threshold current for 100 ms duration. (Bottom) Depiction of the ion channels, pumps and exchangers on the axonal membrane important in threshold electrotonus. Purple depicts the spread of current into the internode which produces the characteristic threshold electrotonus profile.

through the channel pore, and no subsequent action potentials can be generated (6,9,48). The relative refractory period follows the absolute refractory period and lasts a further 3–4 ms during which time it is more difficult to produce an impulse due to the slow recovery of Na^+ channels from inactivation (5).

The relative refractory period is followed by a phase of increased membrane excitability termed superexcitability (peaking at 5–7 ms), when threshold is reduced and action potentials are easier to generate (9,11,47). Superexcitability occurs due to capacitive charging of the internode, which leads to production of a depolarizing after potential and subsequent threshold reduction (49–51). The final phase is subexcitability, when the membrane excitability is again reduced due to activation of slow K^+ channels, which are activated during impulse generation but with slow activation kinetics (11,47,51,52).

Threshold electrotonus

Threshold electrotonus is designed to examine indirectly the changes in membrane potential that occur during prolonged, subthreshold current pulses, which alter the potential difference across the axonal membrane in the internode (9,53). Threshold electrotonus produces a characteristic excitability change profile, plotted as threshold reduction so that increased excitability is plotted upwards and decreased excitability downwards (9). Internodal properties are important determinants of axonal excitability and membrane potential, as the internodal membrane represents up to 99.9% of all axonal membrane area (Fig. 9.5; (9,54,55)). The response to a depolarizing current pulse is an immediate decrease in threshold proportional to the level of current, then a further decrease as current spreads into the internode. This threshold decrease is attenuated by the accommodative action of slow K^+ channels, as in vitro studies have demonstrated via removal of this accommodation by slow K^+ channels blockers (51,56). Once the depolarizing pulse is stopped, threshold undershoots baseline values and slowly returns to baseline, reflecting slow K^+ channel deactivation (9,45).

In response to hyperpolarization, threshold is proportionally increased as the node is polarized. The slow spread of current into the internode produces additional increases in threshold, which is further accentuated by hyperpolarization-mediated closure of K^+ channels. However, the threshold increase is tempered at around 150 ms by the slow activation of I_h (8,39). When the current is terminated, threshold overshoots baseline levels due to the slow kinetics of I_h deactivation and the reactivation of slow K^+ channels (9). The extent of threshold change with hyperpolarizing current pulses is greater than with depolarization, as the hyperpolarization-mediated closure of K^+ channels enables threshold to increase unrectified until I_h is activated.

Current-threshold relationship

The current threshold relationship maintains a constant duration of the polarizing current, while the strength of the current is stepped from +50% to −100% of threshold (Fig. 9.6; (10)). In response to depolarizing current, fast, and slow K^+ channel activation occur as an accommodative response which produces outward rectification (57,58). In response to hyperpolarizing current,

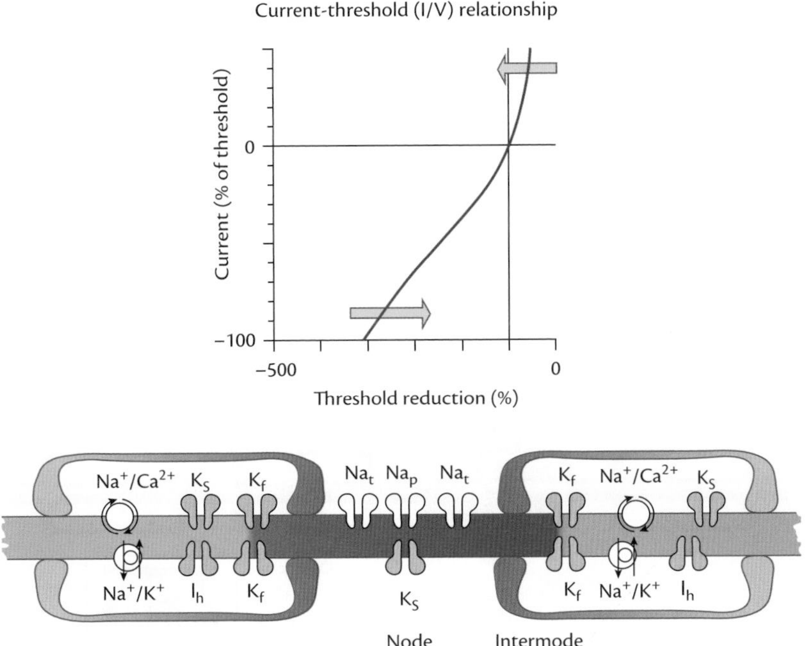

Fig. 9.6 Molecular characterization of the current-threshold relationship (Top) Current–threshold relationship with arrow in the top quadrant indicating outward rectification and the arrow in the lower quadrant depicting inward rectification. (Bottom) Depiction of the ion channels, pumps, and exchangers on the axonal membrane important in current–threshold relationship. Purple depicts the spread of current into the internode and important ion channels are highlighted in purple, including I_h and K^+ channels.

I_h is activated, leading to inward rectification and a reduction in the extent of threshold change (10,39).

Axonal excitability in clinical practice

Motor neuronopathy

Axonal excitability studies have been utilized to provide insights into the pathophysiological processes underlying motor neuron disorders, and have proved useful in dissociating different motor neuronopathies. Axonal excitability studies in patients with amyotrophic lateral sclerosis (ALS) have uncovered potential changes in persistent Na^+ conductances associated with ectopic motor activity and fasciculations, key clinical signs in ALS (59–63). Increased strength–duration time constant is linked to up-regulation of nodal persistent Na^+ conductances, which would depolarize the membrane potential and predispose the axon to fire spontaneously. In addition, evidence of reduced axonal K^+ conductance has been established, with reduced accommodation to depolarization in threshold electrotonus and increased superexcitability in the recovery cycle (60,61,63). Longitudinal studies over 3 months have demonstrated that K^+ channel dysfunction was greater in ALS patients with relatively stable CMAP amplitudes, suggesting that changes were most prominent in axons prior to degeneration (63). Both decreased K^+ conductance and increased persistent Na^+ conductance contribute to produce instability of the axonal membrane, assisting in the development of ectopic activity.

Because of prominent fasciculations and muscle atrophy, multifocal motor neuropathy (MMN) patients may be misdiagnosed with ALS (64). However, despite clinical similarities, axonal excitability studies have demonstrated a different pattern of axonal excitability change in patients with MMN as compared to those with ALS. Axonal excitability studies distal to the site of conduction block demonstrated changes potentially indicative of axonal hyperpolarization, including increased threshold change in threshold electrotonus (both depolarizing and hyperpolarizing) and prominently increased superexcitability (Fig. 9.7; (65)). These abnormalities could be normalized by application of depolarizing current, suggesting that they reflected membrane hyperpolarization. However, proximal to the site of block there were no generalized excitability abnormalities (66). The hypothesis underlying the development of hyperpolarization distal to the site of conduction block has been related to blockade of Na^+/K^+ pump activity at the lesion site, leading to intracellular accumulation of Na^+ (Fig. 9.7). As the excess Na^+ ions diffuse along the axon, the pump distal to the block would tend to over-compensate in an attempt to correct the ionic imbalance. Over-activity of the pump would lead to a net hyperpolarization in membrane potential due to the discrepancy in K^+ and Na^+ transport ratios. Accordingly, areas of depolarization and hyperpolarization would surrounding the site of conduction block along the axon. This pattern of excitability change separates MMN from ALS and provides insights into potential pathophysiological mechanisms underlying axonal dysfunction in MMN.

Evolution of a toxic sensory neuropathy: oxaliplatin-induced neurotoxicity

Chemotherapy-induced peripheral neuropathy (CIPN) is a prominent form of toxic neuropathy that occurs with many different types of chemotherapies including taxanes, platinum compounds, vinca alkaloids, and thalidomide and bortezomib. Oxaliplatin, a third generation platinum analogue chemotherapy, is often used in the treatment of colorectal cancer (67,68). Oxaliplatin produces

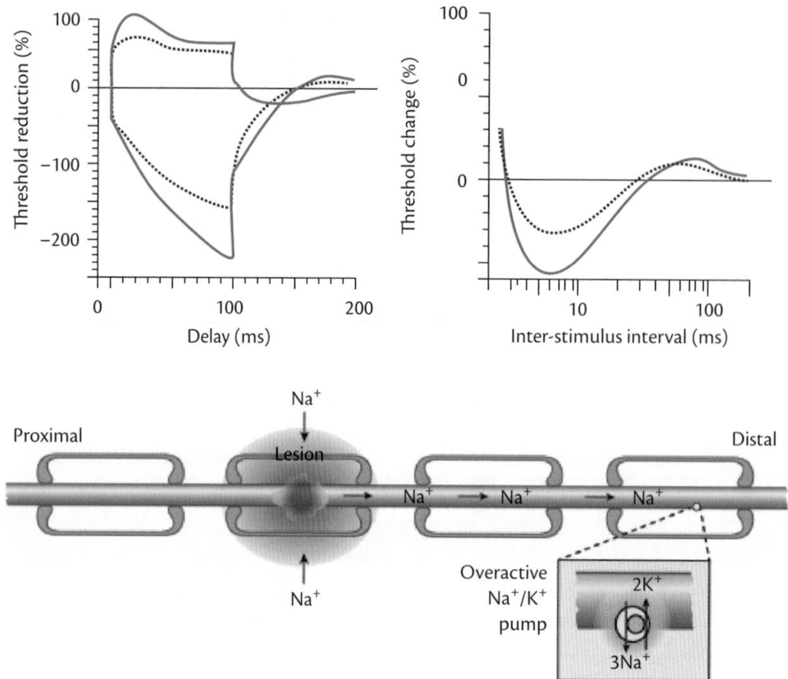

Fig. 9.7 Axonal changes in multifocal motor neuropathy. (Top) Depiction of changes in threshold electrotonus (left) and recovery cycle (right) associated with MMN patients (red) compared with normative data (dotted line). (Bottom) Schematic of the proposed pattern of axonal changes in MMN, with additional Na^+ entry at the lesion site, followed by diffusion of Na^+ along the axon to the distal axon, causing over-activity of the Na^+/K^+ pump.

acute neurotoxicity in virtually all patients—with symptoms of paraesthesia and dysesthesia, fasciculations and cramps in limbs and mouth (69). These symptoms are produced by cold exposure and occur immediately following infusion, lasting up to 7 days. Acute neurotoxicity occurs from the first treatment onwards, however with increasing cumulative exposure a sensory neuropathy develops (70). Chronic oxaliplatin-induced neuropathy is characterized by sensory loss in hands and feet, leading to sensory ataxia and areflexia (Fig. 9.8). These symptoms can lead to substantial disability, significantly affecting quality of life.

Nerve conduction studies and electromyography have been utilized to examine the neurophysiological features of oxaliplatin-induced neurotoxicity. Acute oxaliplatin-induced neurotoxicity is characterized by repetitive motor discharges in response to a single electrical stimuli, observable 1–4 days after infusion, although repetitive compound sensory action potentials (CSAPs) are not observed (69,71,72). Needle EMG studies demonstrated abnormal spontaneous high frequency motor unit activity, identified in all patients at 2–4 days post oxaliplatin infusion and only in 25% two weeks after infusion (69,71,72). In contrast, chronic oxaliplatin-induced neurotoxicity is associated with reduction in CSAP amplitudes in both upper and lower limbs, with conduction velocity unaffected (72). Likewise, motor amplitudes are not affected. Significant reduction in CSAP amplitude can be observed following 3–4 months of oxaliplatin treatment and this reduction may be predictive of the development of severe neurotoxicity (71,72).

Axonal excitability studies have also been undertaken in oxaliplatin-treated patients, both acutely and longitudinally across treatment. Axonal excitability studies have identified acute changes in sensory nerve function. In the recovery cycle of excitability, refractoriness was reduced and superexcitability increased following oxaliplatin treatment (74,75). Interestingly, these changes are associated with alterations in sodium channel function, which has been proposed as a mechanism mediating acute neurotoxicity in several in vitro studies of oxaliplatin on nerve preparations (76–80). In addition, these changes were similar to those found in sensory axons following exposure to the Na^+ channel blocker tetrodotoxin (14), again suggesting a role for voltage-gated Na^+ channels in the etiology of oxaliplatin-induced neurotoxicity.

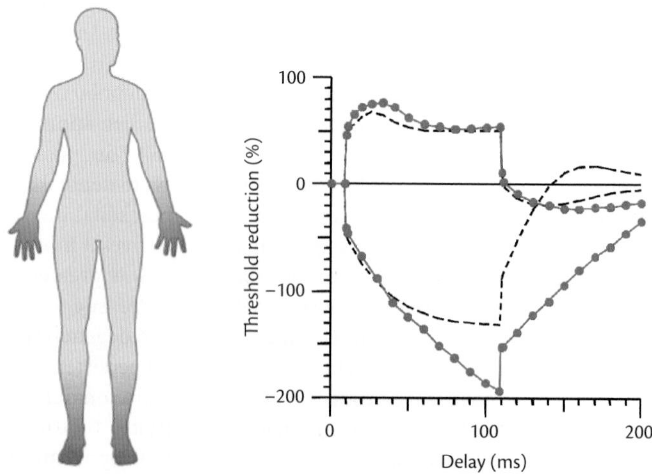

Fig. 9.8 'Glove and stocking' distribution of symptoms in chronic oxaliplatin-induced neurotoxicity (left) and depiction of typical cumulative threshold electrotonus changes in sensory axons with oxaliplatin treatment, illustrating threshold increases in both depolarizing and hyperpolarizing electrotonus (right).

Oxaliplatin-treated patients were also followed longitudinally across treatment to determine the excitability changes that accompany the development of chronic sensory neuropathy. A suite of excitability changes occurred in sensory nerve excitability progressively across treatment (Fig. 9.8). Importantly these changes occurred earlier than reductions in peak CSAP amplitude, by the 2nd month of treatment, suggesting that they may provide a marker of the onset of axonal dysfunction in oxaliplatin treated patients (75). These global changes in sensory axons are similar to those seen in models of Wallerian degeneration (81), suggesting that the early changes in axonal excitability reflect the initial stages of axonal degeneration.

Axonal excitability studies provide important additional information regarding both the pathophysiology and the evolution of oxaliplatin-induced neurotoxicity. Strategies to assist in the early identification of individual patients at-risk of severe neurotoxicity are becoming increasingly important in oncology clinical practice to help in the development of individualized patient treatment approaches.

Monitoring treatment response in chronic inflammatory demyelinating polyneuropathy

Intravenous immunoglobulin (IVIg) is a successful treatment for CIDP and may produce rapid clinical improvement (82). However, many patients require monthly maintenance IVIg treatment that is expensive and there is a need for a marker to identify which patients are responding to treatment and need to continue to require maintenance treatment.

Studies have identified a pattern of axonal excitability change in the motor axons of CIDP patients that corresponds to the timing of their IVIg treatment (83). Patients were studied before and 1-2 weeks after IVIg treatment, which revealed a pattern of axonal excitability change in relation to IVIg. The motor axons of CIDP patients were relatively hyperpolarized at baseline, and after treatment, displayed greater accommodation and reduced superexcitability, suggesting that membrane potential was normalized by IVIg treatment. However, this effect started to return to baseline pre-IVIg levels by two weeks after each IVIg treatment (83). These studies demonstrated that maintenance treatment may be required to rectify this pattern of axonal change, and suggest that axonal excitability studies may provide a novel marker of treatment response in CIDP patients.

Conclusions

Axonal excitability studies provide a means to examine the molecular foundations of the axonal membrane, providing insights into resting membrane potential and ion channel function. These studies provide additional information to assist in understanding the basic physiology of impulse conduction and further provide insights into mechanisms governing pathophysiology in disease states. Used in combination with conventional nerve conduction studies, axonal excitability techniques provide useful information to assist in differential diagnosis, assessment of treatment response, and understanding the mechanisms of nerve dysfunction. From the historical beginnings of the study of nerve electricity, axonal excitability techniques have now developed to represent an important component of the methodology available to clinical neurophysiologists.

References

1. Finger, S. and Wade, N.J. (2002). The neuroscience of Helmholtz and the theories of Johannes Müller. Part 1: nerve cell structure, vitalism, and the nerve impulse. *Journal of Historical Neuroscience*, **11**(2), 136–55.
2. Piccolino, M. (2006). Luigi Galvani's path to animal electricity. *Comptes Rendus Biologies*, **329**(5–6), 303–18.
3. Weiss, G. (1901). Sur la possibilité de render comparables entre eux les appareils servant à l'excitation electrique. *Archives Italiennes de Biologie*, **35**, 413–46.
4. Lapicque, L. (1909). Définition expérimentale de l'excitabilité. *Comptes Rendus de la Société de Biologiques (Paris)*, **67**, 280–3.
5. Hodgkin, A.L. and Huxley, A.F. (1952). Propagation of electrical signals along giant nerve fibers. *Proceedings of the Royal Society, London, Series B Biological Sciences*, **140**(899), 177–83.
6. Catterall, W.A. (2012). Voltage-gated sodium channels at 60: structure, function and pathophysiology. *Journal of Physiology*, **590**(Pt 11), 2577–89.
7. Bergmans, J. (1970). The physiology of single human nerve fibres. Vander, Louvain.
8. Bostock, H., Cikurel, K., and Burke, D. (1998). Threshold tracking techniques in the study of human peripheral nerve. *Muscle & Nerve*, **21**(2), 137–58.
9. Burke, D., Kiernan, M.C., and Bostock, H. (2001). Excitability of human axons. *Clinical Neurophysiology*, **112**(9), 1575–85.
10. Kiernan, M.C., Burke, D., Andersen, K.V., and Bostock, H. (2000). Multiple measures of axonal excitability: a new approach in clinical testing. *Muscle & Nerve*, **23**(3), 399–409.
11. Lin, C.S., Mogyoros, I., and Burke, D. (2000). Recovery of excitability of cutaneous afferents in the median and sural nerves following activity. *Muscle & Nerve*, **23**(5), 763–70.
12. Krishnan, A.V., Lin, C.S., and Kiernan, M.C. (2004). Nerve excitability properties in lower-limb motor axons: evidence for a length-dependent gradient. *Muscle & Nerve*, **29**(5), 645–55.
13. Krishnan, A.V., Hayes, M., and Kiernan, M.C. (2007). Axonal excitability properties in hemifacial spasm. *Movement Disorders*, **22**(9), 1293–8.
14. Kiernan, M.C., Isbister, G.K., Lin, C.S., Burke, D., and Bostock, H. (2005). Acute tetrodotoxin-induced neurotoxicity after ingestion of puffer fish. *Annals of Neurology*, **57**(3), 339–48.
15. Tomlinson, S.E., Tan, S.V., Kullmann, D.M., et al. (2010). Nerve excitability studies characterize Kv1.1 fast potassium channel dysfunction in patients with episodic ataxia type 1. *Brain*, **133**(Pt 12), 3530–40.
16. Krishnan, A.V., Lin, C.S., Park, S.B., and Kiernan, M.C. (2009). Axonal ion channels from bench to bedside: a translational neuroscience perspective. *Progress in Neurobiology*, **89**(3), 288–313.
17. Kiernan, M.C. and Lin, C.S.Y. (2013). Nerve excitability: a clinical translation. In: M. Aminoff (Ed.) *Aminoff's electrodiagnosis in clinical neurology*, pp. 345–65. Amsterdam: Elsevier.
18. Kiernan, M.C. and Kaji, R. (2013). Physiology and pathophysiology of myelinated nerve fibers. *Handbook of Clinical Neurology*, **115**, 43–53.
19. Kiernan, M.C., Cikurel, K., and Bostock, H. (2001). Effects of temperature on the excitability properties of human motor axons. *Brain*, **124**(Pt 4), 816–25.
20. Schwarz, J.R., Reid, G., and Bostock, H. (1995). Action potentials and membrane currents in the human node of Ranvier. *Pflügers Archives*, **430**, 283–92.
21. Dugandzija-Novakovic, S., Koszowski, A.G., Levinson, S.R., and Shrager, P. (1995). Clustering of Na^+ channels and node of Ranvier formation in remyelinating axons. *Journal of Neuroscience*, **15**, 492–503.
22. Arroyo, E.J. and Scherer, S.S. (2000). On the molecular architecture of myelinated fibers. *Histochemistry and Cellular Biology*, **113**, 1–18.
23. Caldwell, J.H., Schaller, K.L., Lasher, R.S., Peles, E., and Levinson, S.R. (2000). Sodium channel $Na_v1.6$ is localized at nodes of Ranvier,

dendrites, and synapses. *Proceedings of the National Academy of Sciences USA*, **97**, 5616–20.

24. Boiko, T., Rasband, M.N., Levinson, S.R., et al. (2001). Compact myelin dictates the differential targeting of two sodium channel isoforms in the same axon. *Neuron*, **30**, 91–104.
25. Crill, W.E. (1996). Persistent sodium current in mammalian central neurons. *Annual Review of Physiology*, **58**, 349–62.
26. Ritchie, J.M. and Rogart, R.B. (1977). Density of sodium channels in mammalian myelinated nerve fibers and nature of the axonal membrane under the myelin sheath. *Proceedings of the National Academy of Sciences USA*, **74**, 211–15.
27. Kiss, T. (2008). Persistent Na-channels: origin and function. A review. *Acta Biologica Hungarica*, **59**, 1–12.
28. Rush, A.M., Dib-Hajj, S.D., and Waxman, S.G. (2005). Electrophysiological properties of two axonal sodium channels, $Na_v1.2$ and $Na_v1.6$, expressed in mouse spinal sensory neurones. *Journal of Physiology*, **564**, 803–15.
29. Baker, M.D. and Bostock, H. (1998). Inactivation of macroscopic late Na^+ current and characteristics of unitary late Na^+ currents in sensory neurons. *Journal of Neurophysiology*, **80**, 2538–49.
30. Roper, J. and Schwarz, J.R. (1989). Heterogeneous distribution of fast and slow potassium channels in myelinated rat nerve fibres. *Journal of Physiology*, **416**, 93–110.
31. Wang, H., Kunkel, D.D., Martin, T.M., Schwartzkroin, P.A., and Tempel, B.L. (1993). Heteromultimeric K^+ channels in terminal and juxtaparanodal regions of neurons. *Nature*, **365**, 75–9.
32. Arroyo, E.J., Xu, Y.T., Zhou, L., et al. (1999). Myelinating Schwann cells determine the internodal localization of Kv1.1, Kv1.2, Kvbeta2, and Caspr. *Journal of Neurocytology*, **28**(4–5), 333–47.
33. Rasband, M.N. and Trimmer, J.S. (2001). Developmental clustering of ion channels at and near the node of Ranvier. *Developmental Biology*, **236**, 5–16.
34. Vacher, H., Mohapatra, D.P., and Trimmer, J.S. (2008). Localization and targeting of voltage-dependent ion channels in mammalian central neurons. *Physiological Review*, **88**(4), 1407–47.
35. Schwarz, J.R., Glassmeier, G., Cooper, E.C., et al. (2006). KCNQ channels mediate I_{Ks}, a slow K^+ current regulating excitability in the rat node of Ranvier. *Journal of Physiology*, **573**, 17–34.
36. Devaux, J.J., Kleopa, K.A., Cooper, E.C., and Scherer, S.S. (2004). KCNQ2 is a nodal K^+ channel. *Journal of Neuroscience*, **24**, 1236–44.
37. Gutman, G.A., Chandy, K.G., Grissmer, S., et al. (2005). International Union of Pharmacology. LIII. Nomenclature and molecular relationships of voltage-gated potassium channels. *Pharmacological Review*, **57**, 473–508.
38. Robinson, R.B. and Siegelbaum, S.A. (2003). Hyperpolarization-activated cation currents: From molecules to physiological function. *Annual Review of Physiology*, **65**, 453–80.
39. Pape, H.C. (1996). Queer current and pacemaker: the hyperpolarization-activated cation current in neurons. *Annual Review of Physiology*, **58**, 299–327.
40. He, C., Chen, F., Li, B., and Hu, Z. (2014). Neurophysiology of HCN channels: from cellular functions to multiple regulations. *Progress in Neurobiology*, **112**, 1–23.
41. Dempski, R.E., Friedrich, T., and Bamberg, E. (2009). Voltage clamp fluorometry: combining fluorescence and electrophysiological methods to examine the structure-function of the Na(+)/K(+)-ATPase. *Biochimica et Biophysica Acta*, **1787**, 714–20.
42. Rakowski, R.F., Gadsby, D.C., and De Weer, P. (1989). Stoichiometry and voltage dependence of the sodium pump in voltage-clamped internally dialyzed squid giant axon. *Journal of General Physiology*, **93**, 903–41.
43. Mogyoros, I., Kiernan, M.C., and Burke, D. (1996). Strength-duration properties of human peripheral nerve. *Brain*, **119**, 439–47.
44. Bostock, H. (1983). The strength-duration relationship for excitation of myelinated nerve: computed dependence on membrane parameters. *Journal of Physiology*, **341**, 59–74.
45. Bostock, H. and Rothwell, J.C. (1997). Latent addition in motor and sensory fibres of human peripheral nerve. *Journal of Physiology*, **498**, 277–94.
46. Nodera, H. and Kaji, R. (2006). Nerve excitability testing and its clinical application to neuromuscular diseases. *Clinical Neurophysiology*, **117**(9), 1902–16.
47. Kiernan, M.C., Mogyoros, I., and Burke, D. (1996). Differences in the recovery of excitability in sensory and motor axons of human median nerve. *Brain*, **119**, 1099–105.
48. Goldfarb, M. (2012). Voltage-gated sodium channel-associated proteins and alternative mechanisms of inactivation and block. *Cell & Molecular Life Sciences*, **69**(7), 1067–76.
49. Adrian, E.D. and Lucas, K. (1912). On the summation of propagated disturbances in nerve and muscle. *Journal of Physiology*, **44**, 68–124.
50. Barrett, E.F. and Barrett, J.N. (1982). Intracellular recording from vertebrate myelinated axons: mechanism of the depolarizing afterpotential. *Journal of Physiology*, **323**, 117–44.
51. Baker, M., Bostock, H., Grafe, P., and Martius, P. (1987). Function and distribution of three types of rectifying channel in rat spinal root myelinated axons. *Journal of Physiology*, **383**, 45–67.
52. Kiernan, M.C., Mogyoros, I., and Burke, D. (1996). Changes in excitability and impulse transmission following prolonged repetitive activity in normal subjects and patients with a focal nerve lesion. *Brain*, **119**(Pt 6), 2029–37.
53. Baker, M.D. (2000). Axonal flip-flops and oscillators. *Trends in Neurosciences*, **23**, 514–19.
54. Abe, I., Ochiai, N., Ichimura, H., Tsujino, A., Sun, J., and Hara, Y. (2004). Internodes can nearly double in length with gradual elongation of the adult rat sciatic nerve. *Journal of Orthopaedic Research*, **22**, 571–7.
55. Salzer, J.L., Brophy, P.J., and Peles, E. (2008). Molecular domains of myelinated axons in the peripheral nervous system. *Glia*, **56**, 1532–40.
56. Bostock, H. and Baker, M. (1988). Evidence for two types of potassium channel in human motor axons in vivo. *Brain Research*, **462**(2), 354–8.
57. Kiernan, M.C. and Bostock, H. (2000). Effects of membrane polarization and ischaemia on the excitability properties of human motor axons. *Brain*, **123**, 2542–51.
58. Lin, C.S., Kiernan, M.C., Burke, D., and Bostock, H. (2006). Nerve excitability measures: biophysical basis and use in the investigation of peripheral nerve disease. In: J. Kimura (Eds) *Handbook of Clinical Neurophysiology, Clinical Neurophysiology of Peripheral Nerve Diseases*, pp. 81–403. Amsterdam: Elsevier.
59. Mogyoros, I., Kiernan, M.C., Burke, D., and Bostock, H. (1998). Strength-duration properties of sensory and motor axons in amyotrophic lateral sclerosis. *Brain*, **121**(Pt 5), 851–9.
60. Vucic, S. and Kiernan, M.C. (2006). Axonal excitability properties in amyotrophic lateral sclerosis. *Clinical Neurophysiology*, **117**(7), 1458–66.
61. Kanai, K., Kuwabara, S., Misawa, S., et al. (2006). Altered axonal excitability properties in amyotrophic lateral sclerosis: impaired potassium channel function related to disease stage. *Brain*, **129**(Pt 4), 953–62.
62. Vucic, S. and Kiernan, M.C. (2010). Upregulation of persistent sodium conductances in familial ALS. *Journal of Neurology, Neurosurgery, & Psychiatry*, **81**(2), 222–7.
63. Cheah, B.C., Lin, C.S., Park, S.B., Vucic, S., Krishnan, A.V., Kiernan, M.C. (2012). Progressive axonal dysfunction and clinical impairment in amyotrophic lateral sclerosis. *Clinical Neurophysiology*, **123**(12), 2460–7.
64. Arcila-Londono, X. and Lewis, R.A. (2013). Multifocal motor neuropathy. *Handbook of Clinical Neurology*, **115**, 429–42.
65. Kiernan, M.C., Guglielmi, J.M., Kaji, R., Murray, N.M., and Bostock, H. (2002). Evidence for axonal membrane hyperpolarization in multifocal motor neuropathy with conduction block. *Brain*, **125**(Pt 3), 664–75.

66. Cappelen-Smith, C., Kuwabara, S., Lin, C.S., and Burke, D. (2002). Abnormalities of axonal excitability are not generalized in early multifocal motor neuropathy. *Muscle & Nerve*, **26**(6), 769–76.
67. de Gramont, A., Figer, A., Seymour, M., et al. (2000). Leucovorin and fluorouracil with or without oxaliplatin as first-line treatment in advanced colorectal cancer. *Journal of Clinical Oncology*, **18**, 2938–47.
68. André, T., Boni, C., Navarro, M., et al. (2009). Improved overall survival with oxaliplatin, fluorouracil, and leucovorin as adjuvant treatment in stage II or III colon cancer in the MOSAIC trial. *Journal of Clinical Oncology*, **27**(19), 3109–16.
69. Wilson, R.H., Lehky, T., Thomas, R.R., Quinn, M.G., Floeter, M.K., and Grem, J.L. (2002). Acute oxaliplatin-induced peripheral nerve hyperexcitability. *Journal of Clinical Oncology*, **20**, 1767–74.
70. Grothey, A. (2003). Oxaliplatin-safety profile: Neurotoxicity. *Seminars in Oncology*, **30**, 5–13.
71. Hill, A., Bergin, P., Hanning, F., et al. (2010). Detecting acute neurotoxicity during platinum chemotherapy by neurophysiological assessment of motor nerve hyperexcitability. *BMC Cancer*, **10**, 451.
72. Lehky, T.J., Leonard, G.D., Wilson, R.H., Grem, J.L., and Floeter, M.K. (2004). Oxaliplatin-induced neurotoxicity: acute hyperexcitability and chronic neuropathy. *Muscle & Nerve*, **29**, 387–92.
73. Velasco, R., Bruna, J., Briani, C., et al. (2014). Early predictors of oxaliplatin-induced cumulative neuropathy in colorectal cancer patients. *Journal of Neurology, Neurosurgery & Psychiatry*, **85**(4), 392–8.
74. Park, S.B., Goldstein, D., Lin, C.S., Krishnan, A.V., Friedlander, M.L., and Kiernan, M.C. (2009). Acute abnormalities of sensory nerve function associated with oxaliplatin-induced neurotoxicity. *Journal of Clinical Oncology*, **27**(8), 1243–9.
75. Park, S.B., Lin, C.S., Krishnan, A.V., Goldstein, D., Friedlander, M.L., and Kiernan, M.C. (2009). Oxaliplatin-induced neurotoxicity: changes in axonal excitability precede development of neuropathy. *Brain*, **132**(Pt 10), 2712–23.
76. Sittl, R., Lampert, A., Huth, T., et al. (2012). Anticancer drug oxaliplatin induces acute cooling-aggravated neuropathy via sodium channel subtype Na(V)1.6-resurgent and persistent current. *Proceedings of the National Academy of Sciences USA*, **109**(17), 6704–9.
77. Grolleau, F., Gamelin, L., Boisdron-Celle, M., Lapied, B., Pelhate, M., and Gamelin, E. (2001). A possible explanation for a neurotoxic effect of the anticancer agent oxaliplatin on neuronal voltage-gated sodium channels. *Journal of Neurophysiology*, **85**(5), 2293–7.
78. Webster, R.G., Brain, K.L., Wilson, R.H., Grem, J.L., and Vincent, A. (2005). Oxaliplatin induces hyperexcitability at motor and autonomic neuromuscular junctions through effects on voltage-gated sodium channels. *British Journal of Pharmacology*, **146**(7), 1027–39.
79. Benoit, E., Brienza, S., and Dubois, J.M. (2006). Oxaliplatin, an anticancer agent that affects both Na+ and K+ channels in frog peripheral myelinated axons. *General Physiology and Biophysics*, **25**(3), 263–76.
80. Adelsberger, H., Quasthoff, S., Grosskreutz, J., Lepier, A., Eckel, F., and Lersch, C. (2000). The chemotherapeutic oxaliplatin alters voltage-gated Na(+) channel kinetics on rat sensory neurons. *European Journal of Pharmacology*, **406**(1), 25–32.
81. Moldovan, M., Alvarez, S., and Krarup, C. (2009). Motor axon excitability during Wallerian degeneration. *Brain*, **132**(Pt 2), 511–23.
82. Eftimov, F., Winer, J.B., Vermeulen, M., de Haan, R., and van Schaik, I.N. (2009). Intravenous immunoglobulin for chronic inflammatory demyelinating polyradiculoneuropathy. *Cochrane Database of Systematic Reviews*, (1), CD001797.
83. Lin, C.S., Krishnan, A.V., Park, S.B., and Kiernan, M.C. (2011). Modulatory effects on axonal function after intravenous immunoglobulin therapy in chronic inflammatory demyelinating polyneuropathy. *Archives of Neurology*, **68**(7), 862–9.

CHAPTER 10

Reflex studies

Josep Valls-Solé

Introduction

The integration of sensory inputs into the motor commands for purposeful voluntary or automatic actions (sensorimotor integration) is a very complex function of the central nervous system. With regard to somatosensory inputs, reflex motor responses to external stimuli range from the simplest tendon jerk to quite elaborate behavioural responses. Probably one of the most important accomplishments of human kind has been the development of a sufficient supraspinal control of reflex behaviour, in such a way that, to a certain extent, we are able to selectively suppress the unwanted reflex responses, or let them fully manifest with no limitations, if necessary. Many clinical neurophysiological studies can be performed to show how the central nervous system exerts its control over the reflexes and, also, to document the loss of such control in certain neurological diseases.

Fig. 10.1 shows a schematic classification of reflexes that are elicitable in humans and can be studied neurophysiologically. Reflexes involving motor responses can be monosynaptic or polysynaptic. Monosynaptic reflexes are of short latency and follow a relatively simple segmental reflex arc. The most widely known monosynaptic reflex is the tendon jerk, which is the muscle response to mechanical activation of muscle stretch receptors. The potential linked to the reflex muscle response is known as the T wave. A similar muscle response can be obtained if the Ia fibres are depolarized by relatively weak electrical shocks applied to the supplying nerve. The response obtained in this way is known as the H reflex. Both mechanical taps and electrical shocks are unnatural stimuli and, consequently, the reflexes induced by these stimuli are likely to be of little functional use. However, they are of great importance for both clinical work in patients with neurological diseases and research in the physiology of sensorimotor circuits. At another level, a stimulus activating

Stimuli	Afferents	REST	CONTRACTION
electrical	muscle	H reflex * (H wave)	Local and distant complex effects
	cutaneous	Blink reflex (R1, R2, R3) Withdrawal reflex (RII, RIII) Anal (bulbocavernous)	Cutaneo-muscular reflexes (E1, I1, E2, I2) (cutaneous silent period) Masseteric inhibitory reflex (MIR I, MIR II)
	mixed	Reflex myoclonus (C wave) Palmomental reflex Somatosensory startle	Exteroceptive reflexes (mixed nerve silent period) Long latency reflexes (LLR I,II,III)
mechanical	muscle joint bone	Tendon jerk * (T wave) Shortening reaction	Stretch reflex (M1, M2, M3) Unloading reflex Postural reflex #
vibration	muscle	Tonic vibration (TVR)	Postural reaction &

Fig. 10.1 Schematic classification of reflex responses consisting of muscle activation, induced by stimulation of somatosensory afferents, obtained at rest and during sustained muscle contraction. Modulatory effects on muscle contraction by electrical stimuli applied on muscle afferents comprise local signs and distant complex effects that can be also obtained by mechanical stimuli and are usually quantified by means of using a probe stimulus, rather than recording a reflex response. Reflexes classified under 'mixed' afferent stimulation at rest may also be induced by just cutaneous nerve stimulation. Some of the polysynaptic reflex responses obtained at rest to cutaneous and mixed nerve afferents may actually require a slight (tonic) preactivation of the target muscle.

*Indicates monosynaptic reflexes; #, Refers to postural reflexes elicited by a push; &, refers to the reflex postural adjustment when having vibration at certain points such as the neck or the Achilles tendon.

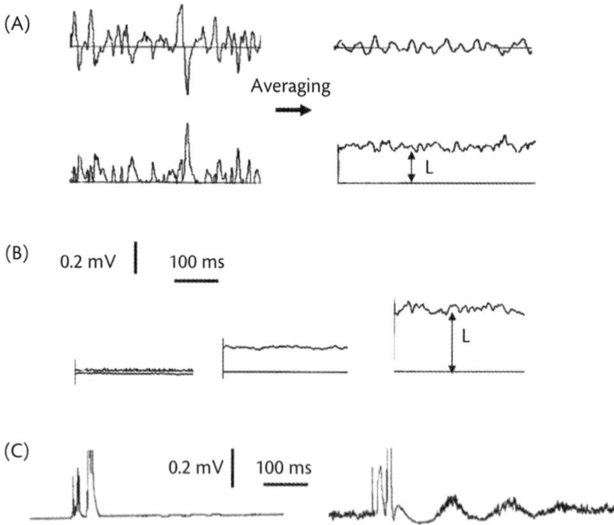

Fig. 10.2 Modulation of EMG activity. (A) Recordings of conventional unrectified (upper traces). and rectified (lower traces) epochs of EMG activity and the results of averaging a number of successive epochs. To facilitate comprehension of the procedure, a theoretical baseline has been drawn that does not exist in most commercially available equipment. Note that, with unrectified EMG activity, the resulting average line around the theoretical baseline (i.e. at zero level), whereas with rectified EMG activity, the resulting line after averaging appears at a certain level from baseline (L). This level of EMG activity will be related to the level of muscle contraction as shown in (B). Calibration is given as an approximation but the same type of recordings apply to any gain and temporal resolution. (C) The application of the technique to the recording of the H reflex in the soleus muscle at rest (left side graph). and during sustained contraction (right side graph). Note the oscillations triggered by synchronization of motor unit firing after elicitation of the H reflex during contraction.

somatosensory afferents of any type is capable of inducing polysynaptic reflexes. These reflexes may be mediated over multi-level spinal and supraspinal pathways, and are of relatively long latency. In contrast to monosynaptic reflexes, which always have a predominant excitatory component, the polysynaptic long latency reflexes can be either excitatory or inhibitory, and most commonly they have both effects combined. To make these effects apparent, the examiner requires the use of specific methodological conditions. For instance, the responses to certain stimuli may be so weak that they will only appear if the motor system is already engaged in a voluntary contraction. Maintaining a background contraction allows the identification of the inhibitory component of the reflex response as a transient decrease of the tonic level of voluntary contraction. The best method to quantify the size of inhibitory and excitatory reflex responses requiring muscle contraction is by averaging several epochs of rectified electromyograph (EMG) activity, i.e. the so-called modulation of EMG activity, as depicted in Fig. 10.2. In other instances, a stimulus may not induce a reflex response on its own, but it may be able to modulate the reflex response to another sensory stimulus in the same reflex circuit or in a distant one. A number of these studies use the H reflex as a probe for the effects of other stimuli.

The following paragraphs review the physiology of monosynaptic and polysynaptic reflexes, and describe methods and technical specifications for performing reflex studies, and using them in the assessment of peripheral nerve disorders. The reader can also find useful information in the literature devoted to technical recommendations published elsewhere (1–4).

Monosynaptic reflexes

The T wave

The T wave is just the electrophysiological counterpart of the tendon jerk. It results from monitoring the muscle response elicited by a mechanical tap to the tendon with a hammer containing an electronic device that triggers the oscilloscopic sweep. As it happens in clinical practice, the tap induces a stretch of muscle spindles, which activate the Ia afferent fibres, sending an afferent volley to the spinal cord. The Ia afferent terminals have monosynaptic excitatory connections with the alpha motoneurons innervating the stretched muscle. Their discharge will produce electrical activation of muscle fibres leading to the response that is picked up by recording electrodes and the physical evidence of muscle contraction that is evaluated clinically.

The electromyograph should be set for external triggering by the built-in hammer switch. Usually, a time window of 50 ms is long enough to record the whole response in any human muscle, and a gain of 0.5 mV per division is enough to reproduce the response recorded within a bandpass frequency range of 10–1000 Hz. The most reliable parameter to measure in the T wave is the onset latency, while its amplitude is less reproducible. The examiner has to be aware of some methodological constraints, which also apply to the clinical assessment of tendon jerks—a variable strength of the tap will give rise to a different response latency and size; a variable degree of relaxation of the agonist and antagonist muscles may inhibit or facilitate the reflex response, and a different position of right and left limbs could be responsible for an asymmetry in the size of the response. Therefore, for the T wave to furnish reliable data, the recording electrodes should be placed symmetrically in muscles of either side, and reflexes should be tested by maintaining the same position of each limb to avoid differences in the degree of tonic muscle contraction. With these precautions, the strength of the tap does not seem to produce significant changes in onset latency since the size of the monosynaptic reflex depends on the spinal excitability state, rather than on the fusimotor drive (5).

The most important feature of the T wave is its apparent simplicity. It can be applied to the study of unilateral lesions of proximal nerve segments, such as radiculopathies and plexopathies in which it may confirm clinical observations of asymmetry. Latency delay may occur in demyelinating polyneuropathies affecting large axons (6). Its absence together with a decrease in the amplitude of the sensory nerve action potentials may be one of the first objective signs of nerve damage in patients with distal axonal neuropathies (7). A particularly interesting muscle for T wave examination is the masseter muscle. The mandibular reflex is elicited by tapping to the chin. However, it is often difficult to discern whether the reflex is normal in both sides, or even whether it is present at all, by inspection only. Therefore, electromyographic monitoring of the masseter or temporalis muscle response is of paramount importance in the evaluation of suspected brainstem lesions (8,9). The mandibular reflex circuit is also unusual regarding the location of the cell bodies. In contrast to those of all other muscles in the body, the proprioceptive neurons of the jaw muscles lie within the neuraxis, protected by the blood–brain barrier from peripheral circulating agents. This is an important piece of information for the diagnosis of some disorders involving immunological attack on sensory neurons of the Gasserian ganglia, such as in sensory neuronopathies in the context of paraneoplastic syndromes or in some patients with Sjögren's syndrome (10). Fig. 10.3 shows the observations in a patient with

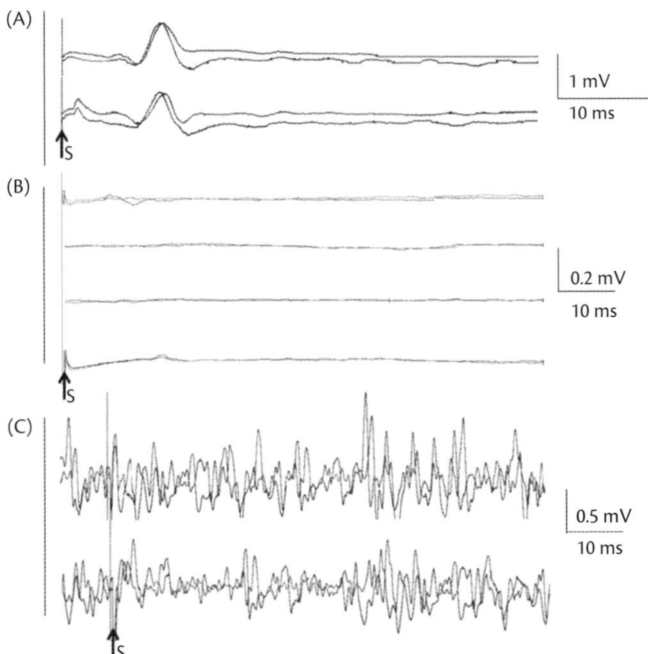

Fig. 10.3 Recording of facial reflexes induced by trigeminal nerve stimuli in a patient with Sjögren's syndrome and trigeminal neuronopathy, who had normal jaw jerk, but absent responses in other trigeminal reflexes. (A) T wave of the masseter muscles to taps to the mandible, showing a normal response. (B) Blink reflex to supraorbital nerve stimuli, showing absent responses in both sides. (C) Electrical stimuli to the mentalis nerve should elicit a masseteric silent period, which was completely absent in this patient.

trigeminal neuronopathy (ganglionopathy). in whom only the jaw jerk was preserved of all reflexes mediated by the trigeminal nerve.

The H reflex

The H reflex, named in the honour of Paul Hoffmann, the physiologist who described it for the first time in 1918, is a monosynaptic reflex that results from activation of alpha motoneurons by electrically induced Ia afferent excitatory volleys. The H reflex is commonly examined in the soleus muscle (Fig. 10.4). However, it is also easily elicitable in other muscles, such as the quadriceps and wrist flexors. It may also be observed in thenar muscles, tibialis anterior, and many other muscles if a slight facilitation is induced with voluntary contraction (11,12). The frequency band pass, gain, and sweep speed of the electromyograph recommended for recording the H reflex are the same as those used in the study of the soleus T wave, i.e. a band pass of 10–1000 Hz, a recording window of at least 50 ms, and a gain of at least 0.5 mV. The main difference in the set up for the study of the T wave and the H reflex is the use of an electrical stimulus, rather than a tendon tap, for elicitation of the H reflex. In the most conventional procedure, the electrical stimulus is applied to the posterior tibial nerve at the level of the popliteal fossa. Stimuli of relatively low intensity and long duration, typically 1 ms, are the most appropriate for selective depolarization of the Ia afferent axons (13). Bipolar stimulation electrodes are perfectly suitable for all purposes in the study of the H reflex. However, the use of the cathode placed over the nerve and the anode placed over the patella, as shown in Fig. 10.4, may be a safer procedure for selective activation of the Ia fibres in some cases (14).

The most common procedure for recording the soleus H reflex is to attach the recording electrode over the soleus muscle at the point where the two gastrocnemii muscles join the Achilles tendon (slightly distal to the midcalf) and the reference electrode about 3 cm distally. The ground electrode is located at a point between the stimulating and the recording electrode. The latency of the H wave should be measured at the point of its initial deflection, whether positive or negative, since this is the point of onset of reflex activity in the soleus. As with the T wave, care should be taken with regard to the position of the knee and ankle joints. Both should be slightly flexed to avoid stretching of the gastrocnemius, which would prevent the stimulating current reaching the nerve, and unnoticed contraction of the tibialis anterior, which would cause reciprocal attenuation of the H reflex. The soleus H reflex has a latency of about 30 ms, while the direct M wave recorded in the same muscle with supramaximal stimulus intensity occurs at a latency of about 5 ms. The latency of the H reflex is similar to that of the F wave, which can be elicited in the same muscle by the same stimulating electrode, using higher intensities. Therefore, the examiner must be aware of the characteristics that distinguish relatively easily the two responses (Table 10.1). Apart from the soleus muscle, the H reflex can be recorded from wrist flexors, albeit not consistently in all healthy subjects. The recording electrodes should be placed over the belly of the flexor carpi radialis (FCR), usually at about one-third of the distance between the medial epicondyle and the radial styloid process (15) with median nerve stimulation at the elbow.

The afferent volley eliciting the H reflex

Although the H reflex is considered to be a monosynaptic reflex, there is no consistent proof for that in humans. Certainly, the onset of the action potential reflects motoneuronal excitation induced by the fastest afferent fibres. In humans, the excitatory post-synaptic potential (EPSP) generating the soleus H reflex by the electrical stimulation of the posterior tibial nerve has a duration of 3.5–5.5 ms (16,17). During this time, other afferent inputs may reach the same motoneurons after a synaptic delay in spinal interneurons and their action potentials contribute to the developing response (17,18). Inhibitory inputs from Ib or type II fibres activated in the posterior tibial nerve also reach the motoneurons some time later, and generate inhibitory post-synaptic potentials (IPSPs) contributing to the termination of the action potential (19,20). One important functional mechanism at play is presynaptic inhibition of excitatory inputs before they reach the motoneurons (21–23). The mechanism of presynaptic inhibition was first demonstrated in the cat motoneurons by showing that the EPSP generated by group Ia fibres was depressed after conditioning stimulation of flexor muscles group I afferents, while there was no change in the size of the antidromic potential. The interneurons responsible for presynaptic inhibition (known as 'primary afferent depolarization' (PAD) neurons) receive inputs from collaterals of the Ia afferents and establish axo-axonal connections to produce depolarization of the same Ia terminals or the dendrites where the Ia axons terminate, causing a reduction of transmitter release at the arrival of excitatory inputs (23). The concept of presynaptic inhibition is very important for understanding physiological mechanisms of motor control over reflex responses (24). Neurons mediating presynaptic inhibition are probably under the control of descending tracts (25,26). and, therefore, the CNS can use this mechanism to modulate the degree of Ia afferent terminal excitation desired at any given moment or with any given condition.

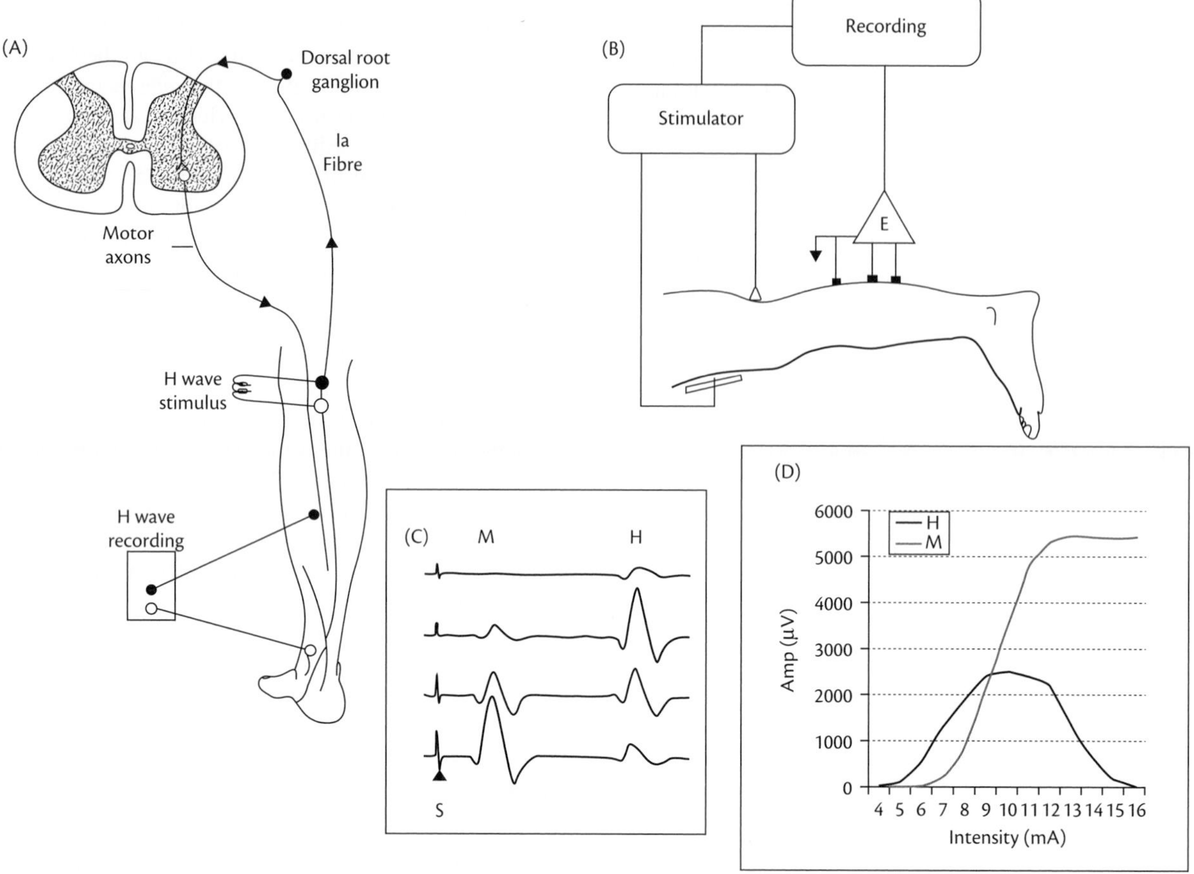

Fig. 10.4 Elicitation of the H reflex in the soleus muscle. (A) Diagram of the circuits of the H reflex. (B) Stimulation and recording electrodes placement. (C) Recordings obtained with progressively increasing the stimulus intensity from the top downwards (recruitment curve). S = stimulus; R = recording. (D) Graphic plot showing the changes in the amplitude of the H reflex and the M wave, obtained with increasing stimulus intensity.

Combining simultaneous presynaptic inhibition and post-synaptic excitation is like pressing the clutch and the accelerator at the same time to prepare the car for a speedy departure. The motoneurons are therefore set in a condition of preparation for very rapid firing (see also Chapter 3).

Table 10.1 Differences between the H reflex and the F wave

Observation	H reflex	F wave
Stimulus intensity (relative to threshold for the M wave)	Subthreshold	Suprathreshold
Largest amplitude (relative to amplitude of the M wave)	About 50%	About 5%
Onset latency	Rather constant	Variable
Shape of the action potential	Smooth and constant	Irregular and variable
Effect of a weak tonic muscle contraction	Increase in amplitude	Increase in persistence
Effect of increasing the stimulus intensity	Decrease in amplitude	Increase in persistence
Effect of vibration	Inhibition	No effect

Tests of H reflex excitability

A stimulus of intensity slightly above the threshold for depolarization of the Ia afferents should induce the H reflex as the only response of the soleus muscle, avoiding concomitant activation of motor axons. Increasing the stimulus intensity causes the H reflex to increase and the response derived from stimulation of motor axons (the M response) to appear. In normal subjects, intensities slightly above threshold for activation of motor axons give rise to H reflexes of larger amplitude than the M wave. As the stimulus intensity increases, the H reflex amplitude continues to increase up to a certain size and then decreases to completely disappear, resulting in the characteristic H reflex recruitment curve (Fig. 10.4). This is due to the fact that, beyond a certain intensity, the stimulus activates not only the Ia afferents, but also a number of motor axons where an antidromic volley will be generated. This antidromic volley will collide with the orthodromic volley generated in the same axons after motoneuronal reflex activation (27), thereby preventing the reflex volley reaching the muscle fibres. Another likely mechanism is the generation of an active inhibition of alpha motoneurons after antidromic excitation of recurrent inhibitory (Renshaw) interneurons (28). The role of the Renshaw cells in the extinction of the H wave will be dealt with in more detail below. Because of its physiological details, the recruitment curve of the H reflex should be kept in mind in all studies in which the size of the H reflex is taken as a measure of motoneuronal excitability.

When two stimuli of the same intensity are applied to the posterior tibial nerve, the second response is conditioned by the response to the first stimulus and the time elapsed between the two stimuli. This technique is known as the paired shock technique and can be used for examining the excitability recovery curve in other reflexes. Taborikova and Sax (29) were the first to describe the changes occurring in the soleus H reflex using this technique. These authors showed an early phase of inhibition, attributed to depletion of neurotransmitter, a phase of relative reduction of inhibition, between 100 and 250 ms, attributed to a long-loop facilitation operating through bulbar and cerebellar centres, and a continuation of the inhibition up to more than 1000 ms. Later studies demonstrated that post-activation depression of excitability had not completely recovered until more than 8 s after the preceding stimulus (30). It should be noted that conditioning and test stimuli may not necessarily activate the same motoneurons, and other factors modifying the excitability of motoneurons, such as recurrent inhibition and after-hyperpolarization, may affect the results. An observation of interest for clinical studies is that the inhibitory effect of repeated stimulation is reduced during voluntary contraction (12). and, therefore, the H reflex may be assessed during voluntary contraction in muscles in which it is not usually obtained at rest.

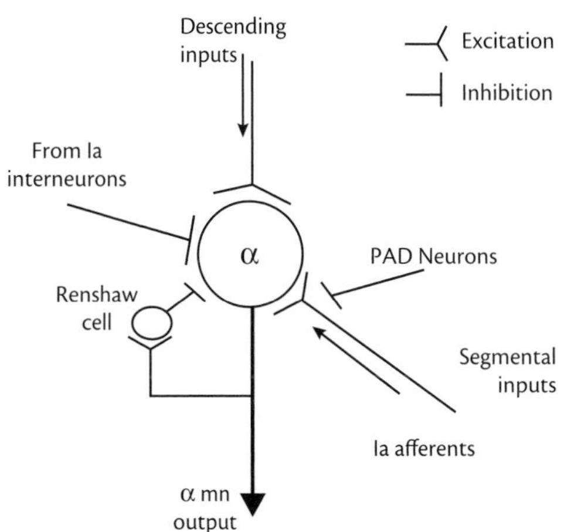

Fig. 10.5 A simplified scheme of some of the excitatory and inhibitory inputs to the alpha motoneuron that can be investigated with the H reflex. The two main sources of excitatory inputs are marked with arrows. Reciprocal inhibition is mediated by Ia afferents from the antagonist muscle. The Ia interneurons may also mediate autogenic Ib inhibition. Recurrent inhibition is mediated by the Renshaw cell. Presynaptic inhibition of the homologous excitatory Ia afferents is represented by the primary afferent depolarization (PAD). neurons, although mechanisms involving dendritic hyperpolarization or other types of control of the Ia inputs are also contributing.

Testing propriospinal circuits with the H reflex

A series of propriospinal interneurons are considered to mediate the excitability of the H reflex arc. These interneuronal circuits, which are summarized in Table 10.2, are available to neurophysiological study using the relatively complex techniques developed by Pierrot-Deseilligny and coworkers (19,31–33). Some of the circuits are schematically shown in Fig. 10.5.

Reciprocal inhibition is mediated by Ia inhibitory interneurons, which receive excitatory inputs from Ia muscle afferents, and project to the motoneurons of the physiological antagonist muscle, in which they induce a disynaptic IPSP. The Ia interneurons impinging on the extensor muscles, and the alpha and gamma motoneurons innervating flexor muscles, are thought to receive parallel inputs from descending pathways ('alpha-gamma linkage with reciprocal Ia interneuron coactivation). The electrophysiological evidence for reciprocal inhibition was first obtained in leg muscles by Mizuno et al. (34). These authors attempted to activate the Ia inhibitory interneuron projecting to soleus alpha motoneurons by stimulating the large Ia afferent fibres from the pretibial muscles. However, they only found the expected disynaptic decrease of the soleus H reflex amplitude to peroneal nerve stimuli when subjects maintained a voluntary dorsal flexion. With muscles at rest, Mizuno et al. (34) found only two late phases of H reflex suppression, suggesting that mechanisms other than reciprocal inhibition were at play. A few years later, Day et al. (35) reported that the circuit in the forearm muscles was different to that of the leg muscles in such a way that reciprocal inhibition of the H reflex of the wrist flexor muscles by a radial nerve shock was indeed present at rest. Interestingly, Day et al reported also two late phases of H reflex suppression following the disynaptic phase. Since then, conditioning of the median nerve H reflex by electrical stimulation of the radial nerve has been considered the test paradigm for reciprocal inhibition. The test has been very useful for the evaluation of patients with forearm dystonia (36–39). Appropriate methodological recommendations have been put forward by Fuhr and Hallett (40).

Table 10.2 Propriospinal inhibitory mechanisms

Mechanism	Interneurons implicated	Neurotransmitter	Tests
Ia inhibition	Ia inhibitory interneuron	Glycine	Reciprocal
Ib inhibition	Ib inhibitory interneuron	Glycine	Autogenetic
Renshaw inhibition	Renshaw cell	Glycine	Recurrent
Presynaptic inhibition	PAD* interneurons	GABA	Presynaptic

PAD = Primary afferent depolarization.

Autogenetic Ib inhibition is mediated by Ib inhibitory interneurons. These interneurons receive excitatory inputs from Ib Golgi tendon organ afferents and project to the homonymous and synergistic motoneurons, in which they induce a disynaptic IPSP. The circuit of Ib inhibition is available to electrophysiological testing with the individual at rest, thanks to the fact that the nerve from the medial gastrocnemius muscle does not carry Ia afferents for soleus motoneurons (41). A low intensity electrical stimulus given to the nerve from the medial gastrocnemius muscle will, therefore, activate the Ib afferents. The subsequent activation of the Ib inhibitory interneuron will cause disynaptic inhibition on the soleus motoneurons. This takes place in the form of a small decrease of the amplitude of the soleus H reflex between 4 and 6 ms after the medial gastrocnemius nerve stimulus. It should be noted that the dorsal reticulospinal system originating in the nucleus reticularis gigantocellularis (NRGC). sends facilitatory projections to the Ib interneurons (42), and inhibitory projections to the Ia interneuron (43). Therefore, dysfunction of the reticulospinal system may responsible in part for the subcortical deficit in preparation for the execution of a motor action observed in patients with Parkinson's disease (43).

Recurrent inhibition is mediated by Renshaw interneurons. These interneurons receive excitatory inputs from the first axonal collateral from alpha motoneurons, and project to the homonymous and neighbouring motoneurons (44), in which they induce a disynaptic IPSP. Pierrot-Deseilligny and Bussel (45) devised an electrophysiological method to assess recurrent inhibition with the soleus H reflex. The method requires a stimulator capable of delivering two stimuli of different intensity through the same pair of electrodes. The first stimulus is of low intensity, capable of inducing a small amplitude H reflex (H1). The second stimulus, delivered 10–15 ms later, should be of supramaximal intensity. Because of that, all motor axons will be depolarized and an antidromic volley will be generated. At that time, the orthodromic volley generated in a few motoneurons by the preceding stimulus, that would give rise to the H1, will be travelling in a few motor axons and collide with the antidromic impulses. The motoneurons which axons become free of any antidromic volley will be again ready to be activated by the Ia afferent axons, producing a new H reflex, termed H'. However, H' is always smaller than H1. If the amplitude of H1 is made to increase by increasing the intensity of the conditioning stimulus, there will be first an increase in the amplitude of H', which will be followed by a progressive decrease when the intensity reaches a certain level. This decrease is due to the effects of Renshaw cell activation on the motoneuron pool involved in the generation of H1. The smaller the H' with respect to H1, the larger the inhibition (46). The beauty of such a method actually lies in the fact that a similar chain of events can be considered partly responsible for silent periods and rebounds induced by stimuli applied to mixed nerves when there is traffic of impulses during voluntary contraction. Renshaw cells inhibit the homonymous motoneuron, other neighbouring motoneurons, and also the Ia inhibitory interneurons projecting to the antagonist muscle (47).

Presynaptic inhibition is mediated by an axo-axonal synapse on Ia afferent terminals themselves. The interneurons mediating such inhibition are known as '*primary afferent depolarization*' neurons. In contrast to the other spinal mechanisms of motor control discussed so far, presynaptic inhibition takes place at a presynaptic level, therefore, leaving the motoneuron uninfluenced. Methods for electrophysiological testing of presynaptic inhibition involve activation of the Ia afferents of the same muscle by means of vibration (46,48). Vibratory stimulation at the Achilles tendon depolarizes Ia afferent fibres, which cause inhibition of the soleus H reflex and, at the same time, tonic activation of motoneurons (the tonic vibration reflex). Vibration-induced presynaptic inhibition may be overcome by muscle contraction, as an example of the control of supraspinal structures on spinal interneurons. However, the exact source of the control is largely unknown. As described in a previous paragraph, presynaptic inhibition is an important strategy of the central nervous system for control of reflexes.

H reflex excitability changes with movement

Various manoeuvres can cause modulation of the soleus H reflex in humans, such as the Jendrassik's manoeuvre. Remote contraction of a muscle induces facilitation of both the tendon jerk and the H reflex (49–51). The temporal profile of the Jendrassik's manoeuvre was accurately studied by Delwaide and Toulouse (49). These authors divided the facilitation in three different phases, the first one due to a general preparatory state, before any EMG activity could be seen, the second one related to the descending motor command, that would activate not only the muscle in which the voluntary contraction was intended but also other muscles in the body, and the third due to a long-loop reflex generated in the re-afferentation during contraction. According to Zher and Stein (51), the increase in the size of the H reflex related to a remote muscle contraction is, in part due to removal of presynaptic inhibition.

Many studies have examined the physiology of the H reflex changes preceding a voluntary movement in reaction time experiments. In these experiments, an imperative (IS). or 'go' signal gives the subject the instruction to begin moving and in some paradigms, the IS is preceded by a warning signal that announces the incoming instruction. Some time after the IS, the size of the H reflex increases progressively until onset of EMG activity in the agonist muscle, and decreases in the antagonist (52–54). The H reflex enhancement prior to a reaction is related to the motor commands issued for the actual movement execution, rather than to any sensory feedback signals. Postural changes also lead to soleus H reflex modulation (55–57). Koceja et al. (55), reported a decrease in the H reflex size in standing compared to sitting, even though the soleus muscle exhibits tonic EMG activity while standing. This is probably the consequence of an increase in the excitability of the soleus motoneurons, with a concomitant increase in the amount of presynaptic inhibition of the Ia afferent terminals. The size of the soleus H reflex is also modulated during walking, with facilitation in the stance phase and inhibition in the swing phase (58–60).

H reflex excitability changes after remote external stimuli

External activation of the reticulospinal tract by auditory stimuli has been reported to cause enhancement of the soleus H reflex by several authors. Rossignol and Melvill-Jones (61). examined the sound-related changes of the H reflex size in volunteers who were hopping to the rhythm of a musical instrument. These authors reported the synchronization of hopping with the strong beats of the musical piece, in such a way that the motor events were timed to make best use of the auditory facilitation of the segmental reflex. Delwaide and Schepens confirmed the finding of Rossignol and Melvill-Jones (61), and reported that such facilitation was abnormal in patients with reticulospinal tract dysfunction (62). Finally, activation of the corticospinal tract with cortical stimulation causes also modulation of the soleus H reflex. Using electrical stimuli, Cowan et al. (11). reported a short latency, short duration facilitation of the H reflex in most muscles, except the soleus, where the H reflex was inhibited. This was attributed to the fact that corticospinal projections to the soleus muscle are mainly inhibitory. However, using transcranial magnetic stimulation (TMS), facilitation of the soleus H reflex has been reported at longer intervals (57): the first phase, occurring between 5 and 30 ms, was assumed to represent the facilitatory effects of a soleus motoneuronal EPSP, and the second occurring between 50 and 100 ms, which may have a reflex origin from reafferentation after activation of other muscles (63).

Electrical or mechanical stimuli applied anywhere in the body are also capable of inducing changes in the size of the soleus H reflex. Gassel and Ott (64) showed that cutaneous stimuli of the dorsal and plantar surfaces of the distal foot in humans induced opposite effects on the soleus H reflex, with short latency facilitation with stimulation of the dorsum, and inhibition with plantar stimulation.

These effects were followed by a prominent late increase of excitability with both stimuli.

Clinical applications of the study of the H reflex

The soleus H reflex can be absent or have an abnormally prolonged latency in 100% of patients with S1 radiculopathy (65–67), and in only up to 26% of patients with L5 radiculopathy (68,69). In patients with neurogenic claudication, conventional electrophysiological tests may not show positive signs at rest, but these may appear after exercise such as walking for 30 min (70). The H reflex may be absent in polyneuropathies affecting large fibres, suggesting conduction block in, or functional loss of, large afferent axons. Different types of polyneuropathies may lead to a different type of abnormality in the H reflex. Hence, a delayed but still present H reflex might be consistent with predominantly demyelinating polyneuropathy, while absence of response or abnormally high threshold would be consistent with a predominantly axonal polyneuropathy (71,72). In Friedreich's ataxia, the involvement of dorsal root ganglia should cause the sensory nerve action potential amplitude to decrease and the H reflex to be absent (73). However, some patients with early onset cerebellar ataxia may have completely normal reflexes even though they show a noticeable sensory loss and absence of response in sensory nerve conduction studies (74). This clinical condition, known as Friedreich's ataxia with retained reflexes (FARR), has a slightly better prognosis than the classical form. The H reflex may be helpful for the assessment of central nervous system disorders such as dystonia and spasticity.

Oligo- and polysynaptic reflexes elicitable at rest

Any electrical stimulus applied over a nerve trunk usually gives rise to a large afferent volley. The effects of that volley in the central nervous system may not be apparent in the majority of nerves when the muscle is at rest. Noteworthy exceptions in healthy human subjects are the orbicularis oculi reflex, the nociceptive withdrawal reflexes, and a few others. Most of these reflexes are known as exteroceptive reflexes.

The blink reflex

Physiological and technical considerations

The blink reflex consists on a stimulus-triggered eyelid closure involving mainly activation of the orbicularis oculi muscle and relaxation of the levator palpebrae muscle. The most commonly used stimulus in clinical practice is an electrical stimulus to the supraorbital nerve (75,76). An appropriate stimulus duration is 0.2 ms, with the stimulus intensity two to four times perception threshold. The reflex response of the orbicularis oculi to supraorbital nerve electrical stimulation consists of two separate components—an early ipsilateral R_1 and a later bilateral R_2 response. R_1 is a pontine reflex, while R_2 is presumably relayed through a more complex route including the pons and lateral medulla (77,78). A mechanical tap over the glabella also elicits a blink reflex. Although the stimulus is a gentle tap, this is a cutaneous, rather than a stretch reflex, probably relayed via the same polysynaptic reflex pathways as the electrically elicited blink reflex (76). The responses are recorded with surface electrodes, the active one placed on the lower eyelid and the reference 2–3 cm lateral. Careful placement of the electrodes may help reduce the stimulus artefact that may sometimes cause difficulties in determining the exact onset latency of R1. A low-frequency cut-off of 200 Hz or more may also be helpful. Although it is recommendable that reference values are obtained at each laboratory for their own comparison with patients, normative values have been published by several authors (see, for instance, ref (79)). The upper limit for R_1 latency is 13.0 ms, and the latency difference between the two sides in the same subject must be less than 1.2 ms. The upper limit of normal latency for the R_2 is 40 ms on the side of stimulation and 41 ms on the contralateral side, and the latency difference between the ipsilateral and the contralateral R_2 evoked simultaneously by stimulation on one side should not exceed 5 ms.

Methods of clinical interest for the study of peripheral nerve disorders with the blink reflex

The R_1 response is relatively stable with repeated trials and is therefore better suited for assessing nerve conduction through the trigeminal and facial nerves. Analysis of R_2, however, is essential in determining whether a lesion involves the afferent or efferent arc of the reflex. With a lesion of the trigeminal nerve, R_2 is slowed or diminished bilaterally when the affected side of the face is stimulated (afferent delay), while stimulation of the unimpaired nerve gives rise to normal responses. With a lesion of the facial nerve, R_2 is abnormal on the affected side (efferent delay), while the responses are normal in the unaffected side, regardless of the side of stimulation (Fig. 10.6). Focal brainstem lesions cause abnormalities of the various components of the blink reflex, as well as of other brainstem reflexes. The relative topodiagnostic value of the neurophysiological study of brainstem reflexes in relation to the brainstem magnetic resonance has been examined by Cruccu et al. (80).

Withdrawal reflexes to nociceptive inputs

Physiological and technical considerations

Fibres of small diameter are not normally accessible to direct electrophysiological tests using conventional methods. The most suitable methods for the study of small fibres are described in Table 10.3. A short train of electrical impulses applied to the sole of the foot, or to the posterior tibial or sural nerves at the ankle causes a reflex response leading to the withdrawal of the leg from the painful stimulus (nociceptive reflex). The response recorded in biceps femoris and tibialis anterior is made of two components, the RII and the RIII (an RI is supposed to exist, but it is not consistent enough in healthy subjects). The RII component is usually obtained at a relatively low, non-painful, stimulus intensity. The RIII is obtained at a high intensity stimulation and is recruited in parallel with an increasingly painful sensation. When recording in the short head of the biceps femoris, the first muscle to be activated after a painful stimulation of the foot, response latency is 50–70 ms for the RII, and 80–130 ms for the RIII (81,82). The RII response is probably due to activation of large group A afferents, while the RIII response should be due to activation of A delta and C fibres. Withdrawal reflex responses have been also recognized in the upper limb to noxious cutaneous stimuli (83).

Reflex responses to nociceptive stimuli can also be generated in pelvic floor and facial muscles. The bulbocavernous reflex (84) is usually obtained by applying electrical stimuli to the dorsal nerve of the penis or clitoris. Recording can be done through needle electrodes inserted in the bulbocavernosus muscle or in the

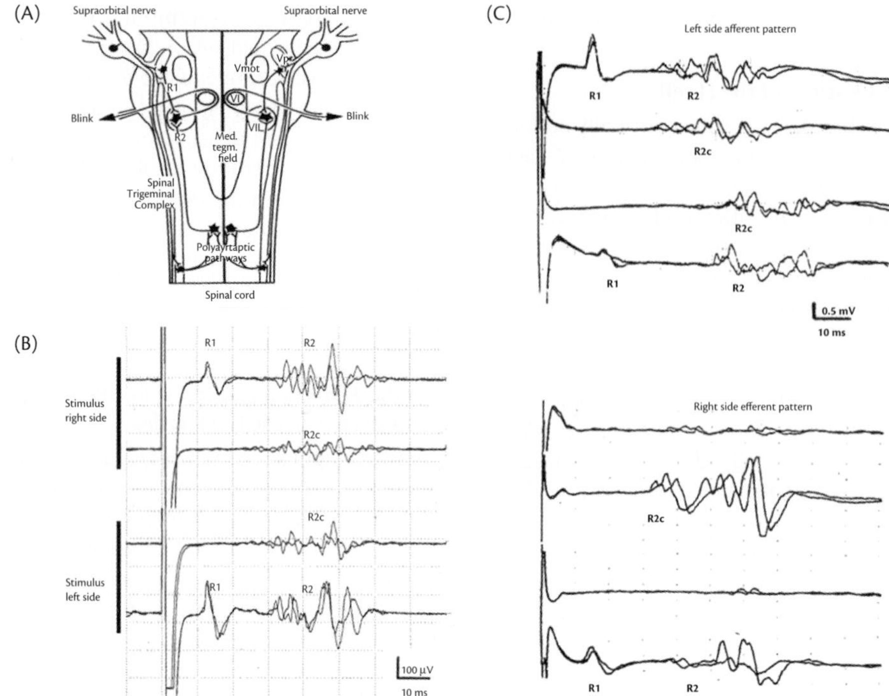

Fig. 10.6 The blink reflex. (A) The schematic pathway followed by R1 and R2 responses to supraorbital nerve stimulation. The R1 is integrated at pontine level and leads to ipsilateral response, while the circuit for the R2 and R2c responses reaches caudal to the medulla and leads to bilateral responses. (B) Typical responses from a healthy subject. Stimuli to the right supraorbital nerve give rise to right-side R1, and bilateral R2 and R2c. Stimuli to the left supraorbital nerve give rise to left side R1 and bilateral R2 and R2c. (C) Abnormal blink reflexes due to lesions in the trigeminal nerve in the upper set of graphs (i.e. an afferent pattern), and of the facial nerve in the lower set of graphs (an efferent pattern).

anal or vesical sphincter muscles. It is also possible to record the response from the anal sphincter using surface electrodes. Reflex responses of facial muscles are obtained with relatively high intensity electrical stimuli to the median or tibial nerves. The clinical correlate of these responses is the palmomental reflex (85). The reflex circuit of the facial responses to peripheral nerve stimulation is not known.

Table 10.3 Methods for neurophysiological evaluation of activity in pain and small fibre systems

Sensory pathways	
Standardized sensory testing	Touch and pain sensations
Vibrometer	Vibratory sensation
Thermotest	Warm, cold, heat nociception, and cold nociception sensations
Algometer	Pressure and pressure nociception sensation
Laser stimulation	Pain sensation
Reflex responses	
Painful electrical stimulation	Reflex withdrawal motor responses
Sudomotor function	Sudomotor skin response. Post-ganglionic axonal reflexes
Blood pressure	Baroreflex functions
Heart rate	Cardiovagal functions

Methods of clinical interest for the study of peripheral nerve disorders with nociceptive reflexes

The RIII response has been utilized in studies on efficacy of analgesic drugs, but it is not applied to clinical evaluation of patients. Nevertheless, the withdrawal reflexes to nociceptive stimuli may be a valuable tool to examine the function of small fibres. The bulbocavernosus reflex is abnormally large and has a shorter latency in patients with neurogenic bladder due to upper motor neuron disease, and is absent or has a prolonged latency in patients with cauda equina or conus medullaris lesions (86). The analysis of facial responses to peripheral nerve stimuli may help in the assessment of several peripheral and central disorders (85,87).

Long latency reflexes as a result of modulation of sustained electromygraph activity

Long latency reflex responses are of very small size and are only elicitable during muscle contraction. The eliciting of long latency reflexes during a sustained contraction allows the recording of both excitatory and inhibitory events, since these would appear as an increase or a decrease of the amount of rectified EMG activity with respect to the background level of activity recorded during the same muscle contraction, but without the interference of the stimuli. Therefore, superimposition or averaging of several traces may be required to demonstrate and quantify excitatory and inhibitory reflex responses. The technique of performing this type of examination using an electronically rectified electromyographic

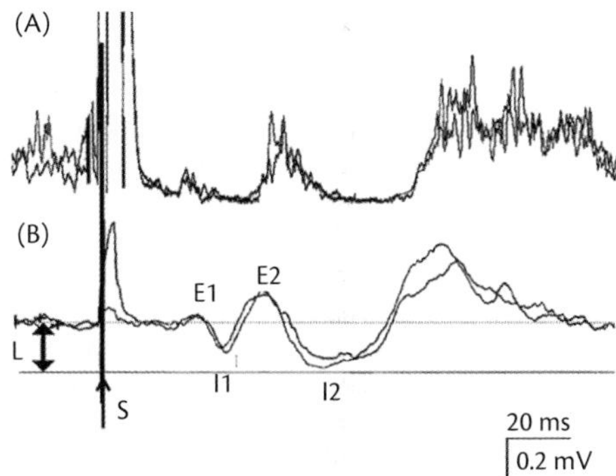

Fig. 10.7 Exteroceptive reflexes elicited in hand muscles during sustained voluntary contraction. Recordings are obtained after averaging 100 consecutive stimuli two times for each graph. The two resulting traces are superimposed to show consistency of the response. (A) The effects induced on thenar muscles by stimulation of mixed afferents from the median nerve at the wrist, while the graph in (B) shows the effects induced on the 1st interosseous by stimulation of cutaneous afferents from the median nerve at the first finger. Traces in (B) have been smoothed. Note the similarity of the effects in both conditions.
S = stimulus; L = level of activity (averaged rectified EMG activity); E1 = first excitatory phase; I1 = first inhibitory phase; E2 = second excitatory phase; I2 = second inhibitory phase.

recording is sometimes referred as to the 'modulation of electromyographic activity'. Using this method, facilitatory and inhibitory reflex phenomena can be quantified with respect to the level of background activity. Methods to obtain reliably the long latency reflexes of hand muscles without sophisticated equipment have been well standardized (88). These involve median nerve or radial superficial nerve stimulation, while recording from thenar muscles during maintenance of a contraction of the opponens pollicis (Fig. 10.7).

The reflex responses recorded in this way are termed long-latency reflex I-III (LLR I-III), following median nerve stimulation, and cLLR I-III following radial superficial nerve stimulation (87). Hand muscle reflexes are mainly used to diagnose alterations within the central nervous system like in myoclonus, multiple sclerosis, Parkinsonian syndromes and choreiform syndromes (88–90). Stimulation of cutaneous nerves gives rise to another type of long latency reflex, the cutaneo-muscular reflex (91,92). Excitatory effects are labelled E waves, and inhibitory effects are labelled I waves. The E and I waves are numbered according to their order of appearance (E1, I1, E2, etc.). Meinck et al. (93) reported on the utility of the cutaneo-muscular reflexes in the assessment of disorders causing changes in the excitability of motoneurons. Abnormalities of leg cutaneo-muscular reflexes have been reported in four patients with the stiff-leg syndrome (94), in whom cutaneous electrical stimuli induced long-lasting spasms.

References

1. Gitter, A.J. and Stolov, W.C. (1995). AAEM minimonograph 16. Instrumentation and measurement in electrodiagnostic medicine. Part I. *Muscle & Nerve*, **18**, 799–811.
2. Gitter, A.J. and Stolov, W.C. (1995). AAEM minimonograph 16. Instrumentation and measurement in electrodiagnostic medicine. Part II. *Muscle & Nerve*, **18**, 812–24.
3. Bischoff, C., Fuglsang-Frederiksen, A., Vendelbo, L., and Sumner, A. (1999). Standards of instrumentation of EMG. *Electroencephalography & Clinical Neurophysiology*, **52**(Suppl.), 199–211.
4. Deuschl, G. and Eisen, A. (1999). Guidelines of the International Federation of Clinical Neurophysiology. *Electroencephalography & Clinical Neurophysiology*, **52**(Suppl.), 263–268.
5. Burke, D., McKeon, B., and Skuse, N.F. (1981). The irrelevance of fusimotor activity to the Achilles tendon jerk of relaxed humans. *Annals of Neurology*, **10**, 547–50.
6. Kuruoglu, H.R. and Oh, S.J. (1994). Tendon-reflex testing in chronic demyelinating polyneuropathy. *Muscle & Nerve*, **17**, 145–50.
7. Sandler, S.G., Tobin, W., and Henderson, E.S. (1969). Vincristine induced neuropathy, a clinical study of fifty leukemic patients. *Neurology*, **19**, 367–74.
8. Hopf, H.C., Thömke, F., and Guttmann, L. (1991). Midbrain vs pontine medial longitudinal fasciculus lesions, the utilization of masseter and blink reflexes. *Muscle & Nerve*, **14**, 326–30.
9. Hopf, H.C. (1994). Topodiagnostic value of brainstem reflexes. *Muscle & Nerve*, **17**, 475–84.
10. Valls-Solé, J., Graus, F., Font, J., Pou, A., and Tolosa, E. (1990). Normal proprioceptive trigeminal afferents in patients with pure sensory neuronopathy and Sjogren's syndrome. *Annals of Neurology*, **28**, 786–90.
11. Cowan, J.M.A., Day, B.L., Marsden, C.D., and Rothwell, J.C. (1986). The effect of percutaneous motor cortex stimulation on H reflexes in muscles of the arm and leg in intact man. *Journal of Physiology*, **377**, 333–47.
12. Burke, D., Adams, R.W., and Skuse, N.F. (1989). The effect of voluntary contraction on the H reflex of various muscles. *Brain*, **112**, 417–33.
13. Panizza, M., Nilsson, J., and Hallett, M. (1989). Optimal stimulus duration for the H reflex. *Muscle & Nerve*, **12**, 576–9.
14. Hugon, M. (1973). Methodology of the Hoffmann reflex in man. In: J. E. Desmedt (Ed.) *New developments in electromyography and clinical neurophysiology*, Vol. 3, pp. 277–93. Basel: Karger.
15. Jabre, J. (1981). Surface recording of the H reflex of the flexor carpi radialis. *Muscle & Nerve*, **4**, 435–8.
16. Ashby, P. and Labelle, K. (1977). Effects of extensor and flexor group I afferent volleys on the excitability of individual soleus motoneurones in man. *Journal of Neurology, Neurosurgery & Psychiatry*, **40**, 910–19.
17. Burke, D., Gandevia, S.C., and McKeon, B. (1984). Monosynaptic and oligosynaptic contributions to the human ankle jerk and H reflex. *Journal of Neurophysiology*, **52**, 435–48.
18. Burke, D., McKeon, B., and Skuse, N.F. (1981). Dependence of the Achilles tendon reflex on the excitability of spinal reflex pathways. *Annals of Neurology*, **10**, 551–6.
19. Pierrot-Deseilligny, E.. Morin, C., Bergego, C., and Tankov, N. (1981). Pattern of group I fibre projections from ankle flexor and extensor muscles in man. *Experimental Brain Research*, **42**, 337–50.
20. Burke, D. (1985). Mechanisms underlying the tendon jerk and H reflex. In P. J. Delwaide and R. R. Young (Eds) *Clinical neurophysiology in spasticity*, pp. 55–62. Amsterdam: Elsevier.
21. Frank, K. and Fuortes, M.G.F. (1957). Presynaptic and postsynaptic inhibition of monosynaptic reflexes. *Federal Proceedings*, **16**, 39–40.
22. Hultborn, H., Meunier, S., Morin, C., and Pierrot-Deseilligny, E. (1987). Assessing changes in presynaptic inhibition of Ia fibers, a study in man and the cat. *Journal of Physiology*, **389**, 729–56.
23. Rudomin, P. (1990). Presynaptic inhibition of muscle spindle and tendon organ afferents in the mammalian spinal cord. *Trends in Neurosciences*, **13**, 499–505.
24. Stein, R.B. (1995). Presynaptic inhibition in humans. *Progress in Neurobiology*, **47**, 533–44.
25. Rymer, W.Z., Houk, J.C., and Crago, P.E. (1979). Mechanisms of the clasp-knife reflex studied in an animal model. *Experimental Brain Research*, **37**, 93–113.
26. Valls-Solé, J., Alvarez, R., and Tolosa, E. (1994). Vibration-induced presynaptic inhibition of the soleus H reflex is temporarily reduced by cortical magnetic stimulation in human subjects. *Neuroscience Letters*, **170**, 149–52.

27. Magladery, J.W. and McDougal, D.B., Jr (1950). Electrophysiological studies of nerve and reflex activity in normal man. I. Identification of certain reflexes in the electromyogram and the conduction velocity of peripheral nerve fibers. *Bulletin of Johns Hopkins Hospital*, **86**, 265–90.
28. Gottlieb, G.L. and Agarwal, G.C. (1976). Extinction of the Hoffmann reflex by antidromic conduction. *Electroencephalography & Clinical Neurophysiology*, **41**, 19–24.
29. Taborikova, H. and Sax, D.S. (1969). Conditioning of H-reflexes by a preceding subthreshold H reflex stimulus. *Brain*, **92**, 203–12.
30. Crone, C. and Nielsen, J. (1989). Methodological implications of the post activation depression of the soleus H-reflex in man. *Experimental Brain Research*, **78**, 28–32.
31. Pierrot-Deseilligny E. and Mazières, L. (1984a). Circuits réflexes de la moelle epinière chez l'homme. Part I. *Reviews in Neurology*, **140**, 605–14.
32. Pierrot-Deseilligny, E. and Mazières, L. (1984b). Circuits réflexes de la moelle epinière chez l'homme. Part II. *Reviews in Neurology*, **140**, 681–94.
33. Pierrot-Deseilligny, E. and Burke D. (2012). *The circuitry of the human spinal cord: its role in motor control and movement disorders.* Cambridge: Cambridge University Press.
34. Mizuno, Y., Tanaka, R., and Yanagisawa, N. (1971). Reciprocal group I inhibition of triceps surae motoneurons in man. *Journal of Neurophysiology*, **34**, 1010–17.
35. Day, B.L., Marsden, C.D., Obeso, J.A., and Rothwell, J.C. (1984). Reciprocal inhibition between the muscles of the human forearm. *Journal of Physiology*, **349**, 519–34.
36. Rothwell, J.C., Day, B.L., Obeso, J.A., Berardelli, A., and Marsden, C.D. (1988). Reciprocal inhibition between muscles of the human forearm in normal subjects and in patients with idiopathic torsion dystonia. *Advances in Neurology*, **50**, 133–40.
37. Panizza, M.E., Hallett, M., and Nilsson, J. (1989). Reciprocal inhibition in patients with hand cramps. *Neurology*, **39**, 85–9.
38. Nakashima, K., Rothwell, J.C., Day, B.L., Thompson, P.D., Shannon, K., and Marsden, C.D. (1989). Reciprocal inhibition between forearm muscles in patients with writer's cramp and other occupational cramps, symptomatic hemidystonia and hemiparesis due to stroke. *Brain*, **112**, 681–97.
39. Deuschl, G., Seifert, C., Heinen, F., Illert, M., and Lucking, C.H. (1992). Reciprocal inhibition of forearm flexor muscles in spasmodic torticollis. *Journal of Neurological Sciences*, **113**, 85–90.
40. Fuhr, P. and Hallett, M. (1993). Reciprocal inhibition of the H reflex in the forearm, methodological aspects. *Electroencephalography & Clinical Neurophysiology*, **89**, 319–27.
41. Pierrot-Deseillgny, E., Katz, R., and Morin, C. (1979). Evidence for Ib inhibition in human subjects. *Brain Research*, **166**, 176–9.
42. Takakusaki, K., Shimoda, N., Matsuyama, K., and Mori, S. (1994). Discharge properties of medullary reticulospinal neurons during postural changes induced by intrapontine injections of carbachol, atropine and serotonin, and their functional linkages to hindlimb motoneurons in cats. *Experimental Brain Research*, **99**, 361–74.
43. Delwaide, P.J., Pepin, J.L., DePasqua, V., and Maertens de Noordhout, A. (2000). Projections from basal ganglia to tegmentum, a subcortical route for explaining the pathophysiology of Parkinson's disease signs? *Journal of Neurology*, **247**(Suppl. 2), 75–81.
44. Renshaw, B. (1941). Influence of discharge of motoneuron pool upon excitation of neighbouring motoneurons. *Journal of Neurophysiology*, **4**, 167–83.
45. Pierrot-Deseilligny, E. and Bussel, B. (1975). Evidence for recurrent inhibition by motoneurons in human subjects. *Brain Research*, **88**, 105–8.
46. Pierrot-Deseilligny, E. and Mazières, L. (1985). Spinal mechanisms underlying spasticity. In: P. J. Delwaide and R. R. Young (Eds), *Clinical neurophysiology in spasticity*, pp. 63–76. Amsterdam: Elsevier.
47. Houk, J.C., Singer, J.J., and Goldman, M.R. (1970). An evaluation of length and force feedback to soleus muscles of decerebrate cats. *Journal of Neurophysiology*, **33**, 784–811.
48. Wong, P.K.H., Verriere, M., and Ashby, P. (1977). The effect of vibration on the F wave in normal man. *Electromyography & Clinical Neurophysiology*, **17**, 319–29.
49. Delwaide, P.J. and Toulouse, P. (1981). Facilitation of monosynaptic reflexes by voluntary contraction of muscle in remote parts of the body. Mechanisms involved in the Jendrassik manoeuvre. *Brain*, **104**, 701–9.
50. Miyahara, T., Hagiya, N., Ohyama, T., and Nakamura, Y. (1996). Modulation of human soleus H reflex in association with voluntary clenching of the teeth. *Journal of Neurophysiology*, **76**, 2033–41.
51. Zher, E.P. and Stein, R.B. (1999). Interaction of the Jendrassik maneuver with segmental presynaptic inhibition. *Experimental Brain Research*, **124**, 474–80.
52. Michie, P.T., Clarke, A.M., Sinden, J., and Glue, L.C.T. (1976). Reaction time and spinal excitability in a simple reaction time task. *Physiology & Behavior*, **16**, 311–15.
53. Eichenberger, A. and Rüegg, D.G. (1983). Facilitation of the H reflex in a simple and choice reaction time situation. *Experimental Brain Research*, **7**, 130–4.
54. Schieppati, M. (1987). The Hoffmann reflex, a means of assessing spinal reflex excitability and its descending control in man. *Progress in Neurobiology*, **28**, 345–76.
55. Koceja, D.M., Trimble, M.H., and Earles, D.R. (1993). Inhibition of the soleus H reflex in standing man. *Brain Research*, **629**, 155–8.
56. Trimble, M.H. (1998). Postural modulation of the segmental reflex, effect of body tilt and postural sway. *International Journal Neurosciences*, **95**, 85–100.
57. Goulart, F., Valls-Solé, J., and Alvarez, R. (2000). Posture-related changes of soleus H-reflex excitability. *Muscle & Nerve*, **23**, 925–32.
58. Dietz, V., Schmidtbleicher, D., and Noth, J. (1979). Neuronal mechanisms of human locomotion. *Journal of Neurophysics*, **42**, 1212–22.
59. Faist, M., Dietz, V., and Pierrot-Deseilligny, E. (1996). Modulation, probably presynaptic in origin, of monosynaptic Ia excitation during human gait. *Experimental Brain Research*, **109**, 441–9.
60. Capaday, C. and Stein, R.B. (1986). Amplitude modulation of the soleus H reflex in the human during walking and standing. *Journal of Neuroscience*, **6**, 1308–13.
61. Rossignol, W. and Melvill-Jones, G. (1976). Audio-spinal influence in man studied by the H-reflex and its possible role on rhythmic movements synchronized to sound. *Electroencephalography & Clinical Neurophysiology*, **41**, 83–92.
62. Delwaide, P.J., Pepin, J.L., and Maertens de Noordhout, A. (1993). The audiospinal reaction in parkinsonian patients reflects functional changes in reticular nuclei. *Annals of Neurology*, **33**, 63–9.
63. Guzmán-López, J., Costa, J., Selvi, A., Barraza, G., Casanova-Molla, J., and Valls-Solé, J. (2012). The effects of transcranial magnetic stimulation on vibratory-induced presynaptic inhibition of the soleus H reflex. *Experimental Brain Research*, **220**, 223–30.
64. Gassel, M.M. and Ott, K.H. (1970). Local sign and late effects on motoneuron excitability of cutaneous stimulation in man. *Brain*, **93**, 95–106.
65. Braddom, R.I. and Johnson, E.W. (1974). Standardization of H reflex and diagnostic use in S1 radiculopathy. *Archives of Physical Medicine and Rehabilitation*, **55**, 161–6.
66. Aiello, I., Rosati, G., Serra, G., and Manca, M. (1981). The diagnostic value of H index in S1 root compression. *Journal of Neurology, Neurosurgery & Psychiatry*, **44**, 171–2.
67. Sabbahi, M.A. and Kahlil, M. (1990). Segmental H reflex studies in upper and lower limbs in patients with radiculopathy. *Archives of Physical Medicine and Rehabilitation*, **71**, 223–7.
68. Schuchmann, J.A. (1978). H-reflex latency in radiculopathy. *Archives of Physical Medicine and Rehabilitation*, **59**, 185–7.
69. Rico, R.E. and Jonkman, E.J. (1982). Measurement of the Achilles tendon reflex for the diagnosis of lumbosacral root compression syndromes. *Journal of Neurology, Neurosurgery & Psychiatry*, **45**, 791–5.

70. Pastor, P. and Valls-Solé, J. (1998). Recruitment curve of the soleus H reflex in patients with neurogenic claudication. *Muscle & Nerve*, **21**, 985–90.
71. Guiheneuc, P. and Bathien, N. (1976). Two patterns of results in polyneuropathies investigated with the H reflex. Correlation between proximal and distal conduction velocities. *Journal of Neurological Sciences*, **30**, 83–94.
72. Albers, J.W. (1993). Clinical neurophysiology of generalized polyneuropathy. *Journal of Clinical Neurophysiology*, **10**, 149–66.
73. Hughes, J.T., Brownell, B., and Hewer, R.L. (1968). The peripheral sensory pathways in Friedreich's ataxia. *Brain*, **91**, 803–18.
74. Harding, A.E. (1981). Early onset cerebellar ataxia with retained tendon reflexes, a clinical and genetic study of a disorder distinct from Friedreich's ataxia. *Journal of Neurology, Neurosurgery, and Psychiatry*, **44**, 503–8.
75. Kimura, J., Powers, J.M., and Van Allen, M.W. (1969). Reflex response of orbicularis oculi muscle to supraorbital nerve stimulation: Study in normal subjects and in peripheral facial paresis. *Archives of Neurology*, **21**,193–9.
76. Shahani, B.T. (1970). The human blink reflex. *Journal of Neurology, Neurosurgery, & Psychiatry*, **33**, 792–800.
77. Kimura, J. and Lyon, L.W. (1972). Orbicularis oculi reflex in the Wallenberg syndrome: alteration of the late reflex by lesions of the spinal tract and nucleus of the trigeminal nerve. *Journal of Neurology, Neurosurgery, & Psychiatry*, **35**, 228–33.
78. Ongerboer de Visser, B.W. and Kuypers, H.G.J.M. (1978). Late blink reflex changes in lateral medullary lesions: an electrophysiological and neuroanatomical study of Wallenberg's syndrome. *Brain*, **101**, 285–94.
79. Kimura, J. (2001). H, T, masseter, and other reflexes. In: J. Kimura (Ed.). *Electrodiagnosis in diseases of nerve and muscle: principles and practice*, 3rd edn, pp. 466–94. New York, NY: Oxford University Press.
80. Cruccu, G., Iannetti, G.D., Marx, J.J., et al. (2005). Brainstem reflex circuits revisited. *Brain*, **128**, 386–94.
81. Willer, J.C. (1983). Nociceptive flexion reflexes as a tool for pain research in man. In: J. E. Desmedt (Ed.), *Motor control mechanisms in health and disease*, pp. 809–27. New York, NY: Raven Press.
82. Sandrini, G., Serrao, M., Rossi, P., Romaniello, A., Cruccu, G., and Willer, J.C. (2005). The lower limb flexion reflex in humans. *Progress in Neurobiology*, 77, 353–95.
83. Floeter, M.K., Valls-Solé, J., Toro, C., Jacobowitz, D., and Hallett, M. (1998). Physiologic studies of spinal inhibitory circuits in patients with stiff-person syndrome. *Neurology*, **51**, 85–93.
84. Vodusek, D.B., Janko, M., and Lokar, J. (1983). Direct and reflex responses in perineal muscles on electrical stimulation. *Journal of Neurology, Neurosurgery, & Psychiatry*, **46**, 67–71.
85. Dehen, H., Bathien, N., and Cambier, J. (1975). The palmo-mental reflex. An electrophysiological study. *European Neurology*, **13**, 395–404.
86. Ertekin, C. and Reel, F. (1976). Bulbocavernosus reflex in normal men and in patients with neurogenic bladder and/or impotence. Journal of Neurological Sciences, **28**, 1–15.
87. Valls-Solé, J., Valldeoriola, F., Tolosa, E., and Martí, M.J. (1997). Distinctive abnormalities of facial reflexes in patients with progressive supranuclear palsy. *Brain*, **120**, 1877–83.
88. Cruccu, G. and Deuschl, G. (2000). The clinical use of brainstem reflexes and hand-muscle reflexes. *Clinical Neurophysiology*, **111**, 371–87.
89. Deuschl, G. and Lucking, C.H. (1990). Physiology and clinical applications of hand muscle reflexes. *Electroencephalography & Clinical Neurophysiology*, **41**(Suppl.), 84–101.
90. Deuschl, G., Strahl, K., Schenck, E., and Lücking, C.H. (1988). The diagnostic significance of long-latency reflexes in multiple sclerosis. *Electroencephalography & Clinical Neurophysiology*, **70**, 56–61.
91. Caccia, M.R., McComas, A.J., Upton, A.R.M., and Blogg, T. (1973). Cutaneous reflexes in small muscles of the hand. *Journal of Neurology, Neurosurgery & Psychiatry*, **36**, 960–7.
92. Jenner, J.R. and Stephens, J.A. (1982). Cutaneous reflex responses and their central nervous pathways studied in man. *Journal of Physiology*, **333**, 405–19.
93. Meinck, H.M., Benecke, R., Küster, S., and Conrad, B. (1983). Cutaneomuscular (flexor). reflex organization in normal man and in patients with motor disorders. In: J. E. Desmedt (Ed.), *Motor control mechanisms in health and disease. Advances in Neurology*, vol **39**, pp. 787–96. New York, NY: Raven Press.
94. Brown, P., Rothwell, J.C., and Marsden, C.D. (1997). The stiff leg syndrome. *Journal of Neurology, Neurosurgery & Psychiatry*, **62**, 31–7.

CHAPTER 11

Electroencephalography

Michalis Koutroumanidis, Dimitrios Sakellariou, and Vasiliki Tsirka

Introduction

Although invented as a tool to explore human psychophysiology (1,2), the electroencephalogram (EEG) was very soon recognized as the principle 'laboratory' investigation for epilepsy, , and has remained so since, as it studies its most relevant functional markers—the ictal and interictal electrical brain signals.

The first part of this chapter discusses the 'basic' technical aspects that are relevant for a clinician interpreting EEGs, recorded with modern, digital technology. The second part examines the role of EEG in the diagnosis and management of patients with epilepsy, and concentrates on the principles and methodology of recording and reporting. The other important uses of video EEG in encephalopathies and coma are described in Chapter 36.

Technical aspects

The EEG records the electrical activity of the brain over time. The electrical signal originates in the extracellular space of the brain.

Brain dipoles, EEG signal, and clinical relevance

The source of the electric activity in the brain can be modelled as dipoles whose orientation (direction) is perpendicular to the cortex with the negative charge at the surface. It follows that the orientation of the dipole in relation to the overlying recording electrodes (which will shape the characteristics of its signal as this eventually appears on our screens) is determined by the actual orientation of the cortical area that generates it.

The EEG electrodes pick up summated electrical signals (dipoles) and the area of the cortex that must be simultaneously activated to produce a distinctive recordable EEG event (for instance a spike) extends over several cm square and consists of gyri and sulci of variable curvature, orientation and depth. This complex concept could be perhaps simplified and understood in its clinical context by assuming that the cortical area responsible for the dipole is flat instead of convoluted. In *lateral (neocortical) temporal lobe epilepsy* a dipole (arising, for instance, from a flat area of the superior temporal gyrus) could be schematically conceived as perpendicular to the overlying electrode surface with maximal negativity over the mid-temporal electrode (Fig. 11.1, left traces); in the example of *childhood epilepsy with centro-temporal spikes* the maximal negativity is again over the same area (as the functional focus is not far in absolute distance), but the orientation of the dipole could be assumed perpendicular to the cortex *within* the Rolandic fissure and therefore almost parallel to the anterior—posterior axis of the brain, producing the classical anterior-posterior dipole (Fig. 11.1, right traces). Indeed, centro-temporal spikes are frequently of maximal amplitude (peak negativity) over the lower central electrodes (C5/C6) positioned halfway between the central (C3/C4) and the mid-temporal (T3/T4) electrodes (3) with tangential orientation of the dipole, whose positivity is recorded over the anterior areas (4).

The relatively large area of cortex that generates a spike and the varying orientation of the generating dipole limit the spatial resolution of EEG, in sharp contrast to its temporal resolution that is high (but see 'Digitization (or analog to digital conversion) and sampling rate').

Electrode placement

To ensure reproducibility of studies and communication between EEG departments/epilepsy units, scalp EEG electrodes are (almost universally) placed at standard positions according to the international 10–20 system (5). This system describes head surface locations (typically 21) as relative distances between stable anatomical points of the skull, namely the nasion, the inion and the pre-auricular position on either side (6). With the advent of multi-channel EEG hardware systems and the development of topographic methods and tomographic signal source localization, higher density electrode systems have been devised as extensions to the original 10/20 system, such as the 10/10 system, with 74 electrodes (7), or even the 10/5 system with up to 320 electrode locations (8).

It is possible that, in the context of presurgical assessment, additional electrodes may provide useful information; for example, adding six electrodes to the 10-20 system in an inferior temporal chain, at the level of the pre-auricular points (F9/F10, T9/T10, and P9/P10 of the 10-10 system) may give a reasonable coverage, particularly of the anterior temporal/inferior frontal areas. The Maudsley system with a lower and more anterior position of the F7/F8 electrodes, which in the 10-20 system are placed over the inferior frontal, rather than the anterior temporal areas, offers an important compromise (9). For most other clinical EEG recordings, the 10-20 is sufficient and devoting extra channels to polygraphy is, in the authors' view, far more rewarding. (For topographic orientation, see explicit graphics at http://www.bci2000.org/wiki/index.php/User_Tutorial:EEG_Measurement_Setup.

EEG channels and amplifier

The 'raw' signal that is picked up by each of these 'active' electrode positions is the voltage difference between the particular electrode

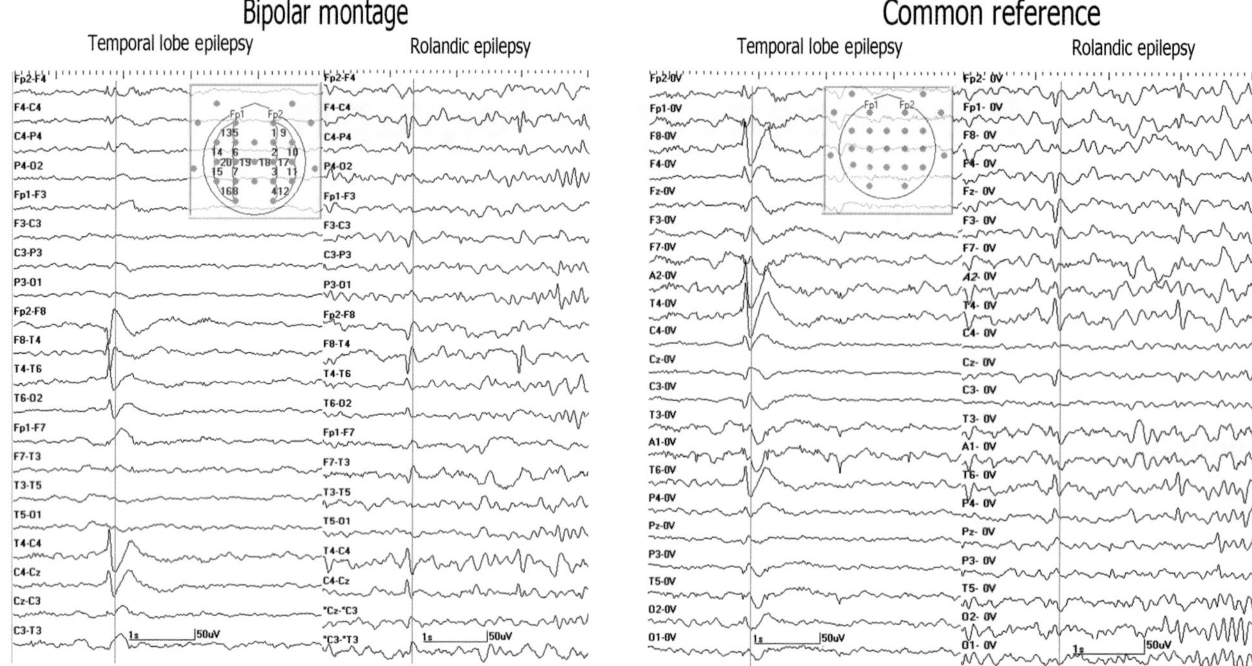

Fig. 11.1 Comparison between the typical mid-temporal focal spike in symptomatic lateral temporal lobe epilepsy (left half of traces) and the also typical rolandic spike in the self-limiting childhood epilepsy with centrotemporal spikes. Note the differences in the spike topography and polarity over the anterior brain areas and the differences between the two montages (bipolar anterior-posterior or 'double banana' and average reference (see text for details).

and a common system reference (CSR), which can be non-cephalic (at the ear A1, A2) separately or conjointly, or a scalp electrode, or a mathematical relationship between electrodes (for instance (A1 + A2)/2 or (C3 + C4)/2), and is specific for a given piece of commercial equipment.

This analogue (raw) voltage signal passes through an isolation (protective) transformer, which ensures that the current flows from the patient to the machine and not the other way round, and via a dedicated channel (one for each active electrode on the head) into its differential amplifier where it is amplified several thousand times. The signal from each channel is the voltage difference between the 'active' electrode that records the signal from the given position, and the 'reference' electrode connected to the CSR. By convention, when voltage difference is negative, the trace deflects upwards; most clinically significant EEG phenomena (i.e. spikes) are negative.

Digitization (or analog to digital conversion) and sampling rate

The amplified voltage signal from each electrode is then 'digitized', i.e. converted into numbers (see also Chapter 5). Sampling rate associates to the temporal representation of an electrical signal by measuring voltage values (samples) at equal time distances (rate). When we refer to the sampling rate, we mean how frequently the analogue to digital converter measures voltage values per unit of time. To define a periodic or quasi-periodic signal, samples must be taken at least twice within its period (Nyquist theorem, Fig. 11.2B); anything less would be inadequate (Fig. 11.2A and C). This is represented by the formula

$$f_s \geq 2 \cdot f_{max}, \quad \text{[eqn 11.1]}$$

where f_s is the sampling rate and f_{max} the highest frequency component for a given analog signal. [eqn 11.1]

Taking the simplified example of a 10-Hz rhythm, the minimum number of samples for meaningful representation should be two within its 100-ms period (1/10 of a second), yielding a sampling rate of 20/s (twice the frequency) with a time-frame of 50 ms (half the period). It follows that for clinical EEG use the *minimum* sampling rate for reasonable representation of biological frequencies up to the high beta range (60–65 Hz) is 128 samples/s. However, for fuller characterization of the composite EEG signal (which is a dynamic mixture of fast and slower frequencies, rather than single periodic frequencies over different brain areas) and better temporal resolution, modern EEG machines can record at a 256-Hz sampling rate or higher.

The commonly used sampling frequencies are 256, 512, 1024, and 2048 Hz. The choice of the sampling rate value depends on the purpose of study—data acquired for research purposes may need to be sampled at high rates, while clinical data are usually sampled at lower rates in an attempt to compromise between clinically useful temporal resolution and minimum possible storage space.

Quantization is the process of making the range of a signal discrete so that it takes on only a discrete, finite set of values (10). The set of input voltage values of the analogue signal is continuous and therefore uncountable. Quantization turns the output values finite or countable infinite. Fig. 11.3 shows a sinusoidal signal quantized where amplitude can take on values between a set of (a) 3 (−1,0,1) (b) 32 (0,1,2…32) amplitude values.

The bigger the set of values (bit depth) of the signal quantization, the higher the precision of the represented amplitude values; to yield the maximum possible information from the acquired signals quantization is fine, usually up to 16 bit depth, resulting in a set of 65 536 values according to the formula:

$$16 \text{ bit} = 2^{16} = 65{,}536 \text{ values} \quad \text{[eqn 11.2]}$$

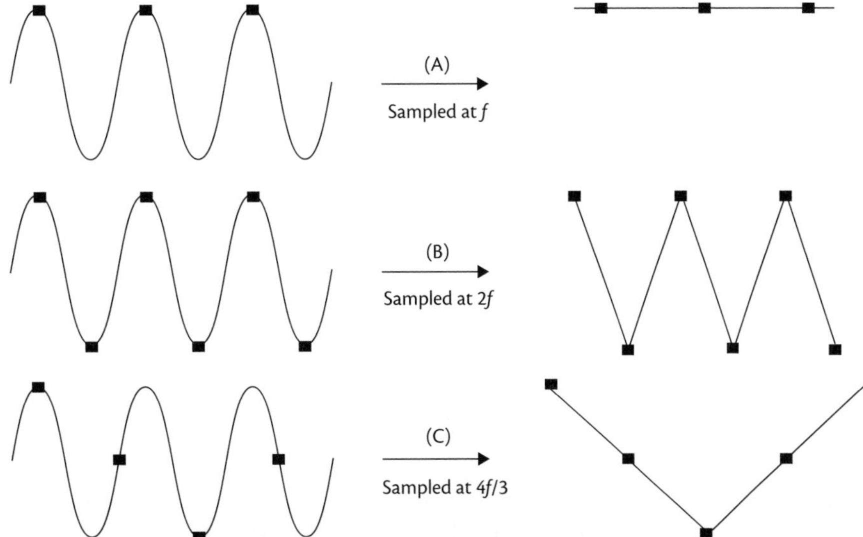

Fig. 11.2 Each dot symbolizes the samples taken per unit of time and f is the frequency of the sinusoid wave shown. The right-hand part of the figure shows the reconstructed signals from different sampling frequencies (f_s) (A) $f_s = f$; (B) $f_s = 2f$; and (C) $f_s = 1.34 \cdot f$). Only the sampling frequency of B is high enough to represent the signal adequately.

Because of the unreasonable increase of *data storage* costs, it is useful to understand that the memory size of a recording depends on the number of electrodes and its duration, the sampling frequency and the quantization bit depth, and can be calculated by the equation below

$$\text{Size}_{\text{recording}}(\text{bits}) = N_{\text{electrodes}} \cdot t_{\text{seconds}} \cdot f_{\text{sampling (Hz)}} \cdot q_{\text{bits}} \quad \text{[eqn 11.3]}$$

where, size is the size of the recorded data in bits, N the number of electrodes, t time in seconds, f the selected sampling rate in Hz and q the bit depth.

For example, 1 h (3600 s) recording from 64-electrode EEG signals sampled at 2000 samples/s, and quantization resolution of 16 bits occupies a memory size of:

$$64 \cdot 360 \text{ sec} \cdot 2000 \frac{1}{\text{sec}} \cdot 16 \text{ bits} \simeq 7.37 \text{ Gbits} \xrightarrow{1\text{Gbit}=0.125\text{Gbyte}} \text{Size} \simeq 0.86 \text{ Gbyte} \quad \text{[eqn 11.4]}$$

Such a recording may be ideal for research purposes, but is untenable for busy clinical EEG departments due to storage limitations and costs; a 24-h (24 × 3600 s) ambulatory recording at f_s = 2 kHz would occupy a memory size of 20.6 Gb, whereas at f_s = 200 Hz would occupy only 2.06 Gb.

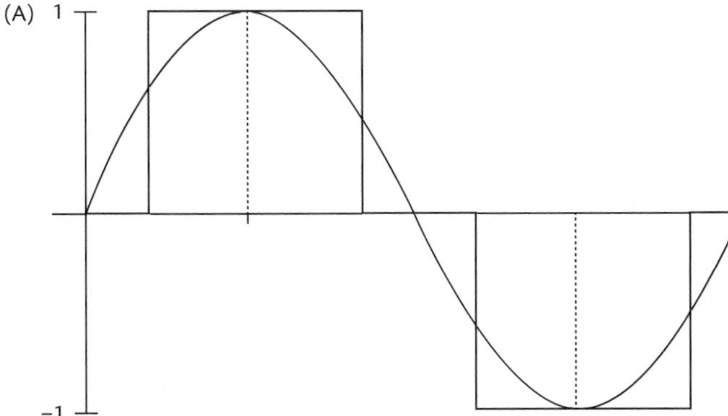

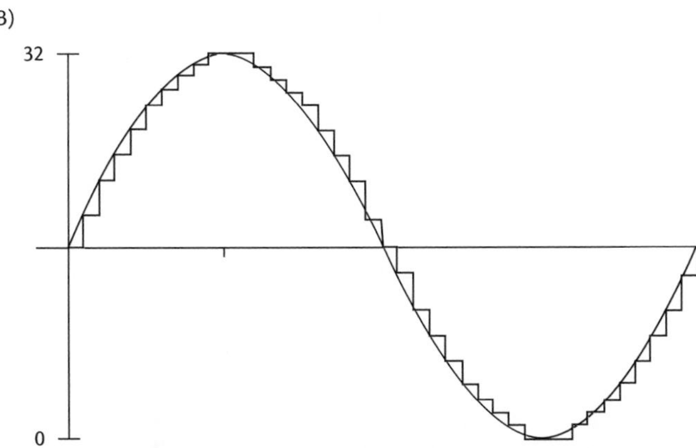

Fig. 11.3 The input analogue sinusoidal wave is shown in black. The output signal (shown in blue) is quantized into a set of three (−1, 0, 1) and thirty-two (0, 1, 2, …, 32) amplitude values for (A) and (B), respectively.

Visualization of the EEG (and video) on the screen

Modern EEG machines allow a number of off-line display modifications of the digital signal for better visualization of the EEG recording on the screen.

Paper-speed

The term *paper-speed* dates from the era of the analogue paper EEG machines and can still describe graphically the time display of the digital EEG on the computer screen. Then and now the paper speed of 30 mm/s is used for most clinical purposes, although larger screens can conveniently accommodate 15 s/page. Faster speeds (fewer seconds of recording displayed on the screen) may help identify details, such as rapid propagation of the signal from one region of the brain to the other, useful in pre-surgical evaluation. Slower paper speeds are used in polysomnography, neonatal EEG, and intensive care unit recordings, as they can better demonstrate EEG patterns and polygraphy variables of long periodicity. Slow speeds can also identify more easily mild background asymmetries or focal slowing, and subtle clinical events against a backdrop of chaotic background, such as infantile spasms in hypsarrhythmia (Fig. 11.4).

Sensitivity

Sensitivity is the ratio of the magnitude of the input electrical potential to the vertical deflection of the graph on the screen (the 'size' of the displayed signal), expressed in μV/mm. Enhancing the signal, for instance, in intensive care unit or in scalp telemetry recordings, may reveal important low voltage rhythms that cannot be reliably identified in the usual 10 μV/mm setting, while its reduction allows full appreciation of high voltage activities (Fig. 11.4).

Filters and their use (and abuse)

High frequency filters (HFF) attenuate components of higher frequency than the set value of the filter while they allow lower frequencies to pass; hence they are also called low pass filters. Low frequency filters (LFF), or high pass filters, do the opposite. It is important to remember that filtering out a component within the EEG range (0.1–70 Hz) also results in loss of EEG data, for instance,

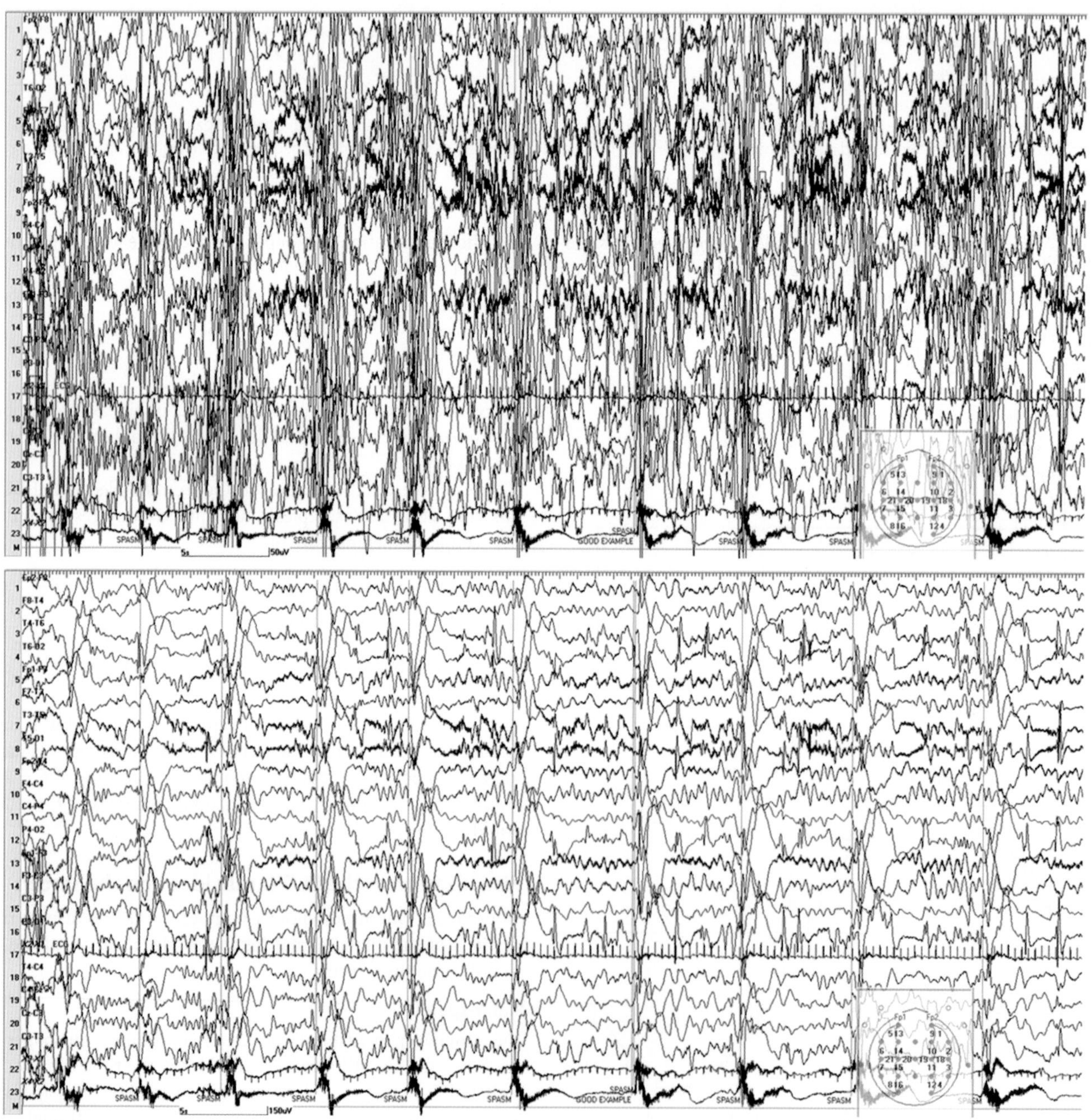

Fig. 11.4 Slow paper speed (upper and lower traces) and reduced sensitivity (lower trace) help appreciate better clusters of seizures, such as mild epileptic spasms against a chaotic high voltage background (hypsarrhythmia). Note the bilateral EMG polygraphy at the bottom of each trace.

clinically important slow activity or low voltage fast polyspike discharges, and that frequency components that are *close* to the filter's cut-off value (but not filtered out) are distorted. The 50 Hz (and 60 Hz in the USA) notch filters that eliminate the electrical noise caused by mains current should be used with extreme caution and only as last resort, as they may 'hide' dysfunctional electrodes.

EEG derivations and montages

Modern digital EEG machines allow re-referencing of the signal obtained using the default common system reference (voltage recorded from each electrode 'minus' voltage of CSR) into four derivations, three monopolar and the bipolar. The monopolar derivations include:

- The *common reference* to a selected, other than the CSR, reference point.
- The *common average reference*, usually with the facility to exclude certain electrodes that record local cerebral or artefactual signal of high voltage that may unevenly distort the mean.
- The *source reference* where localization is by maximal amplitude.

The bipolar derivation displays local voltage differences between adjacent electrodes and localizes by 'phase-reversal' (Table 11.1 and Fig. 11.1). It is useful to emphasize that localization is not always the primary aim of the EEG and that the term 'phase-reversal' is applied to any local phasic event (including the physiological vertex sharp waves that 'phase-reverse' over Cz) and not only to epileptic activities.

The term 'montage' is often used interchangeably with the term derivation and simply denotes the way electrode connections are sequenced and the order they appear on our screen; essentially, it applies to the bipolar derivation and may include a range of longitudinal and transverse arrays. The left halves of the traces in Fig. 11.1 depict four anterior-posterior chains of five electrodes each, connected to four channels, with a transverse array passing through Cz ('double banana'). Use of this montage *and* of a monopolar derivation, such as the common average reference, is sufficient for most clinical purposes. The topic is comprehensively discussed elsewhere (11–13).

EEG in epilepsy

The different diagnostic modalities in the field of the epilepsies afford different views of the same natural phenomenon and are not of higher or lower diagnostic precision or usefulness. As in all diseases, the diagnostic process in epilepsies requires first, description/recognition of the full clinical semiology and second identification of the underlying cause with each diagnostic modality weighting differently between these two levels. The EEG contributes significantly to the clinical, but less so to the aetiological diagnosis.

EEG and clinical diagnosis of epilepsy

Epileptologists and electroencephalographers agree that, apart from confirming the clinical diagnosis of epilepsy, the EEG can contribute to its diagnostic refinement at multiple levels, and by implication to the optimal treatment and management of patients with seizures. It is important to state at the outset that the term 'interictal' scalp EEG is not wholly appropriate, since it does not reflect the overall diagnostic potential across a wide spectrum of epilepsies. Correctly used and activated, the so-called 'interictal' video scalp EEG may record a range of different seizure types (and non-epileptic events) and efficiently contribute to the diagnosis of several epileptic syndromes; examples include typical absences, myoclonic and reflex seizures in adults, and epileptic spasms, typical and atypical absences, tonic–atonic, and myoclonic seizures in children. The use of time-locked video (standard in modern EEG equipment) and polygraphy upgrades the EEG from the most relevant laboratory test for the diagnosis of epilepsy to a decisive extension of the physical examination.

Furthermore, as certain clusters of abnormal features are closely associated with (and sometimes specific to) distinct clinical phenotypes, the EEG has been a decisive tool for the classification of seizures and epilepsies since the very first attempts of the International League Against Epilepsy (ILAE) on classification (14) and the consolidation of the first ILAE classification of 1981/1989 (15), and remains as important today (16). Syndromes may be fewer in adults than in children (as some severe epileptic encephalopathies and inherited disorders lead to early death, and most benign epilepsies remit before the age of 16), but a diagnostic component is always present (and sometimes particularly challenging) because of the dynamic nature of epilepsy, possible earlier missed diagnoses, and of course, the adult-onset epilepsies. The ultimate understanding of the natural history of the various epilepsy syndromes is an additional worthwhile goal.

EEG and aetiological diagnosis of epilepsy

The correlation between EEG and aetiology is relatively poor. A particular distinctive EEG pattern may occur with different

Table 11.1 Relevant advantages and disadvantages of the four derivations

Derivations	Strengths	Weaknesses
Common reference	Good visualization of focal/regional epileptic events (interictal or seizure onset) with good display of background	May be difficult to select the least active reference for each recording; activity such as heartbeat may be seen in all channels
Common average reference	Still good visualization of focal events, essentially obliterating the possibility of an active reference	May be difficult to identify possible faulty electrodes that would affect all channels
Source derivation	Pinpoints discrete foci by suppressing volume conduction and makes a multifocal EEG easier to read	Active areas may look deceptively restricted and regional abnormalities may be underestimated
Bipolar	Good display of inter-hemispheric differences in background rhythms and paroxysmal activities, particularly when the latter are bilateral synchronous or generalized Good localization of focal events. 'Easier' to view, with reasonable localizing capabilities	Solely longitudinal montages may miss or misrepresent transverse dipoles

aetiologies, while epilepsies within a given class of aetiology may have their own individual EEG characteristics.

The EEG is only concerned with aetiology indirectly. In fact, it can only *suggest* an identifiable cause for epilepsy, focal (i.e. a circumscribed brain lesion) or diffuse (i.e. an epileptic encephalopathy), or the absence of such a cause, expressed by the terms 'symptomatic' and 'idiopathic' of the 1989 nomenclature, respectively. As a rule, such suggestion is non-specific, although some EEG patterns may imply that a certain aetiology is more likely than others, introducing a *degree of probability*. for instance, the combination of interictal spikes and slow activity over the temporal lobe in a patient with focal seizures strongly suggests a structural (symptomatic) cause, although it cannot indicate its nature (medial temporal sclerosis, tumour, autoimmune limbic encephalitis, or other). Similar EEG features, however, may occur in patients with non-specific histopathological changes and good outcome after ipsilateral temporal lobectomy (17) and sometimes in familial temporal lobe epilepsy (18). On the other hand, different EEG patterns may reflect a homogenous class of aetiology—syndromes of idiopathic (genetic) generalized epilepsy, such as childhood absence epilepsy (CAE) and juvenile myoclonic epilepsy (JME), Dravet syndrome and familial focal epilepsies, such as the autosomal dominant nocturnal frontal lobe epilepsy and the familial temporal lobe epilepsies, are all genetic and yet have distinctively different electroclinical profiles.

Electroencephalogram recording

Inherent EEG attributes and their relevance to optimal recording methodology

The occurrence of diagnostically important EEG abnormalities is dynamic, subject to a multitude of parameters that include age and stage of natural history (newly manifested drug naïve vs. chronically treated individuals), time of day, and state of arousal, specific and non-specific activation, antiepileptic and other medication, and possibly other environmental factors, all operating in various combinations and at different levels. The full EEG characterization of an epilepsy type or syndrome, or even the mere support of a clinical suspicion of epilepsy, may not be possible from a standard 'snapshot' recording during wakefulness. Such a recording may miss epileptic activity in up to 50% of people with documented epilepsies (19,20), while a sleep EEG may reduce this percentage to 20% (21). The convincing recording of a defining epileptic discharge (for instance, the generalized spike-wave discharge or the spike-wave focus over a temporal lobe) or the occurrence of a defining seizure (for instance, an absence) may well depend on the state of arousal (i.e. following awakening), sleep adequacy (i.e. after partial sleep deprivation), or even on possible seizure triggers, the effect of which can be tested under controlled conditions in the EEG laboratory. Therefore, EEG recordings can be 'personalized' according to the diagnostic hypothesis and the available relevant clinical information to maximize diagnostic yield.

It is perhaps useful to distinguish two levels of EEG, according to the available facilities and equipment that include use of video and polygraphy and the capability for activated and sleep recordings. The basic (or level 1) encountered in most district general hospitals and the advanced (or level 2), expected in tertiary epilepsy centres and teaching hospitals. At both levels, the service is provided by a skilled EEG physiologist who will compile the available electro-clinical information and ensure technical adequacy and the electroencephalographer who will interpret the EEG findings and report to the referring physician. The following discussion concerns level 2 of service with the inference that, notwithstanding differences in facilities and EEG equipment, level 1 could offer almost as much.

In search of the diagnostic hypothesis

A clinically useful EEG recording can only be planned when a good history and a clear diagnostic hypothesis are available. Unfortunately, the EEG request form rarely contains detailed clinical information and the working hypothesis (or the question to be answered) is frequently basic or misguided ('is it epilepsy?', 'exclude epilepsy'). Decisive amendments can be made during the stage of electrode application, when the (trained in epileptology) physiologist has ample time to discuss important aspects of history, including seizure symptoms and semiology (as patients are frequently escorted by relatives or friends who may have witnessed attacks), seizure frequency and timing, and possible triggers (with the view to test during the recording). The relevant information can be entered in a structured template that would generate the EEG report, together with the standard personal data, full list of medications, most recent seizure, handedness, last meal, etc. Physiologists should be prepared to record beyond the expected, ideally after discussion with the consultant electroencephalographer or improvise within established departmental protocols when the latter is not available. For instance, a history of muscle jerks on awakening or during reading should prompt placement of EMG electrodes over the indicated limbs or jaw muscles, and testing of the possible effects of sleep (by letting the patient drowse even when a standard EEG had been requested), or those of reading, talking or listening. Finding past EEG reports is also very important (see section on EEG reporting below).

Activation

Sleep and sleep deprivation

Light slow sleep activates epileptiform discharges in both focal and generalized epilepsies and therefore a sleep EEG should be the recording of choice in the evaluation of patients with a first seizure (22), where accurate diagnosis of the type or syndrome of epilepsy should be achieved as soon as possible. In patients with medically intractable focal epilepsies, sleep EEG studies are essential in the first stage of presurgical evaluation process and have significant lateralization value (17,23–25).

In idiopathic generalized epilepsies (IGEs) (or genetic generalized epilepsies (GGEs)), the spontaneous occurrence of generalized spike-wave discharges (GSWD) and generalized seizures is influenced by the circadian rhythm (26,27). Early morning awakening, particularly when provoked, may activate all types of seizure (generalized tonic clonic seizures (GTCS), absences, and myoclonic). Awakening from daytime naps is also effective showing that, besides circadian susceptibility, the transitional state from sleep to wakefulness is also a principal activator. During slow (non-rapid eye movement (REM)) sleep, GSWD tend to occur more frequently, acquire a polyspike component, and become shorter, incomplete, or fragmented, regardless of the specific sub-syndrome. During REM sleep they tend to attenuate or be inhibited. Their occurrence is particularly associated with phasic EEG arousal events (K-complexes), and at microstructure sleep level with the higher arousal phase of the cyclic alternating pattern (CAP-A), rather than with the lower arousal phase (CAP-B) (28,29). Excessive occurrence of GSWD during CAP-B has been associated with poor seizure control, particularly when discharges directly provoke CAP-A epochs, increasing the arousal rate and therefore destabilizing sleep (30,31). Partial sleep deprivation activates GSWD independently (31–33). It follows that in patients with suspected IGE/GGE sleep EEGs should be pursued after partial sleep deprivation, while in those with suspected focal seizures the choice between sleep deprivation and sedative-induced sleep is less important.

Hyperventilation and assessment of cognition during spike-wave discharges

In absence, IGE syndromes hyperventilation activates GSWD and absence (for the period of the exercise and shortly after), and should be best performed after awakening when hyper excitability is maximal (34,35). It is important to remember that hyperventilation can also activate focal EEG abnormalities and seizures and should not be neglected in diagnostic studies and in candidates for epilepsy surgery (36,37).

GSWD may occur in association with measurable behavioural changes such as impairment of cognition (IoC), motor and autonomic manifestations on video and polygraphy, or may be subclinical. The length of the GSWD is not relevant to the degree of IoC, although the latter is easier to appreciate in lengthy absences; indeed, most patients experience maximal impairment within the first second of the GSWD, while IoC is milder when the associated discharge is incomplete, or asymmetric (38). Traditionally, IoC during GSWD is assessed by simple auditory stimuli (name, numbers, or words) given clearly and loudly as soon as possible after the onset of the discharge. This method detects only absences associated with severe IoC and with discharges longer than 3–4 s, as shorter discharges may be 'fading' by the time the auditory stimulation is given. Briefer (≤2 s) or longer mild absences are diagnosed by demonstrating impaired concentration and motor execution during simple rhythmic motor tasks, such as tapping or counting, performed at a comfortable rate. Counting aloud breaths during hyperventilation may allow clinical assessment of the possible effects of the *provoked* GSWD on cognition, manifested by hesitations, misses, repeating the same number, etc. (39). This method is possible in all subjects older than 5–6 years, and while video monitoring (in level 2 recordings) is desirable, physiologists may accurately annotate performance even on level 1 paper recordings.

Polygraphy

The term refers to the correlation between the EEG signal and physiological signals from the muscles, heart, autonomic system and respiration for clinical diagnostic purposes. Polygraphy can unveil important parts of the seizure semiology that may not be apparent on video, and therefore is a *complementary* diagnostic technique not to be omitted in 'level 2' video recordings. It is technically simple to record and limited versions can be used in 'level 1' recordings, as clinically indicated.

EMG electrodes

EMG electrodes allow detection and study of ictal motor phenomena such as myoclonus, spasms and tonic seizures. The EMG expression of *myoclonus* is a brief (100 ± 50 ms) burst of potentials, followed by an also brief (50–100 ms) interruption of the ongoing muscle activity, called the silent period (SP) and induced by reflex spinal inhibition. A SP without a preceding myoclonus can also occur, called *negative myoclonus* (NM) producing a sudden interruption of the muscle tone with the corresponding pause lasting from 100 to 500 ms. *Epileptic myoclonus (EM)* or *epileptic negative myoclonus (ENM)* are preceded by paroxysmal EEG events and occur in a multitude of epilepsies ranging from early childhood to adulthood and from idiopathic/genetic (benign myoclonic

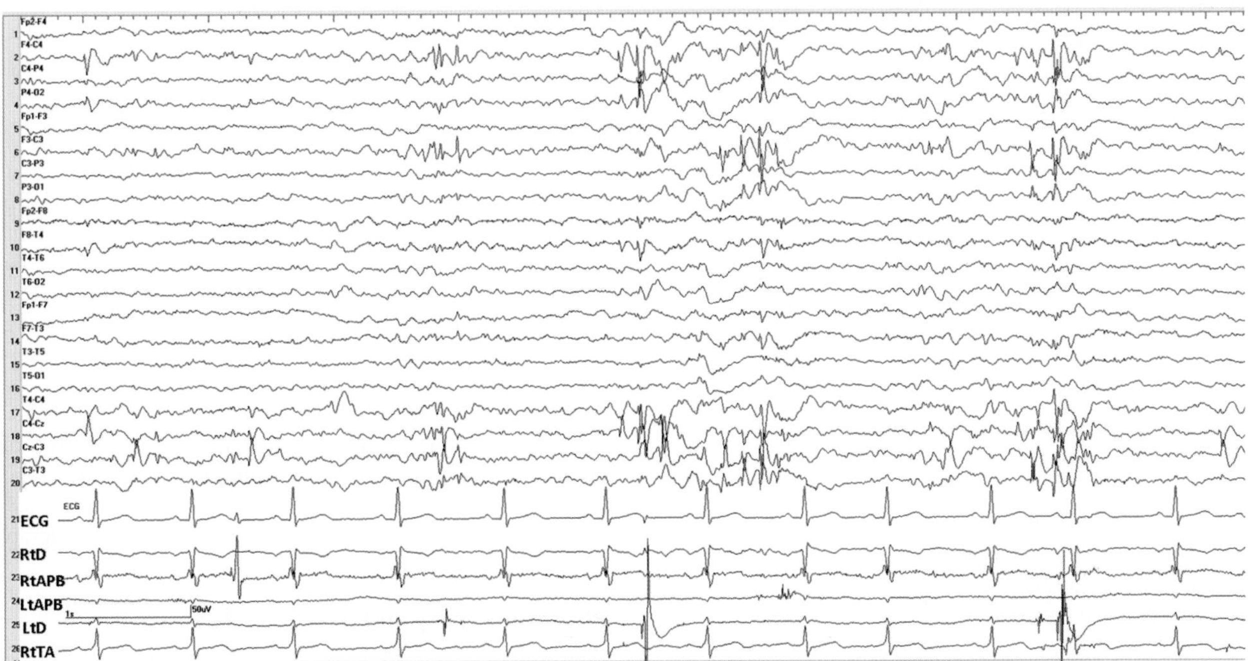

Fig. 11.5 Video EEG on an 18-year-old patient with Lafora disease (LD). Note the bilateral synchronous and independent spike-wave discharges and the associated myoclonic jerks recorded by dedicated proximal and distal EMG electrodes. D, deltoid; APB, abductor pollicis brevis; TA, tibialis anterior.

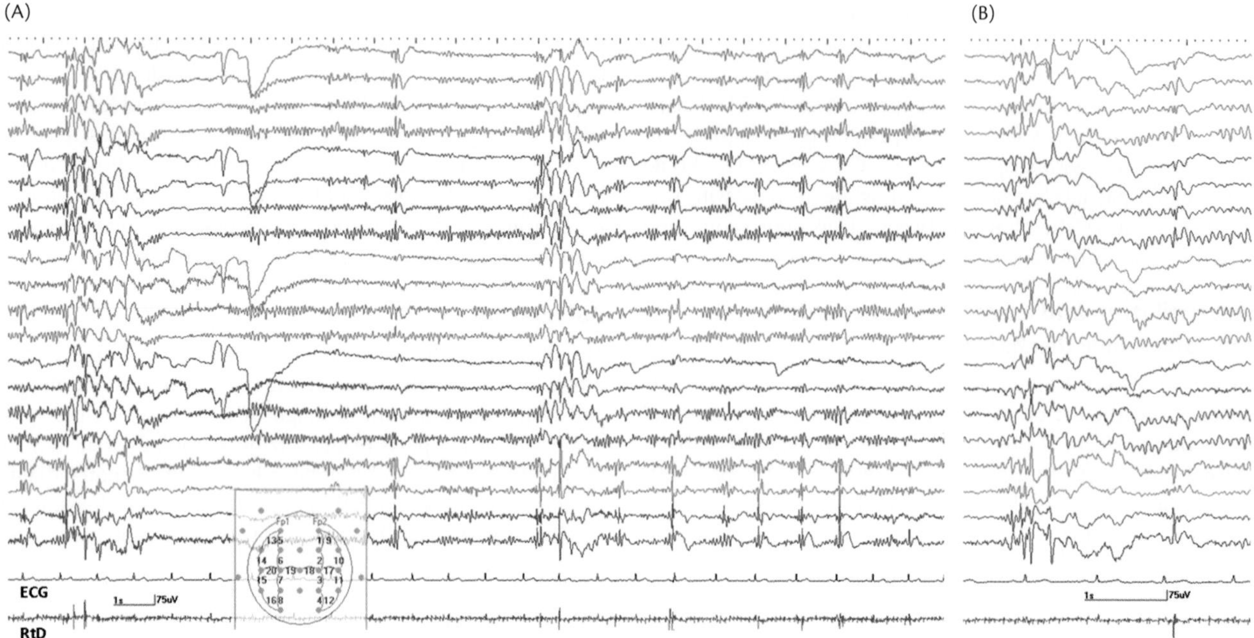

Fig. 11.6 Video EEG on a 20-year-old patient with muscle twitches only when her eyes are closed. She has two types of generalized discharge, regular 3-Hz GSWD for up to 2 s and singular spike-wave, both associated with myoclonic jerks recorded by the EMG electrode (bottom channel). During eyes-open epochs, the EEG is entirely normal and the patient is adamant that she has no symptoms when her eyes are open. (A) Slow paper speed; (B) conventional paper speed; RtD, right deltoid.

epilepsy of infancy, epilepsy with myoclonic absences, JME, Dravet syndrome, progressive myoclonic encephalopathies (PME) and even in benign Rolandic epilepsy) to focal structural epilepsies (Figs 11.5–11.7) and can be spontaneous or reflexive (40). It follows that aspirations to record ENM require the suspected (by history or previous tests) limbs to be in a state of tonic contraction. In other words, the patient not 'lying relaxed with the eyes closed'.

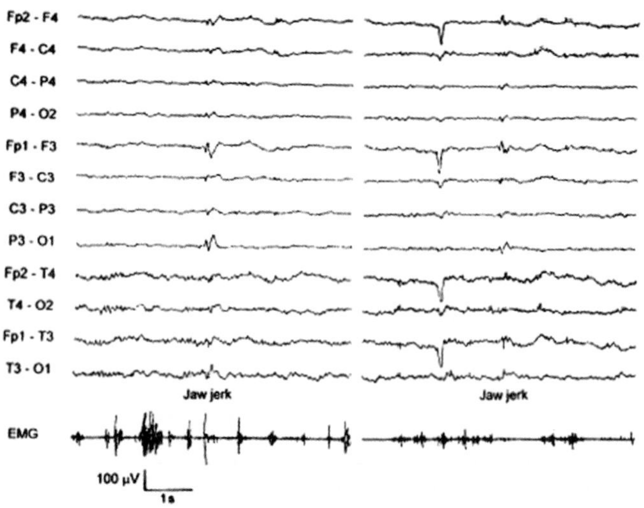

Fp2 - F4
F4 - C4
C4 - P4
P4 - O2
Fp1 - F3
F3 - C3
C3 - P3
P3 - O1
Fp2 - T4
T4 - O2
Fp1 - T3
T3 - O1
Jaw jerk
Jaw jerk
EMG
100 µV
1s

Fig. 11.7 Video EEG activated by reading aloud in an 18-year-old patient with myoclonic reading epilepsy. Note that the EMG from the jaw muscles may not always record a potential associated with the activated by reading, usually bilateral synchronous, spike-wave discharge.

Reproduced from *Brain*, **121**(8), Koutroumanidis M, Koepp M, Richardson MP, et al., The variants of reading epilepsy: a clinical and video-electroencephalographic study of 17 patients with reading-induced seizures, pp. 1409–27, copyright (1998), with permission from Oxford University Press.

Epileptic spasms

Epileptic spasms are the characteristic seizure in West syndrome (but may also occur in neonatal epileptic encephalopathies) and are typically associated with a diffuse high amplitude slow complex, which is synchronous with the burst of EMG activity recorded from proximal muscles, followed by diffuse flattening of the electrical activity for a few seconds, usually with superimposed low amplitude rapid rhythms (Fig. 11.4).

Tonic seizures

Tonic seizures are characterized, clinically by bilateral and prolonged (compared with jerks and spasms) contraction of proximal arm and trunk muscles, and electrographically by a sequence of a slow potentials, a period of desynchronization or flattening (Fig. 11.8A), or low voltage fast rhythms that progressively become slower and of higher voltage (Fig. 11.8B), giving way to postictal slow. The EMG leads demonstrate continuous tonic activity from the proximal muscles and the ictal EEG changes appear bilateral and diffuse.

Further examples of the use of polygraphy in EEG recordings, including plethysmography that may detect seizure-associated autonomic responses, can be found in Figs 36.6 and 36.8–10), while the reader can find more information about polygraphy in the excellent review by Tassinari et al (41).

Photosensitivity and fixation-off sensitivity

Intermittent photic stimulation (IPS) is routinely used to detect abnormal sensitivity to light stimuli. EEG departments may still have their own protocols and methods of assessing photosensitivity,

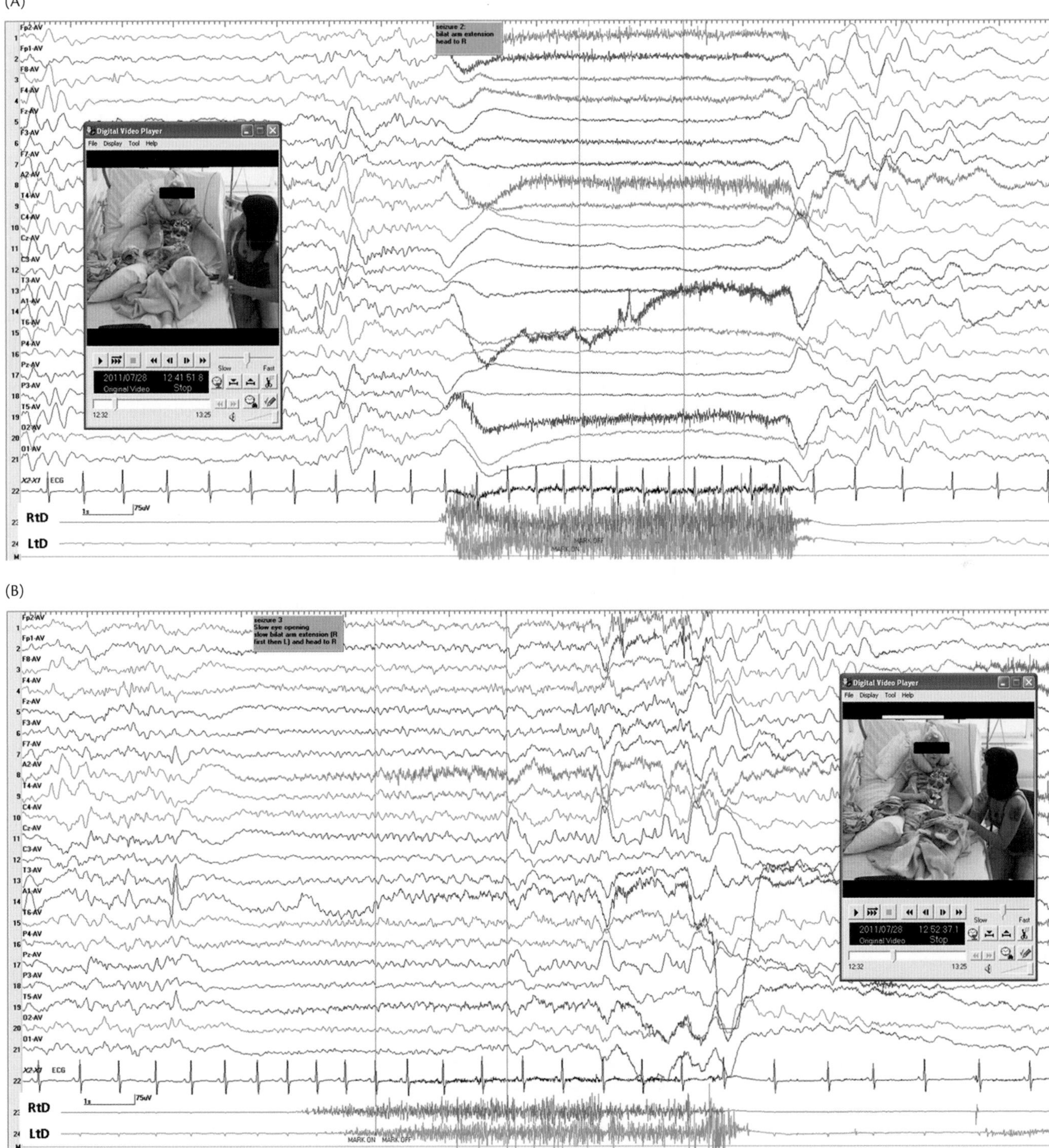

Fig. 11.8 (A,B) Partially sleep-deprived video EEG showing two tonic seizures arising from sleep in a 12-year-old girl with focal and tonic seizures since age 5. Both show generalized onset, although they appear to be preceded by left temporal sharp activity. During this EEG, a focal subclinical seizure occurred over the left posterior temporal area too. Note the different EEG patterns and the associated EMG activity.

but in recent years, efforts have been made to standardize the methods and interpretation of the results of IPS (42).

IPS can elicit several types of EEG response, with different clinical significance: the so-called generalized photoparoxysmal response (GPPR) is strongly associated with clinical photosensitivity and consists of generalized spike-/polyspike-wave discharges, although they may have an occipital onset. Milder spike-wave responses may occur in the temporoparietal-occipital areas, symmetrically or asymmetrically, or be confined within the occipital areas. The latter are also clinically significant when 'stimulus—independent', i.e. when they do not follow the flash frequency or its harmonics. Posterior stimulus–dependent responses can usually occur after suppression of previous GPPR by antiseizure medication, or in neuronal ceroid lipofuscinosis, while time-locked to the flash frequency orbitofrontal photomyoclonus has no epilepsy connotations. The evolution of photosensitivity (and by implication the response to treatment) can be assessed by sequential EEGs, but it is important to keep in mind that the range or the degree of photosensitivity may relate to the overall level of cortical hyperexcitability at the time of the testing; for example, a patient with GPPR after awakening in a sleep deprived EEG may well show only posterior photosensitivity in a routine recording taken only a few days later and vice-versa.

GPPR is far more likely to occur on eye-closure while the train of IPS is being given than when eyes are open or closed, and frequently, photosensitivity can be predicted by eye-closure epileptic activity before photic stimulation. In photosensitive patients it is always useful to test the effect of monocular stimulation with each eye covered by the palm (*not* the fingers), as well as whether common sunglasses can reduce the abnormal responses (43).

While photosensitivity is frequently associated with 'eye-closure' abnormalities, fixation-off sensitivity (*FOS*; the occurrence of EEG epileptic discharges or seizures by elimination of central vision and fixation) can be suspected from the early stages of an EEG recording by paroxysms that constantly occur for as long as the eyes are closed and disappear when the eyes open (44). FOS is confirmed by demonstrating that the same abnormalities also occur (or are activated) by impeding central vision and fixation using translucent spherical lenses, underwater goggles covered with semi-transparent tape, Frenzel lenses, or Ganzfeld stimulation; Fig. 11.9 illustrates the diagnostic sequence.

FOS may occur in children with Panayiotopoulos syndrome, occipital epilepsy of Gastaut, coeliac disease, seizures with posterior cerebral calcification (45) and other types of structural occipital epilepsy, but also in IGE (Fig. 11.6) including photosensitive patients (46).

EEG reporting

EEG interpretation is essentially based on expert visual analysis of the gathered abnormalities and their synthesis into a plausible diagnostic hypothesis, taking into account the available clinical information, physiologist's observations, and comments, and EEG findings from earlier reports or actual traces, if available. Without knowledge of the clinical context, EEG interpretation is less helpful, or even misleading. For instance, focal spikes do not necessarily reflect an underlying irritative structural change (and therefore epilepsy with focal seizures), while generalized spike wave discharges may not always imply epilepsy with generalized seizures of genetic aetiology (idiopathic generalized epilepsy (IGE)). Earlier recordings, particularly those performed shortly after the onset of seizures when patients were off, or early into their first treatment, can frequently hold the key to the diagnosis of a newly referred adult with epilepsy, as robust patterns during childhood and adolescence may become subtle and difficult to recognize with age. Serial recording of the EEG is also important in understanding the natural course of the various epilepsy types and syndromes; significant clinical

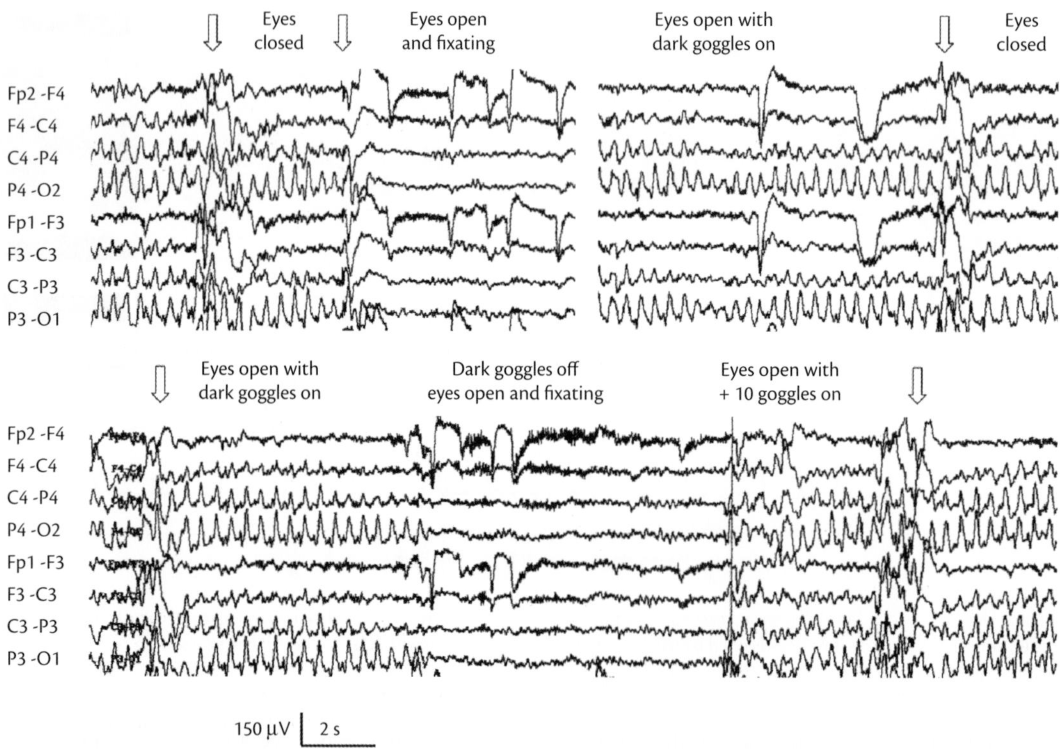

Fig. 11.9 (Top left trace) Bilateral synchronous occipital paroxysms (OP) with eyes closed are promptly inhibited by eye opening and visual fixation. (Top right trace) Ongoing OP irrespective of whether eyes are open or shut, as long as dark goggles are on. (Lower trace) Eyes are open with dark goggles on (left third), dark goggles off and visual fixation (middle third), and +10 translucent goggles on (right third). Sub-clinical generalized polyspike—wave discharges are marked with arrows.

changes that may occur in the first two decades of life (and therefore reflect in the EEG), such as CAE remitting or evolving into JME (47), progressive myoclonic epilepsy imitating epilepsy with myoclonic-astatic seizures or JME (48), idiopathic photosensitive occipital lobe epilepsy progressing into JME (49), are unusual in adult life.

The clinical EEG report should 'translate' the EEG findings into clinically useful information and discuss how they relate to the clinical presentation. Depending on the diagnostic value of the findings, it may be possible to suggest an epilepsy type or syndrome, implicitly with some reference to the possible underlying aetiology. Therefore, the final impression may express a clinical opinion, but should refrain from recommending specific management. Offering further discussion of the findings and their clinical significance would suffice and most likely be welcome by the referring physician. The report should also be written in a clear and simple way, avoiding excessive technical terms particularly when addressed to a non-epilepsy specialist, e.g. to a general physician or medical team (as it is frequently the case).

Compiling a clinical EEG report is not always a smooth and straightforward exercise. The rich visual material of different features and their combinations in different conditions has to be 'rephrased' using a finite lexicon of descriptive terms that, no matter how clear and jargon-free, still have important semantic connotations. For example, a spike wave discharge that occurs in all scalp electrodes, but has a regional onset can be interpreted, either as focal with fast generalization or as generalized with an incomplete initial phase. Similarly, a spike wave discharge that occupies several adjacent (but not all) scalp electrodes can be interpreted as either (diffusing) focal or (incompletely/abortive) generalized. Both examples refer to the distinction between focal epilepsy with secondary bilateral synchrony (SBS) and idiopathic (genetic) generalized epilepsies. When the available EEG evidence is insufficient for electroencephalographers to suggest a diagnosis without being overly subjective in their opinion, it is recommended that the rationale and the EEG evidence are fully explained in the report, other diagnostic possibilities are discussed and further testing is advised as required (sleep or ambulatory EEG, telemetry, etc.). Guidelines for writing EEG reports can be found elsewhere (50,51).

Conclusions

Based on the preceding discussion, this section summarizes the main clinical uses and limitations of the 'interictal' EEG in epileptology (see also 52,53).

Uses

The EEG can:

- Confirm a clinical diagnosis of epilepsy.
- Contribute to refinement of the diagnosis at various levels:
  - Classify the type of epilepsy (i.e. focal versus generalized, or IGE/GNE versus SBS).
  - Delineate the full clinical range with implications for syndrome diagnosis, AED selection and rough prognostication, by recording:
    - — absences/spasms/tonic/atonic/myoclonic/reflex seizures in children;
    - — absences/MJ/absence status/reflex seizures in adults.
  - Confirm/expand on, or rule out triggers/reflex epileptogenesis:
    - — photosensitivity; pattern sensitivity; FOS;
    - — reading–language epilepsy; other high cognitive triggers;
    - — other reflex mechanisms.
  - Detect self-induction:
    - — detect psychogenic non-epileptic seizures (activated by suggestion);
    - — suggest possible antiepileptic drugs intoxication.
- Record previously unrecognized (or new) seizures/triggers in patients who failed initial treatment.
- Guide further testing (repeat video sleep deprivation EEG, diagnostic telemetry, other).
- Acts as a first line investigation for epilepsy surgery.

Limitations

- A normal interictal EEG does not exclude the diagnosis of epilepsy.
- An abnormal interictal EEG does not prove the diagnosis of epilepsy.

References

1. Berger, H. (1929). Über das Elektroenkephalogram des Menschen. *Arch Psychiatr Nervenkr.*, **87**, 527–70.
2. Gloor, P. (1994). Berger lecture. Is Berger's dream coming true? *Electroencephalography & Clinical Neurophysiology*, **90**, 253–66.
3. Legarda, S., Jayakar, P., Duchowny, M., Alvarez, L., and Resnick, T. (1994). Benign rolandic epilepsy: high central and low central subgroups. *Epilepsia*, **35**(6), 1125–9.
4. Gregory, D.L. and Wong, P.K. (1984). Topographical analysis of the centrotemporal discharges in benign rolandic epilepsy of childhood. *Epilepsia*, **25**(6), 705–11.
5. Jasper, H.H. (1958). The ten–twenty electrode system of the International Federation. *Electroencephalography & Clinical Neurophysiology*, **10**, 367–80.
6. No authors listed (1991). American Electroencephalographic Society Guidelines for Standard Electrode Position Nomenclature. *Journal of Clinical Neurophysiology*, **8**, 200–2.
7. Chatrian, G.E., Lettich, E., and Nelson, P.L. (1985). Ten percent electrode system for topographic studies of spontaneous and evoked EEG activity. *American Journal of EEG Technology*, **25**, 83–92.
8. Oostenveld, R. and Praamstra, P. (2001). The five percent electrode system for high-resolution EEG and ERP measurements. *Clinical Neurophysiology*, **112**, 713–19.
9. Fernandez Torre, J.L., Alarcon, G., Binnie, C.D., and Polkey, C.E. (1999). Comparison of sphenoidal, foramen ovale and anterior temporal placements for detecting interictal epileptiform discharges in presurgical assessment for temporal lobe epilepsy. *Clinical Neurophysiology*, **110**, 895–904.
10. Sanei, S. and Chambers, J.A. (2007). *EEG signal processing*. Chichester: John Wiley & Sons Ltd.
11. Binnie, C.D. (2005). EEG technology. In: R. Cooper, C. D. Binnie, and R. Billings (Eds) *Techniques in clinical neurophysiology. A practical manual*, pp. 89–168. Amsterdam: Elsevier.
12. Nunez, P.L. and Srinivasan, R. (2006). *Electric fields of the brain. The neurophysics of EEG*, 2nd edn. New York, NY: Oxford University Press.
13. Pizzagalli, D.A. (2007). Electroencephalography and high density electrophysiological source localization. In: J. Cacioppo,

L. G. Tassinary, and G. G. Berntson (Eds) *The handbook of psychophysiology*, pp. 56–84. New York, NY: Cambridge University Press.
14. Commission on Classification and Terminology of the International League Against Epilepsy. (1969). Clinical and electroencephalographic classification of epileptic seizures *Epilepsia*, **10**(Suppl.), S2–13.
15. Commission on Classification and Terminology of the International League Against Epilepsy. (1989). Proposal for revised classification of epilepsies and epileptic syndromes. *Epilepsia*, **30**, 389–99.
16. Berg, A.T., Berkovic, S.F., Brodie, M.J., et al. (2010). Revised terminology and concepts for organization of seizures and epilepsies: report of the ILAE commission on classification and terminology, 2005–2009. *Epilepsia*, **51**(4), 676–85.
17. Koutroumanidis, M., Martin-Miguel, C., Hennessy, M.J., et al. (2004). Interictal temporal delta activity in temporal lobe epilepsy: correlations with pathology and outcome. *Epilepsia*, **45**, 1351–67.
18. Crompton, D.E., Scheffer, I.E., Taylor, I., et al. (2010).. Familial mesial temporal lobe epilepsy: a benign epilepsy syndrome showing complex inheritance. *Brain*, **133**(11), 3221–31.
19. Pedley, T.A., Mendiratta, A., and Walczack, T.S. (2003). Seizures and epilepsy. In: J. S. Ebersole and T. A. Pedley (Eds) *Current practice of clinical electroencephalography*, 3rd edn, pp. 506–87. New York, NY: Lippincott Williams & Wilkins.
20. Pillai, J. and Sperling, M.R. (2006). Interictal EEG and the diagnosis of epilepsy. *Epilepsia*, **47**(Suppl. 1), 14–22.
21. Binnie, C.D. and Stefan, H. (1999). Modern electroencephalography: its role in epilepsy management. *Clinical Neurophysiology*, **110**, 1671–97.
22. Leach, J.P., Stephen, L.J., Salveta, C., and Brodie, M.J. (2006). Which electroencephalography (EEG) for epilepsy? The relative usefulness of different EEG protocols in patients with possible epilepsy. *Journal of Neurology, Neurosurgery, & Psychiatry*, **77**, 1040–2.
23. Salanova, V., Morris, H.H., Van Ness, P., et al. (1995). Frontal lobe seizures: electroclinical syndromes. *Epilepsia*, **36**(1), 16–24.
24. Adachi, N., Alarcón, G., Binnie, C.D., Elwes, R.D.C., Polkey, C.E., and Reynolds, E.H. (1998). Predictive value of interictal epileptiform discharges during non-REM sleep on scalp EEG recordings for the lateralisation of epileptogenesis. *Epilepsia*, **39**, 628–32.
25. Bautista, R., Spencer, D., and Spencer, S. (1998). EEG findings in frontal lobe epilepsies. *Neurology*, **50**(6), 1765–71.
26. Janz, D. (1953). 'Aufwach'-epilepsien (als ausdruckeinerden 'nacht'- oder 'schlaf'-epilepsien gegenuberzustellenden verlaufsform epileptischer erkrankungen). *Arch Z Neurol*, 73–98.
27. Janz, D. (2000). Epilepsy with grand mal on awakening and sleep-waking cycle. *Clinical Neurophysiology*, **111**(Suppl. 2), S103–10.
28. Terzano, M.G., Parrino, L., Anelli, S., and Halasz, P. (1989). Modulation of generalized spike-and-wave discharges during sleep by cyclic alternating pattern. *Epilepsia*, **30**, 772–81.
29. Gigli, G.L., Calia, E., Marciani, M.G., et al. (1992). Sleep microstructure and EEG epileptiform activity in patients with juvenile myoclonic epilepsy. *Epilepsia*, **33**, 799–804.
30. Bonakis, A. and Koutroumanidis, M. (2009). Epileptic discharges and phasic sleep phenomena in patients with juvenile myoclonic epilepsy. *Epilepsia*, **50**(11), 2434–45.
31. Serafini, A., Rubboli, G., Gigli, G.L., Koutroumanidis, M., and Gelisse, P. (2013). Neurophysiology of juvenile myoclonic epilepsy. *Epilepsy & Behaviour*, **28**(Suppl. 1), S30–9.
32. Fountain, N.B., Kim, J.S., and Lee, S.I. (1998). Sleep deprivation activates epileptiform discharges independent of the activating effects of sleep. *Journal of Clinical Neurophysiology*, **15**, 69–75.
33. Halasz, P., Filakovszky, J., Vargha, A., and Bagdy, G. (2002). Effect of sleep deprivation on spike-wave discharges in idiopathic generalised epilepsy: a 4 × 24 h continuous long term EEG monitoring study. *Epilepsy Research*, **51**, 123–32.
34. Penry, J.K., Porter, R.J., and Dreifuss, R.E. (1975). Simultaneous recording of absence seizures with video tape and electroencephalography. A study of 374 seizures in 48 patients. *Brain*, **98**, 427–40.
35. Panayiotopoulos, C.P., Obeid, T., and Waheed, G. (1989). Differentiation of typical absence seizures in epileptic syndromes. A video EEG study of 224 seizures in 20 patients. *Brain*, **112**, 1039–56.
36. Miley, C.E. and Forster, F.M. (1977). Activation of partial complex seizures by hyperventilation. *Archives of Neurology*, **34**(6), 371–3.
37. Guaranha, M.S.B., Garzon, E., Buchpiguel, C.A., Tazima, S., Yacubian, E.M., and Sakamoto, A.C. (2005). Hyperventilation revisited: physiological effects and efficacy on focal seizure activation in the era of video-EEG monitoring. *Epilepsia*, **46**(1), 69–75.
38. Browne, T.R., Penry, J.K., Porter, R.J., and Dreifuss, F.E. (1974). Responsiveness before, during, and after spike-wave paroxysms. *Neurology*, **24**, 659–65.
39. Panayiotopoulos, C.P., Koutroumanidis, M., Giannakodimos, S., and Agathonikou, A. (1997). Idiopathic generalised epilepsy in adults manifested with phantom absences, generalised tonic-clonic seizures and frequently absence status. *Journal of Neurology, Neurosurgery, & Psychiatry*, **63**(5), 622–7.
40. Rubboli, G., Mai, R., Meletti, S., et al. (2006). Negative myoclonus induced by cortical electrical stimulation in epileptic patients. *Brain*, **129**(1), 65–81.
41. Tassinari, C.A., Cantalupo, G., and Rubboli, G. (2010). Polygraphic recording of epileptic seizures. In: C. P. Panayiotopoulos (Ed.) *The atlas of epilepsies*, pp. 723–40. London: Springer-Verlag.
42. Rubboli, G., Parra, J., Seri, S., Takahashi, T., and Thomas, P. (2004). EEG diagnostic procedures and special investigations in the assessment of photosensitivity. *Epilepsia*, **45**(S1), S35–9.
43. Capovilla, G., Gambardella, A., Rubboli, G., et al. (2006). Suppressive efficacy by a commercially available blue lens on PPR in 610 photosensitive epilepsy patients. *Epilepsia*, **47**(3), 529–33.
44. Panayiotopoulos, C.P. (1998). Fixation-off, scotosensitive, and other visual-related epilepsies. *Advances in Neurology*, **75**, 139–57.
45. Gobbi, G. (2005). Coeliac disease, epilepsy and cerebral calcifications. *Brain Development*, **27**, 189–200.
46. Koutroumanidis, M., Tsatsou, K., Sanders, S., et al. (2009). Fixation-off sensitivity in epilepsies other than the idiopathic epilepsies of childhood with occipital paroxysms: a 12-year clinical-video EEG study. *Epileptic Disorders*, **11**(1), 20–36.
47. Wirrell, E.C., Camfield, C.S., Camfield, P.R., Gordon, K.E., and Dooley, J.M. (1996). Long-term prognosis of typical childhood absence epilepsy: remission or progression to juvenile myoclonic epilepsy. *Neurology*, **47**, 912–18.
48. Oguni, H. (2005). Symptomatic epilepsies imitating idiopathic generalized epilepsies. *Epilepsia*, **46**(Suppl. 9), 84–90.
49. Taylor, I., Marini, C., Johnson, M.R., Turner, S., Berkovic, S.F., and Scheffer, I.E. (2004). Juvenile myoclonic epilepsy and idiopathic photosensitive occipital lobe epilepsy: is there overlap? *Brain*, **127**, 1878–86.
50. American Clinical Neurophysiology Society. (2006). Guideline 7:Guidelines for writing EEG reports. *Journal of Clinical Neurophysiology*, **23**, 118–21.
51. Kaplan, P.W. and Benbadis, S.R. (2013). How to write an EEG report: dos and don'ts. *Neurology*, **80**(Suppl. 1), S43–6.
52. Binnie, C.D. and Prior, P.F. (1994). Electroencephalography. *Journal of Neurology, Neurosurgery, & Psychiatry*, **57**, 1308–19.
53. Koutroumanidis, M. and Smith, S. (2005).Use and abuse of EEG in the Diagnosis of idiopathic generalized epilepsies. *Epilepsia*, **46**, 96–107.

CHAPTER 12

Intracranial electroencephalographic recordings

Gonzalo Alarcón and Antonio Valentín

Introduction

Non-invasive methods of presurgical assessment, such as seizure semiology, neuroimaging findings, neuropsychological testing, scalp inter-ictal, and ictal electroencephalogram (EEG) are the first steps to presurgical evaluation. Assessment with intracranial electrodes can be used in patients where results from non-invasive tests are contradictory or unrevealing (see Chapter 32 and Box 12.1). Before implantation of electrodes, there should be a hypothesis to explain any non-convergence of evidence from different tests. This hypothesis should be testable by the implantation of intracranial electrodes. Broadly speaking, the hypothesis for implantation can be that of localization (the side of seizure onset has been identified, but there is some doubt about localization at lobar or sublobar level) or lateralization (the lobe has been identified, but the hemisphere is in doubt).

Intracranial recordings have the advantage of showing larger EEG signals and less muscle artefact than scalp recordings. They offer higher spatial sampling at the cost of recording from fewer regions, since the number of implanted electrodes is necessarily limited.

In addition to its use during video-telemetry (chronic or subacute recordings), intracranial electrodes can also be used intraoperatively in order to tailor the resection in the operating theatre (acute electrocorticography or ECoG). The latter can be obtained under local or general anaesthesia.

For chronic recordings, after recovering from electrode implantation, video-telemetry is obtained with implanted intracranial electrodes. Video-telemetry not only aims to localize epileptogenic cortex, but also mapping function so as to avoid neurological deficits after surgery.

Box 12.1 Examples of patients requiring intracranial recordings

- The patient with a frontal MRI lesion with generalized interictal discharges and contralateral positron emission tomography (PET) changes.
- The patient with left MTS and focal right temporal seizure onset on scalp telemetry, bilateral independent interictal discharges, and unrevealing neuropsychology and PET.
- The patient with a normal MRI.

Types of intracranial electrodes

Intracranial electrodes usually come as bundles of several electrodes assembled in a variety of shapes, each with its own advantages and disadvantages. Informally, each electrode in a bundle is often referred to as a 'contact', but in fact, each 'contact' is an electrode able to record brain activity independently from the other electrodes. There are different types of intracranial electrodes, including subdural (mats and strips), depth (intracerebral), foramen ovale (FO), and epidural electrodes.

Prior to implantation of electrodes, it is necessary to have identified at least one potential target. The position, type, and combination of electrodes are decided for each patient (Fig. 12.1) based on the location of targets.

Subdural electrodes

Subdural electrodes are bundles of electrodes assembled as either mats (grids) or strips, which can be placed under the dura in contact with the cortical surface. General anaesthesia is required for insertion and removal. Mats are two-dimensional arrays of electrodes, whereas strips are arranged as single rows of electrodes. Subdural electrodes are usually made with stainless steel or platinum/iridium embedded in Silastic® or Teflon® sheets. Platinum and some stainless steel electrodes are magnetic resonance imaging (MRI)-compatible. Each contact typically has 5 mm diameter and contact centres are located 1 cm apart, but new models with high density grids with smaller diameter electrodes are available. Mats are introduced through a craniotomy and can be placed over the cerebral convexity. Mats are ideal for functional mapping with electrical stimulation or with sensory evoked responses. Strips are single rows usually of four or eight electrodes that can be inserted through a burr hole, or through a craniotomy when used in combination with mats. If inserted through a burr hole anterior to the ear, an eight contact strip can be slipped under the temporal lobe and can provide excellent recordings from the parahippocampal gyrus in a similar fashion to those obtained with FO recordings. If inserted in the interhemispheric space, strips, and small mats can provide excellent recordings from the mesial aspect of the cerebral hemispheres.

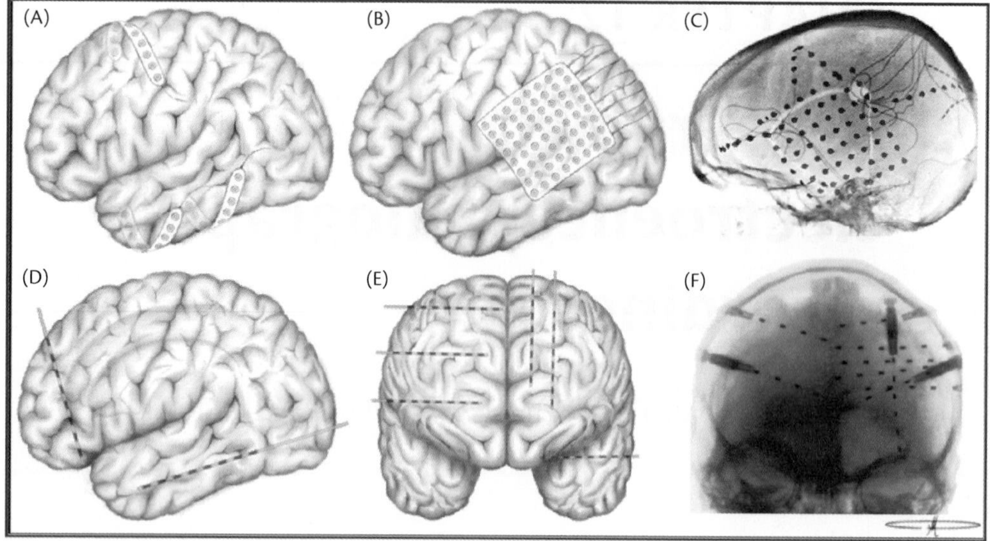

Fig. 12.1 Different types and positioning of intracranial electrodes. (A) Lateral brain view of nine-contact subdural electrodes in subtemporal and frontal structures. (B) Lateral view of a 64-contact mat over the parieto-posterior temporal region. (C) Lateral X-ray from a patient with a 64-contact mat and four 8-contact subdural electrodes. (D) Lateral view with depth electrodes located in orbito-frontal cortex, and one electrode at the hippocampus (posterior approach). (E) Anterior posterior view with depth electrodes over frontal and temporal structures (lateral approach). (F) Anterior posterior X-ray from a patient with depth electrodes in frontal structures.
Courtesy of Dr David Martin Lopez, KCL.

Depth (intracerebral) electrodes

Depth electrodes are thin multicontact electrode bundles that can be inserted stereotactically into the brain under neuroimaging control. Their main advantage is that they can be used for recording from deep brain regions (hippocampus, amygdala, cingulum), while simultaneously sampling some superficial regions. Their main disadvantages are that they can only sample relatively small areas of the brain, and that they need to be inserted into the brain, with some concerns about possible haemorrhage.

Foramen ovale electrodes

These electrodes consist of a thin wire inserted through the FO under fluoroscopic control under local or general anaesthesia (Fig. 12.2). FO recording could be considered as a semi-invasive technique because no skull perforation is required for their implantation, and they can be combined with simultaneous scalp EEG recordings. As the deepest contacts tend to record from medial temporal structures, their main indication is to establish temporal lobe laterality. When combined with peg or scalp electrodes they can be used to distinguish between temporal/extratemporal or medial/lateral temporal seizure onset. Their use has been limited because they are often disagreeable due to irritation of the trigeminal nerve during implantation. Temporary facial pain or hypoesthesia are common side effects. A permanent small area of facial anaesthesia, usually lateral to the chin, can rarely occur.

Epidural peg electrodes

These consist of mushroom-shaped plastic typically positioned in a burr hole to enable epidural recording of brain activity. Peg electrodes are considered semi-invasive, and allow recording from different brain regions in combination with simultaneous scalp EEG. Although they provide better and more localized cortical recordings than scalp EEG electrodes, it has been reported that pegss rarely determine the site of seizure onset or the spread of epileptiform discharges.

Subacute (chronic) recordings in telemetry

Indications and types of electrodes

The type and placement of intracranial electrodes depends on the hypothesis regarding the location of the seizure onset zone. In cases of temporal lobe epilepsy without clear laterality, it would be sufficient to record seizures with FO electrodes or with bilateral eight-contact subtemporal strips inserted through fronto-temporal burr holes. In patients where these procedures provide inconclusive results or in those with suspected bilateral pathology, intracranial recordings with bilateral temporal intracerebral electrodes covering mesial temporal structures are preferred, as they provide higher sensitivity in detecting hippocampal epileptogenesis. In frontal lobe epilepsy with uncertain laterality, bilateral intracerebral electrodes can be used. Subdural recordings with unilateral mats or strips are preferable when seizures are suspected to originate on the temporo-fronto-parietal-occipital convexity, peri-central regions, or in the supplementary motor area.

Interpretation of interictal recordings

Focal slowing

As in the case of scalp recordings, focal slowing of the background activity suggests the presence of underlying pathology. There is a high correlation between focal slowing and the neocortical seizure-onset zone, suggesting that slowing can be considered as an EEG marker of epileptogenic cortex (Fig. 12.3) (1,2).

Interictal epileptiform discharges

When recorded with intracranial electrodes, discharges are larger, sharper, and occur more frequently than in the scalp EEG. Interictal activity recorded with intracranial electrodes should be interpreted cautiously. Each patient usually exhibits several patterns of epileptiform discharge occurring independently at different sites, not necessarily at the site of seizure onset, and often including the hemisphere contralateral to seizure onset (3,4). The incidence of

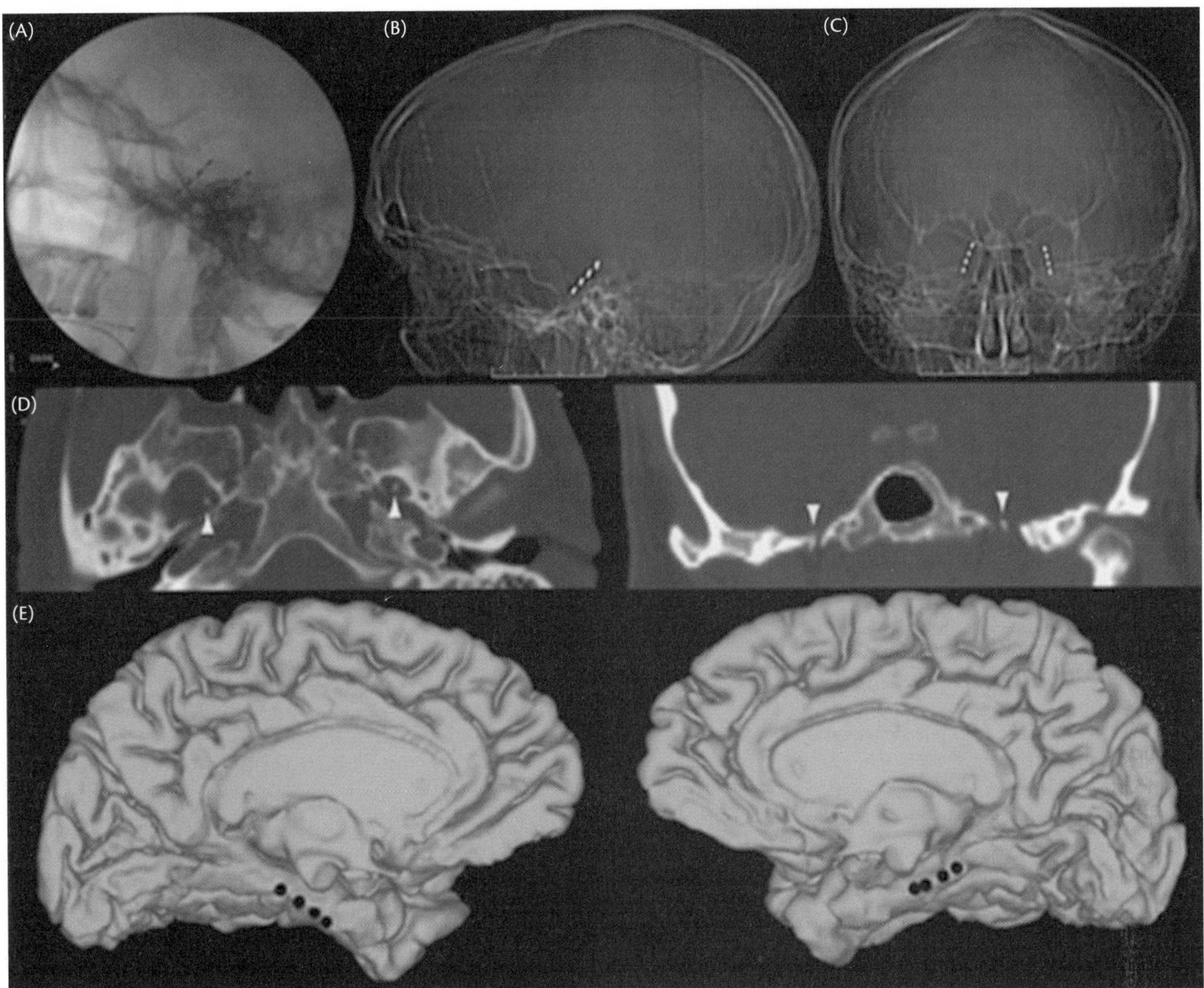

Fig. 12.2 Foramen ovale electrode (FOE) placement and position. (A) Intraoperative lateral oblique fluoroscopic image obtained after placement of bilateral FOE. (B,C) Post-operative CT scan lateral (B) and anteroposterior (C) scout images showing FOE position. (D) Axial and coronal CT scan images at the level of the FO showing the electrode passing through the foramen (arrowheads). (E) A representative patient's post-operative CT scan was co-registered to the preoperative MRI to visualize the position of the FOE contacts using a method described previously. The contacts (black circles) are displayed superimposed on the MRI surface extraction and lie along the mesial temporal lobe.

Reproduced from *Epilepsia*, 55(5), Sheth SA, Aronson JP, Shafi MM, Phillips HW, Velez-Ruiz N, Walcott BP, et al, Utility of foramen ovale electrodes in mesial temporal lobe epilepsy, pp. 713–24, copyright (2014), with permission from John Wiley and Sons.

bilateral independent scalp epileptiform discharges in temporal lobe epilepsy is around 40% (Figs 12.4–12.6). Discharges are usually more prominent on the epileptogenic side, particularly during sleep periods. However, most patients with bilateral independent interictal temporal discharges have unilateral seizures, usually arising where spikes are more frequent (5). Seizure control appears to be better the greater the degree of lateralization of bilateral independent discharges (6,7).

Ictal-like interictal epileptiform discharges

As seen in acute electrocorticographic recordings (see Fig. 12.7), focal cortical dysplasia can show ictal-like discharges in the interictal period, which manifest as focal areas showing bursts of spike-wave activity for a few seconds, or continuous rhythmic spike-wave activity or bursts of fast activity (8,9).

High frequency oscillations or fast ripples

At sampling rates above 2000 Hz, it is possible to study brief high frequency oscillations (HFO) in the 80–250 Hz (ripples) and 250–600 Hz (fast ripples) frequency ranges. Ripples occur over the primary motor cortex and hippocampus, and are thought to be physiological events possibly involved in memory consolidation. On the contrary, fast ripples appear to be pathological, predominately occurring at the seizure onset zone both interictally and at seizure onset (10–13).

Single pulse electrical stimulation

During interictal recordings, the study of cortical excitability with single pulse electrical stimulation (SPES, see Electrical stimulation for the identification of epileptogenic cortex) is emerging as one of the most effective interictal test to identify epileptogenic cortex and predict surgical outcome (2).

Interpretation of ictal recordings

Despite the presence of interictal abnormalities, the interpretation of chronic intracranial recordings relies on the identification of the area where seizures start on intracranial recordings

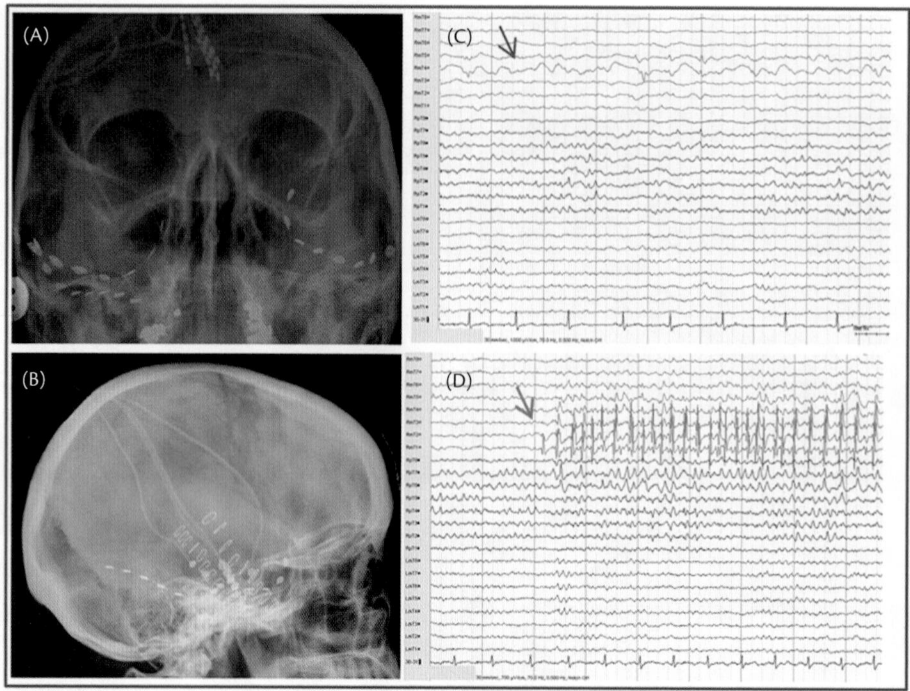

Fig. 12.3 Interictal focal slowing in a patient with two eight-contact subdural strips in the right temporal lobe and one 8-contact subdural strip in the left temporal lobe. Contact 1 is the deepest contact of the subdural electrodes. (A) Anterior posterior skull X-ray. (B) Lateral skull X-ray. (C) Focal slow activity (blue arrow) over the inferior and anterior aspect of the right temporal lobe (RaT 4,5). (D) Typical seizure originated in the deepest contact of the right anterior temporal electrode (red arrow).
RaT = right anterior temporal; RpT = right posterior temporal; LmT = left mid-temporal.

(ictal onset zone). Focal seizures can be associated with a variety of EEG patterns, which are not necessarily focal in distribution (Fig. 12.7). Seizures can start with a run of focal fast activity, sharp waves, spikes, or slow waves lasting for a few seconds; such focal changes are often associated with more widespread patterns, such as a diffuse attenuation of the background activity (diffuse electrodecremental event). These sustained ictal changes are sometimes preceded by a single epileptiform discharge, which often shows a widespread or bilateral distribution, and is associated with a prominent slow wave. The physiological and clinical significance of these widespread changes at seizure onset is unclear, and their presence is often disregarded in clinical practice. However, results

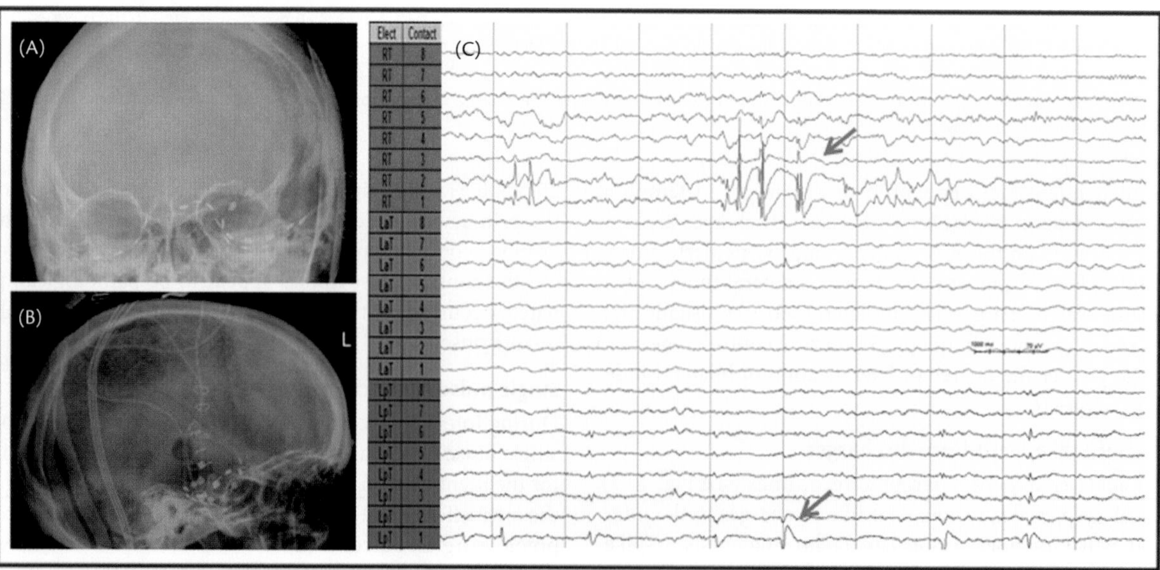

Fig. 12.4 Interictal discharges in a patient with two eight-contact subdural strips in the left temporal lobe and one 8-contact subdural strip in the right temporal lobe. Contact 1 is the deepest contact of the subdural electrodes. (A) Anterior posterior skull X-ray. (B) Lateral skull X-ray. (C) Independent interictal epileptiform discharges (arrows) over the deepest contacts of either hemisphere (Lpt1 and RT 1, 2).
RT = right temporal; LaT = left anterior temporal; LpT = left posterior temporal.

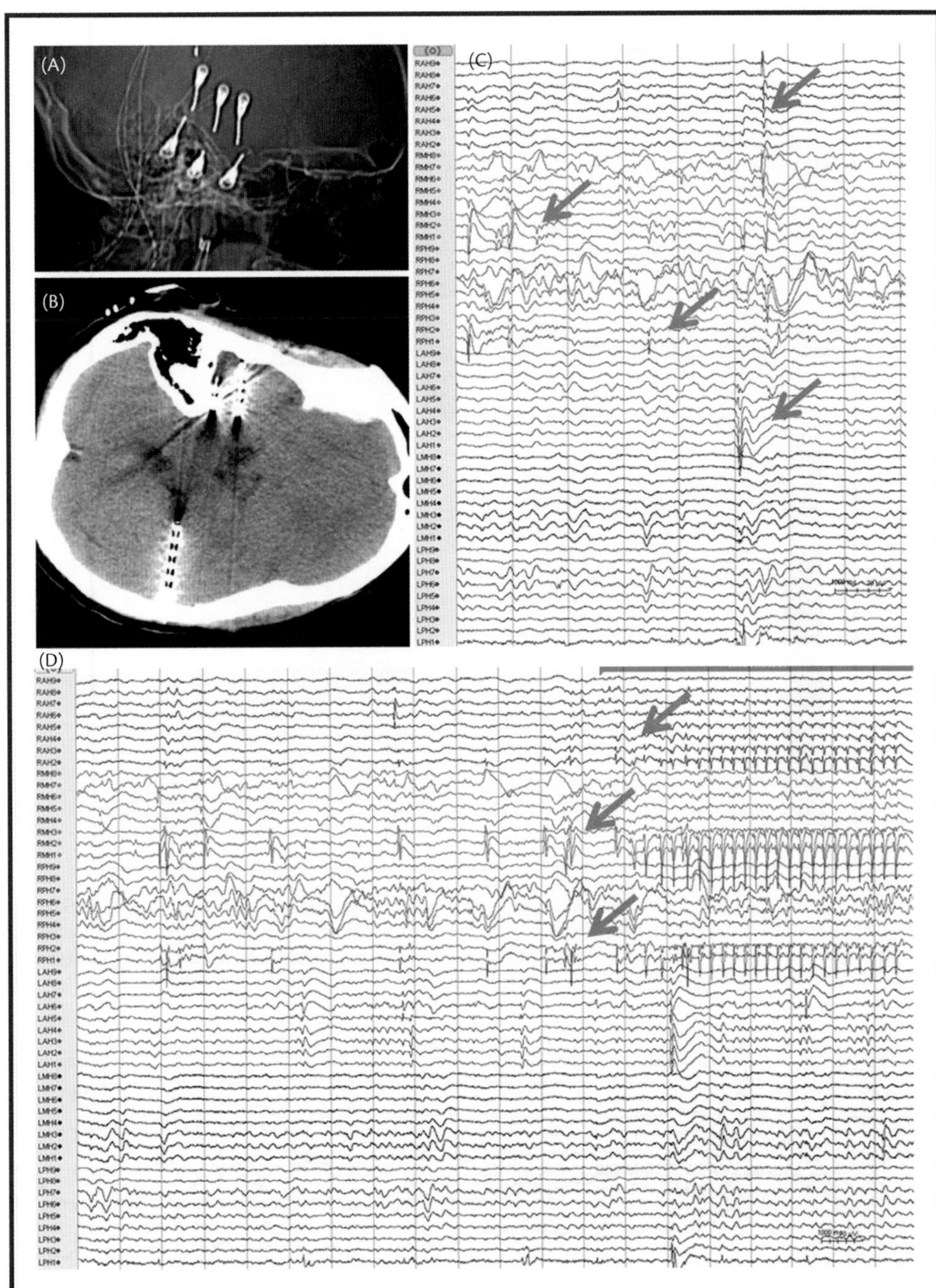

Fig. 12.5 Interictal discharges and seizures in a patient with depth electrodes in both temporal lobes. Contact 1 is the deepest contact. (A) Lateral skull X-Ray. (B) CT showing the posterior depth electrodes. (C) Independent interictal epileptiform discharges (blue arrows) over the deepest contacts of both hemispheres. (D) Typical seizure originated in the deepest contacts of the electrodes located over the right temporal lobe (red arrows).

R = right; L = left; AH = anterior hippocampus; MH = mid hippocampus; PH = posterior hippocampus.

from a large recent series have shown that those patients with seizures starting with focal fast activity tend to become seizure free after surgery, whereas those starting with a diffuse electrodecremental event tend to suffer poor seizure control (14).

The presence of a widespread or bilateral epileptiform discharge preceding focal ictal changes does not seem to affect outcome. This confirms that the best marker of proximity to the epileptogenic zone is focal fast activity.

Examples of seizure onset in patients with intracranial electrodes are shown in Figs 12.3, 12.5–12.16.

Interpretation of intracranial subacute recordings

The results from subacute recordings in telemetry can exclude resective surgery if:

- The EEG shows predominantly generalized interictal discharges in the absence of a discrete lesion on neuroimaging.

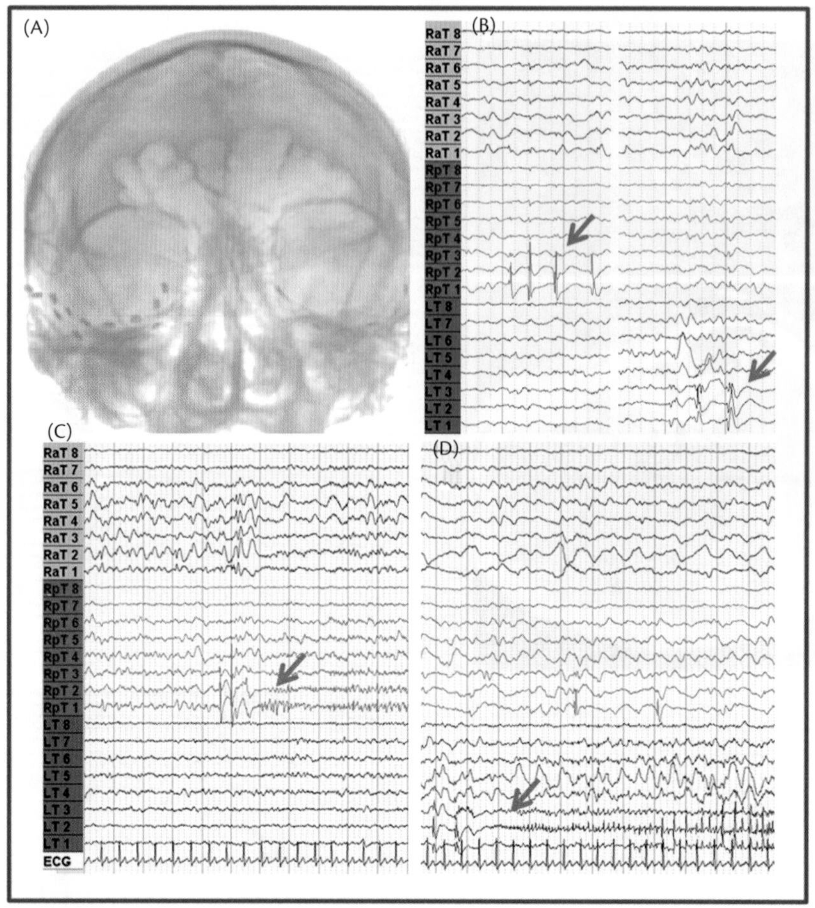

Fig. 12.6 Patient with bilateral independent epileptiform discharges and bilateral independent medial temporal seizures. (A) Anterior posterior skull X-ray showing the position of three subdural temporal electrodes, two under the right and one under the left temporal lobe. (B) EEG recording of interictal activity in the deepest contacts of the RpT electrode and in the deepest contacts of the LT electrode (blue arrows). (C) EEG recording of a typical seizure starting on the deepest contacts of the RpT electrode (red arrow). (D) Recording of a seizure starting on the deepest contacts of the LT electrode (red arrow).

RaT = right anterior temporal; RpT = right posterior temporal; LT = left temporal.

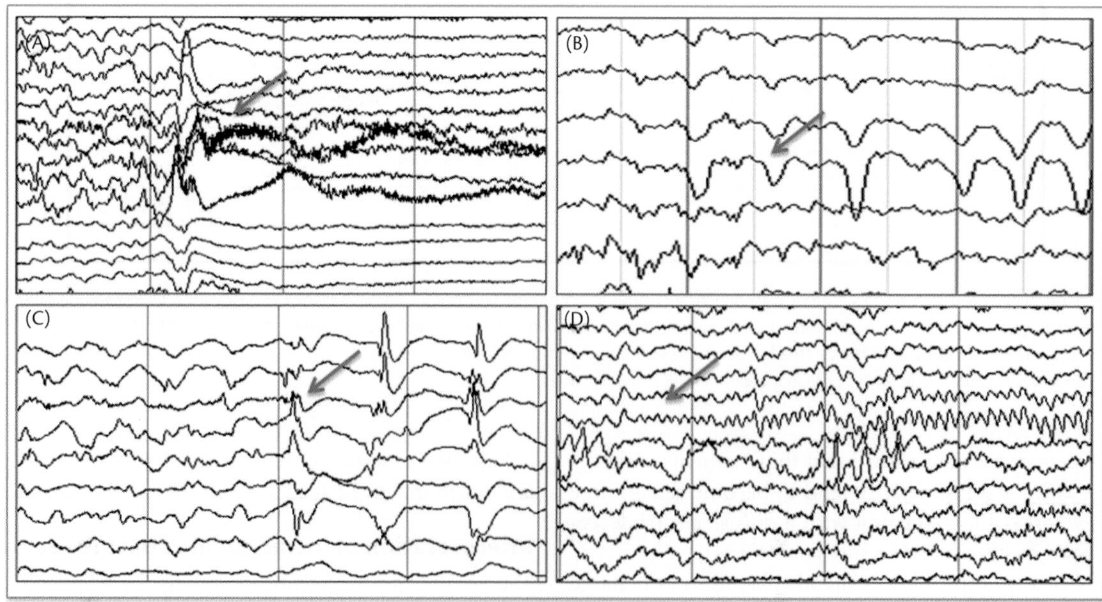

Fig. 12.7 Different types of focal seizure onsets. (A) Preceding epileptiform discharge followed by fast activity. (B) Run of theta waves. (C) Run of epileptiform discharges. (D) Alpha/beta activity. Seizure onset marked by blue arrows.

Courtesy of Dr Diego Jimenez-Jimenez.

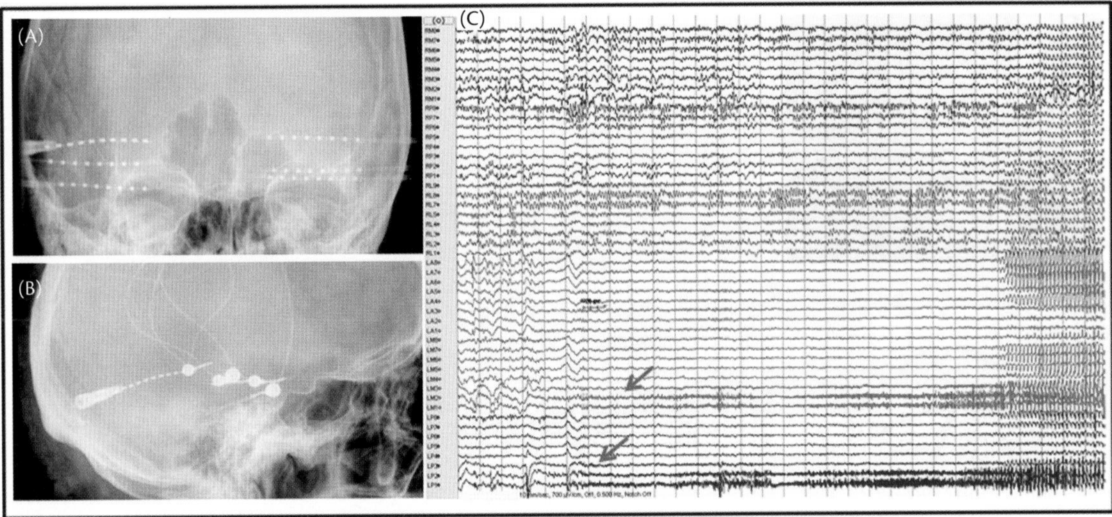

Fig. 12.8 Seizure onset in a patient with depth electrodes implanted in both temporal lobes. (A) Anterior posterior skull X-ray. (B) Lateral X-ray view. (C) Seizure onset at the deepest contacts of LM and LP electrode bundles (arrows).
R = right; L = left; A = anterior; M = medial; P = posterior; L = lateral.

- Seizure onset shows a widespread distribution (Figs 12.10 and 12.12) or independent ictal foci in more than one lobe (Fig. 12.15), and no clear alternative hypothesis exists for further studies with intracranial recordings.
- The site of seizure onset cannot be resected without unacceptable functional deficits (Fig. 12.11).

Acute electrocorticography

Indications

Intraoperative ECoG is performed in patients who are considered suitable for resective surgery or multiple subpial transections. The procedure allows further tailoring of the surgical procedure at the time of surgery and, if performed under local anaesthesia, functional mapping can be carried out. ECoG consists of intraoperative EEG recordings obtained with electrodes covering most of the area around the proposed surgical procedure under general or local anaesthesia.

Local anaesthesia

During the procedure under local anaesthesia, patients are usually sedated with propofol during the craniotomy, and thereafter, propofol is discontinued and the patient is allowed to wake up. Propofol has pharmacokinetic properties that permit quick recovery of the patient on the operating table, allowing functional mapping to

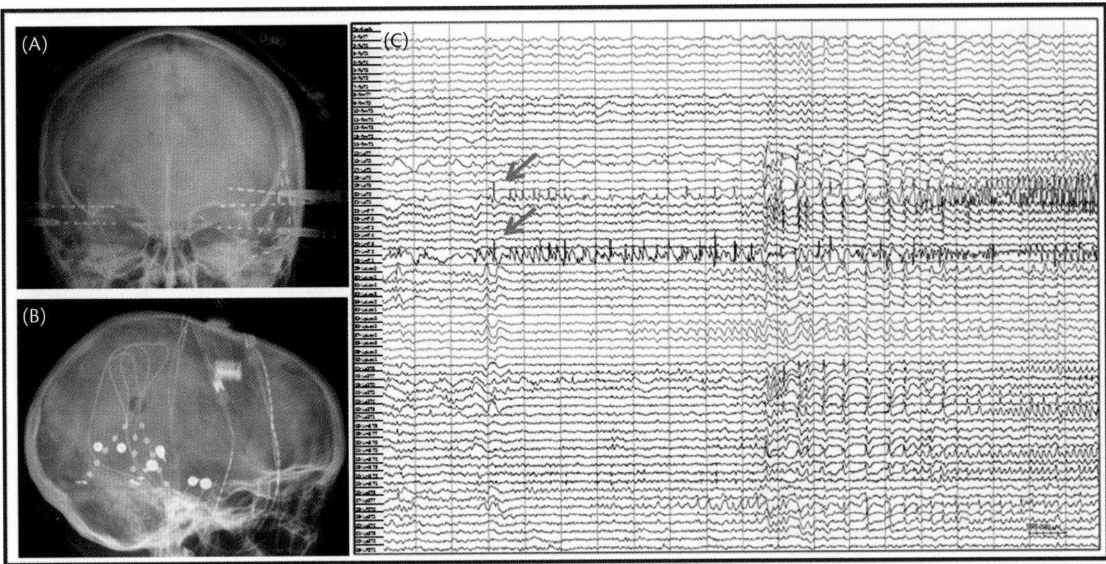

Fig. 12.9 Seizure onset in a patient with a MRI lesion in the lateral aspect of the left temporal lobe, and with subdural and depth electrodes in both temporal lobes. (A) Anterior posterior skull X-ray. (B) Lateral skull X-ray. (C) Seizure onset at the deepest contacts of LmT and LaT depth electrodes (arrows).
R = right; L = left; aT = anterior temporal; mT = mid temporal; aLes = anterior lesion; pLes = posterior lesion; asT = anterior subdural temporal; msT = mid subdural temporal; psT = posterior subdural temporal.

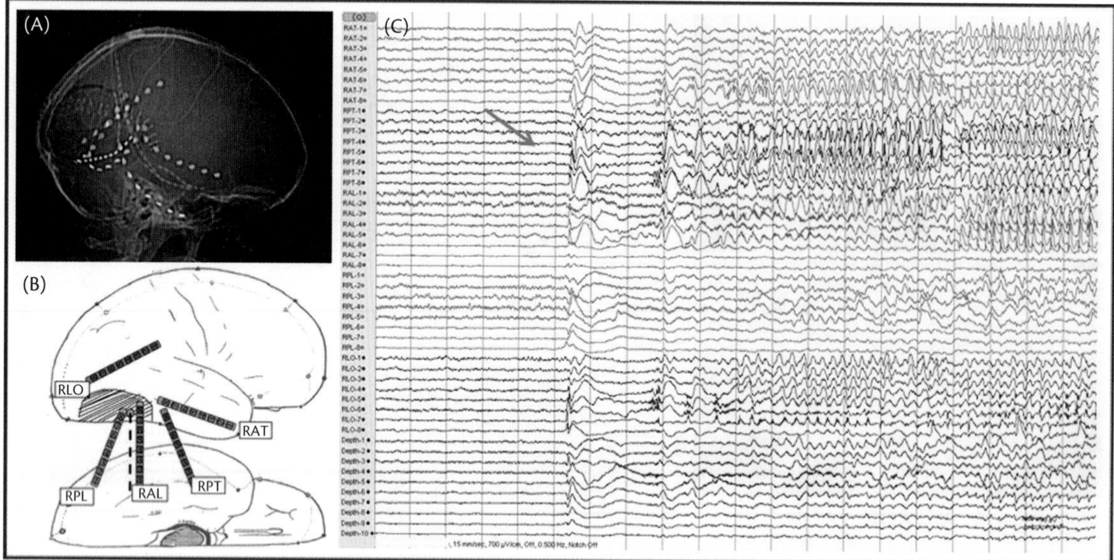

Fig. 12.10 Seizure onset in a patient with subdural strips over the right temporo-parietal lobes. The patient had a previous resection in the posterior temporal region. A) Lateral skull X-ray. B) Drawing showing the electrode's positions. C) Seizure onset showing a 2-sec diffuse electrodecremental event preceded by a single epileptiform discharge (arrow) associated with a prominent slow wave.

RAT = right anterior temporal; RPT = right posterior temporal; RAL = right anterior lesion; RPL = right posterior lesion; RLO = right lateral occipital.

localize motor, sensory, or language areas and avoid functional deficits induced by the surgical procedure. Electrical stimulation during functional mapping carries a significant risk inducing a seizure. This should be avoided at all costs in patients with an open craniotomy. Consequently, it is of primary importance to carry out functional mapping under ECoG control in order to avoid stimulation above after-discharge threshold, where the risk of inducing seizures is significantly increased. The procedure for intraoperative functional mapping is essentially similar to that described below for subacute recordings.

General anaesthesia

Different anaesthetic agents can be used for ECoG during general anaesthesia (alfentanil, fentanyl, remifentanil, sevoflurane, isoflurane, propofol, and methohexitone). Anaesthetic drugs may suppress or exacerbate epileptiform activity, induce EEG slowing, and provoke burst-suppression patterns as anaesthesia deepens. For instance, it has been described that propofol can suppress or induce spontaneous epileptiform activity during seizure surgery. Inhalational anaesthetics, such as isoflurane or sevofluorane, have been reported to have dose-dependent effects in suppressing spike

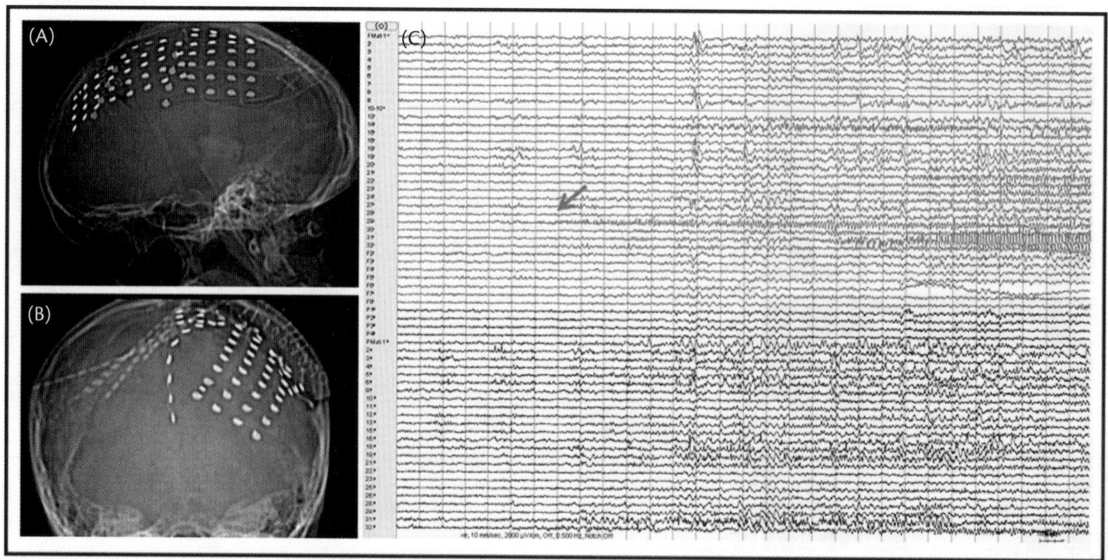

Fig. 12.11 Seizure onset in a patient with subdural mats and strips over fronto-parietal lobes. (A) Lateral skull X-ray. (B) Anterior posterior skull X-ray. (C) Seizure onset mainly over contact 29 of the frontal 32-contact mat (arrow).

F = frontal; P = parietal.

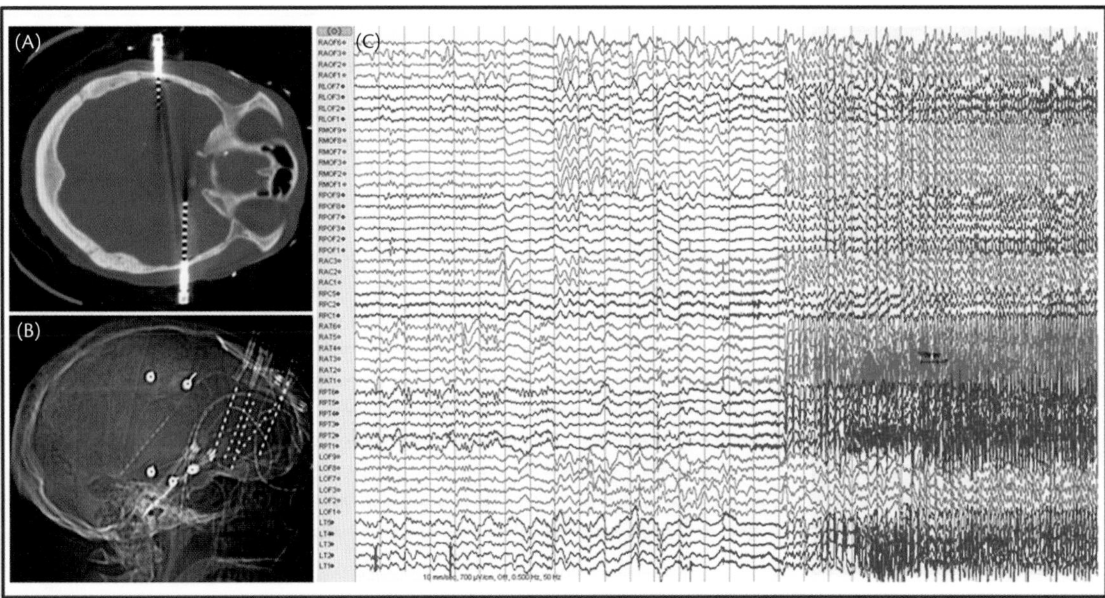

Fig. 12.12 Seizure onset in a patient with depth electrodes over fronto-parietal-temporal lobes. (A) CT of the lateral temporal depth electrodes. (B) Lateral skull X-ray. (C) Seizure onset showing regional-diffuse EEG abnormalities.
R = right; L = left; OF = orbito-frontal; C = central; T = temporal; A = anterior; L = lateral; M = medial; P = posterior.

activity in the ECoG. Experience at King's College Hospital, has shown isoflurane to be the most adequate anaesthetic agent during ECoG, as it allows ECoG recordings similar to those seen in subacute telemetry. If few epileptiform discharges are seen in the baseline record, spike activity can be activated, with progressive administration of thiopentane, methohexital, etomidate, or alfentanil. For instance, during the progressive injection of thiopentane (25 mg every 20 s, up to 250 mg in adults), an initial activation of one or more interictal foci, followed by burst-suppression patterns at higher doses are expected. It is assumed that epileptogenic cortex is located where discharges disappear last and reappear first.

Evaluation of interictal discharges

It is unclear whether the resection should be guided by the extent, incidence, or amplitude of the epileptiform discharges. A distinction between primary and propagated spikes is necessary. Primary discharges are associated with structural lesions and background abnormalities, and they show sharp components,

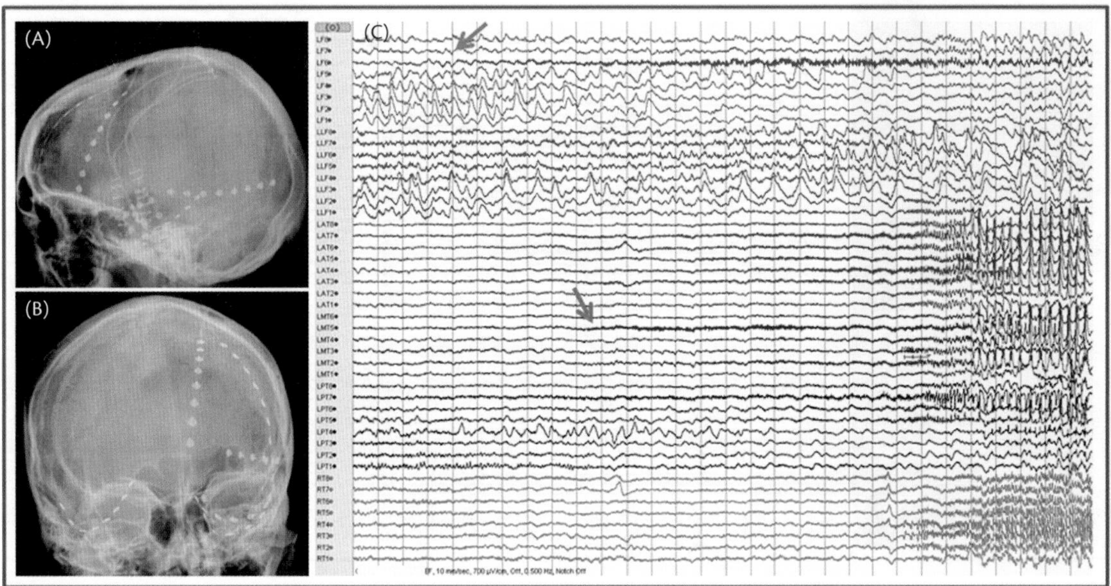

Fig. 12.13 Seizure onset in a patient with subdural electrodes over fronto-temporal-occipital lobes. (A) Lateral skull X-ray. (B) Anterior-posterior skull X-ray. (C) Seizure onset starting with focal fast activity over the left frontal (blue arrow) and spreading over the lateral temporal region (red arrow).
R = right; L = left; F = frontal; T = temporal; A = anterior; L = lateral; M = medial; P = posterior.

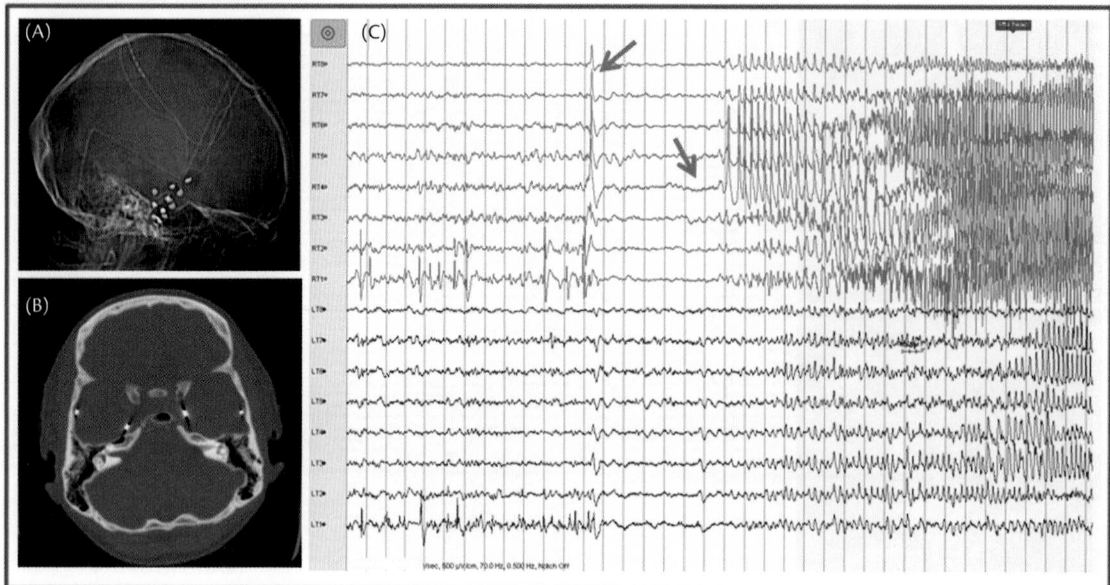

Fig. 12.14 Seizure onset in a patient with subdural electrodes over both temporal lobes. (A) Lateral skull X-ray. (B) CT showing the most lateral and deeper contacts of the subdural strips. (C) Seizure onset starting with a regional epileptiform discharge over the right temporal (blue arrow), followed by a generalized electrodecremental event, and fast activity mainly over the right subtemporal region.
RT = right temporal; LT = left temporal.

that may become blunter with propagation (3,15). Resection of the leading regions correlates with good outcome (16). Excision of all discharging areas has not been considered necessary by the Montreal School. For instance, in temporal lobe epilepsy, amygdalo-hippocampectomy is effective, even though discharges are often recorded outside this region. A reduction of more than 50% in discharges before and after resection correlates with favourable outcome (17). However, epileptiform discharges at the edges of the resection or in the insula region can be ignored.

Electrocorticography in temporal lobe surgery

In mesial temporal sclerosis (MTS), epileptiform discharges are usually seen independently in the hippocampus and in the subtemporal cortex, but can also be seen in the lateral temporal cortex (16). Positive

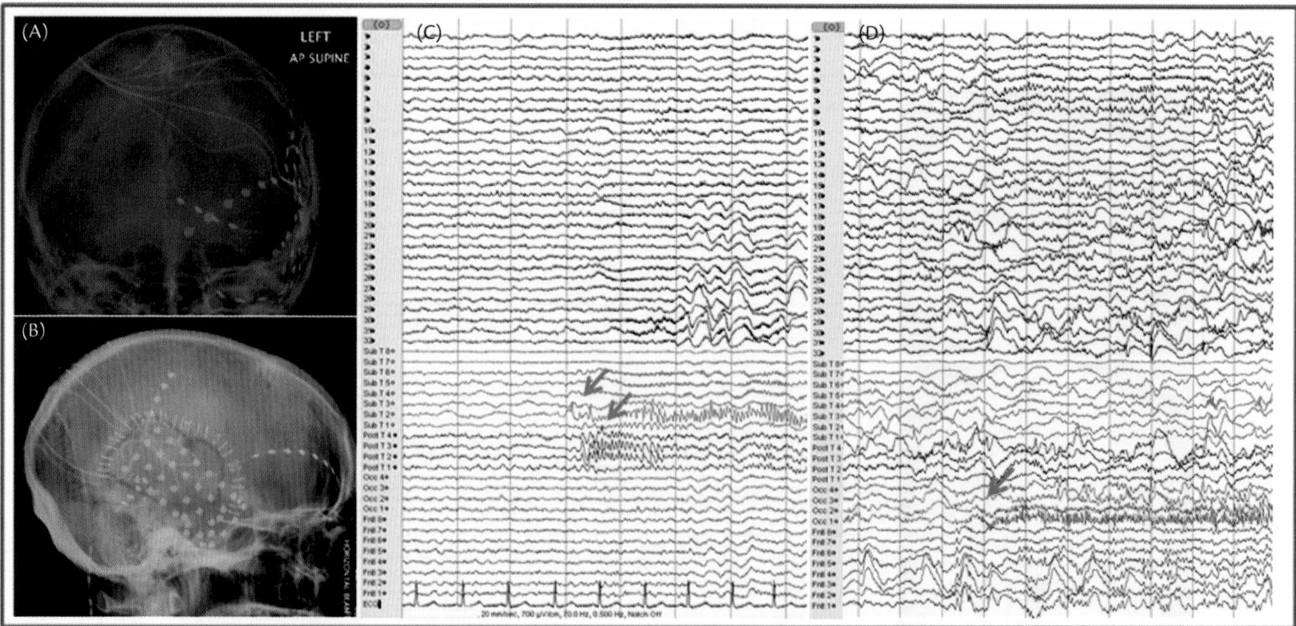

Fig. 12.15 Patient with subdural electrodes over the left temporo-parietal-frontal lobes with two independent seizure onsets. (A) Anterior posterior skull X-ray. (B) Lateral skull X-ray. (C) Seizure onset starting with regional fast activity over the anterior and mid temporal region (blue arrows). (D) Seizure onset starting with focal fast activity at the deepest contact of the posterior temporal/occipital subdural strip (red arrows).
1-32, 32-contact mat; Sub T = subtemporal anterior strip; Post T = posterior temporal strip; Occ = occipital; Frnt = frontal.

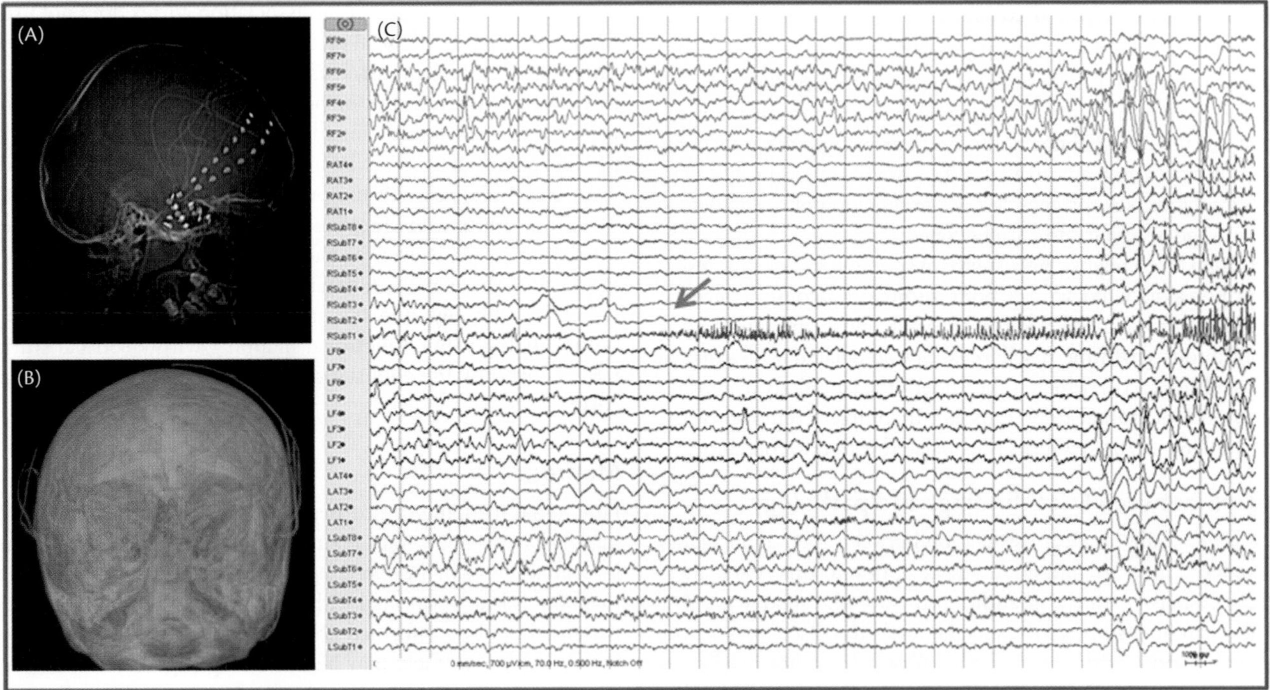

Fig. 12.16 Seizure onset in a patient with subdural electrodes over the left temporo-parietal-frontal lobes. (A) Lateral skull X-ray. (B) Volume rendered CT reconstruction. (C) Seizure onset starting with focal fast activity at the deepest contact of the posterior temporal/occipital subdural strip (blue arrow).
R = right; L = left; F = frontal; AT = anterior temporal; SubT = subtemporal.

spikes can be recorded from the hippocampus with an electrode introduced in the temporal horn of the lateral ventricle (18) (Fig. 12.17).

In many centres, ECoG is standard in non-MTS cases. In MTS cases, some centres consider that the ECoG procedure is unnecessary, as MTS tends to respond to the standard anterior temporal resection. We have found that ECoG can suggest a posterior extension of the resection in a small number of patients with MTS.

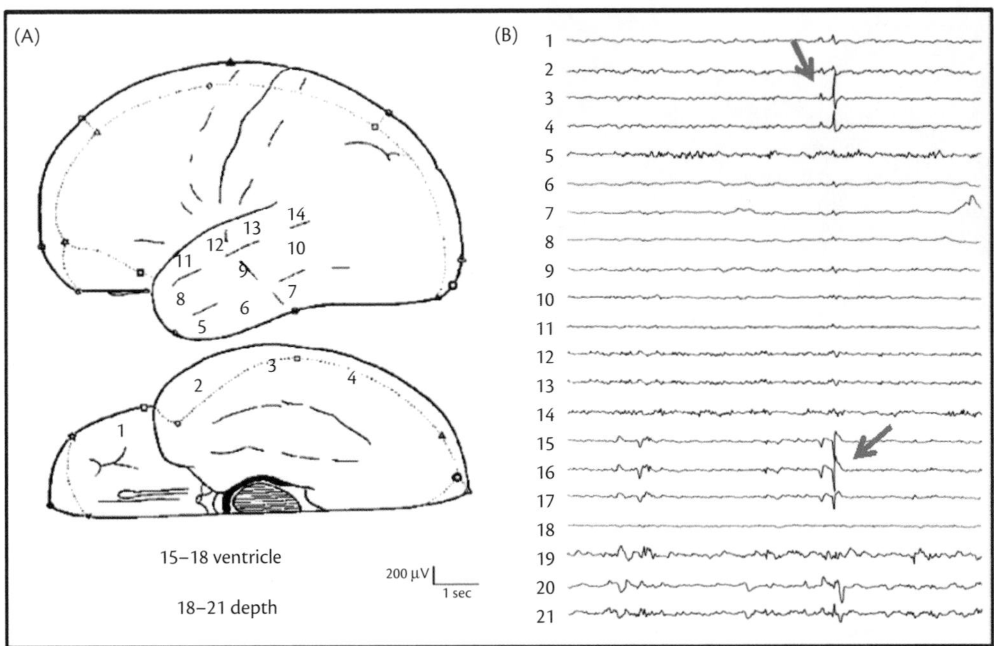

Fig. 12.17 Electrocorticography in a patient with mesial temporal sclerosis (MTS). (A) Drawing showing the position of the electrode recording. (B) Typical recording of MTS, showing positive epileptiform discharges at the hippocampus (blue arrows), associated with negative epileptiform discharges in the temporal lobe.
Adapted from *Epilepsia*, **42**, Ferrier CH, Alarcon G, Engelsman J, Binnie CD, Koutroumanidis M, Polkey CE, et al., Relevance of residual histologic and electrocorticographic abnormalities for surgical outcome in frontal lobe epilepsy, pp. 363–71, copyright (2001), with permission from John Wiley and Sons.

ECoG in extratemporal surgery

Frontal lobe epilepsy is often refractory to medical treatment, and patients are potential candidates for epilepsy surgery. Post-surgical outcome after frontal lobe resections is poorer than for temporal lobe surgery. A focal abnormality on neuroimaging and its complete removal are predictive of a favourable outcome (19–21). ECoG guidance is recommended in the resection of epileptogenic lesions in the occipital and parietal lobes.

In frontal lobe epilepsy, persistence of sporadic spikes after resection is not associated with a poor surgical outcome. However, interictal seizure patterns (bursts of fast activity, rhythmic spikes, or continuous rhythmic discharges) are associated with cortical dysplasia (Fig. 12.18), which is the underlying pathology in approximately 50% of frontal resections (8). Incomplete removal of the areas showing seizure patterns is associated with poor outcome.

Relation between ECoG findings and pathology

There is generally no clear association between EEG abnormality and underlying neuropathology. An exception to this rule is focal cortical dysplasia, which, in addition to sporadic spike wave discharges, can also show fairly specific ictal-like discharges in the interictal period, which manifest as focal areas showing bursts of spike-wave activity for a few seconds, or continuous rhythmic spike-wave activity or bursts of fast activity (8,9).

Functional mapping

Clinical and diagnostic value

To achieve good seizure control, it is often necessary to extend the resection beyond the margins of the lesion, or of the seizure onset zone. A limited resection of the association cortex does not usually induce severe neurological or cognitive deficits. However, a resection in the primary motor, sensory areas, or language areas can be associated with transient or permanent deficits. If the proposed resection is close to functionally eloquent cortex, the risk of major post-operative deficits can be minimized by accurate identification of functional areas with the purpose of avoiding their resection. The initial approach to localize motor, sensory, and language areas is anatomical. However, as there is individual variability in the location of these areas, a detailed localization is required before resection. Although non-invasive functional MRI can have potential localizing value, functional mapping with electrical stimulation is currently the gold standard procedure to identify eloquent cortex.

Methods

Functional mapping can be performed during intracranial recordings in the telemetry ward (extraoperative) or during surgery under local anaesthesia in the operating theatre (intraoperative). Functional mapping is best performed with the patient awake and relaxed. The patient should be informed about the expected clinical responses to stimulation, as these may be distressing (e.g. forced movements, speech arrest), and about the risk of provoking a seizure. Functional mapping is best performed though subdural mats, as they provide excellent spatial sampling. Stimulation of mesial temporal structures through the deepest contacts of subtemporal strips can be painful, presumably due to stimulation of nearby cranial nerves.

Trains of electrical pulses of increasing intensity and duration are applied via pairs of adjacent electrodes. The standard parameters for functional stimulation are 1–10-s trains of 1 ms electrical pulses,

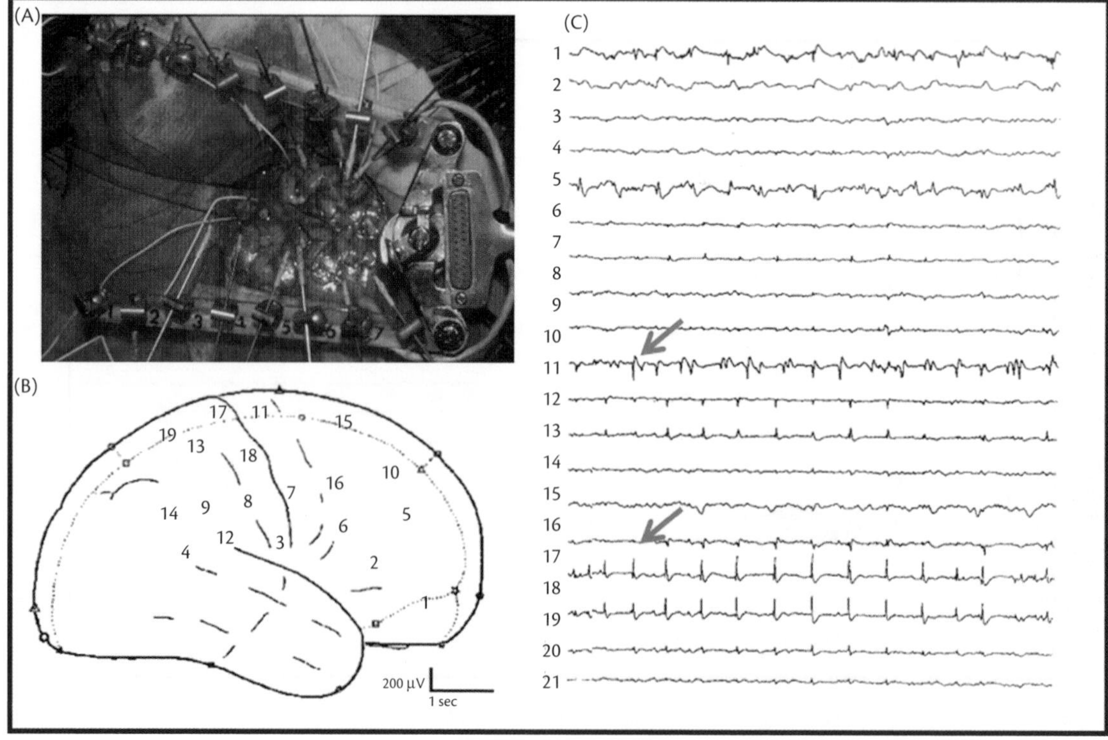

Fig. 12.18 Electrocorticography in a patient with cortico-dysplasia in the central region. (A) Intrasurgical electrode position system. (B) Drawing showing the position of the electrode recording. (C) Typical recording of cortico-dysplasia showing almost continuous epileptiform discharges over the region.

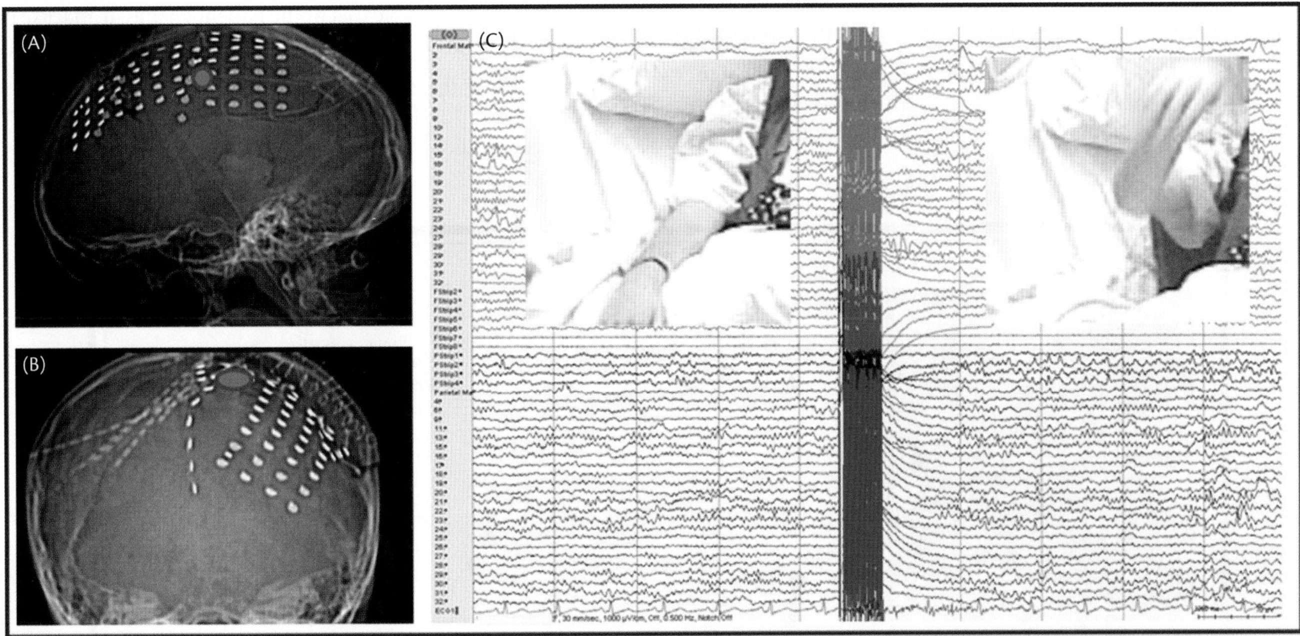

Fig. 12.19 Functional stimulation of the primary motor hand area with subdural mats and strips over frontal and parietal. (A) Lateral X-ray view. (B) Anterior posterior X-ray view. (C) Functional stimulation of contacts 7 and 9 of the subdural strip provoked elevation of the right hand.

with an intensity of 0.5–10 mA, and a frequency of 50 Hz. However, other stimulation parameters have been used with similar results.

Stimulation is carried out with progressive strength, starting at low intensities (0.5–1 mA) and short duration (1 s). If no response is seen, the duration is increased up to 5–10 s, followed by increasing the intensity with a short duration, then increasing duration, and so forth. The duration and intensity of stimulation are progressively increased through each pair of electrode's contacts until at least one of the following circumstances occurs associated to stimulation:

- *Positive clinical signs*: limb movements (Fig. 12.19), brief dystonic postures, or muscle contractions, contralateral paraesthesia (tingling, numbness, burning or sensation of electricity), visual symptoms (flashing lights, lines, colours, moving forms), aversive movements (contralateral head and eye rotation).
- *Negative clinical signs*: they consist of interruptions in normal on-going function, including arrest of motor activity and speech dysfunction. To identify these responses the patient must be engaged in a mental/motor activity.

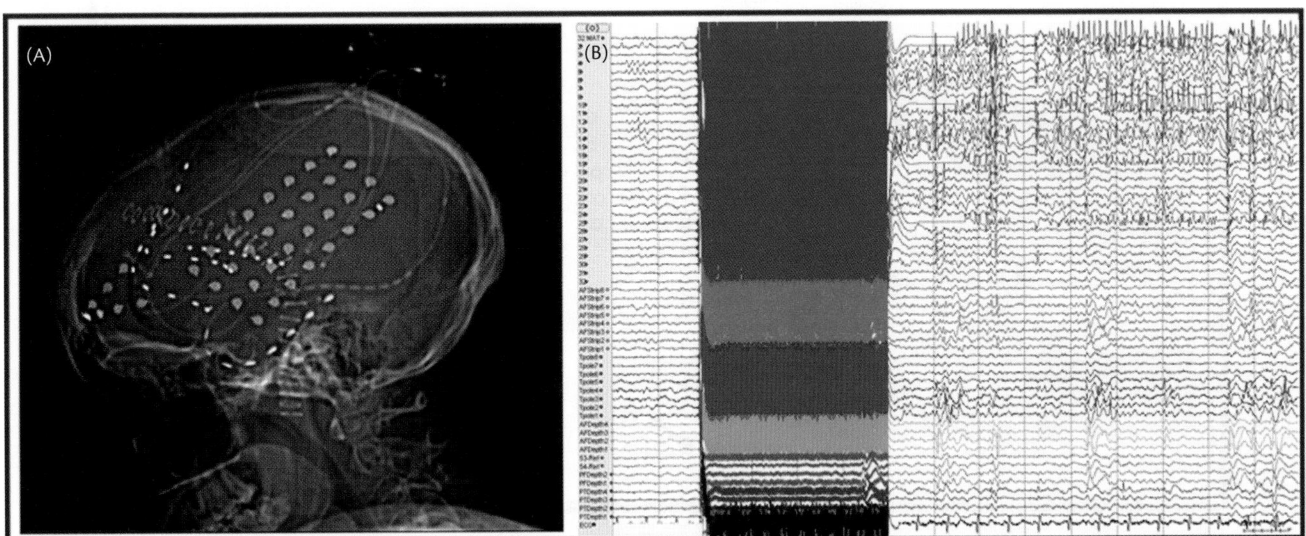

Fig. 12.20 Generation of after discharges during functional stimulation in a patient with subdural strips and a mat over the left fronto-temporo-parietal cortex. (A) Lateral X-ray view. (B) Functional stimulation of contacts 9 and 17 (posterior aspect) of the 32 contact mat provoked after discharges over most contacts located over the parietal lobe.

F = frontal, PT = posterior temporal; PF = posterior frontal.

- *After-discharges*: runs of epileptiform discharges or ictal EEG patterns induced by electrical stimulation (Fig. 12.20). They are usually short and asymptomatic, but stimulation with higher intensities can induce after-discharges, which evolve to seizures with clinical symptoms. The semiology of the induced seizures may or may not resemble that of the patient's habitual attacks. The former have localizing value.
- When reaching the maximal stimulation intensity and duration, usually around 7.5–15 mA for 6–10 sec. High intensities may be necessary in children.

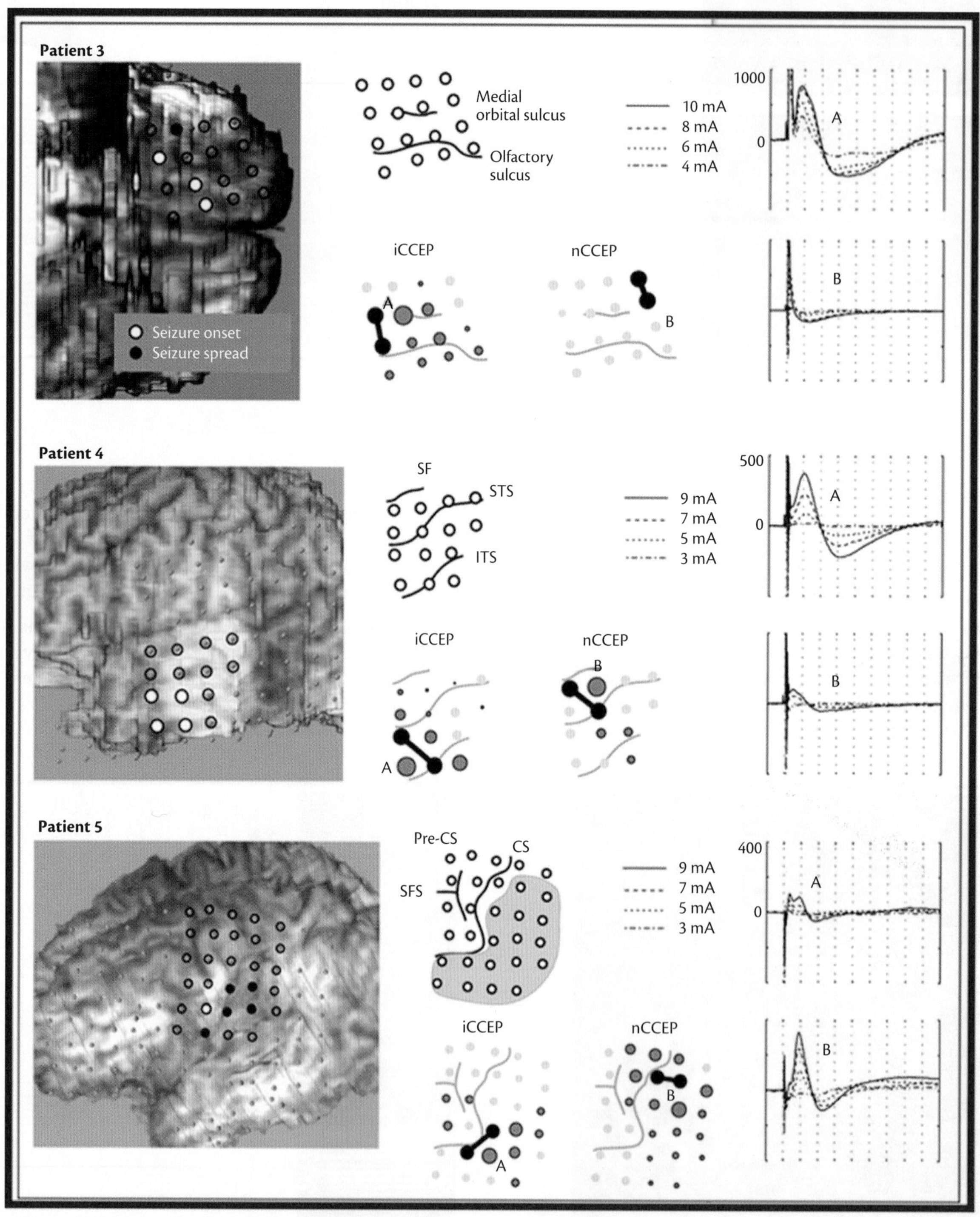

Fig. 12.21 CCEPs recording from the right orbito-frontal cortex (patient 3), left lateral temporal cortex (patient 4), and left peri-Rolandic cortex (patient 5). The latencies and CCEP waveforms are shown on the right with the iCCEP (A) and the nCCEP (B). The stimulation at the ictal onset zone evoked larger N1 in patients 3 and 4. In patient 5, the stimulation outside the ictal onset zone produced larger responses. The grey-painted area in patient 5 shows the area of cortical atrophy.

During functional stimulation, the patient and the EEG should be observed for the occurrence of clinical responses and/or after-discharges. If clinical responses occur, functional stimulation must be repeated at the same position and stimulation parameters after at least 30 s, in order to confirm findings.

Positive or negative clinical responses without after-discharges are the best indicators of correct localization of motor, sensory, or speech areas in the cortex underlying the stimulating electrodes. When clinical responses are associated with after-discharges, they have limited value as clinical responses may be due to cortical activation of relatively distant regions by after-discharges, potentially leading to mislocalization of function.

Clinical responses depending on the stimulated area

- *Motor cortex*: contralateral muscle tonic or clonic contractions and movements of the corresponding limbs and muscle groups.
- *Somatosensory cortex*: contralateral paraesthesia in the corresponding body part, most frequently tingling, numbness, burning or tightening, and less frequently painful sensations or a sense of movement.
- *Supplementary motor cortex* (anterior to the leg area in the interhemispheric fissure and dorsal aspect of the superior frontal gyrus): aversive movements (i.e. this region is also called supplementary eye field) in addition to a variety of movements and dystonic postures (often bilateral, involving aversive head rotation, and posturing of shoulders and elbows), arrest or slowing of voluntary movements and speech, vocalization, sensations (general body sensation, sensation of flush, cephalic, epigastric or indescribable sensations, contralateral or bilateral leg sensations), autonomic changes (pupillary dilation and changes in heart rate) and aphasia (15).
- *Occipital visual cortex*: phosphenes (flickering lights, dancing lights, stars, colours, shades, grey spots) or elementary visual shapes (lines, whirling circles) seen in the contralateral visual field.
- *Parieto-occipital region*: aversive eye movements.
- *Uncus, olfactory bulb, and amygdala*: olfactory and gustatory sensations, disagreeable smells.
- *Insula and Sylvian fissure*: the insular cortex is connected to two different cortical networks, a visceral network extending to the temporomesial structures and a somesthetic network reaching the opercular cortex, which induce visceral (autonomic changes, salivation, taste and abdominal sensations, such as nausea or epigastric sensations), or somatosensory sensations with stimulation of the anterior or the posterior insula, respectively (22).
- *Auditory cortex* (posterior half of the superior temporal gyrus): perception of ringing, humming, buzzing, and other sounds, as well as deafness and distortion of incoming sounds. Sounds are referred to the opposite ear or to both ears. Stimulation of the superior temporal gyrus can also be associated with suppression of hearing (23).

Language mapping

The main two specific areas of the brain responsible for language are the motor speech region (Broca's area) in the inferior frontal

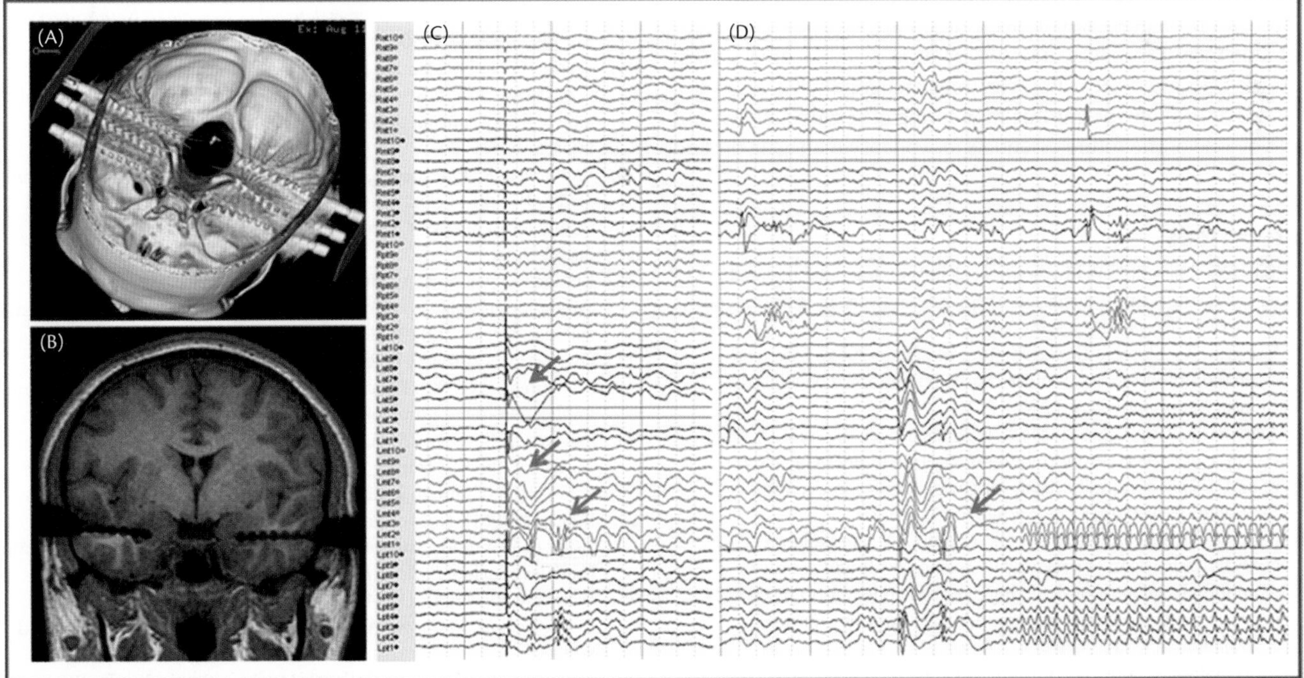

Fig. 12.22 Single pulse electrical stimulation (SPES) through contacts LaT3 and LaT 4 in a patient with bitemporal depth electrodes. (A) Volume rendered CT. (B) T1-MRI sequence. (C) Cortical responses to SPES, including normal early responses (red arrows) and abnormal delayed responses (blue arrow). Delayed responses are mainly seen in hyperexcitable and possibly epileptogenic cortex. (D) Spontaneous seizure. Note the similarities between spontaneous seizures and delayed responses (blue arrow).

L = left; R = right; T = temporal; A = anterior; M = mid; P = posterior.

gyrus (24), and the receptive speech area (Wernicke's area) in the posterior and superior aspect of the lateral temporal lobe (25).

For localization of Broca's area, the traditional task consists of asking the patient to read, count, or recite a known lullaby or song, while checking for pauses in speech associated with stimulation. For identification of Wernicke's area, auditory responsive naming, picture selection, and visual confrontation naming are the methods most commonly used.

Electrical stimulation for the identification of epileptogenic cortex

Stimulation with trains of electrical pulses

The localizing value of after discharges provoked by functional cortical stimulation has long been debated (26–28). After-discharges can be induced in areas other than the seizure onset zone and their presence is not a clear marker for epileptogenic cortex. However, the areas whose stimulation induces the patient's habitual aura or habitual seizures may be a more reliable marker for epileptogenicity (29).

Stimulation with single electrical pulses

SPES can be used to induce early responses or corticocortical evoked responses (CCEPs) (Figs 12.21 and 12.22), which could be useful to identify brain connectivity between cortical regions (30,31). In addition, SPES can provoke delayed responses (DRs), (Fig. 12.22), which are specific of hyperexcitable, and probably epileptogenic, cerebral cortex (32–34).

Recordings from permanently implanted intracranial electrodes

Recent technological advances allow recordings from chronic implantation of intracranial electrodes, either as part of a neuromodulatory therapeutic device (35) or as a monitoring system (chronic unlimited recording electrocorticography, CURE) (36). These devices allow the chronic monitoring of seizure activity, and could be one of the future essentials for epilepsy management and treatment.

References

1. Tao, J.X., Chen, X.J., Baldwin, M., et al. (2011). Interictal regional delta slowing is an EEG marker of epileptic network in temporal lobe epilepsy. *Epilepsia*, **52**(3), 467–76.
2. Valentín, A., Alarcón, G., Barrington, S.F., et al. (2014). Interictal estimation of intracranial seizure onset in temporal lobe epilepsy. *Clinical Neurophysiology*, **125**(2), 231–8.
3. Alarcon, G., Guy, C.N., Binnie, C.D., Walker, S.R., Elwes, R.D., and Polkey, C.E. (1994). Intracerebral propagation of interictal activity in partial epilepsy: implications for source localisation. *Journal of Neurology, Neurosurgery, & Psychiatry*, **57**, 435–49.
4. Fernández Torre, J.L., Alarcón, G., Binnie, C.D., and Polkey, C.E. (1999). Comparison of sphenoidal, foramen ovale and anterior temporal placements for detecting interictal epileptiform discharges in presurgical assessment for temporal lobe surgery. *Clinical Neurophysiology*, **110**, 895–904.
5. So, N., Gloor, P., Quesney, L.F., Jones-Gotman, M., Olivier, A., and Andermann, F. (1989). Depth electrode investigations in patients with bitemporal epileptiform abnormalities. *Annals of Neurology*, **25**(5), 423–31.
6. Chung, M.Y., Walczak, T.S., Lewis, D.V., Dawson, D.V., and Radtke, R. (1991). Temporal lobectomy and independent bitemporal interictal activity: what degree of lateralization is sufficient? *Epilepsia*, **32**(2), 195–201.
7. Blume, W.T., Holloway, G.M., and Wiebe, S. (2001). Temporal epileptogenesis: localizing value of scalp and subdural interictal and ictal EEG data. *Epilepsia*, **42**(4), 508–14.
8. Ferrier, C.H., Alarcon, G., Engelsman, J., et al. (2001). Relevance of residual histologic and electrocorticographic abnormalities for surgical outcome in frontal lobe epilepsy. *Epilepsia*, **42**, 363–71.
9. Palmini, A., Andermann, F., Aicardi, J., et al. (1991). Diffuse cortical dysplasia, or the 'double cortex' syndrome: the clinical and epileptic spectrum in 10 patients. *Neurology*, **41**, 1656–62.
10. Bragin, A., Engel, J., Jr, Wilson, C.L., Fried, I., and Buzsaki, G. (1999). High-frequency oscillations in human brain. *Hippocampus*, **9**, 137–42.
11. Engel, J., Jr, Bragin, A., Staba, R., and Mody, I. (2009). High-frequency oscillations: what is normal and what is not? *Epilepsia*, **50**(4), 598–604.
12. Morris, R., Valentin, A., and Alarcon, G. (2013). The value of high frequency oscillations (HFOs) in pre-surgical assessment of epilepsy. *North African and Middle East Epilepsy Journal*, **2**(2), 25–9.
13. Haegelen, C., Perucca, P., Châtillon, C.E., et al. (2013). High-frequency oscillations, extent of surgical resection, and surgical outcome in drug-resistant focal epilepsy. *Epilepsia*, **54**(5), 848–57.
14. Jiménez-Jiménez, D., Nekkare, R., Flores, L., et al. (2015). Prognostic value of intracranial seizure onset patterns for surgical outcome of the treatment of epilepsy. *Clinical Neurophysiology*, **126**(2), 257–67.
15. Penfield, W. and Little, J.H. (1954). *Epilepsy and the functional anatomy of the human brain*. Boston, MA: Brown.
16. Alarcon, G., Garcia Seoane, J.J., Binnie, C.D., et al. (1997). Origin and propagation of interictal discharges in the acute electrocorticogram. Implications for pathophysiology and surgical treatment of temporal lobe epilepsy. *Brain*, **120**(Pt 12), 2259–82.
17. McBride, M.C., Binnie, C.D., Janota, I., and Polkey, C.E. (1991). Predictive value of intraoperative electrocorticograms in resective epilepsy surgery. *Annals of Neurology*, **30**, 526–32.
18. Polkey, C.E., Binnie, C.D., and Janota, I. (1989). Acute hippocampal recording and pathology at temporal lobe resection and amygdalo-hippocampectomy for epilepsy. *Journal of Neurology, Neurosurgery, & Psychiatry*, **52**(9), 1050–7.
19. Ferrier, C.H., Engelsman, J., Alarcon, G., Binnie, C.D., and Polkey, C.E. (1999). Prognostic factors in presurgical assessment of frontal lobe epilepsy. *Journal of Neurology, Neurosurgery, & Psychiatry*, **66**, 350–6.
20. Wennberg, R., Quesney, L.F., Lozano, A., Olivier, A., and Rasmussen, T. (1999). Role of electrocorticography at surgery for lesion-related frontal lobe epilepsy. *Canadian Journal of Neurological Sciences*, **26**, 33–9.
21. Kumar, A., Valentín, A., Humayon, D., et al. (2013). Preoperative estimation of seizure control after resective surgery for the treatment of epilepsy. *Seizure*, **22**(10), 818–26.
22. Ostrowsky, K., Isnard, J., Ryvlin, P., Guénot, M., Fischer, C., and Mauguière, F. (2000). Functional mapping of the insular cortex: clinical implication in temporal lobe epilepsy. *Epilepsia*, **41**(6), 681–6.
23. Fenoy, A.J., Severson, M.A., Volkov, I.O., Brugge, J.F., and Howard, M.A., 3rd (2006). Hearing suppression induced by electrical stimulation of human auditory cortex. *Brain Research*, **1118**, 75–83.
24. Broca, P. (1961). Nouvelle observation d'aphémie produite par une lésion de la moitié postérieure des deuxième et troisième circonvolution frontales gauches. *Bulletin de la Société Anatomique*, **36**, 398–407.
25. Wernicke, C. (1874). *Der Aphasische Symtmenkomplex. Eine Psychologische Studie auf Anatomischer Basis*. Breslau: M. Cohn und Weigart.
26. Ajmone-Marsan, C. (1972). Focal electrical stimulation. In: Purpura DP, Penry J.K., Woodbury D.M., Tower D.B., Walter RD. (eds) *Experimental Models of Epilepsy – A Manual for the Laboratory*, pp. 147–72. New York: Raven.
27. Bernier, G.P., Richer, F., Giard, N., et al. (1990). Electrical stimulation of the human brain in epilepsy. *Epilepsia*, **31**, 513–20.

28. Gloor, P. (1975). Contributions of electroencephalography and electrocorticography to the neurosurgical treatment of the epilepsies. *Advances in Neurology*, **8**, 59–105.
29. Chauvel, P., Landre, E., Trottier, S., et al. (1993). Electrical stimulation with intracerebral electrodes to evoke seizures. *Advances in Neurology*, **63**, 115–21.
30. Enatsu, R., Piao, Z., O'Connor, T., et al. (2012). Cortical excitability varies upon ictal onset patterns in neocortical epilepsy: a cortico-cortical evoked potential study. *Clinical Neurophysiology*, **123**, 252–60.
31. Iwasaki, M., Enatsu, R., Matsumoto, R., et al. (2010). Accentuated cortico-cortical evoked potentials in neocortical epilepsy in areas of ictal onset. *Epileptic Disorders*, **12**, 292–302.
32. Valentin, A., Alarcon, G., Garcia-Seoane, J.J., et al. (2005). Single-pulse electrical stimulation identifies epileptogenic frontal cortex in the human brain. *Neurology*, **65**, 426–35.
33. Valentin, A., Alarcon, G., Honavar, M., et al. (2005). Single pulse electrical stimulation for identification of structural abnormalities and prediction of seizure outcome after epilepsy surgery: a prospective study. *Lancet Neurology*, **4**, 718–26.
34. Valentin, A., Anderson, M., Alarcon, G., et al. (2002). Responses to single pulse electrical stimulation identify epileptogenesis in the human brain in vivo. *Brain*, **125**, 1709–18.
35. Morrell, M.J. (2011). Responsive cortical stimulation for the treatment of medically intractable partial epilepsy. *Neurology*, **77**, 1295–304.
36. Cook, M.J., O'Brien, T.J., Berkovic, S.F., et al. (2013). Prediction of seizure likelihood with a long-term, implanted seizure advisory system in patients with drug-resistant epilepsy: a first-in-man study. *Lancet Neurology*, **12**, 563–71.

Further reading

Spencer, S., Nguyen, D., and Duckrow, R. (2009). Invasive EEG in presurgical evaluation of epilepsy. In: S. Shorvon, E. Perucca & J. Engel (Eds) *The treatment of epilepsy*, 3rd, pp. 767–99 edn. Oxford: Wiley-Blackwell.

Invasive monitoring techniques. In: J. Engel, T. Pedley, and J. Aicardi. (Eds) *Epilepsy: a comprehensive textbook*, 2nd edn, pp. 1779–832. Philadelphia, PA: Lippincott Williams & Wilkins.

Functional mapping techniques. In: J. Engel, T. Pedley, and J. Aicardi. (Eds) *Epilepsy: a comprehensive textbook,* 2nd edn, pp. 1833–58. Philadelphia, PA: Lippincott Williams & Wilkins.

G. Alarcon (2012). Intraoperative (acute) electrocorticography. In: G. Alarcon and A. Valentin (Eds) *Introduction to epilepsy*, pp. 1777–93. Cambridge: Cambridge University Press.

CHAPTER 13

Magnetoencephalography

Paul L. Furlong, Elaine Foley,
Caroline Witton, and Stefano Seri

Introduction

For presurgical evaluation prior to lesionectomy, non-invasive neuroimaging has played an increasingly important role in the last decade, in planning procedures and evaluating associated risks. In patients with drug-resistant epilepsy, surgical resection of epileptogenic brain tissue affords an important prospect for seizure reduction or prevention, provided that the epileptogenic zone can be accurately defined and that its relationship to functionally important cortex is known. Neuroimaging techniques have aided in the non-invasive pre-surgical evaluation of these questions. Task-based functional magnetic resonance imaging (fMRI) has conventionally been used in surgical planning in epilepsy to identify eloquent cortex that must be spared in a surgical intervention, while other techniques such as electroencephalography (EEG), magnetoencephalography (MEG), and intracranial electrode recordings are used to identify the epileptogenic tissue to be resected, and is most useful when it is combined with structural imaging, most commonly with structural magnetic resonance imaging (MRI) and MR diffusion tensor imaging.

The last decade has seen an exponential growth in studies that evaluate the clinical value of MEG in neurosurgical applications. This chapter attempts to summarize the key findings of these studies and the methodological developments that have improved clinical application.

Magnetoencephalography: background

As with EEG, much of the neural activity measured by MEG originates as the post-synaptic activity in the pyramidal cells of the cerebral cortex. MEG measures the vector sum of post-synaptic potentials, and is therefore a direct measure of neural activity. In contrast, BOLD fMRI, some forms of PET and SPECT imaging reflect neural activity indirectly through changes in blood oxygenation. Magnetic fields are less perturbed than electrical potentials by the scalp, skull, cerebrospinal fluid, and meninges, and are also unperturbed by previous craniotomy or burr holes, which may be problematic for post-surgical EEG assessments.

Modern MEG systems comprise 2–300+ sensors arranged around a fixed cranial helmet (Fig. 13.1). The calculation of the magnetic field is more straightforward than that of the electric field because of the symmetries and conductivity distribution of the human head. MEG therefore affords some practical advantage in comparison with EEG; an important additional consideration is that the non-contact nature of the measurement ensures that 2–300 channels of reference-free and electrode artefact-free data can be measured within a few minutes of the patient arriving for their evaluation.

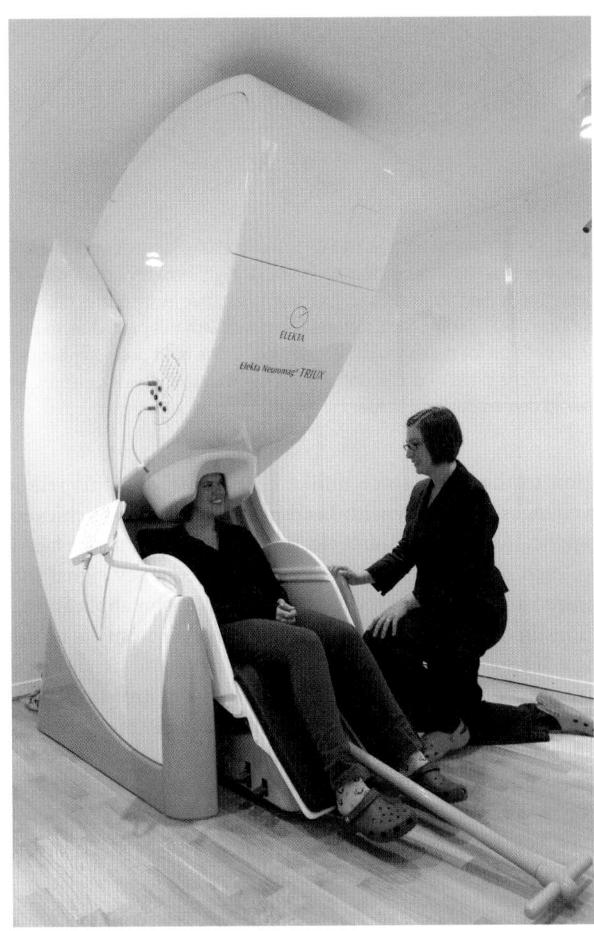

Fig. 13.1 The latest generation of MEG systems comprise over 300 detection coils connected to SQUIDs capable of measuring brain activity in the femtoTesla range (1×10^{-15} T). The patient may also be laid supine as the system pivots through 90° degrees to allow a bed to be inserted level with the helmet. The vessel above the head partly comprises a reservoir that liquid helium is syphoned into keeping the SQUID devices at the required −269°C for superconductivity.

Coregistration

To achieve accurate localization of data, there needs to be a method of establishing the precise position of the measurement helmet in

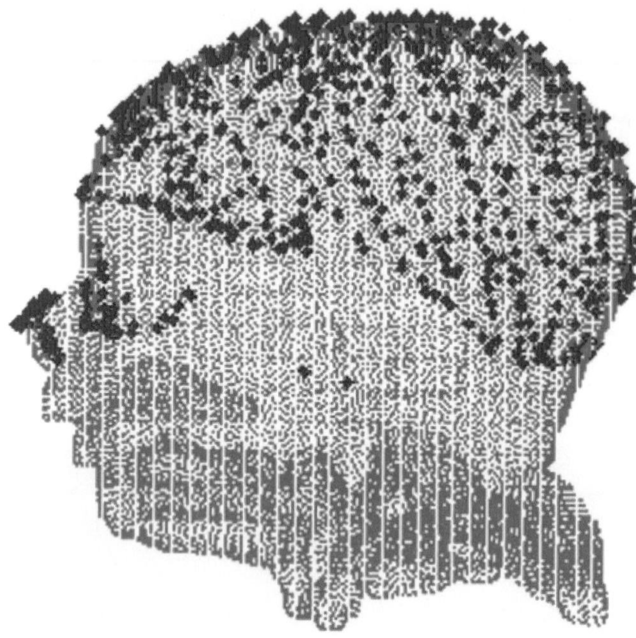

Fig. 13.2 Coregistration between MEG data and MRI volumes is achieved by using devices that allow the construction of a 3D digital image of the shape of the head. This is then optimally aligned using an iterative least-squares fitting algorithm. Including some facial features is essential with this procedure to provide a unique position for the spherical head geometry. The head positioning coils form part of the cloud of darker dots from the 3D tracking device superimposed on an MRI derived 3D head shape.

relation to the head, since unlike EEG electrodes, this relationship changes constantly with head movement. This is achieved by the applications of 3–5 small electromagnetic coils to the face and scalp, each of which emits a unique high frequency signal that is measured and localized by the MEG system. Upon application, the relationship of the coils with respect to head anatomy is achieved using a 3D digitizer (Fig. 13.2). When the magnetic signature emitted by the coils is outside of the normal range of investigated MEG signal, continuous head tracking throughout the measurement is possible.

By combining the information about the location of the coils in the MEG system and their positions with respect to the head surface, we are able to co-register MEG and MRI co-ordinate systems.

Interpretation of magnetoencephalography

Examination of the waveforms recorded by the MEG sensors or superconducting quantum interference devices (SQUIDS) requires the same expertise as the examination of the EEG signal, though critical analysis of over 300 channels of data, sometimes measuring different elements of the same magnetic field, requires some additional advanced visualization tools. Nevertheless, correct identification of artefacts and abnormal transients will be an entirely familiar process to the trained clinical neurophysiologist.

For mapping epileptiform activity, segments of the brain signal are used for source analysis, often without any signal averaging. In order to improve the signal-to-noise ratio of measured signals for evoked activity, the signal is typically averaged based on the timing of the presented stimulus.

In addition to these basic well-established methods, advances in detecting and analysing epileptiform data, as well as evoked activity have been developed and will be discussed in the 'Methodological development' section.

Once the spontaneous or evoked activity has been appropriately identified, the next vital element is to localize the sources of this activity with sufficient robustness to be of clinical value. Source modelling is the one step that falls outside of conventional training for clinical neurophysiologists, and the inherent complexities and choices of the appropriate source model has possibly constrained clinical development of MEG more than any other factor. However, there is now sufficient consensus and numbers of clinical trials to be able to summarize the key approaches.

Source modelling

All currents, both intracellular and extracellular, generate magnetic fields. MEG has the advantage though that the magnetic fields are dependent mainly on the primary currents, whereas the EEG is also influenced by the extracellular volume currents, which are difficult to model accurately. Due to the near-spherical shape of the head, the resultant magnetic fields due to primary currents may be calculated without taking into account the conductivity layers of the head. *This is the primary reason for selecting MEG as the technique of choice over the EEG when source localization is the objective.*

Source modelling determines the origin of the measured neuromagnetic fields, and is the goal of most MEG measurements. The mathematics behind this analysis is known as the inverse solution. The inverse solution of electromagnetic measurements is non-unique and mathematically ill-posed. If proper assumptions are made, however, it becomes soluble. Some of the most commonly used source analysis methods used clinically are discussed below.

Equivalent current dipole

This method is based on the assumption that, if a limited area of cortex is synchronously activated, neural activity can be modelled as a single equivalent-current dipole (ECD). The ECD provides spatial information, magnitude (current dipole moment), and direction of the estimated source. It is typically computed using an iterative least-squares algorithm, which provides a goodness-of-fit measure to match the predicted solution with the actual data. Confidence volumes based on signal to noise estimates provide important measures to evaluate the robustness of localization (1,2).

This approach is particularly effective for localizing evoked-response activity in primary sensory areas and also in focal epilepsies (Fig. 13.3). Routine clinical application of MEG still mostly uses the single ECD model. This has been demonstrated to be generally successful (3,4), although at times it may be limited or inappropriate (see Minimum Norm Estimate (MNE) below and 'Methodological developments in MEG of clinical benefit'). Techniques that may allow a greater range of success for MEG source localization with more advanced spatiotemporal multi-dipole modelling have been developed.

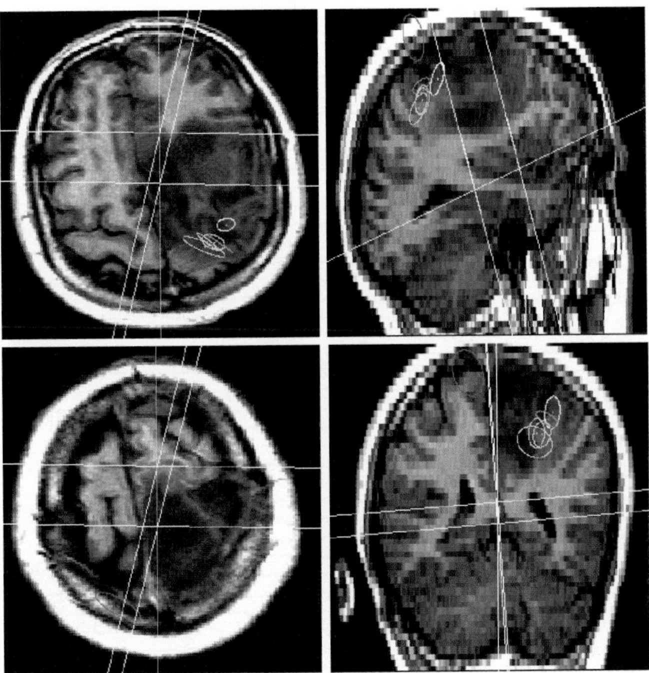

Fig. 13.3 The coloured ellipses superimposed onto the MRI are the 95% confidence volumes of equivalent current dipole fits of somatosensory evoked fields achieved by electrical stimulation of the digits of the hand (yellow, green, and white) and the posterior tibial nerve of the foot (red). The technique is particular valuable in this case since previous surgery distorts the visible gyral anatomy.
Image produced in collaboration with Mr David Sandeman, Frenchay Hospital, Bristol, UK.

Minimum Norm Estimate

If the assumptions that need to be applied for the application of a single ECD are not present (such as a large area, or several areas of simultaneously activated cortex), the single ECD solution may be misleading. Some of the problems of multiple dipole fitting, including the reliable estimation of the dipole location parameters, can be overcome by using a distributed source model that simultaneously models a large number of spatially fixed dipoles whose amplitudes are estimated from the data. The minimum norm estimate (5) is an example of such a distributed source model, and it allows the examination of the time course of activation at the source, which can be valuable in spike source estimations (6). A noise normalized minimum norm estimate is now frequently employed. This method, known as dynamic statistical parametric mapping (dSPM), also has the benefit of allowing hypothesis testing as dSPM provides an F-distribution (7).

Beamformer

Another distributed source solution is the beamformer, which is based on the principal of spatial filters. A minimum variance beamformer approach, also described by some authors as synthetic-aperture magnetometry (SAM) (8), is a beamformer approach that can be applied to both raw data and averaged (evoked) MEG or EEG.

SAM provides an evaluation of the current-dipole amplitude at a given point of interest using the covariance matrix of this activity recorded from all MEG sensors and for the entire brain volume (9). With the SAM method, as with MNE and ECD approaches, an estimate of the continuous trace of the current dipole amplitude at the point of interest can be obtained. This source based time series is sometimes referred to as a 'virtual electrode' output, approximating a time series from an implanted electrode at the source site. Source localization by surface potentials (EEG) or magnetic fields (MEG) are further complicated by the strong correlations that may exist among cortical areas, or between cortical areas and deep nuclei. The contributions of such correlations are minimized by a beamformer.

Dynamic imaging of coherent sources (DICS) (10), uses a spectral analysis with coherence or other correlations. This makes it possible to localize changes in spectral power or functional connectivity that can be used to make evoked, or resting state functional connectivity measures.

Methodological developments in MEG of clinical benefit

Spike detection and localization methodology

The majority of EEG- and MEG-based studies examining localization of epileptogenic activity have used the technique of assessing the equivalent current dipole (ECD) of measured epileptiform spike activity. There are, however, a number of caveats associated with the robust application of the method, some of which have already been discussed under 'Source modelling'. Importantly, the signal amplitude of the spike above the ambient signal needs to be significant. There needs to be an evaluation as to whether the spike activity is best modelled with either a single equivalent dipole source or multiple equivalent sources, and the method requires an initial subjective evaluation about the number and putative location of these sources. Considerable time and expertise is required in ECD analysis to identify interictal epileptiform discharges, select appropriate ECDs, and localize the epileptic zones. Nevertheless, the technique is widely used and appears robust in the hands of skilled practitioners if the preconditions are met (3).

A number of alternative approaches have been proposed for spike localization. In 2004, Robinson and colleagues (11) proposed an automated interictal spike localization technique using adaptive spatial filtering, which was named SAM(g2) (a combination of synthetic aperture magnetometry beamformer approach, and excess kurtosis (g2) method). With this method, the time-course of activity is examined for excess kurtosis and the spatial locations of voxels with high excess kurtosis are assumed to be sources of interictal spikes. The advantage of the technique is that the practitioner is presented with a series of automatically detected cortical locations with associated time series requiring no a-priori estimates of the number or putative location of spikes, and is independent of the spike amplitudes. Skilled practitioners are still required to exclude false positive kurtotic measures that may result from artefacts. The technique was subsequently evaluated on a patient group by the authors and compared with conventional ECD measures, revealing high concordance (12).

Ishii et al. (13) evaluated the excess kurtosis measure as a method to localize interictal epileptogenic activity associated with focal cortical dysplasia and reported high concordance with the ictal onset zone (IOZ) as characterized by electrocorticography (ECoG). Others (14) report a high concordance with MEG- and SAM(g2)-guided

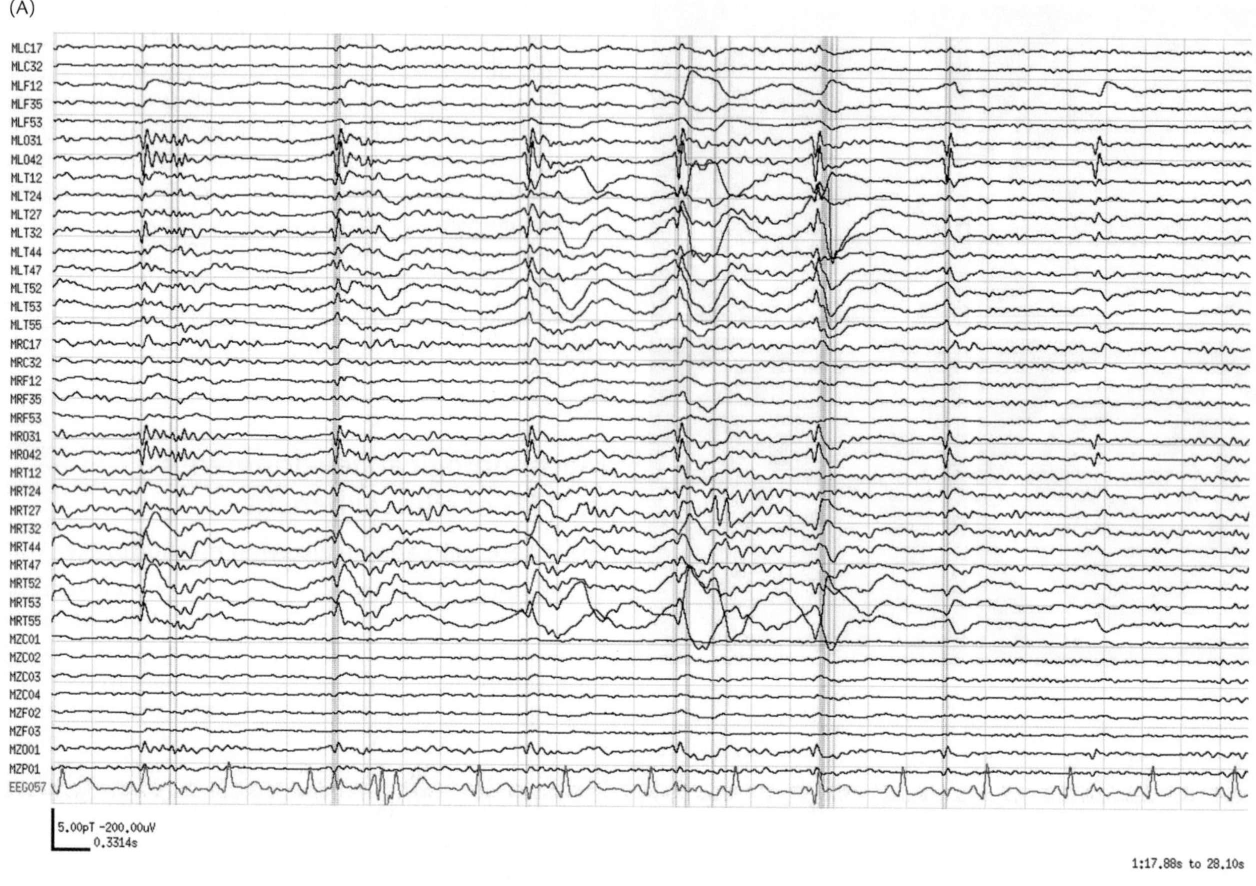

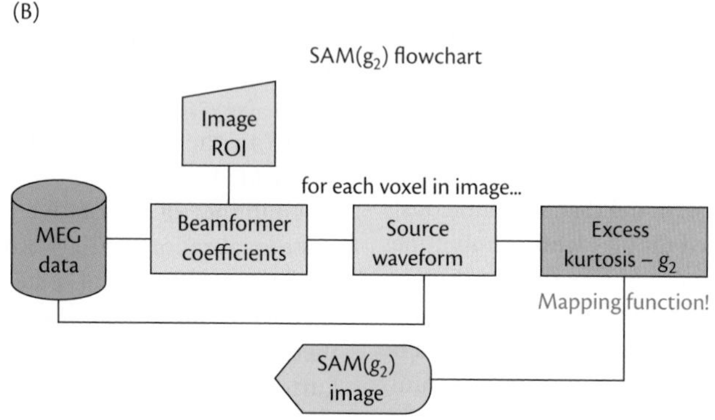

Fig. 13.4 Spike activity localized using a SAM(g2) beamformer approach (11). The technique requires no a-priori assumptions relating to the number or notional position of the dipole sources. Panel A indicates a selection of MEG sensor outputs during an inter-ictal measurement in a pre-surgical candidate. The green vertical lines indicate moments in time when the SAM(g2) algorithm has identified excess kurtosis in the sensor data. Panel B depicts the analysis pipeline employed to produce a source localization from the excess kurtosis defined data shown in Panel A.

intra-cranial electrode placement and intraoperative single pulse electrical stimulation (SPES). This study highlighted the success of the technique in guiding invasive measures when conventional approaches had failed to yield conclusive guidance (Figs 13.4 and 13.5). Other authors have reported the benefits of SAM(g2) for ictal MEG measures, including a reported ictal MEG with SAM(g2) demonstrating an SMA onset of cephalic auras in a case study, where video/EEG recordings showed no recognizable epileptic discharges associated with the onset of the isolated cephalic auras (15).

Other distributed source models have proven valuable in analysis of epileptiform data and the ability to explore the time course of spike propagation, and thus delineate the onset of epileptic activities has been reported (16). The same approach has also been advocated for ictal MEG estimates (17).

(C)

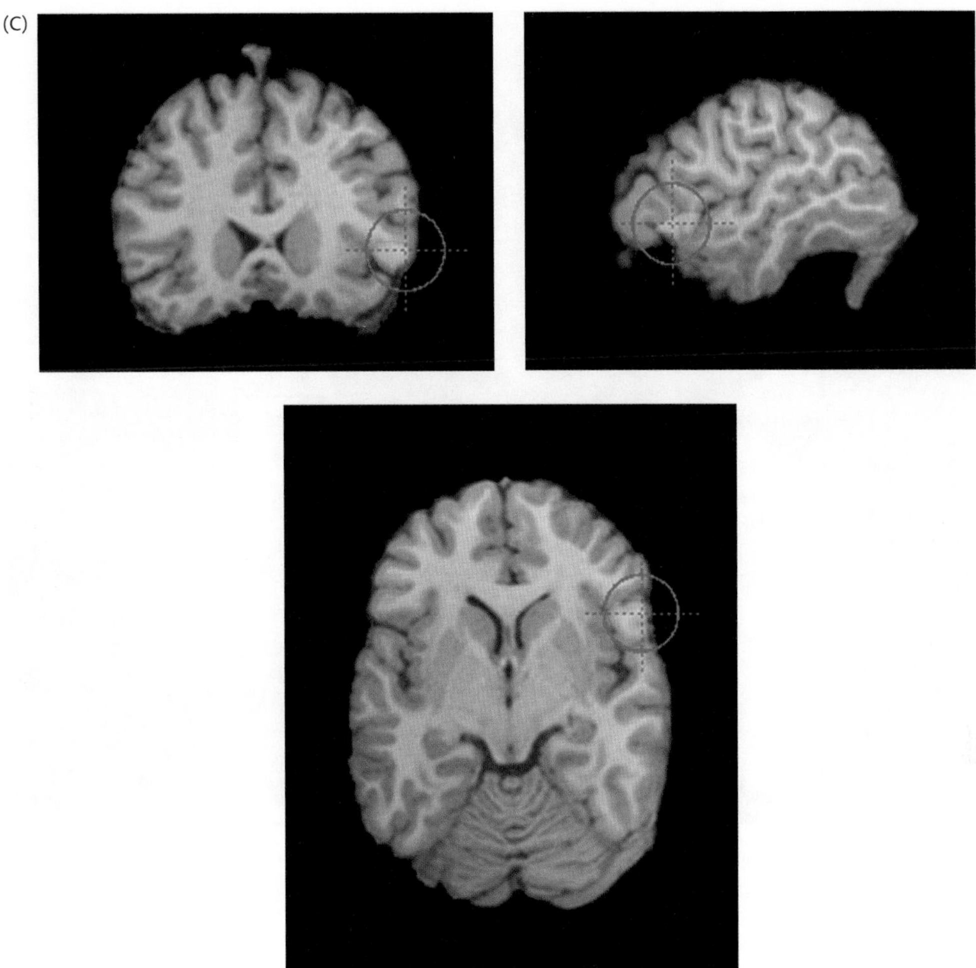

Fig. 13.4 Panel C indicates the determined source localization on the co-registered MRI of the patient, where the cortical area depicted in yellow is the calculated source of the epileptiform activity identified by the kurtosis beamformer method.

Head movement compensation

Methods to compensate for head movements during an MEG recording have been described (18,19) and validated (20). Such motion correction algorithms have enabled the incorporation of standard diagnostic protocols, such as hyperventilation and spontaneous sleep, to become part of routine MEG evaluations and have proven highly valuable for measurements on children (21). Furthermore, reliable measures can now be made at the onset of most convulsive seizures using MEG (22) and the time-locked video-MEG recording of ictal events may well further improve the sensitivity and clinical value of MEG studies (23).

High frequency oscillations

Results from studies in animals with epilepsy and presurgical patients have consistently found a strong association between high frequency oscillations (HFOs) and epileptogenic brain tissue that suggest HFOs could be a potential biomarker of epileptogenicity and epileptogenesis (24). Recent interest has focused on fast ripples (250–600 Hz), which appear to reflect pathological processes related to epileptogenicity (25–28). Ictal HFOs found with intracranial EEG have provided localizing value (29), typically with frequencies up to 500 Hz, though ictal HFOs to 800 Hz have also been reported (30). Interictal HFOs (80–500 Hz) also appear to be an indicator of seizure onset areas, independent of spike activity (31). Recent studies have also reported interictal and ictal HFOs in the medial and mesial temporal lobe that may prove to be a specific marker for mesial temporal lobe epilepsy (MTLE) with hippocampal sclerosis (31–33). All aforementioned studies had utilized intracerebral EEG recordings, although there are some reports of ictal HFOs measured using scalp EEG measures (34).

Recent studies reported non-invasive localization of HFOs (100 Hz–1 KHz) using MEG measures in a cohort of pre-surgically evaluated children (35,36). In 86% (26/30) of the children, HFOs were clearly observed. Furthermore, these authors reported that in 82% (9/11) of these patients, HFOs were concordant with intracranial recordings during surgery.

Head movement compensation, offer an enhanced ability to measure non-invasively HFOs associated with epileptiform activity, representing an important research area of significant potential clinical value.

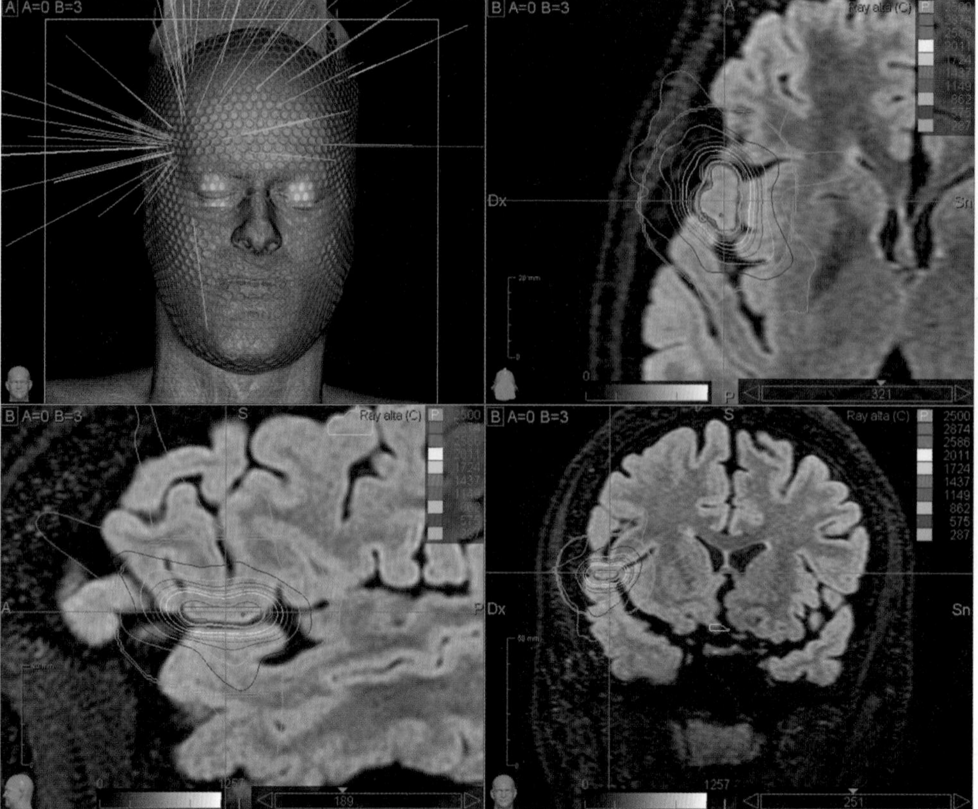

Fig. 13.5 Co-registration of the MEG source localization seen in Fig. 13.4 onto a surgical neuronavigation device allowing cyberknife coregistration.
Images courtesy of Professor Pantaleo Romanelli, Centro Diagnostico Italiano, Brain Radiosurgery, Cyberknife Center, Milano, Italy, and AB Medica, Lainate, Italy.

Functional connectivity

Functional connectivity refers to coupled activity in anatomically distinct regions of the brain, such that if two areas are highly coupled in their activity over time, they are considered functionally connected. The study of brain networks in a relaxed awake state and in the absence of a specific task has gained increasing attention, as spontaneous neural activity has been found to be highly structured at a large scale. This so called 'resting-state' activity has been found to comprise of certain non-random spatiotemporal patterns and fluctuations, and several resting-state networks (RSN) have been found in BOLD-fMRI (37–39) as well as in MEG signal power envelope correlations (40). The underlying anatomical connectivity structure between areas of the brain has been identified as being a key to the observed functional network connectivity.

Functional connectivity analysis in the resting state using fMRI techniques has recently yielded data suggesting that cortical regions are organized differently in epilepsy patients, either as a direct function of the disease or through indirect compensatory responses (41–44). Abnormally high functional connectivity across small distances could influence epileptic spike generation and increased connectivity has been hypothesized to be responsible for the propagation and generalization of epilepsy.

Functional connectivity techniques using MEG is a rapidly developing area (40,45,46) and may help to define epileptogenic cortex (47) and characterize epileptic networks (48).

Magnetoencephalography in clinical epilepsy and brain mapping

MEG accuracy in epilepsy localization

The theoretical advantages of MEG in cortical source localization have been extensively reported (2,49–51), and much resource has been placed into evaluating its practical clinical value with respect to epilepsy, epilepsy surgery and lesionectomy. These validations have broadly taken the form of intracranial electrode co-localization, using phantoms (52), implanted intra-cranial EEG measures (13,16,53–68), and intra-operatively (14). The following sections summarize their key findings.

MEG and simultaneous ICEEG

In three separate studies, Mikuni et al. (59), Shigeto et al. (65), and Oishi et al. (60) measured paroxysmal epileptiform magnetic fields simultaneously with intracranial EEG measures (ICEEG) from superior-lateral and lateral-frontal cerebral cortex. A consensus conclusion was that spikes may be detected using MEG in these regions when the activity extended 3 cm^2 or more across the cerebral cortex, although MEG shows a significantly higher sensitivity to lateral convexity epileptic discharges than to discharges in restricted basal temporal regions. In a similar study by Santiuste et al. (61), MEG detected and localized 95% of the neocortical spikes measured simultaneously with ICEEG, but only 25–60% of spikes from mesial structures. Mesial temporal spikes yielded lower MEG spike amplitudes in comparison with neocortical spikes (see 'Mesial temporal lobe epilepsy').

Comparison of interictal MEG with interictal ICEEG

In common with studies of simultaneous MEG–ICEEG, interictal MEG compared with independently recorded interictal ICEEG report a high concordance, seen independently of epilepsy type (53,57,64,69). Other studies report anatomical site-dependent variance in concordance, with 90% or greater concordance in the inter-hemispheric and frontal orbital region, approximately 75% in the superior frontal, central, and lateral temporal regions, but only approximately 25% in the mesial temporal lobe (53). In a study of anterior temporal spikes (67), concordance with ICEEG was only reliable at the lobar level. In a comparison of ICEEG with MEG and ICEEG with MRI-SPECT (63), sublobar ICEEG-MEG concordance, and complete focus resection significantly increased the chance of seizure freedom after epilepsy surgery (80%). In comparison, in 43% of patients with concordant ICEEG and MRI-SPECT and complete focus resection, only 67% became seizure-free.

Comparison of interictal MEG and ictal-ICEEG

Several authors have reported a positive predictive value of MEG for seizure localization of between 82% to 90%, depending on whether computed against ictal ICEEG alone or in combination with surgical outcome (57,62,68,70). These studies also suggest that MEG can complement ictal semiology in establishing a non-invasive focal localization hypothesis.

ICEEG is used as the 'gold standard' in most comparison/concordance studies of ICEEG-MEG. However Knowlton et al. (57) observed that ICEEG results in seven cases appeared to be either falsely positive (falsely localized; $n = 4$) or falsely negative (falsely non-localized; $n = 3$) compared with early surgical outcome. In two of the seven cases that were false-negative by ICEEG, the MEG appeared to be true-positive. The authors concluded that with more long-term outcome data, the accuracy of MEG independent of ICEEG can be further assessed.

Ictal MEG

Measurement of ictal MEG has conventionally been problematic due to limited time in the MEG recording suite and also patient movement either during prolonged measurements or with the clinical onset of the seizure. Despite this, several studies have demonstrated the value of a few seconds captured at seizure onset (15,54,66,71–73). Three of these describe successful localization of seizures with a single equivalent current dipole (ECD) model (66,71,73). Occasional failures of the ECD approach to localize the seizure onset zone (SOZ) from ictal MEG data have partly been attributed to limitations of the ECD model approach in these cases (66,72,73).

Other authors have reported the benefits of alternative analysis methods for ictal MEG, and these include a kurtosis beamformer measure SAM(g2) (15); minimum norm estimates (17); spectral power change analysis (29,74), and dSPM (75). Medvedovsky (23) using a movement compensation algorithm (18), was able to make lengthy ictal recordings in 34 patients. All MEG ictal onset zone calculations were compatible with overall irritative zone findings and ictal-ICEEG, but in addition, yielded more detailed results than the respective EEG data in two cases.

The value of MEG for the optimal placement of IC electrodes

Although ICEEG is considered a 'gold standard' for localization due to its capability of long-term recording directly from the cortex of real time spontaneous seizures, it has a number of limitations. The most important is limited sampling of a small fraction of cerebral cortex. To overcome this, accurate preoperative indication of which brain region(s) to target with ICEEG is required (56). A resultant goal of non-invasive epilepsy imaging tests is to improve the hypothesis of seizure-onset zone localization that is used to guide ICEEG. Mamelak et al. (58) critically reviewed their experience with whole-head MEG in the management of patients undergoing epilepsy surgery. They describe case studies where MEG was used to guide invasive electrodes to locations that otherwise would not have been targeted, providing unique localization data that strongly influenced the surgical management of these patients.

Agirre-Arrizubieta et al. (14) utilized a case-control study to investigate the role of MEG in the identification of candidate sites for intracranial implantation, where other non-invasive pre-surgical assessments were inconclusive. Primary outcome measures were the identification of focal seizure onset in intracranial electrode recordings together with the application of single pulse electrical stimulation (SPES) and favourable surgical outcome. There was high concordance between seizure onset and SPES findings and between MEG and intracranial seizure onset (76–78). In all MEG patients, at least one virtual MEG electrode generated suitable hypotheses for the location of implantation, supporting the view that MEG can contribute to the identification of implantation sites where standard methods failed. The concordance between seizure onset zone and MEG was high, and in one patient where findings were discordant, the patient did not improve after surgery.

Sutherling et al. (79) determined whether MEG changed the surgical decision in a prospective, blinded, cross-over-controlled, single treatment, observational case series. In 69 sequential patients diagnosed with partial epilepsy of suspected neocortical origin, MEG gave non-redundant information in 23 patients (33%), added ICEEG electrodes in nine (13%), and changed the surgical decision in another 14 (20%). Based on MEG, 16 patients (23%) were scheduled for different ICEEG coverage, and this benefitted 21% (16) who consequently went to surgery. Similarly, Knowlton et al. (56) characterized 62 patients having surgical resection based on ICEEG recording of seizures. Use of MEG indicated additional electrode coverage in 18 of 77 ICEEG cases (23%), of which 39% showed seizure onset ICEEG patterns involving the additional electrodes indicated by MEG. Highly localized MEG activity was also significantly associated with seizure-free outcome (mean = 3.4 years, minimum > 1 year) for the entire surgical population.

Lesion co-localization

Well-delineated lesion co-localization represents strong validation of MEG spike-source localization, particularly with low-grade tumours or lesions that clearly are the single focal cause of a given patient's epilepsy. In several reported cases, MEG played a role in directly determining the functional epileptogenic significance of abnormalities, successfully localizing the perilesional epileptogenic zone in ~86% of cases (80–85).

Focal cortical dysplasia

Focal cortical dysplasia (FCD) is often associated with severe partial epilepsy. A number of studies have explored the effectiveness of MEG in characterizing the epileptogenesis of MRI visible FCD (13,81,83,86–88) indicating that resections with a significant overlap with the MEG source cluster area were associated with better clinical outcome. One significant challenge is to determine the extent of neighbouring tissue that should be included in the resection in addition to the lesion removal. This issue was addressed specifically in an MEG study of 12 children with lesional neocortical epilepsy (81). It was concluded that cortical dysplasia characterized by clusters of MEG spike sources within, and extending from lesions seen on MRI, should be removed to prevent seizures. Other authors concur that when the resected area included the onset zones of averaged EEG and MEG spike activity, these patients had excellent post-operative outcome, whereas the others did not become seizure free (13,86).

Cryptogenic lesions

As previously described, high coverage of MEG dipole sources by the resection area was correlated with a favourable post-operative seizure outcome (80–82), but in these studies nearly all subjects had focal lesions identified on brain MRI. Kim et al. (85) characterized a population in which only seven of 22 cases (31.8%) had localized abnormalities on MRI. In a study of patients with normal MRIs (89), the results showed that cases with multiple clusters required delineation of the epileptogenic zone by ICEEG. In some cases with a single focal cluster (with or without a focal MRI lesion) ICEEG was still required to define the actual epileptogenic zone. Others report good post-surgical outcome in MRI-negative patients when there was good sublobar concordance between MEG and ICEEG and complete resection of both regions (62).

Zhang et al. (90) characterized the value of MEG in providing critical information for the placement of ICEEG electrodes in non-lesional cases. It has also been reported that MEG can contribute to the identification of epileptogenic lesions in MRI-negative cases and can be particularly useful in finding small FCDs not visible on MRI (91).

This includes lesions located in deep anatomical structures, such as peri-insular cortex (92).

MEG predictors of post-surgical freedom in children with refractory epilepsy, and normal or non-focal MRI findings have been identified. These include a study of 22 children by Ramachandran Nair et al. (90) who reported good post-surgical outcomes in the majority of patients where a restricted ictal onset zone predicted post-operative seizure freedom, and was most likely to occur when there was concordance between EEG and MEG localization, and least likely to occur when these results were divergent. Seizure-freedom was also less likely to occur in children with bilateral MEG dipole clusters or only scattered dipoles, multiple seizure types, and incomplete resection of the proposed epileptogenic zone.

Temporal lobe epilepsy

Studies vary in the reported value of MEG for pre-surgical localization of temporal lobe spikes in adults. Several authors report a high yield of accurate spike detection in lateral and basal temporal lobe in comparison with ECoG (53), in comparison with simultaneous EEG (93) and invasive EEG (94). However, mesial temporal lobe epilepsy presents additional challenges.

Mesial temporal lobe epilepsy

MEG helmet design based on adult head sizes are sub-optimal for anterior and inferior temporal lobe, and inferior frontal lobe measures in adults (95), and is significantly more pronounced in children. This raises concerns as to whether MEG is able to provide information in patients with MTLE.

The sensitivity of whole-head MEG to detect and localize epileptic spikes has been evaluated through simultaneous recordings of ICEEG and MEG (61). Interictal MEG and depth electrode activities from the temporal mesial and occipital lobes were simultaneously recorded from four candidates for epilepsy surgery. MEG detected and localized 95% of the neocortical spikes detected by ICEEG, but only 25–60% of spikes from mesial structures. Mesial temporal spikes were reflected by lower MEG spike amplitudes, when compared with neocortical spikes. These data are supported by several studies indicating that MEG detects few, if any, mesial temporal or hippocampal-only spikes (53,59,60,65).

In contrast, others have reported that 60% of patients with MTLE with non-localizing ictal scalp EEG had well-localized spikes on MEG (55). Poor spike yield in MTLE has been attributed to inadequate coverage of the temporal lobes by the MEG helmet (96). There are no comparable systematic comparisons of spike detection and yield in paediatric age in temporal lobe epilepsy (TLE), but given the increased distance of temporal lobe sources from sensors in children, detection rates and yield are likely to be impoverished in comparison with adult evaluation. Overall, the ability of MEG to detect MTLE spikes appears to be determined by the extent of the cortical activation. ICEEG with MEG measures have revealed that at least three subdural electrodes (1 cm apart) are required (57,62).

Ictal HFOs in the medial temporal lobe may be a specific marker for MTLE with hippocampal sclerosis (32), and it has been reported that HFO's are most frequent in mesial temporal structures and are prominent in the SOZ, providing additional information on epileptogenicity independently of spikes (31).

Frontal cortex localization

The benefits of MEG over EEG for epilepsy, particularly in the frontal lobe, are attributed to MEG's significantly improved signal-to-noise-ratio for sources in this area (97). Similar to lateral temporal lobe sources, for MEG to detect the majority of spikes recorded from the dorsolateral frontal lobe, ICEEG activation of at least three subdural electrodes (1cm apart) appears to be required (60,65).

The question as to whether the advantage of MEG over EEG is compromised in paediatric age due to a reduction in signal to noise through increased cortex to sensor distance remains to be evaluated.

Secondary bilateral synchrony

A relatively common epilepsy localization problem in extratemporal epilepsy is that of secondary bilateral synchrony. MEG can be superior to scalp EEG to address this problem (98) and there is significant value of spatiotemporal analysis of spikes from MEG to determine the primary source (99,100). Application of a source space time series, such as the beamformer 'virtual electrode' output may be of particular value in this regard (14).

Spike detection

An issue that is infrequently addressed in outcome studies relates to the number of patients with intractable epilepsy referred for MEG assessment in whom no epileptiform activity is recorded during the session. In a large retrospective study, Cheyne et al. (101) reported that no epileptiform activity was detected in 159 of 460 epilepsy recordings (35%). A similar level of non-epileptic recordings have also been reported in other studies (102,103).

Alkawadri et al. (3) describe their experience in 19 patients (5% of patients over a 3-year period) who underwent a repeat MEG recording due to the absence or paucity of interictal activity in the original recordings. In their repeat cohort, new localizing or complementary information was achieved in 58%.

In a study exploring MEG and TLE, Assaf and colleagues reported an adequate yield of spikes in all 26 patients evaluated and attributed the high success yield to a protocol in which patients were on sub-therapeutic anticonvulsant levels and because they encouraged sleep during MEG recording sessions to facilitate spike recording (104).

Functional mapping

The ability to accurately define functionally critical cortical areas has significant consequences on the extent of resection to avoid potential damage to these areas, and may compromise the eventual outcome. Robust techniques for localizing sensory-motor cortex have been developed, while a number of language and memory paradigms continue to be refined.

Sensory-motor cortex localization

Precise localization of the motor cortex and somatosensory cortex (central sulcus) is commonly required for fronto-parietal lesions or for frontal lobe epilepsy. In presurgical studies, somatosensory mapping is usually confined to a few digits, toes and the lip. Sensory stimulation is often either electrical (105,106), or through tactile/pressure devices (107–109). Motor mapping can be performed by synchronizing movement initiation with a visual or auditory cue, and to recording associated movement electromyographic (EMG) measures.

As with evoked electrical potentials, a premovement 'readiness potential' can be recorded as an MEG field, lasting about 500 ms and occurring 1–2 s before EMG onset (110). Motor evoked fields begin with peaks of 20–50 ms after EMG onset, followed by variable, but typically identified larger amplitude components occurring at around 100s, 200, and 300 ms after movement onset. One problem with mapping motor-activated neural activity for clinical purposes is that, even with simple spontaneous repetitive finger movement, proprioceptive activation of sensory cortex localizes to a post-central gyrus. A second challenge is that motor cortex sources tend to be orientated primarily radially in relation to the detecting coils (101). Therefore, the MEG motor response is a weak signal that is difficult to image with statistical significance using the single ECD technique.

Other methods have been described to improve signal source localization evoked fields generated by motor tasks, and these include an event-related beamformer (ER-beamformer) paradigm (111,112). The ER-beamformer was also successfully applied to pre-surgical functional mapping of sensory and auditory function (113), providing the added advantage of being robust to noise generated by implants and dental braces—a highly valuable benefit when studying paediatric populations (114).

The motor cortex may also be mapped by quantifying the changes in beta rhythm associated with movement. Placing bipolar electrodes over the first interosseous muscle and instructing the patient to make brisk transient abductions 'like scissors opening and closing' using their right or left index finger, generates a large signal in both the EMG and MEG. Calculation of the MEG changes in beta, particularly the so-called 'post-movement beta rebound' observed following the movement, results in source localizations that are known to originate in the motor cortex (115).

Using a similar motor task (116), Cheyne et al. have shown that gamma band activity is observed between 60 and 90Hz, and occurs between 0 and 300 ms following EMG onset, which is robustly localized to M1.These examinations can be completed within 1–2 h with results integrated into frameless stereotaxy systems for intraoperative use. A significant benefit of MEG mapping studies is aiding the preoperative planning of surgery, especially surgical approaches in patients with distorted anatomy from lesions that can preclude accurate predictions of sensorimotor function localization based on anatomy (107,109).

Language cortex evaluation

Surgery for temporal lobe epilepsy is commonly a left anterior temporal lobectomy, making localization of cortex subserving language function an imperative in these cases. Testing whether a temporal lobe is dominant for verbal function is often referred to as determining language 'lateralization'. Language cortex, typically the posterior language cortex, also needs to be precisely *localized* when a lesion is located in the posterior temporal lobe.

Until recently, comprehensive language mapping during surgical planning has relied on the application of invasive diagnostic methods, namely the Wada procedure and direct electrocortical stimulation mapping, often considered as the 'gold standard' techniques for identifying language-specific cortex. Language processes are complex and include phonological, lexical, and also semantic processes. Verbal memory encoding and retrieval occurs concurrently in virtually any task, making it difficult to separate language processing from working memory.

Sequential dipole fitting to both auditory and visual words appear to provide robust lateralization of language function (117). Other approaches that have been compared favourably with Wada include an auditory verb generation task (118), a silent reading task (119), and an auditory stimulus sequence comprised of two one-syllable spoken words presented in an oddball paradigm without subject response (120). Other techniques that have been evaluated include a verb-generation task assessed pre- and post-operatively (121) (see Fig. 13.6), a picture and word verb generation task (122), and an auditorily presented verb generation task (123), both evaluated for concordance with fMRI.

Service evaluation studies

A number of studies have set out to evaluate the value of MEG as part of their service provision for pre-surgical evaluation (79,85,101–104,124,125) and many of these have already been cited under relevant sections. Here, we summarize some general consensus observations.

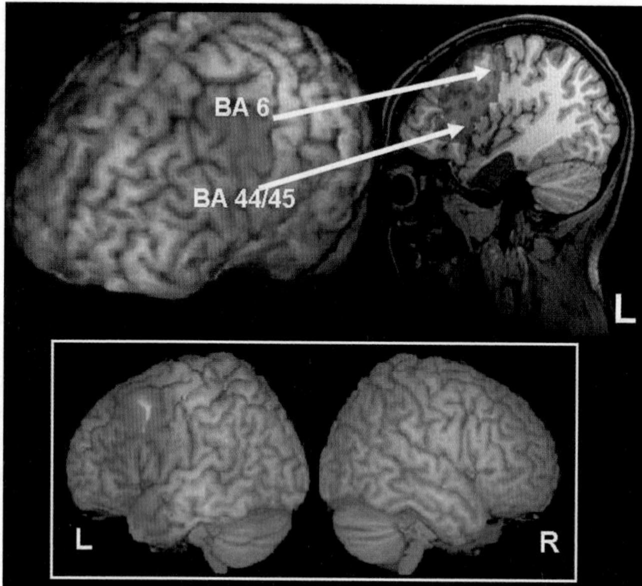

Fig. 13.6 Verb generation tasks are amongst a group of stimuli, which appear to provide valuable lateralization information. In this epilepsy surgery candidate, a dual-state beamformer was used to characterize the pattern of beta power changes in right and left hemispheres. Studies suggest that the beta power drop (indicated by the colour blue on the MRI image) in the left hemisphere corresponds well to left language lateralization. The colour red on the MRI image indicates an increase in beta power during the language task, and is frequently seen, as here, occurring coincidentally in the right hemisphere when language is left lateralised.

Spike detection in comparison to EEG

Seizure freedom is most likely to occur when there is concordance between EEG and MEG localization and least likely to occur when these results are divergent (124). MEG spikes may occur in the absence of spikes on the EEG because of the better signal-to- noise ratio of MEG for sources in superficial, neocortical areas.

Spikes visible on EEG, but undetectable on MEG, are commonly explained by the better sensitivity of EEG to deeper and radial sources.

Post-operative seizure freedom is less likely to occur with bilateral or scattered MEG dipole clusters, multiple seizure types and incomplete resection of the proposed epileptogenic zone.

Detecting activity from deep sources poses challenges for MEG because of the rapid amplitude decrease of magnetic fields with increasing distance between sensor and source. In patients with previous surgery, skull defects, large lesions, and malformations impact on EEG measures, and may result in incorrect localization due to volume conduction effects. In these patients, MEG should be used, especially in patients who are not seizure-free after an initial surgery and require re-evaluation.

Association with lesions

MEG localizations associated with a lesion indicates its epileptogenicity and supports the viability and success of epilepsy surgery. MEG localizations in MRI-negative cases require further supportive investigation. MEG can highlight subtle structural changes when they are accompanied by abnormal activity and scanning with higher field MRI can be beneficial in these cases.

Invasive recordings

Patients with complex results may require invasive recordings. Several studies indicate that MEG yields non-redundant information in around 30% of patients with epilepsy of suspected neocortical origin, aiding in the location and extent of invasive measurements. In some patients, invasive recordings may not be viable or repeatable, due to previous surgery, limited cooperation, young age, or comorbidities. In such cases, MEG localization provides additional information for planning of surgery.

Functional mapping

As a direct measure of neuronal activity, evoked responses using MEG are not influenced by altered perfusion associated with tumours, stroke, or temporal lobe epilepsy as is fMRI. Routine application has been shown to yield good results, complementing other methods, such as intraoperative electrical stimulation or the Wada test in patients with epilepsy.

Evaluation in paediatric age

MEG predictors of post-surgical freedom in children with refractory epilepsy and normal or non-focal MRI findings have also been identified. Restricted ictal onset zone predicts post-operative seizure freedom, and is most likely to occur when there is concordance between EEG and MEG localization. Development of movement compensation technology extends the usefulness of MEG in cases where compliance with functional mapping is limited. There are no systematic comparisons of spike detection and yield in paediatric age in TLE, but given the increased distance of temporal lobe sources from sensors in children, detection rates and yield are likely to be impoverished in comparison with adult evaluation.

An increasing number of studies are demonstrating that MEG adds to the surgical evaluation process and should be applied as part of the clinical routine. Recent developments in technology for movement compensation and enhanced noise reduction, together with the characterization of high frequency oscillations and functional connectivity, provide optimism for continually improving outcomes of MEG-enhanced pre-surgical evaluations.

References

1. Mosher, J.C., Spencer, M.E., Leahy, R.M., and Lewis, P.S. (1993). Error bounds for EEG and MEG dipole source localization. *Electroencephalography & Clinical Neurophysiology*, **86**(5),303–21.
2. Hamalainen, M., Hari, R., and Ilmoniemi, R.J. (1993). Magnetoencephalography—theory, instrumentation, and application to noninvasive studies of the working human brain. *Reviews in Modern Physics*, **65**:413–97.
3. Alkawadri R, Burgess R, Isitan C, Wang IZ, Kakisaka Y, Alexopoulos A V. Yield of repeat routine MEG recordings in clinical practice. *Epilepsy & Behavior*, **27**(2), 416–19.
4. Shiraishi, H. (2011 Source localization in magnetoencephalography to identify epileptogenic foci. *Brain Development*, **33**(3), 276–81.
5. Gramfort, A., Luessi, M., Larson, E., et al. (2014). MNE software for processing MEG and EEG data. *NeuroImage*, **86**, 446–60.
6. Shiraishi, H., Ahlfors, S.P., Stufflebeam, S.M., et al. (2005). Application of magnetoencephalography in epilepsy patients with widespread spike or slow-wave activity. *Epilepsia*, **46**(8), 1264–72.
7. Dale, A.M., Liu, A.K., Fischl, B.R., et al. (2000). Dynamic statistical parametric mapping: combining fMRI and MEG for high-resolution imaging of cortical activity. *Neuron*, **26**(1), 55–67.
8. Vrba, J. and Robinson, S.E. (2001). Signal processing in magnetoencephalography. *Methods*, **25**(2), 249–71.

9. Hillebrand, A., Singh, K.D., Holliday, I.E., Furlong, P.L., and Barnes, G.R. (2005). A new approach to neuroimaging with magnetoencephalography. *Human Brain Mapping*, **25**(2), 199–211.
10. Gross, J., Kujala, J., Hamalainen, M., et al. (2001). Dynamic imaging of coherent sources: Studying neural interactions in the human brain. *Proceedings of the National Academy of Sciences, USA*, **98**(2), 694–9.
11. Robinson, S.E., Nagarajan, S.S., Mantle, M., Gibbons, V., and Kirsch, H. (2004). Localization of interictal spikes using SAM(g2) and dipole fit. *Neurology & Clinical Neurophysiology*, **2004**, 74.
12. Kirsch, H.E., Robinson, S.E., Mantle, M., and Nagarajan, S. (2006). Automated localization of magnetoencephalographic interictal spikes by adaptive spatial filtering. *Clinical Neurophysiology*, **117**(10), 2264–71.
13. Ishii, R., Canuet, L., Ochi, A., et al. (2008). Spatially filtered magnetoencephalography compared with electrocorticography to identify intrinsically epileptogenic focal cortical dysplasia. *Epilepsy Research*, **81**, 228–32.
14. Agirre-Arrizubieta, Z., Thai, N., and Valentín, A. (2014). The value of magnetoencephalography to guide electrode implantation in epilepsy. *Brain Topography*, **27**(1), 197–207.
15. Canuet, L., Ishii, R., Iwase, M., et al. (2008). Cephalic auras of supplementary motor area origin: An ictal MEG and SAM(g2) study. *Epilepsy & Behavior*, **13**(3), 570–4.
16. Kanamori, Y., Shigeto, H., Hironaga, N., et al. (2013). Minimum norm estimates in MEG can delineate the onset of interictal epileptic discharges: a comparison with ECoG findings. *NeuroImage: Clinical*, **2**, 663–9.
17. Alkawadri, R., Krishnan, B., Kakisaka, Y., et al. (2013). Localization of the ictal onset zone with MEG using minimum norm estimate of a narrow band at seizure onset versus standard single current dipole modeling. *Clinical Neurophysiology*, **124**(9), 1915–18.
18. Taulu, S., Simola, J., and Kajola, M. (2004). Clinical applications of the signal space separation method. *International Congress Series*, **1270**, 32–7.
19. Wilson, H.S. (2004). Continuous head-localization and data correction in a whole-cortex MEG sensor. *Neurology & Clinical Neurophysiology*, **2004**, 56.
20. Nenonen, J., Nurminen, J., Kičić, D., et al. (2012). Validation of head movement correction and spatiotemporal signal space separation in magnetoencephalography. *Clinical Neurophysiology*, **123**(11), 2180–91.
21. Wehner, D.T., Hämäläinen, M.S., Mody, M., and Ahlfors, S.P. (2008). Head movements of children in MEG: quantification, effects on source estimation, and compensation. *NeuroImage*, **40**(2), 541–50.
22. Kakisaka, Y., Wang, Z.I., Mosher, J.C., et al. (2012). Clinical evidence for the utility of movement compensation algorithm in magnetoencephalography: Successful localization during focal seizure. *Epilepsy Research*, **101**(1), 191–6.
23. Medvedovsky, M., Taulu, S., Gaily, E., et al. (2012). Sensitivity and specificity of seizure-onset zone estimation by ictal magnetoencephalography. *Epilepsia*, **53**(9), 1649–57.
24. Staba, R.J., Stead, M., and Worrell, G. (2014). Electrophysiological biomarkers of epilepsy. *Neurotherapeutics*, **11**(2), 334–46.
25. Bragin, A., Engel, J., Wilson, C.L., Fried, I., and Buzsáki, G. (1999). High-frequency oscillations in human brain. *Hippocampus*, **9**(2), 137–42.
26. Staba, R. and Wilson, C. (2002). Quantitative analysis of high-frequency oscillations (80–500 Hz) recorded in human epileptic hippocampus and entorhinal cortex. *Journal of Neurophysiology*, **88**, 1743–52.
27. Worrell, G. and Gotman, J. (2011). High-frequency oscillations and other electrophysiological biomarkers of epilepsy: clinical studies. *Biomarkers in Medicine*, **5**(5), 557–66.
28. Jefferys, J.G.R., Menendez de la Prida, L., Wendling, F., et al. (2012). Mechanisms of physiological and epileptic HFO generation. Progress in Neurobiology, **98**(3), 250–64.
29. Fujiwara, H., Greiner, H.M., Lee, K.H., et al. (2012). Resection of ictal high-frequency oscillations leads to favorable surgical outcome in pediatric epilepsy. *Epilepsia*, **53**(9), 1607–17.
30. Kobayashi, K., Agari, T., Oka, M., et al. (2010). Detection of seizure-associated high-frequency oscillations above 500 Hz. *Epilepsy Research*, **88**(2–3), 139–44.
31. Jacobs, J., LeVan, P., Chander, R., Hall, J., Dubeau, F., and Gotman, J. (2008). Interictal high-frequency oscillations (80-500 Hz) are an indicator of seizure onset areas independent of spikes in the human epileptic brain. *Epilepsia*, **49**(11), 1893–907.
32. Usui, N., Terada, K., Baba, K., et al. (2011). Clinical significance of ictal high frequency oscillations in medial temporal lobe epilepsy. *Clinical Neurophysiology*, **122**(9), 1693–700.
33. Mari, F., Zelmann, R., Andrade-Valenca, L., Dubeau, F., and Gotman J. (2012). Continuous high-frequency activity in mesial temporal lobe structures. *Epilepsia*, **53**(5), 797–806.
34. Kobayashi, K., Oka, M., Akiyama, T., et al. (2004). Very fast rhythmic activity on scalp EEG associated with epileptic spasms. *Epilepsia*, **45**(5), 488–96.
35. Xiang, J., Liu, Y., Wang, Y., et al. (2009). Frequency and spatial characteristics of high-frequency neuromagnetic signals in childhood epilepsy. *Epileptic Disorders*, **11**(2), 113–25.
36. Xiang, J., Wang, Y., Chen, Y., et al. (2010). Noninvasive localization of epileptogenic zones with ictal high-frequency neuromagnetic signals. *Journal of Neurosurgery: Pediatrics*, **5**(1), 113–22.
37. Biswal, B., Yetkin, F.Z., Haughton, V.M., and Hyde, J.S. (1995). Functional connectivity in the motor cortex of resting human brain using echo-planar MRI. *Magnetic Resonance Medicine*, **34**(4), 537–41.
38. Fox, M.D. and Raichle, M.E. (2007). Spontaneous fluctuations in brain activity observed with functional magnetic resonance imaging. *Nature Review: Neuroscience*, **8**(9), 700–11.
39. Deco, G. and Corbetta, M. (2011). The dynamical balance of the brain at rest. *Neuroscientist*, **17**(1), 107–23.
40. Brookes, M.J., O'Neill, G.C., Hall, E.L., et al. (2014). Measuring temporal, spectral and spatial changes in electrophysiological brain network connectivity. *NeuroImage*, **91**, 282–99.
41. Pittau, F., Grova, C., Moeller, F., Dubeau, F., and Gotman, J. (2012). Patterns of altered functional connectivity in mesial temporal lobe epilepsy. *Epilepsia*, **53**(6), 1013–23.
42. Negishi, M., Martuzzi, R., Novotny, E.J., Spencer, D.D., and Constable, R.T. (2011). Functional MRI connectivity as a predictor of the surgical outcome of epilepsy. *Epilepsia*, **52**(9), 1733–40.
43. Weaver, K.E., Chaovalitwongse, W.A., Novotny, E.J., Poliakov, A., Grabowski, T.G., and Ojemann, J.G. (2013). Local functional connectivity as a pre-surgical tool for seizure focus identification in non-lesion, focal epilepsy. *Frontiers in Neurology*, **4**(43), 1–14.
44. Cataldi, M., Avoli, M., and de Villers-Sidani, E. (2013). Resting state networks in temporal lobe epilepsy. *Epilepsia*, **54**(12), 2048–59.
45. Hillebrand, A., Barnes, G.R., Bosboom, J.L., Berendse, H.W., and Stam, C.J. (2012). Frequency-dependent functional connectivity within resting-state networks: an atlas-based MEG beamformer solution. *NeuroImage*, **59**(4), 3909–21.
46. Pasquale, F., De, Della Penna, S., Snyder, A.Z., et al. (2010). Temporal dynamics of spontaneous MEG activity in brain networks. *Proceedings of the National Academy of Sciences, USA*, **107**(13), 1–6.
47. Lin, F-H., Hara, K., Solo, V., et al. (2009). Dynamic Granger–Geweke causality modeling with application to interictal spike propagation. *Human Brain Mapping*, **30**(6), 1877–86.
48. Malinowska, U., Badier, J-M., Gavaret, M., Bartolomei, F., Chauvel, P., and Bénar, C-G. (2013). Interictal networks in magnetoencephalography. *Human Brain Mapping*, **35**(6), 2789–805.
49. Lopes da Silva, F.H. and van Rotterdam, A. (2005). Biophysical aspects of EEG and Magnetoencephalographic generation. In: E. Niedermeyer, F. H. Lopes da Silva (Eds) *Electroencephalography, basic principles, clinical applications and related fields*, 5th edn, pp. 1165–98. Philadelphia, PA: Lippincott, Williams and Wilkins.

50. Cohen, D. and Cuffin, B.N. (1991). EEG versus MEG localization accuracy: theory and experiment. *Brain Topography*, **4**(2), 95–103.
51. Cohen, D., Cuffin, B.N., Yunokuchi, K., et al. (1990). MEG versus EEG localization test using implanted sources in the human brain. *Annals of Neurology*, **28**(6), 811–17.
52. Gharib, S., Sutherling, W.W., Nakasato, N., et al. (1995). MEG and ECoG localization accuracy test. *Electroencephalography & Clinical Neurophysiology*, **94**(2), 109–14.
53. Agirre-Arrizubieta, Z., Huiskamp, G.J.M., Ferrier, C.H., van Huffelen A.C., and Leijten, F.S.S. (2009). Interictal magnetoencephalography and the irritative zone in the electrocorticogram. *Brain*, **132**(Pt 11), 3060–71.
54. Fujiwara, H., Greiner, H.M., Hemasilpin, N., et al. (2012). Ictal MEG onset source localization compared to intracranial EEG and outcome: Improved epilepsy presurgical evaluation in pediatrics. *Epilepsy Research*, **99**(3), 214–24.
55. Kaiboriboon, K., Nagarajan, S., Mantle, M., and Kirsch, H.E. (2010). Interictal MEG/MSI in intractable mesial temporal lobe epilepsy: spike yield and characterization. *Clinical Neurophysiology*, **121**(3), 325–31.
56. Knowlton, R. and Razdan, S. (2009). Effect of epilepsy magnetic source imaging on intracranial electrode placement. *Annals of Neurology*, **65**(6), 716–23.
57. Knowlton, R.C., Elgavish, R., Howell, J., et al. (2006). Magnetic source imaging versus intracranial electroencephalogram in epilepsy surgery: a prospective study. *Annals of Neurology*, **59**(5), 835–42.
58. Mamelak, A.N., Lopez, N., Akhtari, M., and Sutherling, W.W. (2002). Magnetoencephalography-directed surgery in patients with neocortical epilepsy. *Journal of Neurosurgery*, **97**(4), 865–73.
59. Mikuni, N., Nagamine, T., Ikeda, A., et al. (1997). Simultaneous recording of epileptiform discharges by MEG and subdural electrodes in temporal lobe epilepsy. *NeuroImage*, **5**(4 Pt 1), 298–306.
60. Oishi, M., Otsubo, H., Kameyama, S., et al. (2002). Epileptic spikes: magnetoencephalography versus simultaneous electrocorticography. *Epilepsia*, **43**(11), 1390–5.
61. Santiuste, M., Nowak, R., Russi, A., et al. (2008). Simultaneous magnetoencephalography and intracranial EEG registration: technical and clinical aspects. *Journal of Clinical Neurophysiology*, **25**(6), 331–9.
62. Schneider, F., Alexopoulos, A.V., Wang, Z., et al. (2012). Magnetic source imaging in non-lesional neocortical epilepsy: additional value and comparison with ICEEG. *Epilepsy & Behavior*, **24**(2), 234–40.
63. Schneider, F., Irene Wang, Z., Alexopoulos, A.V, et al. (2013). Magnetic source imaging and ictal SPECT in MRI-negative neocortical epilepsies: additional value and comparison with intracranial EEG. *Epilepsia*, **54**(2), 359–69.
64. Schwartz, D.P., Badier, J.M., Vignal, J.P., Toulouse, P., Scarabin, J.M., and Chauvel, P. (2003). Non-supervised spatio-temporal analysis of interictal magnetic spikes: comparison with intracerebral recordings. *Clinical Neurophysiology*, **114**(3), 438–49.
65. Shigeto, H., Morioka, T., Hisada, K., et al. (2002). Feasibility and limitations of magnetoencephalographic detection of epileptic discharges: simultaneous recording of magnetic fields and electrocorticography. *Neurological Research*, **24**(6), 531–6.
66. Tilz, C., Hummel, C., Kettenmann, B., and Stefan, H. (2002). Ictal onset localization of epileptic seizures by magnetoencephalography. *Acta Neurologica Scandinavica*, **106**(4), 190–5.
67. Wennberg, R. and Cheyne, D. (2013). Reliability of MEG source imaging of anterior temporal spikes: analysis of an intracranially characterized spike focus. *Clinical Neurophysiology*, **125**(5), 903–18.
68. Wheless, J.W., Willmore, L.J., Breier, J.I., et al. (1999). A comparison of magnetoencephalography, MRI, and V-EEG in patients evaluated for epilepsy surgery. *Epilepsia*, **40**(7), 931–41.
69. Papanicolaou, A.C., Pataraia, E., Billingsley-Marshall, R., et al. (2005). Toward the substitution of invasive electroencephalography in epilepsy surgery. *Journal of Clinical Neurophysiology*, **22**(4), 231–7.
70. Wu, X., Rampp, S., Weigel, D., Kasper, B., Zhou, D., and Stefan, H. (2011). The correlation between ictal semiology and magnetoencephalographic localization in frontal lobe epilepsy. *Epilepsy & Behavior*, **22**(3), 587–91.
71. Assaf, B.A., Karkar, K.M., Laxer, K.D., et al. (2003). Ictal magnetoencephalography in temporal and extratemporal lobe epilepsy. *Epilepsia*, **44**(10), 1320–7.
72. Tanaka, N., Cole, A.J., von Pechmann, D., et al. (2009). Dynamic statistical parametric mapping for analyzing ictal magnetoencephalographic spikes in patients with intractable frontal lobe epilepsy. *Epilepsy Research*, **85**(2), 279–86.
73. Shiraishi, H., Watanabe, Y., Watanabe, M., Inoue, Y., Fujiwara, T., and Yagi, K. (2001). Interictal and ictal magnetoencephalographic study in patients with medial frontal lobe epilepsy. *Epilepsia*, **42**(7), 875–82.
74. Yagyu, K., Takeuchi, F., Shiraishi, H., et al. (2010). The applications of time-frequency analyses to ictal magnetoencephalography in neocortical epilepsy. *Epilepsy Research*, **90**(3), 199–206.
75. Tanaka, N., Hämäläinen, M.S., Ahlfors, S.P., et al. (2010). Propagation of epileptic spikes reconstructed from spatiotemporal magnetoencephalographic and electroencephalographic source analysis. *NeuroImage*, **50**(1), 217–22.
76. Flanagan, D., Valentín, A., García Seoane, J.J., Alarcón, G., Boyd, S.G. (2009). Single-pulse electrical stimulation helps to identify epileptogenic cortex in children. *Epilepsia*, **50**(7), 1793–803.
77. Kokkinos, V., Alarcón, G., Selway, R.P., and Valentín, A. (2013). Role of single pulse electrical stimulation (SPES) to guide electrode implantation under general anaesthesia in presurgical assessment of epilepsy. *Seizure*, **22**(3), 198–204.
78. Valentín, A., Alarcón, G., Honavar, M., et al. (2005). Single pulse electrical stimulation for identification of structural abnormalities and prediction of seizure outcome after epilepsy surgery: a prospective study. *Lancet: Neurology*, **4**(11), 718–26.
79. Sutherling, W.W., Mamelak, A.N., Thyerlei, D., et al. (2008). Influence of magnetic source imaging for planning intracranial EEG in epilepsy. *Neurology*, **71**(13), 990–6.
80. Iwasaki, M., Nakasato, N., Shamoto, H., et al. (2002). Surgical implications of neuromagnetic spike localization in temporal lobe epilepsy. *Epilepsia*, **43**(4), 415–24.
81. Otsubo, H., Ochi, A., Elliott, I., et al. (2001). MEG predicts epileptic zone in lesional extrahippocampal epilepsy: 12 pediatric surgery cases. *Epilepsia*, **42**(12), 1523–30.
82. Genow, A., Hummel, C., Scheler, G., et al. (2004). Epilepsy surgery, resection volume and MSI localization in lesional frontal lobe epilepsy. *NeuroImage*, **21**(1), 444–9.
83. Morioka, T., Nishio, S., Ishibashi, H., et al. (1999). Intrinsic epileptogenicity of focal cortical dysplasia as revealed by magnetoencephalography and electrocorticography. *Epilepsy Research*, **33**(2), 177–87.
84. Burneo, J.G., Bebin, M., Kuzniecky, R.I., and Knowlton, R.C. (2004). Electroclinical and magnetoencephalographic studies in epilepsy patients with polymicrogyria. *Epilepsy Research*, **62**(2), 125–33.
85. Kim, H., Kankirawatana, P., Killen, J., et al. (2013). Magnetic source imaging (MSI) in children with neocortical epilepsy: surgical outcome association with 3D post-resection analysis. *Epilepsy Research*, **106**(1), 164–72.
86. Bast, T., Oezkan, O., Rona, S., et al. (2004). EEG and MEG source analysis of single and averaged interictal spikes reveals intrinsic epileptogenicity in focal cortical dysplasia. *Epilepsia*, **45**(6), 621–31.
87. Jeong, W., Kim, J.S., and Chung, C.K. (2013). Localization of MEG pathologic gamma oscillations in adult epilepsy patients with focal cortical dysplasia. *NeuroImage: Clinical*, **3**, 507–14.
88. Widjaja, E., Otsubo, H., Raybaud, C., et al. (2008). Characteristics of MEG and MRI between Taylor's focal cortical dysplasia (type II) and other cortical dysplasia: surgical outcome after complete resection of MEG spike source and MR lesion in pediatric cortical dysplasia. *Epilepsy Research*, **82**(2–3), 147–55.
89. Oishi, M., Kameyama, S., Masuda, H., et al. (2006). Single and multiple clusters of magnetoencephalographic dipoles in neocortical epilepsy: significance in characterizing the epileptogenic zone. *Epilepsia*, **47**(2), 355–64.

90. Zhang, R., Wu, T., Wang, Y., et al. (2011). Interictal magnetoencephalographic findings related with surgical outcomes in lesional and nonlesional neocortical epilepsy. *Seizure*, **20**(9), 692–700.
91. Wilenius, J., Medvedovsky, M., Gaily, E., et al. (2013). Interictal MEG reveals focal cortical dysplasias: special focus on patients with no visible MRI lesions. *Epilepsy Research*, **105**(3), 337–48.
92. Heers, M., Rampp, S., Stefan, H., et al. (2012). MEG-based identification of the epileptogenic zone in occult peri-insular epilepsy. *Seizure*, **21**(2), 128–33.
93. Lin, Y., Shih, Y., Hsieh, J., et al. (2003). Magnetoencephalographic yield of interictal spikes in temporal lobe epilepsy. *NeuroImage*, **19**(3), 1115–26.
94. Baumgartner, C. and Pataraia, E. (2003). Magnetoencephalography in the definition of the irritative zone. *Handbook of Clinical Neurophysiology*, **3**, 25–47.
95. Hillebrand, A. and Barnes, G.R. (2002). A quantitative assessment of the sensitivity of whole-head MEG to activity in the adult human cortex. *NeuroImage*, **16**(3), 638–50.
96. Leijten, F.S.S., Huiskamp, G-J.M., Hilgersom, I., and Van Huffelen, A.C. (2003). High-resolution source imaging in mesiotemporal lobe epilepsy: a comparison between MEG and simultaneous EEG. *Journal of Clinical Neurophysiology*, **20**(4), 227–38.
97. Ossenblok, P., de Munck, J.C., Colon, A., Drolsbach, W., and Boon, P. (2007). Magnetoencephalography is more successful for screening and localizing frontal lobe epilepsy than electroencephalography. *Epilepsia*, **48**(11), 2139–49.
98. Chang, E., and Nagarajan, S. (2009). Magnetic source imaging for the surgical evaluation of EEG secondary bilateral synchrony in intractable epilepsy. *Journal of Neurosurgery*, **111**(6), 1248–56.
99. Yu, H.Y., Nakasato, N., Iwasaki, M., Shamoto, H., Nagamatsu, K., and Yoshimoto, T. (2004). Neuromagnetic separation of secondarily bilateral synchronized spike foci: report of three cases. *Journal of Clinical Neuroscience*, **11**(6), 644–8.
100. Cheyne, D., Ross, B., Stroink, G., et al. (2007). Intra- and inter-hemispheric propagation of interictal spikes in epilepsy patients as measured by MEG. *International Congress Series*, **1300**, 645–8.
101. Cheyne, D., Ross, B., Stroink, G., et al. (2007). Retrospective review of 733 clinical MEG studies. *International Congress Series*, **1300**, 661–4.
102. Manoharan, A., Bowyer, S.M., Mason, K., et al. (2007). Localizing value of MEG in refractory partial epilepsy: Surgical outcomes. *International Congress Series*, **1300**, 657–60.
103. Paulini, A., Fischer, M., Rampp, S., et al. (2007). Lobar localization information in epilepsy patients: MEG—a useful tool in routine presurgical diagnosis. *Epilepsy Research*, **76**(2), 124–30.
104. Assaf, B.A., Karkar, K.M., Laxer, K.D., et al. (2004). Magnetoencephalography source localization and surgical outcome in temporal lobe epilepsy. *Clinical Neurophysiology*, **115**(9), 2066–76.
105. Sharma, R., Pang, E., Mohamed, I., et al. (2007). Magnetoencephalography in children: Routine clinical protocol for intractable epilepsy at the hospital for sick children. *International Congress Series*, **1300**, 685–8.
106. Vitikainen, A-M., Lioumis, P., Paetau, R., et al. (2009). Combined use of non-invasive techniques for improved functional localization for a selected group of epilepsy surgery candidates. *NeuroImage*, **45**(2), 342–8.
107. Gallen, C.C., Schwartz, B.J., Bucholz, R.D., et al. (1995). Presurgical localization of functional cortex using magnetic source imaging. *Journal of Neurosurgery*, **82**(6), 988–94.
108. Yang, T.T., Gallen, C.C., Schwartz, B.J., and Bloom, F.E. (1993). Noninvasive somatosensory homunculus mapping in humans by using a large-array biomagnetometer. *Proceedings of the National Academy of Sciences, USA*, **90**(7), 3098–102.
109. Roberts, T.P., Zusman, E., McDermott, M., Barbaro, N., and Rowley, H.A. (1995). Correlation of functional magnetic source imaging with intraoperative cortical stimulation in neurosurgical patients. *Journal of Image Guided Surgery*, **1**(6), 339–47.
110. Kristeva, R., Cheyne, D., and Deecke, L. (1991). Neuromagnetic fields accompanying unilateral and bilateral voluntary movements: topography and analysis of cortical sources. *Electroencephalography & Clinical Neurophysiology*, **81**(4), 284–98.
111. Cheyne, D., Bakhtazad, L., and Gaetz, W. (2006). Spatiotemporal mapping of cortical activity accompanying voluntary movements using an event related beamforming approach. *Human Brain Mapping*, **229**, 213–29.
112. Pang, E.W., Drake, J.M., Otsubo, H., et al. (2008). Intraoperative confirmation of hand motor area identified preoperatively by magnetoencephalography. *Pediatric Neurosurgery*, **44**(4), 313–17.
113. Cheyne, D., Bostan, A., Gaetz, W., and Pang, E. (2007). Event-related beamforming: a robust method for presurgical functional mapping using MEG. *Clinical Neurophysiology*, **118**, 1691–704.
114. Hillebrand, A., Fazio, P., de Munck, J.C., and van Dijk, B.W. Feasibility of clinical magnetoencephalography (MEG) functional mapping in the presence of dental artefacts. *Clinical Neurophysiology*, **124**(1), 107–13.
115. Gaetz, W., Macdonald, M., Cheyne, D., and Snead, O.C. (2010). Neuromagnetic imaging of movement-related cortical oscillations in children and adults: age predicts post-movement beta rebound. *NeuroImage*, **51**(2), 792–807.
116. Cheyne, D., Bells, S., Ferrari, P., Gaetz, W., and Bostan, A.C. (2008). Self-paced movements induce high-frequency gamma oscillations in primary motor cortex. *NeuroImage*, **42**(1), 332–42.
117. Merrifield, W.S., Simos, P.G., Papanicolaou, A.C., Philpott, L.M., and Sutherling, W.W. (2007). Hemispheric language dominance in magnetoencephalography: sensitivity, specificity, and data reduction techniques. *Epilepsy & Behavior*, **10**(1), 120–8.
118. Findlay, A.M., Ambrose, J.B., Cahn-Weiner, D.A., et al. (2012). Dynamics of hemispheric dominance for language assessed by magnetoencephalographic imaging. *Annals of Neurology*, **71**(5), 668–86.
119. Hirata, M., Kato, A., Taniguchi, M., et al. (2004). Determination of language dominance with synthetic aperture magnetometry: comparison with the Wada test. *NeuroImage*, **23**(1), 46–53.
120. Kim, J.S. and Chung, C.K. (2008). Language lateralization using MEG beta frequency desynchronization during auditory oddball stimulation with one-syllable words. *NeuroImage*, **42**(4), 1499–507.
121. Fisher, A.E., Furlong, P.L., Seri, S., et al. (2008). Interhemispheric differences of spectral power in expressive language: a MEG study with clinical applications. *International Journal of Psychophysiology*, **68**(2), 111–22.
122. Pang, E.W., Wang, F., Malone, M., Kadis, D.S., and Donner, E.J. (2011). Localization of Broca's area using verb generation tasks in the MEG: validation against fMRI. *Neuroscience Letters*, **490**(3), 215–19.
123. Wang, Y., Holland, S.K., and Vannest, J. (2012). Concordance of MEG and fMRI patterns in adolescents during verb generation. *Brain Research*, **1447**, 79–90.
124. RamachandranNair, R., Otsubo, H., Shroff, M.M., et al. (2007). MEG predicts outcome following surgery for intractable epilepsy in children with normal or nonfocal MRI findings. *Epilepsia*, **48**(1), 149–57.
125. Stefan, H., Rampp, S., and Knowlton, R.C. (2011). Magnetoencephalography adds to the surgical evaluation process. *Epilepsy & Behavior*, **20**(2), 172–7.

CHAPTER 14

Transcranial magnetic stimulation

Kerry R. Mills

Introduction

Development of magnetic devices capable of stimulating elements of the central nervous system began in the 1980s (1–3) and has progressed such that now, the devices are used not only in the diagnosis of central nervous system conditions and has expanded the core of knowledge of motor control in healthy subjects (4), but they also have potential for the treatment of psychiatric and other conditions. The initial devices used a single pulse to excite nerve cells, but devices capable of repetitive stimulation with rates up to 100 Hz are now available (5). These are used principally to modulate cortical excitability in an attempt to treat such conditions as depression. This chapter will be concerned mainly with the diagnostic possibilities of single pulse transcranial magnetic stimulation (TMS).

A magnetic stimulator is essentially a simple device in which a large capacitor is charged up from a power source and then discharged through a copper coil (Fig. 14.1A). This results in a transient intense magnetic field. Magnetic fields pass unattenuated into all surrounding media and if a conductor is within the field, then a current will be induced within it. Thus, if the coil is placed on the scalp, the magnetic field passes through the skull and induces current within the brain. The shape and size of the copper coil into which the charge is sent determines, *inter alia*, the distribution of induced current in the brain.

Currents in nervous tissue excite axons more easily than cell bodies and currents induced by the magnetic field are no different. 'Magnetic stimulation' is strictly speaking a misnomer, since it is the induced electric currents that excite the brain. It is likely that in primary motor cortex, for example, axons in the superficial horizontal layers are excited most easily with deeper axonal elements being excited with increased intensities. Moreover, although the precise sites of stimulation are unclear, it is likely that activation of axons occurs preferentially where they bend or change direction (6).

Physics of magnetic stimulation

It important to understand the spatial and temporal distribution of the induced electric field if we are to understand which structures are excited by TMS. We consider first the simplest situation of a circular coil placed flat on the scalp. The electric field induced has two sources—one is due to magnetic induction and the second is due to the build-up of charge at the interface between the conductor and air. The total electric field is the sum of these two vector quantities (7,8). Modelling of the distribution of these fields (9), which requires a number of assumptions (e.g. that impedance is the same in all directions—isotropy), has shown that there is no current induced in the direction radial to the surface of the conductor where the two components exactly cancel each other (10,11). This holds true for any depth or any angulation of the coil. Secondly, and again assuming isotropy, the induced current below the surface takes the form of concentric circles of reducing diameters with increasing depth. Increasing angulation of the coil has the effect of reducing the induced current. Thus, a flat circular coil placed horizontally on the scalp induces no current in the radial direction, but induces circles of current that are strongest nearest the surface at the circumference of the coil. This explains why a 13-cm diameter circular coil centred at the vertex is most potent at exciting the hand area of the motor cortex since this lies beneath the circumference of the coil.

The situation with coils of other dimensions and shapes has also been modelled. With diminishing sizes of circular coil, in an attempt to improve the focality of stimulation, more and more inducing current is needed to induce the same electric field and a point is soon reached where this becomes prohibitively high. With a figure 8-shaped coil the current flowing in the central segment of the coil is twice that present in each wing of the coil. In the brain, two sets of concentric circles of induced current are produced, which summate below the central segment. Thus a figure 8 coil produces more focal and directional stimulation. The orientation of a figure 8 coil is also important (12). The most effective orientation is with the coil at 45° to the parasagittal plane with the current in the brain flowing in a postero-anterior direction (Fig. 14.2).

The time course of the induced current is also important (Fig. 14.1B). The magnetic field developed by the coil is proportional to the current passing through it, the area of the coil and the number of turns in the coil. The induced electric field is related to the rate of change of magnetic field and is in the opposite sense. Depending on the coil inductance, storage capacitance, and the resistance of the circuit the magnetic field may be monophasic or may oscillate. Strictly speaking, the so-called monophasic pulse is, in fact, large and rapid in one direction and then slowly returns to zero in the opposite direction, the net current flow being zero. This has important implications for which part of the brain will be stimulated. A monophasic pulse of stimulating current in a circular coil will excite predominantly one hemisphere whereas an oscillating pulse will excite first one hemisphere then the other. The direction of current flow in the coil determines which hemisphere is excited by a monophasic pulse—clockwise current flow in the coil excites the left hemisphere; inverting the coil will excite the right hemisphere (Fig. 14.2).

A large number of coils of various dimensions and shapes have been tried in order to improve focality of stimulation (13,14). For most diagnostic purposes, a 13-cm diameter circular coil is most

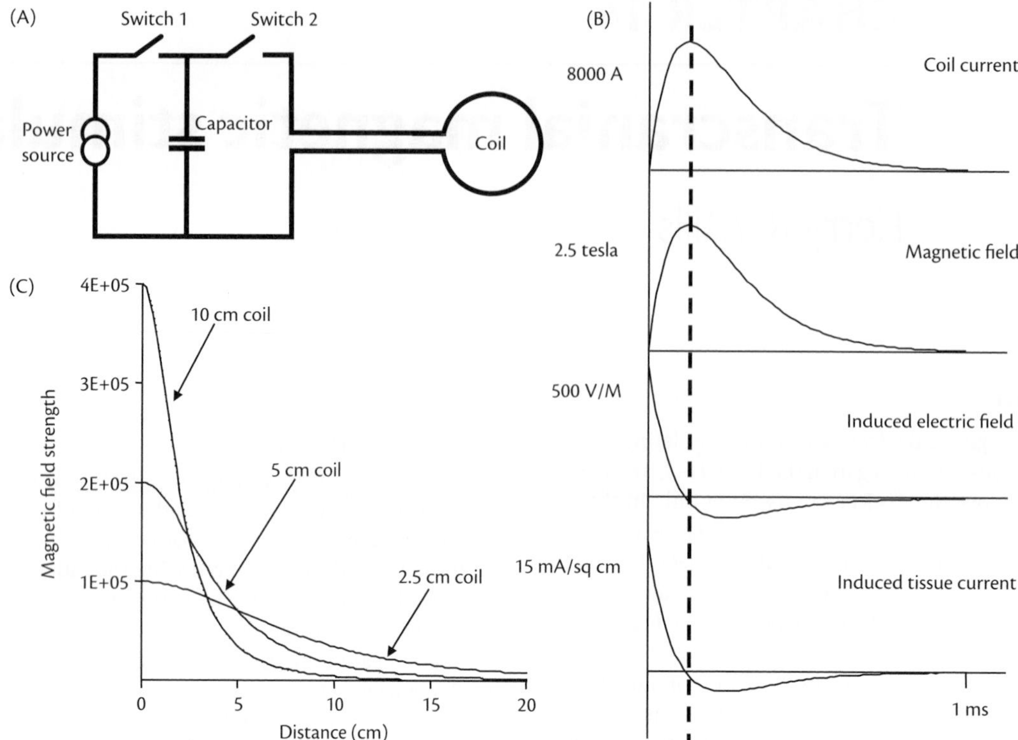

Fig. 14.1 (A) A much simplified circuit of a magnetic stimulator. Closure and then reopening of switch 1 charges the capacitor. Closure of switch 2 discharges the capacitor through the coil. (B) The time courses of coil current, magnetic field, induced electric field and tissue current. Note the time course of the induced tissue current is the differential of the coil current. (C) The decay with distance of the magnetic field produced by coils of 10, 5, and 2.5 cm diameter.

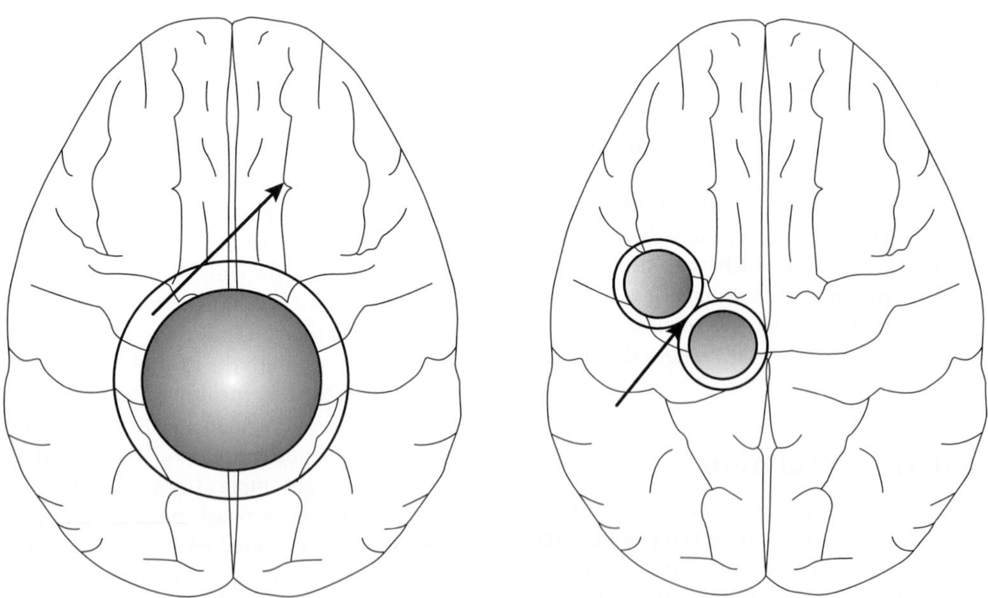

Fig. 14.2 A 13-cm diameter circular coil centred at the vertex induces current maximally at its circumference. The tangent to the coil overlying the hand motor area is approximately at 45° to the parasagittal plane, a position which is optimal for exciting the motor cortex. A figure 8 coil is placed with its axis along the same tangent for optimal excitation of the hand area.

useful for investigating upper limb problems and a double cone coil (a figure 8 coil with the two limbs at 90°) excites the leg area of the motor cortex most successfully.

Safety of transcranial magnetic stimulation

TMS has proved to be a safe procedure if certain guidelines are followed (15). The principal risks relate to the movement of metallic objects in the head or the induction of current in wires such as pacemakers. The magnetic field, however, diminishes rapidly with distance (Fig. 14.1C) reducing risks of current induction in metallic objects outside the head. The risks of inducing a seizure with single pulse TMS is very small even in patients with epilepsy. The majority of reported side effects have been of no great concern even in children (16).

Physiology of transcranial magnetic stimulation

The majority of information relates to the motor effects of TMS. Brain stimulation in other areas *does* have an effect, with phosphenes being reported with stimulation over visual areas (17,18), and more subtle motor effects from stimulation of pre-motor and supplementary motor areas (Fig. 14.3) (19–22).

The intensity of magnetic stimulation is usually expressed as a percentage of the device's maximum output. With different coils, different manufacturers, and so on, it is difficult to compare results using absolute values of magnetic field. A physiological approach is adopted in which the stimulus intensity is expressed as a multiple of the resting threshold (see 'Threshold'). With increasing intensity of stimulation and with the muscle at rest small motor evoked potentials (MEPs) begin to appear. At first, they are not present with each trial, but as intensity increases MEPs become larger and finally appear reliably with each trail. The amplitude of MEPs, however, remains variable. Indeed even with a constant stimulus, MEP amplitude may vary by an order of magnitude. This is in stark contrast to the results of peripheral nerve stimulation where responses are constant once the maximal stimulus level has been reached. If the subject activates the muscle voluntarily, the MEP is considerably larger at the same stimulus intensity: this is referred to as facilitation. The latency of the MEP is also shorter by about 2 ms (23) during voluntary contraction.

It has now been established (24) that a single pulse of TMS to the motor cortex evokes not a single action potential in the corticospinal tract fibres, but a high frequency train of impulses. It has been shown in animal work (25) and from epidural recordings in humans (26–28) that there is a direct (D) wave due to direct excitation of the axon hillock of Betz cells followed by a variable number of indirect (I1, I2, etc.) waves due to trans-synaptic excitation of cells via interneurons (Fig. 14.4). The interval between D and I1 and between I1 and I2 and so on is about 2 ms. The same phenomenon can be seen in single motor unit recordings. Liminal magnetic stimuli cause single motor units to discharge. The first recruited motor unit under voluntary drive is the same one that is fired by a weak magnetic stimulus. By applying many such stimuli during voluntary activation of a single motor unit, the modulation of firing can be documented using a peri-stimulus time histogram (PSTH) (29). TMS drives motor units at several discrete latencies and peaks in the PSTH appear separated by about 2 ms. These peaks correspond to the arrival at the spinal motor neurone of individual I-waves. The shape of peaks in the PSTH also suggests that the corticospinal connection is monosynaptic (29).

The train of impulses descends in the corticospinal tract and reaches the spinal motor neurons. Each pyramidal fibre probably diverges within the motor neurone pool to synapse on many if not all motor neurons. If the muscle is quiescent, then the D wave may be able to fire just a few motoneurons (MNs) to produce a small MEP. If repetitive I waves are present, summation at the MNs may cause more to fire and produce a larger MEP. If the muscle is being activated voluntarily, then more MNs will be nearer the threshold for firing and an even greater MEP will be produced. Indeed, this summation with already active motor neurons results in some

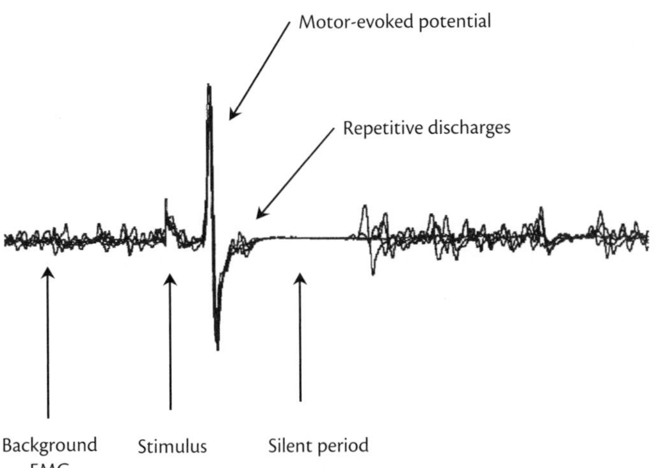

Fig. 14.3 MEPs evoked in a small hand muscle producing ongoing voluntary activation from motor cortex stimulation. The motor evoked potential is followed by small repetitive discharges due to double firing of some motor neurons and then a silent period before resumption of the voluntary EMG.
Reproduced from Mills KR, *Magnetic stimulation of the human nervous system*, copyright (1999), with permission from Oxford University Press.

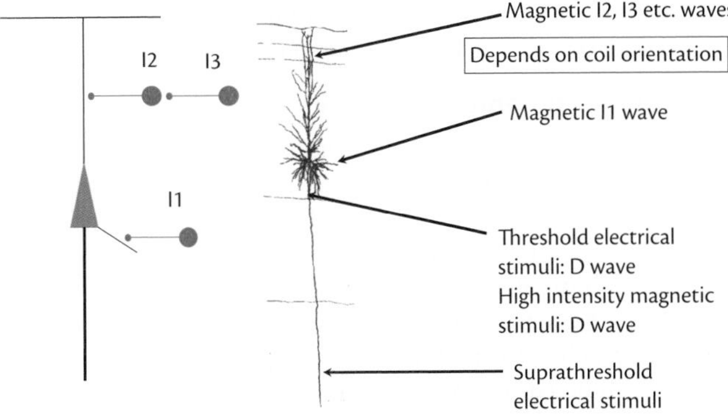

Fig. 14.4 Diagram of a Betz cell showing its apical dendrite receiving inputs from superficial horizontal fibres, basal dendrites and its descending axon. Magnetic stimuli, depending on coil orientation, excite horizontal fibres to produce indirect (I2 and I3 waves, etc.), basal dendrites to produce I1 waves, or if sufficiently intense a direct (D) wave from excitation of the initial segment of the axon. In contrast, high voltage electrical stimuli at threshold excite the initial segment to produce a D wave and if higher in intensity excite descending axons at deeper and deeper locations.

MNs firing on a D wave rather than the first I wave and hence the latency of the MEP in this situation is shorter by about 2 ms.

It is also possible to stimulate the motor cortex through the scalp using high voltage electrical pulses (30–32); this is a rather painful procedure and although it has been used in patients (33), it is not recommended for routine use. However, the MEPs evoked have latencies shorter than magnetically evoked MEPs indicating that stimulation has evoked D-waves. Indeed, it has been shown that in anaesthetized patients high intensity transcranial electrical stimuli can penetrate far into the brain (30).

Threshold

Because of the inherent variability of MEPs, threshold, the minimal intensity to cause an MEP of a given amplitude, must be defined using a statistical procedure. A number of protocols have been developed (34) using the probability with which stimuli evoke an MEP. One such method (35) aims to determine the highest intensity at which no responses are seen in 10 trials and the lowest intensity at which all 10 trials produce an MEP. Resting threshold is then defined as the average of these two levels. A suitable scheme for determining threshold with minimal stimuli is shown in Fig. 14.5. Defining threshold with the muscle active is more problematic because small MEPs may be obscured by the ongoing electromyographic activity. The importance of determining threshold is that the intensity of stimuli can then be scaled to that particular patient's threshold diminishing inter-individual variability.

MEP amplitude

The size of MEPs is related to both stimulus intensity and to the level of ongoing EMG activity, although the latter is less important. As long as the ongoing EMG is above about 5% of the maximum, then facilitation of MEPs remains relatively stable (23). Again, the variability of MEP amplitude must be taken into account and to provide an estimate of amplitude, a number of MEPs should be examined. For clinical purposes, it is sufficient to record the largest of, say, a sequence of 8 MEPs. For research purposes, often up to 10 MEPs are averaged to determine mean amplitude. The MEP amplitude is usually quoted as a ratio to the maximal compound muscle action potential amplitude produced by peripheral nerve stimulation. This reduces variability in the measure by eliminating variations in different individual's muscle size.

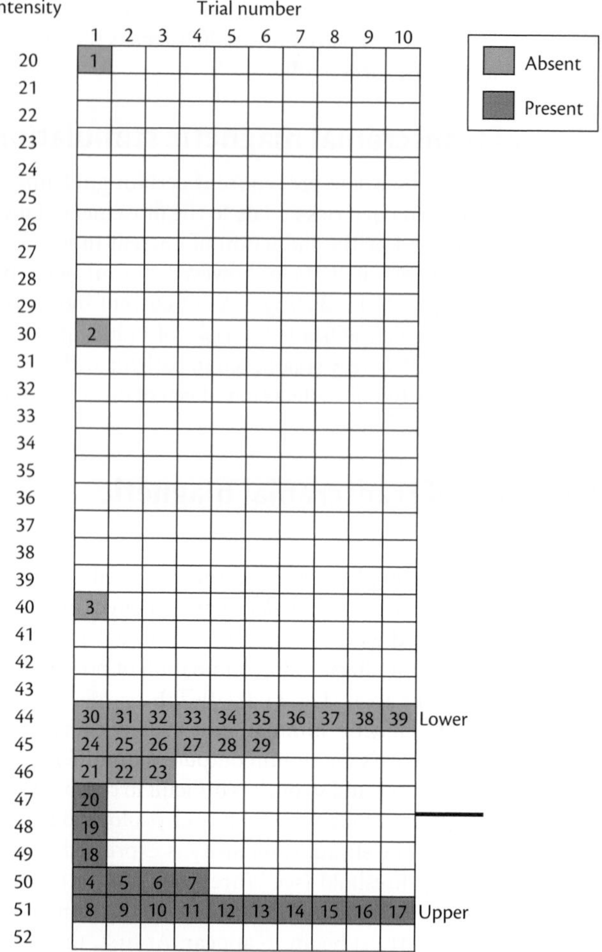

Fig. 14.5 A scheme for determining resting motor threshold. The order of each stimulus is given. No response is shown in red; a present response is shown in blue. Starting at 20% maximum stimulator output (MSO), intensity is increased in 10% steps until a response is present. Stimuli are then given at steps increasing by 1% until 10 consecutive responses are present (upper level). Then beginning at 1% less than the intensity at which a response was initially detected (49% in this example), stimuli are given in steps decreasing by 1% until 10 consecutive trials give no response (lower level). Resting threshold is the mean of upper and lower levels, in this case 47.5% MSO.

Central motor conduction time

The latency of MEPs includes the time for activation of neurons in motor cortex, conduction down the central pathway to the spinal MNs, synaptic relay in the cord, conduction down the peripheral nerve and excitation of the muscle fibres (Fig. 14.6). To estimate the central components of this combined latency, we must subtract the part due to peripheral conduction. This can be estimated in two ways. First, the motor root serving the muscle under study can be stimulated or, secondly the F-wave latency can be used. Root stimulation, either with a needle or with a high voltage surface stimulator (36) excites the motor roots at their exit foramina from the vertebral column and, hence, the calculated central motor conduction time (CMCT) includes a small time for conduction in the motor root within the spinal canal. Root stimulation can also be achieved with low intensity magnetic stimuli. The intensity must be low, otherwise, nerve fibres distant to the coil will be activated resulting in an over-estimation of CMCT. With the F-wave method, allowance must be made for conduction into the spinal motor neurone and its re-excitation. Using this method and if M is the distal latency from say, wrist stimulation, and F is the minimal F latency, CMCT is calculated from:

$$\text{MEP Latency} - (F + M - 1)/2 \qquad \text{[eqn 14.1]}$$

The F-wave method is, however, limited to those muscles in which reliable F-waves can be found; usually the intrinsic hand and foot muscles are used.

Silent period

Ongoing EMG activity during voluntary activation of a muscle is suppressed or silenced after a magnetic stimulus to motor cortex. This is referred to as the silent period and is thought to be due to gamma amino butyric acid A ($GABA_A$) and $GABA_B$ cortical

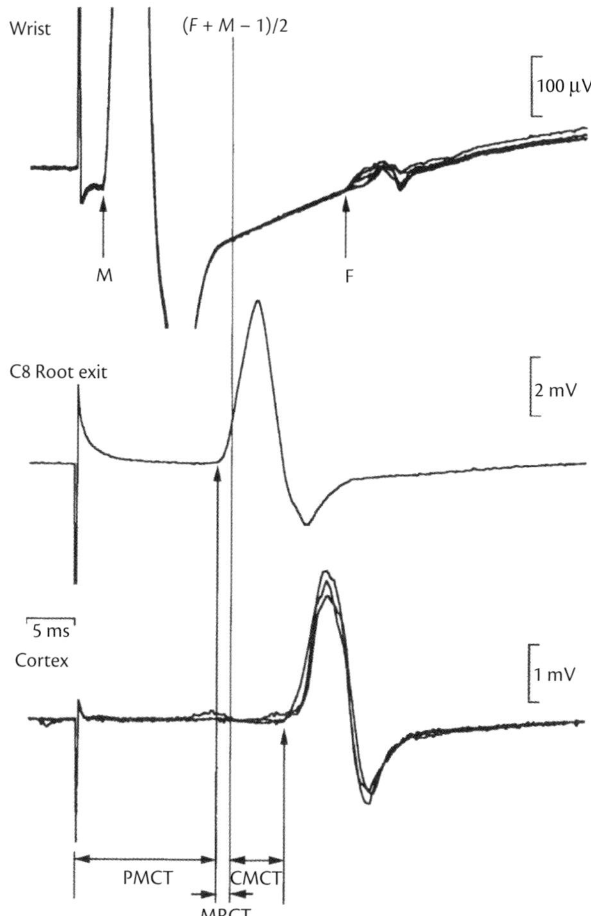

Fig. 14.6 Calculation of central motor conduction time (CMCT). Traces show the responses from a small hand muscle from stimulation of the wrist (above), vertebral column (middle) and cortex (below). The latency of the M wave and F waves are marked as is the line calculated from (*F* + *M*− 1)/2, i.e. the peripheral conduction time CMCT is calculated either by subtracting the latency from root stimulation or the peripheral nerve conduction time from the F wave from the cortex to muscle latency. Since vertebral stimulation excites axons at the exit foramen of the root, the motor root conduction time (MRCT) from motor neuron to root exit can also be calculated.

Reproduced from Mills KR, *Magnetic stimulation of the human nervous system*, copyright (1999), with permission from Oxford University Press.

inhibitory circuits depending on stimulus intensity (37). The threshold for causing such inhibition is, in fact, lower than that to cause an MEP. The duration of the silent period has a sigmoid relation to stimulus intensity and can reach some 400 ms (38); silent period duration, however, has only a weak relation to the degree of ongoing EMG activity (4).

Input:output relations

The size of the MEP, per se, gives no information about the excitability of the motor cortex. A more direct approach to measuring cortical excitability (or inhibitability) is to construct a curve relating stimulus intensity to MEP amplitude (39) or, in the case of inhibition, the silent period duration (38). In both cases there is a sigmoid relationship and the slope of the central almost linear segment of the curve can be used to assess the susceptibility of the cortex to either excitation or inhibition.

Paired-pulse protocols

The effects of pairs of stimuli to the motor cortex, the so-called classical conditioning:test paradigm depends on the intensity of the two pulses and the inter-stimulus interval. If two suprathreshold pulses are of equal strength are applied at very short intervals, say 2 ms, the response to the second pulse is increased. This is thought to be because the second pulse corresponds to the time of I-wave generation evoked by the conditioning pulse. If the first pulse is sub-threshold producing no MEP, at intervals up to 8ms the response to a second supra-threshold pulse is reduced. This is referred to as short interval intracortical inhibition (SICI) (40). This is thought to be mediated over $GABA_A$ pathways. If longer intervals are used (10–50 ms), the response to the second pulse is increased, so called short interval intracortical facilitation (SICF). If conditioning and test pulses are of equal strength and given at longer inter-stimulus intervals of say 80–150 ms, the response to the test pulse is again reduced, referred to as long interval intracortical inhibition (LICI) (41). The mechanism of LICI is also $GABA_B$-mediated and may have the same mechanism as the silent period.

It has also been demonstrated that transcallosal inhibition shown by stimulating both hemispheres independently occurs at latencies of 6–10 ms, corresponding with the transcallosal conduction time (42). Furthermore, the interaction of peripheral afferent inputs with TMS has been investigated, although timing of the afferent volley must be precise for an inhibitory effect to be seen (43).

Triple stimulation technique

The MEP evoked by a single pulse of TMS is mediated over an unknown number of corticospinal fibres each with different conduction velocities. The MEP is therefore desynchronized or dispersed as a result. Comparison with the compound muscle action potential (CMAP) evoked by a peripheral nerve stimulus to give an estimate of the overall corticospinal activation is therefore problematic. This has been ingeniously circumvented with the triple stimulation technique (Fig. 14.7) (44–46). Here, a stimulus to the cortex is followed some 20 ms later by a peripheral nerve stimulus to the wrist. The result is that a collision occurs in the motor axons that had been active as a result of the cortical stimulus. A third shock is now given at Erb's point, the result of which is that a second collision occurs now in the peripheral nerve fibres not blocked by the first collision, i.e. in the fibres originally activated from the cortex. The difference is that now these fibres have not undergone any dispersion in the corticospinal tract. By using a suitable control the amplitude of the response to the Erb's point stimulus now gives an estimate of the number of corticospinal fibres activated by TMS. This technique has been used clinically to estimate corticospinal tract degeneration in conditions like amyotrophic lateral sclerosis (47–49) and MS (50,51).

Magnetic nerve stimulation

The magnetic stimulator was originally developed as a novel way of stimulating peripheral nerves. There are a number of applications where this has proved useful. The problem with limb peripheral nerves is in defining where the actual nerve activation takes place, a necessary requirement if conduction velocity is to be calculated.

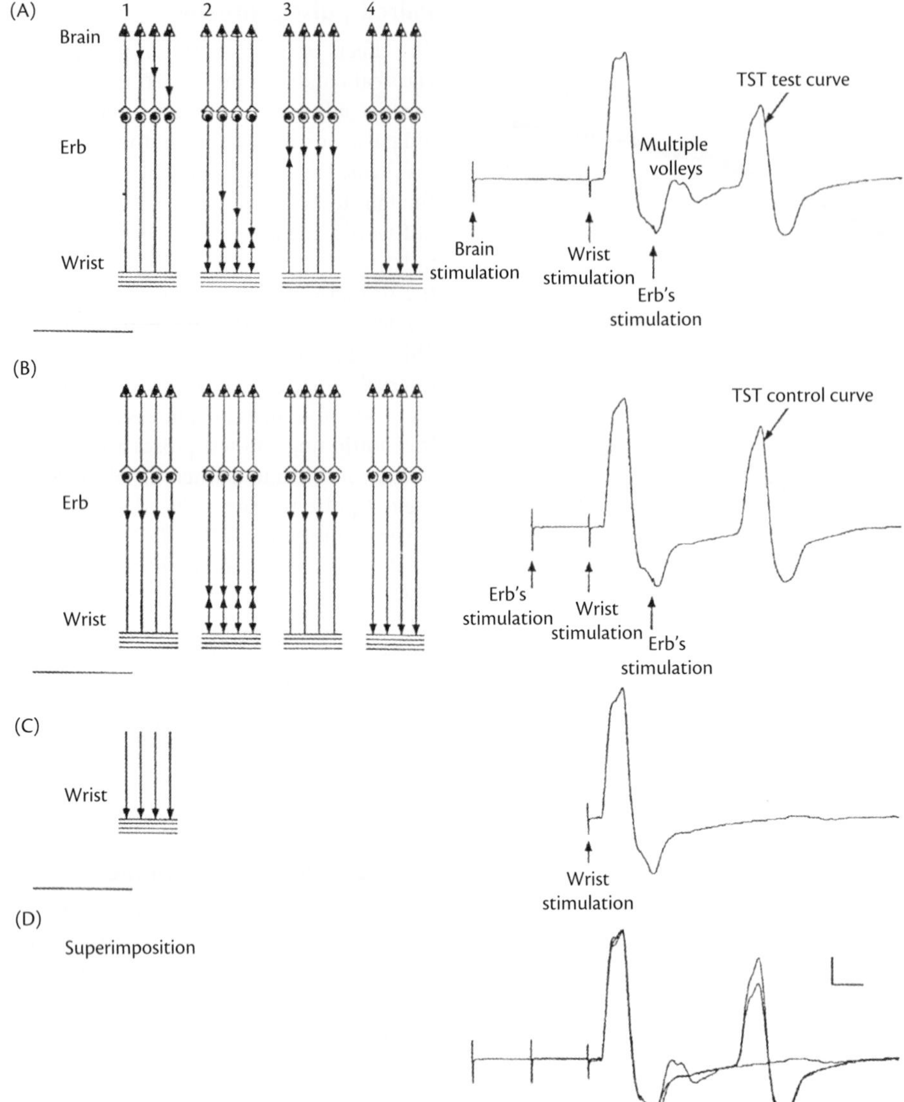

Fig. 14.7 Triple stimulation technique. (A) The test sequence. A cortical stimulus is followed some 20 ms later by a supramaximal wrist stimulus. Collision of descending and ascending impulses occurs in the peripheral nerve. A supramaximal stimulus is then given at Erb's point. Collision of remaining ascending impulses occurs and the response to the uncollided impulses from the Erb's point stimulus now only includes those fibres which were excited by the original cortical stimulus. (B) The control sequence. Here, an Erb's point stimulus is given first, followed by a wrist stimulus and then a final second Erb's point stimulus. This sequence is necessary in order to take account of the desynchronization of the CMAP from Erb's point stimulation due to the peripheral nerve conduction distance (compare the sizes of CMAPs from wrist and Erb's point stimuli in (B)). (C) The CMAP from supramaximal wrist stimulation alone. (D) A superposition of all traces; the difference between the test and control responses represents the proportion of fibres activated by the cortical stimulus.

The circumference of the coil needs to be placed in line with the nerve, which makes coupling of the field to the tissue inefficient.

The situation, however, is different where nerves pass through foramina since here the induced electrical field is channelled into the foramen making nerve stimulation more efficient. This is most evident in the case of the facial nerve. A circular coil placed over the side of the head can easily excite the facial nerve within the facial canal in the temporal bone (52). This has been use to investigate facial nerve lesions such as Bell's palsy (53–55). In addition, magnetic stimulation can be used to excite motor roots in either the cervical (56) or lumbar areas (56–58) in order to calculate root conduction. One problem with this technique is that it is difficult to achieve maximal stimulation and so if the clinical question is one of conduction block then it can be difficult to be certain if this exists if secure supramaximal stimulation cannot be achieved. The same problem arises with magnetic stimulation of the brachial plexus. However, the technique has proved useful in investigating lumbar root and plexus lesions (57).

Repetitive TMS and direct current stimulation

It has been shown that repetitively stimulating the brain or passing a direct current through it can modulate the excitability of the motor cortex (59–62). It follows that the majority of the work on rTMS has been to look for therapies for such conditions as depression (63). rTMS can be applied in a wide variety of patterns: the frequency of stimulation, bursts of rTMS at various intervals and durations and other patterns of stimulation, e.g. theta burst stimulation have all been tried (64). Intensity of stimulation is also an important parameter. Initial studies with rTMS (65) showed the propensity of the motor cortex to develop sustained discharges that occasionally evolved into seizures. A set of guidelines were developed (66) that relate intensity of stimulation (with respect to resting threshold to a single pulse) and frequency. The major physiological finding of these studies is that cortical excitability is either increased or reduced depending on the frequency of stimulation: low frequency (0.2–3 Hz) stimulation down-regulates excitability whereas higher frequency (3–20 Hz) stimulation up-regulates excitability. The changes in excitability may outlast the period of stimulation, but only by a few minutes (64). Nevertheless, workers were encouraged to try rTMS in a variety of conditions where modulation of cortical excitability might be thought to be useful, e.g. depression, schizophrenia, and pain (67–70). Hitherto, results have been mixed and no protocol stands out as the most effective.

Clinical uses of single pulse transcranial magnetic stimulation

TMS has been used in a wide variety of clinical conditions where there is abnormality of central motor fibres. It has been used to monitor therapy (71) and assess prognosis (72), as well as being used diagnostically (73).

It is worth considering the possible causes of central motor conduction abnormality when using the technique for diagnostic purposes.

A raised threshold may be due to a paucity of excitable elements in the upper layers of the cortex as in neurodegenerations but could also be due to excessive ongoing inhibition of these elements. Centrally-acting drugs can also affect resting threshold (74). The silent period suffers from great variability even if stimulus intensity and background force are held constant. Side-to-side variability is, however, small (no more than 20 ms). Shortened or absent silent periods have been found in stroke patients where the lesion is in primary motor cortex (75) and prolonged silent periods have been found in a large number of conditions, including the centrally active drugs (37,76) making their clinical usefulness limited (77). Prolonged central motor conduction time could be due to demyelination of corticospinal fibres as occurs in MS, but could also be due to abnormal excitability of spinal motor neurons such that more summation is required before motor neurons are brought to threshold. This is a possible mechanism for the modest prolongation of CMCT seen in amyotrophic lateral sclerosis (ALS). Short interval intracortical inhibition has been found to be abnormal in a large number of conditions including ALS, dementia, and psychiatric disorders, and may reflect more a loss of cortical inhibitory elements, rather than any specific inhibitory system abnormality (78–85).

Amyotrophic lateral sclerosis (ALS) and other neurodegenerative conditions

Central motor conduction studies can be abnormal in ALS, but only in a minority of cases is the information useful in diagnosis of the individual patient. Threshold in a group of patients with ALS early in their disease was shown to be low and as the disease progressed with the development of mixed upper and lower motor neuron signs, threshold was higher (86–88). Patients with advanced ALS may have no responses recordable to TMS. Central motor conduction time is often normal or mildly prolonged (by a few milliseconds) and the more muscles are examined then the greater is the likelihood of finding a prolongation of CMCT (89). This can be clinically useful where the differential diagnosis is between ALS and a compressive myelopathy (90). In the latter, CMCT tends to be more obviously prolonged since the primary pathology here is demyelination due to the compression (91). Occasionally, in patients with ALS, the finding of a modestly prolonged CMCT can be useful if there are no upper motor neuron signs in that particular limb, since it gives evidence of a central motor abnormality not evident clinically. Similarly, if the differential is with multifocal motor neuropathy (MMN) which can mimic ALS, finding a prolonged CMCT is useful in eliminating MMN. Patients with spinal muscular atrophy have hitherto been reported to have normal CMCT (4).

The triple stimulation technique has given additional information on degeneration of the corticospinal tract in ALS (49,92). The ability to quantify the proportion of the corticospinal tract which is accessible to TMS shows that this falls progressively as upper motor neuron signs become more evident. The technique holds promise as a way to follow upper motor neuron involvement in say, drug trials.

In primary lateral sclerosis, responses to TMS are often not obtainable but where they are, CMCT is markedly prolonged, much beyond that seen in idiopathic ALS (86,93). Familial ALS comprises 5–10% of all ALS cases. In a series of cases with the D90A mutation (94), CMCT was commonly abnormal, but in other mutations CMCT is normal. In the spinocerebellar degenerations, CMCT tends to be normal although thresholds can be raised. However, in Friedreich's ataxia and spinocerebellar atrophy Type 1 (SCA1), CMCT is prolonged to both upper and lower limb muscles and threshold is raised (95).

Demyelinating diseases

Multiple sclerosis (MS) was one of the first conditions to be studied with TMS (Fig. 14.8) and showed that CMCT could be markedly prolonged consistent with reduced conduction velocities in the corticospinal tract (96). Correlation with pyramidal signs is, however poor. In a number of cases, it has been shown that CMCT can be prolonged, even though there is no clinical evidence of a pyramidal lesion giving useful evidence of dissemination in space and analogous to the persistent prolongation of the visual evoked potential (VEP) after optic neuritis. The sensitivity of TMS in MS in detecting silent lesions is similar to the VEP.

With the increased use of MRI scanning with its superior sensitivity, use of TMS is now only indicated where, for example, the patient is unable to tolerate imaging. TMS has been used to monitor treatment in MS (97): with high-dose steroid therapy, CMCT tends to modestly shorten over a few days, but it is unclear whether this represents improvement of central conduction, the effects of steroid on excitatory potentials at spinal motor neurons or a reduction in oedema.

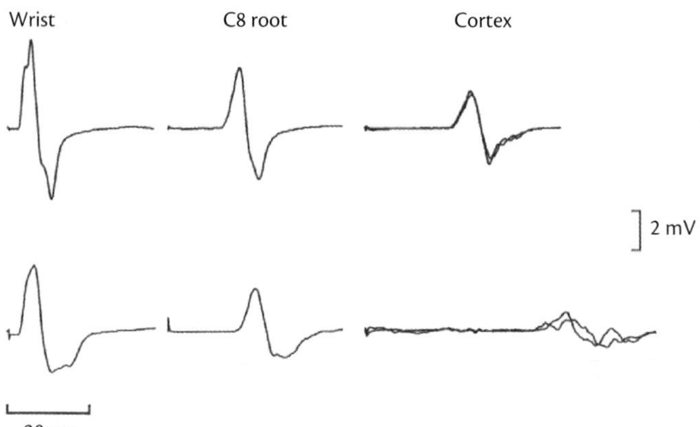

Fig. 14.8 Central motor conduction in a healthy subject (top) and a patient with multiple sclerosis (below). Prolongation of central motor conduction is marked in the patient.

Reproduced from *Ann Neurol*, **18**(5), Mills KR, Murray NM, Corticospinal tract conduction time in multiple sclerosis, pp. 601–5, copyright (1985), with permission from John Wiley and Sons.

Stroke

Patients with a dense hemiplegia following stroke usually have no recordable responses to TMS on the affected side. Less affected patients may have mildly prolonged CMCT or normal CMCT and this group tends to have the better prognosis (72,98–100). Some patients after stroke have noticeable responses on the ipsilateral side of the body; it is unclear whether this represents enhanced activity in ipsilateral pathways, spread of the stimulus to both hemispheres or to transmission over the corpus callosum (101,102).

Spinal cord diseases

Compressive myelopathy is usually determined using imaging of the spine, but several levels of compression are commonly found. By recording from many muscles with different levels of innervation, TMS can narrow down the level of maximal functional compression (103). For example, with lower cervical cord compression, normal CMCT to biceps with prolonged CMCT to EDC and muscles innervated below C7 will localize the compression to the C6 segmental level. The degree of corticospinal tract slowing is of the order seen in demyelinating disease, which can help differentiate compressive myelopathy from ALS.

Complete spinal cord injury naturally leads to absent responses below the site of the lesion and relatively normal responses above. Determining if the injury is complete or not can be assessed with TMS (104). Clearly, any responses below the level of the lesion would indicate partial injury.

In syringomyelia, CMCT may be marginally prolonged or more often, no responses are obtainable (105). Vitamin B12 deficiency sufficiently severe to cause subacute combined degeneration of the cord produces marginally prolonged CMCTs.

Paediatric applications

TMS in children is as in adults considered to be of minimal risk (106,107). It is difficult to elicit muscle responses using TMS in children under the age of 4 years (108). In children between the ages of 4 and 16, CMCT is prolonged such that, taking the peripheral conduction time into account which is related to height, the total time between cortex and muscle is similar to that found in adults. It appears that cortex to muscle time remains constant during the processes of growth and corticospinal tract myelination. Threshold also tends to be higher in children. A number of paediatric conditions have been investigated with TMS. In cerebral palsy, responses are often absent. Where there are unilateral abnormalities, responses may be very prominent on the ipsilateral side suggesting enhancement of the ipsilateral corticospinal projection, TMS has also been used in Rett syndrome where CMCT has been found to be shorter than age matched controls suggesting a heightened cortical excitability (109).

Movement disorders

In general terms, central motor conduction studies are normal in movement disorders such as Parkinson's disease. Some patients with the PARK2 mutation have shown prolonged CMCTs, however. More subtle abnormalities can be revealed by using more complex investigations. For example, cortical excitability as assessed by constructing input:output curves is abnormal as is SICI (80). A summary of the main findings in movement disorders has been reviewed (110).

Functional disorders

Patients with non-organic paralysis of a limb are not uncommonly encountered in clinical practice. TMS has a role in demonstrating the patency of the corticospinal tract in these patients, although care should be exercised in interpreting the results. TMS excites only the fast conducting portion of the corticospinal tract and there may be abnormalities of other corticospinal projections, as well as abnormal inhibitory mechanisms in play. However, there are accounts in the literature of patients with non-organic paralysis being 'cured' by the application of TMS (111).

Use of TMS in neurosurgical monitoring

Monitoring of spinal cord function in situations where there is risk from neurosurgical procedures is clearly important. Responses to TMS are, however, very susceptible to abolition from the anaesthetic. Consequently, motor monitoring during neurosurgery is usually carried out using transcranial electrical stimulation (see Chapter 37). Somatosensory evoked potential monitoring is usually carried out at the same time.

References

1. Barker, A.T., Jalinous, R., and Freeston, I.L. (1985). Non-invasive magnetic stimulation of human motor cortex. *Lancet*, **1**(8437), 1106–7.
2. Barker, A.T., Freeston, I.L., Jabinous, R., and Jarratt, J.A. (1986). Clinical evaluation of conduction time measurements in central motor pathways using magnetic stimulation of human brain. *Lancet*, **1**(8493), 1325–6.
3. Barker, A.T., Freeston, I.L., Jabinous, R., and Jarratt, J.A. (1987). Magnetic stimulation of the human brain and peripheral nervous system: an introduction and the results of an initial clinical evaluation. *Neurosurgery*, **20**(1), 100–9.
4. Mills, K.R. (1999). *Magnetic stimulation of the human nervous system*. Oxford: Oxford University Press.
5. George, M.S., Wassermann, E.M., Williams, W.A., et al. (1996). Changes in mood and hormone levels after rapid-rate transcranial magnetic stimulation (rTMS) of the prefrontal cortex. *Journal of Neuropsychiatry & Clinical Neuroscience*, **8**(2), 172–80.

6. Abdeen, M.A. and Stuchly, M.A. (1994). Modeling of magnetic field stimulation of bent neurons. *IEEE Transactions on Bio-medical Engineering*, **41**(11), 1092–5.
7. Tofts, P.S. (1990). The distribution of induced currents in magnetic stimulation of the nervous system. *Physics in Medicine and Biology*, **35**(8), 1119–28.
8. Tofts, P.S. and Branston, N.M. (1991). The measurement of electric field, and the influence of surface charge, in magnetic stimulation. *Electroencephalography & Clinical Neurophysiology*, **81**(3), 238–9.
9. Davey, K., Epstein, C.M., George, M.S., and Bohning, D.E. (2003). Modeling the effects of electrical conductivity of the head on the induced electric field in the brain during magnetic stimulation. *Clinical Neurophysiology*, **114**(11), 2204–9.
10. Roth, B.J., Saypol, J.M., Hallett, M., and Cohen, L.G. (1991). A theoretical calculation of the electric field induced in the cortex during magnetic stimulation. *Electroencephalography & Clinical Neurophysiology*, **81**(1), 47–56.
11. Saypol, J.M., Roth, B.J., Cohen, L.G., and Hallett, M. (1991). A theoretical comparison of electric and magnetic stimulation of the brain. *Annals of Biomedical Engineering*, **19**(3), 317–28.
12. Mills, K.R., Boniface, S.J., and Schubert, M. (1992). Magnetic brain stimulation with a double coil: the importance of coil orientation. *Electroencephalography & Clinical Neurophysiology*, **85**(1), 17–21.
13. Lin, V.W., Hsiao, I.N., and Dhaka, V. (2000). Magnetic coil design considerations for functional magnetic stimulation. *IEEE Transactions on Bio-medical Engineering*, **47**(5), 600–10.
14. Thielscher, A. and Kammer, T. (2004). Electric field properties of two commercial figure-8 coils in TMS: calculation of focality and efficiency. *Clinical Neurophysiology*, **115**(7), 1697–708.
15. Rossi, S., Hallett, M., Rossini, P.M., and Pascual-Leone, A. (2009). Safety of TMSCG. Safety, ethical considerations, and application guidelines for the use of transcranial magnetic stimulation in clinical practice and research. *Clinical Neurophysiology*, **120**(12), 2008–39.
16. Garvey, M.A. and Gilbert, D.L. (2004). Transcranial magnetic stimulation in children. *European Journal of Paediatric Neurology*, **8**(1), 7–19.
17. Ray, P.G., Meador, K.J., Epstein, C.M., Loring, D.W., and Day, L.J. (1998). Magnetic stimulation of visual cortex: factors influencing the perception of phosphenes. *Journal of Clinical Neurophysiology*, **15**(4), 351–7.
18. Kammer, T., Puls, K., Erb, M., and Grodd, W. (2005). Transcranial magnetic stimulation in the visual system. II. Characterization of induced phosphenes and scotomas. *Experimental Brain Research*, **160**(1), 129–40.
19. Schluter, N.D., Rushworth, M.F., Passingham, R.E., and Mills, K.R. (1998). Temporary interference in human lateral premotor cortex suggests dominance for the selection of movements. A study using transcranial magnetic stimulation. *Brain*, **121**(Pt 5), 785–99.
20. Baumer, T., Rothwell, J.C., and Munchau, A. (2003). Functional connectivity of the human premotor and motor cortex explored with TMS. *Clinical Neurophysiology*, **56**(Suppl.), 160–9.
21. Baumer, T., Bock, F., Koch, G., et al. (2006). Magnetic stimulation of human premotor or motor cortex produces interhemispheric facilitation through distinct pathways. *Journal of Physiology*, **572**(Pt 3), 857–68.
22. Koch, G. and Rothwell, J.C. (2009). TMS investigations into the task-dependent functional interplay between human posterior parietal and motor cortex. *Behavioural Brain Research*, **202**(2), 147–52.
23. Hess, C.W., Mills, K.R., and Murray, N.M. (1987). Responses in small hand muscles from magnetic stimulation of the human brain. *Journal of Physiology*, **388**, 397–419.
24. Patton, H.D. and Amassian, V.E. (1954). Single and multiple-unit analysis of cortical stage of pyramidal tract activation. *Journal of Neurophysiology*, **17**(4), 345–63.
25. Kernell, D. and Chien-Ping, W.U. (1967). Responses of the pyramidal tract to stimulation of the baboon's motor cortex. *Journal of Physiology*, **191**(3), 653–72.
26. Di Lazzaro, V., Oliviero, A., Profice, P., et al. (1998). Comparison of descending volleys evoked by transcranial magnetic and electric stimulation in conscious humans. *Electroencephalography & Clinical Neurophysiology*, **109**(5), 397–401.
27. Di Lazzaro, V., Oliviero, A., Profice, P., et al. (1995). Direct recordings of descending volleys after transcranial magnetic and electric motor cortex stimulation in conscious humans. *Electroencephalography & Clinical Neurophysiology*, **51**(Suppl.), 120–6.
28. Di Lazzaro, V. and Ziemann, U. (2013). The contribution of transcranial magnetic stimulation in the functional evaluation of microcircuits in human motor cortex. *Frontiers in Neural Circuits*, **7**, 18.
29. Mills, K.R. (1991). Magnetic brain stimulation: a tool to explore the action of the motor cortex on single human spinal motoneurones. *Trends in Neuroscience*, **14**(9), 401–5.
30. Burke, D., Hicks, R.G., and Stephen, J.P. (1990). Corticospinal volleys evoked by anodal and cathodal stimulation of the human motor cortex. *Journal of Physiology*, **425**, 283–99.
31. Hicks, R., Burke, D., Stephen, J., Woodforth, I., and Crawford, M. (1992). Corticospinal volleys evoked by electrical stimulation of human motor cortex after withdrawal of volatile anaesthetics. *Journal of Physiology*, **456**, 393–404.
32. Rothwell, J., Burke, D., Hicks, R., Stephen, J., Woodforth, I., and Crawford, M. (1994). Transcranial electrical stimulation of the motor cortex in man: further evidence for the site of activation. *Journal of Physiology*, **481**(Pt 1), 243–50.
33. Mills, K.R. and Murray, N.M. (1985). Corticospinal tract conduction time in multiple sclerosis. *Annals of Neurology*, **18**(5), 601–5.
34. Groppa, S., Oliviero, A., Eisen, A., et al. (2012). A practical guide to diagnostic transcranial magnetic stimulation: report of an IFCN committee. *Clinical Neurophysiology*, **123**(5), 858–82.
35. Mills, K.R. and Nithi, K.A. (1997). Corticomotor threshold to magnetic stimulation: normal values and repeatability. *Muscle & Nerve*, **20**(5), 570–6.
36. Mills, K.R. and Murray, N.M. (1986). Electrical stimulation over the human vertebral column: which neural elements are excited? *Electroencephalography & Clinical Neurophysiology*, **63**(6), 582–9.
37. Kimiskidis, V.K., Papagiannopoulos, S., Kazis, D.A., et al. (2006). Lorazepam-induced effects on silent period and corticomotor excitability. *Experimental Brain Research*, **173**(4), 603–11.
38. Kimiskidis, V.K., Papagiannopoulos, S., Sotirakoglou, K., Kazis, D.A., Kazis, A., and Mills, K.R. (2005). Silent period to transcranial magnetic stimulation: construction and properties of stimulus-response curves in healthy volunteers. *Experimental Brain Research*, **163**(1), 21–31.
39. Carroll, T.J., Riek, S., and Carson, R.G. (2001). Reliability of the input-output properties of the cortico-spinal pathway obtained from transcranial magnetic and electrical stimulation. *Journal of Neuroscience Methods*, **112**(2), 193–202.
40. Kujirai, T., Caramia, M.D., Rothwell, J.C., et al. (1993). Corticocortical inhibition in human motor cortex. *Journal of Physiology*, **471**, 501–19.
41. Daskalakis, Z.J., Farzan, F., Barr, M.S., Maller, J.J., Chen, R., and Fitzgerald, P.B. (2008). Long-interval cortical inhibition from the dorsolateral prefrontal cortex: a TMS-EEG study. *Neuropsychopharmacology*, **33**(12), 2860–9.
42. Meyer, B.U., Roricht, S., Grafin von Einsiedel, H., Kruggel, F., and Weindl, A. (1995). Inhibitory and excitatory interhemispheric transfers between motor cortical areas in normal humans and patients with abnormalities of the corpus callosum. *Brain*, **118**(Pt 2), 429–40.
43. Fischer, M. and Orth, M. (2011). Short-latency sensory afferent inhibition: conditioning stimulus intensity, recording site, and effects of 1 Hz repetitive TMS. *Brain Stimulation*, **4**(4), 202–9.
44. Magistris, M.R., Rosler, K.M., Truffert, A., and Myers, J.P. (1998). Transcranial stimulation excites virtually all motor neurons supplying the target muscle. A demonstration and a method

improving the study of motor evoked potentials. *Brain*, **121**(Pt 3), 437–50.

45. Magistris, M.R. and Rosler, K.M. (2003). The triple stimulation technique to study corticospinal conduction. *Clinical Neurophysiology*, **56**(Suppl.), 24–32.
46. Humm, A.M., Z'Graggen, W.J., von Hornstein, N.E., Magistris, M.R., Rosler, K.M. Assessment of central motor conduction to intrinsic hand muscles using the triple stimulation technique: normal values and repeatability. *Clinical neurophysiology*, **115**(11), 2558–66.
47. Magistris, M.R., Rosler, K.M., Truffert, A., Landis, T., and Hess, C.W. (1999). A clinical study of motor evoked potentials using a triple stimulation technique. *Brain*, **122**(Pt 2), 265–79.
48. Rosler, K.M., Truffert, A., Hess, C.W., and Magistris, M.R. (2000). Quantification of upper motor neuron loss in amyotrophic lateral sclerosis. *Clinical Neurophysiology*, **111**(12), 2208–18.
49. Rosler, K.M. and Magistris, M.R. (2004). Triple stimulation technique (TST) in amyotrophic lateral sclerosis. *Clinical Neurophysiology*, **115**(7), 1715.
50. Humm, A.M., Beer, S., Kool, J., Magistris, M.R., Kesselring, J., and Rosler, K.M. (2004). Quantification of Uhthoff's phenomenon in multiple sclerosis: a magnetic stimulation study. *Clinical Neurophysiology*, **115**(11), 2493–501.
51. Humm, A.M., Z'Graggen, W.J., Buhler, R., Magistris, M.R., and Rosler, K.M. (2006). Quantification of central motor conduction deficits in multiple sclerosis patients before and after treatment of acute exacerbation by methylprednisolone. *Journal of Neurology, Neurosurgery, and Psychiatry*, **77**(3), 345–50.
52. Schmid, U.D., Moller, A.R., and Schmid, J. (1992). Transcranial magnetic stimulation of the facial nerve: intraoperative study on the effect of stimulus parameters on the excitation site in man. *Muscle & Nerve*, **15**(7), 829–36.
53. Schriefer, T.N., Mills, K.R., Murray, N.M., and Hess, C.W. (1988). Evaluation of proximal facial nerve conduction by transcranial magnetic stimulation. *Journal of Neurology, Neurosurgery, and Psychiatry*, **51**(1), 60–6.
54. Nowak, D.A., Linder, S., and Topka, H. (2005). Diagnostic relevance of transcranial magnetic and electric stimulation of the facial nerve in the management of facial palsy. *Clinical Neurophysiology*, **116**(9), 2051–7.
55. Happe, S. and Bunten, S. (2012). Electrical and transcranial magnetic stimulation of the facial nerve: diagnostic relevance in acute isolated facial nerve palsy. *European Neurology*, **68**(5), 304–9.
56. Matsumoto, H., Hanajima, R., Terao, Y., and Ugawa, Y. (2013). Magnetic-motor-root stimulation: review. *Clinical Neurophysiology*, **124**(6), 1055–67.
57. Souayah, N. and Sander, H.W. (2006). Lumbosacral magnetic root stimulation in lumbar plexopathy. *American Journal of Physical Medicine & Rehabilitation*, **85**(10), 858–61.
58. Matsumoto, H., Octaviana, F., Hanajima, R., et al. (2009). Magnetic lumbosacral motor root stimulation with a flat, large round coil. *Clinical Neurophysiology*, **120**(4), 770–5.
59. Nitsche, M.A. and Paulus, W. (2000). Excitability changes induced in the human motor cortex by weak transcranial direct current stimulation. *Journal of Physiology*, **527**(Pt 3), 633–9.
60. Paulus, W. (2003). Transcranial direct current stimulation (tDCS). *Clinical Neurophysiology*, **56**(Suppl.), 249–54.
61. Nitsche, M.A. and Paulus, W. (2011). Transcranial direct current stimulation—update 2011. *Restorative Neurology and Neuroscience*, **29**(6), 463–92.
62. Paulus, W., Peterchev, A.V., and Ridding, M. (2013). Transcranial electric and magnetic stimulation: technique and paradigms. *Handbook of Clinical Neurology*, **116**, 329–42.
63. Schutter, D.J. (2010). Quantitative review of the efficacy of slow-frequency magnetic brain stimulation in major depressive disorder. *Psychological Medicine*, **40**(11), 1789–95.
64. Huang, Y.Z. and Rothwell, J.C. (2004). The effect of short-duration bursts of high-frequency, low-intensity transcranial magnetic stimulation on the human motor cortex. *Clinical Neurophysiology*, **115**(5), 1069–75.
65. Pascual-Leone, A., Valls-Sole, J., Wassermann, E.M., and Hallett, M. (1994). Responses to rapid-rate transcranial magnetic stimulation of the human motor cortex. *Brain*, **117**(Pt 4), 847–58.
66. Wassermann, E.M. (1998). Risk and safety of repetitive transcranial magnetic stimulation: report and suggested guidelines from the International Workshop on the Safety of Repetitive Transcranial Magnetic Stimulation, June 5–7, 1996. *Electroencephalography & Clinical Neurophysiology*, **108**(1), 1–16.
67. Homan, P., Kindler, J., Hauf, M., Hubl, D., and Dierks, T. (2012). Cerebral blood flow identifies responders to transcranial magnetic stimulation in auditory verbal hallucinations. *Translational Psychiatry*, **2**, e189.
68. Bishnoi, R.J. and Jhanwar, V.G. (2011). Extended trial of transcranial magnetic stimulation in a case of treatment-resistant obsessive-compulsive disorder. *Asian Journal of Psychiatry*, **4**(2), 152.
69. Amiaz, R., Levy, D., Vainiger, D., Grunhaus, L., and Zangen, A. (2009). Repeated high-frequency transcranial magnetic stimulation over the dorsolateral prefrontal cortex reduces cigarette craving and consumption. *Addiction*, **104**(4), 653–60.
70. Saba, G., Schurhoff, F., and Leboyer, M. (2006). Therapeutic and neurophysiologic aspects of transcranial magnetic stimulation in schizophrenia. *Neurophysiology Clinics*, **36**(3), 185–94.
71. Ayache, S.S., Creange, A., Farhat, W.H., et al. (2014). Relapses in multiple sclerosis: effects of high-dose steroids on cortical excitability. *European Journal of Neurology*, **21**(4), 630–6.
72. Trompetto, C., Assini, A., Buccolieri, A., Marchese, R., and Abbruzzese, G. (2000). Motor recovery following stroke: a transcranial magnetic stimulation study. *Clinical Neurophysiology*, **111**(10), 1860–7.
73. Chen, R., Cros, D., Curra, A., et al. (2008). The clinical diagnostic utility of transcranial magnetic stimulation: report of an IFCN committee. *Clinical Neurophysiology*, **119**(3), 504–32.
74. Ziemann, U. (2004). TMS and drugs. *Clinical Neurophysiology*, **115**(8), 1717–29.
75. Schnitzler, A. and Benecke, R. (1994). The silent period after transcranial magnetic stimulation is of exclusive cortical origin: evidence from isolated cortical ischemic lesions in man. *Neuroscience Letters*, **180**(1), 41–5.
76. Ziemann, U. (2003). Pharmacology of TMS. *Clinical Neurophysiology*, **56**(Suppl.), 226–31.
77. Cantello, R., Gianelli, M., Civardi, C., and Mutani, R. (1992). Magnetic brain stimulation: the silent period after the motor evoked potential. *Neurology*, **42**(10), 1951–9.
78. Avanzino, L., Martino, D., van de Warrenburg, B.P., et al. (2008). Cortical excitability is abnormal in patients with the 'fixed dystonia' syndrome. *Movement Disorders*, **23**(5), 646–52.
79. Battaglia, F., Quartarone, A., Bagnato, S., et al. (2005). Brain dysfunction in uremia: a question of cortical hyperexcitability? *Clinical Neurophysiology*, **116**(7), 1507–14.
80. Berardelli, A., Abbruzzese, G., Chen, R., et al. (2008). Consensus paper on short-interval intracortical inhibition and other transcranial magnetic stimulation intracortical paradigms in movement disorders. *Brain Stimulation*, **1**(3), 183–91.
81. Daskalakis, Z.J., Christensen, B.K., Fitzgerald, P.B., Moller, B., Fountain, S.I., and Chen, R. (2008). Increased cortical inhibition in persons with schizophrenia treated with clozapine. *Journal of Psychopharmacology*, **22**(2), 203–9.
82. Gilbert, D.L., Bansal, A.S., Sethuraman, G., et al. (2004). Association of cortical disinhibition with tic, ADHD, and OCD severity in Tourette syndrome. *Movement Disorders*, **19**(4), 416–25.
83. Huynh, W., Krishnan, A.V., Vucic, S., Lin, C.S., and Kiernan, M.C. (2012). Motor cortex excitability in acute cerebellar infarct. *Cerebellum*, **12**(6), 826–34.
84. MacKinnon, C.D., Gilley, E.A., Weis-McNulty, A., and Simuni, T. (2005). Pathways mediating abnormal intracortical inhibition in Parkinson's disease. *Annals of Neurology*, **58**(4), 516–24.

85. Rothwell, J.C. and Edwards, M.J. (2013). Parkinson's disease. *Handbook of Clinical Neurology*, **116**, 535–42.
86. Floyd, A.G., Yu, Q.P., Piboolnurak, P., et al. (2009). Transcranial magnetic stimulation in ALS: utility of central motor conduction tests. *Neurology*, **72**(6), 498–504.
87. Mills, K.R. (2003). The natural history of central motor abnormalities in amyotrophic lateral sclerosis. *Brain*, **126**(Pt 11), 2558–66.
88. Vucic, S. and Kiernan, M.C. (2013). Utility of transcranial magnetic stimulation in delineating amyotrophic lateral sclerosis pathophysiology. *Handbook of Clinical Neurology*, **116**, 561–75.
89. Osei-Lah, A.D. and Mills, K.R. (2004). Optimising the detection of upper motor neuron function dysfunction in amyotrophic lateral sclerosis—a transcranial magnetic stimulation study. *Journal of Neurology*, **251**(11), 1364–9.
90. Deftereos, S.N., Kechagias, E.A., Panagopoulos, G., et al. (2009). Localisation of cervical spinal cord compression by TMS and MRI. *Functional Neurology*, **24**(2), 99–105.
91. Chan, Y.C. and Mills, K.R. (2005). The use of transcranial magnetic stimulation in the clinical evaluation of suspected myelopathy. *Journal of Clinical Neuroscience*, **12**(8), 878–81.
92. Attarian, S., Verschueren, A., and Pouget, J. (2007). Magnetic stimulation including the triple-stimulation technique in amyotrophic lateral sclerosis. *Muscle & Nerve*, **36**(1), 55–61.
93. Cruz Martinez, A. and Trejo, J.M. (1999). Transcranial magnetic stimulation in amyotrophic and primary lateral sclerosis. *Electromyography & Clinical Neurophysiology*, **39**(5), 285–8.
94. Turner, M.R., Osei-Lah, A.D., Hammers, A., et al. (2005). Abnormal cortical excitability in sporadic but not homozygous D90A SOD1 ALS. *Journal of Neurology, Neurosurgery, & Psychiatry*, **76**(9), 1279–85.
95. Schwenkreis, P., Tegenthoff, M., Witscher, K., et al. (2002). Motor cortex activation by transcranial magnetic stimulation in ataxia patients depends on the genetic defect. *Brain*, **125**(Pt 2), 301–9.
96. Hess, C.W., Mills, K.R., Murray, N.M., and Schriefer, T.N. (1987). Magnetic brain stimulation: central motor conduction studies in multiple sclerosis. *Annals of Neurology*, **22**(6), 744–52.
97. Fierro, B., Salemi, G., Brighina, F., et al. (2002). A transcranial magnetic stimulation study evaluating methylprednisolone treatment in multiple sclerosis. *Acta Neurologica Scandinavica*, **105**(3), 152–7.
98. Stulin, I.D., Savchenko, A.Y., Smyalovskii, V.E., et al. (2003). Use of transcranial magnetic stimulation with measurement of motor evoked potentials in the acute period of hemispheric ischemic stroke. *Neuroscience and Behavioral Physiology*, **33**(5), 425–9.
99. Timmerhuis, T.P., Hageman, G., Oosterloo, S.J., and Rozeboom, A.R. (1996). The prognostic value of cortical magnetic stimulation in acute middle cerebral artery infarction compared to other parameters. *Clinical Neurology & Neurosurgery*, **98**(3), 231–6.
100. Delvaux, V., Alagona, G., Gerard, P., De Pasqua, V., Pennisi, G., and de Noordhout, A.M. (2003). Post-stroke reorganization of hand motor area: a 1-year prospective follow-up with focal transcranial magnetic stimulation. *Clinical Neurophysiology*, **114**(7), 1217–25.
101. Walther, M., Juenger, H., Kuhnke, N., et al. (2009). Motor cortex plasticity in ischemic perinatal stroke: a transcranial magnetic stimulation and functional MRI study. *Pediatric Neurology*, **41**(3), 171–8.
102. Graziadio, S., Tomasevic, L., Assenza, G., Tecchio, F., and Eyre, J.A. (2012). The myth of the 'unaffected' side after unilateral stroke: is reorganisation of the non-infarcted corticospinal system to re-establish balance the price for recovery? *Experimental Neurology*, **238**(2), 168–75.
103. Lo, Y.L., Chan, L.L., Lim, W., et al. (2006). Transcranial magnetic stimulation screening for cord compression in cervical spondylosis. *Journal of Neurological Sciences*, **244**(1–2), 17–21.
104. Awad, B.I., Carmody, M.A., Zhang, X., Lin, V.W., and Steinmetz, M.P. (2015). Transcranial magnetic stimulation after spinal cord injury. *World Neurosurgery* **83**(2), 232–5.
105. Nogues, M.A., Pardal, A.M., Merello, M., and Miguel, M.A. (1992). SEPs and CNS magnetic stimulation in syringomyelia. *Muscle & Nerve*, **15**(9), 993–1001.
106. Gilbert, D.L., Garvey, M.A., Bansal, A.S., Lipps, T., Zhang, J., and Wassermann, E.M.(2004). Should transcranial magnetic stimulation research in children be considered minimal risk? *Clinical Neurophysiology*, **115**(8), 1730–9.
107. Wu, S.W., Shahana, N., Huddleston, D.A., Lewis, A.N., and Gilbert, D.L. (2012). Safety and tolerability of theta-burst transcranial magnetic stimulation in children. *Developmental Medicine & Child Neurology*, **54**(7), 636–9.
108. Eyre, J.A., Miller, S., and Ramesh, V. (1991). Constancy of central conduction delays during development in man: investigation of motor and somatosensory pathways. *Journal of Physiology*, **434**, 441–52.
109. Nezu, A., Kimura, S., Takeshita, S., and Tanaka, M. (1998). Characteristic response to transcranial magnetic stimulation in Rett syndrome. *Electroencephalography & Clinical Neurophysiology*, **109**(2), 100–3.
110. Cantello, R. (2002). Applications of transcranial magnetic stimulation in movement disorders. *Journal of Clinical Neurophysiology*, **19**(4), 272–93.
111. Chastan, N. and Parain, D. (2010). Psychogenic paralysis and recovery after motor cortex transcranial magnetic stimulation. *Movement Disorders*, **25**(10), 1501–4.

CHAPTER 15

Evoked potentials

Helmut Buchner

Introduction

This chapter will give an overview of the standards and clinical applications of visual, auditory and somatosensory evoked potentials. Therefore, the technical basis of stimulation and recording of evoked potentials will be addressed and the mechanism of generation of the potentials will briefly be discussed.

Evoked potentials (EP) are electric potentials occurring in the peripheral and the central nervous system—time locked to a natural or artificial sensory stimulus.

The scope of this chapter includes the established methods for stimulation and recording for clinical applications; these methods and the interpretation of the results have proven reliable in making clinical decisions. Evoked potentials to natural stimuli tend to be less reproducible and are therefore less useful in the clinical context.

The technical system

There are three main components of the technical system:

- An amplifier amplifying the voltage difference between two recording electrodes.
- An analogue to digital converter.
- An averager—the heart of the system—which is connected via a trigger to the stimulator.

Evoked potentials are recorded with electrodes placed in a suitable location connected to a differential amplifier. The electrodes should have a contact resistance of 5–20 kΩ. Some laboratories prefer to use steel needles because they can be placed very quickly. Amplifiers are designed with a differential input, so that only the difference in the signals applied to the two inputs is amplified. Coupled to the amplifier, a filter limits the signal to the frequency components of interest for the respective measurement (see Tables 15.1–15.7).

Analogue to digital conversion transforms the signal into an amplitude value at a given time point. For exact conversion the sampling frequency must be at least four times of the highest frequency present in the signal and the amplification must be chosen so that the lowest and the highest amplitudes of the signal of interest are within the range of the quantization. Sweeps with high signal amplitudes beyond the upper quantization limit are to be rejected from averaging. The available digitization frequency (in Hz) and the range of the quantization (in bits, e.g. 10 bit, i.e. 2^{10} = 1024 amplitude values) are usually technically fixed by the hardware (see also Chapter 5).

Evoked potentials are usually very low amplitude signals, hidden within the spontaneous electroencephalogram (EEG). The trick is to average sweeps triggered by the stimulus by adding the amplitudes at each digitization point and dividing the sum by the number of sweeps. This leads to reduction of the random noise while the signal, which is time locked to the stimulus, is preserved. The signal to noise ratio improves with the square root of the number of averaged sweeps.

Some residual noise remains and can be estimated from a pre-trigger time period. The best way to measure the noise is plus-minus averaging, whereby every second sweep is subtracted instead of added, resulting in reduction of both noise and evoked potential.

One has to bear in mind that stimulus time locked averaging can miss or modify stimulus dependent signals if the latency varies from trial to trial. This can occur, for example, in demyelinating neuropathy, where the conduction time can change from stimulus to stimulus.

Before starting to measure an EP the technical system should be checked:

- The electrodes should be firmly affixed with a resistance of 5–20 kΩ.
- The amplifier and the filters should be adjusted according to the type of EP.
- The functioning of the trigger and the stimulator should be checked.
- Effective stimulation should be ensured.
- Quality control recommends the recording of two waveforms with good reproducibility of latencies and amplitudes.
- Finally, complete documentation of all parameters and waveforms should be done.

One of the surprising things in neurophysiology is that just one nomenclature for evoked potentials has been established. Peaks are labelled as N for negative and P for positive with a number indicating the peak or alternatively—and most often used—the normal value characterizing the potential.

Although biological data may not be truly normally distributed, 2.5 or 3 standard deviations from the mean is usually used for calculating an upper normal limit, which approximates to a cut-off value of 99% of normal subjects.

Under standard and stable stimulus and recording conditions there are published upper latency limits for evoked potentials. Therefore, it is not necessary to create normative data for each laboratory, but data on at least 10 locally acquired normal subjects should be obtained for confirmation.

Mechanisms of generation of evoked potentials

The mechanism of generation of evoked potentials is of great interest.

The most important facts can be summarized as follows:

- *First*: from intra-cellular sources and sinks, extra-cellular current flow is generated. The extracellular volume current can be summarized as an equivalent dipole. From this, electrical fields on the surface of the head are generated by multiple parallel oriented equivalent dipoles from pyramidal cells either in a predominantly tangential or radial orientation. Since scalp electrodes can be placed close to the generator, this is called near-field recording. A typical near-field potential is the somatosensory evoked potential (SEP) N20 component with a clear tangential orientation between a positivity at a parietal location (CP; international 10–20 system of electrode location) and a frontal negativity (Fz).
- *Second*: the propagation of currents around an axon can be described by a depolarization/repolarization pair of dipoles resulting in a bipolar nerve action potential propagating along an axon. An abrupt change of the characteristics of the volume conductor around the axon either in terms of conductivity or geometry or in the direction of propagation, results in the generation of a potential recordable in the far-field. This means the point of potential generation is located remote from the recording electrode. Typical far-field potentials are the auditory evoked potentials (AEP) peaks arising at the brainstem, but generating a widely distributed positive field on the head with a maximum around the electrode position Cz.

For more details see references (1–4).

Visual evoked potentials

Anatomy and physiology

The physiology underlying the visual system is highly complex, starting with optics at the level of the retina with its organisation of cones, rods and retinal ganglion cells projecting over the magnocellular and the parvocellular pathways for contrast and motion for the first and for colour for the second.[1] For simplicity one can assume that under the conditions of almost normal or corrected vision, pathological processes within the retina itself have only a minor influence on the clinically used checker board visual evoked potential.

Even in the proximal retinal layer structured stimuli evoke the highest amplitudes, since the entire visual system is optimized to detect contours. This explains why contrast stimulation leads to far more efficient and stable evoked potentials than flash stimulation.

In the fovea, a particularly high resolution is achieved by very small, densely packed cones. In the visual cortex the central projection of this high resolution in the fovea is realized by 'cortical magnification'. This means that stimulation of the central 10° of the visual field excites a large part of the visual cortex.

[1] For further reading and for the recommendations of the international federation of clinical neurophysiology see (5–8).

So, visual evoked potentials measure contrast vision in only the central visual field. This explains why VEPs are highly sensitive in diseases with decreased contrast vision, e.g. retrobulbar neuritis but are not able to evaluate a larger part of the visual field.

VEPs have long latencies. Most of this time is consumed by the cones and bipolar cells within the retina due to their graded excitation (not spike coded). The two major peaks of visual evoked potential are the N75 reflecting the input into the striate cortex from the lateral geniculate nucleus (GN) and the P100, which is generated by secondary excitation of the striate cortex.

Stimulus and recording

The stimulus is a pattern reversal black and white alternating stimulus with a very high contrast. This stimulus is a very crude 'supramaximal' stimulation of the visual system. Approximately 80% of the VEP derives from the central 8° of the visual field. It generates a typical waveform which can be easily detected by an Oz–Fz montage or, for more information about waveform changes, by the so called 'Queen Square montage' which includes additional recording electrodes 5 cm right and left from Oz (Fig. 15.1).

The P100 amplitude is, from a checker board size from 17° to 2°, approximately stable. The smaller the selected check size (14–16'), the more the foveal portion of the retina is tested and the VEPs are more sensitive to visual loss, e.g. insufficient refraction.

Quality control recommends two measurements have latencies different by less than 1 ms and amplitudes by +/–20%.

There are two typical sources of error in the recording routine:

- A low black/white contrast results in a reduced amplitude and sharpness of the waveform. Assuming a constant retinal illumination (e.g. cataract) and using a check size of more than 35' a visual acuity of 20/200 still does not prolong the P100.
- Attention to fixation of the checker board´s midpoint is crucial for the amplitude and to some degree for the latency. The lower half of the visual field generates most of the VEP. The amplitude of the P100 depends also on vigilance. With full attention the amplitude can be higher by a factor of up to 2. In sleepy patients α-rhythm in the EEG can also dominate with the eyes open, so that the first positive peak can be superimposed with 'α-driving'. The suspicion should arise when there is no clear linear baseline recorded in the first 70 ms.

There is only minor influence of age, gender, head size, pupil size, and body temperature.

The recommended stimulus and recording parameters are summarized in Table 15.1.

Normal and pathological recordings

Normal and pathological VEPs are defined mainly by the P100 latency. Amplitude and wave shape are less reliable parameters. Sometimes a double P100 waveform can be recorded and can cause uncertainty in defining the latency of the P100. These are rare cases and in most of them the Queen Square montage and/or stimulation of the lower visual field can solve the problem, usually then resulting in a normal potential waveform being recorded. Under standard and stable stimulus conditions there are upper limits of the P100 latency and of its side difference (Table 15.2).

In clinical applications visual evoked potentials measure conduction along the optic nerve and the central visual pathway.

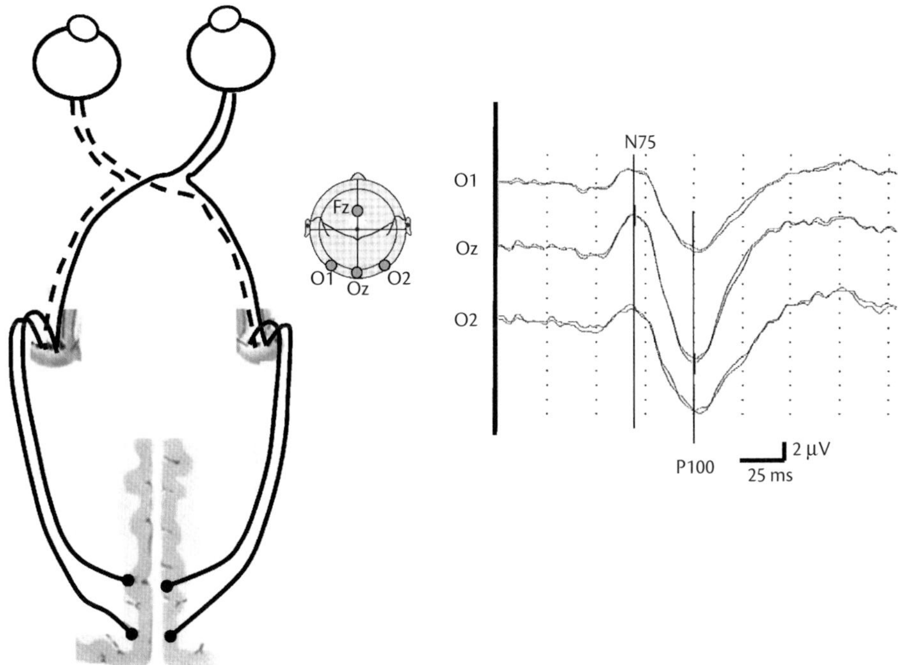

Fig. 15.1 Schematic anatomy of the visual system. Electrode locations (three channel Queen square montage) and normal waveforms (each the average of 100) showing N75 and P100 potentials.

For reasons mentioned above, prechiasmal pathology can be detected very reliably. After the chiasm, the corresponding halves of the field are brought together in the optic tract. Monocular whole visual field stimulation can thus provide no valid information about retrochiasmal deficits, because half of the stimulated axons are routed in the intact hemisphere and thus may mask pathological findings by the normal excitation of the contralateral visual cortex. Although MRI has proved superior in the localization of cerebral foci since its introduction, especially in MS, optic neuritis is not necessarily detectable by MRI.

Therefore pathological processes influencing conduction in general and especially demyelinating diseases like multiple sclerosis are clear indications to perform visual evoked potentials. Fig. 15.2 shows VEPs with a small, but reliable side-difference in latency of the P100 in a subject with an acute optic neuritis with minor loss of visual acuity and colour discrimination.

Table 15.1 Standard stimulus and recording parameters

VEP—Stimulus and recording parameters
One eye full-field stimulation
Stimulus field: ≥12–15°
Check size (15'/50–60')
Contrast black/white: >80%
Reversal rate: 1–2 Hz
Fixation: central
Amplifier filter setting: ≤0.5 Hz, ≥100 Hz
Analysis time: ≥250 ms
Number of averages: ≤100

The main use of the VEP in MS diagnosis is the detection of clinically apparent demyelinating lesions of the optic nerve. Significant latency delay of the P100 (up to 70 ms) with largely maintained amplitude provides strong evidence for demyelination of the nerve. The importance of the VEP in MS diagnosis is based on the frequency of lesions of the optic nerve. Pathological VEP latencies usually remain even after clinically complete resolution of optic neuritis, probably due to incomplete remyelination. Thus, the VEP enables pointers in the history of a prior retrobulbar neuritis to be confirmed. Conversely, delayed VEPs are frequently found in MS patients without visual impairment (60–90% of the eyes studied, 30–40% of the patients studied). These pathological findings are consistent with clinically silent lesions. In patients with clinically definite MS pathological VEPs can be found in 80–95% at least on one side. In acute optic neuritis, the amplitude of the response is reduced or, when vision is reduced below 6/24, may be unrecordable. In severe visual loss, the VEPs are, if still recordable, always markedly delayed in latency and often reduced in amplitude. After improvement of the visual acuity the amplitude recovers incompletely, and in 90% of patients a delayed P100 remains with a well-preserved waveform. An improvement or normalization over the course of many months is rare.

Table 15.2 VEP P100 latency, upper normal limits

	Latency—upper normal limit	Side difference- upper normal limit
P100	111 ms	5 ms

Source data from Paulus W, Elektroretinographie (ERG) und visuell evozierte Potenziale. In: Buchner H. (Ed.), *Praxisbuch—Evozierte Potenziale*, copyright (2014), Thieme Medical Publishers, Inc.

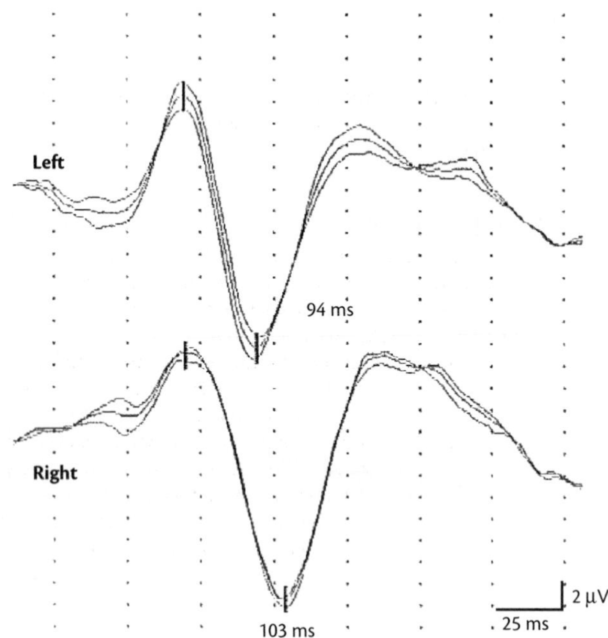

Fig. 15.2 Significant side to side difference in the P100 latency in a patient with acute optic neuritis. Each VEP has been replicated three times with averages of 100 trials.

In clinically mild optic neuritis, the P100 latency is occasionally still within the normal range. Here, a comparison of the sides (side difference ≤5 ms) with stimulation of the central 4° of the visual field can enhance the detection of abnormalities. Overall, with smaller checker sizes (15') the diagnostic sensitivity for the detection of optic nerve lesions is higher than for large (50').

Clinical and advanced use

Three categories (Table 15.3) of clinical utility of visual evoked potential (VEPs) called hard rocks, soft rocks, and beach have been chosen to give clear, but subjective recommendations. This follows (and with compliments) to a presentation by D. L. Jewett at the Evoked Potentials conference in 1986 in Berlin (9).

Table 15.3 Clinical usefulness of the VEP

Hard rocks ◆ Often pathological ◆ Usually technically clear results ◆ Usually unique pathophysiologic interpretation	MS—diagnosis
	Optic neuritis and neuromyelitis optica
	Anterior ischaemic optic neuropathy
	Compressive optic neuropathy
Soft rocks	MS—history
	Degenerative diseases
	Friedreich ataxia, familial spastic paraplegia, hereditary neuropathies
	Toxic neuropathies
Beach ◆ Sometimes pathological ◆ Often technically unclear results ◆ Usually no unique pathophysiologic interpretation	Paraneoplastic disorders
	Visual field defects
	Cortical blindness

Auditory evoked potentials

Anatomy and physiology

For auditory evoked potentials the anatomy and physiology of the system can be divided into three parts[2]. The peripheral ear and the ear canal acting as a compressor followed by the inner ear for sound to mechanical and mechanical to electrical conversion, and the cochlea nerve and the central auditory pathway.

Although mid-latency and long latency auditory EPs can be recorded, for clinical purposes, only the short latency brainstem auditory evoked potentials are of interest.

In the external ear canal, the sound waves are 'compressed' and forwarded to the eardrum. The ear canal behaves physically like a cavity of approximately 2.5 cm in length and a natural frequency of about 3000 Hz. For the measurement of AEP it is essential that this property is preserved. Inflammation and swelling in the ear canal or a larger amount of wax can greatly attenuate the sound leading to absence of all AEP waves. The eardrum with the chain of ossicles converts sound into mechanical vibrations. In addition, a gain is made in the ratio of 1:22 and there is pressure adjustment through the Eustachian tube. From the oval window the mechanical waves are transmitted to the basilar membrane. This is deflected according to the 'travelling-wave' hypothesis, so that a succession of waves of different frequency and amplitude is formed. The point of maximum deflection is located near the end of the cochlea at low frequencies and close to the oval window for high frequencies. The localized displacement of the basilar membrane leads to an excitation of the respective hair cells, which initiate the transmission to the acoustic nerve. Therefore, in diseases of the cochlea, there may be selective conversion of excitation into action potentials.

Direct recording from the cochlear nerve allows the compound action potential of all currently conducting nerve fibres to be measured. At some distance from the auditory nerve, i.e. on the scalp, a volume conducted potential consisting of the waves I and II is recordable.

The central connection of the incoming signal from the cochlear nerve in the brainstem begins at the cochlear nucleus. The anatomy of the acoustic tracts in the brainstem is complicated with a bilateral ascending projection and decussations of fibres at many levels, and efferent systems. For clinical applications of the AEP, a simplified model has proved successful:

- Conduction from the cochlear nucleus via the trapezoid body and the dorsal medial striae acusticae to the superior lateral nuclei of the stimulated and the contralateral side.
- Ascending conduction via the inferior colliculus and the lateral lemniscus (12).

Stimulus and recording

These potentials are evoked using click pulses of short (0.1 ms) duration produced by electrical pulses applied to a headset either first elevating the membrane in the direction towards the tympanum,

[2] For further reading and for the recommendations of the international federation of clinical neurophysiology see (10,11).

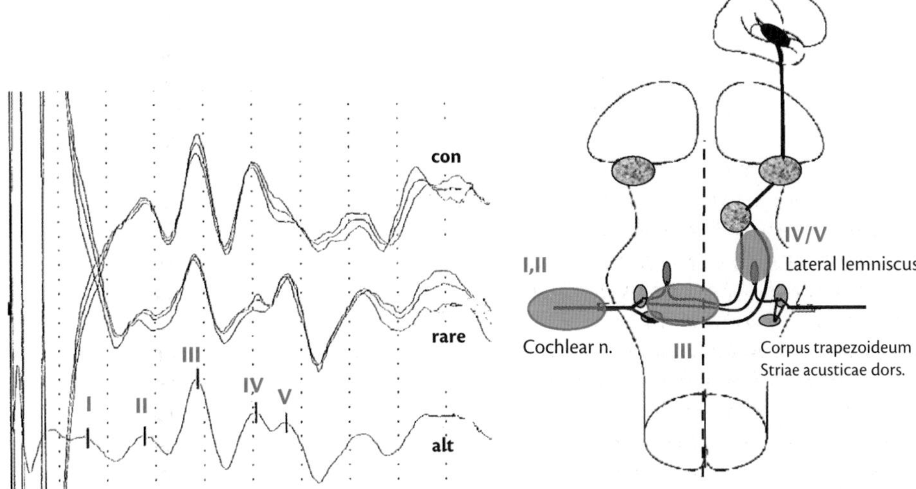

Fig. 15.3 Schematic anatomy of the auditory system. Normal waveforms are shown each replicated three times with averages of 1500 trials. The responses to condensation and rarefaction clicks and to alternating clicks are shown. Note the stimulus artefact obscuring wave I.

called condensation, or away from the tympanum, called rarefaction. To minimize the stimulus artefact an alternating stimulation mode can be used. These condensation/rarefaction clicks make the membrane of the headset vibrate in a wide spectrum of frequencies. Note that the stimulus intensity has to be limited to 70 dB above subjective hearing threshold in order to avoid damaging the cochlea.

This stimulus evokes five positive peaks. In accordance with the model of generation of the peaks, the first two peaks are generated within the intracranial cochlear nerve, the third peak in the brainstem and the fourth and fifth peak in the ascending lateral lemniscus mostly, but not exclusively on the contralateral side.

Brainstem auditory evoked potentials are far-field potentials and following the convention from the first description by Jewett in the 80s (9), AEPs are displayed with positivity upwards and are labelled by roman numerals.

The active electrode is placed at electrode position Cz and the reference, or more accurately the return electrode, usually at the mastoid ipsilateral to the stimulus. The amplitude of wave I is slightly enhanced when a needle electrode is placed within the meatus. Clearer separation of peaks IV/V can be achieved when the reference is placed at the mastoid. Recordings only triggered on the condensation stimulus enhance the amplitude of the wave V while rarefaction triggers enhance the IV/V complex. This can be helpful in identifying central lesions (see Figs 15.3–15.4).

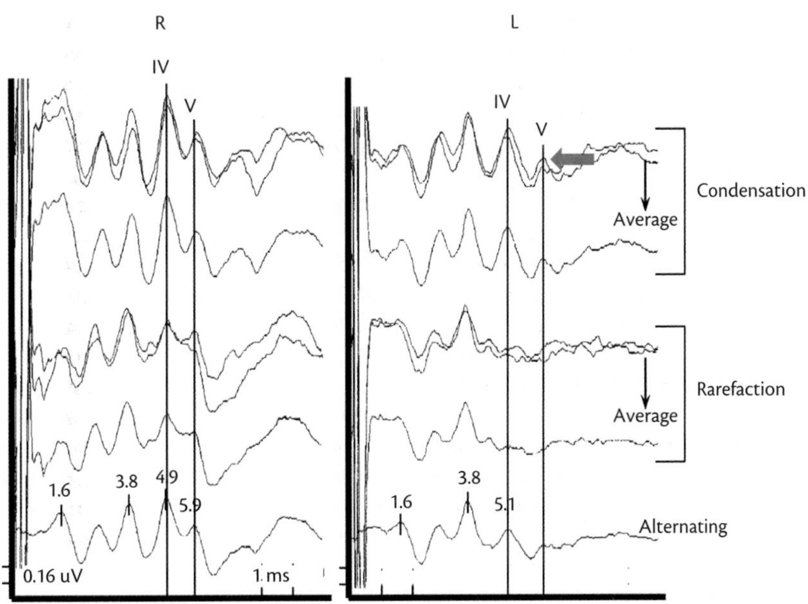

Fig. 15.4 Examples of different potential amplitudes to condensation and rarefaction stimuli. Waves IV and V are preserved in the condensation (red arrow), but lost in the rarefaction condition. Each average of 1500 trials to condensation and rarefaction clicks are shown as is the average to alternating stimuli.

Table 15.4 AEP stimulus and recording parameters

Click duration 0.1 ms
Polarity: condensation – rarefaction = alternating
Frequency: 10–20 Hz (14.7)
Intensity: 70 dBSL max. 90 dBHL
Contralateral—40 dB of stimulus intensity
Amplifier filter setting: ≤100–200 Hz, ≥3000 Hz
Analysis time: 10 ms
Number of averages: ≤2000

dBSL = decibel sensory level. The null value is defined as the stimulus intensity that is just perceived. dBHL = decibel normal hearing level. The null value is defined as the threshold at which a healthy person hears the click of repeated stimuli on 50% of trials.

There are two major sources of errors in recording auditory evoked potentials:

- Body temperature has a major effect on the latency of peak V.
- Peripheral hearing disturbances lead to reduction of wave I and all subsequent potentials.

Thus, auditory evoked potentials are only reliable in subjects with almost normal hearing.

The recommended stimulus and recording parameters are summarized in Table 15.4.

Normal and pathological recordings

Again, for quality control, two measurements should be reproduced with latency differences below 0.1 ms and amplitudes of +/–20%.

The upper normal limits of the latency are given in Table 15.5.

There are common variants of waves IV and V; separate peaks with higher amplitude of wave IV or V is also a variant of a common 'complex' with fused waves IV and V. The best way to identify wave V is to record the AEP with separate condensation and rarefaction stimulation.

It is intuitive to differentiate three patterns of pathological auditory evoked potentials by site of lesion:

- Peripheral ear and cochlear.
- Peripheral auditory nerve.
- Central conduction.

Peripheral and cochlear hearing disturbances cause a delay of waves I to V and when more pronounced reduce amplitudes of all peaks up to wave V (Fig. 15.5A).

Isolated high frequency hearing loss is a special case causing a delayed wave I, but a normal latency of wave V resulting in a shortened I–V peak interval.

In cochlear nerve lesions such as acoustic neuroma a prolonged latency of wave III with a normal III–V interpeak latency is typical. In some cases wave II is included and in rare cases an isolated loss of wave II can be seen.

In central lesions the latency of the wave V is prolonged or the wave is reduced in amplitude or may disappear together with wave IV. In more severe lesions wave III is reduced in amplitude in addition, and finally only wave I remains.

Table 15.5 AEP peak latencies upper limits of normal

	Peak latency: upper normal limit (ms)	Side difference: upper normal limit (ms)
I–III	2.5	0.5
III–V	2.4	0.5
I–V	4.5	0.5
V	5.6	0.5

Source data from Pratt H., Aminoff M., Nuwer M.R., Starr A, Short-latency auditory evoked potentials, In: Deuschl G. and Eisen A. (ed.), *Recommendations for the practice of clinical neurophysiology: Guidelines of the IFCN*, copyright (1999), Elsevier.

These typical patterns show some overlap, see Figs 15.5B and 15.5C.

One unusual phenomenon is the increase of wave I amplitude after loss of all central peaks. This may be caused by loss of central inhibitory impulses to the cochlear. In the early seventies Starr (13) described the loss of all central brainstem auditory evoked potentials in brain death (14). This seems to be valid, but clearly depends on either a preserved wave I or well documented consecutive loss of initially normal potentials. So, follow-up is required to document an irreversible loss of the auditory pathway as an indicator of brain death.

Clinical and advanced use

In summary, indications for clinical application of auditory brainstem evoked potentials are given in Table 15.6.

In multiple sclerosis a small number of patients show pathological auditory evoked potentials at the time of first diagnosis.

The use of auditory evoked potentials in the diagnosis of brain death is problematic because of the early development of peripheral hearing disturbances in intensive care management and in any case, only a very small part of brainstem function is being examined.

Modern MRI has greatly reduced the indications for the AEP. An important application, however, is the monitoring of comatose patients when increased intracranial pressure is suspected. Then the AEP can predict a worsening prognosis indicated by a deteriorating central lesion pattern of findings.

Somatosensory evoked potentials

Anatomy and physiology

Clinically, somatosensory evoked potentials are evoked by electrical pulses to peripheral nerves, not by stimulation of the receptors[3]. This simplifies the physiology and anatomy of SEPs mediated by the lemniscal system. Furthermore, this is the reason why there is no close correlation between SEPs and sensory perception.

The peripheral nerve, the brachial or lumbar plexus, the dorsal horn, and the intramedullary connections, the lemniscal pathway, the thalamus and the cortex are involved in generation of the somatosensory evoked potentials (Fig. 15.6).

[3] For further reading and for the recommendations of the international federation of clinical neurophysiology see (3,15–17).

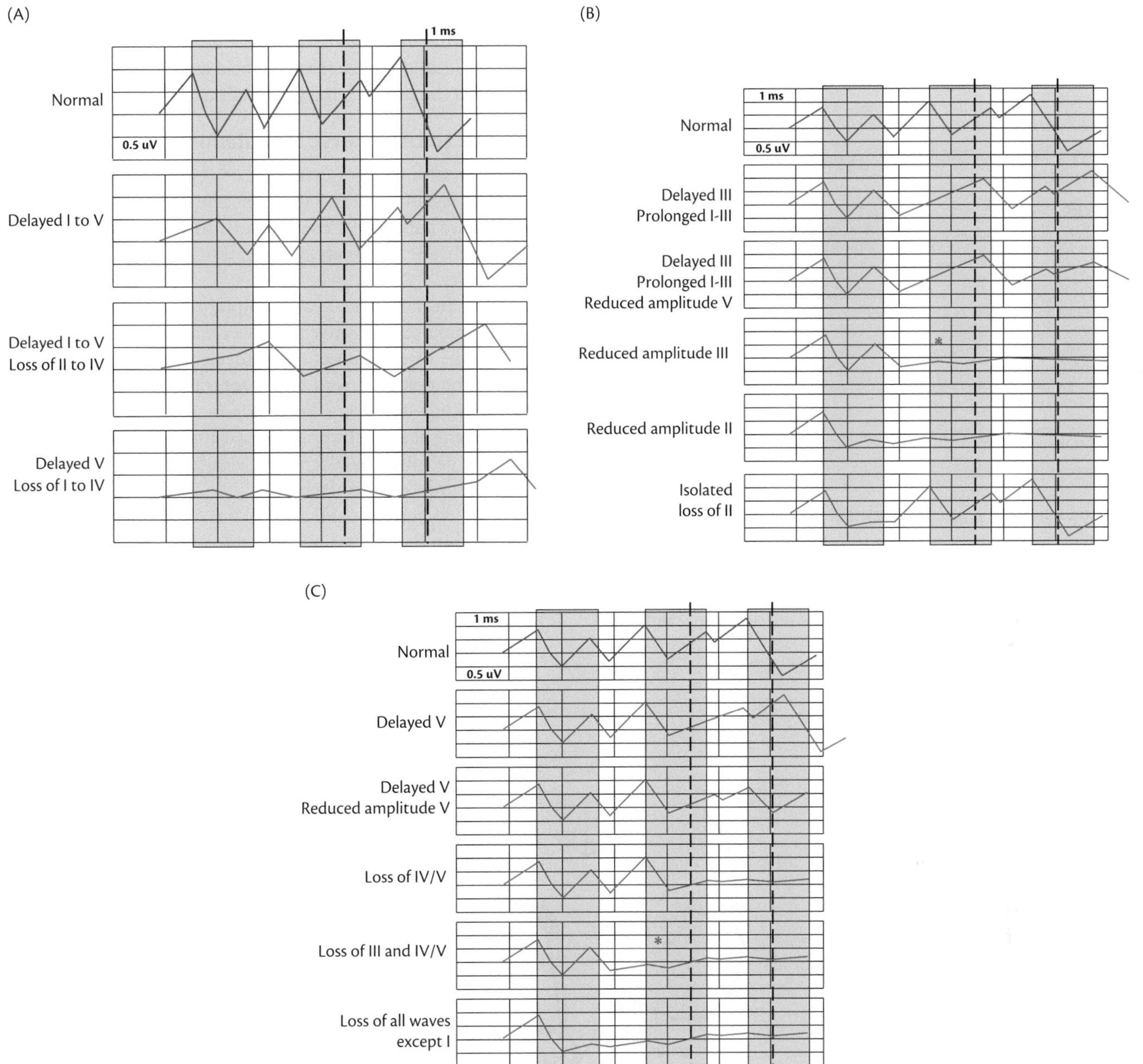

Fig. 15.5 Patterns of pathological auditory evoked potentials by lesion site. (A) Peripheral ear and cochlear nerve lesion showing delay of all waves. (B) Cochlear nerve lesions showing delay and reduced amplitude of wave III and subsequent waves. (C) Central lesions showing delay of wave IV and V and reduced amplitudes.
Overlaps between patterns in cochlear nerve and central lesions are marked by *.

Stimulus and recording

Usually mixed nerves are stimulated, because stimulus intensity can be easily determined by the motor threshold.

The recommended stimulus and recording parameters are given in Table 15.7. Stimulation and recording are straight forward but there are a number of sources of error:

- Low stimulus intensity reduces amplitude and to some degree prolongs latency.
- Suboptimal electrode placement can miss maximum amplitudes.
- The latencies of the somatosensory evoked potential parallel height, but the interpeak latencies from the brachial or lumbar plexus to the cortex are almost unaffected by height.
- Body temperature influences peripheral conduction velocity, but only to a minor extent central conduction times and amplitudes.
- Drugs have almost no effect, but there are rare cases with severe intoxication with carbamazepine inducing prolonged interpeak latency or diminution of amplitude.
- Sleep and attention have no effect on the early SEPs.
- Increasing age tends towards higher amplitudes and longer latencies.

Potentials to median nerve stimulation

Electrical stimulation excites cutaneous afferents travelling via the peripheral nerve and the brachial plexus. The brachial plexus N9 is

Table 15.6 Clinical usefulness of the AEP

Hard rocks ◆ Often pathological ◆ Usually technically clear results ◆ Usually unique pathophysiologic interpretation	Retrocochlear hearing impairment caused by retrocochlear lesions Diagnosis of multiple sclerosis Prognosis in coma
Soft rocks	Brain death Intoxication Peripheral hearing loss
Beach ◆ Sometimes pathological ◆ Often technically unclear results ◆ Usually no unique pathophysiologic interpretation	Dizziness MS—history Diagnosis of ischemic brainstem lesions

generated when the afferent volley leaves the relatively narrow arm and enters the wider thorax. The intramedullary connection from the dorsal horn to the frontal pathway generates the intramedullary N13 by the change in direction of the volley. The N14 is generated in the lemniscal pathway when leaving the narrow spinal canal and entering the posterior fossa. The N20 is a near field potential generated within the pyramidal cells at the dorsal bank of the central sulcus. The location of the maximal field amplitude of the N20 can vary around the contralateral CP position.

Potentials to tibial nerve stimulation

Tibial nerve stimulation generates the very low amplitude N22 in the lumbar plexus and the P40 in the central somatosensory cortex. There are subcortical components, which, however, are very low in amplitude and therefore are difficult to record with sufficient quality.

The P40 can show the so-called paradoxical lateralization. This refers to a maximum positive peak over the ipsilateral hemisphere caused by the orientation of the equivalent dipole standing tangential in the posterior wall of the central sulcus. So, the maximum amplitude of P40 can be missed in a Pz to Fz montage and the P60 might be misinterpreted as a P40. This problem, when it occurs, can be solved by placing the electrode 2 to 3 cm more lateral on the ipsilateral hemisphere.

There is at least some evidence that not only cutaneous afferents are involved in the generation of the SEPs to tibial nerve stimulation. Muscle afferents might cause problems in the interpretation of preserved tibial nerve SEPs in subjects with clearly complete lesions of the dorsal column.

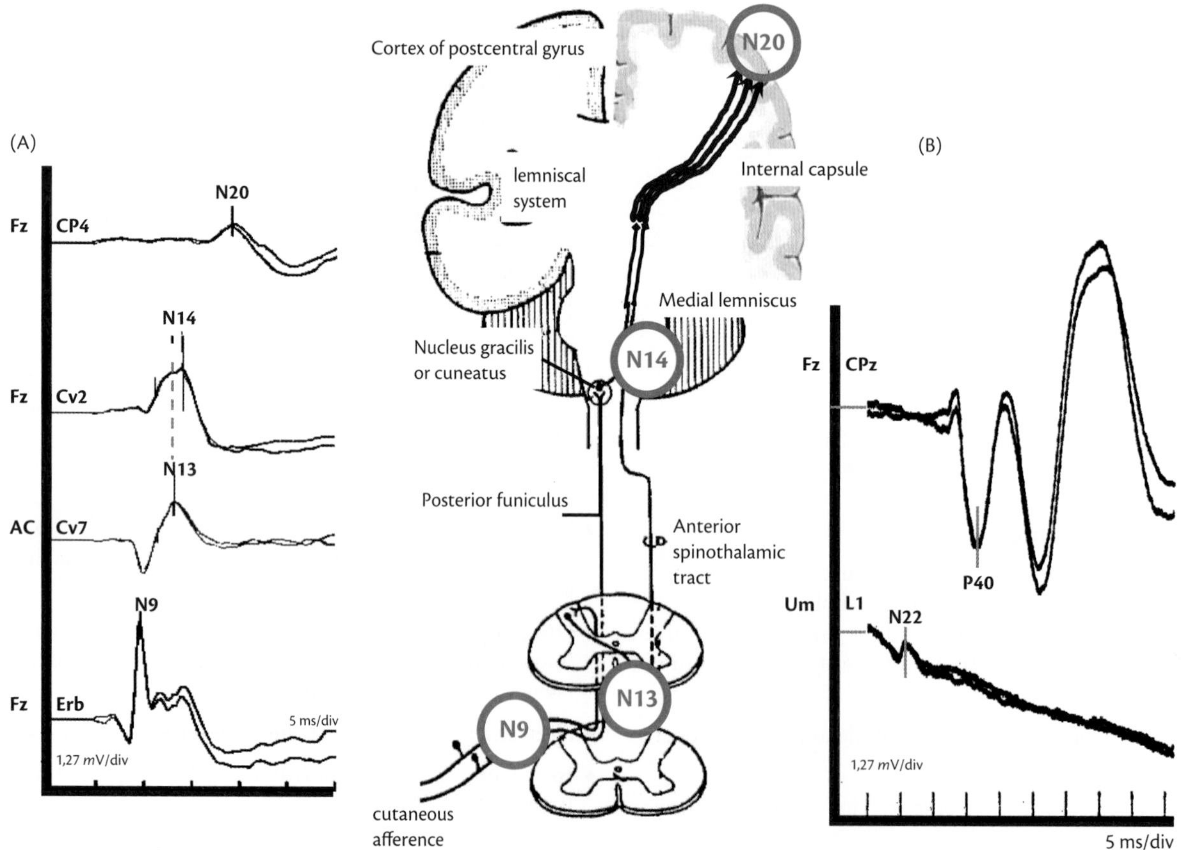

Fig. 15.6 Schematic anatomy of the somatosensory system. Normal waveforms and location of the generators for (A) median nerve stimulation and (B) tibial nerve stimulation.

Table 15.7 SEP—stimulus and recording parameters

SEP—Stimulus and recording parameters
Constant current; duration 0.2 ms
Cathode proximal
1–10/sec; best 4.3/s
Intensity motor threshold plus 4 mA for sensory nerve stimulation 3–4 times sensory threshold
Amplifier filter setting: ≤5 Hz, ≥1500 Hz
Analysis time: (50)100 ms
Number of averages: ≤250–2000

Normal and pathological recordings

For quality control two measurements have to be reproduced with latencies below 0.25 ms to arm nerve stimulated SEPs and 0.5 ms in leg nerve SEPs, respectively. Amplitude values are recommended to be within +/–20%.

The upper normal limits of the latencies and the lower limits of the amplitudes are given in Tables 15.8 and 15.9.

Four patterns of abnormality of median nerve SEPs can be discerned:

- In peripheral nerve lesions the SEPs can be prolonged in latency due to reduction of the peripheral nerve conduction velocity. Most often, the amplitudes of the cortical generated potentials are normal, due to the central amplifier effect.

Table 15.8 SEP from median nerve stimulation. Upper limits of normal for peak latency and amplitude

	Latency—upper normal limit (ms)	Side difference—upper normal limit (ms)
N9	11.5	–
N13	14.5	–
N14	16.7	0.8
N20	23.0	1.4
N9-N13	4.5	1.3
N13-N20	7.2	1.0
N14-N20	6.0	1.1
	Amplitudes—lower normal limit (μV)	**Side difference—upper normal limit (%)**
N9 (baseline to peak)	1.0	50
N13 (baseline to peak)	0.5	–
N20 (baseline to peak)	0.6	50
N20–P25 (peak to peak)	0.8	50

Reproduced from *Clin Neurophysiol*, **113**(11), Facco E, Munari M, Gallo F, Volpin SM, Behr AU, Giron GP, Role of short latency evoked potentials in the diagnosis of brain death, pp. 1855–66, copyright (2002), with permission from Elsevier.

Table 15.9 SEP from tibial nerve stimulation. Upper limits of normal for peak latency and amplitude

	Latency—upper normal limit (ms)	Side difference—upper normal limit (ms)
N22	25.2	1.1
P40	43.9	2.1
N22–P40	20.6	2.1
	Amplitudes—lower normal limit (μV)	**Side difference—upper normal limit (%)**
N22 (baseline to peak)	0.3	–
P40-N50 (peak to peak)	0.5	–

Reproduced from *Clin Neurophysiol.*, **113**(11), Facco E, Munari M, Gallo F, Volpin SM, Behr AU, Giron GP, Role of short latency evoked potentials in the diagnosis of brain death, pp. 1855–66, copyright (2002), with permission from Elsevier.

- In cervical extra medullary lesions the interpeak latency difference between N9 and N14 is prolonged. The typical case is cervical stenosis (Fig. 15.7). In addition to the MRI this can indicate functional disturbance of the sensory pathway.
- In cervical intra medullary lesions an isolated loss of the N13 can be seen (see Fig. 15.8). This is most often described in syringomyelia and is often accompanied by neuropathic pain in the hands. When the lesion is located more laterally, tibial nerve SEPs are affected, while the median nerve SEPs remain normal (see Fig. 15.9).
- In pontine and more rostrally located lesions the peaks up to N14 remain normal, but the cortical N20 can be affected in latency and/or amplitude.

Clinical and advanced use

Somatosensory evoked potentials provide additional evidence in polyneuropathy, especially in the case of prominent proximal demyelinating neuropathies. In the case of normal peripheral nerve conduction velocities (e.g. tibial nerve between ankle and knee) prolonged latency of the plexus potential indicates proximal conduction slowing. In severe demyelinating neuropathies amplitudes of nerve and muscle action potentials are diminished. Then, the cortical SEP can be preserved due to cortical amplification and peripheral nerve conduction can be calculated by evoking the SEP from stimulation at distal and proximal sites.

An important indication for median nerve somatosensory evoked potentials is the prognosis of patients in coma. There is clear evidence that a bilateral loss of N20 indicates a very poor prognosis in patients with severe head trauma or after hypoxia. However, recently it has been reported that some cases with bilateral loss of N20 after cardiac arrest and hypothermia can have a good recovery. In these cases, recordings were done during hypothermia for brain protection (18). To the author's knowledge there are no cases of bilateral loss of N20 recorded during normothermia and after excluding severe intoxications, for instance, with carbamazepine.

Somatosensory evoked potentials can be used to give evidence of the irreversible loss of brain function in the diagnosis of brain death. In the hands of an expert, able to differentiate

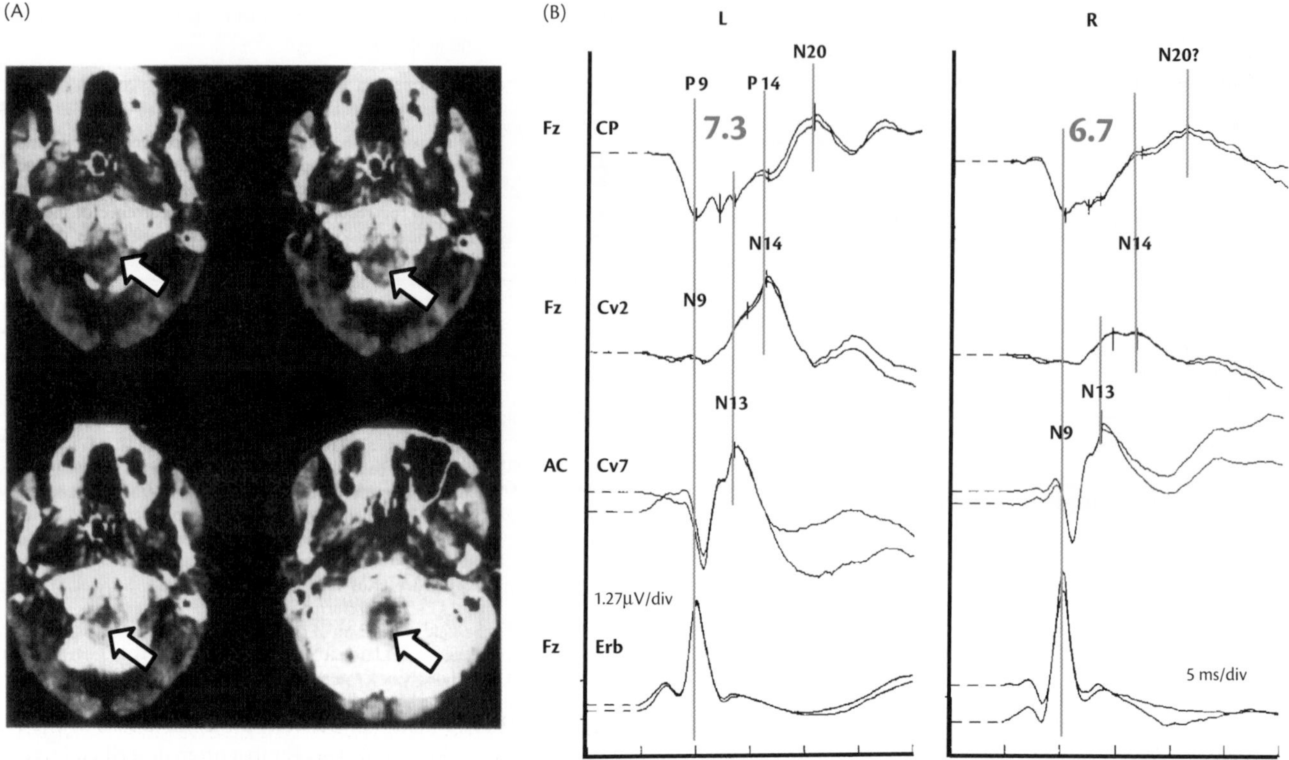

Fig. 15.7 Cervical extramedullary lesion causing prolonged interpeak interval between N9 and N14. (A) CT-scan showing a subarachnoid haemorrhage at the cervico-medullary junction down to C2. (B) Normal N9/P9 and N13 on both sides. Prolonged interpeak latency difference N9–N14/P14 on both sides and reduced amplitude to right median nerve stimulation.

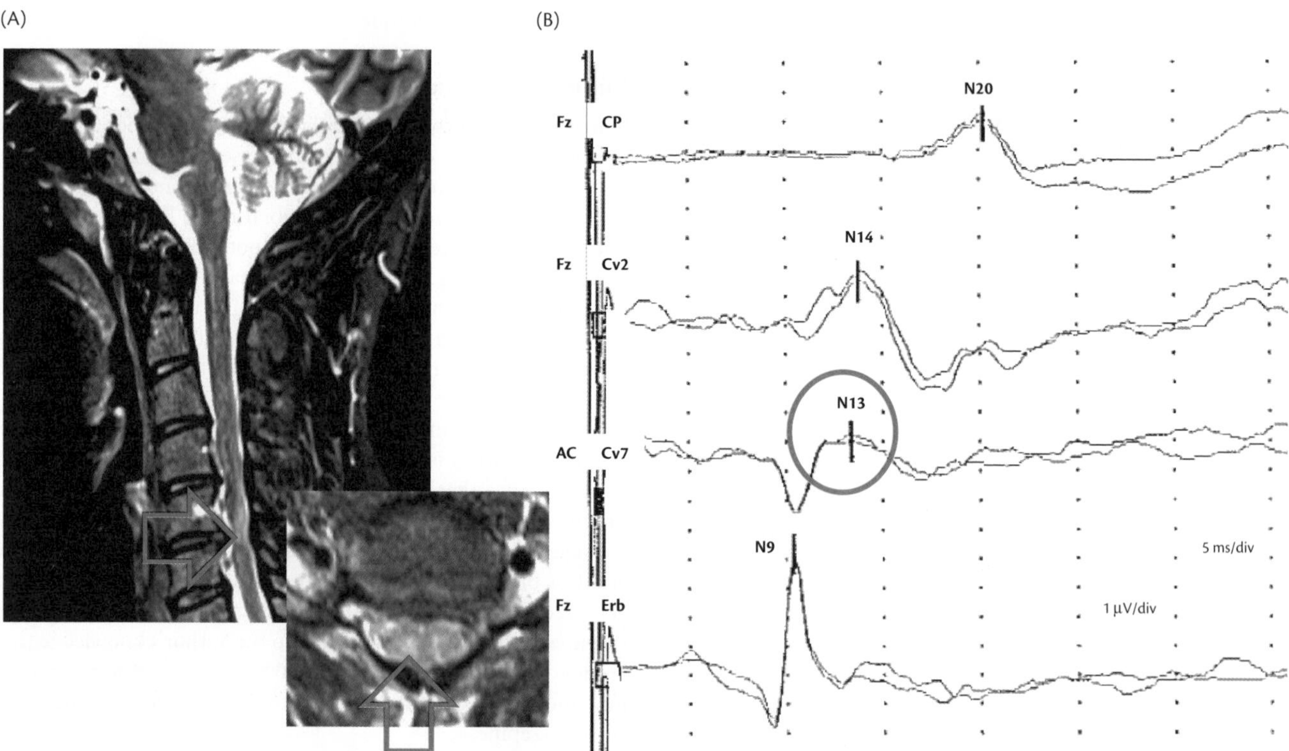

Fig. 15.8 Cervical intramedullary lesion causing an isolated loss of the N13. (A) MRI showing a right-sided intra medullary lesion at C5–6. (B) Median nerve SEP showing a loss of the intramedullary generated N13 (red circle).

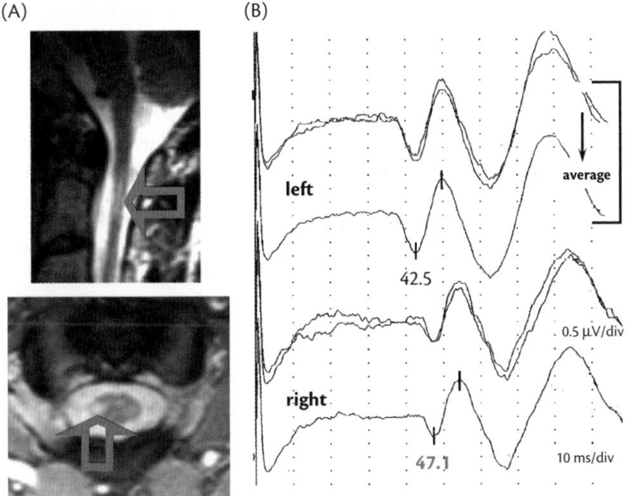

Fig. 15.9 Cervical intramedullary lesion causing a pathological tibial nerve P40. (A) MRI showing a right-sided intra medullary lesion at C2. (red arrows) (B) Tibial nerve SEP showing prolonged latency of P40 to right nerve stimulation.

Table 15.10 Clinical usefulness of the SEP

Hard rocks ◆ Often pathological ◆ Usually technically clear results ◆ Usually unique pathophysiologic interpretation	MS—diagnosis
	Prognosis in coma
	Peripheral nerve lesions—proximal neuropathies
	Intraoperative monitoring/ location of central sulcus
	Cortical myoclonus
Soft rocks	Brain death
	Intoxication
	MS—history
Beach ◆ Sometimes pathological ◆ Often technically unclear results ◆ Usually no unique pathophysiologic interpretation	Stimulation of cutaneous nerves
	Dermatomal stimulation

N13/N14 and its far-field reflection P14, somatosensory evoked potentials are easy to record and very reliable in defining irreversible loss of function of the brainstem somatosensory pathway in an ICU (14). Unfortunately, this can be problematic. The case presented in Fig. 15.10 shows the loss of the N20 and may show loss of the N14 and preserved N13, but in far-field recordings with the reference at the shoulder a preserved P14 is demonstrated. This patient was not brain dead, but in a persistent vegetative state.

In the acute stage of spinal cord lesions the SEP are often more sensitive than MRI. Normal or delayed potentials imply a good prognosis even with a severe functional deficit during the first few days.

Monitoring of comatose patients on the ICU can detect impaired function before a definite lesion occurs.

Monitoring of operations on the spinal cord or during aneurysm surgery is established, because reduction of amplitudes can indicate a critical impairment before it is irreversible.

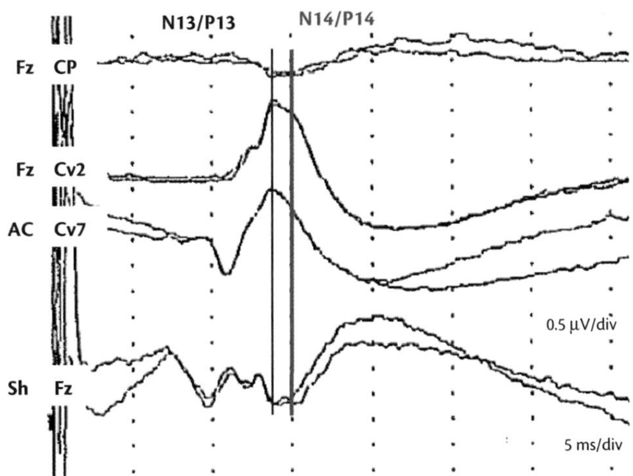

Fig. 15.10 Persistent vegetative state with loss of N20, but preserved N14/P14 and N13.

In post-anoxic myoclonus and myoclonic epilepsy increased amplitudes of the SEPs can be seen.

Within the literature other possible stimulation points for somatosensory evoked potentials are described. Stimulation of the sural nerve in some patients is useful in addition to tibial nerve stimulation. Ulnar nerve stimulation might be useful in questions concerning lesions of the brachial plexus. Unfortunately, surface stimulation or trigeminal nerve stimulation are not reliable.

Hard rocks are peripheral and/or central spinal demyelinating lesions. So, somatosensory evoked potentials are very useful in the diagnosis of multiple sclerosis (see Table 15.10).

Advanced use of evoked potentials

Visual evoked potentials

The methods presented above to record VEPs is not able to detect defects in the visual field. A developing technique is multifocal stimulation, which can detect lesions in the visual field if the subject is able to fixate the stimulus very accurately over several minutes (6).

Auditory evoked potentials

Several auditory stimulus paradigms, e.g. P300 and mismatch negativity, have been used mostly in cognitive research. These more elaborate stimulation and recording techniques might have further impact in the diagnosis of cognitive disturbances (2,4).

Somatosensory evoked potentials

Multichannel recordings clearly increase the chances of identifying and localizing slowed conduction. Therefore, a four channel montage in median nerve recordings is recommended. High frequency (600 Hz) components of the SEP have been the subject of several studies. These low amplitude peaks are shown to be strongly dependent on intact myelination of central pathways and might be an indicator of intracortical inhibitory or excitatory processes (19).

References

1. Buchner, H. and Noth, J. (2005). *Evozierte Potenziale, Neurovegetative Diagnostik, Okulographie.* Stuttgart: Thieme.
2. Buchner, H. (2014). *Praxisbuch—Evozierte Potenziale.* Stuttgart: Thieme.
3. Kimura, J. (2013). *Electrodiagnosis in diseases of nerve and muscle—principles and practice.* Oxford: Oxford University Press.
4. Regan, D. (1988). *Human brain electrophysiology.* New York, NY: Elsevier.
5. Celesia, G.G. and Brigell, M. (1999). Recommended standards for pattern electroretinograms and visual evoked potentials. In: G. Deuschl and A. Eisen (Eds) *Recommendations for the practice of clinical neurophysiology,* EEG Suppl. 52, pp. 53–67. Amsterdam: Elsevier.
6. Green, A.J. (2012). Visual evoked potentials, electroretinography, and other diagnostic approaches to the visual system. In: M. J. Aminoff (Ed.) *Aminoff's electrodiagnosis in clinical neurology,* 6th edn, pp. 477–503. Amsterdam: Elsevier.
7. Holder, G.E., Celesia, G.G., Miyake, Y., et al. (2010). International federation of Clinical neurophysiology: recommendations for visual system testing. *Clinical Neurophysiology,* **121**, 1393–409.
8. Paulus, W. (2014). Elektroretinographie (ERG) und visuell evozierte Potenziale In: H. Buchner (Ed.) *Praxisbuch—Evozierte Potenziale,* pp. 67–74. Stuttgart: Thieme.
9. Jewett, D.L. (1987). Auditory evoked potentials: overview of the field (and shoreline)—1986. In: C. Barber and Th. Blum (Eds) *Evoked potentials III,* pp. 19–30. Oxford: Butterworths.
10. Legatt, A.D. (2012). Brainstem auditory evoked potentials: methodology, interpretation, and clinical application. In: M. J. Aminoff (Ed.) *Aminoff's electrodiagnosis in clinical neurology,* 6th edn, pp. 519–552. Amsterdam: Elsevier.
11. Pratt, H., Aminoff, M., Nuwer, M.R., and Starr, A. (1999). Short-latency auditory evoked potentials. In: G. Deuschl and A. Eisen (Eds) *Recommendations for the practice of clinical neurophysiology,* EEG Suppl. 52, pp. 69–77. Amsterdam: Elsevier.
12. Scherg, M. and von Cramon, D. (1985). A new interpretation of the generators of BEAP waves I–V—results of a spatio-temporal dipole model. *Electroencephalography & Clinical Neurophysiology,* **62**, 290–9.
13. Starr, A. and Achor, L.J. (1975). Auditory brain stem responses in neurological disease. *Archives of Neurology,* **32**, 761–8.
14. Facco, E., Munari, M., Gallo, F., Volpin, S.M., Behr, A.U., and Giron, G.P. (2002). Role of short latency evoked potentials in the diagnosis of brain death *Clinical Neurophysiology,* **113**(11), 1855–66.
15. Aminoff, M.J. and Eisen, A. (2012). Somatosensory evoked potentials. In: M. J. Aminoff (Ed.) *Aminoff's Electrodiagnosis in clinical neurology,* 6th edn, pp. 581–601. Amsterdam: Elsevier.
16. Mauguière, F., Allison, T., Babiloni, C., et al. (1999). Somatosensory evoked potentials. In: G. Deuschl and A. Eisen (Eds) *Recommendations for the Practice of clinical neurophysiology,* EEG Suppl. 5, pp. 79–90. Amsterdam: Elsevier.
17. Cruccu, G., Aminoff, M.J., Curio, G., and Guerit, J.M. (2008). Recommendations for the clinical use of somatosensory-evoked potentials. *Clinical Neurophysiology,* **119**, 1705–19.
18. Leithner, C., Ploner, C.J., Hasper, D., and Storm, C. (2010). Does hypothermia influence the predictive value of bilateral absent N20 after cardiac arrest? *Neurology,* **74**, 965–9.
19. Gobbelé, R., Waberski, T.D., Dieckhofer, A., et al. (2003). Patterns of disturbed impulse propagation in multiple sclerosis identified by low and high frequency somatosensory evoked potential components. *Journal of Clinical Neurophysiology,* **20**, 283–90.

CHAPTER 16

Polysomnography and other investigations for sleep disorders

Zenobia Zaiwalla and Roo Killick

Polysomnography

The term polysomnography (PSG) is used for sleep recordings, reflecting the multiple parameters monitored for sleep analyses when investigating sleep disorders, including analyses and distribution of the various sleep stages, respiratory events, and sleep stage-related normal and abnormal limb movements and behaviours. These studies require additional technical and interpretive skills to those essential for recording and interpreting electroencephalograms (EEG).

Traditionally, clinical and research scientists working in sleep disorders have entered the field from various specialities and may not be trained in neurophysiology. Hence, the parameters included in a PSG study are often modified depending on the investigator's research or clinical interest. While neurophysiologists would consider a PSG recording without including EEG for sleep staging as an incomplete study, respiratory sleep disorder colleagues have considerable experience in interpreting clinically significant respiratory events in sleep using a range of cardiorespiratory parameters, with or without monitoring movements in sleep, and may not routinely include EEG. A clinician or research scientist interested in overnight quantitative distribution of sleep stages and sleep continuity may include only the minimal EEG electrodes with the other parameters required for sleep stage analyses. In contrast, a neurologist/neurophysiologist faced with deciding whether episodic behaviours/movements from sleep are parasomnias or nocturnal epileptic seizures, and whether there are non-epileptic causes for lowering of arousal threshold for the episodic behaviours in sleep, will want to combine the PSG with additional EEG channels to record a video-telemetry/PSG study.

The digitalization of EEG recording and availability of multichannel commercial video-telemetry systems with software for sleep analysis has allowed neurophysiologists to enter the field of sleep disorders, previously avoided by busy clinical departments faced with reels of overnight paper recordings and manual sleep staging. However, while automated sleep staging provides the hypnogram structure and gross analyses, all studies require manual editing by experienced physiologists familiar with sleep staging in children and adults of all ages, and these studies remain time resource heavy.

Most centres investigating sleep disorders admit patients overnight to sleep monitoring rooms, with or without overnight sleep technician attendance. In-patient monitoring has limitations, as while some patients sleep better in monitoring rooms, away from day-to-day life stresses, many do not, especially on the first night. Also the patient's schedule may not match the hospital ward schedule. Even with soundproof monitoring rooms and staff instructed not to wake the patient for meals if still asleep or napping during the day, there is always the risk of interruptions as the hospital rhythm is not conducive to sleep. Home recordings using ambulatory recorders overcome some of these limitations, although absence of simultaneous video recording can diminish the value of the study. Laptop systems with video-EEG/PSG recording systems are now commercially available, but are expensive and their cost can limit the number of available systems compared with the cost of ambulatory EEG recorders. Also their complexity may require a technician to travel to the patient's home to set up the recording. Another limitation of ambulatory recorders is that depending on the number of inputs, some EEG channels in some systems may have to be sacrificed to include the additional parameters for sleep staging, with loss of valuable EEG data in patients with sleep disorders associated with neurological disorders including epilepsy. Of course, some patients do not have the personal resource to manage unsupervised home studies. Finally, if the patient has to return to the department after an overnight home study for the multiple sleep latency test (MSLT), which should start approximately 2 h after habitual wake up time, there is the risk that the patient wakes earlier than usual to allow for travel time. Despite these limitations, home recordings using the ambulatory recorders can be the preferred option in some patients and clinical situations, providing a more representative study, including 24 h recordings allowing situation dependant daytime sleep episodes/behaviours to be recorded over at least a couple of days.

The Rechtschaffen and Kales (R and K) scoring manual for sleep (1), underpins the terminology and technical aspects of sleep scoring today over 40 years after it was published (see Table 16.1). The minimum parameters recorded to allow EEG sleep staging include widely spaced EEG (C4 to A1 or C3 to A2), EOG (electro-oculogram recording right and left eye movements) and chin EMG (electromyograph) (Fig. 16.1).

Normal sleep is cyclical with a periodicity of 90 min (60 min in neonates). Night sleep begins with non-rapid eye movement sleep (NREM) sleep with initial sleep cycles dominated by NREM stages

Table 16.1 Sleep stage characteristics (30 s epoch)

	Wake	Stage 1	Stage 2	Stage 3	Stage 4	REM
EEG	Beta active wake; alpha resting wake	<50% alpha; vertex sharp waves	No alpha, Spindles, & K complexes; some theta, <20% delta	20–50% delta; some theta	>50% delta	Wake-like EEG; saw tooth waves
EOG	Fast, high, amplitude	Slow rolling	No eye movements	No eye movements	No eye movements	Periodic REMs
EMG	High	Moderately low	Lower	Lower	Lower	Dramatic drop
	Delta: 0.5–3.9 Hz (slow wave)	Theta: 4–7.9 Hz	Alpha: 8–12 Hz (relaxed wake)	Sigma: 12–14 Hz (sleep spindles)	Beta: >15 Hz (active wakefulness)	

Reproduced from Rechtschaffen A, Kales A, A manual of standardized terminology, technique and scoring system for sleep stages of human subjects, copyright (1968), with permission from the National Institute of Health.

3 and 4 sleep. Rapid eye movement sleep (REM) sleep occurs later in the night. Normal sleep architecture comprises approximately:

- Stage 1: 2–5%.
- Stage 2: 45–55%.
- Stage 3: 3–8%.
- Stage 4: 10–15%.

REM 20–25% over 4–6 cycles, with wake/arousals up to 5% of the night in the healthy adult (Fig. 16.2) (2).

Over the years there have been attempts to modify and add to the R and K criteria for sleep staging to better correlate sleep analysis with clinical sleep disorders and aid research. The wake after sleep onset and movement times analysed using Rechtschaffen and Kales (R&K) 30-s epochs criteria did not always recognize brief arousals, important in clinical practice. In 1992, the Atlas task force of the American Sleep Disorders Association published rules for EEG arousals (3); scoring an arousal in any sleep stage as an abrupt shift in EEG frequency including alpha, theta and/or frequencies greater than 16 Hz (excluding spindles) that last at least 3 s, with at least 10 s of stable sleep preceding the change. Scoring of arousals in REM requires simultaneous increase in submental EMG lasting at least 1 s. While it is important to analyse arousals, and the facility is available in the automated sleep staging software, reliable normative data for various ages is not available and departments are encouraged to collect their own data. Similarly, clinical usefulness of changes in the microstructure of sleep, as expressed by the cyclical alternating pattern (CAP) reflecting sleep oscillations (4), has been limited in clinical practice. This is in spite of publications reporting changes in CAP in several conditions including NREM arousal parasomnias and nocturnal frontal lobe epilepsy, due to unavailability of normative data and reliable automated analysis

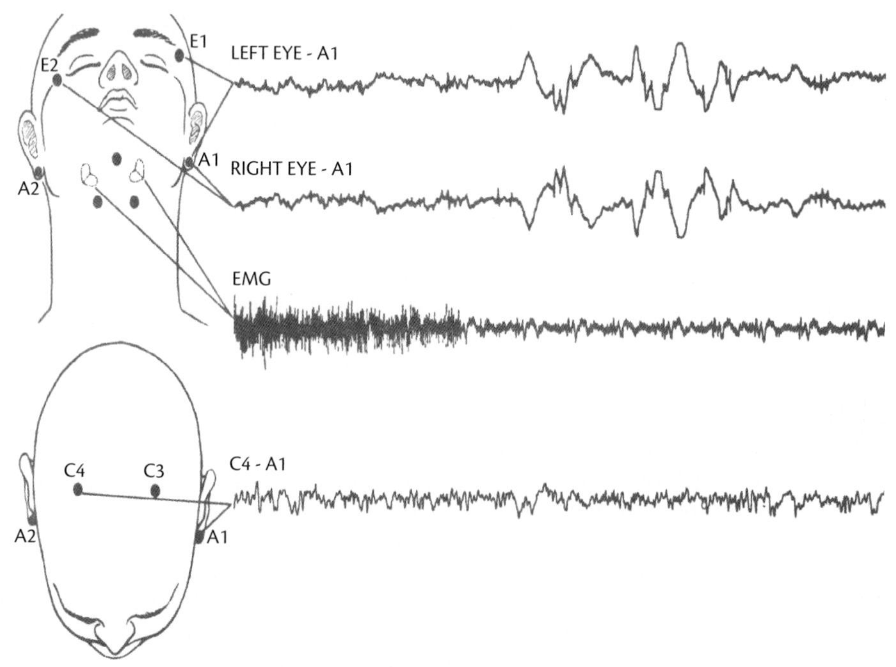

Fig. 16.1 Shows standard placements of EEG, EOG, and EMG electrodes for sleep staging.

Reproduced from Rechtschaffen A, Kales A, *A manual of standardized terminology, technique and scoring system for sleep stages of human subjects*, copyright (1968), with permission from the National Institute of Health.

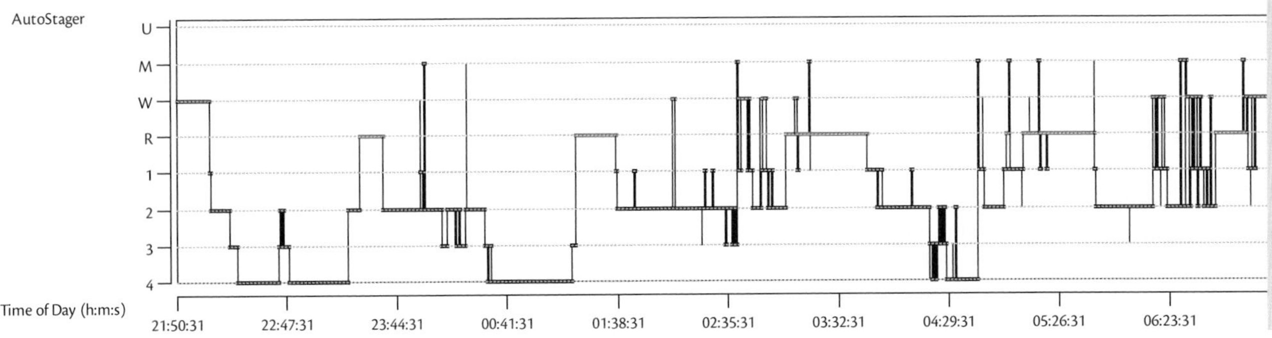

Fig. 16.2 Normal hypnogram.

software, although this may change in the near future. However, identifying changes in the microstructure of sleep has enhanced our understanding of the mechanisms involved in sleep disorders and the patient's complaint, masked with sleep scoring using traditional R and K rules (1) alone.

The manual for Scoring of Sleep and Associated Events was published by the American Academy of Sleep Medicine in 2007 (5) and has been adopted by many centres worldwide. It has recently been updated and an on-line version created (6). For the neurophysiologist, the main change is combining stages 3 and 4 into one sleep stage of slow wave sleep (N3); so NREM sleep stages are classified as N1, N2, and N3. Most sleep software systems provide the facility to select both the old or new nomenclature. The 2007 manual (5) also provides rules for recording movements including periodic leg movements and other limb movements, bruxism, polysomnographic features of rhythmic movement disorder and REM sleep behaviour disorder (RSBD).

Adult respiratory monitoring

Gold standard respiratory PSG involves attended, in-laboratory testing with audio-video recording. However resources and availability of testing means that limited and/or at-home testing with portable monitors has been widely adopted internationally, especially as the prevalence and awareness of sleep disordered breathing (SDB), in particular the global burden of obstructive sleep apnoea (OSA), is increasing. The range of software devices now available has meant that accessibility to simpler and less labour intensive diagnostic testing has improved. However, these must be utilized with the appropriate clinical support of sleep physicians to ensure clinical pathways are established for their use and follow-up. Portable monitors however, have their limitations and recommendations on patient suitability for such testing have been documented in recent guidelines (7).

Respiratory PSGs were described in four levels of complexity by American Academy of Sleep Medicine (AASM) guidelines first established in 1994 (8):

- *Level 1*: in-laboratory full attended PSG (≥7 channels).
- *Level 2*: full unattended PSG (≥7 channels).
- *Level 3*: limited channel devices (4-7 channels).
- *Level 4*: 1 or 2 channels, commonly involving oximetry.

Sleep laboratories may choose which level of monitoring they require for their diagnostic purposes, incorporating available resources, patient referral base and clinical skills available for interpreting more limited testing.

When assessing adult patients for sleep-disordered breathing (SDB) with gold-standard level 1 PSG, the AASM guidelines (6) advise that the following parameters should be monitored in addition to those required for sleep staging/leg movements:

- Airflow signals.
- Respiratory effort signals.
- Oxygen saturations.
- Body position.
- Electrocardiogram (ECG).

All have specific recommendations regarding sampling rates, low and high frequency filters, and maximal electrode impedances to provide optimal signal processing.

Airflow signals

Airflow monitoring is measured using an oronasal thermal airflow sensor or a nasal pressure transducer, or if continuous positive airway pressure (CPAP) or other ventilation is being used during the study, via the ventilator device flow signal. A thermal airflow sensor measures changes in temperature with airflow and calculates airflow accordingly, whereas a nasal pressure transducer measures the nasal cavity pressure relative to atmospheric pressure using a nasal cannula. The pressure difference during respiration is proportional to the magnitude of airflow squared and, hence, an airflow waveform can be created. Respiratory events and snoring can be assessed and scored from this channel.

Snoring can also be objectively monitored using an acoustic sensor (i.e. microphone), in addition to the change in waveform seen on a nasal pressure transducer.

Respiratory effort signals

These are commonly known as thoraco-abdominal 'bands' or 'belts', whose excursion via respiratory inductance plethysmography (RIP) give an estimate of tidal volume and, hence, their waveform is proportional to respiratory effort. Gold standard monitoring requires oesophageal manometry, although this is rarely used in clinical practice as it is invasive and can be uncomfortable for the patient. Respiratory effort is important in differentiating the aetiology of apnoeas and hypopnoeas as either obstructive (with effort) from central (without effort) respiratory events.

Oxygen saturation

Oximetry is essential for respiratory diagnoses. Standard pulse oximeters are used. However, it is recommended they have a maximum signal averaging time of ≤3 s at a heart rate of 80 beats/min to ensure all desaturations are captured.

Electrocardiogram

ECG monitoring provides important information related to SDB events and autonomic arousals from sleep. All arrhythmias and indeed significant bradycardias (<40 beats/min) or tachycardias (>90 beats/min) during sleep should be commented upon by the sleep scientist and reporting physician.

Other cardiac parameters

Other cardiac parameters, for instance pulse transit time (PTT) as a marker of beat-to-beat blood pressure changes, can be extrapolated from the ECG and oximeter data. Blood pressure falls with inspiration and the amount is proportional to respiratory effort; blood pressure also increases with arousals. Therefore, these can be utilized to provide surrogate markers of autonomic arousal and respiratory effort (9). These parameters are utilized in some units and are of particular importance in limited channel PSG, where EEG may not be monitored, providing additional diagnostic information.

Transcutaneous carbon dioxide monitoring

Transcutaneous carbon dioxide monitoring ($tcCO_2$) or end-tidal/arterial partial pressure of carbon dioxide (pCO_2) monitoring (less commonly used) is required to diagnose hypoventilation accurately. This is defined as an increase in overnight pCO_2 ≥ 10 mmHg (1.3 kPa) from awake levels to ≥50 mmHg (6.7 kPa) for ≥10 min; or to a level ≥55 mmHg (7.3 kPa) for ≥10 min. The equipment is expensive and is only available in some centres that deal routinely with respiratory failure patients. It can also be vulnerable to signal drift, and requires regular calibration and careful interpretation.

Body position

Position sensors are important because SDB is often worse in the supine position. If utilized, the report should incorporate a breakdown of respiratory statistics based on body position. The sensor accuracy can be further assessed in conjunction with the video, where available. In order to ensure SDB is not missed by lack of supine sleep during the PSG, some laboratories may ask their night technicians to disturb the patient and ask them to sleep on their back mid-way through the study. This is, however, centre-dependent and certainly not adopted by all, as it involves waking the patient during the study. If no supine sleep is seen, care must be taking in interpreting a 'normal' study and clinical correlation sought.

Definitions of main respiratory outcomes of interest

Apnoea

This is defined as a reduction of ≥90% in airflow for ≥10 s. These can be obstructive (with respiratory effort), central (without effort), or mixed (no effort initially, followed by return of effort during the event). There is no minimum desaturation required to score an apnoea.

Hypopnoea

This is defined as a reduction of ≥30% reduction in airflow for ≥10 s, with an associated ≥3% oxygen desaturation or an arousal. This has been newly defined in the most recent updated scoring guidelines (6), as two different definitions were allowed previously. New optional criteria to allow definition of either obstructive or central hypopnoeas have also been added.

Respiratory effort-related arousal (RERA)

Increased respiratory effort or flattening of inspiratory airflow for ≥10 s, with an associated arousal, when the criteria for hypopnoea or apnoea are not met.

Cheyne–Stokes breathing

Crescendo–decrescendo pattern of breathing, involving ≥3 consecutive central apnoeas and/or central hypopnoeas, a cycle length of ≥40 s, and ≥5 central events/h for at least 2 h of the PSG. This is often associated with heart failure.

Apnoea–hypopnoea index

The combined number of apnoeas and hypopnoeas per hour of sleep.

Oxygen desaturation index

The number of oxygen desaturation events per hour of sleep (note these are often described as 3 or 4% drops, and it is important to note which criteria are used).

Respiratory disturbance index

Refers to the sum of all apnoeas, hypopnoeas, and RERAs per hour of sleep.

As there has yet to be international universal adoption of the AASM guidelines, and there have been alternative scoring rules available for the definition of hypopnoeas in the previous guidelines (5,6), it is very important to state clearly in reports which scoring definitions have been used. The clinician must be aware that apnoea–hypopnoea indexes (AHIs) from different laboratories may not be the same and, hence, the clinical interpretation of the report must be taken in its context. Generally, an AHI of >5/h in an adult is taken as abnormal, so it is vitally important that scoring rules and definitions are taken into account as, for instance 15 RERAs/h may not have the same clinical significance as 15 apnoeas/h. Within a laboratory, it is essential that regular inter-scorer concordance testing, feedback, and education occurs, and there are on-line programmes available to help with this provision, to assess their own scoring consistency against other units.

Multiple sleep latency test

The MSLT measures the propensity to fall asleep in a sleep-inducing environment. The test was developed and standardized by Carskadon, Dement, and Mitler (10) as a measure of sleepiness, and until the option of measuring CSF hypocretin level became available, was the only objective test for reliably diagnosing narcolepsy with cataplexy syndrome. The AASM task force provided guidelines for the clinical use of the MSLT, and the maintenance of wakefulness test (MWT), which is discussed subsequently (11).

The MSLT is a resource heavy test and should never be interpreted in isolation. A detail clinical history, including sleep schedule and drug history in the preceding month is essential, and a sleep diary for 1–2 weeks prior to the study should be obtained.

If the patient cannot be relied upon to provide an accurate sleep diary, recording 1–2 weeks of sleep with actigraphy is recommended (12). It is essential that the patient has sufficient (habitual) sleep on the night preceding the MSLT (at least 7 h in bed for adults) and this should be objectively recorded with PSG or actigraphy. However, 7 h in bed may be insufficient for children, and especially adolescents and some adults, and can lead to false positive MSLT results.

The MSLT should start about 2 h (1½–3 h) after the patient's habitual wake up time. The patient is offered 4 or 5 nap opportunities during the day at 2-h intervals (usually 10.00, 12.00, 14.00, 16.00, and if necessary 18.00 hours), to provide an adequate sampling of the day. The test is performed with the patient in bed lying down and they are instructed to allow themselves to fall asleep if so desired. There are two protocols for the MSLT (13). In the research protocol, the patient is woken from the test nap at sleep onset, which is defined as three consecutive 30-s epochs of NREM stage 1 sleep, or one 30-s epoch of NREM sleep stages 2–4, or REM sleep. In the clinical protocol, the patient is woken 15 min after the first epoch of sleep, or the test is ended after 20 min if no sleep occurs, in which case the sleep latency is recorded as 20 min. Sleep onset in the clinical protocol is defined as the first epoch of greater than 15 s of sleep in a 30-s epoch of any sleep stage, including stage 1. The time from lights off to this epoch is recorded as sleep onset latency (SOL). REM latency is the time from the first epoch of sleep to the beginning of the first epoch of REM sleep. The mean SOL of all the naps is calculated and the number of naps in which REM sleep occurred is reported. Mean SOL in adults without a sleep disorder ranges from 10–20 min; however, the standard deviation in normal controls is large, as the practice parameters outlined by the AASM in 2005 suggest (11), with 16% of normal controls having a mean sleep onset latency of below 5 min. A mean SOL of 8 min or less is abnormal, with patients with narcolepsy usually entering REM sleep in 2 or more naps (sleep-onset REM period (SOREMP) within 15 min of sleep onset).

While the MSLT can be a valuable objective test for sleepiness as it is assumed that patients cannot force themselves to fall asleep, the findings can be compromised if the environment and protocol is not standardized. The sleep room should be dark and quiet during the naps. Smoking should be stopped at least 30 min prior to each nap, and stimulating activity stopped 15 min prior. While a light breakfast is allowed an hour before the first nap and light lunch immediately after the noon trial and before the 14.00 hours trial, the patient should not have caffeinated beverages or exposure to bright light. Also patients must not fall asleep in between the trial naps.

Of more importance is recording subjective quality in addition to objectively recording the quantity of sleep on the preceding night, as sleep disturbance due to PSG recording can reduce the mean SOL. It is not uncommon to find that patients have sufficient sleep on the night preceding the MSLT, but a longer sample of sleep wake schedule recorded with actigraphy over 1–2 weeks shows a very irregular schedule, suggesting that sleepiness is likely to be due to accumulative sleep debt. This is especially common in college/university students, but also seen in older patients reporting sleepiness, with or without shift work employment. In contrast, we also see patients who report sleepiness, but cannot fall asleep in the MSLT test environment, or have a long mean SOL on MSLT, but during a 24-h polysomnography study, several daytime naps are recorded, in spite of sufficient night sleep, supporting their complaint. MSLT should not be attempted in shift workers unless they are able to negotiate with their employers, at least a 2-week stable day shift schedule, to allow night sleep saturation before the test.

The MSLT findings may also be altered by medication. While patients are instructed not to use any prescribed or over the counter sleep aids prior to the MSLT, many patients referred with a complaint of sleepiness may be on treatment for underlying depression or other psychiatric disorders, including drugs for bipolar disorders, which would be unsafe to withdraw prior to the MSLT. The same is true for patients on anticonvulsants for epilepsy or pain. Both tricyclics and serotonin-specific reuptake inhibitors (SSRI)/serotonin–norepinephrine reuptake inhibitors (SNRI) suppress REM sleep at night and increase rebound daytime REM sleep, with SOREMPs during MSLT naps. Drugs prescribed for psychiatric disorders, epilepsy, pain, and other comorbid disorders can be associated with mean SOL on MSLT in the pathological sleepiness range, as even 8 h of habitual night sleep may be insufficient for these patients. Although patients may report that the clinical sleepiness preceded the medication, great caution is advised in interpreting the test findings in these situations. The AASM practice parameters (11) emphasize that the test is indicated mainly to aid diagnosis in patients with suspected narcolepsy or idiopathic CNS hypersomnia, and should not be routinely considered in assessment of patients with obstructive sleep apnoea or other sleep disorders associated with medical, psychiatric, or neurological disorders. In clinical practice, patients with a range of sleep disorders are referred for an MSLT for evaluation of sleepiness, often after OSA has been excluded. These will also include older patients with relatively late onset narcolepsy, and many more with idiopathic hypersomnia may be on treatment for co-morbid depression.

The complexity of interpreting the MSLT makes it mandatory that the physician interpreting the study has a detailed knowledge of the patient's sleep, medical, psychiatric, and social history and lifestyle, and that the standardized environment and protocol for the test is followed (11).

Maintenance of wakefulness test

The MWT measures the propensity to stay awake in a sleep-inducing environment (11,14). The test may be more representative of a patient's complaint that they can stay awake when active, but in a sedentary non-stimulating environment they fall asleep. As in the MSLT, the study is started 1½–3 h after habitual wake up time. The patient is seated in the bed with the back and head supported, in a comfortable dark and quiet room. At the start of each nap the patient is asked to stay awake for as long as possible. The protocol includes four naps with 2 h between each nap and, to avoid the ceiling effect, a 40-min nap protocol is recommended (11). The trial is ended if sleep onset occurs, defined as three consecutive 30 s epochs of stage 1 sleep, or one epoch of any other sleep stage, or if no sleep has occurred after 40 min. The SOL for each test nap is recorded and the mean calculated. A preceding night PSG recording is not mandatory before the MWT as, if the patient can stay awake, the quality or quantity of the preceding night sleep is not as important as for the MSLT. It is left to the physician's clinical discretion (11).

The MSLT and MWT do not correlate well in patients with complaints of sleepiness. While the MSLT is recommended for diagnosis of sleep disorders, such as narcolepsy and idiopathic CNS hypersomnia, the MWT is recommended in the AASM practice parameters when the inability to stay awake by a person is a safety hazard (11). It can also be used in monitoring treatment for narcolepsy or idiopathic CNS hypersomnia or even obstructive sleep apnoea. However, there is little evidence linking MWT latency with real world accidents, and it has to be interpreted in the clinical context and for providing information on treatment adherence. Furthermore, if a patient has fallen asleep at work while employed in a high risk occupation, these patients are usually off work by the time they are seen by a sleep specialist, so their sleep schedule may be different from their work day schedule. This is particularly important where the job requirement includes shift work or very early starts, especially in occupations where start times vary. The patient's morning or evening preference (15) (so-called 'lark' or 'owl' tendency) may also play a role in their inability to stay awake at certain times during the day, when combined with insufficient sleep due to even mild sleep disturbance. The MWT report should acknowledge these limitations.

The lack of normative data limited the clinical use of MWT, until such data was published in 1997 by a consortium of sleep disorder specialists with a range for expected values [16]. Based on statistical analysis of normative data (16), the AASM practice parameters suggest that a mean SOL of 8 min or less or even 10 min is abnormal (11,13). However, while the ability to stay awake for 40 min in all four naps provides the strongest evidence for being able to maintain alertness during the day, a SOL between 10–40 min is therefore of uncertain significance. The mean SOL for presumed normal volunteer subjects was 35.2 +/– 7.9 mins in the original normative dataset (16). It is important for the clinician to recognize the limitation of this test and that normal values do not guarantee safety, especially in occupations that can compromise personal or public safety.

Actigraphy

Actiwatches are movement monitors that allow recording of wake and sleep activity over several days. As the name suggests, they are watch-like devices, usually worn on the non-dominant wrist, although can be worn on either hand. In patients with neurological disorders the arm with more movements is selected and in children, especially babies, the actiwatch can be strapped to the leg. Actiwatches have a movement detector (like an accelerometer) and a battery life long enough to record continuously over 24 h for several days or weeks. The basis of the interpretation is that there is less motor activity during sleep than wake (17). The sampling and epoch rates for analysing the digitized movement data can be set by the investigator, although usually 1-min epochs are used. The patient has to keep a sleep diary recording lights off and getting out of bed times, medication, daytime naps, and times when the actiwatch was removed, as many actiwatches are not waterproof. For analyses of night time sleep, the period for analysis is selected for each night, entering patients recorded lights off and rising times. The software then uses a computer algorithm to analyse sleep, providing information on a range of sleep parameters including time in bed against actual sleep time, sleep efficiency, sleep onset latency, percentage of movement, wake periods after sleep onset, and fragmentation index. In addition to the detail analysis, the plotted data (Fig. 16.3) provides a visual format for quick review of the sleep–wake schedule and sleep disturbance. Daytime data can also be analysed. Many actiwatches include the facility to record environmental light and some can also record sound and skin temperature (17). Actiwatches can also be used to record periodic leg movements in sleep (PLMS). However, this requires two actiwatches, one for each leg, attached either over the tibialis anterior muscles or strapped to the base of the big toe. Algorithms analyse the number of limb movements per hour. As PLMS can vary from night to night, recording limb movements with actigraphy has advantages, but studies demonstrating reliability of frequency and distribution of the periodic leg movements with head-to-head comparison with PSG are limited.

The beauty of actigraphy recording is that it is cost efficient and offers the ability to record non-intrusively over several days, unlike PSG that is traditionally recorded for 1 or 2 nights. Patients often report not sleeping as usual on the PSG nights, with some reporting severely disturbed sleep, making PSG analysis uninterpretable. Patients can go about their day-to-day activity with the actiwatch on. Some patients have skin sensitivity to the strap used and if so the material used for the strap may need to be customized. Children with learning difficulties, especially autism, who have sensory processing problems, and adults with cognitive problems, may not like the continuous skin contact. However, with some creativity, including using locked hospital bands, even young people with learning difficulties can be encouraged to wear an actiwatch for several days, providing valuable data in this population group who may not tolerate a full PSG study. The main problems we have encountered with actigraphy is the occasional patient losing the actiwatch, not returning it in a timely manner, or the actiwatch getting lost in the post, although we discourage return by post for this reason. There can be a tendency to underestimate the cost value of these very small recording systems. Also as the recordings are done over long periods, patients can forget to put the actiwatch on again after a shower or swim, leading to intermittent loss of data, or patients return the actiwatch with incomplete sleep diaries, so estimated lights off and get up times have to be used for analysis, reducing the reliability of the data, especially sleep onset latency. This is more likely to occur in chaotic families, parents juggling the needs of the index child, often with learning difficulties and their other children, but many adults with sleep disorders also struggle with routine and organization.

The main limitation of actigraphy is that it cannot provide information on sleep stage distribution, relying entirely on movements to analyse sleep. Also there are a number of devices available with different algorithms, with no studies comparing data between systems. Most studies have found a high correlation in total sleep time between actigraphy and PSG, in subjects with normal sleep (18). However, in patients with sleep disorders, the coefficient is lower depending on the type of sleep disorders and age and sex of the patient group. In general, actigraphy over-estimates sleep and under-estimates wakefulness. This is partly due to quiet wake/drowsy periods analysed as sleep by actigraphy, especially compromising sleep onset latency. However, actigraphy is superior to sleep logs maintained by patients (or parents).

Actigraphy is the test of choice and is superior to PSG in children and adults with circadian rhythm disorders, including delayed and advanced sleep phase syndromes, shift work sleep

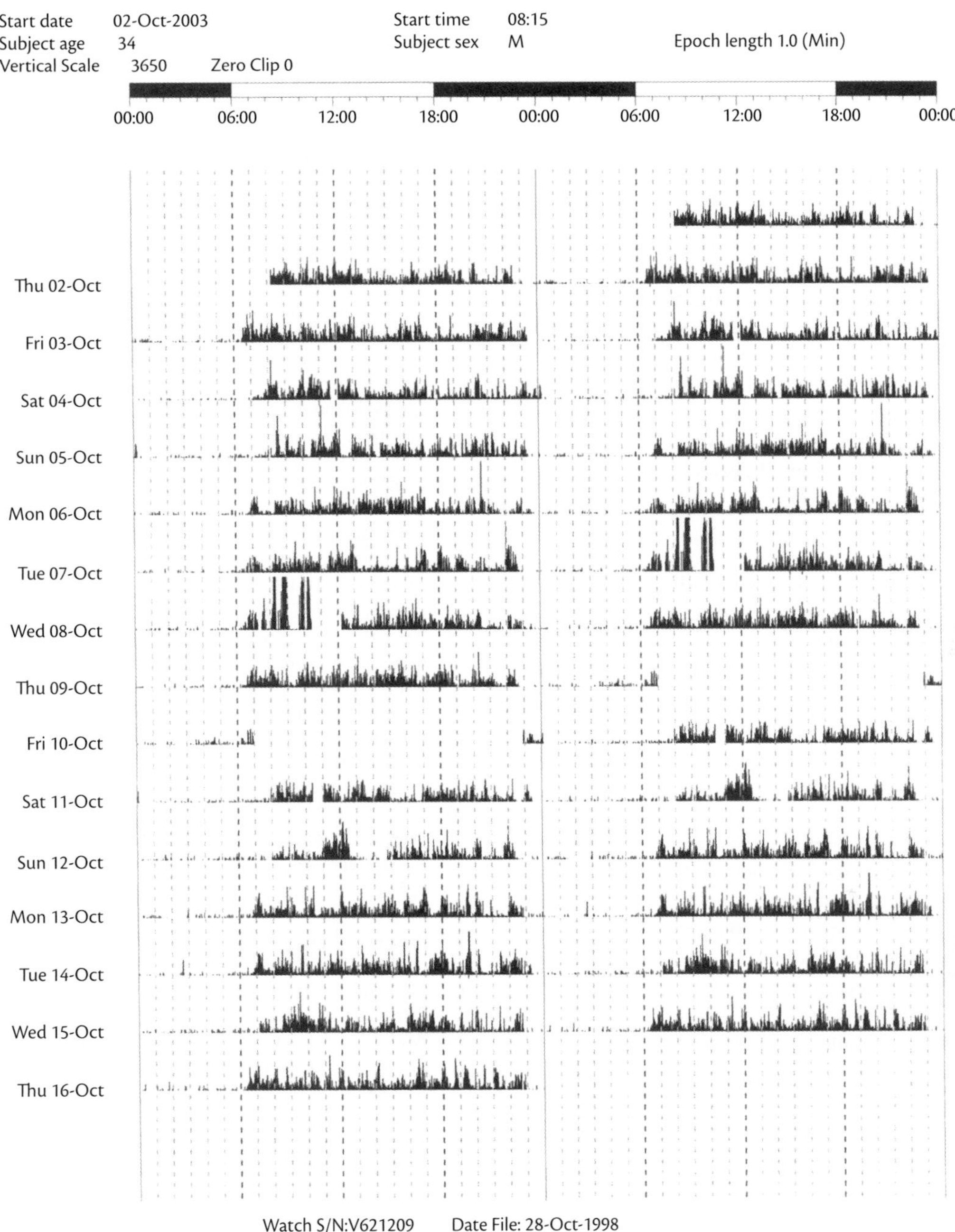

Fig. 16.3 Example of normal actigraphy; data double plotted showing stable bed times and wake up times during working week, later at weekends; the watch was not worn during day 8.

disorders and non-24-h sleep–wake rhythm disorder. In these disorders, actigraphy must be recorded for at least 7 days, but if possible 2 weeks, to include weekends or days off. The recording may need to be extended in some cases for more than 2 weeks, to cover all work shift rotations or allow review of sleep schedules between school/work days and holiday periods. Usually actigraphy alone, with monitoring of environmental light, is sufficient for diagnosing and managing circadian rhythm disorders, and is widely used in sleep and circadian rhythm research. Furthermore, it can be easily repeated to monitor response to treatment.

Actigraphy can sometimes provide a preliminary screen of sleep–wake schedules and quality in patients in whom a reliable sleep history is difficult to obtain, before considering PSG. However, it is best to avoid actigraphy in patients with insomnia unless a circadian rhythm sleep disorder is suspected, as many insomnia patients can get distressed if the actigraphy suggests more sleep than their own perception, which of course may be correct as actigraphy can record quiet wake periods as sleep. Occasionally actigraphy can be helpful in patients with insomnia to monitor hypnotic drug withdrawal, as actigraphy may show that sleep quality is not significantly different on nights when a sedative is taken compared with reduced dose or no medication. Actigraphy can be useful in patients who have not responded to cognitive behaviour therapy for insomnia, a programme that includes sleep restriction, as the actigraphy may show that the patient continues to spend too long in bed at night, maintaining the insomnia.

Inclusion of 1–2 weeks of actigraphy is recommended prior to PSG and MSLT testing, in patients with excessive daytime sleepiness without cataplexy, as sleep diaries are not always reliable. Actigraphy over several weeks can be useful in patients with typical and atypical Kleine–Levin syndrome with episodes of sleepiness over several days or weeks, to include both the sleepy phase and the baseline sleep pattern. The recording may demonstrate the period of long sleeps, and the recovery phase heralded by a night or two of less sleep before the habitual sleep wake pattern returns.

Conclusions

More than with other neurophysiology recordings, sleep changes with life issues including acute or chronic personal stress, lifestyle, medication, and psychological and behavioural changes contributing to its variance. All these factors have to be considered in interpreting studies of patients with sleep disorders. Sleep studies, including PSG and MSLT/MWT, are time and cost resource heavy, and it is essential that the type of study is selected carefully according to the clinical complaint and that strict protocols are followed. Ideally, sleep studies should not be reported in isolation without clinical knowledge of the patient or the neurophysiologist working closely with a sleep medicine specialist. Referrals for individual tests like MSLT in isolation are impossible to interpret and a waste of resource. It is also important that units offering a sleep disorder service regularly update their study protocols and audit the inter-rater reliability between their sleep technologists. This should include PSG sleep staging in patients of different ages and with varying sleep disorders and the reliability of MSLT/MWT mean SOL analysis. Physician reports should include comments on difficulties identified during sleep staging, which may be due to technical problems or change in sleep architecture due to age, neurological disorders or medication, or frequent oscillations/intrusions between sleep stages and wake periods in some sleep disorders. The reports should also include limitations of the study, including if compromised by patient behaviour and discuss the option of repeating the study.

Sleep medicine is a rapidly developing field with recent advances in genetics and basic science research enhancing our understanding of normal and abnormal sleep disorders. It is a fascinating field and in spite of the difficulties in interpretation addressed in this chapter, neurophysiologists are encouraged to engage with Sleep Medicine specialists and contribute and develop this speciality.

References

1. Rechtschaffen, A. and Kales, A. (1968). *A manual of standardized terminology, technique and scoring system for sleep stages of human subjects*. Los Angeles, CA: Brain Information Service/Brain Research Institute, University of California Los Angeles.
2. Carskadon, M. and Dement, W.C. (2011. Normal Human Sleep. In: M. Kryger, T. Roth, and W. C. Dement (Eds) *Principles and Practice of Sleep Medicine*, 5th edn, pp. 16–26. Amsterdam: Elsevier Saunders.
3. Association ASD (1992). EEG arousals: scoring rules and examples: a preliminary report from the Sleep Disorders Atlas Task Force of the ASDA. *Sleep*, **15**, 173–84.
4. Parrino, L., Ferri, R., Bruni, O., and Terzano, M.G. (2012). Cyclic alternating pattern (CAP): the marker of sleep instability. *Sleep Medicine Reviews*, **16**, 27–45.
5. Iber, C., Ancoli-Israel, S., Chesson, A.L., Jr., and Quan, S.F. (2007). *The AASM Manual for the scoring of sleep and associated events: rules, terminology and technical specifications*. Westchester, IL: American Academy of Sleep Medicine.
6. Berry, R.B., Budhiraja, R., Gottlieb, D.J., et al. (2012). Rules for scoring respiratory events in sleep: update of the 2007 AASM Manual for the Scoring of Sleep and Associated Events. Deliberations of the Sleep Apnea Definitions Task Force of the American Academy of Sleep Medicine. *Journal of Clinical Sleep Medicine*, **8**, 597–619.
7. Collop, N.A., Anderson, W.M., Boehlecke, B., et al. (2007). Clinical guidelines for the use of unattended portable monitors in the diagnosis of obstructive sleep apnea in adult patients. Portable Monitoring Task Force of the American Academy of Sleep Medicine. *Journal of Clinical Sleep Medicine*, **3**, 737–47.
8. Association SoPCotASD. (1994). Practice parameters for the use of portable recording in the assessment of obstructive sleep apnea. *Sleep*, **17**, 372–7.
9. Pitson, D.J. and Stradling, J.R. (1998). Value of beat-to-beat blood pressure changes, detected by pulse transit time, in the management of the obstructive sleep apnoea/hypopnoea syndrome. *European Respiratory Journal*, **12**, 685–92.
10. Carskadon, M.A., Dement, W.C., Mitler, M.M., Roth, T., Westbrook, P.R., and Keenan, S. (1986). Guidelines for the multiple sleep latency test (MSLT): a standard measure of sleepiness. *Sleep*, **9**, 519–24.
11. Littner, M.R., Kushida, C., Wise, M., et al. (2005). Practice parameters for clinical use of the multiple sleep latency test and the maintenance of wakefulness test. *Sleep*, **28**, 113–21.
12. Medicine AAoS. (2014). *International classification of sleep disorders*. Darien, IL: AAoS.
13. Arand, D., Bonnet, M., Hurwitz, T., Mitler, M., Rosa, R., and Sangal, R.B. (2005). The clinical use of the MSLT and MWT. *Sleep*, **28**, 123–44.
14. Mitler, M.M., Gujavarty, K.S., and Browman, C.P. (1982). Maintenance of wakefulness test: a polysomnographic technique for evaluation treatment efficacy in patients with excessive somnolence. *Electroencephalography & Clinical Neurophysiology*, **53**, 658–61.
15. Horne, J.A. and Ostberg, O. (1976). A self-assessment questionnaire to determine morningness-eveningness in human circadian rhythms. *International Journal of Chronobiology*, **4**, 97–110.
16. Doghramji, K., Mitler, M.M., Sangal, R.B., et al. (1997). A normative study of the maintenance of wakefulness test (MWT). *Electroencephalography & Clinical Neurophysiology*, **103**, 554–62.
17. Stone, K.L. and Ancoli-Israel, S. (2011). Actigraphy. In: M. H. Kryger, T. Roth, and W. C. Dement (Eds) *Principles and practice of sleep medicine*, 5th edn, pp. 1668–75. Oxford: Elsevier Saunders.
18. Morgenthaler, T., Alessi, C., Friedman, L., et al. (2007). Practice parameters for the use of actigraphy in the assessment of sleep and sleep disorders: an update for 2007. *Sleep*, **30**, 519–29.

CHAPTER 17

Clinical neurophysiology of the pelvic floor

Adrian J. Fowle

Introduction

Few clinical neurophysiologists in the United Kingdom (UK) offer pelvic floor studies, so referrals come from a wide geographical area but despite this an individual department may see only 25–30 cases a year, less than 1% of the workload. This is, therefore, a personal account. Some colleagues in Europe have higher referral rates, and the first suggestion of this chapter is that the rate of referral in the UK is too low for techniques with proven utility. The techniques described here are similar to those used in the limbs. They differ in the details, just as techniques used in the cranial region do. A typical pelvic floor study involves taking a history and clinical examination, followed by some combination of nerve conduction study and electromyography (EMG). Somatosensory evoked potentials (SSEP) and magnetic stimulation can be added as necessary. Any clinical neurophysiologist who has mastered the blink reflex and EMG of facial muscles could just as easily master these techniques. Many practical points are included in the hope that others may be encouraged to learn at least the easier techniques.

Acceptability of pelvic floor studies

I first became interested in these studies as a trainee when my teaching hospital needed them for the investigation of multiple system atrophy (MSA). Although there were some initial reservations on the acceptability of the studies, it was easy to persuade colleagues who saw the studies actually being carried out that they are clinically well tolerated. Indeed, the patients who are referred are often quite desperate for some explanation of their symptoms and for a doctor willing to take them seriously. Patient feedback leads me to the conclusion that EMG studies are a little more painful than EMG of limb muscle, but only a little. In both limb and pelvic floor studies consideration of the patients' perspective is essential to minimize distress. Also it must be recognized that these diagnostic studies, like others, are subject to technical problems, and equivocal, false positive, or false negative results. It is often said of peripheral neurophysiology that it is an extension of the clinical examination with a needle. These studies should be considered in the same light.

Pelvic floor neurophysiology deserves to be developed, bearing in mind, however, that referrals may be infrequent and a sufficient throughput of a range of cases will be needed to maintain skills. It is probably something that only one doctor in a large practice needs to provide. If this chapter serves to remove some of the imagined difficulties around these studies it will have served its purpose.

Approach to the patient

The appointment and consultation are only part of the process of having a clinical neurophysiological study. Patients are often sent a booklet with their appointment letter, which explains something of the test. These booklets were very difficult to write and still cause debate. Patients should have neurophysiological studies explained as fully as they require beforehand, but sometimes this is best done by a sensitive clinician in a face to face consultation, rather than in print and read without support. Too much information beforehand causes fear, sometimes to the point of not attending. Too little information risks being seen as patronizing or not allowing fully informed consent. We have settled on a single booklet, which covers all of nerve conduction and EMG studies, without mentioning particular muscles.

The booklet does not specify which areas are to be tested. Patients who have EMG of tongue muscles for motor neurone disease are not warned and neither are patients having pelvic floor studies. Patients are offered the opportunity to telephone for advice if they wish, but very few do so. The vast majority accept these studies as readily as patients accept routine EMG.

It is preferable to talk to patients while they are clothed and seated, and with an accompanying relative present if desired. They can then be invited to use the toilet while the room and examiner are prepared. The bed should be covered with disposable absorbent towelling, and the examiner should wear a plastic apron. It is essential to have a dedicated assistant present during the procedure. There is, first the need for a chaperone during these intimate procedures. The examiner is unable to see the patient's face and reactions during many of these tests. The assistant should be seated at the far side of the couch facing the patient and able to indicate to the examiner the degree of discomfort that the patient is experiencing. An assistant able to help engage the patient in small talk as distraction therapy is even better. In most routine studies the electromyographer can move around freely while performing them, but with pelvic floor studies it is sometimes necessary to have the assistant pass things to the examiner. Relatives are not normally present during the examination, but occasionally one will comfort the patient.

After the study, when the patient is once again dressed and seated, and the accompanying relatives are present if desired, the patient should be encouraged to ask questions and should be warned that they may be uncomfortable and see small amounts of blood on their underwear if they have had an EMG.

Although opinions differ on the need to discuss results with patients, it is my practice in routine EMG clinics to discuss the results and to give the patient a copy of the report. Patients having pelvic floor studies are treated in the same way. It is preferable to explain complex studies to patients, rather than have them be misinformed by the Internet. Anyone following this approach needs to allow time and have the skills to break bad news to patients. Some patients are distressed to be told that studies are normal when they have symptoms that concern them.

It is important to explain that EMG is only one facet of making a diagnosis. Further investigation and treatment options may need to be discussed. Occasionally, the referral diagnosis is very wide of the mark, in which case referral to another specialty may need to be arranged. Some patients seem not to want to know the result and their wishes should be respected, but usually an honest discussion of unpleasant or unexpected news is generally well received.

Approach to referrers

Clinical neurophysiology should be seen as a continuing service to referrers and not as a series of discrete consultations. It is important to establish a rapport with frequent referrers. Establishment of a multidisciplinary 'pelvic floor team' consisting of a colo-rectal surgeon, uro-gynaecologist, radiologist, specialized nurses and physiotherapists, and neurophysiologist is useful. For example, a service for post-partum cases allows understanding of which conditions the surgeons are trying to distinguish and what the consequences are, as well as allowing surgeons to appreciate the benefits and limitations of neurophysiology. A guide for referrers and an invitation to both junior and senior colleagues to discuss possible patients for referral and to see their patients being examined also fosters a successful multidisciplinary approach.

Selected anatomy and physiology of the pelvic floor

Clinical neurophysiologists often prefer schematic diagrams and function to detailed anatomy. The three-dimensional anatomy of the pelvis is particularly challenging. It is, of course, different in the two genders, and there are those who wish to split the muscles into ever smaller named structures. Access to the external anal sphincter is easy enough, but EMG of other muscles requires understanding their anatomy, and that of the structures to be avoided.

The new digital interactive anatomy tools are very useful to 'move round' a structure in three dimensions, adding and removing structures in virtual dissection. None show all the muscles of interest, however, and there is still a role for old fashioned coronal, sagittal, and transverse sections, and anatomical dissections, for example (1).

The main muscle (Figs 17.3 and 17.4) is the levator ani, a sheet of muscle that takes origin from the pubis, iliacus, and ischium, and inserts into two tendinous masses in the mid-line and into the coccyx. Some regard this as three separate muscles, the pubococcygeus, iliococcygeus, and ischiococcygeus. The levator ani has an anterior margin, leaving a mid-line space through which passes the urethra and vagina. The rectum passes through more posteriorly and, as it does so, receives a sling of muscle passing around it at the ano-rectal junction, a fourth part of levator ani called the puborectalis.

Below the levator ani, the external anal sphincter and external urethral sphincter sit around their respective organs. In strictly anatomical terms, the sphincter muscles and the bulbocavernosus/bulbospongiosus muscles are part of the perineum, rather than the pelvis. The perineum includes other muscles, the external genitalia, and the surrounding skin.

For the purposes of this chapter and access to EMG, the principle differences between the two genders are that in the female the urethra does not run through the clitoris, but below it, and the presence of the vagina and vulva. There are thus two additional mid-line structures running through the pelvis and perineum compared with the male. Loops of muscle surround all these structures and many are named by those who like such detail. In the male, the bulb of the penis and the scrotum pose additional obstacles to needle access. In the male, the bulbocavernosus/bulbospongiosus is a paired mid-line structure lying over the root of the penis. In the female the left and right muscles separate around the vagina.

All these muscles named in the preceding paragraphs receive somatic innervation from S2–S4 roots, via the nerve to levator ani and the pudendal nerve and its branches. Both nerves lie on the surface of levator ani.

The anterior horn cells of the motor nerves to both sphincter muscles are located in Onuf's nucleus at S2 root level. This sits in close proximity to autonomic nerves, which innervate the internal anal sphincter, and has a character between that of most somatic and autonomic nuclei. It is usually spared in motor neurone disease (2), but involved in multiple system atrophy (3).

Spinal roots S2–3 also provide sensation to the peri-anal and genital regions and the parasympathetic outflow. The latter provides motor innervation to the muscular wall of the bladder (detrusor) and an inhibitory supply to the internal urethral sphincter. It is vasomotor to the penis and clitoris and responsible for erections.

The pelvic sympathetic supply arises from lumbar roots L1–3, and is distributed to other roots and vessels via the sympathetic chain. It controls ejaculation amongst other functions.

Integration of parasympathetic, sympathetic, and somatic function occurs at several levels in the spinal cord, brain-stem, and cortex, and may be disturbed in, for example, spinal cord lesions or multiple sclerosis. Conventional clinical neurophysiological techniques provide little access to autonomic or integrative functions, and will not be considered further.

The separation into pelvic and limb functions is, of course, artificial. Several limb muscles share an S2 root level with the sphincters, including the gluteus maximus and sometimes gastrocnemius, and may be affected in the same patient in disorders of the lower spine.

Figs 17.1 and 17.2 give the essential surface anatomy, while figs 17.3 and 17.4 show the important muscles.

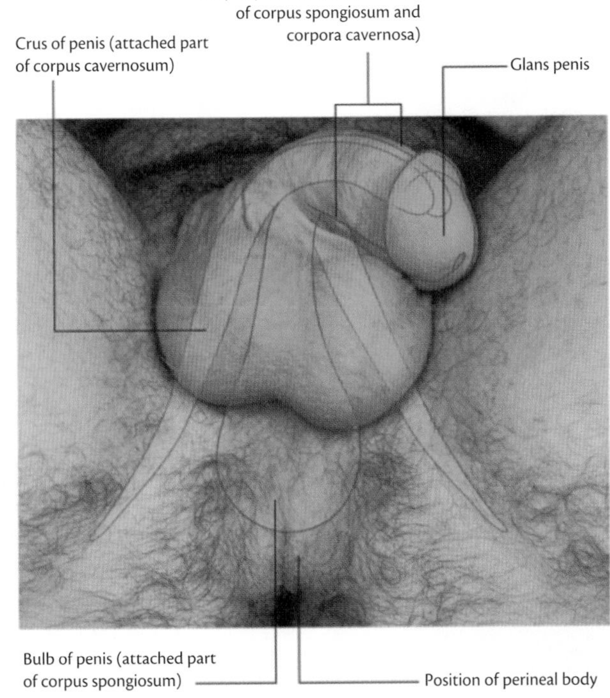

Fig. 17.1 Structures in the male urogenital triangle. Inferior view of the urogenital triangle of a man with the erectile tissues of the penis indicated with overlays.
Reproduced from Drake, Vogl and Mitchell, "*Gray's Anatomy for Students*," copyright (2005), with permission from Elsevier.

History and clinical examination

As with other areas of clinical neurophysiology, pelvic floor investigations have to be understood in the context of the patient's clinical picture. In cases of changes in bladder or bowel habit the previous habit must be understood. Pregnancy-related events may have deteriorated with each previous pregnancy. Changes in sexual function may need to be asked about directly.

Many cases are referred by non-neurologists and the most important examination is that of the nervous system in the legs to look for evidence of extra-pelvic involvement. Peri-anal sensation can be mapped in root or cord lesions and anal tone assessed by rectal examination. Bulbocavernosus reflex is described in the section on sacral reflexes.

Clinical neurophysiological techniques

Nerve conduction studies

Pudendal nerve

The pudendal nerve can be stimulated near the spine of the ischium and a motor response recorded distally. This is accomplished with a 'St Marks Electrode'. This consists of two pairs of electrodes in a disposable unit, which is attached to the examiner's gloved index finger. With the finger inserted into the rectum, the ischial spine can be palpated with the tip of the index finger. A pair of electrodes on the finger tip stimulates the nerve. The other pair of electrodes is at the base of the finger, surrounded by the external anal sphincter from which the response is recorded.

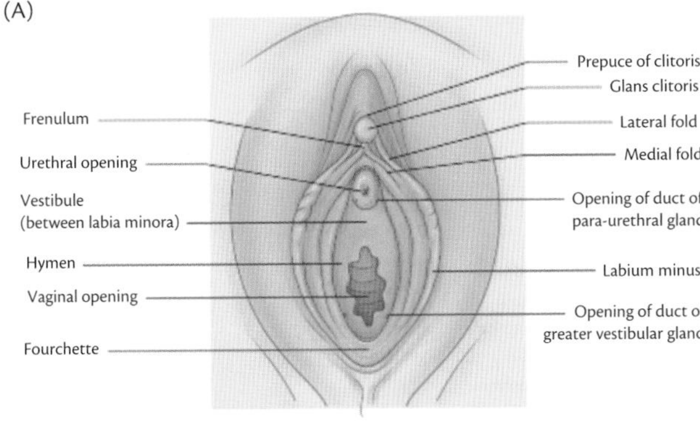

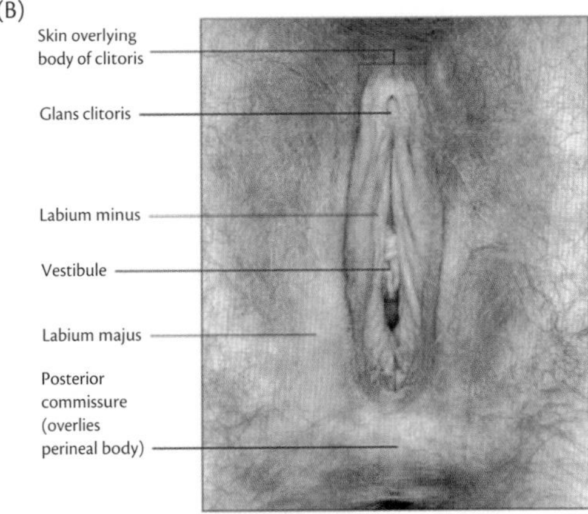

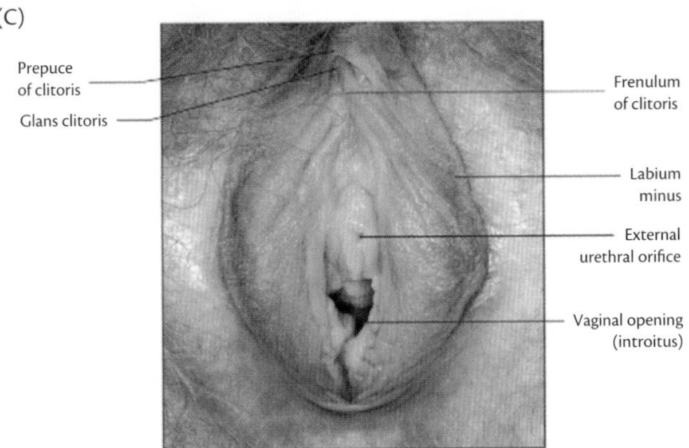

Fig. 17.2 (A) Superficial features of the perineum. (B) The labia majora and surrounding external genitalia. (C) The labia minora.
(A) reproduced from Drake, Vogl and Mitchell, Gray's Anatomy for Students, 2nd edition, copyright (2008), with permission from Elsevier; (B–C) reproduced from Drake, Vogl and Mitchell, "*Gray's Anatomy for Students*," copyright (2005), with permission from Elsevier.

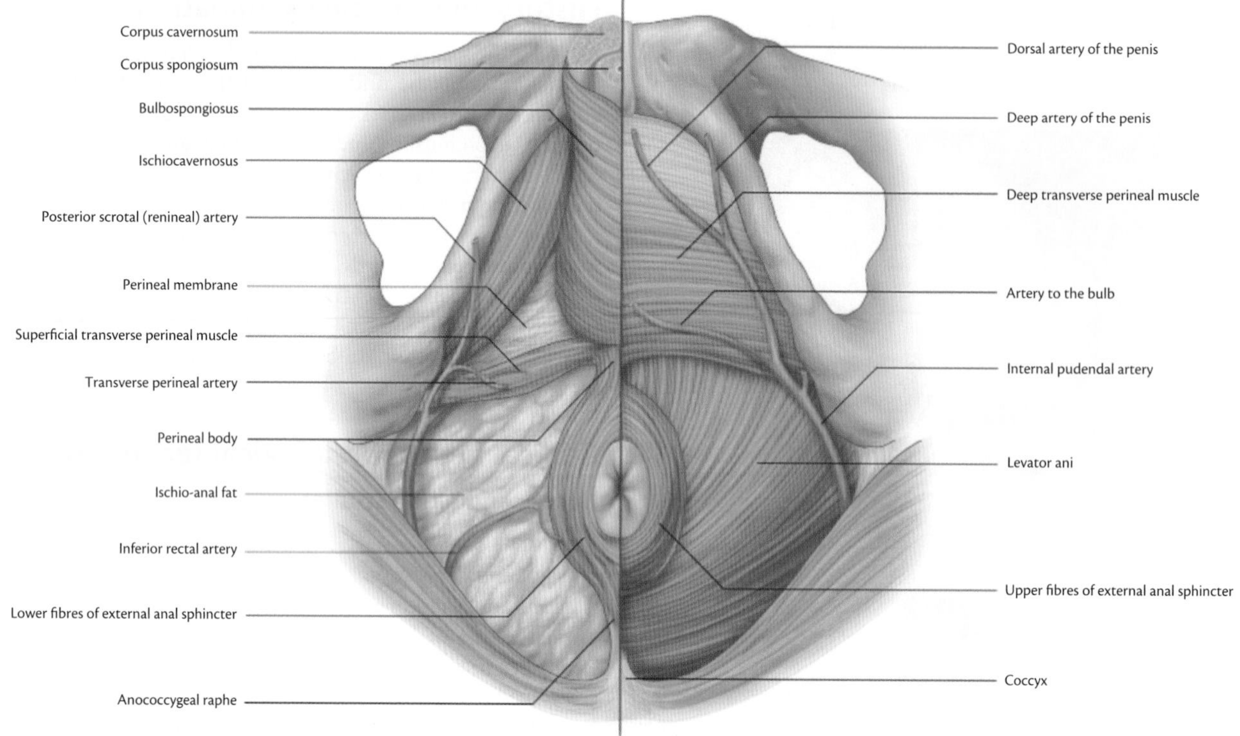

Fig. 17.3 Muscles and fasciae of the male perineum. On the left side the skin and superficial fascia of the perineum only have been removed. The posterior scrotal (perineal) artery has been shown as it runs forward into the scrotal tissues. On the right side, the corpora cavernosa and corpus spongiosum and their associated muscles, the superficial perineal muscles and perineal membrane have been removed to reveal the underlying deep muscles and arteries of the perineum. All veins and nerves have been omitted for clarity.

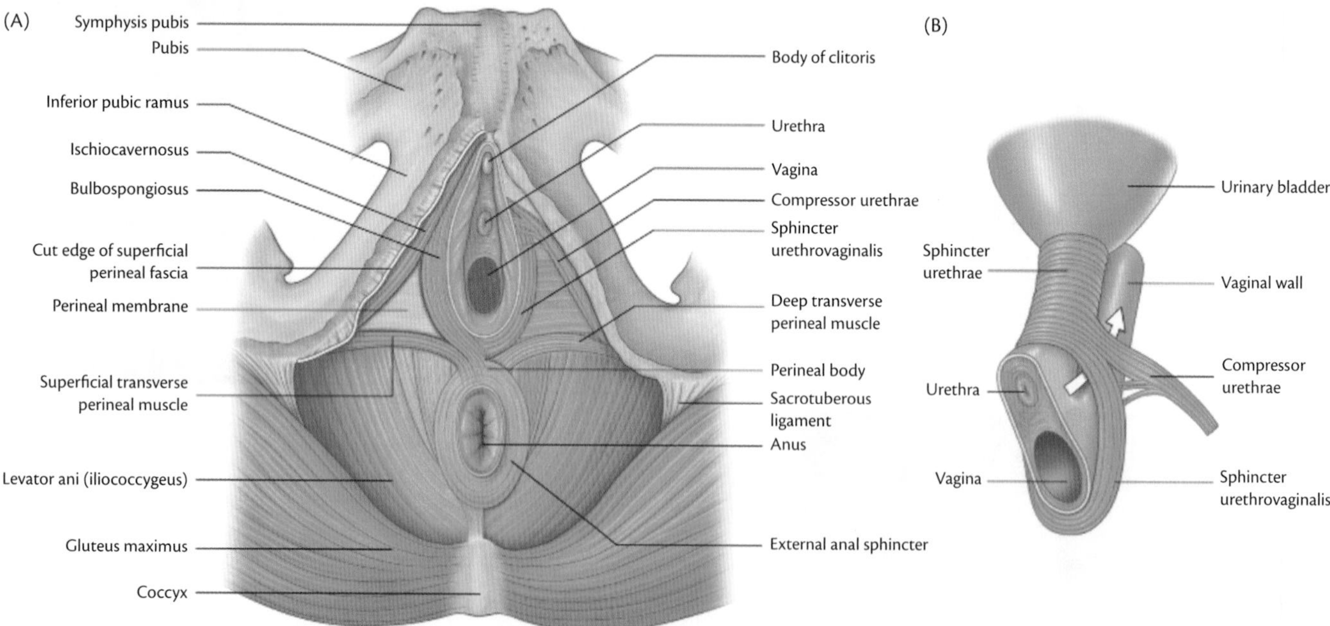

Fig. 17.4 Muscles of the female perineum. (A) On the right side, the membranous layer of superficial fascia has been removed (note the cut edge). On the left side, superficial perineal muscles and overlying fascia have been removed to show the deep perineal muscles. (B) The continuity of the deep perineal muscles with sphincter urethrae.

With this arrangement the responses from one side are inverted compared with the other, which may cause some initial confusion. Once this is understood, it is easy to learn to find the best site of stimulation and the correct response.

The plastic backing to the electrodes can catch in the anus as the examining finger is rotated. A large pelvis or large buttocks may make the procedure uncomfortable, but generally this is a well-tolerated procedure.

There are variations on this theme, with vaginal approaches or recording electrodes mounted on urinary catheters or anal sponges. A conduction velocity cannot be calculated because the length of the nerve pathway is not the length of the finger. The amplitude of the potential is unreliable as a guide to the pathology, even when compared with its homologue. The only measure reported is the distal latency, usually referred to as 'pudendal nerve terminal motor latency' (PNTML) (4). The normal range is of the order of 2.0–2.5 ms.

The utility of this study is disputed. To many neurophysiologists, it has all the attraction of trying to assess an ulnar neuropathy at the elbow using only distal motor latency recorded by stimulation at the wrist. Others point to variations in the latency in different conditions, which presumably depends partly on whether the cause of the dysfunction lies proximal or distal to the point of stimulation. Even when it is abnormal, the clinical neurophysiologist needs needle EMG to assess axonal loss and the prognosis for recovery.

Dorsal nerve of penis

These are left and right nerves which run closely together. They are stimulated together and this is most commonly done with ring electrodes similar to those used for the fingers. There are electrodes made of cloth impregnated with metal, which are more acceptable than the thin wire type.

An orthodromic technique is used and requires five electrodes (5):

- A stimulating pair near the glans.
- A recording pair near the root of the penis.
- The earth lead between.

The reference electrode of the recording pair can be placed on the perineum. That still leaves four electrodes to be placed on a penis that might not be long enough. It is better if this can be assessed quietly without embarrassing the patient with an unsuccessful trial.

Patients can comfortably put these electrodes on themselves. Once the electrodes are in place, the recording can be made like other sensory studies. It appears not to be painful, and it is useful to record limb studies first, without hurting the patient, to reduce apprehension.

The conduction velocity is reduced in some neuropathies (6,7). There is no comparable study in women.

Ilioinguinal, iliohypogastric, and genitofemoral nerves

See the section Post-herniorrhaphy and other inguinal pain.

Sacral root stimulation

Sacral roots can be stimulated with needle electrodes or with magnetic stimulation, with recordings made from many different muscles. Needle electrodes have to be used to localize the response to a single muscle. None of these appears to be clinically useful.

Autonomic and small fibre studies

The clinical neurophysiological assessment of these is much the same in the perineal region as elsewhere. Thermal threshold studies should be performed in the feet, rather than the perineum even for erectile dysfunction. The sympathetic skin response (SSR) can be both elicited in the perineum and recorded there in response to stimulation elsewhere. It is not of great use. There are no tests for the parasympathetic system within the realm of clinical neurophysiology.

Evoked potentials

Pudendal somatosensory evoked potentials

SSEPs can be recorded following stimulation of penile or clitoral nerves. The penile nerve can be stimulated with ring electrodes as in the nerve conduction study, but only one pair of electrodes is needed (8,9). The clitoral nerve can be stimulated with the bipolar stimulator used for nerve conduction studies in limbs. The patient can hold it in place herself and be reassured that she can remove it to stop the stimulation. Some authors state that this is difficult, but I have not found it so.

It is advisable to record these potentials after recording tibial SSEPs. Patients seem quite accepting of the idea of 'right, left, and middle legs'. The recording montage is the same, Cz' referred to Fz. Cz' is posterior by 2–4 cm to Cz. (Cz and Fz are standard electrode positions in the international 10-20 system for recording scalp EEG). Recording of potentials from intermediate points of the path, for instance, the sacral root, are even more difficult than with tibial SSEPs and these are not usually attempted. The peripheral pathway for pudendal stimulation is much shorter than that for tibial nerve at the ankle. The nerves involved are thinner, however, and in practice the latency of the N40 from the pudendal nerve stimulation is within 6 ms of that from the tibial nerve (Fig. 17.5).

The study is of most interest when the tibial SSEPs are normal and the pudendal SSEP is absent or delayed. This suggests dysfunction localized to the sacral roots and cauda equina. This is a rare occurrence, but the technique is easy enough for the examiner and acceptable to patients to warrant doing it. An abnormal tibial SSEP is important and may indicate dysfunction in roots or higher.

Other SSEP and motor evoked potentials

SSEPs can be recorded following mechanical or electrical stimulation of several perineal structures including the bladder. Magnetic stimulation studies of cortex with recording in perineal muscles has also been described. None of these appears to be clinically useful.

Electromyography

General observations

The muscles of the pelvis and perineum are mainly striated, voluntary muscle, and can be studied by standard EMG techniques. Access to some is easier than others. Many of the books of anatomy relevant to placement of EMG needles cover the external anal sphincter (EAS), but none describes all the techniques listed here.

One particular feature of these muscles is that they maintain continence and are, therefore, tonically active, even in sleep. Their activity can or should be suppressed during straining with obvious

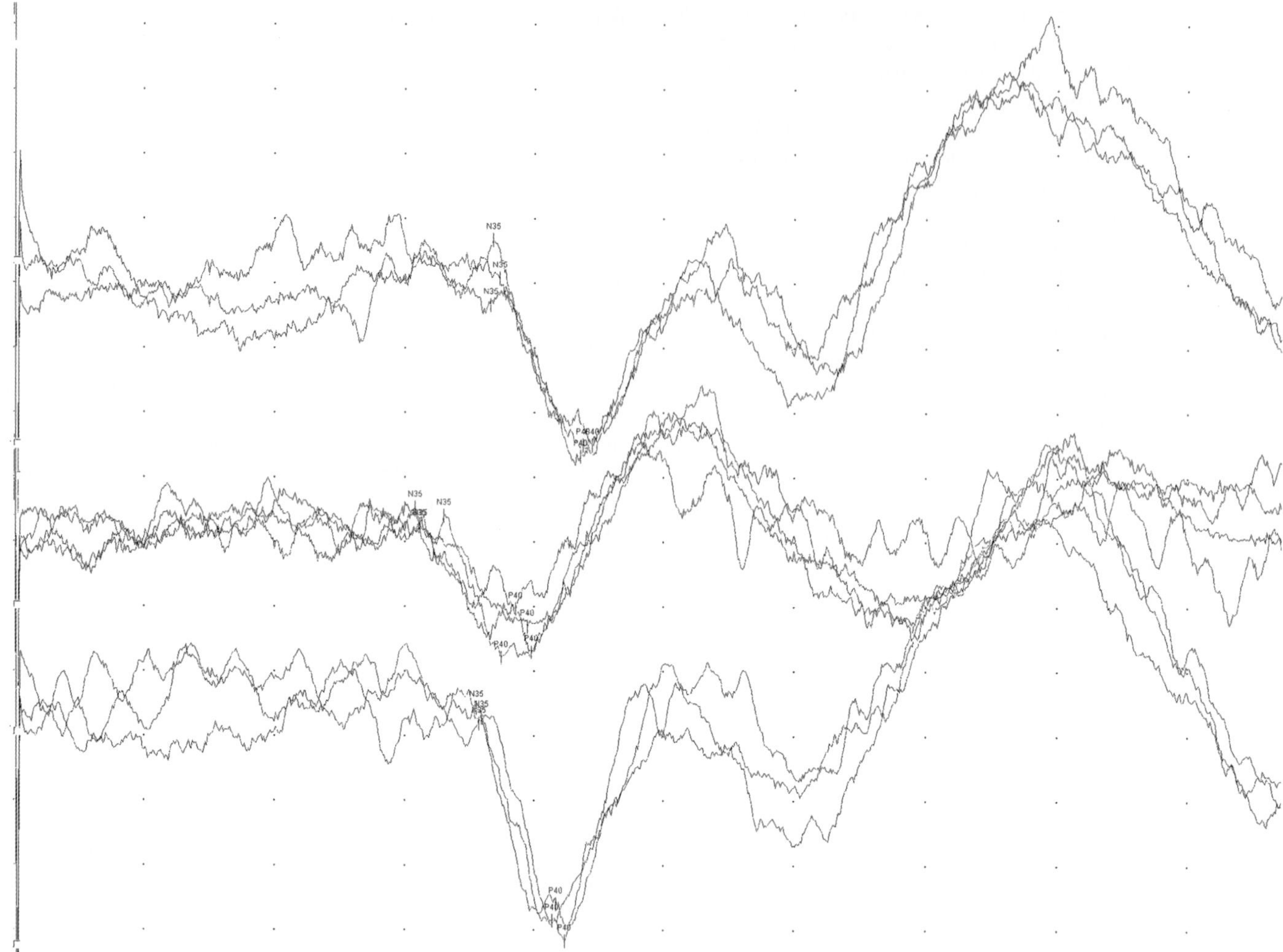

Fig. 17.5 Somatosensory evoked potentials from stimulation of left tibial (upper set), clitoral (middle set), and right tibial (lower set) nerves. Several repetitions of each are shown. Calibration dots represent 10 ms horizontally and 1 microvolt vertically. Recordings were from Cz′ (3–4 cm posterior to Cz) referred to Fz. The patient was a 47-year-old female.

risks during the examination if bowel and bladder are not emptied beforehand. It can be difficult otherwise to distinguish between fibrillations and the motor unit potentials produced by the often small muscle fibres. Voluntary activation can be achieved by asking the patient to clench the muscle concerned, as if to prevent accidental passage of urine or stool. Several techniques require a bi-manual approach to aid needle placement, usually per rectum. To the normal risks of inoculation injury this adds the risk of faecal contamination. The dominant hand is usually employed for rectal examination, but it seems more logical to hold the EMG needle in the dominant hand and use the non-dominant hand for guidance, but this seems to add further to the risks. There are two other consequences of the bi-manual approach. First, it is essential to have an assistant who can operate the EMG machine through the different phases of the examination. Second, the examiner's posture in relation to the patient is unlike either EMG or conventional rectal examination if the non-dominant hand is used. It may be necessary to reposition the screen of the EMG machine with this in mind.

When examining a muscle requiring the bi-manual technique the needle is positioned at the start of its insertion and aimed at the destination, but withdrawal of the guiding finger is suggested rather earlier than the originators of these techniques described. Unless the needle is stable on completion and both hands can be briefly freed, the examiner then needs somewhere to rest the soiled hand while the recording is made.

Abnormalities on needle EMG (2,10–12) alone make it difficult to distinguish lesions of the peripheral nerve, plexus, or root. Electro-clinical assessment of the nerves in the legs is the most useful approach to determine whether the patient has an intra-pelvic problem, or a more general spinal or peripheral nerve disease.

Single fibre EMG and fibre density

These techniques have been used extensively in the past (10,13), but are no longer used.

External anal sphincter

This is the easiest of the EMG techniques listed here, for both patient and examiner, is probably the most commonly used, and is well within the compass of any clinical neurophysiologist who can perform EMG of limb muscles. When both the urinary and anal sphincters are affected by a process, this muscle should be preferentially examined. Many of the EMG atlases recommend the bi-manual technique to avoid penetration of the rectum, but this

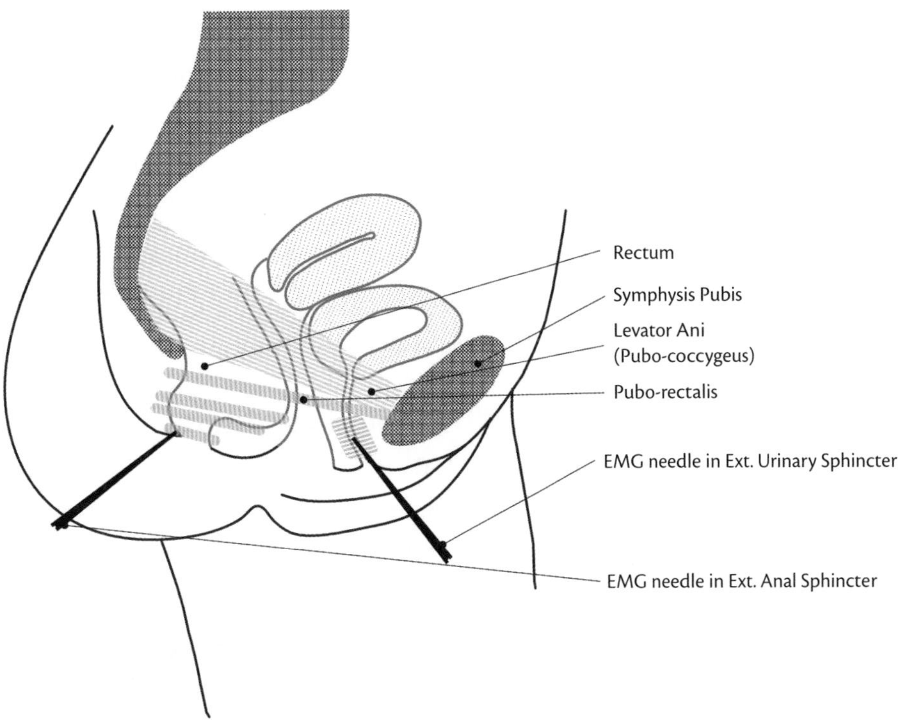

Fig. 17.6 Midline section of female pelvis and perineum to show viscera and left thigh, with overlying muscles of right perineum. EMG approach to subcutaneous part of Ext. Anal Sphincter and Ext. Urethral Sphincter.
© Jeremy Wolfe.

is quite unnecessary. The non-dominant hand is better employed in helping to separate the buttock cheeks to allow access, and in providing counter traction behind the needle entry point.

The EAS has subcutaneous, superficial, and deep parts, which form concentric rings around the anus (Fig 17.6). The superficial and deep parts are wrapped around the internal anal sphincter, which is under autonomic control, and just below the puborectalis sling. The subcutaneous part, which is the most commonly used, is beneath the skin as it slopes in towards the mucocutaneous junction at the anal verge. It is a thin and superficial sheet. The EMG needle needs to be inserted almost parallel to the skin, towards the anus, to avoid going straight through the muscle. The superficial and deep parts can be examined by angling the needle more sharply.

The EAS surrounds the circumference of the anus and different authors recommend different approaches, described in terms of the standard 'laparoscopic clock'. Some prefer four quadrant examinations at 3, 6, 9, and 12 o'clock. For investigation of MSA, 5 and 7 o'clock are preferred.

Positioning is important. For a right-handed examiner the patient should be lying in the left lateral position, with neck and knees tucked up as much as comfortable. The right shoulder should be above the left and similarly for the hips, to avoid a confusing pelvic twist. To examine the subcutaneous part of the left EAS the needle has to slope upward from below. To allow this, the patient has to be positioned with the buttock partly overhanging the edge of the couch towards the examiner with reassurance about not falling. The assistant may need to lift the right buttock to aid access. Insertion is momentarily painful, but the pain subsides quickly to allow the examination to continue. The assistant should signal to the examiner when this has happened.

Examination of the right side seems to be more difficult and more painful in some patients, probably related to support of the buttock. It is not always possible to put the patient in the right lateral position, but the assistant may be able to provide the right amount of lifting and pressing of the right buttock to reduce the pain in the left lateral position.

Fibrillation and interference pattern can be assessed as indicated in the general notes. Motor unit analysis is best undertaken with a multi- motor unit potential (multi-) program on a modern EMG machine. The examination for MSA, in particular, is a quantitative technique and any satellite potentials should be included in the analysis of duration. Only well-recorded units with a short rise time should be included. Normal units have a duration of about 7 ms, and abnormal units are taken to have durations above 10 ms. In MSA the process is patchy, and long duration units and normal units are intermixed. In other disorders the process may be more uniform. This is illustrated in the section on MSA. Several units may be detectable at a single site. These units can have large fields, so the needle may need to be withdrawn some distance before another set of units is recorded, but often several sets can be recorded along one insertion track. Some insertions produce no suitable units.

External urethral sphincter

In women the external urethral sphincter (EUS) can be sampled via a transperineal approach (Fig 17.6). With the patient supine and with feet together, knees flexed and flattened outwards most patients can be studied reasonably easily. This is the only muscle for which a small amount of local anaesthetic (1% lidocaine) is used, injected with a fine gauge hypodermic needle before insertion of the EMG needle. The needle is simply inserted 5 mm lateral to and parallel with the urethral orifice on either side. The female EUS can be thin and easy to miss, which probably accounts for other failures. The procedure results in more blood and serosanguinous exudate than the other muscles and the patient should be warned of this.

Local anaesthetic is not required for examination of the male EUS, but it does require a bi-manual approach—see 'General observations'. The guiding index finger is used to identify the prostate and the needle is aimed at a point just inferior to this. The needle insertion point is between the anterior anus and the bulb of the penis.

Signal analysis is as for the EAS. In young women with urinary retention the decelerating bursts of Fowler's syndrome may also be seen, as described in the clinical section.

Bulbocavernosus/bulbospongiosus

In men this is a thin superficial sheet, which covers the bulb of the penis. A short EMG needle can be inserted a few millimetres lateral to the mid-line, about halfway from the anus to the root of the penis.

In women this is a thin, superficial muscle that surrounds the vagina. It can be approached from the vagina, medial to the labium minorum, or from the skin, lateral to the labium majoram.

These muscles are mainly of use for studying the bulbocavernosus reflex, which is described in a later section.

Levator ani

The levator ani and in particular the puborectalis part of it can be identified by palpation rectally. It surrounds the rectum and forms a large part of the pelvic floor and can be reached in many places. A long needle may be necessary. Usually, the EAS is so much easier to study. However, EMG on a part of the muscle in spasm in levator ani syndrome is feasible where the region in spasm could be palpated to guide the needle.

Transversus perinei superficialis

Perotto (14) describes methods for access to this small and variable muscle in both genders, but it is not clear when this might be useful, although it is suggested that the muscle has only S3–4 innervation which might allow differentiation from S2 root lesions.

Reflexes

Bulbocavernosus reflex

Strictly speaking the bulbocavernosus reflex (BCR) is part of the clinical examination. The reflex is elicited by briskly squeezing the glans penis, and the effects assessed by palpation of the bulbocavernosus muscle or anal sphincter. The clinical study was replaced by an electrical study with the same name. The stimulus can be given in both genders as described for SSEPs. The recordings can be made in bulbocavernosus muscle or the EAS in the course of needle EMG studies (15–17). The increase in firing can be detected audibly and this can be performed in the course of the studies performed above.

A great deal of attention has been focused on these studies. The response can be recorded bilaterally and latencies recorded. It may be necessary to use pulse widths of up to 0.5 ms and paired stimuli with an interstimulus interval of 3 ms. Responses with a latency of less than 40 ms are normal and may have multiple components similar to the blink reflex. Responses from stimulation of different sites use different pathways and may have much longer latencies. Despite this, the abnormalities of BCR that can be detected in a clinical neurophysiology setting do not allow any specific diagnosis to be reached.

Complementary investigations

Urological and ano-rectal physiology departments have techniques for studying pressure, flow, and volume relationships in their respective systems These are sometimes combined with surface EMG or EMG of non-striated muscle. There is surprisingly little overlap with techniques considered to be the natural territory of clinical neurophysiology.

Imaging techniques are the other large group of techniques for studying the pelvis and perineum, and may be combined with the manometric methods, but it must be remembered that the wider neurological system may need imaging too.

Clinical conditions

The majority of cases that are referred fall into one of three groups:

- Conditions local to the pelvic floor, most commonly post-partum.
- Spinal conditions, such as radiculopathy or their treatment.
- Neurological, mainly for Parkinsonism.

Post-partum incontinence

During parturition the baby passes through the anterior space between the two levator ani muscles, which are pushed downwards to allow this. The nerve to levator ani and the pudendal nerve may be stretched and damaged anywhere along their course in the process. The anal sphincter muscles may be torn in third and fourth degree perineal tears too. The anatomy of the sphincter can be assessed by endo-anal ultrasound, and the innervation by needle EMG of the external anal sphincter (18). Repairing the sphincter is a relatively simple process if the innervation is intact. If the sphincter is intact, the uro-gynaecologist may try biofeedback techniques that are thought to be successful. However, if the muscle is denervated and unlikely to recover sufficiently, a more complicated reconstruction may be called for or the use of sacral stimulators. Urinary, as well as faecal incontinence may occur.

Other causes of pudendal neuropathy

Surgical procedures, tumours, and other obvious physical causes within the pelvis can cause lesions of the nerve, and are accompanied by appropriate unilateral sensory loss in the penis or clitoris, erectile dysfunction, but rarely incontinence (19,20).

A well-recognized extra pelvic cause is sitting on a bicycle saddle for long periods, especially a thin one of the kind favoured by athletes. It is thought that these patients develop pressure palsy of the pudendal nerve in Alcock's canal. It can be progressive and associated with impotence. Neurophysiological assessment can be unhelpful in unilateral cases, but normality of the investigation may provide some reassurance to the patient.

The occasional patient with a clear history of trauma to the penis, and others with a convincing history of altered rather than lost sensation on half the penis with difficulties with intercourse are probably also examples of pudendal neuropathy (21).

There is a pain syndrome termed pudendal neuralgia or pudendal nerve entrapment syndrome, characterized by pain in the distribution of the pudendal nerve. There are no signs of neuropathy or objective loss of sensation. It is much discussed by desperate patients with varying symptoms on internet forums. Unfortunately, like many similar pain syndromes the evidence

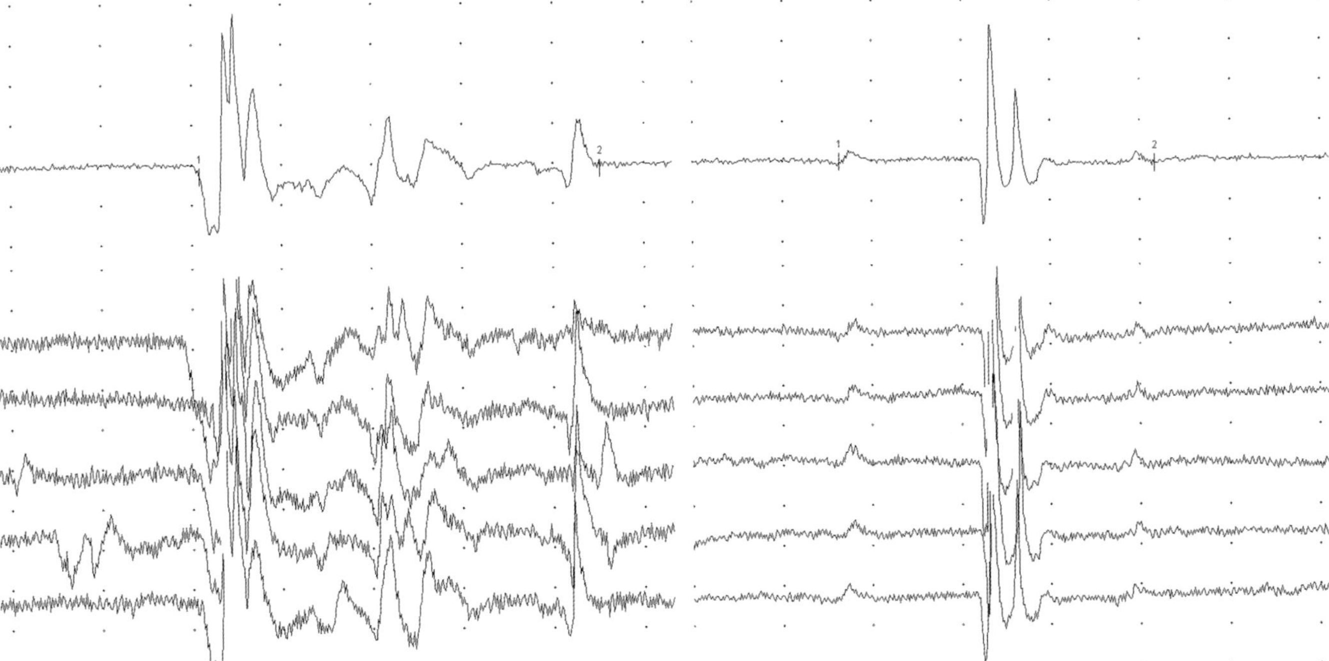

Fig. 17.7 Motor unit potentials (MUP) from EMG of external anal sphincter in two patients with MSA. The two potentials are shown left and right. Upper trace is the 'median' unit produced and analysed by the EMG machine (Synergy, Optima Medical UK). Lower traces are some of the individual recordings from which the median is constructed. Calibration dots are 6 ms horizontally and 50 microvolts vertically in all panels. Note the complexity of the MUP on the left. The MUP on the right appears simpler, but has two small satellite potentials, which define its duration.

that the nerve is the cause of the pain is lacking and there is no rational basis for investigation and treatment.

Levator ani syndrome

This appears to be a syndrome distinct from pudendal neuralgia. A portion of levator ani muscle goes into spasm for a few minutes at a time causing pain or discomfort (22–25). The left side is more commonly affected. Muscle spasm can be both palpated and recorded by EMG, and may respond to botulinum toxin.

Radiculopathy and cauda-equina syndrome

These are covered elsewhere, but it is worth noting that even a single disc lesion in the lumbosacral vertebral column can produce protean manifestations from that level downwards. Tumours, infections, vascular abnormalities, trauma, and surgery can all affect this area, too.

Acute changes in bladder or bowel function are an emergency in the cauda-equina syndrome, but tend not to present to clinical neurophysiology. Patients with more chronic changes may present to urology or colo-rectal surgery and be referred to clinical neurophysiology for investigation of pelvic dysfunction. One of the most important jobs of the clinical neurophysiologist is to make sure that patients referred from outside neuroscience have had a proper clinical and electrophysiological assessment of the nervous system to exclude radiculopathy or cauda-equina syndrome.

The converse is also true. It is important, when assessing the many patients referred to clinical neurophysiology with lumbosacral radiculopathy, to enquire about bowel, bladder, and sexual function.

The sacral roots do not innervate paraspinal muscles so the approach used in the limbs to separate root from plexus do not apply here.

Multiple system atrophy

MSA is a degenerative disorder of the central nervous system, which can present with either Parkinsonian or cerebellar symptoms. In the former case it is one of several diagnoses in the 'Parkinson's Plus group'. Urinary incontinence can precede other symptoms to the extent that prostatectomy may be attempted and fail. EMG of the EAS can help distinguish patients with MSA from those with either idiopathic Parkinson's disease or other Parkinsonian disorders (2,12,26). The criticism is levelled against this method that the separation is incomplete, particularly with regards to progressive supranuclear palsy (PSP) and that it may also be normal in the presence of disease. While this is true, the position seems analogous to the separation of motor neurone disease from either radiculopathy or other motor syndromes.

Both urinary and anal sphincters are affected, but it is easier to study the EAS, even though the main symptom may be urological. In MSA, the motor units are particularly likely to show small, stable satellite potentials, which should be included in the determination of the motor unit duration (Fig. 17.7). A mean duration of over 10 ms or more than 20% of motor units having an individual duration greater than 10 ms is abnormal (3).

Fowler's syndrome

Young women with this syndrome present with urinary retention and in about 50% of cases with polycystic ovaries. It often defies easy diagnosis, to the extent that many patients are thought to have a psychiatric condition (27,28). Correct diagnosis not only relieves them of this stigma, but allows specific treatment with implanted sacral root stimulators.

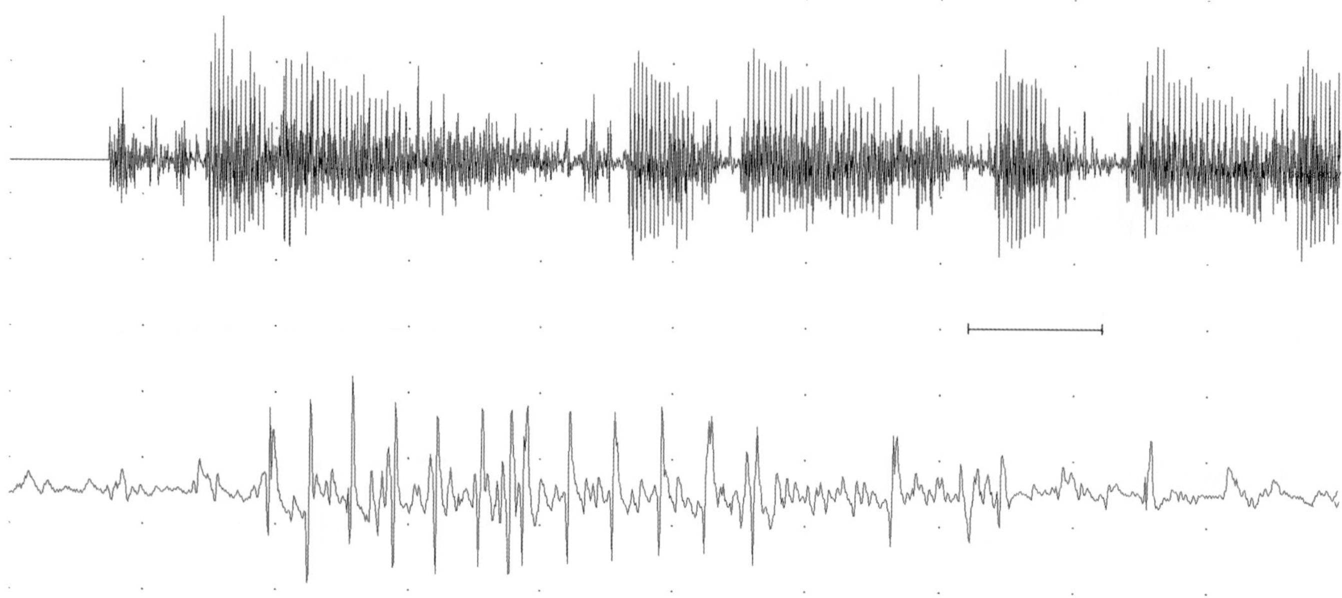

Fig. 17.8 'Whale noises'—decrescendo bursts in EMG recorded from urethral sphincter. The horizontal line between traces represents the portion of the upper trace which is expanded in the lower trace. Calibration dots represent 250 ms horizontally in the upper trace and 25 ms in the lower trace. The patient was a 21-year-old woman.

This does require urethral sphincter study. The abnormality sought is a series of decrescendo and decelerating myotonic bursts with a normal interference pattern. Some of these bursts may be normal, but there is an excess in Fowler's syndrome. In this context the sound of the myotonic bursts on the EMG amplifier has been likened to 'whale noises' (Fig. 17.8).

Imperforate anus

A technique involves searching the perineum to mark out the site of active sphincter muscle, to guide surgical correction in children. It has been replaced by imaging.

Constipation and incontinence in children

Eeg-Olofsson (29) describes neurophysiological investigation for these disorders in children. The methods are essentially those described here with some adaptations for performing them under anaesthesia.

Multiple sclerosis and other central nervous system conditions

The EMG and nerve conduction techniques described here are largely for the assessment of the peripheral nervous system. Spinal conditions affecting pudendal SSEPs are more likely to be discovered with the more familiar tibial SSEPs (30). Nevertheless, techniques such as these can be used to show dysfunction in the higher integration centres. For instance, in the condition called detrusor sphincter dyssnergia there is a failure to suppress the urinary sphincter activity during voiding, and EMG can be used to demonstrate this.

Post-herniorrhaphy and other inguinal pain

The genitofemoral, iliohypogastric and ilioinguinal nerves are branches of the lumbar plexus with root levels T12–L2. All three end up in the inguinal region. These nerves contribute to the sensory innervation of the lower abdominal wall, inguinal area, scrotum or labia, and anterior medial thigh. They overlap with the lateral femoral cutaneous nerve, the lower intercostal nerves, and branches of the pudendal nerve in a variable fashion.

Post-herniorrhaphy pain, which is apparently common in some series and can be debilitating, is occasionally referred for neurophysiological assessment. Conclusive evidence is lacking that the pain is neuropathic, i.e. due to nerve damage, as opposed to nociceptive, i.e. correct signalling by the nerves of pain due to some other cause. Some patients have sensory deficits in this area after surgery, but it can be asymptomatic. Other surgical procedures and non-surgical causes are known. Neurophysiological methods have been described for some, but are not widely used.

Conclusions

The clinical neurophysiological methods available for study of the pelvic floor parallel those used in the limbs, but only a few are in routine use. Most electromyographers could easily master examination of the external anal sphincter and either the bulbocavernosus reflex or pudendal SSEP technique. Assessment of the female urinary sphincter is only a little more difficult.

The clinical neurophysiologist who is armed with these techniques will be able to investigate many patients whose problems are currently unsolved. The cauda-equina syndrome, post-partum problems of continence, pudendal neuropathies, and possibly multiple system atrophy are the main problems seen. Patients have often been ignored or referred to different specialties for some time, so that helping them can be a most gratifying experience.

Acknowledgements

I am grateful to Clare Fowler for introducing these techniques to me, and to past and present members of my team at WSCN for their dedication to our patients.

References

1. http://www.visiblebody.com/
2. Carvalho, M., Schwartz, M.S., and Swash, M. (1995). Involvement of the external anal sphincter in amyotrophic lateral sclerosis. *Muscle & Nerve*, **18**(8), 848–53.
3. Palace, J., Chandiramani, V.A., and Fowler, C.J. (1997). Value of sphincter electromyography in the diagnosis of multiple system atrophy. *Muscle & Nerve*, **20**(11), 1396–403.
4. Bakas, P., Liapis, A., Karandreas, A., and Creatsas, G. (2001). Pudendal nerve terminal motor latency in women with genuine stress incontinence and prolapse. *Gynecologic and Obstetric Investigation*, **51**(3), 187–90.
5. Bradley, W.E., Lin, J.T., and Johnson, B. (1984). Measurement of the conduction velocity of the dorsal nerve of the penis. *Journal of Urology*, **131**(6), 1127–9.
6. Lin, J.T. and Bradley, W.E. (1985). Penile neuropathy in insulin-dependent diabetes mellitus. *Journal of Urology*, **133**(2), 213–15.
7. Vodusek, D.B., Ravnik-Oblak, M., and Oblak, C. (1993). Pudendal versus limb nerve electrophysiological abnormalities in diabetics with erectile dysfunction. *International Journal of Impotence Research*, **5**(1), 37–42.
8. Haldeman, S., Bradley, W.E., and Bhatia, N. (1982). Evoked responses from the pudendal nerve. *Journal of Urology*, **128**(5), 974–80.
9. Opsomer, R.J., Caramia, M.D., Zarola, F., Pesce, F., and Rossini, P.M. (1989). Neurophysiological evaluation of central-peripheral sensory and motor pudendal fibres. *Electroencephalography & Clinical Neurophysiology*, **74**(4), 260–70.
10. Vodusek, D.B., Janko, M., and Lokar, J. (1982). EMG, single fibre EMG and sacral reflexes in assessment of sacral nervous system lesions. *Journal of Neurology, Neurosurgery, and Psychiatry*, **45**(11), 1064–6.
11. Strijers, R.L., Felt-Bersma, R.J., Visser, S.L., and Meuwissen, S.G. (1989). Anal sphincter EMG in anorectal disorders. *Electromyography & Clinical Neurophysiology*, **29**(7–8), 405–8.
12. Pramstaller, P.P., Wenning, G.K., Smith, S.J., Beck, R.O., Quinn, N.P., and Fowler, C.J. (1995). Nerve conduction studies, skeletal muscle EMG, and sphincter EMG in multiple system atrophy. *Journal of Neurology, Neurosurgery, and Psychiatry*, **58**(5), 618–21.
13. Aanestad, O. and Flink, R. (1994). Interference pattern in perineal muscles. A quantitative electromyographic study in patients with faecal incontinence. *European Journal of Surgery: Acta Chirurgica*, **160**(2), 111–18.
14. Perotto, A. (2011). *Anatomical guide for the electromyographer: the limbs and trunk*, 5th edn. Springfield, IL: Charles C Thomas.
15. Kaneko, S. and Bradley, W.E. (1987). Penile electrodiagnosis. Value of bulbocavernosus reflex latency versus nerve conduction velocity of the dorsal nerve of the penis in diagnosis of diabetic impotence. *Journal of Urology*, **137**(5), 933–5.
16. Podnar, S. (2003). Electrodiagnosis of the anorectum: a review of techniques and clinical applications. *Techniques in Coloproctology*, **7**(2), 71–6.
17. Granata, G., Padua, L., Rossi, F., De Franco, P., Coraci, D., and Rossi, V. (2013). Electrophysiological study of the bulbocavernosus reflex: normative data. *Functional Neurology*, **28**(4), 293–5.
18. Torrisi, G., Sampugnaro, E.G., Pappalardo, E.M., D'Urso, E., Vecchio, M., and Mazza, A. (2007). Postpartum urinary stress incontinence: analysis of the associated risk factors and neurophysiological tests. *Minerva Ginecologica*, **59**(5), 491–8.
19. Pfeifer, J., Salanga, V.D., Agachan, F., Weiss, E.G., and Wexner, S.D. (1997). Variation in pudendal nerve terminal motor latency according to disease. *Diseases of the Colon and Rectum*, **40**(1), 79–83.
20. Valles-Antuna, C., Fernandez-Gomez, J., and Fernandez-Gonzalez, F. (2011). Peripheral neuropathy: an underdiagnosed cause of erectile dysfunction. *British Journal of Urology International*, **108**(11), 1855–9.
21. Sangwan, Y.P., Coller, J.A., Barrett, M.S., Murray, J.J., Roberts, P.L., and Schoetz, D.J., Jr. (1996). Unilateral pudendal neuropathy. Significance and implications. *Diseases of the Colon and Rectum*, **39**(3), 249–51.
22. Wright, R.R. (1969). The levator ani spasm syndrome. *American Journal of Proctology*, **20**(6), 447–51.
23. Lilius, H.G. and Valtonen, E.J. (1973). The levator ani spasm syndrome. A clinical analysis of 31 cases. *Annales chirurgiae et gynaecologiae Fenniae*, **62**(2), 93–7.
24. Ng, C.L. (2007). Levator ani syndrome—a case study and literature review. *Australian Family Physician*, **36**(6), 449–52.
25. Hull, M. and Corton, M.M. (2009). Evaluation of the levator ani and pelvic wall muscles in levator ani syndrome. *Urologic Nursing*, **29**(4), 225–31.
26. Cersosimo, M.G. and Benarroch, E.E. (2013). Central control of autonomic function and involvement in neurodegenerative disorders. *Handbook of Clinical Neurology*, **117**, 45–57.
27. Swinn, M.J. and Fowler, C.J. (2001). Isolated urinary retention in young women, or Fowler's syndrome. *Clinical Autonomic Research*, **11**(5), 309–11.
28. Kavia, R.B., Datta, S.N., Dasgupta, R., Elneil, S., and Fowler, C.J. (2006). Urinary retention in women: its causes and management. *British Journal of Urology International*, **97**(2), 281–7.
29. Eeg-Olofsson (2006). *Paediatric clinical neurophysiology.* London: Mac Keith Press
30. Podnar, S. and Vodusek, D.B. (2014). Place of perineal electrophysiologic testing in multiple sclerosis patients. *Annals of Physical and Rehabilitation Medicine*, **57**(5), 288–96.

SECTION 3

Clinical aspects: peripheral nervous system

CHAPTER 18

The clinical approach to neurophysiology

Kerry R. Mills

Introduction

This chapter is concerned with the conduct of the neurophysiological examination, stressing the importance of communication with patients and the subsequent reporting of results. Other important issues, such as definition of normal ranges and the safety of our procedures are also discussed.

Neurophysiological examination

Some neurophysiologists believe that the results of neurophysiological tests should be interpreted without knowledge of the clinical problem so as not to bias the results in any particular direction. Others, in contrast, think all information should be taken into account in reaching a conclusion. I will argue that the latter approach is preferable. Abnormalities encountered during neurophysiological testing must be interpreted within the clinical context; the significance of test results should be weighed with respect to all other available information. Indeed, abnormal test results can sometimes be ignored completely.

The above approach is predicated upon having a complete clinical picture, a situation that is often not the case (see Box 18.1). However, by focusing on the salient primary complaints of patients, whether this is related to a peripheral nerve problem or to a central problem such as epilepsy, the clinical neurophysiologist can provide a far more relevant and useful diagnostic service. The process of taking a focused history, apart from gleaning clinically important information, also serves the important purpose of establishing rapport with the patient and gives an opportunity to explain the nature of the tests to the patient. The focused history should begin with an analysis of the current primary symptom. For example, the patient suffering episodic loss of consciousness should be asked about precipitating factors, duration, orientation after the event, evidence of tongue biting, and, of course, the account of a witness would be very important. The patient with hand weakness should be asked about weakness elsewhere, sensory disturbance or pain, cramps, etc. Supplementary enquiries may be needed to delineate further the likely nature of symptoms. In some cases enquiry as to family or drug history may be needed as will a history of previous medical conditions. At the end of a relatively short interview, the experienced neurophysiologist will have a pretty good idea of the likely differential diagnosis and of the likely pathology.

Box 18.1 Clinical approach

- Introduce yourself to patient.
- Establish rapport and take focused history.
- Conduct a focused examination.
- Plan the investigation.
- Modify the investigation in light of results.
- Reach a conclusion.
- Tell the patient what will happen next.
- Make a report summarizing the findings.
- Make a conclusion.
- Send report to referring physician.

Clinical examination should be equally focused, aimed at further refinement of the differential: does the patient with hand muscle weakness have brisk or absent reflexes; does the patient with loss of consciousness have a cardiac arrhythmia; is the sensory loss in a nerve or root distribution; is the paralysed muscle wasted or not. In neurology, the history is useful in deciding the likely pathology and the examination gives information as to the likely location of the lesion.

Armed with this clinical information, the neurophysiologist is then able to plan the investigation. The aim is to maximize the collection of useful information with the minimum of testing. Similarly, the EEG technologist must be prepared to modify the routine protocol to investigate specific aspects related to the history. For example, if the patient reports loss of consciousness triggered by a specific factor, say, reading of text, it will be useful to try to replicate these conditions during the EEG recording.

Throughout the conduct of the examination the neurophysiologist should maintain rapport with the patient; nothing destroys a patient's trust more than a large electrical shock delivered without warning. Explanations given to the patient go a long way in maintaining rapport. For example, 'It's important for me to decide if your other hand is affected, even though you haven't noticed anything wrong, and so I'm going to do the same tests on the left'. These aspects of the examination are even more important in children (and their parents) and some useful hints are given in Chapter 25 At the end of the initial testing, the neurophysiologist should review the test results along with the clinical findings and decide if any more information is required; it is inefficient (not to say embarrassing) to have to call the patient back because some key test has been omitted.

The above discussion of the importance of communication with patients has relevance to the question of consent. It may appear obvious that if a patient has arrived in the neurophysiology department, then their consent to the procedures has been given. Many departments send out written information on neurophysiological testing before the patient arrives; this is often useful, but can be counter-productive. Often, euphemisms for the sensations of electrical shocks (taps, clicks, pulses, etc.) or needle insertion (scratches, pricks, like a blood test, etc.) will have been used. The patient may have spent hours looking at videos of electromyograph (EMG) or EEG examinations on the internet and may be thoroughly frightened. The principal to be followed is of explanation followed by the obtaining of verbal consent from the patient. At all times the neurophysiologist should give the patient the opportunity to withdraw from further examination. Formal written consent is, of course, required if a photograph or video is to be made for teaching purposes, especially if the patient could be identified.

Test results are usually compared with those from a normal population and abnormality is judged to be present if the test result deviates from the normal mean value by more than two standard deviations, assuming the parameter is normally distributed (see below). This means that 4% of normal individuals will be judged abnormal using these criteria. The problem is compounded, however, when, as is usually the case, multiple parameters are compared with their respective normal values. Thus if, say, 20 nerve conduction parameters are measured in a healthy individual, there is every possibility that one of them will fall outside the normal range. Looked at conversely, if a patient with a putative peripheral neuropathy has an abnormal sural sensory action potential (SAP), then it could be argued that this has occurred merely by chance. However, if the contralateral sural SAP is also small and even more so if upper limb SAPs are also abnormal, then the probability that the sural SAP is abnormal becomes greater.

Before the patient leaves, it is important to tell them what will happen next. A simple 'I will write to your doctor with the test results and you will hear from him in the near future' may be all that is required. Often, patients, quite rightly, will want to know if you have any conclusions; this needs to be managed with care. Clearly, in simple problems like nerve entrapments or diabetic neuropathy it may be appropriate to give the patient a full account including the likely outcome. For example, 'There is a nerve trapped at your wrist and you may well need a small operation to release it'. In other situations, where a prolonged discussion of the likely diagnosis is required, then it is inappropriate to undertake this in the EMG or EEG clinic. The patient with motor neuron disease, for example, deserves a full discussion with a neurologist skilled at giving bad news. An honest statement like 'Your doctor will need to get all his test results back and then he will be able to talk to you more fully about the problem' is a more appropriate approach in these circumstances. Another difficult problem that can arise is with the patient with functional weakness; again it is inappropriate to try to delve into the patient's psychopathology to discover why this is happening. Giving some sort of plausible explanation for the symptom risks perpetuating and solidifying the patient's symptoms. It is probably better to say 'the test results haven't shown any definite abnormality and your doctor may need to explore different avenues to try to get the limb working again'. Transcranial magnetic stimulation (see Chapter 14) can be used to demonstrate patency of the corticospinal tract and this can be useful to demonstrate to patients that 'the pathways are still working'. Very occasional patients experience almost biblical recovery of function after transcranial magnetic stimulation (TMS) (1). Except in specific circumstances, such as to check for progression in motor neuron disease, it is almost never justifiable to recall the patient after an interval to repeat the tests.

The test results, of course, may leave you completely baffled. In this situation, again honesty is required with, for example 'Although the test results have eliminated certain conditions, I'm afraid I don't know what the problem is yet and your doctor may well need to do some other types of tests'.

Once testing has finished and the patient has left the department, the next task is that of providing a report. Clearly, a discussion with the referring physician is preferable, but this is often not available and the report may be the only way of communicating results. The following comments, although illustrated by examples from the EMG clinic, are generalizable to all neurophysiology reports, including EEG and Evoked Potentials. The detail and tenor of the report may differ depending on to whom it is written. For a neurosurgeon or orthopaedic surgeon, a conclusion saying 'This patient has left carpal tunnel syndrome' may suffice. For a neurologist some justification for concluding statements is required: thus, for example, '... the ulnar compound muscle action potential (CMAP) amplitude drops by 50% between elbow and wrist implying conduction block'. Some abnormal findings may be dismissed: so, for example, '... the sural SAP is reduced in amplitude, but this is of no consequence in the clinical context'. The best approach is to provide a summary of the abnormal findings and then to provide a conclusion, always taking into account the clinical context. It goes without saying that the report should be made available to the referring physician as quickly as possible.

Normal values

The above discussion assumes that a decision can been made as to the normality or abnormality of test results. All parameters have a range of normality due to the spread of each parameter determined in healthy individuals and so statements about abnormality are essentially probabilities that the test value could have come from a normal individual. Thus, with a parameter distributed in a Gaussian fashion with a mean value of μ and a standard deviation s, the chance of our test value, x, coming from this normal population is estimated from $(x - \mu)/s$. This is referred to as the z-score and gives the number of standard deviations that our value deviates from the mean; it can readily be converted to a probability, but as a rule of thumb, if the z-score is >2 or <–2 there is a 95% probability that our value did not come from the normal population. Conduction velocity and distal motor latency are distributed in a Gaussian fashion, but amplitudes tend to have skewed distributions, often rendered Gaussian by logarithmic transformation. Tables of normal values are widely available for most nerve conduction (2) and (see Chapter 6), evoked potential (see Chapter 15), motor evoked potential (3) and EMG parameters (see Chapters 7 and 8). Normality of the routine EEG is rather more subjective and depends more on the presence or absence of various features rather than on quantitative measures.

The definition of normal ranges deserves some additional discussion since many factors come into play. Factors related to the particular equipment being used (electrode type and size and inter-electrode distance) amplification, filtering, and so on are under the control of the neurophysiologist. Temperature (nerve, muscle, or skin) may be modified by the neurophysiologist (4,5), whereas external factors such as height, age, gender, body mass index, cannot be controlled (6). The most important factors in relation to nerve conduction are age and temperature. Height is clearly important with respect to F-wave studies.

Because so many factors are involved, it important to compare test results with normal values collected in the same fashion. It is often recommended that neurophysiologists collect their own normal values in sufficient numbers to compare with the published values; at least this gives some confidence in decisions about normality of individual test results.

It is well known that temperature affects nerve conduction parameters. Conduction velocity is slower at lower temperatures whereas amplitudes tend to increase. There are two approaches to dealing with this problem: firstly, the limb can be warmed to a temperature of, say, 32°C; or secondly, temperature may be merely measured and a correction factor based on the linear relation of each parameter to temperature, applied. There are problems with both methods. Warming of the limb is time consuming, requires additional equipment (water baths, electric heating blankets, etc.), there is no consensus as to the target temperature and cooling begins as soon as testing starts. Also it should be born in mind that skin temperature, which is usually measured will not exactly reflect nerve or muscle temperature. Correction factors are often assumed to be the same in normal and diseased nerves; in fact there is some evidence that this is not so (7–10). A third and more pragmatic approach is to only measure temperature and attempt to correct it when the diagnosis critically depends on values of conduction velocity. For example, if a patient with cool extremities with a potential diagnosis of demyelinating neuropathy has a measured conduction velocity in the posterior tibial nerve of 28 m/s, then it would be difficult to decide if the slowing were merely an effect of temperature. Warming should be undertaken and the measurements repeated.

Nerve conduction parameters also change with age. In infants and children below the age of 4, conduction velocity is lower than in adults (see Chapters 6 and 25). Also the amplitude of, predominantly, sensory responses fall with age. In a study of patients over the age of 75 years, 14% of sural and 21% of superficial peroneal sensory responses were absent using the surface recording technique. An absent lower limb sensory potential therefore cannot, in isolation be taken as evidence of a peripheral neuropathy in this age group.

In clinical EEG, 'normality' depends more on the visual recognition of abnormal patterns, a skill that is gained by experience. The patterns of 'normality' in neonates, children, and during sleep also need to be recognized. Furthermore, there are a large number of 'normal variants'; the importance of recognizing these is that they can resemble patterns found in conditions such as epilepsy. Examples include slow or fast alpha rhythms, 6 and 14 Hz positive spikes, lambda and mu rhythms, posterior occipital sharp transients of sleep, benign epileptiform transients of sleep, sleep spindles, and K-complexes, and so on (see Chapters 11 and 34).

There are many methods of quantifying the EEG, but these have not found widespread use in routine work. In research or in specific clinical scenarios, quantitative EEG is increasingly used. For example, spectral analysis has been used in the objective classification of hepatic encephalopathy (11,12), in monitoring depth of anaesthesia (13) and in monitoring brain injury (14). More advanced techniques such as dipole source localization (15) and brain mapping have been used to refine the location of epileptic foci.

Safety of neurophysiological procedures

Neurophysiological procedures are on the whole very safe and there are only a few sporadic accounts of adverse effects. There are, however, a number of situations where particular vigilance is required. By far the commonest complication is vasovagal syncope in the patient (or on-looking relative). This is usually self-limiting and responds to laying the patient flat whilst monitoring pulse and blood pressure.

The risk of introducing infection from EMG needles or from needle EEG electrodes (or scarifying procedures to reduce skin impedance) has largely disappeared with the widespread use of disposable electrodes. It is not recommended that patients at risk for infective endocarditis are given antibiotic cover. It is prudent for electromyographers to wear gloves to mitigate against the risk of blood born infection caused by inadvertent needle stick injury.

Although taken as read that the electromyographer has a sound knowledge of neuroanatomy, puncture of several vessels or nerves has been described: for example, the radial artery or nerve when sampling flexor pollicis longus, the sciatic nerve when sampling gluteus maximus and the median artery or nerve when sampling pronator teres.

A more common adverse effect, which can be severe (16,17), is the patient on heparin or warfarin where it is usually impractical to withdraw the medication. The approach, if diagnostic information is vital, is to use the minimum number of insertions in superficial muscles using the smallest gauge needle with minimal exploration of the muscle. In general, it is better to avoid areas like the neck or the anterior tibial muscles, which could produce a compartment syndrome (18) or cause compression of vessels or nerves, and focus on a small number of large muscles, which can easily be manually compressed after the procedure. In patients with a hereditary or acquired bleeding diathesis, it is recommended that EMG is not performed if the platelet count is less than 50,000 or the prothrombin time is greater than 2 times normal (19), but this is a relative contraindication. Again, the physician has to weigh the need for diagnostic information against the risks of haematoma.

Pneumothorax has been reported after needle EMG of cervical and thoracic paraspinal muscles, intercostals, rhomboids, trapezius, latissimus dorsi, serratus anterior and supraspinatus (20,21).

The question of safety in relation to intravenous lines, implanted pacemakers, and cardioverter-defibrillators should also be addressed, not only in relation to electrical, but also magnetic stimulation (TMS). Despite the potential risks, it appears that electrical stimulation as employed in nerve conduction studies and somatosensory evoked potentials do not cause abnormal triggering of demand pacemakers or cardioverter-defibrillators (22) even when an intravenous line is in place in the same limb that is being stimulated (23). Nevertheless, it would be prudent to check with cardiologists that the particular device implanted in the patient does not pose a particular risk.

With respect to single pulse TMS, a number of issues need to be considered: induction of seizures, induction of current in metallic objects causing movement or heating, effects on hearing, cognitive changes, and so on. These have been exhaustively reviewed (24–27).

Any metallic object in the vicinity of the coil will have current induced in it that could cause it to move. Similarly, currents could be induced in pacemaker wires, deep brain stimulator leads, vagal nerve stimulators, cochlear implants, and so on. Implanted electrodes are currently regarded as absolute contraindications for TMS, but as commented by Rossi et al. (24) and consistent with the author's experience, '... clinical experience suggests that single-pulse TMS in patients with implanted electrodes, including those for vagal nerve stimulation, can be performed safely, provided that the coil is not discharging on the skin overlying the electrical device. The same applies for stimulators, infusion pumps or pace-makers.' It is also considered that single pulse TMS carries a very low risk in children (28). There are a few reports of seizures following single pulse TMS (24), but these are usually (29) in patients with either ongoing spontaneous seizures or a large cerebral haemorrhage.

Repetitive TMS (rTMS), currently being explored in a wide variety of conditions as a means of modulating cortical excitability, has the potential to induce seizures in otherwise healthy individuals, but adherence to published guidelines has meant very few untoward events have occurred [27,30,31].

References

1. Chastan, N. and Parain, D. (2010). Psychogenic paralysis and recovery after motor cortex transcranial magnetic stimulation. *Movement Disorders*, **25**(10), 1501–4.
2. Liveson, J.A. and Ma, D.M. (1992). *Laboratory reference for clinical neurophysiology*. Philadelphia, PA: F. A. Davis.
3. Mills, K.R. (1999). *Magnetic stimulation of the human nervous system*. Oxford: Oxford University Press.
4. Bolton, C.F., Sawa, G.M., and Carter, K. (1981). The effects of temperature on human compound action potentials. *Journal of Neurology, Neurosurgery, and Psychiatry*, **1944**(5), 407–13.
5. Moses, B., Nelson, R.M., Nelson, A.J., Jr., and Cheifetz, P. (2007). The relationship between skin temperature and neuronal characteristics in the median, ulnar and radial nerves of non-impaired individuals. *Electromyography and Clinical Neurophysiology*, **47**(7–8), 351–60.
6. Rivner, M.H., Swift, T.R., and Malik, K. (2001). Influence of age and height on nerve conduction. *Muscle & Nerve*, **24**(9), 1134–41.
7. Franssen, H., Notermans, N.C., and Wieneke, GH. (1999). The influence of temperature on nerve conduction in patients with chronic axonal polyneuropathy. *Clinical Neurophysiology*, **110**(5), 933–40.
8. Franssen, H. and Wieneke, G.H. (1994). Nerve conduction and temperature: necessary warming time. *Muscle & Nerve*, **17**(3), 336–44.
9. Franssen, H., Wieneke, G.H., and Wokke, J.H. (1999). The influence of temperature on conduction block. *Muscle & Nerve*, **22**(2), 166–73.
10. Notermans, N.C., Franssen, H., Wieneke, G.H., and Wokke, J.H. (1994). Temperature dependence of nerve conduction and EMG in neuropathy associated with gammopathy. *Muscle & Nerve*, **17**(5), 516–22.
11. Ciancio, A., Marchet, A., Saracco, G., et al. (2002). Spectral electroencephalogram analysis in hepatic encephalopathy and liver transplantation. *Liver transplantation*, **8**(7), 630–5.
12. Reeves, R.R., Struve, F.A., and Burke, R.S. (2006). Quantitative EEG analysis before and after liver transplantation. *Clinical EEG and Neuroscience*, **37**(1), 34–40.
13. Isley, M.R., Edmonds, H.L., Jr., Stecker, M., and the American Society of Neurophysiological. (2009). Guidelines for intraoperative neuromonitoring using raw (analog or digital waveforms) and quantitative electroencephalography: a position statement by the American Society of Neurophysiological Monitoring. *Journal of Clinical Monitoring and Computing*, **23**(6), 369–90.
14. Haneef Z., Levin H.S., Frost J.D., Jr, and Mizrahi, E.M. (2013). Electroencephalography and quantitative electroencephalography in mild traumatic brain injury. *Journal of Neurotrauma*, **30**(8), 653–6.
15. Plummer, C., Litewka, L., Farish, S., Harvey, A.S., and Cook, M.J. (2007). Clinical utility of current-generation dipole modelling of scalp EEG. *Clinical Neurophysiology*, **118**(11), 2344–61.
16. Butler, M.L. and Dewan, R.W. (1984). Subcutaneous hemorrhage in a patient receiving anticoagulant therapy: an unusual EMG complication. *Archives of Physical Medicine and Rehabilitation*, **65**(11), 733–4.
17. Caress, J.B., Rutkove, S.B., Carlin, M., Khoshbin, S., and Preston, D.C. (1996). Paraspinal muscle hematoma after electromyography. *Neurology*, **47**(1), 269–72.
18. Farrell, C.M., Rubin, D.I., and Haidukewych, G.J. (2003). Acute compartment syndrome of the leg following diagnostic electromyography. *Muscle & Nerve*, **27**(3), 374–7.
19. Al-Shekhlee, A., Shapiro, B.E., and Preston, D.C. (2003). Iatrogenic complications and risks of nerve conduction studies and needle electromyography. *Muscle & Nerve*, **27**(5), 517–26.
20. Honet, J.C. (1988). Pneumothorax and EMG. *Archives of Physical Medicine and Rehabilitation*, **69**(2), 149.
21. Miller, J. (1990). Pneumothorax. Complication of needle EMG of thoracic wall. *New Jersey Medicine*, **87**(8), 653.
22. Derejko, M., Derejko, P., Przybylski, A., et al. (2012). Safety of nerve conduction studies in patients with implantable cardioverter-defibrillators. *Clinical Neurophysiology*, **123**(1), 211–3.
23. Mellion, M.L., Buxton, A.E., Iyer, V., Almahameed, S., Lorvidhaya, P. and Gilchrist, J.M. (2010). Safety of nerve conduction studies in patients with peripheral intravenous lines. *Muscle & Nerve*, **42**(2), 189–91.
24. Rossi, S., Hallett, M., Rossini, P.M., and Pascual-Leone, A. (2009). Safety, ethical considerations, and application guidelines for the use of transcranial magnetic stimulation in clinical practice and research. *Clinical Neurophysiology*, **120**(12), 2008–39.
25. Kandler, R. (1990). Safety of transcranial magnetic stimulation. *Lancet*, **335**(8687), 469–70.
26. Agnew, W.F. and McCreery, D.B. (1987). Considerations for safety in the use of extracranial stimulation for motor evoked potentials. *Neurosurgery*, **20**(1), 143–7.
27. Machii, K., Cohen, D., Ramos-Estebanez, C., and Pascual-Leone, A. (2006). Safety of rTMS to non-motor cortical areas in healthy participants and patients. *Clinical Neurophysiology*, **117**(2), 455–71.
28. Gilbert, D.L., Garvey, M.A., Bansal, A.S., Lipps, T., Zhang, J., and Wassermann, E.M. (2004). Should transcranial magnetic stimulation research in children be considered minimal risk? *Clinical Neurophysiology*, **115**(8), 1730–9.
29. Kratz, O., Studer, P., Barth, W., et al. (2011). Seizure in a nonpredisposed individual induced by single-pulse transcranial magnetic stimulation. *Journal of ECT*, **27**(1), 48–50.
30. Pascual-Leone, A., Houser, C.M., Reese, K., et al. (1993). Safety of rapid-rate transcranial magnetic stimulation in normal volunteers. *Electroencephalography & Clinical Neurophysiology*, **89**(2), 120–30.
31. Wassermann, E.M. (1998). Risk and safety of repetitive transcranial magnetic stimulation: report and suggested guidelines from the International Workshop on the Safety of Repetitive Transcranial Magnetic Stimulation, June 5–7, 1996. *Electroencephalography & Clinical Neurophysiology*, **108**(1), 1–16.

CHAPTER 19

Focal neuropathies

Jeremy D. P. Bland

Introduction

This chapter covers what are often referred to as entrapment neuropathies. What the clinical syndromes under discussion have in common is that they are localized lesions of a single nerve, mostly thought to be of mechanical origin rather than due to auto-immune or vasculitic processes. I am also excluding from consideration major trauma resulting in nerve tears or lacerations, but this still leaves a wide range of pathological variants for consideration not all of which are necessarily entrapment in an anatomical tunnel, hence the choice of focal neuropathy as a generic term. They are worth considering as a group because the neurophysiological approach to them all is essentially similar and the major therapeutic decision to be made in most cases is whether surgical treatment is appropriate. Entire textbooks have been devoted to this group of disorders (1–4) and it is not the purpose of this chapter to enumerate a comprehensive catalogue of all the alleged tunnel syndromes. Some of the disorders are extremely common and well characterized, others are rare, but well documented, and a considerable number have been described only in case reports or short case series with only partial documentation. I will try to describe a general approach to the problem of a focal neuropathy and then illustrate with a few key examples.

What is the purpose of investigation?

It is sometimes useful to stop and think why one is carrying out a 'diagnostic test'. The term suggests that the answer is obvious, to make the diagnosis, but this simple answer deserves closer examination and is in any case not always the primary justification for testing. There are three main reasons for testing:

Diagnosis

Let us assume for the moment that the patient is in the EMG laboratory in order for us to contribute in some way to the process of diagnosis. Diagnosis is not primarily an end in itself. Although patients appreciate being given a name for their problem, the primary function of diagnosis is as a means of grouping together individuals with a medical problem where the result of subsequent actions, including doing nothing, can be predicted to some extent. Diagnosis is thus a guide to subsequent action. If the tests that we carry out have no influence on the subsequent choices made by the patient and doctor then we should question the value of doing them for diagnosis. In the context of focal neuropathies, the nerve involved will often be obvious to anyone with a good knowledge of neuromuscular anatomy and often the site of the lesion along the nerve can be worked out from clinical signs and symptoms with a good deal of precision. In such cases there may be little value in testing merely to show that for example, the median nerve is compromised at the carpal tunnel, but that does not necessarily mean that testing is entirely diagnostically redundant. Neurophysiological methods can reveal other clinically unsuspected problems, which may influence management, such as an underlying generalized polyneuropathy, and consultation with a practitioner who is familiar with peripheral nerve disease will often provide a valuable second opinion on the clinical presentation, which can reduce errors in clinical diagnosis. These important contributions can easily be overlooked where attention is purely focussed on confirming what is already clinically obvious.

There are of course many patients who present with symptoms for which the clinical diagnosis of a focal neuropathy is considered as a possibility in the differential, but in whom there is uncertainty about the site the lesion, or even about whether the problem is nerve related at all. Neurophysiological tests are frequently ordered in an attempt to alleviate the doctor's anxiety over this diagnostic uncertainty. In such circumstances careful thought should be given to the significant incidence of false negative and false positive results in neurophysiological tests for focal neuropathies when evaluating the results.

Prognosis

The clinical outcomes resulting from disease management decisions are generally uncertain, but amenable to statistical prediction. Prognosis is usually dependent on multiple interacting patient, disease, and treatment variables and determining which variables have a big enough effect to be worth considering in advising a patient of likely outcomes and therefore usually requires studies of large numbers of patients. A few of the focal neuropathy syndromes are common enough that such studies have provided good evidence of the prognostic utility of neurophysiological findings, and the neurophysiological results can be used to help guide treatment decisions. For rarer lesions the prognostic value of the neurophysiological results can only be imputed by analogy with the commoner disorders. It is, for instance, generally believed that outcomes are poorer where there is neurophysiological evidence of severe axonal loss. This is demonstrably true in carpal tunnel syndrome, but has not been proven in other focal neuropathies. Predictions made from the neurophysiological findings should be considered as well informed guesses.

Evaluation of progress

A significant proportion of neurophysiological tests for focal neuropathies are carried out, not for diagnosis of the original problem, but to quantify a change in nerve function over time either as a measure of treatment effect or to quantify rate of progression for prognostic purposes.

Both the conduct of the test and the way in which it is reported should be adapted to the nature of the underlying questions being asked.

Syndromic criteria for a focal neuropathy

For a clinical syndrome to be attributed to a focal neuropathy some general requirements can be set out which apply regardless of which nerve or site is being considered.

- The clinical manifestations must be plausibly attributable to a nerve lesion at the proposed site of the lesion on anatomical grounds. There is known to be some variation in sensory and motor innervation between individuals, but most such variants have been documented and should be known to the neurophysiologist. It should be remembered, however, that the symptom of pain often does not respect the boundaries which might be expected from peripheral nerve anatomy.
- There should be objective evidence of a localized abnormality of the nerve at the proposed site of the lesion. This may be:
  - *physiological*—segmentally impaired nerve conduction;
  - *anatomical*—structural change in the nerve or surrounding structures demonstrated on imaging or by direct visualization, though there are pitfalls to the latter;
  - preferably both.
- Treatment directed at the lesion should produce:
  - clinical improvement in the original symptoms and signs;
  - improvement in any impairment of nerve function, which was demonstrable before treatment;
  - improvement in any anatomical change, which was seen before treatment.

Any syndrome which meets all of these criteria can be considered a genuine clinical entity, even as a single case. Many of the conditions described in the literature meet few of them and before attributing a new presentation in a patient to one of these it is well to consider how many of them can be shown to apply.

Mechanisms of nerve injury

A variety of pathological processes can cause localized impairment of a peripheral nerve.

Entrapment (tunnel syndromes)

Several nerves pass through anatomical spaces with more or less rigid anatomical boundaries ranging from the completely unforgiving, the facial nerve passing through the bony facial canal, to the loosely defined and distensible, the median nerve passing between the two heads of pronator teres in the upper forearm. Increased pressure in these confined spaces can impair nerve function, probably at least initially by impairing the microvascular circulation of the nerve itself.

Chronic trauma (stretch/compression/abrasion)

Nerves may be exposed to mechanical stresses imposed by everyday life by virtue of their anatomical location. The classical example of this is the ulnar nerve at the elbow discussed below. Extreme examples of such mechanical stresses can produce acute nerve injury as in 'Saturday night palsy', external compression of the radial nerve.

Other local pathologies

Although these two mechanisms account for most of what are generally thought of as focal neuropathy syndromes it should be remembered that other local processes can also produce or predispose to focal nerve lesions including:

- Inflammatory disorders (focal chronic inflammatory demyelinating polyneuropathy).
- Tumour and other space occupying lesions:
  - primary nerve tumours;
  - compression by mass external to the nerve;
  - mass effect within an anatomical tunnel.
- Ischaemia (nerve infarction).
- Invasion/infiltration (lymphomas).
- Iatrogenic injuries, especially injection injuries and complications of modern anaesthetic techniques.
- Irradiation.
- Thermal injury.

These possibilities should be considered, especially when focal nerve impairment occurs at less common or novel sites.

Anatomical and physiological changes

Depending on the mechanism of nerve injury there may be a variety of focal structural and physiological changes in the nerve at the site of injury and also 'downstream' effects distal to the lesion. Pathologically the principal local effects may include interruption of axons, localized demyelination and changes to the interstitial tissues of nerves including oedema and fibrosis. These pathological effects are reflected in anatomical and physiological phenomena, which can be demonstrated in vivo using imaging and neurophysiological techniques.

Imaging

Ultrasound imaging and MRI scanning can both produce high resolution images of peripheral nerve good enough to be useful for diagnostic purposes. Ultrasound currently has the advantage of low cost, higher in-axis resolution, and the ability to visualize nerves and surrounding structures in movement in real-time. Ultrasound scanners are now inexpensive and portable enough for imaging studies to be performed at the same time as electrophysiological tests by the same operator. MRI can detect signal change within nerve resulting from subtle changes in tissue as well as delineating nerve structure, but is less readily available and more costly. Both technologies are evolving rapidly and techniques such as assessment of tissue vascularity using Doppler ultrasound (5) and measurement of tissue stiffness (elastography) (6) are being actively explored.

At present the most widely documented imaging change seen in focal neuropathies is local enlargement of nerve at the site of the lesion. This appears to be a common response to most forms of nerve injury. Where such enlargement is constrained within an

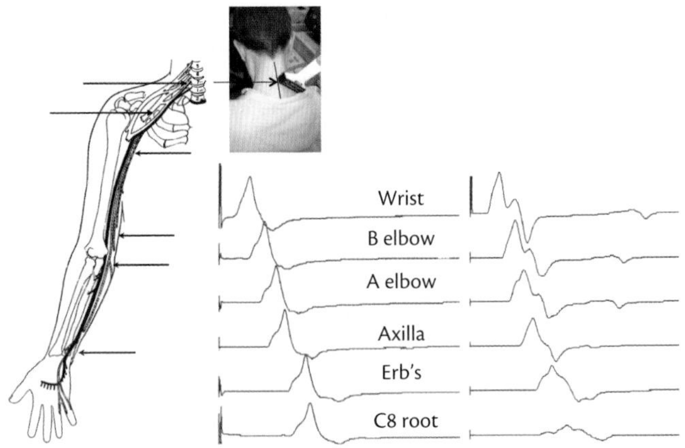

Fig. 19.1 Conduction block in the left ulnar nerve between Erb's point and the C8 root.

anatomical tunnel the greatest swelling may be seen just proximal and distal to the site of entrapment rather than at the site of maximal compression. Caution should be used in interpreting measures of nerve size, however, as a variety of generalized neuropathies can also exhibit either diffuse or multifocal enlargement of nerve (7).

Physiology

The principal physiological markers of focal nerve dysfunction result from demyelination and axonal interruption. Demyelination produces localized slowing of nerve conduction velocity in myelinated sensory and motor axons and the phenomenon of conduction block at the site of the lesion where propagation of axon potentials across the site of the lesion fails even when the axons themselves remain in physical continuity. Conduction block is distinguished from physical interruption of axons by the fact that the distal segment of nerve continues to conduct normally and by the absence of electromyographic evidence of denervation in muscles innervated by the affected nerve distal to the lesion.

Conduction block in some intact axons can co-exist with physical interruption of others at the same lesion. Conduction block can be recognized by an abrupt drop in the amplitude of a surface motor potential on moving the stimulation site proximal to the lesion (Fig. 19.1).

Neurophysiological demonstration of a focal lesion

The mainstay of electrophysiological assessment of focal neuropathies is the measurement of nerve conduction velocity in the affected segment of nerve and its comparison with adjacent segments and other nerves in order to demonstrate that an individual nerve segment is physiologically compromised. This requires that stimulus and recording points either side of the lesion are anatomically accessible and that other nerve segments can also be measured for comparison purposes. Conduction in sensory fibres can be assessed by recording nerve action potentials from purely sensory branches where these are available, motor conduction velocity is assessed by recording muscle action potentials resulting from nerve stimulation at multiple sites. At some sites nerve conduction velocities are measured by stimulating both motor and sensory fibres and recording a mixed nerve action potential. It is widely thought that sensory axons are more susceptible to pressure related demyelination than motor axons, but studies of the lumbrical/interosseous distal motor latency comparison in carpal tunnel syndrome suggest that this may not be strictly true (8). It should also be remembered that conventional nerve conduction velocity studies are measures only of the behaviour of the largest, and fastest conducting, axons. Although it appears to be the case that, in many focal neuropathies, these are amongst the earlier fibres to show physiological abnormalities this is not necessarily true for all types of pathology and all cases and it is often not changes in the function of these large myelinated fibres, which are responsible for most of the patient's symptoms.

The usual measurement of nerve conduction velocity across a lesion involves the assessment of a segment of nerve some 5–10cm long within which the pathology is thought to lie. A more systematic approach to locating a lesion is the technique of 'inching' in which an attempt is made to measure conduction velocity in multiple, short (usually 1–2 cm), adjacent nerve segments of equal length proximal to, across and distal to the suspected lesion. As with ordinary nerve conduction studies both motor and sensory/mixed studies can be devised. To some extent both standard studies and inching methods are limited by the accuracy with which the physical distances involved can be measured (see 'Neurophysiological pitfalls').

The amplitudes of surface recorded nerve and muscle action potentials also convey useful information, but tend to have much wider normal ranges than nerve conduction velocity measurements and are more susceptible to variations produced by anatomical and technical factors such as the distance between the signal generator and recording electrode, the impedance of the electrode to patient connection, and the placement of reference electrodes. Provided these technical constraints are kept in mind then reduction in the amplitude of potentials indicates either axonal loss or conduction block. Amplitude measurements more often contribute to the assessment of the severity of a lesion than to its detection.

Needle EMG sampling, looking for evidence of denervation, is generally of lesser importance than nerve conduction measurements, but can be useful in some specific circumstances. In very severe lesions where no nerve conduction measurements can be obtained the presence of EMG evidence of denervation in muscles not innervated by the suspect nerve may be the only clue to another lesion, as when denervation is seen in tibialis posterior suggesting an L5 radiculopathy or more proximal sciatic lesion in a patient with a suspected common peroneal nerve lesion at the fibular head. The absence of denervation potentials in a muscle from which no surface motor potential is obtained can also be a clue to the presence of conduction block provided that sampling is being undertaken at an appropriate interval after the onset of symptoms.

Neurophysiological pitfalls

When carrying out electrophysiological investigations of focal neuropathies several technical considerations are especially important.

Measures of nerve conduction velocity depend for accuracy on knowing the distance along the nerve between the stimulus and recording points or between two stimulation points. Although the latency of recorded signals can be measured to sub-millisecond precision, the accuracy of distance measurement is compromised by several factors with different relative importance for different nerves and lesion sites:

Mobile nerves

Nerves have to adapt to movement of the body and limbs over a wide range of anatomical positions. In order to do this they can stretch, slide longitudinally in their beds, corrugate when in a greatly shortened position and change position entirely in response to joint movements, as for example when the ulnar nerve subluxes to a position anterior to the medial epicondyle on elbow flexion in some patients. As a result of these movements the relationship of surface stimulation points to underlying nerve is both uncertain and inconstant if the limb is moved during testing, so that the distance measured on the skin surface may differ unpredictably from the actual distance along the underlying nerve.

Non-linear nerve paths

Some important nerves do not follow straight lines through sites where they may be compromised, good examples being the ulnar nerve at the elbow and the common peroneal nerve around the fibular head. Although one can try to approximate the course of the nerve when measuring on the skin surface such efforts are doomed to introduce additional uncertainty in distance measurement.

Stimulus intensity

Careful attention should be paid to stimulus intensity when carrying out nerve conduction studies for focal neuropathies. Two principal problems occur. Firstly, although we may think that we are exciting the nerve at the surface position of the stimulating cathode, high stimulus intensities may in fact trigger action potentials at sites some way along the nerve introducing yet more error in distance measurements. This may especially be a problem when the segment of nerve directly under the stimulator is demyelinated and thus relatively resistant to stimulation at the site. Secondly, stimulation may spread so far that completely different nerves are stimulated, a particular problem when the ulnar or radial nerves are stimulated at the wrist when trying to stimulate the median nerve.

These difficulties in making accurate measurements of distances along nerves are especially relevant in the assessment of focal neuropathies because their effect on the overall accuracy of a nerve conduction velocity measurement is inversely proportional to the length of the nerve segment being studied. In assessing a focal lesion we would prefer to measure conduction velocity only in the affected nerve segment rather than in a larger segment, which includes lengths of normal nerve proximal and distal to the lesion because our measurement shows only the average conduction velocity of the entire segment measured, and if it is slow over only a small proportion of that segment then the average velocity may be within the normal range and the presence of a lesion will be missed. However as we make the measured segment shorter the accuracy of the velocity measurement itself becomes progressively more compromised by the distance measurement errors, again reducing our ability to distinguish an impaired segment from a normally conducting one.

Anatomical variations

Not every patient has neuroanatomy, which follows the textbook illustrations. Examples include the Martin-Gruber anastomosis between the median and ulnar nerves in the forearm, which can lead to mistaken assessments of conduction block in the forearm, ulnar nerve innervation of the muscles of the thenar eminence in the hand, and recurrent motor branches of the median nerve, which do not pass through the carpal tunnel. The examiner should have a good knowledge of the commonly documented anomalies and always be prepared to carry out additional studies to try and elucidate whether an anatomical variant could explain an unusual finding. Most of these variations can be worked out by studying multiple additional stimulus and recording points.

Obesity, oedema, skin lesions, dressings, and casts

Nerve conduction studies for focal lesions are often requested in patients in whom these physical factors either altogether preclude any valid measurement or else significantly reduce the achievable accuracy. Such factors should be documented in reports and the diagnostic conclusions drawn should be cautious and qualified when measurements are compromised in this way.

Specific example syndromes

Carpal tunnel syndrome

As the best known, commonest, and probably the first described tunnel syndrome—a case was reported by Paget in 1854, followed by the description of ulnar neuropathy at the elbow by Panas in 1878—carpal tunnel syndrome is familiar to most doctors and well known to many members of the public. Despite some 80 years of intensive study since the term was coined in 1939, however, there remain considerable gaps in our understanding of many aspects of the condition and much misinformation in standard textbooks, in the ordinary press, and on the internet.

Strictly the term carpal tunnel syndrome applies to the clinical presentation as a whole and the condition is defined as a constellation of symptoms, signs, and laboratory findings, none of which individually define the diagnosis. Various diagnostic criteria have been suggested for use in different circumstances, which rely to varying extents on, for example, the presence of sensory symptoms in median nerve territory, physical signs such as Phalen's test (reproduction of the symptoms by non-forced flexion of the wrist to 90° for 1 min (9)) or electrodiagnostic or imaging abnormalities. In the absence of a gold standard definition for the disorder it is impossible to determine accurately the sensitivity and specificity of any individual feature of the history, examination, or investigations in making the diagnosis.

Pathology

The carpal tunnel is clearly defined anatomically by the bones of the carpus and the transverse carpal ligament and pressure studies have clearly shown that patients with the clinical syndrome of CTS have higher tissue pressures within the carpal tunnel than normal control subjects, sufficient to impair the microvascular circulation of the nerve (10,11). There is thus an entirely plausible physical explanation for suggesting that this is truly an example of injury to a nerve occurring purely because it passes through a confined space in which it can be subjected to pressures sufficient to cause injury. It is therefore the archetypical tunnel syndrome.

Clinical presentation

The classical early presentation of carpal tunnel syndrome is with intermittent paraesthesiae, which most characteristically awaken the patient from sleep in the middle of the night, reported by 77% of patients with CTS confirmed on nerve conduction studies (12). They should be felt in the anatomical territory of the median nerve distal to the wrist, which technically means the thumb, index finger, middle finger, and the adjacent side of the ring finger, but not the palm of the hand for which sensation is supplied by the palmar branch of the median nerve, which passes outside the carpal tunnel.

Although this presentation may be 'classical' and anatomically correct, it is in fact only the minority of patients who eventually have their symptoms attributed to carpal tunnel syndrome who exactly match this distribution of symptoms. Spread of symptoms outside the median nerve territory is common, either because patients are not good at discriminating fine anatomical detail in the middle of the night or because some forms of visceral sensation arising from the tissues of the carpal tunnel are poorly localized at cortical level, unlike cutaneous sensation. It is true, however, that symptoms, which are felt only outside the median nerve territory are usually a pointer to a completely different diagnosis.

Pain in carpal tunnel syndrome is extremely variable. The typical paraesthesiae often have a peculiarly unpleasant subjective quality and patients may confuse the terminology of tingling and pain making it difficult to understand exactly what they are describing. True pain can occur at any time in the course of the evolution of the disease, may be felt almost anywhere in the upper limb, ranges in severity from mild to excruciating, or can be entirely absent.

Loss of sensation in median innervated fingers and wasting and weakness of the thenar muscles, of which the most reliably median innervated is abductor pollicis brevis are signs of relatively advanced CTS and ideally the condition should be recognized and treated before they develop as they often do not recover fully even with surgery.

Many provocative physical tests have been suggested of which the most widely used are Phalen's test, described above, and Tinel's sign, the elicitation of paraesthesiae by percussion over a regenerating neuroma as originally described, though now the term is often applied simply to the elicitation of paraesthesiae by percussion over any nerve whatever the pathology. As with all other diagnostic features and tests much effort has been expended on futile attempts to determine their sensitivity and specificity in CTS. Systematic review of these studies suggest that Phalen's test is better than Tinel's sign in CTS, but neither is by any means definitive (13,14).

Although many medical disorders have been described as predisposing to, or associated with, CTS and should be considered as possible therapeutic targets in preference to treating the CTS directly when present, there is little evidence to suggest that screening tests for conditions, which are not clinically apparent are cost effective in patients presenting with typical CTS (15).

Aims of neurophysiological testing

When presented with a patient with possible CTS the neurophysiologist should first consider how certain that diagnosis is before testing. If the diagnosis appears highly probable on other grounds then even perfectly normal nerve conduction results will not greatly reduce the overall probability that it is correct (16) and one should not persist in performing more and more measurements of median nerve function until one can be coaxed into appearing 'abnormal'. Instead testing should establish two things:

How bad is the objective median nerve dysfunction?

This can range physiologically from no detectable impairment to no detectable function in the nerve. Although it has been argued that nerve conduction study reports should give no indication of the severity of disease (17), one of the main advantages of carrying out physiological measurements of this type is precisely that the degree of nerve dysfunction can be approached in a quantitative rather than qualitative manner. Furthermore, the degree of pre-treatment physiological impairment of nerve function is correlated with the clinical outcome of treatment (18), and is therefore useful in selecting treatment and guiding prognosis. Although individual measurements of sensory and motor conduction velocity and amplitudes of nerve and muscle action potentials are quantitative measures individually they are too complex for practical use by non-specialists in making decisions about treatment and several methods for describing overall neurophysiological severity of CTS have been suggested, notably by Stevens (19), Padua (20), and Bland (21). The system we use is designed so that it can be applied largely independently of the differing normal values of individual laboratories and shows a clear relationship with surgical prognosis (Fig. 19.2).

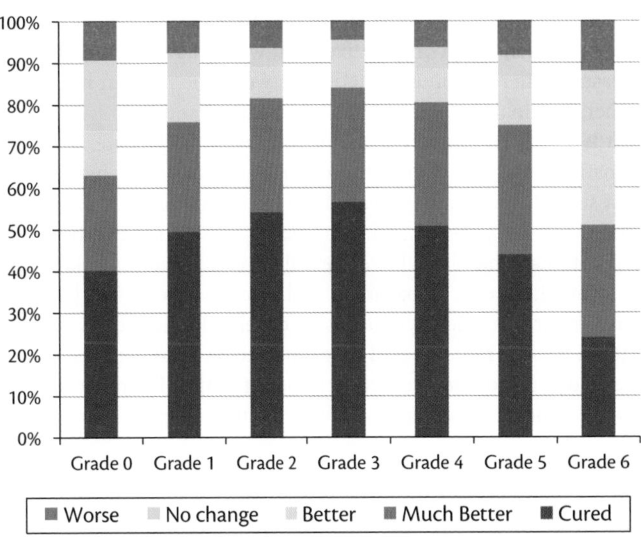

Fig. 19.2 Surgical outcome in 7136 carpal tunnel decompressions.

Is there any other underlying nerve problem?

Carpal tunnel syndrome may be the initial presenting feature of a more widespread polyneuropathy such as that of amyloidosis, or the presence of other nerve problems such as an ulnar lesion at the elbow, a radiculopathy, or an incidental generalized neuropathy may prejudice the outcome of treatment for the CTS. It is therefore important to detect these other pathologies.

When the diagnosis is unlikely on clinical grounds then the neurophysiologist should remember that false positive results on nerve conduction studies become commoner the more tests are performed (22), not concentrate too much on median nerve testing, and ensure that other neurological causes, which might explain the symptoms are tested for when possible. Only when the clinical presentation lies between these extremes, and CTS is a possible, but not obvious contributor to the symptoms, do the nerve conduction studies have an important diagnostic role.

How to test for CTS

Detailed standards for testing for CTS are available (23) and testing should always include:

1. Studies of median sensory conduction between a median innervated digit and the wrist.
2. A study of median motor conduction from wrist to a median innervated muscle, usually abductor pollicis brevis.
3. Studies of enough other nerves to demonstrate that any local median nerve abnormality at the carpal tunnel is not simply part of a wider disorder.

Where (1) and (2) are normal then testing should always include one of the sensitive comparative tests, which have been devised to measure difference in conduction between the median nerve and either the ulnar or radial nerves of the same hand.

It is worth remembering that CTS is eventually bilateral in 80% of patients and that even if the 'other' hand is asymptomatic when first seen the patient has a very high likelihood of returning to the clinic with the second hand some years subsequently. Testing such asymptomatic hands will often reveal nerve conduction evidence of mild median nerve impairment and provides useful baseline data for comparison when symptoms do develop in the second hand. It should also be remembered that ulnar nerve lesions at the elbows are also common and sometimes coexist with CTS making use of the ulnar nerve for comparison with the median problematic.

Imaging

The median nerve at the wrist is very easily visualized with modern ultrasound scanners (Figs 19.3 and 19.4) and a wide variety of measurements of the dimensions of the nerve itself and of surrounding structures can be made as well as assessments of nerve mobility, tissue vascularity, and tissue stiffness.

Evidence-based guidelines have been developed for the use of ultrasound imaging in the assessment of CTS (24). The most extensively studied parameter is the cross sectional area of the median nerve just proximal to the carpal tunnel, either in absolute terms or in comparison to a more proximal section of the nerve. Upper limits of the normal range for the absolute size of the nerve vary from 8.5 to 14.5 square mm. The exact reasons for such variation are not yet known, but probably include differences in the scanners themselves, the transducer frequencies, measurement methods, and operator variables. At present, as with nerve conduction studies, those wishing to use ultrasound imaging as a diagnostic tool for CTS are well advised to determine their own normal ranges for commonly imaged nerves using the scanner and transducers, which they intend to use for diagnostic work. Estimates of the diagnostic performance of ultrasound parameters for CTS range from 43–100% for sensitivity and 55–100% for specificity (25–28).

Imaging studies can show anatomical anomalies, which can either help to account for the occurrence of CTS in a particular patient or which may have relevance for choosing a surgical approach to the problem. Among these findings are unusually proximal branching of the median nerve, persistent median arteries, intracarpal ganglion cysts, anomalous muscle bellies passing through the tunnel or invading it with wrist and finger movement, and median nerve tumours and malformations (Fig. 19.5).

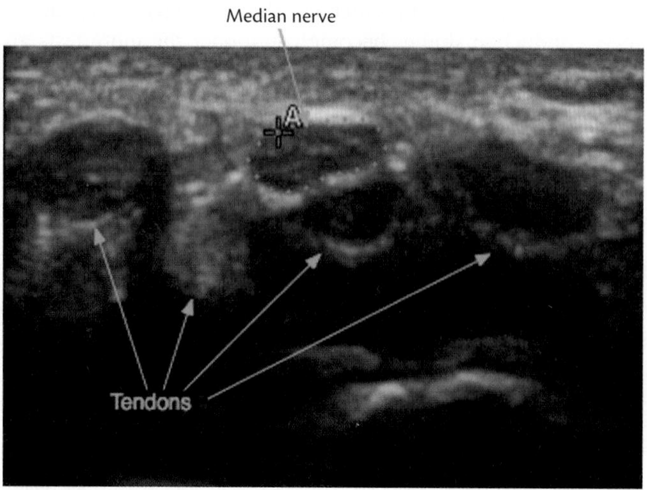

Fig. 19.3 Ultrasound image of normal median nerve in transverse section, cross sectional area of nerve = 9 mm^2.

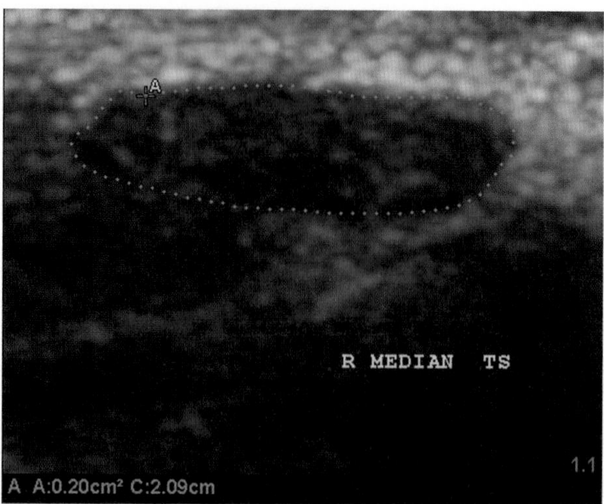

Fig. 19.4 Enlarged median nerve in carpal tunnel syndrome. Cross-sectional area = 20 mm^2.

These findings are analogous to the ability of nerve conduction studies to detect problems such as an underlying neuropathy and full assessment of a case of carpal tunnel syndrome should ideally include both modalities.

Post-treatment testing

A not uncommon clinical problem is the patient who has been treated for CTS, but is no better, or even worse, clinically after treatment. Nerve conduction testing in these patients can be confusing, especially when no pre-treatment results are available, but can sometimes contribute to further clinical management. In the case of failed surgery for carpal tunnel syndrome the main decision to be made is whether to embark on more surgery. This is appropriate in two circumstances, firstly when the surgeon has directly injured a nerve during the operation, and secondly when the attempt to section the transverse carpal ligament has been unsuccessful. Partial division of the ligament is surprisingly common, accounting for about half of all unsatisfactory clinical outcomes in one large study (29). Both iatrogenic nerve injury and incomplete division of the ligament generally manifest as deterioration in nerve conduction study results compared with pre-operative tests. However full release of the ligament without any inadvertent injury, even when wholly clinically successful, is frequently not accompanied by return of either neurophysiological or imaging measurements of the median nerve to the normal range. Thus, when faced only with a moderately abnormal median nerve after failed surgery, it is impossible to be sure whether this represents deterioration from a milder abnormality before surgery, and thus an indication for re-exploration, or improvement from a more severely affected state, in which case it should be left alone and given a chance to recover as improvement from severe CTS after surgery is often slow.

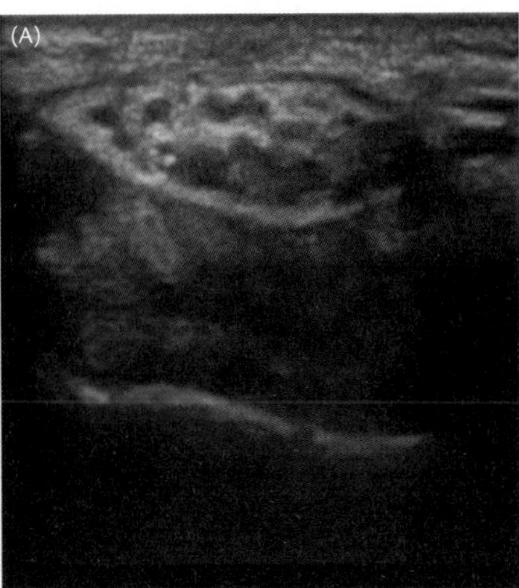

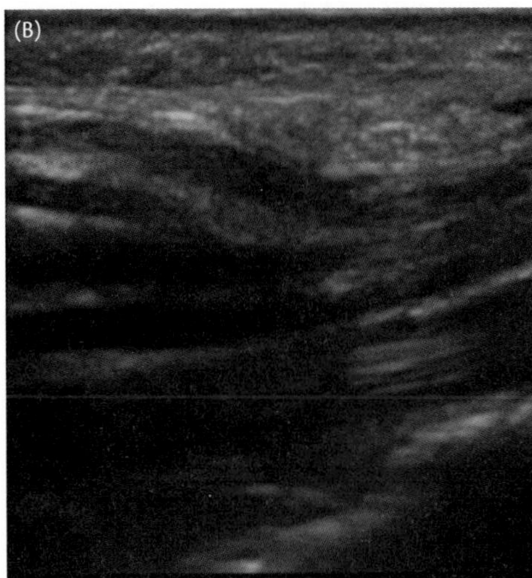

Fig. 19.5 Fibrolipomatous hamartoma of the median nerve (A) transverse view (B) longitudinal view.

The best known iatrogenic nerve injury during carpal tunnel decompression is to the recurrent motor branch of the median nerve leading to weakness and wasting of abductor pollicis brevis. This is usually recognizable clinically by the fact that the sensory symptoms of CTS have improved as expected while thumb abduction has become weaker, and can be confirmed neurophysiologically by demonstrating intact and/or improved motor conduction to the lumbrical muscles and EMG evidence of denervation in APB soon after surgery. Lacerations to the main median nerve trunk and to the palmar and digital branches of the median nerve are also documented (29).

The other two major reasons for surgical failure are extremely severe CTS and incorrect initial diagnosis. Patients with no recordable sensory or motor conduction in the median nerve, fixed sensory loss and marked thenar wasting may have little or no improvement even with correctly performed surgery though many will report some benefit, especially if they had troublesome pain and paraesthesiae before surgery (30). Post-operatively their nerve conduction studies may be improved, but seldom return to normal. Patients in whom the original symptoms were not due to CTS will mostly have had normal nerve conduction results before and after surgery, but some will have had mild median nerve conduction abnormalities, which, when present, will generally have contributed to the decision to operate and if the operation has been done correctly then these abnormalities may well have improved after surgery while symptoms remain unchanged.

Ulnar neuropathy at the elbow

Focal impairment of ulnar nerve function in the region of the elbow is the second commonest localized neuropathy, but is only 1/10th as common as carpal tunnel syndrome, is much more heterogeneous pathologically and clinically, and even less predictable in its clinical course and response to treatment. The incidence has been estimated at approximately 25 per 100,000 person years in Italy (31). Furthermore, the anatomical arrangement of the nerve at the elbow renders the ulnar nerve technically harder to test reliably using neurophysiological methods.

In contrast to the general agreement about the exact site of damage to the median nerve in carpal tunnel syndrome, under the transverse carpal ligament, a variety of anatomical explanations are invoked for ulnar nerve impairment at the elbow with four in particular being generally accepted. From proximal to distal these are:

- *The medial intramuscular septum*. This fibrous layer separates the triceps muscle mass of the upper arm from the biceps and brachialis compartment anteriorly on the medial side of the humerus. It is perforated by the ulnar nerve and thus forms a point at which the nerve is relatively fixed in place and potentially compressed.
- *The condylar groove* between the olecranon process and medial epicondyle of the humerus.
- *The cubital tunnel*—formed by the two heads of flexor carpi ulnaris and the aponeurosis, which joins them superficially and the medial ligaments of the elbow beneath.
- *The exit point* of the nerve from the flexor carpi ulnaris muscle.

Very localized impairments of ulnar nerve conduction at these sites have been demonstrated using intra-operative nerve conduction studies, but the accuracy with which they can be identified using surface studies, even inching techniques, is uncertain (32) and ulnar nerve impairment in any given individual is not necessarily confined to only one of them. It is likely that the mechanical stresses on the nerve at these points are much more varied than the high tissue pressure which is found in the carpal tunnel, consisting of external pressure related to elbow position in daily activities, direct pressure from some of the structures listed above, and stretching of the nerve related to limb position. Nerve conduction studies may fail to demonstrate any abnormality at all in some patients with quite clear ulnar nerve signs and symptoms. Estimates of the sensitivity of NCS for ulnar nerve lesions at the elbow range from 37–86% and as with carpal tunnel syndrome guidelines are available to provide a recommended strategy for testing (33).

Prognosis

Far less work has been done on predicting the response of ulnar neuropathies to surgical or conservative management, but outcomes have been shown to be better in patients whose nerve conduction studies show marked local slowing or conduction block at the elbow and worse in patients with larger nerve diameters on ultrasound imaging (34). No systematic studies are available to help guide the interpretation of nerve conduction studies at the elbow after unsuccessful surgery.

Imaging

Magnetic resonance imaging (MRI) and ultrasound imaging of the ulnar nerve show considerable promise in the diagnostic and prognostic evaluation of patients with ulnar nerve impairment (34,35), but seem unlikely to entirely replace nerve conduction studies as a way of quantifying the physiological impairment of nerve function.

Peroneal neuropathy

Acute or chronic compressive injury to the common peroneal nerve as it winds around the head of the fibula, resulting from prolonged kneeling or habitual sitting with one leg crossed over the other, is generally familiar as perhaps the third most commonly seen focal neuropathy, presenting as a foot drop with a variable sensory disturbance. It should be noted that a similar presentation can occur with a more proximal sciatic nerve lesion in which the peroneal nerve component of the sciatic nerve seems to be more vulnerable to injury in the region of the hip. This is a rare, but well documented event during hip replacement with an incidence of 0.17% in a series of 27,004 operations (36). At the knee however, this nerve provides a good illustration of how a strikingly unusual lesion can lead to a focal neuropathy. The intraperoneal nerve ganglion (Fig. 19.6) is a cyst filled with synovial fluid exactly like those commonly seen around the wrist. However, the weak point in the joint capsule of the tibiofibular joint from which these arise is the point at which the nerve branch from the peroneal nerve to the joint capsule enters it. As a result the cyst tracks up this nerve branch and enters the substance of the common peroneal nerve, sometimes extending all the way up to the sciatic bifurcation and back down the tibial nerve (37). These lesions produce a gradually worsening foot drop with sensory disturbance and sometimes pain and should be suspected in patients who develop a peroneal palsy in the absence of any history of a postural cause. They produce localized impairment of peroneal motor conduction in the region of the fibular head and popliteal fossa and are easily identified on ultrasound imaging

They are very amenable to surgical treatment though it is essential to sever the connection with the tibiofibular joint, sacrificing the nerve branch, or else they recur.

Tarsal Tunnel Syndrome—an example of a debatable syndrome

The general success of the diagnosis of carpal tunnel syndrome as an explanation for median nerve impairment at the wrist leading to successful surgical treatment led logically to thoughts that a similar situation might be found in the foot and the term tarsal-tunnel syndrome was coined by two authors in 1962 (38,39). The anatomical analogue of the median nerve in the wrist at the ankle is the posterior tibial nerve as it passes behind and under the medial malleolus. As with the transverse carpal ligament at the wrist overlying the median nerve the posterior tibial nerve at this point does lie under a recognizable fibrous band, the laciniate ligament, also referred to as the internal angular ligament or the flexor retinaculum of the foot, which connects the medial malleolus to the calcaneum. Unlike the flexor retinaculum in the hand however, the laciniate ligament is not required structurally to resist the sort of forces applied to the flexor retinaculum of the hand by the long flexor tendons during wrist flexion and power grip. The different configuration of the foot

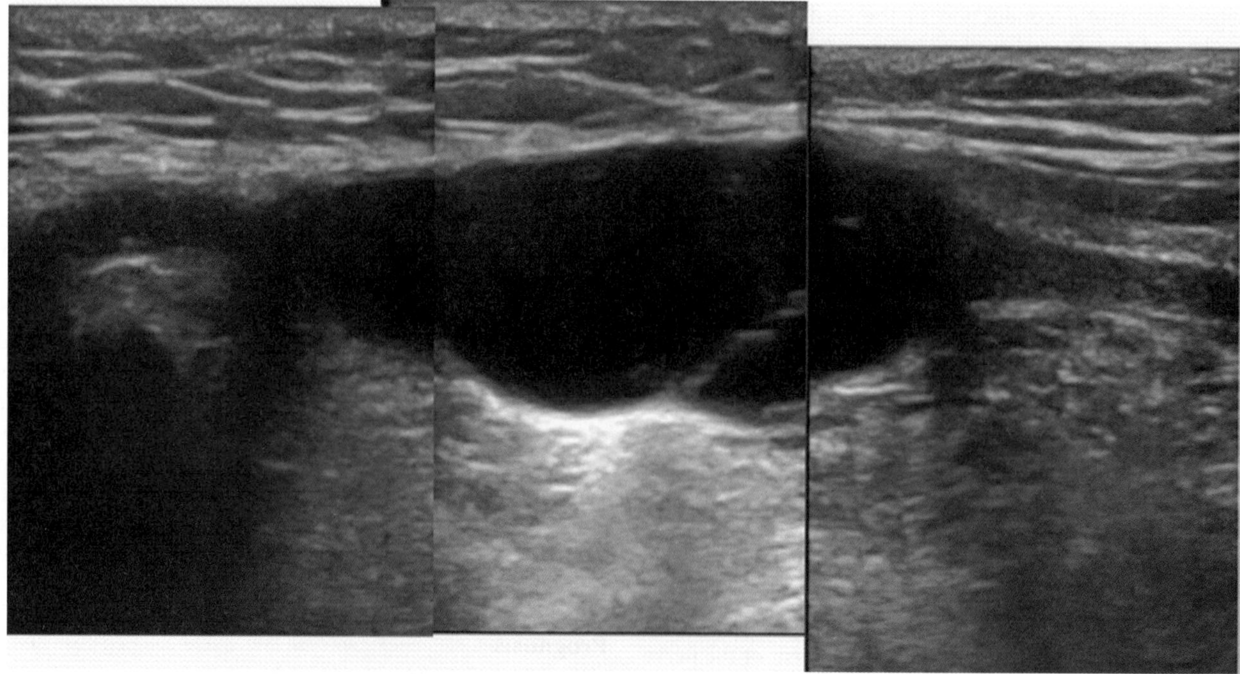

Fig. 19.6 Longitudinal view of peroneal nerve at knee showing intraperoneal ganglion cyst.

means that sideways forces from the tibialis posterior and flexor digitorum longus tendons when they are loaded are mostly exerted against the underlying bony structure rather than the laciniate ligament. As a result it is a comparatively flimsy structure compared to the flexor retinaculum of the hand.

Many case reports and short case series of 'tarsal tunnel syndrome' exist, but in the majority of well documented cases the nerve appears to be impaired as a result of either external pressure or abnormal anatomy, rather than simply as a result of passing through an apparently anatomically normal tarsal tunnel. The situation does not therefore appear to be entirely analogous to carpal tunnel syndrome and the existence of an idiopathic tarsal tunnel syndrome as an exact equivalent of idiopathic CTS is debatable. The general principles of clinical assessment and testing set out in the first part of this chapter can be applied to suspected cases of tarsal tunnel syndrome, but the role of imaging is perhaps more important given the frequency with which anatomical explanations for a lesion are present.

Conclusions

Localized impairment of nerve function challenges the neurophysiologist to identify the site and severity of a lesion and the presence of any underlying or co-existing neuromuscular disorder. Aetiology can sometimes be inferred from the history and recognition of common clinical syndromes, but will often be clarified by imaging studies, especially with ultrasound, which can be performed at the same visit as the nerve conduction studies. The examiner should always adapt their approach to the clinical problem to meet the needs of the patient and any referring clinician in trying to make practical decisions on further management.

References

1. Stewart, J.D. (2010). *Focal peripheral neuropathies*, 4th edn. West Vancouver: JBJ Publishing.
2. Dawson, D.M., Hallett, M., and Wilbourn, AJ. (1999). *Entrapment neuropathies*, 3rd edn. Philadelphia, PA: Lippincott-Raven.
3. Pecina, M.M., Krmpotic-Nemanic, J., and Markiewitz, A.D. (2001). *Tunnel syndromes: peripheral nerve compression syndromes*, 3rd edn. Boca Raton, FL: CRC Press.
4. Strakowski, J.A. (2014). *Ultrasound evaluation of focal neuropathies: correlation with electrodiagnosis*. New York, NY: Demos Medical Publishing.
5. Vanderschueren, G.A., Meys, V.E., and Beekman, R. (2014). Doppler sonography for the diagnosis of carpal tunnel syndrome. A critical review. *Muscle & Nerve*, **50**(2), 159–63.
6. Kantarci, F., Ustabasioglu, F.E., Delil, S., et al. (2014). Median nerve stiffness measurement by shear wave elastography: a potential sonographic method in the diagnosis of carpal tunnel syndrome. *European Radiology*, **24**(2), 434–40.
7. Zaidman, C.M., Al-Lozi, M., and Pestronk, A. (2009). Peripheral nerve size in normals and patients with polyneuropathy: an ultrasound study. *Muscle & Nerve*, **40**(6), 960–6.
8. Loscher, W.N., Auer-Grumbach, M., Trinka, E., et al. (2000). Comparison of second lumbrical and interosseus latencies with standard measures of median nerve function across the carpal tunnel: a prospective study of 450 hands. *Journal of Neurology*, **247**(7), 530–4.
9. Phalen, G.S. and Kendrick, J.I. (1957). Compression neuropathy of the median nerve in the carpal tunnel. *Journal of the American Medical Association*, **164**(5), 524–30.
10. Seradge, H., Jia, Y-C., and Owens, W. (1995). In vivo measurement of carpal tunnel pressure in the functioning hand. *Journal of Hand Surgery*, **20A**, 855–59.
11. Gelberman, R.H., Hergenroeder, P.T., Hargens, A.R., et al. (1981). The carpal tunnel syndrome: a study of carpal canal pressures. *Journal of Bone Joint Surgery*, **63A**, 380–83.
12. Bland, J.D.P. (2000). The value of the history in the diagnosis of carpal tunnel syndrome. *Journal of Hand Surgery*, **25B**(5), 445–50.
13. Hennessey, W.J. and Kuhlman, K.A. (1997). The anatomy, symptoms and signs of carpal tunnel syndrome. *Physical Medicine & Rehabilitation Clinics of North America*, **8**(3), 439–57.
14. Kuschner, S.H. (1999). Reliability and validity of physical examination tests used to examine the upper extremity (letter). *Journal of Hand Surgery*, **24A**(4), 868–69.
15. Bland, J.D.P. (2007). Use of screening blood tests in patients with carpal tunnel syndrome. *Journal of Neurology, Neurosurgery, & Psychiatry*, **78**, 551.
16. Graham, B., Regehr, G., Naglie, G., et al. (2006). Development and validation of diagnostic criteria for carpal tunnel syndrome. *Journal of Hand Surgery*, **31A**(6), 919–24.
17. Robinson, L. (2013). We should not use a modifier to describe the severity of carpal tunnel syndrome. *Muscle & Nerve*, 334–35.
18. Bland, J.D.P. (2001). Do nerve conduction studies predict the outcome of carpal tunnel decompression? *Muscle & Nerve*, **24**(7), 935–40.
19. Stevens, J.C. (1997). The electrodiagnosis of carpal tunnel syndrome. *Muscle & Nerve*, **20**, 1477–86.
20. Padua, L., LoMonaco, M., Padua, R., et al. (1997). Neurophysiological classification of carpal tunnel syndrome assessment of 600 symptomatic hands. *Italian Journal Neurology Science*, **18**(3), 145–50.
21. Bland, J.D.P. (2000). A neurophysiological grading scale for carpal tunnel syndrome. *Muscle & Nerve*, **23**, 1280–83.
22. Redmond, M. and Rivner, H. (1988). False positive electrodiagnostic tests in carpal tunnel syndrome. *Muscle & Nerve*, **11**, 511–17.
23. Jablecki, C.K., Andary, M.T., Floeter, M.K., et al. (2002). Practice parameter: Electrodiagnostic studies in carpal tunnel syndrome. *Neurology*, **58**(11), 1589–92.
24. Cartwright, M.S., Hobson-Webb, L.D., Boon, A.J., et al. (2012). Evidence-based guideline: Neuromuscular ultrasound for diagnosis of carpal tunnel syndrome. *Muscle & Nerve*, **46**, 287–93.
25. Fowler, J.R., Gaughan, J.P., and Ilyas, A.M. (2011). The sensitivity and specificity of ultrasound for the diagnosis of carpal tunnel syndrome: a meta-analysis. *Clinical Orthopaedics and Related Research*, **469**(4), 1089–94.
26. Roll, S.C., Case-Smith, J., and Evans, KD. (2011). Diagnostic accuracy of ultrasonography vs. electromyography in carpal tunnel syndrome: a systematic review of literature review article. *Ultrasound in Medicine & Biology*, **37**(10), 1539–53.
27. Descatha, A., Huard, L., Aubert, F., et al. (2012). Meta-analysis on the performance of sonography for the diagnosis of carpal tunnel syndrome. *Seminars in Arthritis and Rheumatism*, **41**(6), 914–22.
28. Tai, T.W., Wu, C.Y., Su, F.C., et al. (2012). Ultrasonography for diagnosing carpal tunnel syndrome: a meta-analysis of diagnostic test accuracy. *Ultrasound Medicine Biology*, **38**(7), 1121–8.
29. Stutz, N., Gohritz, A., van, S.J., et al. (2006). Revision surgery after carpal tunnel release: analysis of the pathology in 200 cases during a 2-year period. *Journal of Hand Surgery*, **31B**, 68–71.
30. Capasso, M., Manzoli, C., and Uncini, A. (2009). Management of extreme carpal tunnel syndrome: evidence from a long-term follow-up study. *Muscle & Nerve*, **40**(1), 86–93.

31. Mondelli, M., Giannini, F., Ballerini, M., et al. (2005). Incidence of ulnar neuropathy at the elbow in the province of Siena (Italy). *Journal of Neurological Science*, **234**(1–2), 5–10.
32. Campbell, W.W., Pridgeon, R.M., and Sahni, K.S. (1992). Short segment incremental studies in the evaluation of ulnar neuropathy at the elbow. *Muscle & Nerve*, **15**, 1050–54.
33. No authors listed. (1999). Practice parameter: Electrodiagnostic studies in ulnar neuropathy at the elbow. American Association of Electrodiagnostic Medicine, American Academy of Neurology, and American Academy of Physical Medicine and Rehabilitation. *Neurology*, **52**, 688–690.
34. Beekman, R., Wokke, J.H., Schoemaker, M.C., et al. (2004). Ulnar neuropathy at the elbow: follow-up and prognostic factors determining outcome. *Neurology*, **63**, 1675–80.
35. Beekman, R., Visser, L.H., and Verhagen, W.I.M. (2011). Ultrasonography in ulnar neuropathy at the elbow: a critical review. *Muscle & Nerve*, **43**, 627–35.
36. Farrell, C.M., Springer, B.D., Haidukewych, G.J., et al. (2005). Motor nerve palsy following primary total hip arthroplasty. *Journal of Bone and Joint Surgery*, **87A**(12), 2619–25.
37. Spinner, R.J., Hebert-Blouin, M.N., Amrami, K.K., et al. (2010). Peroneal and tibial intraneural ganglion cysts in the knee region: a technical note. *Neurosurgery*, **67**(3), 71–8; discussion 78.
38. Keck, C. (1992). The tarsal-tunnel syndrome. *Journal of Bone and Joint Surgery*, **44A**(1), 180–2.
39. Lam, S.J.S. (1962). A tarsal-tunnel syndrome. *Lancet*, **2**(29), 1354–55.

CHAPTER 20

Generalized peripheral neuropathies

Hessel Franssen

Introduction

The electrophysiological investigation of a patient suspected of polyneuropathy should be preceded by taking a short history related to symptoms of polyneuropathy. Questions should be asked about the date of onset, rate of progression (in days, months, or years), onset in hands or feet, medication, family history, and presence of motor symptoms, sensory symptoms, asymmetry, pain, and autonomic symptoms. This information is helpful for choosing the appropriate electrophysiological protocol and for reassuring the patient that the rather unpleasant electrophysiological examination will be performed by a genuinely interested human being. For the latter reason, it is also essential to take the history and start the examination with the patient supine and covered by a blanket.

The main purposes of the electrophysiological examination are to establish:

- If clinical symptoms and signs can be explained by polyneuropathy.
- Whether the polyneuropathy is related to demyelination of intact axons (demyelinating polyneuropathy), or to loss of axons without preceding demyelination (axonal polyneuropathy).
- To suggest the most likely cause of the polyneuropathy (table 1, (1)). Establishing the cause of polyneuropathy is essential as some axonal polyneuropathies can be treated depending on the underlying cause and as many acquired demyelinating polyneuropathies improve on immunological treatment, for instance, IVIg or corticosteroids.

Methodological considerations

Stimulation

Careful stimulation of a nerve at each site will avoid pain and co-stimulation of nearby nerves. This is important in patients with polyneuropathy because of the increased threshold for stimulation in many neuropathies, including Charcot-Marie-Tooth (CMT) type 1a, chronic inflammatory demyelinating polyneuropathy (CIDP), multifocal motor neuropathy (MMN), and anti-myelin associated glycoprotein (MAG) neuropathy (2). To elicit the largest possible compound muscle action potential (CMAP) using the smallest possible current, place the cathode at the site where the nerve is probably situated (Fig. 20.1). Increase the stimulus current in steps of about 5 mA until a small response is obtained (Fig. 20.1, left upper). Now keep stimulus current at the same value and move the stimulator perpendicular to the nerve until the largest response is obtained (Fig. 1, right upper, left lower, right lower). Next, increase stimulus current until the maximal response is obtained and add 20% for supramaximal stimulation (30% at Erb's point). For sensory nerve action potentials (SNAPs), stimulus current needs subsequently to be reduced until a just maximal SNAP is obtained in order to reduce stimulus artefact.

Recording

For motor nerve conduction studies (NCS), terminal distance (distance between the most distal stimulating site and the active recording electrode) should be standardized since some demyelinating neuropathies preferentially affect distal motor latency (DML). Cut-off values for DML consistent with demyelination were determined using terminal distances of 7 cm for the median, ulnar, and peroneal nerves and 10cm for the tibial nerve (Table 20.1).

For sensory NCS, distal distance must be standardized as well because SNAP amplitude strongly depends on conduction distance. Optimal recording of sural nerve SNAPs is achieved by location of the nerve by gentle palpation behind the lateral malleolus. At this site, the recording electrode is placed and the reference electrode 3cm more distally. Near nerve needle electrodes are useful in cases with ankle oedema and to detect increased temporal dispersion (TD) in a sensory nerve in patients with suspected demyelinating neuropathy; here, TD is reflected by multiple small peaks following the main component of the SNAP.

Temperature

Reducing temperature increases CMAP amplitude, SNAP amplitude, DML, and FM interval. It decreases motor conduction velocity (MCV), sensory conduction velocity (SCV), and the number of blocked axons (3). NCS often assess nerves and muscles in distal limb parts where temperature is usually considerably below 37°C. It is, therefore, necessary to take account of limb temperature in order to avoid attributing abnormal variables to polyneuropathy whereas, in fact, they are due to decreased nerve temperature. Unfortunately, no perfect or simple methods are available.

The best method is to warm limbs in a water bath where temperature is maintained at 37°C (Fig. 20.2). Because nerves are buried in tissue, nerve temperature changes slowly over tens of minutes toward the desired value according to a decaying exponential function. Since in most patients, distal skin temperature is above 27°C and since, with this skin temperature, warming in water at 37°C for

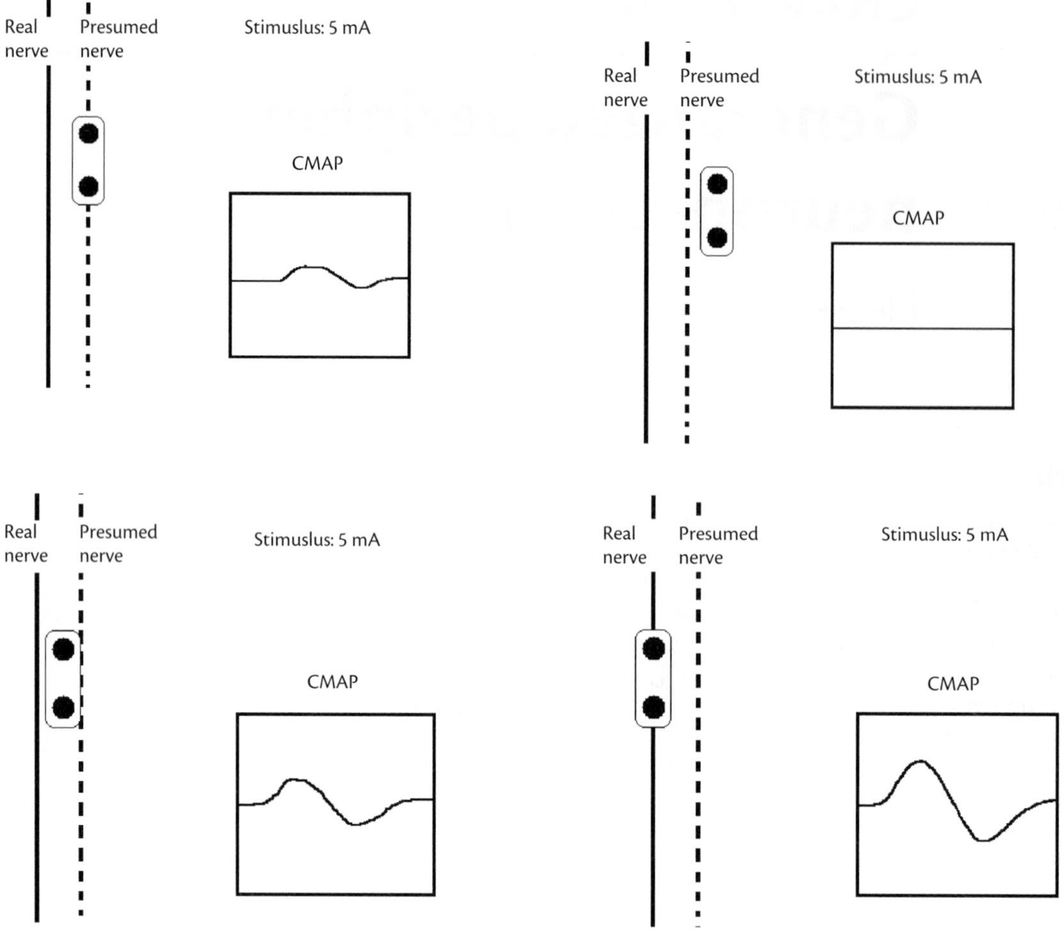

Fig. 20.1 Careful stimulation.

Reproduced from *European Neurological Review*, **7**(2), Franssen H. and van den Berg L. H. Practical Electrophysiology for the Diagnosis of Multifocal Motor Neuropathy, pp. 118–23, copyright (2012), with permission from Touch Medical Media.

Table 20.1 Criteria for demyelinative slowing for nerves investigated at 37°C

	Nerve			
	Median	**Ulnar**	**Peroneal**	**Tibial**
Variable				
DML (ms)	5.8	4.5	6.6	6.4
F–M latency (ms)	38	41	65	64
Distal duration (ms)	9.2	10.5	8.5	8.3
MCV forearm/leg (m/s)	38	40	35	35
MCV upper arm (m/s)	41	43		
MCV shoulder (m/s)	46	46		
Segmental Dur Prol (%)	30 (Sh: 40)	30 (Sh: 40)	100	100

Criteria are only applicable if the CMAP on distal stimulation is at least 1 mV.
CMAP = compound muscle action potential.

DML, distal motor latency; MCV, motor conduction velocity; Segmental Dur Prol, segmental duration prolongation; Sh, shoulder segment.

30 min will achieve a nerve temperature approaching 37°C with an error of less than 1%, a standard warming time of 30 min can be employed (3). Temperature can also be raised by wrapping limbs in sheets through which water at 37°C flows, although this takes a longer time than immersing limbs in water.

Wrapping limbs in sheets with water at 40°C for a short time is incorrect as the nerve temperature at the end of warming will be unknown. Infrared heaters are unsuitable to raise limb temperature as warming by this method takes a very long time. Recalculating CV by correction factors based on the difference between 37°C and actual skin temperature and the increase in CV per °C leads to errors because the change in CV per °C in patients is smaller than that in normal subjects, varying between 2.2 and 0.0m/s/°C (3).

Pathophysiology of single axons

In normal myelinated axons, the *action potential* or *impulse* propagates because, at an active node, transient Na-channels open and Na-ions flow into the axon (Fig. 20.3). This *action current* depolarizes the membrane, resulting in more Na-channel openings and further depolarization. The action current induces a current loop between the active node and the next node that has to be activated. At this latter node, the current leaves the axon as a *driving current*. The driving current deposits positive charges on the inside and draws positive charges from the outside of the node that has to be activated, thereby depolarizing it (this current is capacitative because no ions pass through the axolemma since

Fig. 20.2 Water baths for warming patient's limbs at 37°C prior to electrophysiological studies. The upper small baths are for the arms and the lower baths are for the legs. During warming, the patient can watch TV.

its channels are still closed). When this node is sufficiently depolarized, its Na-channels open and the outward driving current changes into the inward action current of the next action potential. The impulse at the first activated node is terminated because its Na-channels close after 1 ms, a process known as inactivation. Interruption of this current loop at any part will cause conduction block (CB).

Note that, when an ion-channel opens, it is the capacitative return current, which determines membrane potential change! Thus, *Na-channel opening* induces an *inward Na-ion current* along the Na-concentration gradient and an *outward capacitative return current*, which drops positive charges on the inside of the membrane and removes positive charges from its outside, thereby depolarizing the membrane. *K-channel opening* induces an *outward K-ion current* along the K-concentration gradient and an *inward capacitative return current*, which drops positive charges on the outside of the membrane and removes positive charges from its inside, thereby re- or hyperpolarizing the membrane.

Demyelination impairs conduction by several mechanisms (4). Leakage through the damaged myelin decreases driving current, resulting in more time being needed to depolarize a node and, thereby, slowing of internodal conduction time. With more leakage, CB arises because the driving current cannot sufficiently depolarize the node to trigger an impulse. Complete segmental demyelination induces CB because Na-channel density in the internodal axolemma is not sufficient to sustain continuous conduction. Only after upregulation of internodal Na-channel expression will conduction be restored. When paranodal demyelination destroys the attachment of myelin loops, the separation of nodal Na-channels and juxtaparanodal K-channels is lost resulting in dispersion of these ion-channels along the nodal, paranodal and juxtaparanodal axolemma. Nodal Na-channel dispersion produces CB due to diminished action current density. Exposure or dispersion of juxtaparanodal fast K-channels induce CB because their activation by the depolarizing driving current opposes impulse generation.

At critically demyelinated internodes, driving current may just be sufficient for impulse propagation. Adverse physiologic circumstances may then convert slowing into CB. Temperature increase may induce *heat block* because action current is reduced due to the shortened Na-channel open-time and faster fast K-channel opening. Prolonged duration or high frequency of firing of action potentials may induce *rate-dependent* block by several mechanisms of which hyperpolarization induced by increased activity of the electrogenic Na/K-pump is the most important.

Axonal degeneration may occur as a downstream effect of one of the mechanisms causing intra-axonal Na-accumulation (4). These include: increased Na-channel expression along a competely demyelinated internode and Na/K-pump dysfunction due to inflammation, ischaemia, or axonal transport deficits.

Intra-axonal Na-accumulation induces reversed operation of the Na/Ca-exchanger in order to remove the excess Na-ions in exchange for Ca-ions, intra-axonal Ca-accumulation, and Ca-mediated axonal degeneration. In many neuropathies, however, the exact cause of axon loss is unknown.

Demyelination and axonal degeneration on nerve conduction studies

Classification in neurology and pathology

In clinical neurology, demyelinating neuropathy is defined as a neuropathy with features of demyelination on NCS, whether or not NCS or EMG shows features of axon loss (1). Axonal neuropathies are those with signs of axon loss on NCS or EMG, but without signs of demyelination on NCS. Here NCS, rather than pathological studies, are the gold standard for establishing demyelination (Box 20.1). The rather simplified approach in clinical neurology is justified because it serves the practical purpose of establishing diagnosis. It is, therefore, not justified to label a polyneuropathy, with signs of demyelination and axon loss on NCS, as an uncertain type with demyelination and axon loss since this does no help diagnosis: such a polyneuropathy should be classified as demyelinating. Pathological studies, however, may clearly show demyelination in clinically defined axonal neuropathies such as diabetic neuropathy, or show axonal degeneration in clinically defined demyelinating neuropathies.

Conduction slowing

Conduction slowing on NCS is reflected by variables dependent on CV in the fastest conducting axons. These variables include: DML, MCV, shortest F-M interval, and SCV. It is also reflected by variables assessing TD, which is the difference between the slowest and fastest conduction time in a nerve; these include:

- Segmental CMAP duration prolongation, calculated as:

$$[(\text{proximal CMAP duration}) - (\text{distal CMAP duration}) \times 100\%] / [\text{distal CMAP duration}]. \quad \text{[eqn 20.1]}$$

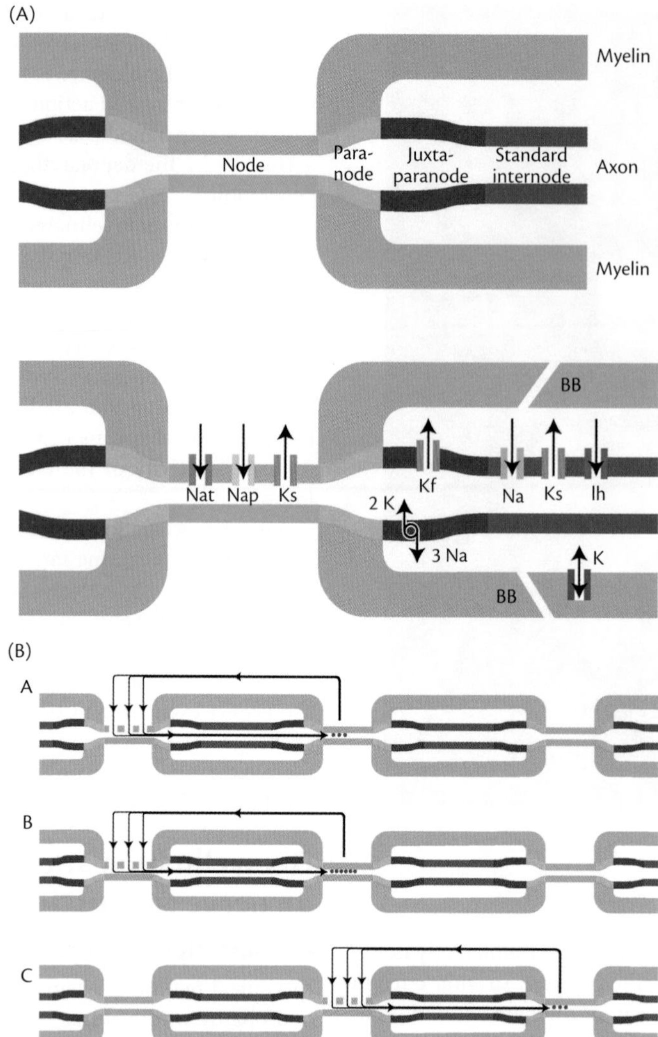

Fig. 20.3 (Upper) schematic drawing of a node and internode of a myelinated axon and the relevant axolemmal ion-channels. The arrows refer to the direction of current through the channel if it is open under normal physiological conditions. (Lower) current loops associated with normal impulse propagation from left to right. (A) At the left node an inward Na-ion current occurs, inducing an impulse; at the right node an outward capacitative current occurs, depolarizing the right node. (B) During the impulse at the left node, the right node is further depolarized by the outward capacitative current. (C) The right node has sufficiently been depolarized for its Na-channels to open, inducing a current-reversal from outward capacitative to an inward current of Na-ions.

Nat, transient Na-current, which sustains the impulse; Nap, persistent Na-current, which facilitates excitation; Ks, slow K-current which protects the axon against extreme depolarization; Kf, fast K-current, which avoids spreading of the impulse to the internode. Ih, hyperpolarization-activated cation current, which protects the axon against extreme hyperpolarization; 2K/3Na, Na/K-pump which maintains ionic concentration gradients.

Reproduced from *Muscle Nerve*, **48**, Franssen H. and Straver D. C. G., Pathophysiology of Immune-mediated Demyelinating Neuropathies—Part I: Neuroscience, pp. 851–864, copyright (2013), with permission from John Wiley and Sons.

- Distal CMAP duration, which reflects distal TD.
- Chronodispersion of F waves.

Criteria for slowing due to demyelination were first formulated by comparing CVs in nerve biopsy proven demyelinating CMT and axonal CMT (5). This revealed a cut-off value that almost perfectly distinguished both CMT types: CV in demyelinating CMT was less than 60% of the mean CV in normal subjects and CV in axonal CMT was above this value. Subsequently, criteria for NCS variables were formulated on the basis of upper or lower normal limits. For instance, MCV was considered to be consistent with demyelination if its value was below 80% of the lower limit of normal and if CMAP amplitude was preserved. The evidence for these criteria cannot be traced back in the literature, but despite this, they feature in current criteria sets for CIDP. Only the diagnostic value of these sets as a whole has been evaluated, an unsuitable procedure to assess criteria for individual variables like MCV. Table 20.1 shows criteria for several NCS variables that are based on the maximal slowing at 37°C found in patients with lower motor neurone disease, the criteria assuming demyelination if slowing is more prominent (6).

Conduction block

CB, which is the cessation of impulse propagation at a site of an axon that is viable, can be detected by NCS if it occurs in a sufficient number of axons of a nerve segment. In that case, the CMAP evoked by proximal stimulation of the segment is smaller than the

Box 20.1 Classification of polyneuropathy

Acute or subacute polyneuropathy

Pure or predominantly motor, symmetric, axonal

- Acute motor axonal neuropathy (AMAN).
- Critical illness neuro-myopathy.
- Porphyria.
- Diphtheria.

Sensorimotor, multifocal, axonal

Vasculitis.

Sensorimotor symmetric, demyelinating

- Acute inflammatory demyelinating polyneuropathy (AIDP).
- Diphtheria.

Pure sensory, symmetric, axonal

Paraneoplastic sensory neuronopathy.

Pure motor, asymmetric, axonal

Diabetic amyotrophy.

Chronic polyneuropathy

Sensorimotor, symmetric, axonal

- Metabolic (diabetes, renal failure, hypothyroidism).
- Intoxication (alcohol, glue, arsenic, medication).
- Deficiency (vitamins).
- Infection (HIV, Lyme disease, leprosy).
- Auto-immune (paraproteinaemia, vasculitis, cryoglobulinaemia, sarcoidosis, amyloidosis).
- Hereditary (Charcot–Marie–Tooth type 2).
- Chronic idiopathic axonal polyneuropathy.
- Paraneoplastic.

Sensorimotor, symmetric, demyelinating

- Chronic inflammatory demyelinating polyneuropathy (CIDP).
- Paraproteinaemia (IgM or IgG).
- HIV.
- Charcot–Marie–Tooth type 1.
- Amiodarone medication.
- Inherited childhood disorders (metachromatic leucodystrofia, adrenomyeloneuropathy, Refsum disease, Tangier disease, Krabbe leucodystrofia, Cockayne syndrome, cerebrotendinous xanthomatosis).

Sensorimotor, multifocal, demyelinating

- Lewis–Sumner syndrome.
- Hereditary neuropathy with pressure palsies (HNPP).
- X-linked Charcot–Marie–Tooth neuropathy.
- Lyme disease.

Pure motor, symmetric, axonal (rare)

- Paraneoplastic.
- Charcot–Marie–Tooth neuropathy type 2.
- Porphyria.
- Lead intoxication.

Pure sensory, symmetric, axonal

- Sjögren disease.
- Paraneoplastic.
- Vasculitis.
- Vitamin E deficiency.
- Vitamin B6 intoxication.
- Idiopathic.
- Biliary cirrhosis.
- Leprosy.
- Polycythaemia.
- Spinocerebellar ataxia
- Hereditary sensory and autonomic neuropathy (HSAN).
- Friedreich ataxia.
- Tangier disease.
- Bassen-Kornzweig disease.

Pure motor, multifocal, demyelinating

Multifocal motor neuropathy.

Painful neuropathy

- Idiopathic small fibre neuropathy.
- Small fibre neuropathy due sodium channel mutations.
- Diabetes mellitus.
- Alcoholism.
- Paraproteinemia.
- HIV.
- Medication.
- Sjögren syndrome.
- Vasculitis.
- Guillain–Barré syndrome.
- Hereditary sensory and autonomic neuropathy (HSAN).
- Leprosy.
- Fabry disease.

CMAP evoked by distal stimulation of the segment (segmental CMAP reduction). This is because part of the nerve action potentials, evoked at the proximal site, will not pass the site where they are blocked, whereas the nerve action potentials, evoked at the distal site, will all reach the muscle. Rate-dependent and heat block are difficult to demonstrate on NCS because both sustained activity and increasing temperature change TD between axons and between muscle fibres. In NCS, CB is usually assessed by:

$$[(\text{distal CMAP area}) - (\text{proximal CMAP area}) \times 100\%] / [\text{distal CMAP area}]. \quad \text{[eqn 20.2]}$$

Because both CB and increased TD give rise to segmental CMAP reduction, criteria are required to distinguish between them. When discussing CB criteria, it is most convenient to describe the CMAP as a summation of surface-recorded motor unit potentials (MUPs), rather than surface-recorded muscle fibre action potentials. TD leads to desynchronized activation of MUPs and cancellation between positive and negative phases of the MUPs that contribute to the CMAP (phase cancellation). Desynchronization results in a lower and broader CMAP and phase cancellation in a lower CMAP. Both desynchronization and phase cancellation are more pronounced on more proximal stimulation, although phase cancellation will be abolished with extreme TD (7). It is unlikely that phase cancellation increases when MUPs are polyphasic after reinnervation by collateral sprouting because surface recorded MUPs are not polyphasic, even if their needle recorded counterpart is (7).

Criteria for CB can be derived neither by comparing different patient groups, nor by comparing patients and normal subjects, because all disorders characterized by CB also feature increased TD. They can be derived, however, by studies in which the MUPs contributing to a CMAP are manipulated to simulate CB and TD. Rhee et al. (8) reconstructed CMAPs from rat MUPs, showing that TD without CB resulted in segmental CMAP reduction of up to 80% for amplitude and up to only 50% for area. Thus, CMAP area reduction exceeding 50% denotes CB in at least a few axons whereas CMAP amplitude reduction is unsuitable to assess CB. Van Asseldonk et al. (7) reconstructed CMAPs from surface MUPs recorded from human hand muscles, showing that less stringent criteria for CB can be applied if TD is limited.

Criteria for axonal and demyelinating polyneuropathy

Demyelinating polyneuropathies are characterized by:

- Slowing of DML, MCV, or F–M interval, fulfilling criteria for demyelination.
- Increased TD fulfilling criteria for demyelination.
- Segmental CMAP reduction fulfilling criteria for CB.

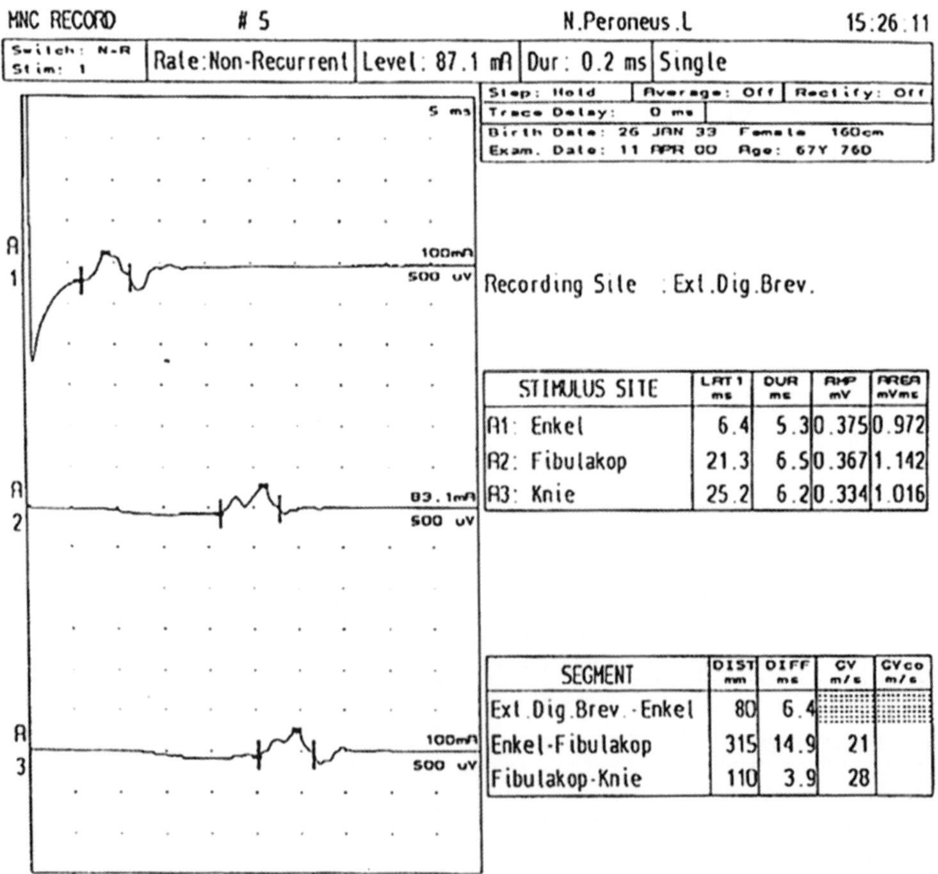

STIMULUS SITE	LAT1 ms	DUR ms	AMP mV	AREA mVms
A1: Enkel	6.4	5.3	0.375	0.972
A2: Fibulakop	21.3	6.5	0.367	1.142
A3: Knie	25.2	6.2	0.334	1.016

SEGMENT	DIST mm	DIFF ms	CV m/s	CVco m/s
Ext.Dig.Brev.-Enkel	80	6.4		
Enkel-Fibulakop	315	14.9	21	
Fibulakop-Knie	110	3.9	28	

Fig. 20.4 Peroneal nerve motor NCS in a patient with moderately severe axonal polyneuropathy. There is marked conduction slowing, but this is insufficient proof of demyelination since all CMAPs are well below 1 mV.

It must be stressed that, in nerves where the CMAP on distal stimulation is smaller than 1 mV, it is not possible to assess CB, TD, and slowing consistent with demyelination reliably. Thus, when lower limb NCS show distal CMAPs smaller than 1 mV accompanied by marked slowing, lower limb SNAPs are decreased, and upper limb NCS are normal, the polyneuropathy should be classified as axonal (Fig. 20.4).

Axonal polyneuropathies are characterized by:

- Signs of motor and sensory axon loss (decreased CMAP and SNAP amplitudes on distal stimulation, neurogenic needle EMG findings).
- No slowing or mild slowing not consistent with demyelination.
- No CB.

These characteristics are based on three assumptions:

- A decreased distal CMAP is most often caused by loss of motor axons and less often by neuromuscular transmission deficits or CB in terminal nerve segments.
- Distal CMAPs are unlikely affected by CB or TD since the distance between the distal stimulation site and recording site is small.
- Partial loss of axons in a nerve leaves the remaining axons intact and these conduct normally so that slowing does not occur.

When these features are not met, diagnostic problems arise. In mild axonal neuropathies, CMAPs can be normal, requiring needle-EMG to demonstrate motor axon loss, since this is more sensitive than distal CMAP amplitude. In mild axonal neuropathies, even SNAPs of distal lower limb nerves can be normal, so that polyneuropathy cannot formally be proven. In moderate to severe polyneuropathies, mild slowing may occur, probably due to loss of the fastest conducting axons, mild demyelination not fulfilling criteria, or increased longitudinal axonal resistance caused by decreased axon diameter or accumulation of intra-axonal structures.

Electrophysiological protocol

The NCS and needle EMG protocols that need to be performed depend on which polyneuropathy is primarily suspected. Since investigators may have to adjust the protocol or repeat an unsatisfactory study, it is important that they are able to assess immediately if a nerve that just has been investigated is normal or shows signs of demyelination or axon loss. Knowledge of normal values is therefore essential. In an adult person, a distal CMAP below 3mV indicates loss of motor axons, a distal SNAP below 7 μV loss of sensory axons, MCV below 38 m/s in an arm nerve and below 30 m/s in a leg nerve indicate demyelination, and MCV below 50 m/s in an arm nerve and below 40 m/s in a leg nerve indicate either slight demyelination not fulfilling criteria, or axonal degeneration. Obviously, the exact cut-off values vary with each nerve (Table 20.1).

Symmetrical length-dependent sensorimotor polyneuropathies associated with diabetes, HIV, renal failure, alcohol abuse, or neurotoxic medication do not always require electrophysiological examination since this will invariably show the expected length-dependent axonal neuropathy or non-specific findings (9). If, however, the clinical picture is not consistent with the expected type of polyneuropathy, electrophysiological investigation is mandatory.

Chronic mild length-dependent sensorimotor neuropathies without established cause, usually require a simple protocol to confirm the presence of polyneuropathy. This may comprise an investigation on one side of the body of motor NCS and F waves in the median (or ulnar), peroneal, and tibial nerves, and distally evoked SNAPs of the median, ulnar, peroneal, and sural nerve SNAPs. If CMAP amplitudes and other motor NCS variables are normal, needle EMG is required to confirm, or exclude, involvement of motor axons. It should be emphasized that polyneuropathy can only be proven if, in several nerves, SNAP amplitudes are decreased or motor NCS are compatible with demyelination. Mild slowing, decreased CMAP amplitudes, and neurogenic needle EMG findings are non-specific as they can also be the result of motor neurone disease or multiple radiculopathy.

Moderate to severe sensorimotor polyneuropathies of unknown cause or with progression over weeks to months require a more extensive examination to establish if the neuropathy is demyelinating or axonal and to suggest a specific cause. The protocol may consist of bilateral investigation of motor NCS and F waves in the median, ulnar, peroneal, and tibial nerves and of distal SNAPs of the median, ulnar, radial, and sural nerves. Upper limb motor NCS should be investigated up to Erb's point. If upper limb motor NCS reveal decreased distal CMAPs below 1 mV, it is important to assess if demyelination can be demonstrated in nerves without signs of axon loss, e.g. the median nerve with recording from the flexor carpi radialis muscle.

Pure or predominantly motor neuropathies without objective sensory loss and asymmetric symptoms and signs, require a very extensive protocol to demonstrate MMN. This should comprise bilateral motor NCS and F waves of the median, ulnar, radial, musculocutaneous, peroneal, and tibial nerves and sensory NCS in arm nerve segments with motor CB. Median nerve motor NCS should be performed with recording from the thenar and flexor carpi radialis muscles.

Pure sensory symmetrical neuropathies require a protocol to demonstrate sensory neuronopathy, axonal sensorimotor neuropathy, or sensory CIDP. This may consist of the above described protocol for moderate to severe neuropathy. Sensory CIDP features objective sensory signs and sensory ataxia. Motor NCS may, however, show signs of demyelination, which then establish the diagnosis (10). In clinically sensory CIDP with demyelination on motor NCS, weakness is absent because there is no definite CB or loss of motor axons. In sensory neuronopathy, electrophysiological signs of motor axon loss are either absent or minimal.

Acute neuropathies require a protocol to demonstrate Guillain–Barré syndrome (GBS) or vasculitic neuropathy. In prototypical GBS, electrophysiology is not required since treatment does not depend on GBS type. If clinical diagnosis is uncertain, NCS are useful to prove that a polyneuropathy is present or to demonstrate signs of demyelination consistent with acute inflammatory demyelinating polyneuropathy (AIDP). If vasculitis is suspected, it is useful to demonstrate that affected nerves show loss of motor and sensory axons and unaffected nerves are normal. Vasculitis, however, may also present as a symmetric axonal polyneuropathy. The protocol for acute neuropathy may comprise unilateral or bilateral motor NCS and F waves of the median, ulnar, peroneal, and tibial nerves and distal median, ulnar, radial, and sural nerve SNAPs. Except if AIDP is suspected, stimulation at Erb´s point is not needed. Needle EMG of affected and unaffected muscles may be useful to demonstrate multifocal motor axon loss. It must be emphasized that, in a nerve recently affected by vasculitis, the distal axon parts are still excitable whereas the proximal parts have become inexcitable. At that stage, NCS reveals normal CMAPs on distal and reduced CMAPs on proximal stimulation. This so-called pseudo-block can be distinguished from true block by repeating NCS 3 or 4 days later: when degeneration has reached distal axon parts, both distal and proximal CMAPs are reduced.

Selected polyneuropathies with characteristic electrophysiological features

Diabetic neuropathy

NCS reveal a non-specific axonal length-dependent sensorimotor polyneuropathy. On the other hand, excitability studies showed rather specific changes suggesting reduced persistent Na-currents and Na/K-pump dysfunction (11). Reduced persistent Na-currents were found by the method of latent addition, a refinement of strength–duration time constant (SDTC) measurement, which enables separate assessment of nodal persistent Na-currents and passive nodal membrane properties (see Chapter 9). The abnormality was more prominent in patients with increased HbA1c levels and reverted to normal following insulin treatment. The reduction in persistent Na-currents in diabetes may be the result of insulin deprivation of Na/K-ATP-ase leading to intra-axonal Na-accumulation, nodal Na-channel dysfunction due to tissue acidosis, or damage to nodal Na-channel clusters.

Neuropathy in renal failure

Neuropathy associated with renal failure is an axonal length-dependent sensorimotor polyneuropathy. MCV may be markedly slowed, but not sufficient to fulfil criteria for demyelination, provided limbs are properly warmed to 37°C. Thresholds may be increased. Patients have an increased incidence of carpal tunnel syndrome and ulnar nerve neuropathy at the elbow. Excitability studies in patients with chronic renal failure, but without clinical or NCS evidence of neuropathy, were shown to be consistent with permanent depolarization of resting membrane potential (11). Remarkably, excitability indices correlated with extracellular serum K-concentration, suggesting that the depolarization was caused by reduced E_K. Excitability-indices improved after dialysis.

Painful neuropathy

Polyneuropathy with known causes may be accompanied by pain, notable examples being GBS, diabetic neuropathy, vasculitic neuropathy, alcoholic neuropathy, and polyneuropathies due to nutritional deficiencies. Another group, small fibre neuropathy (SFN), is characterized by abnormal quantitative sensory testing and reduced density of small unmyelinated axons in skin biopsy specimens. This suggests that SFN is related to loss or dysfunction of small fibres that mediate pain and temperature sense. NCS and needle EMG are normal, indicating that myelinated axons are not involved. Approximately 30% of patients with SFN have gain of function mutations in Na-channel subtypes that are expressed in dorsal root ganglion cell bodies, which may explain the occurrence of spontaneous pain (12). No cause has been established in the remaining SFN patients.

Demyelinating CMT neuropathy

CMT1a is caused by duplication of the peripheral myelin protein 22 (*PMP22*) gene, resulting in demyelination, remyelination, and axon loss. MCVs and SCVs are uniformly slowed and are consistent with demyelination. NCS usually reveals considerably increased thresholds. Because of these features, NCS in these patients require a high stimulus current, prolonged stimulus duration, and an increased time-base to record responses. With proper stimulation, CB is not found. TD is usually minimal, indicating that all axons in a nerve conduct uniformly slowly. The clinical motor deficit is correlated with decreased CMAPs, indicating that this is determined by axon loss (13).

CMT1b is related to several possible mutations in the myelin protein zero (P0) gene. In early-onset forms, severe demyelination and remyelination with short internodes and extremely slow NCS (10 m/s) may occur (14). In late-onset forms, NCS is consistent with axon loss.

HNPP

HNPP is characterized by transient palsies, related to an increased sensitivity of nerves to pressure at common entrapment sites and is due to a deletion of one *PMP22* allele. NCS over entrapment sites show MCV consistent with demyelination or CB. Generalized NCS features include DML slowing (except for tibial nerve DML, which is normal), distal SCV slowing, and decreased SNAP amplitudes (15).

GBS

GBS is a self-limiting acute neuropathy, with a nadir within 4 weeks and complete or partial recovery. Subtypes include AIDP, acute motor axonal neuropathy (AMAN), acute motor and sensory axonal neuropathy (AMSAN), and Miller–Fisher syndrome. AIDP occurs mainly in Europe and North America and AMAN mainly in northern China and Japan (16).

NCS are considered important for distinguishing demyelinating and axonal subtypes, but their interpretation in GBS is not straightforward (17). First, in GBS criteria, CB is considered supportive of demyelination and AIDP. CB resolving without TD was, however, also observed in early stages of AMAN, suggesting that it was due to nodal Na-channel dysfunction rather than demyelination. Therefore, CB cannot be attributed to a particular subtype. Secondly, follow-up NCS led to reclassification in 40% of patients, especially from AIDP to axonal GBS. Thirdly, different cut-off values for demyelinative slowing are used in GBS, including 95, 90, 80, 75, and 70% of the lower limit of normal, as well as 60% of the normal mean. Except for the latter, evidence for the other values cannot be traced in the literature.

AIDP is related to cellular immunity, but its exact mechanisms are unclear. Autopsy shows T-lymphocyte infiltration, invasion of myelin sheaths by macrophages, segmental demyelination, and axonal degeneration (16). NCS performed within 2–15 days after onset, may show motor CB or slowing consistent with evidence-based criteria for demyelination in 60% of patients (17). NCS abnormalities are most prominent 4–8 weeks after onset and recover after 6–10 weeks; CB usually resolves with signs of increased TD (18). In severe cases, distal CMAPs are permanently decreased, indicating loss of motor axons.

AMAN is related to antibodies against peripheral nerve gangliosides. One of its main causes is the development, after a gastro-enteritis, of antibodies against the wall of *C. jejuni* that cross-react with peripheral nerve GM1 ganglioside. Autopsy shows nodal complement depositions, perinodal macrophage invasion, and axonal degeneration (16). AMAN-rabbits showed disruption of nodal Na-channel clusters, associated with CB on whole nerve recordings (19). In human AMAN, NCS shows reduced CMAPs on distal stimulation, consistent with axon loss and associated with poor prognosis. Serial studies, however, showed two patterns (18). In patients with poor prognosis, distal and proximal CMAPs remained permanently decreased, indicating persistent axon loss. In patients with a better prognosis, distal CMAP amplitudes normalized within days after onset (indicating resolution of distal CB) and proximal CMAPs normalized (indicating resolution of more proximal CB). The CB in forearm segments disappeared without signs of increased TD. This fast recovery possibly reflects temporary nodal Na-channel dysfunction because remyelination or regeneration of motor axons would have taken longer.

Excitability studies in AMAN showed abnormal recovery cycles with an abrupt threshold increase at inter-stimulus intervals of about 2.0 ms without an increase in refractory period (20). These findings may indicate Na-channel dysfunction at the wrist or conduction failure in distal axon branches.

Chronic inflammatory demyelinating polyneuropathy

CIDP is characterized by progressive sensorimotor, mainly motor, or purely sensory deficits progressing over more than 2 months. Biopsy studies show demyelination, remyelination, axonal degeneration, and nerve oedema. Immunohistochemistry reveals loss and diffusion towards the internode of nodal Na-channels and K-channels and, in a minority of patients, loss of molecules, which attaches terminal myelin loops to the paranodal axolemma (21). The pathogenesis is poorly understood.

NCS typically shows symmetrical, but multifocal features of demyelination, including CB, increased TD, and slowing of DML, MCV, and F-M interval compatible with demyelination (Fig. 20.5). CIDP may present with pure sensory deficits with sensory ataxia

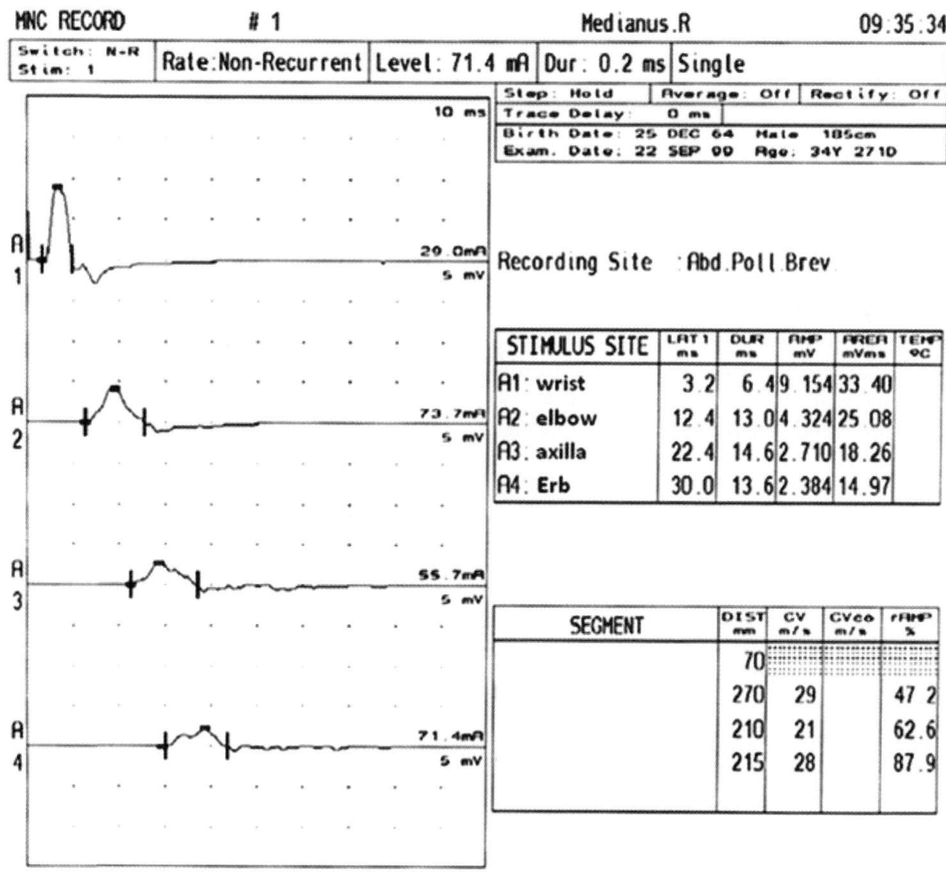

Fig. 20.5 Median nerve motor NCS in a patient with CIDP. The distal CMAP is way above the lower limit of normal, indicating that there is no evidence of motor axon loss. MCVs fulfil criteria for demyelination. Comparing CMAPs on wrist and elbow stimulation shows increased temporal dispersion, but no block. Segmental CMAP duration increase in the forearm is [(13.0 – 6.4 ms) × 100%]/6.4 ms = 103%; this fulfils the criterion for increased temporal dispersion (segmental CMAP duration increase >30% in an arm nerve). Segmental CMAP area reduction in the forearm is [(33.40 – 25.08 mVms)] × 100%/33.40 mVms = 25%; this is does not fulfil the criterion for possible block (segmental CMAP area reduction >30% in an arm nerve), nor that for definite block (segmental CMAP area reduction in any nerve >50%).

on clinical examination and subclinical demyelination of motor axons on NCS (10). To assist in diagnosing CIDP, criteria-sets were developed, which require several NCS variables to be consistent with demyelination. These sets are mainly based on expert opinion and have a high specificity (95%), but moderate sensitivity (60%), which limits identification of patients. Nevertheless, when compared with other tests for CIDP, including sural nerve biopsy, some sets had the highest odds-ratio and likelihood ratio for predicting CIDP (22).

Excitability studies showed different changes among studies (21). One of these was a decreased strength–duration time constant (SDTC), which was unexpected since paranodal demyelination should have increased nodal capacitance and, therefore, the passive component of SDTC.

Multifocal motor neuropathy

MMN is a slowly progressive motor neuropathy in adults selectively affecting nerves at randomly distributed sites and leading to asymmetrical muscle weakness and wasting predominantly in the arms and hands. Pathological studies show demyelination, remyelination, and axonal degeneration. Half of MMN patients have high-titre antibodies against ganglioside GM1, which is expressed on the nodal axolemma and perinodal Schwann cell membranes, suggesting that they cause CB by nodal Na-channel damage. The selective involvement of motor axons in mixed nerves may be due to differences in immunologic or ion-channel properties between motor and sensory axons (21).

Motor NCS show normal nerves as well as randomly distributed nerves affected by CB, slowing consistent with demyelination (DML, MCV, TD, or F wave latency), or axon loss (23). CB, increased TD, and MCV slowing consistent with demyelination may occur together in single nerve segments. Needle-EMG shows marked spontaneous muscle fibre activity and neurogenic MUPs, also in non-atrophic muscles. Sensory NCS is normal in segments with motor CB (Fig. 20.6). The mechanism of CB in MMN is unclear and may be related to demyelination, persistent changes in resting membrane potential and damage to nodal Na-channel clusters (21).

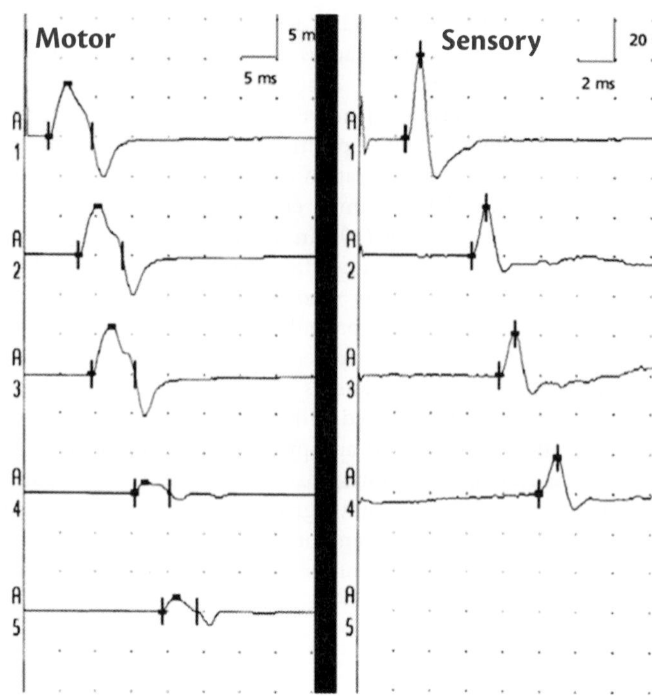

Fig. 20.6 Left ulnar nerve NCS in a patient with MMN. Between the stimulation sites A3 (just proximal to the sulcus) and A4 (axilla) motor NCS reveals conduction block (segmental area reduction 80%) and marked slowing (23 m/s) whereas sensory conduction in the same segment is normal (SCV 64 m/s).

Reproduced from *Brain*, **126**, Van Asseldonk J. T. H., Van den Berg L. H., Van den Berg-Vos R. M., Wieneke G. H, Wokke J. H. J., and Franssen H., Demyelination and axonal loss in multifocal motor neuropathy: distribution and relation to weakness, pp. 186–198, copyright (2003), with permission from Oxford University Press.

IgM neuropathy

IgM neuropathy with signs of demyelination and anti-MAG neuropathy are slowly progressing symmetrical sensorimotor polyneuropathies with sensory ataxia and serum IgM antibodies. The IgM is directed against an extracellular sugar residue shared by P0, PMP22, and sulphated glycolipids. These molecules play a role in adhesion and spacing of myelin lamellae and maintenance of axon diameter by promoting axonal neurofilament side-arm phosphorylation. Pathological studies reveal IgM-deposits in the myelin sheath, demyelination, axon loss, widening between myelin lamellae, and a decreased distance between axonal neurofilaments (24,25). The latter is consistent with impairment of MAG-function in maintaining neurofilament phosphorylation.

Typical cases show prolonged DMLs consistent with demyelination, less pronounced MCV slowing in adjacent nerve segments, and decreased CMAPs and SNAPs consistent with axon loss; conduction block is rare (Fig. 20.7). In more severe cases, however, both DMLs and MCVs are consistent with demyelination. Motor and sensory NCS in nerves with short, medium-length and long axons, revealed that DMLs were more prolonged and signs of axon loss more prominent in nerves with longer axons. The unusual combination of length-dependent axon loss and distal demyelination was not present in disease controls with CIDP and normal controls and its mechanism is unclear (21). To assess the disproportionate distal slowing in IgM neuropathy, slowing in the terminal segment (expressed by DML in ms) can be compared with slowing in the adjacent segment (expressed by MCV in m/s), by recalculation of MCV into a latency difference. Terminal latency index (TLI) is the ratio between both latencies and is given by:

$$\text{(terminal distance)} / (\text{MCV} \times \text{DML}). \qquad \text{[eqn 20.3]}$$

Residual latency (RL) is the difference between both latencies and is given by:

$$(\text{DML}) - [(\text{distal distance}) / \text{MCV}]. \qquad \text{[eqn 20.4]}$$

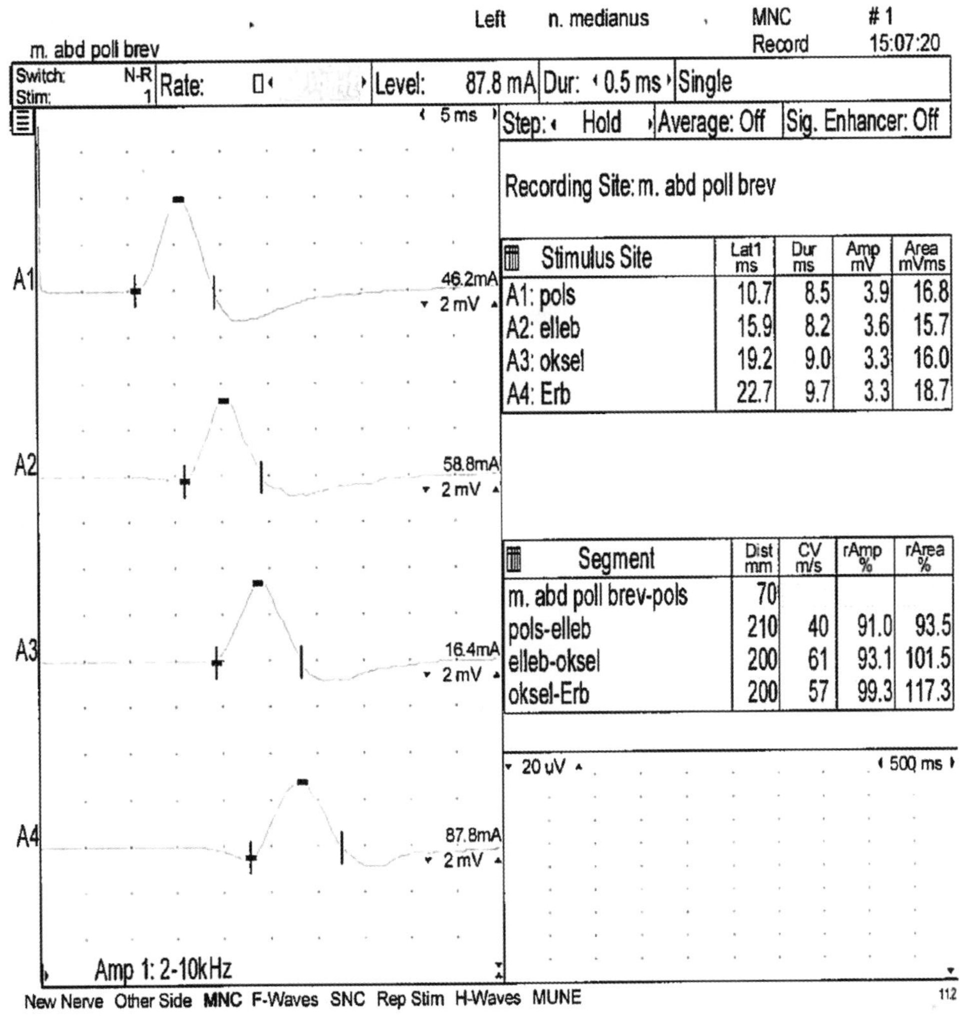

Fig. 20.7 Left median nerve motor NCS in a patient with neuropathy associated with IgM anti-MAG antibodies. DML is markedly prolonged and fulfils criteria for demyelination. Forearm MCV is moderately slowed and does not fulfil criteria for demyelination. Terminal latency index (TLI), calculated as 70 mm/(40 m/s × 10.7 ms) = 0.16, is markedly decreased (normal limit 0.27), indicating disproportional distal slowing.

References

1. Donofrio, P.D. and Albers, J.W. (1990). AAEM minimonograph #34: polyneuropathy: classification by nerve conduction studies and electromyography. *Muscle & Nerve*, **13**, 889–903.
2. Meulstee, J., Darbas, A., Van Doorn, P.A., Van Briemen, L., and Van der Meché, F.G.A. (1997). Decreased electrical excitability of peripheral nerves in demyelinating polyneuropathies. *Journal of Neurology, Neurosurgery, and Psychiatry*, **62**, 398–400.
3. Franssen, H., Notermans, N.C., and Wieneke, G.H. (1999). The influence of temperature on nerve conduction in patients with chronic axonal polyneuropathy. *Clinical Neurophysiology*, **110**, 933–40.
4. Franssen, H. and Straver, D.C. (2013). Pathophysiology of immune-mediated demyelinating neuropathies—part I: Neuroscience. *Muscle & Nerve*, **48**, 851–64.
5. Buchtal, F. and Behse, F. (1977). Peroneal muscular atrophy (PMA) and related disorders. I. Clinical manifestations as related to biopsy findings, nerve conduction and electromyography. *Brain*, **100**, 41–66.
6. Van Asseldonk, J.T., Van den Berg, L.H., Kalmijn, S., Wokke, J.H., and Franssen, H. (2005). Criteria for demyelination based on the maximum slowing due to axonal degeneration, determined after warming in water at 37°C: diagnostic yield in chronic inflammatory demyelinating polyneuropathy. *Brain*, **128**, 880–91.
7. Van Asseldonk, J.T.H., Van den Berg, L.H., Wieneke, G.H., Wokke, J.H.J., and Franssen H. (2006). Criteria for conduction block based on computer simulation studies of nerve conduction with human data obtained in the forearm segment of the median nerve. *Brain*, **129**, 2447–60.
8. Rhee, E.K., England, J.D., and Sumner, A.J. (1990). A computer simulation of conduction block: effects produced by actual block versus interphase cancellation. *Annuals of Neurology*, **28**, 146–56.
9. Rosenberg, N.R., Portegies, P., De Visser, M., and Vermeulen, M. (2001). Diagnostic investigation of patients with chronic polyneuropathy: evaluation of a clinical guideline. *Journal of Neurology, Neurosurgery, and Psychiatry*, **71**, 205–9.
10. Van Dijk, G.W., Notermans, N.C., Franssen, H., and Wokke, J.H.J. (1999). Development of weakness in patients with chronic inflammatory demyelinating polyneuropathy and only sensory symptoms at presentation: a long-term follow-up study. *Journal of Neurology*, **246**, 1134–9.
11. Krishnan, A.V., Lin, C.S-Y., Park, S.B., and Kiernan, M.C. (2008). Assessment of nerve excitability in toxic and metabolic neuropathies. *Journal of the Peripheral Nervous System*, **13**, 7–26.

12. Faber, C.G., Lauria, G., Merkies, I.S.J., et al. (2012). Gain-of-function Nav1.8 mutations in painful neuropathy. *Proceedings of the National Academy of Sciences of the United States of America*, **109**, 19444–9.
13. Krajewski, K.M., Lewis, R.A., Fuerst, D.R., et al. (2000). Neurological dysfunction and axonal degeneration in Charcot-Marie-Tooth disease type 1A. *Brain*, **123**, 1516–27.
14. Bai, Y., Ianokova, E., Pu, Q., et al. (2006). Effect of an R69C mutation in the myelin protein zero gene on myelination and ion channel subtypes. *Archives in Neurology*, **63**, 1787–94.
15. Li J., Krajewski, K., Shy, M.E., and Lewis, R.A.(2002). Hereditary neuropathy with liability to pressure palsy. The electrophysiology fits the name. *Neurology*, **58**, 1769–73.
16. Hughes, R.A. and Cornblath, D.R. (2005). Guillain-Barré syndrome. *Lancet*, **366**, 1653–66.
17. Franssen, H. (2012). Towards international agreement on criteria for Guillain-Barré syndrome. *Clinical Neurophysiology*, **123**, 1483–4.
18. Kuwabara, S., Yuki, N., Koga, M., et al. (1998). IgG anti-GM1 antibody is associated with reversible conduction failure and axonal degeneration in Guillain-Barre syndrome. *Annuals of Neurology*, **44**, 202–8.
19. Susuki, K., Yuki, N., Schafer, D.P., et al. (2012). Dysfunction of nodes of Ranvier: a mechanism for anti-ganglioside antibody-mediated neuropathies. *Experimental Neurology*, **233**, 534–42.
20. Kuwabara, S., Bostock, H., Ogawara, K., et al. (2003). The refractory period of transmission is impaired in axonal Guillain-Barre syndrome. *Muscle & Nerve*, **28**, 683–9.
21. Franssen, H. and Straver, D.C. (2014). Pathophysiology of immune-mediated demyelinating neuropathies—part II: neurology. *Muscle & Nerve*, **49**, 4–20.
22. Molenaar, D.S.M., Vermeulen, M., and De Haan, R.J. (2002). Comparison of electrodiagnostic criteria for demyelination in patients with chronic inflammatory demyelinating polyneuropathy (CIDP). *Journal of Neurology*, **249**, 400–3.
23. Van Asseldonk, J.T.H., Van den Berg, L.H., Van den Berg-Vos, R.M., Wieneke, G.H., and Franssen, H. (2003). Demyelination and axonal loss in multifocal motor neuropathy: distribution and relation to weakness. *Brain*, **126**, 186–98.
24. Steck, A.J., Stalder, A.K., and Renaud S. (2006). Anti-myelin-associated glycoprotein neuropathy. *Current Opinion in Neurology*, **19**, 458–63.
25. Lunn, M.P.T., Crawford, T.O., Hughes, R.A.C., Griffin, J.W., and Sheikh, K.A. (2002). Anti-myelin associated glycoprotein antibodies alter neurofilament spacing. *Brain*, **125**, 904–11.

CHAPTER 21

Disorders of single nerves, roots, and plexuses

Kerry R. Mills

Introduction

This chapter is concerned with conditions affecting individual nerves, nerve roots, and plexuses. Diagnosis depends on the history and an examination that indicates sensory loss and/or motor weakness in a distribution that can only be explained by a lesion at a defined site. Because muscles have multiple nerve root innervation, it may not be possible clinically to decide whether, for instance, a lesion is due to a single root or peripheral nerve. Electromyography, then, since it can demonstrate denervation that may not be evident as clinically definite weakness, can help in this differentiation. Weakness, for instance of the abductor pollicis brevis muscle (APB) could be due to a median nerve lesion or a C8 root lesion, and, in the absence of definite sensory clues, the finding, for example, of EMG changes of denervation in the first dorsal interosseous (FDI) muscle clearly indicates a C8 root origin since the muscle is supplied by the ulnar nerve. Similar logical exercises using clinical, EMG and nerve conduction information, can be used to differentiate most lesions of peripheral nerve, plexuses, or roots. Such exercises, can of course, only be conducted with a thorough knowledge of muscle innervation (root, plexus, and nerve) and the sensory distributions of nerves and roots Figs 21.1–21.3; a scheme setting out the minimal information for upper and lower limbs is given in Tables 21.1 and 21.2, using signature muscles chosen for the purity of their innervation and ease of examination and the sensory distributions are shown in Figs 21.1 and 21.2.

A complicating factor that must be born in mind is that some nerves have fascicles, destined to become branches on the nerve, present quite proximally. This is particularly evident in the sciatic where fascicles destined to become the tibial and common peroneal components may be present in the pelvis. Thus a partial sciatic nerve lesion can resemble a common peroneal nerve lesion.

Mononeuritides

Focal nerve lesions due to compression at common entrapment sites are considered in Chapter 19. Here, we are concerned with those lesions, usually associated with systemic disease such as vasculitis, that affect single nerves, either in isolation or as several separate lesions (mononeuritis multiplex)(1). Although, many conditions have been reported as being associated with mononeuritis (2–13), the important causes of such lesions are summarized in Box 21.1. Nerve conduction studies and EMG are helpful in defining mononeuropathies; often a comparison of nerve conduction findings with the corresponding contralateral nerve is sufficient, but if multiple nerves are involved a widespread search for involved nerves may be required. In the acute stage, in which pain over the nerve is associated with motor weakness without wasting, conduction block may be suspected. Indeed, the compound muscle action potential (CMAP) amplitude may be reduced to proximal stimulation compared with that from a distal site, also suggesting block. However, this is so-called pseudo-block and if follow up is undertaken after a few days it will be found that wasting is under way, acute denervation is found on EMG and the CMAP amplitude is now reduced to both proximal and distal stimulation. Other mononeuropathies involve demyelinating lesions with acute or subacute development of loss of nerve function with, in the initial stages at least, weakness without wasting. Examples include multifocal motor neuropathy with conduction block (MMN) (14,15), chronic demyelinating inflammatory polyneuropathy (CIDP) (16) and the Lewis–Sumner syndrome (17).

Diabetic mononeuritis

Mononeuropathies in diabetes often, but not exclusively, occur in the context of a more generalized symmetrical sensory polyneuropathy (18,19). They occur at common entrapment sites such as the carpal tunnel or ulnar groove, in which case the onset is often gradual, or in nerves not commonly entrapped such as the oculomotor, truncal nerves, or lateral cutaneous nerve of the thigh, in which case the onset is usually acute and painful (20–22). If the associated polyneuropathy is severe, then nerve conduction studies may struggle to define an additional mononeuropathy. Reliance must then be placed on asymmetries in nerve conduction findings. For instance, in the case of carpal tunnel syndrome in a diabetic, the median and ulnar sensory potentials may be absent bilaterally, but if the distal motor latency to APB from wrist stimulation is, say, 12 ms on one side, but only 5 ms on the contralateral side, then an additional median nerve lesion may be inferred.

Vasculitic mononeuritis

The patient typically recounts sequential, acute, painful episodes with loss of function in isolated nerves or branches thereof. There are often other stigmata of vasculitis: e.g. skin lesions, renal impairment, polyarthritis and so on depending on the condition (23–25). The important causes of vasculitic mononeuropathy or mononeuropathy multiplex are shown in Box 21.1 (25–27). Nerve conduction studies and EMG are used to document peripheral nerve lesions and to show that these conform to the innervation of isolated nerves. A wide search is often justified in these patients to detect clinically unsuspected lesions. If

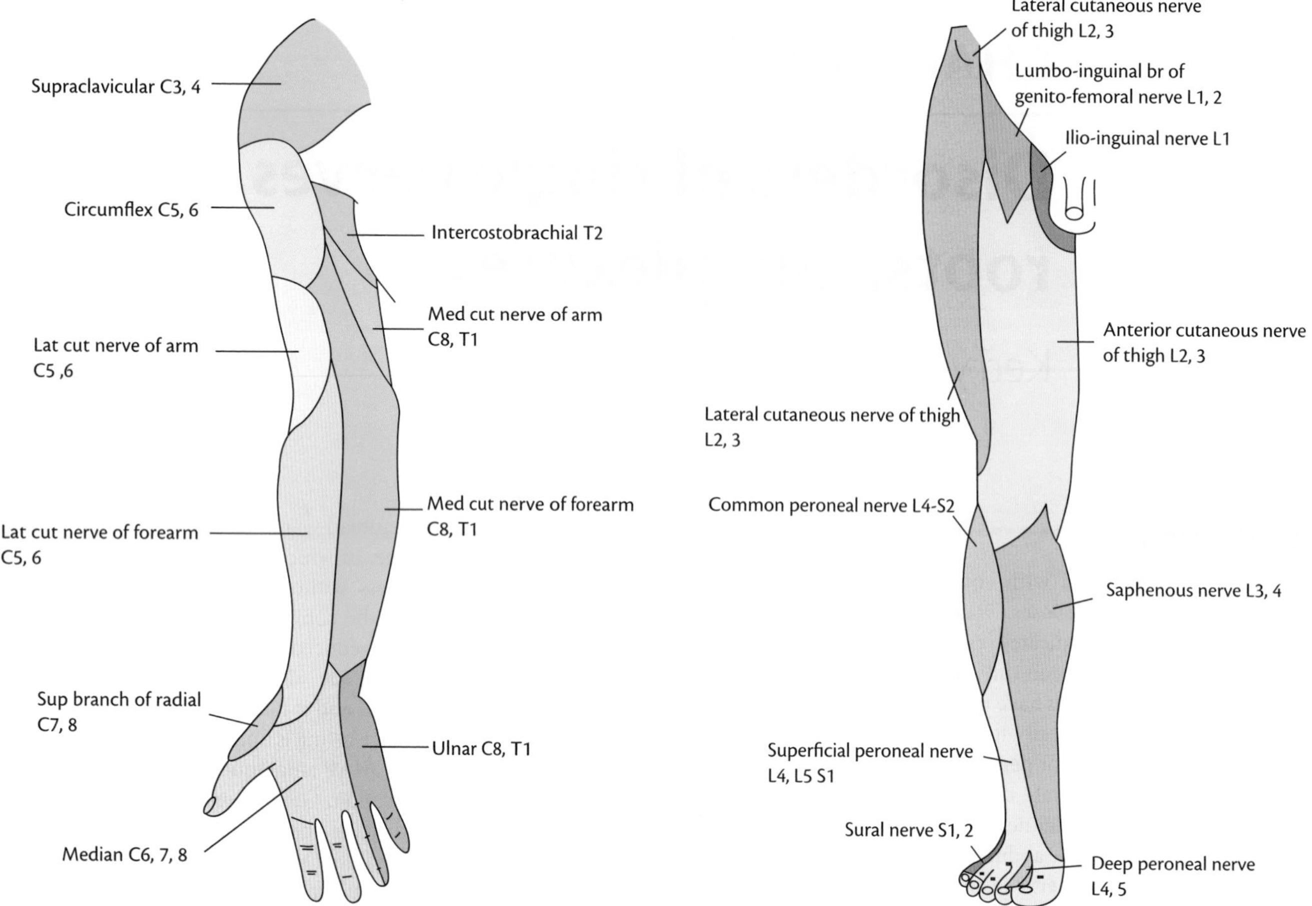

Fig. 21.1 Upper limb dermatomes and nerve supply.
Reproduced from Standring S, *Gray's Anatomy: The Anatomical Basis of Clinical Practice*, 40th edn, copyright (2008), with permission from Elsevier.

Fig. 21.2 Lower limb dermatomes and nerve supply.
Reproduced from Standring S, *Gray's Anatomy: The Anatomical Basis of Clinical Practice*, 40th edn, copyright (2008), with permission from Elsevier.

Table 21.1 Signature muscles for EMG to help differentiating nerve, plexus, and root lesions in the upper limb

	FCU	Triceps	BR	APB	FDI	EDC	Biceps	Deltoid
T1	■	■			■			
C8	■	■		■	■	■		
C7		■		■		■		
C6			■				■	■
C5			■				■	■
Inferior trunk	■	■		■	■	■		
Middle trunk		■				■	■	
Superior trunk			■	■				■
Medial cord					■			
Dorsal cord		■	■			■		■
Lateral cord							■	
Ulnar nerve	■				■			
Radial Nerve		■	■			■		
Median nerve				■				
Musculoctaneous nerve							■	
Axillary nerve								■

FCU = flexor carpi ulnaris; BR = brachioradialis; APB = abductor pollicis brevis; FDI = first dorsal interosseous; EDC = extensor digitorum communis.

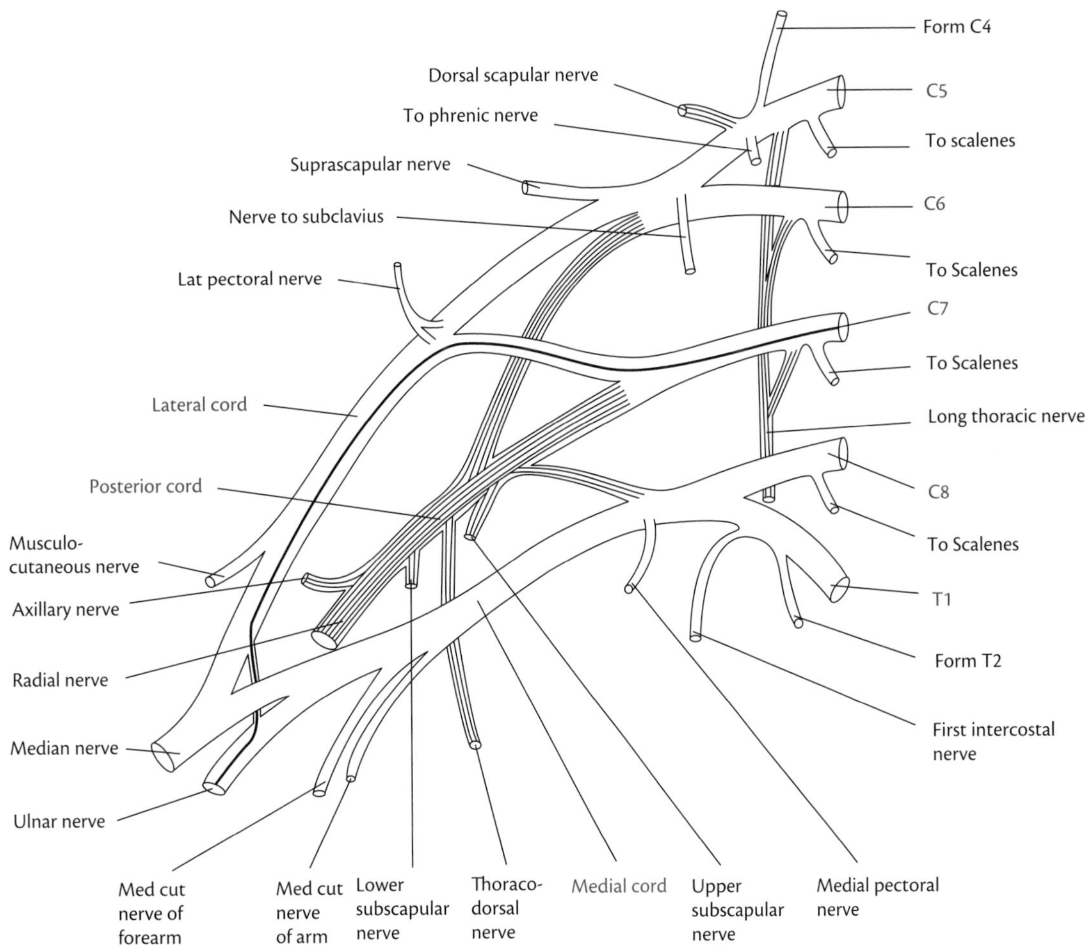

Fig. 21.3 Brachial plexus and upper limb nerve supply.
Reproduced from Standring S, *Gray's Anatomy: The Anatomical Basis of Clinical Practice*, 40th edn, copyright (2008), with permission from Elsevier.

Table 21.2 Signature muscles for EMG to help differentiating nerve and root lesions in the lower limb

	HF	Quads	TA	TP	PL	Gast	G med	G max	LH bic
L1	■	■							
L2	■	■							
L3	■	■							
L4		■	■	■			■		
L5			■	■	■		■	■	
S1				■	■	■	■	■	■
S2				■	■	■			■
Superior gluteal							■		
Inferior gluteal								■	
Sciatic			■	■	■	■			■
Femoral	■	■							
Common peroneal			■		■				
Posterior tibial				■		■			

HF = hip flexors; Quads = quadriceps; TA = tibialis anterior; PL = peroneus longus; Gast = gastrocnemius/soleus; G med = gluteus medius; G max = gluteus maximus; LH bic = long head of biceps.

Box 21.1 Causes of mononeuritis multiplex

Multiple axonal degeneration lesions

- Diabetes.
- Vasculitis:
  - Polyarteritis nodosa.
  - Systemic lupus erythematosus.
  - Rheumatoid arthritis.
  - Churge–Strauss syndrome.
- Waldenstrom's macroglobulinaemia.
- Cryoglobulinaemia.

Infections

- Hepatitis A and C.
- Lyme disease.
- Cytomegalovirus.
- Leprosy.
- Sarcoidosis.
- Neoplasm, e.g. lymphoma.
- Amyloidosis.

Multiple demelinative lesions

- Multifocal motor neuropathy.
- Lewis–Sumner syndrome.
- Chronic inflammatory demyelinating neuropathy.
- Hereditary neuropathy with pressure palsies.

the condition is chronic, then multiple isolated nerve lesions may become confluent and produce a picture more akin to a severe generalized peripheral neuropathy; the diagnosis may then depend of the history of multiple episodes of sequential loss of function.

Demyelinating types of mononeuropathy

These conditions are important to recognize because of their potential for treatment and reversal. They may be inherited or acquired.

Multifocal motor neuropathy

MMN is considered in Chapter 20. It usually presents as a subacute pure motor mononeuropathy in the upper limb and in the early stages has the clinical hallmark of conduction block, i.e. weakness without wasting. Blocks tend not to occur at entrapment sites; in a series of 23 patients with MMN, most blocks were found in the upper arm (42%), the forearm (26%), the brachial plexus (16%), and proximally in the motor root (16%) (28). Neurophysiologically, conduction block where CMAP amplitudes evoked by proximal stimulation are at least 50% smaller than those evoked by distal stimulation, is seen. Conventional nerve conduction studies will detect about 85% of blocks, but if MMN is suspected and no block has been found then stimulation of motor roots either by a deeply inserted needle electrode (29) or by high voltage stimulation over the cervical area (28) should be undertaken (see Chapter 19, Fig 19.1). Absence of F-waves is a clue to proximal block, but cannot be used to verify the block. Sensory studies typically show normal sensory conduction through the region of motor block in a mixed nerve. EMG can also be useful in suggesting conduction block. For example, a patient with block in the proximal segments of the musculocutaneous nerve may show severe weakness of biceps. Nerve conduction studies may not be able to stimulate the nerve proximal to the focal block, but EMG will show severely reduced recruitment of normal motor units in the weakened muscle.

Lewis–Sumner syndrome

This is also known as multifocal acquired demyelinating sensory and motor neuropathy (MADSAM) and appears to be a form of multifocal CIDP with both motor and sensory involvement (30–33). Again multiple foci of conduction block are found. Sensory conduction block is difficult to verify neurophysiologically. This arises from the short duration of sensory action potentials and the consequent marked phase cancellation, which has the effect of markedly reducing the amplitude of sensory nerve action potentials the greater is the distance between stimulus and recording sites (see Chapter 6). In Lewis–Sumner syndrome the presentation is usually of an upper limb multifocal sensori-motor dysfunction with conduction block demonstrable on motor nerve conduction studies usually in the context of a more generalized demyelinating peripheral neuropathy.

Hereditary neuropathy with pressure palsies

An unsuspected demyelinating hereditary neuropathy, particularly Hereditary neuropathy with pressure palsies (HNPP), can present as a mononeuropathy multiplex (34–40). In HNPP, multiple focal nerve lesions at entrapment sites may be found, commonly the carpal tunnel or ulnar nerve at the elbow, in the context of a mild demyelinating generalized neuropathy. Thus typically, for example, carpal tunnel syndrome with delayed sensory responses and delayed distal motor latency are found, but on further investigation, prolonged tibial F-wave latencies for example will also be found and conduction velocities in upper and lower limb nerves will be just below the lower limit of normal. HNPP can also present in a similar way to brachial neuritis with painful onset of weakness in proximal arm muscles and again evidence of a mild generalized demyelinating neuropathy.

Radiculopathy

Conditions affecting specifically the nerve roots are inflammatory, traumatic, or compressive.

Compressive lesions at the nerve root level are very common. Extruded material from the intervertebral disc, narrowing of intervertebral foramina, or osteophyte formation compress nerve roots in either cervical or lumbosacral regions and give rise to pain in the distribution of the root. Often multiple roots may be involved and there may be an additional myelopathy due to compression of the cord. The diagnosis of radiculopathy is often achieved by a combination of history, physical findings, and imaging. Neurophysiology has a limited and confirmatory role in diagnosis (41–46). The patient with brachialgia only with no physical signs in the arm is very unlikely to have any abnormal neurophysiological test. However, if there is weakness in the arm or leg or reduced or absent reflexes, then EMG can be useful in confirming denervation changes limited to a single nerve root distribution. The scheme set out in Tables 21.1 and 21.2 is useful in determining root, plexus, or nerve distribution. Although it might be thought that H-reflexes (see Chapters 3 and 10) or F-waves would be useful in investigating radiculopathy, in practice, this is rarely the case. With F-waves, for instance, the slowing of conduction at the level of the compressed nerve root may be severe, but is present over such a short distance that the inclusion of a large section of normal nerve in the measurement effectively dilutes the abnormality and renders it undetectable.

Cervical radiculopathy

Cervical radiculopathy most commonly affects the C7 root producing neck pain radiating into the arm and loss of the triceps jerk with or without weakness of C7 innervated muscles. EMG can be useful in finding chronic denervation changes limited to the C7 innervated muscles (see Table 21.1). C6 radiculopathy produces arm pain with loss of the brachioradialis jerk and/or biceps jerk with sometimes weakness of the biceps. Similarly, C5 radiculopathy produces neck pain radiating into the shoulder with possible weakness of deltoid and other C5 innervated shoulder muscles. Myelopathy may also be present in which case, long tract signs (e.g. brisk reflexes) will be found in segments below the level of compression. An important differential diagnosis in this clinical scenario is amyotrophic lateral sclerosis (ALS), which also typically produces a mixture of upper and lower motoneuron signs (47–50). EMG may show denervation changes or fasciculation in muscles innervated by cranial nerves or in muscles below the site of cord compression both of which suggests ALS. Transcranial magnetic stimulation (see Chapter 14) may also be helpful in the differentiation; by recording from, say, trapezius, deltoid, biceps, and a hand muscle, the central motor conduction time (CMCT) to sequentially lower segments can be found and this may not only help to localize the lesion, but may also be useful in deciding if the myelopathy is demyelinative, as in compressive lesions, or degenerative, as in ALS. In the former, central motor conduction time tends to be markedly prolonged whereas in the latter, prolongation of CMCT is usually modest.

Cervical plexopathy

Traumatic brachial plexus injuries commonly arise from motor cycle accidents (51–54). If the driver lands on the point of the shoulder then upper cervical roots will be stretched, whereas if the landing is on an outstretched arm, lower cervical roots will bear the brunt of the trauma. From a treatment and prognostic point of view it is useful to distinguish between root avulsion and stretching injury to nerves of the brachial plexus. Sensory action potentials are useful here: in root avulsion, since the sensory nerve is damaged proximal to the dorsal root ganglion, the integrity of sensory fibres is maintained and sensory action potentials persist (despite the patient being anaesthetic in the relevant area). In contrast, stretching lesions of the plexus will cause degeneration of axons and subsequent loss of sensory action potentials. Both lesions may, of course, be present with avulsion of some roots and stretch lesions of others. EMG is again useful in deciding, which nerve roots are affected and also on prognostication.

Anterior or inferior shoulder dislocation can also result in traumatic brachial plexus palsy. It may be complicated by an additional fracture or rotator cuff tear (55). An axillary nerve lesion may also be present (56,57). In uncomplicated cases, even when there is a total brachial plexus palsy, prognosis is relatively good with nearly full recovery after 2 years especially in younger patients (58). Inferior dislocation predominantly affects lower cervical roots affecting hand function. EMG is used to confirm the distribution of abnormality; nerve conduction studies confirm the expected reduction in sensory action potential amplitude and reduction in CMAP amplitude in the relevant dermatomes and myotomes.

Obstetric brachial plexus palsy

Brachial plexus palsies due to birth trauma follow two main patterns. The commonest (80%) is the so called Erb-Duchenne palsy affecting C5 and C6 roots followed by the Dejerine-Klumpke palsy (10%) affecting the C8 and T1 roots often accompanied by a Horner syndrome (59–64). Again EMG can be useful in assessing the distribution of affected roots and sensory studies are used to detect root avulsion. Prognosis is in general good with conservative treatment with according to one series 85% making a good functional recovery by 2 years, although neurophysiological abnormalities may persist (62). Good improvement by 3 months, in effect presages complete recovery.

Brachial neuritis

Also known as neuralgic amyotrophy or Parsonage–Turner syndrome (65), brachial neuritis presents with severe shoulder pain, often after a prodrome of trivial infection. The pain persists for a few days and then weakness of shoulder girdle muscles develops. The weakness typically is in a patchy distribution not conforming to a single nerve or root distribution. Commonly and as originally described by Parsonage and Turner, the anterior interosseous nerve is involved with weakness of the long flexors of the thumb and index finger. Weakness of shoulder girdle muscles, hand muscles, trapezius, diaphragm, and cranial nerves have all been described. The condition runs a self-limiting course with improvement of muscle strength over a matter of months although this is rarely complete. Diagnosis is usually clinical with neurophysiology providing confirmation of the patchy distribution of denervation. Interestingly, there is some evidence that early in the course, proximal roots show conduction block before long-standing denervation changes develop. Nerve conduction studies are usually normal unless there is an underlying condition such as HNPP (see above).

A hereditary form of brachial neuritis, hereditary neuralgic amyotrophy (HNA) also exists and in some cases is associated with mutations in the septin gene (*SEPT9*). In one large Japanese family (66), painful episodes were all asymmetrically located in the upper limbs with motor weakness in 88% and sensory signs in 59%. All patients had hypertelorism. Nerve conduction studies revealed focal demyelinating lesions as well as prominent axonal loss and MRI showed gadolinium enhancement of the brachial plexus.

Cervical rib syndrome

A rudimentary ectopic rib attached to the 7th cervical vertebra and associated with a band of fibrous tissue, which connects to the first rib, may compress the C8 and T1 roots. Some rudimentary ribs may also compress the subclavian vessels causing a vascular steal syndrome in the arm. The condition may be bilateral and predominantly affects young females. A complication in diagnosis is the fact that such rudimentary ribs are present in about 1% of the population and yet very rarely cause the neurological or vascular syndrome. Typically a young woman complains of pain and paraesthesiae in the arm and hand and may develop weakness; carpal tunnel syndrome is commonly suspected and indeed may co-exist. Neurophysiologically, there is evidence of C8 and T1 compression: classically the combination of a small ulnar sensory action potential with evidence of denervation on EMG of the APB muscle and slightly prolonged upper limb F-waves suggests the diagnosis and should trigger imaging studies to look for cervical rib.

Lumbosacral radiculopathy

Lumbosacral radiculopathy is also very common predominantly affecting the L5 and S1 roots. Here back pain radiating down the back of the leg with an absent ankle jerk and possible weakness of L5 and S1 innervated muscles is the usual clinical picture. EMG can again be useful in confirming the distribution of abnormality to a single or pair of nerve roots (see Table 21.2). Similarly, L3 and L4 root lesions typically cause back and anterior thigh pain with an absent knee jerk. The lumbar and sacral roots, of course, lie together within the spinal canal well below the termination of the spinal cord (at L2 segmental level). Thus lumbosacral radiculopathies are less associated with cord compression that cervical radiculopathy. Equally, it is more common in the lumbar region for multiple roots to be affected.

EMG, as stated above is usually used as a confirmatory test, along with imaging, to delineate the distribution of denervation changes (67–72). The usefulness of EMG, of course, is dependent on the experience of the physician (73,74). Posterior tibial or common peroneal F-wave latency, especially where the root compression is unilateral allowing comparison of the two sides. A side to side difference of greater than 4ms suggests L5 and/or S1 root compression. The H-reflex (75), elicited in gastrocnemius or soleus can also be useful in this context, although if the ankle jerk is absent then the H-reflex is likely to be unelicitable. A side to side difference in H-reflex latency of more than 1ms is said to be significant. Magnetic stimulation can be used to investigate root conduction in the lumbar area (76–80). The problem that arises is of knowing precisely where the root has been stimulated (81). Whereas with stimulation over the cervical vertebral column, current is focused into intervertebral foramina to excite nerve roots at this location (82), over the lumbar area roots may be excited in their parallel course within the spinal canal. Nevertheless, there are reports of magnetic lumbar root stimulation having clinical utility in lumbar radiculopathy by showing slowing of conduction in the relevant motor roots (76). The use of dermatomal somatosensory evoked potentials in lumbosacral radiculopathy (41,71,83) has shown inconsistent results and is currently not regarded as useful.

Lumbosacral plexopathy

Lumbosacral plexus lesions are much less common than those of the brachial plexus and are most commonly due to tumours, trauma, or haematomas due to coagulopathies.

Radiation plexopathy

Radiotherapy can cause fibrotic change in the tissues around the brachial or lumbar plexus and can produce weakness and/or dysaesthsiae, sometimes years after the radiation therapy. The principal differential diagnosis is from recurrence of the original tumour or metastasis in, for example, the axillary nodes. Pain, occurring with tumour recurrence in the most reliable sign for differentiation. EMG will show denervation changes in the muscles innervated by the affected roots, but in radiation induced changes, myokymic discharges are also seen and this can be a useful pointer (84–86).

References

1. Parry, G.J. (1985). Mononeuropathy multiplex (AAEE case report #11). *Muscle & Nerve*, **8**(6), 493–8.
2. Cucurachi, L. and Sperber, S.A. (2012). Meningococcemia presenting as acute painful mononeuritis multiplex. *Clinical Neurology and Neurosurgery*, **114**(3), 284–6.
3. Ekiz, E., Ozkok, A. and Ertugrul, N.K. (2013). Paraneoplastic mononeuritis multiplex as a presenting feature of adenocarcinoma of the lung. *Case Reports in Oncological Medicine*, **2013**, 457346.
4. Garg, S., Wright, A., Reichwein, R., Boyer, P., Towfighi, J., and Kothari, M.J. (2005). Mononeuritis multiplex secondary to sarcoidosis. *Clinical Neurology and Neurosurgery*, **107**(2), 140–3.
5. Kilpatrick, T.J., Hjorth, R.J., and Gonzales, M.F. (1992). A case of neurofibromatosis 2 presenting with a mononeuritis multiplex. *Journal of Neurology, Neurosurgery, and Psychiatry*, **55**(5), 391–3.
6. Kumar, S. (2005). Painful mononeuritis multiplex in idiopathic thrombocytopenic purpura. *Indian Paediatrics*, **42**(6), 621–2.
7. Liveson, J.A. and Goodgold, J. (1974). Mononeuritis multiplex associated with infectious mononucleosis. *Canadian Journal of Neurological Sciences*, **1**(3), 203–5.
8. Massey, E.W. (1980). Sjogren's syndrome and mononeuritis multiplex. *Annals of Internal Medicine*, **92**(1), 130.
9. Reid, A.C. and Bone, I. (1981). Lymphoma presenting as mononeuritis multiplex. *Postgraduate Medical Journal*, **57**(665), 176–7.
10. Rosenberg, R.N. and Lovelace, R.E. (1968). Mononeuritis multiplex in lepromatous leprosy. *Archives of Neurology*, **19**(3), 310–4.
11. Takeuchi, A., Kodama, M., Takatsu, M., Hashimoto, T., and Miyashita, H. (1989). Mononeuritis multiplex in incomplete Behcet's disease: a case report and the review of the literature. *Clinical Rheumatology*, **8**(3), 375–80.
12. Tezzon, F., Corradini, C., Huber, R., et al. (1991). Vasculitic mononeuritis multiplex in patient with Lyme disease. *Italian Journal of Neurological Sciences*, **12**(2), 229–32.
13. Yaqoob, M. and Ahmad, R. (1989). Mononeuritis multiplex as an unusual complication of leptospirosis. *Journal of Infection*, **19**(2), 188–9.
14. Magistris, M. and Roth, G. (1992). Motor neuropathy with multifocal persistent conduction blocks. *Muscle & Nerve*, **15**(9), 1056–7.
15. Slee, M., Selvan, A., and Donaghy, M. (2007). Multifocal motor neuropathy: the diagnostic spectrum and response to treatment. *Neurology*, **69**(17), 1680–7.
16. Rajabally, Y.A., Sarasamma, P., and Abbott, R.J. (2004). Chronic inflammatory demyelinating polyneuropathy after *Campylobacter jejuni* infection mimicking vasculitic mononeuritis multiplex in a diabetic. *Journal of the Peripheral Nervous System*, **9**(2), 98–103.
17. Viala, K., Renie, L., Maisonobe, T., et al. (2004). Follow-up study and response to treatment in 23 patients with Lewis–Sumner syndrome. *Brain*, **127**(9), 2010–7.
18. Raff, M.C., Sangalang, V., and Asbury, A.K. (1968). Ischemic mononeuropathy multiplex in association with diabetes mellitus. *Neurology*, **18**(3), 284.
19. Stamboulis, E., Voumvourakis, K., Andrikopoulou, A., et al. (2009). Association between asymptomatic median mononeuropathy and diabetic polyneuropathy severity in patients with diabetes mellitus. *Journal of Neurological Sciences*, **278**(1–2), 41–3.
20. Fraser, D.M., Campbell, I.W., Ewing, D.J., and Clarke, B.F. (1979). Mononeuropathy in diabetes mellitus. *Diabetes*, **28**(2), 96–101.
21. Massey, E.W. (1980). Diabetic truncal mononeuropathy: electromyographic evaluation. *Acta Diabetologica Latina*, **17**(3–4), 269–72.

22. Naghmi, R., and Subuhi, R. (1990). Diabetic oculomotor mononeuropathy: involvement of pupillomotor fibres with slow resolution. *Hormone and Metabolic Research*, **22**(1), 38–40.
23. Lovelace, R.E. (1964). Mononeuritis multiplex in polyarteritis nodosa. *Neurology*, **14**, 434–42.
24. Brune, M.B., Walton, S.P., and Varga, D. (1995). Extensive mononeuritis multiplex as a solitary presentation of polyarteritis nodosa. *Journal of the Kentucky Medical Association*, **93**(1), 15–8.
25. Steidl, C., Baumgaertel, M.W., Neuen-Jacob, E., and Berlit, P. (2012). Vasculitic multiplex mononeuritis: polyarteritis nodosa versus cryoglobulinemic vasculitis. *Rheumatology International*, **32**(8), 2543–6.
26. Said, G., Lacroix, C., Plante-Bordeneuve, V., et al. (2002). Nerve granulomas and vasculitis in sarcoid peripheral neuropathy: a clinicopathological study of 11 patients. *Brain*, **125**(2), 264–75.
27. Agarwal, V., Singh, R., Wiclaf, et al. (2008). A clinical, electrophysiological, and pathological study of neuropathy in rheumatoid arthritis. *Clinical Rheumatology*, **27**(7), 841–4.
28. Arunachalam, R., Osei-Lah, A., and Mills, K.R. (2003). Transcutaneous cervical root stimulation in the diagnosis of multifocal motor neuropathy with conduction block. *Journal of Neurology, Neurosurgery, and Psychiatry*. **74**(9), 1329–31.
29. Berger, A.R., Busis, N.A., Logigian, E.L., Wierzbicka, M., and Shahani, B.T. (1987). Cervical root stimulation in the diagnosis of radiculopathy. *Neurology*, **37**(2), 329–32.
30. Mezaki, T., Kaji, R., and Kimura, J. (1999). Multifocal motor neuropathy and Lewis Sumner syndrome: a clinical spectrum. *Muscle & Nerve*, **22**(12), 1739–40.
31. Saperstein, D.S., Amato, A.A., Wolfe, G.I., et al. (1999). Multifocal acquired demyelinating sensory and motor neuropathy: the Lewis–Sumner syndrome. *Muscle & Nerve*, **22**(5), 560–6.
32. Katz, J.S. and Saperstein, D.S. (2001). Asymmetric acquired demyelinating polyneuropathies: MMN and MADSAM. *Current Treatment Options in Neurology*, **3**(2), 119–25.
33. Heckmann, J.G., Pawlowski, M., Seifert, F., Dutsch, M., and Bickel, A. 2006). Bilateral ulnar neuropathy due to multifocal acquired demyelinating sensory and motor neuropathy (MADSAM). *Journal of Hand Surgery*, **31**(5), 583.
34. Andreadou, E., Yapijakis, C., Paraskevas, G.P., et al. (1996). Hereditary neuropathy with liability to pressure palsies: the same molecular defect can result in diverse clinical presentation. *Journal of Neurology*, **243**(3), 225–30.
35. Chance, P.F. (2006). Inherited focal, episodic neuropathies: hereditary neuropathy with liability to pressure palsies and hereditary neuralgic amyotrophy. *Neuromolecular Medicine*, **8**(1–2), 159–74.
36. Gouider, R., LeGuern, E., Gugenheim, M., et al. (1995). Clinical, electrophysiologic, and molecular correlations in 13 families with hereditary neuropathy with liability to pressure palsies and a chromosome 17p11.2 deletion. *Neurology*, **45**(11), 2018–23.
37. Hong, Y.H., Kim, M., Kim, H.J., Sung, J.J., Kim, S.H., and Lee, K.W. (2003). Clinical and electrophysiologic features of HNPP patients with 17p11.2 deletion. *Acta Neurologica Scandinavica*, **108**(5), 352–8.
38. Luigetti, M., Del Grande, A., Conte, A., et al. (2014). Clinical, neurophysiological and pathological findings of HNPP patients with 17p12 deletion: a single-centre experience. *Journal of the Neurological Sciences*. **341**(1–2), 46–50.
39. Potulska-Chromik, A., Sinkiewicz-Darol, E., Ryniewicz, B., et al. (2014). Clinical, electrophysiological, and molecular findings in early onset hereditary neuropathy with liability to pressure palsy. *Muscle & Nerve*, **50**(6), 914–18
40. Yurrebaso, I., Casado, O.L., Barcena, J., Perez de Nanclares, G., and Aguirre, U. (2014). Clinical, electrophysiological and magnetic resonance findings in a family with hereditary neuropathy with liability to pressure palsies caused by a novel PMP22 mutation. *Neuromuscular disorders*, **24**(1), 56–62.
41. Aminoff, M.J., Goodin, D.S., Barbaro, N.M., Weinstein, P.R., and Rosenblum, M.L. (1985). Dermatomal somatosensory evoked potentials in unilateral lumbosacral radiculopathy. *Annals of Neurology*, **17**(2), 171–6.
42. Dillingham, T.R. and Lauder, T. (2005). Electromyographic evaluation of cervical radiculopathy. *Archives of Physical Medicine and Rehabilitation*, **86**(11), 2224; author reply 5.
43. Hakimi, K. and Spanier, D. (2013). Electrodiagnosis of cervical radiculopathy. *Physical Medicine and Rehabilitation Clinics of North America*, **24**(1), 1–12.
44. Leis, A.A., Kofler, M., Stetkarova, I., and Stokic, D.S. (2011). The cutaneous silent period is preserved in cervical radiculopathy: significance for the diagnosis of cervical myelopathy. *European Spine Journal*, **20**(2), 236–9.
45. Lo, Y.L., Chan, L.L., Leoh, T., et al. (2008). Diagnostic utility of F waves in cervical radiculopathy: electrophysiological and magnetic resonance imaging correlation. *Clinical Neurology and Neurosurgery*, **110**(1), 58–61.
46. Pawar, S., Kashikar, A., Shende, V., and Waghmare, S. (2013). The study of diagnostic efficacy of nerve conduction study parameters in cervical radiculopathy. *Journal of Clinical and Diagnostic Research*, **7**(12), 2680–2.
47. Chan, K.M., Nasathurai, S., Chavin, J.M., and Brown, W.F. (1998). The usefulness of central motor conduction studies in the localization of cord involvement in cervical spondylytic myelopathy. *Muscle & Nerve*, **21**(9), 1220–3.
48. Chan, Y.C. and Mills, K.R. (2005). The use of transcranial magnetic stimulation in the clinical evaluation of suspected myelopathy. *Journal of Clinical Neuroscience*, **12**(8), 878–81.
49. Lo, Y.L., Chan, L.L., Lim, W., et al. (2004). Systematic correlation of transcranial magnetic stimulation and magnetic resonance imaging in cervical spondylotic myelopathy. *Spine*, **29**(10), 1137–45.
50. Truffert, A., Rosler, K.M., and Magistris, M.R. (2000). Amyotrophic lateral sclerosis versus cervical spondylotic myelopathy: a study using transcranial magnetic stimulation with recordings from the trapezius and limb muscles. *Clinical Neurophysiology*, **111**(6), 1031–8.
51. Trojaborg, W. (1994). Clinical, electrophysiological, and myelographic studies of 9 patients with cervical spinal root avulsions: discrepancies between EMG and X-ray findings. *Muscle & Nerve*, **17**(8), 913–22.
52. Wilbourn, A.J. (2007). Plexopathies. *Neurologic Clinics*, **25**(1), 139–71.
53. Barman, A., Chatterjee, A., Prakash, H., Viswanathan, A., Tharion, G., and Thomas, R. (2012). Traumatic brachial plexus injury: electrodiagnostic findings from 111 patients in a tertiary care hospital in India. *Injury*, **43**(11), 1943–8.
54. Hulser, P.J. (1988). Combined brachial root and plexus lesions—typical sequelae of motor-bike accidents. *European Archives of Psychiatry and Neurological Sciences*. **237**(5), 304–6.
55. Groh, G.I. and Rockwood, C.A., Jr. (1995). The terrible triad: anterior dislocation of the shoulder associated with rupture of the rotator cuff and injury to the brachial plexus. *Journal of Shoulder and Elbow Surgery*, **4**(1 Pt 1), 51–3.
56. Gumina, S., Bertino, A., Di Giorgio, G., and Postacchini, F. (2005). Injury of the axillary nerve subsequent to recurrence of shoulder dislocation. Clinical and electromyographic study. *La Chirurgia degli organi di movimento*, **90**(2), 153–8.
57. Liveson, J.A. (1984). Nerve lesions associated with shoulder dislocation; an electrodiagnostic study of 11 cases. *Journal of Neurology, Neurosurgery, and Psychiatry*, **47**(7), 742–4.
58. Kosiyatrakul, A., Jitprapaikulsarn, S., Durand, S., and Oberlin, C. (2009). Recovery of brachial plexus injury after shoulder dislocation. *Injury*, **40**(12), 1327–9.
59. Hardy, A.E. (1981). Birth injuries of the brachial plexus: incidence and prognosis. *Journal of Bone and Joint Surgery British*, **63**–B(1), 98–101.
60. Wilbourn, A,J. (1985). Electrodiagnosis of plexopathies. *Neurologic Clinics*, **3**(3), 511–29.
61. Bisinella, G.L., Birch, R., and Smith, S.J. (2003). Neurophysiological prediction of outcome in obstetric lesions of the brachial plexus. *Journal of Hand Surgery*, **28**(2), 148–52.

62. Strombeck, C., Remahl, S., Krumlinde-Sundholm, L., and Sejersen, T. (2007). Long-term follow-up of children with obstetric brachial plexus palsy II: neurophysiological aspects. *Developmental Medicine and Child Neurology*, **49**(3), 204–9.
63. Van Dijk, J.G., Pondaag, W., Buitenhuis, S.M., Van Zwet, E.W., and Malessy, M.J. (2012). Needle electromyography at 1 month predicts paralysis of elbow flexion at 3 months in obstetric brachial plexus lesions. *Developmental Medicine and Child Neurology*, **54**(8), 753–8.
64. Pitt, M. and Vredeveld, J.W. (2005). The role of electromyography in the management of the brachial plexus palsy of the newborn. *Clinical Neurophysiology*, **116**(8), 1756–61.
65. van Alfen, N. and van Engelen, B.G. (2006). The clinical spectrum of neuralgic amyotrophy in 246 cases. *Brain*, **129**(Pt 2), 438–50.
66. Ueda, M., Kawamura, N., Tateishi, T., et al. (2010). Phenotypic spectrum of hereditary neuralgic amyotrophy caused by the SEPT9 R88W mutation. *Journal of Neurology, Neurosurgery, and Psychiatry*, **81**(1), 94–6.
67. Cho, S.C., Ferrante, M.A., Levin, K.H., Harmon, R.L., and So Y.T. (2010). Utility of electrodiagnostic testing in evaluating patients with lumbosacral radiculopathy: an evidence-based review. *Muscle & Nerve*, **42**(2), 276–82.
68. Coster, S., de Bruijn, S.F., and Tavy, D.L. (2010). Diagnostic value of history, physical examination and needle electromyography in diagnosing lumbosacral radiculopathy. *Journal of Neurology*, **257**(3), 332–7.
69. Mondelli, M., Aretini, A., Arrigucci, U., Ginanneschi, F., Greco, G., and Sicurelli, F. (2013). Clinical findings and electrodiagnostic testing in 108 consecutive cases of lumbosacral radiculopathy due to herniated disc. *Clinical Neurophysiology*, **43**(4), 205–15.
70. Pastore-Olmedo, C., Gonzalez, O., and Geijo-Barrientos, E. (2009). A study of F-waves in patients with unilateral lumbosacral radiculopathy. *European Journal of Neurology*, **16**(11), 1233–9.
71. Rodriquez, A.A., Kanis, L., Rodriquez, A.A., and Lane, D. (1987). Somatosensory evoked potentials from dermatomal stimulation as an indicator of L5 and S1 radiculopathy. *Archives of Physical Medicine and Rehabilitation*, **68**(6), 366–8.
72. Tong, H.C., Haig, A.J., Yamakawa, K.S., and Miner, J.A. (2006). Specificity of needle electromyography for lumbar radiculopathy and plexopathy in 55- to 79-year-old asymptomatic subjects. *American Journal of Physical Medicine & Rehabilitation*, **85**(11), 908–12; quiz 13–5, 34.
73. Chouteau, W.L., Annaswamy, T.M., Bierner, S.M, Elliott, A.C., and Figueroa, I. (2010). Interrater reliability of needle electromyographic findings in lumbar radiculopathy. *American Journal of Physical Medicine & Rehabilitation*, **89**(7), 561–9.
74. Kendall, R. and Werner, R.A. (2006). Interrater reliability of the needle examination in lumbosacral radiculopathy. *Muscle & Nerve*, **34**(2), 238–41.
75. Misiaszek, J.E. (2003). The H-reflex as a tool in neurophysiology: its limitations and uses in understanding nervous system function. *Muscle & Nerve*, **28**(2), 144–60.
76. Banerjee, T.K., Mostofi, M.S., Us, O., Weerasinghe, V., and Sedgwick, E.M. (1993). Magnetic stimulation in the determination of lumbosacral motor radiculopathy. *Electroencephalography & Clinical Neurophysiology*, **89**(4), 221–6.
77. Bischoff, C., Meyer, B.U., Machetanz, J., and Conrad, B. (1993). The value of magnetic stimulation in the diagnosis of radiculopathies. *Muscle & Nerve*, **16**(2), 154–61.
78. Chokroverty, S., Sachdeo, R., Dilullo, J., and Duvoisin, R.C. (1989). Magnetic stimulation in the diagnosis of lumbosacral radiculopathy. *Journal of Neurology, Neurosurgery, and Psychiatry*, **52**(6), 767–72.
79. Ertekin, C., Nejat, R.S., Sirin, H., et al. (1994). Comparison of magnetic coil stimulation and needle electrical stimulation in the diagnosis of lumbosacral radiculopathy. *Clinical Neurology and Neurosurgery*. **96**(2),124–9.
80. Souayah, N. and Sander, H.W. (2006). Lumbosacral magnetic root stimulation in lumbar plexopathy. *American Journal of Physical Medicine & Rehabilitation*, **85**(10), 858–61.
81. Maccabee, P.J., Amassian, V.E., Eberle, L.P., et al. (1991). Measurement of the electric field induced into inhomogeneous volume conductors by magnetic coils: application to human spinal neurogeometry. *Electroencephalography & Clinical Neurophysiology*, **81**(3), 224–37.
82. Mills, K.R. and Murray, N.M. (1986). Electrical stimulation over the human vertebral column: which neural elements are excited? *Electroencephalography & Clinical Neurophysiology*, **63**(6), 582–9.
83. Dvonch, V., Scarff, T., Bunch, W.H., et al. (1984). Dermatomal somatosensory evoked potentials: their use in lumbar radiculopathy. *Spine*, **9**(3), 291–3.
84. Fardin, P., Lelli, S., Negrin, P., and Maluta, S. (1990). Radiation-induced brachial plexopathy: clinical and electromyographical (EMG) considerations in 13 cases. *Electromyography & Clinical Neurophysiology*, **30**(5), 277–82.
85. Harper, C.M., Jr, Thomas, J.E., Cascino, T.L., and Litchy, W.J. (1989). Distinction between neoplastic and radiation-induced brachial plexopathy, with emphasis on the role of EMG. *Neurology*, **39**(4), 502–6.
86. Lederman, R.J. and Wilbourn, A.J. (1984). Brachial plexopathy: recurrent cancer or radiation? *Neurology*, **34**(10), 1331–5.

CHAPTER 22

Neurophysiology in amyotrophic lateral sclerosis and other motor degenerations

Mamede de Carvalho and Michael Swash

Introduction

Degenerative disorders of the motor system usually require neurophysiological investigations for definitive diagnosis, since these tests reinforce and complement clinical examination. Neurophysiological tests, however, are not a substitute for clinical assessment and must always be designed and interpreted in relation to the clinical syndrome. In this chapter we will describe the features of lower motor neuron (LMN) change and show how they are used in diagnosis of motor degenerations, such as amyotrophic lateral sclerosis (ALS) and spinal muscle atrophy (SMA) (Box 22.1). We shall also address the assessment of cortical motor (UMN) changes that are a feature of ALS, and discuss the use of electrophysiological changes in the diagnosis of ALS.

Neurophysiological investigation in ALS has several roles. First, it is essential to confirm and test the clinical diagnosis by showing signs of widespread on-going LMN loss, in particular in clinically unaffected muscles, and by excluding other LMN conditions that resemble ALS. Secondly, it may be helpful in staging the disorder and, thirdly, it can help in establishing the prognosis. Lastly, it is a useful tool for monitoring disease progression.

Needle electromyographic investigation in motor degenerations

Up to 80% of the normal pool of MUs may be lost before wasting and weakness become clinically apparent (1–3). This reflects the capacity of MUs to expand their innervation field (4) by collateral sprouting which, at first, preserves muscle strength despite progressive loss of anterior horn cells. Reinnervated muscle fibres assume the same histological characteristics as those of the reinnervating motor units. Denervated muscle fibres are reinnervated by axonal sprouting from nearby nerve terminals, which causes clustering of muscles fibres innervated by the same MU, with the appearance of fibre-type grouping on histological examination.

Needle electromyography (EMG) is essentially selective, due to the small 'uptake area' characteristic of the electrode. However, needle EMG is a sensitive method to assess the characteristics of MU micro-anatomy. Reinnervation results in larger and more complex MUPs. However, the anatomical territory of the reinnervated MU, although larger (5), does not increase very much, as shown by scanning EMG studies (6), as a result of anatomical constraints within the muscle (7). Loss of LMNs causes reduced electrical activity on maximal muscle contraction, as evaluated by the interference pattern, although this is also modified in UMN dysfunction. UMN lesions or dysfunction reduce the normal variability in the firing rate of early recruited MUs (8). The interference pattern is useful in differentiating MUP loss from poor central recruitment, e.g. poor cooperation or spasticity. In the latter only a few MUPs are recruited, firing at a slow rate (3).

Box 22.1 Motor degenerations

- Amyotrophic lateral sclerosis.
- ALS variants:
  - progressive muscular atrophy;
  - progressive bulbar palsy, primary lateral sclerosis;
  - spinal muscular atrophy.
- Bulbospinal muscular atrophy (Kennedy's disease).
- Post-polio syndrome.

Motor unit potential analysis

The morphology of MUPs should be studied during slight voluntary contraction in terms of amplitude, duration, area, and number of phases, as well as the firing pattern (recruitment ratio), and MUP stability. The recruitment ratio is the ratio between the firing rate of the recruited MUP and the number of the other MUPs that discharge at the same time; the normal value is between 3 and 5. This measurement is generally increased in ALS, (3). In active denervation and reinnervation, MUPs are unstable, as a result of slowed distal motor nerve conduction in partially-myelinated regenerating nerve fibres, immature end-plates and atrophic muscle fibres (6). Newly-formed end-plates have immature acetyl choline receptor subunits, and a lower safety factor for neuromuscular transmission (6). MUP stability, the 'jiggle', is best evaluated using the 500 Hz or 1 KHz high-pass filter (9), giving important qualitative information (Fig. 22.1). Increased MUP amplitude results from clustering of

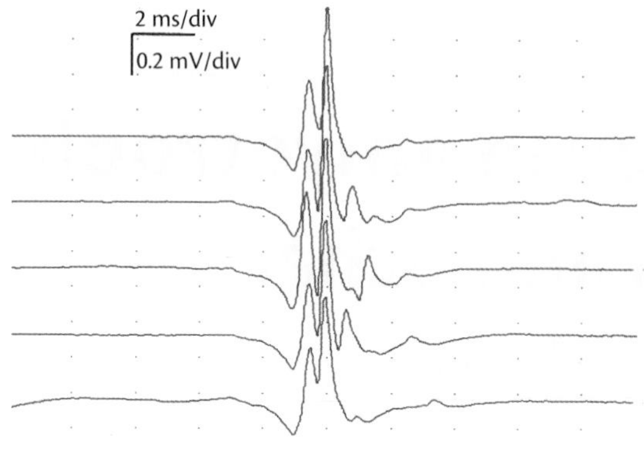

Fig. 22.1 Complex and very unstable motor unit with blocking observed in an ALS patient (high-pass filter of 500 Hz).

reinnervated muscle fibres. Increased MUP duration results from several different factors, including the number of innervated muscle fibres, end-plate dispersion, end-plate delay, and slow conduction through terminal branches, as well as slow conduction along small muscle fibres. The latter results in decreased synchronicity of muscle fibre activation within a motor unit, and increased MUP duration (3). Changes in amplitude and area are probably more sensitive as an index of neurogenic change than increased duration. Giant MUPs (above 10 mV) are rarely observed in ALS, although they can sometimes be found in slowly progressive cases, or other slowly progressive spinal disorders (3). In ALS, at an end-stage, small MUPs may be observed, possibly resulting from degeneration of distal nerve terminals through axonal dying-back pathology. An increased number of satellite potentials are also often observed in early affected muscles (10).

Single-fibre EMG

With this technique the number of muscles fibres belonging to a single MU, within a radius of 300μm from the tip of a single-fibre needle (fibre density), can be quantified. This number represents a correlate of fibre grouping observed in histological samples (6). This measurement is very sensitive in quantifying reinnervation (11,12). Although, the FD increases greatly in affected muscles, at an end-stage it decreases, representing a failure of the reinnervation process. This may not occur until less than 5% of the original neuronal pool survive (2).

SFEMG is also used to determine the 'jitter', that is the variation in the interval between the onset of the action potentials of two repeatedly firing muscle fibres belonging to the same MU. The neuromuscular jitter gives information about the stability of neuromuscular transmission (6). Increased MU instability anticipates neurogenic MU potentials as shown by concentric needle EMG (13) and 'jiggle' increases linearly over the first 6 months in association with developing reinnervation changes in muscles with normal MU potentials at the start of a prospective series of studies (14).

Macro-EMG

In this technique a modified needle electrode is used to record non-selectively from all muscles fibres of the same MU (15). Enlargement of the macro-EMG area is found in reinnervation, in general associated with an increased FD. This may be recorded at a very early stage of muscle involvement, although the FD may be abnormal before the macro-area has changed (6). In end-stage wasted muscles, both FD and macro-EMG may show a lower than normal value. In diseases with a slower course than ALS, the FD and the Macro area can reach much higher values than observed in ALS.

Multi-electrode surface EMG

For surface EMG (sEMG) studies, motor units are recorded via a multi-electrode pad placed on a forearm or thigh muscle. A single muscle fibre action potential is used as the triggering source for studies of firing pattern, as well as the physical aspects of volume conduction. Using the patterns of spatial spread of MUPs recorded over the skin, MU classification and the determination of current density and polarity can be done non-invasively. With high-density surface EMG, it is possible to analyse MUPs, permitting the detection of neurogenic MUPs, both by analysis of the raw signal itself and by analysis of extracted single MUPs (16). Moreover, FPs may also be analysed using this technique (17).

Active denervation

Fibrillation and positive sharp-waves at rest, which are considered a cardinal sign of denervation (18), are non-specific findings that can be recorded in myopathies, as well as in any neurogenic disorder. They represent action potentials generated by individual muscle fibres that have lost their nerve supply, either by axonal damage or by direct muscle fibre damage. Both fibrillation (fib) and positive sharp-waves (p-sw) appear to be generated from a biphasic intracellular action potential with a long hyperpolarization phase (19).

In ALS these spontaneous potentials are more frequently observed in muscles that are moderately or severely weak, with signs of LMN loss (20–22). In ALS fibs-sw persist because reinnervation is ineffective, although they may be sparse in very wasted muscles. They tend to decrease in cold muscles (23). The presence of abundant and diffuse fibs-sw is considered a poor prognostic sign (3). Patients with bulbospinal neuronopathy (Kennedy´s disease), Guamanian ALS, and spinal muscular atrophy type III have less prominent fibs-sw (3,24), consistent with the slow progression profile of these conditions.

Fasciculation potentials

Fasciculations are typically observed in ALS and may be its presenting clinical feature (25). Fasciculation potentials (FPs) are evident from the earliest clinical stages of the disease, even in muscles of normal strength, but generally become less evident in end-stage, wasted muscles (13). They are not absolutely essential to the electrophysiological diagnosis, but FPs are so regularly observed that one rarely accepts the diagnosis if no FPs are recorded (3). It is important to recognize that FPs may arise in other neurogenic disorders, for example in root lesions associated with cervical spondylosis. Needle EMG and ultrasound studies (26) may disclose FPs coming from deep layers of the muscle, although visual inspection detects only those arising in superficial layers. Electrophysiological studies of ALS have shown that FPs are present in clinically unaffected muscles in which MUPs are normal and stable, in the absence of fibs-sw (13), thus suggesting that increased segmental excitability is a very early feature of the disease—the excitotoxic hypothesis.

Although, fasciculations are a characteristic feature of ALS (27) their significance is unknown. Nonetheless, they probably represent motor neuronal hyperexcitability. Fasciculations were once

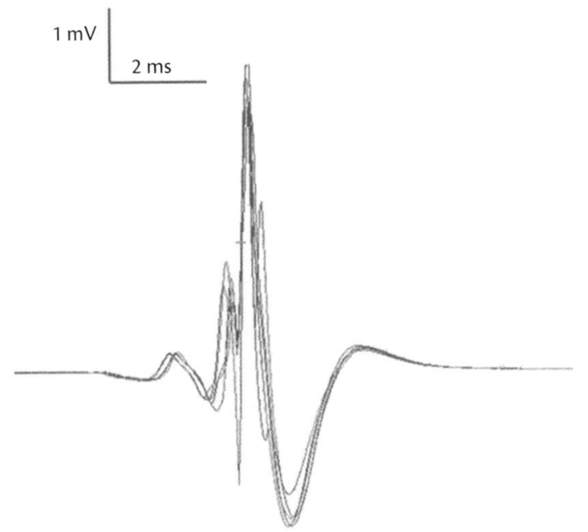

Fig. 22.2 An unstable fasciculation potential in ALS.

thought always to originate in the LMN, perhaps near their soma (18). However, Norris (28) noted that FPs could sometimes be recorded synchronously in different muscles innervated by the same myotome, but by different motor nerves, and suggested that these FPs had a central, perhaps cortical, origin. Wettstein (29) and Roth (30) used a collision technique to record FP F-waves to determine the origin of FPs and concluded that fasciculations can arise both in distal (about 80%) and proximal parts of the LMN. Kleine et al. have suggested predominantly distal origin of fasciculation potentials in ALS using high-density surface EMG (17). Conradi et al. (31) noted that FPs observed in severely affected muscles generally persisted after nerve block, suggesting a distal origin, and confirmed Denny-Brown and Pennybacker's (18) observation that some FPs could be voluntarily driven implying that these had an origin proximal to the distal LMN arborization. Voluntarily driven FPs have also been described by other authors (32,33). In addition, some FPs in ALS can be driven by cortical magnetic stimulation (34). This, and voluntary activation of a fasciculation potential, implies simply that a distal axonal generator can be over-ridden by more proximally generated neuronal activity, not that the FP has necessarily a motoneuronal or even an upper motor neuron origin.

Contrary to the normal morphology of the fasciculation potentials observed in muscles of subjects with benign fasciculations (13,33) and in ALS patients without signs of reinnervation, complex FPs are more frequently observed in ALS (Fig. 22.2). They can arise in abnormal, reinnervated, surviving motor units, or from a generator localized in the distal axonal branches (35,36).

Ephaptic crosstalk between adjacent distal motor axons from different motor units, or antidromic conduction from a distal generator toward a more proximal axonal branch point and then orthodromic transmission in non-refractory axon branches in this axonal arborization, could both lead to complex FPs in non-reinnervated units (33,37). Previous evidence (33,37) and findings using double-needle recording to detect time-locked FPs originating from different axons (38) suggest that the generator of these spontaneous potentials change during the course of muscle denervation in ALS.

An important practical issue in the electrodiagnosis of ALS is how long the EMG must be observed before a muscle can be identified as presenting fasciculation potentials with some degree of certainty. Mills (39) suggests that 90s of observation is required in each tested concentric needle insertion. It should be underlined that fasciculation potentials sometimes occur in healthy subjects (13,33), as well as in multiple motor neuropathy with conduction block (MMNCB), radiculopathies, cramp-fasciculation syndrome, electrolyte depletion, and in thyroid or parathyroid disorders.

Other forms of spontaneous activity

Cramp, a repetitive 200–300 Hz MUP discharge, accompanied by involuntary muscle contraction, are a common early symptom in ALS, frequently induced by voluntary contraction. In our experience, sometimes they can be so disturbing that full contraction for EMG investigation of the interference pattern may be impossible. Myokymic discharges may be associated with demyelination, but are rare in ALS, although they have been observed more frequently in facial muscles (40).

Complex repetitive discharges, an abrupt train of simple or complex spikes between 5–150Hz, are one example of non-specific spontaneous activity that occurs in situations of chronic denervation-reinnervation, or in myopathies. Its origin seems to be ephaptic transmission through an excitatory loop between adjacent muscle fibres (41). It may be observed in ALS, in particular in patients with slower progression. It is often found in the tongue (21).

Discharges resembling *myotonic discharges* occur extremely rarely in ALS. Such discharges, characteristic of membrane disturbance, may also be observed in patients with other long-lasting neurogenic disorders, such as radiculopathies and radiation plexopathy.

Which muscles should be studied by EMG?

The main purpose of the EMG is to show widespread loss of MUs. Therefore, a number of muscles should be assessed in the routine EMG investigation. The pattern of the EMG study should be related to the clinical problem. It is not necessary to examine muscles showing clear clinical features of weakness and/or wasting by EMG, but very important to assess the extent and distribution of neurogenic change in *clinically less affected, or clinically normal muscles*, in order to rule out other patterns of neurogenic change for example those characteristic of multiple root lesions or peripheral neuropathy. This issue is particularly critical in cranial-innervated muscles since fibs-sw is not so frequently observed in these muscles (21). Involvement of cranial muscles is very important in the diagnosis of ALS, since it excludes the diagnosis of cervical and lumbar spondylosis.

Orbicularis oris, masseter, trapezius, tongue (genioglossus), and sternomastoid muscles have been recommended for study in ALS, since these muscles are rostral to the cervical myotomes. However, it is very difficult to relax the tongue, in order to show fib-sw, and performing many needle insertions of the tongue is too invasive in clinical practice. However, the tongue is probably a more sensitive muscle to study for showing fibs-sw than the masseter, temporalis, frontalis, and mentalis muscles (42,43). Sternomastoid or trapezius muscles seem more sensitive than the frontalis and masseter (44–46). Qualitative EMG analysis showed that the sternomastoid muscle is abnormal in half of bulbar-onset patients and two-thirds of spinal onset patients, in which it is more sensitive than the tongue, although fibs-sw are not frequent in this muscle (45). As the entire extramedullary pathway of the spinal accessory nerve is protected from cervical bony compression, neurogenic change in the sternocleidomastoid muscle is of major diagnostic importance.

Intrinsic hand muscles are often involved early in ALS, and should always be studied (47). In particular the first dorsal interosseous is sensitive (21,48), easily investigated in all patients, and usually more severely affected than the abductor digiti minimi. In ALS the thenar hand muscles seems more severely affected than the hypothenar muscles, the 'split hand phenomenon' (49). Proximal muscles, such as the biceps, deltoid, or the extensor digitorum communis are useful for obtaining information about more proximal myotomes, especially for demonstrating FPs. In the lower limbs, distal muscles such as tibialis anterior regularly show fibs-sw in ALS (21,48); but the more proximal vastus medialis or the gastronemius should also be assessed.

Paraspinal muscles are affected early in ALS, at the same time as limb muscles innervated by the same spinal roots, and often show FPs, although fibs-sw are less frequent in paraspinal muscles than in distal limb muscles (50). Study of thoracic paraspinal muscles (51) may be helpful in confirming ALS, but cervical and lumbar spinal muscles may be misleading as degenerative changes in the elderly can cause abnormal MUPs in these muscles.

Abdominal muscles, internal intercostal muscles activated during expiration, and diaphragm activated during inspiration, are other useful muscles for testing. The superior part of rectus abdominis has also been indicated as a useful muscle to study in the diagnosis of ALS (T8-12) (52).

ALS may sometimes present with acute ventilatory failure, caused by severe loss of MUs in the respiratory muscles (53). Denervation of the respiratory muscles indicates impending respiratory failure (54). Involvement of the respiratory muscles and diaphragm is associated with denervation of paraspinal muscles (55), but diaphragmatic involvement cannot be predicted from involvement of limb muscles in the same myotome (56). Thus, EMG of the diaphragm can be useful in patients with probable respiratory involvement.

In our experience (21) patients with lower-limb onset always show fibs-sw in lower limb muscles, and the same is almost always true for patients in whom the weakness commences in upper limbs, when fibs-sw are easily found in first dorsal interosseous muscle. Regionally limited disease, in particular in the bulbar region, can raise problems regarding the confirmation of the diagnosis of ALS. The presence of widespread fibs-sw in lower limb or upper limbs, associated with diffuse fasciculation supports the diagnosis of ALS (21). It is important to understand that when one limb is affected, the next to be involved is often the homologous contralateral muscle, indicating the characteristic mode of spread of disease to contiguous motor neuronal cell columns within the cord (57).

Motor nerve conduction studies

Motor nerve conduction studies are essential in the diagnosis of ALS, in order to exclude axonal or demyelinating neuropathy and motor neuropathies with conduction block (58–61). Note that it is of the first importance to warm the limbs when studying nerve conduction in ALS patients, because wasted limbs tend to be cold and a satisfactory temperature may be difficult to maintain.

Certain motor conduction abnormalities are well recognized in ALS. The distal motor latency may be increased, a feature that has been correlated with the degree of weakness, and linked to slowly conducting, thin, distal, regenerated motor axons (2). Mild to moderate slowing of conduction velocity may be found in nerves innervating wasted muscles in ALS, due to random loss of myelinated motor nerve fibres, atrophy of axons, secondary changes in myelin or to the predominance of thin regenerated axons in the nerves innervating wasted muscles (2,58,62). Histological evidence of selective loss of large myelinated nerve fibres has been noted (63), but in other studies no preferential involvement of fast conducting fibres was found (64). Both the maximal and minimal motor conduction velocities are slowed (64), suggesting that faster and slower conducting motor fibres are similarly susceptible. This finding is consistent with the histological finding of involvement of somatic extrafusal and intrafusal (gamma) motor fibres in the disease (65). In ALS, significant conduction block or temporal dispersion is not observed, even when the motor responses are of very small amplitude (58).

In ALS, reduced CMAP amplitude represents the combined effect of denervation, muscle fibre atrophy, and compensatory reinnervation. In ALS, there is a significant correlation between CMAP amplitude and the degree of muscle weakness and atrophy (58). However, in very chronic conditions, such as old polio, the presence of a few giant remaining MUs may preserve normal mean CMAP amplitude in spite of moderate weakness.

In routine conduction studies only the fastest-conducting, large myelinated fibres are evaluated. The *collision technique* allows assessment of both the fast and slow conducting motor fibres in a motor nerve (66). This approach has shown that the minimum conduction velocity is also decreased in ALS (67), probably as a result of the presence of thin regenerating axons.

Late responses

F-waves are of increased amplitude, duration, and possibly, latency in upper motor neuron (UMN) syndromes (68). Although it is often stated that F-wave frequency is also increased, this assertion is poorly documented (68,69). Using single-fibre recording it seems that in conditions with spasticity, but no LMN dysfunction an increased number of F-waves can be detected from responding neurons, but there is no increase in the proportion of responding motor neurons (70). In ALS progressive loss of MUs is correlated with a decreased F-wave frequency (58), this is more important in determining F-wave persistence than the clinical signs of UMN involvement, at least in upper limbs (69). It is a common observation that, in severely weakened muscles in ALS, F-waves cannot be recorded. F-wave amplitude is frequently increased (70), which may be due to the presence of large remaining reinnervated MUs. F-waves show increased latency and dispersion, and there may be an increased frequency of repeater F-waves (69).

The *H reflex (Hoffmann reflex)* is a monosynaptic response, which assesses the integrity of both motor and sensory nerve fibres, and the nerve excitability at the spinal level. The most noticeable change in the presence of UMN lesion is the presence of H-reflex responses in muscles other than the calf muscles (Fig. 22.3). However, other changes may be observed, such as an increased H/M ratio, diminished inhibition of the H-reflex by simultaneous vibratory stimulation of the soleus muscle or its tendon, and lesser reduction of the amplitude of the H-reflex in a paired-stimulation paradigm (71). In ALS patients, dis-inhibition of the LMN may be observed, as revealed by the last two tests above, whether or not there are clinical signs of UMN lesion (71,72), suggesting a reduction of postsynaptic inhibition by Renshaw cells.

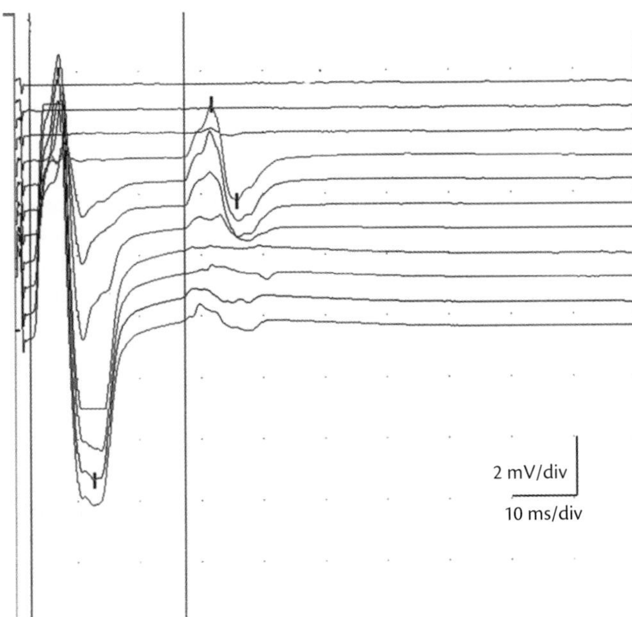

Fig. 22.3 The H reflex in a spastic patient with ALS recorded in hand muscle. Normally H reflexes are undetectable in this muscle.

Repetitive nerve stimulation

Abnormal decrement of the motor response with Repetitive Nerve Stimulation (RNS) was documented in ALS many years ago (73), in particular in wasted muscles and in patients with rapidly progressive disease. The decrement is usually less than 10% and, as in myasthenia gravis, it is enhanced after brief exercise, becoming maximal after 3–5 stimuli. Probably, this abnormal decrement results from disturbed electrical conduction through immature nerve terminals and neuromuscular junctions (74). The rate of decay of CMAP amplitude with RNS correlated with rate of change in CMAP over time (75).

Electromechanical coupling

In reinnervation, of any cause, an increased twitch tension may be observed, as a consequence of the increased MUP area. However, in the later stages of ALS the twitch tension decreases in spite of an enlarged macro EMG area, representing disturbed electromechanical coupling (76,77). This could be caused by mitochondrial abnormalities, which have been described in muscles in ALS, causing abnormal muscle metabolism and associated abnormalities of calcium release and re-uptake in the sarcoplasmic reticulum. The dissociation between electrical signal and force output of motor units in ALS contributes to muscle weakness in the later stages of the disease, in addition to loss of motor units and corticospinal dysfunction. In selected cases, combined studies of MUP amplitude and motor unit force output could be helpful in monitoring the course of the disease and in studying the therapeutic effects of drugs (78).

Table 22.1 summarizes the most important findings in ALS.

Respiratory muscle studies

Since the survival of the ALS patient is dependent on ventilatory function, neurophysiological studies of the respiratory muscles are potentially useful. The thinness of the diaphragm (3–4mm) and its movement during respiration creates some difficulty in using a needle EMG recording electrode. Nevertheless it is a useful technique, readily mastered, and can detect abnormal spontaneous activity in the diaphragm as well as abnormalities of motor unit potentials recorded during inspiration (54).

Percutaneous electrical stimulation of the phrenic nerve in the neck is simple and non-invasive (79,80). The diaphragmatic motor response amplitudes in ALS are symmetric (56) and correlate highly with transdiaphragmatic pressure (81). In ALS abnormally small motor responses suggest a weak diaphragm and therefore predict hypoventilation (80) and also survival in ALS patients (82). The response declines significantly in a three-month period (83). These electrophysiological studies should be considered in conjunction with functional pulmonary tests.

Table 22.1 A plan for a search for diagnostic abnormalities in a selection of muscles in a patient with suspected ALS

Changes	Relative importance
Cranial muscles (at least one)	
Fibs-sw	+/–
Fasciculations	+
Loss of MU	+
Complex MU	+
Instability	+
Upper and lower limb muscles (at least two muscles innervated by different nerves and roots, in each limb)	
Fibs-sw	+
Fasciculations	+
Loss of MU	+
Complex MU	+
Instability	+
Trunk muscles (at least one)	
Fibs-sw	+/–
Fasciculations	+
Loss of MU	+
Complex MU	+
Instability	+
Motor studies (at least one in each limb)	
No segmental demyelination	+
CMAP amplitude > 50% of normal - normal DML	+
CMAP amplitude > 50% of normal - normal F-waves	+
CMAP amplitude > 50% of normal - normal CV	+
Sensory studies (at least two in the most affected limb)	
Normal SNAPs	+

Sensory conduction studies in amyotrophic lateral sclerosis

Sensory potentials should always be studied to confirm that the disease process does not involve sensory neurons. However, very minor changes in sensory studies have been reported in at least one nerve in 23% of ALS patients, consistent with an associated mild axonal sensory neuropathy in 13% (84). Nevertheless, these minor changes are not proportional to the degree of involvement of motor nerves (4), and when these studies are abnormal, another diagnosis should be suspected. It should be remembered that sensory potentials decline in amplitude with age, and ALS often affects older subjects. Two exceptions are Kennedy´s disease, in which abnormal sensory potentials are frequently observed (85), and type 1 SMA (86).

More refined tests may discern some changes, as described using the near-nerve stimulation technique (87), measuring resistance to ischaemia (88), or testing quantitatively the threshold to vibration (89). Morphometry of the dorsal root ganglia showed some changes in sensory neurons (90) and loss of axons has been reported in the sural nerve (63), which may explain the abnormalities observed in some neurophysiological investigations. The strength-duration time constant using a threshold tracking system has been used to analyse the function of sensory fibres in ALS; no abnormality was found (91).

Autonomic function tests

In ALS autonomic nervous system involvement is rarely clinically evident. In Guamanian ALS (92) and in ventilated end-stage patients (93) clinical autonomic manifestations are more frequently observed, and in this last group of patients autonomic dysfunction can contribute to sudden death. Laboratory tests may reveal mild autonomic dysfunction in ALS (94,95). Cardiovascular sympathetic hyperactivity has been reported in ALS patients (96). Decreased heart rate variability has been associated with sudden death in ALS (97). Sympathetic skin response studies have revealed abnormalities in ALS (98).

Threshold tracking

Strength-duration time constant studies using a threshold tracking system have confirmed abnormal function of motor fibres in ALS (99), indicating that motor axons are hyperexcitable and show upregulation of persistent sodium channel conduction, and reduced potassium currents. This could be important in the generation of FPs and cramps in ALS. Threshold tracking is a complex technique of undetermined diagnostic utility although it may be useful in detecting axonal or neural hyperexcitability. This technique is discussed in Chapter 10.

Transcranial corticomotor stimulation studies

There is evidence for increased excitability due to reduced inhibition in the motor cortex. The cortical motor threshold is reduced in ALS (100), and hyperexcitability has been shown using peristimulus time histograms (101). Subsequent studies using cortical threshold tracking disclosed reduced short interval cortical inhibition (SICI) not only in patients with ALS (102), but also in familial SOD-1 ALS families before clinical onset of the disease (103). This test distinguished ALS from 'mimic disorders' (104). The technique is discussed in Chapter 17.

Differential diagnosis

Demyelinating neuropathies

Focal demyelination with partial conduction block causes weakness in demyelinating motor neuropathies. It causes segmental slowing of conduction, increased threshold, and ectopic discharges at the site of demyelination (105). Focal paranodal demyelination leads to dissipation of the action potential as a result of decreased impedance in the pathological region, so that depolarization of the next node fails to reach threshold, and conduction block occurs. Antibodies against myelin, such as the anti-GM1 antibody, leading to paranodal demyelination, or block of sodium channels, is associated with focal demyelination. Focal demyelination in motor nerves can be difficult to demonstrate, since it may occur proximal to the most proximal stimulation site, or more distal than the distal stimulation point, and may therefore resemble axonal loss. Proximal focal demyelination should be suspected when a normal CMAP amplitude is recorded from a weak, non-atrophic muscle (105) and, in particular, when the F-waves are very abnormal. Proximal conduction can be reliably detected in only a few nerves, and is more difficult to assess when atrophy and small CMAPs introduce further technical difficulties. Root stimulation can be performed using high-voltage electrical stimulation, needle stimulation, or magnetic stimulation over the spine. However, the first two are poorly tolerated, and the stimulation obtained by the last method is not always maximal. Proximal stimulation of lower limb nerves is not technically satisfactory. Pure motor neuropathies without conduction block, responsive to IVIg might perhaps be associated with very proximal conduction block (106), although in our experience this is a very rare finding.

In *multifocal motor neuropathy with conduction block* (59,60) only the motor nerve fibres are affected. This probably results from the different ceramide composition of GM1 in motor and sensory nerve fibres (107). In *motor and sensory demyelinating neuropathy with conduction block* (58) both the motor and sensory fibres are affected, although with motor predominance. *Pure motor neuropathies* have been described in patients with monoclonal gammopathy (108), a rare disease.

A number of criteria have been proposed for the definition of conduction block. This is particularly relevant, since, a 20% physiological (non-diagnostic) 'conduction block' in motor fibres may occur due to phase cancellation (105). The latter is likely to occur when there are small and dispersed motor responses. Ideally, conduction block should be identified by stimulation over short nerve segments. A consensus definition of conduction block has been agreed (109). Although such carefully defined criteria are useful, when conduction block is clinically relevant it is almost always very marked. It is important to establish an early diagnosis of focal demyelination, because patients with very wasted muscles are less likely to benefit from treatment with intravenous immunoglobulin (110).

Reports of several patients diagnosed as multifocal motor neuropathy with conduction block who have been found at autopsy to have loss of anterior horn cells and Bunina bodies in motor neurons has increased the difficulties of diagnosis and understanding in this field (111).

Cervical and lumbar canal stenosis

ALS can be difficult to recognize when there is spinal canal stenosis. Nonetheless, the presence of diffuse fibs-sw and very unstable MUPs in spondylosis is rare, unless there is also severe pain and with sensory loss. In these circumstances the clinical neurophysiologist should consider the clinical picture before reaching any conclusion. In addition, in ALS a second EMG investigation performed a few months later will show progression, with more widespread involvement of additional myotomes. Neuro-imaging is generally regarded as essential in ruling out clinically significant treatable spinal cord disease in all cases of suspected ALS.

Spinal muscular atrophy

SMA is a genetic, restricted LMN degeneration generally causing proximal, slowly progressive weakness, usually starting at a young age; the most difficult differential diagnosis is with muscular dystrophy, not ALS. However, differential diagnosis between ALS and distal spinal muscular atrophy, a rare condition (112), can be more difficult. In these cases it can be rewarding to study less affected muscles, to detect the very chronic, giant MUPs.

Old polio and post-polio syndrome

In patients with old polio some very wasted muscles may continue to show sparse fibs-sw, but in these patients almost all muscles show chronic changes, and giant MUPs are typically seen. Post-polio syndrome is a condition in which there is late clinical progression, as a result of a further loss of MUs, probably caused by the normal aging process in a limited functional pool of MUs. The differential diagnosis between post-polio and old polio with no progression, based on neurophysiological grounds is virtually impossible, since fibs-sw and increased jitter are common in residual polio, not indicating progressive disease (113). However, it is believed by some neurologists that the continuing motor unit loss with aging is more rapid in patients with previous polio than in healthy subjects (114).

Polymyositis and inclusion body myositis

These inflammatory myopathies can be mistaken clinically for ALS, but motor unit potential analysis will disclose myopathic potentials. It was previously thought that neurogenic potentials were found in inclusion body myositis, but this was not confirmed in newer studies using macro-EMG and quantitative motor unit analysis (115). However, MUPs may be of increased amplitude and to a lesser extent of increased duration, due to a combination of hypertrophy and atrophy of muscle fibres in the condition, accounting for this difficulty.

Electrodiagnostic criteria for amyotrophic lateral sclerosis

The importance of EMG in the diagnosis of this clinical disorder was recognized many years ago, in particular as a consequence of the work of Edward Lambert, who proposed a number of critical findings in the EMG investigation to support the clinical diagnosis of ALS (Lambert criteria) (20,116). These are described below:

1. Fibrillation and FPs in muscles of the upper and lower extremities (three different muscles with different root and peripheral nerve innervation), or in extremities as well as in muscles innervated by cranial nerves.
2. Reduction of the number and increase in the amplitude and duration of the motor unit potentials.
3. Motor nerve conduction velocities, which are normal when recorded from relatively unaffected muscles, and not less than 70% of average normal value, for age, when recorded from severely affected muscles.
4. Normal sensory nerve conduction responses.

These criteria are still applicable. The relative preservation of motor conduction velocities has been confirmed more recently (117,118). In addition, recognition of motor neuropathies mimicking ALS, some with conduction block (58–61), implies that these entities should be excluded in the EMG work-up of people with suspected ALS.

Another issue is the need for early diagnosis in the management of the disease, to alleviate diagnostic uncertainty and allow timely inclusion in clinical trials (119). In the future, as more effective treatments become available, it may be expected that early diagnosis will become increasingly important (119). The role of EMG in early diagnosis has been systematically evaluated. It is the only available method to support early diagnosis; since LMN abnormalities are typically evident in EMG before they are clinically recognizable (3,48). Distal muscles in any myotome tend to show more fibrillation and sharp-waves potentials (fibs-sw). Cranial innervated muscles do not often show fibs-sw (21,35), which might imply a later diagnosis for those patients presenting with bulbar features, if the strict Lambert criteria are used. On the other hand, weak and atrophic muscles show fibs-sw more consistently and in a greater number than strong muscles (20–22). As a result, a large proportion of ALS patients fail to show classical findings in the initial EMG study, mainly because needle investigation may not show widespread active denervation early in the disease (120). For these reasons, about one-third of ALS patient followed in one centre did not meet the Lambert criteria (120) when investigated. Nonetheless, these criteria are thought to have a high specificity for ALS.

Consensus criteria for diagnosis of amyotrophic lateral sclerosis

The El Escorial criteria for diagnosis of ALS were formulated to facilitate and standardize diagnosis in the context of research, especially clinical trials (121). They were not intended for use in clinical practice. They were revised in 1998 (Airlie House criteria Box 22.2) (122). However, these revised criteria did not explicitly encourage clinical and electrophysiological data to be used concordantly. Thus, the clinical diagnosis of ALS was defined as a separate process from electrophysiological diagnosis, perpetuating a separation derived from the seminal contributions of Lambert and Mulder at the Mayo Clinic (116). In most clinical trials definite ALS, probable ALS and probable-laboratory supported ALS have been used as entry criteria (121,122) Unfortunately, the sensitivity of the El Escorial criteria has proved to be low in clinical practice. For example, Traynor et al. (123) found that only 56% of 388 patients diagnosed clinically as suffering from ALS met the 'definite ALS' or 'probable ALS' criteria set out in the Airlie House recommendations, and as many as 10% of patients in their series died without achieving a level of diagnostic certainty greater than 'possible ALS'. Clearly, it is difficult for a physician to tell a patient with progressive and severe disability that their diagnosis is only probable or possible.

A consensus meeting held at Awaji-shima in late 2006 (Box 22.3) resolved these issues by recognizing the equivalence of clinical and EMG data in detecting chronic neurogenic change, and thus integrating EMG and clinical neurophysiological data into a single diagnostic algorithm (124). In addition, in this formulation of the diagnostic process, when a diagnosis of ALS is under consideration the importance of complex unstable MUPs as representing chronic neurogenic change with ongoing denervation was recognized and, in the presence of neurogenic MUPs, unstable and complex FPs were given equivalent importance to fibs-sw in recognition of progressive partial denervation. It was expected that this revised algorithm

Box 22.2 Revised El Escorial criteria

The diagnosis of amyotrophic lateral sclerosis [ALS] requires:

A—the presence of:

(A:1) Evidence of *lower motor neuron (LMN) degeneration* by clinical, electrophysiological, or neuropathological examination;

(A:2) evidence of *upper motor neuron (UMN) degeneration* by clinical examination;

and

(A:3) *progressive spread of symptoms or signs* within a region or to other regions, as determined by history or examination

Together with:

B—the absence of:

(B:1) Electrophysiological and pathological evidence of other disease processes that might explain the signs of LMN and/or UMN degeneration; and

(B:2) neuroimaging evidence of other disease processes that might explain the observed clinical and electrophysiological signs.

Diagnostic categories

Clinically definite ALS

Defined on clinical evidence alone by the presence of UMN, as well as LMN signs, in three regions.

Clinically probable ALS

Defined on clinical evidence alone by UMN and LMN signs in at least two regions with some UMN signs necessarily rostral to the above LMN signs.

The terms clinically probable ALS—laboratory-supported and clinically possible ALS are used to describe these categories of clinical certainty on clinical and investigative criteria or on clinical criteria alone:

Clinically probable—laboratory-supported ALS

Defined when clinical signs of UMN and LMN dysfunction are present in only one region, or when UMN signs alone are present in one region and LMN signs defined by EMG criteria are present in at least two limbs, with proper application of neuroimaging and clinical laboratory protocols to exclude other causes.

Clinically possible ALS

- UMN and LMN signs are found together in only one region; or
- UMN signs are found alone in two or more regions; or
- LMN signs are found rostral to UMN signs and the diagnosis of clinically probable—laboratory-supported ALS cannot be proven by evidence on clinical grounds in conjunction with electrodiagnostic, neurophysiologic, neuroimaging or clinical laboratory studies.
- Other diagnoses must have been excluded to accept a diagnosis of clinically possible ALS.

Reproduced from *Amyotroph Lat Scler*, **1**(5), Brooks BR, Miller RG, Swash M, Munsat TL, for the World Federation of Neurology Research Group on Motor Neuron Diseases, El Escorial revisited: revised criteria for the diagnosis of amyotrophic lateral sclerosis, pp. 293–300, copyright (2000), with permission from Taylor & Francis.

Box 22.3 Awaji consensus electrodiagnostic criteria for ALS

These criteria must be applied *in conjunction* with clinical findings, and other investigative findings, e.g. neuroimaging.

Nerve conduction studies

- Normal SNAP amplitude and sensory conduction velocities in the absence of concomitant entrapment or other neuropathies. Mildly reduced SNAP amplitudes and CVs in the presence of neuropathy of identified etiology are acceptable.
- If clinical signs of UMN lesion except for exaggerated deep tendon reflex are present, go to needle EMG criteria.
- Motor CV >75% of the lower limit of normal or the minimum F latency <130% of the upper limit of normal.
- Distal CMAP latency and its duration <150% of upper limit (UL) of normal.
- Absence of conduction block (CB). CB is defined by negative CMAP area reduction on proximal vs. distal stimulation > 50%, and distal baseline-negative peak CMAP amplitude must be >1 mV and an increase of proximal negative peak CMAP duration must be <30%.

Needle EMG

For the evaluation of LMN disease in any given body region clinical and electrophysiological abnormalities have equal weight and can be used together:

- In the presence of chronic neurogenic change, fasciculation potentials (preferably complex) are equivalent to fibrillations-sharp waves (fibs-sw) and complex repetitive discharges in the recognition of ongoing denervation.
- fibs-sw must be present in weak, wasted limb muscles.
- EMG features of chronic neurogenic change are defined as follows:
  - Motor units of increased amplitude, increased duration and usually increased number of phases, as assessed by qualitative or quantitative studies.

- Decreased motor unit recruitment, defined by rapid firing of a reduced number of motor units. In limbs affected by clinical features of significant UMN abnormalities rapid firing may not be achieved.
- Using a narrow band-pass filter (500 Hz to 5 KHz) unstable and complex motor units (Fig. 22.1) should be observed.
- Increased fibre density can also serve as an indicator of chronic neurogenic change.
- Chronic neurogenic change can be documented by the presence of 1 and 2, or by the presence of either 3 or 4.

Tests for UMN lesion

In the absence of clinical features of UMN lesion electrophysiological tests may be useful in detecting abnormality in the UMN pathway. Other causes of UMN lesion must be excluded by appropriate clinical and imaging investigation. Transcranial magnetic stimulation (TMS) requires that the target muscle has a detectable CMAP. The following features may be useful:

- Increased central motor conduction time (CMCT).
- Increased absolute latency to a tested muscle, provided that distal conduction slowing can be excluded.
- In patients with bulbar onset, the finding of an absent response to TMS is supportive of UMN lesion.
- The triple stimulation technique.
- F-wave studies, including F/M amplitude ratios.

NB: these suggested criteria for detection of UMN lesion by transcortical motor stimulation will need modification according to the newer studies discussed in the text of this chapter.

would facilitate early diagnostic categorization since it more closely accords with established clinical practice in the diagnosis of ALS. Nonetheless, there remains insensitivity in the detection of UMN abnormality in ALS (125), a problem that may yet be resolved by ongoing studies on corticomotor excitability in the disease (126).

The application of the Awaji algorithm to the diagnosis of ALS (Box 22.3) has been evaluated in many retrospective and prospective studies, and validated in a meta-analysis of the reported results in 8 studies (5 retrospective and 3 prospective studies) involving 1187 patients (127). There was a 23% increase in the proportion of patients diagnosed as definite or probable ALS as compared with the revised El Escorial criteria. The Airlie House/Awaji criteria are particularly useful in facilitating in the diagnosis of bulbar onset ALS. This process should therefore be applied to research studies, although in clinical practice the exercise of clinical judgement will always take precedence in a disease such as ALS for which there is no distinct diagnostic test.

Clinical neurophysiological assessment in clinical trials

Quantitative evaluation of the LMN pool is a feasible measure, and several methods are available.

CMAP amplitude

CMAP amplitude reflects the combined effect of denervation, muscle atrophy, and compensatory reinnervation, and is an indirect measure of the number of innervated fibres (2). In ALS, CMAP amplitude correlates well with muscle strength as determined by maximum voluntary isometric contraction (117,128), or as determined by electrical stimulation (129). There is also a significant correlation between CMAP amplitude and the motor unit number estimation (MUNE) (128,130), as expected since CMAP is a dependent variable for MUNE calculation. For these reasons CMAP amplitude has been included as a variable in several ALS clinical trials (128). However, CMAP amplitude measures do not enable detection of the effects of collateral reinnervation in masking MU loss, since successful reinnervation may maintain CMAP amplitude despite ongoing motor unit death (2,128). This effect will be most marked when there is relatively extensive reinnervation, as in slowly progressive ALS. Reinnervated MUPs often contain late components, causing temporal dispersion. Temporal dispersion, and phase cancellation, may cause a reduction in CMAP amplitude, independent of muscle fibre denervation, reducing the significance of any change in amplitude of the CMAP. Although the effect of temporal dispersion can be at least partially overcome if area measurement is used, the latter variable is no more sensitive than CMAP amplitude (131).

Neurophysiological index

De Carvalho and Swash developed a multimetric index, the neurophysiological index (NI) derived from the CMAP amplitude, F-wave frequency and distal motor latency of the ulnar nerve-abductor digiti minimi (ADM) system (117). This is a sensitive measure in evaluating progression in ALS patients (131). The NI is sensitive to LMN loss and weakness in ALS, whether the disease is rapidly or slowly progressive (131). A major advantage of the NI is that it is calculated from standard neurophysiological measurements. It can therefore be used in any neurophysiological laboratory, and requires no special equipment or expertise. It has been shown that it is more sensitive than CMAP amplitude (132) and with a rate of change similar to MUNE (132). It is reproducible in healthy controls (131) and in ALS patients (131,132) is a useful measure in detecting progression. Cheah et al. (133) showed that in ALS the NI reduced by 0.04 per week ($p < 0.0001$), indicating great sensitivity in all ALS types, including bulbar onset disease.

Motor unit number estimation

Motor unit number estimation, first described by McComas et al. in 1971 (1) is a quantitative electrophysiological method for calculating the number of functional motor units in a muscle, requiring division of the amplitude or area of the CMAP by the mean amplitude or area of the single motor unit potential derived from incremental nerve stimulation. Accurate determination of these two measures is essential, but difficult. Several techniques have been proposed for measurement of the mean single motor unit potential amplitude, the most critical measurement. MUNIX is a motor unit number index comparing different degrees of muscle contraction with CMAP amplitude, both measures expressed as power functions. This technique appears reliable and sensitive in controls and ALS patients (134,135). It is easy and quick to perform and does not require needle insertions.

Other techniques

In general, the F-wave frequency is reduced in weak muscles in ALS (4), but more studies are needed to apply them as a measurement of disease progression. SFEMG fibre density requires a period of training and experience to achieve competence. Macro-EMG is a relatively complex and invasive. This technique was studied longitudinally for 6 months in one study, but was not sensitive in detecting change (130). Successful longitudinal single motor unit tracking in patients with ALS, measuring change in functional innervation in a single motor unit, is possible. However, this is a technically challenging technique and is not recommended as a standard measure in sequential studies of ALS (136).

Related ALS syndromes

In primary lateral sclerosis the clinical syndrome is initially restricted to UMN disorder, although later, some five years after the onset more prominent LMN signs, first detected by EMG may become apparent. In some patients the syndrome evolves slowly into an ALS-like picture, but unlike typical ALS the syndrome does not usually progress rapidly. Diagnosis in the early stages depends on absence of LMN features (137).

Primary muscular atrophy (PMA) is characterized by progressive LMN signs without clinically evident UMN involvement, although at autopsy there are invariably UMN abnormalities, which were masked by the extent of the LMN change in life. The disorder may progress only slowly in some cases (138).

Monomelic syndrome is a rare disorder that presents with involvement of only one limb. The prognosis is variable. EMG may disclose subtle abnormalities in other limbs, suggesting the diagnosis of an ALS syndrome (138).

Progressive bulbar palsy, a syndrome commoner in women, progresses with development of more prominent LMN features in limb muscles. The latter are often present at first assessment, suggesting the diagnosis.

References

1. McComas, A.J., Sica, R.E., Campbell, M.J., and Upton, A.R. (1971). Functional compensation in partially denervated muscles. *Journal of Neurology, Neurosurgery, and Psychiatry*, **34**, 453–60.
2. Hansen, S. and Balantyne, J.P. (1978). A quantitative study of motor neurone disease. *Journal of Neurology, Neurosurgery, and Psychiatry*, **41**, 773–83.
3. Daube, J.R. (2000). Electrodiagnostic studies in amyotrophic lateral sclerosis and other motor neuron disorders. *Muscle & Nerve*, **23**, 1488–502.
4. Wohfart, G.L. (1957). Collateral regeneration from residual motor nerve fibres in ALS. *Neurology*, **7**, 124–34.
5. Erminio, F., Buchtal, F., and Rosenfalk, P. (1957). Motor unit territory and muscle fiber concentration in paresis due to peripheral nerve injury and anterior horn cell involvement. *Neurology*, **9**, 657–71.
6. Stalberg, E. (1982). Electrophysiological studies of reinnervation in amyotrophic lateral sclerosis. In: L. P. Rowland (Ed.) *Human motor neuron diseases*, pp. 47–57. New York, NY: Raven Press.
7. Kugelberg, E., Edstrom, L., and Abbruzzese, M. (1970). Mapping of motor units in experimentally reinnervated rat muscle. Interpretation of histochemical and atrophic fibre patterns in neurogenic lesions. *Journal of Neurology, Neurosurgery, and Psychiatry*, **33**, 319–29.
8. de Carvalho, M., Turkman, A., and Swash, M. (2012). Motor unit firing in amyotrophic lateral sclerosis and other upper and lower motor neurone disorders. *Clinical Neurophysiology*, **123**, 2312–8.
9. Stalberg, E. and Sonoo, M. (1994). Assessment of variability in the shape of the motor unit action potential, the 'jiggle,' at consecutive discharges. *Muscle & Nerve*, **17**, 1135–44.
10. Partanen, J. and Nousiainen, U. (1990). Motor unit potentials in a mildly affected muscle in amyotrophic lateral sclerosis. *Journal of Neurological Sciences*, **95**, 193–9.
11. Stalberg, E., Schwartz, M.S., and Trontelj, J.V. (1975). Single fibre electromyography in various processes affecting the anterior horn cell. *Journal of Neurological Sciences*, **24**, 403–15.
12. Swash, M. and Schwartz, M.S. (1982). A longitudinal study of changes in motor units in motor neuron disease. *Journal of Neurological Sciences*, **56**, 185–97.
13. de Carvalho, M. and Swash, M. (2013). Fasciculation potentials and earliest changes in motor unit physiology in ALS. *Journal of Neurology, Neurosurgery, and Psychiatry*, **84**, 963–8.
14. de Carvalho, M., Turkman, A., and Swash, M. (2014). Sensitivity of MUP parameters in detecting change in early ALS. *Clinical Neurophysiology*, **125**, 166–9.
15. Stalberg, E. (1980). Macro EMG, a new recording technique. *Journal of Neurology, Neurosurgery, and Psychiatry*, **43**, 469–74.
16. Drost, G., Stegeman, D.F., Schillings, M.L., et al. (2004). Motor unit characteristics in healthy subjects and those with postpoliomyelitis syndrome: a high-density surface EMG study. *Muscle & Nerve*, **30**, 269–76.
17. Kleine, B.U., Stegeman, D.F., Schelhaas, H.J., and Zwarts, M.J. (2008). Firing pattern of fasciculations in ALS: evidence for axonal and neuronal origin. *Neurology*, 353–9.
18. Denny-Brown, D. and Pennybacker, J.B. (1983). Fibrillation and fasciculation in voluntary muscles. *Brain*, **61**, 311–32.
19. Dumitru, D. (1999). Electrophysiologic basis for single muscle fiber discharge morphology. Thesis, University of Nijmegen.
20. Lambert, E.H. (1969). Electromyography in amyotrophic lateral sclerosis. In: F. H. Norris and L. T. Kurland (Eds) *Motor neuron diseases: research in amyotrophic lateral sclerosis and related disorders*, 135–53. New York, NY: Grune and Stratton.
21. de Carvalho, M., Bentes, C., Evangelista, T., and Sales-Luís, M. (1999). Fibrillation and sharp-waves: do we need them to diagnose ALS? *Amyotrophic Lateral Sclerosis*, **1**, 29–32.
22. Cappellari, A., Brioschi, A., Barbieri, S., Braga, M., Scarlato, G., and Silani V. (1999). A tentative interpretation of electromyographic regional differences in bulbar- and limb-onset ALS. *Neurology*, **52**, 644–6.
23. Denys, E.H. (1991). The influence of temperature in clinical neurophysiology. *Muscle & Nerve*, **14**, 795–811.
24. Ahlskog, J.E., Litchy, W.J., Peterson, R.C., et al. (1999). Guamanian neurodegenerative disease: electrophysiologic findings. *Journal of Neurological Sciences*, **15**, 28–35.
25. de Carvalho, M. and Swash, M. (2004). Cramps, muscle pain and fasciculations—not always benign? *Neurology*, **63**, 721–3.
26. Misawa, S., Noto, Y., Shibuya, K., et al. (2011). Ultrasonographic detection of fasciculations markedly increases diagnostic sensitivity of ALS. *Neurology*, **77**, 1532–7.
27. Li, T.M., Swash, M., Alberman, E., and Day, S.J. (1991). Diagnosis of motor neuron disease by neurologists: a study in three countries. *Journal of Neurology, Neurosurgery, and Psychiatry*, **54**, 980–3.
28. Norris, F.H. (1965). Synchronous fasciculations in motor neurone disease, *Archives on Neurology*, **13**, 495–500.
29. Wettstein, A. (1979).The origin of fasciculations in motor neuron disease. *Annuals in Neurology*, **5**, 295–300.
30. Roth, G. (1982). The origin of fasciculations, *Annuals in Neurology*, **12**, 542–54.
31. Conradi, S., Grimby, L., and Lundemo, G. (1982). Pathophysiology of fasciculations in ALS as studied by electromyography of single motor units. *Muscle & Nerve*, **5**, 202–8.

32. Guiloff, R.J. and Modarres-Sadeghi, H. (1992). Voluntary activation and fiber density of fasciculations in motor neuron disease. *Annuals in Neurology*, **31**, 416–24.
33. de Carvalho, M. and Swash, M. (1997). Fasciculation potentials: a study of amyotrophic lateral sclerosis and other neurogenic disorders. *Muscle & Nerve*, **21**, 336–44.
34. de Carvalho, M., Miranda, P.C., Lourdes Sales Luís, M., and Ducla-Soares, E. (2000). Neurophysiological features of fasciculation potentials evoked by transcranial magnetic stimulation in amyotrophic lateral sclerosis. *Journal of Neurology*, 247, 189–94.
35. de Carvalho, M. (2000). Pathophysiological significance of fasciculations in the early diagnosis of ALS. *Amyotrophic Lateral Sclerosis*, **1**(1), S43–6.
36. Mills, K.R. (2010). Characteristics of fasciculations in amyotrophic lateral sclerosis and the benign fasciculation syndrome. *Brain*, **133**, 3458–69.
37. Janko, M., Trontelj, J.V., and Gersak, K. (1989). Fasciculations in motor neuron disease: discharge rate reflects extent and recency of collateral sprouting. *Journal of Neurology, Neurosurgery, and Psychiatry*, **52**, 1375–81.
38. de Carvalho, M. and Swash, M. (2013). Origin of fasciculations in amyotrophic lateral sclerosis and benign fasciculations syndrome. *Journal of the American Medical Association: Neurology*, **70**, 1582–5.
39. Mills, K.R. (2011). Detecting fasciculations in amyotrophic lateral sclerosis: duration of observation required. *Journal of Neurology, Neurosurgery, and Psychiatry*, **82**, 549–51.
40. Whaley, NR. and Rubin, DI. (2010). Myokymic discharges in amyotrophic lateral sclerosis (ALS): a rare electrophysiologic finding? *Muscle & Nerve*, **41**, 107–9.
41. Trontelj, J.V. and Stalberg, E. (1983). Bizarre repetitive discharges recorded with single fibre EMG. *Journal of Neurology, Neurosurgery, and Psychiatry*, **46**, 305–9.
42. Preston, D.C., Shapiro, B.E., Raynor, E.M., and Kothari, M.J. (1997). The relative value of facial, glossal, and masticatory muscles in the electrodiagnosis of amyotrophic lateral sclerosis. *Muscle & Nerve*, **20**, 370–2.
43. Tankisi, H., Otto, M., Pugdahl, K., and Fuglsang-Frederiksen, A. (2013). Spontaneous electromyographic activity of the tongue in amyotrophic lateral sclerosis. *Muscle Nerve*, **48**, 296–8.
44. Finsterer, J., Erdorf, M., Mamoli, B. and Fuglsang-Frederiksen, A. (1998). Needle electromyography of bulbar muscles in patients with amyotrophic lateral sclerosis: evidence of subclinical involvement. *Neurology*, **51**, 1417–22.
45. Li, J., Petajan, J., Smith, G., and Bromberg, M. (2002). Electromyography of sternocleidomastoid muscle in ALS: a prospective study. *Muscle & Nerve*, **25**, 725–8.
46. Xu, Y.S., Zheng, J.Y., Zhang, S., and Fan, D.S. (2011). Upper trapezius electromyography aids in the early diagnosis of bulbar involvement in amyotrophic lateral sclerosis. *Amyotrophic Lateral Sclerosis*, **12**, 345–8.
47. Swash, M. (1980). Vulnerability of lower brachial myotomes in motor neurone disease: a clinical and single fibre EMG study. *Journal of Neurological Sciences*, **47**, 59–68.
48. Troger, M. and Dengler, R. (2000). The role of electromyography (EMG) in the diagnosis of ALS. *Amyotrophic Lateral Sclerosis*, **1**(1) S33–40.
49. Wilbourn AJ, Seveeney PJ. (1994). Dissociated wasting of the medial and lateral hand muscles with motor neuron disease. *Canadian Journal of Neurology Science*, **21**(2), 59.
50. de Carvalho, M., Pinto, S., and Swash, M. (2008). Paraspinal and limb motor neuron involvement within homologous spinal segments in ALS. *Clinical Neurophysiology*, **119**, 1607–13.
51. de Carvalho, M., Pinto, S., and Swash, M. (2009). Motor units changes in thoracic muscles in amyotrophic lateral sclerosis. *Muscle & Nerve*, **119**, 1607–13.
52. Xu, YS., Zheng, JY., Zhang, S., Kang, D., Zhang, J., and Fan, D. (2007). Needle electromyography of the rectus abdominis in patients with amyotrophic lateral sclerosis. *Muscle & Nerve*, **35**, 383–5.
53. de Carvalho, M., Matias, T., Coelho, F., Evangelista, T., Pinto, A., and Luis, M.L. (1996). Motor neuron disease presenting with respiratory failure. *Journal of Neurological Sciences*, **139**, 117–22.
54. Stewart, H, Eisen, A., Road, J., Mezei, M., and Weber, M. (2001). Electromyography of respiratory muscles in amyotrophic lateral sclerosis. *Journal of Neurological Sciences*, **191**, 67–73.
55. de Carvalho, M., Pinto, S., and Swash, M. (2010). Association of paraspinal and diaphragm denervation in ALS. *Amyotrophic Lateral Sclerosis*, **11**, 63–6.
56. Pinto, S. and de Carvalho, M. (2010). Symmetry of phrenic nerve motor response in amyotrophic lateral sclerosis. *Muscle & Nerve*, **52**, 822–5.
57. Simon, N.G., Lomem-Hoerth, C., and Kiernan, M.C. (2014). Patterns of clinical and electrodiagnostic abnormalities in early amyotrophic lateral sclerosis. *Muscle & Nerve*, **50**(6), 894–9.
58. Lewis, R.A., Sumner, A.J., Brown, M.J., and Asbury, A.K. (1982). Multifocal demyelinating neuropathy with persistent conduction block. *Neurology*, **32**, 958–64.
59. Parry, G.J. and Clarke, S. (1988). Multifocal acquired demyelinating neuropathy masquerading as motor neuron disease. *Muscle & Nerve*, **11**, 103–7.
60. Pestronk, A., Cornblath, D.R., Ilyas, A.A., et al. (1988). A treatable multifocal motor neuropathy with antibodies to GM1 ganglioside. *Annual in Neurology*, **24**, 73–8.
61. Evangelista, T., de Carvalho, M., Conceição, I., Pinto, A., and Sales-Luís, M. (1996). Motor neuropathies mimicking amyotrophic lateral sclerosis/motor neuron disease, *Journal of Neurology Science*, **139**, 95–8.
62. Borg, J. (1984). Conduction velocity and refractory period of single motor nerve fibers in motor neuron disease. *Journal of Neurology, Neurosurgery, & Psychiatry*, **47**, 349–53.
63. Bradley, W.G., Good, P., Rasool, C.G., and Adelman, L.S. (1983). Morphometric and biochemical studies of peripheral nerves in amyotrophic lateral sclerosis. *Archives of Neurology*, **41**, 267–77.
64. Ijima, M., Arasaki, K., Iwamoto, H., and Nakanishi, T. (1991). Maximal and minimal motor nerve conduction velocities in patients with motor neuron diseases: correlation with age of onset and duration of illness, *Muscle & Nerve*, **14**, 1110–15.
65. Swash, M. and Fox, K.P. (1974). Pathology of the muscle spindle-effect of denervation, *Journal of Neurology Science*, **22**, 785–9.
66. Ingram, D.A., Davis, G.R., and Swash, M. (1987). The double collision technique: a new method for measurement of the motor nerve refractory period distribution in man. *Electroencephalography & Clinical Neurophysiology*, **66**, 225–34.
67. Nakanishi, T., Tamaki, M., and Arasaki, K. (1989). Maximal and minimal motor nerve conduction velocities in amyotrophic lateral sclerosis, *Neurology*, **39**, 580–3.
68. Milanov, I.G. (1992). F-wave for assessment of segmental motoneurone excitability, *Electromyography Clinical Neurophysiology*, **32**, 11–15.
69. de Carvalho, M. and Swash, M. (2002). The F-waves and the corticospinal lesion in amyotrophic lateral sclerosis. *Amyotrophic Lateral Sclerosis*, **3**, 131–6.
70. Schiller, H.H. and Stalberg, E. (1978). F responses studied with single fibre EMG in normal subjects and spastic patients, *Journal of Neurology, Neurosurgery, & Psychiatry*, **41**, 45–53.
71. Argyropoulos, C.J., Panayiotopoulos, C.P., and Scarpalezos, S. (1978). F- and M-wave conduction velocity in amyotrophic lateral sclerosis, *Muscle & Nerve*, **1**, 479–85.
72. Drory, V.E., Kovach, I., and Groozman, G.B. (2001). Electrophysiologic evaluation of the upper motor neuron involvement in amyotrophic lateral sclerosis. *Amyotrophic Lateral Sclerosis*, **2**, 147–52.
73. Raynor, E.M. and Shefner, J.M. (1994). Recurrent inhibition is decreased in patients with amyotrophic lateral sclerosis. *Neurology*, **44**, 2148–53.
74. Mulder, D.W., Lambert, E.H., and Eaton, L.M. (1959). Myasthenic syndrome in patients with amyotrophic lateral sclerosis, *Neurology*, **9**, 627–31.

75. Bernstein, L.P. and Antel, J.P. (1981). Motor neuron disease decremental responses to repetitive nerve stimulation. *Neurology*, **31**, 202–4.
76. Wang, F.C., De Pasqua, V., Gerard, P., and Delwaide, P.J. (2001). Prognostic value of decremental responses to repetitive nerve stimulation in ALS patients. *Neurology*, **11**, 897–9.
77. Dengler, R., Konstanzer, A., Küther, G., Wolf, W., Hesse, S., and Struppler, A. (1990). Amyotrophic lateral sclerosis: macro-EMG and twitch forces of single motor units. *Muscle & Nerve*, **13**, 545–50.
78. Schmied, A., Pouget, J., and Vedel, J.P. (1999). Electromechanical coupling and synchronous firing of single wrist extensor motor units in sporadic amyotrophic lateral sclerosis. *Clinical Neurophysiology*, **110**, 960–74.
79. Bolton, C.F. (1993). AAEM Minimonograph # 40: clinical neurophysiology of the respiratory system, *Muscle & Nerve*, **16**, 809–18.
80. Pinto, S., Turkman, A., Pinto, A., Swash, M. and de Carvalho, M. (2009). Predicting respiratory insufficiency in amyotrophic lateral sclerosis: the role of phrenic nerve studies. *Clinical Neurophysiology*, **120**, 941–6.
81. Luo, Y.M., Lyall, R.A., Harris, M.L., Rafferti, G.F., Polkey, M.I., and Moxham, J. (1999). Quantification of the esophageal diaphragm electromyogram with magnetic phrenic nerve stimulation. *American Journal of Respiratory Critical Care Medicine*, **160**, 1629–34.
82. Pinto, S. and de Carvalho, M. (2012). Phrenic nerve studies predict survival in amyotrophic lateral sclerosis. *Clinical Neurophysiology*, **123**, 2454–9.
83. Pinto, S., Geraldes, R., Vaz, N., Pinto, A., and de Carvalho, M. (2009). Changes of the phrenic nerve motor response in amyotrophic lateral sclerosis: longitudinal study. *Clinical Neurophysiology*, **120**, 2082–5.
84. Pugdahl, K., Fuglsang-Frederiksen, A., de Carvalho, M., et al. (2007). Generalised sensory system abnormalities in amyotrophic lateral sclerosis: a European multicentre study. *Journal of Neurology, Neurosurgery, & Psychiatry*, **78**, 746–9.
85. Ferrante, M.A. and Wilbourn, A.J. (1997). The characteristic electrodiagnostic features of Kennedy's disease, *Muscle & Nerve*, **20**, 323–9.
86. Duman, O., Uysal, H., Skjei, K.L., Kizilay, F., Karauzum, S., and Haspolat, S. (2013). Sensorimotor polyneuropathy in patients with SMA type-1: electroneuromyographic findings. *Muscle & Nerve*, **48**, 117–21.
87. Shefner, J.M., Tyler, R.H., and Krarup, C. (1991). Abnormalities in the sensory action potential in patients with amyotrophic lateral sclerosis. *Muscle & Nerve*, **14**, 1242–6.
88. Shahani, B., Davies-Jones, G.A., and Russell, W.R. (1972). Motor neurone disease. Further evidence for an abnormality of nerve metabolism. *Journal of Neurology, Neurosurgery, & Psychiatry*, **34**, 185–91.
89. Mulder, D.W., Bushek, W., Spring, E., Karnes, J., and Dyck, P.J. (1983). Motor neuron disease (ALS): evaluation of detection thresholds of cutaneous sensation. *Neurology*, **33**, 1625–7.
90. Kawamura, Y., Dyck, P.J., Shimura, M., Okazaki, H., Tateishi, J., and Doi, H. (1981). Morphometric comparison of the vulnerability of motor and sensory neurons in amyotrophic lateral sclerosis. *Journal of Neuropathology and Experimental Neurology*, **40**, 667–75.
91. Mogyoros, I., Kiernan, M.C., Burke, D., and Bostock, H. (1998). Strength-duration properties of sensory and motor axons in amyotrophic lateral sclerosis, *Brain*, **121**, 851–9.
92. Low, P.A., Ahlskog, J.E., Petersen, R.C., Waring, S.C., Santillan, E.C., and Kurland, L.T. (1997). Autonomic failure in Guamanian neurodegenerative disease. *Neurology*, **49**, 1031–4.
93. Sato, K., Namba, R., Hayabara, T., Kashihara, K., and Morimoto, K. (1995). Autonomic nervous disorder in motor neuron disease: a study of advanced stage patients. *Internal Medicine*, **34**, 972–5.
94. Murata, Y., Harada, T., Ishizaki, F., Izumi, Y., and Nakamura, A. (1997). An abnormal relationship between blood pressure and pulse rate in amyotrophic lateral sclerosis. *Acta Neurologica Scandinavica*, **96**, 118–22.
95. Pisano, F., Miscio, G., Mazzuero, G., Lanfranchi, P., Colombo, R., and Pinelli, P. (1995). Decreased heart rate variability in amyotrophic lateral sclerosis. *Muscle & Nerve*, **18**, 1225–31.
96. Tanaka, Y., Yamada, M., Koumura, A., et al. (2013). Cardiac sympathetic function in the patients with amyotrophic lateral sclerosis: analysis using cardiac [123I] MIBG scintigraphy. *Journal of Neurology*, **260**, 2380–6.
97. Pinto, S., Pinto, A., and de Carvalho, M. (2012). Decreased heart rate variability predicts death in amyotrophic lateral sclerosis. *Muscle & Nerve*, **46**, 341–5.
98. Dettmers, C., Fatepour, D., Faust, H., and Jerusalem, F. (1993). Sympathetic skin response abnormalities in amyotrophic lateral sclerosis. *Muscle & Nerve*, **16**, 930–4.
99. Bostock, H., Sharief, MK., Reid, G. and Murray, NM. (1995). Axonal ion channel dysfunction in amyotrophic lateral sclerosis. *Brain*, **118**, 217–25.
100. Caramia, MD., Cicinelli, P., Paradiso, C., et al. (1991). 'Excitability' changes of muscular responses to magnetic brain stimulation in patients with central motor disorders. *Electroencephalography & Clinical Neurophysiology*, **81**, 243–50.
101. Weber, M. and Eisen, A. (2000). Peristimulus time histograms (PSTHs)—a marker for upper motor neuron involvement in ALS? *Amyotrophic Lateral Sclerosis and Other Motor Neuron Disorders*, **1**(2), S51–6.
102. Vucic, S. and Kiernan, M.C. (2008). Cortical excitability testing distinguishes Kennedy's disease from amyotrophic lateral sclerosis. *Clinical Neurophysiology*, **119**, 1088–96.
103. Vucic, S., Nicholson, G., and Kiernan, M.C. (2008). Cortical excitability may precede the onset of familial amyotrophic lateral sclerosis. *Brain*, **131**, 1540–50.
104. Vucic, S., Cheah, B.C., Yiannikas, C., and Kiernan, M.C. (2011). Cortical excitability may distinguish ALS from mimic disorders. *Clinical Neurophysiology*, **122**, 1860–86.
105. Kimura, J. (1997). Multifocal motor neuropathy and conduction block. In: J. Kimura and R. Kaji (Eds) *Physiology of ALS and related diseases*, pp. 57–72. Amsterdam: Elsevier Science.
106. Pakiam, A.S. and Parry, G.J. (1998). Multifocal motor neuropathy without overt conduction block. *Muscle & Nerve*, **21**, 243–5.
107. Ogawa-Goto, K., Funamoto, N, Abe, T., and Nagashima, K. (1990). Different ceramide compositions of gangliosides between human motor and sensory nerves. *Journal of Neurochemistry*, **55**, 1486–93.
108. Parry, G.J., Holtz, S.J., Ben-Zeev, D., and Drori, J.B. (1986). Gammopathy with proximal motor axonopathy simulating motor neuron disease. *Neurology*, **36**, 273–6.
109. Olney, R.K., Lewis, R.A., Putnam, T.D., and Campellone, J.V. (2003). Consensus criteria for the diagnosis of multifocal motor neuropathy. *Muscle & Nerve*, **27**, 117–21.
110. Bouche, P., Moullonguet, A., and Younes-Chennoufi, A.B. (1995). Multifocal motor neuropathy with conduction block—a study of 24 patients. *Journal of Neurology, Neurosurgery, & Psychiatry*, **59**, 38–44.
111. Veugelers, B., Theys, P., Lammens, M., Van Hees, J., and Robberecht, W. (1996). Pathological findings in a patient with amyotrophic lateral sclerosis and multifocal motor neuropathy with conduction block. *Journal of Neurology Science*, **136**, 64–70.
112. McLeod, J.G. and Prineas, J.W. (1971). Distal type of chronic spinal muscular atrophy. Clinical, electrophysiological and pathological studies. *Brain*, **94**, 703–14.
113. Ravits, J., Hallett, M., Baker, M., Nilsson, J., and Dalakas, M. (1990). Clinical and electromyographic studies of postpoliomyelitis muscular atrophy. *Muscle & Nerve*, **13**, 667–74.
114. McComas, A.J., Quartly, C., and Griggs, R.C. (1997). Early and late losses of motor units after poliomyelitis. *Brain*, **120**, 1415–21.

115. Luciano, C.A. and Dalakas, M.C. (1997). Inclusion body myositis: no evidence for a neurogenic component. *Neurology*, **48**, 29–33.
116. Lambert, E.H. and Mulder, D.W. (1957). Electromyographic studies in amyotrophic lateral sclerosis. *Mayo Clinic Proceedings*, **32**, 441–6.
117. Cornblath, D.R., Kuncl, R.W., Mellits, E.D., et al. (1992). Nerve conduction studies in amyotrophic lateral sclerosis. *Muscle & Nerve*, **15**, 1111–15.
118. De Carvalho, M. and Swash, M. (2000). Nerve conduction studies in amyotrophic lateral sclerosis. *Muscle and Nerve*, **23**, 344–52.
119. Swash M. (2000). Shortening the time to diagnosis in ALS: the role of electrodiagnostic studies. *Amyotrophic Lateral Sclerosis*, **1**(1), S67–72.
120. Behnia, M. and Kelly, J. (1991). Role of electromyography in amyotrophic lateral sclerosis. *Muscle & Nerve*, **14**, 1236–41.
121. Brooks, B.R. and World Federation of Neurology Sub-Committee on Motor Neuron Diseases. (1994). El Escorial WFN criteria for the diagnosis of amyotrophic lateral sclerosis. *Journal of Neurological Sciences*, **124**, 965–1085.
122. Brooks, B.R., Miller, R.G., Swash, M., and Munsat, T.L. (2000). For the World Federation of Neurology Research Group on Motor Neuron Diseases. El Escorial revisited: revised criteria for the diagnosis of amyotrophic lateral sclerosis. *Amyotrophic Lateral Sclerosis*, **1**, 293–300.
123. Traynor, B.J., Codd, MB., Corr, B., et al. (1997). Clinical features of amyotrophic lateral sclerosis according to the El Escorial and Airlie House diagnostic criteria. *Archives of Neurology*, **49**, 1292–8.
124. De Carvalho, M., Dengler, R., Eisen, A., et al. (2008). Electrodiagnostic criteria for diagnosis of ALS: Consensus of an International Symposium sponsored by IFCN, December 3–5 2006, Awaji-shima, Japan. *Clinical Neurophysiology*, **119**, 407–503.
125. Swash, M. (2012). Why are upper motor neuron signs difficult to elicit in amyotrophic lateral sclerosis? *Journal of Neurology, Neurosurgery, & Psychiatry*, **83**, 659–62.
126. Geevasinga, N., Menon, P., Yiannikas, C., Kiernan, M.C., and Vucic, S. (2014). Diagnostic utility of cortical excitability studies in amyotrophic lateral sclerosis. *European Journal of Neurology*, **21**(12), 1451-7.
127. Costa, J., Swash, M., and de Carvalho, M. (2012). Awaji criteria for the diagnosis of amyotrophic lateral sclerosis: a systematic review. *Archives in Neurology*, **69**, 1410–16.
128. de Carvalho, M., Chio, A., Dengler, R., Hecht, M., Weber, M., and Swash, M. (2005). Neurophysiologic measures in amyotrophic lateral sclerosis: markers of progression in clinical trials. *Amyotrophic Lateral Sclerosis*, **6**, 17–28.
129. Kelly, J.J., Thibodeau, L., Andres, P.L., and Finison, L.J. (1990). Use of electrophysiologic tests to measure disease progression in ALS therapeutic trials. *Muscle & Nerve*, **13**, 471–9.
130. Bromberg, M., Fries, T., Forshew, D., and Tandan R. (2001). Electrophysiological endpoint measures in a multicenter ALS drug trials. *Journal of Neurology Science*, **184**, 51–4.
131. de Carvalho, M., Lopes, A., Scotto, M., and Swash, M. (2001). Reproducibility of neurophysiological and myometric measurement in the ulnar nerve-abductor digiti minimi system, *Muscle & Nerve*, **24**, 1391–5.
132. de Carvalho, M., Scotto, M., Lopes, A., and Swash, A. (2005). Quantitating progression in ALS. *Neurology*, **24**, 1783–5.
133. Cheah, B.C., Vucic, S., Krishnan, A.V., Boland, R.A., and Kiernan, M.C. (2011). Neurophysiological index as a biomarker for ALS progression: validity of mixed effects models. *Amyotrophic Lateral Sclerosis*, **12**, 33–8.
134. Neuwirth, C., Nandedkar, S., Stalberg, E., et al. (2011). Motor Unit Number Index (MUNIX): a novel neurophysiological marker for neuromuscular disorders; test-retest reliability in healthy volunteers. *Clinical Neurophysiology*, **122**, 1867–72.
135. Boekestein, W.A., Schelhass, H.J., van Putten, M.J., Stegeman, D.F., Zwarts, M.J., and van Dijk, J.P. (2012). Motor unit number index (MUNIX) versus motor unit number estimation (MUNE): a direct comparison in a longitudinal study of ALS patients. *Clinical Neurophysiology*, **123**, 1644–9.
136. Gooch, C.L. and Harati, Y. (1997). Longitudinal tracking of the same single motor unit in amyotrophic lateral sclerosis. *Muscle & Nerve*, **20**, 511–3.
137. Pringle, C.E., Hudson, A.J., Muniz, D.G., Kiernan, J.A., Brown, W.R., and Ebers, G.C. (1992). Primary lateral sclerosis. Clinical features, neuropathology and diagnostic criteria. *Brain*, **115**, 495–520.
138. de Carvalho, M. and Swash, M. (2007). Monomelic neurogenic syndromes: a prospective study. *Journal of Neurology Science*, **263**, 26–34.

CHAPTER 23

Clinical aspects of neuromuscular junction disorders

Donald B. Sanders

Neuromuscular junction disorders

Myasthenia gravis

Acquired myasthenia gravis (MG) is an autoimmune disease in which weakness results from an immunologic attack against the neuromuscular junction (NMJ). MG is not common, affecting approximately 140 people per million in the USA. MG most often begins in the 3rd or 4th decade of life in women and after age 50 in men, but it can affect either sex at any age. As the population has aged, MG now affects more men than women and begins after age 50 in most patients.

Weakness in MG typically varies during the day, usually being least in the morning and becoming worse as the day progresses, especially after prolonged use of affected muscles. For example, ocular symptoms typically become worse while reading, watching television, or driving, especially in bright sunlight. Jaw muscle weakness typically becomes worse during prolonged chewing.

In most patients symptoms initially fluctuate over the short term, but become progressively more severe during the first few years. Periods of spontaneous improvement, even remission, are common, especially early in the disease, but are rarely permanent. Weakness becomes maximal during the first year in two-thirds of patients. Weakness remains limited to ocular muscles in about 10% of patients (ocular myasthenia—OM). Clinical worsening in MG has been described following administration of aminoglycoside, macrolide, and fluoroquinolone antibiotics, magnesium, calcium-channel blockers, and iodinated intravenous contrast agents.

The diagnosis of MG may be elusive, particularly in mild cases, and is frequently delayed for months or even years. The most important factor in arriving at the diagnosis is thinking of MG when the patient complains of muscle dysfunction that varies from time to time, especially if symptoms involve eyelid and extraocular muscle weakness.

The examination of patients suspected of having MG must be performed in a way to detect variable weakness in specific muscle groups. Strength should be assessed repetitively during maximum effort and again after rest. Fluctuating strength in MG is most easily appreciated by observing ocular and oropharyngeal muscle function.

Almost all patients with MG have weakness of ocular muscles, although close examination may be necessary to demonstrate this in some patients. Asymmetrical weakness of several muscles in both eyes is typical. The pattern of weakness doesn't usually fit that of lesions affecting one or more individual nerves, and pupillary reactions are normal. Eyelid ptosis is usually asymmetrical and varies during sustained activity.

Palatal weakness produces nasal speech. Weakness of the laryngeal muscles causes hoarseness and difficulty making a high-pitched 'eeeee' sound. If jaw muscles are weak, the patient may have to close the jaw manually.

Any trunk or limb muscle can be weak, but some are more often affected than others: neck flexors are usually weaker than neck extensors, and the deltoids, triceps, extensors of the wrist and fingers and dorsiflexors of the ankles are frequently weaker than other limb muscles. In about 10% of patients, symptoms begin in one or more of these limb muscle groups, producing a clinical pattern more suggestive of neuropathy or focal myopathy.

The differential diagnosis of muscle weakness or oculomotor symptoms is broad, but in most patients the correct diagnosis is usually apparent from the history and the examination. This is particularly true if eyelid ptosis clearly fluctuates or alternates from side to side. In patients with less typical manifestations, the differential includes motor neuron disease, primary muscle diseases, central nervous system lesions affecting the brainstem nuclei, cavernous sinus thrombosis, various toxins, botulism, diphtheritic neuropathies, and other rare conditions.

The diagnosis is confirmed by demonstrating abnormal neuromuscular transmission by electrodiagnostic testing or by observing improvement after administration of cholinesterase inhibitors; and, in most patients, finding serum antibodies to the acetylcholine receptor (AChR) or muscle specific kinase (MuSK).

Pathophysiology of MG

In MG associated with AChR antibodies the physiologic abnormality results from reduction in the concentration of AChR on the muscle endplate, complement-mediated damage that distorts and simplifies the postsynaptic muscle membrane, and blockade of AChR by antibodies. Acetylcholine (ACh) is released normally from the nerve, but its effect on the muscle is reduced as a consequence of these endplate changes and the probability that a nerve impulse will generate a muscle action potential is reduced.

Diagnostic testing in MG

The edrophonium (Tensilon©) test

Weakness caused by abnormal neuromuscular transmission characteristically improves after administration of cholinesterase inhibitors, such as edrophonium chloride. The edrophonium test is most reliable when it produces dramatic improvement in eyelid ptosis, ocular muscle weakness, or dysarthria. Changes in strength of other muscles must be interpreted cautiously, especially in a suggestible patient. The edrophonium test is positive in more than 90% of patients with MG (1). However, a response to edrophonium is not unique to MG and may also be seen in motor neuron disease, with lesions of the oculomotor nerves and in muscle disease affecting the ocular muscles.

The ideal dose to be administered cannot be predetermined. A single fixed dose, such as 10mg, is too much for many patients, thus incremental injections of small doses is recommended. Two milligrams should be injected initially and the response monitored for 60 s. Subsequent injections of 3 and 5 mg are then given. If improvement is seen within 60 s after any dose, no further injections are necessary. Weakness that develops or worsens after injection of less than 10mg edrophonium also indicates a neuromuscular transmission defect, as this dose will not weaken normal muscle.

Some patients who do not respond to intravenous edrophonium may respond to intramuscular neostigmine, because of its longer duration of action. Intramuscular neostigmine is particularly useful in infants and children whose response to edrophonium may be too brief for accurate observation. In some patients, a therapeutic trial of oral pyridostigmine or neostigmine for several days may produce improvement that can't be appreciated after a single dose of edrophonium or neostigmine.

Electrodiagnostic testing in MG

Abnormal neuromuscular transmission can be confirmed by electrodiagnostic testing. These tests are particularly valuable when the clinical findings, antibody testing, and response to cholinesterase inhibitors do not give conclusive diagnostic information. See chapter 7 for detailed descriptions of these techniques.

Repetitive nerve stimulation (RNS) is the most commonly used electrodiagnostic test in MG. A conclusive decrementing response to RNS confirms that neuromuscular transmission is abnormal in the tested muscle. If the patient has clinical features consistent with MG and there is no other nerve or muscle disease to produce abnormal neuromuscular transmission, this is strong confirmation of the diagnosis. The major limitation to RNS testing in MG is its relative insensitivity.

To obtain the maximum diagnostic yield, it may be necessary to examine several muscles, including those that are most involved in the patient. Facial and proximal muscles are usually affected earlier and more severely than distal muscles. If several muscles are examined, selected on the basis of the clinical findings, RNS demonstrates a decrementing response in most MG patients with weakness in limb muscles. RNS in proximal or facial muscles detect abnormalities in almost all MG patients with moderate or severe weakness.

In a comparison study, we found an abnormal decrement in a hand or shoulder muscle in 75% of patients with generalized MG, and in 50% of those with ocular myasthenia (2). The results from a proximal shoulder muscle and a hand muscle were concordant in most patients, but occasionally a decrement was seen only in a proximal muscle; only rarely was the hand muscle abnormal if the shoulder was normal.

Measurement of the neuromuscular jitter is the most sensitive clinical test of neuromuscular transmission and shows increased jitter in some muscles in almost all patients with MG. Jitter is greatest in weak muscles, but is usually increased even in muscles with normal strength. In most patients with MG, the jitter abnormality is greater and found more often in facial than in limb muscles (2). This distribution pattern is not invariable, however: rarely jitter is normal in the face and increased in the arm.

When myasthenic weakness is limited to ocular muscles, 60% of examinations show increased jitter in the arm (2). In the rare patient with weakness limited to a few limb muscles, it may be necessary to examine a weak muscle to demonstrate increased jitter—this is particularly true in some patients with MuSK MG (3).

Jitter is more often increased in any given muscle in patients with more severe disease. However, jitter varies markedly among patients with similar weakness and overall disease severity cannot be inferred from the amount of jitter alone.

Jitter is usually abnormal even when the patient is taking cholinesterase inhibitors. However, in rare patients with purely ocular or only mild limb weakness, jitter is increased only after cholinesterase inhibitors have been discontinued (4). Although it is not necessary to withhold cholinesterase inhibitors before jitter studies in all patients, the diagnostic yield will be higher if this is done in patients with mild disease. If jitter is normal while the patient is taking these medications, testing should be repeated after they have been withheld for at least 24 hours.

Jitter is also increased in diseases of nerve and muscle; these must be excluded by other electrophysiological and clinical examinations before concluding that the patient has MG. If neuronal or myopathic disease is present, increased jitter does not indicate that MG is also present.

The most efficient sequence of electrodiagnostic testing in patients who are being investigated for possible MG depends on the clinical findings, and on the techniques that are readily available in the testing laboratory.

If there is weakness in limb muscles, testing can begin with RNS in a shoulder muscle. If there is an unequivocal decrement in this muscle, this is adequate to conclude that neuromuscular transmission is abnormal. If the decrement is normal or only slightly increased, a facial muscle is then tested. A significant decrement in this muscle confirms that neuromuscular transmission is abnormal. If RNS is normal in both of these muscles, the patient still may have MG.

If jitter measurements are readily available, an alternative testing procedure may be more efficient. One approach is to measure jitter if RNS testing is normal in a shoulder muscle. Alternatively, considering its greater sensitivity, jitter may be measured as the initial test, particularly in patients with focal or mild disease in whom RNS testing is unlikely to be abnormal. If symptoms are mild or limited to the ocular muscles, it is most efficient to begin jitter testing with a facial muscle, either the frontalis or orbicularis oculi. If the first facial muscle tested is abnormal, another facial muscle should then be tested. If jitter is normal in all tested muscles in a patient with mild symptoms, the examination should be repeated after a few weeks. If jitter is normal in a muscle with definite weakness, the weakness is not due to MG.

When abnormal neuromuscular transmission has been demonstrated by RNS, it is not necessary to measure jitter to confirm the diagnosis. However, it may be useful to obtain baseline jitter values for comparison if subsequent studies will be performed to monitor the response to treatment. Jitter in the orbicularis may not correlate well with clinical improvement; the frontalis or extensor digitorum may provide better correlation. Even when jitter is increased, it still may be useful to perform the RNS test in a distal muscle to exclude the Lambert–Eaton myasthenic syndrome (LEMS) or to provide a baseline to follow the effects of treatment.

Antibody testing in MG

Serum antibodies that bind human AChR are found in 80% of patients with acquired generalized MG and 55% with ocular myasthenia. The AChR antibody level varies widely among patients with MG of similar severity and does not predict the severity of disease in individual patients. The antibody level may be normal or low at symptom onset and become elevated later, thus repeat testing after several months is appropriate when the initial level is normal.

Antibodies to MuSK are found in up to 50% of patients with generalized MG who lack AChR antibodies (5) and in some patients with ocular myasthenia as well (6,7) (MuSK MG).

An elevated AChR or MuSK antibody level in a patient with compatible clinical features confirms the diagnosis of MG, but normal antibody levels do not exclude the diagnosis.

Muscle specific kinase myasthenia gravis

MuSK MG predominantly affects females, and may begin from childhood through middle age. In some patients, the clinical findings are indistinguishable from MuSK-negative MG, with fluctuating ocular, bulbar, and limb weakness. However, many Musk MG patients have predominant weakness in cranial and bulbar muscles, frequently with marked atrophy of these muscles (8). Others have prominent neck, shoulder, and respiratory muscle weakness, with little or no involvement of ocular or bulbar muscles. Electrodiagnostic abnormalities may not be as widespread as in other forms of MG and it may be necessary to examine different muscles to demonstrate abnormal NMT (8).

Many MuSK MG patients do not improve with cholinesterase inhibitors—some actually become worse, and many have profuse fasciculations with these medications (9). The diagnosis of MuSK MG may be particularly challenging when the clinical features, electrodiagnostic findings, and response to cholinesterase inhibitors differ from typical MG.

Treatment of MG

The best treatment for each patient with MG is determined by many factors, including the severity, distribution and rate of progression of weakness, the age and sex of the patient, the degree of functional impairment and comorbidities.

In patients with ocular myasthenia, pyridostigmine is used until or unless the disease spreads to oropharyngeal or limb muscles—if this is going to occur, it will practically always do so within two years after onset of symptoms. If the ocular weakness is disabling despite optimum pyridostigmine treatment, treat with prednisone, and in some patients, consider immunosuppression or thymectomy.

In patients with generalized MG and onset of symptoms before age 50, thymectomy should be considered. If there is oropharyngeal or respiratory muscle weakness, treat with therapeutic plasma exchange (TPE) or immune immunoglobulin (IVIg) before thymectomy. If weakness, particularly of oropharyngeal muscles, remains consider treating with steroids before thymectomy. If thymectomy is performed without immunosuppression and significant weakness persists 12 months after surgery, steroids or other immunosuppression should then be considered.

If steroids fail to produce improvement, consider TPE and the addition of azathioprine. Azathioprine may be given as the initial immunosuppressant for relatively mild disease or if steroids are contraindicated. Azathioprine may also be used as a steroid-sparing agent in patients who have improved on steroids. Cyclosporine or mycophenolate mofetil may be used instead of azathioprine if the latter is ineffective or produces unacceptable side-effects.

In patients over 50 years of age, concurrent illness becomes a major factor in determining the best therapeutic approach. If there is no evidence of thymoma and pyridostigmine does not produce a satisfactory response, treat with azathioprine, steroids, TPE, IVIg, or a combination of these, tapering the pyridostigmine and steroid to the minimal maintenance dose as rapidly as feasible. Cyclosporine or mycophenolate mofetil may be used as an alternative to azathioprine, as above. Consider thymectomy in these patients if this approach has not produced a satisfactory response. When patients have achieved their optimal response to medical therapy, medications should be reduced gradually and sequentially to the fewest and the smallest dose of each necessary to maintain this response. As a general rule, medications with the shortest duration of action are reduced first.

Surgery should be planned in virtually all patients with a thymoma to remove the tumour and all thymus tissue. Thymomas in stage I and II with a WHO classification type A, AB, and B1 have a low risk of recurrence. Thymomas in stage II and WHO type B2 and B3 and all stages III and IV should be treated with an interdisciplinary approach after standard radiation therapy.

Ocular myasthenia

Ptosis and/or diplopia are the initial symptoms of MG in up to 85% of patients (10), and almost all patients have both symptoms within two years of disease onset. Myasthenic weakness that remains limited to the ocular muscles is termed ocular myasthenia, and comprises approximately 10–15% of all MG in Caucasian populations. If weakness remains limited to the ocular muscles after two years, there is a 90% likelihood that the disease will not generalize. Ocular myasthenia is more common in Asian populations (up to 58% of all MG patients) (11).

Confirmation of the diagnosis of ocular myasthenia may be a challenge as RNS studies and AChR antibodies are often negative, in which case jitter measurements are particularly useful in confirming the diagnosis.

In patients with purely ocular weakness, pyridostigmine is used until or unless the disease spreads to oropharyngeal or limb muscles. If the ocular muscle weakness is disabling despite optimum pyridostigmine treatment, steroids, immune-suppression and in some patients, thymectomy should be considered.

Childhood MG

The onset of immune-mediated MG before age 18 is referred to as juvenile MG (JMG) (12). Twenty percent of JMG and almost 50% of those with onset before puberty are seronegative—the distinction from a congenital myasthenic syndrome is most challenging in the latter group (see Congenital Myasthenic Syndromes, later in this chapter). A beneficial response to TPE or IVIg may help to establish the diagnosis of autoimmune MG. Many children who are initially seronegative later develop AChR antibodies. Thymomas are rare, but not unheard of in this age group.

Cholinesterase inhibitors are used alone in prepubertal children not disabled by weakness. Steroids are indicated for patients who remain symptomatic despite optimal dosing with cholinesterase inhibitors, although chronic steroid side effects potentially have a long-term impact in children. Steroid-sparing immunosuppressive drugs are used in more severe or refractory cases, as in adult MG. TPE and IVIg are effective short-term therapies in JMG. Thymectomy has been reported to produce favourable results in JMG, even in patients less than 5 years of age, although the frequency of spontaneous remission in JMG makes it difficult to assess the benefit from thymectomy.

Foetal and transient neonatal myasthenia

About 10–20% of infants born to mothers with autoimmune MG develop transient neonatal myasthenia gravis (TNMG), which results from transplacental passage of maternal auto-antibodies (13–16). The maternal antibody level correlates with the frequency and severity of TNMG (17). An affected mother who delivers an infant with TNMG is likely to have subsequent similarly-affected infants.

Symptoms of TNMG usually develop a few hours after birth, but may be delayed for up to three days. These include hypotonia, generalized weakness, facial diplegia, poor sucking, weak cry, intermittent cyanosis (especially during feeds), respiratory weakness, and respiratory failure. Ptosis and external ophthalmoplegia are usually not as marked as in autoimmune MG. TNMG is transient, typically lasting about three weeks, but may persist for up to three months. AChR antibody levels fall as the weakness improves. The differential diagnosis of TNMG includes congenital myasthenia, congenital myotonic dystrophy, Moebius syndrome, and neonatal hypotonia.

Weakness may manifest in utero, particularly when maternal antibodies are directed against the foetal AChR, and may lead to arthrogryposis multiplex congenita (18). Birth of a child with arthrogryposis should prompt a search for MG in the mother.

If the mother is known to have MG, the diagnosis of TNMG in the affected infant is straight-forward. Detection of AChR antibodies in the mother and child confirms the diagnosis, but seronegative mothers have delivered seronegative infants with TNMG (19,20). Improvement following injection of edrophonium supports the diagnosis, although it may be difficult to demonstrate a clear response in an intubated neonate and not all affected infants respond to edrophonium. RNS testing can be of great value in demonstrating a decremental response in the rare TNMG patient in whom there is no clear response to edrophonium, the mother is not known to have myasthenia and AChR testing does not confirm the diagnosis (21,22). RNS may also be of value in following the response to treatment (22).

Congenital (genetic) myasthenic syndromes

These constitute a heterogeneous group of disorders produced by genetic abnormalities that affect different proteins at the neuromuscular junction. Different congenital myasthenic syndrome (CMS) subtypes have been identified based on detailed analysis of cDNA sequences of the acetylcholine receptor and acetylcholinesterase subunits (23,24).

Neuromuscular symptoms are typically present at birth or early childhood, but can be delayed until young adult life in some forms of CMS, in which case it is a challenge to distinguish them from acquired MG. Defined syndromes are due to specific functional and structural defects that result from mutations of the genes coding for proteins that regulate synthesis, aggregation, attachment and stability of the components of the neuromuscular junction. They can be classified according to the site of the defect—presynaptic, synaptic, or post-synaptic—by ultrastructural and histochemical examination, microphysiological techniques, and molecular genetic studies.

Diagnosing CMS in newborns is relatively straight-forward if other family members have the diagnosis. Otherwise, the diagnosis largely depends on a high index of suspicion. Recognizing extraocular, facial, and bulbar muscle weakness is particularly difficult in infants who have been intubated. The clinical assessment becomes more reliable with age. A careful examination can exclude most differential diagnoses. Once myasthenia has been confirmed by clinical or electrodiagnostic findings or response to cholinesterase inhibitors, the major differential is between congenital and acquired autoimmune myasthenia. If AChR or MuSK antibodies are present, the diagnosis of acquired MG is assured; however, 50% of children with acquired MG are seronegative (25,26), thus the absence of antibodies is not informative. The sophisticated techniques that enable precise diagnosis of the different forms of CMS are only available at a few centres. A trial of TPE or IVIg may be necessary in questionable cases—clear improvement after these treatments confirms the diagnosis of autoimmune MG. Failure to improve, however, does not exclude acquired MG.

Acetylcholine receptor deficiency

The most common form of congenital myasthenia results from one of several genetic neuromuscular defects associated with a deficiency or kinetic abnormality of the AChR. These can only be distinguished by DNA analysis. Most patients present with feeding problems and ptosis at birth or later in infancy. Ophthalmoplegia is usually present at birth, but may not be noticed until later.

In most cases RNS findings are the same as in autoimmune MG: normal CMAP amplitude and a decrementing response to low frequency RNS that is corrected by edrophonium. In severe cases, the RNS findings may resemble LEMS, with low amplitude CMAPs that facilitate during high frequency stimulation, although the microphysiological findings confirm a post-synaptic abnormality (27). Most respond to cholinesterase inhibitors and 3,4-diaminopyridine (3,4-DAP).

Rapsyn deficiency

Patients with deficiency of the AChR clustering protein rapsyn are usually hypotonic at birth, with breathing and feeding problems, ptosis and strabismus. Mild arthrogryposis is common. RNS shows a decremental pattern in most, but not all patients, and may be seen only after prolonged low-frequency stimulation. Most improve with cholinesterase inhibitors and 3,4-DAP (28) and improve spontaneously in later years.

DOK7 deficiency

These patients typically develop proximal muscle weakness in childhood or adolescence with relative sparing of ocular muscles. RNS demonstrates a decremental pattern, especially in shoulder and facial muscles, and there are EMG and biopsy findings of myopathy. Cholinesterase inhibitors are of limited benefit in most patients, but 3,4-DAP, albuterol, and ephedrine benefit many (29,30).

Choline acetyltransferase deficiency

This condition has distinctive clinical and electrophysiological characteristics. Affected infants are hypotonic at birth and have severe and repeated bouts of respiratory insufficiency and feeding difficulty (31). They may also have fluctuating eyelid ptosis, but other ocular muscle function is usually normal. Cholinesterase inhibitors improve the weakness and respiratory distress. Strength improves spontaneously within weeks, but episodes of weakness and life-threatening apnoea recur throughout infancy and childhood, sometimes even into adult life (32). Microphysiological findings indicate a defect of choline re-uptake into the nerve terminal or deficient ACh resynthesis or mobilization (33). Mutations in choline acetyltransferase have been identified in several families (34).

These patients have a characteristic EMG pattern: there is a decremental response to RNS in weak muscles, but in strong muscles a decrement may be seen only after sustained voluntary muscle activity or prolonged low frequency nerve stimulation. The CMAP amplitude begins to fall progressively after 2–3 min of continuous low frequency RNS. The progressive abnormality of neuromuscular transmission during sustained activity that characterizes this condition can also be demonstrated by measuring jitter during continuous axonal stimulation (2). We have seen symptomatic improvement from 3,4-DAP in patients from several families with this condition.

Slow channel syndrome

The onset of symptoms in this condition may be delayed until adult life, which can make it difficult to distinguish from acquired MG (35,36). Transmission is by autosomal dominant inheritance, but a corroborative family history is not obtained in all cases. Symptoms begin after infancy, sometimes as late as the third decade, with slowly progressive weakness of arm, leg, neck, and facial muscles, which may be atrophic. Mutations of the AChR produce delayed closure of the ACh ion channel once it has been opened (37), which prolongs the endplate potential beyond the refractory period of the muscle fibre action potential, resulting in repetitive CMAPs. Defective neuromuscular transmission results from loss of AChR, altered endplate geometry and conduction block from the summated prolonged endplate potentials (38). RNS produces a decrementing response. Characteristic repetitive CMAPs are seen after single nerve stimulation in most, but not all muscles (39), and have also been described in some asymptomatic relatives (36). Cholinesterase inhibitors increase the size and number of these repetitive CMAPs (40) and may worsen symptoms. Fluoxetine (41) and quinidine sulphate (42), which is contraindicated in other forms of myasthenia, are effective in treating slow channel syndrome.

Congenital endplate acetylcholinesterase deficiency

In this condition, weakness of facial, oropharyngeal, neck, and limb muscles is usually noted in the neonatal period or shortly thereafter, but rarely may not be apparent until childhood or early adulthood. The pupillary light response is sluggish (43) and there is variable ptosis and ophthalmoparesis. Weakness progresses slowly and patients develop postural, then fixed, spinal column deformity, a characteristic finding in this condition (44). Staining for AChE in the muscle biopsy helps in making the diagnosis. Ultrastructural and microphysiological studies demonstrate that the asymmetric form of cholinesterase is missing from the endplate (43). Single nerve stimulation produces repetitive CMAPs that fade quickly during repetitive stimulation, similar to findings in slow channel syndrome. Unlike slow channel syndrome, the repetitive CMAPs are not affected by cholinesterase inhibitors (40). RNS produces a decrement that does not improve with cholinesterase inhibitors, which produce clinical worsening. Treatment is with albuterol or ephedrine (45,46).

Fast channel syndromes

These autosomal recessive disorders present at birth or early infancy with hypotonia, ptosis, and oropharyngeal and limb weakness (47). Several specific kinetic mutations of the AChR have been identified in these patients (48), most of whom improve with pyridostigmine and 3,4-DAP (49). Microphysiological studies are necessary to distinguish these conditions from AChR deficiency, as they have the same clinical and electrodiagnostic findings.

Genetic abnormality of glycosylation pathways

Recently described CMS subtypes due to mutations in genes encoding glycosylation pathway enzymes typically have onset of limb girdle weakness in early childhood, with normal oculomotor function (50–54). The clinical picture is that of a myopathy, and muscle biopsy shows tubular aggregates. EMG shows myopathic findings, there is a decrement on RNS testing and increased jitter. Cholinesterase inhibitors, 3,4-DAP, albuterol, and ephedrine have all been reported to produce clinical improvement in these patients.

Paucity of synaptic vesicles and reduced quantal release

Only one case with this condition has been reported (55). The patient had feeding difficulty and a weak cry at birth and later developed fatigable eyelid ptosis, ocular and bulbar muscle weakness, and delayed motor development. Weakness improved somewhat with cholinesterase inhibitors. Microphysiological studies demonstrated reduced quantal content and a reduced number of quanta available for release. RNS studies at age 23 showed a pattern identical to acquired MG: normal CMAP amplitude, a decrementing response to 2/sec stimulation in multiple muscles, and no facilitation after exercise or 50/sec stimulation.

Congenital MuSK deficiency

MuSK acts together with DOK7 and rapsyn to aggregate AChR on the post-synaptic muscle membrane. A patient with both a frameshift mutation and a missense mutation of the MuSK gene had respiratory distress and lid ptosis at birth (56). Symptoms improved with age, and at age 22 she had ptosis, fatigability, and oropharyngeal symptoms. RNS produced a decrement and she improved with 3,4-DAP, but not pyridostigmine.

Congenital Lambert–Eaton syndrome

There have been rare reports of a congenital syndrome in which RNS studies show a pattern similar to LEMS (57,58). The precise abnormality in these patients has not been determined by microphysiological studies. A presynaptic defect similar to LEMS has also been demonstrated in some children whose RNS studies suggest a post-synaptic abnormality (27), indicating limitations in interpreting the clinical and EMG findings in some cases.

The Lambert–Eaton myasthenic syndrome

LEMS is a presynaptic abnormality of ACh release that is frequently associated with malignancy, usually small cell lung cancer (SCLC). LEMS results from an antibody-mediated attack against the voltage-gated calcium channels (VGCC) on presynaptic nerve terminals. LEMS usually begins after age 40, but has been reported in children. Males and females are equally affected. About half the patients with LEMS have an underlying malignancy (CA-LEMS), most of these have SCLC (59). The cancer may be discovered years before or after the symptoms of LEMS begin.

Weakness of proximal muscles, especially the legs, is the major symptom. Oropharyngeal and ocular muscles may be mildly affected. The weakness demonstrated on examination is usually relatively mild compared to the severity of symptoms. Strength may improve initially after exercise and then weaken with sustained activity. Edrophonium chloride does not improve strength to the degree seen in MG. Tendon reflexes are reduced or absent, but may be normalized by repeated muscle contraction or tapping the tendon repeatedly. Dry mouth is a common symptom of autonomic dysfunction; other features are impotence and postural hypotension.

LEMS may be first discovered when prolonged paralysis follows the use of neuromuscular blocking agents during surgery. Although LEMS and MG are both immune-mediated disorders of neuromuscular transmission, their clinical features are usually quite distinct. As in MG, clinical worsening in LEMS has been described following administration of aminoglycoside, macrolide, and fluoroquinolone antibiotics, magnesium, calcium-channel blockers. and iodinated intravenous contrast agents.

Immunopathology of LEMS

LEMS patients who do not have cancer frequently have organ specific serum auto-antibodies and other coincidental autoimmune diseases. SCLC cells contain high concentrations of VGCC and antibodies to the VGCC are found in the sera of most LEMS patients with SCLC, many of those without cancer, and some patients with SCLC who do not have LEMS (60). SCLC cells induce VGCC antibodies that react with the VGCC of peripheral nerves and cause LEMS. In LEMS patients who do not have SCLC, the VGCC antibodies are produced as part of a more general autoimmune diathesis. VGCC antibody titres do not correlate with disease severity among individuals, but the antibody levels may fall as the disease improves in patients receiving immunosuppression (61).

Diagnostic procedures for LEMS

The diagnosis of LEMS is confirmed by EMG. The characteristic findings are: decreased size of CMAPs with further reduction in response to RNS at frequencies between 1 and 5 Hz; increased CMAP size in response to RNS at 20–50 Hz; and a transitory increase in CMAP size after brief maximum voluntary contraction. Virtually all patients with LEMS have a decrementing response to low-frequency RNS in a hand or foot muscle and almost all have low amplitude CMAPs in some muscle (62).

Because the clinical features of LEMS frequently suggest a myopathy, the diagnosis may initially be suspected when needle EMG performed to assess muscle disease demonstrates markedly unstable MUAPs, a typical finding in LEMS.

Jitter is markedly increased with frequent blocking. At many NMJs the jitter and blocking decreases as the firing rate increases (63-65), but this is not seen in all endplates nor in all LEMS patients (65).

Immunoprecipitation assays demonstrate VGCC antibodies in almost all patients with CA-LEMS and in more than 90% without cancer (66). Antibodies may not be detectable early in the disease and repeat antibody testing may be useful.

In a recent report, antibodies to SOX1, a transcription factor involved in neural development, were found in 64% of LEMS patients with SCLC and in none without cancer (67).

Treatment of LEMS

Once the diagnosis of LEMS is established, an extensive search for underlying malignancy, especially SCLC, is mandatory. Chronic smokers should undergo bronchoscopy and/or PET scan if chest-imaging studies are normal. The target of initial treatment is the underlying malignancy. Weakness may improve after effective cancer therapy and some patients require no further treatment. The search for occult malignancy should be repeated periodically, especially during the first 2 years after symptom onset. The frequency of re-evaluation is determined by the patient's risk for cancer.

Cholinesterase inhibitors improve strength significantly only in occasional LEMS patients. Randomized controlled trials have shown that 3,4-DAP improves muscle strength in patients with LEMS (68-72). Other treatments, such as TPE, IVIg, corticosteroids, and immunosuppressive agents, including rituximab (73) may be of benefit in some patients.

In CA-LEMS patients, the prognosis is determined by the response to cancer therapy. In patients without cancer, treatment with immunosuppression produces improvement in many patients, but most require substantial and continuing doses of immunosuppressive medications (74).

MG/LEMS overlap syndrome

No single clinical or electrodiagnostic feature distinguishes MG from LEMS in all patients. There have been rare cases in which AChR and VGCC antibodies are both present, giving evidence for the co-existence of LEMS and MG in the same patient (75). Facilitation more than 100% may be seen in some muscles in patients with clinically typical MG if high stimulation rates are used (63,76-81). Conversely, patients with LEMS may have electrodiagnostic features more characteristic of MG early in their illness (82). In other patients, mixed clinical and electrodiagnostic features make it impossible to distinguish between the two conditions (83,84). In patients with such mixed features, the clinical characteristics and the presence of elevated AChR or VGCC antibodies or lung cancer define the most likely diagnosis (85).

Botulism

Botulism results from toxin produced by an anaerobic bacterium, *Clostridium botulinum*. Food-borne botulism results from ingestion of toxin produced in foods that have been incompletely sterilized. Neuromuscular symptoms usually begin 12–36 hours after ingestion of the contaminated food.

Wound botulism results when *C. botulinum* colonizes devitalized tissue and releases toxin that produces local and patchy systemic weakness. Major symptoms of food-borne and wound botulism include blurred vision, dysphagia, and dysarthria. Pupillary responses to light are impaired and tendon reflexes are variably reduced. The weakness progresses for several days and then reaches a plateau. Fatal respiratory paralysis may occur rapidly. Most patients have evidence of autonomic dysfunction, such as dry mouth, constipation, or urinary retention. In patients who survive, recovery may take many months, but is usually complete. The edrophonium test is positive in only about 1/3 of patients and does not distinguish botulism from other causes of neuromuscular blockade (86).

The diagnosis of wound botulism is confirmed by wound cultures and serum assay for botulinum toxin.

Infantile botulism results from the growth of *C. botulinum* in the infant gastrointestinal tract and the elaboration of small quantities of toxin over a prolonged period. Symptoms of constipation, lethargy, poor suck, and weak cry usually begin at about four months of age. Patients have weakness of the limb and oropharyngeal muscles, poorly reactive pupils, and hypoactive tendon reflexes. Most patients require ventilatory support. The diagnosis of infant botulism is confirmed by demonstrating botulinum toxin in the stool or by isolating *C. botulinum* from stool culture.

Other conditions with disturbed neuromuscular transmission

Drugs

Muscle relaxants block neuromuscular transmission, either by interfering with the interaction of ACh with receptors on the muscle or by directly depolarizing the muscle membrane. Many other medications also have blocking effects on neuromuscular transmission. The most frequently encountered are aminoglycoside, macrolide and fluoroquinolone antibiotics. For the most part, the neuromuscular blocking effects of these drugs are clinically apparent only when the safety factor of neuromuscular transmission has been lowered by disease or concomitant administrations of other drugs. Abnormal neuromuscular transmission may be confirmed in such cases by EMG tests or administration of cholinesterase inhibitors, but the diagnosis can usually be made by withdrawing the offending medication and observing improvement in muscle function.

Magnesium

Disturbed neuromuscular transmission from hypermagnesaemia occurs in patients with renal insufficiency who receive oral magnesium such as laxatives, and in women who receive magnesium for pre-eclampsia. Hypermagnesaemia resembles LEMS clinically: there is proximal muscle weakness, which may progress to respiratory insufficiency in severe cases. The ocular muscles are spared and tendon reflexes are depressed. The diagnosis is made by demonstrating elevated serum magnesium levels and observing the return of tendon reflexes as the serum magnesium level falls. Edrophonium may improve strength in some patients. The response to RNS resembles that in LEMS or botulism, with low amplitude CMAPs, a decremental response to low frequency nerve stimulation and marked facilitation after muscle activation.

Organophosphates

These agents irreversibly inhibit cholinesterase, producing neuromuscular blockade as well as autonomic and central nervous system dysfunction. RNS testing can be of great value in making a diagnosis and in following the course of intoxication by these agents (87). When acetylcholinesterase has been blocked, excess ACh accumulates at the NMJ and impairs neuromuscular transmission by depolarizing the post-junctional muscle membrane. Receptors on the pre-synaptic nerve ending are also activated by the excess ACh, producing repetitive discharges when the nerve is activated. Single nerve stimuli produce repetitive muscle discharges that follow the CMAP. RNS produces a decrementing pattern. This combination of findings is seen within hours after ingestion of organophosphates (87). These electrodiagnostic findings are similar to those seen in congenital endplate acetylcholinesterase deficiency or slow-channel syndrome, and after administration of high doses of cholinesterase inhibitors. A distinctive 'decrement-increment' pattern of response to RNS is seen in the early stages of organophosphate intoxication and again later as the intoxication resolves (87). At the peak of intoxication, the decremental response is so severe that no response is seen after the first few stimuli in a train. The evolution of these electrodiagnostic patterns has been used to assess the severity and progression of intoxication by these agents.

Nerve or muscle disease

Neuromuscular transmission is abnormal in many motor unit diseases other than those that primarily affect the NMJ. Patients with amyotrophic lateral sclerosis (ALS) may have fluctuating weakness that responds to cholinesterase inhibitors (88). An abnormal decrement on RNS has been described in two-thirds of patients with ALS so studied (89). Increased jitter and impulse blocking are seen in most patients with ALS, presumably resulting from collateral reinnervation. Various manifestations of abnormal NMT have also been reported in syringomyelia, poliomyelitis, peripheral neuropathies, and inflammatory myopathies (90). Jitter is increased, especially in facial muscles, in patients with oculocraniosomatic myopathy (91), most of whom have myopathic features on EMG of shoulder muscles and characteristic 'ragged-red fibres' on muscle biopsy.

Animal venoms and toxins

Neuromuscular block is the primary effect of envenomation by cobras, kraits, and some other poisonous snakes. Snake toxins that act post-synaptically and bind competitively to the receptor produce patterns of weakness similar to that of MG (92). In such cases, RNS demonstrates a decrementing response and weakness is reversed by cholinesterase inhibitors. Such toxins include those of cobras and the death adder.

References

1. Pascuzzi, R.M. (2003). The edrophonium test. *Seminars in Neurology*, **23**, 83–8.
2. Stålberg, E.V., Trontelj, J.V., and Sanders, D.B. (2010). Myasthenia gravis and other disorders of neuromuscular transmission. In: *Single fiber EMG*, , 3rd edn, pp. 218–66. Fiskebäckskil: Edshagen Publishing House.
3. Sanders, D.B., el-Salem, K., Massey, J.M., McConville, J., and Vincent, A. (2003). Clinical aspects of MuSK antibody positive seronegative myasthenia gravis. *Neurology*, **60**(12), 1978–80.
4. Massey, J.M., Sanders, D.B., and Howard, J.F, Jr. (1989). The effect of cholinesterase inhibitors on SFEMG in myasthenia gravis. *Muscle & Nerve*, **12**, 154–5.
5. Guptill, J.T. and Sanders, D.B. (2010). Update on MuSK antibody positive myasthenia gravis. *Current Opinions in Neurology*, **23**, 530–5.
6. Caress, J.B., Hunt, C.H., and Batish, S.D. (2005). Anti-MuSK myasthenia gravis presenting with purely ocular findings. *Archives of Neurology*, **62**(6),1002–3.
7. Bau, V., Hanisch, F., Hain, B., and Zierz, S. (2006). Ocular involvement in MuSK antibody-positive myasthenia gravis. *Klinische Monatsblätter für Augenheilkunde*, **223**(1), 81–3.
8. Stickler, D.E., Massey, J.M., and Sanders, D.B. (2005). MuSK-antibody positive myasthenia gravis: clinical and electrodiagnostic patterns. *Clinical Neurophysiology*, **116**, 2065–8.
9. Hatanaka, Y., Claussen, G.C., and Oh, S.J. (2005). Anticholinesterase hypersensitivity or intolerance is common in MuSK antibody positive myasthenia gravis. *Neurology*, **64**, 6(l), A79.
10. Grob, D., Brunner, N.G., Namba, T., and Pagala, M. (2008). Lifetime course of myasthenia gravis. *Muscle & Nerve*, **37**,141–9.

11. Zang, X., Yang, M., Xu, J., et al. (2007). Clinical and serological study of myasthenia gravis in HuBei Province, China. *Journal of Neurology, Neurosurgery, & Psychiatry*, **78**, 386–90.
12. Andrews, P.I. and Sanders, D.B. (2002). Juvenile myasthenia gravis. In: H. R. Jones, D. C. DeVivo and B. T. Darras (Eds) *Neuromuscular disorders of infancy, childhood, and adolescence*, pp. 575–97. Boston: Butterworth Heinemann.
13. Strickroot, F.L., Schaeffer, R.L., and Bergo, H.E. (1942). Myasthenia gravis occurring in an infant born of a myasthenic mother. *Journal of the American Medical Association*, **120**, 1207–9.
14. Greer, M. and Schotland, M. (1960). Myasthenia gravis in the newborn. *Pediatrics*, **26**, 101–8.
15. Namba, T., Brown, S,B., and Grob, D. (1970). Neonatal myasthenia gravis: a report of two cases and review of the literature. *Pediatrics*, **45**, 488–504.
16. Papazian, O. (1992). Transient neonatal myasthenia gravis. *Journal of Child Neurology*, **7**, 135–41.
17. Eymard, B., Morel, E., Dulac, O., et al. (1989). Myasthenia and pregnancy: a clinical and immunologic study of 42 cases (21 neonatal myasthenia cases). *Reviews of Neurology*, **145**, 696–701.
18. Barnes, P.R.J., Kanabar, D.J., Brueton, L., et al. (1995). Recurrent congenital arthrogryposis leading to a diagnosis of myasthenia gravis in an initially asymptomatic mother. *Neuromuscular Disorders*, **5**, 59–65.
19. Oteiza Orradre, C., Navarro Serrano, E., Rebage Moises, V., et al. (1996). Seronegative transient neonatal myasthenia: report of a case. *Annales Espanoles de Pediatria*, **45**, 651–2.
20. Lefvert, A.K. and Osterman, P.O. (1983). Newborn infants to myasthenic mothers: a clinical study and an investigation of acetylcholine receptor antibodies in 17 children. *Neurology*, **33**, 133–8.
21. Wise G.A and McQuillen, M.P. (1970). Transient neonatal myasthenia. *Archives of Neurology*, **22**, 556–65.
22. Hays, R.M. and Michaud, L.J. (1988). Neonatal myasthenia gravis: specific advantages of repetitive stimulation over edrophonium testing. *Pediatric Neurology*, **4**, 245–7.
23. Engel, A.G. (2008). Congenital myasthenic syndromes. In: A. G. Engel (Ed.) *Neuromuscular junction disorders*, Series 3, pp. 285–331. Amsterdam: Elsevier.
24. Finlayson, S., Beeson, D., and Palace, J. (2013). Congenital myasthenic syndromes: an update. *Practical Neurology*, **13**(2), 80–91.
25. Andrews, P.I., Massey, J.M., and Sanders, D.B. (1993). Acetylcholine receptor antibodies in juvenile myasthenia gravis. *Neurology*, **43**, 977–82.
26. Sanders, D.B., Andrews, P.I., Howard, J.F., Jr, and Massey, J.M. (1997). Seronegative myasthenia gravis. *Neurology*, **48**(5), S40–51.
27. Maselli, R., Kong, D.Z., Bowe, C.M., et al. (2001). Presynaptic congenital myasthenic syndrome due to quantal release deficiency. *Neurology*, **57**, 279–89.
28. Banwell, B.L., Ohno, K., Sieb, J.P., and Engel, A.G. (2004). Novel truncating RAPSN mutations causing congenital myasthenic syndrome responsive to 3,4-diaminopyridine. *Neuromuscular Disorders*, **14**(3), 202–7.
29. Lorenzoni, P.J., Scola, R.H., Kay, C.S., et al. (2013). Salbutamol therapy in congenital myasthenic syndrome due to DOK7 mutation. *Journal of Neurological Sciences*, **331**, 155–7.
30. Lashley, D., Palace, J., Jayawant, S., Robb, S., and Beeson, D. (2010). Ephedrine treatment in congenital myasthenic syndrome due to mutations in DOK7. *Neurology*, **74**, 1517–23.
31. Robertson, W.C., Chun, R.W.M., and Kornguth, S.E. (1980). Familial infantile myasthenia. *Archives of Neurology*, **37**, 117–9.
32. Gieron, M.A. and Korthals, J.K. (1985). Familial infantile myasthenia gravis. Report of three cases with follow-up until adult life. *Archives of Neurology*, **42**, 143–4.
33. Mora, M., Lambert, E.H., and Engel, A.G. (1987). Synaptic vesicle abnormality in familial myasthenia. *Neurology*, **37**, 206–14.
34. Ohno, K., Tsujino, A., Brengman, J.M., et al. (2001). Choline acetyltransferase mutations cause myasthenic syndrome associated with episodic apnea. *Proceedings of the National Academy Science*, **98**, 2017–22.
35. Engel, A.G., Lambert, E.H., Mulder, D.M., et al. (1982). A newly recognized congenital myasthenic syndrome attributed to a prolonged open time of the acetylcholine-induced ion channel. *Annals of Neurology*, **11**, 553–69.
36. Oosterhuis, H.J.G.H., Newsom-Davis, J., and Wokke, J.H.J. (1987). The slow channel syndrome. Two new cases. *Brain*, **110**, 1061–79.
37. Hutchinson, D.O., Nakano, S., Sieb, J.P., and Engel, A. (1993). Patch clamp analysis of the slow channel myasthenic syndrome. *Neurology*, **4**, A180.
38. Zhou, M., Engel, A.G., and Auerbach, A. (1999). Serum choline activates mutant acetylcholine receptors that cause slow channel congenital myasthenic syndrome. *Proceedings of the National Academy of Science*, **96**, 10466–71.
39. Bedlack, R.S., Bertorini, T.E., and Sanders, D.B. (2000). Hidden after discharges in slow channel congenital myasthenic syndrome. *Journal of Clinical Neuromuscular Disease*, **1**, 186–9.
40. Harper, C.M. (2002). Congenital myasthenic syndromes. In: W. F. Brown, C. F. Bolton, and M. J. Aminoff (Eds) *Neuromuscular function and disease*, pp. 1687–95. Philadelphia, PA: W.B. Saunders Company.
41. Harper, C.M., Fukudome, T., and Engel, A.G. (2003). Treatment of slow-channel congenital myasthenic syndrome with fluoxetine. *Neurology*, **60**, 1710–13.
42. Harper, C.M. and Engel, A.G. (1998). Quinidine sulfate therapy for the slow-channel congenital myasthenic syndrome. *Annals of Neurology*, **43**, 480–4.
43. Hutchinson, D.O., Walls, T.J., Nakano, S., et al. (1993). Congenital endplate acetylcholinesterase deficiency. *Brain*, **116**, 633–53.
44. Hutchinson, D.O., Engel, A.G., Walls, T.J., et al. (1993). The spectrum of congenital end-plate acetylcholinesterase deficiency. *Annals of the New York Academy of Sciences*, **681**, 469–86.
45. Liewluck, T., Selcen, D., and Engel, A.G. (2011). Beneficial effects of albuterol in congenital endplate acetylcholinesterase deficiency and Dok-7 myasthenia. *Muscle & Nerve*, **44**, 789–94.
46. Mihaylova, V., Muller, J.S., Vilchez, J.J., et al. (2008). Clinical and molecular genetic findings in COLQ-mutant congenital myasthenic syndromes. *Brain*, **131**(3), 747–59.
47. Ohno, K., Wang, H.L., Milone, M., et al. (1996). Congenital myasthenic syndrome caused by decreased agonist binding affinity due to a mutation in the acetylcholine receptor e subunit. *Neuron*, **17**, 157–70.
48. Ohno, K. and Engel, A.G. (2002). Congenital myasthenic syndromes. *Current Neurology & Neuroscience Reports*, **2**, 79–88.
49. Harper, C.M. and Engel, A.G. (2000). Treatment of 31 congenital myasthenic syndrome patients with 3,4-diaminopyridine. *Neurology*, **54**(3), A395.
50. Belaya, K., Finlayson, S., Slater, C.R., et al. (2012). Mutations in DPAGT1 cause a limb-girdle congenital myasthenic syndrome with tubular aggregates. *American Journal of Human Genetics*, **91**(1), 193–201.
51. Finlayson, S., Palace, J., Belaya, K., et al. (2013). Clinical features of congenital myasthenic syndrome due to mutations in DPAGT1. *Journal of Neurology, Neurosurgery, & Psychiatry*, **84**(10), 1119–25.
52. Guergueltcheva, V., Muller, J.S., Dusl, M., et al. (2012). Congenital myasthenic syndrome with tubular aggregates caused by GFPT1 mutations. *Journal of Neurology*, **2259**, 838–50.
53. Huh, S.Y., Kim, H.S., Jang, H.J., Park, Y.E., and Kim, D.S. (2012). Limb-girdle myasthenia with tubular aggregates associated with novel GFPT1 mutations. *Muscle & Nerve*, **46**, 600–4.
54. Selcen, D., Shen, X.M., Milone, M., et al. (2013). GFPT1-myasthenia: clinical, structural, and electrophysiologic heterogeneity. *Neurology*, **81**(4), 370–8.
55. Walls, T.J., Engel, A.G., Nagel, A.S., Harper, C.M., and Trastek, V.F. (1993). Congenital myasthenic syndrome associated with paucity of synaptic vesicles and reduced quantal release. *Annals of the New York Academy of Sciences*, **681**, 461–8.

56. Chevessier, F., Faraut, B., Ravel-Chapuis, A., et al. (2004). MUSK, a new target for mutations causing congenital myasthenic syndrome. *Human Molecular Genetics*, **13**(24), 3229–40.
57. Albers, J.W., Faulkner, J., Dorovini-Zis, K., Barald, K.F., Must, R.E., and Ball, R.D. (1984). Abnormal neuromuscular transmission in an infantile myasthenic syndrome. *Annals of Neurology*, **16**, 28–34.
58. Bady, B., Chauplannaz, G., and Carrier, H. (1987). Congenital Lambert–Eaton myasthenic syndrome. *Journal of Neurology, Neurosurgery, & Psychiatry*, **50**, 476–8.
59. Sanders, D.B. (1995). Lambert–Eaton myasthenic syndrome: clinical diagnosis, immune-mediated mechanisms, and update on therapy. *Annals of Neurology*, **37**(1), S63–73.
60. Lennon, V.A. and Lambert, E.H. (1989). Autoantibodies bind solubilized calcium channel-omega-conotoxin complexes from small cell lung carcinoma: a diagnostic aid for Lambert–Eaton myasthenic syndrome. *Mayo Clinic Proceedings*, **64**, 1498–504.
61. Leys, K., Lang, B., Johnston, I., and Newsom-Davis, J. (1991). Calcium channel autoantibodies in the Lambert–Eaton myasthenic syndrome. *Annals of Neurology*, **29**, 307–14.
62. Tim, R.W., Massey, J.M., and Sanders, D.B. (2000). Lambert–Eaton myasthenic syndrome. Electrodiagnostic findings and response to treatment. *Neurology*, **54**, 2176–8.
63. Schwartz, M.S. and Stålberg, E. (1975). Myasthenia gravis with features of the myasthenic syndrome. *Neurology*, **25**, 80–4.
64. Trontelj, J.V. and Stålberg, E. (1991). Single motor end-plates in myasthenia gravis and LEMS at different firing rates. *Muscle & Nerve*, **14**, 226–32.
65. Sanders, D.B. (1992). The effect of firing rate on neuromuscular jitter in Lambert–Eaton myasthenic syndrome. *Muscle & Nerve*, **15**, 256–8.
66. Harper, C.M. and Lennon, V.A. (2002). The Lambert–Eaton myasthenic syndrome. In: H. J. Kaminski (Ed.) *Current clinical neurology: myasthenia gravis and related disorders*, pp. 269–91. Totowa, NJ: Humana Press.
67. Sabater, L., Titulaer, M., Saiz, A., Verschuuren, J., Gure, A.O., and Graus, F. (2008). SOX1 antibodies are markers of paraneoplastic Lambert–Eaton myasthenic syndrome. *Neurology*, **70**, 924–8.
68. Sanders, D.B., Massey, J.M., Sanders, L.L., and Edwards, L.J. (2000). A randomized trial of 3,4–diaminopyridine in Lambert–Eaton myasthenic syndrome. *Neurology*, **54**, 603–7.
69. McEvoy, K.M., Windebank, A.J., Daube, J.R., and Low, P.A. (1989). 3,4-Diaminopyridine in the treatment of Lambert–Eaton myasthenic syndrome. *New England Journal of Medicine*, **321**, 1567–71.
70. Maddison, P. and Newsom-Davis, J. (2005). Treatment for Lambert–Eaton myasthenic syndrome. *Cochrane Database of Systematic Reviews*, CD003279.
71. Oh, S.J., Claussen, G.G., Hatanaka, Y., and Morgan, M.B. (2009). 3,4-Diaminopyridine is more effective than placebo in a randomized, double-blind, cross-over drug study in LEMS. *Muscle & Nerve*, **40**, 795–800.
72. Wirtz, P.W., Verschuuren, J.J., van Kijk, J.G., et al. (2009). Efficacy of 3,4-diaminopyridine and pyridostigmine in the treatment of Lambert–Eaton myasthenic syndrome: a randomized, double-blind, placebo-controlled, crossover study. *Clinical Pharmacology & Therapeutics*, **86**, 44–8.
73. Maddison, P., McConville, J., Farrugia, M.E., et al. (2011). The use of rituximab in myasthenia gravis and Lambert–Eaton myasthenic syndrome. *Journal of Neurology, Neurosurgery, & Psychiatry*, **82**, 671–3.
74. Maddison, P., Lang, B., Mills, K., and Newsom-Davis, J. (2001). Long term outcome in Lambert–Eaton myasthenic syndrome without lung cancer. *Journal of Neurology, Neurosurgery, & Psychiatry*, **70**, 212–17.
75. Newsom-Davis, J., Leys, K., Vincent, A., Ferguson, I., Modi, G., and Mills, K. (1991). Immunological evidence for the co-existence of the Lambert–Eaton myasthenic syndrome and myasthenia gravis in two patients. *Journal of Neurology, Neurosurgery, & Psychiatry*, **54**, 452–3.
76. Simpson, JA. (1966). Disorders of neuromuscular transmission. *Proceedings of the Royal Society for Medicine*, **59**, 993–8.
77. Mayer, R.F. and Williams, I.R. (1974). Incrementing responses in myasthenia gravis. *Archives of Neurology*, **731**, 24–6.
78. Brown, J.C. and Johns, R.J. (1974). Diagnostic difficulties encountered in the myasthenic syndrome sometimes associated with carcinoma. *Journal of Neurology, Neurosurgery, & Psychiatry*, **37**, 1214–24.
79. Singer, P., Smith, L., Ziegler, D.K., and Festoff, B.W. (1981). Posttetanic potentiation in a patient with myasthenia gravis. *Neurology*, **31**, 1345–7.
80. Dahl, D.S. and Sato, S. (1974). Unusual myasthenic state in a teen-age boy. *Neurology*, **24**, 897–901.
81. Lambert, E.H. and Rooke, E.D. (1965). Myasthenic state and lung cancer. In: W. R. Brain and F. H. Norris (Eds) *Contemporary neurology symposia: the remote effects of cancer on the nervous system*, pp. 67–80. New York, NY: Grune & Stratton.
82. Scoppetta, C., Casali, C., Vaccario, M.L., and Provenzano, C. (1984). Difficult diagnosis of Eaton Lambert myasthenic syndrome. *Muscle & Nerve*, **7**, 680–1.
83. Boiardi, A., Bussone, G., and Negri, S. (1979). Alternating myasthenia and myastheniform syndrome in the same subject. *Journal of Neurology*, **220**, 57–64.
84. Fettel, M.R., Shin, H.S., and Penn, A.S. (1978). Combined Eaton–Lambert syndrome and myasthenia gravis. *Neurology*, **28**, 398.
85. Sanders, D.B. and Stålberg, E. (1987). The overlap between myasthenia gravis and Lambert–Eaton myasthenic syndrome. *Annals of the New York Academy of Sciences*, **505**, 864–5.
86. Burningham, M.D., Walter, F.G., Mechem, C., Haber, J., and Ekins, B.R. (1994). Wound botulism. *Annals of Emergency Medicine*, **24**, 1184–7.
87. Besser, R., Gutmann, L., Dillmann, U., Weilemann, L.S., and Hopf H.C. (1989). End-plate dysfunction in acute organophosphate intoxication. *Neurology*, **39**, 561–7.
88. Mulder, D.W., Lambert, E.H., and Eaton, L.M. (1959). Myasthenic syndrome in patients with amyotrophic lateral sclerosis. *Neurology*, **9**, 627–31.
89. Denys, E.H. and Norris, F.H. (1979). Amyotrophic lateral sclerosis impairment of neuromuscular transmission. *Archives of Neurology*, **36**, 202–5.
90. Stålberg, E., Schwartz, M.S., and Trontelj, J.V. (1975). Single fibre electromyography in various processes affecting the anterior horn cell. *Journal of Neurological Sciences*, **24**, 403–15.
91. Krendel, D.A., Sanders, D.B., and Massey, J.M. (1985). Single fiber electromyography in chronic progressive external ophthalmoplegia. *Muscle & Nerve*, **8**, 624.
92. Kumar, S.M. and Usgaonkar, R.S. (1968). Myasthenia gravis like picture resulting from snake bite. *Journal of the Indian Medicine Association*, **50**, 428–9.

CHAPTER 24

Primary muscle diseases

Robin P. Kennett and Sidra Aurangzeb

Introduction

The clinical neurophysiologist receiving a request to investigate a patient with suspected myopathy faces the daunting prospect of a bewilderingly large differential diagnosis of rare and obscure conditions. The rapid recent improvement in genetics, immunology, and histology make the task easier and now the role of the neurophysiologist is to determine if there are electrophysiological features of primary muscle disease and if so to use their nature and distribution in order to refine the differential diagnosis to one that can be tackled by these complimentary investigatory techniques. The neurophysiologist's ability to sample many sites helps direct the muscle biopsy to a region that is most likely to define pathological changes (1,2). In this chapter we outline the neurophysiological finding expected in muscle disease and go on to describe the combinations that are likely to be found in the more commonly encountered disorders.

Nerve conduction studies are often of limited value except for excluding alternative or additional peripheral nerve disease. In muscle disease sensory nerve conduction and motor velocity are expected to be normal, and compound muscle action potentials are rarely reduced in amplitude, except in distal disease. The mainstay of neurophysiological investigation is needle electromyography (EMG) and the prime consideration is to determine whether the muscle disease is associated with abnormal insertional or spontaneous activity or not. Needle insertion into normal muscle is expected to evoke a brief burst of electrical discharges: only end-stage muscle disease would be incapable of showing this activity. In some muscles, particularly the calves, insertional activity may continue at low frequency often with small positive sharp wave appearance, until the electrode is moved. End-plate noise is at a higher frequency and is similarly localized to one region where the patient may find the needle painful. Pathological persisting insertional activity takes the form of fibrillation potentials and positive sharp waves, shown in Fig. 24.1, or recurrent myotonic or complex discharges, which may appear as spontaneous activity if occurring without needle movement (1–8). It is generally agreed that fibrillation potentials and positive sharp waves both represent the same phenomenon of spontaneous depolarization along the muscle membrane, and here the term 'fibrillation' is used to infer both types of activity. It is less certain what causes this increased 'instability', but the strong association between fibrillations and muscle fibre necrosis on biopsy suggests that separation of portions of membrane from the neuromuscular junction is important, perhaps leading to up-regulation of acetyl choline receptors on the disconnected membrane that react to circulating acetyl or butyl choline. This cannot be the only mechanism for membrane instability as can be seen from Box 24.1, fibrillations are seen in disease where necrosis does not occur (1,9–11).

In true myotonia repetitive muscle fibre depolarization demonstrates an accelerating and decelerating pattern of discharge giving rise to the unmistakable waxing and waning sound when amplified through a speaker. This activity is characteristic of myotonic dystrophy and channelopathy (see below). Repetitive muscle fibre discharge without this pattern is termed 'pseudomyotonia' or myotonia-like activity. This may occur at low frequency when individual components have an appearance similar to positive sharp waves, or at higher frequency with a decelerating pattern. Pseudomyotonia may be seen in a wide range of muscle disorders with membrane instability, particularly those with muscle fibre necrosis (see Box 24.2) (1,2,4,7,11). Complex repetitive discharges occur with more extreme membrane instability where spontaneously generated action potentials are transmitted

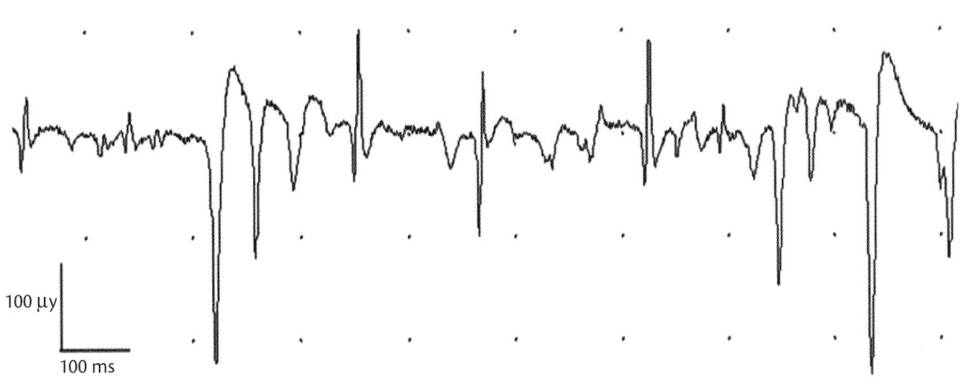

Fig. 24.1 Spontaneous EMG activity consisting of fibrillation potentials and positive sharp waves.

Box 24.1 Myopathies associated with fibrillations potentials and positive sharp waves

- *Inflammatory myopathies*:
  - Polymyositis (PM)/dermatomyositis (DM).
  - Inclusion body myositis (IBM).
  - Immune-mediated necrotizing myopathy (with or without association with cholesterol lowering agents).
- Critical illness myopathy.
- *Toxic/necrotic myopathies*:
  - Alcohol.
  - Cholesterol-lowering agent myopathies (e.g. statin myopathies).
  - Chloroquine/hydroxychloroquine.
  - Colchicine.
  - Amiodarone.
- *Muscular dystrophies*: Duchenne, Becker, limb-girdle, Emery–Dreifuss muscular dystrophy, distal muscular dystrophies, myofibrillar myopathies.
- *Congenital myopathies*:
  - Nemaline rod myopathy.
  - Centronuclear/myotubular myopathy.
- *Myopathies associated with selected infectious agents*:
  - HIV and HTLV- 1 associated myositis.
  - Trichinosis.
  - Toxoplasmosis.
- Infiltrative disorders (e.g. sarcoidosis, amyloid).
- *Metabolic myopathies*:
  - Pompe disease (acid maltase deficiency).
  - McArdle's disease.
  - Glycogen storage disease III and IV.
  - Carnitine palmityl transferase (CPT) deficiency.
  - Muscle carnitine deficiency.
- Hyperkalemic periodic paralysis (in acute stage).

Box 24.2 Myopathies with myotonic discharges

With clinical myotonia

- Myotonic dystrophy type I, II and proximal myotonic myopathy.
- myotonia congenita.
- Paramyotonia congenita.

Without clinical myotonia

- Hyperkalemic periodic paralysis.
- *Metabolic myopathy:*
  - Acid maltase deficiency.
  - Debrancher deficiency.
  - McArdle's disease (myophosphorylase deficiency).
- *Endocrine myopathies*: hypothyroidism.
- *Toxic myopathies*:
  - Chloroquine/hydroxychloroquine myopathy.
  - Statin myopathy.
  - Colchicine myopathy.
- Centronuclear myopathy.
- Inflammatory myopathies: PM and DM.

to adjacent fibres ephaptically, producing a recurrent circuit. On EMG this appears as stable polyphasic complexes that terminate when one action potential in the circuit falls out (1,2,4,7,11,12). Myopathies that may show complex repetitive discharges are listed in Box 24.3.

The morphology of the voluntarily activated motor unit potential can vary greatly between myopathies of different aetiology. EMG studies usually examine fatigue resistant units that recruit early and fire at steady frequency. Consequently disease affecting type 2 motor units preferentially may show no abnormality on standard testing. Disorders that produce uniform reduction in muscle fibre diameter may cause an overall decrease in motor unit size without morphological change, which also may be difficult to detect unless quantitative techniques are used. More often muscle disease causes a variation of muscle fibre diameters and an increased range of muscle fibre conduction velocities that manifest at an increase in polyphasia of the motor unit action potential (MUAP). If there is a general reduction in muscle fibre diameter the MUAP amplitude is reduced, but compensatory hypertrophy has the opposite effect. If muscle fibres are lost from the unit, for instance in the acute phase of inflammatory myopathy, the duration of the MUAP becomes shorter, but compensatory mechanisms such as muscle fibre splitting (giving 2 action potential for each fibre), and collateral reinnervation may increase the overall duration with satellite potentials separated from the main part of the unit. These processes can also increase polyphasicity. It can be seen from this that concluding long-duration/high amplitude MUAPs to be 'neurogenic' and small, short duration units to be 'myopathic' is an over simplification that not always holds, and in some cases it may be impossible

Box 24.3 Myopathies with complex repetitive discharges

- Chronic inflammatory myopathies (PM, DM, IBM).
- Muscular dystrophies (especially Duchenne's).
- Schwartz–Jampel syndrome.
- Metabolic myopathy:
  - Acid maltase deficiency.
  - Debrancher deficiency.
- Hypothyroid myopathy.
- Myofibrillar/centronuclear myopathy.

to decide if the pathological process is in the motor neuron or muscle fibre by the EMG appearance of MUAPs alone (1,2,4,7,13).

Because diseased muscle fibres generate less force, the interference pattern on voluntary contraction appears to contain more motor units than in neurogenic disorders giving the same strength. Prolonged polyphasic motor units with multiple spike components of similar amplitude may also give the appearance of a complete interference pattern on early recruitment, whereas use of a trigger and delay line will confirm that the screen is actually filled with few motor units (1,2,4,7,13).

It will be appreciated that an understanding of the underlying pathophysiological process leading to concentric needle EMG findings will allow the neurophysiologist to make a judgement of the nature and duration of the disease under examination, and the response to treatment. These findings may be enhanced by specialized techniques including quantitative analysis of the motor units of recruitment patterns, or using adapted electrodes as in macro and single fibre EMG. The advantage of concentric needle EMG is that many muscles may be rapidly assessed, reducing the sampling error inherent in muscle biopsy and some of the quantitative EMG methods for which normal values are available for only a small number of muscles (1,2). Nonetheless, quantitative EMG may be essential when the disease is restricted to sites such as facial muscles where the routine EMG assessment of motor unit parameters can be misleading (14). In the next sections we explore the combinations of EMG findings that may be expected in specific disease entities.

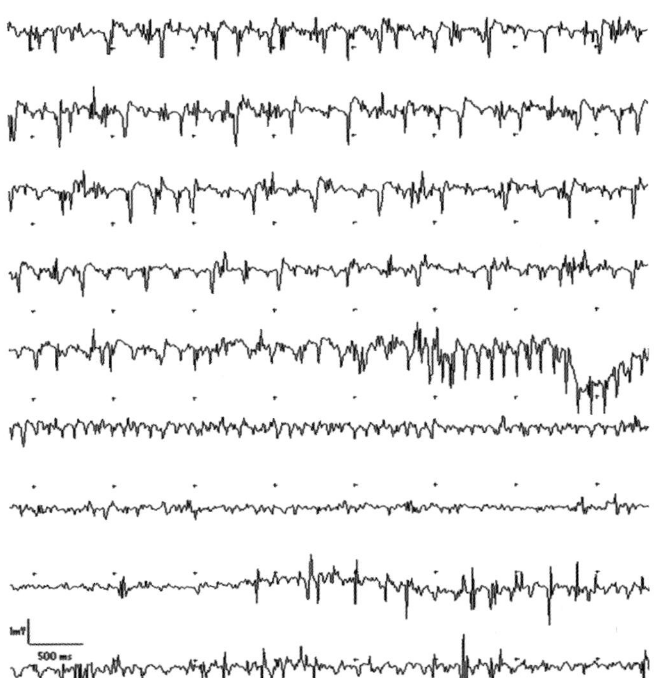

Fig. 24.2 Example of florid spontaneous EMG activity recorded in myositis, including fibrillations and myotonic-like discharges.

Inflammatory myopathy

The different causes of acquired inflammatory disease of muscle produce similar neurophysiological findings and may be considered together. These disorders affect all ages and include poly- and dermatomyositis, and overlap syndromes of connective tissue disease characterized by their autoantibody profiles (for instance, systemic lupus erythmatosis and antisynthetase syndrome). This category is enlarging as more autoantibodies are detected, examples being HMG CoA reductase antibody found in myositis caused by statin medication (15). Other inflammatory disorders such as sarcoidosis may produce similar neurophysiological findings (16).

The 'classical triad' of needle EMG findings in myositis described by Bohan et al. (13) is the presence of fibrillation potentials and positive sharp waves, complex repetitive discharges, and small short-duration polyphasic MUAPs. These changes are most readily apparent in the acute phase with severe muscle fibre necrosis, and the spontaneous activity may be less pronounced in milder disease or after treatment, and in the duration of MUAPs may be increased rather than short, particularly in the more chronic phase. An example of florid spontaneous activity recorded in the active stage of myositis is shown in Fig. 24.2. Typically the EMG abnormalities show a proximal to distal gradient with changes more pronounced towards the axis, although this may be less pronounced in the legs and in the elderly (17). It is therefore essential to examine the paraspinal muscles if myositis is suspected, but the EMG examination of the limb muscles is normal (5,6,11,17,18). The sensitivity of EMG for the diagnosis of acute myositis thus depends on the extent and completeness of the examination: membrane instability with fibrillations and positive sharp waves may only be seen in 1 or 2 muscles, particularly in mild disease, in about 10% of patients (6), but more usually is widespread giving an overall sensitivity of about 80% (4,19). The degree and distribution of instability does not appear to improve the probability of finding myositis (6). In the acute phase fibrillation potentials have higher amplitude than later, when muscle fibre atrophy has developed (7). Membrane instability producing complex repetitive discharges is seen in the sub-acute phase and less often, in about 40% of patients with poly- or dermatomyositis. This is probably because the conditions necessary to initiate ephaptic transmission between muscle fibres requires extensive muscle fibre necrosis (6).

With acute muscle fibre necrosis, drop out of muscle fibres from the motor unit produces MUAPs with reduced amplitude and duration. Within a few weeks of onset compensatory mechanism start, including hypertrophy and splitting of surviving muscle fibres, increased variation of fibre diameters with regeneration, and collateral reinnervation where segmental necrosis has occurred giving increased fibre type grouping. These changes affect the morphology of MUAPs and increased polyphasia, increased amplitude and duration, including the presence of satellite potentials, may be seen in addition to classical small MUAPs. Quantitative analysis giving only the mean MUAP duration may thus lose sensitivity as the excess of short duration units is counterbalanced by those with increased duration (7).

With treatment of myositis the degree of muscle fibre instability is expected to reduce and the compensatory mechanisms increase. With successful treatment fewer fibrillations and positive sharp waves will be recorded and MUAP complexity is increased (4,7,13,17). Quantitative EMG techniques are needed to track these changes, but often assessment of spontaneous activity is qualitative, introducing uncertainty unless the difference is marked. Increasing weakness on treatment could be due to either disease reactivation or steroid induced myopathy, but both qualitative and quantitative analysis may be insensitive in deciding which of these is responsible as steroid myopathy affects type 2 muscle fibres that are not usually assessed by EMG. Nonetheless, a return of fibrillations would suggest recurrence of necrosis (2,7,11).

In young patients the differential diagnosis between juvenile polymyositis and muscular dystrophy is a major concern because of the implications for prognosis and treatment. The needle EMG findings may be similar in these conditions, but myositis is more often associated with complex repetitive discharges and variation of muscle fibre diameter is greater in dystrophy. If serial studies are performed the response to treatment will be apparent only in myositis, but more often diagnosis of these conditions is based on other features such as the presence of atrophy (in dystrophy), autoantibodies (in myositis) and the biopsy appearance (20).

In the elderly, difficulty with the diagnosis of myopathy occurs because of type 2 muscle fibre atrophy and age-related neurogenic change, particularly in the lower limbs (5,17). Moreover the main differential diagnosis of inflammatory muscle disease in the elderly is inclusion body myopathy (IBM), which is the most common idiopathic inflammatory myopathy after the age of 50 years (21). Correct diagnosis is important because the response to treatment is poor, and muscle biopsy confirmation is essential for suspected cases. Nerve conduction is often abnormal in IBM with loss of sural sensory nerve action potentials and mild slowing of conduction velocity (2,22). The needle EMG examination shows fibrillation potentials with low-amplitude, short duration motor unit potentials, as in other inflammatory myopathies, but long-duration high amplitude polyphasic MUAPs are also encountered, as shown in Fig. 24.3. Recruitment patterns may be reduced and this latter combination of findings may erroneously suggest a neurogenic process. The correct diagnosis is usually reached on the basis of the long duration of symptoms (unlike motor neuron disease) and the typical distribution of particular involvement of the quadriceps and forearm flexor muscles (2,23–25). A further important differential diagnosis of inflammatory myositis is adult onset acid maltase deficiency (Pompe's disease) (4).

Critical illness myopathy

Over the last 15 years neuromuscular disease in critically ill patients has become an increasingly recognized complication leading to failure to wean from ventilation and prolonged admission to intensive therapy units. It is especially likely to develop in patients with multiple organ failure or sepsis with changes usually becoming apparent during the first 2 weeks of hospital admission and often persisting after discharge, leading to chronic disability (26–29). The pathophysiological mechanisms of acute quadriplegia may either involve peripheral nerves (axon degeneration or loss of excitability) or muscle fibres where necrosis or loss of myosin thick filaments are typically seen, the latter particularly occurring when high dose corticosteroid medication is given in conjunction with neuromuscular blockade. Neurophysiological evaluation thus requires a combination of motor and sensory nerve conduction with needle electromyography, with reduction of evoked compound muscle action potential amplitude and increase in duration being the first signs of the condition (28–31). Although these changes may be apparent in the extensor digitorum brevis muscle with peroneal nerve stimulation (32), multiple muscle recordings are recommended to confirm the illness (33). The needle EMG examination typically shows florid fibrillation activity. Assessment of motor units can be difficult in the ITU setting because patients are often comatose or sedated and they are unable to perform contraction to command, but if able, short duration low amplitude MUAPs are recorded (28,29,31). In the myopathic form of critical illness quadriplegia muscle fibres lose excitability (34), and a useful technique to distinguish between neuropathy and myopathy is to record muscle potentials following both peripheral nerve and direct muscle stimulation: in myopathy equally small responses are obtained by both methods of stimulation, but in neuropathy the response to direct muscle stimulation is larger than that evoked from the nerve (29,31,35,36). This technique may be used to follow recovery as return of muscle fibre excitability mirrors clinical improvement in strength (34). Use of the technique has shown that primary muscle disease is more

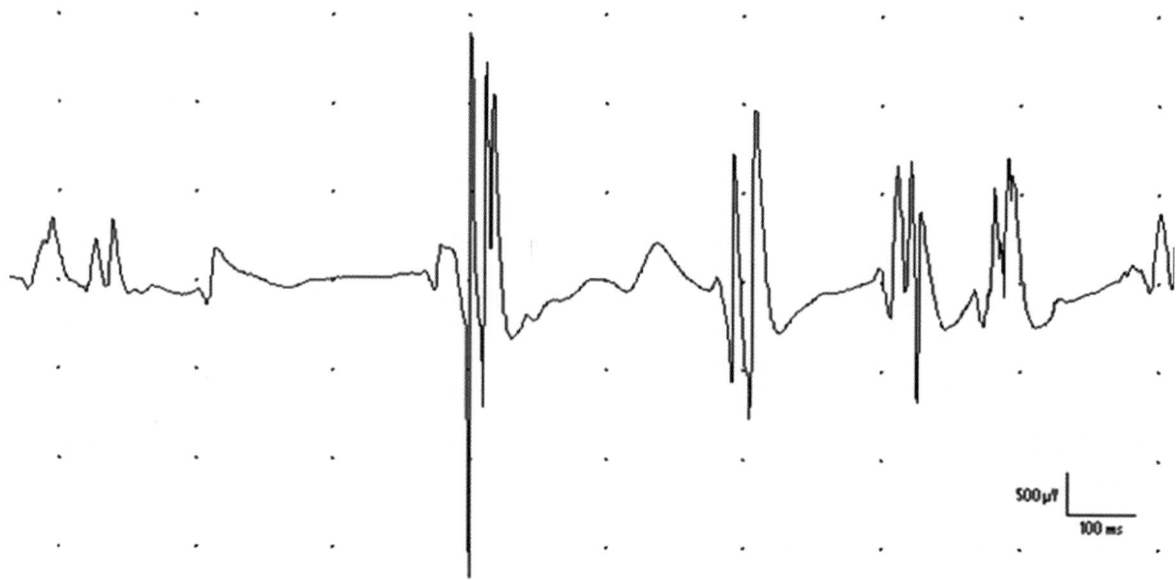

Fig. 24.3 Motor unit action potentials recorded from the quadriceps in a patient with inclusion body myopathy.

common than neuropathy in critical illness quadriplegia (36). Research studies have further defined the pathophysiological process in the myopathic form: slowing of muscle fibre conduction velocity correlates with the prolongation of CMAP duration (37), and studies of velocity recovery cycles indicate that inexcitability is associated with either depolarization or increased sodium channel inactivation, or both (38).

The finding of inexcitable muscles with profuse fibrillation potentials in the acute phase of critical illness may lead the neurophysiologist to predict a gloomy prognosis, but recovery may be surprisingly good as muscle fibre excitability returns, especially in children (26,39). Nonetheless, long term studies usually show persistent abnormality after critical illness disease (40).

Disorders with myotonia

Myotonia is one of the most unmistakable findings in needle electromyography. Repetitive muscle fibre discharges with waveforms similar to fibrillation potentials or positive sharp waves firing between 20 and 80 Hz with waxing and waning frequency and amplitude give a characteristic sound when amplified through a loud-speaker (41). Although a wide range of inherited and acquired muscle disease may be associated with needle myotonia, the finding is characteristic of myotonic dystrophy, and a group of inherited muscle membrane ion channelopathies, which will be discussed in this section.

Electrical myotonia is profuse in myotonic dystrophy type 1 (MD-1, dystrophia myotonica) when found at rest or with needle movement in both distal and proximal muscles including the paraspinals, although more prominent in the periphery (42). The diagnosis of MD-1 is usually made clinically because of the multisystem features (43), but if these are not all present the condition may be unsuspected until the electrophysiologist, testing for distal weakness, especially of the finger flexor muscles, finds electrical myotonia. In addition to the myotonic activity, the EMG examination shows abnormalities to motor unit potentials with an excess of short-duration, low amplitude polyphasic MUAPs. Examples of EMG recordings in myotonic dystrophy type 1 are shown in Fig. 24.4. Although rarely needed for diagnosis, a short exercise test shows immediate reduction in CMAP amplitude with rapid recovery (41,44–46). Unlike other conditions associated with myotonia, nerve conduction may be abnormal, showing features of a peripheral neuropathy that may be sub-clinical or associated with mild sensory symptoms (41,47,48).

More difficulty is encountered with the diagnosis of myotonic dystrophy type 2 (DM-2, proximal myotonic dystrophy). These patients tend to be older at presentation and proximal weakness, which can be focal, is more prominent than myotonia (49). These patients show less florid electrical myotonia, and indeed this may not be observed at all in some patients (42,43,50,51). A careful search of proximal and distal muscles is therefore required, including the quadriceps where myotonia is more prevalent than in MD-1. The myotonia itself may not have waxing and waning characteristic and can be only of the waning or positive sharp wave shape in some patients. The MUAPs in weak muscles show an excess of short-duration, low amplitude polyphasic units, but this myopathic feature is not specific in the absence of myotonia, and a high index of suspicion is needed if MD-2 is to be diagnosed (42,51).

The non-dystrophic myotonias are a group of muscle ion channelopathies in which membrane electrical potential is altered, resulting in either hyperexcitability with myotonia, or inexcitability with weakness. Stiffness due to myotonia is the predominant symptom in chloride (myotonia congenita) or sodium channelopathy (paramyotonia congenita, and Group 2 sodium channel myotonia), whereas weakness is the main symptom in periodic paralysis, which is caused by mutations in sodium, potassium, or calcium channel genes (52–54). Although due to mutations in different sodium channel genes, there is clinical overlap between hyperkalaemic periodic paralysis and myotonia congenita, which both show myotonia and episodes of skeletal weakness. The needle EMG examination in all these conditions shows features similar to MD-2: the electrical myotonia may not be pronounced and can be atypical, but at other times can be prolific and at unusually high frequency; although an excess of short-duration, low amplitude polyphasic potentials is seen, especially in later stages with persistent weakness, the MUAP changes are often not very pronounced. Taking these factors into consideration, the electrophysiological diagnosis of non-dystrophic myotonia can be difficult and often relies on a combined consideration of electrophysiological and clinical findings supported by genetic analysis (54). Rather than relying on electrical myotonia for diagnosis, a better neurophysiological approach is to quantify the membrane inexcitability following exercise. The abductor digiti minimi response to supra-maximal ulnar nerve stimulation is recorded to obtain a stable baseline value and recordings are repeated after either 10 s of maximum voluntary contraction (the short exercise test) or after 3–5 min of repeated 15s bursts of contraction with brief rest intervals (the long exercise test). For the short exercise test, CMAP recordings of amplitude and area are made at 10-s intervals for 60 s following contraction, and the exercise is repeated 3 times at normal temperature and after limb cooling. Patients with paramyotonia congenita typically show a decrement lasting 60 s that increases with subsequent trials and cooling. Patients with myotonia congenita typically show immediate decrement that recovers during the first 60 s and is less pronounced in subsequent trials, but in some this pattern is only apparent after cooling, and at normal temperature there is slight increment immediately after contraction with return to base line, but no decrement. This latter pattern may be seen in normal individuals, showing the need to repeat the test after limb cooling (2,41,44–46,54).

The long exercise test is used for the diagnosis of periodic paralysis. After the 3–5 min of contraction the CMAP amplitude and area are measured immediately and at 1-min intervals for 10 min and then at 5-min intervals for 30–45 min (protocols vary in detail) (2,55,56). Both normal controls and patients with periodic paralysis may show an increment in amplitude immediately after exercise followed by decrement. This is not more than 30% in controls and greater decrement is diagnostic for periodic paralysis. Patients with sodium channelopathy (hyperkalaemic and some hypokalaemic periodic paralysis) show increment followed by decrement that is maximal 30–45 min later. Calcium channel periodic paralysis (the majority of patients with hypokalaemic PP) show decrement at 20–40 min with or without initial increment. This pattern is also seen in the Andersen-Tawil syndrome (a potassium channelopathy associated with ventricular arrhythmias and prolonged QT interval, and congenital anomalies of the face and digits with short

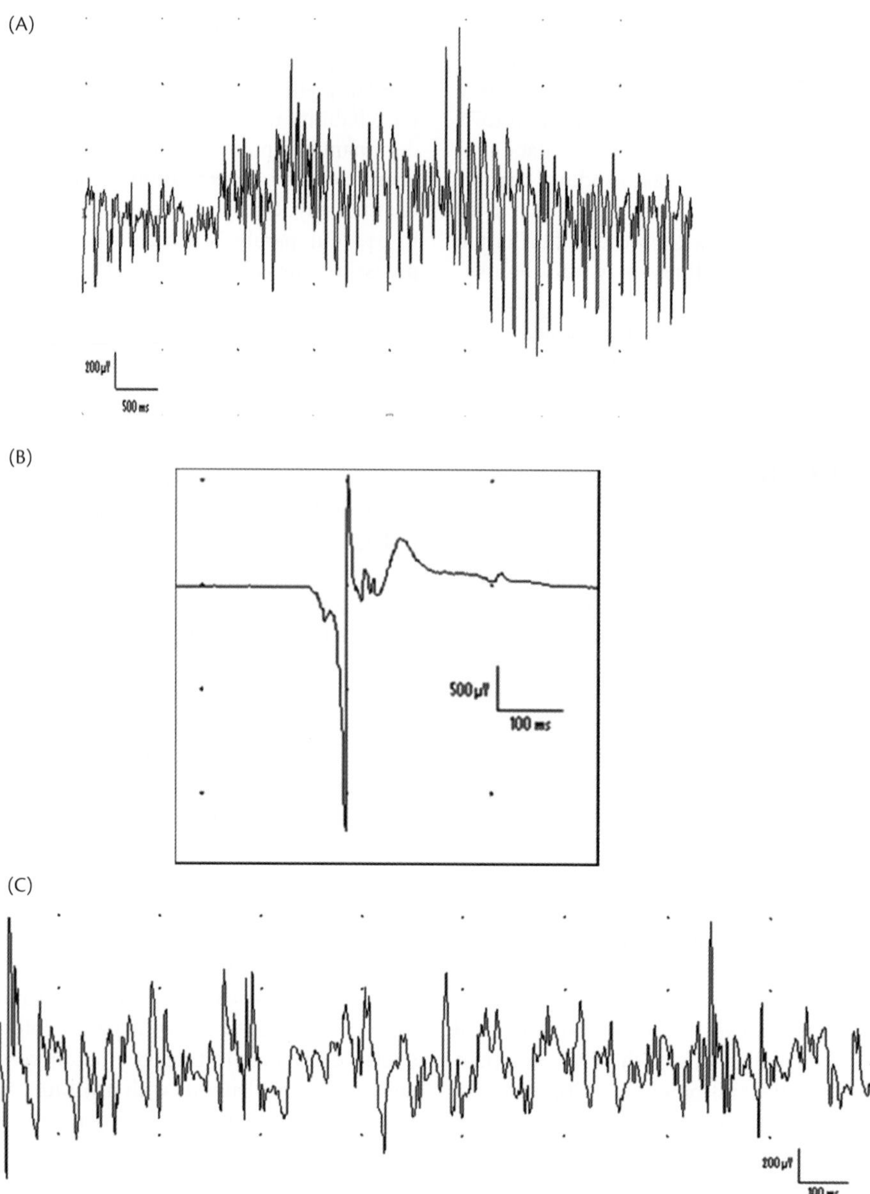

Fig. 24.4 EMG activity recorded in myotonic dystrophy type 1. (A) Myotonic discharge at rest, showing waxing and waning of amplitude and frequency. (B) Polyphasic motor unit action potential with a satellite component. (C) Full interference pattern recorded in a muscle that does not have strength to move the limb against gravity.

stature and scoliosis), but this and hypokalaemic periodic paralysis do not show needle myotonia (41,44–46,54,57).

Inherited myopathy

Advances in molecular genetics have greatly diminished the role of neurophysiology in the management of the dystrophinopathies (Duchenne and Becker muscular dystrophy), and the patients likely to be encountered are sporadic presentation with a differential diagnosis of acquired myositis, mild clinical presentations of Becker MD, and in manifesting female carriers. Nerve conduction studies are expected to be normal, except where severe distal disease causes reduction in evoked CMAP amplitude. The needle EMG findings are dependent on the severity and duration of the illness, but typically fibrillations are detected in proximal and distal muscles and a combination of short-duration, low amplitude polyphasic MUAPs (2,58) with longer duration units that may contain satellite potentials (59) and high amplitude spike components (60). In the later stages, muscle becomes extensively replaced by fibrous tissue causing resistance to needle insertion and reduced interference patterns. These are consequences of muscle disease and should not be interpreted as 'neurogenic' change (2,58).

In the authors' experience, fascioscapulohumeral is the most common of the other muscular dystrophies requiring neurophysiological testing. Asymmetrical shoulder girdle and facial weakness at onset may lead to an erroneous clinical diagnosis of peripheral nerve disease (for instance long thoracic nerve palsy), and unusual presentations with foot-drop, lower limb, or axial weakness are recognized (61,62). The EMG examination often shows a patchy distribution of abnormality and a number of periscapular, arm, and facial muscles should be examined. Apparently asymptomatic abnormalities can be found in the peroneal muscles. Fibrillations may or may not be detected and are usually scanty. The typical short-duration, low amplitude MUAPs confirm a primary myopathic disorder (58), but this may be difficult to assess in the face, where potentials are normally smaller than in the limbs with a greater degree of polyphasia. Quantitative motor unit analysis can be helpful in these circumstances (14). The same holds for the detection of dystrophic changes to MUAPs in the face of patients with oculopharyngeal muscular dystrophy.

The neurophysiological examination is of limited value in the diagnosis of the inherited limb-girdle, scapuloperonal, and distal myopathy syndromes as the findings are similar in all, with only variation in distribution. Typically nerve conduction is normal, but some conditions such as myofibrillar myopathy, may be associated with peripheral neuropathy (63). Needle EMG shows restricted fibrillations that are less prominent than in the dystrophinopathies, and a combination of short-duration or long duration MUAPs of varying amplitude. Recruitment patterns are usually complete, but may decline in late-stage disease (2,5,11,58).

Defects in the genes coding for the mitochondrial respiratory chain lead to multi-system disease including both central and peripheral nervous system (64). Mitochondrial proteins are encoded by both mitochondrial and nuclear genome and in recent years a large number of mutations have been identified (for review see Pferrer & Chinnery, 2013) (65). These genetics studies have made it clear that only weak correlation exists between clinical phenotype and genetic defects, and that variation within phenotypes is large. Mitochondrial myopathy is a differential diagnosis of many clinical presentations and for the clinical neurophysiologist a high index of suspicion is required if a correct diagnosis is to be reached. Tissues with high metabolic requirements typically show clinical abnormality, and myopathy is often, but not always a component of the syndromes of mitochondrial disease (64,65). The most common clinical presentation is with ocular myopathy producing ptosis and ophthalmoplegia (progressive external ophthalmoplegia, PEO). This may appear at any age, but is often more severe with younger onset, and although it may be isolated with or without proximal myopathy, it can be associated with other syndromes, including Kearns-Sayre (with pigmentary retinopathy, cardiac defect, ataxia, diabetes, deafness), ataxia neuropathy, and myopathy neurogastrointestinal encephalopathy (MNGIE) (64,66). The neurophysiologist is often requested to differentiate PEO from acquired or congenital myasthenia or oculopharyngeal muscular dystrophy (67,68). Single fibre EMG studies in myasthenia usually show greater abnormality, particularly the presence of impulse blocking, but mild increase in jitter values can be found in all 3 conditions (69,70). In the authors' experience, the appearance of motor unit potentials may be helpful as in mitochondrial disease they may show less polyphasia than in oculopharyngeal muscular dystrophy, and the presence of low amplitude, short-duration MUAPs in proximal limb muscles is more in keeping with myopathy than myasthenia. These are general guidelines and there is overlap in the neurophysiological findings in the 3 diseases.

Isolated mitochondrial myopathy may present without PEO, as may other syndromes that have myopathy as a component (such as the MELAS and MERRF syndromes). The neurophysiological features of the myopathy are not specific and there is wide clinical variation, both in the age of presentation and in the EMG findings. Typically, abnormal spontaneous activity is minimal, and although fibrillations and myotonia have been reported they are not conspicuous. Motor unit potentials may show an excess of short-duration low amplitude polyphasic potentials, but again this is less prominent than in the muscular dystrophies (66,71). The ataxic neuropathy syndromes (ANS) also show variable association with PEO, but the characteristic feature of this group is a sensory axonal neuropathy, usually with complete loss of sensory nerve action potentials on nerve conduction testing. Peripheral neuropathy on formal nerve conduction testing is reported in about a third of patients with mitochondrial disease, but this may not be clinically apparent (71). Although slowing of conduction velocity has been reported, the consensus opinion is that the neuropathy associated with mitochondrial disease is axonal, being either pure motor, pure sensory or mixed. Neuropathy may be secondary to diabetes in some of the mitochondrial syndromes (72).

Inherited disorders of muscle metabolism are rare and the more severe conditions, such as debrancher enzyme deficiency or carnitine deficiency, present with weakness in childhood. Milder diseases may first appear in adult life, and the common presenting complaints in addition to weakness are exercise intolerance with cramping, exercise induced muscle pain, or myoglobinuria. The main differential diagnosis for these symptoms is deficiency of carnitine palmitoyltransferase, myophosphorylase (McArdle's disease), and phospho-fructokinase (Tarui disease). In all three the EMG examination may be normal in the early stages with intermittent symptoms, but short-duration low amplitude polyphasic MUAPs are seen when permanent weakness has developed. In McArdle and Tarui disease it may be possible to induce electrically silent hand muscle cramps by exercise during the EMG examination (58).

Adult-onset acid maltase deficiency (Pompe's disease) is important as the clinical presentation is with axial muscle weakness with respiratory failure, and the neurophysiologist must therefore differentiate the condition from motor neuron disease and inflammatory myositis. The needle EMG examination shows very pronounced insertional and spontaneous activity that may be restricted to the paraspinal muscles. Fibrillations, myotonic, and complex repetitive discharges in this distribution give a clue to the correct diagnosis (2,73,74).

Endocrine, metabolic, and toxic disorders

Patients with neuromuscular disorders are often treated with prolonged courses of high dose corticosteroid immunosuppression therapy. Should they have increasing weakness, the question often asked is whether this is due to recurrence of the underlying disease or an independent steroid myopathy. Unfortunately, this cannot be answered by neurophysiology as steroid myopathy causes type 2 muscle fibre atrophy, and these units are not assessed by conventional EMG. The same holds for the proximal weakness of Cushing's disease, which is not associated with EMG abnormality. Both chronic alcoholism and paraneoplastic myopathy can also be associated with type 2 muscle fibre atrophy, explaining why the EMG examination of these patients shows no abnormality despite the presence of proximal weakness (2,5,7).

In contrast, thyroid disease may be complicated by muscle disease that can be addressed neurophysiologically. The symptoms of clinical hypothyroidism often include proximal muscle weakness, pain, and stiffness, and biochemical analysis may show elevated muscle enzymes (the serum creatine kinase can be raised in hypothyroidism without muscle symptoms), leading to the clinical suspicion of myositis. In fact the EMG studies of these patients do not show abnormal spontaneous activity (fibrillation potentials) and muscle necrosis is not found on muscle biopsy (75,76). The needle EMG examination may show an increase of small polyphasic MUAPs in proximal muscles, but this is not universally the case. There is a correlation between the presence of myopathy and carpal tunnel syndrome in hypothyroid patients, but not with peripheral neuropathy (77). It is worth remembering that myopathic symptoms may persist after successful biochemical treatment of hypothyroidism, as these perplexing patients may be referred for neurophysiological assessment (78). Hypothyroidism is commonly caused by Hashimoto's thyroiditis, which may be associated with other autoimmune neurological diseases (79,80).

Hyperthyroidism may also be associated with symptoms of proximal myopathy. As with hypothyroidism the EMG examination does not show evidence of myositis, and while low-amplitude, short-duration polyphasic MUAPs may be detected in proximal muscles (81), it often requires quantitative motor unit analysis to differentiate thyrotoxic patients from normal controls (82). A small proportion of patients with thyrotoxicosis present with symptoms of periodic paralysis, with typical changes during a long exercise test, but without myotonia. This association is higher in Chinese people (83,84).

Transient biochemical disturbance in metabolic disease, especially hypokalaemia, is associated with weakness, but no neurophysiological abnormality unless causing peripheral nerve hyperexcitability, when extra-discharges are seen following the M-response of compound muscle action potentials evoked by nerve stimulation (85).

Of the drug associated muscle disorders, statin myopathy has received the most recent attention. Huge numbers of patients are prescribed these lipid-lowering drugs world-wide, and myalgia is reported in up to 25%. No neurophysiological abnormalities are seen in these people and testing is not required. Patients on medication may develop weakness, but there is no close temporal association between taking medication and onset, leading to the opinion that statin use may expose underlying myopathy of other aetiology in these people. The more serious condition of rhabdomyolysis associated with HMG-CoA reductase antibodies is rare, and presents with weakness and high serum creatine kinase levels (86,87). The needle EMG examination shows changes akin to myositis with very prominent abnormal spontaneous activity including myotonia and fibrillations (88).

Other drugs may rarely cause myopathy. D-penicillamine and interferon-α are known to produce immune mediated disease manifesting as polymyositis. Colchicine inhibits microtubules and can lead to a vacuolar proximal myopathy, particularly in the elderly with renal impairment. The EMG findings are not specific, but may be associated with a mild sensory peripheral neuropathy. Nucleoside analogue reverse transcriptase inhibitors used in the treatment of HIV infection may induce proximal muscle weakness, fatigue and myalgia with elevated lactate and CK levels. Zidovudin is particularly responsible for this complication, which presents similarly to mitochondrial myopathy (89-91).

Conclusions

This chapter has described the neurophysiological findings that may be expected in the more commonly encountered muscle diseases. It is not intended to be exhaustive, and indeed one of the reasons that clinical neurophysiology is a continually interesting career is that patients will be seen who do not fit straightforwardly into clinical or electrographic patterns, prompting a literature search to find whether similar presentations have been reported. It will be apparent that a blanket approach to 'myopathy' is unlikely to give the best value from the neurophysiological examination, and that testing tailored to the patient's presenting symptoms and signs will be more valuable. To be able to do this the neurophysiologist needs a working knowledge of the clinical presentation of muscle diseases and be able to generate a list of differential diagnoses that can be explored or rejected by the neurophysiological examination. As in other diseases, clinical neurophysiology is an extension of the neurological examination and must be seen as a component of the process of investigation rather than a definitive test in its own right. Nonetheless, we recommend that a starting point for all patients with suspected muscle disease is to perform at least one motor and sensory nerve conduction study with the objective of eliminating alternative conditions that may present in a similar way (for instance demyelinating polyneuropathy and Eaton Lambert syndrome) and defining conditions with combined peripheral neuropathy. If this is normal, proceed to needle EMG examination of proximal and distal muscles of the arm and leg. If this appears normal always remember the importance of studying the paraspinal muscles before concluding the test. The information gathered by this process will direct the need for further specialized studies such as exercise testing or single fibre EMG.

References

1. Wilbourn, A.J. (1993). The electrodiagnostic examination with myopathies. *Journal of Clinical Neurophysiology*, **10**(2), 132–48.
2. Paganoni, S. and Amato, A. (2013). Electrodiagnostic Evaluation of myopathies. *Physical Medicine and Rehabilitation Clinics of North America*, **24**(1), 193–207.
3. Bohan, A. and Peter, J.B. (1975). Polymyositis and dermatomyositis. part 1. *New England Journal of Medicine*, **292**, 344–7.
4. Gutiérrez-Gutiérrez, G., Barbosa López, C., Navacerrada, F., et al. (2012). 'Use of electromyography in the diagnosis of inflammatory myopathie. *Rheumatology Clinic*, **8**(4), 195–200.
5. Lacomis, D. (2012). Electrodiagnostic approach to the patient with suspected myopathy. *Neurology Clinic*, **30**(2), 641–60.
6. Lyu, R.K., Cornblath, D.R., and Chaudhry, V. (1999). Incidence of irritable electromyography in inflammatory myopathy. *Journal of Clinical Neuromuscular Disease*, **1**(2), 64–7.
7. Amato, A.A. and Dumitru, D. (2002). Acquired myopathies. In: D. Dumitru, A. A. Amato, and M. J. Zwarts (Eds) *Electrodiagnostic Medicine*, 2nd edn, pp. 1371–432. Philadelphia, PA: Hanley and Belfus Medical Publishers.
8. Hanisch, F., Kronenberger, C., Zierz, S., and Kornhuber, M. (2013). The significance of pathological spontaneous activity in various myopathies. *Clinical Neurophysiology*, **125**(7), 1485–90.
9. Purves, D. and Sakmann, B. (1974). Membrane properties underlying spontaneous activity of denervated muscle fibres. *Journal of Physiology*, **239**(1), 125–53.
10. Thesleff, S. and Ward, M.R. (1975). Studies on the mechanism of fibrillation potentials in denervated muscle. *Journal of Physiology*, **244**(2), 313–23.

11. Lynch, M.C. and Cohen, J.A. (2011). A primer on electrophysiologic studies in myopathy. *Rheumatic Disease Clinics of North America*, **37**(2), 253–68.
12. Stoehr, M. (1978). Low frequency bizarre discharges. A particular type of electromyographical spontaneous activity in paretic skeletal muscle. *Electromyography & Clinical Neurophysiology*, **18**(2), 147–56.
13. Bohan, A., Peter, J.B., Bowman, RL., et al. (1977). Computer-assisted analysis of 153 patients with polymyositis and dermatomyositis. *Medicine (Baltimore)*, **56**(4), 255–86.
14. Farrugia, M.E. and Kennett, R.P. (2005). Turns amplitude analysis of the orbicularis oculi and oris muscles. *Clinical Neurophysiology*, **116**(11), 2550–9.
15. Mastaglia, F.L., Garlepp, M.J., Phillips, B.A., et al. (2003). Inflammatory myopathies: clinical, diagnostic and therapeutic aspects. *Muscle Nerve*, **27**(4), 407–25.
16. al-Saffar, Z.S., Kelsey, C.R., Kennet, R.P., et al. (1994). Myositis and eosinophilia in a patient with sarcoidosis. *Postgrad Medicine Journal*, **70**(829), 833–5.
17. Blijham, P.J., Hengstman, G.J., Hama-Amin, A.D., et al. (2006). Needle electromyographic findings in 98 patients with myositis. *European Neurology*, **55**(4). 183–8.
18. Mitz, M., Change, B.S., Albers, J.W., et al. (1981). Electromyographic and histologic paraspinal abnormalities in polymyositis/dermatomyositis. *Archives of Physical Medicine Rehabilitation*, **62**(3), 118–21.
19. Buchthal, F. and Kamieniecka, Z. (1982). The diagnostic yield of quantified electromyography and quantified muscle biopsy in neuromuscular disorder. *Muscle Nerve*, **5**(4), 265–80.
20. Mamyrova, G., Katz, J.D., Jones, R.V., et al. (2013). Childhood Myositis Heterogeneity Collaborative Study Group. Clinical and laboratory features distinguishing juvenile polymyositis and muscular dystrophy. *Arthritis Care & Research (Hoboken)*, **65**(12),1969–75.
21. Dimachkie, M.M. and Barohn, R.J. (2013). Inclusion body myositis. *Current Neurology Neuroscience Report*, **13**(1), 321.
22. Estephan, B., Barohn, R.J., Dimachkie, M.M., et al. (2011). Sporadic IBM: a case cohort. *Journal of Clinical Neuromuscular Disease*, **12**(3), 18–9.
23. Hatanaka, Y. and Oh, S.J. (2007). Single-fiber electromyography in sporadic inclusion body myopathy. *Clinical Neurophysiology*, **118**(7), 1563–8.
24. Barkhaus, P.E., Periquet, M.S., and Nandedkar, S.D. (1999). Quantitative electrophysiologic studies in sporadic inclusion body myositis. *Muscle & Nerve*, **22**(4), 480–7.
25. Hokkoku, K., Sonoo, M., Higashihara, M., et al. (2012). Electromyographs of the flexor digitorum profundus muscle are useful for the diagnosis of inclusion body myositis. *Muscle & Nerve*, **46**(2), 181–6.
26. Vondracek, P. and Bednarik, J. (2006). Clinical and electrophysiological findings and long-term outcomes in paediatric patients with critical illness polyneuromyopathy. *European Journal of Paediatric Neurology*, **10**(4), 176–81.
27. Visser, L.H. (2006). Critical illness polyneuropathy and myopathy: clinical features, risk factors and prognosis. *European Journal of Neurology*, **13**(11), 1203–12.
28. Goodman, B.P. and Boon, A.J. (2008). Critical illness neuromyopathy. *Physical Medicine and Rehabilitation Clinics of North America*, **19**(1), 97–110.
29. Trojaborg, W. (2006). Electrophysiologic techniques in critical illness-associated weakness. *Journal of Neurology Science*, **242**(1–2), 83–5.
30. Park, E., Nishida, T., Sufit, R., et al. (2004). Prolonged compound muscle action potential duration in critical illness myopathy: report of nine cases. *Clin Neuromuscular Disease*, **5**, 176–83.
31. Lacomis, D. (2013). Electrophysiology of neuromuscular disorders in critical illness. *Muscle & Nerve*, **47**(3), 452–63.
32. Latronico, N., Bertolini, G., Guarneri, B., et al. (2007). Simplified electrophysiological evaluation of peripheral nerves in critically ill patients: the Italian multi-centre CRIMYNE study. *Critical Care*, **11**(1), R11.
33. Goodman, B.P., Harper, C.M., and Boon A.J. (2009). Prolonged compound muscle action potential duration in critical illness myopathy. *Muscle & Nerve*, **40**(6), 1040–2.
34. Rich, M.M., Teener, J.W., Raps, E.C., et al. (1996). Muscle is electrically inexcitable in acute quadriplegic myopathy. *Neurology*, **46**(3), 731–6.
35. Rich, M.M., Bird, S.J., Raps, E.C., et al. (1997). Direct muscle stimulation in acute quadriplegic myopathy. *Muscle & Nerve*, **20**(6), 665–73.
36. Lefaucheur, J.P., Nordine, T., Rodriguez, P., et al. (2006). Origin of ICU acquired paresis determined by direct muscle stimulation. *Journal of Neurology, Neurosurgery, & Psychiatry*, **77**(4), 500–6.
37. Allen, D.C., Arunachalam, R., and Mills, K.R. (2008). Critical illness myopathy: further evidence from muscle-fiber excitability studies of an acquired channelopathy. *Muscle & Nerve*, **37**(1), 14–22.
38. Z'Graggen, W.J., Brander, L., Tuchscherer, D., et al. (2011). Muscle membrane dysfunction in critical illness myopathy assessed by velocity recovery cycles. *Clinical Neurophysiology*, **122**, 834–41.
39. Guarneri, B., Bertolini, G., and Latronico, N. (2008). Long-term outcome in patients with critical illness myopathy or neuropathy: the Italian multicentre CRIMYNE study. *Journal of Neurology, Neurosurgery, & Psychiatry*, **79**(7), 838–41.
40. Fletcher, S.N., Kennedy, D.D., Ghosh, I.R., et al. (2003). Persistent neuromuscular and neurophysiologic abnormalities in long-term survivors of prolonged critical illness. *Critical Care Medicine*, **31**(4), 1012–16.
41. Heatwole, C.R., Statland, J.M., and Logigian, E.L. (2013). The diagnosis and treatment of myotonic disorders. *Muscle & Nerve*, **47**(5), 632–48.
42. Logigian, E.L., Ciafaloni, E., Quinn, L.C., et al. (2007). Severity, type, and distribution of myotonic discharges are different in type 1 and type 2 myotonic dystrophy. *Muscle & Nerve*, **35**(4), 479–85.
43. Machuca-Tzili, L., Brook, D., and Hilton-Jones, D. (2005). Clinical and molecular aspects of the myotonic dystrophies: a review. *Muscle & Nerve*, **32**(1), 1–18.
44. Fournier, E., Viala, K., Gervais, H., et al. (2006). Cold extends electromyography distinction between ion channel mutations causing myotonia. *Annals of Neurology*, **60**, 356–65.
45. Fournier, E., Arzel, M., Sternberg, D., et al. (2004). Electromyography guides toward subgroups of mutations in muscle channelopathies. *Annals of Neurology*, **56**, 650–61.
46. Tan, S.V., Matthews, E., Barber, M., et al. (2011). Refined exercise testing can aid DNA-based diagnosis in muscle channelopathies. *Annals of Neurology*, **69**, 328–40.
47. Hermans, M.C., Faber, C.G., Vanhoutte, E.K., et al. (2011). Peripheral neuropathy in myotonic dystrophy type 1. *Journal of the Peripheral Nervous System*, **16**(1), 24–9.
48. Bae, J.S., Kim, O.K., Kim, S.J., et al. (2008). Abnormalities of nerve conduction studies in myotonic dystrophy type 1: primary involvement of nerves or incidental coexistence? *Journal of Clinical Neuroscience*, **15**(10), 1120–4.
49. Milone, M., Batish, S.D., and Daube, J.R. (2009). Myotonic dystrophy type 2 with focal asymmetric muscle weakness and no electrical myotonia. *Muscle & Nerve*, **39**(3), 383–5.
50. Day, J.W., Ricker, K., Jacobsen, J.F., et al. (2003). Myotonic dystrophy type 2: molecular, diagnostic and clinical spectrum. *Neurology*, **60**, 657–64.
51. Young, N.P., Daube, J.R., Sorenson, E.J., et al. (2010). Absent, unrecognized, and minimal myotonic discharges in myotonic dystrophy type 2. *Muscle & Nerve*, **41**(6), 758–62.
52. Hehir, M.K. and Logigian, E.L. (2013). Electrodiagnosis of myotonic disorders. *Physical Medicine and Rehabilitation Clinics of North America*, **24**(1), 209–20.
53. Miller, T.M. (2008). Differential diagnosis of myotonic disorders. *Muscle & Nerve*, **37**(3), 293–9.
54. Heatwole, C.R. and Moxley, R.T., 3rd. (2007). The nondystrophic myotonias. *Neurotherapeutics*, **4**(2), 238–51.
55. McManis, P.G., Lambert, E.H., and Daube, J.R. (1986). The exercise test in periodic paralysis. *Muscle & Nerve*, **9**, 704–10.

56. Kuntzer, T., Flocard, F., Vial, C., et al. (2000). Exercise test in muscle channelopathies and other muscle disorders. *Muscle & Nerve*, **23**, 1089–94.
57. Venance, S.L., Cannon, S.C., Fialho, D., et al. (2006). The primary periodic paralyses: diagnosis, pathogenesis and treatment. *Brain*, **129**(Pt 1), 8–17.
58. Amato, A.A. and Dumitru, D. (2002). Hereditary myopathies. In: D. Dumitru, A. A. Amato, and M. J. Zwarts (Eds) *Electrodiagnostic medicine*, 2nd edn, pp. 1265–370. Philadelphia, PA: Hanley and Belfus Medical Publishers.
59. Zalewska, E., Szmidt-Salkowska, E., Rowinska-Marcinska, K., at al. (2013). Motor unit potentials with satellites in dystrophinopathies. *Journal of Electromyography and Kinesiology*, **23**(3), 580–6.
60. Willison, R.G. (1964). Analysis of electrical activity in healthy and dystrophic muscle in man. *Journal of Neurology, Neurosurgery, & Psychiatry*, **27**, 386–94.
61. Hassan, A., Jones, L.K., Jr., Milone, M., et al. (2012). Focal and other unusual presentations of facioscapulohumeral muscular dystrophy. *Muscle & Nerve*, **46**(3), 421–5.
62. Pastorello, E., Cao, M., and Trevisan, CP. (2012). Atypical onset in a series of 122 cases with facioscapulohumeral muscular dystrophy. *Clinical Neurology and Neurosurgery*, **114**(3), 230–4.
63. Selcen, D., Ohno, K., and Engel, A.G. (2004). Myofibrillar myopathy: clinical, morphological and genetic studies in 63 patients. *Brain*, **127**(Pt 2), 439–51.
64. Nardin, R.A. and Johns D.R. (2001). Mitochondrial dysfunction and neuromuscular disease. *Muscle & Nerve*, **24**(2), 170–91.
65. Pfeffer, G. and Chinnery, P.F. (2013). Diagnosis and treatment of mitochondrial myopathies. *Annals of Medicine*, **45**(1), 4–16.
66. Arpa, J., Cruz-Martínez, A., Campos, Y., et al. (2003). Prevalence and progression of mitochondrial diseases: a study of 50 patients. *Muscle & Nerve*, **28**(6), 690–5.
67. Schoser, B.G. and Pongratz, D. (2006). Extraocular mitochondrial myopathies and their differential diagnoses. *Strabismus*, **14**, 107–13.
68. Ben Yaou, R., Laforêt, P., Bécane, H.M., et al. (2006). Misdiagnosis of mitochondrial myopathies: a study of 12 thymectomized patients. *Reviews in Neurology (Paris)*, **162**(3), 339–46.
69. Girlanda, P., Toscano, A., Nicolosi, C., et al. (1999). Electrophysiological study of neuromuscular system involvement in mitochondrial cytopathy. *Clinical Neurophysiology*, **110**(7), 1284–9.
70. Fawcett, P.R., Mastaglia, F.L., and Mechler, F. (1982). Electrophysiological findings including single fibre EMG in a family with mitochondrial myopathy. *Journal Neurological Sciences*, **53**(2), 397–410.
71. Mancuso, M., Piazza, S., Volpi, L., et al. (2012). Nerve and muscle involvement in mitochondrial disorders: an electrophysiological study. *Neurological Sciences*, **33**(2), 449–52.
72. Pareyson, D., Piscosquito, G., Moroni, I., et al. (2013). Peripheral neuropathy in mitochondrial disorders. *Lancet: Neurology*, **12**(10), 1011–24.
73. Müller-Felber, W., Horvath, R., Gempel, K., et al. (2007). Late onset Pompe disease: clinical and neurophysiological spectrum of 38 patients including long-term follow-up in 18 patients. *Neuromuscular Disorder*, **17**(9–10), 698–706.
74. Hobson-Webb, L.D., Dearmey, S., and Kishnani, P.S. (2011). The clinical and electrodiagnostic characteristics of Pompe disease with post-enzyme replacement therapy findings. *Clinical Neurophysiology*, **122**(11), 2312–7.
75. Madariaga, M.G. (2002). Polymyositis-like syndrome in hypothyroidism: review of cases reported over the past twenty-five years. *Thyroid*, **12**(4), 331–6.
76. Toscano, A., Bartolone, S., Rodolico, C., et al. (1996). Onset of hypothyroidism with polymyositis-like clinical features in elderly patients. *Archives of Gerontology and Geriatrics*, **22**(1), 573–6.
77. Eslamian, F., Bahrami, A., Aghamohammadzadeh, N., et al. (2011). Electrophysiologic changes in patients with untreated primary hypothyroidism. *Journal of Clinical Neurophysiology*, **28**(3), 323–8.
78. Duyff, R.F., Van den Bosch, J., Laman, DM., et al. (2000). Neuromuscular findings in thyroid dysfunction: a prospective clinical and electrodiagnostic study. *Journal of Neurology and Neurosurgery Psychiatry*, **68**(6), 750–5.
79. Wang, H., Li, H., Kai, C., et al. (2011). Polymyositis associated with hypothyroidism or hyperthyroidism: two cases and review of the literature. *Clinical Rheumatology*, **30**(4), 449–58.
80. Turker, H., Bayrak, O., Gungor, L., et al. (2008). Hypothyroid myopathy with manifestations of Hoffman's syndrome and myasthenia gravis. *Thyroid*, **18**(2), 259–62.
81. Mussa, K.I., Mahmoud, H.N., and Abdul-Zehra, I.K. (2003). Electromyographic changes in thyrotoxicosis. *Neurosciences (Riyadh)*, **8**(3), 173–6.
82. Puvanendran, K., Cheah, J.S., Naganathan, N., et al. (1979). Thyrotoxic myopathy: a clinical and quantitative analytic electromyographic study. *Journal of Neurology Science*, **42**(3), 441–51.
83. Kelley, D.E., Gharib, H., Kennedy, F.P., et al. (1989). Thyrotoxic periodic paralysis. Report of 10 cases and review of electromyographic findings. *Archives of Internal Medicine*, **149**(11), 2597–600.
84. Barahona, M.J., Vinagre, I., Sojo, L., et al. (2009). Thyrotoxic periodic paralysis: a case report and literature review. *Clinical Medicine & Research*, **7**(3), 96–8.
85. Seelig, M.S., Berger, A.R., and Spielholz, N. (1975). Latent tetany and anxiety, marginal magnesium deficit, and normocalcemia. *Diseases of the Nervous System*, **36**(8), 461–5.
86. Mohassel, P. and Mammen. A.L. (2013). The spectrum of statin myopathy. *Current Opinions in Rheumatology*, **25**(6), 747–52.
87. Sathasivam, S. (2012). Statin induced myotoxicity. *European Journal of Internal Medicine*, **23**(4), 317–24.
88. de Almeida, D.F., Lissa, T.V., and Melo, A.C., Jr. (2008). Myotonic potentials in statin-induced rhabdomyolysis. *Arquivos de Neuro-psiquiatria*, **66**(4), 891–3.
89. Valiyil, R. and Christopher-Stine, L. (2010). Drug-related myopathies of which the clinician should be aware. *Current Opinions in Rheumatology Reports*, **12**, 213–20.
90. Dalakas, M.C. (2009). Toxic and drug-induced myopathies. *Journal of Neurology, Neurosurgery, & Psychiatry*, **80**(8), 832–8.
91. Sieb, J.P. and Gillessen, T. (2003). Iatrogenic and toxic myopathies. *Muscle & Nerve*, **27**(2), 142–56.

CHAPTER 25

Paediatric conditions

Matthew Pitt

Introduction

The objective of this chapter is to equip its readers with a competence in paediatric electromyography. While other chapters in this textbook cover the technical and theoretical details of the different elements of the EMG and the diseases themselves, this chapter is not going to be devoted only to the diseases, but will introduce modifications to standard techniques, which will be required to be successful in doing paediatric EMG.

Before one goes onto the body of the chapter it is necessary to devote a little time to justify the very existence of paediatric EMG. Paediatric EMG perhaps unlike any other particular element of neurophysiology has had a somewhat chequered history. It has laboured under several misconceptions amongst which three stand out. These are that it is too painful, too difficult, and no longer needed in the era of molecular genetics.

The problem of being perceived as too painful certainly needs to be addressed. It is unclear why this perception has come about, but historically it may have been influenced by such things as the size and quality of the needles used. Also the fact that they were not single use meant that insertion was uncomfortable, because the needles were blunt, and children were justifiably alarmed by this. Also over the time that paediatric EMG has evolved there has been a clear change in the perception of children by not only the society as a whole, but by doctors as well. The child being 'seen, but not heard' is an outdated perception of their needs. Another element which is really not under the control of doctors is that in some societies, because of the funding methods in existence, doctors may feel obliged to investigate a child more than seems warranted. In reality, sometimes a limited number of investigations including needle EMG is all that is needed accurately to point the direction of subsequent investigation. Getting the maximum information with the minimum number of tests must be the overriding aim of any paediatric electromyographer.

With regard to the test being difficult, some of this problem arises from the previous discussion and certainly if the child is finding the experience painful then it is likely that the practitioner will as well. There are, however, undoubted difficulties doing studies on children particularly looking at the smallest babies and unless changes are made to one's techniques it will prove to be a challenge. This is discussed later when the technical aspects of paediatric EMG are highlighted.

The final criticism that the tests themselves are redundant in the era of molecular genetics is suggested by some of the developments in genetics, but perhaps more accurately the role of EMG will change rather than it become redundant. It is almost certain that as the technology increases and price of the exome screening drops a reversal of the normal way of proceeding will likely occur. Instead of phenotypic characterization, which might include EMG, followed by search for genetic markers it is perfectly feasible that we will see the child coming as an outpatient, having blood taken at the entry into the outpatients and then the genetic screening will be available to the clinician when seen. This will, however, not make the EMG or any other investigation strategy completely redundant because there are very few genes that have an individual phenotype and instead of the 'phenotype up' approach, investigation will be 'genotype down' in the future. EMG will also remain central to the diagnostic evaluation of children and adolescents with acquired neuromuscular disorders, including inflammatory/autoimmune, toxic, and traumatic insults.

The technical aspects

Recording electrodes have changed immeasurably certainly since this author started nerve conduction studies and now single use, self-adhesive pre-gelled electrodes are the norm, which can be cut down to any size to suit almost any limb. These are invaluable for the tiny arm of the baby. While the electrodes are more professionally produced and of better quality, the cables connecting them are also often shielded, which allows really excellent recordings to be made even in electronically hostile environments like ITU. Stimulators are still a slight problem. There are a lot of companies, which produce what can be described as 'cattle prod' electrodes, which are perfectly fine if the subject is prepared to stay still, but in many instances this is not possible in children. The author's favourite stimulator is one that allows you to encircle the limb at the same time as giving the stimulation (Fig. 25.1).

The needles for EMGs have also seen significant changes. First and foremost we now have to use in the United Kingdom disposable electrodes because of the fears about CJ disease and also as a consequence of the 'mad cow or bovine encephalitis' scare. This has meant that the needle does not become blunted and bent, which regularly occurred with the needles, which were reused and sterilized. Also some important developments with the coating of the needle have meant that their passage through the skin can be exceptionally easy and practically painless. Finally the EMG machines themselves have made the transition from analogue technology to digital successfully and also have decreased in size. One of the most important consequences of this change is that it is now possible to store the EMG in the computer for later analysis. This is invaluable and almost a requirement for the paediatric EMG.

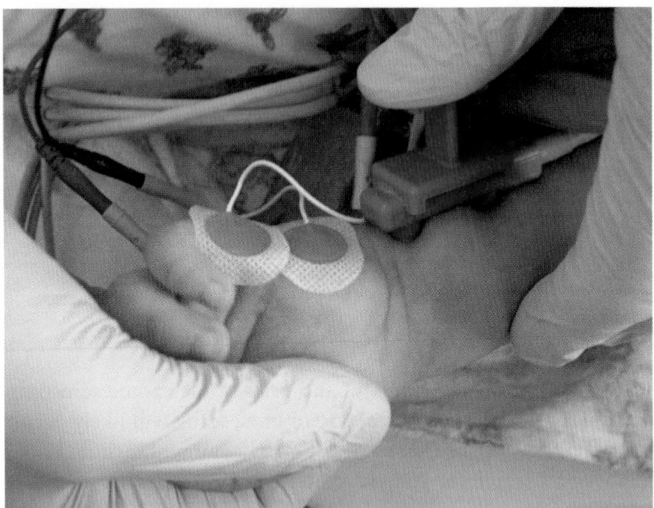

Fig. 25.1 The use of stimulating electrodes to stimulate the ulnar nerve showing how stimulation can be achieved at the same time as circling the limb.

Normative data

Normative data is a problem for all neurophysiological tests, but it is particularly challenging in children because of the difficulty obtaining data from a test that is perceived as uncomfortable and invasive in well children. It is lucky therefore that over the years data had been collected on some of the parameters that we use in paediatric EMG. In particular there is a wealth of data on nerve conduction velocity, which confirms the changes of maturation in the first five years of life (1–4). Several examples are present and while there may be slight variations in the upper and lower limits they define, by using them it is possible to estimate the normality of tests.

Fortunately the data that were collected on motor unit duration by Buchthal and co-workers over many years at the Rigs-hospital in Copenhagen are now available to be used (5). Initially, because these data were collected using standard concentric needle electrodes, which are larger in diameter and have a larger recording surface than the facial needles used in paediatrics, it was felt that it was unlikely that these normative data would be relevant with this different means of collection. It was fortunate therefore that Barkhaus and co-authors (6) were able to demonstrate the duration measurements taken not only with our facial needles and concentric needles, but also monopolar needles, used a great deal in the United States, all share similar motor unit durations. Amplitudes are a different matter, but these are not so important. Therefore this wealth of data is available, which is invaluable for paediatric electromyography. Other normative data is either non-existent or very laboratory specific. It is one of the mysteries to all people working in electromyography, not just in paediatrics, that there has never been a standardization of the recording size of electrodes and therefore, not surprisingly, the amplitudes obtained with these different electrodes for such things as sensory nerve action potentials or the amplitude of the compound muscle action potentials have not been transferable across different laboratories and even sometimes within laboratories if the electrodes used were different.

However, it is very worthwhile remembering that even if one could ideally collect normal controls on every parameter measured in EMG and nerve conduction generally as well as in paediatrics there are multiple variables that affect these recorded potentials. It is salutary to remember that nerve velocity as one example, may be affected by such variables as skin thickness, handedness, height, sex, and we have also seen significant differences between different races (7–15). So, even with the number of variables known with some possibly still unknown, but contributing factors, it would be a lifetime's work to collect for every patient the exact age-matched control recordings. This, however, does not mean that it is impossible to make judgements on the results. What is important is that you should never use a single measure as a determinant of abnormality, but always seek corroboration. So for example it is essential that a sensory neuropathy will be diagnosed not simply on finding one nerve's response reduced in amplitude, but only move to that conclusion if other nerves are tested and also after denervation has been excluded on EMG. It is worth comparing the situation in paediatric EMG with that in MRI. It is impossible to perform MRI in many children unless they are anesthetized. There is no ethics committee that would permit an anaesthetic in a normal child simply to obtain MRI data on a normal brain. This however, does not prevent experienced interpreters of MRI making judgements of the normality or otherwise of MRI even when using new sequences about, which little is known. A similar situation exists in paediatric EMG. Experience is a key factor.

Personal characteristics

It is not easy to do paediatric EMG and quite a few people fail when they attempt it, to neither their nor the child's benefit. It is important to realize that everybody is not suited to paediatric EMG. Some neurophysiologists simply find children too terrifying to work with, in the sense that they react bizarrely, or they may simply not be interested in this part of their work. Children over the age of around eight years are in many ways simply small adults. A humane approach will see an effective examination. Below that age the skills become increasingly difficult and specialized and so when doing neonates on an intensive care ward a whole new set of skills is needed compared with the adult.

To increase the likelihood of success it is important to follow a clear strategy. The first is not to give much information before the child comes to the appointment. It is very hard to put into words a description of a test, which involves electrocution and needles, without causing anxiety. When the child and parents come to the department it is very important to be honest about the likely discomfort. We always compare the test to a blood test. An unpublished audit of our cases suggested that the discomfort experienced during an EMG was roughly the same level as a blood test and certainly a great deal less than that experienced during venous cannulation. It is very important also to make the parents and the patient, if old enough, complicit in what follows by explaining why you are doing it and, particularly with a test such as stimulation single fibre EMG, the real importance to the management and care of the child that this momentary discomfort will possibly result in. Next it is very important to engage with the child as soon as possible. As soon as the child is able to communicate various distraction techniques can be used, but one of the most effective is to ask them to speak about their favourite thing. This is an extraordinary powerful tool and the quietest, shyest child will beam with enjoyment when they start to expand on their particular favourite interest. During the

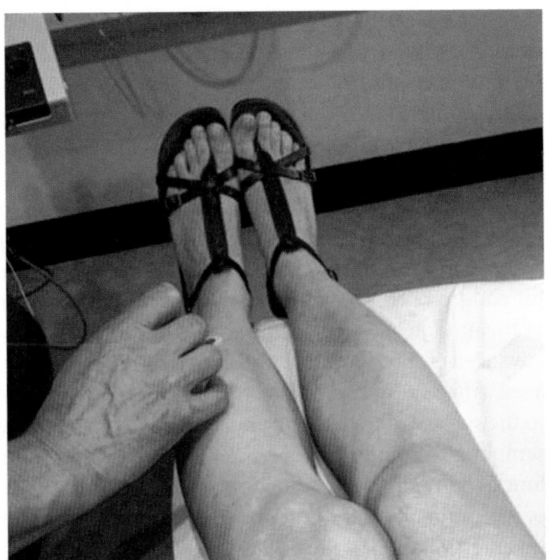

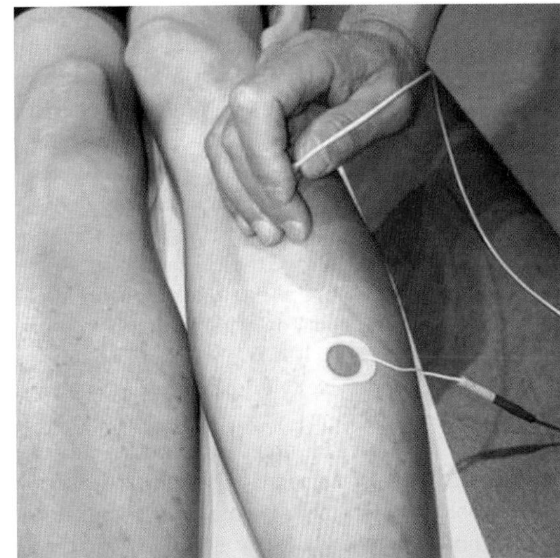

Fig. 25.2 The method used to hide the needle from the child during EMG.

test itself other distractions can be used, using other devices such as iPads, phone games, or any other things that are enjoyed by the child. The actual tests themselves can be used to distract children. Use of language is very important, so for example, when coming to the actual placement of electrodes you do not talk about electric shocks, instead tickling the feet. When doing the motor stimulation you look for movement of the feet, which always causes amusement not only in the child, but in the accompanying adults. When we do the EMG it is essential that you never mention needles. Our preferred description is that a 'microphone' is being used and this is a very important because we use the loudspeaker on the EMG machine as feedback and then a discussion as to what the noise sounds like can be undertaken. It is very important in placing the needle into the muscle that it is hidden in the hand and no glimpse of it is given (see Fig. 25.2).

Play specialists are available in many centres and they often do a very good job. However, with the importance of personal engagement with the child, if you use a play specialist you lose the close interaction between the electromyographer and the child, which in many ways is the key to success. The environment in which you do the test is crucial. It is not fair on children, whatever their age, to be rushed into an appointment, perhaps in a busy adult clinic. The atmosphere is frightening enough for them anyway and the sight of many older people can be very disturbing. Finally, you must always allow yourself time to do the studies in children. Even if you do not have time you must always give the impression that you do. If you allow yourself 5 min the test will take an hour if you give yourself an hour the test will take much less.

Finally it is important to look at yourself and determine whether you have the characteristics that will result in successful electromyography in children. There are three that are most crucial. Clearly compassion is essential, but should be a prerequisite of being a doctor. The next is somewhat surprising, but the practitioner must not be frightened of children. This is not in the sense of fear that the child is likely to attack the electromyographer rather that they are comfortable with children who perhaps cry when becoming distressed or suddenly move unexpectedly. At this point the value of having explained to the parents that you recognize very clearly the difference between agitation and temper in a child and pain is realized. Parents, who know their children will identify very clearly with this and often will have no concerns when the child appears to be acting up and to an outside observer possibly suffering some kind of considerable pain. When this is noted, they often say 'they are always like this'.

The final characteristic is resolve. It is very important even when it appears that many things are not going to plan, perhaps with the child distressed, that you realize that the test you are doing is an essential examination and its results will have a significant bearing on future investigations for the child. It is possible that if you do not do your job properly and work to a predefined conclusion the child may experience tests that are either more painful such as a lumbar puncture or muscle biopsy or perhaps a little more dangerous requiring a general anaesthetic. Once embarked on, paediatric EMG must be continued until it reaches a point when you know you have sufficient information to direct the future investigations.

Investigation strategies

Many of the technicalities are discussed in other chapters. What follows here is a description of some of the nuances of paediatric EMG that will help interventions to be successful.

With regard to nerve conduction studies in reality in children often you do not need to conduct an exhaustive examination. If one considers children as those under 8 years of age, even if the exact definition might include children under 18 years of age, this group do not commonly suffer from entrapment neuropathies. Certainly there is a population of children with mucopolysaccharidosis (16–22) who have carpal tunnel syndrome, but such conditions as anterior or posterior interosseous palsies, for example, or even peroneal palsies are vanishingly rare in the paediatric population, in our experience. More commonly you are confronted with the question of whether this is a generalized neuropathy. In these circumstances testing perhaps one or perhaps two sensory nerves in the leg with a motor nerve study is sufficient screen for this.

When doing this, the preference for sensory testing is to use the medial plantar or superficial peroneal nerve. The former is an excellent nerve to use in children particularly so in the neonates (Fig. 25.3). The difficulties that you encounter in adults do not exist in children. The sural nerve because it is a composite variously produced by different combinations of the peroneal and tibial branch of the sciatic nerve (23,24) perhaps is not the best even though historically it has been used. Another factor against its use is that, to do it successfully, you often have to have the children turn away from you. This means that you are doing a test, which is uncomfortable without them seeing what is going on. This is not always the best. In the hand, palm to wrist study is easy once you have learnt the technique and even in small babies with distances lower than 3 cm it is still possible to test effectively.

For motor nerve studies in the leg the tibial nerve is often chosen in the very young as stimulation of the nerve proximally at the knee is not a significant problem, but as they become older stimulation proximally becomes more painful and the peroneal nerve can then be used. The ulnar nerve is much better than the median in the arm because the proximal stimulation is much less painful. F waves are rarely used in routine examination, but are particularly valuable when looking at Guillain-Barré syndrome, where they may be the only abnormality (25–27).

There are not many small fibre studies that can successfully be done in children. The importance of cooperation and collaboration makes thermal thresholds a daunting prospect in many children. The sympathetic skin response is a valuable test although it is so variable as to make it of little value in practice. The cold and hot evoked potentials (CHEPs) presented a very exciting prospect for testing small fibres without cooperation, but its efficacy is still being evaluated (28,29).

When doing EMG it is very much better in children to study one muscle in detail than many only briefly. This is quite different from the situation when studying adults where the differentiation between peripheral nerve entrapments and multiple root entrapment makes it mandatory to sample widely. Often the changes are quite prominent and therefore the detailed examination recommended in children is not necessary. Fortunately for the paediatric electromyographer this problem rarely occurs in children. Most children will be suffering a generalized abnormality. Of pre-eminent importance is that the muscle used is easy to activate. Tibialis anterior (TA) is exceptional in this respect because even in the youngest children tickling the sole of the foot will cause a dorsiflexion of the foot and activation in TA. Even though weakness may be distributed proximally it is almost inevitable that TA when examined in enough detail will show abnormalities. The approach pioneered by Buchthal, but adapted to be more suitable for children is the way forward. A needle is inserted in the middle of the muscle between the insertion and the origin and the needle advanced in the muscle with isometric contraction being encouraged. After study in one direction, the needle is withdrawn to close to the surface of the muscle and then angled in a different direction sampling from another quadrant. It is perfectly routine to take as long as 5–6 min when doing this in a child. Along with this method of recruitment, motor unit potential analysis is absolutely crucial in the interpretation of the findings. As a routine, the last examination is with a full contraction either evoked by tickling the foot or if the child is old enough, by pressing against the hand. The information so collected can then be subjected to an interference pattern automatic program, available on many EMG machines, which might include turns analysis, envelope analysis, and the number of short segments (30–32). While it is important to attempt to quantify the findings on EMG, one should not diminish the importance of visual assessment of the interference pattern, which will allow you to give information about the recruitment of motor units, more information about polyphasia as sometimes this is not picked up clearly by the motor unit duration programs, and also spontaneous activity including such phenomena as myotonia.

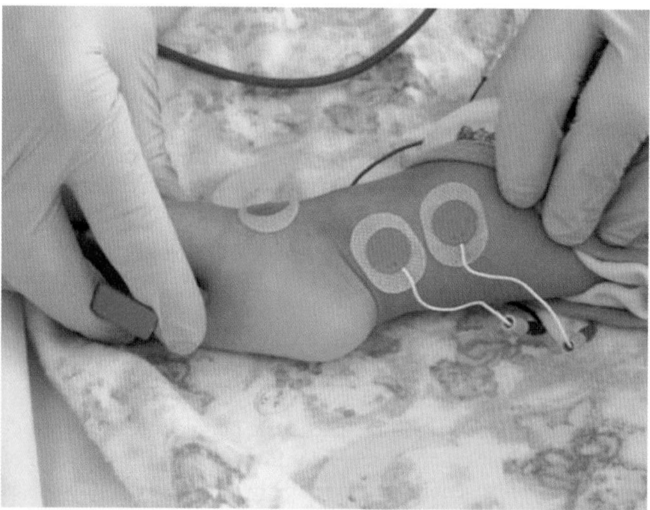

Fig. 25.3 Sensory study of the medial plantar nerve in a neonate.

The neuromuscular junction is a very important part of the peripheral nervous system in children with the possibility of diagnosing and initiating treatment in the congenital myasthenic syndromes and less commonly those caused by autoimmune abnormalities. Whatever has been said, repetitive nerve stimulation is painful to perform in children; it is also insensitive and well recognized to be negative in a proportion of children or adults with proven myasthenic syndromes (33–38). Single fibre EMG performed voluntarily, even if performed by an expert, is virtually impossible under eight years of age and maybe not possible even over that age. Stimulation single fibre EMG is therefore the answer and a technique that has been developed into a routine investigation in our department (39,40). The orbicularis oculi is chosen and a monopolar needle placed near the facial nerve as it crosses the zygomatic arch provides the stimulation. Activity is recorded by a concentric needle facial electrode. The increase of the high band pass to 3 kHz is a throwback from the blanket principle pioneered by Peter Payan (41) and allows an approximation of a single fibre assessment to be made. Concentric needle electrodes have been used for long time in our department not simply because of the difficulty of obtaining reusable single fibre needles, even though they are beginning to come on the market, but for other reasons. First among these is that the concentric needle electrode, by having the recording surface at its tip, allows certainty when assessing whether you are in the muscle itself when you are trying to recruit units. With a single fibre electrode, with its recording area set back from the point, it may on occasion not be possible to place it in the muscle, particularly in a baby where the muscle is very thin. There is evidence that, while not ideal, concentric facial needle electrodes can give a fair approximation of a single fibre EMG assessment (42–45).

Amongst the other specialized techniques that may be important to have in the armament of the paediatric electromyographer, diaphragmatic EMG is one of the more important. The diaphragm is easily approached, even in the youngest babies, by the technique pioneered by Bolton using a trans-costal method (46). Despite concerns from studies of ultrasound of the diaphragm (47), from anecdotal experience this appears to be a safe procedure in children, with over 50 having been performed at all ages with none having the misfortune to suffer a pneumothorax (48). The diaphragm is sampled most commonly in the intensive care ward, where a complication such as pneumothorax is perhaps less worrisome than in an outpatient setting. Combined with phrenic nerve stimulation an effective assessment of the nerve integrity to the diaphragm can be made.

The blink reflex is one of these tests like the sympathetic skin response, which theoretically should have an enormous contribution to the assessment not only of the fifth and seventh cranial nerves, but also the multiple pathways, which serve it in the brainstem. It has been extensively studied in children (49,50). The reality is otherwise; in our experience we have found it really very difficult reliably to obtain the later components of the blink reflex. Whether this is because we commonly use this technique when assessing bulbar palsy or whether this really is the situation in children is hard to determine. This needs to be the subject of a detailed investigation.

Of the other tests available motor unit number estimation (51–53) is perfectly feasible to do in children and is well tolerated. Exercise tests also are easily tolerated by children and will be discussed in the section on myotonia. While we have no experience of nerve excitability studies (54) these should be possible to do in children, as reported by Michelle Farrar in babies (personal communication), and will feature more in the future.

Findings in different conditions

In this section we are going to discuss the different EMG findings in paediatric neuromuscular disorders. The structure of this chapter will follow anatomy; so start with the anterior horn cell, passing from the nerve to the neuromuscular junction, before finally considering conditions affecting the muscle. Within this anatomical description the aetiologies will follow routine assessment namely separating into congenital and acquired conditions. Within the latter, discussion will cover the different aetiologies that may be relevant from infective, autoimmune, metabolic, vascular, iatrogenic, trauma, drugs and toxins, and finally unknown causes. There are many more neuromuscular disorders, which affect children than will be discussed in this chapter. The aim of this chapter is to highlight the particular uses and unique contribution of paediatric EMG in determining the aetiology of different afflictions of the peripheral neuromuscular system.

Anterior horn cell disease

Congenital forms

While faraway the most common cause of anterior horn cell disease in children is SMA related to mutations on chromosome 5q, in practice these are an uncommon reason for children to present in EMG Departments. Fortunately in many parts of the world, even in less developed areas, the availability of the mutation testing is such that one is rarely presented with a suspected anterior horn cell disease without SMA having already been excluded. This has not always been the case and there are reports documenting the EMG findings in this condition (55–59). However, when one used to do this test, in a similar way to the problems neuropathology suffered, it was not always an easy task to determine that anterior horn cell disease was the cause of the findings on EMG. The presence of fibrillation potentials was certainly a common feature although this does not allow distinction between a myopathy and a neuropathy, but the major concern was that the established changes of chronic reinnervation often were not present.

More commonly nowadays one is encountering anterior horn cell disease without any known aetiology. With the increasing access and use of the EMG in our hospital because it is perceived as an important investigation without undue trauma to the children we are encountering anterior horn cell disease as perhaps the most common finding in early onset hypotonia. Certainly we have demonstrated that in some conditions, previously considered to be involving only the central nervous system such as the neuronal form of Gaucher's disease, there are clear indications that the anterior horn cells are also affected. Also we have seen such changes in children suffering definite hypoxic ischaemic events causing cerebral palsy. Previously their weakness was explained entirely on the basis of the upper motor neuron lesion.

It is important at this point to emphasize the importance of examination of the tongue in the investigation of these children. Unlike in adults the differential diagnosis does not include concomitant cervical and lumbar spondylosis with resulting radiculopathy, one of the drivers behind the different EMG criteria for diagnosis of ALS (60–63). Instead in children the differential diagnosis lies between segmental involvement of the spinal cord either in the cervical or lumbar region only, or a more generalized abnormality. To this end when evidence of anterior horn cell disease is seen in tibialis anterior the next step is not exhaustively to demonstrate bilateral involvement and involvement of other segments, but to go straight away to examination of the tongue approached by the submental route (Fig. 25.4). If there is denervation, it is safe to assume that there is denervation in all intervening segments

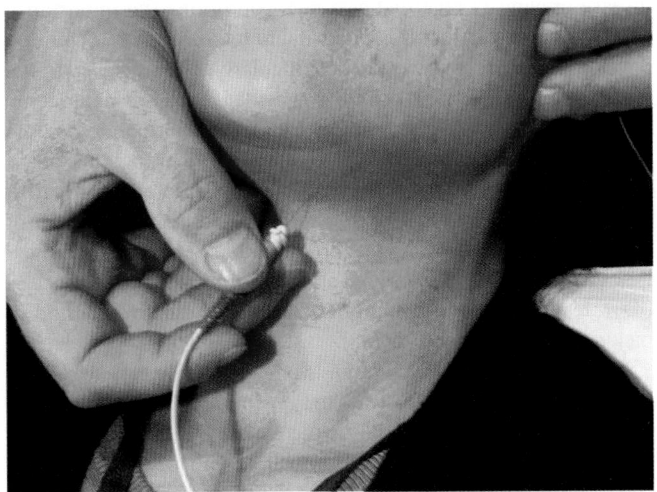

Fig. 25.4 The sub-mental route to the tongue.

without further EMG. If the tongue is normal it is important then to identify how far up the reinnervation occurs so examination in the arm follows, which, if normal, is then followed by the examination of the contralateral leg. If only the legs have been involved a lumbar segmental involvement can be confidently diagnosed.

Acquired anterior horn cell disease

Segmental myelopathy has already been discussed and this may be seen particularly in the neonate where it is likely to be the result of vascular involvement of the anterior spinal arteries. Amongst the most devastating manifestation is the so-called 'baby in a barrel' where the cervical bulb is involved, presumably from infarction of the anterior spinal artery (64–66). The children have completely flaccid arms with preserved sensory responses, but often little or no motor response. This is compatible with a normal lifespan and normal development otherwise. We have seen the infarction expand further down the spinal cord to involve the thoracic region and in these circumstances the nerve supply to the intercostals is affected (67). The situation is far more serious and the children are unable to breathe without ventilation. This is a completely different prognostic situation to those with an isolated cervical involvement.

Amongst the acquired forms of anterior horn cell disease it is important to consider the acute flaccid paralysis (AFP), which occurs in epidemics in areas in the tropics and in Australia and is caused by a variety of bacteria and other infective organisms (68–81). Enteroviruses feature commonly, but there are many different causes. The problem in United Kingdom is that the cases tend to be sporadic rather than epidemic, since the only common reason for epidemics, the polio virus, has been eliminated. For a time, when polio had been nearly eradicated, the commonest reason for AFP was the vaccination rather than the underlying condition. However, it is really important to consider these conditions when presented with the child who has a patchy weakness particularly bulbar, following an infective condition.

In this context is perhaps one of the most important recent developments in paediatric neurology, with the discovery that a proportion of children suffering Brown Vialetto van Laere disease have been proven to have a disorder of their riboflavin metabolism (82–87). More remarkably treatment with large doses of riboflavin can reverse the inexorable and relentless progression of this disease previously regarded as a fatal condition. The importance of this discovery is that it is now vital to look in detail at children developing progressive weakness particularly when involved in the bulbar region in order to confirm whether there is a neurogenic abnormality. If found a search for riboflavin abnormalities is warranted. Some of the children passed very rapidly through this phase of motor neuronopathy to develop a sensorimotor neuropathy. While it is counselled that hearing tests will be the first indicator of the condition with hearing loss an important associated finding, a bulbar EMG is easily performed and well within the competence of most neurophysiologists and offers another alternative method of picking these cases up early (88).

Neurogenic change in children is often relatively easy to be certain of, but it is important always to keep in mind that if neurogenic changes are present on EMG in the context of other myopathic changes, the underlying pathology will be a myopathy with fibre splitting and re-innervation causing the neurogenic changes. The most well recognized cause of this phenomenon is in muscular dystrophy, and particularly facioscapular humeral dystrophy. Most muscular dystrophies should not come to the attention of the electromyographer, as diagnosis is clinical, backed by genetic and muscle biopsy analysis.

Disorders of peripheral nerve

Hereditary disorders of nerves are an ever expanding and important part of paediatric electromyography. It is almost impossible for those not working continuously in this field to have any chance of keeping up with the number of chromosome abnormalities that have been identified. With the increasing number of genes being identified has also come a realization of the expanding phenotype of hereditary neuropathies. The paediatric electromyographer confronted with such a child will have their primary duty to distinguish between an axonal or demyelinating neuropathy. While some practitioners using nerve conduction studies feel comfortable only if an extensive examination of almost every nerve in the body is made sometimes this is not possible without some discomfort in a child. The paediatric electromyographer has to judge what is possible and what contributes most to subsequent genetic localization of the defect. Undoubtedly conduction studies both in the upper and lower limb are essential as some may vary in the extent of their involvement in the upper and lower limbs. Also comparing the velocity throughout the length of the nerve is useful expecting to see an increase in the velocity in the most cranial segments, which is seen in hereditary neuropathies, but not in acquired neuropathies. Similar velocities in similar limb segments of the two sides are another feature strongly in favour of a hereditary neuropathy.

In acquired disorders of the nerve chronic inflammatory demyelinating polyneuropathy (CIDP) can present significant challenges sharing features in common with the hereditary neuropathies. While it has been reported that certain features are strongly suggestive of the CIDP and these include temporal dispersion and conduction block, unfortunately it is becoming increasingly our experience that these are not pathognomonic of CIDP. In some families with CMT1A proximal swelling of the nerve may be the reason for this finding, with the nerve more difficult to stimulate, and it has recently been reported as a distinctive feature of certain identified genetic defects in CMT (89,90).

Guillain-Barré syndrome is well recognized to occur in children although less frequently than in adults. It is, on the whole, a more benign condition than that seen in adults. Most commonly it is demyelinating rather than axonal in type (91–94). The characteristic neurophysiological findings are often varied with some children only having abnormalities of the F waves while others demonstrate the whole range of abnormalities described such as significant prolongation of the distal motor latency, prominent slowing of the main nerve with marked dispersion on proximal stimulation (Fig. 25.5).

Focal neuropathy may be seen in conditions, which affect the vascular supply to the peripheral nerve and in our experience we have seen cases of Behcet's disease and polyarteritis nodosa, as well as other vasculitic conditions. The nerve conduction studies show focal slowing likely due to the infarction of the nerve.

An increasing part of our work has been to monitor the effects of neurotoxic drugs, in particular thalidomide, which has made a re-appearance as an effective therapy particularly in disorders of the skin as well as gastrointestinal disorders (95–100). Nearly all

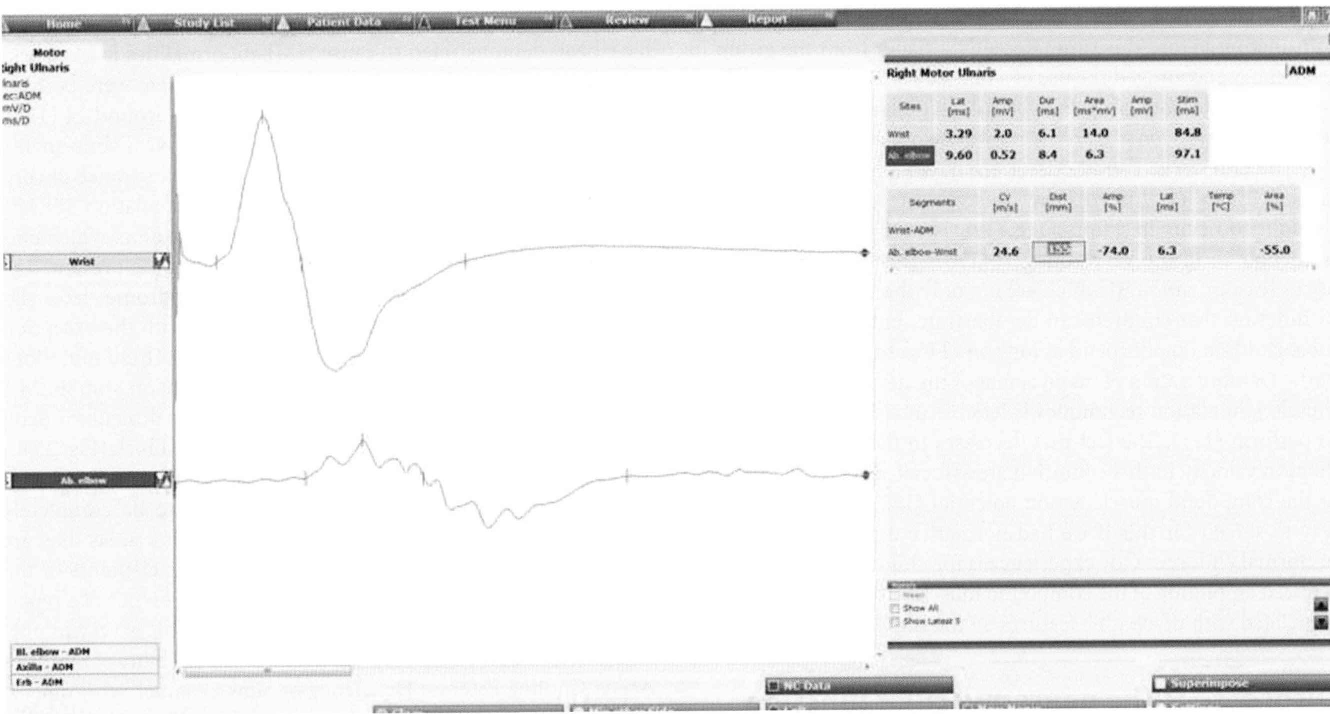

Fig. 25.5 Motor nerve conduction study of the ulnar nerve showing proximal temporal dispersion in a child with Guillain-Barre syndrome.

these children will eventually develop a peripheral neuropathy (101,102). While one is recommended to examine many nerves, the examination of the sensory nerves in the legs is sufficient to alert the clinicians to the neuropathic change. If the nerves in the leg become affected, the next study can incorporate the arms as well. Despite the demonstration of clear neuropathic changes on the conduction study children do not seem to develop the same sensory disturbances as adults. This makes the decision by the parents and children to stop the treatment difficult, particularly as in many instances, it has been the most effective treatment for their condition that they have encountered.

Metabolic conditions such as leucodystrophies are now a rare indication for peripheral nerve studies, having usually been diagnosed by metabolic means. In the past before such screening became commonplace, the demonstration of a significant demyelinating neuropathy in a child showing developmental regression was an important pointer to these diagnoses.

The mononeuropathies on the whole are uncommon in children. In some studies such as the recent report from Boston (103), where they report that they have a relatively high incidence of mononeuropathies it is important to realize that their practice is on the whole mainly concerned with children over eight years of age. The only exception to this rule is nerve injury as a result of trauma, particularly of the upper limb.

That said, if you have a practice, which includes care of the mucopolysaccharidoses carpal tunnel syndrome (CTS) remains a very serious concern (104–106). Prior to the introduction of enzyme replacement and bone marrow transplant it was our experience that all children with Hurler's disease by the age of two years had CTS. Worryingly the situation was that the progression of the disorder from one showing only sensory abnormalities to complete loss of motor responses was very rapid indeed often occurring in less than six months. When we realized this, the demonstration of any neurophysiological indications of CTS prompted immediate operation. The situation has changed completely and with effective therapy it is uncommon to see CTS and, even if found, one can follow an expectant policy.

In the lower limb our experience is that peroneal palsy is vanishingly rare. Much more common, even though still rare, is sciatic nerve palsy. These can occur for a variety of reasons, but most commonly as the result of surgical intervention (107). The nerve is particularly vulnerable in operations around the pelvis, especially if involving the lithotomy position (108–110). Also less easy to explain are those that have occurred in the context of operations were no intervention either for vascular access or positioning has been made in the pelvic region. For instance, we have seen sciatic nerve palsy occurring after cardiac surgery. Possibly this is testimony to the perilous blood supply to the nerve. Follow-up data is difficult to obtain on these children because of the nature of our referral pattern, but from the few that have been seen again they seem to recover much better than adults.

Plexopathies can occur although very rarely. Thoracic outlet syndrome must always be sought in situations with numbness in the medial aspect of the arm and hand, but is very rare. The presence of cervical ribs, a known risk factor for thoracic outlet syndrome, may make the baby more susceptible to obstetric brachial plexus injury (111). Obstetric brachial plexus palsy (OBPP) continues to be an important cause of plexopathy in paediatrics. The strategy to be employed in the investigation of this condition is highly controversial and despite many years of investigation there is still no real clear guideline. For a long time that it was the rule to investigate around three months of age, but this was driven by the surgical strategies, which would encourage surgery if satisfactory biceps function had not been achieved by that date (112–114). However, this has meant that parents have been denied important prognostic information before this time. Recently a paper from the group in Lieden has encouraged the use of a limited EMG study focusing on biceps at one month of age, which will give a clear indication, if no motor units are seen, that recovery will be poor (115–117).

Peripheral nerve conditions of unknown aetiology include the critical illness neuromyopathy. This, despite the high incidence thought to occur in adults (118,119), is rare in children and in particular so in the very youngest (120). This may be for a variety of reasons, amongst which are not only the very good recovery of function that children can demonstrate, but also possibly that most children do not spend as long on ITUs as their adult counterparts. Demonstration of involvement of the nerve or muscle by the muscle stimulation techniques is feasible in children, but difficult to perform (121). The fact that decreases in the muscle fibre conduction velocity in this condition are associated with prolongation of the compound muscle action potential (122,123) offers an easy way to screen for this if we had normative data on the duration in normal children. Our experience is for children to demonstrate reduced amplitude of the compound muscle action potentials often associated with myopathic features on the EMG.

Disorders of the neuromuscular junction

The neuromuscular junction in children is affected by either the congenital myasthenic syndromes or the autoimmune form with antibodies against acetyl choline receptors or, less commonly, those against the MuSK receptor. The expanding number of genes that have been demonstrated to cause NMJ abnormalities is a remarkable feature of this field. Around 10 years ago there were perhaps only one or two, while at the last count there are around 14 (124). This means that for some time demonstration of NMJ abnormality by neurophysiological means did not have the reassurance of confirmation by genetic studies. The findings on stimulation SFEMG (stimSFEMG) themselves do not allow identification of which particular form of myasthenic syndrome is likely to be present. The common finding is that most of myasthenic syndromes have significant abnormalities on stimSFEMG perhaps with the exception of those associated with Dok7 mutations (125). These mutations produce a very interesting pattern of abnormality on stimSFEMG with normal studies mixed with other areas of the orbicularis oculi that have very significant abnormalities including block (Fig. 25.6). This parallels nicely the histological findings in this condition with the neuromuscular junctions often appearing to be completely normal in certain parts of muscle whereas in other areas they are completely destroyed. The contribution of other elements of the neurophysiological armament demonstrate the presence of a repetitive CMAP on motor nerve stimulation (126). This is very important to look for and will, if found, indicate the possibility of Endplate Acetyl Choline Esterase Deficiency or Slow Channel Syndrome. It is important to recognize these two conditions because patients deteriorate when given pyridostigmine.

StimSFEMG can produce positive results, which are not caused by myasthenia, but still indicate abnormalities of neuromuscular

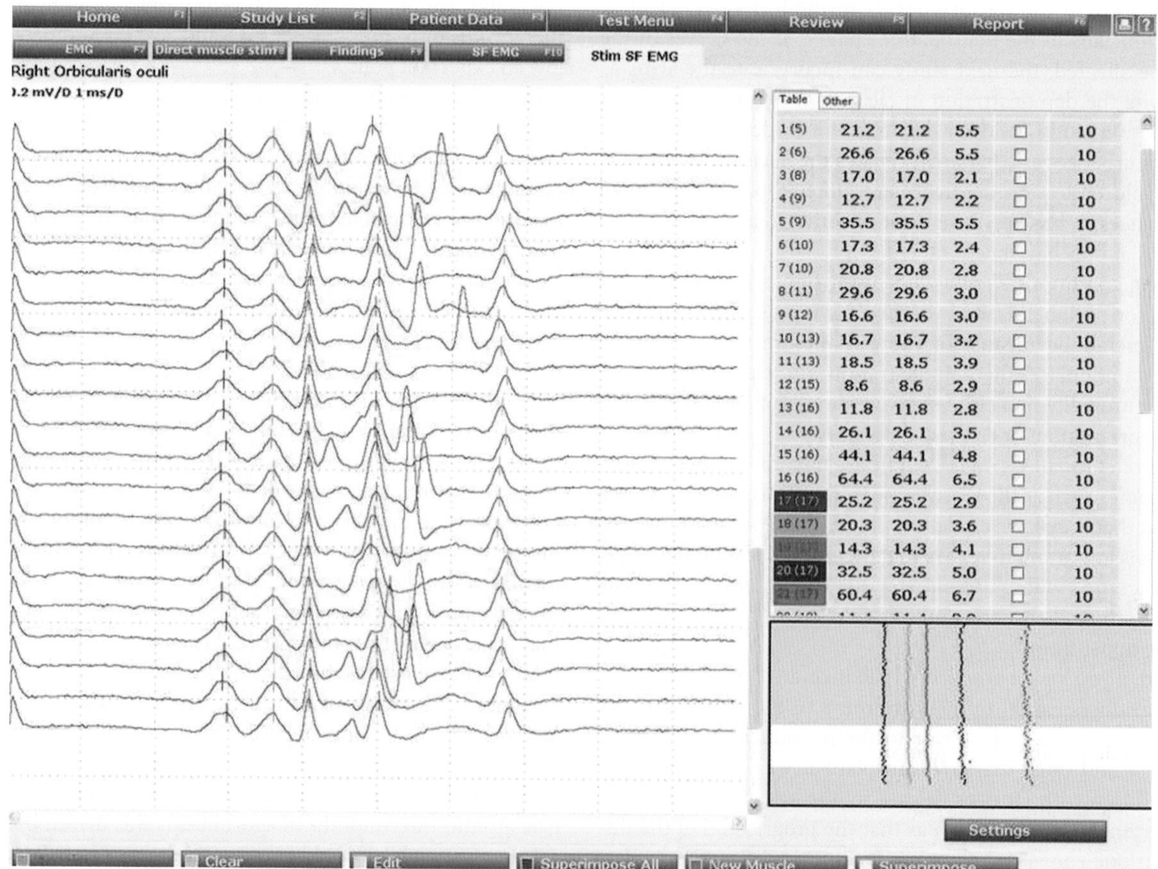

Fig. 25.6 StimSFEMG findings in Dok7 congenital myasthenic syndrome showing highly abnormal units with blocking in the presence of other normal potentials.

transmission. The situation is quite different from that in adults where once the common conditions in the differential diagnosis are excluded the finding of an abnormality on the test equates to a diagnosis of myasthenia. In children the clinical presentations of myasthenia are protean with varying presentations such as, feeding difficulty, stridor, arthrogryposis, apnoea, to list just a few examples. It is therefore very difficult to exclude other conditions that might influence the differential diagnosis before doing the test. We would counsel using stimSFEMG within an investigative strategy rather than a test that stands on its own. We have evolved a strategy, which begins with stimSFEMG, being the most sensitive test and also the one most likely to be disturbed by lack of cooperation. If an abnormal result is found we will leave the needle in the orbicularis oculi looking very carefully for neurogenic change. A bulbar palsy is the most common reason for an abnormal single fibre in children under one year of age. If any doubt exists the tongue is examined by the submental route. Once that has been excluded the next stage is to perform repetitive nerve stimulation (RNS). The reason this is done here is that RNS is uncomfortable and if it is done first may make any further investigations very difficult, possibly jeopardizing a successful conclusion to the StimSFEMG examination. If RNS is abnormal this gives a very strong support of myasthenia. If it is normal it will allow the subsequent EMG of a peripheral muscle if showing a myopathy to be interpreted as more likely indicating the presence of a myopathy with an associated disorder of neuromuscular transmission (127,128) and not the apparent myopathy seen in myasthenia. The latter is only commonly seen when there is block of neuromuscular transmission, which should be detected by RNS. This strategy, which is in the provisional phase of its development is presented here to stress the importance of using the test in the context of a complete EMG examination when used in exploring myasthenia as a cause of weakness.

Other conditions, which may affect the neuromuscular junction include botulism. The common form encountered in paediatrics is infantile botulism (129–131). This occurs at around about three months, often contemporaneous with weaning. In the areas of the world where this condition is endemic, with 30–40 cases seen per annum, neurophysiology in reality has no role as clinicians pick it up almost immediately when the child is seen. Early diagnosis has become of considerable importance because of the use of botulinum immunoglobulin in the treatment of these children (132–134). If this is given early, before the botulinum is irreversibly bound onto the neuromuscular junctions, there may be very significant reductions in the time the child spends in the ITU. Unfortunately the window for its use is very short, and may be less than 10 days. The problem therefore in this country, where the condition is very rare, is that it may take longer than that to make the diagnosis (135). The classic EMG findings in botulism is the small CMAP with a significant increment of repetitive nerve stimulation at high rates of stimulation (136–138), while seen in some cases this does not always appear. It very much depends on how much of the neuromuscular junction pool has been affected for these classic findings to be demonstrated. Perhaps more useful is the recognition that a reduction of the compound muscle action potential and a demonstrable abnormality of NMJ, whether or not associated with RNS changes, should always raise the possibility of botulism. The key to the diagnosis is a recent onset of constipation (139–141).

Myopathies

The role of EMG in myopathies is simply to demonstrate this is the likely cause of the clinical presentation, directing further investigations to characterize more precisely what kind of myopathy is present. Another way to look at it is that EMG in a child who is hypotonic will be one way to determine whether a muscle biopsy is not needed. If the EMG shows a peripheral neuropathy, neuromuscular junction abnormality or motor neurone disease most of these would not warrant a muscle biopsy. The practice of the EMG in the investigation of myopathy has changed over time. Very few of the well recognized myopathies need to have nerve conduction or EMG as many clinicians, particularly those with neuromuscular expertise, are exceptionally good at determining the likely underlying myopathy from the clinical presentation. The next step is either genetic screening or muscle biopsy. The more common ways for a myopathy to be discovered by EMG is for the child referred with an unexpected and unexplained hypotonia in which the clinicians are not sure whether there is any abnormality at all and if so whether that is a myopathy or another cause. In this circumstance a detailed analysis, particularly using motor unit duration analysis is the key to determining whether there is an abnormality. Then the next step will often be a muscle biopsy. Infective myopathies are uncommon. Usually they have been clinically diagnosed, and do not need an EMG confirmation. The same situation applies to dermatomyositis as an example of an autoimmune condition, also rarely needing an EMG.

Perhaps the most important condition in terms of numbers of these children who are referred with these clinical concerns, are those who may be harbouring a disorder of the mitochondria. EMG can be very helpful here as some will have peripheral neuropathy and others myopathy. This will lead to a completely different direction in their subsequent investigation.

Conclusions

Despite the negative aspects of some of the comments made at the beginning of this chapter this author firmly believes that paediatric EMG is far from in decline rather that it is evolving. This is supported by the month on month increase in the cases referred for EMG in his hospital. While around 30 was commonplace 7 or 8 years ago as many as 100 can be seen in one month. Certainly it is not needed for many conditions, which are diagnosed by other means and should never be encouraged when an easier and less uncomfortable route to the final diagnosis exists. However, with its more general acceptance as not only a humane examination, but also one, which can very quickly give information more easily obtained than by any other means, the range of clinical presentations for which it should be considered within the investigation strategy has expanded. Even though exome sequencing will alter the way it is employed, it will still, like clinical examination itself, have a role.

References

1. Garcia, A., Calleja, J., Antolin, F.M. and Berciano, J. (2000). Peripheral motor and sensory nerve conduction studies in normal infants and children. *Clinical Neurophysiology*, **111**(3), 513–20.
2. Hyllienmark, L., Ludvigsson, J., and Brismar, T. (1995). Normal values of nerve conduction in children and adolescents. *Electroencephalography Clinical Neurophysiology*, **97**(5), 208–14.

3. Smit, B.J., Kok, J.H., De Vries, L.S., Dekker, F.W., and Ongerboer, de Visser, B.W. (1999). Motor nerve conduction velocity in very preterm infants. *Muscle & Nerve*, **22**(3), 372–7.
4. Vecchierini-Blineau, M.F. and Guiheneuc, P. (1979). [Maturation of nerve conduction velocity from birth to 5 years of age]. *Archives of French Pediatrics*, **36**(6), 563–72.
5. Ludin, H.P. (2013). *Normal values. Electromyography in practice*. New York: Georg Thieme Verlag Thieme-Stratton Inc.
6. Barkhaus, P.E., Roberts, M.M., and Nandedkar, S.D. (2006). 'Facial' and standard concentric needle electrodes are not interchangeable. *Electromyography & Clinical Neurophysiology*, **46**(5), 259–61.
7. Bodofsky, E., Tomaio, A., and Campellone J. (2009). The mathematical relationship between height and nerve conduction velocity. *Electromyography & Clinical Neurophysiology*, **49**(4), 155–60.
8. Campbell, W.W., Jr., Ward, L.C., and Swift, T.R. (1981). Nerve conduction velocity varies inversely with height. *Muscle & Nerve*, **4**(6), 520–3.
9. Gupta, N., Sanyal, S., and Babbar, R. (2008). Sensory nerve conduction velocity is greater in left handed persons. *Indian Journal of Physiology & Pharmacology*, **52**(2), 189–92.
10. Hays, R.M., Hackworth, S.R., Speltz, M.L., and Weinstein, P. (1992). Exploration of variables related to children's behavioral distress during electrodiagnosis. *Archives on Physical Medical Rehabilitation*, **73**(12), 1160–2.
11. Kommalage, M. and Gunawardena, S. (2013). Influence of age, gender, and sidedness on ulnar nerve conduction. *Journal of Clinical Neurophysiology*, **30**(1), 98–101.
12. Rivner, M.H., Swift, T.R., and Malik, K. (2001). Influence of age and height on nerve conduction. *Muscle & Nerve*, **24**(9), 1134–41.
13. Robinson, L.R., Stolov, W.C., Rubner, D.E., Wahl, P.W., Leonetti, D.L., and Fujimoto, W.Y. (1992). Height is an independent risk factor for neuropathy in diabetic men. *Diabetes Research & Clinical Practice*, **16**(2), 97–102.
14. Stetson, D.S., Albers, J.W., Silverstein, B.A., and Wolfe, R.A. (1992). Effects of age, sex, and anthropometric factors on nerve conduction measures. *Muscle & Nerve*, **15**(10), 1095–4.
15. Todnem, K., Knudsen, G., Riise, T., Nyland, H., and Aarli, JA. (1989). The non-linear relationship between nerve conduction velocity and skin temperature. *Journal of Neurology, Neurosurgery & Psychiatry*, **52**(4), 497–501.
16. Van, M.N. and De, S.L. (2003). Carpal tunnel syndrome in children. *Acta Orthopedic Belgium*, **69**(5), 387–95.
17. Gschwind, C. and Tonkin, M.A. (1992). Carpal tunnel syndrome in children with mucopolysaccharidosis and related disorders. *Journal of Hand Surgery in America*, **17**(1), 44–7.
18. Karpati, G., Carpenter, S., Eisen, A.A., Wolfe, L.S., and Feindel, W. (1974). Multiple peripheral nerve entrapments. An unusual phenotypical variant of the Hunter syndrome (mucopolysaccharidosis II) in a family. *Archives in Neurology*, **31**(6), 418–22.
19. Taori, G.M., Iyer, G.V., Abraham, J., and Mammen, K.C. (1971). Electrodiagnostic studies in lipidoses, mucopolysaccharidoses, and leukodystrophies. I. Nerve conduction and needle electromyographic studies. *Neurology*, **21**(3), 303–6.
20. Sri-Ram, K., Vellodi, A., Pitt, M., and Eastwood, D.M. (2007). Carpal tunnel syndrome in lysosomal storage disorders: simple decompression or external neurolysis? *Journal of Pediatric Orthopedic B*, **16**(3), 225–8.
21. Haddad, F.S., Jones, D.H., Vellodi, A., Kane, N., and Pitt, M.C. (1997). Carpal tunnel syndrome in the mucopolysaccharidoses and mucolipidoses. *Journal of Bone Joint Surgery*, **79**(4), 576–82.
22. Haddad, F.S., Jones, D.H., Vellodi, A., Kane, N., and Pitt, M. (1996). Review of carpal tunnel syndrome in children. *Journal of Hand Surgery*, **21**(4), 565–6.
23. Huelke, D.F. (1958). The origin of the peroneal communicating nerve in adult man. *Anatomical Records*, **132**(1), 81–92.
24. Huelke, D.F. (1957). A study of the formation of the sural nerve in adult man. *American Journal of Physical Anthropology*, **15**(1), 137–47.
25. Gordon, P.H. and Wilbourn, A.J. (2001). Early electrodiagnostic findings in Guillain-Barre syndrome. *Archives in Neurology*, **58**(6), 913–7.
26. Kuwabara, S., Ogawara, K., Mizobuchi, K., et al. (2000). Isolated absence of F waves and proximal axonal dysfunction in Guillain–Barré syndrome with antiganglioside antibodies. *Journal of Neurology, Neurosurgery & Psychiatry*, **68**(2), 191–5.
27. Fraser, J.L. and Olney, R.K. (1992). The relative diagnostic sensitivity of different F-wave parameters in various polyneuropathies. *Muscle & Nerve*, **15**(8), 912–8.
28. Ruscheweyh, R., Emptmeyer, K., Putzer, D., Kropp, P., and Marziniak, M. (2013). Reproducibility of contact heat evoked potentials (CHEPs) over a 6 months interval. *Clinical Neurophysiology*, **124**(11), 2242–7.
29. Kramer, J.L., Taylor, P., Haefeli, J., et al. (2012). Test-retest reliability of contact heat-evoked potentials from cervical dermatomes. *Journal of Clinical Neurophysiology*, **29**(1), 70–5.
30. Nandedkar, S.D., Sanders, D.B., and Stalberg, E.V. (1986). Simulation and analysis of the electromyographic interference pattern in normal muscle. Part II: Activity, upper centile amplitude, and number of small segments. *Muscle & Nerve*, **9**(6), 486–90.
31. Nandedkar, S.D., Sanders, D.B., and Stalberg, E.V. (1986). Automatic analysis of the electromyographic interference pattern. Part II: Findings in control subjects and in some neuromuscular diseases. *Muscle & Nerve*, **9**(6), 491–500.
32. Nandedkar, S.D., Sanders, D.B., and Stalberg, E.V. (1986). Automatic analysis of the electromyographic interference pattern. Part I: development of quantitative features. *Muscle & Nerve*, **9**(5), 431–9.
33. Benatar, M. (2006). A systematic review of diagnostic studies in myasthenia gravis. *Neuromuscular Disorder*, **16**(7), 459–67.
34. Zivkovic, S.A. and Shipe, C. (2005). Use of repetitive nerve stimulation in the evaluation of neuromuscular junction disorders. *American Journal of Electroneurodiagnostic Technology*, **45**(4), 248–61.
35. Witoonpanich, R., Dejthevaporn, C., Sriphrapradang, A., and Pulkes, T. (2011). Electrophysiological and immunological study in myasthenia gravis: Diagnostic sensitivity and correlation. *Clinical Neurophysiology*, **122**(9), 1873–7.
36. Srivastava, A., Kalita, J., and Misra, U.K. (2007). A comparative study of single fiber electromyography and repetitive nerve stimulation in consecutive patients with myasthenia gravis. *Electromyography & Clinical Neurophysiology*, **47**(2), 93–6.
37. AAEM Quality Assurance Committee. American Association of Electrodiagnostic Medicine. (2001). Literature review of the usefulness of repetitive nerve stimulation and single fiber EMG in the electrodiagnostic evaluation of patients with suspected myasthenia gravis or Lambert-Eaton myasthenic syndrome. *Muscle & Nerve*, **24**(9), 1239–47.
38. AAEM Quality Assurance Committee. American Association of Electrodiagnostic Medicine. (2001). Practice parameter for repetitive nerve stimulation and single fiber EMG evaluation of adults with suspected myasthenia gravis or Lambert-Eaton myasthenic syndrome: summary statement. *Muscle & Nerve*, **24**(9), 1236–8.
39. Pitt, M. (2008). Neurophysiological strategies for the diagnosis of disorders of the neuromuscular junction in children. *Developmental Medicine & Child Neurology*, **50**(5), 328–33.
40. Pitt, M. (2009). Workshop on the use of stimulation single fibre electromyography for the diagnosis of myasthenic syndromes in children held in the Institute of Child Health and Great Ormond Street Hospital for Children in London on April 24th, 2009. *Neuromuscular Disorder*, **19**(10), 730–2.
41. Payan J. (1978). The blanket principle: a technical note. *Muscle & Nerve*, **1**(5), 423–6.
42. Kouyoumdjian, J.A. and Stalberg, E.V. (2011). Concentric needle jitter on stimulated Orbicularis Oculi in 50 healthy subjects. *Clinical Neurophysiology*, **122**(3), 617–22.
43. Orhan, E.K., Deymeer, F., Oflazer, P., Parman, Y., and Baslo, M.B. (2013). Jitter analysis with concentric needle electrode in the masseter

muscle for the diagnosis of generalised myasthenia gravis. *Clinical Neurophysiology*, **124**(11), 2277–82.

44. Sanders D.B. (2013). Measuring jitter with concentric needle electrodes. *Muscle & Nerve*, **47**(3), 317–8.
45. Stalberg, E. (2012). Jitter analysis with concentric needle electrodes. *Annals of the New York Academy of Sciences*, **1274**, 77–85.
46. Bolton, C.F., Grand'Maison, F., Parkes, A., and Shkrum, M. (1992). Needle electromyography of the diaphragm. *Muscle & Nerve*, **15**(6), 678–81.
47. Amirjani, N., Hudson, A.L., Butler, J.E., and Gandevia, S.C. (2012). An algorithm for the safety of costal diaphragm electromyography derived from ultrasound. *Muscle & Nerve*, **46**(6), 856–60.
48. Pitt, M.C. (2013). An algorithm for the safety of costal diaphragm electromyography derived from ultrasound. *Muscle & Nerve*, **48**(6), 996–7.
49. Tomita, Y., Shichida, K., Takeshita, K., and Takashima, S. (1989). Maturation of blink reflex in children. *Brain Development*, **11**(6), 389–93.
50. Vecchierini-Blineau, M.F, and Guiheneuc, P. (1984). Maturation of the blink reflex in infants. *European Neurology*, **23**(6), 449–58.
51. Bromberg, M.B. (2004). Motor unit number estimation: new techniques and new uses. *Supplement in Clinical Neurophysiology*, **57**, 120–36.
52. Shefner, J.M. and Gooch, C.L. (2003). Motor unit number estimation. *Physical Medicine & Rehabilitation Clinics of North America*, **14**(2), 243–60.
53. Daube, J.R., Gooch, C., Shefner, J., Olney, R., Felice, K., and Bromberg, M. (2000). Motor unit number estimation (MUNE) with nerve conduction studies. *Supplement in Clinical Neurophysiology*, **53**, 112–5.
54. Bostock, H. (2004). Nerve excitability studies: past, present, future? *Supplement in Clinical Neurophysiology*, **57**, 85–90.
55. Renault, F., Chartier, J.P., and Harpey, J.P. (1996). [Contribution of the electromyogram in the diagnosis of infantile spinal muscular atrophy in the neonatal period]. *Archives on Pediatrics*, **3**(4), 319–23.
56. Rudnik-Schoneborn, S., Forkert, R., Hahnen, E., Wirth, B. and Zerres, K. (1996). Clinical spectrum and diagnostic criteria of infantile spinal muscular atrophy: further delineation on the basis of SMN gene deletion findings. *Neuropediatrics*, **27**(1), 8–15.
57. Ignatius, J. (1996). Nerve conduction velocity in severe childhood spinal muscular atrophy (SMA type I). *Polish Journal of Neurology & Neurosurgery*, **30**(3), 91–4.
58. Hausmanowa-Petrusewicz, I. (1988). Electrophysiological findings in childhood spinal muscular atrophies. *Reviews on Neurology (Paris)*, **144**(11), 716–20.
59. Hausmanowa-Petrusewicz, I. and Karwanska, A. (1986). Electromyographic findings in different forms of infantile and juvenile proximal spinal muscular atrophy. *Muscle & Nerve*, **9**(1), 37–46.
60. Costa, J., Swash, M., and de Carvalho M. (2012). Awaji criteria for the diagnosis of amyotrophic lateral sclerosis: a systematic review. *Archives of Neurology*, **69**(11), 1410–16.
61. de Carvalho M., Dengler, R., Eisen, A., et al. (2011). The Awaji criteria for diagnosis of ALS. *Muscle & Nerve*, **44**(3), 456–7.
62. Dengler, R. (2012). El Escorial or Awaji Criteria in ALS diagnosis, what should we take? *Clinical Neurophysiology*, **123**(2), 217–18.
63. Okita, T., Nodera, H., Shibuta, Y., et al. (2011). Can Awaji ALS criteria provide earlier diagnosis than the revised El Escorial criteria? *Journal of Neurological Science*, **302**(1–2), 29–32.
64. Ruggieri, M., Smarason, A.K., and Pike, M. (1999). Spinal cord insults in the prenatal, perinatal, and neonatal periods. *Development Medicine & Child Neurology*, **41**(5), 311–7.
65. de Leon, G.A., Radkowski, M.A., Crawford, S.E., Swisher, C.N., Uzoaru, I., and de León W. (1995). Persistent respiratory failure due to low cervical cord infarction in newborn babies. *Journal of Child Neurology*, **10**(3), 200–4.
66. Rousseau, S., Metral, S., Lacroix, C., Cahusac, C., Nocton, F., and Landrieu, P. (1993). Anterior spinal artery syndrome mimicking infantile spinal muscular atrophy. *American Journal of Perinatology*, **10**(4), 316–18.
67. Gorigolzarri, I., Pitt, M., Molyneux, A., and De Pablos, C. (2010). Spinal cord injury at birth as a consequence of postulated prenatal anterior spinal artery ischaemic infarct: the value of electromyographic studies. *Clinical Neurophysiology*, **121**(253), 25.
68. Douglas, M.W., Stephens, D.P., Burrow, JN., Anstey, N.M., Talbot, K., and Currie, B.J. (2007). Murray Valley encephalitis in an adult traveller complicated by long-term flaccid paralysis: case report and review of the literature. *Transactions of the Royal Society for Tropical Medicine and Hygiene*, **101**(3), 284–8.
69. Gould E.A. and Solomon, T. (2008). Pathogenic flaviviruses. *Lancet*, **371**(9611), 500–9.
70. Gyure, K.A. (2009). West Nile virus infections. *Journal of Neuropathology & Experimental Neurology*, **68**(10),1053–60.
71. Kleinschmidt-DeMasters, B.K., Marder, B.A., Levi, M.E., et al. (2004). Naturally acquired West Nile virus encephalomyelitis in transplant recipients: clinical, laboratory, diagnostic, and neuropathological features. *Archives of Neurology*, **61**(8), 1210–20.
72. Marx, A., Glass, J.D., and Sutter, R.W. (2000). Differential diagnosis of acute flaccid paralysis and its role in poliomyelitis surveillance. *Epidemiology Reviews*, **22**(2), 298–316.
73. Misra, U.K. and Kalita, J. (2010). Overview: Japanese encephalitis. *Progress in Neurobiology*, **91**(2), 108–20.
74. Ooi, M.H., Wong, S.C., Lewthwaite, P., Cardosa, M.J., and Solomon, T. (2010). Clinical features, diagnosis, and management of enterovirus 71. *Lancet Neurology*, **9**(11), 1097–105.
75. Russell, R.C. (1998). Vectors vs. humans in Australia—who is on top down under? An update on vector-borne disease and research on vectors in Australia. *Journal of Vector Ecology*, **23**(1), 1–46.
76. Saad, M., Youssef, S., Kirschke, D., et al. (2005). Acute flaccid paralysis: the spectrum of a newly recognized complication of West Nile virus infection. *Journal of Infections*, **51**(2), 120–7.
77. Solomon, T. and Willison, H. (2003). Infectious causes of acute flaccid paralysis. *Current Opinions in Infectious Diseases*, **16**(5), 375–81.
78. Torno, M., Vollmer, M., and Beck, C.K. (2007). West Nile virus infection presenting as acute flaccid paralysis in an HIV-infected patient: a case report and review of the literature. *Neurology*, **68**(7), E5–7.
79. Wang, S.M., Ho, T.S., Shen, C.F., and Liu, C.C. (2008). Enterovirus 71, one virus and many stories. *Pediatric Neonatology*, **49**(4):113–5.
80. Centers for Disease Control and Prevention (2010). Nonpolio enterovirus and human parechovirus surveillance. United States, 2006–2008. *Morbidity & Mortality Weekly Report*, **59**(48), 1577–80.
81. Watkins, R.E., Martin, P.A., Kelly, H., Madin, B., and Watson, C. (2009). An evaluation of the sensitivity of acute flaccid paralysis surveillance for poliovirus infection in Australia. *BioMedical Centre for Infectious Diseases*, **9**, 162.
82. Bosch, A.M., Abeling, N.G., Ijlst, L., et al. (2011). Brown-Vialetto-Van Laere and Fazio Londe syndrome is associated with a riboflavin transporter defect mimicking mild MADD: a new inborn error of metabolism with potential treatment. *Journal of Inherited Metabolic Diseases*, **34**(1), 159–64.
83. Bosch, A.M., Stroek, K., Abeling, N.G., Waterham, H.R., Ijlst, L., and Wanders, R.J. (2012). The Brown-Vialetto-Van Laere and Fazio Londe syndrome revisited: natural history, genetics, treatment and future perspectives. *Orphanet Journal of Rare Diseases*, **7**, 83.
84. Ciccolella, M., Catteruccia, M., Benedetti, S., et al. (2012). Brown-Vialetto-van Laere and Fazio-Londe overlap syndromes: a clinical, biochemical and genetic study. *Neuromuscular Disorder*, **22**(12), 1075–82.
85. Ciccolella, M., Corti, S., Catteruccia, M., et al. (2013). Riboflavin transporter 3 involvement in infantile Brown-Vialetto-Van Laere disease: two novel mutations. *Journal of Medical Genetics*, **50**(2), 104–7.
86. Spagnoli, C., Pitt, M.C., Rahman, S., and DeSousa, C. (2014). Brown-Vialetto-van Laere syndrome: A riboflavin responsive neuronopathy of infancy with singular features. *European Journal of Paediatric Neurology*, **18**(2), 231–4.

87. Spagnoli, C. and DeSousa, C. (2012). Brown-Vialetto-Van Laere syndrome and Fazio-Londe disease—treatable motor neuron diseases of childhood. *Developmental Medicine & Child Neurology*, **54**(4), 292–3.
88. Pitt, M. (2012). Support your local paediatric neurophysiologist. *Developmental Medicine & Child Neurology*, **54**(12), 1164.
89. Cottenie, E., Menezes, M.P., Rossor, A.M., et al. (2013). Rapidly progressive asymmetrical weakness in Charcot–Marie–Tooth disease type 4J resembles chronic inflammatory demyelinating polyneuropathy. *Neuromuscular Disorder*, **23**(5), 399–403.
90. Yger, M., Stojkovic, T., Tardieu, S., et al. (2012). Characteristics of clinical and electrophysiological pattern of Charcot–Marie–Tooth 4C. *Journal of Peripheral Nervous Systems*, **17**(1), 112–22.
91. Rosen, B.A. (2012). Guillain-Barre syndrome. *Pediatric Reviews*, **33**(4), 164–70.
92. Rabie, M. and Nevo, Y. (2009). Childhood acute and chronic immune-mediated polyradiculoneuropathies. *European Journal of Paediatric Neurology*, **13**(3), 209–18.
93. Ryan, M.M. (2005). Guillain–Barré syndrome in childhood. *Journal of Paediatric Child Health*, **41**(5–6), 237–41.
94. Evans, O.B. and Vedanarayanan, V. (1997). Guillain–Barré syndrome. *Pediatric Reviews*, **18**(1), 10–16.
95. Lazzerini, M., Martelossi, S., Magazzu, G., et al. (2013). Effect of thalidomide on clinical remission in children and adolescents with refractory Crohn disease: a randomized clinical trial. *Journal of the American Medical Association*, **310**(20), 2164–73.
96. Noel, N., Mahlaoui, N., Blanche, S., et al. (2013). Efficacy and safety of thalidomide in patients with inflammatory manifestations of chronic granulomatous disease: a retrospective case series. *Journal of Allergy Clinic Immunology*, **132**(4), 997–1000.
97. Kawai, T., Watanabe, N., Yokoyama, M., et al. (2013). Thalidomide attenuates excessive inflammation without interrupting lipopolysaccharide-driven inflammatory cytokine production in chronic granulomatous disease. *Clinical Immunology*, **147**(2), 122–8.
98. Bailey, K.M., Castle, V.P., Hummel, J.M., Piert, M., Moyer, J., and McAllister-Lucas, L.M. (2012). Thalidomide therapy for aggressive histiocytic lesions in the pediatric population. *Journal of Pediatric Hematology Oncology*, **34**(6), 480–3.
99. Rigante, D., La, T.F., Calcagno, G., and Falcini, F. (2012). Clinical response to thalidomide and colchicine in two siblings with Behcet's disease carrying a single mutated MEFV allele. *Rheumatology International*, **32**(6), 1859–60.
100. Zheng, C.F., Xu, J.H., Huang, Y., and Leung, Y.K. (2011). Treatment of pediatric refractory Crohn's disease with thalidomide. *World Journal of Gastroenterology*, **17**(10), 1286–91.
101. Priolo, T., Lamba, L.D., Giribaldi, G., et al. (2008). Childhood thalidomide neuropathy: a clinical and neurophysiologic study. *Pediatric Neurology*, **38**(3), 196–9.
102. Fleming, F.J., Vytopil, M., Chaitow, J., Jones, H.R, Jr, Darras, B.T., and Ryan, M.M. (2005). Thalidomide neuropathy in childhood. *Neuromuscular Disorder*, **15**(2), 172–6.
103. Karakis, I., Liew, W., Darras, B.T., Jones, H.R., and Kang, P.B. (2013). Referral and diagnostic trends in pediatric electromyography in the molecular era. *Muscle & Nerve*, **50**(2), 244–9.
104. Haddad, F.S., Jones, D.H., Vellodi, A., Kane, N., and Pitt, M.C. (1997). Carpal tunnel syndrome in the mucopolysaccharidoses and mucolipidoses. *Journal of Bone Joint Surgery*, **79**(4), 576–82.
105. Haddad, F.S., Jones, D.H., Vellodi, A., Kane, N., and Pitt, M. (1996). Review of carpal tunnel syndrome in children. *Journal of Hand Surgery*, **21**(4), 565–6.
106. Sri-Ram, K., Vellodi, A., Pitt, M., and Eastwood, D.M. (2007). Carpal tunnel syndrome in lysosomal storage disorders: simple decompression or external neurolysis? *Journal of Pediatric Orthopedics Belgium*, **16**(3), 225–8.
107. Srinivasan, J., Ryan, M.M., Escolar, D.M., Darras, B., and Jones, H.R. (2011). Pediatric sciatic neuropathies: a 30-year prospective study. *Neurology*, **76**(11), 976–80.
108. Cruccetti, A., Kiely, E.M., Spitz, L., Drake, D.P., Pritchard, J., and Pierro, A. (2000). Pelvic neuroblastoma: low mortality and high morbidity. *Journal of Pediatric Surgery*, **35**(5), 724–8.
109. Kubiak, R., Wilcox, D.T., Spitz, L., and Kiely, E.M. (1998). Neurovascular morbidity from the lithotomy position. *Journal of Pediatric Surgery*, **33**(12), 1808–10.
110. Malone, P.S., Spitz, L., Kiely, E.M., Brereton, R.J., Duffy, P.G., and Ransley, P.G. (1990). The functional sequelae of sacrococcygeal teratoma. *Journal of Pediatric Surgery*, **25**(6), 679–80.
111. Desurkar, A., Mills, K., Pitt, M., et al. (2011). Congenital lower brachial plexus palsy due to cervical ribs. *Developmental Medicine & Child Neurology*, **53**(2), 188–90.
112. Pitt, M. and Vredeveld, J.W. (2004). The role of electromyography in the management of obstetric brachial plexus palsies 1. *Clinical Neurophysiology*, **57**(Suppl.), 272–9.
113. Pitt, M. and Vredeveld, J.W. (2005). The role of electromyography in the management of the brachial plexus palsy of the newborn. *Clinical Neurophysiology*, **116**(8), 1756–61.
114. Pitt, M. (2013). Update in electromyography. *Current Opinions in Pediatrics*, **25**(6), 676–81.
115. Pitt, M. (2012). Why wait 3 months before doing electromyography in obstetric brachial plexus lesions? Challenging the norm. *Developmental Medicine & Child Neurology*, **54**(8), 682.
116. van Dijk J.G, Pondaag W., Buitenhuis S.M, Van Zwet E.W, and Malessy M.J. (2012). Needle electromyography at 1 month predicts paralysis of elbow flexion at 3 months in obstetric brachial plexus lesions. *Developmental Medicine & Child Neurology*, **54**(8), 753–8.
117. Malessy, M.J., Pondaag, W., Yang, L.J., Hofstede-Buitenhuis, S.M., le Cessie, S., and van Dijk, J.G. (2011). Severe obstetric brachial plexus palsies can be identified at one month of age. *PLoS One*, **6**(10), e26193.
118. Semmler, A., Okulla, T., Kaiser, M., Seifert, B., and Heneka, M.T. (2013). Long-term neuromuscular sequelae of critical illness. *Journal of Neurology*, **260**(1), 151–7.
119. Khan, J., Harrison, T.B., Rich, M.M., and Moss, M. (2006). Early development of critical illness myopathy and neuropathy in patients with severe sepsis. *Neurology*, **67**(8), 1421–5.
120. Williams, S., Horrocks, I.A., Ouvrier, R.A., Gillis, J., and Ryan, M.M. (2007). Critical illness polyneuropathy and myopathy in pediatric intensive care: a review. *Pediatric Critical Care Medicine*, **8**(1), 18–22.
121. Lefaucheur, J.P., Nordine, T., Rodriguez, P., and Brochard, L. (2006). Origin of ICU acquired paresis determined by direct muscle stimulation. *Journal of Neurology, Neurosurgery, & Psychiatry*, **77**(4), 500–6.
122. Allen, D.C., Arunachalam, R., and Mills, K.R. (2008). Critical illness myopathy: further evidence from muscle-fiber excitability studies of an acquired channelopathy. *Muscle & Nerve*, **37**(1), 14–22.
123. Goodman, B.P., Harper, C.M., and Boon, A.J. (2009). Prolonged compound muscle action potential duration in critical illness myopathy. *Muscle & Nerve*, **40**(6), 1040–2.
124. Finlayson, S., Beeson, D., and Palace, J. (2013). Congenital myasthenic syndromes: an update. *Practical Neurology*, **13**(2), 80–91.
125. Klein, A., Pitt, M.C., McHugh, J.C., et al. (2013). DOK7 congenital myasthenic syndrome in childhood: early diagnostic clues in 23 children. *Neuromuscular Disorder*, **23**(11), 883–91.
126. Kinali, M., Beeson, D., Pitt, M.C., et al. (2008). Congenital myasthenic syndromes in childhood: diagnostic and management challenges. *Journal of Neuroimmunology*, 201–2, 206–12.
127. Robb, S.A., Sewry, C.A., Dowling, J.J., et al. (2011). Impaired neuromuscular transmission and response to acetylcholinesterase inhibitors in centronuclear myopathies. *Neuromuscular Disorder*, **21**(6), 379–86.
128. Munot, P., Lashley, D., Jungbluth, H., et al. (2010). Congenital fibre type disproportion associated with mutations in the tropomyosin

3 (TPM3) gene mimicking congenital myasthenia. *Neuromuscular Disorders*, **20**(12), 796–800.

129. Koepke, R., Sobel, J., and Arnon, S.S. (2008). Global occurrence of infant botulism, 1976–2006. *Pediatrics*, **122**(1), e73–82.
130. Brook, I. (2007). Infant botulism. *Journal of Perinatology*, **27**(3), 175–80.
131. Brook, I. (2006). Botulism: the challenge of diagnosis and treatment. *Reviews in Neurological Diseases*, **3**(4), 182–9.
132. Long, S.S. (2007). Infant botulism and treatment with BIG-IV (BabyBIG). *Pediatric Infectious Diseases Journal*, **26**(3), 261–2.
133. Arnon, S.S., Schechter, R., Maslanka, S.E., Jewell, N.P., and Hatheway, C.L. (2006). Human botulism immune globulin for the treatment of infant botulism. *New England Journal of Medicine,* **354**(5), 462–71.
134. Thompson, J.A., Filloux, F.M., Van Orman, C.B., et al. (2005). Infant botulism in the age of botulism immune globulin. *Neurology*, **64**(12), 2029–32.
135. Grant, K.A., Nwarfor, I., Mpamugo, O., et al. (2009). Report of two unlinked cases of infant botulism in the UK in October 2007. *Journal of Medicine Microbiology*, **58**(12), 1601–6.
136. Gutmann, L., Gutierrez, A., and Bodensteiner, J. (2000). Electrodiagnosis of infantile botulism. *Journal of Child Neurology*, **15**(9), 630.
137. Jones, H.R, Jr, and Darras, B.T. (2000). Acute care pediatric electromyography. *Muscle & Nerve*, **9**(Suppl.), 53–62.
138. Gutierrez, A.R., Bodensteiner, J., and Gutmann, L. (1994). Electrodiagnosis of infantile botulism. *Journal of Child Neurology*, **9**(4), 362–5.
139. Mygrant, B.I. and Renaud, M.T. (1994). Infant botulism. *Heart and Lung*, **23**(2), 164–8.
140. Ravid, S., Maytal, J., and Eviatar, L. (2000). Biphasic course of infant botulism. *Pediatric Neurology*, **23**(4), 338–9.
141. Thompson, J.A., Glasgow, L.A., Warpinski, J.R., and Olson, C. (1980). Infant botulism: clinical spectrum and epidemiology. *Pediatrics*, **66**(6), 936–42.

CHAPTER 26

EMG-guided botulinum toxin therapy

V. Peter Misra and Santiago Catania

Introduction

Botulinum neurotoxin (BoNT) has been the subject of a makeover. Notorious in the past only as a deadly poison causing fatal botulism or as an agent for biological warfare, the image of BoNT has gradually evolved over the last 50 years. Today, it is better recognized as an increasingly popular prescription medication with an ever expanding list of indications (1).

In previous years oral pharmaceutical agents including anticholinergic drugs, dopamine modulators, muscle relaxants, and other pharmacologic agents had been used to treat dystonia. These treatments were only partially effective and often associated with unwanted side effects. Similarly the mainstay of treatment for spasticity utilized oral pharmacological drugs such as baclofen and dantrolene.

Over the last quarter of a century BoNT treatment has truly revolutionized the treatment of dystonia and impacted upon the treatment of spasticity. In the 1970s there was an interest in the development of BoNT for therapeutic purposes and this became reality in the 1980s. Over the next decade there was an explosion in the usage of BoNT for dystonia and its use increased exponentially over the subsequent decades (1). Building on the initial success and experience of BoNT treatment for dystonia, clinicians then turned their attention to the potential use of BoNT for spasticity. Studies to demonstrate efficacy of BoNT for the treatment of stroke, multiple sclerosis (MS), and cerebral palsy (CP) were initiated and by 2010 BoNT was a standard mainstream treatment for all these conditions.

In the early days BoNT for therapeutic purposes was delivered by straightforward intramuscular injections into affected muscles. These injections were placed based on anatomical landmarks and palpation. As clinical experience and the indications for its use grew, it became clear that some muscles were easier than others to satisfactorily inject using anatomical landmarks alone. For example, in cervical dystonia (CD) injecting a superficial, easily palpable muscle, such as the sternomastoid, was found to be a relatively easy task, but a deeper and less easily palpable muscle, such as the levator scapulae, proved to be a greater challenge to inject correctly. The electromyographer with his/her experience in accurately recording from specific muscles for diagnostic purposes was therefore recruited to help with placing accurately targeted BoNT injections. (In the early days of BoNT treatment we were approached by a Movement disorder Neurologist who said 'The other day you were able to tell me that a patient had selective denervation of just the fascicles of flexor digitorum profundus affecting only the index and middle finger confirming an anterior interosseous nerve lesion. Since you are able so accurately to place an EMG needle would you be able to help me with a patient who needs BoNT injections for a very focal hand dystonia affecting his index finger flexors only?' Thus began our EMG guided BoNT service).

EMG guidance of BoNT injections are based on the electromyographer's knowledge of muscle anatomy, on his/her special EMG training and experience and the technical ability to reliably insert a needle into a specific muscle. The electromyographer can also confirm that this is the correct positioning by utilizing the EMG signal evoked by selective muscle activation. A hollow needle electrode, which incorporated simultaneous recording of the EMG signal, as well as BoNT delivery was developed to allow simultaneous EMG recording and BoNT injection. As experience and expertise grew, patients with conditions such as laryngeal dystonia, oromandibular dystonia, and writer's cramp became regular attenders of Neurophysiology clinics devoted to EMG guided BoNT injections. In these focal dystonic conditions, all of which involve deep muscle groups, BoNT treatment without EMG guidance would have been practically impossible. Other targeting techniques such as ultrasound imaging to guide needle placement have also become popular more recently, and in some instances EMG and ultrasound guidance techniques have been combined with success.

Botulinum neurotoxin

BoNT is a potent neurotoxin produced by Clostridium (Cl.) botulinum. The connection between food contaminated by Cl. botulinum and botulism was first postulated by Justinus Kerner (1786—1862), a German poet and district medical officer. He did not succeed in defining the suspected 'biological poison', which he called 'sausage poison' or 'fatty poison', but he insightfully suggested that there may be a possible therapeutic use of the toxin when it was discovered. In 1895, Emile Pierre van Ermengem, Professor of bacteriology at the University of Ghent investigated a botulism outbreak after a funeral dinner with contaminated smoked ham in the small Belgian village of Ellezelles. This led to the discovery of the pathogen Cl. botulinum. The bacterium was so called because of its pathological association with the sausage (Sausage in Latin = 'botulus'). These and other historical issues are discussed by Ergburth (2).

After its discovery, research for the next 100 years mainly centred around understanding the structure and mechanism of action of BoNT. Scientists were able to show that there were 7 subtypes of

BoNT (Type A, B, C, through to G). All of these shared the same basic structure of 2 heavy and 2 light chains and it was demonstrated that the heavy chain was responsible for the entry of the BoNT inside the axon while the light chain cleaved the protein responsible for the release of Acetylcholine from the presynaptic plate. The cleaved protein distinguished the different subtypes. BoNT was shown to produce an irreversible presynaptic cholinergic blockade causing chemodenervation. Recovery of function was only possible by axonal sprouting and reconnection. These aspects of BoNT structure and function are reviewed by Jankovic (3).

In the 1960 and 70s there was interest in the use of BoNT therapeutically and in 1981, Alan Scott a Canadian Ophthalmologist, after years of animal experimentation, successfully used BoNT type A (BoNTA) to treat strabismus in children (4). Over the next 25 years BoNT treatment became increasingly common place and there are now 4 commercially available preparations of BoNT in the UK.

All of the commercially available preparations have different production techniques and therefore have unique and different units of dosage, which are not interchangeable. The different commercial products have been designated by the following non-proprietary names: onabotulinumtoxin A (Botox [Type A]), rimabotulinumtoxin B (neurobloc [Type B]), abobotulinumtoxin A (Dysport [Type A]), and incobotulinumtoxin A (Xeomin [Type A]). Botox, Dysport, and Xeomin come in powder form, which needs to be reconstituted with normal saline while Neurobloc comes as a liquid ready for injection.

The potency per ml of BoNT therefore depends upon the commercial preparation used and the volume of normal saline used for reconstitution. Switching patients from one product to another is normally not required nor desirable (except for the exceptional circumstance of switching from BoNTA to BoNT type B (BoNTB) or vice versa in case of resistance due to antibody formation to either BoNTA or BoNTB). However, if for some practical consideration a patient has to be switched from one product to another it is worth noting that Botox and Xeomin are considered to have a dose equivalence of 1:1 (5) while the dose equivalence for Dysport to Botox/Xeomin has been reported in several studies to vary from between 2:1 and 4:1 (6–9) and the dose equivalence for Botox to Neurobloc has been shown to be 1:70 (10). Common to all of these products, the BoNT effects begin within a few days after injection with a peak effect at about 4 weeks followed by a gradual reduction in effect over the next 3 or 4 months.

The most frequent adverse effects associated with BoNT treatment are related to unwanted muscle weakness. This may be related to BoNT overtreatment of the injected muscle or due to unwanted inadvertent spread of BoNT into muscles adjacent to those treated. EMG targeted injection can potentially reduce both of these adverse effects because the accurate placement of BoNT should allow a reduction in the dose needed to get a good effect as well as minimizing spread. Dry mouth often accompanies injections with BoNTB. Despite the introduction of foreign protein, true allergic reactions are extremely rare in patients treated with BoNT. It is not clear whether some of the systemic side effects, such as generalized malaise, fever, and other flu-like symptoms, or a skin rash, occurring in about 2% to 20% of all patients, represent allergic reactions (3).

The high cost of BoNT is a major limitation of therapy. However, studies analysing the cost-effectiveness of BoNT treatment have shown that the loss of productivity as a result of untreated dystonia and the cost of medications or surgery more than justify the expense of BoNT treatments (11). It is likely that future research, which will include not only today's commercially available serotypes BoNTA and BoNTB, but possibly other serotypes such as BoNT-type C and BoNT-type F, will result in the development of new, more effective neuromuscular blocking agents that provide therapeutic chemodenervation with long-term benefits and at lower cost.

EMG-guided delivery of botulinum neurotoxin

EMG guided BoNT injections are made using a special Teflon coated hollow monopolar EMG needle, which has a port where a syringe can be attached. Different lengths and gauges of needles are available (Fig. 26.1). A surface reference electrode is placed close to the site of needle insertion and an earth electrode nearby. The needle is inserted into the desired muscle using standard anatomical landmarks and EMG techniques. The accurate positioning of the needle can be fine-tuned and confirmed by recording the EMG signal when the muscle is voluntarily activated. If a muscle cannot be voluntarily activated or if mass movement prevents individual muscle activation (as may be the case in patients with spasticity) then the EMG needle can be used as an electrical stimulator. In this case a small electrical stimulation is delivered through the EMG needle, which is then carefully positioned to evoke an isolated twitch and/or movement of the desired muscle. This confirms that the needle is accurately placed before the injection if made.

Once the EMG needle has been inserted the patient with dystonia is often encouraged to 'let the muscles do what they like'. A strong EMG signal associated with the involuntary movement allows confirmation that the muscle being examined is involved in the dystonic movement. Careful positioning of the needle to achieve the strongest EMG signal allows the physician to further optimize the most accurate needle placement. Once the appropriate needle placement has been made the BoNT injection is injected

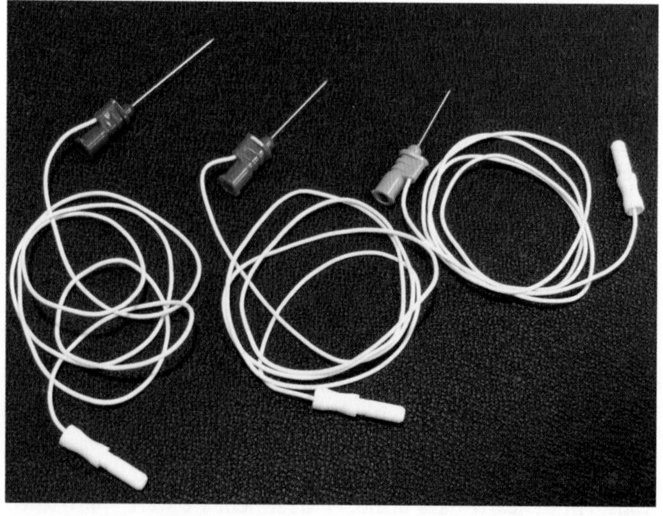

Fig. 26.1 EMG needles of different lengths and gauges, especially designed to allow simultaneous EMG recordings and BoNT injection.

from the syringe connected to the port on the needle. In this way the EMG signal is used to guide the injection. As will be discussed later, in routine practice the EMG signal does not discriminate between muscles affected by dystonia and those, which may be active due to compensatory movements—all it indicates is that the muscle is active.

In those patients who have been receiving repeated BoNT injections it may be possible to identify neurogenic changes in the previously injected muscle. If the previous injections had been associated with a good clinical effect then this EMG denervation signal increases the electromyographer's confidence that the current injection, because it is similarly placed as in the past, should once again produce a similar good result. Furthermore, in a patient who has developed secondary non-responsiveness to BoNT (which will be discussed later) such denervation changes, since they reflect chemodenervation due to BoNT effect, indicate that the non-responsiveness is unlikely to be due to ineffectiveness of BoNT as a consequence of development of BoNT antibodies.

Conditions which benefit from botulinum neurotoxin EMG targeting

Dystonia

According to a recent consensus update (12) dystonia is defined as persistent or intermittent muscle contractions causing abnormal, often repetitive movements, postures, or both. The movements are usually patterned and twisting and may resemble a tremor and it is often initiated or worsened by voluntary movements and symptoms may 'overflow' into adjacent muscles. It can be classified by clinical characteristics and cause.

Dystonia has multiple aetiologies, which for ease of understanding can be divided into primary and secondary causes. Primary causes are either idiopathic or related to specific genes (named DYT genes), which have been identified in families with dystonia. The list of DYT genes is rapidly increasing (DYT1 through to DYT 25 currently) of which 11 have been shown unequivocally to cause different forms of dystonia (13). Some patients with genetically determined dystonia manifest with a generalized form of dystonia although within this there is often focal emphasis in one or more body areas, while in others the dystonia may be entirely focal. Secondary dystonia may be the consequence of structural brain abnormalities from a variety of causes (cerebral infarction, tumours, etc.) or from the side effects of drugs. (See 12, for a review).

Careful observation of the patient while seated, standing, walking, lying down, and while performing various activities helps the physician to understand the various dystonic movements. A video recording, apart from acting as a record for the future, allows for a more careful analysis to be made. Movement patterns and the muscles responsible may be better identified by playing the video in slow motion or by studying individual frames using a video stop-start function. Based upon these clinical observations a plan regarding, which muscles require injection and the BoNT doses required for each can be drawn up.

The patient's input into the clinical decision making is essential. Clinical decision making is often influenced by the patient's insights. The patient often will volunteer that some involuntary movements are more intrusive than others, or that certain movements may be associated with an element of pain while others are not.

Cervical dystonia

This is the commonest form of focal dystonia (14) with a prevalence of 89 per million in parts of the US (Rochester Minnesota) (15). It affects women more than men. Due to the dystonic effect on the muscles of the neck and head the commonest predominant pattern is of neck rotation (torticollis) followed by a sideways head tilt (laterocollis) and associated head tremor or jerk may be common. (16).

The main muscles involved in CD are the splenius capitis (which rotates the head to the same side while pulling the head backwards), the sternomastoid (which rotates the head to the opposite side and pulls the head forward to the ipsilateral side), the trapezius (which pulls the head laterally to the same side and elevates the shoulder) and the levator scapulae (which elevates the shoulder and turns the head to the same side). Deeper muscles, which cannot be palpated (such as the suboccipital group of muscles, comprising Rectus capitis and Caput obliques) also affect head movements.

There are several scales, which have been designed to assess the pattern and severity of CD. (Toronto Western Spasmodic Torticollis Scale, the Tsui Scale, and CD Impact Profile-58 are examples). These are also useful to study serially treatment efficacy and/or disease progression.

Injections of superficial and easily palpable neck muscles such as the sternomastoid, splenius capitis, or trapezius are usually easily made by unguided injections although there is evidence that the greatest benefit is obtained by EMG guided injections as blindly placed injections may often be incorrectly placed (17,18). Unguided injections can be especially tricky in patients with thick set necks or in those cases where muscles have become atrophic following repeated BoNT injections. Table 26.1 shows suggested doses for the commonly injected neck muscles.

At times, especially in complex cases, it may be necessary to inject deeper muscles such as the levator scapulae or muscles from the sub-occipital group. Fig. 26.2 shows a patient with a complex CD showing lateral shift and lateral head tilt. In the case of such complex cases the patients often also require injections into deeply situated muscles. Carefully planned EMG guided injections of various neck muscles would be mandatory.

Unwanted side effects mainly relate to excess temporary weakness or dysphagia. The former is related to the dose of BoNT injected and the latter to unwanted spread of BoNT into pharyngeal or tongue muscles. Dysphagia is usually related to injections of the sternomastoid or other anterior placed neck muscles. Bilateral injections for these muscles groups using high doses are therefore to be avoided if at all possible.

Oromandibular dystonia

This is a relatively uncommon form of focal dystonia affecting women more than men (19). The jaw and tongue are commonly involved and it is often also associated with dystonia of the upper part of the face (especially blepharospasm).

Oromandibular dystonia (OMD) may cause involuntary jaw closing, jaw opening, or jaw side to side deviation. The tongue and/or lips may also be involved. The masseter, temporalis, and medial pterygoid muscles are the main muscles of jaw closure while the

Table 26.1 Suggested BoNTA doses (in Dysport units) for commonly injected muscles in dystonia

Muscle	Dose (Dysport units)	Muscle	Dose (Dysport units)
Cervical dystonia			
Splenius capitis	150–350	Levator scapulae	50–200
Sternomastoid*	75–200	Trapezius	100–300
Oromandibular dystonia			
Masseter	50–150	Lateral pterygoid	50–125
Temporalis	50–150	Medial pterygoid	10–25
Laryngeal dystonia			
Thyroarytenoid	2.5–10	Posterior cricothyroid*	1.5–10
Upper limb task specific dystonia			
Flex carpi ulnaris	20–60	Flex digitorum sup	5–15 per fascicle
Flex carpi radialis	20–60	Flex digitorum prof	5–15 per fascicle
Ext carpi ulnaris	20–60	Ext digitorum comm	5–15
Ext carpi radialis	20-60	Ext indicis	5–10
Pronator teres	20–60	Flex pollicis longus	5–15
Pronator quadratus	10–40	Supinator	10–40

*Avoid bilateral injections if possible.

Fig. 26.2 Complex CD showing lateral shift of the neck to the left and head tilt to the right.

lateral pterygoid is the main muscle for jaw opening. The medial and lateral pterygoids also act to deviate the jaw to the opposite side. As with all dystonic conditions the assessment of the abnormal movements and an analysis of which muscles are affected is extremely important. Patients with OMD should be observed during various activities, especially including talking and eating and a video recording helps with the analysis of the movements and as a record for the future.

Injection of the masseters and temporalis are relatively straightforward and do not usually need EMG guidance. On the other hand EMG guidance is invariably required for medial or lateral pterygoid injections. In our practice we prefer to inject these muscles using an external percutaneous approach although some practitioners prefer an intra-oral approach. On occasions the tongue may need injecting, we usually prefer to inject the genioglossus using a submental percutaneous approach, but at times direct injections into the tongue are also given. Injection of the tongue commonly may cause dysphagia as a side effect and so the doses of BoNT should be carefully titrated. The patient should be warned in advance of this possible side effect and advised of how to modify their eating and drinking should dysphagia occur.

Table 26.1 shows suggested doses for the commonly injected muscles in OMD.

Side effects are mainly related to temporary difficulty with swallowing and weakness of masticatory muscles.

Laryngeal dystonia

Laryngeal dystonia is a relatively uncommon form of focal dystonia affecting women more often than men (20). It may affect the laryngeal adductor or abductor muscles. Laryngeal adductor muscle dystonia is the more common form and the patient's voice typically has a croaky or strangled quality. In the laryngeal abductor dystonia the patient's voice is typically whispery and breathy. Indirect laryngoscopy demonstrates the dystonic muscles. The thyroarytenoid (adductor muscles) and the posterior cricothyroid (abductor) muscles can be conveniently injected percutaneous through the neck; although a per-oral injection approach has also been described. Once the needle has been inserted a burst of EMG activity when the patient phonates confirms accurate needle position and the injection can then be given (20). Table 26.1 shows suggested doses for commonly injected muscles. The main side effects are temporary post injection dysphonia and or dysphagia.

Focal limb dystonia

Focal limb dystonia may affect the upper or the lower limbs and may be task or non-task specific. EMG or other guidance techniques are essential to ensure correct delivery of BoNT to the appropriate muscles (21).

Task specific focal limb dystonia

In this form of focal limb dystonia the abnormal movements only occur when the patient performs certain specific tasks. For instance it may be apparent only when writing or only while typing, but not for other tasks or activities. The task may be related to occupational or recreational activities and as already mentioned may be extremely task specific. For example, in our clinic we have seen a musician who develops dystonia when she plays the piano, but not when she plays the keyboard on an accordion! These forms of dystonia are invariable primary (idiopathic).

These patients should be carefully assessed while performing the provoking task and a video recording made. It is often useful to observe the patient carrying out the task in different positions, including while seated at different heights and while standing. Patients with musician's dystonia will need to bring their musical instrument when they attend the clinic. To properly assess patients with writing dystonia the clinic should have a variety of types of pens, pencils and writing surfaces available. Patients often will say that the thickness of the pen, the type of nib, the relative position of the paper, the writing surface, or the way they grip the pen can all affect the dystonia. These observations can also lead to very practical insights and recommendations some of which were not previously known to the patient and this may positively impact on how they cope with their disability. Using the non-dominant hand to perform the task may evoke involuntary dystonic mirror movements in the opposite dominant hand and these may give a clue regarding the main muscles involved in a dystonic movement.

Following a careful assessment of the movement pattern a plan of the muscles to be treated and the doses to be applied to each is made. Table 26.1 shows the suggested doses for commonly injected forearm muscles in upper limb focal dystonia. Very often it may be necessary to specifically target certain fascicles of a muscle.

As the dystonia is only apparent for certain tasks (and not others) it is important not to compromise normal function by overly weakening muscles. We find that there is often a trade-off between effectively treating the dystonia at the expense of causing unwanted weakness. The patient's view of what is an acceptable balance is often an important factor here.

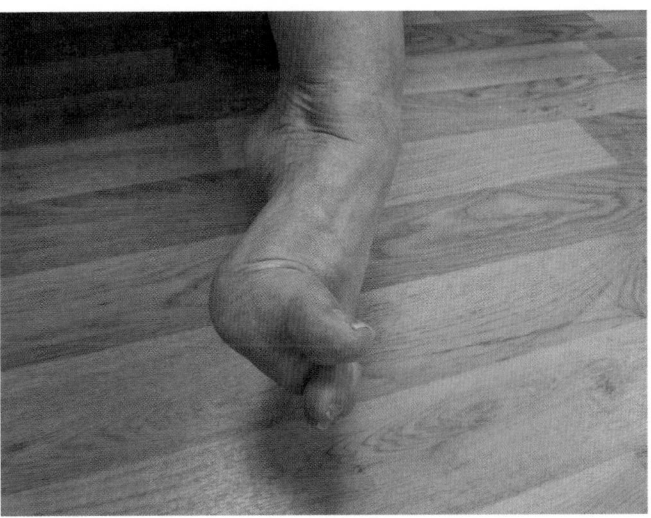

Fig. 26.3 Patient with Parkinsonism demonstrating dystonic focal toe extension and ankle inversion.

Non-task specific limb dystonia

As the name suggest in this case the dystonia affects the limb consistently regardless of the task. The dystonia may be isolated or as part of a more generalized dystonia. It may be primary or secondary in aetiology. The clinical assessment is similar to that made for task specific dystonia with the patient being carefully observed while using the limb for various multiple activities in different positions. Upper or lower limbs may be affected. The selection of muscles to be injected once again requires careful consideration as some of the movements may be compensatory and will reduce once the main dystonic component has been treated. Fig. 26.3 shows a patient with dystonic toe extension with ankle inversion related to Parkinsonism. Fig. 26.4 shows a patient with Parkinsonism who shows prominent wrist and finger dystonia, in such cases it may be necessary to also inject the lumbricals in addition to the finger and wrist flexor muscles.

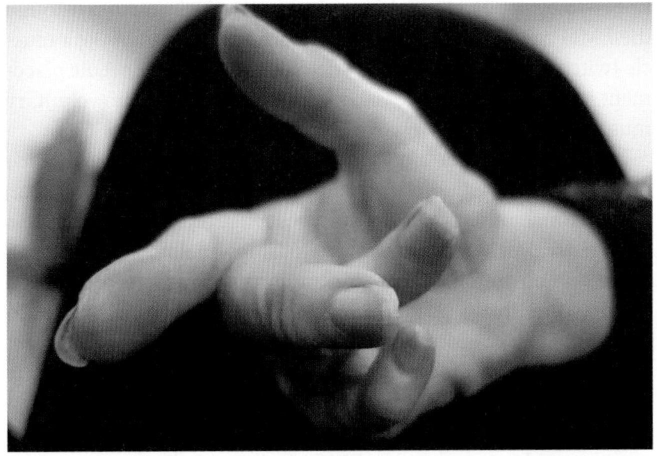

Fig. 26.4 Patient with Parkinsonism demonstrating complex focal hand dystonia.

Spasticity

Spasticity has been defined as a motor disorder characterised by a velocity dependent increase in the tonic stretch reflex (muscle tone) with exaggerated tendon jerks, resulting from hyperexcitabily of the stretch reflex (22). It is just one positive symptom component of the upper motor neurone syndrome. Spasticity as a consequence of stroke, head injury, cerebral anoxia, MS, and CP account for the bulk of patients with spasticity who are referred to us for BoNT injections. A comprehensive review of spasticity and its treatment with botulinum toxin is beyond the scope of this chapter and the reader is directed to national guidelines for the use of BoNT in management of spasticity (23).

Upper limb spasticity affecting elbow, wrist, and finger flexors have been shown to respond well to BoNT injections and have been best studied in the context of post-stroke spasticity. Benefits include reduction in deformity and pain, improvement in washing and dressing the upper limb and a reduction in caregiver burden. Some patients also show improvement in active limb movement and function (24). BoNT treatment is just one tool of the rehabilitation process and the general approach to management begins with assessment, goal setting, and outcome measurement, all to be developed in a multidisciplinary team environment including, amongst others, physiotherapists and occupation therapists.

Spasticity affects muscle groups rather than individual muscles and therefore the accurate placement of BoNT injection is usually not as crucial as it is in focal dystonia. For example, upper limb spasticity may affect all the finger and wrist flexor muscle groups resulting in a clenched fist and flexed wrist, which in association with pain affects limb function, dressing, caring, and in extreme cases hand hygiene. In situations such as this it is desirable to inject all affected muscle groups and any spread of toxin due to inaccurate needle placements is not of such a major concern as with task specific dystonia. Despite this, the results of EMG guided BoNT injections in patients with post stroke spasticity have still been shown to be superior to injections based on anatomical landmarks alone (25). This is probably because EMG guidance ensures that the BoNT is applied to contractile muscle tissue where it will be most effective. The relative intensity of the EMG signal in different muscles may also act as a guide to the differential degrees of muscle spasticity and this can also be of help in determining the dose of BoNT to be applied to each muscle. Electrical stimulation through the EMG needle (as discussed earlier) can be used to evoke a muscle twitch and/or movement, which helps to confirm needle placement in patients who have uncoordinated, limited, or no voluntary muscle movement.

In the case of lower limb spasticity the common muscles, which need injection are the gastrocnemius, soleus, and tibialis posterior, but proximal muscles such as the hamstrings or iliopsoas often also require treatment. At times more focal lower limb spasticity can cause very obvious foot inversion and/or clawing of the toes, which requires the tibialis posterior and/or flexor digitorum muscles to be specifically targeted by EMG guidance. Fig. 26.5 shows a post stroke patient with focal toe flexion. Children with CP frequently benefit from focal BoNT injections into the iliopsoas and, since that muscle is deep and cannot be easily palpated, it is essential to inject it under guidance with either EMG or ultrasound.

The intensity of the EMG signal, either occurring continuously due to increased muscle tone, during volition or when evoked by triggering a reflex by muscle stretch can also help distinguish between severe muscle contraction due to spasticity and contracture. In the former case the EMG signal will be strong while in contracture due to irreversible tissue changes the EMG signal will be difficult to obtain and needle insertion may have had a 'woody' feel. These observations may also be helpful for the clinical team to decide on appropriate patient treatment strategies.

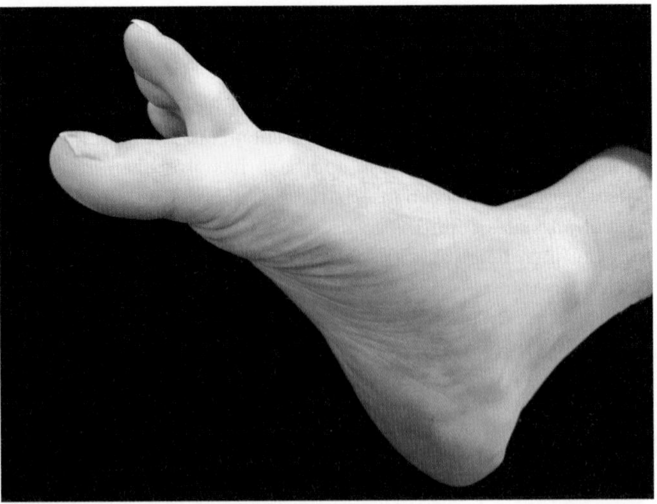

Fig. 26.5 Patient with focal toe flexion following a stroke.

Side effects of BoNT treatment mainly relate to iatrogenic production of excess weakness, which may interfere with function. A careful co-ordinated analysis with the therapy team regarding the short and long term goals and an integrated treatment plan is essential. It is also important in reducing unwanted side effects.

Other uses of the EMG in a botulinum toxin clinic

Movement disorder neurophysiology is an entire discipline and a discussion of this field is outside the scope of this chapter. However, we touch upon a few situations where techniques available in a routine EMG laboratory may be used to assist the management of patients who are receiving BoNT treatment.

Multichannel EMG video telemetry

On occasions the assessment of patients with complex dystonic movements can be aided by multichannel EMG recordings, which are recorded and time locked to simultaneous video telemetry (26). Multiple surface EMG recordings may be adequate for large superficial neck muscles, but smaller and deeper muscles may require needle EMG recordings. This analysis can inform targeted treatment with BoNT injections and can improve treatment outcome in selected patients with CD (27).

EMG signal markers for the diagnosis of dystonia

Co-contraction of agonist and antagonist muscles is one of the hallmarks of dystonia and therefore researchers have tried to use neurophysiological studies to identify these patterns of co-contraction (28). However co-contraction has also been observed in normal muscles performing isometric contractions (29) and in some reports the levels of co-contraction were actually found to be decreased in muscles of individuals affected by upper extremity dystonia compared with normal healthy controls (30). Thus, the sensitivity of co-contraction for the diagnosis of dystonia is limited. Spectral analysis of EMG signals (31) has been recently utilized

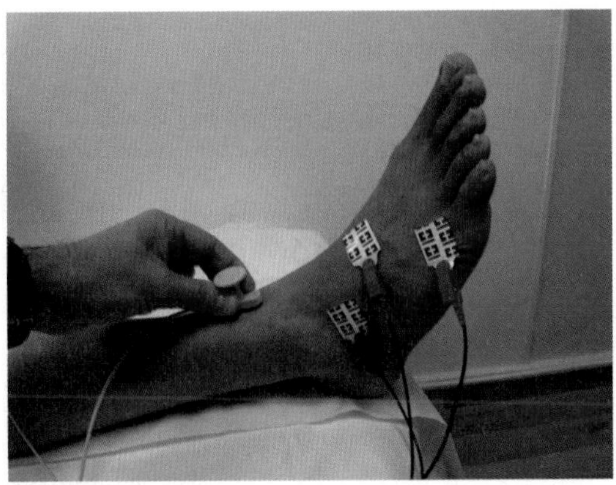

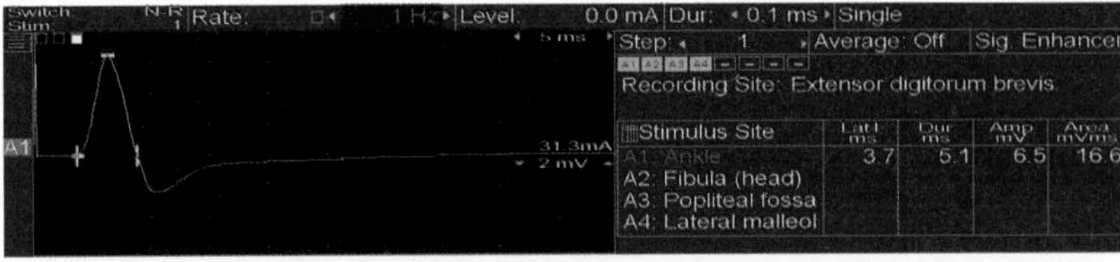

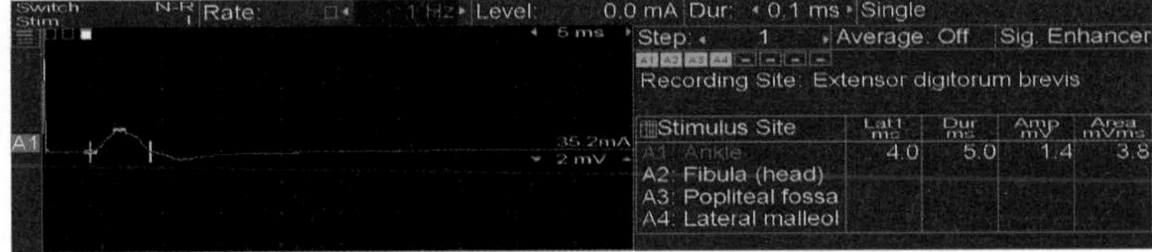

Fig. 26.6 EDB test. Stimulation of the common peroneal nerve at the ankle recording from EDB. Upper recording (EDB CMAP 6.5 mV) performed before injection of 90 units of BoNTA (Dysport). Lower recording (EDB CMAP 1.4 mV), 2 weeks after injection of BoNTA. Note that the amplitude of the CMAP is now less than 50% of the original, demonstrating the clinical effectiveness of BoNTA (i.e. the patient has not developed antibodies against BoNTA).

for the diagnosis of dystonia and studies have shown that muscles exhibiting dystonic behaviour have a characteristic shift of their power spectra to lower frequencies. Further research to increase the specificity and sensitivity of such investigations is on-going.

Secondary non-responders to BoNT and the extensor digitorum brevis test

Less than 5% of patients who were previously responsive to BoNT treatment may subsequently become non-responsive (secondary non-responders). Often this apparent non-responsiveness may just be due to a change in pattern of the dystonia, which has not been recognized (26) and improvement follows a reassessment of the dystonia and review of the muscles, which need injection. However, in some cases this may be related to the development of neutralizing antibodies to BoNT. From the 1990s onwards a reduction in the complexing protein load in the commercially available BoNT preparations coincided with a fall in the incidence of antibody development from 10% to only 1–2%. This suggests that antibody formation may be related to the load of complexing protein in the BoNT manufacturing process (32).

The extensor digitorum brevis (EDB) test is a convenient functional test, which can be easily performed in the EMG laboratory to check for immunological resistance to BoNT (33). A surface electrode is placed over the EDB muscle and the maximum compound muscle action potential (CMAP) to peroneal nerve stimulation is recorded. The EDB is then injected with BoNT. Two weeks later when the BoNT effect is maximum the recording is repeated. A decrement in the CMAP amplitude indicates that there has been a BoNT effect and that the patient is still responsive to BoNT (Fig. 26.6). Sloop et al. (34) have demonstrated a dose response curve of EDB muscle function to BoNT, which may be useful to document partial responsiveness. Partial responsiveness is difficult to assess in other functional tests, e.g. in the frontalis test the frontalis muscle is injected on one side and the ipsilateral disappearance of forehead wrinkles after 10 days determines responsiveness to BoNT in a yes or no fashion.

Conclusions

In this chapter we have highlighted the expanding therapeutic role of BoNT in the treatment of neurological disorders. We have specifically focussed on the conditions of dystonia and spasticity. We have highlighted the importance of specific intramuscular placement of BoNT injection and have concentrated on those conditions where it is especially desirable and on those instances where it is essential. We have discussed how accurately placed injections by EMG guidance improves the response to treatment, utilizes the smallest necessary dose and helps reduce side effects. While EMG guided injections are universally accepted as a standard method to deliver targeted injections other image guided techniques, such as ultrasound, can also be very useful. Combining the two techniques should lead to even better results. Finally we have briefly highlighted how other EMG techniques may be incorporated into the management of patients who receive BoNT treatment.

References

1. Misra, V.P. (2002). The changed image of botulinum toxin. *British Medical Journal*, **325**(7374), 1188.
2. Erbguth, F.J. (2004). Historical notes on botulism, *Clostridium botulinum*, botulinum toxin, and the idea of the therapeutic use of the toxin. *Movement Disorder*, **19**(8), S2–6.
3. Jankovic, J. (2013). Medical treatment of dystonia. *Movement Disorders*, **28**(7), 1001–12.
4. Scott, A.B. (1981). Botulinum toxin injection of eye muscles to correct strabismus. *Trans American Ophthalmology Society*, **79**, 734–70.
5. Pagan, F.L. and Harrison, A. (2012). A guide to dosing in the treatment of cervical dystonia and blepharospasm with Xeomin®: a new botulinum neurotoxin A. *Parkinsonism and Related Disorders*, **18**(5), 441–5.
6. Odergren, T., Hjaltason, H., and Kaakkola, S. (1998). A double blind, randomised, parallel group study to investigate the dose equivalence of Dysport and Botox in the treatment of cervical dystonia. *Journal of Neurology and Neurosurgery Psychiatry*, **64**(1), 6–12.
7. Karsai, S. and Raulin, C. (2009). Current evidence on the unit equivalence of different botulinum neurotoxin A formulations and recommendations for clinical practice in dermatology. *Dermatology Surgery*, **35**(1), 1–8.
8. Wohlfarth, K., Sycha, T., Ranoux, D., Naver, H., and Caird, D. (2009). Dose equivalence of two commercial preparations of botulinum neurotoxin type A: time for a reassessment? *Current Medicine Result Opinions*, **25**(7), 1573–84.
9. Rystedt, A., Nyholm, D., and Naver, H. (2012). Clinical experience of dose conversion ratios between 2 botulinum toxin products in the treatment of cervical dystonia. *Clinical Neuropharmacology*, **35**(6), 278–82.

10. Lee, D.H., Jin, S.P., Cho, S., et al. (2013). RimabotulinumtoxinB versus OnabotulinumtoxinA in the treatment of masseter hypertrophy: a 24-week double-blind randomized split-face study. *Dermatology*, **226**(3), 227–32.
11. Burbaud, P., Ducerf, C., Cugy, E., et al. (2011). Botulinum toxin treatment in neurological practice: how much does it really cost? A prospective cost-effectiveness study. *Journal of Neurology*, **258**, 1670–75.
12. Albanese, A., Bhatia, K., Bressman, SB., et al. (2013). Phenomenology and classification of dystonia: A consensus update. *Movement Disorder*, **28**, 863–73.
13. Klein, C. (2014). Genetics in dystonia. *Parkinsonism and Related Disorders*, **20**(1), 137–42.
14. Bhidayasiri, R. and Tarsy, D. (2006). Treatment of dystonia. *Expert Reviews on Neurotherapy*, **6**, 863–86.
15. Nutt, J.G., Muenter, M.D., Aronson, A., Kurlan, L.T., and Melton, L.J., 3rd. (1988). Epidemiology of focal and generalized dystonia in Rochester, Minnesota. *Movement Disorder*, **3**, 188–94.
16. Misra, V.P., Ehler, E., Zakine, B., Maisonobe, P., and Simonetta-Moreau, M. (2012). Factors influencing response to botulinum toxin type A in patients with idiopathic cervical dystonia: results from an international observational study. *British Medical Journal*, **2**(3), e000881.
17. Comella, C., Buchman, A.S., Tanner, C.M., Brown-Toms, N.C., and Goetz C.G. (1992). Botulinum toxin injection for spasmodic torticollis: increased magnitude of benefit with electromyographic assistance. *Neurology*, **42**, 878–82.
18. Speelman, J.D. and Brans, J.WM. (1995). Cervical dystonia and botulinum treatment: is electromyographic guidance necessary? *Movement Disorders*, **10**, 802.
19. Sinclair, C.F., Gurey, L.E., and Blitzer, A. (2013). Oromandibular dystonia: Long-term management with botulinum toxin. *Laryngoscope*, **123**(12), 3078–83.
20. Blitzer A. (2010). Spasmodic dysphonia and botulinum toxin: experience from the largest treatment series. *European Journal of Neurology*, **17**(1), 28–30.
21. Molloy, F.M., Shill, HA., Kaelin-Lang, A., and Karp, B.I. (2002). Accuracy of muscle localization without EMG: implications for treatment of limb dystonia. *Neurology*, **58**(5), 805–7.
22. Lance, J.W. (1980). Symposium synopsis. In: R. G. Feldman, R. R. Young, and K. P. Koella (Eds) *Spasticity: disorder of motor control*, pp. 17–24. Chicago, IL: Year Book.
23. Royal College of Physicians, British Society of Rehabilitation Medicine, Chartered Society of Physiotherapy, and the Association of Chartered Physiotherapists Interested in Neurology. (2009). *Spasticity in adults: management using botulinum toxin*, National guidelines. London: RCP, BSRM, ACP.
24. Sheean, G., Lannin, N.A., Turner-Stokes, L., Rawicki, B., and Snow, B.J. (2010). Botulinum toxin assessment, intervention and after-care for upper limb hypertonicity in adults: international consensus statement. *European Journal of Neurology*, **17**(2), 74–93.
25. Ploumis, A., Varvarousis, D., Konitsiotis, S., and Beris, A. (2014). Effectiveness of botulinum toxin injection with and without needle electromyographic guidance for the treatment of spasticity in hemiplegic patients: a randomized controlled trial. *Disability Rehabilitation*. **36**(4), 313–8.
26. Cordivari, C., Misra, V.P., Vincent, A., Catania, S., Bhatia, K.P., and Lees, A.J. (2006). Secondary non-responsiveness to botulinum toxin A in cervical dystonia. The role of electromyogram guided injections, botulinum toxin A antibody assay and the extensor digitorum brevis test. *Movement Disorder*, **21**(10), 1737–41.
27. Nijmeijer, S.W.R., Koelman, J.H.T.M., Standaar, T.S.M., Postma, M., and Tijssen, M.A.J. (2013). Cervical dystonia: improved treatment response to botulinum toxin after referral to a tertiary centre and the use of polymyography *Parkinsonism and Related Disorders*, **9**(5), 533–8.
28. Tijssen, M.A, Marsden, J.F, and Brown, P. (2000). Frequency analysis of EMG activity in patients with idiopathic torticollis. *Brain*, **123**(4), 677–86.
29. Carolan, B. and Cafarelli, E. (1992). Adaptations in coactivation after isometric resistance training. *Journal of Applied Physiology*, **73**, 911–17.
30. Malfait, N. and Sanger, T.D. (2007). Does dystonia always include co-contraction? A study of unconstrained reaching in children with primary and secondary dystonia. *Experimental Brain Research*, **176**, 206–16.
31. Go, S.A., Coleman-Wood, K., and Kaufman, K.R. (2014). Frequency analysis of lower extremity electromyography signals for the quantitative diagnosis of dystonia. *Journal of Electromyography and Kinesiology*, **24**(1), 31–6.
32. Jankovic, J., Vuong, K.D., and Ahsan, J. (2003). Comparison of efficacy and immunogenicity of original versus current botulinum toxin in cervical dystonia. *Neurology*, **60**, 1186–8.
33. Hamjian, J.A. and Walker, F.O. (1993). Serial neurophysiologic studies of intramuscular botulinum A toxin in humans. *Muscle & Nerve*, **16**, 181–7.
34. Sloop, R.R., Escutin, R.O., Matus, J.A., Cole, B.A., and Peterson, G.W. (1996). Dose–response curve of human extensor digitorum brevis muscle function to intramuscularly injected botulinum toxin type A. *Neurology*, **46**, 1382–86.

SECTION 4

Clinical aspects: central nervous system

CHAPTER 27

Genetic generalized epilepsy

Friederike Moeller, Ronit M. Pressler, and J. Helen Cross

Introduction

The genetic generalized epilepsies, previously termed the idiopathic generalized epilepsies, are a group of disorders characterized by generalized seizures, and on EEG by generalized spike-wave activity occurring from normal background activity. Seizure types usually consist of absence seizures, myoclonic seizures, and generalized tonic-clonic seizures. Photosensitivity is a common feature in patients and according to some studies may be observed in 40–90% of the patients (1,2). MRI does not show any abnormality and the patients usually show normal development. According to the electroclinical picture with age specific onset, seizure types, and EEG patterns different subgroups can be differentiated.

In 2010, the International League Against Epilepsy (ILAE) commission on Classification and Terminology revised its terminology and replaced the term idiopathic generalized epilepsy (IGE) (3) with the term genetic generalized epilepsy (GGE) (4). The new term describes, that the GGEs are caused by a known or presumed genetic defect, implied from twin and family studies. However, while several genes have been identified in different GGEs in the majority of patients, the underlying gene defect remains unclear (5). Microdeletions in the chromosomal region 15q13.3 have been identified in 1% of all GGE patients and constitute the most common risk factor for GGE (6). GGE accounts for approximately 30% of epilepsies in adults and children (7).

In this chapter, the clinical characteristics and electroencephalogram (EEG) features of the GGEs will be described in order of their age of onset: myoclonic astatic epilepsy (MAE) childhood absence epilepsy (CAE), epilepsy with myoclonic absences (EMA), eyelid myoclonia with absences (ELMA), juvenile absence epilepsy (JAE), juvenile myoclonic epilepsy (JME) and epilepsy with generalized tonic seizures on awakening (see Table 27.1).

Myoclonic astatic epilepsy

Clinical presentation

MAE, first described by Doose and colleagues accounts for approximately 1–2% of childhood-onset epilepsies (8). It is a difficult to treat subgroup of GGE characterized by age of onset mostly between 2 and 6 years, myoclonic and astatic drop attacks, tonic seizures, generalized tonic clonic seizure, and short absences (9). Non-convulsive status epilepticus and myoclonic status epilepticus with erratic myoclonic jerks may occur. Prior to the onset of epilepsy the patients are normally developed, but may show psychomotor retardation as the epilepsy progresses.

Electroencephalogram

At onset the background EEG may be normal or show rhythmic theta activity predominantly over the parietal regions (Fig. 27.1). Interictal discharges consist of bursts of 2–3 Hz generalized spike and wave or polyspike wave discharges (8,10,11). Myoclonic jerks are associated with spike wave or polyspike wave discharges with time locked EMG changes followed by a post-myoclonic silent period (12). Atonic seizures show generalized polyspike wave or spike wave discharges associated with an inhibition in the EMG (Fig. 27.2). Polyspike and wave discharges are seen during tonic seizures. During myoclonic status irregular and independent spike wave discharges are seen, at times resembling hypsarrhythmia (11).

Treatment

No randomized clinical trials are available regarding antiepileptic treatment in MAE. Valproate is often used as the first line antiepileptic drug (AED). Other antiepileptic drugs, such as ethosuximide, topiramate, levetiracetam, and lamotrigine, as well as ketogenic diet and adrenocorticotropic therapy have been reported to be effective (13–16). A combination of valproate, lamotrigine, and a benzodiazepine seems to be effective to treat drop attacks (17). Inappropriate treatment with carbamazepine and vigabatrin may lead to myoclonic status epilepticus (18).

Prognosis

The course of the disorder is variable and its outcome unpredictable (19). Many are self-limiting with seizures abating in 50–89%, but in a proportion seizures may persist and become intractable for a long period of time (19). Although up to 58% have normal cognitive outcome, some may develop an epileptic encephalopathy with up to 22% developing severe mental retardation (14). Tonic seizures and myoclonic status epilepticus (if not provoked by inappropriate AED treatment) are associated with a poor prognosis (11).

Genetics

In families with generalized epilepsy with febrile seizures plus and mutations in the sodium channel SCN1A patients with MAE have been reported (20). In rare cases, mutations in the glucose transporter 1 deficiency have been reported as a treatable cause of MAE (21), but this does not totally explain the favourable response to the ketogenic diet.

Childhood absence epilepsy

Clinical presentation

CAE is the most common type of GGE and accounts for 15% of childhood epilepsy (22). CAE occurs in normally developed children between the ages of 4 and 10, and is characterized by numerous absences per day (up to several hundreds per day). Absences are characterized by an abrupt onset and offset of reduced responsiveness. Most absences last between 4–20 s with a range from 1

Table 27.1 Overview of genetic generalized epilepsy syndromes

Syndrome	Onset (range)	Primary seizure type	Other seizure types	EEG background	EEG interictal/ictal	Photosensitivity
Myoclonic astatic epilepsy	2–6 years	Myoclonic and astatic	GTCS, tonic, absences, non-convulsive status epilepticus	Normal Theta at 4–7 Hz max. over parietal	Irregular 2–3 Hz (poly) spike wave or polyspike wave	>50%
Childhood absence epilepsy	6–7 years (3–12 years)	Absences, simple or complex, lasting	GTCS in <20%	Normal ORIDA in 1/3 of children	*Interictal and ictal*: bilateral synchronous and symmetrical 3/s spike wave (2.5–5 Hz), focal discharges in 15%	18–40%
Epilepsy with myoclonic absences	5–10 years	Absences with marked myoclonus of upper limbs	GTCS, typical absences, drop attacks	Normal	Bilateral synchronous and symmetrical 3/s (poly) spike wave (2.5–5 Hz),	14%
Eyelid myoclonia with absences	2–14 years	Myoclonic jerks of the eyelids, often with absences	GTCS	Normal	Bilateral synchronous 3–6 Hz and polyspike wave discharges	100%
Juvenile absence epilepsy	10–14 years (7–16 years)	Absences	GTCS common, myoclonic jerks	Normal	Bilateral synchronous, fairly irregular 3.5–4 Hz spike wave discharges	20%
Juvenile myoclonic epilepsy	12–18 years (8–26 years)	Irregular bilateral myoclonic jerks	GTCS in 90%, absences in 30%	Normal	Irregular, 3.5–6 Hz polyspike and wave discharges, focal discharges in 30%	30–50% (up to 90%)
Epilepsy with generalized tonic-clonic seizures on awakening	9–25 years	GTCS especially within of 2 h of waking		Normal	Bilateral synchronous regular or irregular 3.5–6 Hz spike and wave activity, focal discharges in 5%	30%

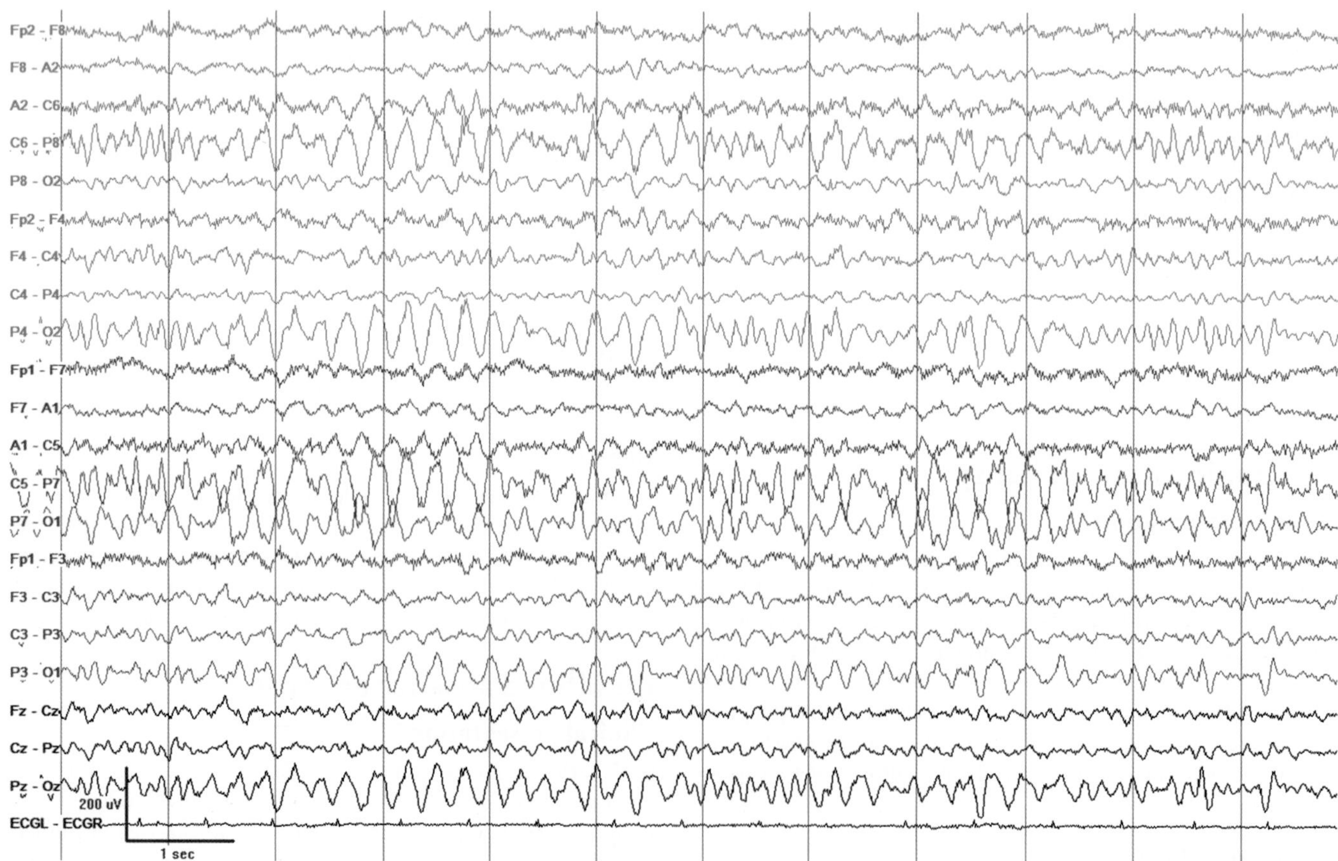

Fig. 27.1 Rhythmic theta activity most prominent over the posterior regions in a 4-year-old boy with MAE. The colour red represents right-sided electrodes and blue left-sided electrodes, whereas midline electrodes are displayed in black.

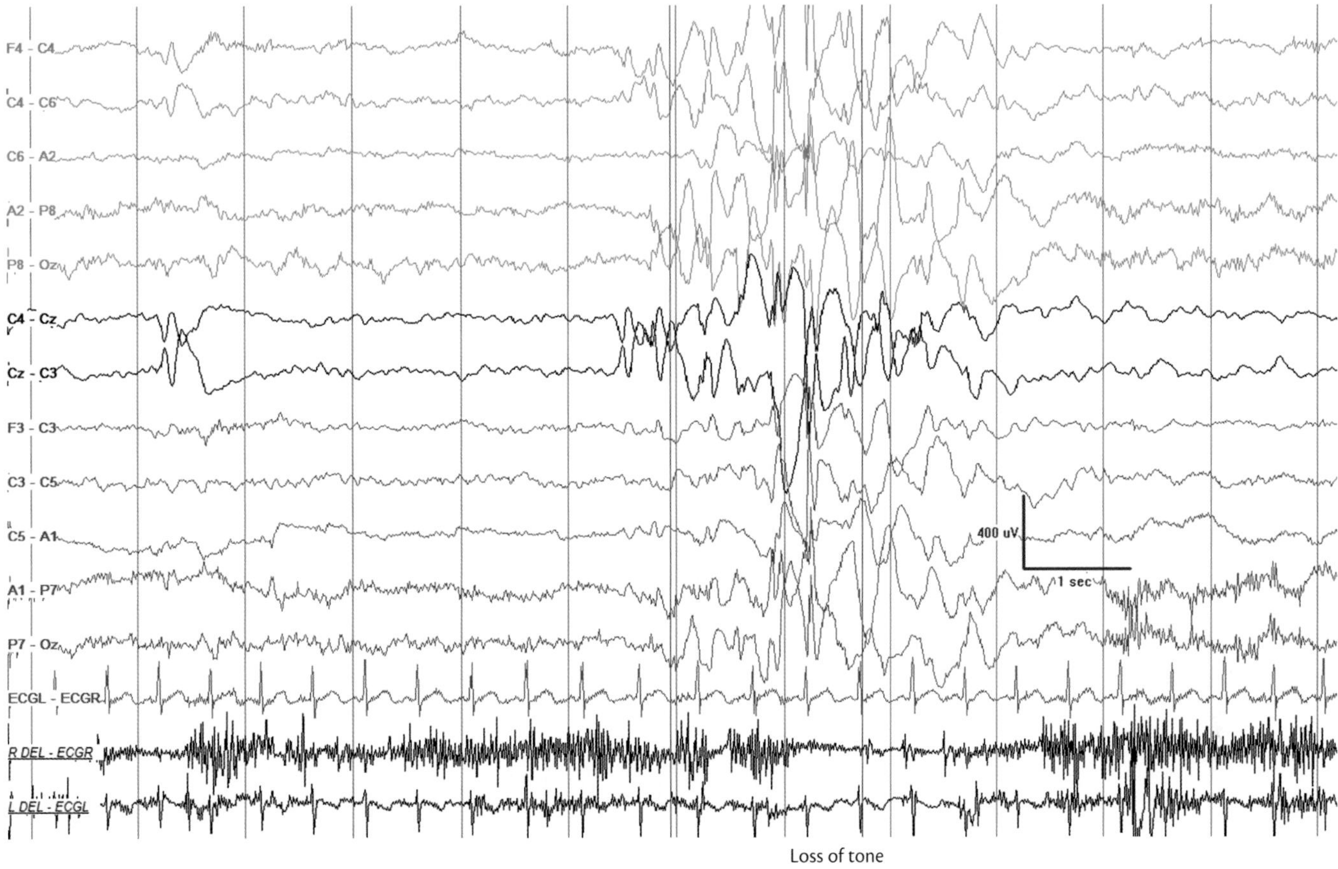

Fig. 27.2 Generalized polyspike and wave discharges associated with loss of tone (see EMG) in a 3-year old girl with MAE. The colour red represents right-sided electrodes and blue left-sided electrodes, whereas midline electrodes are displayed in black.

to 45 s (23). The degree of unresponsiveness may vary; especially in short events specialized testing is needed to detect clinical deficits. Subtle motor phenomena (especially eyelid movements and occasionally also mild myoclonic movements in the lips or limbs) may be seen in some patients. Automatisms occur in around 25% of children. Generalized tonic clonic seizures are not common, but were described in 16% of the patient with CAE (24).

Electroencephalogram

The ictal EEG shows, high amplitude (200–300 μV) generalized spike and wave discharges (GSW), with an amplitude maximum in the frontal midline (Fig. 27.3). The frequency is typically 3 Hz, but can range from 2.5 to 5 Hz. The initial frequency of the absences is slightly faster and slows towards the end of the absence. Typically, absence seizures arise from a normal EEG background. During drowsiness and sleep polyspike wave discharges are seen. Lateralized onset of discharges is not uncommon, but this should not be consistently seen from one side. Focal discharges may occur in 15% of the patients (23). Occipital intermittent rhythmic delta activity (OIRDA) consisting of bursts of rhythmic delta activity most prominent over the occipital region is seen in around one-third of children with CAE, typically in drowsiness and diminishing in sleep. However, OIRDA is not specific for CAE and may rarely be also seen in other GGE syndromes. Hyperventilation induces absences in almost 70% of the children (25). Photosensitivity is reported in 18–40% of CAE (2,26). This discrepancy could be due to population differences, AED therapy, sleep deprivation, intermittent photic stimulation (IPS) technique, and the classification of photo paroxysmal response (PPR) (27).

Treatment

A randomized double blind multicentre study compared ethosuximide, valproic acid, and lamotrigine as initial monotherapy in CAE. Ethosuximide and valproic acid were more effective in controlling seizures compared with lamotrigine. Valproic acid was associated with a higher rate of adverse events leading to drug discontinuation, as well as significant higher rate of negative effects on attention, which was not seen in the cohort treated with ethosuximide (28).

Prognosis

Children with CAE have a good prognosis; 65% become seizure-free (22). Patients with a history of longer lasting absences pretreatment are more likely to achieve seizure freedom (29). Generalized tonic clonic seizures (GTCS) seem to be associated with less favourable prognosis (25).

Genetics

In rare families CAE has been associated with $GABA_A$ receptor genes *GABRA1* (30) and *GABRG2* (31), the calcium channel gene *CACNA1H* (32) and the nicotinic acetylcholine gene *CHRNA4* (33).

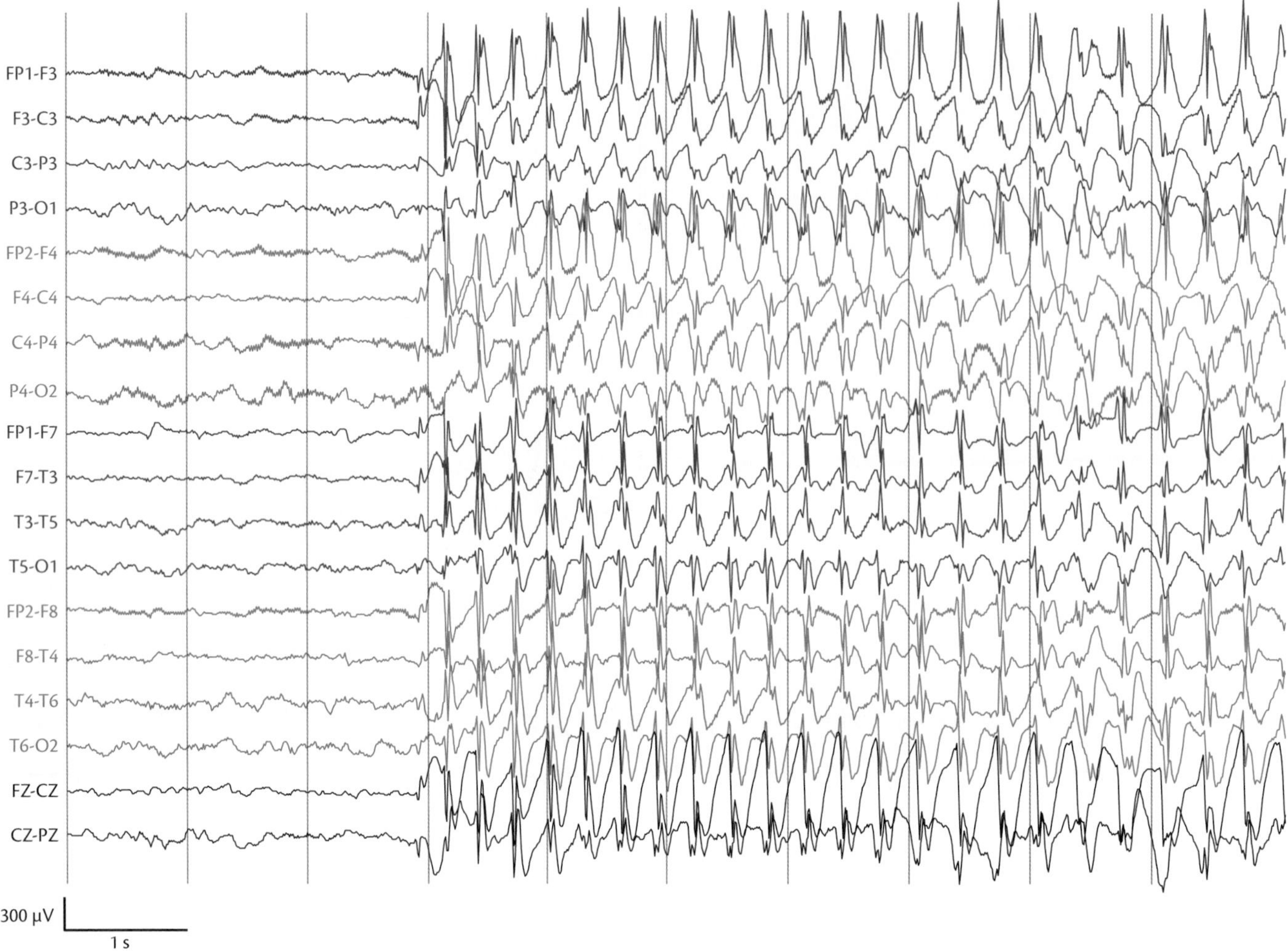

Fig. 27.3 Ictal recording of an absence seizure in a 7-year-old boy with CAE. The spike and wave discharges show a frontal emphasis. The colour red represents right-sided electrodes and blue left-sided electrodes, whereas midline electrodes are displayed in black.

In patients with early onset absence epilepsy (onset before the age of 4 years) mutations in the *SLC2A1* gene, encoding the glucose transporter (GLUT1), are found in approximately 10% of cases (34). These children will benefit from treatment with ketogenic diet.

Epilepsy with myoclonic absences

Clinical presentation

EMA is a rare GGE syndrome with onset usually around the age of 7 years (range 1–12 years). The patients show frequent absences a day lasting up to 60 s. These absences are accompanied by myoclonic jerks of the shoulders and arms, often with a tonic stiffening of the upper arm and progressive elevation of the arms. In two-thirds of the patients other seizures occur, most often GTCS, but typical absences or drop attacks have also been described. Intellectual disability is noted in 45% of children before seizure onset, and deterioration of cognitive function after seizure onset is common (35).

Electroencephalogram

As for the CAE, the EEG shows a normal background. Interictally, GSW and polyspike and wave discharges are seen. Similar to CAE the ictal EEG is characterized by symmetrical and regular 3 Hz GSW (ranging from 2.5 to 5 Hz), which occur in strict relation with myoclonia recorded on EMG (Fig. 27.4). Photic stimulation triggers myoclonic absences in 14% of patients (35).

Treatment

There are no controlled trials reported on the treatment of myoclonic absences. Standard first line therapy (valproate, ethosuximide) is less effective than in CAE patients (36). Lamotrigine seem to be effective in some children and does not aggravate the myoclonia. Second line treatment includes topiramate, levetiracetam, benzodiazepam, or zonisamide.

Prognosis

Seizure freedom is often more difficult to achieve than in CAE. Approximately 40% will become seizure free. Patients in whom myoclonic seizures are the only seizure type seem to have a better prognosis. Patients with therapy-refractory seizures may develop psychomotor retardation and show features of Lennox Gastaut syndrome (36).

Genetics

The genetics of EMA is currently unknown. In about 20% there is a family history of epilepsy for GGE. Chromosomal

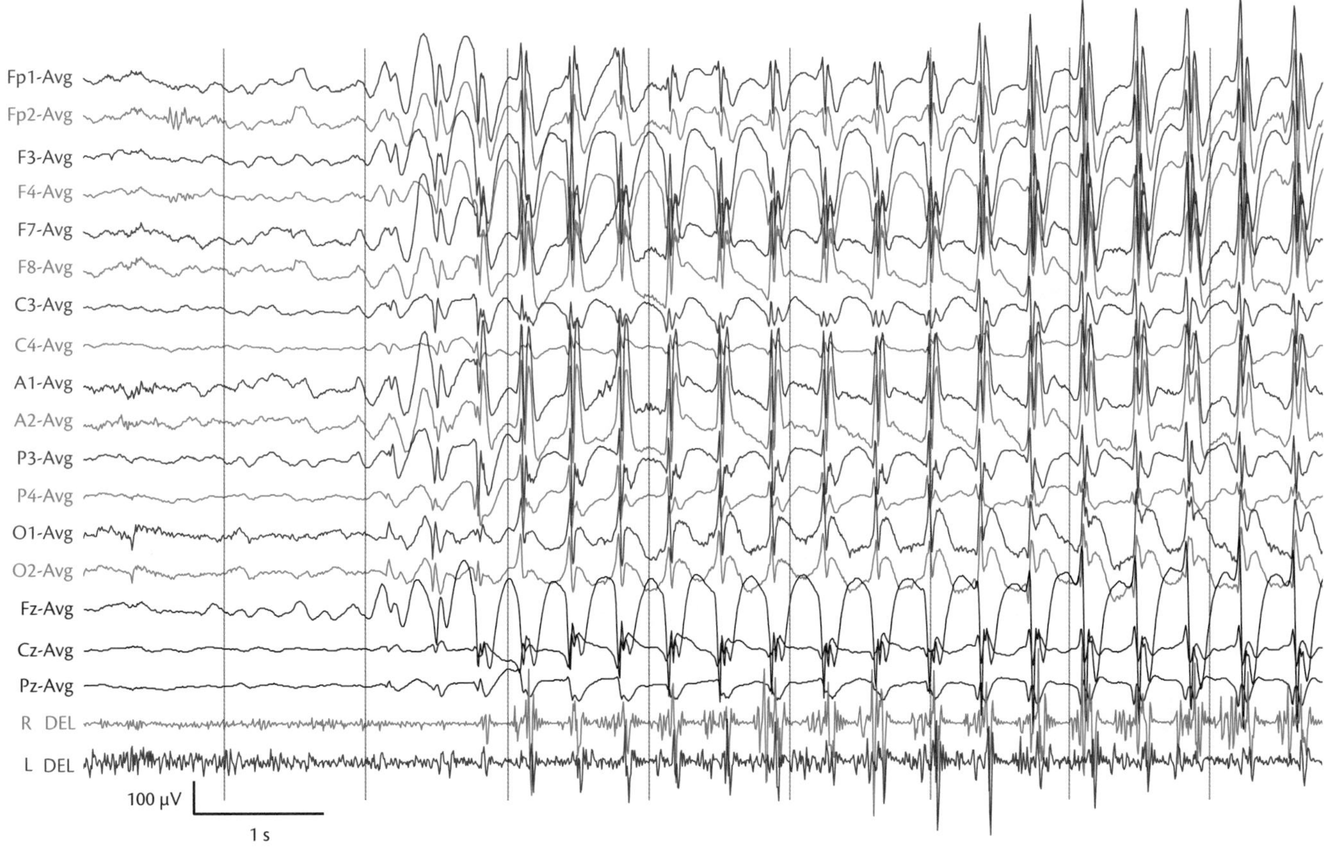

Fig. 27.4 Ictal recording of myoclonic absence seizure in a 6-year-old boy with EMA. The absence seizure is associated with rhythmical shoulder abduction. Deltoid EMG shows strict correlation with the 3 Hz GSW. The colour red represents right-sided electrodes and blue left-sided electrodes, whereas midline electrodes are displayed in black.

abnormalities have been described in single patients with partial trisomy 12p (37,38).

Eyelid myoclonia with absences

Clinical presentation

ELMA is another rare GGE syndrome that has not been formally recognized as a subgroup in the ILAE classification. It usually occurs in children aged 2–14 years and is characterized by brief (1–6 s) events with rhythmic fast (4–6 Hz) myoclonic jerks of the eyelids and simultaneous upward deviation of the eyeballs with extension of the head (39,40). These events may or may not be associated with impairment of consciousness (41). All patients show photosensitivity and many patients also develop GTCS, which may be induced by light (42). Most patients show normal development, but borderline intellectual functioning may occur (43).

Electroencephalogram

The EEG shows brief paroxysms of GSW usually 3–6 Hz and polyspike wave discharges accompanied by eyelid myoclonia (Fig. 27.5). The events are triggered by eye closure, photic stimulation, and hyperventilation. Photosensitivity is seen in all young children but decreases with age and antiepileptic treatment (42).

Treatment

Patients with ELMA are often resistant to treatment. Despite adequate antiepileptic treatment 83% of the patient continued to have eyelid myoclonia with or without absences (44). Valproate, ethosuximide, benzodiazepines are commonly used. A pilot trial multicentre, open label prospective study showed that levetiracetam is effective and well tolerated. Seizure freedom was seen in 17% of the patients, seizure reduction of 50%, or more was reported in 62% of the patients (45).

Prognosis

ELMA is considered a lifelong condition (42,43).

Genetics

Studies reporting on concordant monozygotic twins and epilepsy in relatives of patients suggest a genetic aetiology to ELMA (40,46). However, no mutations associated with ELMA have been described to date.

Juvenile absence epilepsy

Clinical presentation

JAE is less common than CAE. In patients with JAE seizure onset usually is at the age of 10 or later. Absence seizures are less

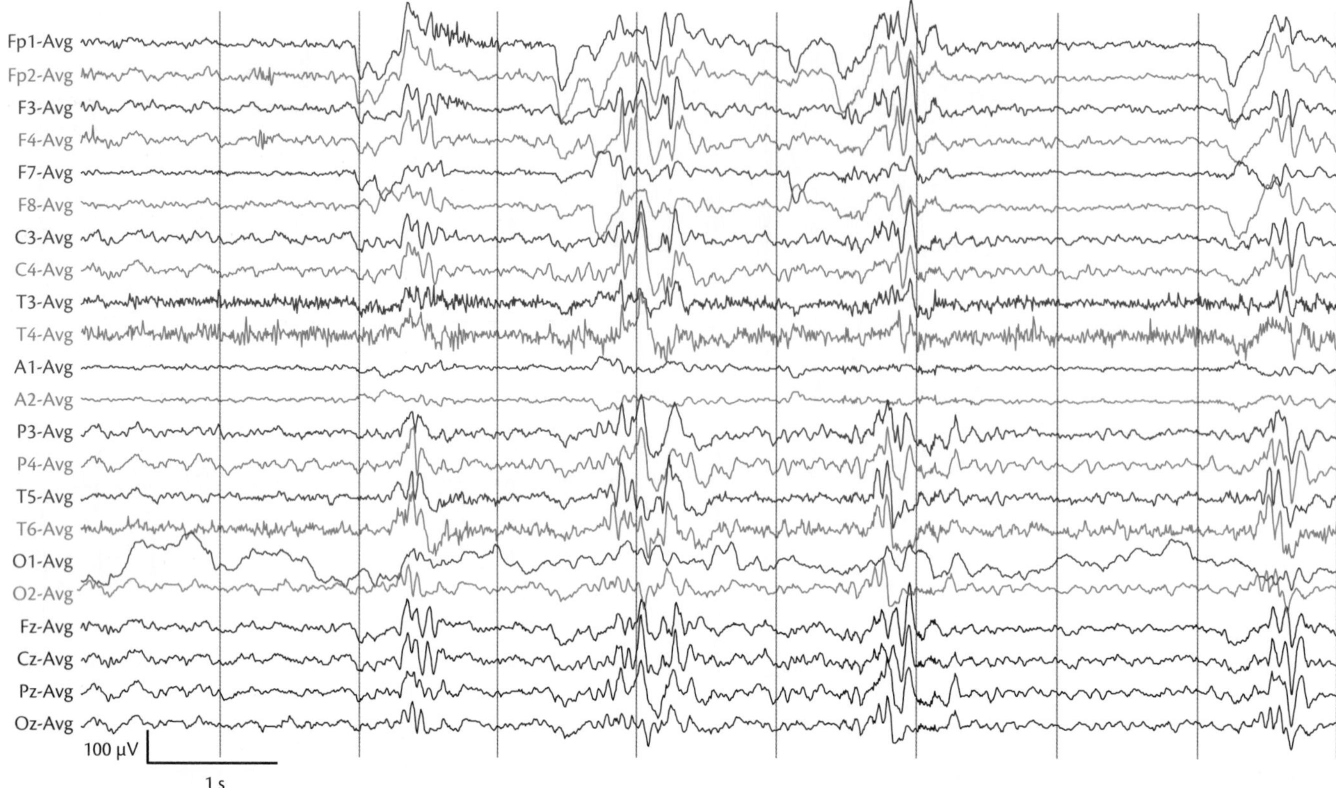

Fig. 27.5 Generalized polyspike wave discharges accompanied by eyelid myoclonia in 8-year-old girl with ELMA. The colour red represents right-sided electrodes and blue left-sided electrodes, whereas midline electrodes are displayed in black.

frequently seen than in CAE and may not occur daily. Generalized tonic clonic seizures are more common than in CAE and also sporadic myoclonic seizure may be seen (47).

Electroencephalogram

The background activity is normal. During absence seizures the frequency of generalized spike wave discharges is slightly higher (3.5–4 Hz) than in CAE and often less regular. During drowsiness and sleep polyspike wave discharges are seen (48). The discharges tend to be longer than in childhood absence epilepsy, yet loss of consciousness is often less complete.

Treatment

There are no randomized clinical trials regarding antiepileptic treatment in JAE. As first line treatment ethosuximide, valproate, or lamotrigine are used. Second line treatments include levetiracetam, topiramate, and zonisamide.

Prognosis

While Wolf and Inoue reported that 84% of patients with JAE responded to treatment (49) only around 40% of the patients become seizure free in the study of Tovia and colleagues. Patients without GTCS show better seizure control (50).

Genetics

Mutations in the *EFHC1* gene and the chloride channel gene *CLCN2* gene have been described in JAE, but also in other GGE syndromes (51,52). A genetic overlap exists between JAE, CAE, JME, and epilepsy with generalized tonic clonic seizures on awakening (EGTCSA) since these different phenotypes of GGE may exist in one family (53).

Juvenile myoclonic epilepsy

Clinical presentation

JME was first described by Janz and is therefore also known as Janz syndrome (54). It accounts for 5–10% of all epilepsies and usually starts around puberty (54,55). It is characterized by seizures with bilateral jerks, predominantly in the arms, which may cause some patients to fall suddenly. Consciousness is often not impaired during the jerks. While about 5% of the patients only experience myoclonic jerks more than 90% of the patients also show (GTCSs), and in about 30% of the patients absence seizures are present (56). The seizures usually occur shortly after awakening and are often precipitated by sleep deprivation. Further triggering factors may be stress, alcohol as well as photic stimulation, reading, or cognitive effort in some patients (55–58). Neurological examination in patients with JME is normal. However, detailed neuropsychological testing often shows impaired frontal lobe functions and psychiatric comorbidities (59,60).

Electroencephalogram

The interictal EEG shows generalized spike and polyspike wave discharge (3–6 Hz) arising from normal background, often high

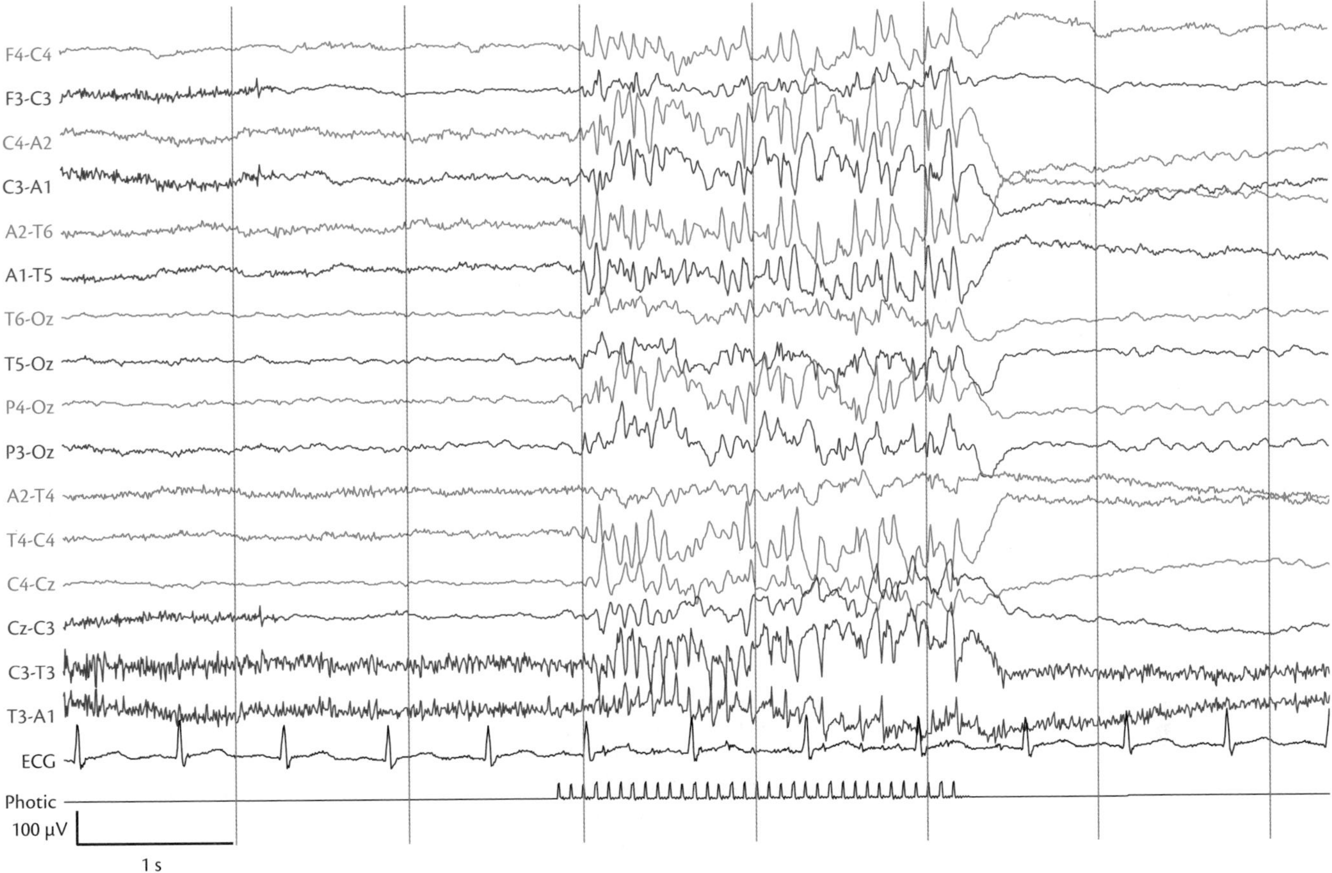

Fig. 27.6 Generalized polyspike discharges evoked by photic stimulation in a 13-year-old girl with JME. The colour red represents right-sided electrodes and, blue left-sided electrodes, whereas midline electrodes, ECG electrode, and photic stimulation are displayed in black.

voltage alpha activity. The generalized discharges usually have a frontal maximum. In comparison to CAE the discharges are more irregular and may be asymmetrical (56). Discharges may be provoked by sleep, hyperventilation and photic stimulation. Photosensitivity (Fig. 27.6) is common and has been described in 30% (26) up to 90% of the patients (1). The authors discussed that the discrepancy could be due to differences in age and design of photic stimulation (1). Brief focal discharges, mainly over the frontal areas, are seen in 30% of the patients (55). Myoclonic jerks are associated with high amplitude generalized spike or polyspike wave activity of 10–16 Hz with frontocentral emphasis, which may be followed by 1–3 Hz slow waves (Fig. 27.7) (56).

Treatment

Valproate is effective in about 90% of the patients (56), but should be avoided where possible in young women of child-bearing age due to weight gain, teratogenicity, and specifically the risk of impaired cognitive function in children exposed to valproate during pregnancy (61). Lamotrigine is less effective than valproate, but is a good alternative for young women (62). Levetiracetam is effective in controlling GTCS and myoclonic jerks (63), and is therefore a further suitable alternative. Topiramate seems to be effective for GTCS, but less effective for myoclonic jerks and absences (64). Carbamazepine, phenytoin, and vigabatrin can provoke seizures and should be avoided (65).

Prognosis

JME is considered a lifelong condition. Adequate antiepileptic treatment leads to seizure freedom in most patients. However, if antiepileptic medication is stopped almost 90% of the patients will have recurrence of seizures even after long seizure-free periods (56). A recent study suggests that about 20% of the patients may remain seizure free without medication (66).

Genetics

About 50% of the patients have a family history of epilepsy (56). In a large family with patients with JME a mutation in the $GABA_A$ receptor gene *GABRA1* has been described, which was inherited in an autosomal dominant fashion (67). However, in the majority of the patients the inheritance seems to be polygenic. Mutations in the *EFHC1* gene and the *BRD2* gene have been described in patients with JME (68,69).

Epilepsy with generalized tonic clonic seizures on awakening

Clinical presentation

EGTCSA usually starts in the second decade of life and is characterized by GTCS seizures occurring exclusively or predominantly within 2 h after awakening. Seizures are often provoked by sleep

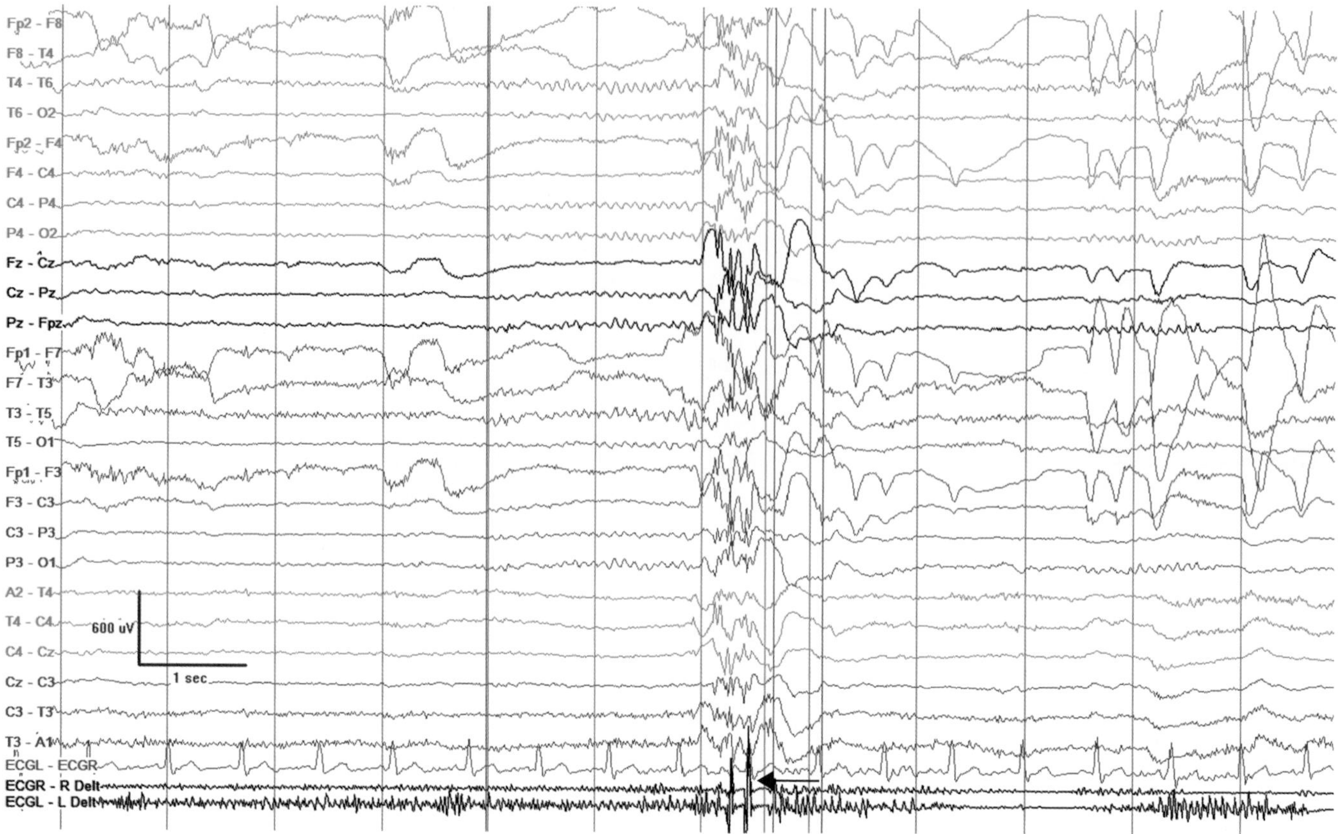

Fig. 27.7 Generalized polyspike wave discharges associated with a bilateral jerk of the upper limb (seen as activation of the deltoid EMG indicated by the black arrow) in a 17-year-old boy with JME. The colour red represents right, and blue left-sided electrodes whereas midline electrodes and EMG electrodes are displayed in black.

deprivation and alcohol consumption. Absences or myoclonic seizures may occur (70). In the revised ILAE classification from 1989 the syndrome was expanded as 'epilepsy with generalized tonic clonic seizures only' excluding patients with other seizure types (3).

Electroencephalogram

The EEG shows generalized fast spike and polyspike wave discharges similar to JME. Generalized discharges are reported in 40–90% of the patients (70,71). Focal discharges are rare and are seen in approximately 3% of the patients (70). About 30% of the patients are photosensitive (71).

Treatment

Valproate is often first line therapy, however, especially in young women lamotrigine or levetiratcetam may be preferred due to fewer side effects. Alternative treatment options are topiramate or zonisamide.

Prognosis

Like JME, EGTCSA is considered a lifelong condition.

Genetics

In single patients with mutations in the EFHC1 gene have been described (72). This mutation, however, was not specific for this syndrome but was also found in patients with CAE, JAE, or JME.

Pathophysiology of genetic generalized epilepsy

Studies in humans and animals have shown that, unlike generalized discharges might suggest, not the entire brain but specific networks are involved in GGE, while other networks are spared. Animal models showed that absences involve a network consisting of cortex, thalamus and basal ganglia (73), but are triggered by a cortical focus (74). In humans combined recordings of EEG and functional MRI (EEG-fMRI) allow to study structures involved in GGE non-invasively. Previous EEG-fMRI studies in absences showed that thalamus, areas of the default mode resting network and caudate nuclei are associated with absence seizures in children with newly diagnosed CAE (Fig. 27.8) (75). GSW-associated decrease in BOLD signal in default mode areas may indicate a disturbance of this physiological resting activity and may contribute to impaired consciousness during absence seizures (76). Early patient-specific blood oxygenation level dependent (BOLD) signal changes could mirror a cortical focus (77). Although the traditional concept of generalized seizures implies whole brain involvement, data from such studies have led to the revised definition of generalized seizures; namely, that they originate at some point within and rapidly engage bilaterally distributed networks and can include cortical and subcortical structures but not necessarily the entire cortex (4).

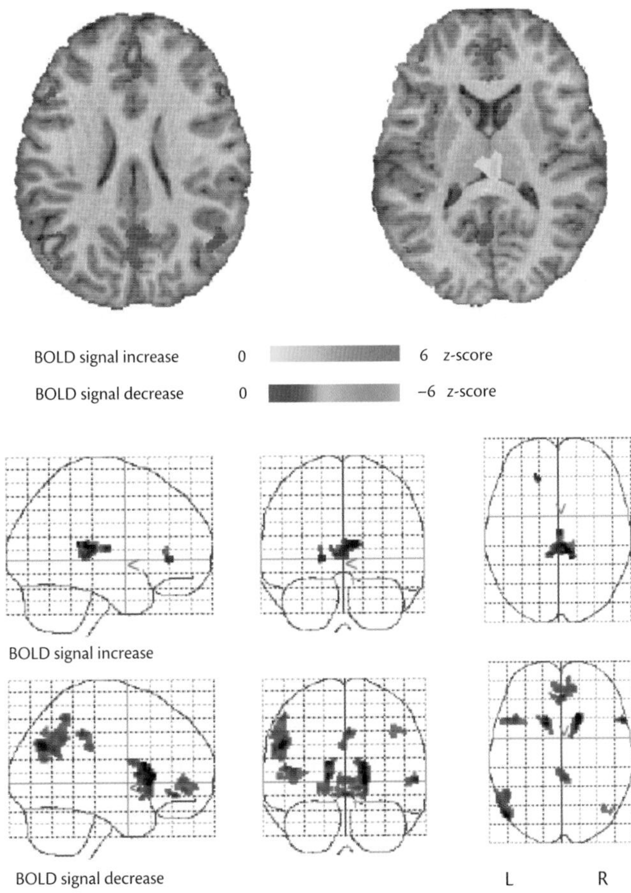

Fig. 27.8 EEG-fMRI investigation in children with absence seizures: positive BOLD signal changes (yellow-red) in the thalamus, and decreases in BOLD signal (green-blue) in parietal areas, precuneus, and in the caudate nucleus are found during absence seizures.
Reproduced from *Epilepsia*, **49**(9), Moeller F, Siebner HR, Wolff S, Muhle H, Granert O, Jansen O, Stephani U, Siniatchkin M, Simultaneous EEG-fMRI in drug-naive children with newly diagnosed absence epilepsy, pp. 1510–19, copyright (2008), with permission from John Wiley and Sons.

Alteration within the motor network might predispose to myoclonic seizures. Patients with JME show an increased excitability of the motor cortex especially in the morning when myoclonic jerks occur more often (78). An EEG-fMRI study in patients with MAE showed a network in the motor system and putamen associated with generalized spike wave discharges (Fig. 27.9) (79). This may indicate that a dysfunction in the motor network may lead to myoclonic seizures observed in these groups of patients. Avanzini and colleagues proposed the term 'system epilepsy', suggesting that some types of epilepsy may depend on the dysfunction of a specific brain network (80).

References

1. Appleton, R., Beirne, M., and Acomb, B. (2000). Photosensitivity in juvenile myoclonic epilepsy. *Seizure*, **9**(2), 108–11.
2. Lu, Y., Waltz, S., Stenzel, K., Muhle, H., and Stephani, U. (2008). Photosensitivity in epileptic syndromes of childhood and adolescence. *Epileptic Disorders*, **10**(2), 136–43.
3. Engel, J., Jr, and the International League Against Epilepsy (ILAE). (2001). A proposed diagnostic scheme for people with epileptic seizures and with epilepsy: report of the ILAE Task Force on Classification and Terminology. *Epilepsia*, **42**(6), 796–803.
4. Berg, A.T., Berkovic, S.F., Brodie, M.J., et al. (2010). Revised terminology and concepts for organization of seizures and epilepsies: report of the ILAE Commission on Classification and Terminology, 2005–2009. *Epilepsia*, **51**(4), 676–85.
5. Gallentine, W.B. and Mikati, M.A. (2012). Genetic generalized epilepsies. *Journal of Clinical Neurophysiology*, **29**(5), 408–19.
6. Helbig, I., Mefford, H.C., Sharp, A.J., et al. (2009). 15q13.3 microdeletions increase risk of idiopathic generalized epilepsy. *Nature: Genetics*, **41**(2), 160–2.
7. Jallon, P., Loiseau, P., and Loiseau, J. (2001). Newly diagnosed unprovoked epileptic seizures: presentation at diagnosis in CAROLE study. Coordination Active du Réseau Observatoire Longitudinal de l' Epilepsie. *Epilepsia*, **42**(4), 464–75.
8. Kelley, S.A. and Kossoff, E.H. (2010). Doose syndrome (myoclonic-astatic epilepsy): 40 years of progress. *Developmental Medicine and Child Neurology*, **52**, 988–93.

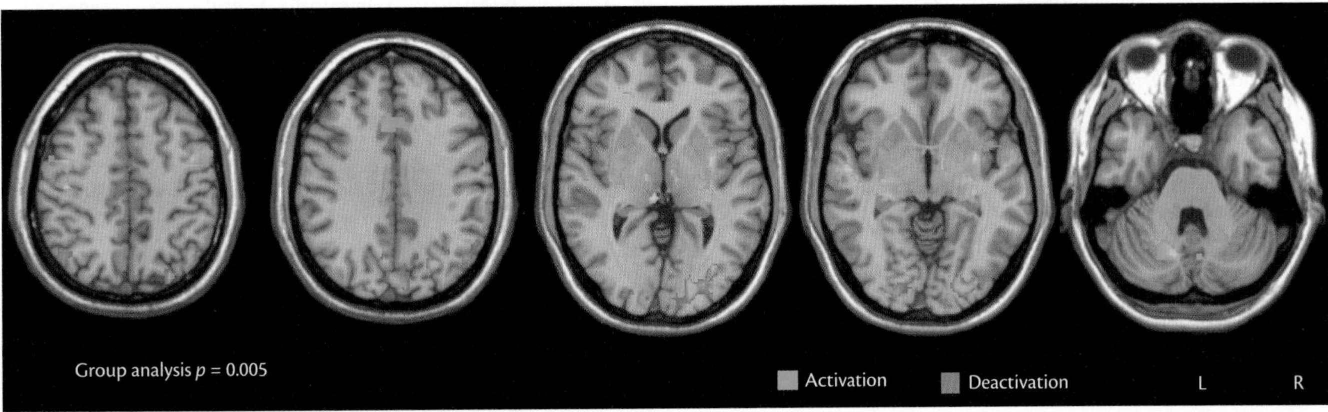

Fig. 27.9 EEG-fMRI study of generalized discharges in patients with MAE. Positive BOLD signal changes (orange) are found in thalamus, premotor cortex, putamen, and cerebellum, whereas deactivations (blue) are found in in the default mode network.
Reproduced from *Neurology*, **82**(17), Moeller F, Groening K, Moehring J, Muhle H, Wolff S, Jansen O, Stephani U, Siniatchkin M, EEG-fMRI in myoclonic astatic epilepsy (Doose syndrome), pp. 1508–13, copyright (2014), with permission from Wolters Kluwer Health, Inc.

9. Doose, H., Gerken, H., Leonhardt, R., Volzke, E., and Volz, C. (1970). Centrencephalic myoclonic astatic petit mal. Clinical and genetic investigation. *Neuropadiatrie*, **2**, 59–78.
10. Doose, H. and Baier, W.K. (1987). Epilepsy with primarily generalized myoclonic-astatic seizures: a genetically determined disease. *European Journal of Pediatrics*, **146**, 550e4.
11. Doose, H. (1992). Myoclonic-astatic epilepsy. *Epilepsy Research*, **6**(Suppl.), 163–8.
12. Dravet, C., Guerrini, R., Bureau, M. (1997): Epileptic syndromes with drop seizures in children. In: Beaumanoir, A., Andermann, F., Avanzini, G., Mira, L. (Eds) Falls in epileptic and non-epileptic seizures during childhood, pp. 95–111. London: John Libbey.
13. Panayiotopoulos, C.P., Ferrie, C.D., Knott, C., and Robinson, R.O. (1993). Interaction of lamotrigine with sodium valproate. *Lancet*, **341**, 445.
14. Oguni, H., Tanaka, T., Hayashi, K., et al. (2002). Treatment and long-term prognosis of myoclonic-astatic epilepsy of early childhood. *Neuropediatrics*, **33**, 122e32.
15. Kilaru, S. and Bergqvist, A.G. (2007). Current treatment of myoclonic astatic epilepsy: clinical experience at the Children's Hospital of Philadelphia. *Epilepsia*, **48**, 1703e7.
16. Doege, C., May, T.W., Siniatchkin, M., von Spiczak, S., Stephani, U., and Boor, R. (2013). Myoclonic astatic epilepsy (Doose syndrome)—a lamotrigine responsive epilepsy? *European Journal of Paediatric Neurology*, **17**(1), 29–35.
17. Machado, V.H., Palmini, A., Bastos, F.A., and Rotert, R. (2011). Long-term control of epileptic drop attacks with the combination of valproate, lamotrigine, and a benzodiazepine: a 'proof of concept', open label study. *Epilepsia*, **52**, 1303e10.
18. Guerrini, R., Bonanni, P., Rothwell, J., Hallett, M. (2002): Myoclonus and epilepsy. In: Guerrini, R., Aicardi, J., Andermann, F., Hallett, M. (eds) Epilepsy and movement disorder, pp. 165–210. Cambridge University Press.
19. Stephani, U. (2006). The natural history of myoclonic astatic epilepsy (Doose syndrome) and Lennox Gastaut syndrome. *Epilepsia*, **47**, 53e5.
20. Scheffer, I.E. and Berkovic, S.F. (1997). Generalized epilepsy with febrile seizures plus. A genetic disorder with heterogeneous clinical phenotypes. *Brain*, **120**(Pt 3), 479–90.
21. Mullen, S.A., Marini, C., Suls, A., et al. (2011). Glucose transporter 1 deficiency as a treatable cause of myoclonic astatic epilepsy. *Archives of Neurology*, **68**, 1152e5.
22. Callenbach, P.M., Bouma, P.A., Geerts, A.T., et al. (2009). Long-term outcome of childhood absence epilepsy: Dutch study of epilepsy in childhood. *Epilepsy Research*, **83**(2–3), 249–56.
23. Sadleir, L.G., Farrell, K., Smith, S., Connolly, M.B., and Scheffer, I.E. (2006). Electroclinical features of absence seizures in childhood absence epilepsy. *Neurology*, **67**(3), 413–18.
24. Loiseau, P., Duché, B., and Pédespan, J.M. (1995). Absence epilepsies. *Epilepsia*, **36**(12), 1182–86.
25. Wirrell, E.C., Camfield, C.S., Camfield, P.R., Gordon, K.E., and Dooley, J.M. (1996). Long-term prognosis of typical childhood absence epilepsy: remission or progression to juvenile myoclonic epilepsy. *Neurology*, **47**(4), 912–18.
26. Wolf, P. and Goosses, R. (1986). Relation of photosensitivity to epileptic syndromes. *Journal of Neurology, Neurosurgery, & Psychiatry*, **49**(12), 1386–91.
27. Seneviratne, U., Cook, M., and D'Souza, W. (2012). The electroencephalogram of idiopathic generalized epilepsy. *Epilepsia*, **53**(2), 234–48.
28. Glauser, T.A., Cnaan, A., Shinnar, S., et al. (2010). Ethosuximide, valproic acid, and lamotrigine in childhood absence epilepsy. *New England Journal of Medicine*, **362**(9), 790–9.
29. Dlugos, D., Shinnar, S., Cnaan, A., et al. (2013). Pretreatment EEG in childhood absence epilepsy: associations with attention and treatment outcome. *Neurology*, **81**(2), 150–6.
30. Maljevic, S., Krampfl, K., Cobilanschi, J., et al. (2006). A mutation in the GABA(A) receptor alpha(1)-subunit is associated with absence epilepsy. *Annals of Neurology*, **59**(6), 983–7.
31. Wallace, R.H., Marini, C., Petrou, S., et al. (2001). Mutant GABA(A) receptor gamma2-subunit in childhood absence epilepsy and febrile seizures. *Nature: Genetics*, **28**(1), 49–52.
32. Chen, Y., Lu, J., Pan, H., et al. (2003). Association between genetic variation of CACNA1H and childhood absence epilepsy. *Annals of Neurology*, **54**(2), 239–43.
33. Steinlein, O., Sander, T., Stoodt, J., Kretz, R., Janz, D., and Propping, P. (1997). Possible association of a silent polymorphism in the neuronal nicotinic acetylcholine receptor subunit alpha4 with common idiopathic generalized epilepsies. *American Journal of Medical Genetics*, **74**(4), 445–9.
34. Suls, A., Mullen, S.A., Weber, Y.G., et al. (2009). Early-onset absence epilepsy caused by mutations in the glucose transporter GLUT1. *Annals of Neurology*, **66**(3), 415–19.
35. Bureau, M. and Tassinari, C.A. (2005). Epilepsy with myoclonic absences. *Brain Development*, **27**(3), 178–84. [Review.]
36. Genton, P. and Bureau, M. (2006). Epilepsy with myoclonic absences. *Central Nervous System Drugs*, **20**(11), 911–16. [Review.]
37. Guerrini, R., Bureau, M., Mattei, M.G., Battaglia, A., Galland, M.C., and Roger, J. (1990). Trisomy 12p syndrome: a chromosomal disorder associated with generalized 3-Hz spike and wave discharges. *Epilepsia*, **31**(5), 557–66.
38. Elia, M., Musumeci, S.A., Ferri, R., and Cammarata, M. (1998). Trisomy 12p and epilepsy with myoclonic absences. *Brain Development*, **20**(2), 127–30.
39. Jeavons, P.M. (1977). Nosological problems of myoclonic epilepsies in childhood and adolescence. *Developmental Medicine and Child Neurology*, **19**(1), 3–8.
40. Giannakodimos, S. and Panayiotopoulos, C.P. (1996). Eyelid myoclonia with absences in adults: a clinical and video-EEG study. *Epilepsia*, **37**(1), 36–44.
41. Appleton, R.E., Panayiotopoulos, C.P., Acomb, B.A., and Beirne, M. (1993). Eyelid myoclonia with typical absences: an epilepsy syndrome. *Journal of Neurology, Neurosurgery, & Psychiatry*, **56**(12), 1312–16.
42. Panayiotopoulos, C.P. (2005). Syndromes of idiopathic generalized epilepsies not recognized by the International League Against Epilepsy. *Epilepsia*, **46**(Suppl. 9), 57–66. [Review.]
43. Striano, S., Capovilla, G., Sofia, V., et al. (2009). Eyelid myoclonia with absences (Jeavons syndrome): a well-defined idiopathic generalized epilepsy syndrome or a spectrum of photosensitive conditions? *Epilepsia*, **50**(Suppl. 5), 15–19.
44. Capovilla, G., Striano, P., Gambardella, A., et al. (2009). Eyelid fluttering, typical EEG pattern, and impaired intellectual function: a homogeneous epileptic condition among the patients presenting with eyelid myoclonia. *Epilepsia*, **50**(6), 1536–41.
45. Striano, P., Sofia, V., Capovilla, G., et al. (2008). A pilot trial of levetiracetam in eyelid myoclonia with absences (Jeavons syndrome). *Epilepsia*, **49**(3), 425–30.
46. Sadleir, L.G., Vears, D., Regan, B., Redshaw, N., Bleasel, A., and Scheffer, I.E. (2012). Family studies of individuals with eyelid myoclonia with absences. *Epilepsia*, **53**(12), 2141–8.
47. Janz, D. (1997). The idiopathic generalized epilepsies of adolescence with childhood and juvenile age of onset. *Epilepsia*, **38**(1), 4–11. [Review.]
48. Sadleir, L.G., Scheffer, I.E., Smith, S., Carstensen, B., Farrell, K., and Connolly, M.B. (2009). EEG features of absence seizures in idiopathic generalized epilepsy: impact of syndrome, age, and state. *Epilepsia*, **50**(6), 1572–8.
49. Wolf, P. and Inoue, Y. (1984). Therapeutic response of absence seizures in patients of an epilepsy clinic for adolescents and adults. *Journal of Neurology*, **231**(4), 225–9.
50. Tovia, E., Goldberg-Stern, H., Shahar, E., and Kramer, U. (2006). Outcome of children with juvenile absence epilepsy. *Journal of Child Neurology*, **21**(9), 766–8.

51. Stogmann, E., Lichtner, P., Baumgartner, C., et al. (2006). Idiopathic generalized epilepsy phenotypes associated with different EFHC1 mutations. *Neurology*, **67**(11), 2029–31.
52. Stogmann, E., Lichtner, P., Baumgartner, C., et al. (2006). Mutations in the CLCN2 gene are a rare cause of idiopathic generalized epilepsy syndromes. *Neurogenetics*, **7**(4), 265–8.
53. Delgado-Escueta, A.V., Greenberg, D., Weissbecker, K., et al. (1990). Gene mapping in the idiopathic generalized epilepsies: juvenile myoclonic epilepsy, childhood absence epilepsy, epilepsy with grand mal seizures, and early childhood myoclonic epilepsy. *Epilepsia*, **31**(Suppl. 3), S19–29. [Review.]
54. Janz, D. (1985). Epilepsy with impulsive petit mal (juvenile myoclonic epilepsy). *Acta Neurologica Scandinavica*, **72**(5), 449–59. [Review.]
55. Panayiotopoulos, C.P., Obeid, T., and Tahan, A.R. (1994). Juvenile myoclonic epilepsy: a 5-year prospective study. *Epilepsia*, **35**(2), 285–96.
56. Delgado-Escueta, A.V. and Enrile-Bacsal, F. (1984). Juvenile myoclonic epilepsy of Janz. *Neurology*, **34**(3), 285–94.
57. Matsuoka, H., Takahashi, T., Sasaki, M., et al. (2000). Neuropsychological EEG activation in patients with epilepsy. *Brain*, **123**(Pt 2), 318–30.
58. Guaranha, M.S., da Silva Sousa, P., de Araújo-Filho, G.M., et al. (2009). Provocative and inhibitory effects of a video-EEG neuropsychologic protocol in juvenile myoclonic epilepsy. *Epilepsia*, **50**(11), 2446–55.
59. Trinka, E., Kienpointner, G., Unterberger, I., et al. (2006). Psychiatric comorbidity in juvenile myoclonic epilepsy. *Epilepsia*, **47**(12), 2086–91.
60. Piazzini, A., Turner, K., Vignoli, A., Canger, R., and Canevini, M.P. (2008). Frontal cognitive dysfunction in juvenile myoclonic epilepsy. *Epilepsia*, **49**(4), 657–62.
61. Meador, K.J., Penovich, P., Baker, G.A., et al. (2009). Antiepileptic drug use in women of childbearing age. *Epilepsy & Behaviour*, **15**(3), 339–43.
62. Marson, A.G., Al-Kharusi, A.M., Alwaidh, M., et al. (2007). The SANAD study of effectiveness of valproate, lamotrigine, or topiramate for generalised and unclassifiable epilepsy: an unblinded randomised controlled trial. *Lancet*, **369**(9566), 1016–26.
63. Berkovic, S.F., Knowlton, R.C., Leroy, R.F., Schiemann, J., Falter, U., and Levetiracetam N01057 Study Group. (2007). Placebo-controlled study of levetiracetam in idiopathic generalized epilepsy. *Neurology*, **69**(18), 1751–60.
64. Biton, V. and Bourgeois, B.F., and the YTC/YTCE Study Investigators. (2005). Topiramate in patients with juvenile myoclonic epilepsy. *Archives of Neurology*, **62**(11), 1705–8.
65. Thomas, P., Valton, L., and Genton, P. (2006). Absence and myoclonic status epilepticus precipitated by antiepileptic drugs in idiopathic generalized epilepsy. *Brain*, **129**(Pt 5), 1281–92.
66. Senf, P., Schmitz, B., Holtkamp, M., and Janz, D. (2013). Prognosis of juvenile myoclonic epilepsy 45 years after onset: seizure outcome and predictors. *Neurology*, **81**(24), 2128–33.
67. Cossette, P., Liu, L., Brisebois, K., et al. (2002). Mutation of GABRA1 in an autosomal dominant form of juvenile myoclonic epilepsy. *Nature: Genetics*, **31**(2), 184–9.
68. Pal, D.K., Evgrafov, O.V., Tabares, P., Zhang, F., Durner, M., and Greenberg, D.A. (2003). BRD2 (RING3) is a probable major susceptibility gene for common juvenile myoclonic epilepsy. *American Journal of Human Genetics*, **73**(2), 261–70.
69. Cavalleri, G.L., Walley, N.M., Soranzo, N., et al. (2007). A multicenter study of BRD2 as a risk factor for juvenile myoclonic epilepsy. *Epilepsia*, **48**(4), 706–12.
70. Janz, D. (2000). Epilepsy with grand mal on awakening and sleep-waking cycle. *Clinical Neurophysiology*, **111**(Suppl. 2), S103–10. [Review.]
71. Koutroumanidis, M., Aggelakis, K., and Panayiotopoulos, C.P. (2008). Idiopathic epilepsy with generalized tonic-clonic seizures only versus idiopathic epilepsy with phantom absences and generalized tonic-clonic seizures: one or two syndromes? *Epilepsia*, **49**(12), 2050–62.
72. Greenberg, D.A., Durner, M., Resor, S., Rosenbaum, D., and Shinnar, S. (1995). The genetics of idiopathic generalized epilepsies of adolescent onset: differences between juvenile myoclonic epilepsy and epilepsy with random grand mal and with awakening grand mal. *Neurology*, **45**(5), 942–6.
73. Blumenfeld, H. (2005). Cellular and network mechanisms of spike-wave seizures. *Epilepsia*, **46**(Suppl. 9), 21–33. [Review.]
74. Meeren, H.K., Pijn, J.P., Van Luijtelaar, E.L., Coenen, A.M., and Lopes da Silva, F.H. (2002). Cortical focus drives widespread corticothalamic networks during spontaneous absence seizures in rats. *Journal of Neuroscience*, **22**(4), 1480–95.
75. Moeller, F., Siebner, H.R., Wolff, S., et al. (2008). Simultaneous EEG-fMRI in drug-naive children with newly diagnosed absence epilepsy. *Epilepsia*, **49**(9), 1510–19.
76. Gotman, J., Grova, C., Bagshaw, A., Kobayashi, E., Aghakhani, Y., and Dubeau, F. (2005). Generalized epileptic discharges show thalamocortical activation and suspension of the default state of the brain. *Proceedings of the National Academy of Sciences, USA*, **102**(42), 15236–40.
77. Moeller, F., LeVan, P., Muhle, H., et al. (2010). Absence seizures: individual patterns revealed by EEG-fMRI. *Epilepsia*, **51**(10), 2000–10.
78. Badawy, R.A., Macdonell, R.A., Jackson, G.D., and Berkovic, S.F. (2009). Why do seizures in generalized epilepsy often occur in the morning? *Neurology*, **73**(3), 218–22.
79. Moeller, F., Groening, K., Moehring, J., et al. (2014). EEG-fMRI in myoclonic astatic epilepsy (Doose syndrome). *Neurology*, **82**(17), 1508–13.
80. Avanzini, G., Manganotti, P., Meletti, S., et al. (2012). The system epilepsies: a pathophysiological hypothesis. *Epilepsia*, **53**(5), 771–8.

CHAPTER 28

Focal epilepsy

Tim Wehner, Kanjana Unnwongse, and Beate Diehl

Introduction

Focal (partial, localization-related) epilepsy accounts for about two-thirds of epilepsies. As primary generalized epilepsies present commonly in the first two decades of life, the relative proportion of focal epilepsies increases in later life.

Diagnostic neurophysiological modalities in the evaluation of people with focal epilepsy are standard outpatient electroencephalography (EEG), ambulatory EEG, video EEG telemetry, intracranial EEG recordings, and magnetoencephalography (MEG). Although the latter two are used in a small minority of patients only when surgical candidacy is considered, these modalities have contributed much to our understanding of the neurophysiology of epilepsy.

In this chapter, we will discuss principles of electrophysiological source localization, and present interictal and ictal video EEG findings in various focal epilepsy syndromes. We will also review those EEG patterns of unclear significance ('normal variants') that are important when considering the differential diagnosis of epileptiform EEG abnormalities.

Electrophysiological source localization

The electrical activity measured by scalp EEG is thought to be the summated activity of post-synaptic potentials in the superficial cortical layers. The magnitude of an electrical potential at a given location is inversely related to the square of the distance from its generator. Thus, the activity reflected on scalp EEG is largely generated in the outermost layers of the neocortex. The size of the potential is further determined by the orientation of the source. A source that is orientated perpendicular to the scalp generates the highest potential, whereas a source of the same strength that is orientated parallel to the scalp generates no potential on scalp EEG. Thus, currents that flow perpendicular to the scalp surface in the gyral crowns contribute relatively more to the scalp EEG signal than currents of similar strength in the gyral sulci. Consequently, the scalp EEG signal is biased towards electrical activity of the lateral convexity (and against the activity of the interhemispheric fissure and brain surface), and within the convexity, towards the activity in gyral crowns (and against the signal in the gyral sulci). The EEG 'signal of interest' is attenuated by a factor of 5–10 by the anatomical structures between the scalp sensor and its generator—the meninges, cerebrospinal fluid, skull, cutis, and epidermis. It may be contaminated by internal and external artefacts. Internal artefacts are generated within the body. Examples are EMG activity of scalp muscles, ECG signal, pulse artefact, movements of the eyeball, eye lids, tongue, and the entire body. Examples of external artefacts, those generated in the recording equipment and the environment, are momentary impedance changes in the recording electrodes ('pops'), movement of the recording leads, and alternating current artefact (50 Hz in the UK).

The potential difference between two electrodes is amplified. One *channel* of digital scalp EEG displays the potential difference between two electrodes over time, with a resolution directly proportional to the sampling frequency (every ~4 ms at a sampling frequency of 256 Hz). By convention, an upwards deflection refers to a relatively higher negativity at the electrode on the left hand side of the channel annotation (or a relatively higher positivity at the electrode on the right-hand side of the channel annotation). A *montage* refers to the combination of several channels. A *bipolar montage* refers to a montage where adjacent channels share consecutive electrodes. Examples are the standard longitudinal bipolar montage ('double banana'), and the coronal transverse montage. A bipolar montage directly displays the potential difference between adjacent electrodes. In a *referential montage*, several channels share one common electrode (the reference electrode). The activity at the reference electrode is thus reflected in every channel of that montage. Digital EEG allows the post-acquisition formatting of montages and the generation of weighted average references (e.g. Laplacian montage, where each electrode is referred to the average of its surrounding neighbours).

Source localization in EEG attempts to solve the 'inverse problem'—what kind of electrical source generates a signal on a limited number of channels (e.g. on scalp EEG)? A single dipole is an idealized solution that allows the mapping of the peak electrical negativity (and sometimes positivity) on scalp EEG at a given time point. This is a useful concept in the mapping of sharp waves and spikes. Nonetheless, it is important to emphasize that a single dipole solution does not equate to the anatomical location of the epileptic 'focus'. It is rather an abstract simplification to visualize the distribution of the field potential. A *phase reversal* refers to a situation where there are opposite deflections in adjacent channels. In a *negative phase reversal*, the deflections point to each other. In a *positive phase reversal*, they point away from each other. A simple set of rules allows determination of the peak negativity (and positivity) of a single dipole solution to a given scalp EEG signal (Figs 28.7A, 28.8A, 28.10A, 28.12).

In a longitudinal bipolar montage, a negative phase reversal indicates the electrode location of the peak negativity, and a positive phase reversal indicates the electrode location of the peak positivity. If there is no phase reversal in a bipolar montage, the maximum negativity and maximum positivity are located at the electrodes at the end of the chain.

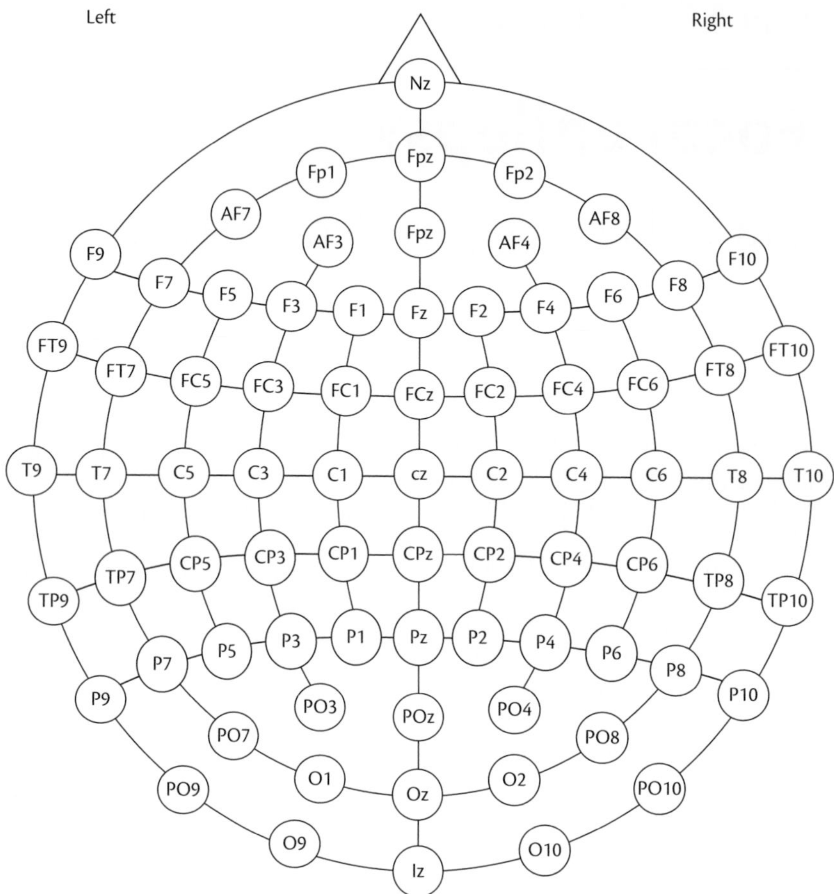

Fig. 28.1 EEG electrode positions according to international 10-10 nomenclature.

In a referential montage, the highest amplitude deflection indicates the electrode location of the peak negativity (upward deflection) or positivity (downward deflection), provided that there is no phase reversal. A phase reversal indicates that the reference electrode is not the relative electrical minimum and is thus potentially part of the field distribution that one is attempting to map.

It is important to recognize that *phase reversal* is a descriptive term indicative of where an EEG activity is maximum, and therefore does not per se imply any EEG pathology. For example, physiological vertex waves of stage I sleep typically display a negative phase reversal over the midline channels Fz and Cz.

Physiological EEG activity varies with age, level of awareness, and anatomical location. EEG findings in people with epilepsy thus can be compared to the patient's own baseline, as well as to healthy controls.

Fig. 28.1 provides a schematic of the international 10-10 system of EEG electrode positions that is used in the EEG samples of this chapter.

Non-specific interictal EEG abnormalities

Focal slowing can indicate a functional or structural abnormality in the respective location. As a general rule, the slower the activity, and the more rhythmic, and the higher the proportion of the record it occupies, the more likely it implies an underlying abnormality (Fig. 28.6A). Focal slowing is not specific for epilepsy and can be seen in underlying structural lesions, as well as in patients with migraine, concussion, neuropsychiatric disorders, and even in the absence of clinically evident neuropsychiatric disease. Intermittent temporal slow in the theta range is considered normal in people above the age of 40 years. Focal slowing can easily be mimicked by tongue or chewing movements.

A persistent amplitude asymmetry of >50% in a referential montage is considered abnormal. A relatively reduced amplitude may indicate an underlying dysfunction, for example due to the presence of an acute pathology. A relative increase in amplitude may be seen in the presence of a bone defect, e.g. if one or more electrodes are placed adjacent to a craniotomy scar or a burr hole. In this situation, faster EEG frequencies may be relatively less attenuated. This combination of relative amplitude increase and faster frequencies is also known as the 'breach effect' (Fig. 28.2, further discussed under 'Electrophysiological differential diagnosis of epileptiform abnormalities'). A relatively reduced amplitude and/or focal slowing may also be seen in the post-ictal period.

Interictal epileptiform abnormalities and ictal EEG patterns

These abnormalities indicate cortical irritability ('an increased liability to generate seizures') and are thus highly specific for epilepsy (~99% in young adults without structural neurological pathology, slightly lower in children and the elderly, (1)). Interictal epileptiform discharges reflect a sudden widespread change in neuronal

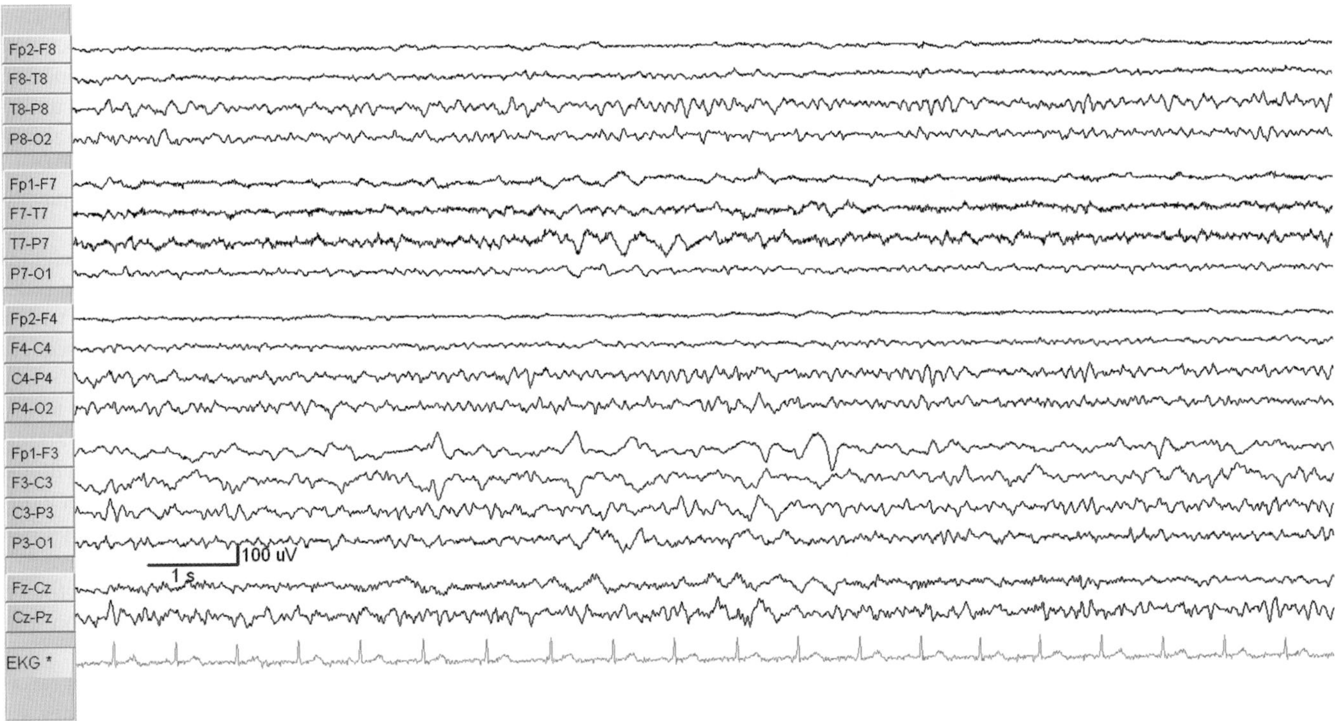

Fig. 28.2 Left frontotemporal breach activity in a 69-year-old woman 5 weeks following resection of a left frontal osseous meningioma. There was no bone flap over the left frontocentral area.

current distribution and thus mirror the fundamental pathophysiological mechanism in epilepsy. Nonetheless, it is important to emphasize that epilepsy is a clinical diagnosis that should not be made based on interictal EEG findings alone. The updated ILAE practical definition of epilepsy (2) has summarized currently accepted criteria:

- At least two unprovoked (or reflex) seizures occurring >24 h apart.
- One unprovoked (or reflex) seizure and a probability of further seizures similar to the general recurrence risk (at least 60%) after two unprovoked seizures, occurring over the next 10 years.
- Diagnosis of an epilepsy syndrome.

EEG plays an important part in assessing seizure recurrence risk, as abnormalities, especially epileptiform ones, will contribute to predict seizure recurrence risk and definition of epilepsy syndrome.

Ictal EEG patterns in focal epilepsy reflect the recruitment of cortical regions to participate in the seizure. The hallmark criteria for an EEG change to be considered ictal are thus evolution (i.e. progressive change) in frequency, amplitude, and distribution. A focal seizure by definition originates in a circumscribed cortical area. If only a small cortical area is recruited, no change may be detectable on scalp EEG, even though the patient may experience an altered sensation (aura) or exhibit objective behavioural changes (for example clonic jerks). 70–90% of focal seizures without impairment of awareness (auras or focal motor seizures) have no scalp EEG correlate. Simultaneous recordings of subdural and scalp EEG suggest that a cortical area of 6–10 cm^2 needs to be activated for a spike to show up on scalp EEG (3). The clinical behavioural change in a seizure thus may precede the detection of an ictal EEG change by many seconds or even minutes. This is particularly common in seizures originating in the hippocampus (Case 28.1). Ictal activity may spread contiguously along the cortical surface, or discontiguously via white matter tracts intra- and interhemispherically. In a study of nearly 500 seizures in 72 patients who became seizure-free following resective epilepsy surgery, the ictal onset pattern consisted of rhythmic delta activity in 29%, rhythmic theta in 25%, and rhythmic alpha in 3%. Repetitive epileptiform activity was observed in 16% of seizures, and paroxysmal fast (>13 Hz) activity in 14%. Suppression of background activity occurred in 8%, and arrhythmic activity in 3%. Rhythmic theta was more common in temporal than extratemporal lobe epilepsy, whereas paroxysmal fast and repetitive epileptiform activity were more common in extratemporal than temporal lobe epilepsy. Correctly localized seizures were more common in temporal than in extratemporal epilepsy. Generalized ictal patterns from the onset were more common in extratemporal epilepsy, in particular in mesial frontal lobe epilepsy, compared with temporal, parietal, and lateral frontal lobe epilepsy. False localization was seen in 28% of occipital lobe seizures and in 16% of parietal lobe seizures (4). Ictal scalp EEG patterns may rarely consistently lateralize the seizure onset to the opposite side. This phenomenon has been described in temporal lobe epilepsy with gross hemispheric lesions (5) and hippocampal sclerosis (6), as well as focal epilepsy with parasagittal lesions (7). In the former scenario, the ictal discharge may propagate to the contralateral hippocampus and evolve further over the contralateral temporal area, without manifesting over the ipsilateral temporal scalp electrodes. In patients with parasagittal lesions, paradoxical lateralization is thought to occur if the cortical source is largely situated on the mesial surface. In this scenario, its negative electrical activity projects obliquely to the contralateral parasagittal scalp electrodes (7).

Normal variants and patterns of unclear significance to be considered in the differential diagnosis of interictal and ictal EEG findings of focal epilepsy

Breach effect (Fig. 28.2) refers to a focal increase in amplitude of alpha, beta and mu activity over or near a skull defect due to craniotomy, head trauma, or osteolytic bone disease. Bone is the major resistive element between the cortex and the scalp electrode, acting as a filter attenuating higher frequencies. Absence of bone results in an increase in current from cortex to scalp. Breach activity thus represents less filtered brain activity through a skull defect. It often includes arch-like or spiky waveforms. The amplitude of the scalp EEG signal over a skull defect may increase up to five times. Physiological breach activity may attenuate with eye-opening (in the posterior regions), movement of the contralateral limbs (in the frontocentral regions) or sleep. In pure breach activity, there is no slow activity. However, a breach effect following resective neurosurgery or head injury is often associated with focal slowing, indicating an underlying area of structural damage or dysfunction. In pure breach activity, a sharpened or spiky waveform is restricted to the area of skull defect (8).

Small sharp spikes (Fig. 28.3), also confusingly called 'benign epileptiform transients of sleep' refer to a low-voltage (usually <50 μV), brief duration (usually <50 ms) waveform with a monophasic or diphasic spike with a steep ascending and descending limb. They may have a subsequent slow wave of lower amplitude. However, they are not associated with focal slowing. Small sharp spikes appear in stage II sleep and disappear in slow wave sleep. They are seen unilaterally over the anterior to mid-temporal derivations, but are almost always seen bilaterally independent. They occur in 20-25% of adults and are thus an important differential diagnosis of genuine epileptiform discharges in the temporal areas (9).

Wicket spikes (Fig. 28.3) are seen over the temporal regions during drowsiness in adults. They are usually bilateral and independent. Wicket spikes occur sporadically or in trains with a crescendo-decrescendo envelope. They do not have a subsequent slow wave and are not associated with focal slowing (10).

Phantom spike and wave discharges at 6 Hz include two forms, described by the acronyms WHAM (wakefulness, high amplitude, anterior, male) and FOLD (female, occipital, low amplitude, drowsy, Fig. 28.4). This is the only EEG phenomenon with a different manifestation in females and males. The spike component is usually of low amplitude, with a slow wave repetition rate of 5–7 Hz lasting for 1–2 s. A higher amplitude, frequency of less than 6 Hz, and persistence in slow wave sleep suggest an association with seizures (9).

Positive bursts of 14 and 6 Hz, also called ctenoids, appear as comb-like spindles over the posterior temporal regions. They are seen in bursts lasting up to 1 s, usually unilaterally or independently bilaterally. Ctenoids are best seen in a contralateral ear reference. They are most common in adolescence and decreasingly seen with age (9,11).

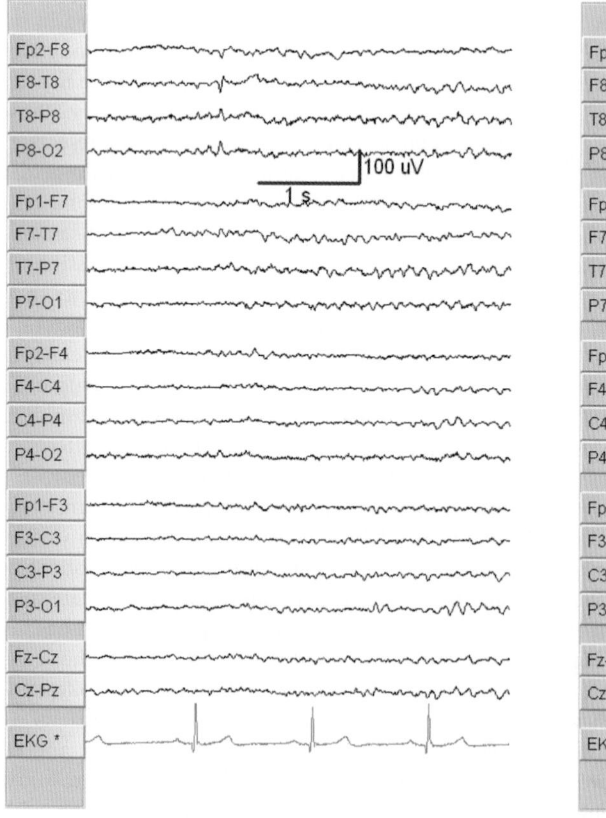

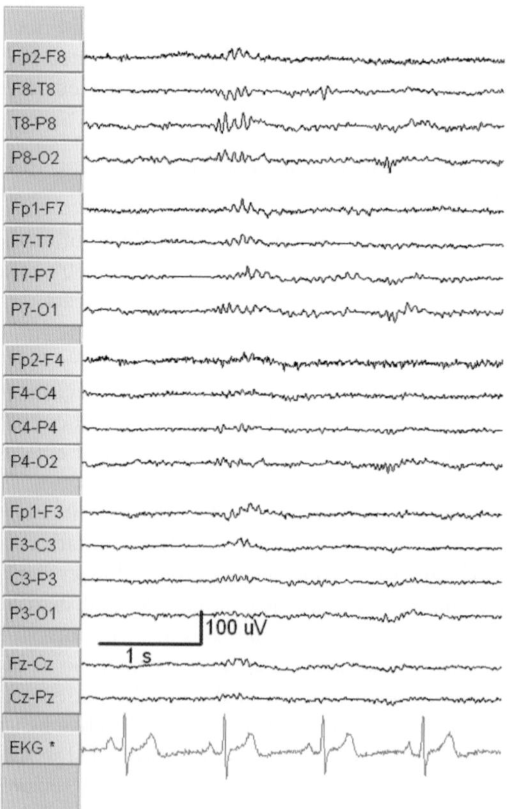

Fig. 28.3 Left: Right temporal small sharp spike in a 62-year-old woman referred for excessive hypnic jerks. Right: Wicket spikes in a 26-year-old woman seen after a first seizure.

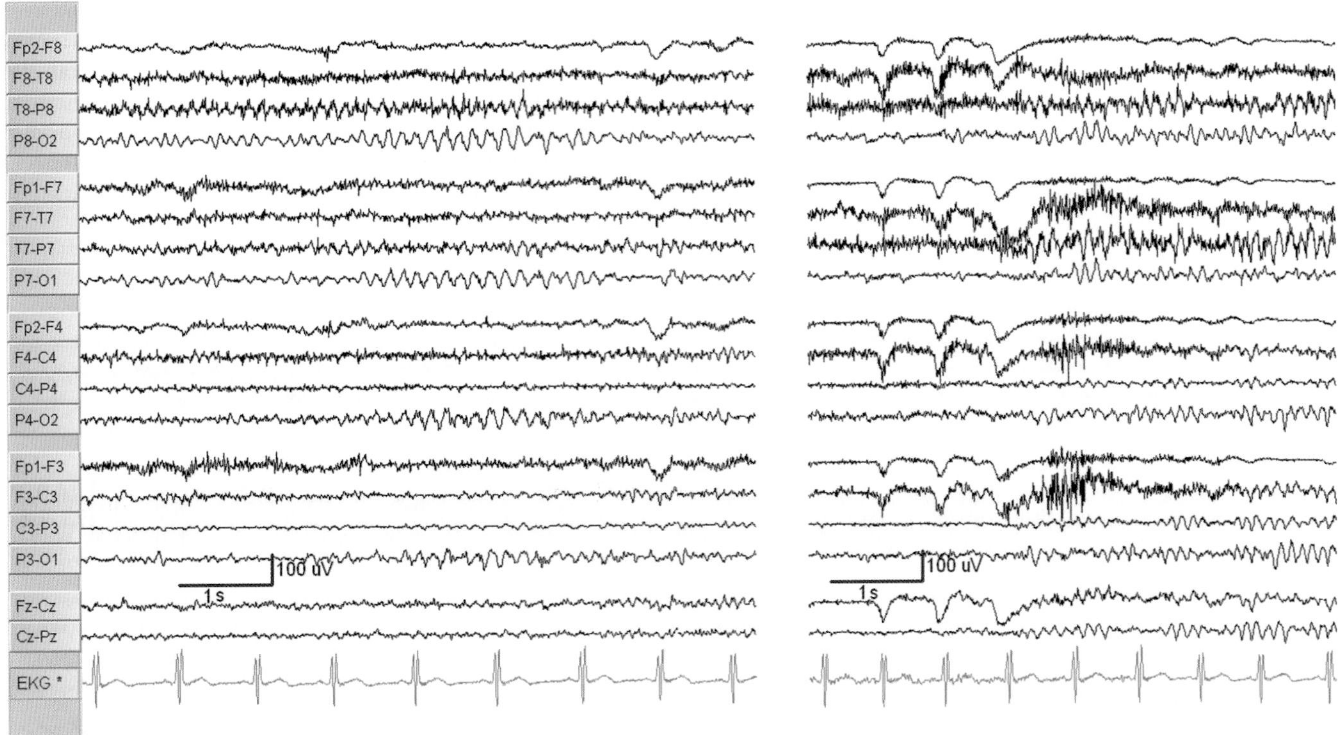

Fig. 28.4 FOLD (*f*emale, *o*ccipital, *l*ow amplitude, *d*rowsy) variant of 6 Hz phantom spike (left) in a 40-year-old woman with psychogenic nonepileptic seizures. 10 Hz posterior background activity (right).

Rhythmic mid-temporal theta bursts of drowsiness (RMTD, previously also described as psychomotor variant, Fig. 28.5) consists of bursts of 5–7 Hz waves that may appear sharp, flat, or notched. It does not evolve spatially or temporally. It may be seen unilaterally, and (usually independently) bilaterally. It is seen in up to 2% of normal adults and adolescents in relaxed wakefulness (9).

Subclincial rhythmical electrographic discharges in adults (SREDA, Fig. 28.6A,B) is a paroxysmal pattern in the theta or delta

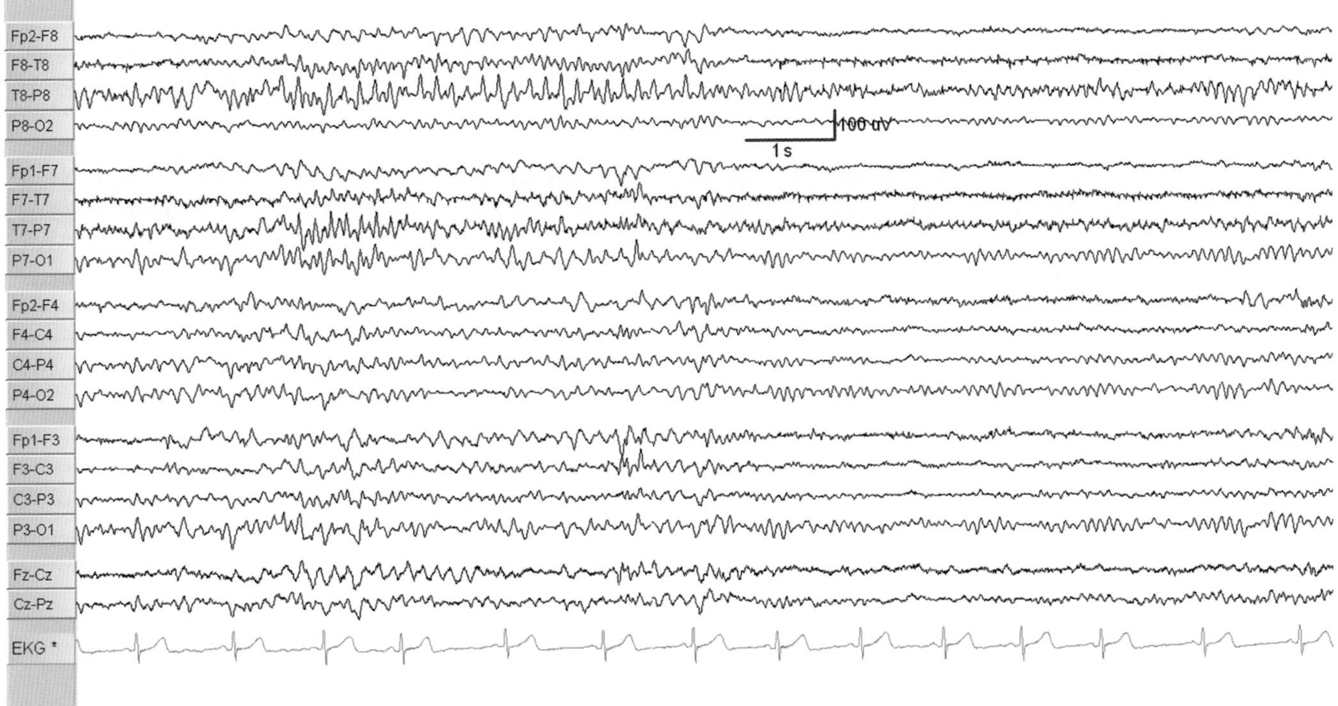

Fig. 28.5 Rhythmic mid-temporal theta in drowsiness. 17-year-old woman with absence epilepsy in childhood, now seizure-free, no longer taking anticonvulsants.

range that lasts on average 40–80 s (range 10–300 s). It is seen in about 1 in 2000 people above the age of 40 years, i.e. even a busy EEG department may see it less than once per year. The onset may have an evolving appearance for a few seconds, or the continuous pattern may be preceded by a single or few 'discharges', making it prone to misdiagnosis as an electrographic seizure. The offset may be abrupt or gradual, but without subsequent slow activity. Physiological activity such as posterior background rhythms, photic driving, frontocentral beta activity, or vertex waves and sleep spindles may be preserved during SREDA. Its distribution is usually generalized with a temporoparietal maximum, however it can also be seen unilaterally or bilaterally asymmetrical and/or restricted to the frontal, temporal or central regions. SREDA is not associated with changes in subjective experience, objective behaviour, or cognitive dysfunction, and is thus considered a benign EEG variant of no diagnostic significance (12,13).

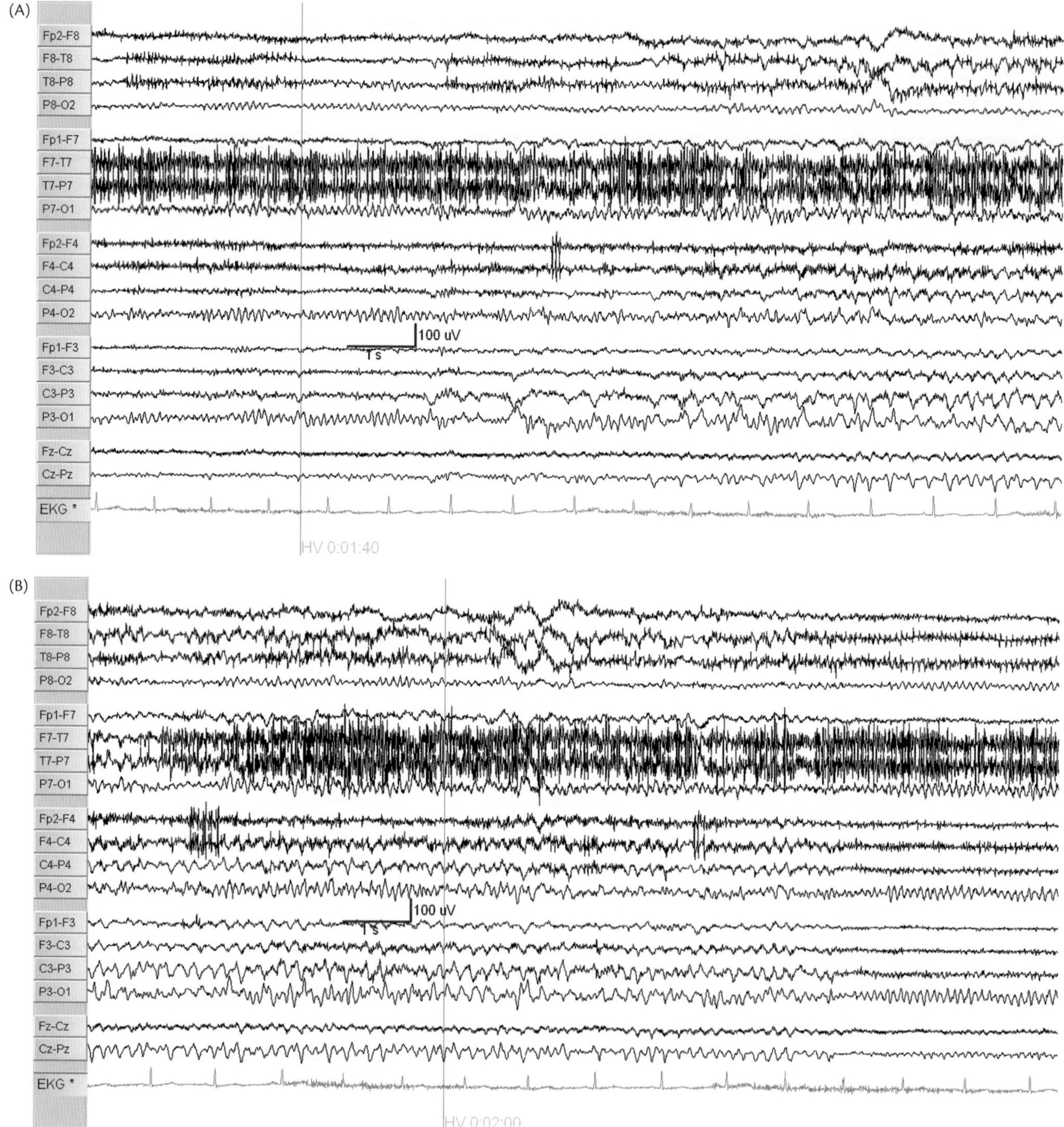

Fig. 28.6 (A) Subclinical rhythmic electrographic discharge (SREDA) in a 63-year-old woman with memory complaints. Note the 'evolving' onset over the left centroparietal area. (B) Subsequent EEG page. Note the abrupt offset with immediate return of posterior background activity. This pattern was seen in several recordings over 3 years and never associated with behavioural or cognitive abnormalities.

Electroclinical findings of various focal epilepsy syndromes are traditionally divided by anatomical location. This concept has limitations—ictal symptoms and signs reflect the amount and distribution of cortical and subcortical activation, and disinhibition generated by the ictal discharge. Even the initial ictal symptoms may be generated in cortical areas remote from the electrographic seizure onset.

Interictal epileptiform discharges and ictal scalp EEG patterns may also propagate within the brain. In a study of nearly 400 people with unilobar MRI lesions evaluated with prolonged video EEG monitoring, localization of interictal epileptiform discharges and ictal patterns were seen restricted to the lobe of the lesion in about 60% of people with temporal lobe lesions, and less than 30% of people with extratemporal lesions. Interictal epileptiform discharges

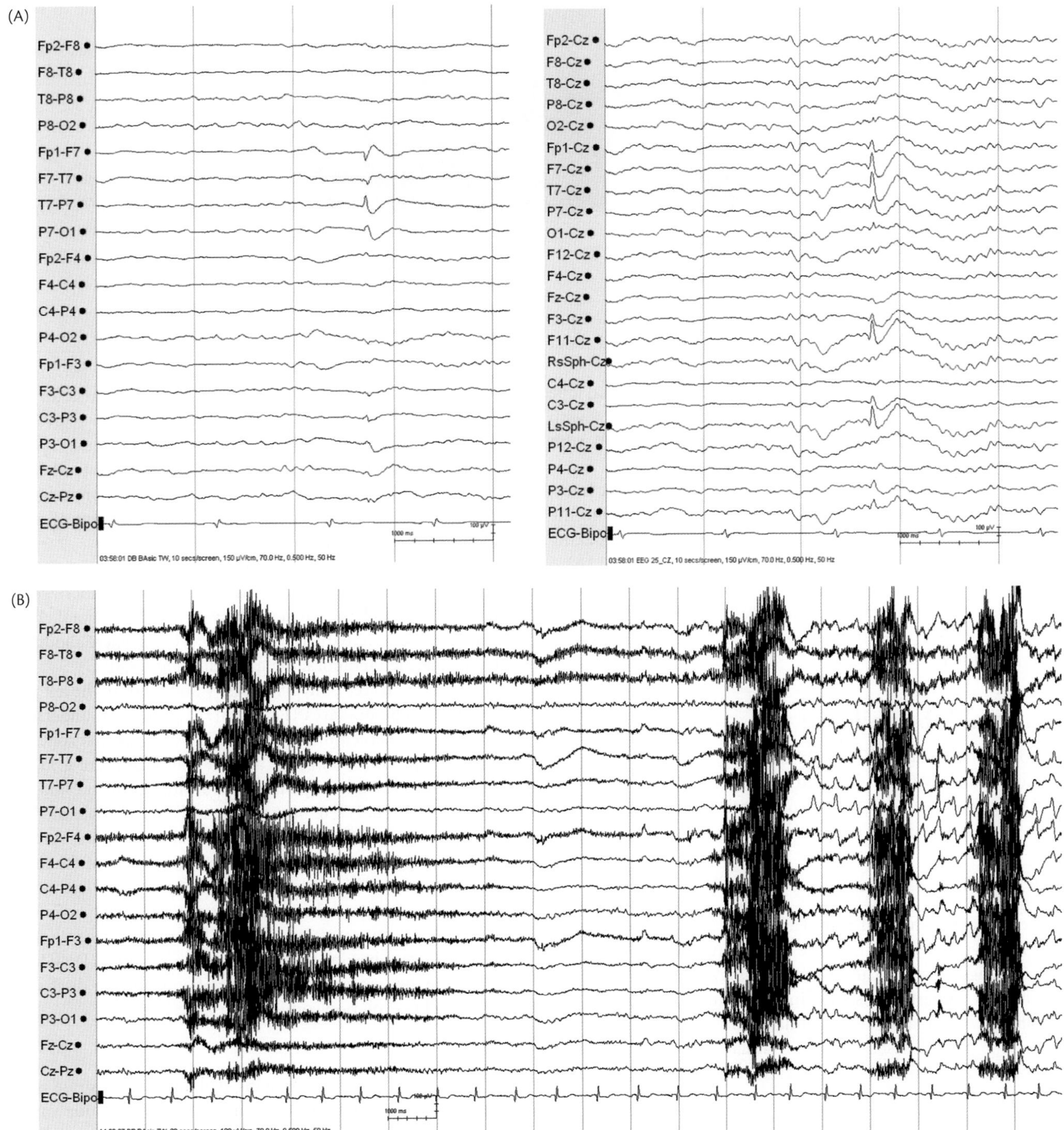

Fig. 28.7 (A) Left temporal sharp wave, longitudinal bipolar montage (left), referential montage (right). (B) Ictal EEG. Rhythmic delta activity over the left anterior temporal contacts appears ~12 s after the behavioural change (orofacial and right hand automatisms, EMG artefact). (C) Coronal T2-weighted MRI shows decreased volume and loss of internal architecture of the left hippocampus, indicating left hippocampal sclerosis. This was confirmed on histopathology.

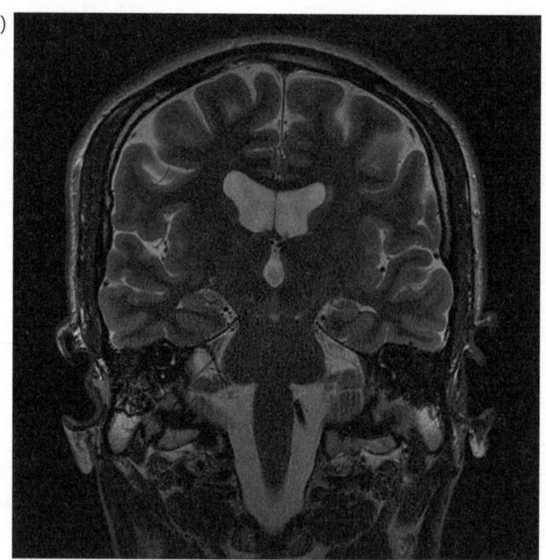

Fig. 28.7 Continued.

were seen exclusively outside the lesional lobe in 2% with temporal lesions, but in nearly 50% of parieto-occipital lesions, and about 20% of frontal or central lesions.

Ictal EEG patterns were seen exclusively outside the lesional lobe in the majority of people (64%) with central lesions, in 21% with parieto-occipital lesions, 13% with frontal lesions, and 1% with temporal lesions (14).

Secondary bilaterally synchrony refers to a situation where interictal or ictal electrical activity spreads within milliseconds to the corresponding area in the contralateral hemisphere, mimicking scalp EEG findings of primary generalized epilepsy (Fig. 28.12A).

The illustrative cases discussed in this section are derived from patients who either became seizure free following lobar or sublobar resections, and/or had seizure onset zones defined by intracranial EEG.

Temporal lobe epilepsy

Temporal lobe epilepsy (Cases 28.1 and 28.2) is the most common focal epilepsy syndrome, and mesial temporal lobe epilepsy (MTLE) accounts for ~80% of people with temporal lobe onset (15). The most common underlying pathological substrate is hippocampal sclerosis. Other common aetiologies include cavernous angioma, low grade brain tumours, especially low grade glial or glioneuroal tumours such as dysembryoplastic neuroepithelial tumour (DNET), focal cortical dysplasia, encephalomalacia, and hamartomas. However, about 30–40% of patients have no identifiable pathology on MR imaging (15). In MTLE, there is often a

Case 28.1 Mesial temporal epilepsy

33-year-old right-handed man. Parents recalled one prolonged febrile convulsion in infancy. The first habitual seizure was witnessed at age 25 years. Current seizures may include a '*déjà-vu*' feeling and 'butterfly sensation' in the stomach and last 30–60 s, followed by orofacial and right hand automatisms, and humming. The patient describes post-ictal language difficulties for several minutes. Seizures occur in clusters of up to three per day on 2–3 days per month.

Scalp EEG shows sharp waves in the left temporal area (Fig. 28.7A), with equipotential maximum at contacts F7 and T7, and a left temporal ictal pattern (Fig. 28.7B). MRI revealed left hippocampal sclerosis (Fig. 28.7C), confirmed on histopathology post- resection.

Case 28.2 Neocortical temporal epilepsy

31-year-old right-handed man. At age 18 months, he fell and hit his head at a banister, resulting in loss of consciousness for 2 h, and hospital admission for 3 weeks, with temporary regression of motor development. Habitual seizures started at age 4 ½ years, described as vacant staring with eye blinking, bilateral hand fidgeting, and heavy sighing. They are sometimes preceded by an aura of a 'strange feeling'. Seizures occur in clusters of up to three per day on 2–3 days per month.

EEG shows left temporal sharp waves with maximum at contacts T7 and TP7 (Fig. 28.8A). Ictal EEG is characterized by attenuation of background activity over the left temporal chain, followed by delta activity over the right (!) hemisphere about 20 s later, and late left temporal theta activity (Fig. 28.8B,C).

MRI shows an area of hyperintensity on FLAIR in the left posterior basal temporal area (Fig. 28.8D). The patient underwent intracranial EEG recordings to define the seizure onset zone and to map the posterior language area. A combination of grid and depth electrodes was implanted, recording from the imaging abnormality, the lateral temporal lobe, and the amygdala and hippocampus on the left. Electrographic seizures were first seen on the contacts overlying the anterior or posterior rim of the imaging abnormality. Electrocortical stimulation of the lateral and basal temporal areas did not interfere with language tasks. The patient underwent resection of the MRI lesion. Histopathology revealed a dysembryoplastic neuroepithelial tumour (DNET).

Note the difference in distribution of epileptiform discharges and ictal EEG evolution compared to Case 28.1.

history of febrile convulsions, central nervous system (CNS) infection, or precipitating CNS injury in early childhood. Habitual seizures typically start with a latency period of 5–20 years, but the latency period may be longer. Many patients report an aura of a rising epigastric sensation, nausea, or unnatural and usually unpleasant olfactory or gustatory hallucinations, the latter are fairly specific for MTLE (16). Patients may also experience a sensation of familiarity (*déjà vu, déjà entendu*), or estrangement from the environment (*jamais vu, jamais entendu*), complex visual or auditory hallucinations, affective symptoms, or fear. These symptoms have no lateralizing value (17).

Compared with MTLE, patients with neocortical temporal lobe epilepsy (NTLE) tend to have seizure onset after the second decade of life. A history of febrile seizures in infancy is uncommon (16,18). Auras are reported by about 50–75% of patients in NTLE. Typical symptoms are simple auditory hallucinations (ringing or buzzing noises), muffled or distorted auditory perception (e.g. echoes), and a sensation of dizziness, unsteadiness, or vertigo.

Common objective symptoms of seizures arising in the temporal lobe are loss of awareness, semi-purposeful oral and manual automatisms, exploring head movements (initially ipsilateral to the side of seizure onset (19), eye blinking, and dystonic posturing of contralateral limbs (20), as well as autonomic features such as mydriasis, hypersalivation, vomiting, piloerection, and tachycardia (or, less frequent, bradycardia and even ictal asystole). The onset of symptoms tends to be gradual and evolving over the course of tens of seconds. Post-ictally, patients are usually confused and amnestic for the seizure, with a variable period of anterograde amnesia. Some patients exhibit more complex behaviours such as wandering, drinking, or kissing during or immediately after a seizure, the latter two associated with non-dominant onset (21). Naming difficulties are often seen after dominant temporal seizures. Ictal speech, defined as formed words out of situational context, suggests that the seizure arises from the non-dominant hemisphere (22). Staring with behavioural arrest and lack of motor features has been linked to posterior neocortical seizure onset (23).

Interictal epileptiform discharges (IIEDs) in MTLE are typically seen with a maximum over the anterior/inferior or mid-temporal contacts (contacts F7/F8 and T7/T8 in the international 10-20 system). IIEDs are often seen over the temporal areas bilaterally, in particular during long-term ambulatory or video EEG recordings. In some patients with MTLE, however, no IIEDs are seen in prolonged video EEG recordings, including post-ictally. The typical ictal scalp EEG pattern in MTLE consists of rhythmic high amplitude activity in the theta to alpha range that is localized to the anterior/inferior or mid-temporal contacts. Clinical seizure manifestations may precede the emergence of an ictal scalp EEG pattern by up to 60 s. In NTLE, IIEDs may have a similar peak negativity as those seen in MTLE, but may also be seen over the posterior temporal derivations (contacts P7/P8 in the 10-20 system). The orientation of the underlying dipole may predict the underlying anatomical substrate better than the location of the absolute maximum. Spikes with an inferior temporal peak negativity and a vertex positivity are generated by the inferior basal cortex and typically seen in MTLE. Spikes with a negative maximum over the lateral temporal contacts and a radial or horizontal orientation are associated with neocortical generators (24). Ictal EEG activity in NTLE is often described as polymorphic delta activity that is lateralized rather than localized. The ictal onset and offset may be less clear, and bilateral involvement may be seen earlier compared to MTLE (25).

In clinical practice, features of mesial and neocortical temporal lobe epilepsy often overlap (15,26). A study combining scalp and intracranial EEG recordings demonstrated that the rhythmic 5–9 Hz temporal scalp EEG pattern commonly associated with MTLE requires the recruitment of adjacent inferolateral temporal neocortex. This suggests that ictal scalp EEG patterns in temporal seizures reflect not only cortical activity at ictal onset, but also ictal evolution, propagation and synchronization (27). Ictal scalp EEG activity is thus only moderately specific for the ictal onset zone.

The term 'temporal plus' epilepsy has been coined for patients in whom the seizures are generated in an epileptogenic network that extends beyond the temporal lobe to include neighbouring structures such as the orbitofrontal area, frontal and parietal opercula, insula, and parieto-occipital junction. Compared to patients with pure TLE, such patients are more likely to present with gustatory hallucinations, vertigo, or auditory illusions, and they may manifest versive head movements to the contralateral side, piloerection, and ipsilateral dystonia (28). Electrophysiological features to suggest a 'temporal plus' epilepsy are precentral IIEDs, and ictal patterns with maxima in the anterior frontal, precentral or temporoparietal areas (28). Secondary generalization may be seen in seizures originating in any brain area.

Frontal lobe epilepsy

Frontal lobe epilepsy (FLE; Cases 28.3 and 28.4)) can be subdivided into orbitofrontal, dorsolateral, and mesial frontal lobe compartments. The distinction of these syndromes can be challenging due to the complexity of functionally interconnected frontal and extrafrontal areas and the variability of epileptic propagation patterns. The orbitofrontal lobe and the cingulate are connected to the temporal lobe and limbic system, and the mesial and dorsolateral frontal areas are connected to parietal areas. While ictal symptoms and signs reflect the activation or disinhibition of a specific symptomatogenic area, the ictal discharge may originate in a different frontal compartment or even outside the frontal lobe. Auras in frontal lobe seizures are often non-specific. Patients may report or exhibit a facial expression of anxiety or fear, epigastric auras, and *déjà vu* (29), or report an ill-defined sensation of movement in the head ('cephalic aura', 30), sometimes associated with blurred vision (31). Movements often involve the proximal extremities and may appear 'ballistic' or unnatural. Activation or disinhibition of mesial premotor and prefrontal areas may produce bizarre gestures and repetitive movements such as bicycling or scissoring leg automatisms, shuddering, pelvic thrusting, jumping, kicking, thrashing, crawling, and unformed vocalizations (screaming, mirthless laughter, crying, singing, howling, or barking (32,33). These prominent features have been linked to the ventromesial frontal cortex, whereas more subtle horizontal rocking or rotational body movements with or without tonic contractions have been associated with the mesial premotor cortex, mesial intermediate frontal cortex, and dorsal anterior cingulate cortex (34). Seizures involving the supplementary sensorimotor area (SSMA) feature symmetrical or asymmetrical tonic contractions of face, throat, and limb muscles. Classic manifestations of SSMA seizures are:

- 'Fencing posture' referred to a position where the contralateral upper extremity is extended, the ipsilateral arm flexed and abducted at the shoulder, and the head rotated contralateral to the seizure focus.

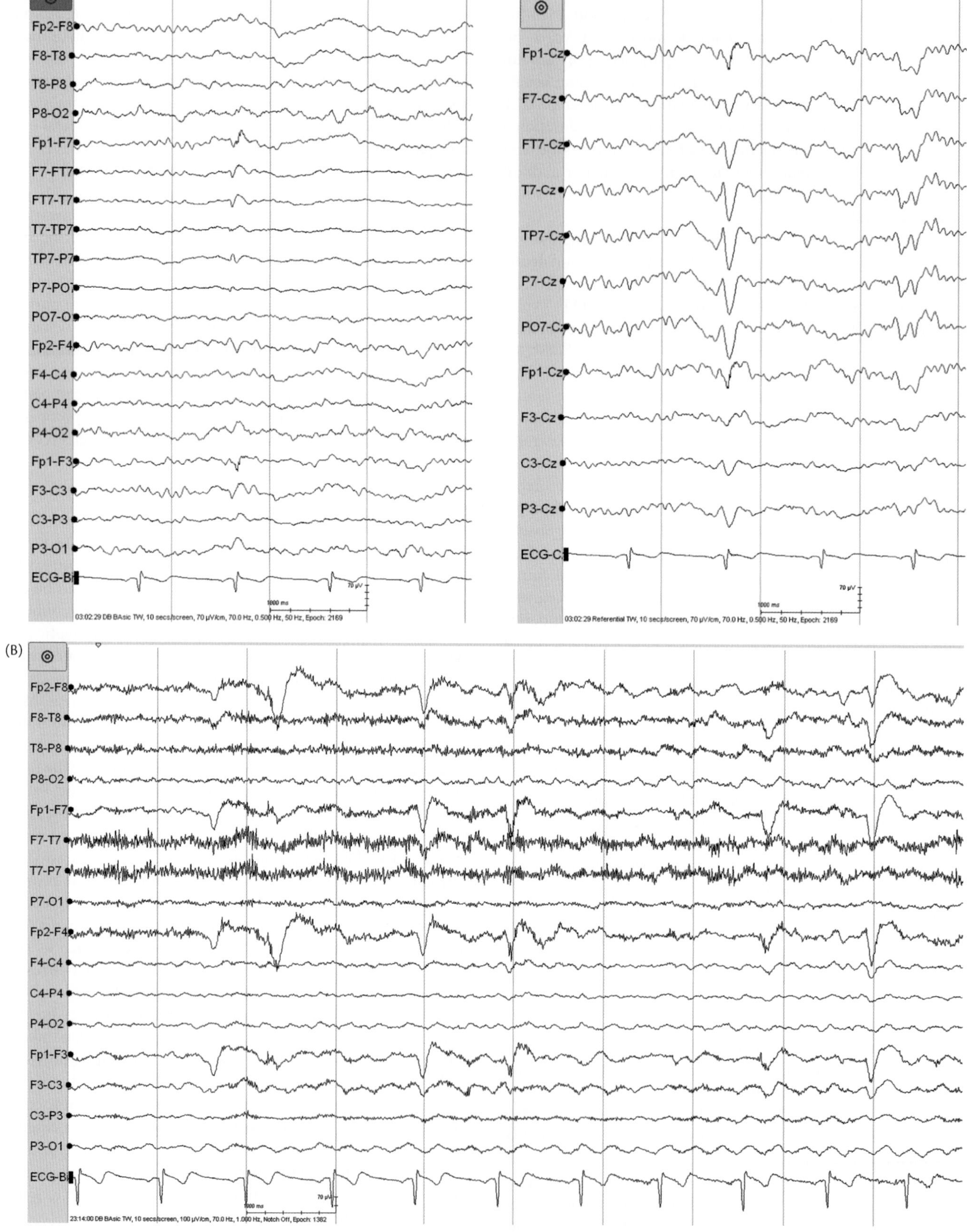

Fig. 28.8 (A) Left posterior temporal sharp wave, longitudinal bipolar montage (left), referential montage (right). Note additional 10/10 electrodes in the left temporal chain. (B) Ictal EEG onset. Subtle rhythmic delta activity appears over the right>left (!). (C) Ictal EEG evolution. Nearly 1 min later, rhythmic activity at T7. (D) Coronal FLAIR MRI shows hyperintensity in the left posterior basal temporal lobe. Intracranial EEG confirmed diffuse onset from this lesion. Patient became seizure free following resection. Pathology revealed dysembrioplastic neuroepithelial tumour (DNET).

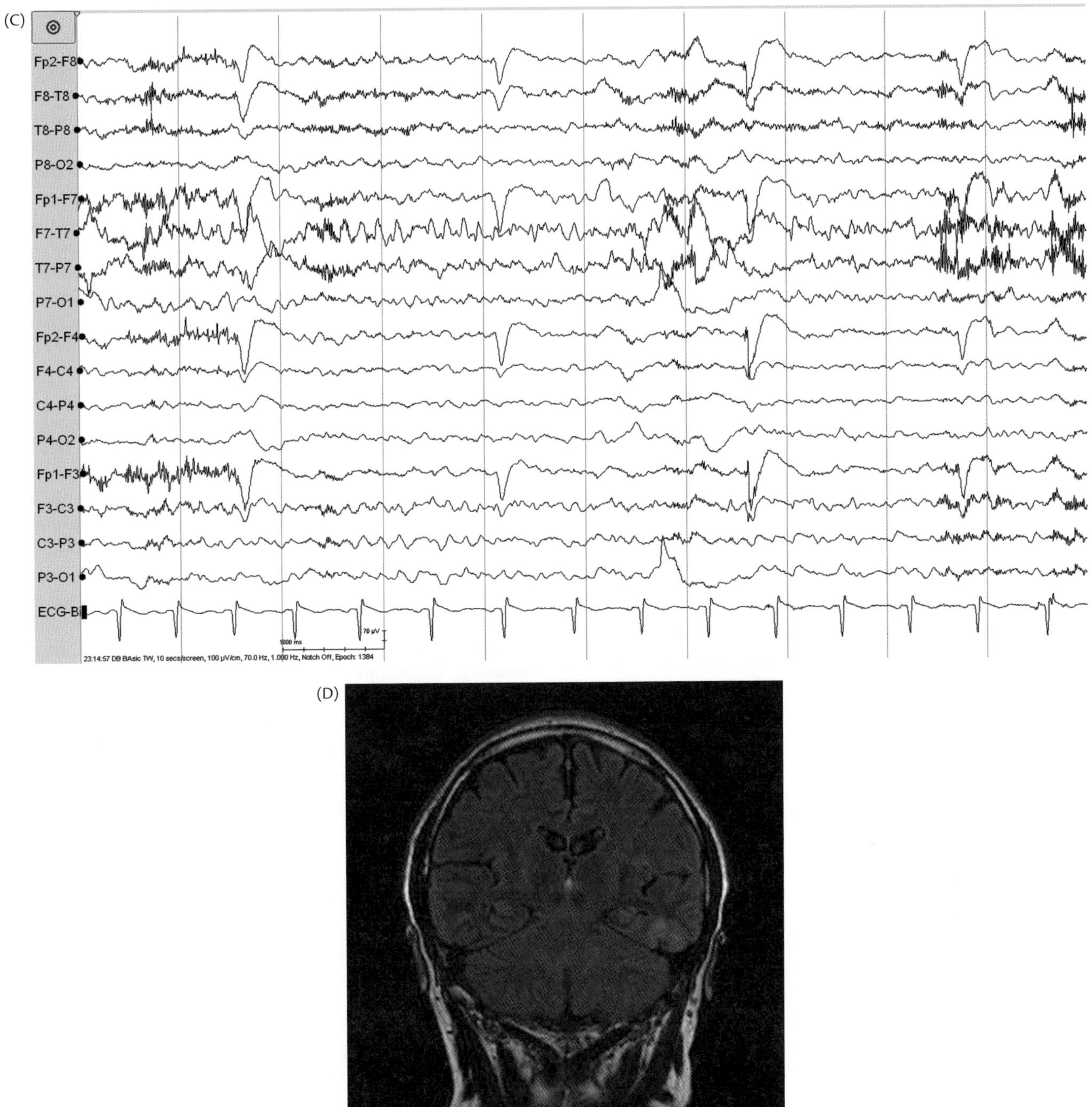

Fig. 28.8 Continued.

- 'M2e' posture, consisting of contralateral shoulder abduction, elbow flexion, and head deviation towards the affected arm.
- 'Figure of four' extension of the contralateral upper extremity across the chest and ipsilateral arm flexion at the elbow (35).

These postures are often accompanied by vocalizations and preserved awareness. Principal component and cluster analysis of semiological features in an intracranial EEG study of FLE have linked elementary motor manifestations (clonic or tonic contractions) to the precentral and premotor areas, whereas more complex gestural motor behaviours involved anterior lateral and medial prefrontal regions, and fearful behaviour mesial temporal with or without ventromesial frontal cortex activation (36). Negative motor phenomena, such as inability to speak or move, are thought to reflect involvement of the negative motor area, rostral to the SSMA (37). Loss of postural tone (ictal atonia) may lead to head drop or falls.

In contrast to seizures seen in TLE, seizures seen in FLE, especially those involving the SSMA, often have an abrupt onset and offset, last on the order of 30 ± 15 s, and tend to occur repeatedly during non-rapid eye movement sleep stages I and II.

Few semiological features in FLE have strong lateralizing value: unilateral clonic or tonic/dystonic contractions of the face

Case 28.3 Inferior lateral frontal lobe epilepsy

34-year-old right-handed woman. Few febrile seizures in infancy. Onset of staring spells with or without elevation of the right arm at age 7 years. Episodes of status epilepticus at age 13 and 17 years. Currently, three different seizure types:

- Blank staring with unresponsiveness.
- Tonic elevation of the right upper and/or right lower extremity.
- Inability to speak with retained awareness.

Interictal scalp EEG demonstrates sharp waves in the left frontocentral and left inferior frontal areas, with occasional propagation (secondary bilateral synchrony) to the homotopic areas on the right (Fig. 28.9A).

During video EEG, the patient experienced a cluster of 24 seizures over 30 min. The only EEG correlate of individual seizures were broad sharp waves at electrode F3, followed by diffuse EMG artefact without discernible underlying EEG pattern (Fig. 28.9B). Following administration of intravenous diazepam, clinical seizures resolved, while sharp waves at F3 persisted in the post-ictal period for >1 h (Fig. 28.9C).

MRI demonstrated a subtle area of hyperintensity surrounding the inferior frontal sulcus (Fig. 28.9D), suggestive of focal cortical dysplasia. The patient became seizure-free without language decline following intracranial EEG and awake resection of the anatomical abnormality, and surrounding middle and inferior frontal gyrus, sparing the anterior language area. Histopathology confirmed a focal cortical dysplasia type IIb.

Case 28.4 Mesial/orbitofrontal epilepsy

29-year-old right-handed man. Onset of seizures at age 13 years, remembered as sudden loss of awareness while at school. Subsequent seizures characterized by a tingling sensation in his chest and arms, then loss of awareness with violent limb movements and flight-like behaviour—he may hide under a blanket during a seizure.

Interictal scalp EEG showed left frontopolar spikes and left anterior temporal sharp waves (Fig. 28.10A). Ictal EEG onset characterized by a low amplitude left fronto-temporal sharp wave, followed by evolving rhythmic bifrontal delta activity (Fig. 28.10B). About 25 s later, a rhythmic theta pattern appears confined to the left frontocentral/temporal contacts (Fig. 28.10C).

MRI was repeatedly normal at 1.5 and 3T. Interictal FDG-PET demonstrated an area of reduced glucose uptake in the left mesial orbitofrontal/frontopolar area (Fig. 28.10D). Intracranial EEG confirmed seizure onset from the anterior mesio-frontal area. The patient experienced significant decline in seizure frequency following a resection of the seizure onset zone. No specific abnormality was found on histopathology. Note that ictal (and some interictal) EEG changes are not seen on the scalp electrodes closest to the seizure onset zone as defined by intracranial EEG, but in a more widespread and relatively remote area, reflecting ictal propagation and recruitment of a larger cortical area.

and extremities, with or without post-ictal Todd's paresis, lateralize to the contralateral primary motor areas. Forced deviation of the eyes and head, often in combination with hyperextension in the neck suggests activation of the contralateral frontal eye field (Brodman area 8), in particular if followed by secondary generalization. If a secondarily generalized tonic clonic seizure ends in an asymmetrical fashion, the side of the last clonic jerk(s) lateralizes to the hemisphere of seizure origin (38). Frontal lobe seizures are often sudden in onset and offset. Awareness may be preserved even when bilateral motor features are present.

Depending on the ictal propagation pathway, seizures arising in the orbitofrontal areas may present with 'temporal' or 'frontal' features. In surgical series, patients reported auras of choking, *déjà vu* sensation, or 'butterflies' in the stomach. Objective features are behavioural arrest, staring, head deviation to either side, and other features of temporal lobe seizures, including affective or autonomic auras, and oral or gestural automatisms. Propagation to mesial or lateral frontal areas reflects the hyperkinetic features described above (39). These features may, however, also be the presenting manifestations of seizures arising in the temporal lobe (40).

Interictal scalp EEG features in FLE are often non-diagnostic or may be misleading (41). Interictal rhythmic midline theta activity during clear wakefulness, excluding periods of mental activation and drowsiness, is seen in approximately half of patients with FLE, but rarely in TLE and not in patients with psychogenic non-epileptic seizures (42). This activity is often the only interictal EEG abnormality seen in people with FLE even in prolonged video EEG recordings.

Dorsolateral foci more commonly have concordant IIEDs than seizures arising in the mesial or orbitofrontal areas (43). The latter may present with IIEDs localizing to the frontopolar, inferior frontal, or anterior temporal electrodes, or IIEDs in a widespread distribution. In mesial FLE, IIEDs may be seen in the parasagittal or frontocentral electrodes, ipsi- and/or contralateral to the seizure focus (44). Often, IIEDs escape detection by scalp EEG in mesial FLE.

Ictal patterns in lateral FLE are localizing or lateralizing in about 70% of seizures and most often consist of rhythmic spiking, rhythmic delta activity (45), and rhythmic fast activity (41,46). In mesial FLE, ictal patterns are often marred or completely obscured by movement or EMG artefact, in particular in seizures with vigorous motor manifestations. Mesial frontal lobe seizures may present as rhythmic activity in the delta to beta range, repetitive epileptic discharges, diffuse suppression, or arrhythmical activity distinct from background activity. Ictal patterns occur in a generalized distribution in the majority of seizures and significantly more commonly than in any other focal epilepsy syndrome (4). The post-ictal EEG tends to be uninformative as well in mesial FLE. In orbitofrontal epilepsy, ictal patterns may localize to the ipsilateral frontal or temporal contacts, with or without preceding background attenuation (39).

Insular epilepsy

Seizures arising in the insula may present with throat constriction or the sensation thereof, and/or unpleasant paraesthesias in the orofacial area, the contralateral half of the body, or bilateral lower extremities (47). Further symptoms may include ictal dysarthria and widespread dystonia. Insula seizures may also present

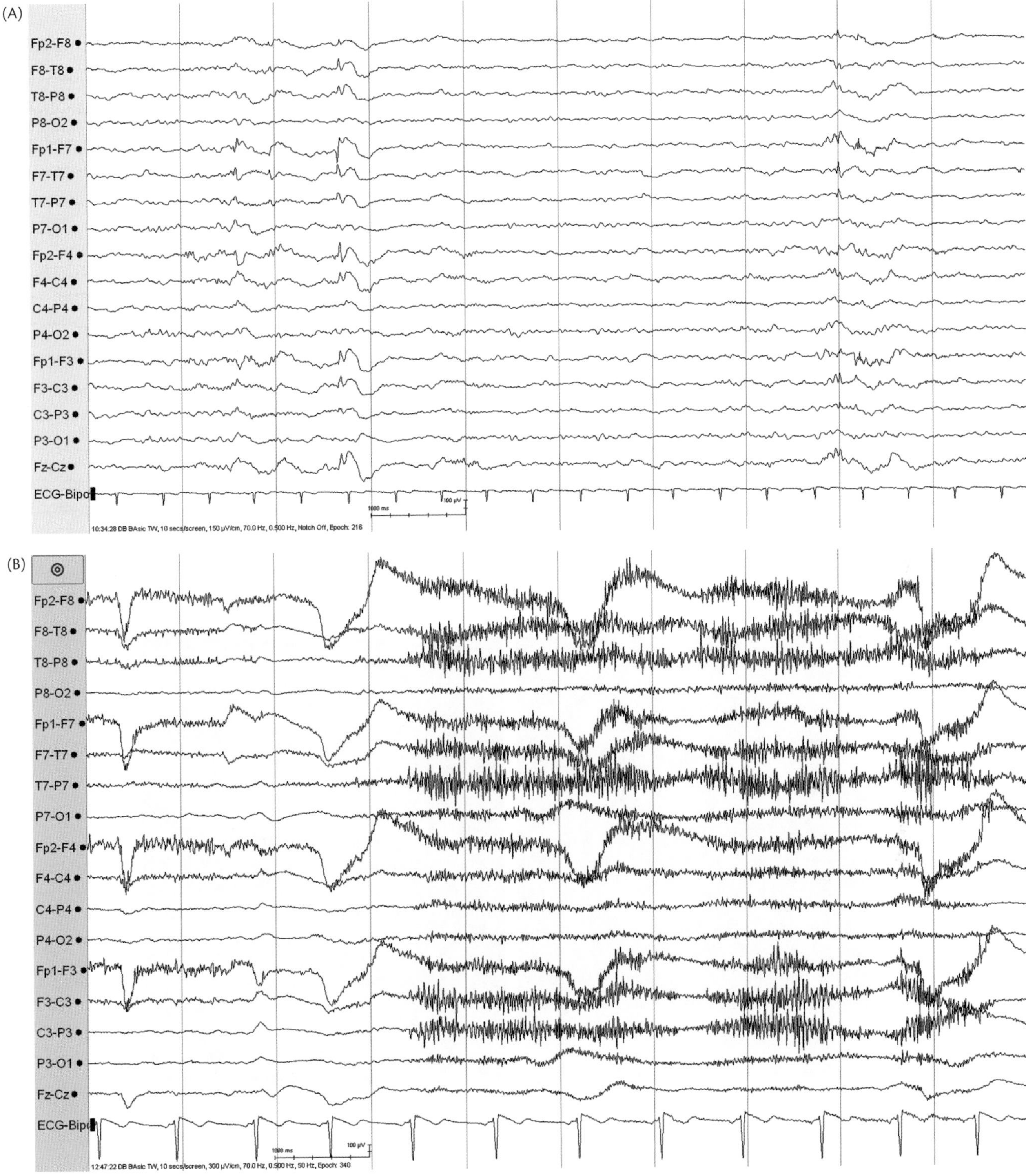

Fig. 28.9 (A) Frontal spikes in longitudinal bipolar montage. Note the spikes at the beginning and the end of the page are confined to contacts F7 and F3, whereas the second spike propagates to involve the frontopolar area bilaterally. (B) Ictal EEG. Note the broad sharp wave at contact F3 at the onset of the seizure (arrow). There is diffuse EMG artefact ~1.5 s later, with no further EEG change. (C) Post-ictal sharp waves persist at contact F3. (D) Coronal FLAIR (left) and axial T2 MRI reveal an area of signal hyperintensity around the left inferior frontal sulcus, suggestive of focal cortical dysplasia. This was confirmed on histopathology (Taylor type 2b). The patient became seizure free following resection of this area.

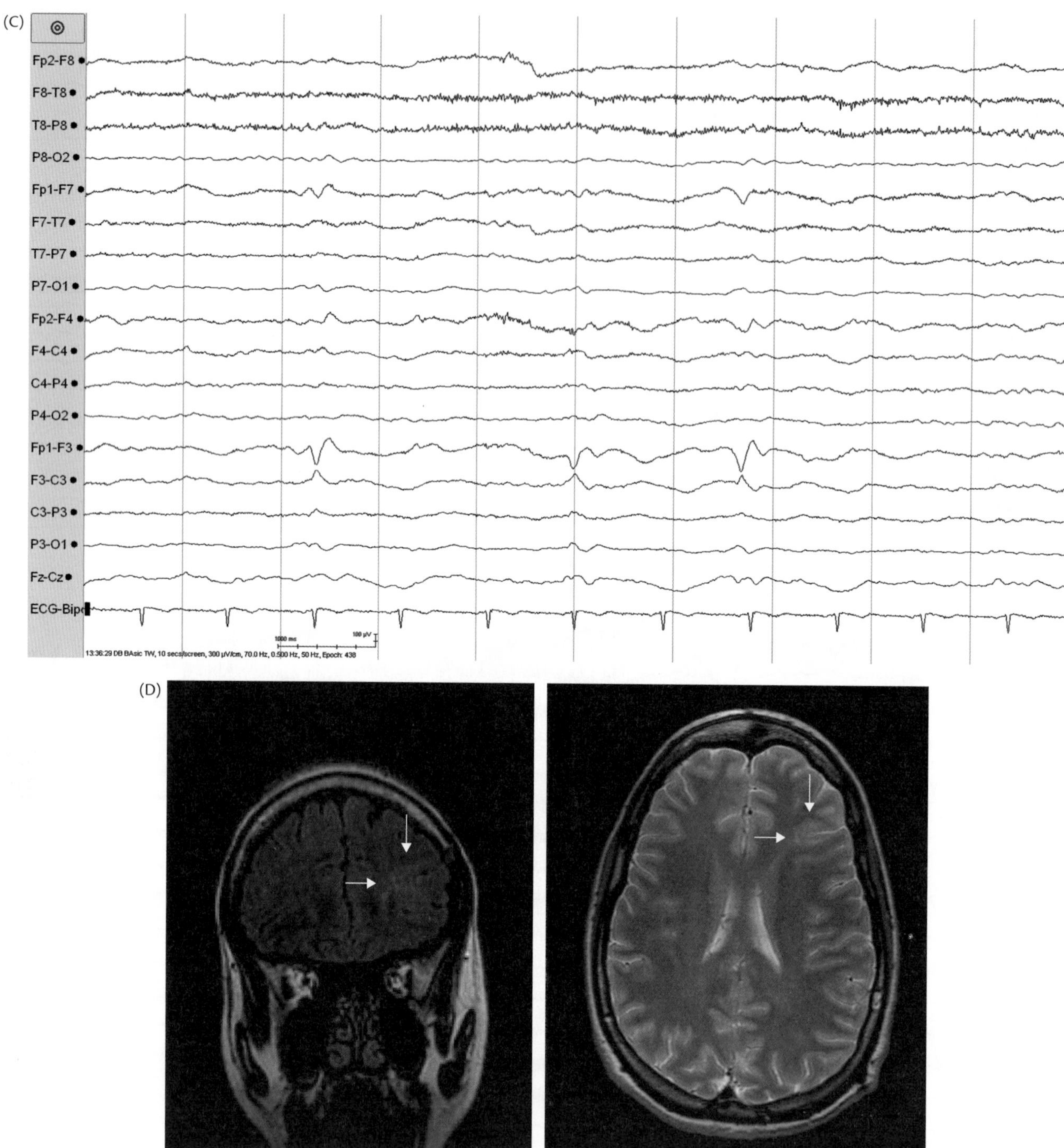

Fig. 28.9 Continued.

with oro-alimentary automatisms (48) or hyperkinetic movements (49,50). Interictal and ictal scalp EEG may imitate those of TLE or FLE (reflecting spread of activity), or can be uninformative, especially when the seizures remain restricted to the insula.

Occipital lobe epilepsy

Patients with occipital lobe epilepsy (OLE, Case 28.5) may report elementary visual auras (flickering or flashing lights, stars, dots or shapes) or complex visual auras (visual illusions or hallucinations, distortions of proportions or colours), and visual field defects, as well as paraesthesias, epigastric sensations, and experiential symptoms (51–53). The latter may reflect propagation and ictal activation of temporal-limbic, or frontoparietal systems. Visual auras do not distinguish mesial from lateral occipital seizure onsets (54). Careful visual field examination during or after the aura may allow lateralization or localization, or demonstrate a visual field deficit, which localizes the seizure onset to the contralateral occipital lobe. Epileptic visual auras are shorter (seconds to a few minutes) than visual auras reported by migraneurs (4–20 min, 55). Objective

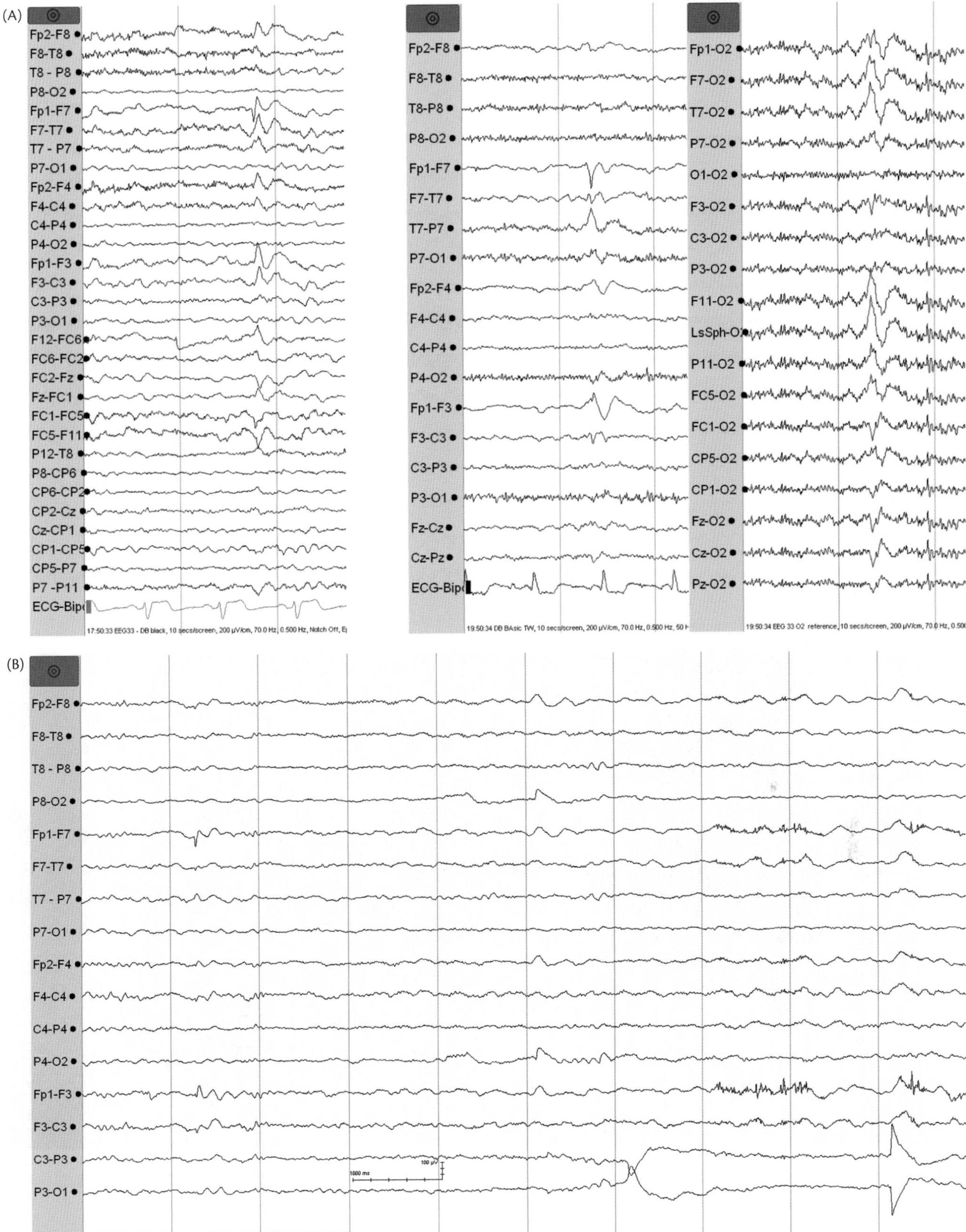

Fig. 28.10 (A) Left frontopolar spike (left, longitudinal bipolar montage) and left fronto-temporal sharp wave in longitudinal bipolar and referential montage (middle, right). Note the negative phase reversal across F7 and T7 and the positive phase reversal across F3, compare with the sharp wave in Fig 28.7A. (B) Ictal EEG onset. Low amplitude left frontotemporal sharp wave, followed by bifrontal slow activity. (C) 25 s later emergent left frontocentral/temporal theta pattern. Selected channels only. (D) FDG-PET shows a focal area of reduced glucose uptake in the left mesial frontopolar/orbitofrontal area (bright green area). Surface-rendered view of normalized FDG-distribution in comparison to a pooled normal control population. No specific histopathology was identified.

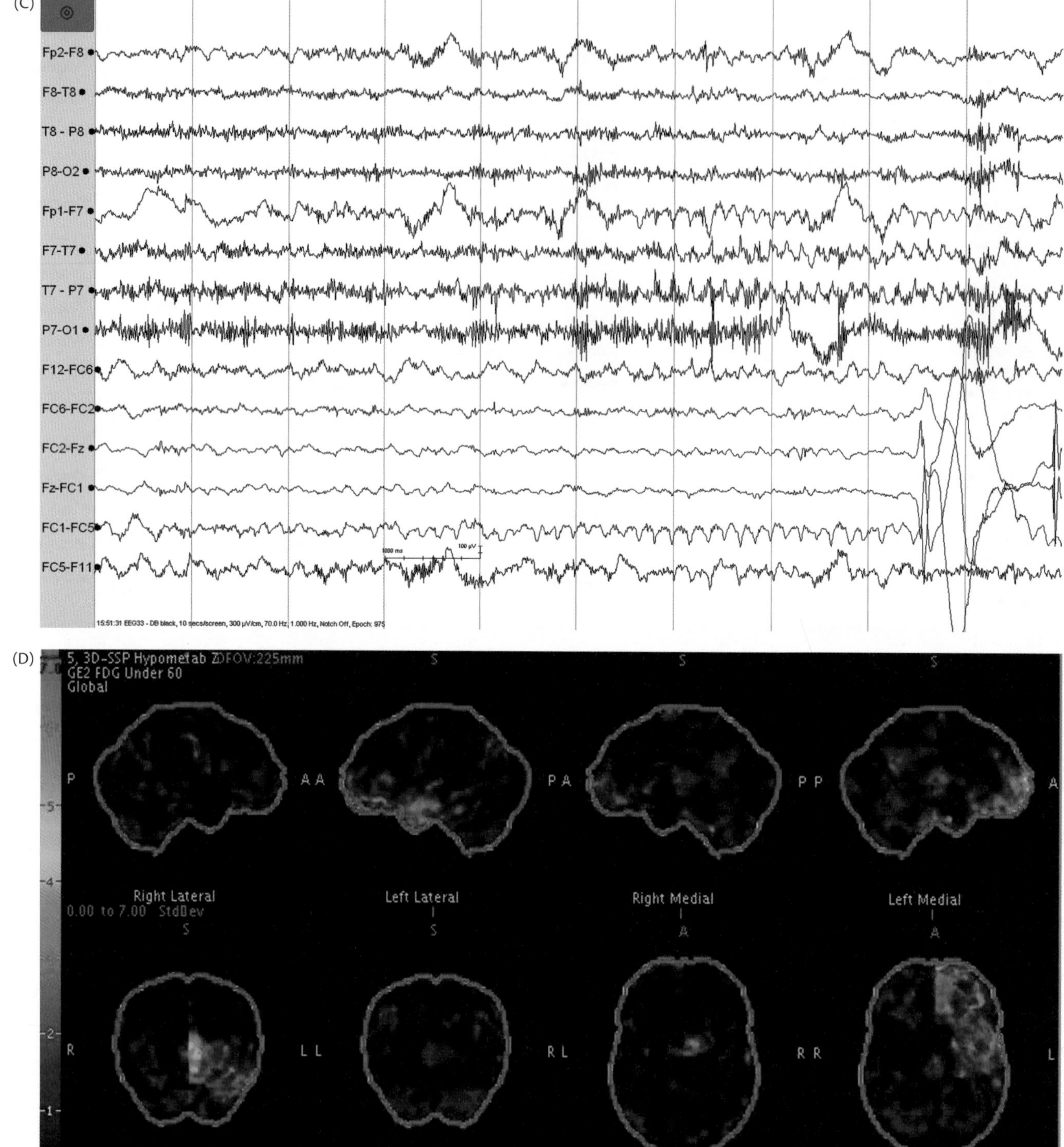

Fig. 28.10 Continued.

ictal manifestations seen in occipital lobe epilepsy are rapid eye blinking or eyelid flutter (56,57), and pursuit-like involuntary eye movements or nystagmus (55). Further semiological signs reflect seizure spread into the temporolimbic or frontoparietal areas (loss of contact, oral and gestural automatisms in the former, dystonic posturing and hyperkinetic automatisms in the latter (52).

Interictal EEG abnormalities may include asymmetry in frequency or amplitude of physiological activities (posterior background rhythm, lambda activity, and positive occipital sharp transients), and focal posterior slowing. IIEDs (spikes or spike wave complexes) may be restricted to the occipital electrodes, or occur together with posterior temporal sharp waves. The initial ictal EEG

Case 28.5 Occipital epilepsy

22 year old right handed man, born via Caesarean section due to foetal distress during prolonged labour. Few seizures noted during the neonatal period, normal development thereafter. Onset of habitual seizures at age 11 years, described by the patient as a 'strange feeling' with altered visual perception and blurred vision, with subsequent loss of awareness, automatisms and occasional forced head version to the right and generalized tonic clonic convulsions. Neurological exam revealed inferior quadrantanopia to the right.

Scalp EEG showed continuous slow activity in the left temporal chain in keeping with the structural lesion (Fig. 28.11A). No interictal epileptiform abnormalities were seen. Ictal EEG changes manifested as repetitive spiking in left parieto-occipital contacts P7, PO7 and PO3 (Fig. 28.11B), with propagation to the left frontal and temporal leads (Fig. 28.11C). MRI brain showed left parieto-occipital signal hyperintensity with volume loss and increased sulcation (Fig. 28.11D). The patient underwent intracranial EEG with multiple stereotactically placed depth electrodes, in order to define ictal onset and propagation, and to define the posterior language area. First ictal changes were seen over the left mesial parietal area. Seizures decreased significantly following resection of the imaging abnormality and the ictal onset zone. Histopathology demonstrated focal cortical dysplasia with ulegyria and remote hypoxic-ischaemic injury.

As in Case 28.4, ictal scalp EEG findings reflect propagation / recruitment of cortical areas remote from the ictal onset found on intracranial EEG.

Case 28.6 Parietal epilepsy

38-year-old right-handed man. Seizure onset, described as generalized convulsions, from age 3 years, on occasion followed by right sided weakness for few days. Currently describes several seizure types:

- Growling vocalization from sleep, with extension of the right and flexion of the left arm, as well as hand automatisms and cycling movements of the legs.
- Violent movements of all limbs that may sometimes result in him falling out of bed.
- Staring with loss of awareness.
- Left eyelid twitching.
- Episodes of right arm weakness for 10–20 min.

Interictal scalp EEG revealed bilateral parieto-occipital spikes with amplitude maximum at contact PO3 (Fig. 28.12A).

Ictal EEG characterized by bilateral posterior attenuation with parieto-occipital fast activity, and subsequent repetitive spiking at electrode P3 (Fig. 28.12B).

MRI showed a cavernous haemangioma in the left parietal area (Fig. 28.12C). The patient underwent an extended lesionectomy following intracranial EEG to map seizure onset zone and eloquent sensory cortex. He became free from disabling seizures.

Histopathology confirmed a cavernous haemangioma with adjacent chronic gliosis.

patterns often consist of repetitive bilateral occipital spiking or paroxysmal fast (55). Paradoxical lateralization of interictal and ictal EEG findings is most common in occipital lobe epilepsy (4).

Parietal lobe epilepsy

Clinical features suggesting parietal seizure onset (Case 28.6) are a somatosensory aura in contralateral hemibody areas, in particular if presenting as a Jacksonian march. The aura may also involve pain or a thermal sensation (58–60). Other subjective symptoms are a disturbance of body image, illusions of movement, vertigo, and complex visual or auditory illusions. Affective symptoms and waxing and waning of sensory symptoms have also been described (61). Objective behavioural manifestations are non-specific and may reflect ictal propagation into primary motor areas or SSMA (focal clonic activity contralateral to the side of seizure onset, tonic posturing), or the temporolimbic system (behavioural arrest, automatisms). Patients with parietal lobe epilepsy may have several seizure types (59,62). IIEDs may have a widespread distribution (fronto-centro-parietal, parieto-posterior-temporal, parieto-occipital, fronto-centro-temporal, fronto-temporoparietal or entire hemispheric distribution, with frequent secondary bilateral synchrony (63). In another series, they were actually least(!) localizing to the parietal electrodes (64). Ictal patterns may also be multiregional and/or falsely localizing, reflecting the widespread connectivity of the parietal cortex with occipital, temporal, frontal, and insula areas (59,64). Parietal lobe epilepsy therefore appears to be characterized by the absence of specific features, and has been named the 'great imitator' amongst focal epilepsy syndromes (64).

Treatment and prognosis of focal epilepsy

The ultimate goals in the treatment of people with focal epilepsy are freedom from disabling seizures with no disabling adverse effects of medication. Importantly, all anticonvulsant medications are a symptomatic treatment, i.e. they reduce the risk of seizures, but they do not affect the underlying pathological or pathophysiological substrate in the brain that generates seizures. In this sense, the commonly used term 'antiepileptic medication' is a misnomer. First line treatments for focal epilepsy recommended by the National Institute for Clinical Excellence (NICE, 65) are controlled-release carbamazepine and lamotrigine, based on the SANAD trial (66). NICE recommends levetiracetam, oxcarbazepine, and sodium valproate as further first-line treatment options if carbamazepine extended release and lamotrigine are unsuitable or not tolerated. Subsequently, zonisamide has been shown to be non-inferior to carbamazepine extended release in newly diagnosed focal epilepsy (67). Importantly, most patients require long-term or life-long treatment. Patient-specific factors that may influence the choice for or against a specific anticonvulsant are potential effects of an anticonvulsant on comorbidities, potential interactions with other medications and oral contraceptives, anticipated side effects and long-term effects, convenience of use, and cost. Fifteen anticonvulsants and vagus nerve stimulation have been licensed since 1990 for the add-on treatment of focal epilepsy in adults. To date, there are no patient-inherent factors known that predict the individual response (in terms of efficacy and tolerability) to any anticonvulsant.

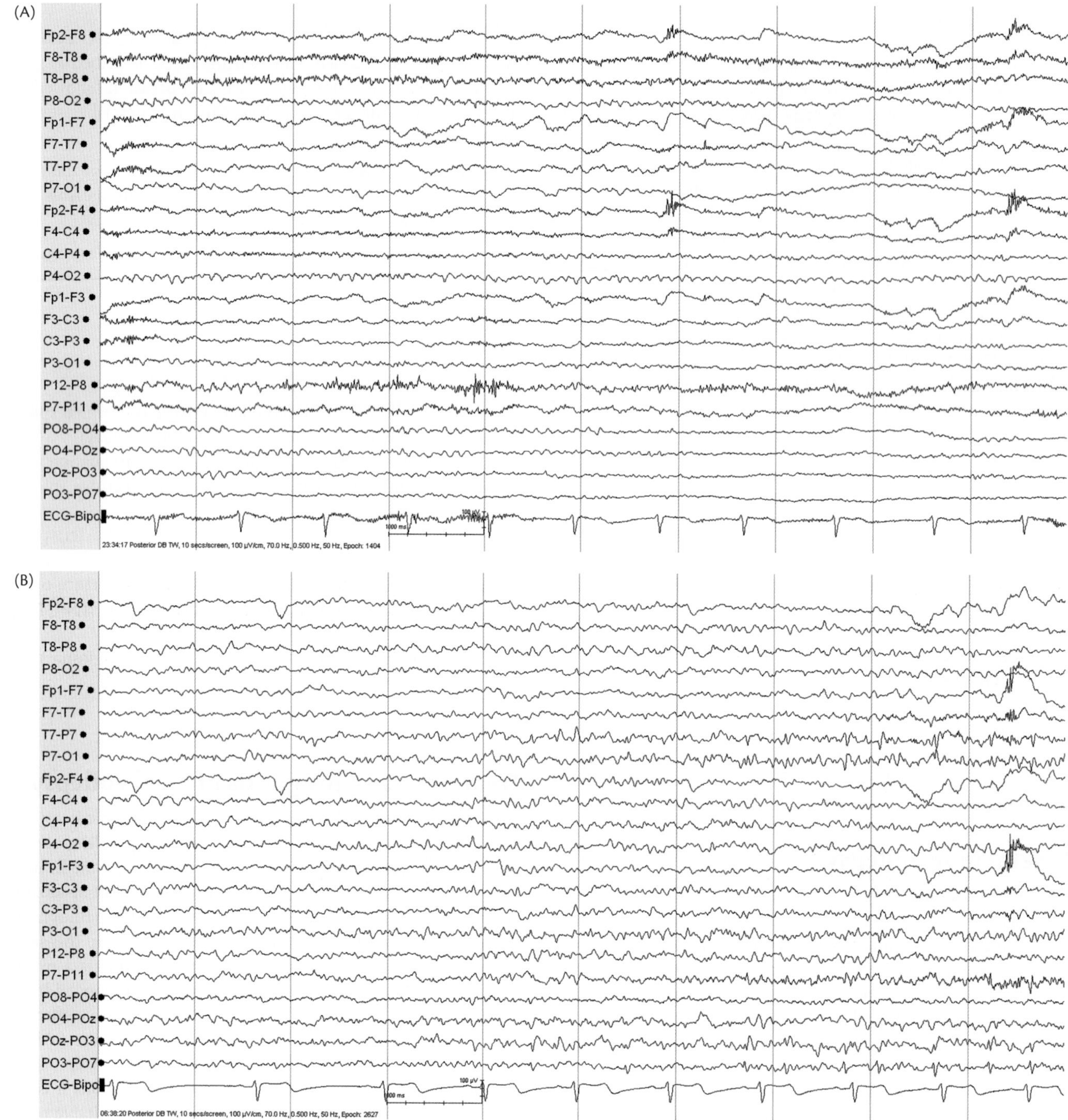

Fig. 28.11 (A) Interictal EEG shows continuous delta slow activity with no discernible posterior background activity over the left hemispheric leads. No interictal spikes were recorded. (B) Ictal EEG onset shows repetitive spikes at contacts PO7, P7, and PO3. (C) ~50 s later, a theta pattern emerges in the left frontotemporal leads. Note contacts P7, PO7 and PO3 are no longer involved. (D) Coronal FLAIR (left) and sagittal T1 (right) MRI demonstrate left occipital signal hyperintensity and volume loss with increased sulcation. Histopathology revealed a focal cortical dysplasia with ulegyria on the background of neonatal hypoxic-ischaemic injury.

A large observational study of more than one thousand patients identified four categories of treatment response: About 37% of people became seizure free on their first medication. A further 22% achieved seizure freedom within six months of treatment. About 25% of people never became seizure free on anticonvulsants. The remaining 16% alternated between periods of seizure freedom and periods of seizure relapses (68). Importantly, the chance of sustained seizure freedom is largely predicted by the initial response to anticonvulsant therapy and decays with each unsuccessful treatment attempt (69). Accordingly, the ILAE has defined drug resistant epilepsy 'as failure of adequate trials of two tolerated, appropriately chosen and used antiepileptic drug schedules (whether as monotherapies or in combination) to achieve sustained seizure freedom.' (70). In selected patients with drug-resistant focal epilepsy, resective neurosurgery offers an up to 70% chance of seizure freedom over 2–5 years, provided the

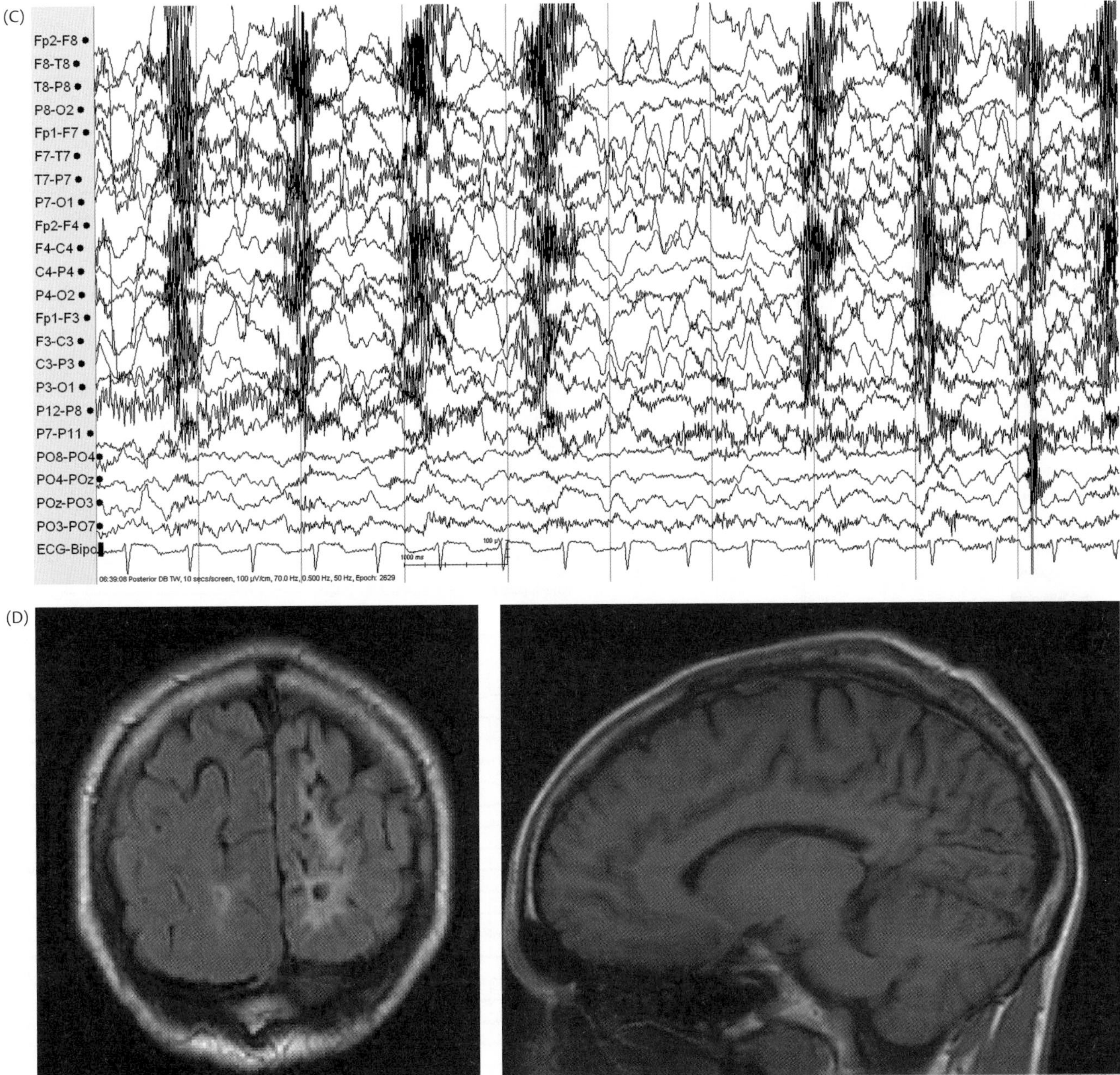

Fig. 28.11 Continued.

epileptogenic zone can be precisely identified and safely resected. Evaluation for surgical candidacy should thus be considered in patients who failed two or more adequate anticonvulsant trials (71). At a minimum, this consists of inpatient video EEG monitoring to record habitual seizures, dedicated MR imaging, and psychometric and psychiatric evaluation prior to a multidisciplinary team review. In selected cases, further diagnostic tests may include FDG-PET and ictal and interictal SPECT imaging, MEG, and intracranial EEG recordings (72).

Irrespective of surgical candidacy, NICE (65) recommends referral to specialist (tertiary) epilepsy centres if one or more of the following criteria are met:

- The epilepsy is not controlled with medication within 2 years.
- Management is unsuccessful after two drugs.
- The child is aged under 2 years.
- The patient experiences, or is at risk of, unacceptable side effects from medication.
- There is a unilateral structural lesion.
- There is psychological and/or psychiatric co-morbidity.
- There is diagnostic doubt as to the nature of the seizures and/or epilepsy syndrome.

Conclusions

Clinical and EEG findings in focal epilepsies reflect interictal and ictal activation or inhibition of restricted cortical areas. Although no single clinical or EEG feature allows precise anatomical classification of a focal epilepsy, a careful history and detailed observation

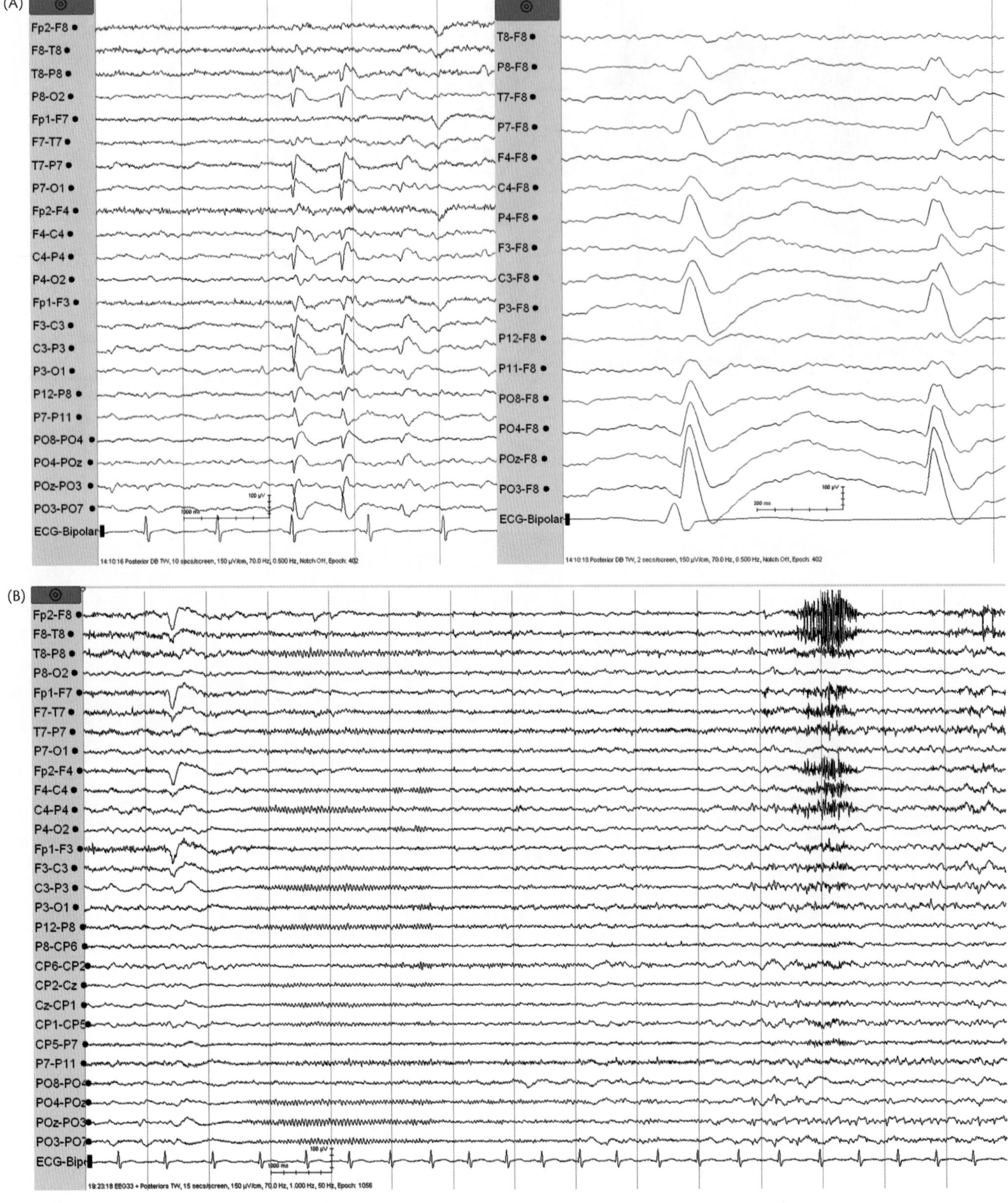

Fig. 28.12 (A) Run of parieto-occipital spikes. Longitudinal bipolar montage (left), referential montage of selected channels with reduced time base (right). Note the phase reversal at PO3 on the left, and the amplitude maximum at PO3, the minimal delay in peak negativity between PO3 and other ipsi- and contralateral parieto-occipital contacts, and the slightly blunter configuration of the spikes in the latter electrodes, all suggesting spike propagation from PO3. (B) Ictal EEG shows bilateral parieto-occipital fast activity, followed by repetitive spiking at PO3. (C) Axial T2 MRI revealed a cavernoma abuting the left post-central gyrus (left). Surface rendered view of the cavernoma (red), and foot and hand motor areas identified by tapping paradigms on functional MRI (green, right).

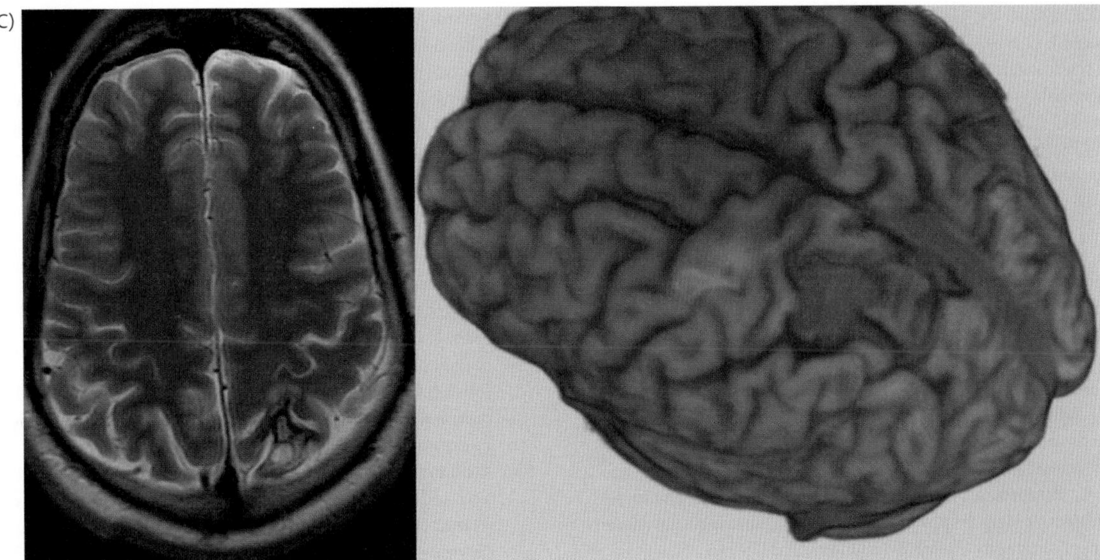

Fig. 28.12 Continued.

of seizure semiology in combination with interictal and ictal EEG analysis provide relevant diagnostic information that helps to define the patient's epilepsy.

References

1. Pedley, T.A., Mendiratta, A., and Walczak, TS. (2003). Seizures and epilepsy. In: Ebersole JS and Pedley TA (Eds) *Current Practice of Clinical Electroencephalography* 3rd edn. Philadelphia, PA: Lippincott Williams & Wilkins. 506–77.
2. Fisher, R.S., Acevedo, C., Arzimanoglou, A., et al. (2014). ILAE official report: a practical clinical definition of epilepsy. *Epilepsia*, **55**, 475–8.
3. Tao, J.X., Ray, A., Hawes-Ebersole, S., and Ebersole, J.S. (2008). Intracranial EEG substrates of scalp interictal spikes. *Epilepsy Research*, **82**, 190–3.
4. Foldvary, N., Klem, G., Hammel, J., Bingaman, W., Najm, I., and Lüders, H. (2001). The localizing value of ictal EEG in focal epilepsy. *Neurology*, **57**, 2022–8.
5. Sammaritano, M., de Lotbinière, A., Andermann, F., Olivier, A., Gloor, P., and Quesney, L.F. (1987). False lateralization by surface EEG of seizure onset in patients with temporal lobe epilepsy and gross focal cerebral lesions. *Annals of Neurology*, **21**, 361–9.
6. Mintzer, S., Cendes, F., Soss, J., et al. (2004). Unilateral hippocampal sclerosis with contralateral temporal scalp ictal onset. *Epilepsia*, **45**, 792–802.
7. Catarino, C.B., Vollmar, C., and Noachtar, S. (2012). Paradoxical lateralization of non-invasive electroencephalographic ictal patterns in extra-temporal epilepsies. *Epilepsy Research*, **99**, 147–55.
8. Brigo, F., Cicero, R., Fiaschi, A., and Bongiovanni, L.G. (2011). The breach rhythm. *Clinical Neurophysiology*, **122**, 2116–20.
9. Tatum, I.V.W.O., Husain, A.M., Benbadis, S.R., and Kaplan, P.W. (2006). Normal adult EEG and patterns of uncertain significance. *Journal of Clinical Neurophysiology*, **23**, 194–207.
10. Krauss, G.L., Abdallah, A., Lesser, R., Thompson, R.E., and Niedermeyer, E. (2005). Clinical and EEG features of patients diagnosed with EEG wicket rhythms misdiagnosed with epilepsy. *Neurology*, **64**, 1879–83.
11. Blume, W.T., Kaibara, M., and Young, G.B. (Eds) (2002). *Atlas of adult electroencephalography*, 2nd edn. Philadelphia, PA: Lippincott Williams & Wilkins.
12. Westmoreland, B.F. and Klaas, D.W. (1981). A distinctive rhythmic EEG discharge of adults. *Electroencephalography & Clinical Neurophysiology*, **51**, 186–91.
13. Westmoreland, B.F. and Klaas, D.W. (1997). Unusual variants of subclinical rhythmic electrographic discharge of adults (SREDA). *Electroencephalography & Clinical Neurophysiology*, **102**, 1–4.
14. Rémi, J., Vollmar, C., de Marinis, A., Heinlin, J., Peraud, A., and Noachtar, S. (2011). Congruence and discrepancy of interictal and ictal EEG with MRI lesions in focal epilepsies. *Neurology*, **77**, 1383–90.
15. Tatum, I.V.W.O. (2012). Mesial temporal lobe epilepsy. *Journal of Clinical Neurophysiology*, **29**, 356–65.
16. Gil-Nagel, A. and Risinger, M.W. (1997). Ictal semiology in hippocampal versus extrahippocampal temporal lobe epilepsy. *Brain*, **120**, 183–92.
17. Ferrari-Marinho, T., Caboclo, L.O., Marinho, M.M., et al. (2012). Auras in temporal lobe epilepsy with hippocampal sclerosis: relation to seizure focus laterality and post surgical outcome. *Epilepsy & Behaviour*, **24**, 120–5.
18. Foldvary, N., Lee, N., Thwaites, G., et al. (1997). Clinical and electrographic manifestations of lesional neocortical temporal lobe epilepsy. *Neurology*, **49**, 757–63.
19. Rémi, J., Wagner, P., O'Dwyer, R., et al. (2011). Ictal head turning in frontal and temporal lobe epilepsy. *Epilepsia*, **52**, 1447–51.
20. Kuba, R., Tyrlíková, I., Brázdil, M., and Rektor, I. (2010). Lateralized ictal dystonia of upper and lower limbs in patients with temporal lobe epilepsy. *Epileptic Disorders*, **12**, 109–15.
21. Rashid, R.M., Eder, K., Rosenow, J., Macken, M.P., and Schuele, S.U. (2010). Ictal kissing: a release phenomenon in non-dominant temporal lobe epilepsy. *Epileptic Disorders*, **12**, 262–9.
22. Loddenkemper, T. and Kotagal, P. (2005). Lateralising signs during seizures in focal epilepsy. *Epilepsy & Behaviour*, **7**, 1–17.
23. Lee, S.Y., Lee, S.K., Yun, C.H., Kim, K.K., and Chung, C.K. (2006). Clinico-electrical characteristics of lateral temporal lobe epilepsy; anterior and posterior lateral temporal lobe epilepsy. *Journal of Clinical Neurology*, **2**, 118–25.
24. Ebersole, J.S. (2000). Sublobar localization of temporal neocortical epileptogenic foci by source modeling. *Advances in Neurology*, **84**, 353–63.

25. O'Brien, T.J., Kilpatrick, C., Murrie, V., Vogrin, S., Morris, K., and Cook, M.J. (1996). Temporal lobe epilepsy caused by mesial temporal sclerosis and temporal neocortical lesions. A clinical and electroencephalographic study of 46 pathologically proven cases. *Brain*, **119**, 2133–41.
26. Kennedy, J.D. and Schuele, S.U. (2012). Neocortical temporal lobe epilepsy. *Journal of Clinical Neurology*, **29**, 366–70.
27. Ebersole, J.S. and Pacia, S.V. (1996). Localization of temporal lobe foci by ictal EEG patterns. *Epilepsia*, **37**, 386–99.
28. Barba, C., Barbati, G., Minotti, L., Hoffmann, D., and Kahane, P. (2007). Ictal clinical and scalp-EEG findings differentiating temporal lobe epilepsies from temporal 'plus' epilepsies. *Brain*, **130**, 1957–67.
29. von Lehe, M., Wagner, J., Wellmer, J., Clusmann, H., and Kral, T. (2012). Epilepsy surgery of the cingulate gyrus and the fronto-mesial cortex. *Neurosurgery*, **70**, 900–10.
30. Canuet, L., Ishii, R., Iwase, M., et al. (2008). Cephalic auras of supplementary motor area origin: an ictal MEG and SAM(g2) study. *Epilepsy & Behaviour*, **13**, 570–4.
31. Beauvais, K., Biraben, A., Seigneuret, E., Saïkali, S., and Scarabin, J.M. (2005). Subjective signs in premotor epilepsy: confirmation by stereo-electroencephalography. *Epileptic Disorders*, **7**, 347–54.
32. Unnwongse, K., Wehner, T., Bingaman, W., and Foldvary-Schaefer, N. (2010). Gelastic seizures and the anteromesial frontal lobe: a case report and review of intracranial EEG recording and electrocortical stimulation case studies. *Epilepsia*, **51**, 2195–8.
33. Leung, H., Schindler, K., Clusmann, H., et al. (2008). Mesial frontal epilepsy and ictal body turning along the horizontal body axis. *Archives of Neurology*, **65**, 71–7.
34. Rheims, S., Ryvlin, P., Scherer, C., et al. (2008). Analysis of clinical patterns and underlying epileptogenic zones of hypermotor seizures (HMS). *Epilepsia*, **49**:2030–40.
35. Unnwongse, K., Wehner, T., and Foldvary-Schaefer, N. (2012). Mesial frontal lobe epilepsy. *Journal of Clinical Neurophysiology*, **29**, 371–8.
36. Bonini, F., McGonigal, A., Trebuchon, A., et al. (2014). Frontal lobe seizures: From clinical semiology to localization. *Epilepsia*, **55**, 264–77.
37. Ikeda, A., Hirasawa, K., Kinoshita, M., et al. (2005). Negative motor seizure arising from the negative motor area: is it ictal apraxia? *Epilepsia*, **5**, 2072–84.
38. Bonelli, S.B., Lurger, S., Zimprich, F., Stogmann, E., Assem-Hilger, E., and Baumgartner, C. (2007). Clinical seizure lateralisation in frontal lobe epilepsy. *Epilepsia*, **48**, 517–23.
39. Kriegel, M.F., Roberts, D.W., and Jobst, B.C. (2012). Orbitofrontal and insular epilepsy. *Journal of Clinical Neurophysiology*, **29**, 385–91.
40. Staack, A.M., Bilic, S., Wendling, A.S., et al. (2011). Hyperkinetic seizures in patients with temporal lobe epilepsy: clinical features and outcome after temporal lobe resection. *Epilepsia*, **52**, 1439–46.
41. Lee, R.W. and Worrell, G.A. (2012). Dorsolateral frontal lobe epilepsy. *Journal of Clinical Neurophysiology*, **29**, 379–84.
42. Beleza, P., Bilgin, O., and Noachtar, S. (2009). Interictal rhythmical midline theta differentiates frontal from temporal lobe epilepsies. *Epilepsia*, **50**, 550–5.
43. Vadlamudi, L., So, E.L., Worrell, G.A., et al. (2004). Factors underlying scalp-EEG interictal epileptiform discharges in intractable frontal lobe epilepsy. *Epileptic Disorders*, **6**, 89–95.
44. Blume, W.T. and Oliver, L.M. (1996). Noninvasive electroencephalography in supplementary sensorimotor area epilepsy. *Advances in Neurology*, **70**, 309–17.
45. Foldvary-Schaefer, N. and Unnwongse, K. (2011). Localizing and lateralizing features of auras and seizures. *Epilepsy & Behaviour*, **20**, 160–6.
46. Bautista, R., Spencer, D., and Spencer, S. (1998). EEG findings in frontal lobe epilepsies. *Neurology*, **50**, 1765–71.
47. Isnard, J., Guénot, M., Sindou, M., and Mauguiere, F. (2004). Clinical manifestations of insular lobe seizures: a stereo-electroencephalographic study. *Epilepsia*, **45**, 1079–90.
48. Isnard, J., Guénot, M., Ostrowsky, K., Sindou, M., and Maguière, F. (2000). The role of the insular cortex in temporal lobe epilepsy. *Annals of Neurology*, **48**, 614–23.
49. Ryvlin, P., Minotti, L., Demarquay, G., et al. (2006). Nocturnal hypermotor seizures, suggesting frontal lobe epilepsy, can originate in the insula. *Epilepsia*, **47**, 755–65.
50. Dobesberger, J., Ortler, M., et al. (2008). Successful surgical treatment of insular epilepsy with nocturnal hypermotor seizures. *Epilepsia*, **49**, 159–62.
51. Bien, C.G., Benninger, F.O., Urbach, H., et al. (2000). Localizing value of epileptic visual auras. *Brain*, **123**, 244–53.
52. Jobst, B.C., Williamson, P.D., Thadani, V.M., et al. (2010). Intractable occipital lobe epilepsy: clinical characteristics and surgical treatment. *Epilepsia*, **51**, 2334–7.
53. Williamson, P.D., Thadani, V.M., Darcey, T.M., Spencer, D.D., Spencer, S.S., and Mattson, R.H. (1992). Occipital lobe epilepsy: clinical characteristics, seizure spread patterns, and results of surgery. *Annals of Neurology*, **31**, 3–13.
54. Blume, W.T., Wiebe, S., and Tapsell, L.M. (2005). Occipital epilepsy: lateral versus mesial. *Brain*, **128**, 1209–25.
55. Adcock, J.E. and Panayiotopoulos, C.P. (2012). Occipital lobe seizures and epilepsies. *Journal of Clinical Neurophysiology*, **29**, 397–407.
56 Kun Lee, S., Young Lee, S., Kim, D.W., Soo Lee, D., and Chung, C.K. (2005). Occipital lobe epilepsy: clinical characteristics, surgical outcome, and role of diagnostic modalities. *Epilepsia*, **46**, 688–95.
57. Tandon, N., Alexopoulos, A.V., Warbel, A., Najm, I.M., and Bingaman, W.E. (2009). Occipital epilepsy: spatial categorization and surgical management. *Journal of Neurosurgery*, **110**, 306–18.
58. Kim, D.W., Lee, S.K., Yun, C.H., et al. (2004). Parietal lobe epilepsy: the semiology, yield of diagnostic workup, and surgical outcome. *Epilepsia*, **45**, 641–9.
59. Salanova, V. (2012). Parietal lobe epilepsy. *Journal of Clinical Neurophysiology*, **29**, 392–6.
60. Williamson, P.D., Boon, P.A., Thadani, V.M., et al. (1992). Parietal lobe epilepsy: diagnostic considerations and results of surgery. *Annals of Neurology*, **31**, 193–201.
61. McGonigal, A. and Bartolomei, F. (2014). Parietal lobe seizures mimicking psychogenic nonepileptic seizures. *Epilepsia*, **55**, 196–7.
62. Salanova, V., Andermann, F., Rasmussen, T., Olivier, A., and Quesney, L.F. (1995). Parietal lobe epilepsy: clinical manifestations and outcome in 82 patients treated surgically between 1929–1988. *Brain*, **118**, 607–27.
63. Salanova, V., Andermann, F., Olivier, A., Rasmussen, T., and Quesney, L.F. (1992). Occipital lobe epilepsy: electroclinical manifestations, electrocorticography, cortical stimulation and outcome in 42 patients treated between 1930 and 1991. Surgery of occipital lobe epilepsy. *Brain*, **115**, 1655–80.
64. Ristic, A.J., Alexopoulos, A.V., So, N., Wong, C., and Najm, I.M. (2012). Parietal lobe epilepsy: the great imitator among focal epilepsies. *Epileptic Disorders*, **14**, 22–31.
65. National Institute for Health and Clinical Excellence. (2012). The epilepsies: the diagnosis and management of the epilepsies in adults and children in primary and secondary care. London: NICE. Available at: www.nice.org.uk/cg137 (accessed 3 April 2016).
66. Marson, A.G, Al-Kharusi, A.M., Alwaidh, M., et al. (2007). The SANAD study of effectiveness of carbamazepine, gabapentin, lamotrigine, oxcarbazepine, or topiramate for treatment of partial epilepsy: an unblinded randomised controlled trial. *Lancet*, **369**, 1000–15.
67. Baulac, M., Brodie, M.J., Patten, A., Segieth, J., and Giorgi L. (2012). Efficacy and tolerability of zonisamide versus controlled-release

carbamazepine for newly diagnosed partial epilepsy: a phase 3, randomised, double-blind non-inferiority trial. *Lancet: Neurology*, **11**, 579–88.

68. Brodie, M.J., Barry, S.J.E., Bamagous, J.A., Norrie, J.D., and Kwan, P. (2012). Patterns of treatment response in newly diagnosed epilepsy. *Neurology*, **78**, 1548–54.
69. Schiller, Y. and Najjar, Y. (2008). Quantifying the response to antiepileptic drugs: effect of past treatment history. *Neurology*, **70**, 54–65.
70. Kwan, P., Arzimanoglou, A., Berg, A.T., et al. (2010). Definition of drug resistant epilepsy: consensus proposal by the ad hoc Task Force of the ILAE Commission on Therapeutic Strategies. *Epilepsia*, **51**, 1069–77.
71. Wiebe, S. and Jette, N. (2012). Pharmacoresistance and the role of surgery in difficult to treat epilepsy. *Nature Reviews: Neurology*, **8**, 669–77.
72. Lüders, H.O. (Ed) (2008). *Textbook of epilepsy surgery*. London: Informa Healthcare.

CHAPTER 29

Syncope

Shane Delamont

Definition of syncope: background

The word *syncope* comes to us from the Greek συγκοπή meaning to cut down, with its implication of sudden alteration in functioning. It has been in use from at least the 16th century as a synonym for a loss of consciousness with a weak or thready pulse, and has been used synonymously with fainting ever since. From this definition, it is evident that it is a disorder that has been described from the outside, in other words from the bystander's perspective, rather than the sufferer's. While a faint is a common experience with a cumulative experience in females over the age of 80 years of 50%; (1) most popular literature has pictured it from the observer's perspective. This concentration on the dramatic appearances, the suddenness of onset, and the potential for injury has made detailed studies difficult.

Definition of syncope: current

The abrupt failure of effective vascular global cerebral circulation with secondary loss of awareness is the current working definition for syncope. It is of short duration and there is spontaneous recovery (2). The failure to lose awareness is often described as presyncope and suggests a milder form of the disorder, although both terms are frequently lumped together in discussion and pre-syncope acts as an aura for syncope where there is loss of consciousness.

Epidemiology

With such a broad definition, it has become increasingly clear that determining the incidence of syncope is fraught with difficulties as a result of varying interpretations of its definition. This is true even for serial studies carried out by the same institution, as illustrated by the Framingham studies in 1985 (3), 1999 (4), and 2002 (5). The years of patient follow-up were 26, 4, and 17, respectively. Age ranges studied varied from 30 to 62 years, 26 to 84 years, and 20 to 96 years. In the last study with 7814 participants, the incidence of first syncopal event was 6.2 per 1000 person years, but it was age dependent rising from 5.7 per 1000 person years in men aged 60–69 years to 11.1 per 1000 person years in men aged 70–79 years. The most common form of syncope was reflex syncope with 21.2% of episodes followed by cardiac syncope 9.5% and orthostatic hypotension in 9.4%. Of great interest was the finding that 44% of the cohort was not evaluated at all. They report that there was no difference in survival in those with neurocardiogenic syncope and those without evaluation. The authors included seizures, strokes, and transient ischaemic attacks in their population studied, making interpretation of these data even more difficult.

In contrast, data from a number of studies suggest that the cumulative incidence is higher and has at least two age peaks, one at 15 years and the second at 70 years (3–7).

Specific groups

Young people have a higher cumulative incidence of syncope in several studies with up to 47% of females and 24% of males reporting at least one syncopal event by age 24 years (6). These findings are supported by several other studies suggesting these are likely to represent a truer picture of the incidence of syncope in this part of the population (8,9).

Amongst the older age group, the effects of syncope are associated with a greater morbidity and mortality (10) with similar impairment to quality of life as seen in other chronic disorders, such as epilepsy (11). The data for the elderly is not as clear cut as for the young because of the overlap with falls (12). This limited data suggests an incidence of syncope of at least 6% per year, with a 10% prevalence and a 30% 2-year recurrence rate in those over 70 years of age (12). A more recent study using the Irish Longitudinal Study on Aging (TILDA) suggests that by combining the history of faints and non-accidental falls in the previous year the true prevalence ranges from 4.4 to 8.8% (1).

Syncope and co-morbidities

In a large prospective Danish study identifying all admissions to emergency departments (ED) (13) it was noted that 0.9% of total hospital admissions and 0.6% of ED visits were due to syncope, and from 1997 to 2009 there was an increase in the incidence rates of syncope from 13.8 to 19.4 per 1000 person years similar to the UK and the USA (14). The study showed that 28% of the syncopal population had cardiovascular co-morbidity compared with 14% of the control population, and 66% of the 50–79-year-old age group with syncope were taking at least one cardiovascular medication compared with 51% without syncope. The authors emphasize the importance of assessing for co-morbidities and medication, particularly cardiovascular disease and medication. This has been recognized by National Institute for Clinical Excellence (NICE) (15), which released guideline CG109 for managing transient loss of consciousness in August 2010 with guidelines in the initial assessment commencing with a cardiac assessment followed by orthostatic hypotension and epilepsy (Sections 1.1.2.1–1.1.4.6, 1.2.1, and 1.2.2). The importance of selecting patients for further assessment is further emphasized by a Dutch study using a GP database (16). The prevalence of syncope is estimated to lie between 18.1 and 39.7 per 1000 patient years. 9.3 will attend their general practice and 0.7

will present to an ED. This further emphasizes the importance of risk stratification of these patients and was the reasoning behind the NICE guidance, which was directed primarily at managing patients in the ED. The causes for syncope have been shown to be reflex syncope (up to 40%), orthostatic hypotension (6–24%), cardiac syncope (10–20%), and psychogenic causes (1–5%). A number of studies have been undertaken in order to introduce order into a complex field without as yet definite answers being found. There is currently a Canadian study that is trying to address this issue (17). There is an age effect with more cardiac and orthostatic causes seen in older people, but reflex syncope remains the commonest cause throughout the ages, although it has a less benign course in the elderly compared with the young (18).

Syncope: the clinical syndrome

The very word, syncope suggests a sudden event, but the clinical spectrum, in fact, is quite wide. In the emergency room, outpatient setting or clinical decision unit, it is useful to divide the history into three components:

- *The time period prior to loss of consciousness* (LOC): the prodrome.
- *The time period of LOC*: the attack.
- *The time period of recovery following LOC*: post-attack.

The time period prior to LOC should be looked upon conceptually the same way that an aura is in the field of epilepsy. A detailed history needs to be taken from the patient who can often describe the prodrome and post-event features and any eyewitness who can also describe the three components and may note external triggers, such as stress not recognized by the patient (19). Furthermore, the 'eyewitness' should include videos of any attacks, increasingly available via mobile phones, CCTV, and cameras. Relatives often cite a desire not to embarrass the patient by videoing their attacks, but the overall benefits of an accurate diagnosis need to be emphasized. This is particularly important for recurrent and frequent attacks. Careful questioning can often identify similar symptoms without LOC occurring prior to the first LOC bringing the patient to medical attention. Furthermore, when repeated episodes of LOC occur, the prodromes can both evolve differently between attacks and at variable speeds between attacks. The underlying aetiology can be suggested by the details of the prodrome, attack, and post-attack features, but they are by no means diagnostic. Certain features can point in a particular direction. Neurocardiogenic syncope or reflex syncope is most closely associated with physiological triggers. Cardiac syncope may have a cardiac stressor, but can be unheralded. Orthostatic syncope because of its name has an orthostatic trigger, while psychogenic syncope may have any of the preceding triggers, but there tends to be more variability in the history, and the attacks can be of longer duration and may have some features of psychogenic seizures. The commonest cause is reflex neurocardiogenic syncope and a detailed understanding of the clinical features is invaluable for accurate diagnosis.

Reflex neurocardiogenic syncope

The duration of the three components (prodrome, event, post-event) can vary between different attacks in the same person and between different patients. Prodromal symptoms can be present for many hours in a mild form in patients with a significant tendency to collapse, such as is seen in those with autonomic failure or severe postural orthostatic tachycardia syndrome (PoTS). The severity will fluctuate and can be influenced by posture being worse on prolonged standing, with exercise or post-prandially. These symptoms are often seen without LOC and are described as presyncopal in nature and should be sort diligently. They include a non-specific sense of disequilibrium, muzzy heads, poor concentration, a fear of collapse, weak legs, and there may be active manoeuvres to abort the symptoms, such as sitting down and resting. There may be sweating, nausea, and a desire to escape the situation ('get some fresh air'). There may be specific triggers for LOC superimposed on this background of presyncope. They vary with age with prolonged standing being more important in older populations with up to 54.9% reporting this trigger beyond age 30 years (20) compared with between 11.9 and 27% reporting this in studies with a mean age of 21–22 years (6,8). Syncope in older age is associated with the act of standing, but can be seen in the sitting position and post-prandially, and is more commonly seen when cardiovascular active drugs, such as anti-hypertensives are used. Younger patients report the effects of pain, venesection, viscous pain, concomitant illnesses, emotion, menstruation, drugs, fatigue, and fasting (1). Multiple triggers can be reported either on separate occasions or contributing to one event. The psychological impact of the physical triggers is often neglected in taking the history from the patient. Their understanding of the potential importance of the physical symptoms needs to be recognized. Sudden knee injury with pain following knee surgery may bring forth the belief that they may crippled as a result and is likely to be a more potent trigger for syncope than if they have no fear of that outcome. This illustrates the importance of the beliefs and patient's psychological state in assessing the significance of triggers including physical ones.

The immediate pre-LOC symptoms reported include overwhelming fatigue, hunger, sweating, breathlessness, a narrowing of vision, and general darkening of vision, sounds becoming more distant with a ringing in the ears, which can be high or low pitched. A rising sensation in the abdomen often accompanied by a sense of nausea without vomiting can lead to confusion with an epileptic aura, although the presyncopal symptoms are less intense as an experience. It should be noted that complex partial seizures particularly of fronto-temporal origin with secondary seizure-induced bradycardia or asystole is a well-recognized syndrome, which can be a trap for the clinician (21). In the classification of syncope, situational syncope is often separated from classical syncope. Examples include carotid sinus syndrome, swallow, defaecation, micturition, post-exercise, post-prandial, cough, sneeze, laughing, brass instrument playing, weight lifting, medical instrumentation, and prankster valsalva manoeuvres. This is artificial from a physiological perspective, but clinically useful. A detailed assessment of the triggering organ looking for local pathology should be undertaken. If related to direct intervention, a history of previous neurocardiogenic syncope or presyncope should be elucidated (22).

Orthostatic syncope

The definition includes syncope associated only with orthostatic changes. Syncope in this clinical situation can be due to a failure to maintain adequate cerebral perfusion due to blood volume deficiencies, reduced cardiac output, or impaired venous return with or without reflex neurocardiogenic syncope. The working definition is a reduction in systolic blood pressure (BP) by 20 mmHg or a reduction in diastolic BP by 10 mmHg at 3 min with associated syncope not confined to occurring within 3 min, however. Presyncopal symptoms are similar to those previously described, but low pressures can be surprisingly well tolerated by the older patient with a loss of the 'aura' before LOC. Slowly developing autonomic failure as is seen in primary autonomic failure can also be associated with remarkably well-tolerated low blood pressures.

Cardiac syncope

This is syncope which is associated with primary cardiac disease. This can be structural or arrhythmogenic and is described as being triggered by increasing cardiac workload particularly during exercise or with cardiac ischaemia. There is a greater probability with increasing age. In young populations, it is rarer, but can be seen particularly as part of a genetic disorder, so enquiries about a family history of sudden death is necessary. In a large preparticipation screening programme of 7568 athletes, only 57 had experienced pre-exertion syncope in the preceding 5 years and six had exertion related syncope. Of the latter group, one had hypertrophic obstructive cardiomyopathy, one had right ventricular outflow tachycardia, and the remaining four had neurocardiogenic syncope during the head-up tilt (HUT) testing, which reproduced their symptoms. Thus, in the young, exercise-induced syncope is rare and is not associated with a poor prognosis if there is no evidence of cardiac disease (23). What is not so clear is the role for genetic studies in those athletes who may have a family history of sudden death if there is no evidence for cardiac disease on formal assessment.

Other types of syncope

Much of the literature on syncope, presumes that transient loss of conciousness (TLOC) has been firmly established. In practice, in the ED or outpatients, the exact nature of the attack is not always quite so clear. Seizures and psychogenic attacks are the main differential diagnoses that must be considered. The importance of establishing the mechanism of the syncopal attack has received attention (14) and the authors draw attention to the variability in diagnosis and management of syncope with the implications for reduced quality of life and excess costs. Well-developed pathways with dedicated units and experienced clinicians are needed to manage these patients. Given the variability in the prevalence of syncope, being dependent on the population being studied, there are difficulties in determining accurately the incidence of psychogenic attacks. This is more difficult given the importance of psychological factors in contributing to syncopal events, and the recognition that syncope and psychogenic attacks can co-exist in the same patient. Therefore, the gold standard in identifying psychogenic attacks is triggering an event that shows normal heart rate (HR), BP, and electroencephalogram (EEG) patterns during an attack. Moreover, it is possible to trigger neurocardiogenic syncope (NCS) with physiological recovery but no behavioural recovery after the event where both processes are present in the same patient. This makes accurate assessment of incidence and prevalence of psychogenic attacks difficult. Previous studies suggest an incidence varying between 3.7% during tilt testing (24), 6% in possible or recurrent syncope (25), and up to 90% during tilt testing with recurrent syncope (26). This reflects the populations being studied and the importance of detailed history taking and assessment in this population.

Pathophysiology of syncope

The one thing in common with the different types of syncope is the reduction in cerebral perfusion, which underpins the LOC. The brain is highly dependent on cardiac output to meet its metabolic needs and it takes a disproportionate amount of the cardiac output. At rest despite constituting about 2% of body mass at 1400 g, it takes 15–20% of cardiac output (27). Cessation of blood flow by neck compression leads to loss of consciousness within 6 or 7 s (28). In addition to the central role for cardiac output, the baroreflex, which maintains blood pressure and cerebral autoregulation, which endeavours to maintain consistent cerebral blood flow despite variations in systemic blood pressure are the other major controllers of cerebral perfusion.

The baroreflex system

The baroreflex arc consists of afferent loops taking sensory information from receptors in carotid sinuses via the glossopharangeal nerves and from the aortic arch via the vagus to the nuclei of the tractus solitarius in the medulla. Integration with other inputs from throughout the nervous system takes place at this point before the efferent signals pass through the vagus to the heart, controlling HR rate and sympathetic nerves to influence heart muscle contractility, and vascular peripheral resistance to affect systemic blood pressure. An increase in BP at the carotid sinus will lead to an immediate slowing in HR (fast reflex response) and a reduction in sympathetic output which takes 2–3 s to be seen (slow response) (29,30) with a reduction in peripheral resistance and cardiac contractility.

Cerebral blood flow

Cerebral blood flow (CBF) in contrast to other vascular bed systems is best thought of as a high flow low resistance system (31). This means that when the system is working, there is flow during both systole and diastole. When resistance to flow rises, the effects are initially seen with a reduction in diastolic flow. CBF depends on the difference between mean arterial pressures and intracranial pressures (32,33). Intracranial pressures are influenced by venous pressures, tissue pressures, and the tendency for vessel walls to collapse. Venous drainage is influenced by position. While supine, it is largely by the jugular veins, while upright it is via the vertebral complex (34,35). Raised central venous pressures due to obstructive disease or during valsalva manoeuvres, e.g. cough can be associated with reflex syncope. Tissue pressures can be significant because the skull is a rigid limit on space for brain expansion as is seen with tumours or bleeding.

Cerebral circulation

CBF is controlled by a number of mechanisms, some of which are unique. At a cerebral vascular level, the humoral effects are thought to be attenuated by the blood–brain barrier, although differential effects of humoral mechanisms on large versus small vessels may play a part (36). There is a dense autonomic and sensory network on cerebral blood vessels. Sympathetic innervation is associated with vasoconstriction of larger vessels, but this is compensated by cerebral autoregulation with downstream vasodilation (37), so there is no net change in flow. While this has been shown in mammals, it has not been specifically shown in humans. The function of parasympathetic output is less clear, but supraphysiological stimulation of the sphenopalantine ganglion is associated with increased cerebral blood flow (38). Sensory fibres are associated with modulating cerebral blood flow and can be associated with increased blood flow as is seen in cortical spreading depression (39), seizures (40), and reactive hyperaemia (41). Metabolic stimuli such as tissue adenosine, lactate, pCO_2, pH, and hypoxaemia are thought to act via a coupling mechanism to modulate cerebral blood flow. This mechanism involves nitric oxide released from neurons (42–44). Autoregulation of blood flow is most effective in cerebral, renal, and mesenteric vessels (45), and the mechanisms involved include myogenic, metabolic, neural, and activation of potassium channels (46–48). Hypercapnia and hypoxaemia of the cerebral circulation are both potent vasodilators of cerebral blood vessels, although the precise mechanisms may vary between the two (49,50). In terms of syncope, the regulation of cerebral blood flow by large arteries compensating for steal type phenomena is of particular interest (51,52). Finally, autoregulation has both static and dynamic aspects with the latter having a delay of between 2 and 10 s, which is the timescale seen for most forms of syncope and is the timescale for Mayer type BP waves seen during presyncope (32).

Physiology of neurocardiogenic syncope

There is an understanding of aspects of the mechanisms involved in NCS, but the ultimate controlling mechanism remains to be established. There are two arms to the process, the afferent and the efferent, with the brainstem acting as the integrating mechanism. The ultimate mechanism for all syncope is loss of arterial pressure due to either failure of entry of blood into the circulation or loss of peripheral resistance. In NCS, there is loss of BP and there may be slowing of HR. These two events are related, but independent and variable in manifestation. The efferent pathway has been divided into four phases during orthostatic induced NCS using tilt and lower body suction (53). Each phase can vary in duration. Phase 1 immediately following tilt is a period of stability with reduced cardiac vagal tone and increased peripheral resistance. Phase 2 is associated with rising HR, further reduction in cardiac vagal tone and little change in peripheral resistance. Phase 3 is associated with rhythmical oscillations in BP and to a lesser extent HR at 0.1 Hz, which is thought to represent the intrinsic frequency of the cardiovascular system (54). Cardiac vagal tone and baroreflex sensitivity as measured by the cardiac sensitivity to the baroreflex (53) are maximally withdrawn with loss of effective cardiovascular buffering and the appearance of Mayer BP waves. Such changes are seen in patients on standing and can be associated with symptoms of presyncope, and they can be modified by techniques to increase BP, such as squeezing lower limbs or buttocks. Phase 4 follows with a fall in systolic BP initially and increasing oscillations in BP with diastolic BP affected and a spiral of decreasing BP. The overall fall in BP is associated with a fall in peripheral resistance due to withdrawal of sympathetic function as assessed indirectly via blood flow studies (53) or by direct microneurographic techniques (55). Furthermore, baroreflex sensitivity has been shown to decrease in subjects undergoing orthostasis (53,56). This is not always shown when other methods are used, such as neck suction and blood volume manipulation (57). Despite this, the consistent finding of reduced baroreflex sensitivity with orthostasis alone suggests that in this posture, bipedal animals such as humans are physiologically under stress. Any further contributors to that stress will therefore tend to bring on symptoms when in that posture.

Clinical approach to syncope

The approach to syncope varies depending on the circumstances where the patient is seen, such as the emergency department, outpatients, short stay wards in hospital, and geriatric departments. The underlying theme is to establish whether this discrete event represents a disruption in normal brain electrical activity, cerebral perfusion, a psychological functioning, a combination of any of the preceding, or something altogether different, such as primary cardiac disease or a sleep-related disorder. The index event should be described by the patient who will often be able to describe the pre-event circumstances, symptoms leading up to the loss of awareness and post-event symptoms. Any eye-witnesses should be interrogated to cover the same period and they should be encouraged to act out what they have seen, despite the potential for embarrassment for both eye-witness and patient. Identifying the eye-witness is also useful in case the history needs to be reviewed in the future and establishing their experience in assessing episodic behavioural events. If there is any photographic evidence such as a home video, it should be examined. This last aspect is invaluable for recurrent attacks and it forms an essential component of the clinical assessment of patients with recurrent attacks.

The NICE approach

NICE released a guideline in August 2010 for the management of TLOC in adults and young people. The guideline was designed for aiding the assessment, diagnosis and specialist referral for people with TLOC. The idea was to introduce some order into a condition that is inherently complex, and where clinical decisions are often made with inadequate data leading to variability in management and outcomes with the potential for serious morbidity and mortality affecting a substantial component of the population, at least 50% of whom will have a TLOC at some time in their life.

Initial assessment

- Circumstances of the event.
- Person's posture immediately before loss of consciousness.
- Prodromal symptoms (such as sweating or feeling warm/hot).
- Appearance (for example, whether eyes were open or shut) and colour of the person during the event.
- Presence or absence of movement during the event (for example, limb-jerking and its duration).
- Any tongue-biting (record whether the side or the tip of the tongue was bitten).
- Injury occurring during the event (record site and severity).
- Duration of the event (onset to regaining consciousness).
- Presence or absence of confusion during the recovery period.
- Weakness down one side during the recovery period.
- Electrocardiogram (ECG) looking for a conduction abnormality (for example, complete right or left bundle branch block or any degree of heart block) evidence of a long or short QT interval, or any ST segment or T wave abnormalities.

With this basic information it is possible to send the patient down the most appropriate initial pathway. If there is an ECG abnormality, evidence for heart failure (history or physical signs), TLOC during exertion, a family history of sudden cardiac death in people aged younger than 40 years and/or an inherited cardiac condition, new or unexplained breathlessness, or a heart murmur, then they should be seen within 24 h at a cardiac clinic. One should also consider referring within 24 h for cardiovascular assessment for anyone aged older than 65 years who has experienced TLOC without prodromal symptoms.

If there are features of neurocardiogenic syncope then they should be reviewed through primary care. If there are features of seizure then they should be assessed through the epilepsy service or the first seizure service.

Syncope: singular or repeated

While a single episode of syncope can be the presentation for significant disease particularly cardiac, it may not reach medical attention. It is repeated events, particularly if associated with injury that will bring patients to medical attention consistently. If confident from the history that these events are TLOC, then their frequency determines how to record ECG and this is the next step. If the TLOCs occur several times a week, external Holter monitoring up to 48 h is the next step. If TLOC occurs every 1–2 weeks then offer an external recorder with the ability to note when the events occur. If the events are less frequent, then offer an implantable event recorder (15). These approaches are recommended before considering HUT. This advice applies if there is certainty as to the presence of TLOC. Where there is doubt as to the nature of the events, then an attempt to trigger an event is often necessary. It is important to realize that clinically this is the most important indication for performing HUT. There remains some debate as to how to undertake this procedure. A balance must be struck between sensitivity and specificity for the induced event being clinically relevant.

The role for neurophysiology in syncope

Neurophysiology is used to assess autonomic cardiovascular regulation and identify any other relevant abnormalities, such as an abnormal EEG or for signs of a generalized neuropathy. The general approach is to determine if there is a fixed deficit in autonomic regulation. This requires assessing autonomic function, while resting after the body has established a steady state and then during a series of manoeuvres. A basic screening test can include 24-h HR and BP recordings to identify basic circadian regulation. This information is useful, but does not provide any great detail on the different components of cardiovascular autonomic control. To elucidate this requires an assessment in an autonomic laboratory. Access to autonomic laboratories is often difficult or limited. The laboratories are expensive to maintain and need dedicated well trained staff to carry out investigations in a reproducible manner in a controlled environment. Neuroscience centres have well-developed physiology departments, which have well trained staff used to reproducibly performing physiological tests on patients. This is an ideal environment to develop physiological assessment of autonomic function in close collaboration with cardiology departments.

The approach to assessment of autonomic function

Within the autonomic laboratory, it is necessary to assess cardiorespiratory parameters at rest and during certain manoeuvres. The manoeuvres are designed to stress the regulatory system and identify the functioning of different components of the autonomic nervous system. The components include cardiac vagal control, cardiac sympathetic, vascular sympathetic, and splanchnic sympathetic control. These simple measures enable one to approach the complexities of cardiovascular autonomic regulation systematically. There are a number of ancillary measures that can be assessed including HR changes, and baroreflex sensitivity under different conditions. The manoeuvres used include carotid stimulation to assess cardiac vagal and baroreflex blood pressure responses, the effects of hyperventilation on cardiac vagal control, the act of standing and remaining standing for a period of time. To assess for PoTS, the commonest cause for recurrent neurocardiogenic syncope, standing for a minimum of 10 min is needed. The effects of a valsalva manoeuvre can be used to assess cardiovagal baroreflex sensitivity during the phase II_E and the adrenergic component (58). It can also be used to assess the splanchnic vascular response to reduced cardiac filling (59). Furthermore, pressor responses can be elicited by isometric grip, mental arithmetic and cold stimulation (60). These stimuli induce changes in both HR and BP, and give an idea about sympathetic responses mediated through $\beta1$ (heart) and $\alpha1$ (resistance blood vessels) receptors (61).

Management

There is a perception in clinical medicine that a syncopal event is not as serious as an epileptic seizure and perhaps this reflects a widespread neurophobia in clinical medicine (62). Nonetheless, recurrent syncope can be profoundly debilitating. This dichotomy may be partly a reflection of the literature, an essentially benign prognosis suggesting that there is less to be concerned about (63), while patient perceptions of the impact of recurrent episodes of loss of consciousness may not be fully taken into account (64). Publication of NICE 109 guidance addresses a component of risk associated from syncope, largely the cardiac component, but leaves unaddressed the considerable psychological impact, physical consequences, and restrictions that may be imposed on people with recurrent syncope. While some of these restrictions are external, such as driving restrictions, many more are self- or family-imposed. This is particularly so in the older population where concerns relating to the effects of falling are particularly acute and severe (12).

Despite the commonest final common pathway for syncope being neurocardiogenic in origin, determining the best treatment for an individual remains challenging. Precise characterization of the cause remains the most important step for deciding on a management pathway. This largely means determining if it is a primary cardiac process and treating according to the best guidelines. The second group to consider is the group with pyschogenic attacks, which largely correspond to the hypomotor form of non-epileptic attack disorder. The third, and numerically largest group, are those with neurocardiogenic syncope. A fourth and often overlooked group are those with mixed attacks, most frequently neurocardiogenic and psychogenic attacks. Amongst this population, numbers are difficult to establish, but they were under-recognized for some time in the epilepsy population with up to 60% of patients having psychogenic attacks in addition to epileptic attacks (65). Furthermore, patients with panic attacks can hyperventilate and this can lead to secondary syncope, making it difficult to disentangle the primary process.

Approach to management

Using NICE 109 guidance as a means for identifying cardiac causes with a low threshold for asking for a neurological opinion on whether the event/s could represent seizures enables the identification of those with these disorders to be directed down the appropriate pathways. Thereafter, the management question hinges on the number and frequency of attacks.

Clinical interpretation of physiology

It is the approach to untangling the multiple potential causes for syncope that is key to using neurophysiology in the most efficient manner. If the history suggests that a cardiac cause is the likely candidate, then investigations at identifying a cardiac cause, such as structural heart disease using ECG, cardiac echocardiography are essential with early referral to cardiology. If a rhythm disturbance is suspected, then monitoring cardiac rhythms is the way forward. The precise means depends on the frequency of attacks. Daily episodes are most easily assessed via cardiac monitoring in a cardiac ward, if the events occur up to twice a week; ambulatory ECG for 48 h is a minimum, while some laboratories can monitor for 7 days. If the frequency is less than twice a week, give serious thought to an implantable device to identify any serious arrhythmia.

If the history raises the question of epileptic seizures, then a sleep EEG would be most appropriate along with MRI scanning of the intracranial structures, a referral to the neurology department, and a first seizure clinic if it is available.

If neurocardiogenic syncope is the likely diagnosis, procedures to trigger an episode of syncope with neurophysiological monitoring are undertaken. This can be done with HUT and glycerine trinitrate stresses. There are a number of protocols in use around the world in different laboratories. It is important to establish experience of normal and abnormal responses. An ideal arrangement includes continuous BP monitoring, which can be invaluable in identifying rhythmic swings in BP with a duration of between 5–12 s, so-called Mayer type waves, which are a marker for presyncope and can be associated with symptoms of presyncope, chest tightness, and brain fogging, which may lead to syncope if the person is kept upright in that circumstance. It also enables one to identify the characteristic rise in HR that can be seen in postural tachycardia syndrome, the commonest cause for recurrent syncope.

The other component of neurophysiological assessment is to establish whether there is evidence of autonomic failure, which may be affecting the cardiovascular system. There are a number of means by which autonomic function can be assessed. In general, people follow the consensus statement of San Antonio (66). In this, the diagnosis of autonomic failure is not based on a single abnormal parameter, but the failure of multiple parameters. These parameters are a set of special manoeuvres set out in this statement and which include isometric exercise, deep breathing, carotid massage, response to standing, and the valsalva breath. HR, mean BPs, cardiac vagal tone, cardiac sympathetic tone, sympathetic tone to skeletal muscles, and to the splanchnic vasculature, baroreflex responsiveness can all be assessed. How extensive the investigations are depends on the experience of both the referring clinicians and the laboratory serving them.

Targeted management

The most effective way of managing syncope is being able to direct specific treatments at the disability. This may mean a simple cardiovascular fitness programme with dietary modifications increasing salt intake and water intake, avoiding diuretics such as caffeine and alcohol, and patient education to understand their condition and the role of counter-manoeuvres and small frequent meals in managing the symptoms. The importance of exercise lies in improved aerobic fitness, blood volume, increased heart size, and quality of life (67).

The role for medication is less clear though has an established a role in symptomatic treatment in preparation for longer-term lifestyle and dietary changes. Such agents can include fludrocortisone, and the sympathomimetic midodrine, though there are anecdotal reports that other medications can be helpful including serotonin re-uptake inhibitors, bisoprolol, pyridostigmine, and ivabradine.

References

1. Kenny, R.A., Bhangu, J., and King-Kallimanis, B.L. (2013). Epidemiology of syncope/collapse in younger and older Western patient populations. *Progress in Cardiovascular Disease*, **55**(4), 357–63.
2. Moya, A., Sutton, R., Ammirati, F., et al. (2009). Guidelines for the diagnosis and management of syncope (version 2009): the Task Force for the Diagnosis and Management of Syncope of the European Society of Cardiology (ESC) 2009. *European Heart Journal*, **30**(21) 2631–71.
3. Savage, D.D., Corwin, L., McGee, D.L., Kannel, W.B., Wolf, P.A. (1085). Epidemiologic features of isolated syncope: the Framingham Study. *Stroke*, **16**(4), 626–9.
4. Freed, L.A., Levy, D., Levine, R.A., et al. (1999). Prevalence and clinical outcome of mitral-valve prolapse *New England Journal of Medicine*, **341**(1), 1–7.
5. Soteriades, E.S., Evans, J.C., Larson, M.G., et al. (2002). Incidence and prognosis of syncope. *New England Journal of Medicine*, **347**(12), 878–85.
6. Ganzeboom, K.S., Colman, N., Reitsma, J.B., Shen, W.K., and Wieling, W. (2003). Prevalence and triggers of syncope in medical students. *American Journal of Cardiology*, **91**(8), 1006–8, A8.
7. Colman, N., Nahm, K., Ganzeboom, K.S., et al. (2004). Epidemiology of reflex syncope. *Clinical Autonomic Research*, **14**(Suppl. 1), 9–17.
8. Providencia, R., Silva, J., Mota, P., Nascimento, J., and Leitao-Marques, A. (2011). Transient loss of consciousness in young adults. *International Journal of Cardiology*, **152**(1), 139–43.
9. Lamb, L.E., Green, H. C., Combs, J.J., Cheeseman, S.A., and Hammond, J. (1960). Incidence of loss of consciousness in 1,980 Air Force personnel. *Aerospace Medicine*, **31**, 973–88.
10. Marrison, V.K., Fletcher, A., and Parry, S.W. (2012). The older patient with syncope: practicalities and controversies. *International Journal of Cardiology*, **155**(1), 9–13.
11. Santhouse, J., Carrier, C., Arya, S., Fowler, H., and Duncan, S. (2007). A comparison of self-reported quality of life between patients with epilepsy and neurocardiogenic syncope. *Epilepsia*, **48**(5), 1019–22.
12. Kenny, R.A. (2003). Syncope in the elderly: diagnosis, evaluation, and treatment. *Journal of Cardiovascular Electrophysiology*, **14**(Suppl. 9), S74–7.
13. Ruwald, M.H., Hansen, M.L., Lamberts, M., et al. (2012). The relation between age, sex, comorbidity, and pharmacotherapy and the risk of syncope: a Danish nationwide study. *Europace*, **14**(10), 1506–14.
14. Brignole, M. and Hamdan, M.H. (2012). New concepts in the assessment of syncope. *Journal of the American College of Cardiology*, **59**(18), 1583–91.
15. NICE. (2010). Transient loss of consciousness ('blackouts') in over 16s, NICE guideline CG109. London: NICE. Available at: https://www.nice.org.uk/guidance/cg109 (accessed 5 April 2016).
16. Olde Nordkamp, L.R.A., van Dijk, N., Ganzeboom, K.S., et al. (2009). Syncope prevalence in the ED compared to general practice and population: a strong selection process. *American Journal of Emergency Medicine*, **27**(3), 271–9.
17. Thiruganasambandamoorthy, V., Stiell, I.G., Sivilotti, M.L.A., et al. (2014). Risk stratification in emergency department syncope patients

to predict short term serious outcomes after discharge (RiSEDS) study. *BMC Emergency Medicine*, **14**(8), DOI: 10.1186/1471-227X-14-8.
18. Duncan, W., Tan, M.P., Newton, J.L., Reeve, P., and Parry, S.W. (2010). Vasovagal syncope in the older person: differences in presentation between older and younger patients. *Age and Ageing*, **39**(4), 465–70.
19. Reuber, M., Jamnadas-Khoda, J., Broadhurst, M., et al. (2011). Psychogenic nonepileptic seizure manifestations reported by patients and witnesses. *Epilepsia*, **11**, 2028–35.
20. O'Dwyer, C. and Kenny, R.A. (2010). Prodrome and characteristics of young and old patients presenting with vasovagal syncope to a teaching hospital. *Age and Ageing*, **39**(Suppl. 1), 20.
21. Stokes, M.B., Palmer, S., Moneghetti, K.J., Mariani, J.A., and Wilson, A.M. (2012). Asystole following complex partial seizures. *Heart Lung and Circulation*, **22**(2), 146–8.
22. Ryo Wakita, Yuka Ohno, Saori Yamazaki, Hikaru Kohase, and Masahiro Umino. (2007). Vasovagal syncope with asystole associated with intravenous access. *Oral Surgery, Oral Medicine, Oral Pathology, Oral Radiology, and Endodontology*, **102**(6), e28–32.
23. Colivicchi, F., Ammirati, F., and Santini, M. (2004). Epidemiology and prognostic implications of syncope in young competing athletes *European Heart Journal*, **25**(19), 1749–53.
24. Van Dijk, N., Sprangers, M.A., Boer, K.R. Colman, N., Wieling, W., and Linzer, M. (2007). Quality of life within one year following presentation after transient loss of consciousness. *American Journal of Cardiology*, **100**, 672–6.
25. Petersen, M.E., Williams, T.R., and Sutton, R. (1995). Psychogenic syncope diagnosed by prolonged head-up tilt testing. *Quarterly Journal of Medicine*, **88**, 209–13.
26. Benbadis, S.R. and Chichkova, R. (2006). Psychogenic pseudosyncope: an underestimated and provable diagnosis. *Epilepsy & Behaviour*, **9**, 106–10.
27. Van Lieshout, J., Wieling, W., Karemaker, J.M., and Secher, N.H. (2003). Syncope, cerebral perfusion, and oxygenation. *Journal of Applied Physiology*, **94**, 833–48.
28. Rossen, R., Kabat, H., and Anderson, J.P. (1943). Acute arrest of the cerebral circulation in man. *Archives of Neurology & Psychiatry*, **50**, 510–28.
29. Eckberg, D.L. and Sleight, P. (1992). *Human baroreflexes in health and disease*. Monograph of the Physiological Society. Oxford: Clarendon Press.
30. Wieling, W. and Karemaker, J.M. (2012). Measurement of heart rate and blood pressure. In: C. Mathias and R. Bannister (Eds) *Autonomic Failure*, 5th edn, pp. 290–306. Oxford: Oxford University Press.
31. Folino, A.F. (2007). Cerebral autoregulation and syncope. *Progress in Cardiovascular Disease*, **50**, 49–80.
32. Bor-Seng-Shu, E., Kita, W.S., Figueiredo, E.G., et al. (2012). Cerebral hemodynamics: concepts of clinical importance. *Arquivos de Neuro-Psiquiatria*, **70**, 352–6.
33. Panerai, R.B. (2003). The critical closing pressure of the cerebral circulation *Medical Engineering & Physics*, **25**, 621–32.
34. Valdueza, J.M., von Münster, T., Hoffman, O., Schreiber, S., and Einhäupl, K.M. (2000). Postural dependency of the cerebral venous outflow. *Lancet*, **355**, 200–1.
35. Gisolf, J., van Lieshout, J.J., van Heusden, K., Pott, F., Stok, W.J., and Karemak, J.M. (2004). Human cerebral venous outflow pathway depends on posture and central venous pressure. *Journal of Physiology*, **560**, 317–27.
36. Faraci, F.M., Mayhan, W.G., Schmid, P.G., and Heistad, D.D. (1988). Effects of arginine vasopressin on cerebral microvascular pressure. *American Journal of Physiology*, **255**(*Heart & Circulation Physiology* 24), H70–6.
37. Baumbach, G.L. and Heistad, D.D. (1983). Effects of sympathetic stimulation and changes in arterial pressure on segmental resistance of cerebral vessels in rabbits and cats. *Circulation Research*, **52**, 527–33.
38. Mugge, A., Brandes, R.P., Boger, R.H., et al. (1994). Vascular release of superoxide radicals is enhanced in hypercholesterolemic rabbits. *Journal of Cardiovascular Pharmacology*, **24**, 994–8.
39. Colonna, D.M., Meng, W., Deal, D.D., and Busija, D.W. (1994). Calcitonin gene-related peptide promotes cerebrovascular dilation during cortical spreading depression in rabbits. *American Journal of Physiology*, **266**(*Heart & Circulation Physiology* 35), H1095–102.
40. Sakas, D.E., Moskowitz, M.A., Wei, E.P., Kontos, H.A., Kano, M., and Ogilvy, C.S. (1989). Trigeminovascular fibers increase blood flow in cortical gray matter by axon reflex-like mechanisms during acute severe hypertension or seizures. *Proceedings of the National Academy of Sciences, USA*, **86**, 1401–5.
41. Moskowitz, M.A., Sakas, D.E., Wei, E.P., et al. (1989). Postocclusive cerebral hyperemia is markedly attenuated by chronic trigeminal ganglionectomy. *American Journal of Physiology*, **257**(*Heart & Circulation Physiology* 26), H1736–9.
42. Faraci, F.M. and Bresse, K.R. (1993). Nitric oxide mediates vasodilatation in response to activation of *N*-methyl-D-aspartate receptors in brain. *Circulation Research*, **72**, 476–80.
43. Faraci, F.M., Breese, K.R., and Heistad, D.D. (1993). Nitric oxide contributes to dilatation of cerebral arterioles during seizures. *American Journal of Physiology*, **265**(*Heart & Circulation Physiology* 34), H2209–12.
44. Pelligrino, D.A., Gay, R.L., Baughman, V.L., and Wang, Q. (1996). NO synthase inhibition modulates NMDA-induced changes in cerebral blood flow and EEG activity. *American Journal of Physiology*, **271**(*Heart & Circulation Physiology* 40), H990–5.
45. Heistad, D.D., and Kontos, H.A. (1983). Cerebral circulation. In: J. T. Shephard and F. M. Abbound (Eds) *Handbook of physiology. The cardiovascular system. peripheral circulation and organ blood flow*, pp. 137–82. Bethesda, MD: American Physiological Society.
46. Busija, D.W. (1993). Cerebral autoregulation. In: J. W. Phillis (Ed.) *The Regulation of Cerebral Blood Flow*, pp. 45–61. Boca Raton, FL: CRC.
47. Lee, W.S., Kwon, Y.J., Yu, S.S., Rhim, B.Y., and Hong, K.W. (1993). Disturbances in autoregulatory responses of rat pial arteries by sulfonylureas. *Life Sciences*, **52**, 1527–34.
48. Paulson, O.B., Strandgaard, S., and Edvinsson, L. (1990). Cerebral autoregulation. *Cerebrovascular Brain Metabolic Review*, **2**, 161–92.
49. Armstead, W.M. (1997). Role of nitric oxide, cyclic nucleotides, and the activation of ATP-sensitive K^+ channels in the contribution of adenosine to hypoxia-induced pial artery dilation. *Journal of Cerebral Blood Flow and Metabolism*, **17**, 100–8.
50. Fabricius, M. and Lauritzen, M. (1994). Examination of the role of nitric oxide for the hypercapnic rise of cerebral blood flow in rats. *American Journal of Physiology*, **266**(*Heart & Circulation Physiology* 35), H1457–64.
51. Faraci, F.M. and Heistad, D.D. (1990). Regulation of large cerebral arteries and cerebral microvascular pressure. *Circulation Research*, **66**, 8–17.
52. Fujii, K., Heistad, D.D., and Faraci F.M. (1991). Flow-mediated dilatation of the basilar artery in vivo. *Circulation Research*, **69**, 697–705.
53. Julu, P.O.O., Cooper, V.L., Hansen, S., and Hainsworth, R. (2003). Cardiovascular regulation in the period preceding vasovagal syncope in conscious humans. *Journal Physiology*, **549**(1), 299–311.
54. Deboer, R.W., Karemaker, J.M., and Strackee, J. (1987). Hemodynamic fluctuations and baroreflex sensitivity in humans—a beat-to-beat model. *American Journal Physiology*, **253**, H680–9.
55. Sanders, J.S. and Ferguson. D.W. (1989). Diastolic pressure determines autonomic responses to pressure perturbation in humans. *Journal of Applied Physiology*, **66**, 800–7.
56. Youde, J., Panerai, R.B., Gillies, C., and Potter, J.F. (2002). Continuous cardiac baroreceptor measurement during tilt in healthy elderly subjects. *Clinical Autonomic Research*, **12**, 379–84.
57. Thompson, C.A., Tatro, D.L., Ludwig, D.A., & Convertino, V.A. (1990). Baroreflex responses to acute changes in blood volume in humans. *American Journal of Physiology*, **259**, R792–8.
58. Huang, C.C., Sandroni, P., Slette, D., Weigand, S., and Low, P.A. (2007). Effect of age on adrenergic and vagal baroreflex sensitivity in normal subjects. *Muscle & Nerve*, **36**, 637–42.

59. Bang, D-H., Son, Y., Lee, Y.H., and Yoon, K-H. (2014). Doppler ultrasonography measurement of hepatic dynamics during Valsalva manoeuvre: healthy volunteer study. *Ultrasonography*, **34**(1), 32–8.
60. Ewing, D.J., Irving, J.B., Kerr, F., Wildsmith, J.A., and Clarke, B.F. (1974). Cardiovascular responses to sustained hand grip in normal subjects and in patients with diabetes mellitus: a test of autonomic function. *Clinical Science and Molecular Medicine*, **46**, 295–306.
61. Guyenet, P.G. (2006). The sympathetic control of blood pressure. *Nature Reviews: Neuroscience*, **7**, 335–46.
62. Ridsdale, L. (2009). No more neurophobia: welcome neurology in general practice. *British Journal of General Practice*, **59**(565), 567–9.
63. Soteriades, E.S., Evans, J.C., Larson, M.G., et al. (2002). Incidence and prognosis of syncope. *New England Journal of Medicine*, **347**, 878–85.
64. Linzer, M., Pontinon, M., and Gold, D.T. (1991). Impairment of physical and psychosocial health in recurrent syncope. *Journal of Clinical Epidemiology*, **44**, 1037–43.
65. Gates, J.R. (2000). Epidemiology and classification of non-epileptic events. In: J. R. Gates and A. J. Rowan (Eds) *Non-epileptic seizures*, 2nd edn, pp. 3–14. Boston: Butterworth-Heinemann.
66. Anonymous Consensus statement (1988). Report and recommendations of the San Antonio conference on diabetic neuropathy. American Diabetes Association, Academy of Neurology. *Diabetes Care*, **11**, 592–7.
67. Fu, Q., Tiffany, B. VanGundy, M.S., et al. (2010). Levine cardiac origins of the postural orthostatic tachycardia syndrome. *Journal of the American College of Cardiology*, **55**(25), 2858–68.

CHAPTER 30

Convulsive and non-convulsive status epilepticus

Matthew C. Walker

Introduction and definitions

Seizures are mostly self-terminating, usually lasting less than 5–10 min. However, occasionally, seizure activity can continue or repeat at such frequent intervals so as to constitute a persistent epileptic state, status epilepticus (SE). Although the initial definitions of SE were confined to tonic-clonic seizures (convulsive SE), in 1964 SE was redefined at a meeting in Marseille, so that SE depended on the persistence of the seizure rather than its form (1). The definition was 'a condition characterized by epileptic seizures that are sufficiently prolonged or repeated at sufficiently brief intervals so as to produce an unvarying and enduring epileptic condition'. Because of the existence of non-convulsive forms of SE that can mimic other conditions (e.g. acute confusional states), the detection of SE has become critically dependent upon EEG.

However, this definition introduces two aspects that remain contentious, first the duration of a seizure before it becomes 'enduring' and second the concept of SE as 'unvarying'. In addition, the definition does not distinguish between the different forms of SE. Conventionally, SE was defined as seizures lasting longer than 30 min, but this definition is unhelpful when it comes to treatment, because most would treat convulsive seizures that last longer than 5 min (2). The definition of SE has, therefore, been recently revisited by the International League against Epilepsy in 2015 (3) and a new definition has been proposed:

> SE is a condition resulting either from the failure of the mechanisms responsible for seizure termination or from the initiation of mechanisms which lead to abnormally prolonged seizures (after time point t_1). It is a condition that can have long-term consequences (after time point t_2), including neuronal death, neuronal injury, and alteration of neuronal networks, depending on the type and duration of seizures.

This definition encompasses two important concepts, first that there is a time beyond which a seizure will become persistent and second, that there is a separate time point at which the seizure results in long-lasting disruption to normal brain function. These time points may be different for different forms of SE, so, for example, convulsive SE has been proposed to be defined by $t_1 = 5$ min and $t_2 = 30$ min, while focal SE with impaired consciousness has $t_1 = 10$ min and $t_2 > 60$ min (3).

Epidemiology

Status epilepticus, defined as seizures lasting more than 30 min, is a common condition with an incidence of 10–60 per 100,000 person-years, depending on the population studied (the higher incidence is in people from poorer socioeconomic backgrounds) (4). In people with epilepsy, SE is mostly precipitated by drug withdrawal (reintroduction of the withdrawn drug can lead to a rapid resolution of the episode) (5). It has been estimated that 4–16% of people with epilepsy (over 25% of all children with epilepsy) will have at least one episode of SE (6,7).

Most SE, however, occurs in people who have no prior history of epilepsy, usually as a result of an acute symptomatic cause. Non-central nervous system infections with fever are the most common cause of SE in children, while in adults cerebrovascular accidents, hypoxia, metabolic causes and alcohol are the main acute causes (5). Such causes are usually identified by a combination of neuro-imaging, appropriate blood tests and, if necessary, cerebrospinal fluid (CSF) examination. For those people who prove refractory to treatment and are transferred to the intensive care unit, a different spectrum of aetiologies is apparent with central nervous system infection, autoimmune encephalitis, and mitochondrial disorders playing a prominent role (8,9).

SE is recurrent in over 10% of people (5), emphasizing the need to have a protocol in place to prevent further episodes of status epilepticus. The prognosis of SE is related to aetiology; however, the prognosis of certain conditions such as stroke is worse if associated with status epilepticus. The mortality for SE is 10–20% and is higher in the elderly (10). For those who prove resistant to treatment and require transfer to intensive care, the mortality is high, approximately 20–50%, and those who survive are often left with permanent neurological and cognitive deficits (8,9,11). Those who have acute symptomatic SE *de novo* have a 40% chance of developing chronic epilepsy (this is approximately three-fold greater than those who have acute symptomatic seizures) (12).

Classification

The early classifications of SE classified by seizure type (e.g. simple partial status epilepticus, complex partial status epilepticus), but this failed to include many other forms of SE such as electrical SE during slow wave sleep. The recent ILAE classification of SE proposes to define SE by semiology, aetiology, EEG correlates, and age (3). The semiology is defined by the presence or absence of prominent motor symptoms and the degree of impaired consciousness. Thus, there is SE with prominent motor symptoms including convulsive SE and focal motor SE, and SE without motor symptoms (non-convulsive SE) with and without coma. Non-convulsive SE without coma can be further subdivided into generalized and focal forms (3).

Status epilepticus as a staged phenomenon

SE can be staged by EEG/clinical findings, progressive systemic physiological compromise, progressive neuronal damage and progressive drug resistance.

In animal models of untreated status epilepticus, there is a clear evolution of EEG patterns (13). First there are discrete seizures (stage 1), then the seizures merge (stage 2) and form continuous seizure activity (stage 3), and eventually the seizure is punctuated by flat periods (stage 4), evolving to periodic epileptiform discharges (stage 5). Whether such an EEG progression occurs in human SE has been debated, but the progression has been recorded in one patient (Fig. 30.1) (14).

One confounder is that there may be the emergence of periodic epileptiform discharges early with some pathologies such as viral encephalitis. Nonetheless, these stages may be relevant to human SE because often there is a premonitory phase of increasing numbers of seizures before progression to seizures without full recovery between and then continuous seizure activity. Finally, the end-stages of SE are often marked by periodic epileptiform discharges. Whether in the end stages of convulsive SE such discharges represent underlying neuronal damage or ongoing ictal activity is unclear. A fluorodeoxyglucose PET scan during such activity, however, has clearly indicated hypermetabolism associated with the periodic epileptiform discharges (15), suggesting that they should be considered pathogenic. More recently, it has been recognized that, in people with acute brain injury, ictal activity evident on intracortical recordings may not be apparent on surface EEG, possibly because such activity in such clinical circumstances represents multifocal, but poorly synchronized, 'mini-seizures' (16). Together, these findings suggest that

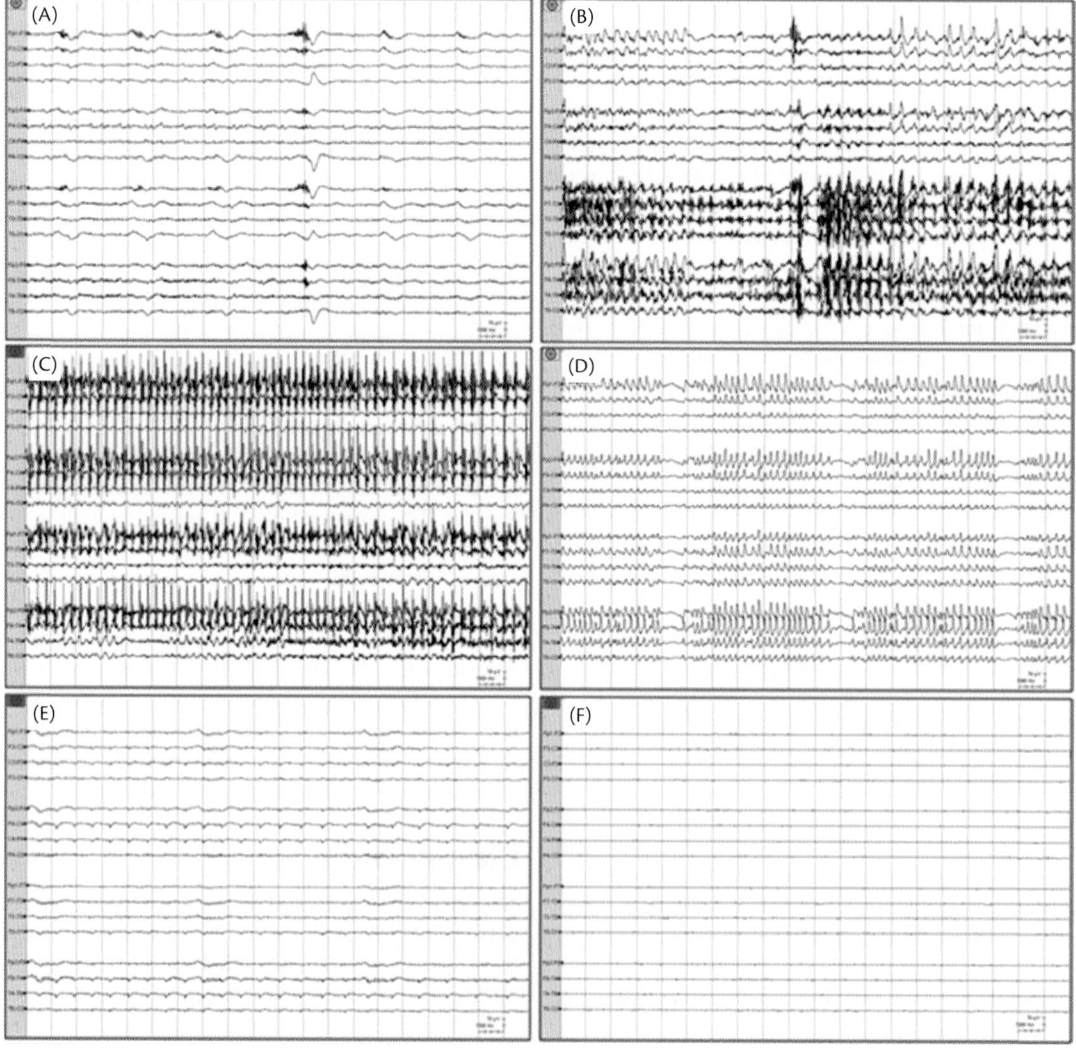

Fig. 30.1 EEG progression in a patient with status epilepticus. (A) Stage 1: Discrete seizures with interictal slowing. (B) Merging seizures with waxing and waning amplitude and frequency of EEG rhythm. (C) Continuous ictal activity. (D) Continuous ictal activity punctuated by low-voltage flat periods. (E) Periodic epileptiform discharges on a flat background. (F) Complete suppression of the background consistent with electrocerebral silence.

the later stages of SE warrant, at least initially, aggressive treatment, but studies in this area to determine how aggressive and how long treatment should continue are sorely lacking. The systemic effects of convulsive SE can be divided into early and late stages (17). The initial consequence of a prolonged convulsion is a massive release of plasma catecholamines, which results in an increase in heart rate, blood pressure (BP), and plasma glucose. During this stage the increase in cerebral blood flow (CBF) compensates for the increased brain metabolism. After 60–90 min, but earlier in the elderly, the SE enters a second phase in which there is a loss of cerebral autoregulation, so that CBF depends on systemic BP, which together with the hypotension that occurs at this stage, can result in significant cerebral ischaemia. In these patients, even after the SE has stopped, the EEG may demonstrate evidence of an ischaemic encephalopathy, and the EEG may help predict prognosis; a reactive EEG background, in my experience, predicts good recovery, especially if faster rhythms (theta range) are present, while an unreactive background with periodic epileptiform discharges, a burst-suppression EEG pattern or unreactive alpha (alpha coma) usually predicts a poor outcome. It is also worth noting that severe metabolic derangements can also occur in the later stages of the status epilepticus, in particular hypoglycaemia, renal failure, hyponatraemia and acidosis, and these may contribute both to resistance of the SE to treatment, as well as the encephalopathy, thus influencing the EEG pattern. Until such derangements are corrected, the prognostic value of the EEG should be treated with caution.

Independent of the physiological compromise, SE can result in neuronal death (18). This is secondary to the excessive activation of glutamate receptors and the consequent accumulation of calcium within neurons that can lead to mitochondrial failure, free radical generation, and also programmed cell death (19). This cell death is time dependent (i.e. the longer SE continues the more cell death occurs) and occurs even if there are no clinical signs of seizure activity (i.e. it can occur with electrical SE alone). In animal models, significant neuronal death is associated with progression of the EEG pattern to periodic epileptiform discharges (13).

Finally, there is progressive resistance of SE to antiepileptic drugs. This is due to not only rapid upregulation of multidrug transporter proteins (20), but also internalization of synaptic GABA(A) receptors (so profoundly decreasing the response to benzodiazepines) (21,22).

Diagnosis

Although the diagnosis of SE would appear to be straightforward, in an audit of patients transferred in apparently refractory status epilepticus, almost half did not have SE, but rather had pseudostatus epilepticus (psychogenic non-epileptic seizures) (23). Psychogenic attacks (dissociative seizures) can be differentiated by their semiology (such as poorly co-ordinated thrashing, back-arching, eyes held shut and head rolling), by clinical signs (responsive pupils, flexor plantar responses) and, if necessary, by EEG. Serum prolactin plays no role, as it can increase in a variety of circumstances and can normalize during prolonged seizure activity (17).

Conversely non-convulsive SE can be difficult to diagnose. As seizures continue, the clinical manifestation can become more subtle, so that, for example, people with absence SE may complain of 'feeling strange' and complex partial SE may manifest as a change in personality, varying degrees of confusion or an apparent prolonged (>20 min) post-ictal period (24). Possibly up to 8% of people in coma in an intensive therapy unit without a prior history of epilepsy or seizures are in non-convulsive SE (25). Clues that someone is in non-convulsive SE include fluctuations in pupillary size, nystagmus, and stereotypical motor manifestations, but an EEG is almost always necessary to confirm or refute the diagnosis. Even with EEG the diagnosis may not be straightforward (see 'Is it non-convulsive status epilepticus?').

Electroencephalogram in diagnosis and treatment

Convulsive status epilepticus

In the early stages, EEG is rarely necessary to make the diagnosis of convulsive SE as the clinical appearances are typical. However, in uncertain cases, electroencephalogram (EEG) can help distinguish cases of psychogenic seizures from frontal lobe SE. The EEG in pseudostatus epilepticus (psychogenic SE) is usually masked by movement artefact, but often the movements will slow sufficiently to enable reactive posterior alpha rhythms to become apparent.

Late-on in convulsive SE with transfer to the intensive care unit, the EEG becomes critical.

The eventual progression of convulsive SE is to a form of electromechanical dissociation, so that there may be no overt clinical activity evident despite on-going electrographic SE (Fig. 30.2A). It is therefore imperative that people in the later stages of SE (i.e. on the intensive care unit) are monitored with (ideally continuous) EEG (26). Although brief EEGs can be helpful, at least 24 h of recording is necessary to be confident that there are no seizures (27). An alternative to EEG, for example, when EEG is not easily available, is to use cerebral function monitors or bispectral index (BIS) monitors, which can be used to monitor burst suppression (i.e. depth of anaesthesia), but cannot identify regional epileptic activity (28,29). The EEG endpoint on the intensive care unit is subject to debate. Conventionally burst suppression patterns have been used as the anaesthetic endpoint (Fig. 30.2B). However, there is growing evidence that such an endpoint may be unnecessary and that seizure suppression may suffice. It is certainly our experience that seizures can occur despite a burst suppression pattern (i.e. it may not be sufficient) and that often less anaesthesia is required to obtain successful seizure suppression than is necessary to obtain burst suppression. There has been a suggestion that seizure recurrence is more likely with seizure suppression than burst suppression (11), but the eventual effect on outcome is unclear.

During anaesthesia, the focality of the SE may become apparent. This is critical because, in such cases, resective surgery may be possible even with a normal MRI with the extent of the operation guided by intracranial electrodes (30). In many such cases, an occult dysplasia has been identified.

Certain anaesthetic agents, in particular barbiturates, have a great propensity to accumulate and so a persistent burst suppression pattern, even after anaesthesia has been stopped could be due to residual anaesthesia and, in such cases, drug levels can be helpful.

Focal motor status epilepticus

In focal motor SE with preserved consciousness, such as epilepsia partialis continua, the EEG can be normal or can show focal abnormalities (31). Even with back-averaging the EEG may not reveal a cortical abnormality. EMG can sometimes be useful when it is

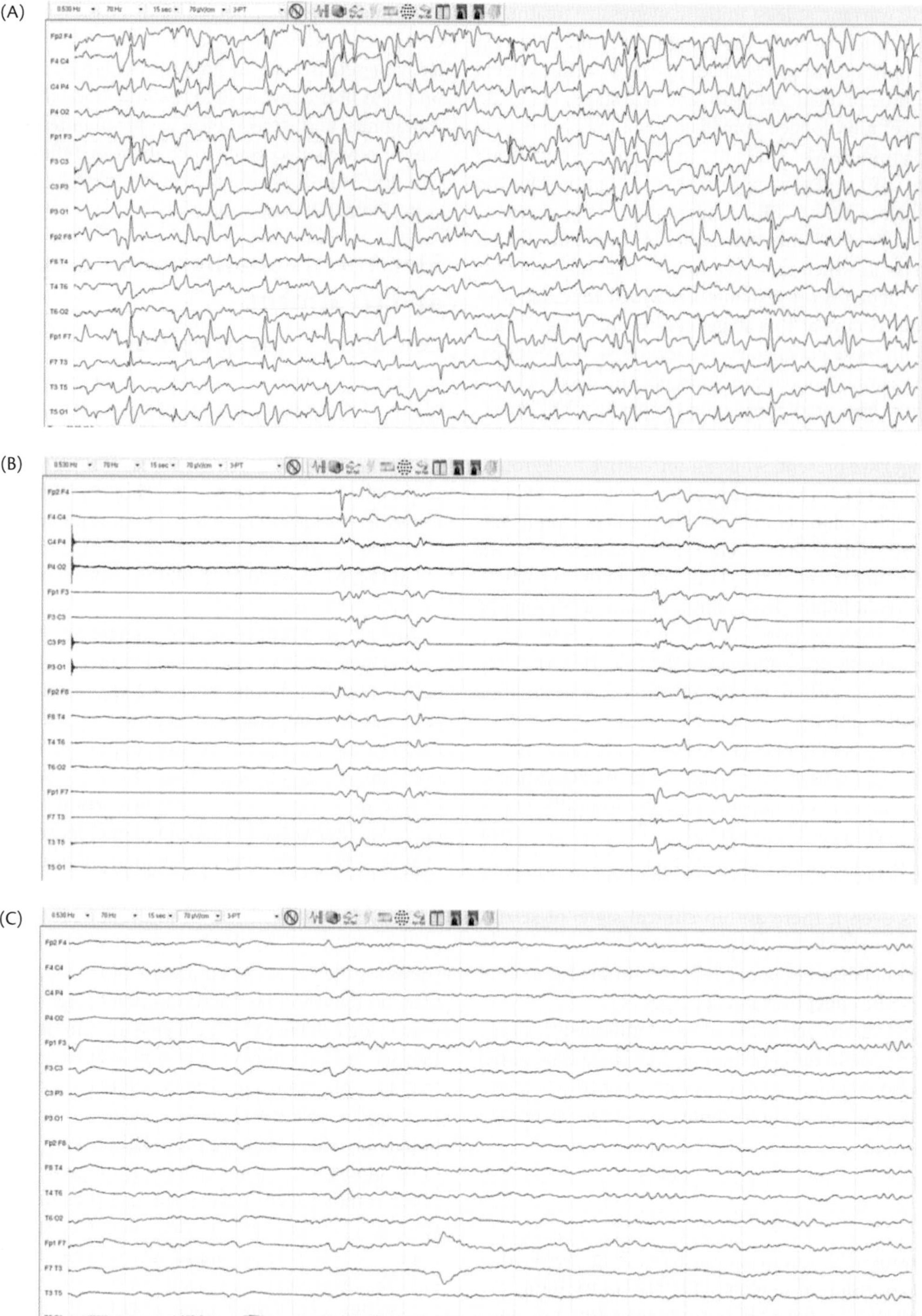

Fig. 30.2 Patient is admitted into ICU following convulsive status epilepticus. (A) On-going seizure activity is apparent even though there were no clinical indicators of seizure activity. (B) Burst suppression was achieved with anaesthesia. (C) Coming off anaesthesia, there is no recurrence of seizures and the faster background rhythms are a good prognostic indicator.

uncertain whether there is a cortical or subcortical focus, as the cortical jerks are usually brief (<50 ms) and are synchronous in antagonist muscles.

Myoclonic status epilepticus in coma

This usually occurs in patients who have suffered anoxic or ischaemic brain injury. Such patients are in deep coma, sometimes with subtle myoclonic twitching. The prognosis is invariably poor with poor response to antiepileptic drug treatment (32–34). The EEG can demonstrate bursts of generalized spikes and polyspikes that are usually time-locked to the jerks. Other patterns including burst suppression, periodic complexes and alpha coma have also been described, and probably have a poorer prognosis, representing an agonal pattern. Survival can occur, however, especially if the initial

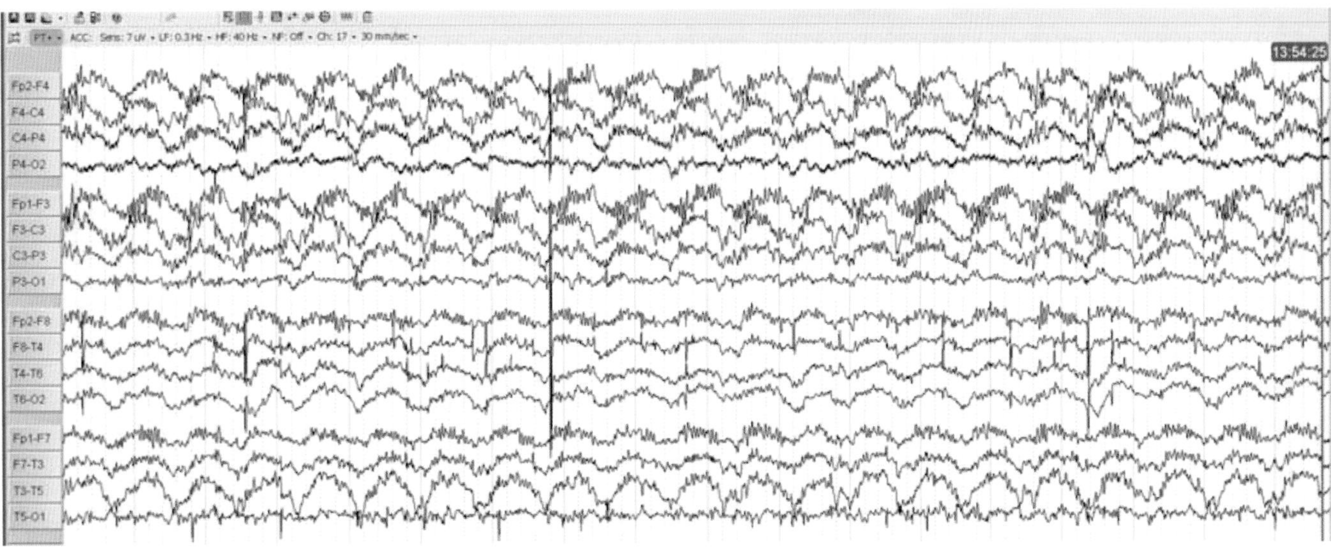

Fig. 30.3 Extreme delta brush in a person with NMDA receptor encephalitis.

insult was primarily hypoxia related (35). Survivors are often left with Lance–Adams-type action myoclonus or in a vegetative state. Whether the EEG changes represent electrographic SE is contentious, and in my view these usually indicate underlying widespread cortical damage or dysfunction.

Non-convulsive status epilepticus

The EEG is often critical for the diagnosis of non-convulsive status epilepticus, yet there is still no universally accepted EEG definition. Moreover, after deciding that the EEG is diagnostic of non-convulsive status epilepticus, can we use the EEG to subdivide the non-convulsive SE into specific groups (e.g. focal and generalized)?

Is it non-convulsive status epilepticus?

In many instances the EEG interpretation is straight forward (e.g. when there are clear evolving EEG patterns), but problems in interpretation occur when slow repetitive discharges or rhythmic slow are present. Another challenge is interpreting the EEG in people with epileptic encephalopathies, whose background EEG is grossly abnormal.

It is therefore helpful to subdivide the EEG clinically into those with and without an epileptic encephalopathy. In those without an epileptic encephalopathy, epileptiform discharges occurring at a frequency of 3 Hz represent status epilepticus (36,37). Most would also concur that epileptiform discharges at <3 Hz or rhythmic slow that show clear evolving spatiotemporal patterns with changes in frequency and amplitude and spread to additional electrodes would also be diagnostic of SE. When there is rhythmic slow without obvious spatiotemporal evolution, great care needs to be taken as associated clinical signs may not always represent seizure activity, e.g. the oral dyskinesia associated with N-methyl-D-aspartate (NMDA) receptor encephalopathy in which the EEG frequently shows high amplitude frontal delta, often with a delta brush, which is pathognomic (Fig. 30.3) (38).

However, if there is associated clinical seizure activity or the emergence of normal EEG rhythms/clinical improvement following treatment, then such patterns can be interpreted as seizure activity.

Repetitive epileptiform discharges (sharp waves, spikes or polyspikes) at less than 3 Hz, 50–300 μV in amplitude and occurring in a 10-min epoch are termed period epileptiform discharges and can be subdivided into periodic lateralized epileptiform discharges (PLEDs), bilateral (independent) periodic epileptiform discharges (PEDs) and generalized periodic epileptiform discharges (GPEDs) (39). Such discharges may become more prominent following stimulation and these are then termed stimulus-induced rhythmic periodic or ictal discharges (SIRPIDs) (Fig. 30.4) (40). The clinical relevance of SIRPIDs is unclear, especially since some studies have not shown an increase in blood flow during SIRPIDs with SPECT, suggesting that they may not be of any metabolic consequence (41).

The controversy surrounding periodic patterns is evident in the early descriptions. For example Cobb and Hill in 1950 described periodic discharges in association with subcortical pathology (42), while Chatrian, in 1964, described periodic epileptiform discharges in association with epilepsia partialis continua (43). Indeed, such patterns can be generated by both subcortical and cortical pathology, with subcortical generators resulting in more monomorphic and briefer discharges (44). There are also a range of aetiologies that can result in PEDs (including stroke, tumours, infections, and metabolic causes) and PEDs are not specific to any aetiology (45). Indeed, it can be difficult to distinguish periodic complexes in the latter stages of SE from those observed in metabolic encephalopathies (Fig. 30.5).

Often in such instances PEDs can be considered interictal, as seizures will occur in approximately 70% of patients with PLEDs and 30–40% of patients with BIPLEDs or GPEDs. The prognosis for people with PEDs is usually poor and probably relates to the underlying aetiology. Overall it is my opinion that PEDs should only be treated as epileptic if there is other evidence of ictal activity; PEDs can be ictal, most obviously when time-locked to jerks in epilepsia partialis continua, but there have also been descriptions of PEDs in association with complex partial seizures (46) and, recently, it has been shown that PEDs may be observed on surface electrodes while intracranial electrodes demonstrate seizure activity. We are, therefore, left in the difficult position that PEDs can be ictal and, therefore, in some circumstances could represent SE.

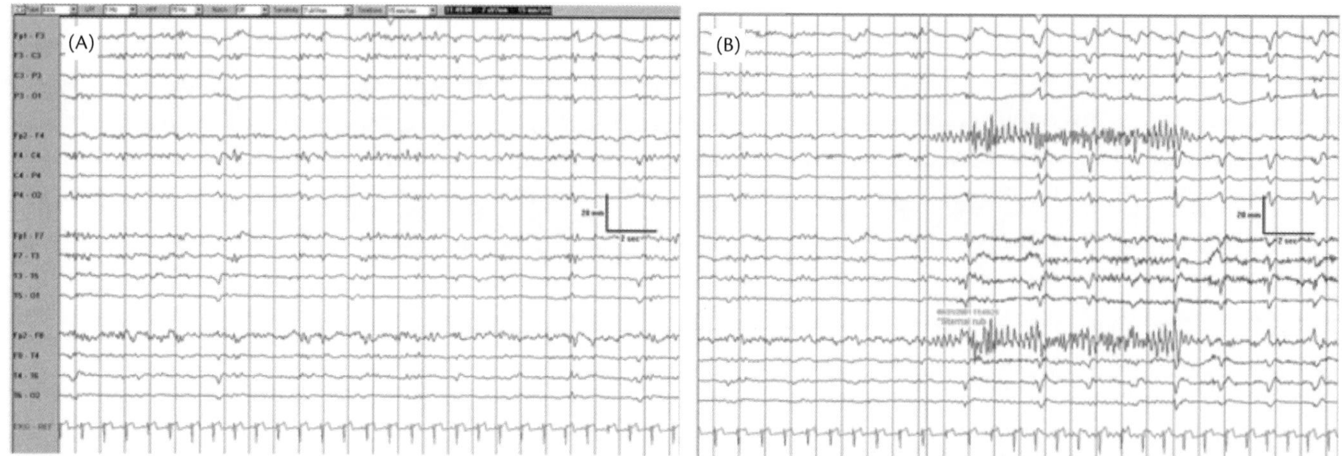

Fig. 30.4 The occurrence of stimulus induced rhythmic periodic discharges(SIRPIDs) following sternal rub in the second panel.
Reproduced from *Epilepsia*, 45, Hirsch LJ, Claassen J, Mayer SA, Emerson RG, Stimulus-induced rhythmic, periodic, or ictal discharges (SIRPIDs): a common EEG phenomenon in the critically ill, pp. 109–23, copyright (2004), with permission from John Wiley and Sons.

Certain additional criteria can help determine this. First, if there is a clinical manifestation that is time-locked to the PEDs, then they should be considered ictal. An important catch is that PEDs can occur in situations in which subcortical myoclonus can occur (e.g. following hypoxic brain injury) and, therefore, the presence of myoclonus alone is not sufficient—there should be a close temporal association between the discharges and the myoclonus. Secondly, if there is clinical improvement on administration of antiepileptic medication such as benzodiazepines. This, however, is more likely to occur in ambulatory, but confused patients. In people in coma (usual in people with PEDs), the clinical improvement may not be immediately apparent. In such patients, there should be improvement in the EEG. However, many EEG patterns including triphasic waves can improve with benzodiazepines (47). Therefore, there should be not only resolution of the PEDs but also the appearance of normal EEG patterns.

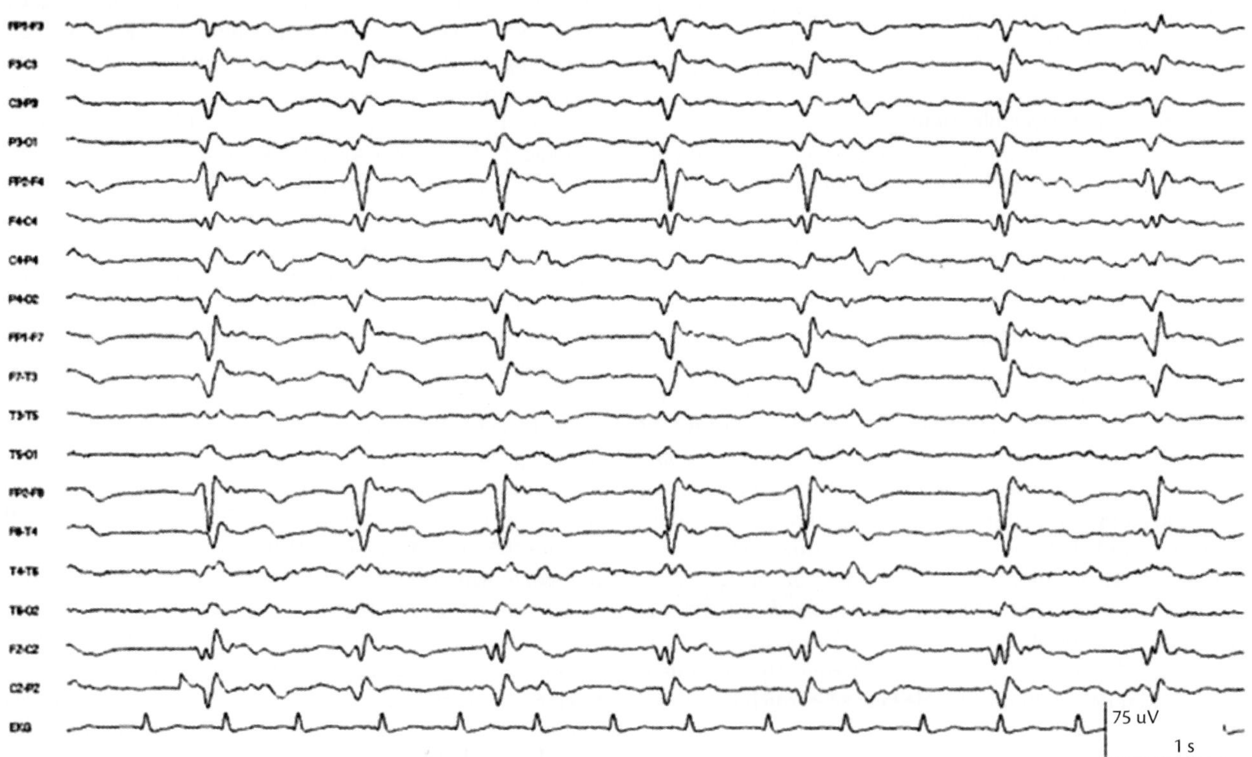

Fig. 30.5 Periodic epileptiform discharges in a 75-year-old woman in status epilepticus. Such periodic discharges are indistinguishable from other periodic discharges that can occur with metabolic derangements.
Reproduced from *Epilepsia*, **43**(Suppl. 3), Brenner RP, Is it status?, pp. 103–13, copyright (2002), with permission from John Wiley and Sons.

What type of SE is it?

Having decided that the EEG pattern represents non-convulsive SE, does the pattern help determine the type of SE?

Certain EEG patterns are pathognomonic. Epileptic SE in slow wave sleep is an epileptic encephalopathy characterized by the presence of generalized 1–3 Hz spike-wave discharges occupying 85% or more of the EEG of non-REM sleep (48). Typical absence SE is associated with trains of 3 Hz spike and wave, and atypical absence *SE* slower 2–3 Hz spike-wave (49). The EEG disturbance in complex partial SE typically has some focality. Unfortunately, the EEG changes are not always so clear-cut, and there is overlap between these syndromes. In typical absence status epilepticus, the EEG discharges can vary in frequency from 1 to 4 Hz, and consist of spike-wave or polyspike-wave complexes, so that electrographic differentiation from atypical absence SE can sometimes be difficult (49). Complex partial SE can also manifest with generalized spike-wave discharges. However, focal EEG features often become apparent following treatment. Furthermore, there is often a more prolonged post-ictal state associated with EEG slowing in cases of treated partial SE compared with absence SE in which there is more rapid recovery (49). Simple partial SE may not be associated with any change in the scalp EEG.

Treatment of status epilepticus

Randomized control trials in SE

There have been at least 14 randomized studies of intravenous (iv) drug treatment in SE (50–63). These studies have methodological problems including different definitions of SE, different doses of drugs and varying outcome measures. These studies have variably compared lidocaine with placebo, lorazepam with diazepam, phenobarbital with diazepam and phenytoin, intramuscular midazolam with iv diazepam, four different iv treatment regimens (lorazepam, phenytoin alone, diazepam and phenytoin, and phenobarbital), valproate with phenytoin, valproate with iv diazepam in refractory status epilepticus, levetiracetam with valproate, and phenytoin and levetiracetam with lorazepam. Very few conclusions can be drawn from these studies:

- Lidocaine is effective in the treatment of status epilepticus.
- Lorazepam and diazepam are equally effective, although more patients required additional antiepileptic drugs if given diazepam.
- Lorazepam is more effective than phenytoin alone.
- Intramuscular midazolam is as effective and quicker to give than iv lorazepam.
- No particular drug or drug combination has significantly more side-effects, including respiratory depression.

Overall drug choice is perhaps not as important as having a protocol so that adequate doses of appropriate anti-epileptic drugs are given rapidly.

Drug treatment of convulsive status epilepticus

Premonitory phase

Prior to SE becoming established (people with repeated or prolonged seizures), there is a wealth of data indicating that rectal diazepam can prevent the progression to status epilepticus, and can be used safely in the community (64). More recent evidence in children indicates that buccal midazolam is a superior alternative (65–67).

Established SE

Once SE has become established rapid administration of an iv therapy is required. At this stage, iv access is necessary, bloods should be taken, fluids should be administered and glucose/thiamine may be given if hypoglycaemia is suspected. Oxygen should be administered because of the respiratory depression that occurs in SE.

At present iv lorazepam is the drug of choice as it has a long redistribution half-life (3–10 h) and so is less likely to result in rebound seizures (17). If the first bolus (usually 4 mg in an adult) fails then a second bolus can be given followed by a loading dose of either phenytoin (fosphenytoin), levetiracetam or valproate (68). If rapid iv access is not possible then intramuscular midazolam is an alternative (62).

Refractory SE

If convulsive SE has not responded to first-line treatments above and has continued for 60–90 min, then people will be entering a phase of physiological compromise, neuronal damage, and increasing drug resistance. At this stage, the patient should be transferred to an intensive therapy unit and given anaesthetic (propofol, thiopentone/pentobarbital, or midazolam) to stop the seizure activity (11,26). Pressor therapy may be necessary at this stage to maintain BP. Phenytoin administration should be continued. The patient should be loaded with phenobarbitone. Once all seizure activity has stopped for more than 24 h and provided there are adequate blood levels of anti-epileptic drugs then the anaesthetic can be slowly withdrawn. If SE should recur then other drugs can be tried, but at this stage good clinical trials are lacking and a variety of treatments have been suggested (69).

Drug treatment of non-convulsive status epilepticus

There is some controversy concerning the degree of neuronal damage that occurs with non-convulsive status epilepticus, and consequently how aggressively it should be treated (70). There is no evidence of neuronal damage in typical absence status epilepticus, which responds very well to oral or iv benzodiazepines. Complex partial status epilepticus (CPSE) is more difficult, and the outcome depends upon circumstance. Patients with epilepsy who have CPSE usually respond well to treatment with an iv benzodiazepine and if they have repeated episodes, then oral clobazam (or oral diazepam) is an alternative in the community. Benzodiazepines can, however, increase mortality in the critically-ill elderly who have CPSE and, in this population (71), an alternative, such as iv valproate or levetiracetam, could be considered. In both these instances, there is rarely justification to progress to anaesthesia and intensive therapy unit.

Predictive value of electroencephalogram after status epilepticus

The EEG also plays an independent role in predicting outcome after status epilepticus (72). Those whose EEG returns to normal have a very good prognosis and a very low mortality rate. However, the following EEG patterns are associated with a very high mortality (40–60% within 30 days of the cessation of status epilepticus):

- Periodic lateralized epileptiform discharges.
- Burst suppression pattern with bursts of activity (spikes, sharp waves, or slow) punctuated by periods of EEG flattening.
- After SE ictal discharges defined as ictal EEG activity lasting at least 10 s and up to a few minutes, starts and stops abruptly, and is not associated with a clinical change.

Conclusions

SE is the maximal expression of epilepsy and is associated with a high morbidity and mortality. Treatments need to be given in staged fashion. In the later stages, on-going electrographic seizure activity may not be accompanied by overt clinical signs. EEG plays a critical role in SE including its diagnosis, monitoring of treatment and prognostication. The clinical importance of certain EEG patterns in status epilepticus, in particular periodic discharges in the later stages, remains uncertain and requires further research.

References

1. Gastaut, H. (1970). Clinical and electroencephalographical classification of epileptic seizures. *Epilepsia*, **11**, 102–12.
2. Lowenstein, D.H., Bleck, T., and Macdonald, R.L. (1999). It's time to revise the definition of status epilepticus. *Epilepsia*, **40**, 120–2.
3. Trinka, E., Cock, H., Hesdorffer, D., et al. (2015). A definition and classification of status epilepticus. Report of the ILAE Task Force on Classification of Status Epilepticus. *Epilepsia*, **56**, 1515–23.
4. Walker, M.C. (1998). The epidemiology and management of status epilepticus. *Current Opinions in Neurology*, **11**, 149–54.
5. DeLorenzo, R.J., Hauser, W.A., Towne, A.R., et al. (1996). A prospective, population-based epidemiologic study of status epilepticus in Richmond, Virginia. *Neurology*, **46**, 1029–35.
6. Hauser, W.A. (1990). Status epilepticus: epidemiologic considerations. *Neurology*, **40**, 9–13.
7. Sillanpää, M. and Shinnar, S. (2002). Status epilepticus in a population-based cohort with childhood-onset epilepsy in Finland. *Annals of Neurology*, **52**, 303–10.
8. Mayer, S.A., Claassen, J., Lokin, J., Mendelsohn, F., Dennis, L.J., and Fitzsimmons, B-F. (2002). Refractory status epilepticus: frequency, risk factors, and impact on outcome. *Archives of Neurology*, **59**, 205–10.
9. Holtkamp, M. (2005). Predictors and prognosis of refractory status epilepticus treated in a neurological intensive care unit. *Journal of Neurology, Neurosurgery, & Psychiatry*, **76**, 534–9.
10. Neligan, A. and Shorvon, S.D. (2011). Prognostic factors, morbidity and mortality in tonic-clonic status epilepticus: a review. *Epilepsy Research*, **93**, 1–10.
11. Claassen, J., Hirsch, L.J., Emerson, R.G., and Mayer, S.A. (2002). Treatment of refractory status epilepticus with pentobarbital, propofol, or midazolam: a systematic review. *Epilepsia*, **43**, 146–53.
12. Hesdorffer, D.C., Logroscino, G., Cascino, G., Annegers, J.F., and Hauser, W.A. (1998). Risk of unprovoked seizure after acute symptomatic seizure: effect of status epilepticus. *Annals of Neurology*, **44**, 908–12.
13. Treiman, D.M., Walton, N.Y., and Kendrick, C. (1990). A progressive sequence of electroencephalographic changes during generalized convulsive status epilepticus. *Epilepsy Research*, **5**, 49–60.
14. Pender, R.A. and Losey T.E. (2012). A rapid course through the five electrographic stages of status epilepticus. *Epilepsia*, **53**, e193–5.
15. Handforth, A., Cheng, J.T., Mandelkern, M.A., and Treiman, D.M. (1994) Markedly increased mesiotemporal lobe metabolism in a case with PLEDs: further evidence that PLEDs are a manifestation of partial status epilepticus. *Epilepsia*, **35**, 876–81.
16. Waziri, A., Claassen, J., Stuart, R.M., et al. (2009). Intracortical electroencephalography in acute brain injury. *Annals of Neurology*, **66**, 366–77.
17. Walker, M. (2005). Status epilepticus: an evidence based guide. *British Medical Journal*, **331**, 673–7.
18. Meldrum, B. (1991). Excitotoxicity and epileptic brain damage. *Epilepsy Research*, **10**, 55–61.
19. Walker, M. (2007). Neuroprotection in epilepsy. *Epilepsia*, **48**(Suppl. 8), 66–8.
20. Bankstahl, J.P. and Löscher, W. (2008). Resistance to antiepileptic drugs and expression of P-glycoprotein in two rat models of status epilepticus. *Epilepsy Research*, **82**, 70–85.
21. Naylor, D.E., Liu, H., and Wasterlain, C.G. (2005). Trafficking of GABA(A) receptors, loss of inhibition, and a mechanism for pharmacoresistance in status epilepticus. *Journal of Neuroscience*, **25**, 7724–33.
22. Goodkin, H.P., Yeh, J-L., and Kapur, J. (2005). Status epilepticus increases the intracellular accumulation of GABAA receptors. *Journal of Neuroscience*, **25**, 5511–20.
23. Walker, M.C., Howard, R.S., Smith, S.J., Miller, D.H., Shorvon, S.D., and Hirsch, N.P. (1996). Diagnosis and treatment of status epilepticus on a neurological intensive care unit. *Quarterly Journal of Medicine*, **89**, 913–20.
24. Walker, M.C. (2007). Treatment of nonconvulsive status epilepticus. *International review of neurobiology* **81**, 287–97
25. Towne, A.R., Waterhouse, E.J., Boggs, J.G., et al. (2000). Prevalence of nonconvulsive status epilepticus in comatose patients. *Neurology*, **54**, 340–5.
26. Walker, M.C. (2003). Status epilepticus on the intensive care unit. *Journal of Neurology*, **250**, 401–6.
27. Claassen, J., Mayer, S.A., Kowalski, R.G., Emerson, R.G., and Hirsch, L.J. (2004). Detection of electrographic seizures with continuous EEG monitoring in critically ill patients. *Neurology*, **62**, 1743–8.
28. Altafullah, I., Asaikar, S., and Torres, F. (1991). Status epilepticus: clinical experience with two special devices for continuous cerebral monitoring. *Acta Neurologica Scandinavica*, **84**, 374–81.
29. Musialowicz, T., Mervaala, E., Kälviäinen, R., Uusaro, A., Ruokonen, E., and Parviainen, I. (2010). Can BIS monitoring be used to assess the depth of propofol anesthesia in the treatment of refractory status epilepticus? *Epilepsia*, **51**, 1580–6.
30. Lhatoo, S.D. and Alexopoulos, A.V. (2007). The surgical treatment of status epilepticus. *Epilepsia*, **48**(Suppl. 8), 61–5.
31. Cockerell, O.C., Rothwell, J., Thompson, P.D., Marsden, C.D., and Shorvon, S.D. (1996). Clinical and physiological features of epilepsia partialis continua. Cases ascertained in the UK. *Brain*, **119**(Pt 2), 393–407.
32. Hui, A.C.F., Cheng, C., Lam, A., Mok, V., and Joynt, G.M. (2005). Prognosis following postanoxic myoclonus status epilepticus. *European Neurology*, **54**, 10–3.
33. Jumao-as, A. and Brenner, R.P. (1990). Myoclonic status epilepticus: a clinical and electroencephalographic study. *Neurology*, **40**, 1199–202.
34. Young, G.B., Gilbert, J.J., and Zochodne, D.W. (1990). The significance of myoclonic status epilepticus in postanoxic coma. *Neurology*, **40**, 1843–8.
35. Morris, H.R., Howard, R.S., and Brown, P. (1998). Early myoclonic status and outcome after cardiorespiratory arrest. *Journal of Neurology, Neurosurgery, & Psychiatry*, **64**, 267–8.
36. Kaplan, P.W. (2007). EEG criteria for nonconvulsive status epilepticus. *Epilepsia*, **48**, 39–41.
37. Beniczky, S., Hirsch, L.J., Kaplan, P.W., et al. (2013). Unified EEG terminology and criteria for nonconvulsive status epilepticus. *Epilepsia*, **54**(Suppl. 6), 28–9.
38. Schmitt, S.E., Pargeon, K., Frechette, E.S., Hirsch, L.J., Dalmau, J., and Friedman, D. (2012). Extreme delta brush: a unique EEG pattern in adults with anti-NMDA receptor encephalitis. *Neurology*, **79**, 1094–100.
39. Brenner, R.P. (2002). Is it status? *Epilepsia*, **43**(Suppl. 3), 103–13.

40. Hirsch, L.J., Claassen, J., Mayer, S.A., and Emerson, R.G. (2004). Stimulus-induced rhythmic, periodic, or ictal discharges (SIRPIDs): a common EEG phenomenon in the critically ill. *Epilepsia*, **45**, 109–23.
41. Zeiler, S.R., Turtzo, L.C., and Kaplan, P.W. (2011). SPECT-negative SIRPIDs argues against treatment as seizures. *Journal of Clinical Neurophysiology*, **28**, 493–6.
42. Cobb, W. and Hill, D. (1950). Electroencephalogram in subacute progressive encephalitis. *Brain*, **73**, 392–404.
43. Chatrian, G.E., Shaw, C.M., and Plum, F. (1964). Focal periodic slow transients in epilepsia partialis continua: clinical and pathological correlations in two cases. *Electroencephalography & Clinical Neurophysiology*, **16**, 387–93.
44. Kalamangalam, G.P., Diehl, B., and Burgess, R.C. (2007). Neuroimaging and neurophysiology of periodic lateralized epileptiform discharges: observations and hypotheses. *Epilepsia*, **48**, 1396–405.
45. Orta, D.S.J., Chiappa, K.H., Quiroz, A.Z., Costello, D.J., and Cole, A.J. (2009). Prognostic implications of periodic epileptiform discharges. *Archives of Neurology*, **66**, 985–91.
46. Singh, G., Wright, M-A., Sander, J.W., and Walker, M.C. (2005). Periodic lateralized epileptiform discharges (PLEDs) as the sole electrographic correlate of a complex partial seizure. *Epileptic Disorders*, **7**, 37–41.
47. Fountain, N.B. and Waldman, W.A. (2001). Effects of benzodiazepines on triphasic waves: implications for nonconvulsive status epilepticus. *Journal of Clinical Neurophysiology*, **18**, 345–52.
48. Nickels, K. and Wirrell, E. (2008). Electrical status epilepticus in sleep. *Seminars in Pediatric Neurology*, **15**, 50–60.
49. Shorvon, S. and Walker, M. (2005). Status epilepticus in idiopathic generalized epilepsy. *Epilepsia*, **46**(Suppl. 9), 73–9.
50. Taverner, D. and Bain, W.A. (1958). Intravenous lignocaine as an anticonvulsant in status epilepticus and serial epilepsy. *Lancet*, **2**, 1145–7.
51. Leppik, I.E., Derivan, A.T., Homan, R.W., Walker, J., Ramsay, R.E., and Patrick, B. (1983). Double-blind study of lorazepam and diazepam in status epilepticus. *Journal of the American Medical Association*, **249**, 1452–4.
52. Chamberlain, J.M., Okada, P., Holsti, M., et al. (2014). Lorazepam vs diazepam for pediatric status epilepticus. *Journal of the American Medical Association*, **311**, 1652.
53. Sreenath, T.G., Gupta, P., Sharma, K.K., and Krishnamurthy, S. (2010). Lorazepam versus diazepam-phenytoin combination in the treatment of convulsive status epilepticus in children: a randomized controlled trial. *European Journal of Paediatric Neurology*, **14**, 162–8.
54. Shaner, D.M., McCurdy, S.A., Herring, M.O., and Gabor, A.J. (1988). Treatment of status epilepticus: a prospective comparison of diazepam and phenytoin versus phenobarbital and optional phenytoin. *Neurology*, **38**, 202–7.
55. Treiman, D.M., Meyers, P.D., Walton, N.Y., et al. (1998). A comparison of four treatments for generalized convulsive status epilepticus. Veterans Affairs Status Epilepticus Cooperative Study Group. *New England Journal of Medicine*, **339**, 792–8.
56. Misra, U.K., Kalita, J., and Patel, R. (2006). Sodium valproate vs phenytoin in status epilepticus: a pilot study. *Neurology*, **67**, 340–2.
57. Mehta, V., Singhi, P., and Singhi, S. (2007). Intravenous sodium valproate versus diazepam infusion for the control of refractory status epilepticus in children: a randomized controlled trial. *Journal of Child Neurology*, **22**, 1191–7.
58. Misra, U.K., Kalita, J., and Maurya, P.K. (2012). Levetiracetam versus lorazepam in status epilepticus: a randomized, open labeled pilot study. *Journal of Neurology*, **259**, 645–8.
59. Mundlamuri, R.C., Sinha, S., Subbakrishna, D.K., et al. (2015). Management of generalised convulsive status epilepticus (SE): a prospective randomised controlled study of combined treatment with intravenous lorazepam with either phenytoin, sodium valproate or levetiracetam—pilot study. *Epilepsy Research*, **114**, 52–8.
60. Agarwal, P., Kumar, N., Chandra, R., Gupta, G., Antony, A.R., and Garg, N. (2007). Randomized study of intravenous valproate and phenytoin in status epilepticus. *Seizure*, **16**, 527–32.
61. Appleton, R., Sweeney, A., Choonara, I., Robson, J., and Molyneux, E. (1995). Lorazepam versus diazepam in the acute treatment of epileptic seizures and status epilepticus. *Developmental Medicine & Child Neurology*, **37**, 682–8.
62. Silbergleit, R., Durkalski, V., Lowenstein, D., et al. (2012). Intramuscular versus intravenous therapy for prehospital status epilepticus. *New England Journal of Medicine*, **366**, 591–600.
63. Alldredge, B.K., Gelb, A.M., Isaacs, S.M., et al. (2001). A comparison of lorazepam, diazepam, and placebo for the treatment of out-of-hospital status epilepticus. *New England Journal of Medicine*, **345**, 631–7.
64. Dreifuss, F.E., Rosman, N.P., Cloyd, J.C., et al. (1998). A comparison of rectal diazepam gel and placebo for acute repetitive seizures. *New England Journal of Medicine*, **338**, 1869–75.
65. Scott, R.C., Besag, F.M., and Neville, B.G. (1999). Buccal midazolam and rectal diazepam for treatment of prolonged seizures in childhood and adolescence: a randomised trial. *Lancet*, **353**, 623–6.
66. McIntyre, J., Robertson, S., Norris, E., et al. (2005) Safety and efficacy of buccal midazolam versus rectal diazepam for emergency treatment of seizures in children: a randomised controlled trial. *Lancet*, **366**, 205–10.
67. Mpimbaza, A., Ndeezi, G., Staedke, S., Rosenthal, P.J., and Byarugaba, J. (2008). Comparison of buccal midazolam with rectal diazepam in the treatment of prolonged seizures in Ugandan children: a randomized clinical trial. *Pediatrics*, **121**, e58–64.
68. Shorvon, S., Baulac, M., Cross, H., Trinka, E., and Walker, M. (2008). The drug treatment of status epilepticus in Europe: consensus document from a workshop at the first London Colloquium on Status Epilepticus. *Epilepsia*, **49**, 1277–85.
69. Shorvon, S. and Ferlisi, M. (2012). The outcome of therapies in refractory and super-refractory convulsive status epilepticus and recommendations for therapy. *Brain*, **135**(Pt 8), 2314–28.
70. Walker, M.C. (2001). Diagnosis and treatment of nonconvulsive status epilepticus. *CNS Drugs*, **15**, 931–9.
71. Litt, B., Wityk, R.J., Hertz, S.H., et al. (1998). Nonconvulsive status epilepticus in the critically ill elderly. *Epilepsia*, **39**, 1194–202.
72. Jaitly, R., Sgro, J.A., Towne, A.R., Ko, D., and DeLorenzo, R.J. (1997). Prognostic value of EEG monitoring after status epilepticus: a prospective adult study. *Journal of Clinical Neurophysiology*, **14**, 326–34.

CHAPTER 31

Presurgical evaluation for epilepsy surgery

Robert Elwes

Historical development of epilepsy surgery

Hughlings Jackson is generally acknowledged as providing the intellectual basis for the modern surgical treatment of epilepsy. He defined epilepsy as 'the name for occasional, sudden, excessive, rapid, and local discharge of grey matter'. This was many years before the electrical properties of the cortex were described. Berger demonstrated the electroencephalogram (EEG) in humans and shortly after this Foerster in Germany pioneered direct recordings of the EEG from the exposed brain and also the intraoperative localization of cortical function by stimulation (1). Gibbs and Gibbs in Boston described the anterior temporal lobe focus in patients with psychomotor seizures and with Bailey carried out anterior temporal resections for what we now term medial temporal lobe epilepsy (2). An important advance occurred in Paris in the middle of the 20th century. Talairach developed the stereotactic atlas and, with Bancaud at the Hôpital St Anne, inserted EEG electrodes into deep brain structures (3). The EEG was recorded in three dimensions, so-called stereo EEG. At first, acute interictal recordings were used. Chronic recordings from depth electrodes using telemetry were developed, particularly by Crandall at University of California, Los Angeles (4). The Montreal Neurological Institute was the dominant centre for epilepsy surgery in the middle of the 20th century. Penfield, who worked with Foerster, and was succeeded by Rasmussen, carried out a large series of resections of epileptogenic lesions (5), where the extent of the surrounding cortical resection was guided by intraoperative electrocorticography (6). This work largely established the specialty of epilepsy surgery as we now know it. Falconer at the Maudsley Hospital in London developed the en bloc temporal lobectomy and this group went on to show that hippocampal sclerosis, indolent glioneuronal tumours and cortical dysplasia are the commonest pathological substrates in temporal lobe epilepsy (7,8).

Selection of candidates for epilepsy surgery

The electroclinical syndromes open to epilepsy surgery are shown in Table 31.1. In practice medial temporal epilepsy and frontal epilepsy will form some 80–90% of cases selected for evaluation. Most will have seizures on a weekly or monthly basis, and will have failed on both first and second line anti-epileptic medications (Box 31.1). Identification of a well-defined electroclinical syndrome prior to presurgical evaluation can usually be done on the basis of clinical

Table 31.1 Electroclinical syndromes, common aetiologies and operative procedures

Electroclinical syndrome	Common pathology or imaging	Operation
Medial temporal lobe epilepsy	Hippocampal sclerosis Indolent glioneuronal tumour Normal MR	En bloc temporal lobectomy Selective amygdalo-hippocampectomy
Lateral temporal lobe epilepsy	Indolent glioneronal tumour Cortical dysplasia Cavernoma Normal MR	Lesionectomy and corticectomy guided by electrocorticography En bloc temporal lobectomy
Frontal lobe epilepsy Primary motor Secondary motor Hypermotor seizures Orbitofrontal Frontal absences Opercular	Cortical dysplasia Indolent glioneronal tumour Cavernoma Traumatic scar Normal MR	Lesionectomy and corticectomy guided by electrocorticography
Focal motor/central epilepsy	Perinatal stroke Rasmussen's encephalitis Hemimegalencephaly Cortical dysplasia Sturge Weber syndrome Focal encephalitis	Hemisperectomy Hemispherotomy
Landau–Kleffner syndrome	None	Multiple subpial transection
Symptomatic generalized epilepsy or drop attacks	Multiple aetiologies	Anterior two-thirds callosotomy Vagal nerve stimulation
Primary generalized or secondary generalized epilepsy	Multiple aetiologies	Vagal nerve stimulation Deep brain stimulation
All syndromes after exclusion of resective surgery	Multiple aetiologies	Vagal nerve stimulation

Box 31.1 Selection criteria for consideration of resective epilepsy surgery

General Selection Criteria

- Active epilepsy for at least 2 years.
- Have tried first line and second line drugs.
- Most will have seizures every month, often more frequently.
- Sufficiently disabled by seizures to warrant surgery.
- A well-defined electroclinical syndrome suggestive of focal epilepsy.
- No general contraindications to brain surgery.

Relative contraindications

- Multiple seizure types or multifocal EEG.
- Prominent generalized epileptiform discharges.
- Moderate or severe learning difficulties.
- Electroclinical syndromes likely to involve eloquent cortex with normal MR.
- Fixed schizophreniform psychosis.

features, interictal EEG, and structural imaging. Multifocal or generalized epileptiform discharges, as with learning difficulties or a history of developmental delay, may be indicative of a more widespread or pervasive disorder, possibly less amenable to localized resective surgery. Frequent generalized convulsions have a similar implication. There are, however, rare instances of the coexistence of idiopathic generalized epilepsy and focal lesional epilepsy which are not an overlap of the two disorders and open to surgery (9). Importantly, hippocampal sclerosis with bilateral interictal EEG changes or a normal MR with a clear electroclinical diagnosis of temporal lobe epilepsy (TLE) or frontal lobe epilepsy (FLE), are not contraindications to presurgical evaluation.

General principles of electroclinical evaluation

Multidisciplinary assessment

Successful epilepsy surgery requires input from multiple specialties. The initial assessment is usually led by a physician with training in EEG. Interpretation of ictal EEG, high resolution structural magnetic resonance imaging (MRI) with appropriate epilepsy protocols and neuropsychological evaluation requires considerable experience. Functional imaging, post-processing of structural imaging and specialist EEG techniques, such as single pulse electrical stimulation may require input from research based groups. The team should meet to discuss each case. They should have a clear protocol (see Fig. 31.1) and avoid ad hoc decisions. The surgeon should see each case before any invasive procedure, including carotid amytal; the patient must understand and accept the risks and benefits of an operation, and agree in principle to surgery before invasive tests. In order to gain the appropriate skills and maintain a complex series of facilities around 20 operative procedures should be performed per year in the setting of a tertiary referral centre. Biopsy or treatment of malignant tumours is in general not considered epilepsy surgery. Access to neuropsychiatry is crucial. Some centres may wish to operate only on cases using only non-invasive tests. They should collaborate with an advanced centre willing to take on complex cases requiring intracranial EEG and not exclude potentially operable patients. Complex epilepsy surgery with major intracranial EEG evaluations usually requires the back-up of university or research based groups to undertake EEG signal analysis and advanced imaging.

Congruence of non-invasive tests

Surgery should be offered wherever possible on the basis of congruence of non-invasive tests. This occurs in around 80% of cases. If the patient has a lesion known to be associated with epilepsy, the electroclinical features are congruent, and the lesion can be removed with a low risk of cognitive or neurological deficits, invasive evaluation is not needed, and is unlikely to improve outcome. It has to be appreciated, however, that the relationship of lesion to seizures can be complex. Strokes and traumatic lesions often have extensive epileptogenic areas and there are few operative series, other than hemispherectomy for perinatal middle cerebral artery strokes with dense hemiplegia. The lesions in CD often extend well beyond the imaging abnormality (10). Lesions can occur incidentally in the context of generalized epilepsy. Lesions may cause secondary epileptogenesis at distant sites (9,11) and in some cases good outcome still occurs if the lesion is not fully removed (12).

The principle indication for intracranial EEG is discordance among baseline tests. If it is accepted that, in many cases, removal of pathology is the most important predictor of outcome, the commonest indication for intracranial EEG is to assess whether the lesion is the cause of the epilepsy in the face of atypical electroclinical or other discordant features. In this respect intracranial EEG is used to test a hypothesis generated from the non-invasive evaluation. Speculative evaluations or 'fishing expeditions' are to be avoided. Sometimes intracranial EEG is done to test the hypothesis that a lesion is not the cause of the seizures thus avoiding an unhelpful therapeutic operation that may carry added risk.

Methods of intracranial EEG evaluation

The type and proposed site of electrode placement (Box 31.2) (13,14) should be decided at the preoperative multidisciplinary meeting, which reviews all the clinical and non-invasive tests. In general the testing of a hypothesis relating to pathology or delineation of functional cortex is often done with electrodes in the subdural space, often strips. Depth is used in selective cases such as bitemporal disease, deep or inaccessible lesions or for stereo-EEG. If a subdural mat is being inserted, it is best that the neurophysiologist attends theatre. The orientation and relationship of the mat to the putative lesion, gyral anatomy, probable eloquent areas, and the margins of the craniotomy can be quite complex. Intraoperative electrocorticography may help in the placing of electrodes.

Most patients require a period of 24 h in a high dependency unit to check for secondary deterioration due to bleeding, most often subdural if mats are used. Post-operative imaging with skull X-rays and X-ray computed tomography (CT) can be done, which can be used to fuse the position of the electrodes with other imaging data. If the precise position of the tip of a depth electrode has to be found then magnetic resonance imaging (MRI) can be performed. The

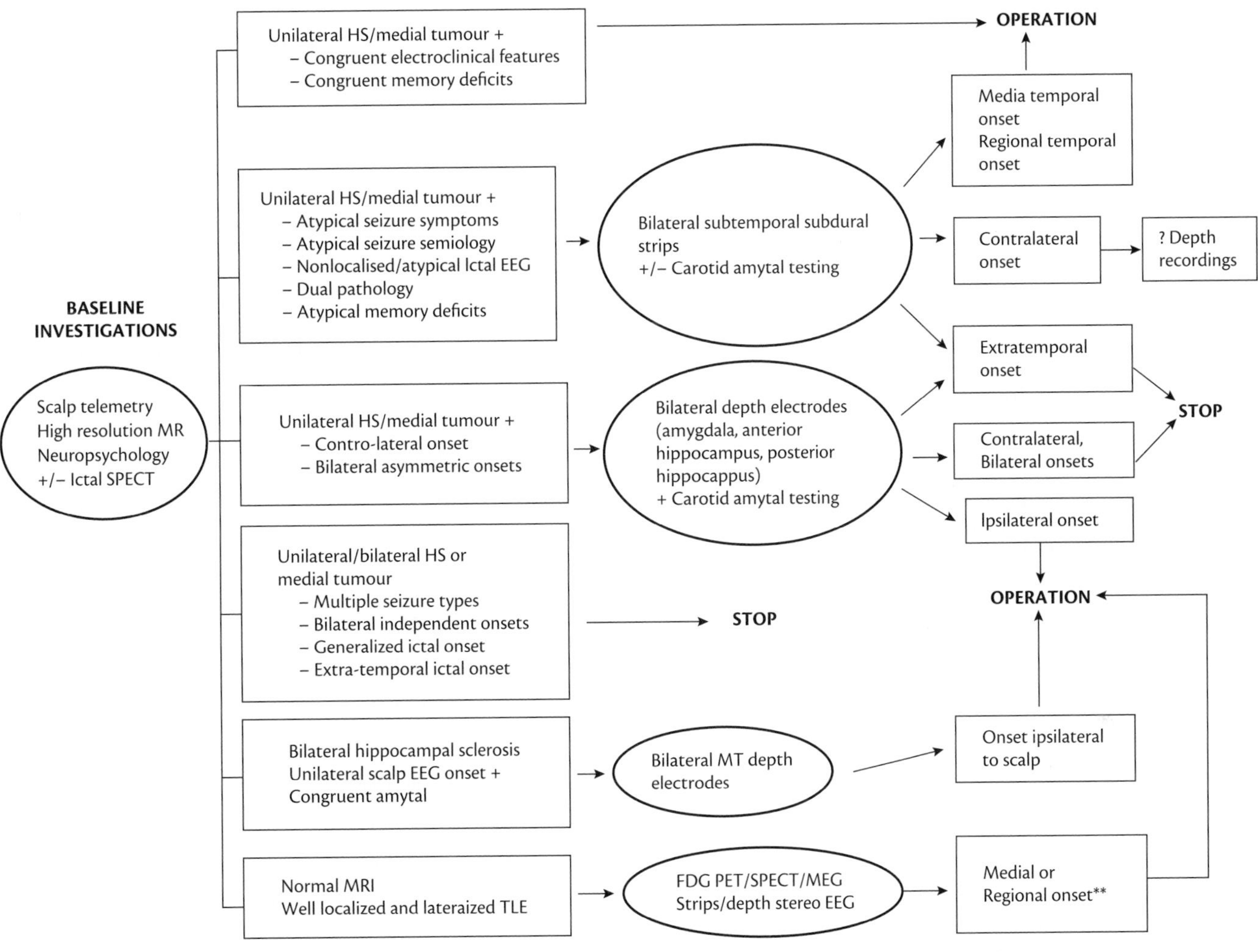

Fig. 31.1 Electroclinical evaluation of temporal lobe epilepsy; flow diagram based on hypothesis testing.
**Focal medial onsets on IC EEG are those confined to one or two contacts, typically the deepest contacts of the subtemporal strip or depth electrode. Regional onsets involve lateral structures, but are confined to the temporal lobe.

chief risk with non-ferrous electrodes is the presence of loops that produce heating and the test is probably safe if the electrodes do not come into close apposition (15).

Most intracranial recording will require 64 or more channels, and multiple preamplifier boxes. Artefacts from non-functioning electrodes are common. The site of the problem needs to be identified by switching cables and connections. It is most often due to internal disruption of the recording electrode, which can also produce the same electrical signal from all contacts. If located in cerebrospinal fluid (CSF) or blood, and not in apposition to the brain the recordings may be flat. For ease of interpretation montages should have the electrodes on a mat sequentially numbered (Fig. 31.6A). Brief abbreviated names can be given to electrodes that indicate their position such as amygdala, anterior and posterior hippocampus (Amg, AH, PH, etc.) or they can be given a letter if a large number of closely spaced depths are being used for stereo EEG. Typically, the contacts on a depth electrode or subdural strip are numbered from the highest contact (most superficial) to the lowest (deepest contact). Once the montage has been drawn up, and the electrodes connected and recording, a biological check should be done by an independent person to check the labelling is correct. Simple montages with a single left and right electrode are the ones most open to error. It is our practice to choose an internal reference from an electrode that appears inactive, but many use bipolar recordings. If the ictal onset consists of decrements and fast activity high sampling frequency may be needed to show ictal onsets in the gamma range. Interictal recordings often have many spikes in diverse places and compared with scalp EEG more emphasis is placed on ictal onsets. It is often best to have one day on standard drugs and then reduce; on rare occasions drug reduction may precipitate a new focus or seizure type (16). All the EEG should be scanned electronically by eye. Daily files of interictal EEG awake and asleep should be stored. Many commercial systems are now available for seizure detection that can help reduce data volume, but visual analysis is always needed. It is surprising how often brief ictal events with subtle behavioural changes that were missed on scalp recording are identified. Electrical seizures may be seen and if different from the site of onset of habitual seizures are an adverse prognostic sign (17). Usually a minimum of three or four habitual seizures are needed, but more if there is a suggestion of multiple seizure types. Whilst the ictal pattern may be very complex it is usually remarkably stereotyped for each case. A single habitual

Box 31.2 Specialist and intracranial electrode placements

Sphenoidal electrodes

- Provide stable anterior temporal recordings on telemetry.
- Surface electrode at same site provides similar information.

Foramen ovale electrodes (Fig. 31.2B) (13)

- Intermediate invasiveness.
- Can be combined with scalp EEG.
- Inserted through the foramen ovale and trigeminal ganglion.
- Pain and swelling.
- Lie in the middle cranial fossa and along free edge of tentorium.
- Recording restricted to medial temporal area.
- Helpful if doing selective amygdalo-hippocampal resections.

Bilateral temporal subdural strips (see Fig. 31.2B)

- Commonest IC EEG, usually 8-contact strip.
- Bilateral burr holes.
- Covers lateral and medial temporal lobes.
- Always inserted bilaterally.
- If frontal or posterior seizure features multiple electrodes may be inserted (Fig. 31.3A).
- May be difficult to position and can pass into brain substance.
- Used to test if epilepsy is of temporal origin prior to temporal lobectomy.
- May give incorrect lateralization in small proportion of cases.

Depth electrodes (Fig. 31.5A)

- Placed using MR coordinated and stereotactic head frame.
- Commonly bilateral mesial temporal: amygdala, anterior and posterior hippocampus.
- Deep seated inaccessible lesions, orbitofrontal, medial hemisphere.
- Small risk or bleeding, avoid primary motor and speech areas.

Stereo-EEG (14)

- Multiple orthogonal depth electrodes, usually closely spaced in one hemisphere.
- Site depends on scalp electroclinical and functional imaging results.
- Assesses three-dimensional neural network involved in seizures.
- May be combined with functional stimulation.
- Often in cases with normal MRI.
- Tailored resection based on EEG findings rather than pathology.

Subdural mats (Fig. 31.6A)

- Most complex and invasive.
- Requires full craniotomy.
- CSF leaks, deficits due to disordered blood flow, risk of subdural.
- Often with strips to the side.
- Focus close to eloquent cortex.
- 64-contact mat straddling motor strip.
- 32/20 contact mat across speech areas.
- Combined with functional stimulation.

seizure with other electrical or more minor seizures (Fig. 31.5B) may provide sufficient information to localize the focus.

Intracranial EEG gives highly detailed information on the spread of seizures and latencies of fraction of a second may be important in determining the site of onset. Conversely, if the electrodes are not in the correct position the results can be highly misleading. Considerable time and care must be taken to establish the first change that indicates the onset of the seizure. There may be a subtle movement or arousal on the video, a decrement in the EEG or even the disappearance of interictal spikes (Fig. 31.6B) that indicate the timing of seizure onset. Repeated and careful examination of seizures is needed. If the onset to the seizure precedes the first ictal EEG rhythm this implies that the changes may be due to spread from another site. This is a major difference from scalp EEG where ictal onsets after behavioural changes are common (Fig. 31.2D). The commonest rhythm at onset is low voltage fast activity. It is usual to divide onsets into focal (one or two contacts or a single electrode), regional (multiple electrodes or contacts, but confined to a lobe) or multilobar/hemispheric. Low voltage fast activity may be difficult to identify on visual inspection. The use of filters to remove slower rhythms may make them easier to see, especially if the gain is then is increased.

Risks of intracranial EEG

Subdural grids or mats in general carry the greatest risk of complication and probably occur in some form in around 5% of cases (18,19). The insertion of grids requires a craniotomy and then sewing of the dura. CSF leaks and low pressure headaches are very common. The high risks of central nervous system (CNS) infection shown in Box 31.3 are probably less now that the electrode wires are usually tunnelled out of through a separate wound and prophylactic antibiotics given. The greatest risk arises from subdural collections. Close monitoring in a high dependency unit is needed for the first 24 h. After transfer to the EEG monitoring unit there may be reducing conscious level, appearance of a hemiparesis and gradual loss of EEG signal as blood collects below the mat. Deficits may also arise from interference with blood supply. Most studies imply that the deficits are temporary, although it is difficult to extract clear statistics on this. Brain swelling and raised intracranial pressure are important adverse events, presumably either due to interference of blood supply or possibly as a direct reaction to the implants. The risk of complications was double in those with greater that 67 electrode contacts. The complication rate was said to rise by 4% per day after the first week of recording, although the major ones relating to bleeding

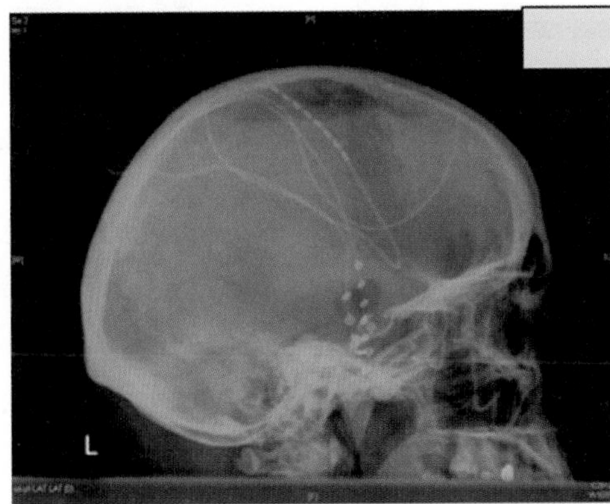

Chan	Subdural	Contact
1	RT	8
2	RT	7
3	RT	6
4	RT	5
5	RT	4
6	RT	3
7	RT	2
8	RT	1
9	LT	8
10	LT	7
11	LT	6
12	LT	5
13	LT	4
14	LT	3
15	LT	2
16	LT	1
17	ECG	
Ref	Scalp	
GND	Scalp	

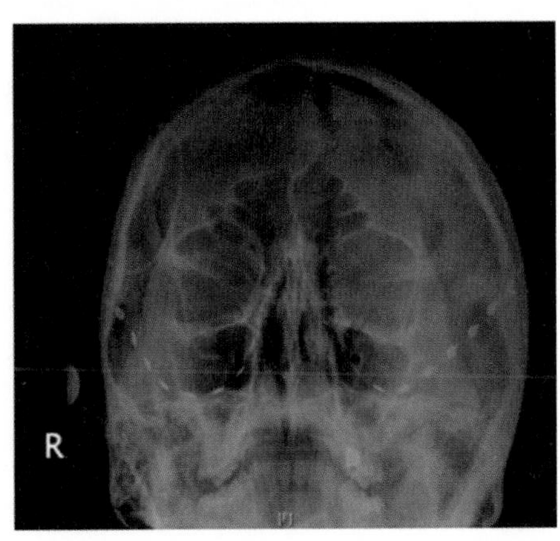

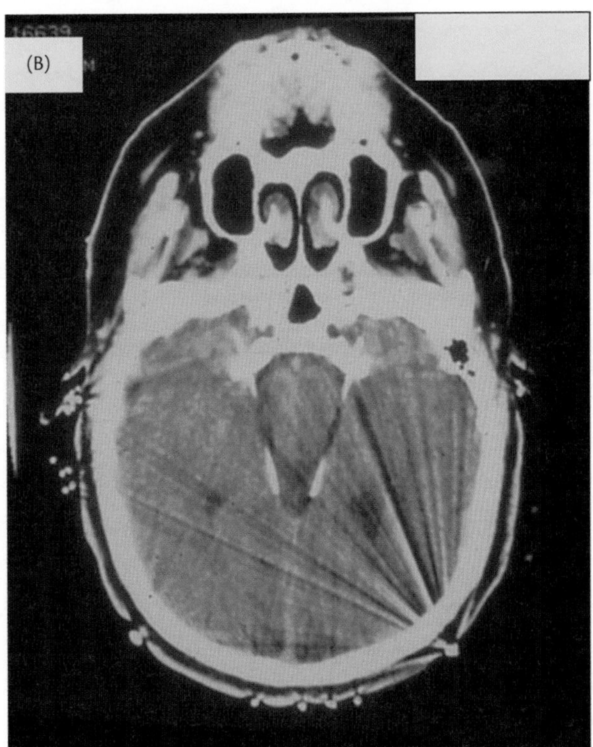

Fig. 31.2 Intracranial EEG recordings from subtemporal strips and foramen ovale electrodes in medial temporal lobe epilepsy. 29-year-old right-handed lady. Complicated febrile convulsions followed by epilepsy from the age of 2½ years of age. Initial imaging thought to be normal, but subsequently suggestive of left hippocampal sclerosis. Scalp EEG was left temporal, but there was probable bilateral hypometabolism on FDG PET and memory was poor on the right on amytal. Bilateral subtemporal strips were inserted. (A) Lateral and anteroposterior skull X-ray showing bilateral subtemporal strips. The deepest contacts 1 and 2 lying medially in the middle cranial fossa, near the parahippocampal gyri with proximal contacts 6-8 over the lateral temporal cortex. (B) Axial CT scan of foramen ovale electrodes with contacts near the parahippocampal gyrus. The electrodes enter through the floor of the middle cranial fossa and lie along the free edge of the tentorium. Note there is no coverage of lateral temporal cortex. (C) Intracranial EEG of bilateral subtemporal strips. Note widespread decrement of all left-sided channels with fast activity in deepest contacts of left temporal strip. (D) + 20 s. Rhythmic activity at 7 Hz has now spread over all lateral contacts on the left. At this time the first surface localized rhythm with ictal transformation is seen. (E) + 30 s, mid-seizure. Ictal activity now in lateral temporal contacts on the left. New rhythm appears on the right. (F) + 110 s, seizure end. Seizure ends asymmetrically, now more on the right. Note muscle jerk artefacts on the ECG as the seizure becomes secondarily generalized.

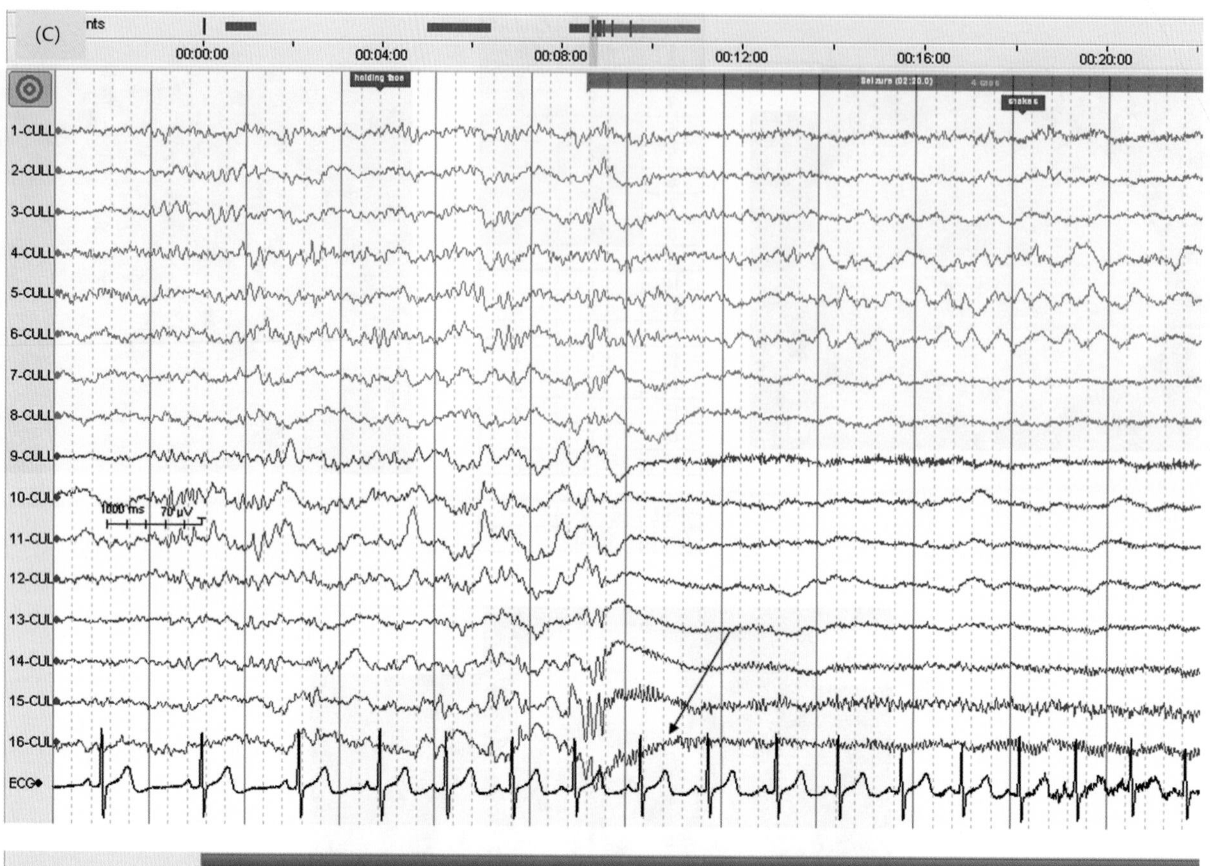

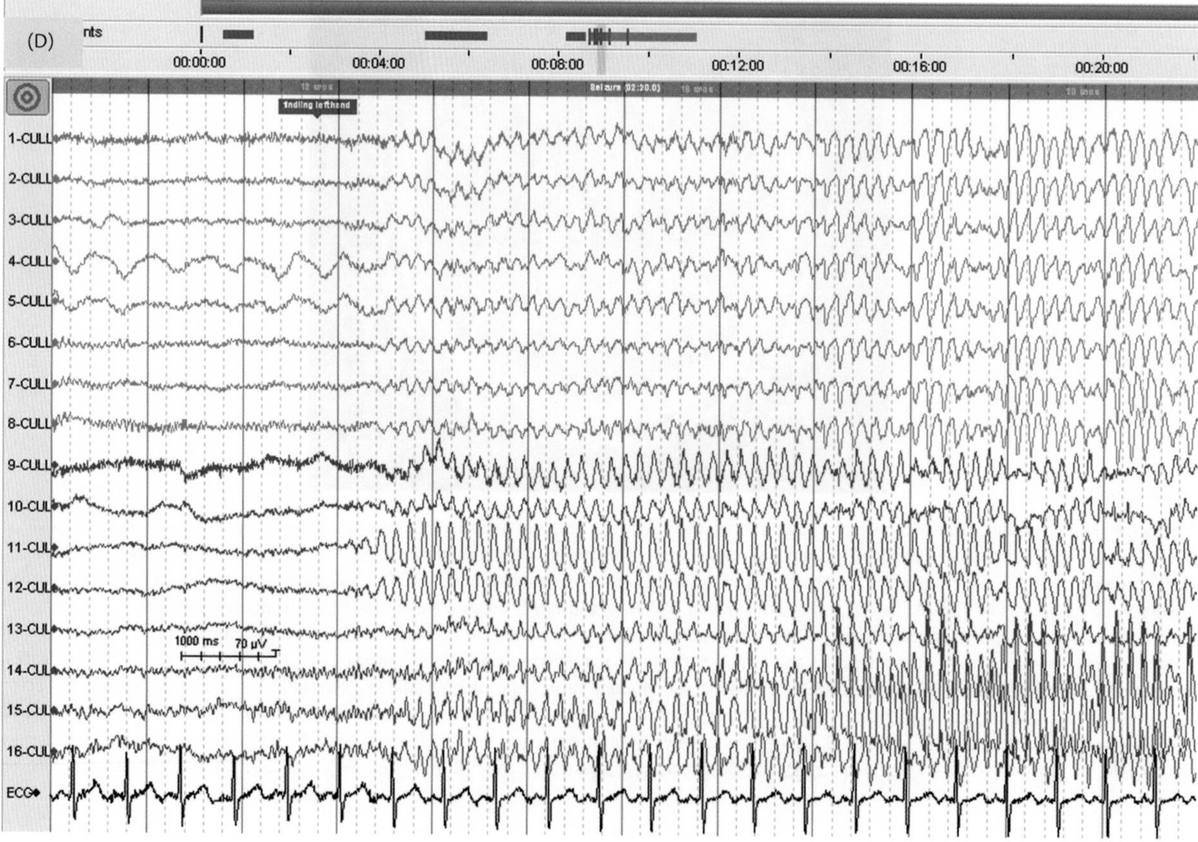

Fig. 31.2 Continued.

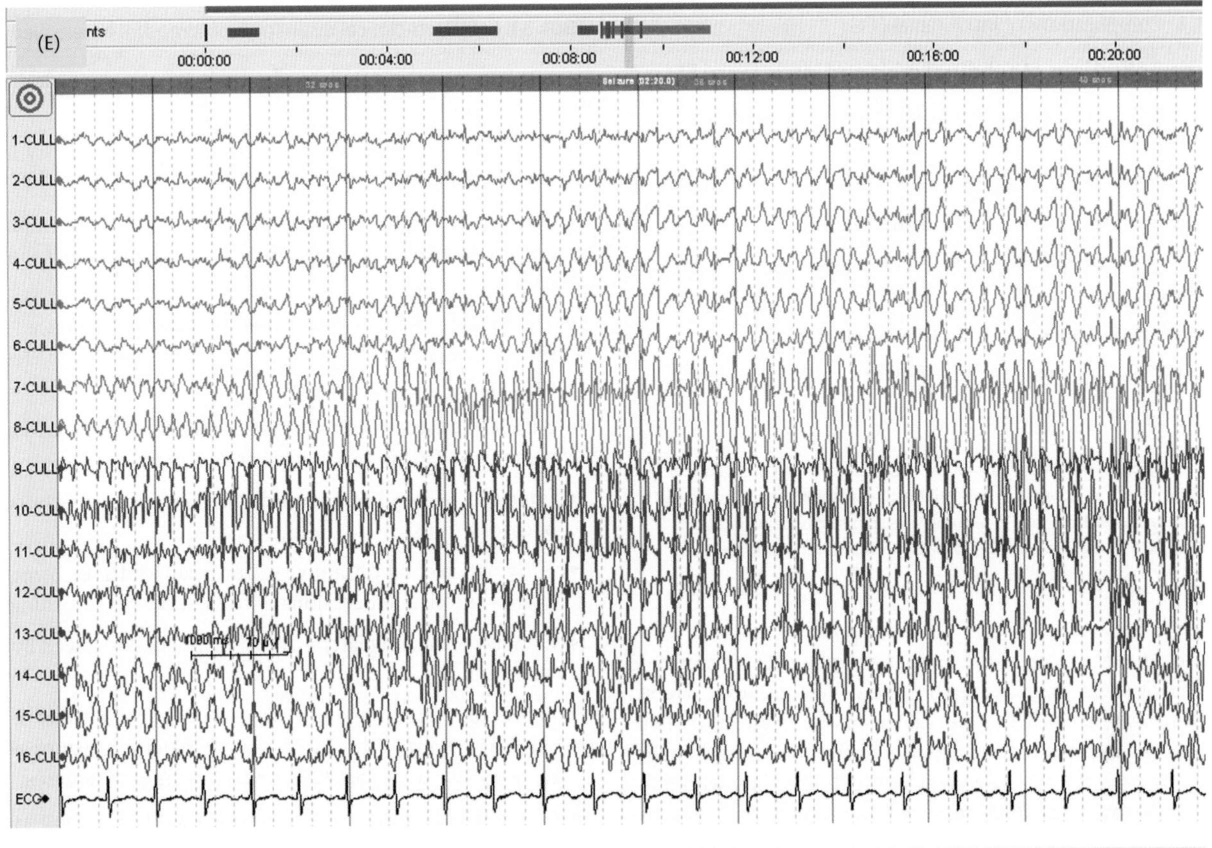

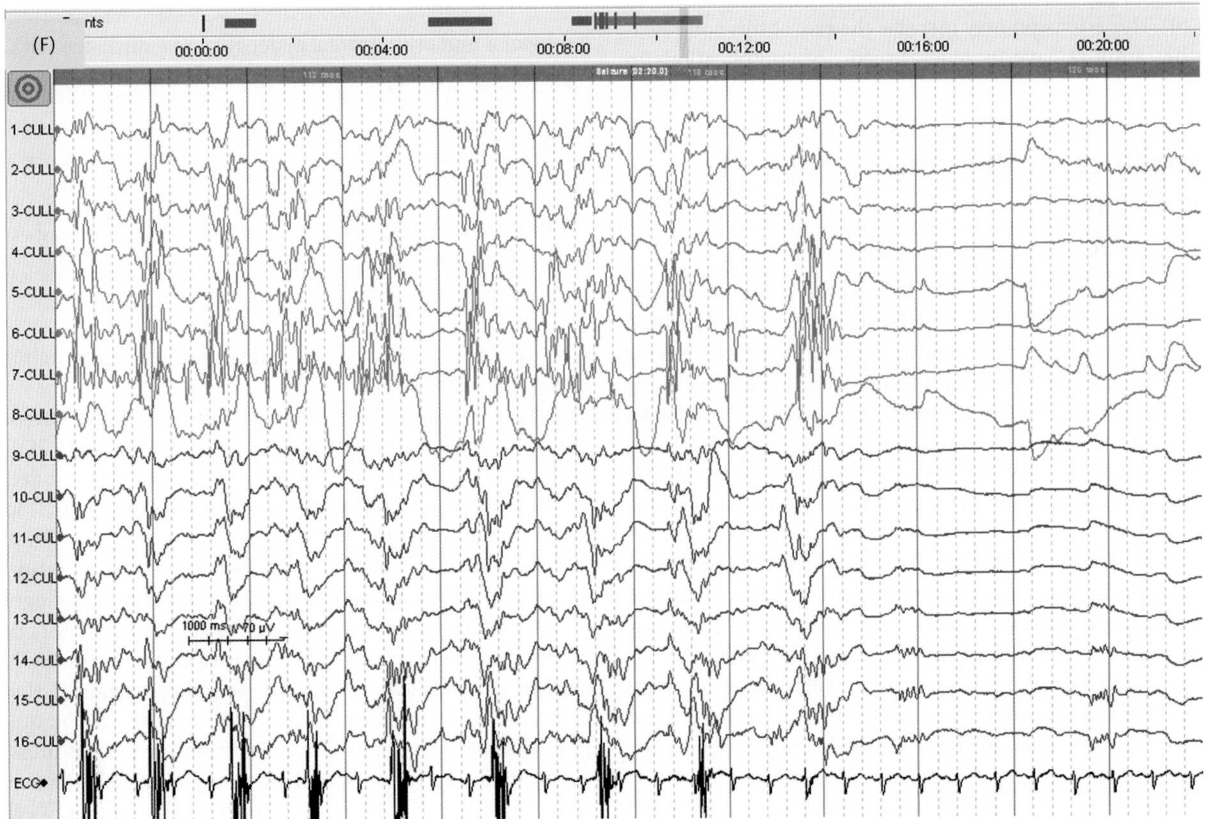

Fig. 31.2 Continued.

Box 31.3 Complications of intracranial EEG

Major complications of subdural grids and intracranial EEG

Review of 2542 cases showing complication percentage with 95% confidence intervals

- *Infection of CNS 2.3% (1.5–3.1%)*: most bacterial meningitis.
- *IC haemorrhage 4% (3.2–4.8)*: mostly subdural.
- *Neurological deficits 4.6% (3.2–6%)*: most hemiparesis or dysphasia.
- Surgical intervention 3% (2.5–3.5%) required.
- Raised IC pressure 2.4% (1.5–3.3%).
- Deaths ($n = 5$).

Adapted from *Epilepsia*, 54(5), Arya R, Mangano FT, Horn PS, Holland KD, Rose DF, Glauser TA, Adverse events related to extraoperative invasive EEG monitoring with subdural grid electrodes: a systematic review and meta-analysis, pp. 828–39, copyright (2013), with permission from John Wiley and Sons.

usually occur early. Despite the lowering of platelets by valproate this does not seem to increase the risk of haematomas. Depth electrodes pass through the substance of the brain and there is a risk of parenchymal as opposed to subdural haemorrhage. Overall, however, the risks seem to be much less than with subdural mats. Sansur et al. 2007 reported one intracranial haemorrhage after 274 depth electrode implantations with no permanent deficit (20). Hypertension may increase the risk of this complication. Other authors have in general recorded complications in about 1% of procedures (21–23).

Electroclinical evaluation of temporal lobe epilepsy

Ictal symptoms, semiology, and surface EEG

Medial temporal lobe epilepsy (MTLE) is the commonest syndrome seen in epilepsy surgery programs. Historically, it formed some 80% of cases, although worldwide the disorder now seems less common. The electroclinical features have been extensively reviewed (24) and only those issues of particular relevance to presurgical evaluation are discussed here (see Box 31.4). In medial temporal epilepsy the focus is basal, maximal on the surface EEG in the area between the eye and the ear, over the zygoma. Unfortunately the lowest line of electrodes in the 10-20 system is higher, running from the inion and nasion. Basal electrodes, such as the inferior frontal in the Maudsley system, or the superficial or deep sphenoidal have been widely used to record from this area. All spikes in MTLE are, therefore, generally biggest at the lowest line of electrodes, if referential recordings are used. The scalp EEG is a rather insensitive instrument for precise localization, but if a spike is bigger at any higher electrode, for example the superior frontal or sylvian, it is more likely to be extra temporal. The spikes are often biggest at the temporal pole, i.e. on F7 (25). However, they characteristically have a wide field spreading from the prefrontal to the occipital contacts (26). The occurrence of a mid- or posterior temporal maximum is more commonly associated with a lateral focus. Because of the wide

Box 31.4 Electro-clinical evaluation of temporal lobe epilepsy

- 80% of cases operated on the basis of congruent lesion and scalp EEG.
- Distinguishing medial from lateral temporal epilepsy is difficult on scalp EEG.
- Selective operations to remove only the hippocampus and amygdala require invasive EEG to confirm medial temporal epilepsy.
- The focus in MTLE is maximal over the zygoma, well below the placement of electrodes in the standard 10-20 electrode system.
- Basal electrodes, such as the sphenoidal or modified Maudsley system are used to provide lower coverage.
- Interictal bilateral temporal discharges occur in around 30% of cases.
- Early switch of hippocampal seizures can lead to incorrect scalp lateralization, and occasionally from subdural recordings.
- Doubts concerning laterality of TLE are best resolved by depth recording.
- Poorer operative outcomes are seen in the presence of multifocal or generalized discharges, rapid switch of seizure sides or the presence of moderate to severe learning difficulties.
- Lesional TLE with poorly localized scalp EEG is best assessed with bilateral sub temporal strips, or ictal SPECT in some cases.
- Insular and orbitofrontal epilepsy can be mistaken for TLE.
- Children with focal epilepsy often have more widespread EEG changes.
- Extra temporal EEG onset can occur in the presence of hippocampal sclerosis.
- Frontal and posterior epilepsies often spread to the temporal lobe.

electrical field, reference recordings may be more helpful than bipolar derivations.

During the aura, intracranial recording shows the onset of low voltage fast activity in the hippocampus (Fig. 31.2C) and/or amygdala. The surface EEG is often normal or shows a decrement or non-specific anterior slowing (27). The seizure may stop at this point or the patient then loses awareness and has a motionless stare. Around the time of loss of awareness the hippocampal fast activity increases in amplitude and slows to the theta range, taking on a rhythmic sharpened appearance (Fig. 31.2D). This becomes visible on the surface EEG as neocortical structures on the basal surface are recruited to form the so-called sustained rhythm or rhythmic ictal transformation (28). The aura can be quite prolonged and a delay of 30 s or more from the patient pressing the event marker to the appearance of rhythmic ictal transformation is not uncommon. In the overall assessment of ictal EEG changes, however, the later a feature occurs in the seizure the more likely it may be due to spread from another site. Ictal single photon emission computed tomography (SPECT) studies suggest that contralateral dystonia or ictal paralysis may be due to basal ganglia spread. At mid-seizure temporal rhythmic theta becomes more widespread and irregular (Fig. 31.2E). Spikes and slowing are seen, although often remaining maximal in the temporal lobe. If the ictal discharges spread over the hemisphere the contralateral arm may come up with the head turned towards it and clonic jerks appear as the seizure becomes secondarily generalized.

A striking feature of medial temporal epilepsy is that the seizures may switch sides. Low voltage fast activity may disappear from one hippocampus and reappear on the other side, such that the seizure is incorrectly lateralized on the surface EEG. This may even occur with subdural recordings. Post-ictal dysphasia is a strong lateralizing sign for MTLE and should be specifically examined for.

Indications for intracranial EEG in MTLE

A flow chart summarizing an approach to invasive evaluation in presumed medial temporal epilepsy is shown in Fig. 31.1. Cases in whom there is unilateral hippocampal sclerosis or a medial tumour, concordant surface EEG, and ipsilateral memory deficits, can proceed directly to resective surgery. For amydalohippocampectomy intracranial EEG, often with foramen ovale electrodes, is needed to demonstrate that the seizures are indeed of medial temporal origin (29). Most centres perform an en bloc anterior temporal resection as advocated by Falconer, or a modified resection after the work of Spencer, which leaves more lateral temporal neocortex (30).

Atypical electroclinical features

The commonest indication for intracranial EEG (IC EEG) is the presence of atypical or discordant clinical, EEG, or imaging features. Lateralized somatosensory or visual auras may indicate other extratemporal pathologies, such as dysplasia that may not be visible on MR (Fig. 31.3). Generalized epileptiform discharges, predominantly tonic clinic seizures, and a low IQ all suggest a more pervasive disorder and are independent predictors of a poor surgical outcome (31). A non-localized EEG onset with a lack of rhythmic ictal transformation is unusual in MTLE. Fast activity at onset on the scalp EEG suggests a neocortical onset. Pathology outside of the temporal lobe, so called dual pathology, occurs in around 10% of cases (32). In practice, there is often some degree of combination of these features that lead to intracranial EEG. Bilateral subdural strips are commonly used. More strips may be directed towards other sites depending on imaging and electroclinical findings (Fig. 31.3A).

Bilateral temporal lobe epilepsy

Unilateral hippocampal sclerosis with contralateral scalp EEG onset is best resolved using symmetrical depth electrodes in the amygdala, and anterior and posterior hippocampus (Fig. 31.4). In most cases, there is onset in the lesion and rapid early switch, but exceptions with contralateral onsets are sometimes seen. The only operation that will be performed is removal of the scarred side so that before depth recording an amytal test is performed to show that the side opposite the lesion is able to support memory.

Bilateral hippocampal sclerosis is also open to surgery in some cases (33–35). Post-mortem studies and evidence from imaging suggests the pathology is bilateral in around 10% of patients with MTLE (36). Careful screening for limbic encephalitis is needed. Herpes simplex encephalitis often leads to bilateral onsets. Neuropsychology and amytal testing are needed to show that the temporal lobes support memory. If the scalp EEG shows a well lateralized electroclinical syndrome and there are memory reserves on the contralateral side, then depth recordings may be performed to confirm side of seizure onset. Small operative series suggest that, if these criteria are rigorously followed, early outcomes appear not to be not dissimilar from unilateral hippocampal sclerosis. Some have reported an increased incidence of memory deficits after left-sided operations, again as seen with unilateral pathology.

Medial temporal lobe epilepsy with normal MRI

Operative outcomes may not be so good for temporal lobe epilepsy with normal MRI with around 40–60% seizure free (37). Removal of non-scarred structures, especially from the dominant hemisphere, may have greater risks for memory. Others, however, have shown that resections guided by invasive EEG recordings may give better results (38). Patients should have a classic electroclinical syndrome and a well-localized temporal onset on scalp EEG. Functional imaging may also be helpful. Intracranial EEG with sub-temporal strips or stereo EEG with depth recording is needed to exclude extratemporal onsets, establish medial or lateral onsets, and to determine the extent of the resection.

Lateral temporal epilepsy, insular, and opercular seizures

The presence of unformed auditory auras or early ictal dysphasia, suggests involvement of the superior temporal gyrus. Psychic auras with illusions or hallucinations imply involvement of posterior association areas, while excessive salivation, epigastric, and other autonomic auras are seen with insular seizures. The frontal operculum covering the sylvian fissure may produce motor seizures of the face and tongue, and speech arrest while posteriorly the second sensory area may give rise to extensive hemisensory disturbances and possibly pain. EEG changes are often largest at the mid- and posterior electrodes, as opposed to the typical anterior temporal focus, or larger at the sylvian, rather than the lower temporal electrodes

EEG monitoring during carotid amytal testing

Amylobarbitone, a short-acting barbiturate, is injected into the carotid artery in the angiography suite. Usually, diffuse hemisphere slowing occurs a few seconds after injection. During this time a series of objects and words are presented and memorized over about 5 min. EEG monitoring is performed to confirm unilateral changes, as there may be aberrant circulation or technical difficulties during the angiography. The EEG is then monitored to show that the slowing outlasts the period of presentation of the memory items. Neurological recovery of the hemiparesis and disappearance of the EEG slowing usually have a similar time course. About 10 min after the EEG and hemiparesis has resolved, recall of the objects is tested. It is important to emphasize that the memory function and integrity of the hippocampus contralateral to the side of injection is tested. The procedure is then repeated on the other side. The occurrence of seizures during the test is also assessed.

Ictal SPECT

This investigation is performed during scalp telemetry. A technetium labelled pharmaceutical is injected usually within around 30 s of seizure onset, which is taken up in the brain in proportion to blood flow. It remains fixed for a period of hours and imaging is done with a rotating gamma camera producing tomographic

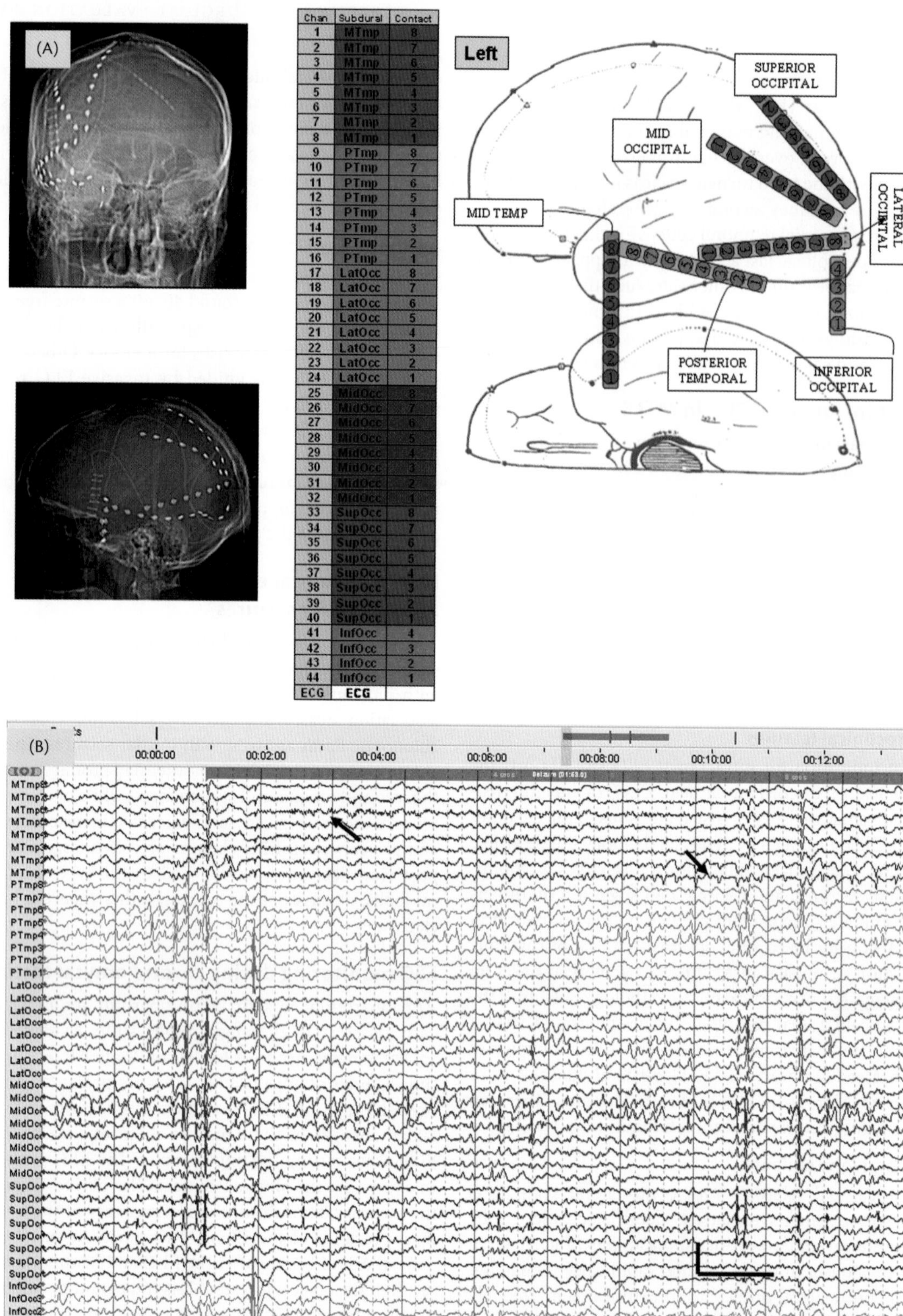

Chan	Subdural	Contact
1	MTmp	8
2	MTmp	7
3	MTmp	6
4	MTmp	5
5	MTmp	4
6	MTmp	3
7	MTmp	2
8	MTmp	1
9	PTmp	8
10	PTmp	7
11	PTmp	6
12	PTmp	5
13	PTmp	4
14	PTmp	3
15	PTmp	2
16	PTmp	1
17	LatOcc	8
18	LatOcc	7
19	LatOcc	6
20	LatOcc	5
21	LatOcc	4
22	LatOcc	3
23	LatOcc	2
24	LatOcc	1
25	MidOcc	8
26	MidOcc	7
27	MidOcc	6
28	MidOcc	5
29	MidOcc	4
30	MidOcc	3
31	MidOcc	2
32	MidOcc	1
33	SupOcc	8
34	SupOcc	7
35	SupOcc	6
36	SupOcc	5
37	SupOcc	4
38	SupOcc	3
39	SupOcc	2
40	SupOcc	1
41	InfOcc	4
42	InfOcc	3
43	InfOcc	2
44	InfOcc	1
ECG	ECG	

Fig. 31.3 Hippocampal sclerosis with extratemporal seizure onsets. Temporal and posterior subdural strip recordings. 38-year-old right-handed man who has mild learning difficulties, epilepsy since infancy. Complex partial seizures with a visual aura and left hippocampal sclerosis and no other lesion. Scalp telemetry showed left regional occipital and temporal onset. Intracranial EEG showed independent onsets in the left medial temporal (B–D), and occipital lobe(E,F). Left hippocampectomy and occipital resection. Histology shows cortical dysplasia type 1 in the occipital cortex. Seizure-free following surgery. (A) Cartoon of placement of subdural strips placed through temporal and occipital burr holes. Note the strips radiating form the site of insertion. The deepest contacts (Channel 8, contact 1) of the strip labelled mid temporal lies near the left hippocampus. (B) Seizure onset, complex partial seizure of temporal lobe origin. First arrow, decrement in the EEG and loss of interictal spikes, possible fast activity. Second arrow, build-up of rhythmic spikes in the deepest contacts of the mid temporal electrode, near to hippocampus. (C) + 40 s. Extensive build-up of ictal activity over the mid-temporal strip. Note little involvement of posterior temporal or occipital strips. (D) Seizure end. Ictal activity is now in the lateral, superficial contacts of the mid- and posterior temporal strips, again with little involvement of the occipital electrodes. (E) Complex partial seizure with visual aura. Diffuse spiking at posterior contacts at seizure onset. (F) + 30 s. Ictal activity now maximal at proximal contacts of the mid-occipital electrode. Note gain reduced. Note some sharp waves over the deepest mid temporal strip, but no prominent hippocampal spread.

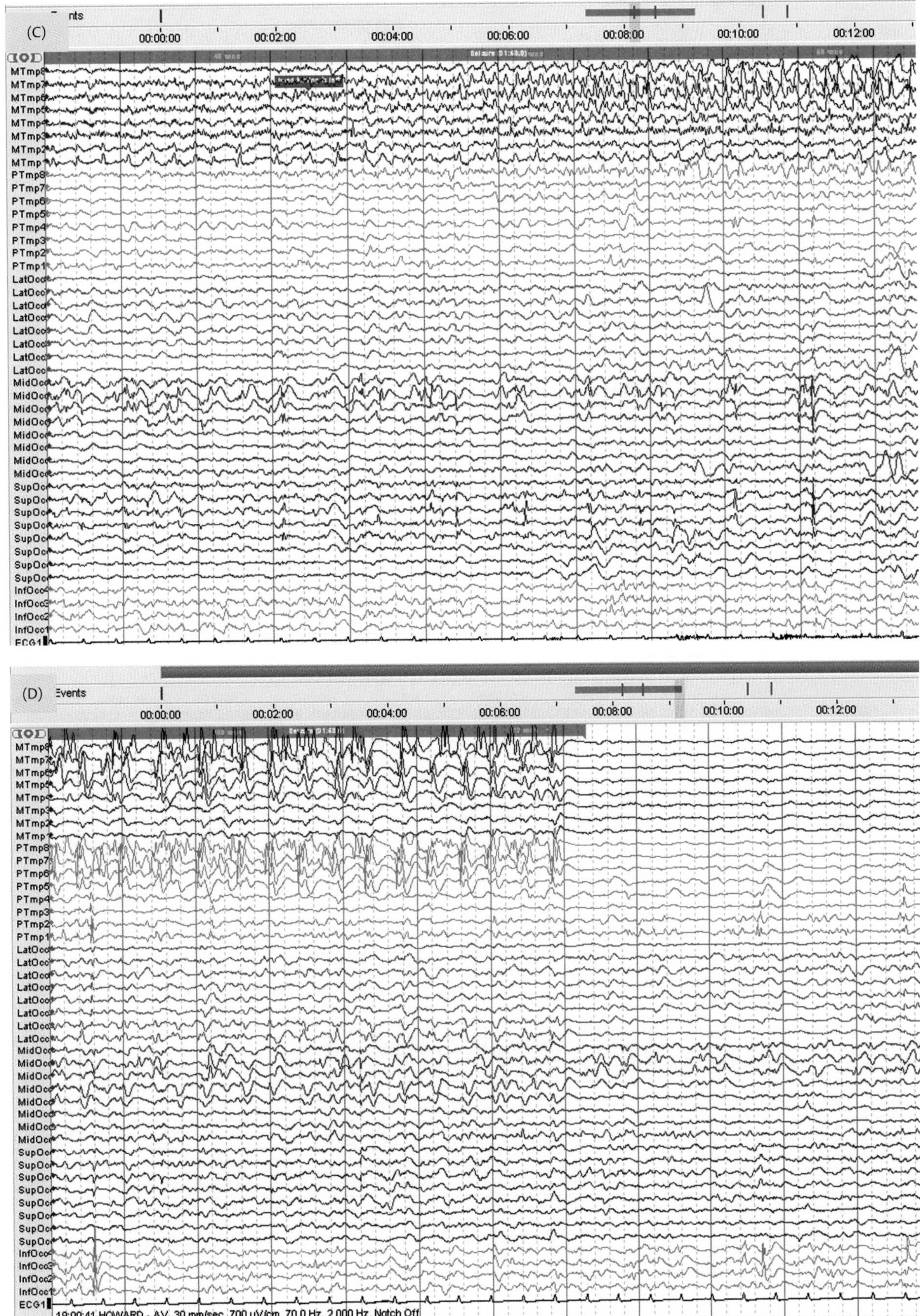

Fig. 31.3 Continued.

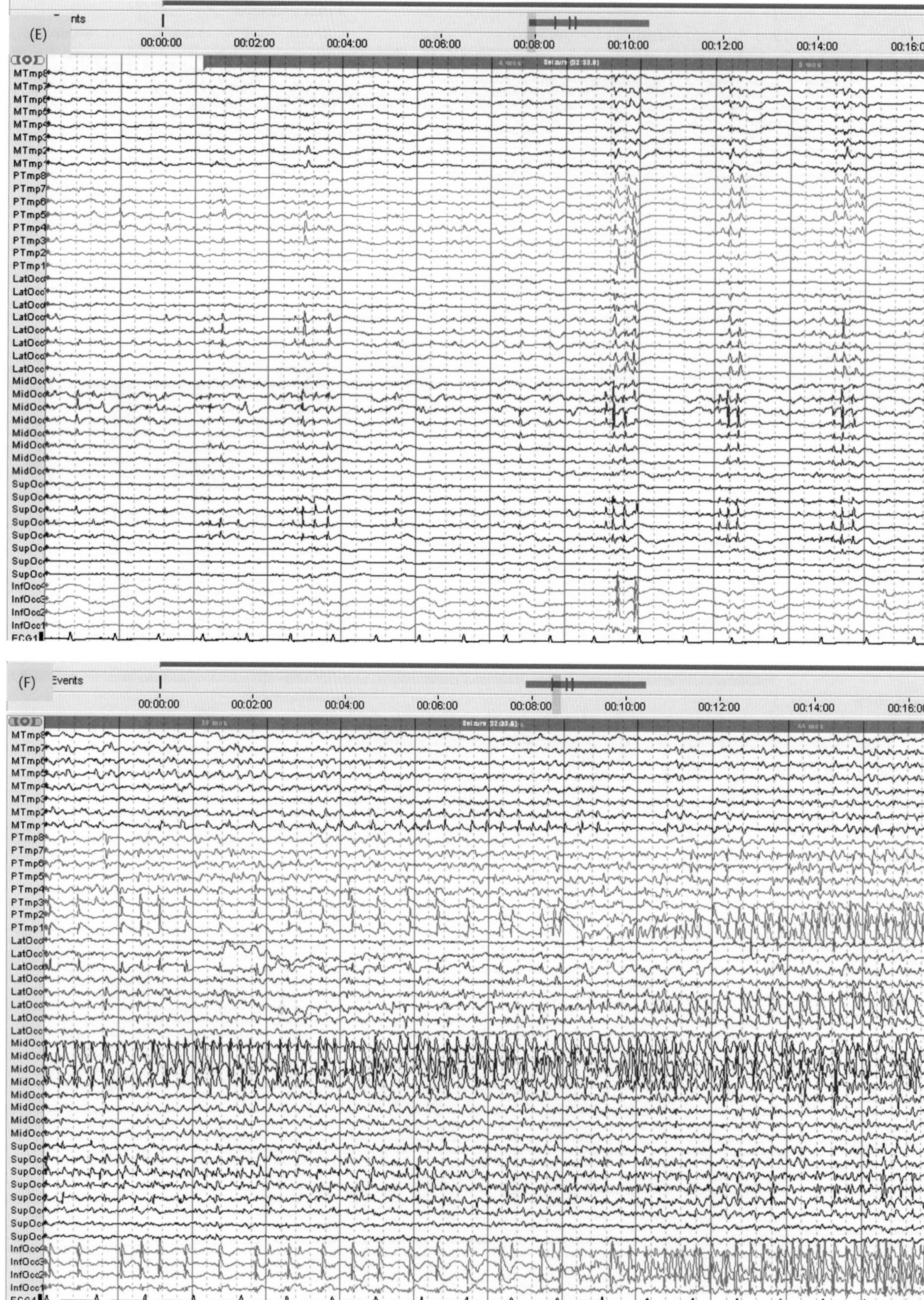

Fig. 31.3 Continued.

images. Appropriate changes are seen in some 70–80% of cases of temporal lobe epilepsy (39). It produces information useful to presurgical evaluation in around 30% of patients (40,41). Cases with a lesion, but poorly localized scalp EEG and those with dual pathology may avoid the need for intracranial EEG. It requires considerable organization to set up and late injections can be misleading due to seizure spread. It could be argued that SPECT should be available to all cases being considered for intracranial EEG. It may also help in the placement of intracranial electrodes in those with normal MRI (42). Although it aids the process of evaluation there is no clear evidence that it effects outcomes in those coming to surgery (43).

Electroclinical evaluation of frontal lobe epilepsy

Focal cortical dysplasia (FCD) is the commonest cause of frontal lobe epilepsy (FLE) in surgical material and may not be visible on MRI in 40–50% of cases (44,45). It is thought to be a brain malformation occurring during the first trimester. There is loss of the normal layered architecture, thickening of the cortex and loss of the grey white junction. Histology shows giant malformed neurons. Unlike tumours or scars, that presumably produce seizures by an effect on surrounding neurons, cortical dysplasia is inherently epileptogenic (46). Direct intralesional

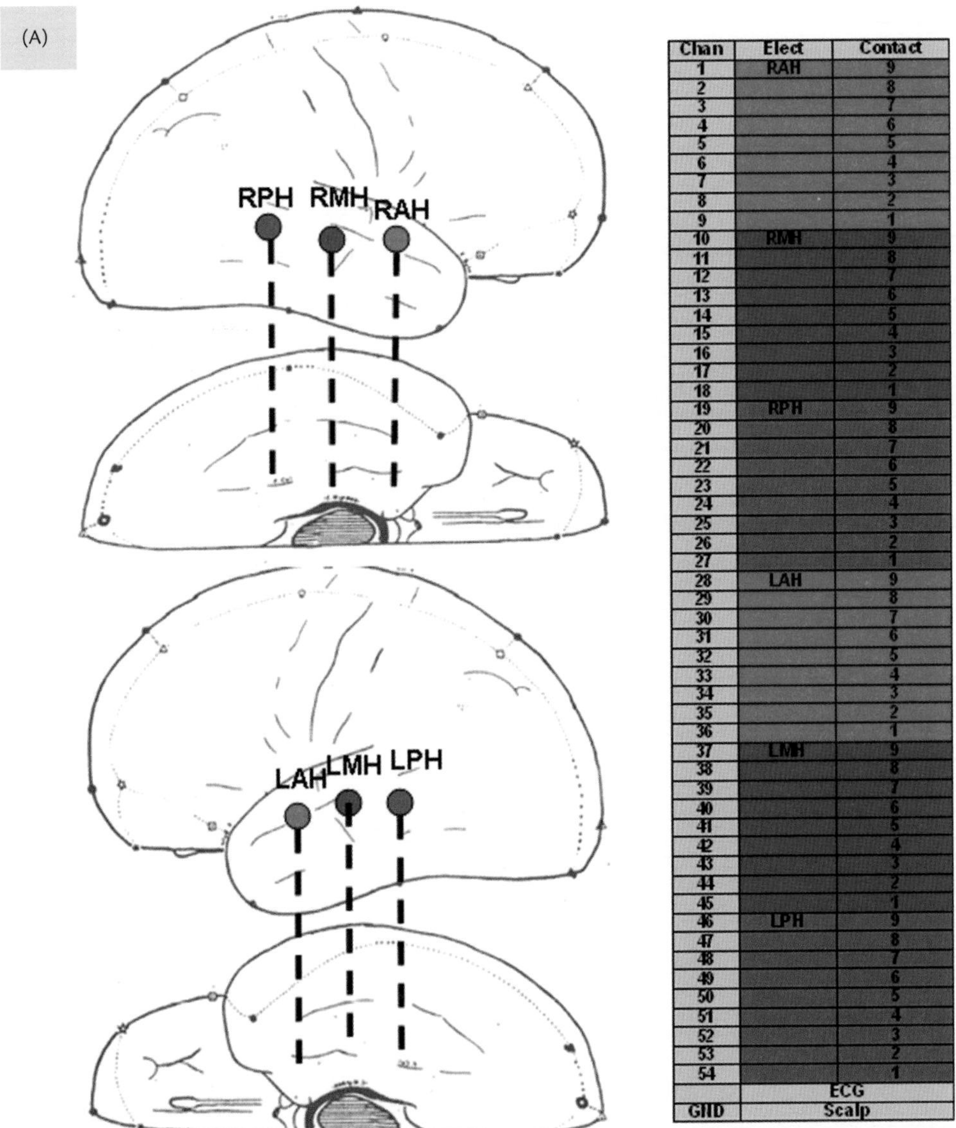

Chan	Elect	Contact
1	RAH	9
2		8
3		7
4		6
5		5
6		4
7		3
8		2
9		1
10	RMH	9
11		8
12		7
13		6
14		5
15		4
16		3
17		2
18		1
19	RPH	9
20		8
21		7
22		6
23		5
24		4
25		3
26		2
27		1
28	LAH	9
29		8
30		7
31		6
32		5
33		4
34		3
35		2
36		1
37	LMH	9
38		8
39		7
40		6
41		5
42		4
43		3
44		2
45		1
46	LPH	9
47		8
48		7
49		6
50		5
51		4
52		3
53		2
54		1
		ECG
GND		Scalp

Fig. 31.4 Depth electrode exploration of left hippocampal sclerosis with non-lateralized scalp EEG. Depth electrodes show contralateral onset.20-year-old left-handed man, epilepsy from age 1 year 6 months. No febrile convulsions, normal development and cognition. Daily auras of fear and strange smell. Monthly complex partial seizures. Imaging shows left hippocampal sclerosis, right normal. Scalp telemetry, CPS of temporal origin, not clearly lateralized? Independent onsets. Amytal, left hemisphere dominant, both sides support memory. (A) Cartoon of bilateral medial temporal depth electrode exploration. (B) Simple partial seizure with onset in deepest contacts of left posterior hippocampal electrode with early involvement of the mid hippocampal electrode. (C) Complex partial seizure. Arrow shows brief burst of fast activity and a decrement in the deepest contact of right anterior hippocampal/amygdala electrode followed by activity in the next hippocampal depth (blue channels). (D) Mid seizure at + 30 s. Spread to mid- and posterior hippocampal electrodes. Note the major seizure type, complex partial seizures arises from the non-scarred hippocampus. Vagal nerve stimulator (VNS) inserted.

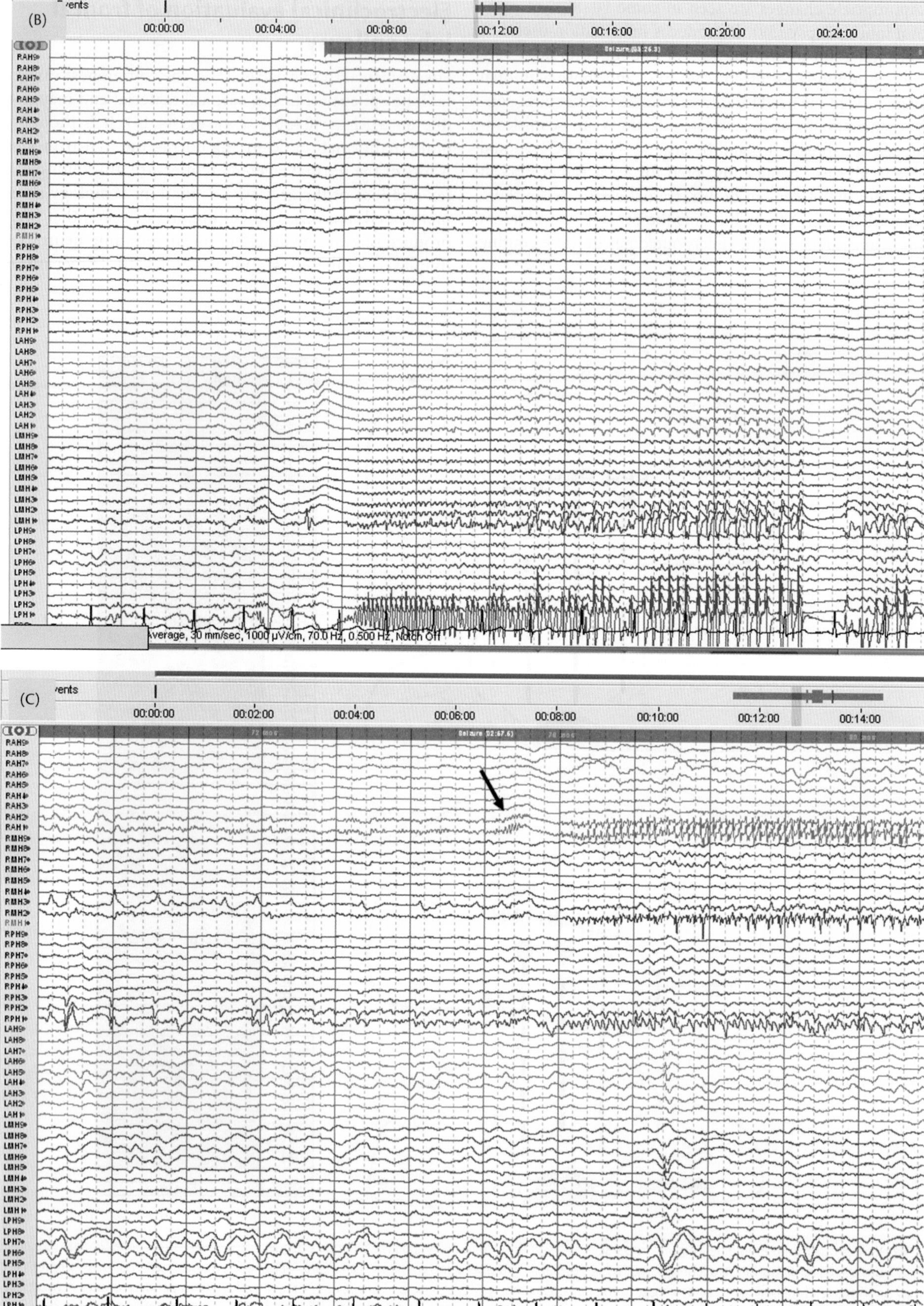

Fig. 31.4 Continued.

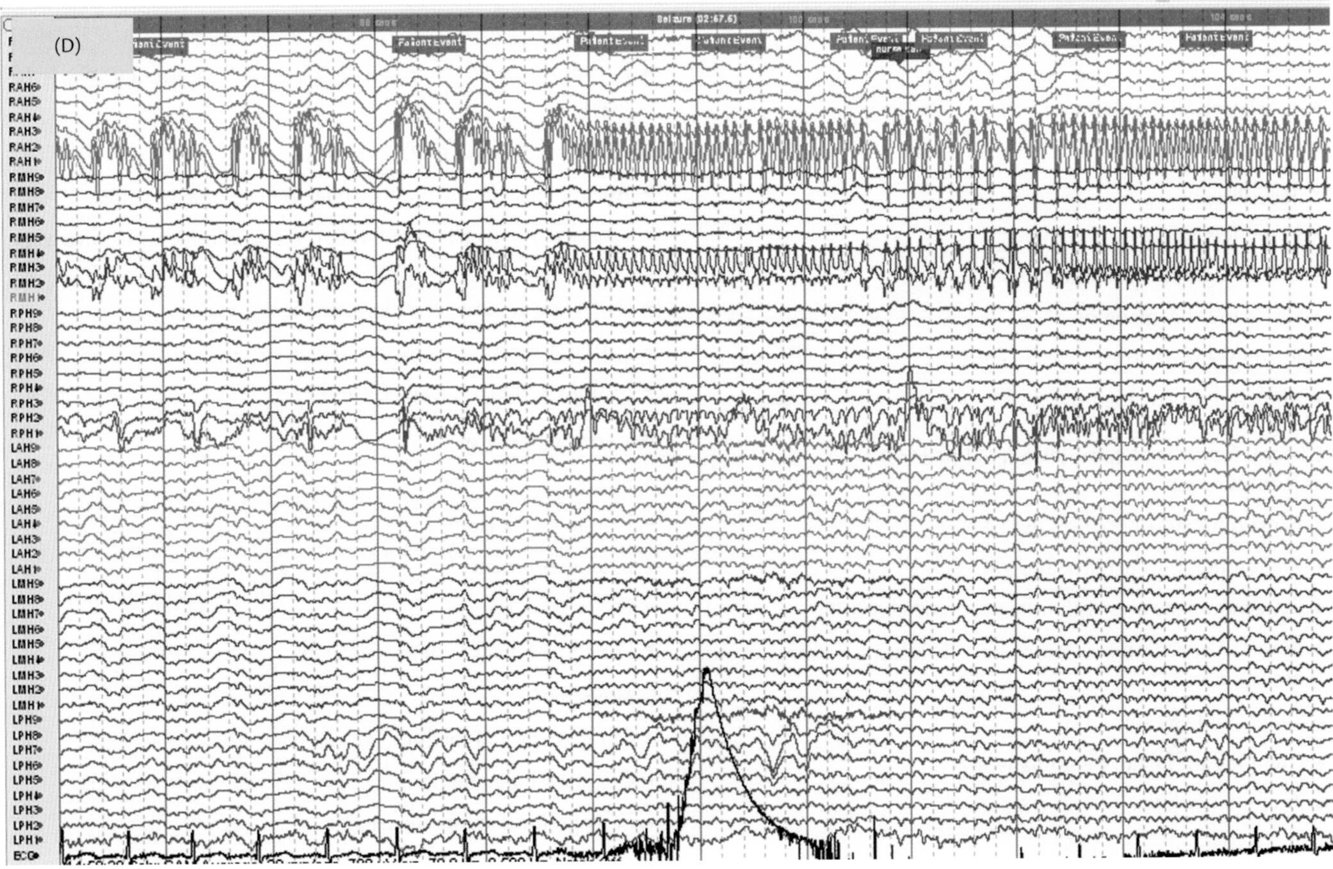

Fig. 31.4 Continued.

recordings show repeated and continuous spiking (47). This may explain the high seizure frequency and tendency to bouts of serial seizures and status epilepticus seen in frontal epilepsy. The lesions are often quite circumscribed, which makes them open to resection and they often occur in people with normal development and well preserved cognition. They also seem to have a predilection for the depths of cerebral sulci and for areas around the central and sylvian fissures, overlapping with eloquent cortex (see Box 31.5).

Ictal symptoms, semiology, and interictal EEG

The extent to which FLE can be subdivided remains controversial. Important subtypes (48) are summarized in Box 31.6. Pure forms of the disorder are rare and most patients have a mixture of features, in keeping with the concept of a network of neural substrates in focal epilepsy (49). Lesions at the temporal pole or the insula can give rise to classic 'frontal' seizures.

As there are extensive orbital and medial surfaces in the frontal lobe the interictal and ictal EEG may be normal. In the 10-20 system only six electrodes cover the frontal lobes. Precise localization is difficult, including distinguishing a lateral temporal from a frontal focus. Modelling of dipoles using closely spaced electrodes or magneto encephalography, may be of help (50). In focal cortical dysplasia continuous runs of rhythmic spike discharges are often seen (Fig. 31.5B). On intracranial EEG, brief runs or bursts of fast activity are common, which are probably electrical seizures. Ictal fast activity may be seen on the scalp EEG, as in other neocortical seizures. Frontal seizures are said to have a rapid spread and a tendency to become quickly generalized; cases due to focal cortical dysplasia, however, often have a long aura and more prolonged onsets.

Indications for intracranial EEG

A higher proportion of frontal lobe epilepsy cases need intracranial EEG. Identifying cortical dysplasia on current MRI can be difficult. Not infrequently functional tests localize the focus and detailed directed re-examination of the MRI reveals a possible lesion. The epileptogenic area is often bigger than the lesion and resection must be guided by intra or extraoperative corticography. Most commonly therefore a depth electrode is placed in the area thought to represent the primary lesion and electrodes placed around this to determine the extent of the abnormality. The electroclinical syndromes are not as well delineated in frontal lobe epilepsy. Unless the primary motor area is involved even lateralization can be difficult. A curious feature of cortical dysplasia is that when localized it often lies at the bottom of the sulcus and subdural recording may be misleading. It also has a predilection for peri-Rolandic and -sylvian cortex and often the medial hemisphere. The focus often abuts eloquent cortex or is inaccessible. Functional mapping of the cortex is best done through subdural mats and those with high density spacing of electrodes give better results (Fig. 31.6).

Box 31.5 Electroclinical features of frontal lobe epilepsy

Aetiology

- Focal cortical dysplasia.
- Stroke.
- Trauma.
- Malignant glioma.
- Indolent glioneuronal tumours.
- Cavernous haemangioma.
- Rasmussen's or other focal encephalitis.
- Mitochondrial cytopathy.

General features of complex partial seizures of frontal origin

- Large numbers of seizures per 24 h.
- Predominant occurrence at night.
- Brief motor seizures with little post-ictal confusion.
- Bizarre hypermotor automatisms, thrashing, kicking, cycling movements.
- Shouts and screams.
- Complex ambulatory automatisms.
- Asymmetric bilateral postures.
- Contra-lateral clonic jerks.
- Incontinence.
- Frequent occurrence of serial seizures.
- Partial retention of awareness.
- Spikes maximal at the prefrontal, superior frontal, midline and central electrodes.
- Rhythmic spike discharges/bursts of fast activity if pathology is cortical dysplasia.

Primary motor (M1) seizures

- Faciobrachial maximum.
- Jacksonian spread.
- Often tonic posturing with ictal fast activity.
- Version of eyes in frontal eye field.
- Spike, maximal at C3 or C4, or Cz if leg involved, with clonic jerks.
- *Epilepsia partialis continua*: jerk locked back averaging may be needed.

Supplementary motor (M2) seizures

- Abrupt onset of asymmetric bilateral dystonic postures.
- Classically adversive, but other more complex postures very common.
- Often late contralateral clonic jerks.
- Partial preservation of consciousness.
- Midline spikes, typically Cz and Fz.

Anterior cingulate

- Fear.
- Non mirthful laughter.
- Ictal incontinence.
- Other autonomic.
- Possibly Fz spikes, may be normal.

Orbitofrontal

- Complex partial seizures.
- Olfactory hallucinations.
- Early gestural automatisms.
- Autonomic signs.
- Similar to MTLE.
- Prefrontal and inferior frontal spikes or normal.

Frontal operculum

- Jerking of tongue and face.
- Speech arrest, especially if Broca's area involved.

Frontal absences

- May have generalized spike and wave and medial lesion.

Ictal EEG

- Focal rhythmic spikes in localization as described above.
- Often rapid spread of ictal discharges.
- May be normal with medial or inferior focus, or focus deep in sulcus.
- Late temporal spread.

Electroclinical evaluation of other syndromes or pathology

Extratemporal epilepsy with normal MRI

This is one of the most challenging areas of epilepsy surgery (51,52) and should only be approached in centres with extensive experience of intracranial EEG using depth electrodes and functional imaging. Our practice is that those evaluated should have a phenotype suggestive of focal cortical dysplasia. This includes strikingly focal seizures, often with multiple daily or nocturnal attacks in the context of normal development and cognition with a frontal, parietal, or perisylvian semiology. Scalp telemetry must give a clear indication of the localization and lateralization. Atypical features such as multifocal or generalized EEG changes, non-localized scalp EEG, or rapid spread of ictal discharges should not be present. Placement of depth electrodes should be guided by other functional tests, particularly ictal SPECT. Co-registration of fluorodeoxyglucose (FDG) positron emission tomography (PET) and MRI allows the identification of small areas of hypometabolism that may be missed if the gyral anatomy is not assessed (47). Dipole modelling using magnetoencephalography

Box 31.6 Surgery for frontal lobe epilepsy

- Second most common form of epilepsy coming to surgery.
- Focal cortical dysplasia commonest pathology.
- FCD is inherently epileptogenic causing rhythmic spike discharges, high seizure frequency, and bouts of serial seizures and status.
- Frontal seizures are classically divided into primary and secondary motor, hypermotor, orbitofrontal, and anterior cingulate.
- Most have mixed features and localization on semiology is difficult.
- Lateralization of seizures usually requires involvement of primary motor areas which can occur late.
- Some cases may have normal interictal and ictal scalp EEGs.
- Intracranial EEG is more commonly performed, as FCD may be poorly seen on MRI and often lies near or on eloquent cortex.
- More resections are tailored and guided by intra- or extraoperative electrocorticography.
- Assessment of cases with normal MRI requires functional assessment with ictal scalp EEG/SPECT and FDG PET, followed by stereo EEG.

may give helpful information in localizing a focus on the convexity, but has not been convincingly shown to identify deep-seated foci. Many of these techniques require the presence of an expert research team to produce reliable results and to quantify imaging results. When a clear hypothesis has been generated stereo-EEG with multiple depth electrodes is needed. Around eight to ten, are placed in a localized area on one hemisphere with the localization determined by seizure semiology, surface EEG, and functional imaging.

Major hemisphere disorders

Infantile hemiplegia

The commonest pathology in this group of patients is an intrauterine or perinatal stroke affecting the middle cerebral artery. There is a dense hemiplegia and loss of all fine movements in the upper limb. A major resection is unlikely to produce a functional deterioration and indeed it is one of the most effective operations. Because of extensive loss of brain tissue the spikes are commonly generated by the lesion, but are propagated and appear bilateral or bigger on the contralateral good side (Fig. 31.7). If the paper speed is increased and careful analysis of the phase undertaken it may be seen that they are, in fact, being propagated from the damaged hemisphere. Ictal EEG changes may also be misleading. Careful visual assessment of the semiology however may show early contralateral posturing, aided by placing EMG or movement sensors on homologous areas of the limbs. As described by Gastaut (53,54) this group of patients may have startle seizures where contra-lateral posturing is stimulus sensitive. There is often prominent sweating and other autonomic features with surprisingly little surface EEG change. More localized strokes with no hemiplegia rarely come to surgery, possibly because the area of epileptogenic brain is often shown to be quite extensive and often overlaps with eloquent brain.

Rasmussen's syndrome

Rasmussen's syndrome classically presents in children aged around 5 years of age. There may be hemispheric slowing on the EEG. Epilepsia partialis continua or bouts of focal motor status are the commonest electroclinical syndromes. In adults, more localized indolent forms are occasionally seen and can present with temporal lobe epilepsy. As the disease progresses other focal seizures may occur. Progressing hemiatrophy on MR and hemiparesis, sometimes with chorea, is seen. Immune modulatory treatments are used that may slow progression and the severe end-stage hemiatrophy now seems less common. There are no specific EEG features that establish the diagnosis, which can be difficult. Over many years the EEG may show contralateral changes and these may correlate with cognitive decline (55). Bilateral disease, however, seems to be extremely rare in surgical cases.

Landau–Kleffner syndrome

This disorder can be diagnosed when there is normal cognitive and language development for the first 2 years with subsequent speech loss due to an acquired auditory agnosia, typically occurring between the ages of 3 and 5 (56). There are often prominent behaviour disturbances. Seizures occur, but are often not severe. The syndrome is thought to be part of the spectrum of the benign partial epilepsies of childhood and the EEG is central in establishing the diagnosis (57). The interictal EEG shows a dominant hemisphere focus with a maximum in the Sylvian area. Many atypical forms are seen and the EEG abnormality may overlap with the centrotemporal spikes seen in Rolandic epilepsy (58). Dipole modelling using magnetoencephalography can be particularly effective in localizing the focus. Similar to other idiopathic partial epilepsies there is marked activation of the spikes in sleep to produce virtually continuous discharges in slow wave sleep or electrical status epilepticus in slow wave sleep (ESES). The aetiology of the cognitive changes is uncertain. Localized interictal spikes can produce transient cognitive impairment, which is specific for the location of the discharge. Alternatively, the greatly disrupted sleep architecture may lead to unravelling at night of all that was learned in the day, the so called Penelope syndrome (59). It is important to distinguish this disorder from the cognitive deficits and autistic features of symptomatic generalized epilepsy, where spike and wave discharges can also be very prominent in sleep. The discharges in Landau–Kleffner syndrome can be shown on phase analysis to be driven from one hemisphere, not always the left (60). At surgery, the focus on electrocorticography is seen to lie across the sylvian fissure and increasing doses of barbiturates can be given until the last remaining spikes are seen in the sylvian area. The fissure is opened and multiple subpial transections, using the method of Morrell, are performed under electrocorticography until the spikes disappear. In some cases there is marked postoperative improvement in language and behaviour, but the overall impact of surgery on the long-,term prognosis is less certain.

Symptomatic generalized epilepsy and section of the corpus callosum

Callosotomy for epilepsy was developed when it was noted that if gliomas invaded the corpus callosum the epilepsy could get better. Early operations were complete sections, sometimes including the anterior and hippocampal commissures and the

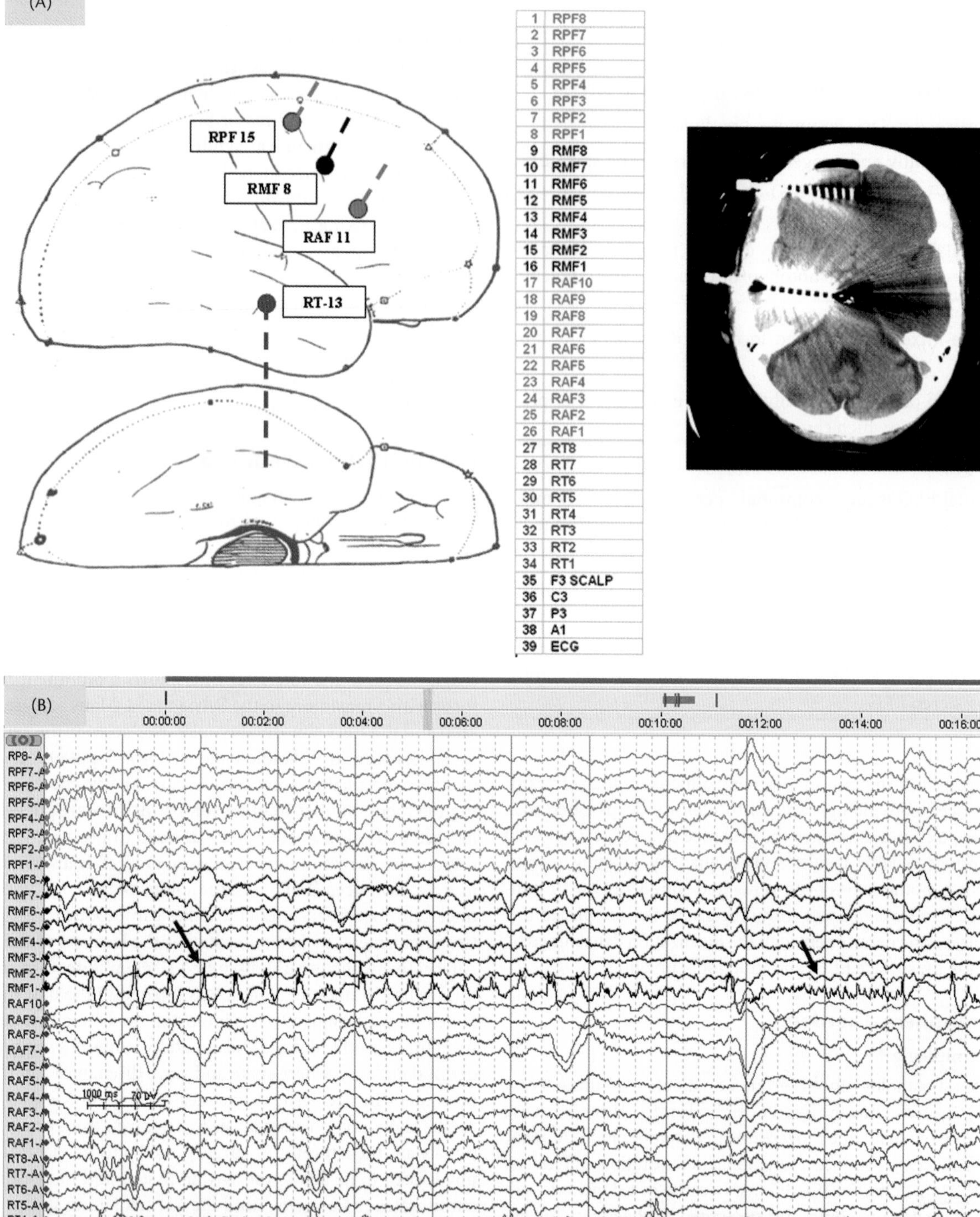

Fig. 31.5 Medial frontal and temporal depth electrode exploration for focal cortical dysplasia. 42-year-old left-handed female accountant. No past history. Seizures from the age of 8, five per day. Anxious, then incontinent of urine. Unable to talk during an episode (not definitely dysphasic). Face goes red, may cough or laugh. Rapid movement of arms and feet. Quick recovery. Interictal and ictal scalp EEG not localized. Imaging showed probable cortical dysplasia in anterior cingulate on the right. Intracranial EEG performed as unusual seizure type, scalp EEG not localized, and lesion indistinct. Seizure free following resection. (A) Cartoon of depth electrode positions. RMF (black) deepest contact pass through the lesion in the anterior medial cingulate. This is seen in the axial CT as the most anterior electrode. RPF and RAF lie superior and anterior to the lesion. There is also a depth electrode in the temporal lobe, seen posteriorly in the axial CT. Note the skull bolts securing the depth electrodes. (B) Pre-ictal recording. Note rhythmic spike discharges at RMF1 (first arrow). There is a brief burst of fast activity in the same electrode (second arrow) which has similar features to those at seizure onset. These occurred frequently and were probably electrical seizures. (C) Seizure onset. Rhythmic spike discharges give way to 14 Hz ictal changes in RMF1, the deepest contact. (D) Mid-seizure. + 30 s. Spread to RAF contacts 2 and 1 (mauve channels). At this point incontinence is the main clinical accompaniment.

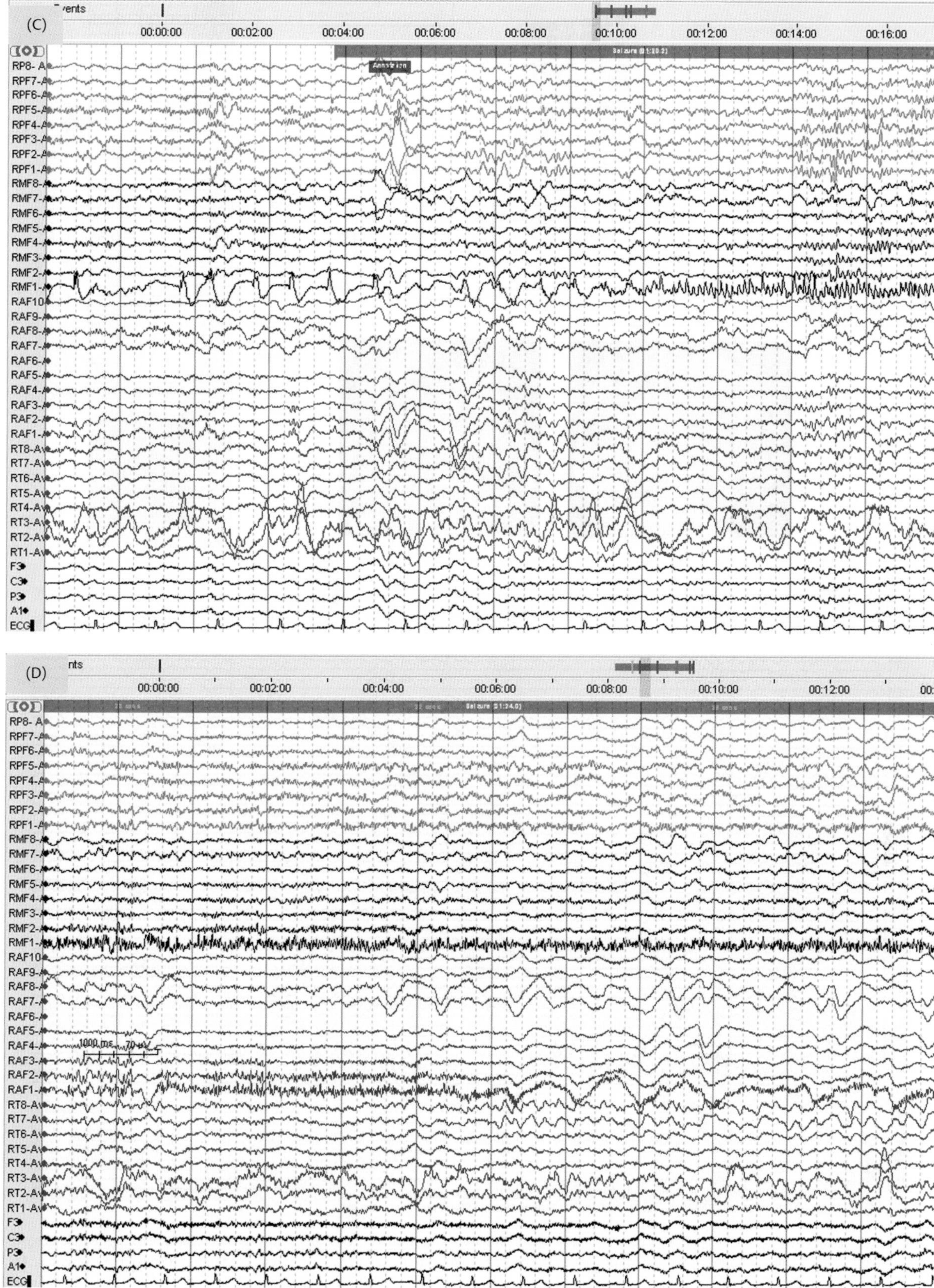

Fig. 31.5 Continued.

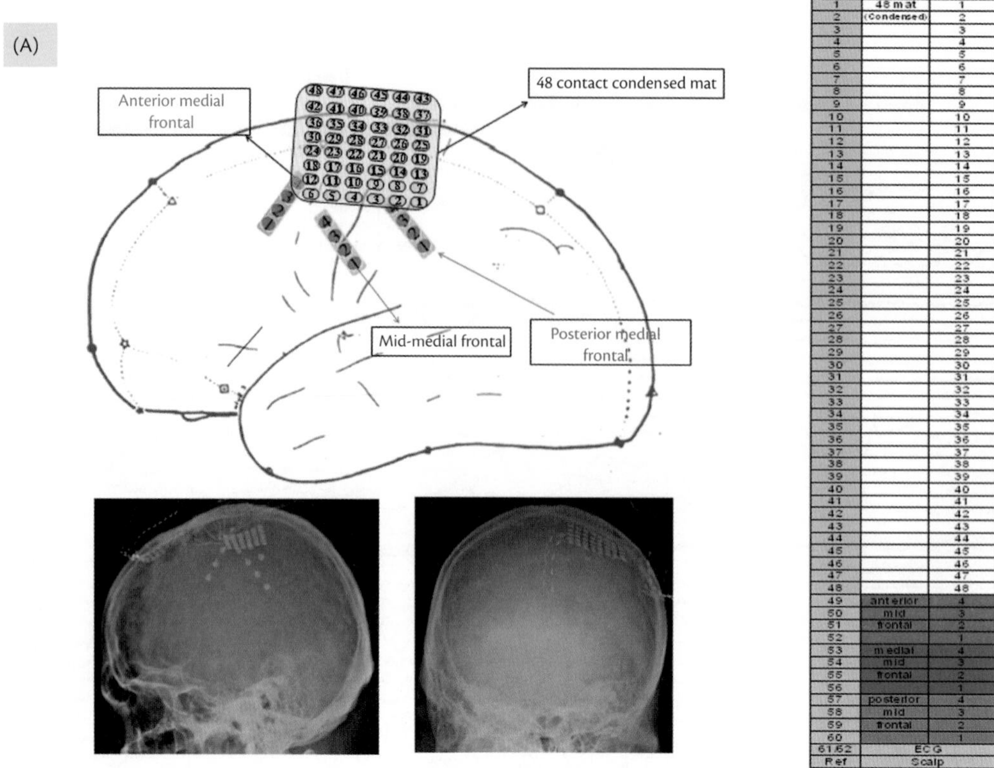

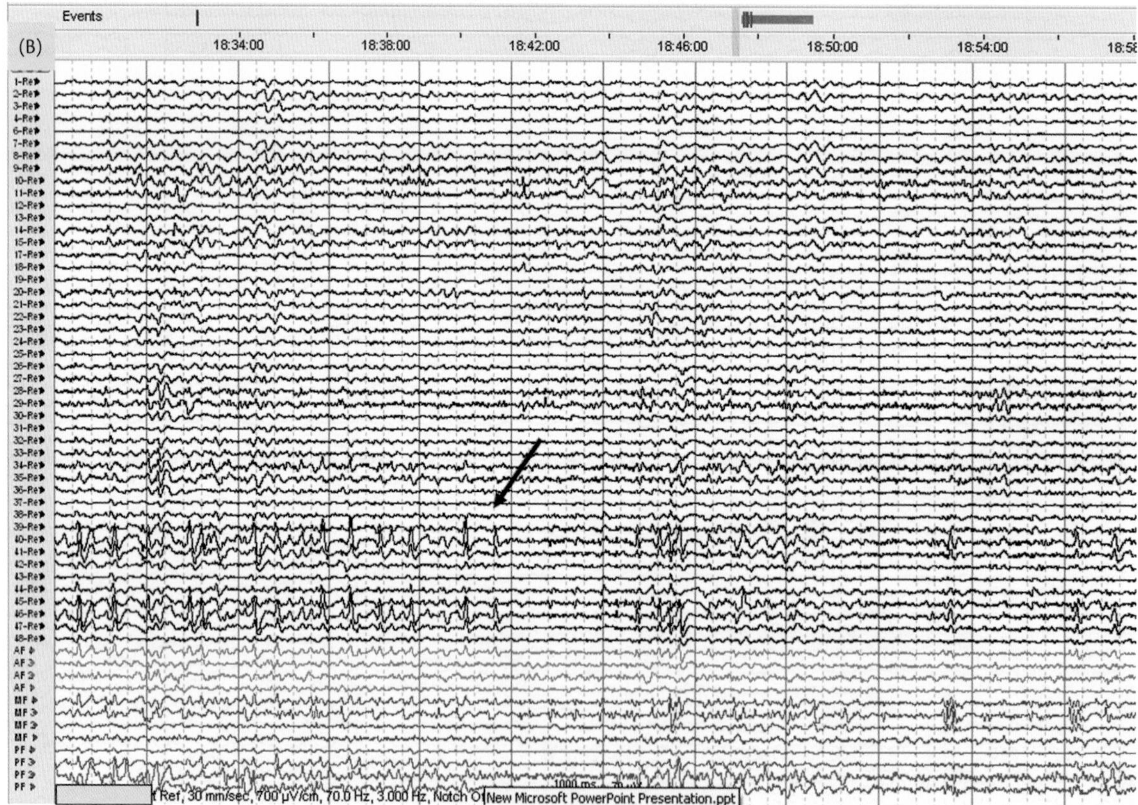

Fig. 31.6 Use of high density mat to map out the areas of ictal onset and the primary and secondary motor areas. 20-year-old, right-handed student. Seizures for 5 years, up to 5 a day. Stiffening in right thigh, whole leg goes stiff, some jerks at the end. May spread to right arm. Also convulsive seizures that start in the same way. MRI shows lesion in left paracentral lobule just anterior to M1 leg area, probably dysplastic. (A) Cartoon and SXR showing position of 58 contact high density mat and medial 4 contact subdural strips. The mat cannot be placed medially due to possibility of damaging veins. The lesion is thought to lie deep in the middle of the mat superiorly. A high density mat was used as the lesion was thought to involve the motor area on anatomical grounds. (B) Interictal EEG. Note rhythmic spikes in the superior part of the mat, contacts 40 and 41, and also 46 and 47, and also blue and green medial strips. The spikes disappear some 10 s before first ictal EEG change, possible related to ictal onset deep in a sulcus not samples with subdural electrodes. (C) Ictal onset with low voltage fast activity. The distribution is broadly similar to the interictal spikes. (D) Cartoon of site of ictal onset and motor map in the high density subdural mat. The medial strips also returned motor responses, the anterior probably in M2, supplementary motor. Limited resection and only partial response to surgery.

(C)

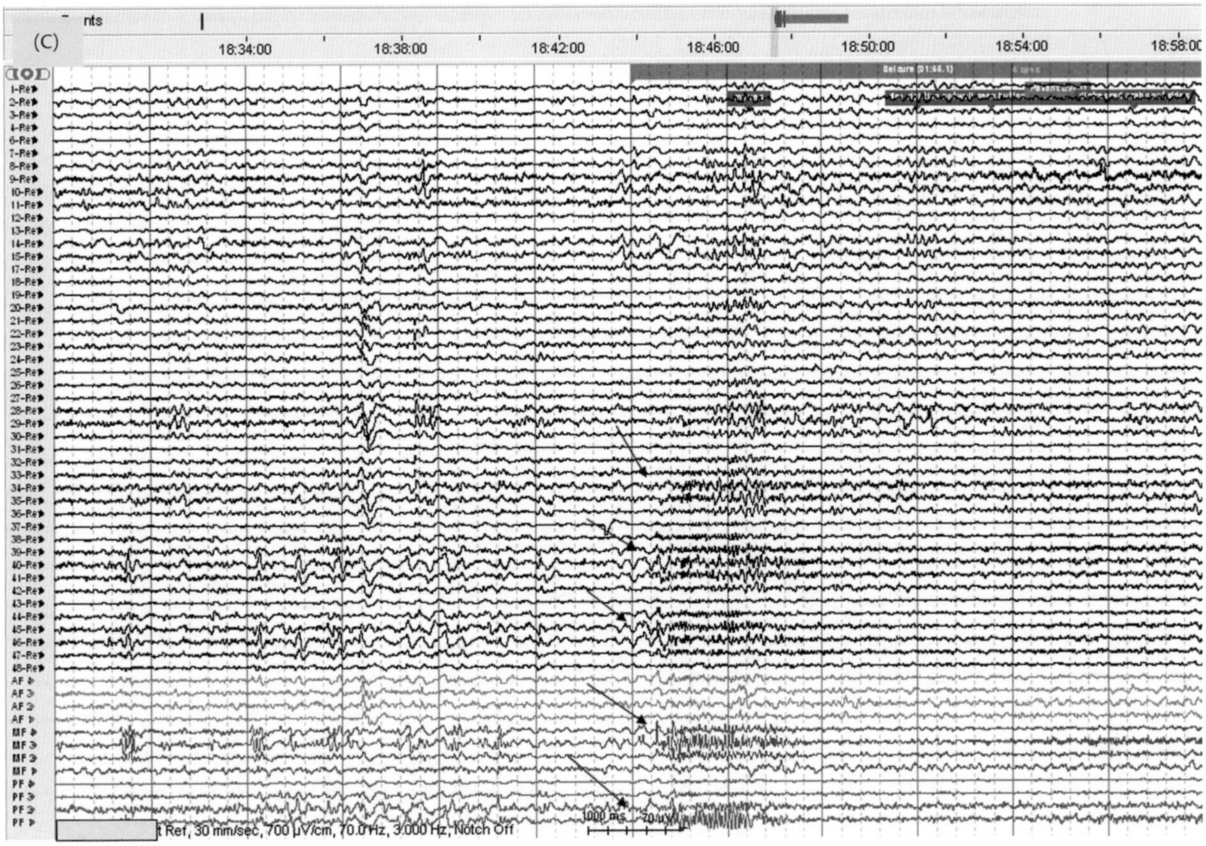

(D)

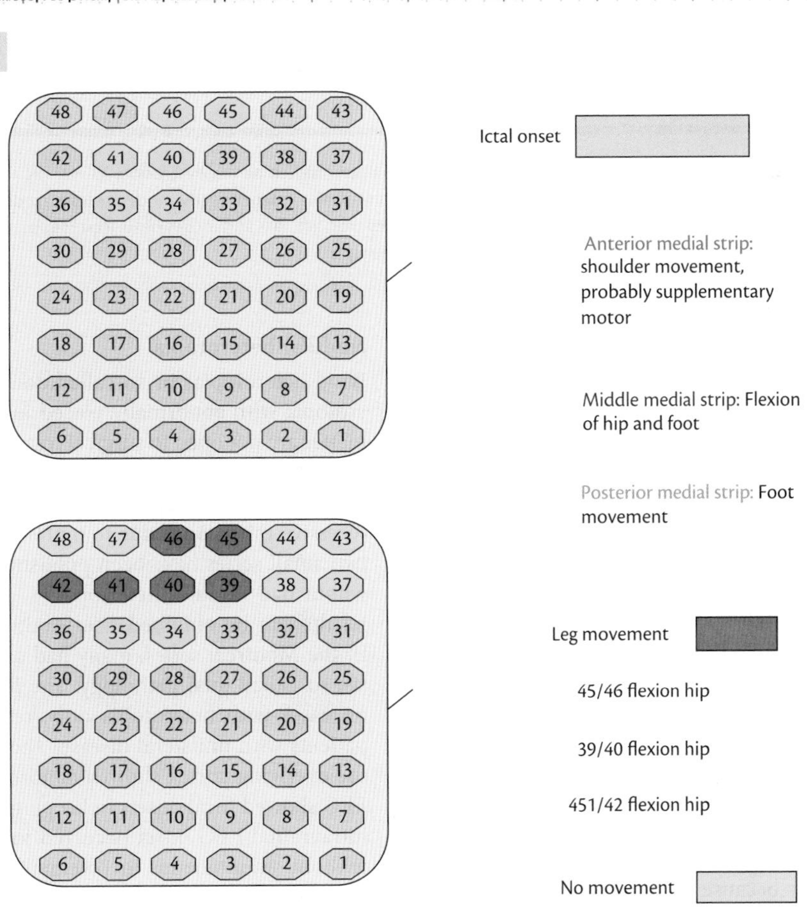

Fig. 31.6 Continued.

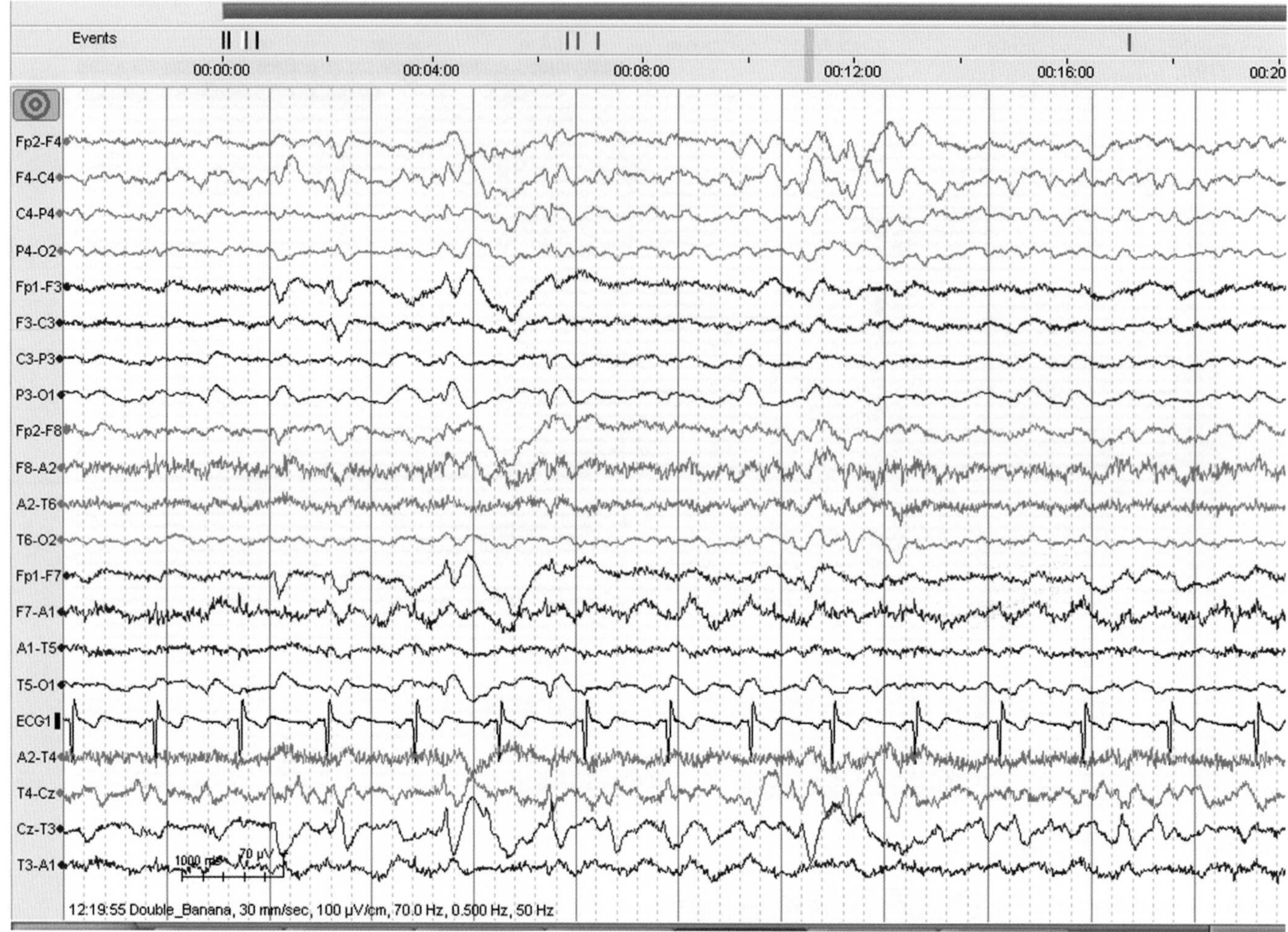

Fig. 31.7 Inappropriately lateralized epileptiform discharges in the presence of a major atrophic left hemisphere lesion. 30-year-old man, intractable epilepsy. Infantile right hemiplegia with left middle cerebral artery infarct. No useful function in upper limb, walks independently, moderate learning difficulties, behavioural disturbance. Fitting at birth. Daily seizures. Tonic seizures, posturing of upper limbs right much more than left, falls and injures head, ictal EEG bilateral fast activity.
Note flatter EEG and reduced rhythms over the damaged left hemisphere with spikes bigger on the right.

fornix. Neuropsychological sequelae were reduced with anterior two thirds callosotomy, which is now the most common operation (61).

The patients who do best appear to be those with atonic seizures and an ictal decrement in the EEG, tonic seizures with diffuse fast activity, and tonic clonic seizures generalized at onset (62). Many studies do not include detailed EEG evaluation and the rather non-specific term of 'drop attacks' is used, which remains the main indication for this operation. This spectrum of seizures occurs in symptomatic generalized epilepsies, particularly in children with Lennox–Gastaut syndrome. They have severe epilepsy, daily seizures with frequent falls and injuries, and need to wear helmets for protection. The operation can be performed in adults (63). The operation is therefore palliative, and reduces or occasionally abolishes one or more of the generalized seizures types. Post-operatively the discharges often appear to occur independently over both hemispheres and this can be monitored intraoperatively to guide the surgeon. Partial seizures do not improve and in some cases deteriorate, possibly because the contralateral hemisphere exerts an inhibitory effect.

Nodular heterotopia

Band heterotopias or so-called double cortex, and neuronal heterotopia with abnormally placed individual neurons in the white matter are not open to surgical treatment. The third group, nodular heterotopia consists of nodules of gray matter, either subependymal or subcortical. Subependymal heterotopia may have an x-linked recessive inheritance due to mutations of the *filamin 1* gene. Only two small series have been published of surgical outcomes, all of whom had nodular hetertopia (12,64). These showed that good results can be obtained after evaluation using depth electrodes. In the Montreal series, many had associated hippocampal sclerosis, so-called dual pathology. The Milan series, which was based on cases with heterotopia only, reported good results in seven patients with unilateral disease. They emphasized that in all cases the overlying cortex showed abnormal gyration. The surface ictal EEG showed focal onsets in keeping with the site of the lesion. On intracranial EEG onset occurred in the overlying cortex, with or without simultaneous involvement of the subependymal or subcortical nodule. The initial ictal changes consisted of low voltage fast activity. Extensive sampling of the overlying cortex is therefore needed. Interestingly, good outcomes were not dependent on the complete removal of the nodule. Although none had hippocampal sclerosis this structure was removed in most of the patients.

Neural stimulation

Classic physiological experiments on the alerting effects of stimulation of mid brain grey matter led to our concept of the reticular activating system. It was noted that in some animal species, stimulation of the lower cranial nerves led to desynchronization of the EEG. Vagal stimulation was found to acutely abort strychnine induced seizures in dogs (65). In humans, the stimulation is applied intermittently for safety reasons, classically with a 30 Hz frequency and a duty cycle of 30 s on and 5 min off. The stimulus artefact can often be seen by careful examining of the ECG signal. Adults and children considered for vagal stimulation should be assessed as part of a formal surgery programme, undergo imaging and scalp telemetry, and only implanted after exclusion of resective surgery. All forms of partial epilepsy, and idiopathic and symptomatic generalized epilepsy, have been reported to respond, although the treatment is only palliative and helps around 30–40% of cases. Curiously, improvement occurs over periods of many months and EEGs may show similar reductions in epileptiform discharges (66).

Re-operation

Cases with frequent ongoing seizures after resection should have repeat imaging and scalp telemetry. In medial temporal epilepsy operative failure is often unexpected and occurs despite a clear electroclinical syndrome and removal of pathology. Commonly, there is no change in seizure type and an ipsilateral mid temporal scalp EEG onset is seen (67). If a full temporal lobectomy has been performed, there is no scope to extend the resection back. Extratemporal or contralateral ictal onsets are much less common and, again, not open to further surgery. Failure to remove the medial temporal structures and completion of the operation the second time leads to good outcomes in around 50% of cases. Re-operations for extratemporal epilepsy again are usually based on removal of residual pathology. If intracranial recordings are needed these may be challenging because of adhesions and distorted anatomy.

References

1. Wolf, P. (1992). The history of surgical treatment of epilepsy. In: H. O. Lüders (Ed.) *Epilepsy surgery*, pp. 9–17. New York, NY: Raven Press.
2. Bailey, P., Green, J.R., Amador, L., and Gibbs, F.A. (1953). Treatment of psychomotor states by anterior temporal lobectomy. *Research publications—Association for Research in Nervous and Mental Disease*, **31**, 341–6.
3. Bancaud, J., Angelergues, R., Bernouilli, C., et al. (1970). Functional stereotaxic exploration (SEEG) of epilepsy. *Electroencephalography & Clinical Neurophysiology*, **28**(1), 85–6.
4. Crandall, P.H., Walter, R.D., and Rand, R.W. (1963). Clinical applications of studies on stereotactically implanted electrodes in temporal-lobe epilepsy. *Journal of Neurosurgery*, **20**, 827–40.
5. Penfield, W. and Steelman, H. (1947). The treatment of focal epilepsy by cortical excision. *Annals of Surgery*, **126**(5), 740–62.
6. Jasper, H.H., Arfel-Capdeville, G., and Rasmussen, T. (1961). Evaluation of EEG and cortical electrographic studies for prognosis of seizures following surgical excision of epileptogenic lesions. *Epilepsia*, **2**, 130–7.
7. Falconer, M.A., Meyer, A., Hill, D., Mitchell, W., and Pond, D.A. (1955). Treatment of temporal-lobe epilepsy by temporal lobectomy; a survey of findings and results. *Lancet*, **268**(6869), 827–35.
8. Taylor, D.C., Falconer, M.A., Bruton, C.J., and Corsellis, J.A. (1971). Focal dysplasia of the cerebral cortex in epilepsy. *Journal of Neurology, Neurosurgery, & Psychiatry*, **34**(4), 369–87.
9. Koutroumanidis, M., Hennessy, M.J., Elwes, R.D., Binnie, C.D., and Polkey, C.E. (1999). Coexistence of temporal lobe and idiopathic generalized epilepsies. *Neurology*, **53**(3), 490–5.
10. Palmini, A., Andermann, F., Olivier, A., et al. (1991). Focal neuronal migration disorders and intractable partial epilepsy: a study of 30 patients. *Annals of Neurology*, **30**(6), 741–9.
11. Morrell, F. and deToledo-Morrell, L. (1999). From mirror focus to secondary epileptogenesis in man: an historical review. *Advances in Neurology*, **81**, 11–23.
12. Tassi, L., Colombo, N., Cossu, M., et al. (2005). Electroclinical, MRI and neuropathological study of 10 patients with nodular heterotopia, with surgical outcomes. *Brain*, **128**(Pt 2), 321–37.
13. Wieser, H.G., Elger, C.E., and Stodieck, S.R. (1985). The 'foramen ovale electrode': a new recording method for the preoperative evaluation of patients suffering from mesio-basal temporal lobe epilepsy. *Electroencephalography & Clinical Neurophysiology*, **61**(4), 314–22.
14. Munari, C., Hoffmann, D., Francione, S., et al. (1994). Stereo-electroencephalography methodology: advantages and limits. *Acta Neurologica Scandinavica*, **152**(Suppl.), 56–67, discussion.
15. Carmichael, D.W., Thornton, J.S., Rodionov, R., et al. (2008). Safety of localizing epilepsy monitoring intracranial electroencephalograph electrodes using MRI: radiofrequency-induced heating. *Journal of Magnetic Resonance Imaging*, **28**(5), 1233–44.
16. Engel, J., Jr and Crandall, P.H. (1983). Falsely localizing ictal onsets with depth EEG telemetry during anticonvulsant withdrawal. *Epilepsia*, **24**(3), 344–55.
17. Zangaladze, A., Nei, M., Liporace, J.D., and Sperling, M.R. (2008). Characteristics and clinical significance of subclinical seizures. *Epilepsia*, **49**(12), 2016–21.
18. Arya, R., Mangano, F.T., Horn, P.S., Holland, K.D., Rose, D.F., and Glauser, T.A. (2013). Adverse events related to extraoperative invasive EEG monitoring with subdural grid electrodes: a systematic review and meta-analysis. *Epilepsia*, **54**(5), 828–39.
19. Hedegard, E., Bjellvi, J., Edelvik, A., Rydenhag, B., Flink, R., and Malmgren, K. (2013). Complications to invasive epilepsy surgery workup with subdural and depth electrodes: a prospective population-based observational study. *Journal of Neurology, Neurosurgery, & Psychiatry*, **85**(7), 716–20.
20. Sansur, C.A., Frysinger, R.C., Pouratian, N., et al. (2007). Incidence of symptomatic hemorrhage after stereotactic electrode placement. *Journal of Neurosurgery*, **107**(5), 998–1003.
21. Fernandez, G., Hufnagel, A., Van, R.D., et al. (1997). Safety of intrahippocampal depth electrodes for presurgical evaluation of patients with intractable epilepsy. *Epilepsia*, **38**(8), 922–9.
22. Guenot, M., Isnard, J., Ryvlin, P., et al. (2001). Neurophysiological monitoring for epilepsy surgery: the Talairach SEEG method. Stereoelectroencephalography. Indications, results, complications and therapeutic applications in a series of 100 consecutive cases. *Stereotactic and Functional Neurosurgery*, **77**(1–4), 29–32.
23. Bekelis, K., Desai, A., Kotlyar, A., et al. (2013). Occipitotemporal hippocampal depth electrodes in intracranial epilepsy monitoring: safety and utility. *Journal of Neurosurgery*, **118**(2), 345–52.
24. Wieser, H.G. (2004). ILAE Commission Report. Mesial temporal lobe epilepsy with hippocampal sclerosis. *Epilepsia*, **45**(6), 695–714.
25. Sparkes, M., Valentin, A., and Alarcon, G. (2009). Mechanisms involved in the conduction of anterior temporal epileptiform discharges to the scalp. *Clinical Neurophysiology*, **120**(12), 2063–70.
26. Gibbs, E.L., Gibbs, F.A., and Fuster, B. (1948). Psychomotor epilepsy. *Archives of Neurology & Psychiatry*, **60**(4), 331–9.
27. Alarcon, G., Kissani, N., Dad, M., et al. (2001). Lateralizing and localizing values of ictal onset recorded on the scalp: evidence from

simultaneous recordings with intracranial foramen ovale electrodes. *Epilepsia*, **42**(11), 1426–37.
28. Alarcon, G., Muthinji, P., Kissani, N., Polkey, C.E., and Valentin, A. (2012). Value of scalp delayed rhythmic ictal transformation (DRIT) in presurgical assessment of temporal lobe epilepsy. *Clinical Neurophysiology*, **123**(7), 1269–74.
29. Wieser, H.G. and Yasargil, M.G. (1982). Selective amygdalohippocampectomy as a surgical treatment of mesiobasal limbic epilepsy. *Surgical Neurology*, **17**(6), 445–57.
30. Spencer, D.D., Spencer, S.S., Mattson, R.H., Williamson, P.D., and Novelly, R.A. (1984). Access to the posterior medial temporal lobe structures in the surgical treatment of temporal lobe epilepsy. *Neurosurgery*, **15**(5), 667–71.
31. Hennessy, M.J., Elwes, R.D., Rabe-Hesketh, S., Binnie, C.D., and Polkey, C.E. (2001). Prognostic factors in the surgical treatment of medically intractable epilepsy associated with mesial temporal sclerosis. *Acta Neurologica Scandinavica*, **103**(6), 344–50.
32. Cendes, F., Li, L.M., Andermann, F., et al. (1999). Dual pathology and its clinical relevance. *Advances in Neurology*, **81**, 153–64.
33. King, D., Spencer, S.S., McCarthy, G., Luby, M., and Spencer, D.D. (1995). Bilateral hippocampal atrophy in medial temporal lobe epilepsy. *Epilepsia*, **36**(9), 905–10.
34. Cukiert, A., Cukiert, C.M., Argentoni, M., et al. (2009). Outcome after cortico-amygdalo-hippocampectomy in patients with severe bilateral mesial temporal sclerosis submitted to invasive recording. *Seizure*, **18**(7), 515–18.
35. Li, L.M., Cendes, F., Antel, S.B., et al. (2000). Prognostic value of proton magnetic resonance spectroscopic imaging for surgical outcome in patients with intractable temporal lobe epilepsy and bilateral hippocampal atrophy. *Annals of Neurology*, **47**(2), 195–200.
36. Margerison, J.H. and Corsellis, J.A. (1966). Epilepsy and the temporal lobes. A clinical, electroencephalographic and neuropathological study of the brain in epilepsy, with particular reference to the temporal lobes. *Brain*, **89**(3), 499–530.
37. Vale, F.L., Effio, E., Arredondo, N., et al. (2012). Efficacy of temporal lobe surgery for epilepsy in patients with negative MRI for mesial temporal lobe sclerosis. *Journal of Clinical Neuroscience*, **19**(1), 101–6.
38. Lee, R.W., Hoogs, M.M., Burkholder, D.B., et al. (2014). Outcome of intracranial electroencephalography monitoring and surgery in magnetic resonance imaging-negative temporal lobe epilepsy. *Epilepsy Research*, **108**(5), 937–44.
39. Rowe, C.C., Berkovic, S.F., Austin, M.C., McKay, W.J., and Bladin, P.F. (1991). Patterns of postictal cerebral blood flow in temporal lobe epilepsy: qualitative and quantitative analysis. *Neurology*, **41**(7), 1096–103.
40. Rathore, C., Kesavadas, C., Ajith, J., Sasikala, A., Sarma, S.P., and Radhakrishnan, K. (2011). Cost-effective utilization of single photon emission computed tomography (SPECT) in decision making for epilepsy surgery. *Seizure*, **20**(2), 107–14.
41. Ta, K.M., Britton, J.W., Buchhalter, J.R., et al. (2008). Influence of subtraction ictal SPECT on surgical management in focal epilepsy of indeterminate localization: a prospective study. *Epilepsy Research*, **82**(2–3), 190–3.
42. Van, P.W., Dupont, P., Sunaert, S., Goffin, K., and Van, L.K. (2007). The use of SPECT and PET in routine clinical practice in epilepsy. *Current Opinions in Neurology*, **20**(2), 194–202.
43. von Oertzen, T.J., Mormann, F., Urbach, H., et al. (2011). Prospective use of subtraction ictal SPECT coregistered to MRI (SISCOM) in presurgical evaluation of epilepsy. *Epilepsia*, **52**(12), 2239–48.
44. Chassoux, F., Devaux, B., Landre, E., et al. (2000). Stereoelectroencephalography in focal cortical dysplasia: a 3D approach to delineating the dysplastic cortex. *Brain*, **123**(Pt 8), 1733–51.
45. Nobili, L., Francione, S., Mai, R., et al. (2007). Surgical treatment of drug-resistant nocturnal frontal lobe epilepsy. *Brain*, **130**(Pt 2), 561–73.
46. Avoli, M., Bernasconi, A., Mattia, D., Olivier, A., and Hwa, G.G. (1999). Epileptiform discharges in the human dysplastic neocortex: in vitro physiology and pharmacology. *Annals of Neurology*, **46**(6), 816–26.
47. Chassoux, F., Rodrigo, S., Semah, F., et al. (2010). FDG-PET improves surgical outcome in negative MRI Taylor-type focal cortical dysplasias. *Neurology*, **75**(24), 2168–75.
48. No authors cited. (1989). Proposal for revised classification of epilepsies and epileptic syndromes. Commission on Classification and Terminology of the International League Against Epilepsy. *Epilepsia*, **30**(4), 389–99.
49. Bonini, F., McGonigal, A., Trebuchon, A., et al. (2014). Frontal lobe seizures: from clinical semiology to localization. *Epilepsia*, **55**(2), 264–77.
50. Agirre-Arrizubieta, Z., Thai, N.J., Valentin, A., et al. (2014). The value of Magnetoencephalography to guide electrode implantation in epilepsy. *Brain Topography*, **27**(1), 197–207.
51. Alarcon, G., Valentin, A., Watt, C., et al. (2006). Is it worth pursuing surgery for epilepsy in patients with normal neuroimaging? *Journal of Neurology, Neurosurgery, & Psychiatry*, **77**(4), 474–80.
52. Elwes, R.D., Binnie, C.D., and Polkey, C.E. (1999). Normal magnetic resonance imaging and epilepsy surgery. *Journal of Neurology, Neurosurgery, & Psychiatry*, **66**(1), 3.
53. Gastaut, H., Poirier, F., Payan, H., Salamon, G., Toga, M., and Vigouroux, M. (1960). H.H.E. syndrome; hemiconvulsions, hemiplegia, epilepsy. *Epilepsia*, **1**, 418–47.
54. Aguglia, U., Tinuper, P., and Gastaut, H. (1984). Startle-induced epileptic seizures. *Epilepsia*, **25**(6), 712–20.
55. Varadkar, S., Bien, C.G., Kruse, C.A., et al. (2014). Rasmussen's encephalitis: clinical features, pathobiology, and treatment advances. *Lancet: Neurology*, **13**(2), 195–205.
56. Cross, J.H. and Neville, B.G. (2009). The surgical treatment of Landau-Kleffner syndrome. *Epilepsia*, **50**(Suppl. 7), 63–7.
57. Scheltens-de, B.M. (2009). Guidelines for EEG in encephalopathy related to ESES/CSWS in children. *Epilepsia*, **50**(Suppl. 7), 13–7.
58. Fejerman, N. (2009). Atypical rolandic epilepsy. *Epilepsia*, **50**(Suppl. 7), 9–12.
59. Tassinari, C.A., Cantalupo, G., Rios-Pohl, L., Giustina, E.D., and Rubboli, G. (2009). Encephalopathy with status epilepticus during slow sleep: 'the Penelope syndrome'. *Epilepsia*, **50**(Suppl. 7), 4–8.
60. Martin Miguel, M.C., Garcia Seoane, J.J., Valentin, A., et al. (2011). EEG latency analysis for hemispheric lateralisation in Landau-Kleffner syndrome. *Clinical Neurophysiology*, **122**(2), 244–52.
61. Gates, J.R., Leppik, I.E., Yap, J., and Gumnit, R.J. (1984). Corpus callosotomy: clinical and electroencephalographic effects. *Epilepsia*, **25**(3), 308–16.
62. Hanson, R.R., Risinger, M., and Maxwell, R. (2002). The ictal EEG as a predictive factor for outcome following corpus callosum section in adults. *Epilepsy Research*, **49**(2), 89–97.
63. Park, M.S., Nakagawa, E., Schoenberg, M.R., Benbadis, S.R., and Vale, F.L. (2013). Outcome of corpus callosotomy in adults. *Epilepsy & Behaviour*, **28**(2), 181–4.
64. Aghakhani, Y., Kinay, D., Gotman, J., et al. (2005). The role of periventricular nodular heterotopia in epileptogenesis. *Brain*, **128**(Pt 3), 641–51.
65. Zabara, J. (1992). Inhibition of experimental seizures in canines by repetitive vagal stimulation. *Epilepsia*, **33**(6), 1005–12.
66. Koo, B. (2001). EEG changes with vagus nerve stimulation. *Journal of Clinical Neurophysiology*, **18**(5), 434–41.
67. Hennessy, M.J., Elwes, R.D., Binnie, C.D., and Polkey, C.E. (2000). Failed surgery for epilepsy. A study of persistence and recurrence of seizures following temporal resection. *Brain*, **123**(Pt 12), 2445–66.

CHAPTER 32

Encephalopathy, central nervous system infections, and coma

Michalis Koutroumanidis and Robin Howard

Introduction

Despite advances in intensive medical care and new diagnostic procedures, encephalopathy remains a frequent and under-recognized critical medical condition with high morbidity and mortality. Electroencephalography (EEG) enables rapid bedside electrophysiological measurements of brain dysfunction and complements clinical and neuroimaging assessment of patients with altered mental status. Importantly, the EEG is the only test that can demonstrate the occurrence of non-convulsive seizures (NCSz) and non-convulsive status epilepticus (NCSE) that frequently complicate acute cerebral insults and can contribute to coma. This chapter discusses the multiple uses and limitations of the acute video EEG and continuous EEG monitoring (cEEG) in the diagnosis and prognostication of the various encephalopathies, encephalitides, and coma. It also examines the role of EEG in the diagnosis of NCSz and NCSE with emphasis on post-cardiac arrest hypoxic-ischaemic brain injury (HIBI). Other types of NCSE in the various epilepsy syndromes and types are discussed in Chapter 34. Box 32.1 summarizes indications for continuous Video EEG monitoring (cEEG) or acute video EEG (vEEG) in the assessment of the critically ill patients.

These indications partially overlap. For instance, management of convulsive status incorporates detection of NCSE after the definitive treatment (and resolution) of convulsive status epilepticus, while identifying and successfully treating NCSE that is associated with periodic epileptic activity against a depressed background, may allow the emergence of physiological cerebral biological rhythms and permit prognostication.

Practical considerations of recording and interpreting EEG in the encephalopathic and critically-ill patient: artefacts

The EEG remains the single most important bedside test that can provide information about the function of the brain during states of impaired consciousness and coma, detect occult epileptic activity and prognosticate. The entire diagnostic process is far from simple. In contrast to the EEG department, recording on the medical ward and particularly in the ICU environment presents unique challenges to the physiologist and the electroencephalographer. The acutely, or subacutely-ill patient is variably uncooperative and often confused and restless or agitated causing muscle, movement and sweat artefacts, and secondary electrode dysfunction. Patients are unable to give a history and describe their symptoms, or even respond, and clinical information to assist EEG interpretation may only rely on provisional summary charts. Important history of past insults, such as cerebrovascular disorders or trauma, or active epilepsy may be missing at the time of the recording or reporting, and the premorbid state is frequently unknown. In comatose post cardiac arrest patients, concurrent sedation, confounding sepsis or metabolic derangements have their own impact on cerebral rhythms blurring the clinical and EEG picture to an uncertain degree. A number of abnormal EEG patterns are still of uncertain clinical or prognostic significance, repetitive EEG patterns may reflect cerebral hypoxia, NCSE or both, and responses to acute administration of a rapidly acting antiepileptic are often impossible to interpret for several reasons.

From the technical viewpoint, a multitude of *artefacts* have to be identified and dealt with as they mar the recording and may mimic seizure activity and misdirect treatment. Artefacts are signals that originate from the body of the patient, the monitoring equipment, life support systems, and ICU personnel. The ideal time to identify and correct them is during the recording; if this is not possible (for example in prolonged monitoring) reviewing the EEG *with* the physiologist and with the assistance of the simultaneous video and audio recordings will help their recognition.

Of the *physiological* artefacts, ocular movements such as vertical nystagmus and bobbing in brainstem lesions can simulate anterior bursts of delta, while glossokinetic artefact (from tongue movements) can mimic frontal intermittent delta activity (FIRDA), a typical finding in deep midline lesions and encephalopathies. Additional electrodes, placed above and below the eyes or the mouth and slightly off-centre, can show the origin of the activity. Pulse and cardioballistic (head moving with each pulse) artefacts can be confused with focal or more diffuse delta activity. The slow pulse artefact is time locked to the QRS, about 200ms from the R. Muscle artefacts including chewing movements, tremor, and shivering (for example due to therapeutic hypothermia) typically spare midline electrodes (Cz, Pz). Respiration can be monitored by a dedicated lead.

Of the *non-physiological* artefacts, electrode 'pops' can mimic a sharp wave focus, while ventilators can produce movement artefacts that, coupled to the respiratory rate, may mimic periodic bursts and even a burst-suppression pattern. Finally, mobile phones (1), vibrating bed artefacts and sternal rub may imitate rhythmic

Box 32.1 Indications for acute video EEG (vEEG) or continuous EEG monitoring (cEEG) in the ICU

Indications

Detection of NCSE/NCSz

- Acute focal (i.e. traumatic, vascular) or diffuse (i.e. HIBI) cerebral insults.
- Following definitive treatment (DT) of refractory status epilepticus.
- In the context of acute or sub-acute encephalopathies.
- Characterization of possible ictal clinical signs.

Management of convulsive status

Induction and maintenance of and emergence from DT.

Prognostication in encephalopathies and anoxic coma

Other uses

- Detection of early reversible vasospasm after aneurismal SAH or ICH.
- Detection of other acute or subacute events such as rising of ICP by automatic quantitative EEG analysis that can identify changes that are difficult to appreciate by viewing row EEG.
- Monitor depth of sedation in anaesthetic coma for managing increased ICP for example after traumatic brain injury.

NCSE, non-convulsive status epilepticus; NCSz, non-convulsive seizures; CA, cardiac arrest; SAH, sub-arachnoid haemorrhage; ICH, intracerebral haemorrhage; ICP, intracranial pressure.

ictal patterns; the latter starts and stops abruptly, is rhythmical and may even appear to evolve in frequency and voltage, and may show a physiological field as it is occurs, or is maximal, in the electrodes that contact the bed.

Electroencephalography in encephalopathies and encephalitides

The term refers to acute or sub-acute global cerebral insults that may be reversible depending on the cause and an early diagnosis. Clinical symptoms and signs are 'non-localizing' (such as altered consciousness, generalized seizures or myoclonus, psychiatric and sleep-wake cycle disturbances, abnormal reflexes), but can suggest certain diagnostic possibilities when focal, such as temporal lobe seizures as in herpes simplex encephalitis (HSE), or facial-brachial seizures as in autoimmune limbic encephalitis (LE). Similarly, EEG findings are typically diffuse or non-localizing and aetiologically non-specific; importantly however, their sequential changes parallel clinical severity. The earliest EEG change is a slowed alpha rhythm with intermixed theta components; theta activities progressively enhance and diffuse and, as severity increases, they gradually give way to delta frequencies, which become dominant and less reactive. In advance stages, amplitudes drop and periods of diffuse attenuation appear, becoming increasingly longer and leading to burst-suppression. In the appropriate clinical context some focal or paroxysmal EEG patterns may suggest more specific aetiologies, although they are *never* pathognomonic. Such electrographic patterns include *temporal lobe epileptiform discharges/PLEDs or seizures* in HSE or autoimmune LE, *triphasic waves* in the various metabolic encephalopathies, such as hepatic and uremic (Fig. 32.1), but also in anoxic encephalopathy, Alzheimer's disease (2) and sporadic CJD, or the recently described delta brushes in anti-NMDA receptor encephalitis.

Fig. 32.2 illustrates a range of non-specific EEG features in a rare case of steroid-responsive amyloid encephalopathy.

In summary, the EEG can:

- Provide objective evidence for organic aetiology and differentiate from purely psychiatric acute or subacute states.
- Diagnose non-convulsive seizures (NCSzs) or non-convulsive status epilepticus (NCSE).
- Objectively gauge the severity of the encephalopathy.
- Follow up course and assess the effectiveness of treatment.
- Assist in determining prognosis when the cause is known.

On the other hand the EEG cannot:

- Specify aetiology (with the exception of the few patterns that in the appropriate clinical setting can point to certain diagnostic possibilities).
- Differentiate between acute or chronic conditions (hence the need to repeat the EEG in order to assess for rapid evolution).
- Exclude epileptic seizures (as in the investigation of epilepsy—see Chapter 11), particularly when using standard video EEG as opposed to cEEG (3).

Most *toxic encephalopathies* are typically associated with 'generic' EEG slowing with the notable exception of BZD and barbiturate intoxication that (also) enhance fast rhythms, clozapine that (also) induces epileptiform activity, and lithium intoxication that is (also) associated with triphasic waves (4). Phenytoin toxicity is initially associated with slow alpha and may progress to diffuse slowing and worsening of seizures (5). EEG findings can alert physicians to the possibility and to dose reduction than further increase. Valproate-induced hyperammonaemic encephalopathy is a rare, but life-threatening adverse event, associated with EEG slowing and even triphasic waves and normal liver function. Co-medication with topiramate or phenobarbital increases the risk (6,7). Of the substances and drugs of abuse, alcohol intoxication produces generalized slowing of the EEG background, though minimal if any in chronic abusers who have developed tolerance; in acute alcohol withdrawal and seizures the interictal EEG shows no epileptic activity. Focal background abnormalities and paroxysmal activity may relate (and alert physicians) to brain injuries, to which alcoholics are prone. Intoxication with amphetamines and cocaine may result in non-specific EEG slowing while cocaine can provoke, or exacerbate, epileptic seizures (8). Cannabis and ecstasy (MDMA) do not visibly alter the EEG, but the latter produces slowing of the alpha and increase of theta rhythms in users with moderate to high life-time dosages, who are also liable to cognitive impairment including memory and attention (9).

The EEG changes in *metabolic encephalopathies* generally lack specificity, but parallel clinical severity; for example, triphasic sharp waves occur equally in advanced hepatic and renal disorders. Dialysis encephalopathy is associated with diffuse slowing maximal frontally and bursts of FIRDA, and with bilateral spike-wave

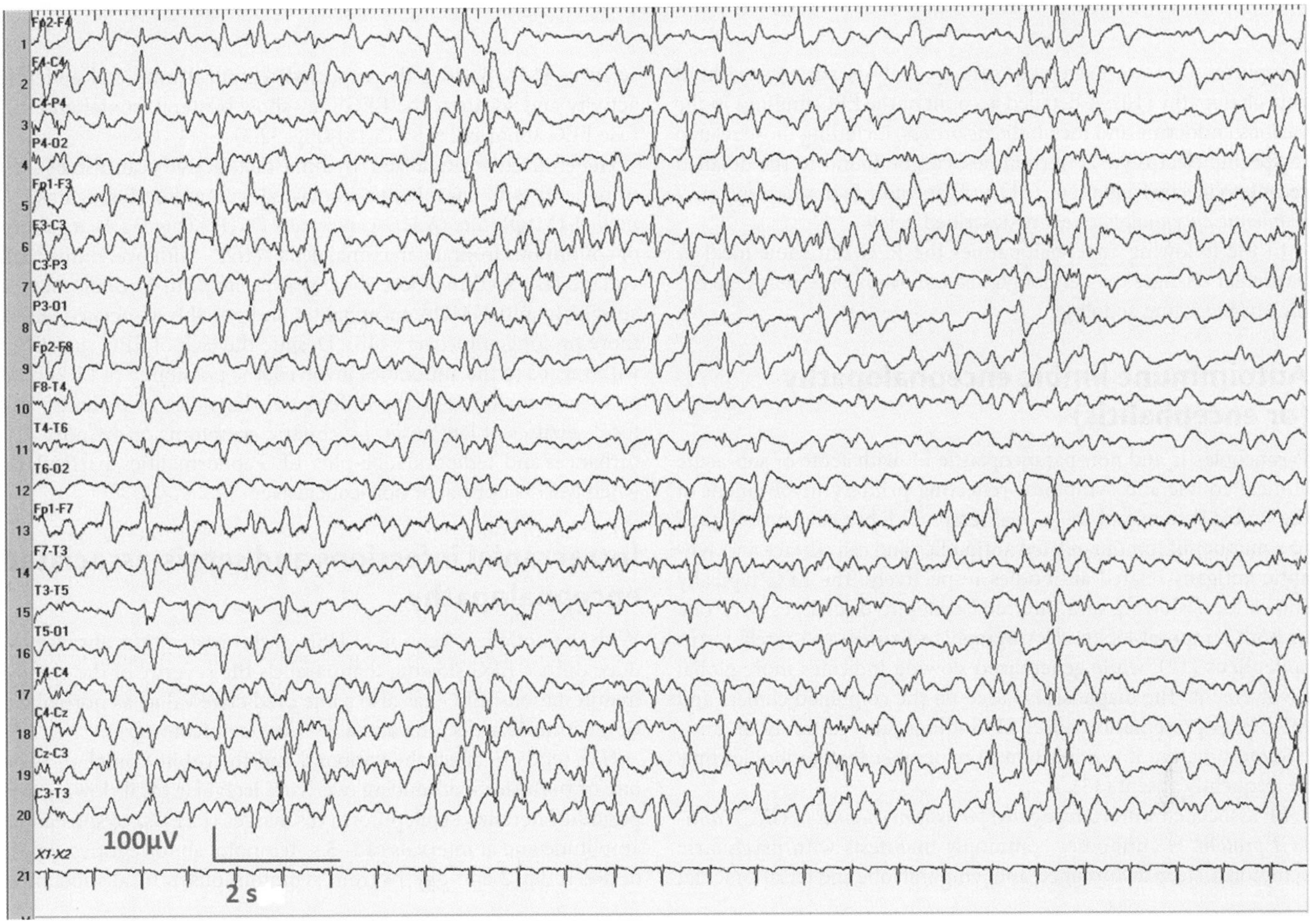

Fig. 32.1 Triphasic waves in hepatic encephalopathy.

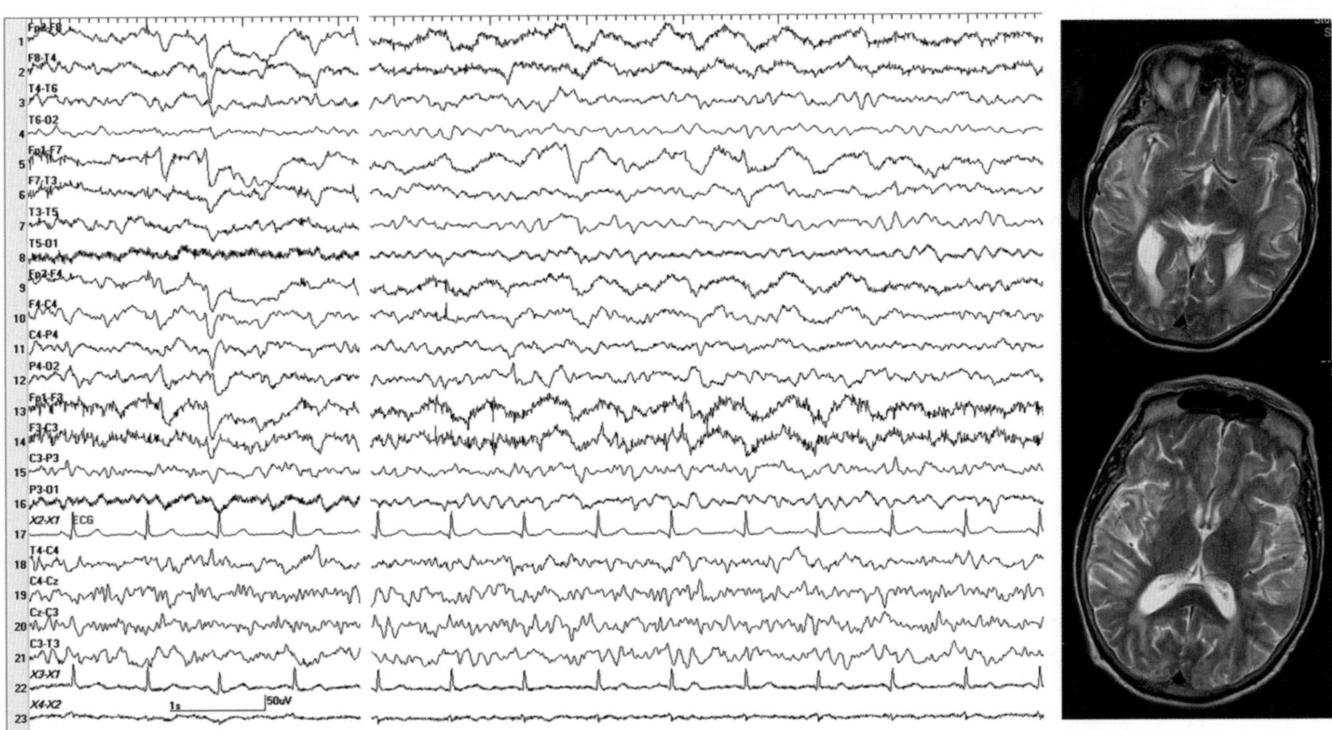

Fig. 32.2 Non-specific encephalopathic features in a 60-year-old woman with amyloid encephalopathy and well controlled JME, followed up in our epilepsy clinic for 15 years. Bilateral synchronous triphasic waves (left trace) and bi-frontal delta activity (right trace), alpha rhythm has slowed down in comparison with her previous recordings.

complexes that are extremely rare in dialysis patients without encephalopathy (10). A detailed account of the EEG findings in the various endocrine and metabolic disorders, including those related to specific electrolyte abnormalities, can be found in the detailed recent review by Faigle et al. (11) and the findings in *severe hypoxic/ischaemic encephalopathies* are described below.

In the following encephalopathies the EEG can show focal or particular changes that, although *not pathognomonic*, may help clinicians regarding aetiology.

Autoimmune limbic encephalopathy (or encephalitis)

Paraneoplastic and non-paraneoplastic LE with acute or sub-acute clinical course and symptoms reflecting primary involvement of the limbic system and the mesial temporal lobes have been linked to a number of tumour-related antibodies and cell surface and synaptic antigens-related antibodies respectively. The EEG typically shows focal slowing and interictal epileptic discharges over one or both temporal lobes, electrographic seizures, and rarely status epilepticus (12), while generalized slowing indicates more global involvement. The diagnosis is based on the combined clinical and MRI/EEG presentation and CSF findings, and certainly on antibody testing, but it is important to remember that antibodies may be frequently absent (13).

LE associated with *voltage gated potassium channel (VGKC) (anti-LGI protein 1)* antibodies commonly manifests with psychiatric symptoms, sleep disturbance, and temporal lobe and facial-brachial seizures (14). Besides the expected temporal interictal epileptic activity and seizures, the EEG may show bilateral frontal and diffuse EEG abnormalities (15,16) (Fig. 32.3).

Generalized or lateralized rhythmic delta activity can also be seen in patients with encephalitis associated with antibodies against *N-methyl-D-aspartate (NMDA) receptors* (17,12) (Fig. 32.4); a pattern of continuous frontally maximal delta activity with overriding high voltage fast rhythms, described as *extreme delta brushes*, may be specific to anti-NMDA encephalitis, and possibly associated with a more prolonged recovery (18). Despite the lack of EEG specificity with regard to the antibodies involved, the possibility of LE should be always included in the differential diagnosis of patients with focal seizures of late onset, psychiatric symptoms or memory disturbances and 'temporal lobe-plus' EEG abnormalities, particularly when MRI is normal or non-conclusive.

Intracranial infections and sepsis-associated encephalopathy

With the notable exception of HSE *viral encephalitides* always produce diffuse EEG slowing that parallels the severity of the clinical manifestations. EEG has also some predictive value, as normalization within 5 weeks indicates a favourable outcome (19).

HSE tends to affect the temporal and the orbitofrontal areas on one or both sides, producing overlying irregular focal slowing and periodic lateralized epileptiform discharges (PLEDs) of substantial amplitude and at intervals of 1–5 s. Temporal abnormalities appear between day 2 and day 14 from symptom onset, focal subclinical

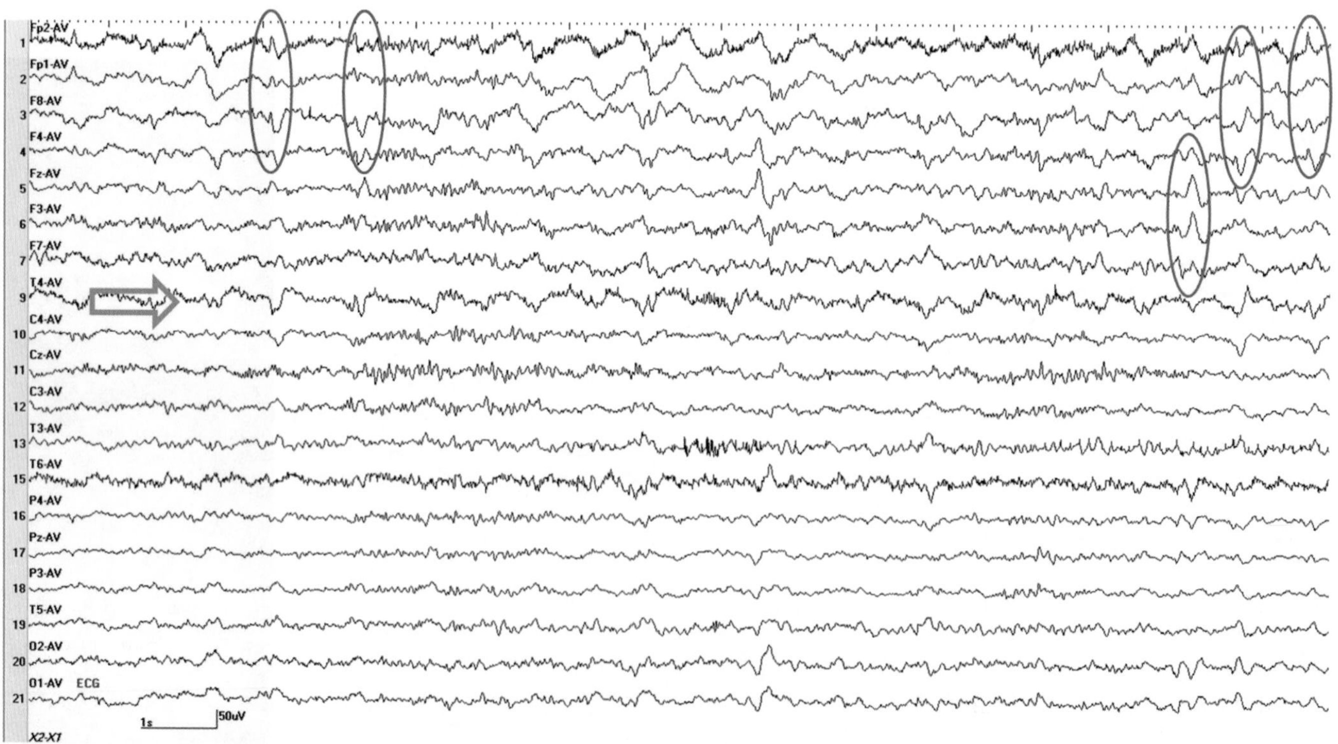

Fig. 32.3 Limbic encephalitis associated with increased title of anti-LGI protein 1 antibodies. Urgent EEG on a 52-year-old man admitted with new onset generalized convulsions, following a brief period of predominantly psychiatric symptoms. Note the almost continuous spike-wave activity over the right mid-temporal area (arrow), diffusing to the ipsilateral frontal areas (oval marks); independent sharp waves also occurred on the left. The EEG normalized in a few weeks, while brain MRI remained normal.

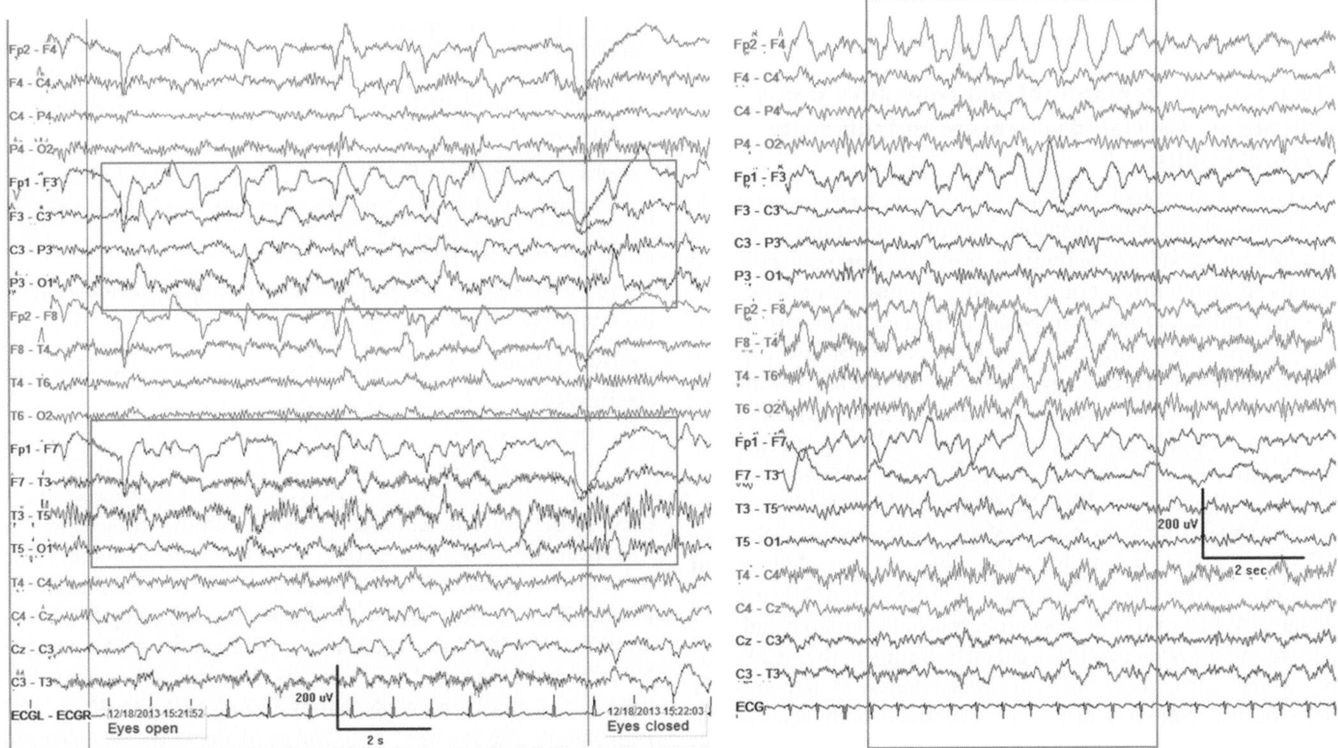

Fig. 32.4 Video EEG on a 20-year-old woman with predominantly psychiatric symptomatology of subacute onset associated with increased titre of anti-NMDA receptor antibodies. In the first outpatient recording the left hemisphere appears more affected (left trace), while a month later bilateral bursts of high voltage sharp rhythmic delta activity showed right frontotemporal emphasis (right trace). The EEG was the only abnormal test; MRI scans and an FDG-PET were normal.

seizure discharges and seizures frequently occur, sometimes emanating from the PLEDs, while focal slowing soon diffuses although usually retaining frontal-temporal emphasis. Although not pathognomonic, in the appropriate clinical setting such early changes can suggest HSE with high degree of certainty and prompt immediate treatment. Certainly, their absence does not rule out HSE, and it is also useful to remember that PLEDs can be very small and hardly noticeable amidst background rhythms, or that they may appear later. Compared to most viral encephalitides, sequential follow up recordings in patients with HSE are less useful in the prediction of neurological outcome, because EEG findings tend to lag clinical changes.

EEG in *HIV encephalopathy* shows mild generalized slowing, while focal/multifocal changes occur in patients with opportunistic infections such as toxoplasmosis, or progressive multifocal leukencephalopathy; around 6% of these patients develop seizures or epilepsy (20). The EEG is expected to be normal in asymptomatic HIV-seropositive patients (21).

The term sepsis-associated encephalopathy (SAE) refers to diffuse brain dysfunction, secondary to infection outside of the CNS (22). It is the commonest type of encephalopathy encountered in up to 70% of septic ICU patients; in the most severely affected SAE is usually combined with critical illness peripheral neuropathy or myopathy. EEG findings lack specificity, but correlate well with the severity of encephalopathy (23): mild degrees show only diffuse slowing in the theta frequency range while the occurrence of triphasic waves usually coincides with renal impairment. The advanced stage of burst-suppression is associated with 70% mortality due to multi-organ failure (24). Therefore, apart from establishing the presence of encephalopathy, the EEG can grade the severity and monitor the course of the disease. In contrast to metabolic encephalopathies, in SAE epileptic EEG activity and seizures are uncommon, although they may occur in severe cases (25).

The diagnosis of sporadic (classical) Creutzfeldt-Jacob disease (sCJD) can be made with high degree of certainty when the typical periodic complexes (PC) occur in the appropriate clinical context. PC show a repetition rate between 0.5 and 2 Hz and are typically bilateral or diffuse, but can also be unilateral or focal. They tend to develop relatively late in the course of the disease, usually in conjunction with the myoclonic jerks, while at early stages the EEG shows only non-specific encephalopathic changes and progressive slowing (3). A further limitation of the diagnostic value of the EEG, even in advanced stages, is that PC occur only in a subset of patients, depending on the molecular subtype (usually MM1 and MV1) (26). Furthermore, PCs are not pathognomonic of CJD, and their morphology can overlap with that of triphasic waves and periodic complexes that occur in other encephalopathies and even in cases of non-convulsive status (see below). The degree of diagnostic certainty increases in the presence of the protein 14-3-3 in the CSF and of characteristic changes on the brain MRI, and indeed a recent multicentre international study on 436 sCJD patients has updated the clinical MRI, CSF, and EEG diagnostic criteria, yielding an overall combined sensitivity of almost 99%; the sensitivity of the positive EEG in all molecular subtypes was less than 40% in that study (27). PCs are not found in other transmissible spongiform encephalopathies, such as Kuru, familial fatal insomnia,

Gerstmann–Staussler–Scheinker syndrome (with the exception of very rare cases), or in variant CJD (3).

Electroencephalography of non-convulsive status epilepticus, associated with acute and sub-acute cerebral insults

The term NCSE describes a range of electroclinically pleomorphic and aetiologically heterogeneous epileptic conditions, in which electrographic seizure activity is prolonged or recurrent and results in unremitting non-convulsive clinical symptoms (28). On pure clinical grounds, in NCSE patients with history of epilepsy, knowledge or identification of the particular epilepsy type or syndrome and its aetiology will define the optimal therapeutic approach and prognosis (chapter 34). In NCSE patients with acute brain insults and largely unknown history including epilepsy, the first diagnostic step is to distinguish between *focal (simple or complex partial)* and *generalized* NCSE. In the former, EEG seizure activity appears focal or multifocal and, in the appropriate clinical setting, may suggest distinctive diagnostic possibilities and prompt aetiological treatment (such as in HSV) and early focused imaging. When ictal clinical semiology is not lateralized and EEG paroxysms appear generalized, characterization of the NCSE simply as generalized is of little clinical usefulness as aetiologically non-specific. Non-convulsive epileptic states with 'generalized' EEG features include truly generalized forms, such as idiopathic (typical) and non-idiopathic (atypical) absence status (29,30) and *de novo* absence—like status (31,32), and secondary generalized states, such as those related to acute symptomatic cerebral pathology, some forms of focally driven rapidly generalized NCSE (for example frontal lobe status) (33), and even frank complex partial NCSE, the localized EEG onset of which was missed just because the recording started late when ictal changes had already become bilateral.

Although most of the forms of NCSE can be encountered in the ICU, typical and atypical absence, absence-like status, and the various forms of complex partial NCSE tend to occur in ambulatory patients who would be treated in medical wards, while the most frequent types of NCSE in the critically ill patients are focal and mainly secondary generalized states following acute focal or diffuse (i.e. hypoxic after cardiac arrest) cerebral insults (34); indeed, an operational distinction between *proper* and *comatose* forms of NCSE has recently been proposed (35). The following section focuses on the critically-ill patients and discusses the usefulness of the acute video EEG and cEEG in the various primary diagnoses that are associated with a high risk of NCSE, the relevant diagnostic criteria and pitfalls and some practical methodological aspects.

Primary diagnoses associated with increased risk of non-convulsive seizures (NCSz) and NCSE

Box 32.2 lists the main cerebral disorders that are associated with increased risk of non-convulsive seizure activity, particularly in comatose patients. Other independent risk factors include younger age (36) and a history of epilepsy or past cerebral insults, such as stroke, trauma, or tumour (37).

Box 32.2 Primary diagnoses associated with increased risk of NCSz and NCSE

- Hypoxic ischaemic brain injury after out of hospital cardiac arrest.
- Severe infections, either systemic (septic encephalopathy) or primarily cerebral (encephalitis or meningitis).
- Metabolic encephalopathies.
- Cerebrovascular disorders (subarachnoid haemorrhage, intracerebral haemorrhage and infarcts).
- Head trauma.
- Tumours.

Hypoxic ischaemic encephalopathy after cardiac arrest

There is great variability in the frequency of the reported seizures after cardiac arrest that, apart from the different definitions of seizures and methodology (cEEG vs. acute EEG), may be due to early sedation to allow tracheal intubation and ventilation, the masking effect of the neuromuscular blockade used in the management of therapeutic hypothermia (TH), or even to a possible direct effect of TH itself against seizures. In 101 comatose post-cardiac arrest patients treated with TH, the incidence of NCSE using cEEG was around 12% (38). In a large international study of 765 patients Nielsen et al (39) found NCSz in 24%, but only 316 patients had EEG. Using cEEG, a smaller study found NCSz in 20% of 34 patients with cardiac arrest (40). The incidence of NCSE appears to be higher in younger patients: cEEG monitoring during hypothermia (24 h), rewarming (12–24 h), and then an additional 24 h of normothermia disclosed electrographic seizures in 47% and NCSE in 32% of 19 children with cardiac arrest (41).

Ischaemic stoke

Acute post-stroke clinical seizures have been reported to occur in 5.4% of patients (42) and have been associated with increased mortality (43), while the incidence of post-stroke epilepsy increases with time reaching 12.4% after a follow-up of 10 years, particularly in patients with total anterior circulation infarcts (44). A cEEG study demonstrated NCSzs in 11% of patients with acute ischaemic stroke (37), while NCSE also appears to be more frequent than convulsive status in the early post-stroke phase (45). The significance of NCSz for the outcome after stroke is still uncertain (46).

Subarachnoid haemorrhage

cEEG studies have shown that non-convulsive seizures and NCSE are more common in patients with subarachnoid haemorrhage (SAH) than clinically appreciated (47). Persisting NCSE after SAH predicts poor outcome (48,49), and adequate anti-seizure treatment is needed. However, this should be given to seizing patients only, as prophylactic treatment has been shown to relate to overall worse outcome (50).

Intracerebral haemorrhage

Acute, immediate (within the first 24 h from intracerebral haemorrhage (ICH)) clinical seizures have been reported to occur in 7–12% of patients (51,52) and in up to 15% in the first 48 h (53). As in the other stroke subtypes, cEEG has shown that the actual occurrence of seizures is higher: NCSz were found in 18–21% (54,55) with periodic epileptiform discharges appearing as an independent predictor of poor outcome (55). As with SAH though, treatment should be limited to the patients with clinical or NCSzs as prophylaxis with phenytoin has been associated with more fever and worse outcomes (51).

Following convulsive status epilepticus

In patients who did not awake shortly after treatment of convulsive status, EEG monitoring showed subsequent NCSzs in 48% and NCSE in 14% (56). The incidence and treatment of convulsive status epilepticus is discussed in Chapter 34.

Traumatic brain injury

Traumatic brain injury (TBI) is the leading cause of death and disability in patients younger than 45 years with a mortality of 30-40% after withdrawal of support in ICU (57). TBI is associated with seizures in up to 22% of patients with blunt trauma and in up to 50% of those with penetrating trauma (58), while the widespread use of anti-seizure prophylaxis after TBI (59) has indeed resulted in clinical seizures occurring less commonly (60). NCSz and NCSE still occur, and they too are associated with poor outcome (61,62). cEEG studies have demonstrated a 10–18% occurrence of NCSzs (63,37) and an 8% occurrence of NCSE (37). The incidence of seizures appears to be even higher in children. In a recent study of 87 children with TBI requiring ICU admission, 42.5% had seizures, with subclinical seizures occurring in 16.7% and subclinical status in 13.8% of children; solely subclinical seizures occurred in 6.9% (6/87 children). Children with subclinical seizures and status demonstrated worse outcomes (64). These studies emphasize the need of early seizure detection, although it is not clear yet whether or not effective treatment of subclinical seizures can improve short-term and long-term outcomes, and how aggressively such treatment should be pursued.

Because of the considerable equipment and staffing costs, cEEG is available only in a few centres and often sparsely; in the authors' major tertiary hospital there is only one such machine without quantitative EEG trending algorithms that time compress and simplify EEG evaluation. Consequently, diagnosis relies on acute video EEG recordings, the inferior diagnostic yield of which can be redressed by enhanced clinical awareness. This is discussed in the next section.

Diagnosis of NCSE

Detection of NCSE *depends* on the clinical suspicion and *relies* on the correct interpretation of the obtained EEG evidence. The possibility of NCSE must always be in the differential diagnosis of the ambulatory and the ICU patient whose behavioural or cognitive state changes compared to the baseline. Box 32.3 contains a list of the behavioural patterns or states that may reflect underlying NCSE and a number of subtle signs that in the appropriate clinical setting (see Box 32.2) may alert to possible NCSz and prompt video-EEG recordings.

The diagnosis can be missed because of the subtlety of the clinical signs (Box 32.3) or delayed by coexisting systemic or CNS illnesses that may contribute to altered mental status and blur the clinical picture in the medical ward or the ICU environment. Examples here include concurrent infections or metabolic derangements, bilateral frontal pathology associated with akinetic mutism, sedating medications and psychiatric co morbidity (including catatonia) or even dementia in elderly patients. Some of them, such as septic or metabolic encephalopathies, may also precipitate NCSE on their own right.

On the EEG front, it is relatively easy to diagnose NCSE when there are frequent focal electrographic seizures with change in frequency, field of distribution, voltage, and morphology of the discharge (65). However, EEG diagnosis of NCSE may be difficult or sometimes even impossible to ascertain, particularly in the comatose patient with generalized periodic epileptiform discharges (GPEDs). Despite numerous proposals based on fairly robust

Box 32.3 Clinical clues alerting to the possibility of NCSz/NCSE

- Fluctuating mental status.
- Unexplained alteration of mental status (agitation or depression).
- Ongoing mental depression after lightening of sedation (i.e. after definitive treatment of refractory convulsive status epilepticus).
- Prolonged encephalopathy following an operation, or a known neurological insult, or in the course of systemic infection (septic encephalopathy).

The degree of suspicion increases when the above clinical states are associated with any of the following episodic behavioural changes that may or may not reflect NCSz activity

- Blinking, nystagmus, eye deviation.
- Mouthing movements, lip smacking, facial twitching.
- Mutism, staring.
- Repetitive, stereotyped, episodic motor phenomena such as limb or axial tonic posturing or myoclonus*; hiccups and tremors are less likely to reflect epileptic (cortical) dysfunction, rigors are frequently seen because of therapeutic hypothermia.
- Episodic autonomic dysfunction (blood pressure, heart rate changes).

Clinical and EEG features that suggest epileptic myoclonus include

- A topography that involves body parts with greater cortical representation (face, hand or arm); can be bilateral synchronous, lateralized, or indeed focal.
- Synchrony of jerks in all affected parts.
- Fast frequency (≥2 Hz) (see also EEG criteria for NCSE (Box 32.4)).

Non-epileptic myoclonus is suggested by midline topography of jerks (trunk or neck), lack of EEG correlates, or low frequency (<1.5 Hz) of EEG discharges (when present) and inconsistent temporal association between them and EMG artefacts.

Note 1: Inability to suppress jerking by manual restriction has been described as a primary criterion of epileptic myoclonus. Although there is some truth in this statement, epileptic myoclonus may appear to attenuate or pause when feeble, while non-epileptic myoclonus due to severe diffuse anoxic subcortical damage (status myoclonus—Fig. 32.6) cannot be manually suppressed or in any way affected by change of posture.

Note 2: Asynchrony of multifocal myoclonus does not necessarily indicate a non-epileptic nature because there may be multiple cortical epileptic foci.

Note 3: In the post-anoxic state, usually epileptic and non-epileptic myoclonus co-exist.

**Myoclonus* in the comatose patient can be cortical (epileptic) or subcortical (non-epileptic), secondary to diffuse anoxia, brainstem pathology, metabolic encephalopathy or chemical exposure (128).

evidence (Box 32.4), there are no universally agreed diagnostic criteria and interpretation of what is and what is not NCSE can be subjective. The clinical significance of several EEG patterns is still uncertain while, in contrast to the ictal patterns recorded in the diagnostic or the epilepsy surgery telemetry unit, ictal EEG changes in the ICU patients may appear understated in direct analogy to the generally subtle clinical seizure signs. For instance, possible ictal periodic electrographic patterns may show slower frequencies and lower voltage (particularly when arising from a depressed brain), hesitant evolution, or both. These uncertainties may explain the relatively low inter-observer agreement of the EEG diagnosis of NCSz/NCSE, particularly when periodic discharges are concerned (66), and also the wide variation in the reported incidence of NCSzs and NCSE in the ICU environment.

Neurophysiological criteria of NCSE

While there are patterns that are certain to reflect a non-convulsive epileptic state (67) (Fig. 32.5) and others that clearly do not (Fig. 32.6), there are intermediate forms that arouse diagnostic uncertainty.

The greatest concern in comatose post cardiac arrest patients is the misinterpretation of a potentially epileptic generalized pattern (that falls short of the above proposed criteria) as anoxia, leading to under treatment with anti-seizure agents or even their withdrawal when they are most needed; there is good evidence that ongoing NCSz and NCSE are associated with poor outcome, both in traumatic head injury leading to delayed increase of intracranial pressure and hippocampal atrophy (61,62) and in post-anoxic coma after cardiac arrest where they are associated with increased mortality (38–40).

For example, *non-evolving* GPEDs at low (≤2 Hz) frequencies against an attenuated or suppressed background after hypoxic—ischaemic brain insults are almost invariably associated with severe and irreversible cortical damage (68,69). From Box 32.4, which outlines the diagnostic criteria for NCSE as these have been proposed by leading experts in the field, the reader will sense that the frequency of the repetitive epileptiform discharges in *non-evolving* generalized or focal patterns is a foremost criterion; ≥2 Hz argues for NCSE rather than cortical anoxic damage. Yet, quasi-periodic discharges at 1.5–2.5 Hz without clearly evolving tendencies may still reflect epileptic activity for the reasons stated in the previous paragraph, and this is supported by the similar pattern that may occur in clearly epileptic states, such as late in the transition from the overt generalized convulsive status to subtle status (70,71).

Evolution of the periodic pattern is (correctly) another major criterion (Box 32.4, Fig. 32.5) and most experts appear to accept that generalized patterns indicative of epileptic state should not be invariant, nor should changes, if present, be only state related. However, a *non-evolving* pattern may not necessarily be '*invariant*' and Fig. 32.7 shows an example of an epileptic (as it was proven in the end) GPED pattern consisting of two different frequencies, a fast and a slow, without the required gradual transition of one to the other according to the criteria in Box 32.4 (72).

As for the secondary criterion of Box 32.4, the clinical endpoint of the test may not occur even in patients with documented NCSE; while a test dose of lorazepam may clear most EEG patterns, including the triphasic sharp waves of metabolic encephalopathies that are clearly non-epileptic (73), a clear clinical improvement is rare in comatose patients (74) (Fig. 32.8). Reasons include a co-existent confounding metabolic or septic encephalopathy (in which case the emerging background activity after the resolution of the epileptic discharges would be diffusely slow, rather than a well-formed alpha), or simply that the status may be highly resistant to anti-seizure drugs.

In post-anoxic coma, epileptic activity is secondary to the hypoxic neuronal dysfunction, but the relatively limited expression of EEG phenomenology does not allow quantitative assessment of the severity of the hypoxic ischaemic damage and distinction from its epiphenomenon. In case of doubt, clinical decisions should rely on a combination of EEG and other neurophysiological indicators, such as SSEPs and BAEPs as well as on clinical signs (pupillary reflexes), rather than exclusively on EEG criteria. It is also important to note that diagnostic criteria proposed before the widespread use of therapeutic hypothermia (or controlled normothermia) need to be revised and become more flexible and versatile. Recognizing the complexity and the clinical importance of NCSE diagnosis, the American Clinical Neurophysiology Society recently published a

Box 32.4 Proposed criteria for the diagnosis of NCSE

An EEG pattern lasting for at least 30 min and satisfying any one of the following three primary criteria.

Primary criteria

- Repetitive generalized or focal epileptiform discharges at ≥3.0 Hz (≥2.5 Hz*).
- Repetitive Generalised or focal epileptiform discharges at <3.0 Hz (or <2.5 Hz*) *and* the secondary criterion.
- Sequential rhythmic, periodic or quasi-periodic waves at ≥1 Hz with unequivocal evolution in:
  - Frequency, gradually increasing or decreasing by at least 1 Hz (i.e. from 2 to 3 Hz or vice-versa).
  - Location, gradually spread into or out of a region involving at least two electrodes.
  - Morphology.

  Change of amplitude or sharpness only is not sufficient to satisfy evolution in morphology.

Secondary criterion

Significant improvement in clinical state or appearance of previously absent normal EEG patterns temporary coupled to acute administration of a rapidly acting antiepileptic drug (i.e. lorazepam or levetiracetam). Resolution of the 'epileptiform' discharges leaving diffuse slowing, without normal EEG patterns or clinical improvement does not satisfy this criterion.

Source data from Jirsch J and Hirsch LJ, Nonconvulsive status epilepticus in critically ill and comatose intensive care unit patients. In: Kaplan PW, Drislane FW (Eds), *Nonconvulsive status epilepticus*, pp. 175–86, copyright (2008), Demos Medical Publishing.

*Modifications on frequency criteria proposed by Drislane FW and Kaplan PW, Non convulsive seizures and status epilepticus. In: Fisch BJ, (Ed), *Epilepsy and intensive care monitoring*, pp. 287–307, copyright (2009), Demos Medical Publishing); the same criteria apply for the diagnosis of non-convulsive seizures, setting a minimum duration of 10 s for the suspected EEG pattern.

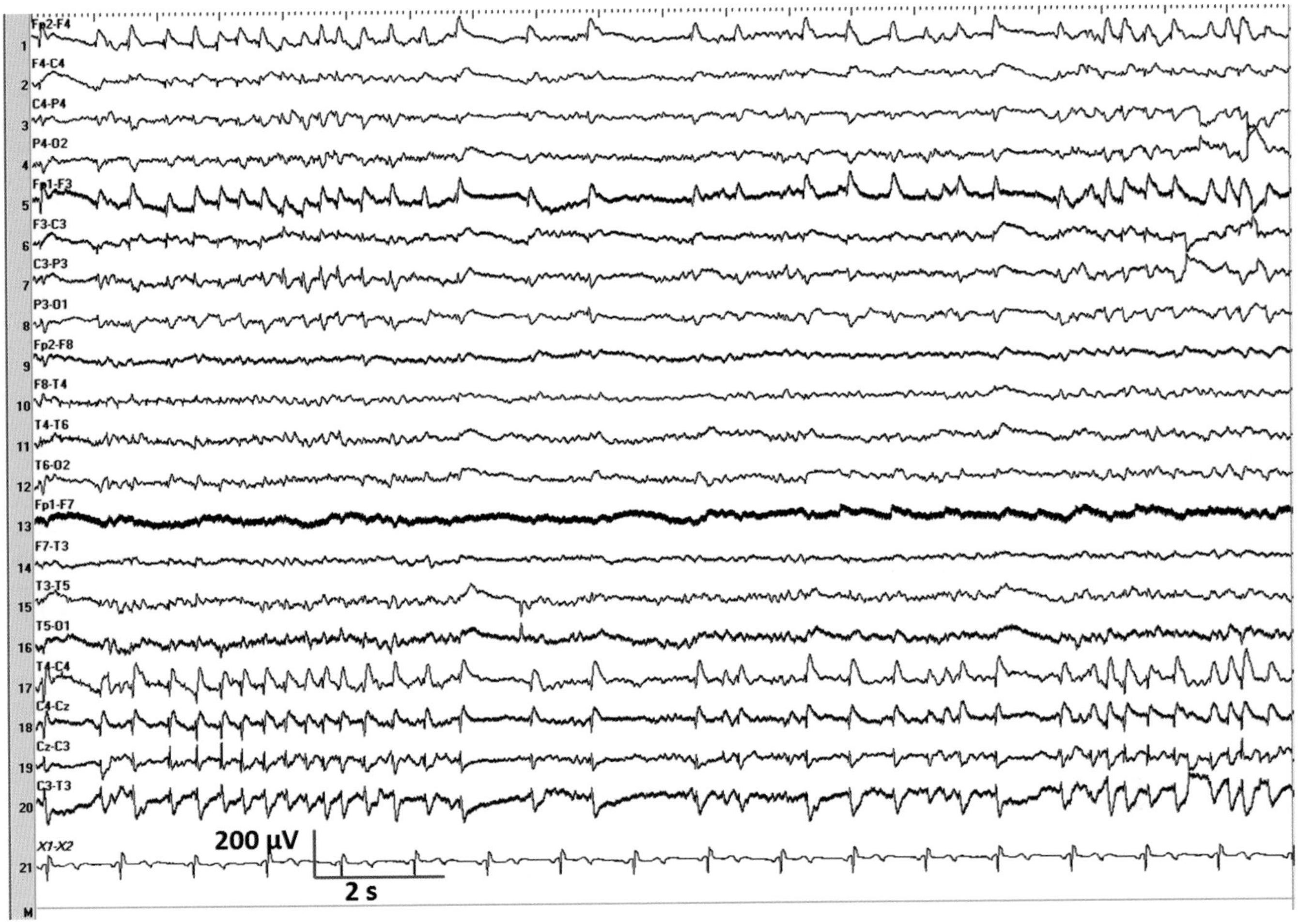

Fig. 32.5 GPEDs showing spontaneous evolution in frequency; the pattern involves the parasagittal areas with some left sided emphasis and reflects NCSE (see proposed criteria of status in Box 32.4).

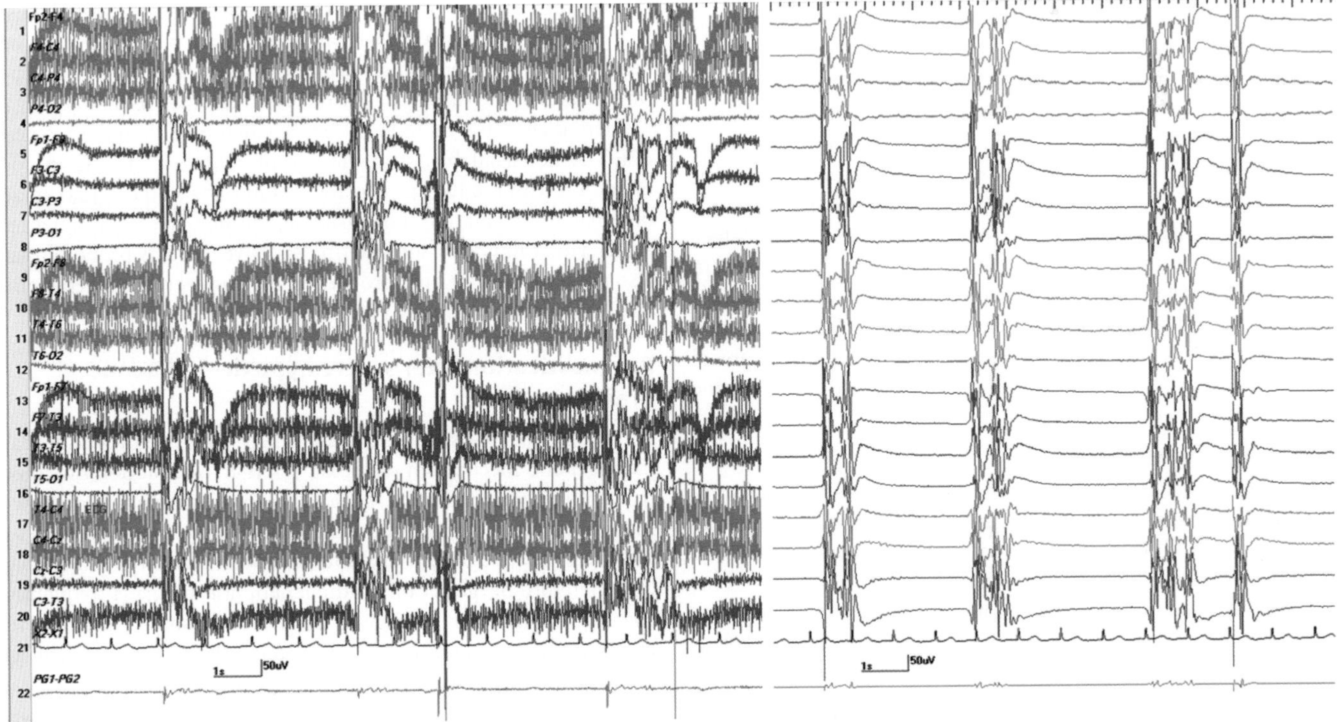

Fig. 32.6 Anoxic status myoclonus 24 h after OOHCA. Note the high voltage polyspike discharges that occur irregularly and at low frequency, and are associated with axial myoclonus (bottom channel). Administration of muscle relaxant leaves no doubt that there is no biological activity between the bursts. SSEPs, performed one hour later recorded no N20.

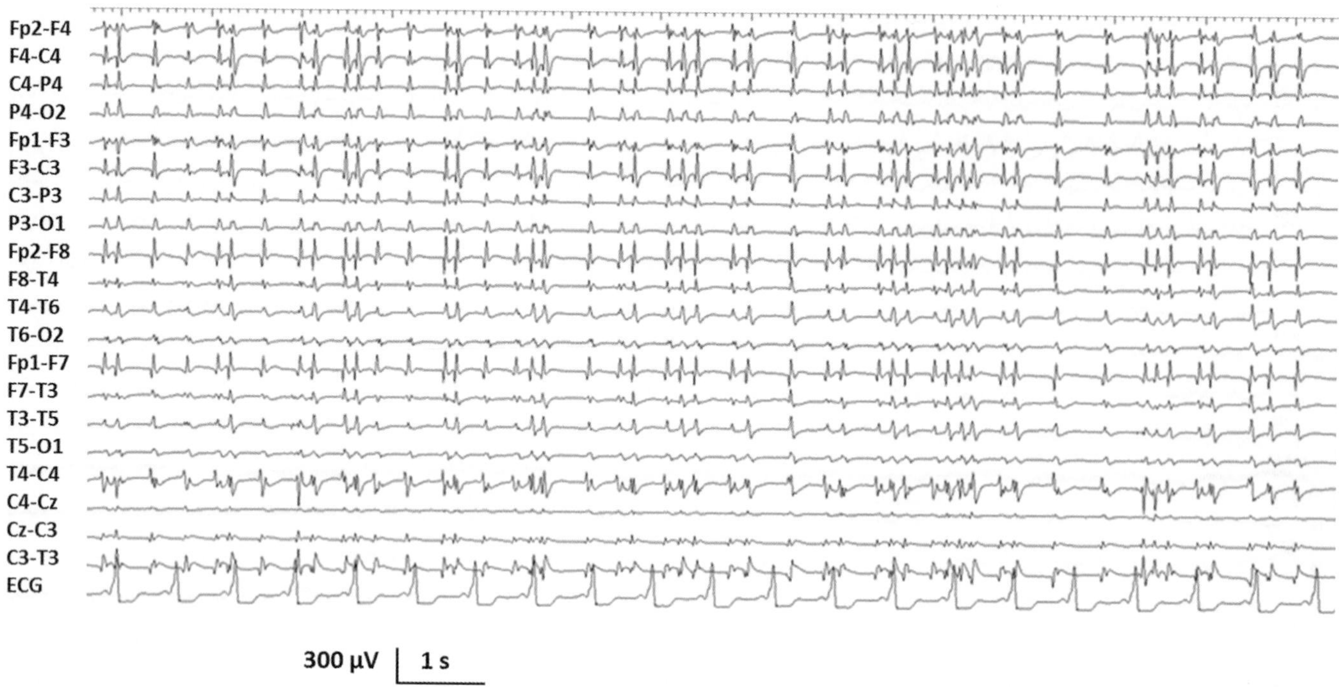

Fig. 32.7 Unremitting, non-evolving, unresponsive 2-6Hz high voltage GPED—isoelectric pattern from 36 to 96 h post insult in a comatose 53-year-old woman with OOHCA; she was completely off sedation, on triple antiepileptic treatment and without systemic confounding disorders. She eventually made meaningful functional recovery and the last EEG 6 months later showed satisfactory background rhythms and stimulus-induced epileptiform discharges without seizures.

comprehensive list of EEG descriptors that include important modifiers, such as 'fluctuating' and 'plus (+)', aiming for better identification of NCSE through clinical research.

Duration of cEEG in ICU patients (and the timing of the evaluation)

Studies using cEEG have shown that of patients with electrographic seizures during the monitoring period, 80% of comatose patients and 95% of non-comatose patients had the first seizure within 24 h of monitoring, and 87% of comatose patients and 98% of non-comatose patients experienced seizures within 48 h of monitoring. Consequently, a period of 24 h has been suggested as a reasonable duration of monitoring to screen for seizure activity in non-comatose patients, while longer periods are advisable for comatose patients or patients with periodic discharges (37). A more recent study demonstrated that most epileptic states start within the first 12–18 h after resuscitation from cardiac arrest (75). Acute video EEG recordings lasting for 30–60 min are usually guided by clinical suspicion and should be carried out as soon post-insult as possible and repeated as required.

cEEG or acute video recordings?

The video-EEG confirms the diagnosis of NCSE/NCSz by showing epileptiform seizure patterns in the appropriate clinical setting. The choice between cEEG monitoring (with video) and acute 30–60 min video EEG depends on the clinical question.

If the purpose is to exclude NCSz as cause of persistent stupor or coma after discontinuation of sedation, cEEG is clearly superior to the acute EEG recordings because:

- It captures significantly more seizures/clinical events that may not happen during a standard acute recording. It is estimated that a standard video EEG of 30–60 min misses NCSz in 50% of patients with NCSz recorded by cEEG for a period of 2–3 days (76).
- It diagnoses NCSE faster and can monitor response to antiepileptic treatment.

If instead the EEG is requested to evaluate specific clinical signs, suspected as seizure manifestations (eye deviation or nystagmus, or other body movements and postures) a finite recording may capture several of the suspected episodes and elucidate their nature.

There is certainly more to the acute video EEG than that. When cEEG is not available, strategic use of acute video EEG in co-operation with the ICU physicians and nursing staff can maximize obtainable information. For instance, temporary discontinuation under video EEG of a short-acting sedative, such as propofol that also possesses anti-seizure properties, may reveal rebound epileptic EEG activity and therefore guide management.

Prognostication after severe hypoxic–ischaemic brain injury

The term hypoxic–ischaemic brain injury (HIBI) indicates a dual pathogenetic mechanism. Hypoxia refers to a reduction of either oxygen supply or utilization, such as in low haemoglobin or following poisoning of the mitochondrial cytochrome enzymes (such as in CO exposure), while ischaemia to a reduction in blood supply and therefore oxygen. Both pathogenetic mechanisms frequently coexist, but the brain damage in ischaemia is more severe than in hypoxia alone because there is also accumulation of toxic cellular metabolites, such as lactate, H + , and glutamate.

Coming after out of hospital cardiac arrest is associated with poor outcome. The rate of survival to hospital discharge is 7–10% with only a few patients making a satisfactory neurological recovery (77).

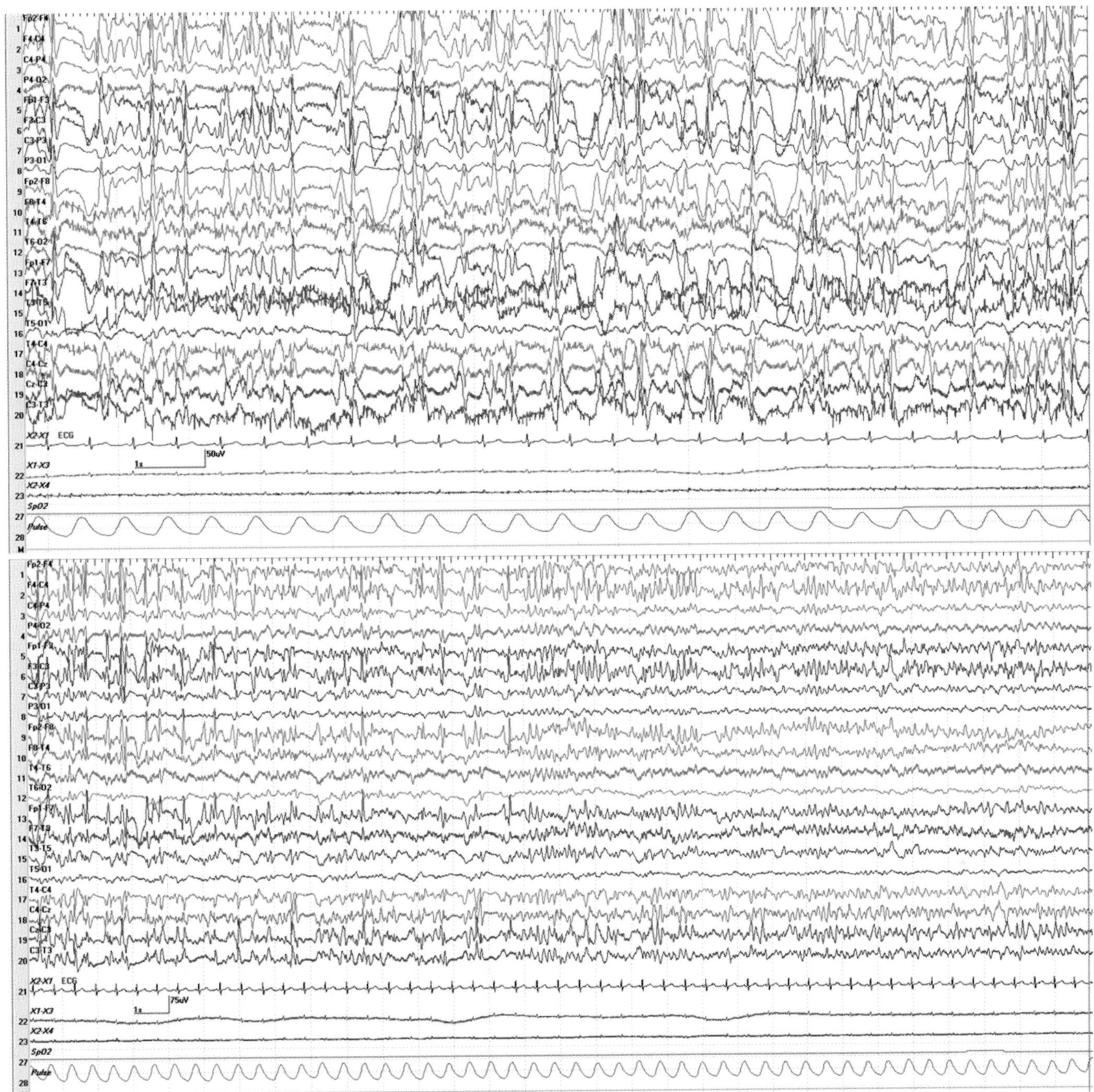

Fig. 32.8 Resolution of bilateral epileptic activity without clinical response after administration of lorazepam 2mg IV in NCSE. The patient made good neurological recovery. Note the fast biological rhythms in the second half of the lower trace, but also the lack of EMG activity in the respective channels and the unchanged plethysmography.

Early prediction of meaningful recovery is important to guide decisions on patients' management, particularly given the ethical, social, and economic implications of continuing intensive support of patients with severe and irreversible anoxic brain damage. The assessment of the degree of hypoxic injury remains primarily clinical, but is decisively assisted by neurophysiological and imaging investigations (78).

The EEG has been widely used over many years to guide prognosis after HIBI. EEG appearances are independent of the primary cause (ischaemia or anoxia) and can be graded according to the degree of damage, but may be, and often are, heavily influenced by a multitude of confounding factors that include sedation and coexistent metabolic or septic encephalopathies and secondary to the hypoxic damage NCSE. EEG findings further depend on the interval between resuscitation and actual recording, as malignant patterns, such electrical silence, or burst—suppression, recorded early after resuscitation may gradually give way to distinct rhythms, which might indicate a better prognosis.

The next section examines the effects of the various procedures and agents commonly used in the ICU on the EEG. The effects of possibly co-existent confounding disorders and NCSE have been discussed in the preceding sections.

Therapeutic hypothermia

Patients who remain unconscious after resuscitation from cardiac arrest tend to show significant improvement in neurological function and survival when treated with TH (32–34°C for 12–24 h) compared with others with standard treatment (79,80). Relative contraindications include bleeding or high risk of bleeding, such as post-surgery or post-trauma, but most patients are eligible and need to remain heavily sedated and possibly on neuromuscular blockade (to prevent breakthrough of shivering), intubated and mechanically ventilated (38). From the EEG perspective, various patterns that have previously been considered as 'malignant' may perhaps be 'less predictive' of poor outcome in comatose post-cardiac arrest patients treated with TH, and therefore EEG interpretation during or after TH may need to be more cautious. Current evidence is limited (81) and prospective multicenter studies using widely accepted terminology are needed to explore the effects of TH on the various EEG patterns. It is perhaps worth mentioning here that the usefulness of the actual cooling is currently under revision: a recently published international multicenter study showed that TH has no advantage compared to therapeutic normothermia, which targets 36°C and aims to simply prevent fever (82).

Effects of commonly used ICU medications on EEG

Sedation decreases cerebral oxygen consumption, reduces intracranial pressure and has potent anti-convulsive properties. It is also used along with neuromuscular blockade to prevent discomfort, agitation, and awareness, while it is an absolute requirement for the induction and maintenance of TH, as discussed in the preceding paragraph. Most anaesthetic medications produce similar EEG changes. At high doses, a burst-suppression pattern occurs, followed by electrocerebral inactivity.

Propofol is a short-acting anaesthetic frequently used in the intensive care unit for sedation and anaesthesia, as well as for termination of status epilepticus, and its mechanism of action involves activation of GABA-A receptors, inhibition of NMDA receptors and modulation of calcium influx through slow calcium ion channels (83). Hypotension and other adverse effects of high dose anaesthetics are not more severe or frequent than with midazolam or barbiturates (84), but high doses may induce the so-called 'propofol infusion syndrome', characterized by cardiac failure, rhabdomyolysis, severe metabolic acidosis and renal failure. To minimize the risk, the upper limit of 4mg/kg/h has been suggested in the sedation of critically ill patients for longer than 48 h, and slightly higher infusion rates have been safely used for shorter periods (85). Irrespective of this, propofol is considered the best initial choice as its major advantages include fast seizure control and fast recovery from anaesthesia, both due to its pharmacokinetic properties. The former is important for the optimal management of convulsive status epilepticus (Chapter 34), and the latter, due to its rapid metabolism, allows meaningful neurological assessment soon after the agent is stopped (86) and can be used in the diagnostic process of NCSE (see earlier section). Caution is needed in the interpretation of the so-called 'seizure-like phenomena' that can occur either during escalation or during withdrawal of the drug (87) and may not be epileptic (88).

Apart from their *analgesic action*, short-acting synthetic opiates, such as fentanyl or remifentanyl, are also used to suppress shivering and can also produce diffuse and often synchronized delta EEG activity. *Neuromuscular paralysis* is sometimes required in critically ill patients for facilitation of ventilation, control of intracranial pressure or muscle spasms, and is also used in patients treated with TH to abate shivering. Although it cannot influence cerebral rhythms, it is mentioned here because it can obliterate any motor phenomena, including signs that can alert physicians to possible seizure activity and EEG request. From the EEG standpoint, short neuromuscular blockade can be used to clear muscle artefacts when stopping propofol and allow meaningful EEG interpretation (Fig. 32.6).

Prediction of outcome

Since the early work of Hockaday et al. (89) there have been several attempts to grade the early post-anoxic EEG changes in order to predict outcome and improve inter-observer agreement (90,91). Higher grades reflect greater severity. The first grade is defined by regular alpha and some theta and is associated with good outcome, while the highest grades indicate poor outcome and include non-reactive alpha/theta coma, diffuse reduction of amplitude to ≤20 μV, or burst suppression, and finally electro-cerebral silence (<2 μV). Grades in-between the two ends of the spectrum correspond to progressive increment of the delta/theta ratio, decrement of the overall voltage and loss of reactivity, and the appearance of specific features, such as generalized periodic patterns, or intermittent periods of diffuse suppression, but are less accurate in predicting outcome. It is perhaps characteristic that a systematic meta-analysis of studies that explored predictors of death or unconsciousness after 1 month or unconsciousness or severe disability after 6 months between 1966 and 2006 concluded that even 'malignant' EEG patterns, such as 'generalized suppression to ≤20 μV, burst-suppression pattern with generalized epileptiform activity, or generalized periodic complexes on a flat background are strongly, *but not invariably* associated with poor outcome' (92).

The timing of the EEG recording post-insult and the level and type of sedation should also be taken into account, alone or in combination, when attempting prognostication. Comatose patients with 'malignant' patterns early after insult may recover, as opposed to those in whom such patterns persist after the second or third day (see discussion on alpha-theta coma below), and a burst-suppression pattern is associated with extremely poor prognosis in post cardiac arrest comatose patients who are not under sedation (93), although the same pattern is the desired target in the treatment of refractory status by depressants, such as barbiturates, propofol, and midazolam (84,94).

Amid the EEG patterns encountered in severe anoxic encephalopathies, *repetitive generalized epileptiform discharges occurring against a suppressed background* is considered particularly 'malignant', as it is almost invariably associated with a fatal outcome or persistent vegetative state (68,69) even in patients who were treated with therapeutic hypothermia (95). The term '*status myoclonus*' refers to the occurrence of epileptiform discharges (usually polyspikes at low and regular or irregular repetition rate) in association with time-locked axial myoclonus (Fig. 32.6). Recovery has been sporadically reported only when the pattern is present early and transiently (usually for 24 h) after resuscitation. There are only a handful of patients described with such a pattern persisting after the first day post-insult and subsequent meaningful recovery (96), but these might have been patients with treatable NCSE and only moderate hypoxic brain damage (72). Similar electroclinical features also occur in subtle status (70) (see discussion on NCSE above).

Caution is recommended when using the term '*generalized periodic pattern*' without further specifying the morphology and the frequency of the periodic activity, and whether the periods in-between the discharges are flat or contain other rhythms, to mention only a few variables. Even in the recent literature, the term appears to embrace, beyond the above described pattern, a continuum of any bilateral, synchronous and symmetric pattern of brief, usually sharp, complex discharge that repeats itself at quasi-regular intervals, including also triphasic waves as they appear in severe metabolic derangements (97). Although a 'generalized periodic pattern' following cardiac arrest has also been associated with poor outcome (69,92,97–99), lumping all types of discharge together, particularly in retrospective studies based on chart review, offers little insight into the complex relationship between various EEG patterns and outcome.

The EEG terms *alpha coma (AC), theta coma (TC)*, or a *combination of alpha-theta coma (ATC)* were initially coined to characterize persistent, unreactive, alpha, and/or theta frequencies as the principal electroencephalographic features in comatose patients, mostly following cardiac arrest and invariably with bad outcome (100). It has since become clear that a few AC patterns may show some reactivity (101) while some of the patients with AC recovered consciousness (102). Subsequently, Young et al. (103) showed that these patterns are transient and not reliably predictive of poor outcome, regardless of coma aetiology, before follow up EEGs are taken within 5 days to show a change to a more definitive pattern. Kaplan et al. (104) emphasized that in coma associated with alpha-theta activities, mortality depends on aetiology (being very high in patients with cardiorespiratory arrest and minimal in those with drug-induced coma), but also on reactivity independently of aetiology. The debate on the prognostic value of ATC in post-anoxic coma after cardiac arrest seems to have concluded by the study of Berkhoff et al. (105) (and their review of the literature); they distinguished two types of ATC, the *complete*, which is monotonous, frontally accentuated and non-reactive, and invariably associated with a poor outcome, and the *incomplete*, which is either not monotonous, partially reactive or posteriorly dominant and could be associated with full recovery. Patients with alpha coma need to be clinically differentiated from those with 'locked-in' syndrome, as in the latter EEG reactivity may not be always present (106) (see also section on chronic disorders of consciousness below).

On EEG reactivity

Comatose patients typically lie with eyes closed and cannot be roused to respond appropriately to vigorous stimulation, i.e. they may grimace or show stereotyped withdrawal responses, but they make no localizing or discrete defensive movements, nor do they volitionally respond to visual, auditory, or tactile stimuli. As coma deepens, clinical responsiveness may diminish or disappear, even to noxious, but electrographic cerebral responses may persist. Electrophysiological testing can then provide an objective measure of brain reactivity to external stimuli, which could be considered as a surrogate measure of minimal 'consciousness'. EEG and evoked potentials can complement (and extend) bedside clinical assessment, and also provide significant prognostic information as preserved normal responses (in the absence of clinical response) are positively correlated with favourable neurological outcome. A recent study showed that in coma following cardiac arrest, 10 out of 11 patients with EEG reactivity evolved positively while only one out of 18 patients with non-reactive EEG had good outcome (95).

Demonstrating reactivity is an important task for the physiologist and the electroencephalographer, but requires meticulous testing and careful interpretation. A normal response is a *reproducible* change to a pattern that reflects higher vigilance, i.e. towards lower voltage 'desynchronized' faster rhythms. There are a few caveats and factors of error here. It is important to recognize that, as with clinical responsiveness, any EEG responses to external auditory and tactile stimuli may be intermittent or vary with time because of alternating changes of the level of vigilance. Therefore, the recording has to be long enough to allow identification of possible different patterns that may reflect cyclical changes and stimulations should be carefully planned and executed in a timely fashion. Auditory stimuli (calling patients by their name for instance) may be more effective when given by a family member, such as a parent or spouse (see also cognitive potentials below). Similar responses may occur spontaneously, presumably in response to internal stimuli, such as pain, or external factors including bleeping monitors or medical and nursing interventions; these should be noted by both the physiologist at the time of the recording and the electroencephalographer at the time of video-EEG reviewing as they can confuse interpretation. All four limbs must be stimulated in turn, each one at least twice. Stimuli must be intense and prolonged (for example nail bed pressure across the proximal edge of the finger or toe nail sustained for a few seconds) to ensure temporal summation. Central stimulation (sternal rub or supraorbital pressure) and finally endotracheal suctioning must follow, particularly when auditory or peripheral tactile stimuli fail to show clear and reproducible responses. As hinted above, stimulation must be delivered when the patient appears quietly resting and during a period of constant EEG pattern that is likely to correspond to the lower level of vigilance. For example if the EEG shows alternating periods of slower and faster rhythms stimuli should be given during the slower epoch. Ideally, physiologists should work in pairs, or a nurse can stimulate the patient under the instructions of the physiologist. Corroborative evidence for arousal is worth seeking from polygraphy channels, such as the ECG (tachycardia) or the plethysmogram (amplitude reduction), obtained from a pulse oximetry sensor (Figs 32.8–32.10). There is an important pitfall in the rule of stimulation being given during a slow EEG epoch: the physiologist must be aware of the phenomenon of 'paradoxical slow wave arousal' that may occur in brainstem lesions, but also in metabolic or other encephalopathies, reflects 'higher vigilance state' and can last for several minutes. Stimulation during this phase (which might have started before the EEG recording begun and be erroneously mistaken for reflecting low vigilance) will yield no further electrographic change and the EEG will be falsely scored as non-reactive (Fig. 32.9).

The 'paradoxical slow wave arousal' does not appear to affect prognosis, and one has to remember that a 'paradoxical' response may be better than getting no response at all; after all, a type of paradoxical reactivity can be seen in normal (or non-encephalopathic) subjects in whom alpha activity reappears upon arousal with open eyes. An extreme example of abnormal paradoxical response

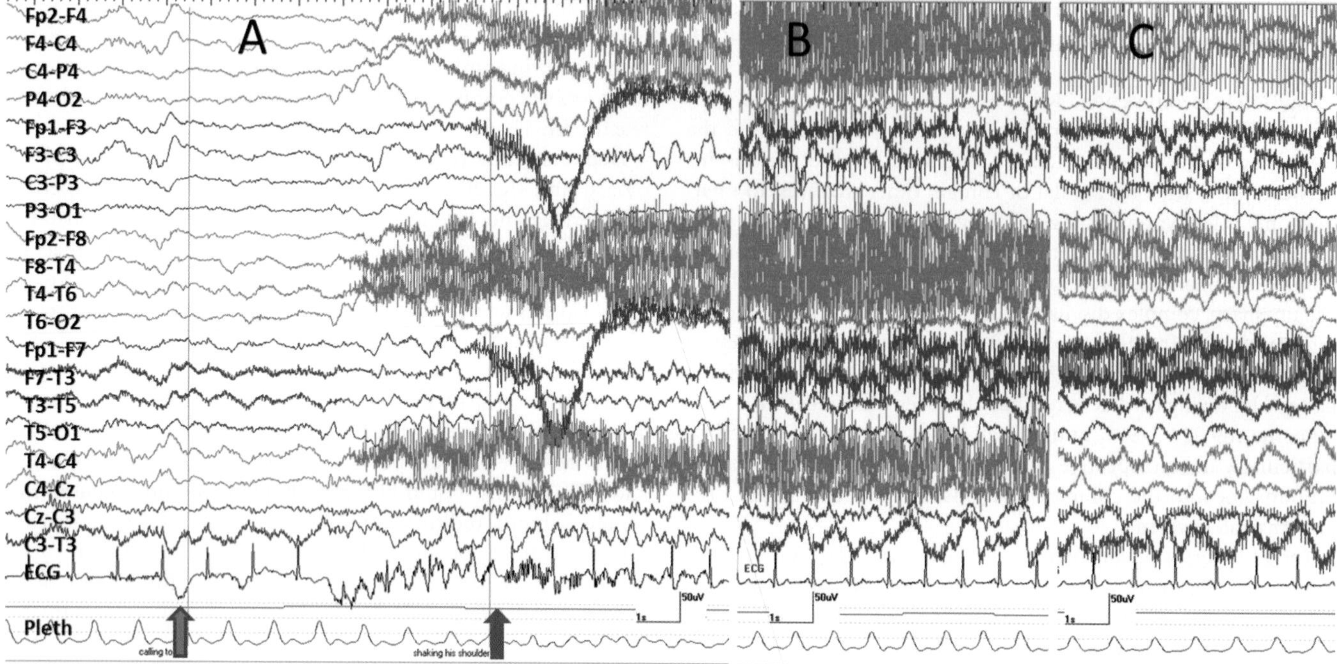

Fig. 32.9 A: Paradoxical 'slow wave' arousal follows a 3 second period of diffuse synchronization after the physiologist calls a 58-year-old man by his name (blue arrow). The patient stiffens and appears to shrug his shoulders (grey arrow). The pattern is ongoing 1 (B) and 4 min (C) after onset and in fact lasted for more than 15 min. Note the amplitude reduction of the plethysmogram that reflects increased sympathetic tone and therefore corroborates higher vigilance. Any further stimulation during this prolonged period would produce no further EEG change (see text).

of the comatose brain is the enigmatic phenomenon of SIRPIDs (stimulus induced rhythmic, periodic or ictal discharges), which consist of clearly paroxysmal patterns that occur in response to alerting stimuli, and in most cases also represent arousal responses (Fig. 32.10) (107).

Indeed, 'paradoxical delta arousal' may occasionally acquire sharp appearance suggesting that there might be a continuum between this arousal pattern and SIRPIDs (Fig. 32.11).

However, in some patients SIRPIDs may be accompanied by clinical ictal activity, such as focal motor (108) or bilateral epileptic

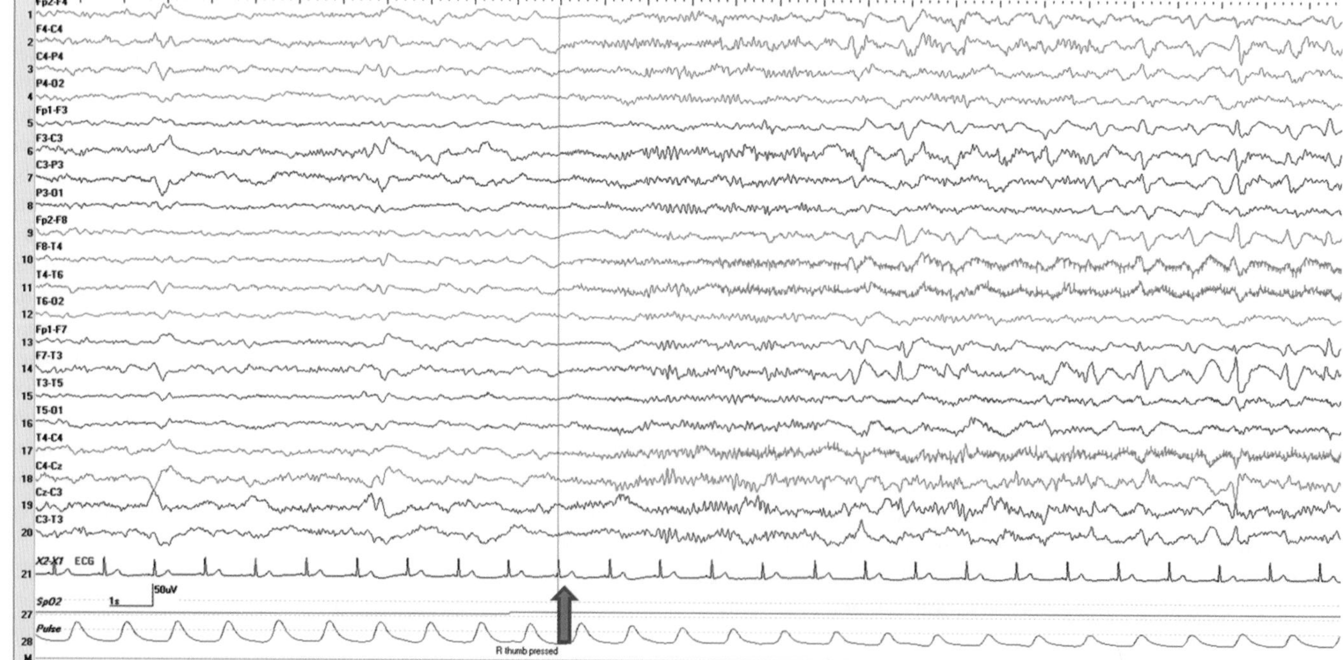

Fig. 32.10 Bilateral synchronous triphasic activity following right thumb press (blue arrow). Note the amplitude reduction of the plethysmogram in the bottom channel as in Fig. 32.9.

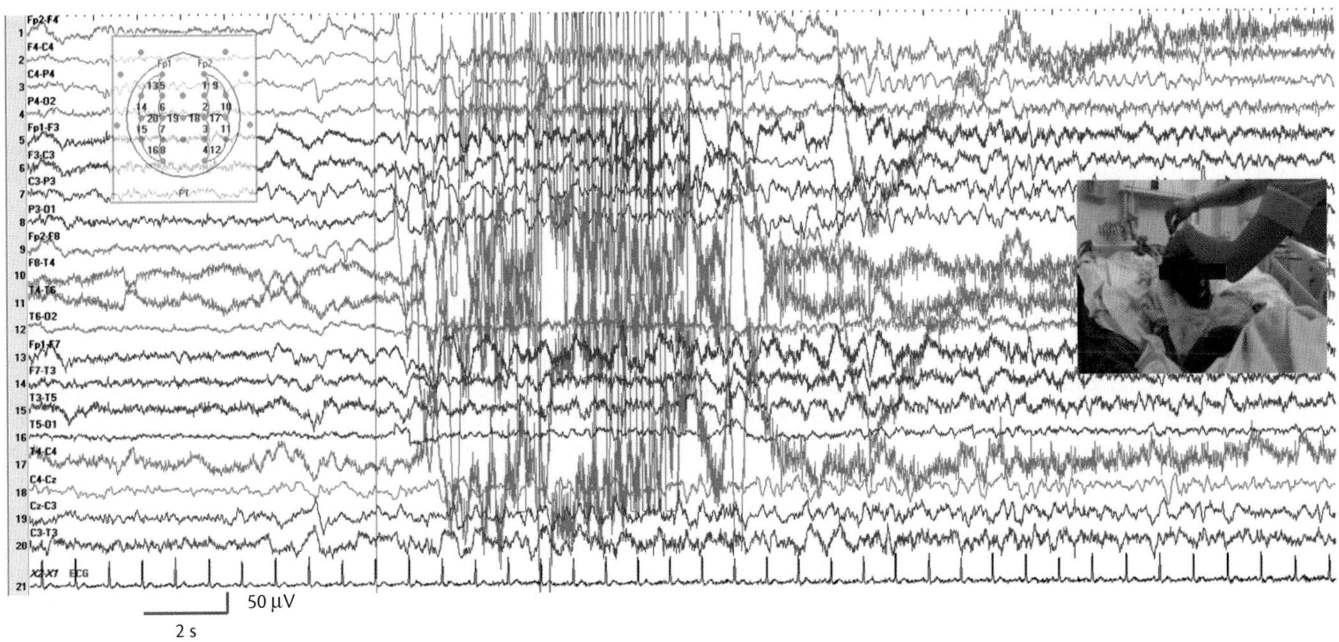

Fig. 32.11 'Spiky' paradoxical arousal caused by the physiologist who attempts to correct the malfunctioning T4 electrode (the green vertical line indicates the exact time she touched the head of the patient, shown in the time-locked video shot).

myoclonus (32) and therefore they *can* reflect arousal-induced seizures, warranting perhaps a trial of antiepileptic medication; even so, SIRPIDs have not been shown to directly affect outcome, which depends on the severity of the coma and the underlying cause rather than on their presence or absence.

Relevant to the detection of arousal responses following external stimuli and of similarly clinical and prognostic significance is the identification of rudimentary sleep features, such as midline potentials akin to K complexes or fast oscillations akin to sleep spindles, and the recognition of periodicities of concurrent EEG and autonomic variables that define what we now know as cyclical alternating pattern (Fig. 32.12) (109).

Long EEG recordings have shown that in post-traumatic coma, the presence of organized sleep patterns rather than the Glasgow

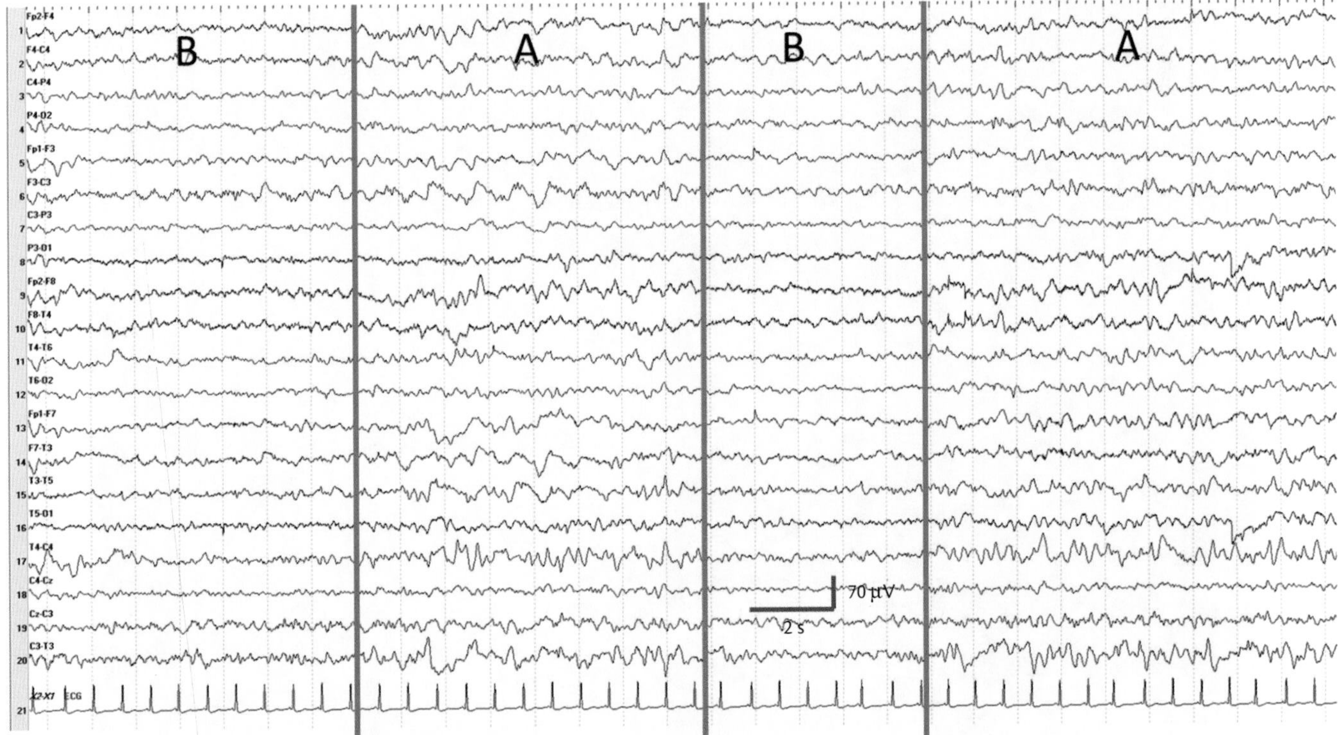

Fig. 32.12 Spontaneous ongoing cycling alternating pattern consisting of phases of reduced vigilance (CAP B) alternating with phases of increased vigilance (CAP A).

Coma Scale, is highly predictive of better outcome (110), while preservation of electrophysiological characteristics typical of normal sleep, such as an alternating non-rapid eye movement/rapid eye movement sleep pattern clearly differentiates minimally conscious patients from those in persistent vegetative state (111). Long overnight ICU recordings may not be possible for many EEG departments, but particular effort should still be made to determine variability, cyclic changes and sleep-related patterns in long enough daytime recordings.

Evoked potentials in outcome prediction

Being less influenced by sedative medication and metabolic derangements or other confounding conditions that commonly occur in comatose survivors of cardiac arrest, short-latency somatosensory evoked potentials (SSEPs) are an important tool in outcome prediction, alongside EEG and clinical indicators. Many studies have shown that the absence of the 'cortical' N20 potential to median nerve stimulation (in the presence of earlier potentials) implies severe damage at the thalamocortical level and invariably indicates poor neurological outcome (Fig. 32.13).

On the other hand, bilateral preservation of cortical responses does not reliably predict good outcome (92); about half of patients with present N20 die or remain in persistent vegetative state (PVS), while from the remainder, better chance for good neurological outcome without significant deficits have those with normal rather than delayed central conduction times (112). Longer latency SSEPs can help prediction of poor outcome in patients with present N20 when N70 potential is absent or prolonged (>130 ms), but, again, normal N70 cannot reliably predict a good outcome (113).

The predictive value of the middle latency auditory evoked potentials (MLAEPs) is similar: abolition of cortical MLAEPs precludes post-anoxic comatose patients from returning to consciousness with 100% specificity (114).

How early after cardiopulmonary resuscitation should neurophysiological studies be performed to assist prognostication? From the systematic meta-analysis of almost 400 articles from January 1966 to January 2006, the American Academy of Neurology Practice Parameter report of 2006 could not recommend a minimum time for the first EEG recording (in most studies EEGs were recorded within the first 3 days from insult, but with considerable

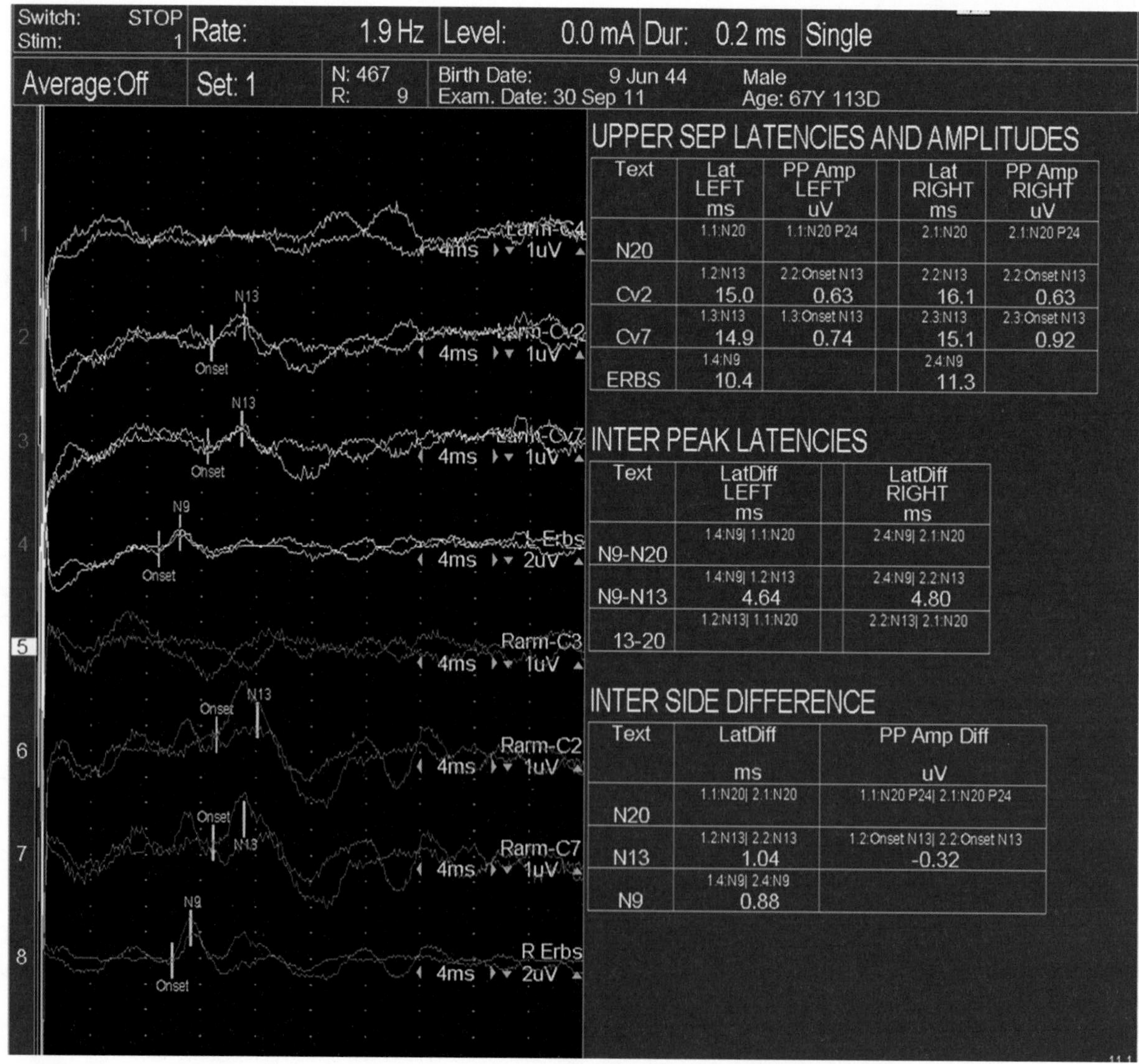

Fig. 32.13 Absent cortical response (N20) in a patient with out of hospital cardiac arrest. The test was performed on the second day post insult.

variation), but was able to conclude that the assessment of poor prognosis can be guided by the bilateral absence of cortical SSEPs (N20 response) within 1–3 days (recommendation level B) (92). A recent prospective cohort study of 60 patients treated with hypothermia and continuously monitored with EEG during the first 24 h after resuscitation showed that continuous patterns within 12 h predicted good outcome, while the sensitivity for predicting poor neurological outcome of low-voltage and isoelectric EEG patterns 24 h after resuscitation was 40% with a 100% specificity. Burst-suppression patterns after 24 h were also associated with poor neurological outcome, but not inevitably so. The sensitivity and specificity of absent somatosensory evoked potential responses during the first 24 h were 24 and 100%, respectively (115).

Acute video EEG and short and longer latency SSEPs and MLAEPs should be best performed soon after the patient is rewarmed from hypothermia, with the EEG repeated as required and the evoked potentials perhaps repeated only in case of unexplained clinical worsening. Bilateral preservation of cortical evoked responses cannot predict outcome, but may shape a positive therapeutic attitude, for instance, encourage anti-seizure treatment in those comatose patients in whom EEG features stand astride anoxia and NCSE.

It is possible that the future experience from the use of therapeutic hypothermia, or normothermia (82), may modify the prognostic significance of the various neurophysiological findings and therefore modify the rules and overall diagnostic strategies of recording EEG and evoked potentials in the ICU. A recent meta-analysis of 492 adult patients treated with TH after cardiopulmonary resuscitation showed that the reliability of the SSEP to predict poor outcome is comparable to that in patients not treated with hypothermia (116).

Cognitive evoked potentials and quantitative EEG in chronic disorders of consciousness

Most of the survivors following severe hypoxic ischaemic brain insult either remain in PVS (117) recently renamed as *state of unresponsive wakefulness* (UW) (118), or gradually improve to *minimally conscious state (MCS)*, in which they remain unable to communicate, but appear to retain minimal consciousness, deduced by observed behavioural responses (119). Further improvement is evidenced by gradually increased complexity of behavioural responses (MCS + state), while emergence from MCS is marked by recovery of communication usually in the presence of severe disability (for full discussion see reference 120).

Bedside clinical assessment of consciousness in the unresponsive patients following coma defines short-term management and guides prediction of functional outcome, but is particularly challenging (120) and prone to errors (121). The standard EEG can differentiate chronic disorders of consciousness from the 'locked-in' syndrome, in which cerebral rhythms are normal or minimally slow and reactivity is usually (but not invariably) preserved (122,106), and successfully predict outcome in the early neurorehabilitation stage (123,124), using the simple Synek scale (90), but cannot detect signs of consciousness. As mentioned earlier, sleep EEG studies can differentiate MCS from PVS/UW (111), while clinical behavioural examination can be assisted by more sophisticated neurophysiological tools that include quantitative EEG and long latency responses to cognitive evoked potentials. Using spectral analysis, the EEG can identify residual cognitive function in clinically unresponsive patients by demonstrating task-oriented responses to acoustic oddball paradigms (125) or to command to move a part of the body (126), as well as by showing deviant brain resting-state connectivity (127).

Long latency (cognitive) evoked responses depend on the psychological connotations of the stimuli and are thought to reflect more complex activation of subcortical areas and thalamus, primary cortex and associative areas. Passive paradigms (such as mismatched negativity) use oddball acoustic—are used to assess residual cognition while active paradigms may identify conscious responses.

Conclusions

Recording, reading, and reporting an ICU EEG (and evoked potentials) is a complex and demanding process that requires specific skills and can be flawed by a number of technical, clinical, and conceptual pitfalls. The main concern is false positive prediction of a poor outcome that may influence decision about continuation of treatment or withdrawal of life support. As in epileptology (Chapter 12), correct interpretation of the EEG findings requires that comprehensive information is available; in the comatose patient (for example, after out-of-hospital cardiac arrest) this should include time of the insult and estimated downtime, neurological examination on first assessment (prior to ICU admission) and upon EEG recording, time hypothermia was started/ended, type, and dosage of medication at the start of the recording (propofol, fentanyl, midazolam muscle paralysers) and discontinuation during recording (propofol) and possible confounding factors. Liaising with the responsible for the patient intensivist(s) should be actively pursued to plan further diagnostic strategies.

References

1. Sethi, P.K., Sethi, N.K., and Torgovnick, J. (2006). Mobile phone artifact. Clinical *Neurophysiology*, **117**(8), 1876–8.
2. Sundaram, M.B. and Blume, W.T. (1987). Triphasic waves: clinical correlates and morphology. *Canadian Journal of Neurology Science*, **14**(2), 136–40.
3. Smith, S.J.M. (2005). EEG in neurological conditions other than epilepsy: when does it help, what does it add? *Journal of Neurology, Neurosurgery, & Psychiatry*, **76**(II), ii8–12.
4. Blatt, I. and Brenner, R.P. (1996). Triphasic waves in a psychiatric population: a retrospective study. *Journal of Clinical Neurophysiology*. **13**(4), 324–9.
5. Levy, L.L. and Fenichel, G.M. (1965). Diphenylhydantoin activated seizures. *Neurology*, **15**, 716–22.
6. Hamer, H.M., Knake, S., Schomburg, U., and Osenow, F. (2000). Valproate-induced hyperammonemic encephalopathy in the presence of topiramate. *Neurology*, **54**(1), 230–2.
7. Segura-Bruna, N., Rodriguez-Campello, A., Puente, V., and Roquer, J. (2006). Valproate-induced hyperammonemic encephalopathy. *Acta Neurologica Scandinavica*, **114**(1), 1–7.
8. Pascual-Leone, A., Dhuna, A., Altafullah, I., and Anderson, D.C. (1990). Cocaine-induced seizures. *Neurology*, **40**(3 Pt 1), 404–7.
9. Adamaszek, M., Khaw, A.V., Buck, U., Andresen, B., and Thomasius, R. (2010). Evidence of neurotoxicity of ecstasy: sustained effects on electroencephalographic activity in polydrug users. *PLoS One*, **5**(11), e14097.
10. Hughes, J.R. and Schreeder, M.T. (1980). EEG in dialysis encephalopathy. *Neurology*, **30**(11), 1148–54.
11. Faigle, R., Sutter, R., and Kaplan, P.W. (2013). Electroencephalography of encephalopathy in patients with endocrine and metabolic disorders. *Journal of Clinical Neurophysiology*, **30**, 505–16.

12. Kirkpatrick, M.P., Clarke, C.D., Sonmezturk, H.H., and Abou-Khalil, B. (2011). Rhythmic delta activity represents a form of nonconvulsive status epilepticus in anti-NMDA receptor antibody encephalitis. *Epilepsy Behaviour*, **20**, 392–4.
13. Bataller, L., Kleopa, K.A., Wu, G.F., et al. (2007). Autoimmune limbic encephalitis in 39 patients: immunophenotypes and outcomes. *Journal of Neurology, Neurosurgery, & Psychiatry*, **78**, 381–5.
14. Irani, S.R., Michell, A.W., Lang, B., et al. (2011). Faciobrachial dystonic seizures precede Lgi1 antibody limbic encephalitis. *Annals of Neurology*, **69**, 892–900.
15. Kaplan, P.W., Rossetti, A.O., Kaplan, E.H., and Wieser, H.G. (2012). Proposition: limbic encephalitis may represent limbic status epilepticus. A review of clinical and EEG characteristics. *Epilepsy Behaviour*, **24**, 1–6.
16. van Vliet, J., Mulleners, W., and Meulstee, J. (2012). EEG leading to the diagnosis of limbic encephalitis. *Clinical EEG Neuroscience*, **43**, 161–4.
17. Iizuka, T., Sakai, F., Ide, T., et al. (2008). Anti-NMDA receptor encephalitis in Japan: longterm outcome without tumor removal. *Neurology*, **70**, 504–11.
18. Schmitt, S.E., Pargeon, K., Frechette, E.S., et al. (2012). Extreme delta brush: a unique EEG pattern in adults with anti-NMDA receptor encephalitis. *Neurology*, **79**, 1094–100.
19. Vas, G.A. and Cracco, J.B. (1990). Diffuse encephalopathies. In: D. D. Daly and T. A. Pedley (Eds) *Current practice of clinical electroencephalography*, 2nd edn, pp. 371–99. New York, NY: Raven Press Ltd.
20. Kellinghaus, C., Engbring, C., Kovac, S., et al. (2008). Frequency of seizures and epilepsy in neurological HIV-infected patients. *Seizure*, **17**(1), 27–33.
21. Nuwer, M.R., Miller, E.N., Visscher, B.R., et al. (1992). Asymptomatic HIV infection does not cause EEG abnormalities: results from the Multicenter AIDS Cohort Study (MACS). *Neurology*, **42**(6), 1214–19.
22. Gofton, T.E. and Young, G.B. (2012). Sepsis-associated encephalopathy. *National Reviews of Neurology*, **8**, 557–66.
23. Young, G.B., Bolton, C.F., Archibald, Y.M., et al. (1990). The encephalopathy associated with sepsis. *Clinical Investigations in Medicine*, **13**, 297–304.
24. Young, G.B. (2013). Encephalopathy of infection and systemic inflammation. *Journal of Clinical Neurophysiology*, **30**, 454–61.
25. Kaplan, P.W. (2004). The EEG in metabolic encephalopathy and coma. *Journal of Clinical Neurophysiology*, **21**, 307–18.
26. Collins, S.J., Sanchez-Juan, P., Masters, C.L., et al. (2006). Determinants of diagnostic investigation sensitivities across the clinical spectrum of sporadic Creutzfeldt–Jakob disease. *Brain*, **129**, 2278–87.
27. Zerr, I., Kallenberg, K., Summers, D.M., et al. (2009). Updated clinical diagnostic criteria for sporadic Creutzfeldt–Jakob disease. *Brain*, **132**, 2659–68.
28. Walker, M., Cross, H., Smith, S., et al. (2005). Nonconvulsive status epilepticus: Epilepsy Research Foundation workshop reports. *Epileptic Disorder*, **7**(3), 253–96.
29. Andermann, F. and Robb, J.P. (1972). Absence status. A reappraisal following review of thirty-eight patients. *Epilepsia*, **13**, 177–87.
30. Agathonikou, A., Panayiotopoulos, C.P., Giannakodimos, S., and Koutroumanidis, M. (1998). Typical absence status in adults: diagnostic and syndromic considerations. *Epilepsia*, **39**, 1265–76.
31. Thomas, P., Beaumanoir, A., Genton, P., Dolisi, C., and Chatel, M. (1992). 'De novo' absence status of late onset: report of 11 cases. *Neurology*, **42**, 104–10.
32. Koutroumanidis, M. (2009). Absence status epilepticus In: P. W. Kaplan and F. W. Drislane (Eds) *Nonconvulsive status epilepticus*, pp. 153–73. New York, NY: Demos Medical Publishing.
33. Thomas, P., Zifkin, B., Migneco, O., Lebrun, C., Darcourt, J., and Andermann, F. (1999). Nonconvulsive status epilepticus of frontal origin. *Neurology*, **52**, 1174–83.
34. Drislane, F.W. and Kaplan, P.W. (2009). Nonconvulsive seizures and status epilepticus. in: B. J. Fisch (Ed.) Epilepsy and intensive care monitoring, pp. 287–307. New York, NY: Demos Medical Publishing.
35. Fernández-Torre, J.L., Rebollo, M., Gutiérrez, A., López-Espadas, F., and Hernández-Hernández, M.A. (2012). Nonconvulsive status epilepticus in adults: electroclinical differences between proper and comatose forms. *Clinical Neurophysiology*, **123**(2), 244–51.
36. Jette, N., Claassen, J., Emerson, R.G., and Hirsch, L.J. (2006). Frequency and predictors of nonconvulsive seizures during continuous electroencephalographic monitoring in critically ill children. *Archives in Neurology*, **63**, 1750–5.
37. Claassen, J., Mayer, S.A., Kowalski, R.G., Emerson, R.G., and Hirsch, L.J. (2004). Detection of electrographic seizures with continuous EEG monitoring in critically ill patients. *Neurology*, **62**, 1743–8.
38. Rittenberger, J.C., Popescu, A., Brenner, R.P., Guyette, F.X., and Callaway, C.W. (2012). Frequency and timing of nonconvulsive status epilepticus in comatose post-cardiac arrest subjects treated with hypothermia. *Neurocritical Care*, **16**, 114–22.
39. Nielsen, N., Sunde, K., Hovdenes, J., et al. (2011). Hypothermia network. Adverse events and their relation to mortality in out-of-hospital cardiac arrest patients treated with therapeutic hypothermia. *Critical Care Medicine*, **39**, 57–64.
40. Rossetti, A.O., Urbano, L.A., Delodder, F., Kaplan, P.W., and Oddo, M. (2010). Prognostic value of continuous EEG monitoring during therapeutic hypothermia after cardiac arrest. *Critical Care*, **14**, R173.
41. Abend, N.S., Topjian, A., Ichord, R., et al. (2009). Electroencephalographic monitoring during hypothermia after pediatric cardiac arrest. *Neurology*, **72**, 1931–40.
42. Giroud, M., Gras, P., Fayolle, H., et al. (1994). Early seizures after acute stroke: A study of 1,640 cases. *Epilepsia*, **35**, 959–64.
43. Vernino, S., Brown, R.D., Jr, Sejvar, J.J., Sicks, J.D., Petty, G.W., and O'Fallon, W.M. (2003). Cause-specific mortality after first cerebral infarction: a population-based study. *Stroke*, **34**, 1828–32.
44. Graham, N.S., Crichton, S., Koutroumanidis, M., Wolfe, C.D., and Rudd, A.G. (2013). Incidence and associations of poststroke epilepsy: the prospective South London Stroke Register. *Stroke*, **44**(3), 605–11.
45. Afsar, N., Kaya, D., Aktan, S., et al. (2003). Stroke and status epilepticus: stroke type, type of status epilepticus, and prognosis. *Seizure*, **12**, 23–27.
46. Sutter, R., Stevens, R.D., and Kaplan, P.W. (2013). Continuous electroencephalographic monitoring in critically ill patients: indications, limitations, and strategies. *Critical Care Medicine*, **41**, 1124–32.
47. Claassen, J., Hirsch, L.J., Frontera, J.A., et al. (2006). Prognostic significance of continuous EEG monitoring in patients with poor-grade subarachnoid hemorrhage. *Neurocritical Care*, **4**, 103–12.
48. Butzkueven, H., Evans, A.H., Pitman, A., et al. (2000). Onset seizures independently predict poor outcome after subarachnoid hemorrhage. *Neurology*, **55**, 1315–20.
49. Dennis, L.J., Claassen, J., Hirsch, L.J., Emerson, R.G., Connolly, E.S., and Mayer, S.A. (2002). Nonconvulsive status epilepticus after subarachnoid hemorrhage. *Neurosurgery*, **51**, 1136–43, discussion 1144.
50. Rosengart, A.J., Huo, J.D., Tolentino, J., et al. (2007). Outcome in patients with subarachnoid hemorrhage treated with antiepileptic drugs. *Journal of Neurosurgery*, **107**, 253–60.
51. Naidech, A.M., Garg, R.K., Liebling, S., et al. (2009). Anticonvulsant use and outcomes after intracerebral hemorrhage. *Stroke*, **40**, 3810–5.
52. Faught, E., Peters, D., Bartolucci, A., et al. (1989). Seizures after primary intracerebral hemorrhage. *Neurology*, **39**, 1089–93.
53. Kilpatrick, C.J., Davis, S.M., Tress, B.M., et al. (1990). Epileptic seizures in acute stroke. *Archives of Neurology*, **47**, 157–60.
54. Vespa, P.M., O'Phelan, K., Shah, M., et al. (2003). Acute seizures after intracerebral hemorrhage: a factor in progressive midline shift and outcome. *Neurology*, **60**, 1441–6.
55. Claassen, J., Jette, N., Chum, F., et al. (2007). Electrographic seizures and periodic discharges after intracerebral hemorrhage. *Neurology*, **69**, 1356–65.

56. Friedman, D., Claassen, J., and Hirsch, L.J. (2009). Continuous electroencephalogram monitoring in the intensive care unit. *Anesthetic & Analgesia*, **109**(2), 506–23.
57. Turgeon, A.F., Lauzier, F., Simard, J.F., et al. (2011). Mortality associated with withdrawal of life-sustaining therapy in patients with severe traumatic brain injury: a Canadian multi-centre cohort study. *Canadian Medical Association Journal*, **183**, 1581–8.
58. Yablon, S.A. (1993). Posttraumatic seizures. *Archives of Physical Medical Rehabilitation*, **74**, 983–1001.
59. Brain Trauma Foundation. (2007). American Association of Neurological Surgeons, Congress of Neurological Surgeons, et al: Guidelines for the management of severe traumatic brain injury. XIII. Antiseizure prophylaxis. *Journal of Neurotrauma*, **24**(1), S83–6.
60. Temkin, N.R., Anderson, G.D., Winn, H.R., et al. (2007). Magnesium sulfate for neuroprotection after traumatic brain injury: a randomised controlled trial. *Lancet: Neurology*, **6**, 29–38.
61. Vespa, P.M., Miller, C., McArthur, D., et al. (2007). Nonconvulsive electrographic seizures after traumatic brain injury result in a delayed, prolonged increase in intracranial pressure and metabolic crisis. *Critical Care Medicine*, **35**, 2830–6.
62. Vespa, P.M., McArthur, D.L., Xu, Y., et al. (2010). Nonconvulsive seizures after traumatic brain injury are associated with hippocampal atrophy. *Neurology*, **75**, 792–8.
63. Vespa, P.M., Nuwer, M.R., Nenov, V., et al. (1999). Increased incidence and impact of nonconvulsive and convulsive seizures after traumatic brain injury as detected by continuous electroencephalographic monitoring. *Journal of Neurosurgery*, **91**, 750–60.
64. Arndt, D.H., Lerner, J.T., Matsumoto, J.H., et al. (2013). Subclinical early posttraumatic seizures detected by continuous EEG monitoring in a consecutive pediatric cohort. *Epilepsia*, **54**(10), 1780–8.
65. Brenner, R.P. (2004). EEG in convulsive and nonconvulsive status epilepticus. *Journal of Clinical Neurophysiology*, **21**(5), 319–31.
66. Ronner, H.E., Ponten, S.C., Stam, C.J., and Uitdehaag, B.M. (2009). Inter-observer variability of the EEG diagnosis of seizures in comatose patients. *Seizure*, **18**(4), 257–63.
67. Hirsch, L.J., LaRoche, S.M., Gaspard, N., et al. (2013). American Clinical Neurophysiology Society's Standardized Critical Care EEG Terminology: *Journal of Clinical Neurophysiology*, **30**, 1–27.
68. Zandbergen, E.G.J., de Haan, R.J., Stoutenbeek, C.P., Koelman, J.H., and Hijdra. A. (1998). Systematic review of early prediction of poor outcome in anoxic ischaemic coma. *Lancet*, **352**, 1808–12.
69. Koenig, M.A., Kaplan, P.W., and Thakor, N.V. (2006). Clinical neurophysiologic monitoring and brain injury from cardiac arrest. *Neurology Clinics*, **24**, 89–106.
70. Treiman, D.M., Walton, N.Y., and Kendrick, C. (1990). A progressive sequence of electroencephalographic changes during generalized convulsive status epilepticus. *Epilepsy Results*, **5**, 49–60.
71. Sutter, R. and Kaplan, P.W. (2013). The neurophysiologic types of nonconvulsive status epilepticus: EEG patterns of different phenotypes. *Epilepsia*, **54**(6), 23–7.
72. Sreedharan, J., Gourlay, E., Evans, M.R., and Koutroumanidis, M. (2012). Falsely pessimistic prognosis by EEG in post-anoxic coma after cardiac arrest: the borderland of nonconvulsive status epilepticus. *Epileptic Disorder*, **14**(3), 340–4.
73. Fountain, N.B. and Waldman, W. (2001) Effects of benzodiazepines on triphasic waves: implications for nonconvulsive status epilepticus. *Journal of Clinical Neurophysiology*, **18**, 345–53.
74. Treiman, D.M., Meyers, P.D., Walton, N.Y., et al. (1998) A comparison of four treatments for generalized convulsive status epilepticus. *New England Journal of Medicine*, **339**, 792–8.
75. Rittenberger, J.C., Polderman, K.H., Smith, W.S., and Weingart, S.D. (2012). Emergency neurological life support: resuscitation following cardiac arrest. *Neurocritical Care*, **17**(1), S21–8.
76. Jirsch, J. and Hirsch, L.J. (2008). Nonconvulsive status epilepticus in critically ill and comatose intensive care unit patients. In: P. W. Kaplan and F. W. Drislane (Eds) *Nonconvulsive status epilepticus*, pp. 175–86. New York, NY: Demos Medical Publishing.
77. Howard, R.S., Holmes, P.A., and Koutroumanidis, M.A. (2011). Hypoxic-ischaemic brain injury. *Practical Neurology*, **11**, 4–18.
78. Howard, R.S., Holmes, P.A., Siddiqui, A., Treacher, D., Tsiropoulos, I., and Koutroumanidis, M. (2012). Hypoxic-ischaemic brain injury: imaging and neurophysiology abnormalities related to outcome. *Quarterly Journal of Medicine*, **105**(6), 551–61.
79. Bernard, S.A., Gray, T.W., Buist, M.D., et al. (2002). Treatment of comatose survivors of out-of hospital cardiac arrest with induced hypothermia. *New England Journal of Medicine*, **346**, 557–63.
80. The Hypothermia after Cardiac Arrest Study Group. (2002). Mild therapeutic hypothermia to improve the neurologic outcome after cardiac arrest. *New England Journal of Medicine*, **346**, 549–56. [Erratum, *New England Journal of Medicine*, 2002, 346, 1756.]
81. Rossetti, A.O., Oddo, M., Liaudet, L., and Kaplan, P.W. (2009). Predictors of awakening from postanoxic status epilepticus after therapeutic hypothermia. *Neurology*, **72**(8), 744–9.
82. Nielsen, N., Wetterslev, J., Cronberg, T., et al. (2013). TTM Trial Investigators. Targeted temperature management at 33°C versus 36°C after cardiac arrest. *New England Journal of Medicine*, **369**(23), 2197–206.
83. Kotani, Y., Shimazawa, M., Yoshimura, S., Iwama, T., and Hara, H. (2008). The experimental and clinical pharmacology of propofol, an anesthetic agent with neuroprotective properties. *CNS Neuroscience Therapy*, **14**, 95–106.
84. Claassen, J., Hirsch, L.J., Emerson, R.G., and Mayer, S.A. (2002). Treatment of refractory status epilepticus with pentobarbital, propofol, or midazolam: a systematic review. *Epilepsia*, **43**(2), 146–53.
85. Parviainen, I., Uusaro, A., Kälviäinen, R., Mervaala, E., and Ruokonen, E. (2006). Propofol in the treatment of refractory status epilepticus. *Intensive Care Medicine*, **32**, 1075–9.
86. Marik, P.E. (2004). Propofol: therapeutic indications and side-effects. *Current Pharmaceutical Design*, **10**(29), 3639–49.
87. Walder, B., Tramer, M.R., and Seeck, M. (2002). Seizure-like phenomena and propofol: a systematic review. *Neurology*, **58**, 1327–32.
88. Zubair, S., Patton, T., Smithson, K., Sonmezturk, H.H., Arain, A., and Abou-Khalil, B. (2011). Propofol withdrawal seizures: non-epileptic nature of seizures in a patient with recently controlled status epilepticus. *Epileptic Disorder*, **13**(1), 107–10.
89. Hockaday, J.M., Potts, F., Epstein, E., Bonazzi, A., and Schwab, R.S. (1965). Electroencephalographic changes in acute cerebral anoxia from cardiac or respiratory arrest. *Electroencephalography & Clinical Neurophysiology*, **18**, 575–86.
90. Synek, V.M. (1988). Prognostically important EEG coma patterns in diffuse anoxic and traumatic encephalopathies in adults. *Journal of Clinical Neurophysiology*, **5**, 161–74.
91. Young, G.B., McLachlan, R.S., Kreeft, J.H., and Demelo, J.D. (1997). An electroencephalographic classification for coma. *Canadian Journal of Neurological Science*, **24**, 320–5.
92. Wijdicks, E.F.M., Hijdra, A., Young, G.B., Bassetti, C.L., and Wiebe, S. (2006). Practice Parameter: Prediction of outcome in comatose survivors after cardiopulmonary resuscitation (an evidence-based review): report of the Quality Standards Subcommittee of the American Academy of Neurology. *Neurology*, **67**, 203–10.
93. Kuroiwa, Y. and Celesia, G. (1980). Clinical significance of periodic EEG patterns. *Archives on Neurology*, **37**(1), 15–20.
94. Herkes, G.K., Wszolek, Z.K., Westmoreland, B.F. and Klass, D.W. (1992). Effects of midazolam on electroencephalograms of seriously ill patients. *Mayo Clinic Proceedings*, **67**(4), 334–8.
95. Thenayan, E.A.L., Savard, M., Sharpe, M.D., Norton, L., and Young, B. (2010). Electroencephalogram for prognosis after cardiac arrest. *Journal of Critical Care*, **25**, 300–4.
96. Chen, R., Bolton, C.F., and Young, G.B. (1996). Prediction of outcome in patients with anoxic coma: s clinical and electrophysiologic study. *Critical Care Medicine*, **24**, 672–8.
97. Foreman, B., Claassen, J., Abou Khaled, K., et al. (2012). Generalized periodic discharges in the critically ill. A case-control study of 200 patients. *Neurology*, **79**, 1951–60.

98. Young, G.B., Doig, G., and Ragazzoni, A. (2005). Anoxic-ischemic encephalopathy: clinical and electrophysiological associations with outcome. *Neurocritical Care*, **2**(2), 159–64.
99. Koenig, M.A. and Kaplan, P.W. (2013). Clinical neurophysiology in acute coma and disorders of consciousness. *Seminar in Neurology*, **33**, 121–32.
100. Westmoreland, B.F., Klass, D.W., Sharbrough, F.W., and Reagan, T.J. (1975). Alpha-coma. Electroencephalographic, clinical, pathologic and etiologic correlations. *Archives in Neurology*, **32**, 713–18.
101. Tomassen, W. and Kamphuisen, H.A.C. (1986). Alpha coma. *Journal of Neurological Science*, **76**, 1–11.
102. Austin, E.J., Wilkus, R.J., and Longstreth, W.T. Jr, (1988). Etiology and prognosis of alpha coma. *Neurology*, **38**, 773–7.
103. Young, G.B., Blume, W.T., Campbell, V.M., et al. (1994). Alpha, theta and alpha-theta coma: a clinical outcome study utilizing serial recordings *Electroencephalography & Clinical Neurophysiology*, **91**, 93–9.
104. Kaplan, P.W., Genoud, D., Ho, T.W., and Jallon, P. (1999). Etiology, neurologic correlations, and prognosis in alpha coma. *Clinical Neurophysiology*, **110**(2), 205–13.
105. Berkhoff, M., Donati, F., and Bassetti, C. (2000). Postanoxic alpha (theta) coma: a reappraisal of its prognostic significance. *Clinical Neurophysiology*, **111**, 297–304.
106. Gütling, E., Isenmann, S., and Wichmann, W. (1996). Electrophysiology in the locked-in-syndrome. *Neurology*, **46**(4), 1092–101.
107. Hirsch, L.J., Claassen, J., Mayer, S.A., and Emerson, R.G. (2004). Stimulus-induced rhythmic, periodic, or ictal discharges (SIRPIDs): a common EEG phenomenon in the critically ill. *Epilepsia*, **45**, 109–23.
108. Hirsch, L.J., Pang, T., Claassen, J., et al. (2008). Focal motor seizures induced by alerting stimuli in critically ill patients. *Epilepsia*, **49**, 968–73.
109. Evans, B.M. (1992). Periodic activity in cerebral arousal mechanisms—the relationship to sleep and brain damage. *Electroencephalography & Clinical Neurophysiology*, **83**(2), 130–7.
110. Valente, M., Placidi, F., Oliveira, A.J., et al. (2002). Sleep organization pattern as a prognostic marker at the subacute stage of post-traumatic coma. *Clinical Neurophysiology*, **113**, 1798–805.
111. Landsness, E., Bruno, M.A., Noirhomme, Q., et al. (2011). Electrophysiological correlates of behavioural changes in vigilance in vegetative state and minimally conscious state. *Brain*, **134**, 2222–32.
112. Rothstein, T.L. (2000). The role of evoked potentials in anoxic-ischemic coma and severe brain trauma. *Journal of Clinical Neurophysiology*, **17**(5), 486–97.
113. Zandbergen, E.G., Koelman, J.H., de Haan, R.J., and Ijdra, A. (2006). PROPAC-Study Group. SSEPs and prognosis in postanoxic coma: only short or also long latency responses? *Neurology*, **67**(4), 583–6.
114. Logi, F., Fischer, C., Murri, L., and Mauguière, F. (2003). The prognostic value of evoked responses from primary somatosensory and auditory cortex in comatose patients. *Clinical Neurophysiology*, **114**(9), 1615–27.
115. Cloostermans, M.C., van Meulen, F.B., Eertman, C.J., Hom, H.W., and van Putten, M.J. (2012). Continuous electroencephalography monitoring for early prediction of neurological outcome in postanoxic patients after cardiac arrest: a prospective cohort study. *Critical Care Medicine*, **40**(10), 2867–75.
116. Kamps, M.J., Horn, J., Oddo, M., et al. (2013). Prognostication of neurologic outcome in cardiac arrest patients after mild therapeutic hypothermia: a meta-analysis of the current literature. *Intensive Care Medicine*, **39**(10), 1671–82.
117. The Multi-Society Task Force on PVS. (1994) Medical aspects of the persistent vegetative state (1). *New England Journal of Medicine*, **330**, 1499–508.
118. Bruno, M.A., Vanhaudenhuyse, A., Thibaut, A., Moonen, G., and Laureys, S. (2011). From unresponsive wakefulness to minimally conscious PLUS and functional locked-in syndromes: recent advances in our understanding of disorders of consciousness. *Journal of Neurology*, **258**(7), 1373–84.
119. Giacino, J.T., Ashwal, S., Childs, N., et al. (2002). The minimally conscious state: definition and diagnostic criteria. *Neurology*, **58**, 349–53.
120. Andrews, K., Murphy, L., Munday, R., and Littlewood, C. (1996). Misdiagnosis of the vegetative state: Retrospective study in a rehabilitation unit. *British Medical Journal*, **313** (7048), 13–16.
121. Childs, N.L., Mercer, W.N., and Childs, H.W. (1993). Accuracy of diagnosis of persistent vegetative state. *Neurology*, **43**(8), 1465–7.
122. Markand, O.N. (1976). Electroencephalogram in 'locked-in' syndrome. *Electroencephalography & Clinical Neurophysiology*, **40**(5), 529–34.
123. Bagnato, S., Boccagni, C., Prestandrea, C., Sant'Angelo, A., Castiglione, A., and Galardi, G. (2010). Prognostic value of standard EEG in traumatic and nontraumatic disorders of consciousness following coma. *Clinical Neurophysiology*, **121**(3), 274–80.
124. Boccagni, C., Bagnato, S., Sant Angelo, A., Prestandrea, C., and Galardi, G. (2011). Usefulness of standard EEG in predicting the outcome of patients with disorders of consciousness after anoxic coma. *Journal of Clinical Neurophysiology*, **28**(5), 489–92.
125. Fellinger, R., Klimesch, W., Schnakers, C., et al. (2011). Cognitive processes in disorders of consciousness as revealed by EEG time-frequency analyses. *Clinical Neurophysiology*, **122**(11), 2177–84.
126. Cruse, D., Chennu, S., Chatelle, C., et al. (2011). Bedside detection of awareness in the vegetative state: a cohort study. *Lancet*, **378**(9809), 2088–94.
127. Lehembre, R., Marie-Aurélie, B., Vanhaudenhuyse, A., et al. (2012). Restingstate EEG study of comatose patients: a connectivity and frequency analysis to find differences between vegetative and minimally conscious states. *Functions of Neurology*, **27**(1), 41–7.
128. Shibasaki, H. and Hallett, M. (2005). Electrophysiological studies of myoclonus. *Muscle & Nerve*, **31**(2), 157–74.

CHAPTER 33

Migraine, stroke, and cerebral ischaemia

Gonzalo Alarcón, Marian Lazaro, and Antonio Valentín

Migraine and headache

Episodes of disabling headache affect approximately 10% of the population. In a minority, headaches are due to organic structural pathology. Before the advent of modern neuroimaging, the EEG was often requested as a non-invasive method to identify intracranial pathology in these patients. In subjects with migraine or tension headache, the EEG between attacks is usually normal, or may contain minor, non-specific abnormalities in adults (1) and children (2), or may show those findings often seen in anxious subjects such as low amplitude EEG, excessive muscle artefacts, blinking, or beta activity. Other minor abnormalities of unclear significance have also been described, such as excessive slowing on hyperventilation and unusually sharp photic following responses, which is probably due to the excess of beta activity (3). During migranous attacks, sharp waves and sharpened theta activity have been described over temporal regions (4). More severe abnormalities, such as unilateral mild slowing or reduction in alpha activity, have been described during attacks of hemiplegic migraine associated with hemiparesis or hemianopia. EEG changes during migranous attacks may be due to cerebrovascular disturbances and associated cortical spreading depression (5,6).

A different problem arises in patients with epilepsy whose seizure semiology includes headache. Post-ictal headache is exceedingly common, especially after convulsive seizures, but its diagnosis is usually straight forward, as they are immediately preceded by a severe seizure. Pre-ictal or ictal headaches may be more difficult to diagnose. A variety of cephalic sensations are described by patients with seizures during the auras. Most are characterized as strange or 'indescribable', rather than painful. An exception is the syndrome of Childhood Epilepsy with Occipital Paroxysms, where seizures may occur associated with headache, visual symptoms, and vomiting, resembling migraine. The interictal EEG will show normal background activity and focal occipital, in addition to extraoccipital, epileptiform discharges, and fixation-off phenomenon (activation of discharges on eye closure or darkness, Fig. 33.1). A family history of migraine is common in this syndrome.

Cerebrovascular disease (stroke)

Cerebrovascular disease is defined as a pathological process affecting the blood vessels supplying or removing blood from parts of the brain. Essentially, cerebrovascular disease is due to occlusion of or haemorrhage from one of the vessels supplying the brain, or their branches. This process may lead to damage of the brain substance and it is this process that provides the substrate for EEG abnormalities. Cerebrovascular disease differs from situations where there is a sudden and massive drop in oxygen supply to the complete brain, such as in cardiac arrest, drowning, asphyxiation, or hanging, which will be described in the next section (hypoxia).

Within western countries, the incidence of stroke disease is reducing, but because of the aging population, the total burden for society is increasing. The mortality rate of stroke increases with aging, more so for men than women. The underlying causes vary between populations. For example, haemorrhagic stroke is more commonly seen in East Asians than in Europeans.

The mechanisms causing brain damage consist of two basic processes with either too little blood supplied (ischaemic) or too much blood present usually outside the vascular space (haemorrhagic). Both, the mechanisms and the location of the damage, are important in causing EEG changes.

The EEG is less precise or specific in localizing cerebrovascular disease than CT or MRI, but may show functional abnormalities at an earlier stage in the disease. Consequently, the EEG has a role in early detection of patients with transient or minor cerebrovascular disease (7). Transient ischaemia generates less marked EEG abnormalities than complete stroke. In patients with transient ischaemic attack (TIA), the EEG is either normal or shows mild non-specific changes, whereas the EEG is seldom normal after frank stroke (8).

Depending on the severity, age, depth, topography, and extension of the disease, cerebrovascular accidents can be associated with:

- No EEG change.
- Attenuation in amplitude or reactivity, or frequency decrease in the alpha, mu rhythm and beta activities, which may lead to inter-hemispheric asymmetries in these rhythms.
- Appearance of mu rhythm in older subjects.
- Slowing of the background activity in the theta or delta ranges (Fig. 33.2).
- Frontal intermittent rhythmic delta activity (FIRDA) (Fig. 33.2).
- Attenuation of all EEG activity.
- Sharp waves or, less frequently, epileptiform discharges, either occurring sporadically or in the form of periodic lateralized epileptiform discharges (PLEDs (9)).

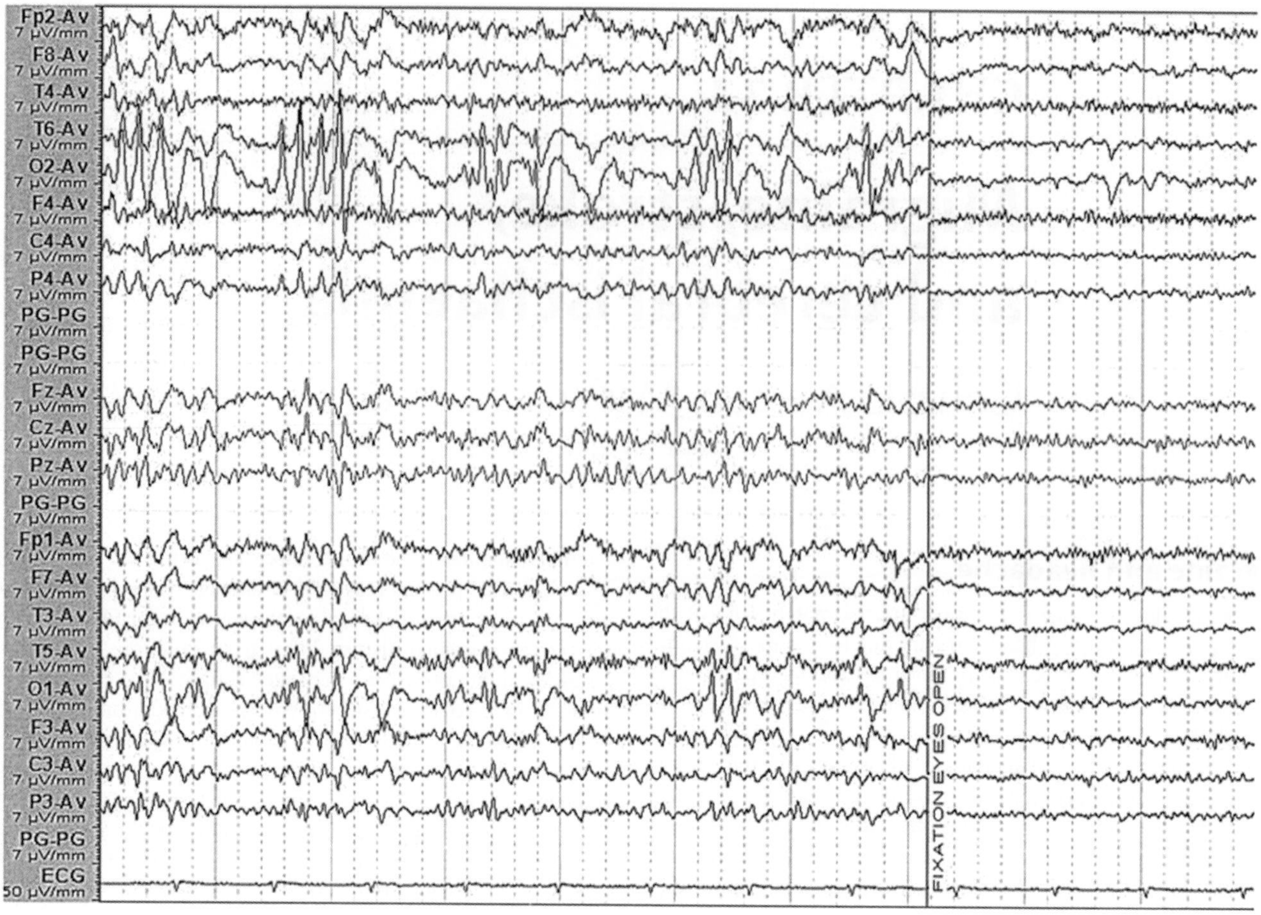

Fp2-Av 7 µV/mm
F8-Av 7 µV/mm
T4-Av 7 µV/mm
T6-Av 7 µV/mm
O2-Av 7 µV/mm
F4-Av 7 µV/mm
C4-Av 7 µV/mm
P4-Av 7 µV/mm
PG-PG 7 µV/mm
PG-PG 7 µV/mm
Fz-Av 7 µV/mm
Cz-Av 7 µV/mm
Pz-Av 7 µV/mm
PG-PG 7 µV/mm
Fp1-Av 7 µV/mm
F7-Av 7 µV/mm
T3-Av 7 µV/mm
T5-Av 7 µV/mm
O1-Av 7 µV/mm
F3-Av 7 µV/mm
C3-Av 7 µV/mm
P3-Av 7 µV/mm
PG-PG 7 µV/mm
ECG 50 µV/mm
FIXATION EYES OPEN

Fig. 33.1 EEG recording demonstrating fixation–off sensitivity; high amplitude right occipital paroxysms are seen when fixation is lost, with corresponding termination of epileptiform activity when fixation is resumed.

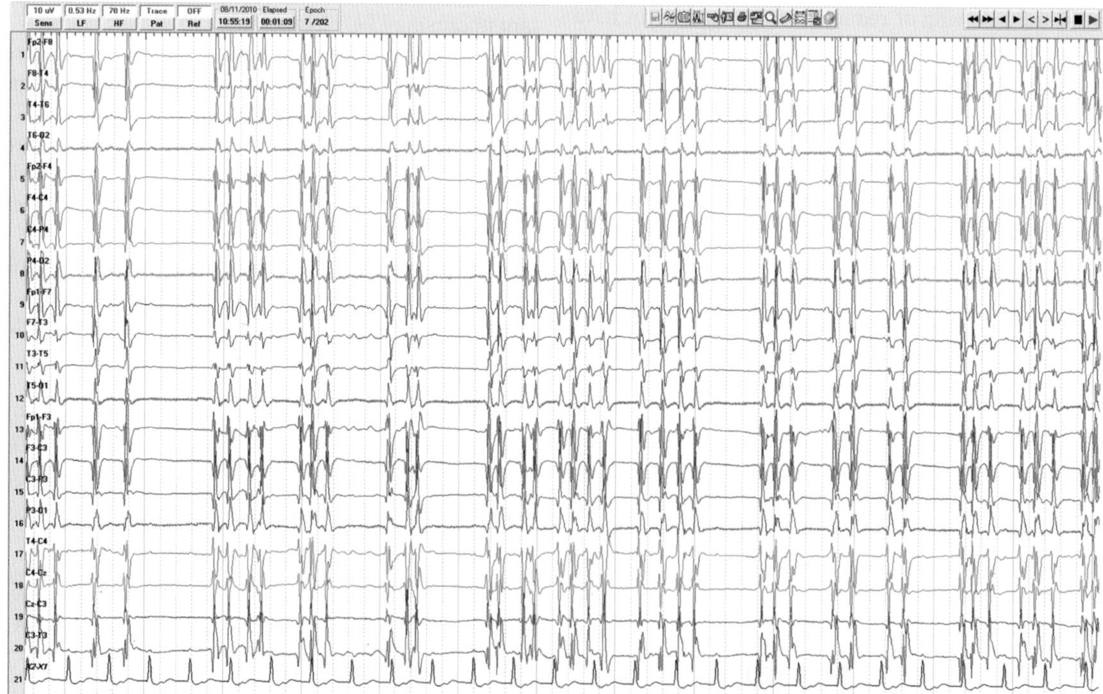

Fig. 33.2 Background activity is symmetrical with variable biological rhythms that cycle spontaneously and appeared also to be responsive to external stimuli. There were brief bursts of FIRDA in keeping with encephalopathy. Right hemisphere channels are shown in red and left hemisphere channels appear in blue.

Quantitative EEGs may show abnormalities in 50–70% of patients with normal standard EEGs (10,11).

The EEG may show changes following a stroke depending on the location and extent of the stroke. If extensive, slowing of background rhythms may be seen acutely down to delta range, but with time there is often an improvement into the theta range. The presence of periodic lateralized epileptic discharges (commonly called PLEDs) acutely is a marker for early seizures. However, in a patient with post-stroke epilepsy, the inter-ictal EEG may be remarkably bland, with few epileptiform discharges seen. Ictal EEG changes usually show slowing of background rhythms whereas a focal spike onset with rhythmic transformation is rarely seen.

Changes in physiological rhythms tend to persist whereas slowing of the background activity gradually decreases and may disappear. Extensive cortical strokes are associated with widespread EEG abnormalities whereas small deep (lacunar) lesions may not show EEG changes on the scalp. Strokes affecting the carotid artery territory induce more severe changes than strokes involving the vertebrobasilar system. Changes are most pronounced immediately after stroke, and gradually decrease thereafter. This is in contrast with CT/MRI abnormalities that generally appear a few hours after stroke. However, in contrast to neuroimaging, EEG changes are not sufficiently specific to differentiate between intracranial haemorrhage and infarct. Around 75% of cerebrovascular accidents involve the middle cerebral artery. Delta activity tends to be most pronounced in the anterior temporal region, FIRDA over frontal regions symmetrically and synchronously and PLEDs are usually predominant over posterior temporal regions. The EEG over the non-affected hemisphere tends to be normal unless there is increased intracranial pressure due to oedema or haemorrhagic infarction.

The following EEG changes would be seen according to the topography of the stroke:

- *Ischaemia in the territory of the posterior cerebral artery*: this is very rare and should be suspected in the presence of an abnormality of the alpha rhythm with posterior delta activity in the same region.
- *Watershed or arterial boundary zone ischaemia*: changes in the background activity and epileptiform discharges or PLEDs are seen with a parietal emphasis unilaterally or bilaterally (if ischaemia is bilateral).
- *Ischaemia in the territory of the anterior cerebral artery*: this is rare, but should be suspected in the presence of unilateral frontal delta activity and bilateral FIRDA.
- *Stenosis or complete occlusion of the internal carotid artery:* depending on collateral circulation, this can induce no EEG changes or pronounced EEG changes similar to those seen in watershed or middle cerebral artery infarcts.
- *Vertebrobasilar artery ischaemia*: the EEG may be normal or show bilateral EEG abnormalities with posterior predominance in addition to paroxysmal frontal delta. Patients with 'locked-in' syndrome, who are unable to react except with vertical eye movements, may falsely appear comatose. However, the EEG may be normal and reactive to external stimuli, including speech.
- *Cerebral venous thrombosis*: this is uncommon, but represents a very severe condition, which can present papilloedema, headache, hemiplegia, focal motor seizures, coma, and may lead to death. Diagnosis is confirmed with angiography and/or MRI. The EEG is abnormal in over 73% of patients, showing focal or bilateral asymmetrical changes, with epileptiform discharges in 21% (12,13).
- *Cerebral vasculitis*: EEG changes will depend on the severity and distribution of the disease, but the EEG is abnormal in over 75% of patients, showing diffuse, focal, or multifocal slowing, attenuation of physiological rhythms, sharp waves, or epileptiform discharges.
- *Cerebral haemorrhage*: EEG changes may be similar to ischaemia unless intracerebral pressure increases, inducing bilateral abnormalities. Diagnosis should be confirmed with CT/MRI.
- *Subarachnoid haemorrhage*: the role of the EEG is limited, as diagnosis is provided by CT and/or lumbar puncture. EEG changes may be present due ischaemia secondary to vasospasm.
- *Arteriovenous malformation*: the EEG may show signs of ischaemia and/or focal epilepsy.
- *Subdural haematoma*: the acute presentation is an emergency best diagnosed by neuroimaging. The EEG will show unilateral attenuation of normal rhythms. The chronic form, sometimes seen in the elderly, in association with headache and mental deterioration can show a normal EEG or unilateral attenuation of normal rhythms, slowing of the background activity and FIRDA. If normal, the EEG cannot exclude the diagnosis, but the presence of abnormalities can suggest the diagnosis and prompt the need for a CT in an elderly patient with fluctuating confusion.

Acute brain hypoxia

Acute brain hypoxia and anoxia are caused by any event that severely reduces oxygen supply to the brain. Such events may be internal or external. Internal causes include stroke, cardiovascular shock, and cardiac arrest. External causes include situations where blood flow may be normal, but blood is not carrying or delivering sufficient oxygen to the brain. This includes poor oxygen content in the air, exposure to certain poisons such as carbon monoxide, or any event that prevents breathing such as choking, asthma attack, suffocation, or a near drowning.

As described above, ischaemia and hypoxia can be associated with a variety of EEG changes depending on the severity of ischaemia. This would range from diminution or abolition of normal rhythms, slowing of the background activity, presence of epileptiform discharges to the abolition of EEG activity (Fig. 33.3). Changes can be focal or unilateral (unilateral infarct) or bilateral (cardiac arrest, drowning). Reactivity of EEG patterns to peripheral stimulation (pain, loud noises) is usually considered as a favourable prognostic sign. EEG reactivity can vary from EEG flattening to increment in slow activity. Evaluation of sedation is of paramount importance because hypoxia and some pharmacological agents used in intensive care can induce similar EEG patterns, such as slowing and burst suppression (see below). Perhaps, the most characteristic pattern is the presence of rhythmic epileptiform discharges (spike and wave) occurring at around 1 Hz. If unilateral, the pattern is called periodic lateralized epileptiform discharges (PLEDs). If bilateral independent or generalized, the patterns are called bilateral or generalized periodic epileptiform discharges (BiPLEDs or GPEDs, respectively; Figs 33.4–33.9) (14–18).

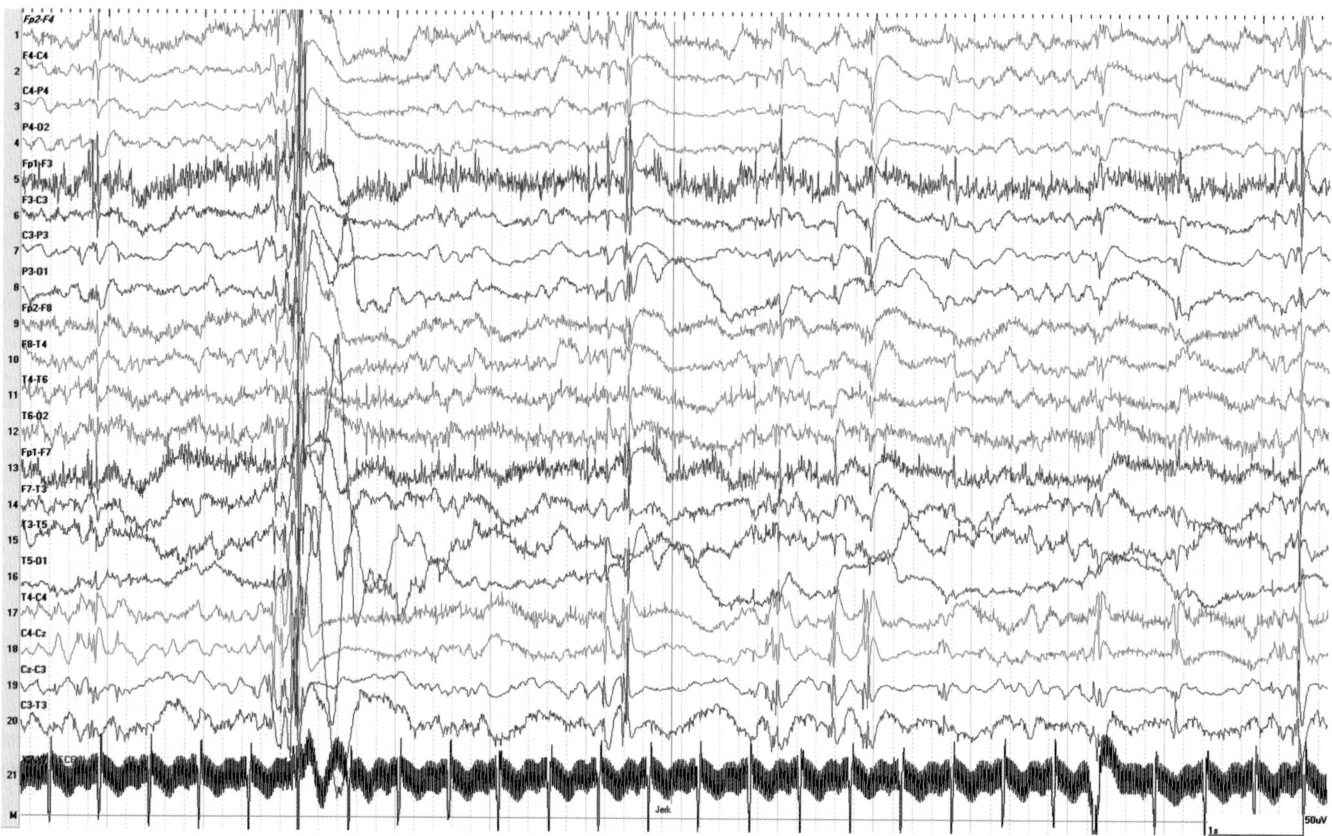

Fig. 33.3 Collapsed with immediate resuscitation with six cycles of cardiopulmonary resuscitation (CPR), six shocks and 1 mg of adrenaline, leading to return of spontaneous circulation. The background activity consists of fast reactive rhythms, interspersed by brief bursts of frontal intermittent rhythmic delta activity (FIRDA). The findings indicate a relatively mild to moderate encephalopathy, of apparently hypoxic nature. There are no features to suggest a neurologically unfavourable outcome.

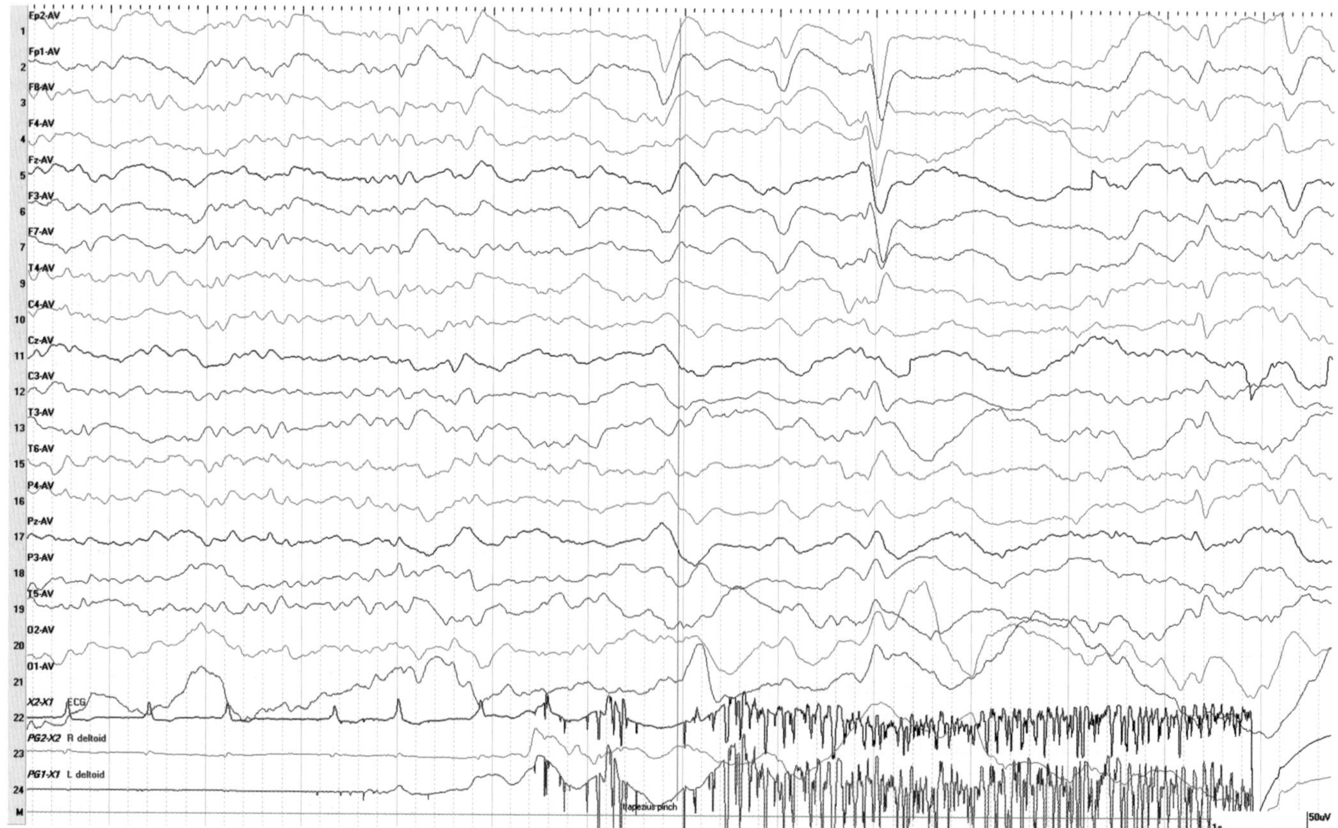

Fig. 33.4 EEG recording demonstrating bursts of diffuse synchronous generalized periodic epileptiform discharges (GPEDs) occurring in clusters.

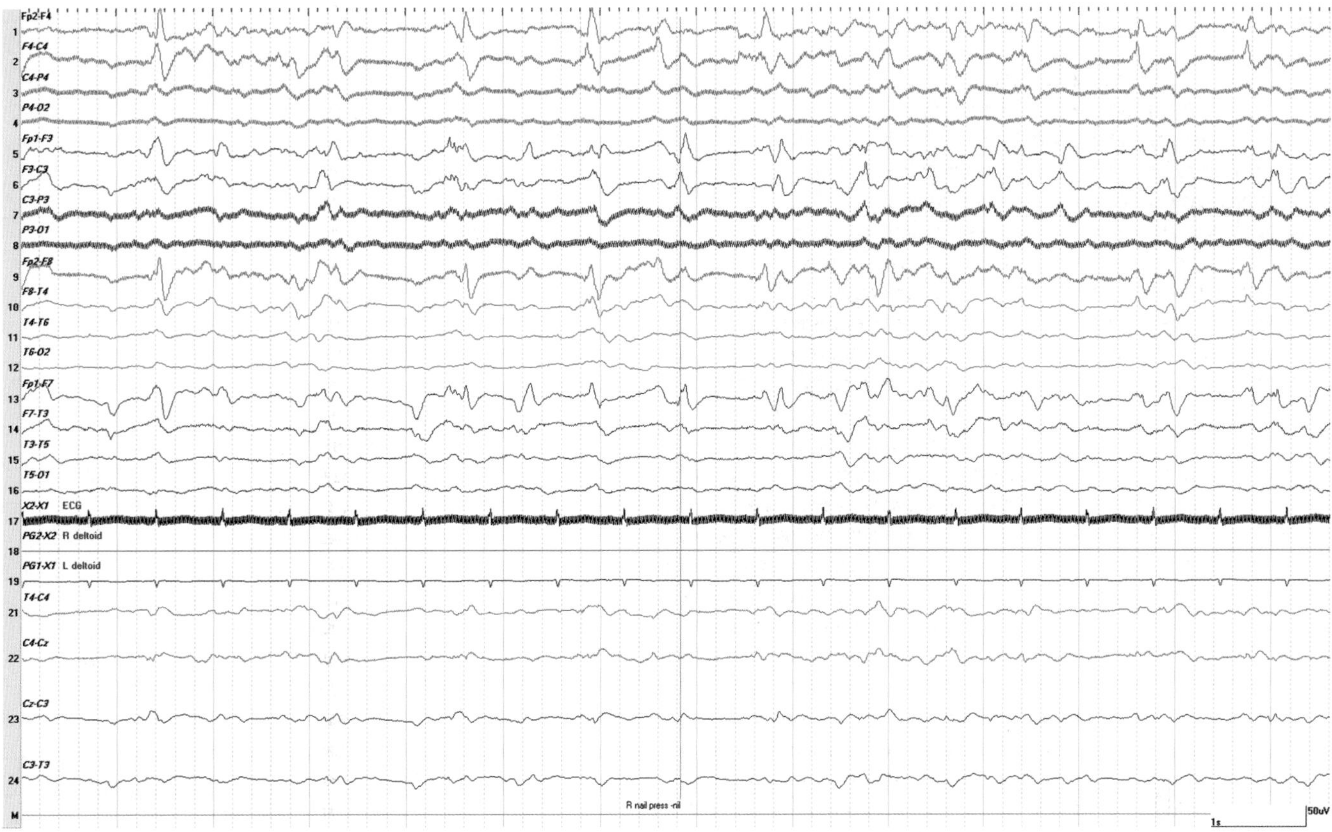

Fig. 33.5 EEG of a patient suffering from cerebral hypoxia after a near drowning incident. The EEG shows high-voltage bursts of slow waves with sharp or spike transients occurring against a depressed background.

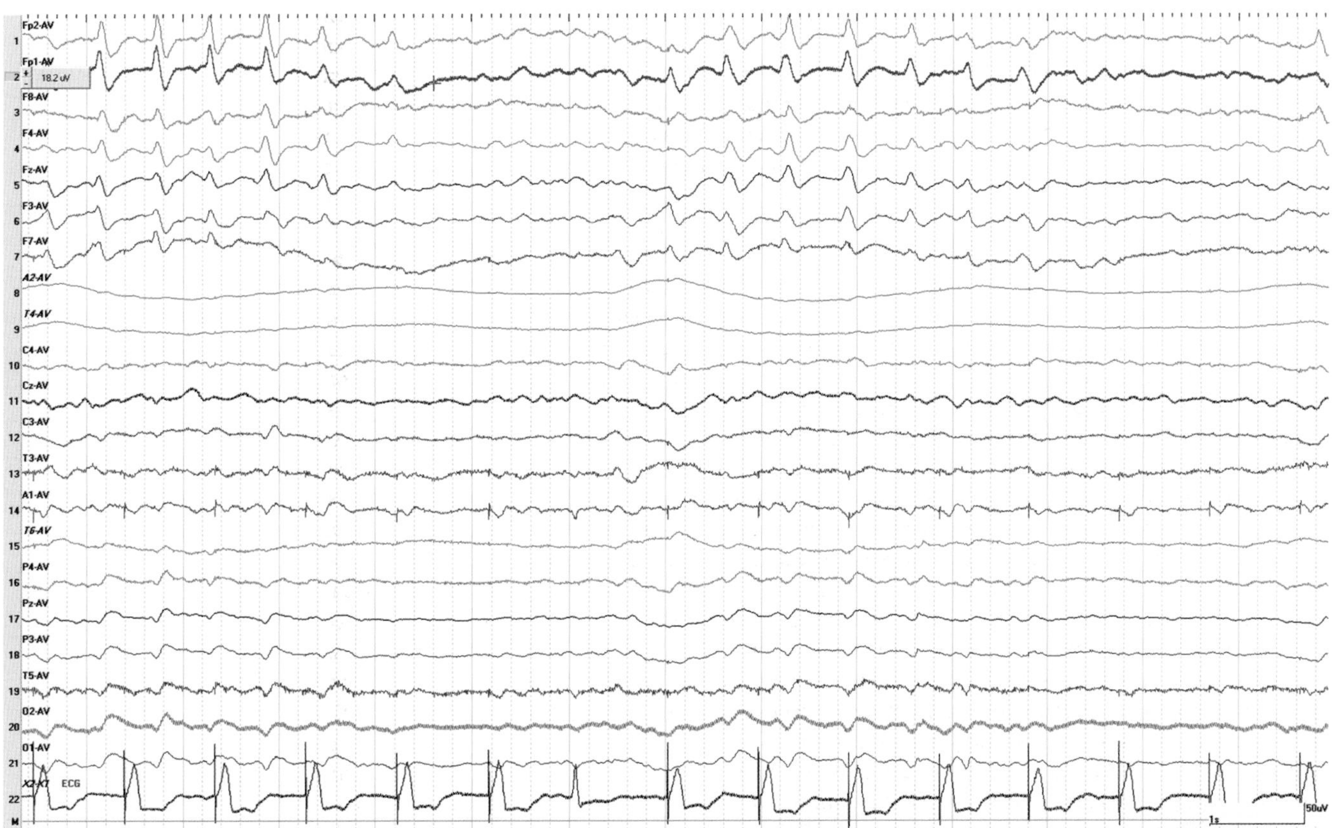

Fig. 33.6 EEG after a brief pulseless electrical activity (PEA) cardiac arrest. Sedation was stopped 22 h before the recording. Against a diffuse slow background, repetitive triphasic sharp wave discharges occurred mainly in response to external stimuli (stimulus induced rhythmic periodic ictal discharges or SIRPIDs). The diffuse theta slowing reflects a degree of hypoxic encephalopathy. As for the SIRPIDs, there is no good evidence from the literature that they do necessarily represent epilepsy-related rhythms and their prognostic value is uncertain.

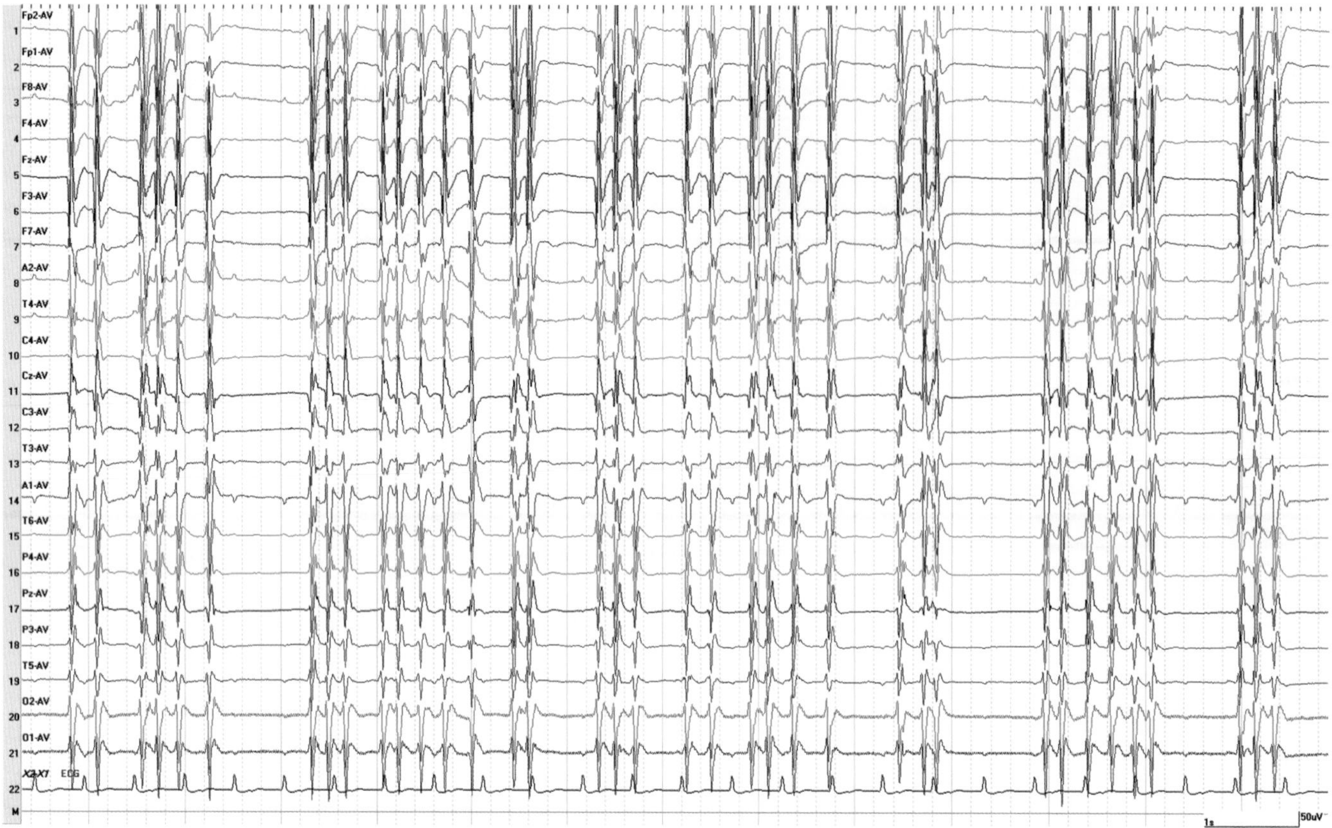

Fig. 33.7 Out of hospital cardiac arrest following ethanol and cocaine abuse, sedation stopped 5 h before test. The EEG shows bilateral independent or synchronous subclinical frontal BiPEDs, punctuating a diffusely slow background with brief periods of diffuse suppression. The BiPEDs are non-evolving in frequency, topography or amplitude and therefore have no characteristics of epileptic activity. This pattern is consistent with hypoxic encephalopathy.

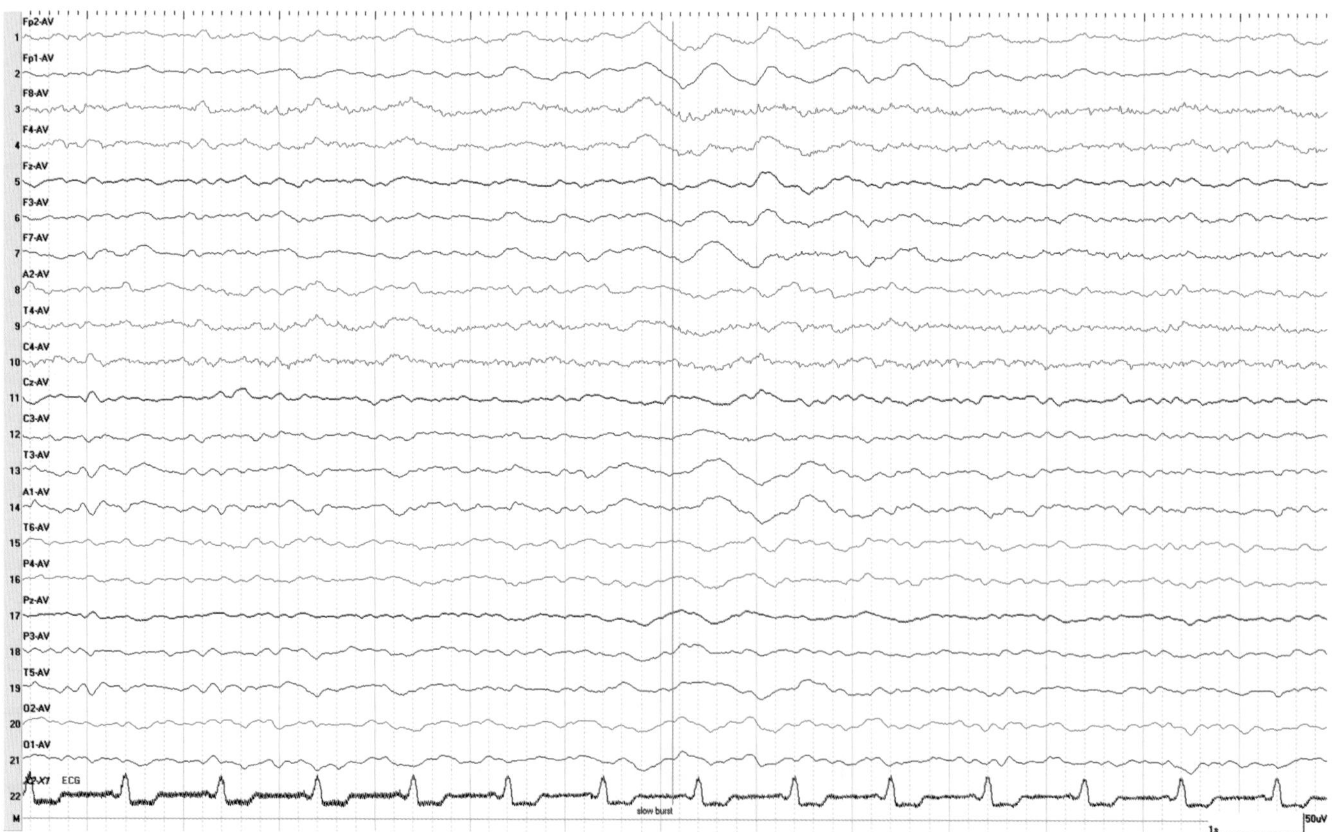

Fig. 33.8 Ventricular fibrillation with pulseless electrical activity, cardiopulmonary resuscitation for 35 min. Bifrontal sharp waves occurred semi-rhythmically in brief runs (BiPEDs) at relatively low frequency (<2 Hz). There are also brief periods of diffuse suppression; these features reflect severe ischaemic damage. The overall picture is that of a diffuse hypoxic encephalopathy with little chance of satisfactory neurological recovery.

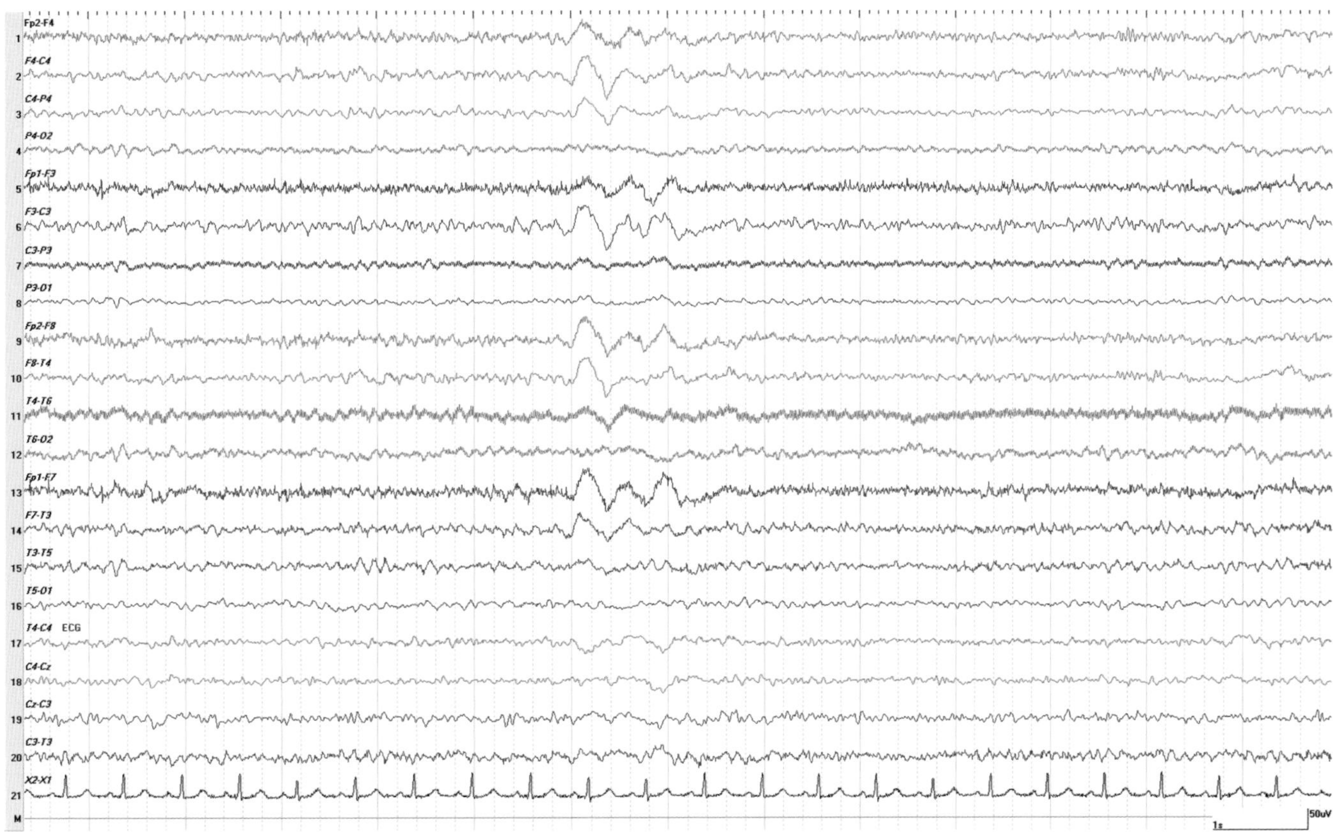

Fig. 33.9 Ventricular fibrillation secondary to probable hypokalaemia. Cardiopulmonary resuscitation at home, response team arrived within 3 min, patient was in ventricular fibrillation. Patient was off midazolam for an hour before recording. The EEG shows an unremitting pattern of repetitive burst–suppression; the bursts consist of diffuse synchronous GPEDs and occur in clusters while the periods of (complete) flattening become shorter as midazolam wears off and disappear towards the end of the recording. These features are indicative of severe cerebral anoxia and predict very poor neurological outcome.

PLEDs are spike-wave or sharp-wave complexes occurring periodically, typically at around 1 Hz. The most common condition associated with PLEDs is cerebrovascular disease, but can also be seen in herpes virus encephalitis and in other forms of focal pathology, such as brain tumours, head injury, and cerebral abscess (Fig. 33.5) (19). If the slow wave is sufficiently long to merge two successive complexes, they may resemble triphasic waves (20). The presence of PLEDs tends to be a self-limiting phenomenon even in the presence of progressive pathology. BiPEDs (Fig. 33.7) indicate a more severe condition (21).

One practical difficulty is the distinction between status epilepticus and rhythmic discharge patterns (PLEDs, BiPLEDs, or GPEDs). Not only can the EEG patterns be distinctly similar, but PLEDs, BiPLEDs, and GPEDs can be abolished by anticonvulsants and are commonly associated with clinical seizures or status epilepticus. The question often arises of whether to treat with anticonvulsants a patient with such periodic patterns in the absence of clear clinical seizures. If the EEG patterns evolve in frequency, topography, or amplitude, it is assumed that they are epileptic seizures. Otherwise, the relation between such rhythmic patterns and status epilepticus is unclear. It could be claimed that PLEDs, BiPLEDs, and GPEDs can be considered as a form of electrical status epilepticus occurring in the context of a cortex, which is so severely dysfunctional that is unable to generate symptoms. In this scenario, it is fundamental to consider the whole clinical situation: presence of clinical seizures other than the occasional myoclonic jerk, the high risk of suffering seizures in these patients even if none have yet been observed, the state of the underlying cortex and reversibility of the cause.

Bust-suppression (Figs 33.10–33.13) is another EEG pattern seen after severe brain hypoxia and, in this context, suggests poor prognosis in the absence of sedation (22). Burst-suppression may be the expression of reduced cortical metabolism, as suggested by the fact that it can also be induced by deep sedation (thiopenotone, propofol) and hypothermia below 20°C. This should be distinguished from similar, but normal EEG patterns seen in the newborn (tracé discontinue and tracé alternant) (Figs. 33.14 and 33.15).

In the absence of sedation, low-voltage EEG and isoelectric EEG patterns represent the most severe stages of hypoxic encephalopathy. Recording criteria and prognostic implications are the same as in other terminal conditions. Absence of EEG activity in an unresponsive non-sedated patient (electroclinical silence) would suggest brain death. In the diagnosis of brain death, it is important to follow national guidelines to minimize false positives.

In patients with coma resulting from hypoxic encephalopathy, the EEG may reflect the severity of brain dysfunction, although the exact relationship among the EEG changes, the extent of neuronal damage, and consequent prognosis is still under study.

Many prognostications are based on particular EEG patterns at a time point, such as burst suppression or generalized periodic discharges, but with sequential, repeated, or with prolonged or

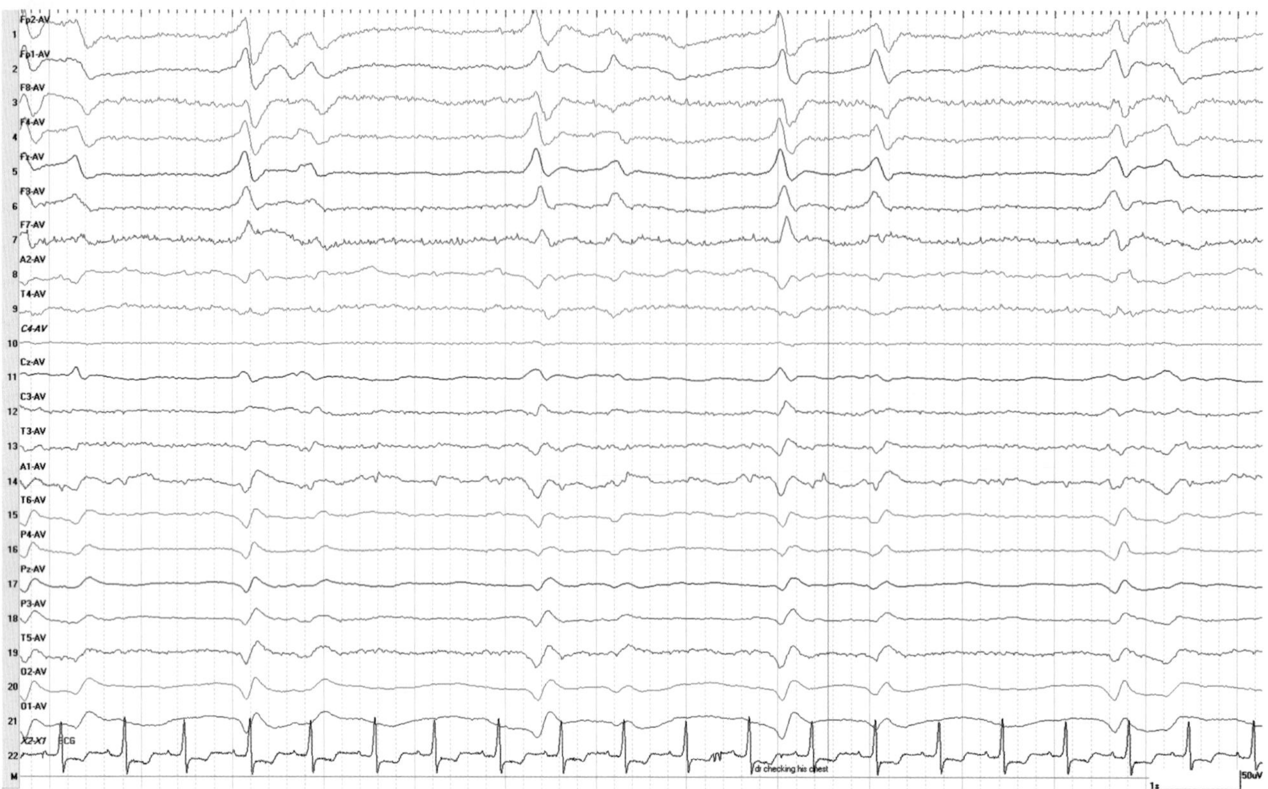

Fig. 33.10 EEG performed 3 days post-arrest due to ventricular fibrillation while in the recovery room after haemodialysis. Off sedation for 48 h and normothermic, the patient is comatose with an EEG showing a burst suppression pattern; suppression is complete and bursts consist of solitary sharp waves at slow frequency (from <1.5 to 0.3 Hz). The pattern did not show any spontaneous change and remained unresponsive to all types of external stimulus. The findings indicate severe anoxic encephalopathy and suggest that meaningful neurological outcome is unlikely.

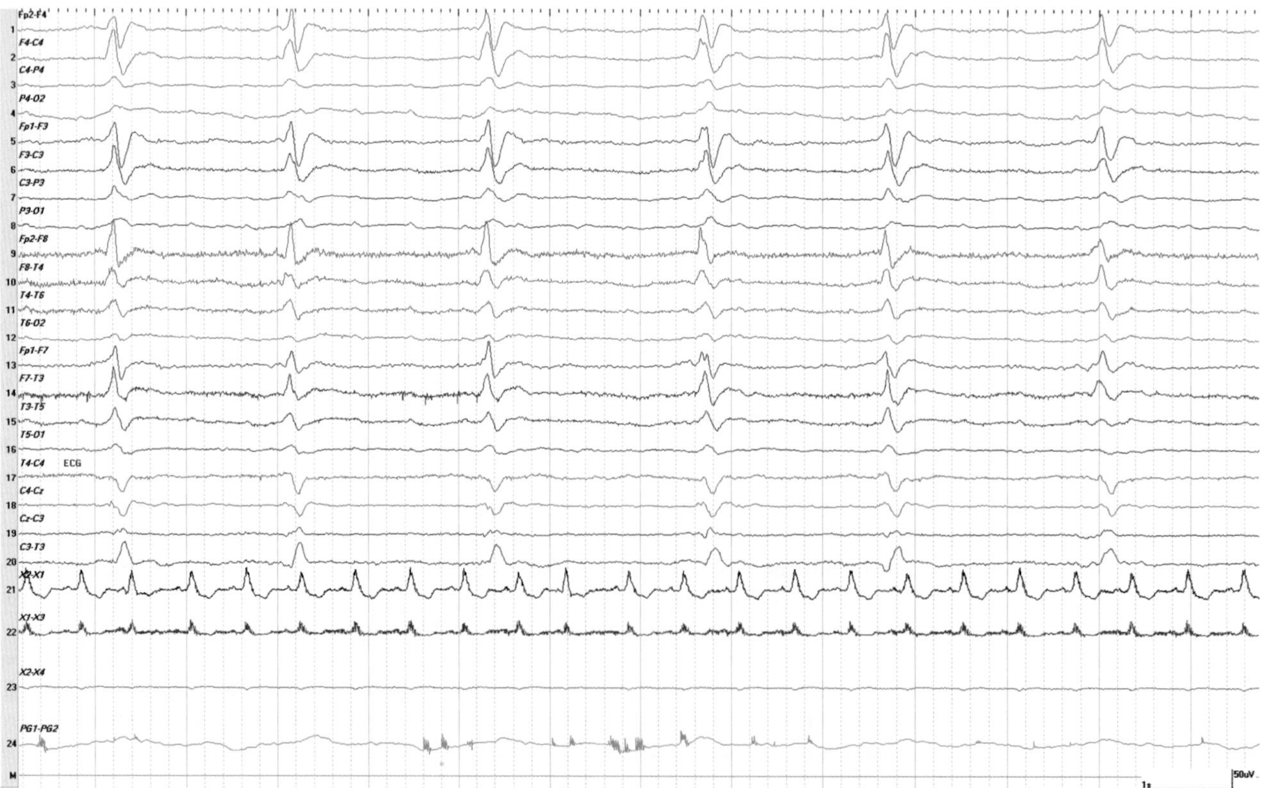

Fig. 33.11 Ventricular fibrillation with return of spontaneous circulation after 30 min. EEG recoding performed four days after the insult and two days off sedation. Continuous high voltage GPEDs at 0.4–0.7 Hz occurring throughout the recording with the intervals between the bursts containing very little, if any, biological activity. The pattern was unresponsive to external stimulation. This pattern of burst suppression is 'malignant' and reflects extensive, most likely irreversible cortical damage, predicting a poor neurological outcome.

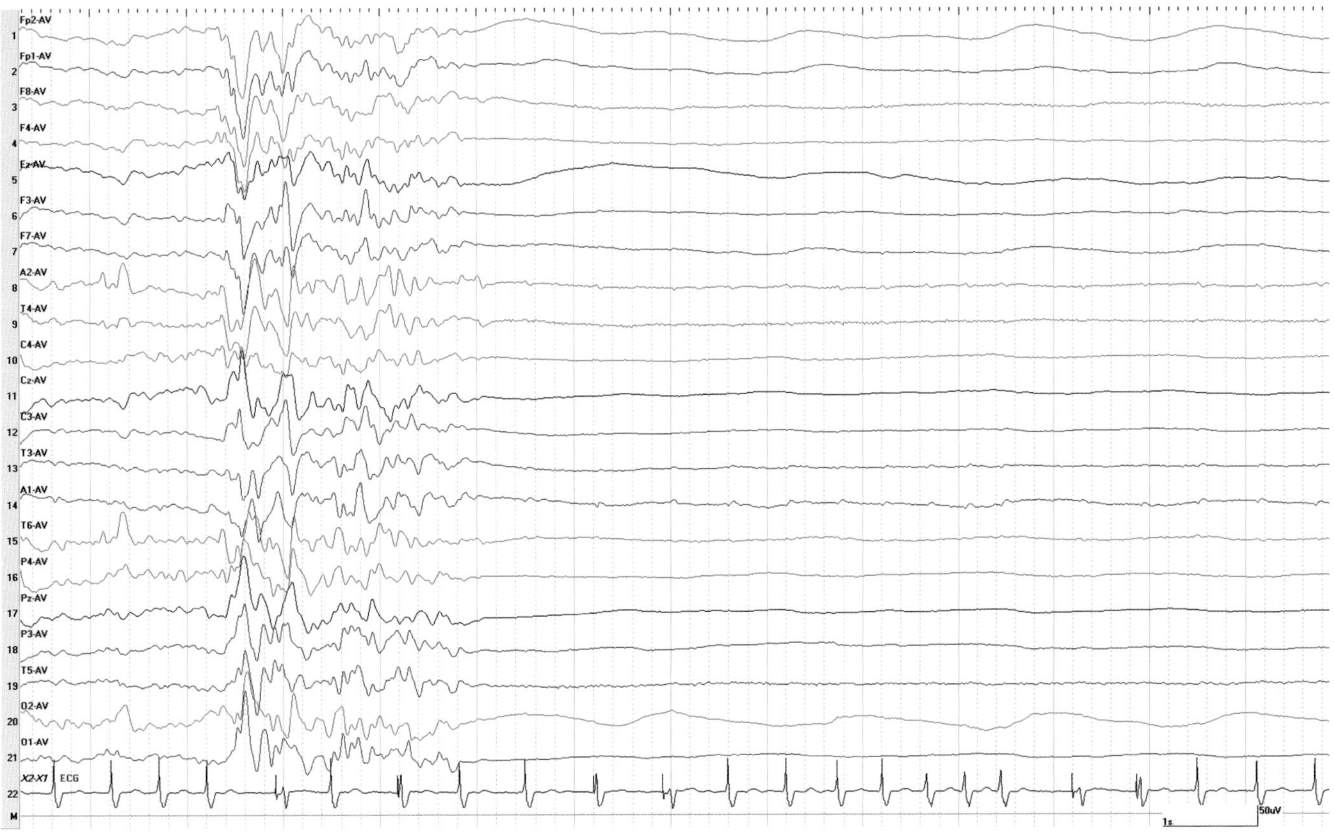

Fig. 33.12 Out of hospital arrest with approximately 30 min downtime. 24 h after sedation stopped the EEG pattern is that of severe burst suppression with long periods of complete flatness. The pattern reflects severe irreversible damage and predicts grave neurological outcome.

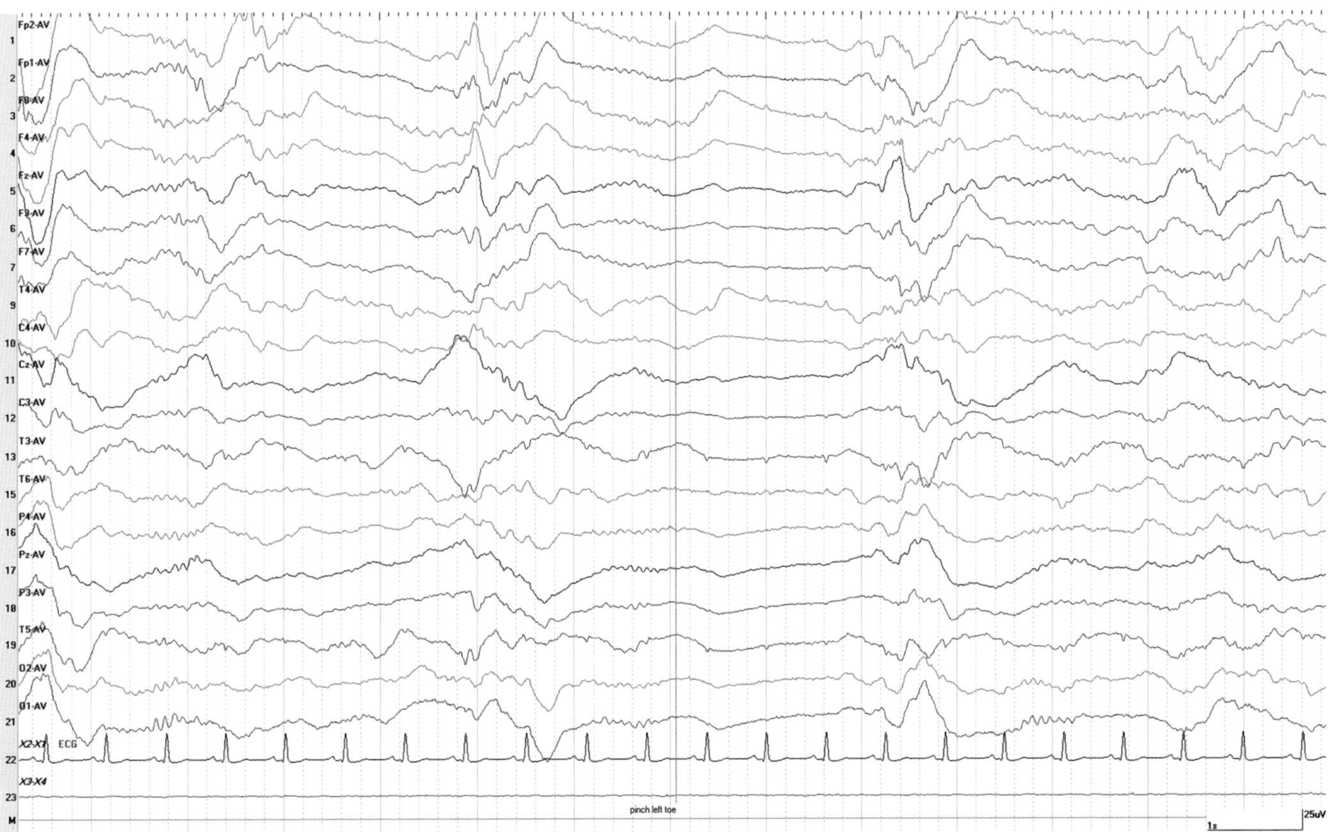

Fig. 33.13 Pneumococcal bacteraemia/pneumonia. Acute coronary syndrome in context of left ventricular hypertrophy and acute illness. Severe respiratory failure with profound hypoxia. Now on extracorporeal membrane oxygenation. Failure to wake after 60 h off sedation. Glasgow coma scale of 3. The pattern is that of burst-suppression with the periods of activity extending over up to 20–25 s and the periods of suppression up to 2–3 s. If the patient has been off sedation for 60 h, the present pattern leaves little hope for meaningful recovery.

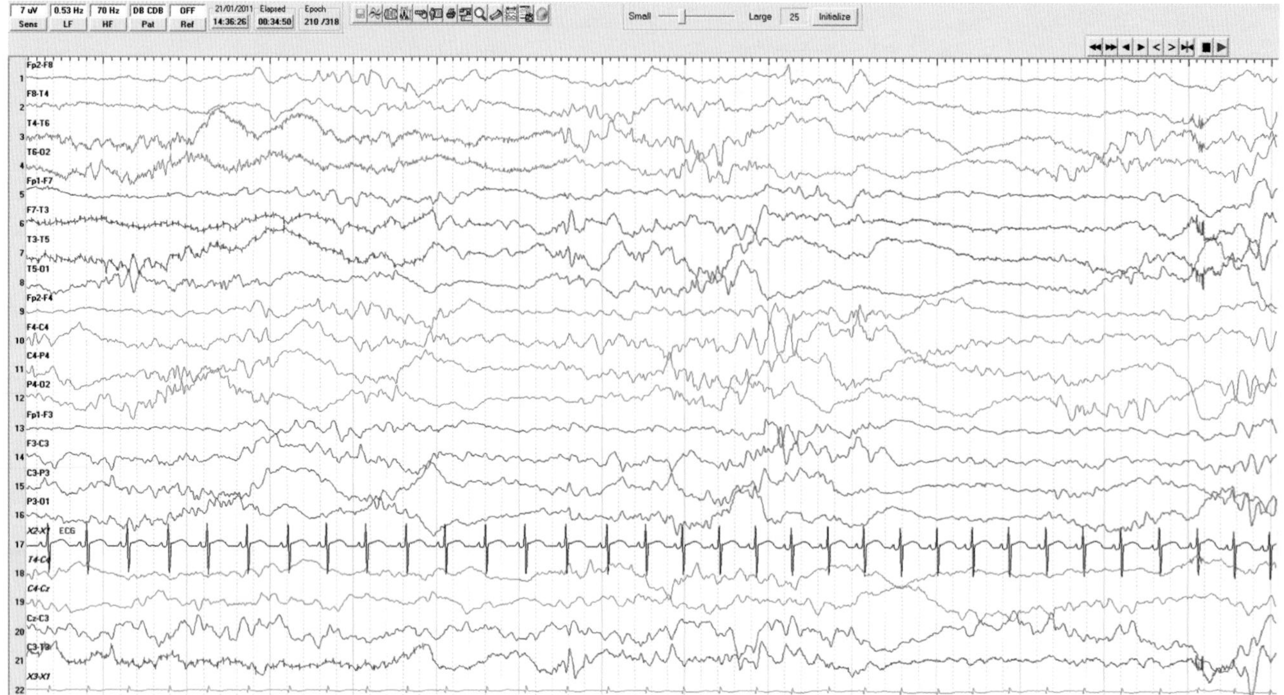

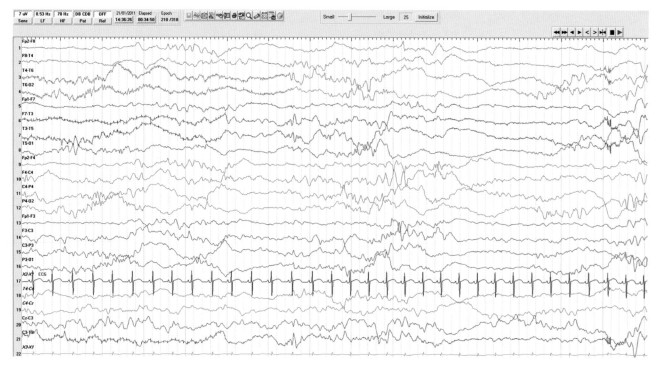

Fig. 33.14 EEG at 40-week conceptional age showing normal trace alternant of quiet sleep. Activity in the inter-burst intervals consist of mixed frequencies with amplitudes between 25 and 50 µV. The duration of bursts is 2–4 s. Bursts consist of delta activity with superimposed fast frequencies and amplitude as high as 300 µV.

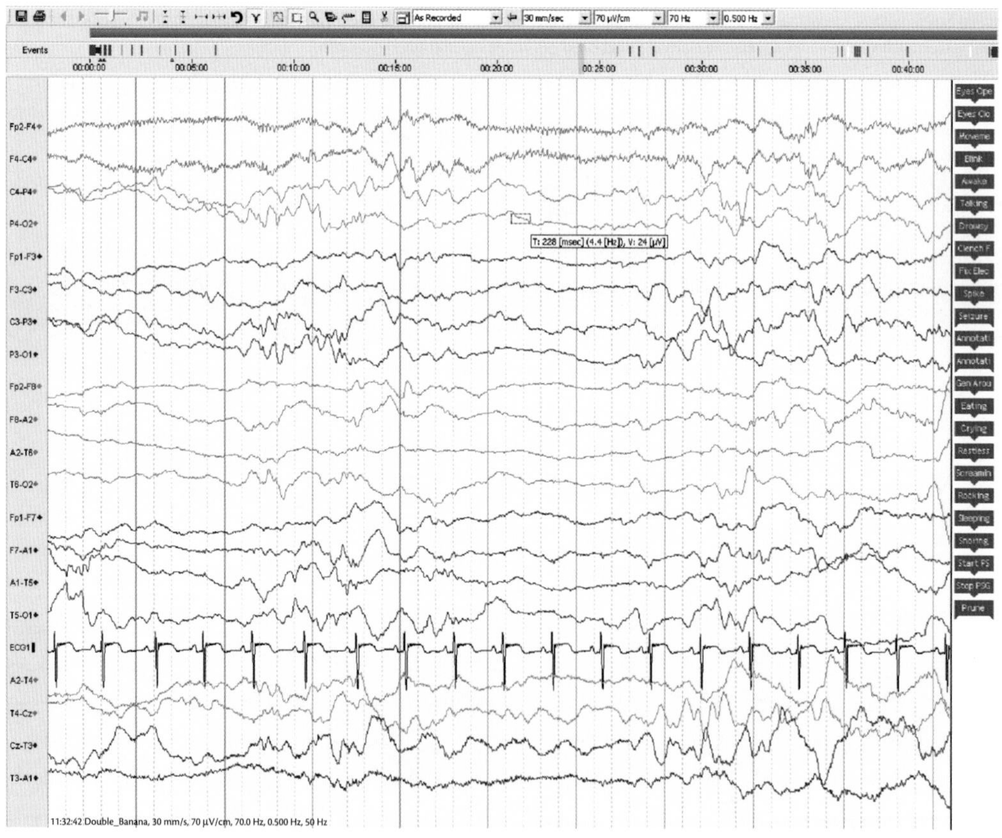

Fig. 33.15 Normal quiet sleep (37–44 weeks of gestational age) consisting of continuous theta (4–7 Hz, 25–50 µV) interspersed with bilateral bursts of delta each lasting for 3–8 s.

continuous EEG monitoring, it has become increasingly clear that more information might be gleaned from EEG pattern changes over time. Temporal dynamic changes (as opposed to permanent transitions), or preserved reactions to exogenous stimuli, have to be differentiated (23).

In some patients suffering coma after hypoxia, peripheral stimulation can induce EEG patterns resembling seizures, which are not associated with ictal clinical manifestations. This phenomenon has been called stimulus induced rhythmic periodic ictal discharges (SIRPIDs) (Fig. 33.6). There is no evidence that SIRPIDs necessarily represent true epileptic seizures, and their prognostic value is unclear.

References

1. Ulett, G.A., Evans, D., and O'Leary, J.L. (1952). Survey of EEG findings in 1,000 patients with chief complaint of headache. *Electroencephalography & Clinical Neurophysiology*, **4**(4), 463–70.
2. Gamstorp, I. (1985). *Paediatric neurology*, 2nd edn. London: Butterworth.
3. Golla, F.L. and Winter, A.L. (1959). Analysis of cerebral responses to flicker in patients complaining of episodic headache. *Electroencephalography & Clinical Neurophysiology*, **11**(3), 539–49.
4. Hockaday, J.M. and Whitty, C.W. (1969). Factors determining the electroencephalogram in migraine: a study of 560 patients, according to clinical type of migraine. *Brain*, **92**(4), 769–88.
5. Tfelt-Hansen, P.C. and Koehler, P.J. (2011). One hundred years of migraine research: major clinical and scientific observations from 1910 to 2010. *Headache*, **51**, 752–78.
6. Smith, J.M., Bradley, D.P., James, M.F., and Huang, C.L.H. (2006). Physiological studies of cortical spreading depression. *Biology Reviews*, **81**, pp. 457–81.
7. Faught, E. (1993). Current role of electroencephalography in cerebral ischemia. *Stroke*, **24**(4), 609–13.
8. Schaul, N., Green, L., Peyster, R., and Gotman, J. (1986). Structural determinants of electroencephalographic findings in acute hemispheric lesions. *Annals of Neurology*, **20**(6), 703–11.
9. Chatrian, G.E., Shaw, C.M., and Leffman, H. (1964). The significance of periodic lateralized epileptiform discharges in EEG: an electrographic, clinical and pathological study. *Electroencephalography & Clinical Neurophysiology*, **17**, 177–93.
10. Van Huffelen, A.C., Poortvliet, D.C., and Van der Wulp, C.J. (1984). Quantitative electroencephalography in cerebral ischemia. Detection of abnormalities in 'normal' EEGs. *Progressive Brain Results*, **62**, 3–28.
11. Veering, M.M., Jonkman, E.J., Poortvliet, D.C., De Weerd, A.W., Tans, J.T., and John, E.R. (1986). The effect of reconstructive vascular surgery on clinical status, quantitative EEG and cerebral blood flow in patients with cerebral ischaemia. A three month follow-up study in operated and unoperated stroke patients. *Electroencephalography & Clinical Neurophysiology*, **64**(5), 383–93.
12. Bousser, M.G., Chiras, J., Bories, J., and Castaigne, P. (1985). Cerebral venous thrombosis: a review of 38 cases. *Stroke*, **16**(2), 199–13.
13. Rousseaux, P., Vieillart, A., Scherpereel, B., Bernard, M.H., Motte, J., and Guyot, J.F. (1985). Benign intracranial hypertension (17 cases) and cerebral venous thromboses (49 cases). Comparative study. *Neurochirurgie*, **31**(5), 381–9.
14. Howard, R.S., Holmes, P.A., Siddiqui, A., et al. (2012). Hypoxic-ischaemic brain injury: imaging and neurophysiology abnormalities related to outcome. *Quarterly Journal of Medicine*, **105**, 551–61.
15. Kaplan, P.W. (2004). The EEG in metabolic encephalopathy and coma. *Journal of Clinical Neurophysiology*, **21**, 307–18.
16. Kaplan, P.W. (2006). Electrophysiological prognostication and brain injury from cardiac arrest. *Seminars in Neurology*, **26**, 403–12.
17. Kaplan, P.W. and Rossetti, A.O. (2011). EEG patterns and imaging correlations in encephalopathy: encephalopathy part II. *Journal of Clinical Neurophysiology*, **28**, 233–51.
18. Sutter, R., Stevens, R.D., and Kaplan, P.W. (2013). Clinical and imaging correlates of EEG patterns in hospitalized patients with encephalopathy. *Journal of Neurology*, **260**, 1087–98.
19. Pohlmann-Eden, B., Hoch, D.B., Cochius, J.I., and Chiappa, K.H. (1996). Periodic lateralized epileptiform discharges critical review. *Journal of Clinical Neurophysiology*, **13**, 519–30.
20. Hirsch, L.J., LaRoche, S.M., Gaspard, N., et al. (2013). American Clinical Neurophysiology Society's standardized critical care EEG terminology: 2012 version. *Journal of Clinical Neurophysiology*, **30**, 1–27.
21. Bauer, G. and Trinka, E. (2010). Nonconvulsive status epilepticus and coma. *Epilepsia*, **51**, 177–90.
22. Niedermeyer, E. (2009). The burst-suppression electroencephalogram. *American Journal of Electroneurodiagnostic Technology*, **49**, 333–41.
23. Bauer, G., Trinka, E., and Kaplan, P.W. (2013). EEG Patterns in hypoxic encephalopathies (post–cardiac arrest syndrome): fluctuations, transitions, and reactions. *Clinical Neurophysiology*, **30**, 477–89.

CHAPTER 34

Electroclinical features of paediatric conditions

Sushma Goyal

Basic concepts of neonatal and paediatric electroencephalogram

Commonly used neonatal terminology

- GA = gestational age, the number of completed weeks and days from the date of the first day of the last menstrual period.
- CGA = corrected gestational age, GA + number of weeks post-partum.
- Term = neonate born between 37–42 weeks GA.
- Preterm = GA less than 37 weeks.
- Discontinuity = periods of hypoactivity (<10 μV) of variable duration. This is the normal for preterm electroencephalograms (EEGs) and abnormal in term EEGs.
- Inter-burst interval = length of period of hypoactivity. Normal for preterm.
- HIE = hypoxic ischaemic encephalopathy. Failure to initiate and sustain breathing immediately after delivery. This is associated with hypoxic ischemic injury to the central nervous system, the clinical manifestations of which are termed as HIE.

Normal neonatal EEG at term (1)

Awake state—Activité moyenne

Continuous, irregular, diffuse activity with a Rolandic predominance mainly in the theta band (4–7 Hz), with a voltage of 25–50 μV, occipital delta of similar voltage.

Quiet sleep

- *Slow continuous tracing or high voltage pattern*: continuous delta-wave activity (1–3 Hz) with occipital predominance and variable voltage (50–150 μV).
- *Tracé alternant pattern*: bilateral bursts of delta waves occurring on a background of continuous theta activity (4–7 Hz and 25–50 μV). This pattern is characterized by:
  - Bursts (1–3 Hz, 50–150 μV) varying in duration from 3 to 8 s.
  - Intervening periods of low voltage activity (4–7 Hz, 25–50 μV) of similar duration.
  - Appears between 37 and 44 weeks.
  - Only in quiet sleep.

Normal paediatric EEG

Important variables to consider when interpreting a paediatric EEG are age, state, and medication. There is considerable intersubject variability due to different rates of maturation. Artefact recognition and an awareness of normal variants are equally important. The paediatric EEG should ideally include a representation of all states of awareness.

Awake state

The most important aspect of determining a normal paediatric EEG is the posterior background rhythm. Hans Berger was the first to illustrate that the frequency of background activity in childhood increases with age. Passive eye closure is used to demonstrate this as the background can be attenuated by eye opening as early as 3 months of age. There is considerable variability in the appearance of the posterior alpha rhythms though it is generally accepted that it should appear by 4 years. Numerous studies (2–5) have provided the frequency milestones outlined below:

- 3 month, 3–4 Hz.
- 6 months, 5–6 Hz.
- 9–18 months, 6–7 Hz.
- 2 years, 7–8 Hz.
- 7 years, 9 Hz.
- 15 years, 10 Hz.

During passive eye closure, considerable wave-to-wave variability in the amplitude is seen in the first year of life, usually from 30 to 100 μV, with occasional waves reaching 200 μV in the latter part of the first year. Pampiglione (3) reported amplitudes of 50–80 μV at age 2 years. The alpha amplitude in the study of Petersen and Eeg-Olofsson (4) increased to a maximum at 6–9 years and then declined.

Sleep state

The reliable interpretation of paediatric EEG requires a thorough appreciation of drowsy, sleep, and arousal phenomena, as normal patterns occuring in these states can be mistakenly interpreted. In addition, the morphology of common sleep patterns may differ between children and adults. Lack of a clear correlation between clinical and EEG manifestations of state reflects a diffuse encephalopathy in children.

Electroencephalographic features precede the clinical state of drowsiness. The background activity appears about 1–2 Hz slower in comparison to wakefulness. In active children this may be the best period to ascertain posterior rhythms as they are relaxed with their eyes are closed. Diffuse rhythmic to sinusoidal theta appears, which gradually replaces the awake pattern and may persist for several minutes. This may be maximal centrally, more posteriorly, or frontocentrally.

Hypnogogic hypersynchrony (6) is a well-recognized normal variant of drowsiness in children aged 3 months to 13 years. These are paroxysmal bursts at 3–5 Hz of high voltage (up to 350 μV) sinusoidal waves, maximally expressed in the prefrontal and central areas. They do not persist in stage 2 sleep.

Rudimentary sleep spindles appear by 6–8 weeks of age after term (7). Spindles are most prolific between 3–9 months. The mean length of individual spindles varies from 0.5–5 seconds in infancy. Spindle wave frequency remains relatively consistent at 13–14 Hz from infancy to 4–5 years. Interhemispheric asynchrony of spindles occurs more commonly in children less than 2 years of age (8). The spindles of infants and young children may also be comb-shaped.

Rudimentary vertex waves (V waves) appear in light sleep as early as 3–4 months of age and are usually well developed by 5 months (7).

Variants

Normal variants including posterior slow waves of youth, lambda waves, slow alpha variant, Mu rhythms, and central theta rhythms, can mimic epileptiform discharges. 6 and 14 positive spikes are notorious for being misinterpreted (see Fig. 34.1).

Classification of epileptic seizures

Epileptic seizures are defined as localized, generalized, or unknown. Generalized seizures are conceptualized as originating at some point within, and rapidly engaging bilaterally distributed networks. The bilateral networks may include cortical and sub-cortical structures, but may not necessarily involve the whole cortex. Focal seizures are conceptualized as originating within networks limited to one hemisphere. Ictal onset is consistent from one seizure to the next with preferential propagation patterns that can involve the contralateral hemisphere. If there is uncertainty regarding classification then the seizures should be considered unclassified until further information allows accurate diagnosis. The controversy in defining whether epileptic spasms are focal or unknown has led them to remain in the unknown category. It is important to acknowledge this uncertainty particularly in children to prevent misdiagnosis.

Generalized seizures

- Tonic–clonic (in any combination).
- *Absence*:
  - Typical.
  - Atypical.
- *Absences with special features*:
  - Myoclonic absence.
  - Eyelid myoclonia.
- Myoclonic:
  - Myoclonic atonic.
  - Myoclonic tonic.
- Clonic.
- Tonic.
- Atonic.

Focal

- Without impairment of consciousness.
- With autonomic, motor components.

Subjective or psychic phenomena

- With impairment of consciousness.
- Evolving to a bilateral convulsive seizure.

Unknown

Epileptic spasms.

Electroclinical syndromes

An electroclinical syndrome is a cluster of signs and symptoms that customarily occur together. It is defined by its age of onset, seizure type, EEG patterns, imaging features, co-morbidities, provoking or triggering factors, and patterns of seizure occurrence with respect to sleep (see Box 34.1). They may not necessarily have the same aetiology. Children who frequently defy syndromic diagnosis are infants and children with recurrent febrile and afebrile seizures, neurologically abnormal children with polymorphic seizures who do not fulfil the criteria for Lennox–Gastaut syndrome or other myoclonic epilepsies and children who present with focal seizures with normal or non-specific EEG and magnetic resonance imaging (MRI) findings (9).

Conditions in neonates and infancy

Hypoxic ischaemic encephalopathy

Early EEG within 72 h after birth remains a useful bedside diagnostic tool to inform prognosis in babies with Hypoxic ischemic encephalopathy (HIE) (11). Normal or mildly abnormal results have a 100% positive predictive value at 6, 12, or 24 h of age. At 48 h of age the positive predictive value of abnormal results is 93% and negative predictive value is 71% (12). EEG features, which help to prognosticate are background amplitude <30 μV, interburst intervals >30 s, electrographic seizures and absence of sleep wave cycling at 48hrs. Studies on amplitude integrated EEG and EEG during controlled hypothermia have reported a time window of 48 h after the hypoxic ischaemic event for optimal prognosis of outcome (13).

Infantile epileptic encephalopathies

Epileptic encephalopathies are conditions in which cognitive, sensory, and/or motor function deterioration results from epileptic activity (14). This concept is now widely accepted as it embodies the notion that the epileptic activity itself may contribute to severe cognitive and behavioural impairments above and beyond what might be expected from the underlying aetiology alone. Included in this syndrome are:

- Early infantile epileptic encephalopathy (EIEE) or Ohtahara's syndrome.
- Myoclonic encephalopathy (ME or EME).
- Infantile spasms (IS).

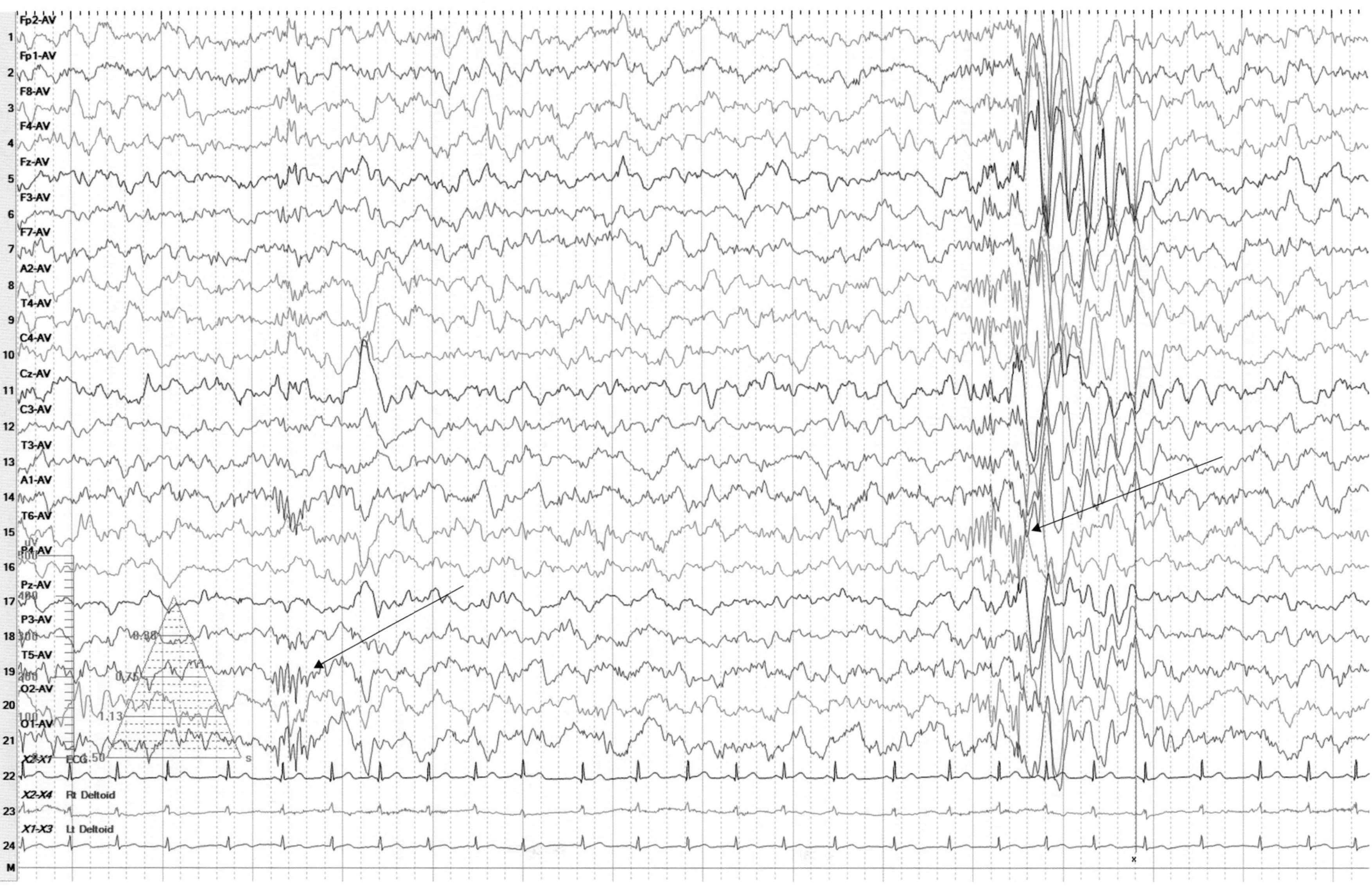

Fig. 34.1 Normal EEG variants in sleep. 6 & 14 positive spikes seen independently and preceding 6 Hz phantom spike-and-wave.

Box 34.1 Electroclinical syndromes according to the age of presentation

Neonates (<44 weeks of gestational age)

- Benign familial neonatal epilepsy.
- EME.
- Ohtahara syndrome.

Infancy (<1 year)

- Epilepsy in infancy with migrating focal seizures.
- West syndrome.
- Benign infantile epilepsy.
- Benign familial infantile epilepsy.
- Dravet syndrome.
- Myoclonic encephalopathy in non-progressive disorders.

Childhood (1–12 years)

- Febrile seizures plus (FS +): can start in infancy.
- Panayiotopoulos syndrome.
- BECTS.
- Late onset childhood occipital epilepsy (Gastaut type).
- Lennox-Gastaut syndrome.
- CSWS or ESES.
- Landau–Kleffner syndrome.
- Autosomal dominant nocturnal frontal lobe epilepsy.
- Childhood absences epilepsy.
- Epilepsy with myoclonic atonic (previously astatic) seizures.
- Epilepsy with myoclonic absences.

Adolescence–adult (12–18 years)

- JAE.
- JME.
- Epilepsy with generalized tonic-clonic seizures.
- Progressive myoclonic epilepsies.
- Autosomal epilepsy with auditory features.
- Other familial temporal lobe epilepsies.

Less specific age relationship

- Familial focal epilepsy with variable foci (childhood to adult).
- Reflex epilepsies.

Distinctive constellations

- Mesial temporal lobe epilepsy with hippocampal sclerosis.
- Rasmussen syndrome.
- Gelastic seizures with hypothalamic hamartoma.
- Hemiconvulsion-hemiplegia-epilepsy.

Epilepsies attributed to structural: metabolic causes

- Malformation of cortical development (tuberous sclerosis complex, Sturge–Weber, etc.).
- Tumour.
- Infection.
- Trauma.
- Angioma.
- Perinatal insults.
- Stroke.

Epilepsies of unknown cause

- Conditions that are traditionally not diagnosed as a form of epilepsy per se.
- Benign neonatal seizures (BNS).
- Febrile seizures (FS) (10).

- Partial migrating epilepsy in infancy (PMEI).
- Dravet syndrome.

Early onset epileptic encephalopathy

Early (neonatal) myoclonic encephalopathy (EME or ME) and Ohtahara's syndrome (see Fig. 34.2) both present with early onset epileptic encephalopathy with burst suppression EEG. This EEG abnormality is a stable or invariant pattern, which persists longer than 2 weeks and has been traditionally associated with these 2 conditions (15). They both have an early onset with seizure intractability and poor prognosis. The main differentiating features are:

- Tonic spasms in Ohtahara's syndrome (OS) versus partial seizures and erratic myoclonias in EME.
- Continuous burst suppression pattern in OS whereas this pattern is augmented in sleep in EME.
- Aetiology is usually static structural brain damage in OS versus genetic or metabolic disorders in EME.
- Evolution to West syndrome and thereafter Lennox–Gastaut in OS.
- EME shows no unique evolution. It may continue as such or evolve to partial epilepsy or severe epilepsy with multiple independent spike foci.

Infantile epileptic encephalopathies (**IEE**) are early onset syndromic or non-syndromic epilepsies associated with a poor developmental outcome. Emerging genetic causes for these include mutations in *KCNQ2*, aristalsis-related homeobox (*ARX*), *CDLK5, SLC25A22, SPTAN1, STXBP1, SCN1A, SCN8A, PCDH19*, and *GR1N2 A*.

KCNQ2 encephalopathy

Benign familial neonatal seizures (BFNC) is associated with *KCNQ2* and *KCNQ3* mutations in 60–70% of families and presents with

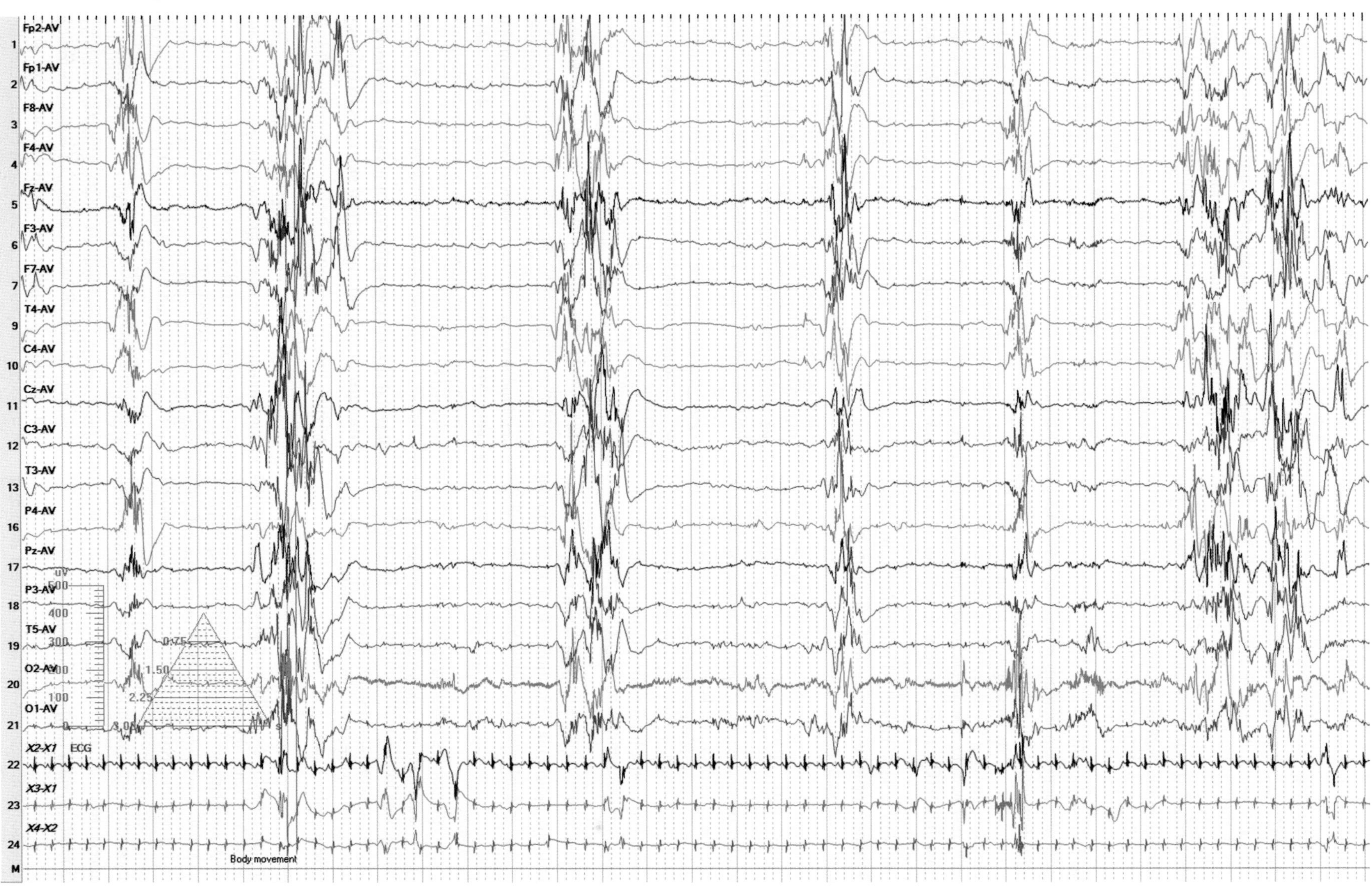

Fig. 34.2 Early (neonatal) myoclonic encephalopathy with burst-supression. Note myoclonic jerks in association with some of the bursts.

seizure typically between days 2 and 8 of life remitting by 16 months. Seizures begin with a tonic component followed by a range of autonomic and motor features, which may be unilateral, bilateral, or symmetrical. They are often accompanied with apnoea. Development is usually normal. *KCNQ2* mutation is also associated with a more severe phenotype presenting with intractable tonic seizures with autonomic signs in the first week of life with burst suppression or multifocal epileptiform activity (16). Seizures are intractable in the first few months to the first year of life with subsequent intellectual and motor impairment in the majority. Early magnetic resonance imaging (MRI) of the brain shows characteristic hyperintensities in the basal ganglia and thalamus that resolve later.

STXBP1- related encephalopathy

STXBP1 encodes the syntaxin binding protein 1, a regulatory component of the soluble NSF protein attachment protein receptor (SNARE) complex, with its binding to syntaxin required for fusion of membranes allowing for exocytosis of synaptic vesicles. Mutations in *STXBP1* have been reported in neonates with EIEE with suppression burst (Ohtahara syndrome) without a structural brain abnormality. It may also present with drug responsive infantile spasms with focal or lateralized epileptiform discharges. A generalized tremor appearing after the first year of life has been described (17).

Proctahedrin19

This presents as an early onset epilepsy exclusively in females with variable severity, with or without mental retardation. PCDH19 encodes proctahedrin 19 on chromosome Xq22.3, which plays a significant role in neuronal migration and synaptic connections. Clinical features include early onset (mean 8–10months) recurrent clusters of brief seizures, fever sensitivity, tonic seizures (including focal tonic), tonic–clonic seizures, focal seizures, which can generalize, intellectual disability and autistic traits. Focal seizures with ictal screaming can present in infancy within the context of hypomotor semiology or later on with prominent motor manifestations. Ictal EEGs during the focal seizures show a prominent involvement of the frontotemporal regions (18).

SPTAN encephalopathy

This presents with epileptic encephalopathy with early onset hypsarryhthmia, no visual attention, acquired microcephaly, spastic quadriplegia, and severe intellectual disability (19). Brainstem and cerebellar atrophy and cerebral hypomyelination are specific hallmarks. *SPTAN* encodes α-II spectrin, which is essential for proper myelination in zebra fish.

ARX *mutations*

These were first identified in three different phenotypes including X-linked lissencephaly with abnormal genitalia, non-syndromic X linked mental retardation, and X-linked infantile spasms. In other families Ohtahara syndrome, X- linked myoclonic epilepsy, Partington syndrome (intellectual disability with dystonic movements, ataxia, and seizures), mental retardation with tonic seizures and dystonia and infantile epileptic dyskinetic encephalopathy have also been reported.

Epilepsy due to PNPO (pyridox (am) ine 5′-phosphate oxidase) mutations

These can also present with seizures with onset under 1 month or after 3 months with an epileptic encephalopathy, which responds to pyridoxal 5′-phosphate. Seizure types range from clonic seizures, myoclonic jerks, and tonic seizures, generalized tonic clonic, and focal seizures. EEG abnormalities are also varied and include burst-suppression and hypsarrythmia (20).

Infantile spasms (epileptic spasms)

A spasm is a sudden brief flexion, extension, or mixed flexion-extension of predominantly proximal and trunchal muscles, which is usually more sustained than a myoclonic jerk, but not as sustained as a tonic seizure. The initial component is less the 2 s, which can be followed by the second tonic component, which lasts up to 5 s. Subtle spasms with grimacing or head nodding can also occur.

Spasms were first reported by Dr W. J. West in 1841 in his own 4-month-old son. They can present soon after birth up to 5 years of age; but in the majority of children occur within the first year of life, usually between 3 and 8 months. The triad of infantile spasms, hypsarryhthmia on EEG with or without developmental regression is termed as West syndrome. The aetiology of this age specific syndrome is diverse ranging from Idiopathic/Genetic causes to Symptomatic/Structural brain malformations and metabolic disorders. Hypsarryhthmia the pathognomic EEG pattern consists of disorganized and chaotic high voltage (>300 μV) random irregular slow waves interspersed with multifocal spikes and sharp waves (Fig. 34.3). During sleep, there is an increase in the frequency of epileptiform activity with discharges at times appearing more synchronous. Atypical hypsarryhthmia occurs when the discharges appear in synchrony and are not so chaotic or when there are intermittent periods of attenuation. Unilateral or asymmetric hypsarryhthmia is seen in structural malformations such as hemimegancephaly. Spasms can occur singly or in clusters and infantile spasms can occur without hypsarryhthmia and vice versa. Clinically spasms can be flexor, extensor, mixed, or subtle. They typically cluster in drowsiness or on awakening. Asymmetrical spasms indicate a structural aetiology. The most common electrophysiological ictal correlate is an electrodecrement, consisting of generalized attenuation with superimposed fast activity or a high amplitude slow-wave burst followed by attenuation. Factors associated with a poor prognosis include age of onset, symptomatic aetiology and other seizure types. Normal neurological outcome is reported in up to 10–15% patients.

Partial migrating epilepsy in infancy formerly malignant migrating partial seizures in infancy

Seizures usually begin in the first 6 months of life. A first phase can start soon after birth or even on the first day of life with sporadic seizures. Seizures are manly focal with rapid secondary generalization. Autonomic manifestation such as apnoea, gushing, or cyanosis frequently occur. Seizures may begin with the second 'stormy phase' occurring from 1–12 months with almost continuous migrating polymorphous focal seizures, in clusters of 5–30, several times a day or being continuous for days. Based on the origin of focal ictal discharges, semiology varies and may include lateral deviation of head and eyes, twitches of eyelids, clonic or tonic jerks of one or both limbs, apnoea, flushing and/or cyanosis of the face, chewing movements, mastication, and secondary tonic generalization. Multifocal ictal EEG discharges are seen due to involvement of different independent areas of both hemispheres. In sequential seizures the area of ictal onset may shift from one region to another

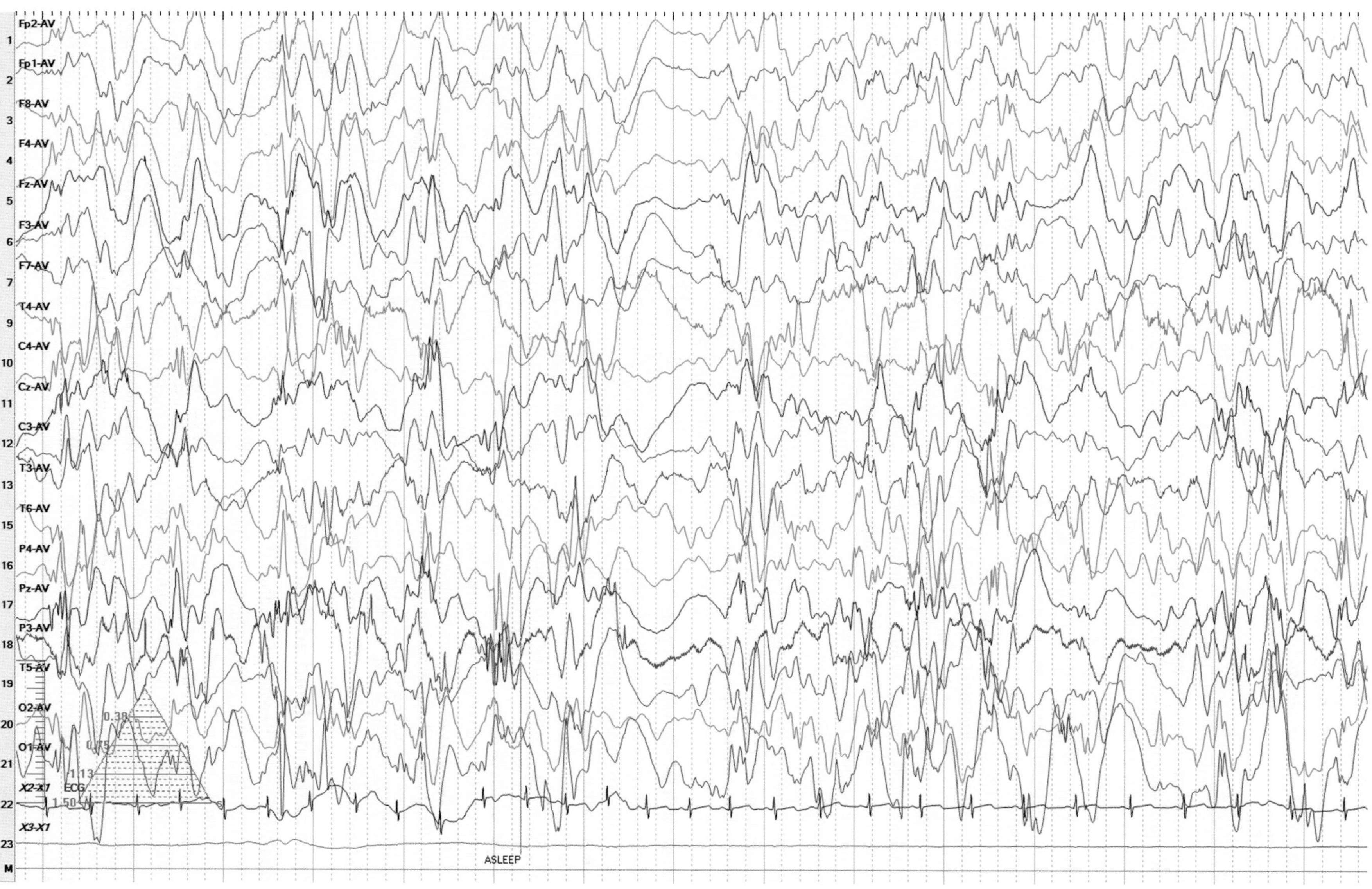

Fig. 34.3 Hypsarryhthmia interspersed with brief periods of generalized attenuation.

and from hemisphere to another (21). There is progressive deterioration of psychomotor development and acquired microcephaly. With increasing age, the amplitude of the ictal discharges tends to increase and frontal areas are more affected (22).

Dravet syndrome

This early onset epileptic encephalopathy was first described by Charlotte Dravet in 1978. It presents with generalized or unilateral clonic/tonic-clonic seizures in infancy in normal infants.

These are usually prolonged febrile seizures (23). This is followed by intractable epilepsy with polymorphic seizures, psychomotor delay, and ataxia. It is also known as severe myoclonic epilepsy of infancy (SMEI) due to emergence of myoclonic seizures in the second year of life. During the evolution from 1–4 years of age, children may also develop atypical absences, atonic, and focal seizures. The EEG is often normal at first or may show some non-specific posterior focal slowing. With the onset of myoclonic jerks generalized spike/polyspike-and-wave discharges may appear on an otherwise preserved background. Later on the epileptiform discharges can become focal with or without background slowing. SMEI-B (borderline) is a phenotype lacking one or more features of SMEI, such as myoclonus or generalized spike-and-wave discharges on EEG with less severe psychomotor impairment.

The voltage gated Na channel alpha subunit (SCN1A) is the most clinically relevant epilepsy gene and has been linked to several epilepsy syndromes. More than 70–80% of patients with Dravet's syndrome have mutations of *SCN1A*. About 95% of the mutations are de novo with overlapping clinical characteristics, which vary in severity (24).

Genetic epilepsy with febrile seizures plus

Genetic epilepsy with febrile seizures plus (GEFS +) is a familial epilepsy syndrome with heterogeneous phenotypes and autosomal dominant inheritance with incomplete penetrance, which can also be caused by *SCN1A* mutations. GEFS + has also been associated with mutations of genes encoding the sodium channel β1 subunit SCN1B, and the GABA (A) receptor gamma 2 subunit, GABRG2. The most common phenotypes are febrile seizures FS and febrile seizures plus FS +, when seizures persist beyond the age 6 years or they are associated with afebrile tonic-clonic seizures. Other phenotypes included mild generalized epilepsies, myoclonic astatic epilepsy, and SMEI, which represents the severe end of the spectrum. Temporal lobe epilepsy with or without hippocampal sclerosis has been increasingly recognized in the GEFS + spectrum with *SCN1A mutation* (25).

Conditions in childhood and adolescence

Genetic/idiopathic focal epilepsies of childhood

Idiopathic focal epilepsies is a term used to emphasize the absence of underlying structural abnormality and a genetic propensity to seizures. The term 'benign' implies that the seizures are easily treated or require no treatment, show remission without sequelae with ultimate or definitive remission before adulthood. Children usually do not have severe or exceedingly disturbing seizures, and have no associated serious intellectual or behavioural disturbance (26). However, the current evidence suggests that these assumptions are rapidly crumbling, but because of widespread traditional use of the term benign by clinicians, it continues to be accepted, with its limitations.

- Benign epilepsy with centrotemporal spikes (BECTS)/Benign Rolandic epilepsy (BRE).
- Panayiotopoulos syndrome.
- Late onset childhood occipital epilepsy (Gastaut type).

Benign epilepsy with centrotemporal spikes/Benign rolandic epilepsy

This is the most common epilepsy in school-going children, accounting for 15% of all childhood epilepsies. The seizure onset is usually between 3 and 10 years classically with rolandic type focal nocturnal seizures. Most have motor seizures, but sensory symptoms often occur. The focal seizures involve preferentially one side of the face (37%) with tonic contraction of one side of the face, clonic jerks of the cheek and eyelids, or both. There is also involvement of the oropharyngeal muscles (57%) and to a lesser extent the upper limb (20%). Lower limb involvement may occur less commonly. Oropharyngeal signs include guttural sounds, mouth movement, and profuse salivation. Sensory symptoms involve the corner of the mouth, the inside of one cheek, the tongue, and gums. Speech arrest often occurs. As seizures are mainly nocturnal in more than half of patients, the focal onset may be missed and they may be reported as generalized convulsions. Seizure frequency is generally low, with one-quarter of patients having only one attack, half have fewer than five and only 8% have 20 seizures or more. When seizures are frequent they often occur in clusters. The total duration of epilepsy is relatively short with only 7% children having epilepsy lasting 3–8 years. In 92% patients' remission occurs by 12 years of age (27). Normalization of the EEG occurs after the clinical remission.

The background EEG is normal, with variability in the topography and frequency of epileptiform discharges. There is no relationship between the frequency and extent of epileptiform abnormalities and the seizure frequency or duration. The characteristic interictal EEG abnormality is high voltage sharp waves, which are often followed by slow waves, and are focal or multifocal over the centro-temporal/sylvian, mid temporal, centro-parietal, fronto-central, or centro-occipital regions. Bilateral discharges, which may be synchronous or asynchronous can also occur. These focal discharges occur frequently in bursts of 1.5–3 Hz and activate in slow wave sleep diffusing to the ipsilateral and sometimes contralateral hemisphere (see Fig. 34.4). The topography of the paroxysms may change in sequential recordings. The main negative component of Rolandic spikes can be modelled by a tangential dipole with the negative pole in the centro-temporal region and the positive pole in the frontal region (28), which may have a bearing on prognosis. Occasionally generalized discharges can occur concomitantly.

This epilepsy is no longer considered to be entirely 'benign' as there are reported stable or intermittent neuropsychological and specific cognitive deficits such as that of language, attention, and visual-motor co-ordination. Dyspraxia and mild intellectual and behavioural disorders are also reported. Hommet et al. (29) found no significant difference in memory, language function, or executive function in adults who had recovered from BECTS.

Panayiotopoulos syndrome

This is characterized by early onset (3–6 years) autonomic seizures, which are often prolonged and infrequent. Seizures are mostly nocturnal and autonomic symptoms and signs with a feeling of nausea and ictal vomiting occur from the onset in 80% of the seizures.

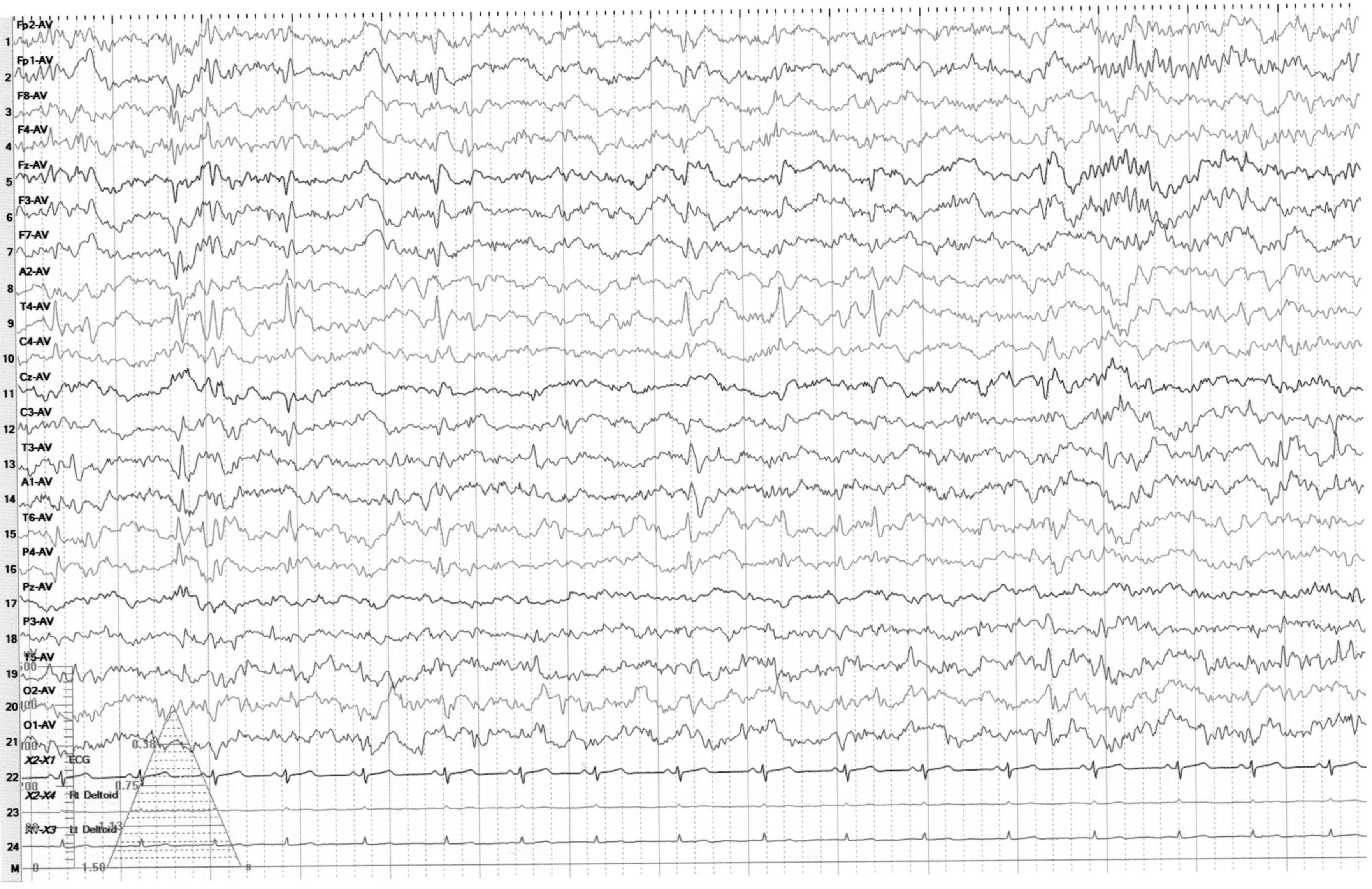

Fig. 34.4 Benign childhood epilepsy with centro-temporal spikes. Activation of focal sharp waves over the centro-temporal regions independently and bilaterally.

Pallor, cyanosis, pupillary dilatation, thermoregulatory changes, and hypersalivation are noted. These autonomic features are followed in the majority of seizures by impairment of consciousness (94%), deviation of eyes (60–80%), hemi clonic (20%), or generalized convulsions (20%). Visual symptoms are infrequent. Seizure duration can vary from 30 min to hours. The child is back to a normal state after a few hours of sleep. The clinical presentation is often confused with encephalitis, syncope, cyclical vomiting, or atypical migraine. The interictal EEG shows multifocal spikes/sharp waves or sharp and slow waves at various locations. Though occipital discharges, which can be similar in morphology to rolandic discharges, predominate they are neither prerequisite nor specific to this syndrome (30).

Late onset childhood occipital epilepsy (Gastaut type)

Age of onset of this syndrome is 3–15 years and often children have a family history of epilepsy or migraine. The characteristic clinical feature is occipital seizures with elementary or formed visual hallucinations or visual illusions and at times transient amaurosis. This can be followed by deviation of the eyes with ipsilateral turning of the head. Gastaut reported 41% of the patients developed hemiclonic seizures, 19% became unresponsive 8% had generalized tonic-clonic seizures. Seizures are usually diurnal and brief lasting 1–3 min. Post-ictal headache is seen in half of the patients and in a minority this can be associated with nausea and vomiting. The interictal EEG shows a normal background with unilateral or bilateral occipital paroxysms seen on eye closure with fixation- off sensitivity.

Idiopathic/genetic generalized epilepsies of childhood

These are probably genetic syndromes in which children are normal prior to epilepsy onset and have generalized spike-and-wave discharges against a normal background EEG with normal neuroimaging. The prognosis is considered favourable. There is now emerging evidence that these syndromes are not as drug responsive as previously thought. There are associated cognitive deficits such as attention problems and deficits in executive functioning. Individuals with Primary generalized or Idiopathic generalized epilepsy and other so called benign epilepsies have worse social and educational outcomes as compared to their peers (31). The ILAE recommends the use of Genetic generalized epilepsies to classify these syndromes.

- Childhood absence epilepsy.
- Epilepsy with myoclonic atonic (previously astatic) seizures.
- Epilepsy with myoclonic absences.

Childhood absence epilepsy

This is a common epilepsy in school age children presenting between 4–10 years of age, usually with normal neurology and development. Absences are brief (4–20 s, exceptionally longer) and occur very frequently in tens to hundreds per day. Abrupt impairment of consciousness is the rule with automatisms often seen in accompaniment (32). The ictal EEG shows bilateral, synchronous, and symmetrical 3 Hz generalized spike-and-wave discharges on a normal background. Absences are produced via reciprocally connected thalamocortcial neurons. Hyperventilation can activate typical absences in 70–90% children (33,34). Absence seizures per se may occur within different epilepsy syndromes. Therefore the stricter exclusion criteria as proposed by Panayiotopoulos (35) allows for definition of CAE within a homogenous group, which is helpful for prognostication (36).

The exclusion criteria are:

- Presence of seizures such as GTCS, or myoclonic jerks before or during the active stage of absences.
- Absences with marked eyelid or peri-oral myoclonus, single or rhythmic massive limb jerking and single or arrhythmic myoclonic jerks of the head, trunk, or limbs.
- Absences with mild or clinically undetectable impairment of consciousness during the 3–4 Hz discharges.
- Stimulus sensitive absences (photosensitive, pattern sensitive).
- Irregular arrhythmic spike and multiple spike and -slow wave discharges with marked variations of the inter-discharge frequency or of the spike and multiple spike-and-slow wave relations.
- Predominant brief discharges 3–4 Hz lasting <4 s.
- Fixed 'lead in' anomaly in the frontal region.

Behavioural, psychiatric, language, and cognitive comorbidities have been reported in children with CAE. In particular attention deficits are associated with reduced academic performance and persist despite successful treatment. Remission rates range from 21 to 74% with GTCS developing in about 40% children.

Epilepsy with myoclonic atonic (previously astatic) seizures

Also known as Doose syndrome it is a rare generalized epilepsy regarded as having a genetic aetiology with multiple seizure types in which the major seizure types are myoclonic and astatic/atonic seizures. Onset occurs within the first five years of life, usually within 3–4 years of age in children who are normal with mild speech delay. However, in 24% children the initial presentation may occur within the first year of life. The ILAE 1989 inclusion criteria are:

- Normal development until the onset of seizures.
- No organic or other obvious cause for seizures.
- Onset of myoclonic-astatic seizure between 7 months and 6 years.
- Often a hereditary predisposition.
- *Seizure types*: myoclonic, astatic, myoclonic-astatic, absence, tonic-rarely during sleep (debatable inclusion), clonic, generalized tonic-clonic.
- Status epilepticus is common.

The EEG usually shows preserved posterior background rhythms and sleep architecture. Central-parietal theta rhythms may be seen and with progression of the disease, brief 2–5 Hz spike/polyspike-and-wave discharges appear. Photosensitivity may also be seen. In younger children the EEG may show continuous irregular activity similar to hypsarryhthmia. Slow spike-and-wave discharges (≤2.5 Hz) may be seen in the later course of the disease and reflects the severe end of this epilepsy spectrum, which is now regarded as an epileptic encephalopathy. (see Fig. 34.5).

Prognosis is variable in children who appear similar at presentation and Doose identified risk factors for unfavourable

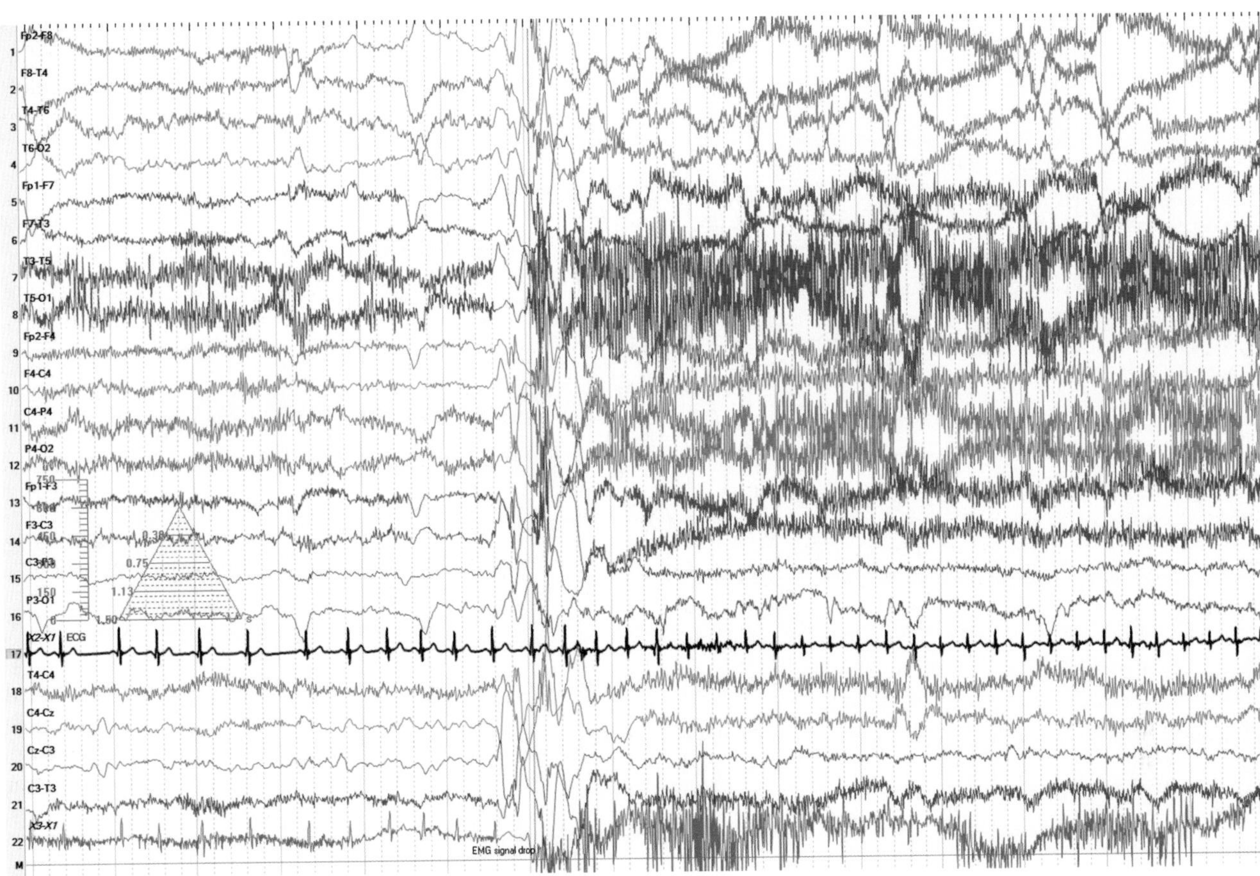

Fig. 34.5 Myoclonic-atonic seizure associated with a generalised burst of spike-and-wave discharges with a drop in the EMG polygraphy signal from left deltoid (Channel X3-X1).

prognosis; onset with febrile and afebrile GTCS during the first 18 months of life, status of minor seizures, persistent of 4–7 Hz rhythms and failure to develop an occipital alpha rhythm. Presences of tonic seizures, minor or absence status, and slow spike-and-wave discharges on EEG also indicate an unfavourable prognosis. Mullen et al. (37) reported Glucose transporter deficiency as a treatable cause of myoclonic astatic epilepsy in 5% of their cohort.

Epilepsy with myoclonic absences

This is a rare childhood epilepsy, which can occur as early as infancy peaking at 7 years with a male predominance. It occurs in two subgroups of children: an Idiopathic subset who are normal, and a symptomatic group with pre-existing developmental and neurological abnormalities. The Idiopathic subset is what really constitutes this syndrome though it is less frequently seen. Frequent typical absences, which have an abrupt onset and offset are seen in association with rhythmical myoclonias and generalized 3 Hz spike-and-wave discharges on the EEG lasting 10–60 s (see Fig. 34.6). There is axial hypertonia as arms are raised with myoclonic jerks involving the muscles of the shoulder, arms, and legs. This leads to rhythmic jerking, which can be symmetrical or asymmetrical. Polygraphic EMG leads show the myoclonic jerks occur concurrently with the spike component of the generalized spike-and-wave discharges. The background EEG is normal in the Idiopathic group and may be slowed in the Symptomatic group. GTCS or atonic seizures are present in the symptomatic subgroup and indicate an unfavourable prognosis. The evolution is variable with half of those with normal development prior to epilepsy onset developing cognitive and behavioural impairment.

Genetic generalized epilepsies in adolescence (12–18 years)

- Juvenile absence epilepsy (JAE).
- Juvenile myoclonic epilepsy (JME).
- Epilepsy with generalized tonic–clonic seizures.

Juvenile absence epilepsy

JAE presents between 10–17 years. The absences in JAE are also associated with an impairment of consciousness though they may be milder than those seen in CAE and their duration is often longer (4–30 s). The ictal discharges may have a faster frequency at 3–4 Hz of spike/polyspike-and-wave discharges. Absences also occur less frequently up to 1–10/day and GTCS occur in about 80% of patients, mostly on awakening. Less frequently myoclonic jerks may also occur in association.

Juvenile myoclonic epilepsy

This the most common genetic generalized epilepsy of childhood with the majority of children presenting between 12 and 18 years (range of 5–34 years). It is clinically and genetically a heterogeneous

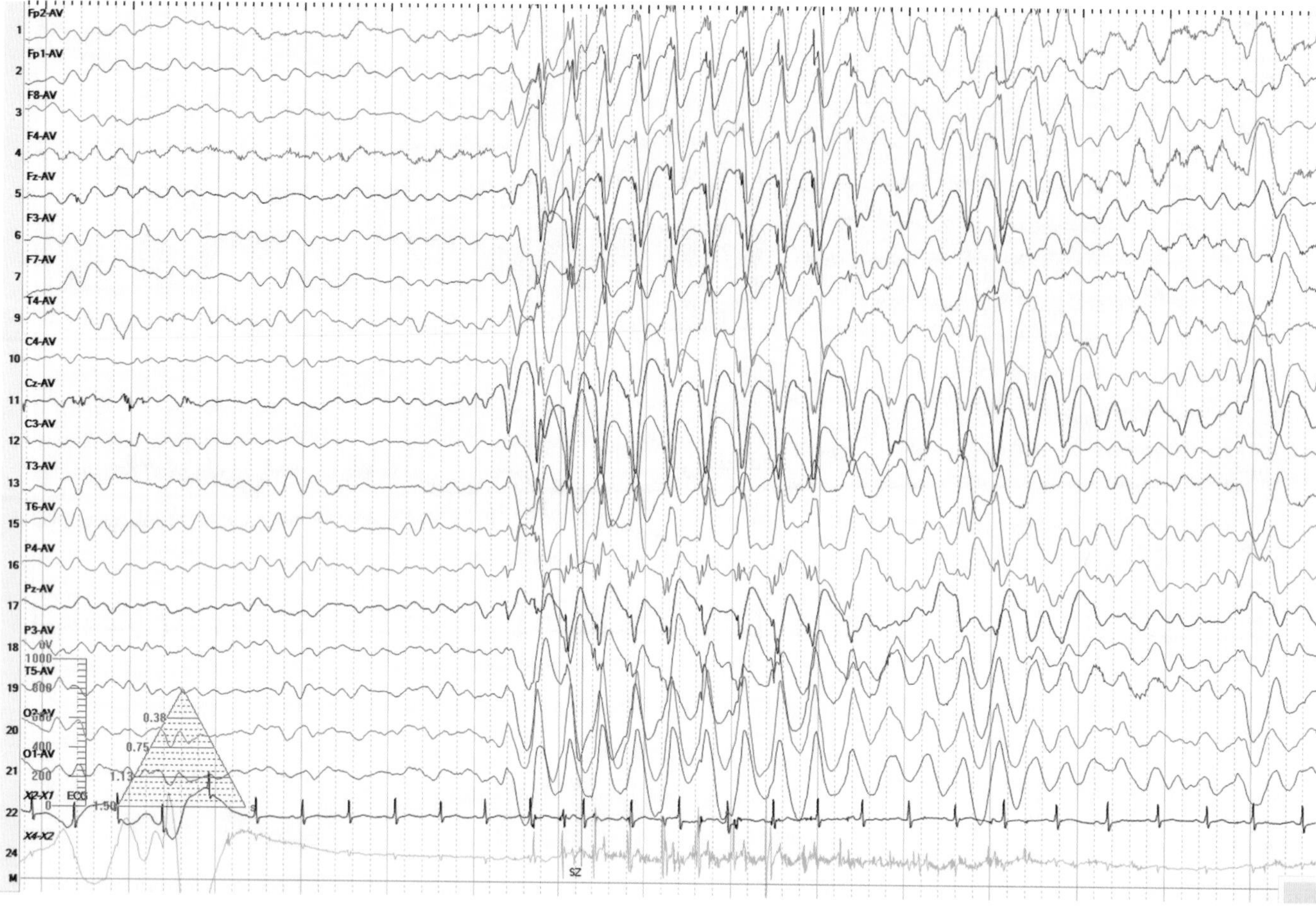

Fig. 34.6 Myoclonic absences with a series of myoclonic jerks.
Note: EMG channel (X4-X2) jerks correlating with the spike component of the generalised spike-and-wave discharges.

epilepsy caused by a dysfunction in the frontocortical-subcortical neuronal network. It is characterized by arrhythmic myoclonus, which can be bilateral, single, or repetitive, with the irregular jerks predominantly affecting the arms. Some jerks occur unilaterally. Absences and generalized tonic clonic seizures (GTCS) usually occur in association.

Absences first present typically between 5 and 16 years. Myoclonic jerks follow between 1 and 9 years later and GTCS occur in 90–95% of patients, a few months after the myoclonus. The seizures usually occur after awakening and are precipitated by sleep deprivation, emotional stress, menstruation, and alcohol. Inter-ictal EEG shows a normal background with brief (0.5–10 s) bursts of 4–6 Hz generalized spike/polyspike-and-wave discharges. Photosensitivity is seen in 30–40% (38).

Epilepsy with generalized tonic-clonic seizures

This presents between 6–47 years of age with a peak at 16–17 years with predominantly generalized tonic clonic seizures. In the majority, these occur a few hours after awakening from nocturnal or diurnal sleep. It has similar electroclinical features to JME such as precipitation by sleep deprivation, stress, and alcohol. EEG features are also similar and often unreported absences are found to occur concurrently. Photosensitivity is seen in 15% of patients.

Conditions associated with electrical status during slow wave sleep

This is the EEG pattern of continuous bilateral and diffuse spike-and-wave discharges occupying ≥ 85% NREM sleep; with a varied phenotype. It is postulated that the topography of the discharges and the duration of electrical status during slow wave sleep (ESES) account for varying degree of cognitive dysfunction (see Table 34.1).

Lennox–Gastaut syndrome

This polymorphic syndrome is characterized by Tonic axial, Atonic, and Atypical absence seizures in association with an inter-ictal EEG, which shows diffuse slow spike-and-wave discharges during wakefulness (<2.5 Hz). In sleep bursts of fast rhythmic waves and slow polyspikes are seen along with characteristic generalized fast rhythms at about 10 Hz. This is clinically associated with intellectual disability either before or at the time of diagnosis. Frequent or prolonged Atypical absences can result in period of non-convulsive status. Clinically, atypical absences are often associated with a loss in tone, which may be localized to the head or neck muscles with excessive drooling from hypersalivation. The aetiology is diverse, but often symptomatic/structural secondary to trauma, cerebral malformation, tumour, encephalitis, or as a sequelae to West syndrome.

Table 34.1 Clinical syndromes associated with ESES (Electrical status in slow wave sleep) in the absence of a structural lesion

Syndrome	Features	Awake EEG	Sleep EEG SWI
CSWS/ESES syndrome, or Encephalopathy with ESES or Penelope syndrome (38)	Global neuropsychological decline Behavioural disturbance Motor impairment Nocturnal focal seizures Generalized minor seizures (atonic) seizures No tonic seizures	Florid/diffuse 2–3 Hz spike-and-wave discharges, which have a FT/CT emphasis. Long runs can have clinical correlate (see Fig. 34.7) Posterior background rhythms preserved Clinically child may appear to be in non-convulsive status	Non-REM ESES > 85%
Atypical BECTS	Presents like BECTS, but may have earlier onset with normal development Nocturnal focal seizures Atypical absences/atonic seizures in the day Outcome better than CSWS syndrome	Spike-and-wave discharges more florid than what is seen in BECTS with bilateral runs of 2–3 Hz spike-and-wave discharges, which have a CT maximum	Non-REM SWI > 50%
Landau-Kleffner Syndrome	Acquired auditory agnosia Receptive/mixed aphasia, verbal agnosia Behavioural disturbance ± cognitive impairment Infrequent seizures	CT/TO discharges	Variable Any % BTESES/ UnilateralESES (39)
Opercular syndrome	Resembling BECTS with fluctuating periods of sialorrhoea, drooling, and oromotor dyspraxia	Marked activation of CT discharges (bilateral)	Non-REM ± ESES

CSWS/ESES = continuous spike-and-wave during slow wave sleep/electrical status during slow wave sleep; BECTS = benign childhood epilepsy with centrotemporal spikes; REM = rapid eye movement; BTESES = bitemporal ESES; SWI = spike wave index; CT = centrotemporal; FT fronto-temporal; TO = temporo-occipital.

Source data from *Seizure*, **9**(2), Appleton R, Beirne M, Acomb B, Photosensitivity in juvenile myoclonic epilepsy, pp. 108–11, copyright (2000), Elsevier.

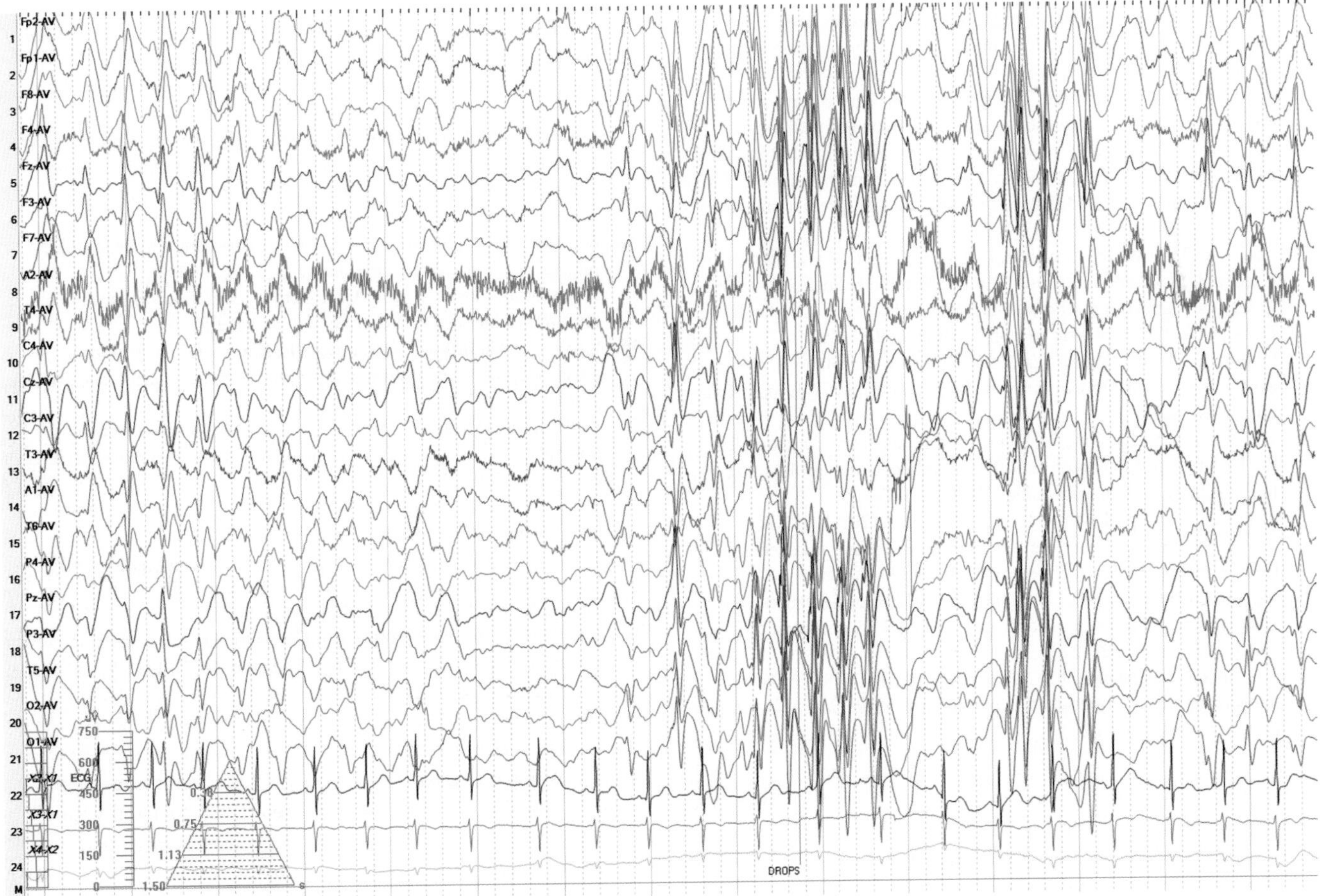

Fig. 34.7 CSWS/ESES syndrome. The awake EEG shows florid 2–2.5 Hz bilateral diffuse spike-and-wave discharges, at times clinically associated with intermittent head drops (atonia). There is also global neuropsychological decline. In sleep ESES is seen.

References

1. André, M., Lamblin, M.D., d'Allest, A.M., et al. (2010). Electroencephalography in premature and full-term infants. Developmental features and glossary *Neurophysiology Clinics*, **40**(2), 59–124.
2. Dreyfus-Brisac, C. (1975). Neurophysiological studies in human premature and full-term newborns. *Biological Psychiatry*, **10**(5), 485–96.
3. Pampiglione, G. (1972). Some criteria of maturation in the EEG of children up to the age of 3 years. *Electroencephalography & Clinical Neurophysiology*, **32**, 463(P).
4. Petersen, I. and Eeg-Olofsson, O. (1971). The development of the electroencephalogram in normal children from the age of 1 through 15 years. Nonparoxysmal activity. *Neuropaediatrie*, **2**, 247–304.
5. Samson-Dollfus, S. and Goldberg, P. (1979). Electroencephalographic quantification by time domain analysis in normal 7–15-year-old children. *Electroencephalography Clinical Neurophysiology*, **46**, 147–54.
6. Gibbs, F.A. and Gibbs, E.L. (1950). *Atlas of electroencephalography, Vol 1: Normal controls*. Cambridge MA: Addison-Wesley.
7. Kellaway, P. and Fox, B.J. (1952). Electroencephalographic diagnosis of cerebral pathology in infants during sleep. I. Rationale, technique, and the characteristics of normal sleep in infants. *Journal of Paediatrics*, **41**(3), 262–87.
8. Tanguay, P.E., Ornitz, E.M., Kaplan, A., and Bozzo, E.S. (1975). Evolution of sleep spindles in childhood. *Electroencephalography & Clinical Neurophysiology*, **38**, 175–81.
9. Duchowny, M. and Harvey, A.S. (1996). Pediatric epilepsy syndromes: an update and critical review. *Epilepsia*, **37**(1), S26–40.
10. Berg, A.T., Berkovic, S.F., Brodie, M.J., et al. (2010). Revised terminology and concepts for organization of seizures and epilepsies: report of the ILAE Commission on Classification and Terminology, 2005–2009. *Epilepsia*, **51**(4), 676–85.
11. van Laerhoven, H., de Haan, T.R., Offringa, M., Post, B., and van der Lee, J.H. (2013). Prognostic tests in term neonates with hypoxic-ischemic encephalopathy: a systematic review. *Pediatrics*, **131**(1), 88–98.
12. Murray, D.M., Boylan, G.B., Ryan, C.A., and Connolly, S. (2009). Early EEG findings in hypoxic-ischemic encephalopathy predict outcomes at 2 years. *Pediatrics*, **124**(3), e459–67.
13. Thoresen, M., Hellström-Westas, L., Liu, X., and de Vries, L.S. (2010). Effect of hypothermia on amplitude-integrated electroencephalogram in infants with asphyxia. *Pediatrics*, **126**(1), e131.
14. Dulac, O. (2001). Epileptic encephalopathy. *Epilepsia*, **42**(3), 23–6.
15. Ohtahara, S. and Yamatogi, Y. (2003). Epileptic encephalopathies in early infancy with supression-burst. *Journal of Clinical Neurophysiology*, **20**(6), 398–407.
16. Weckhuysen, S., Ivanovic, V., Hendrickx, R., et al. (2013). Extending the KCNQ2 encephalopathy spectrum: clinical and neuroimaging findings in 17 patients. *Neurology*, **81**(19), 1697–703.
17. Mignot, C., Moutard, M.L., Trouillard, O., et al. (2011). STXBP1-related encephalopathy presenting as infantile spasms and generalized tremor in three patients. *Epilepsia*, **52**(10), 1820–7.
18. Marini, C., Darra, F., Specchio, N., et al. (2012). Focal seizures with affective symptoms are a major feature of PCDH19 gene-related epilepsy. *Epilepsia*, **53**(12), 2111–19.
19. Tohyama, J., Nakashima, M., Nabatame, S., et al. (2015). SPTAN1 encephalopathy: distinct phenotypes and genotypes. *Journal of Human Genetics*, **60**(4), 167–73.
20. Mills, P.B., Camuzeaux, S.S., Footitt, E.J., et al. (2014). Epilepsy due to PNPO mutations: genotype, environment and treatment affect presentation and outcome *Brain*, **137**(Pt 5), 1350–60.
21. Coppola, G. (2009). Malignant migrating partial seizures in infancy: an epilepsy syndrome of unknown etiology. *Epilepsia*, **50**(5), 49–51.
22. Dulac, O. (2005). Malignant partial seizure in infancy. In: J. Roger, M. Bureau, Ch. Dravet, P. Genton, C.A. Tassinari, and P. Wolf (Eds) *Epileptic syndromes in infancy, childhood and adolescence*, 4th edn, pp. 73–6. Montrouge: John Libbey Eurotext Ltd.
23. Dravet, C. and Oguni, H. (2013). Dravet syndrome (severe myoclonic epilepsy in infancy). *Handbook of Clinical Neurology*, **111**, 627–33.
24. Gambardella, A. (2009). Marini Clinical spectrum of SCN1A mutations. *Epilepsia*, **50**(5), 20–3.
25. Mantegazza, M., Gambardella, A., Rusconi, R., et al. (2005). Identification of an Nav1.1 sodium channel (SCN1A) loss-of-function mutation-associated with familial simple febrile seizures. *Proceedings of the National Academy of Science United States of America*, **102**(50), 18177–82.
26. Guerrini, R. and Pellacani, S. (2012). Benign childhood focal epilepsies. *Epilepsia*, **53**(4), 9–18.
27. Bouma, P.A., Bovenkerk, A.C., and Westendorp, R.G. (1997). Brouwer of the course of benign partial epilepsy of childhood with centrotemporal spikes: a meta-analysis. *Neurology*, **48**(2), 430–7.
28. Gregory, D.L. and Wong, P.K. (1992). Clinical relevance of a dipole field in rolandic spikes. *Epilepsia*, **33**(1), 36–44.
29. Hommet, C., Billard, C., Motte, J., et al. (2001). Cognitive function in adolescents and young adults in complete remission from benign childhood epilepsy with centro-temporal spikes. *Epileptic Disorder*, **3**(4), 207–16.
30. Koutroumanidis, M. (2002). Panayiotopoulos syndrome. *British Medical Journal*, **324**(7348), 1228–9.
31. Sillanpää, M., Jalava, M., Kaleva, O., and Shinnar, S. (1998). Long-term prognosis of seizures with onset in childhood. *New England Journal of Medicine*, **338**(24), 1715–22.
32. Panayiotopoulos, C.P. (1999). Typical absence seizures and their treatment. *Archives of Disease in Childhood*, **81**, 351–5.
33. Drury, I. (2000). Activation of seizures by hyperventilation. In: H. O. Luders, S. Noachtars (Eds) *Epileptic seizures: pathophysiology and clinical semiology*, pp. 575–9. Philadelphia, PA: Churchill & Livingstone.
34. Mendez, O.E. and Brenner, R.P. (2006). Increasing the Yield of EEG. *Journal of Clinical Neurophysiology*, **23**(4), 282–93.
35. Panayiotopoulos, C.P. (2005). Idiopathic generalized epilepsies: a review and modern approach. *Epilepsia*, **46**(9), 1–6.
36. Grosso, S., Galimberti, D., Vezzosi, P., et al. (2005). Childhood absence epilepsy: evolution and prognostic factors. *Epilepsia*, **46**(11), 1796–801.
37. Mullen, S.A., Marini, C., Suls, A., et al. (2011). Glucose transporter 1 deficiency as a treatable cause of myoclonic astatic epilepsy. *Archives of Neurology*, **68**(9), 1152–5.
38. Appleton, R., Beirne, M., and Acomb, B. (2000). Photosensitivity in juvenile myoclonic epilepsy. *Seizure*, **9**(2), 108–11.
38. Tassinari, C.A., Cantalupo, G., Rios-Pohl, L., Giustina, E.D., and Rubboli, G. (2009). Encephalopathy with status epilepticus during slow sleep: 'the Penelope syndrome'. *Epilepsia*, **50**(7), 4–8.
39. Scheltens-de Boer, M. (2007). Guidelines for EEG in encephalopathy related to ESES/CSWS in children. *Epilepsia*, **50**(7), 13–17.

CHAPTER 35

Sleep disorders

Zenobia Zaiwalla and Roo Killick

The *International Classification of Sleep Disorders* published by the American Academy of Sleep Medicine (AASM) (1) recognizes a long list of sleep disorders in both children and adults (Box 35.1). However, in practice patients are referred mainly for the complaint of excessive daytime sleepiness (too much sleep), insomnia (too little sleep), circadian rhythm disorders (sleeping at wrong times), or parasomnias (episodic behaviours including motor activity in sleep). Patients may present with one or more of these complaints and it is the clinician's role to identify the primary sleep disorder and target investigations to aid diagnosis and management.

Excessive daytime sleepiness

Excessive daytime sleepiness (EDS) is the inability to stay awake and alert during the major waking episodes of the day (1), to be distinguished from fatigue and tiredness, which denote a lack of energy, motivation, and strength (2). Inability to stay awake can have major implications for personal and public safety, is a symptom of many disorders and can have multifactorial origins (see Box 35.2).

Sleepiness can be persistent and daily, or episodic. It is the most frequent complaint prompting referral for diagnostic sleep studies, including for the two best known sleep disorders; obstructive sleep apnoea (OSA) and narcolepsy syndrome. While sleepiness due to sleep-related breathing disorder can be diagnosed with confidence following a respiratory sleep study often without EEG, the diagnosis of non-respiratory sleep disorders can be challenging, especially in adults who may be on medication for comorbid depression, or other medical/psychiatric disorders including potentially sedative drug combinations for epilepsy or chronic pain. Also patients may have more than one sleep disorder with accumulative effect, for example, OSA and narcolepsy or Restless Leg Syndrome (RLS).

The Epworth Sleepiness Scale (ESS) (3) is a subjective standardized measure of sleepiness. It includes eight situations for which patients indicate their likelihood of dozing ('never', 'slight', 'moderate', or 'high' chance). Primary care physicians often use the ESS of ≥10/24, along with the clinical history and examination, as an aid to deciding whether the patient should be referred for sleep studies, though 10% of population have scores of 11 or more (4,5). Hirshkowitz et al. (6), based on published studies and their work,

Box 35.1 Sleep disorders: ICSD-3 (1)

- Insomnia.
- Sleep-related breathing disorders.
- Central disorders of hypersomnolence.
- Circadian rhythm sleep disorders.
- Parasomnias.
- Sleep-related movement disorders.
- Other sleep disorders when cannot be classified elsewhere and includes environmental sleep disorder.
- Sleep-related medical and neurological disorders including fatal familial insomnia, sleep-related epilepsy, sleep-related headaches, sleep-related laryngospasm, sleep-related gastro-oesophageal reflux, sleep-related myocardial ischaemia.

Source data from *International Classification of Sleep Disorders*, Third Edition (ICSD-3), copyright (2014), American Academy of Sleep Medicine.

Box 35.2 Excessive daytime sleepiness (EDS)

Common causes of EDS

- Sleep-disordered breathing.
- Narcolepsy types 1 and 2.
- Idiopathic hypersomnia.
- Insufficient sleep.
- Circadian rhythm sleep–wake disorders including shift work.
- Interrupted sleep due to a primary sleep disorder (e.g. RLS/PLMD) or other medical disorders (e.g. with pain).
- Prescribed and non-prescribed sedative medication/substances.
- Psychiatric disorders including depression and bipolar disorder.
- Chronic fatigue syndrome/fibromyalgia.

Less common causes

- *Other medical disorders*: e.g. myotonic dystrophy, Prader–Willi syndrome.
- Head injury.
- Tumours or autoimmune disorders involving the hypothalamus.
- Endocrine disorders.
- Epilepsy.
- Malingering/psychogenic sleep like periods.

suggest a score of 0–8 as normal, 9–12 as mild sleepiness, 13–16 as moderate, and more than 16 as severe sleepiness. The 'STOP-BANG' questionnaire is a shorter questionnaire, specifically used to screen for OSA that takes into account snoring, tiredness, observed pauses in breathing, and common risk factors for OSA (7).

Sleep-related breathing disorders

Sleep-related breathing disorders (often referred to as sleep disordered breathing (SDB)) refer to a group of disorders characterized by disturbed nocturnal respiration. The commonest conditions are OSA, central sleep apnoea (CSA), and sleep-related hypoventilation. There can be an overlap of conditions in one individual.

Obstructive sleep apnoea

OSA syndrome is a common disorder affecting up to 4% of middle aged men and 2% of women (8). OSA refers to the objective abnormalities during sleep, whereas OSA 'syndrome' includes the symptoms associated with it. It is characterized by repetitive obstructions of the upper airway with associated oxygen desaturations, cognitive arousals and subsequent sleep fragmentation. The obstructions can either be full (apnoeas, defined as cessation of airflow for at least 10 s) or partial (hypopnoeas), but both lead to reduction in oxygen saturations and/or arousals from sleep.

OSA is strongly associated with obesity, body mass index (BMI), and neck circumference being important predictors of disease (9). Although OSA is most commonly seen in overweight individuals, certain anatomical features of the upper airway (for instance, retrognathia, significant tonsillar hypertrophy, and other craniofacial abnormalities) can also predispose to disease in normal weight individuals, along with alterations in ventilatory control (10) and there is an increased prevalence seen in certain conditions including Marfan's disease (11).

OSA is commoner in men than women and incidence increases with age. It occurs in all ethnic groups. It can run in families and although the exact genetic marker remains elusive at this time, several polymorphisms have been proposed requiring further investigation with genome wide studies (12). Often patients will present following a family member's diagnosis, having become informed of the condition and the improvement in symptoms following treatment.

There is a significant mortality risk associated with untreated significant OSA (13–15), as well as important associated comorbidities including cardiovascular disease, hypertension, atrial fibrillation, stroke, type 2 diabetes mellitus, depression, and erectile dysfunction. There is also an increased risk of motor vehicle accidents, especially in heavy goods vehicle drivers (16).

Common symptoms of OSA include daytime sleepiness, snoring, pauses in breathing heard by the bed partner (witnessed apnoeas), gasping or choking during sleep, dry mouth, sore throat, unrefreshing sleep, restless sleep, gastro-oesophageal reflux, and nocturia. Sleepiness is due to sleep fragmentation and can be highly subjective. Often it is first noticed in passive situations, but can occur at all times in more severe cases. The ESS is widely used in both primary and specialist care settings, which assesses recent subjective daytime sleepiness (3). It is useful to seek an objective history from the partner as well, with regards to sleep quality and sleepiness in the individual.

Clinical examination should include BMI calculation, neck circumference measurement, daytime oxygen saturations, cardiovascular examination and assessment of the upper airway and dentition. The Mallampati score, commonly used by anaesthetists, is used to describe visualization of the airway, with a narrowed, crowded airway and prominent tongue being associated with higher risk of OSA. Assessment of the jaw structure and dentition should be noted, included any significant micrognathia, under- or over-bite, and state of dentition, as good dentition is required for treatment with a mandibular advancement device to be considered.

A sleep study may show a range of obstructive events during sleep, from full apnoeas, hypopnoeas, or indeed respiratory effort related arousals (RERAs)—whereby partial airway narrowing, that may not cause significant oxygen desaturations, does however, still lead to arousals and sleep fragmentation. Those patients with predominantly the latter may be referred to as having upper airway resistance syndrome (UARS). The spectrum of disease can be thought of as a continuum; from snoring to UARS to more significant degrees of OSA.

There are published diagnostic criteria for OSA by the AASM for scoring sleep studies performed with EEG, airflow, and oximeter parameters, whereby the apnoea-hypopnoea index (AHI) indicates the degree of OSA:

- Mild: 5–15 events/h.
- Moderate: 15–30 events/h.
- Severe: >30 events/h.

There is wide variability in the UK, and indeed worldwide, regarding what type of sleep study system is utilized in different centres. For instance, if using a single channel device measuring oximetry, this will not provide a scored AHI value, but will calculate the oxygen desaturation index (ODI). Assessing what type of study has been performed is important in its interpretation and skilled sleep physicians locally should report each study with a clinical interpretation based on the information available by that system. Clinical correlation is always required and if there is any doubt in the diagnosis with a limited channel device, then a more detailed study should be recommended at a centre providing that service.

Central sleep apnoea

CSA occurs when there is cessation in breathing without associated respiratory effort, unlike OSA. CSA is often associated with alterations in the ventilatory control system in the brainstem, leading to changes in ventilatory drive and unstable breathing due to pCO_2 dropping below the apnoeic threshold. CSA is most often associated with cardiac failure where a waxing-waning pattern of breathing called Cheyne–Stokes breathing can occur. It is important that patients with heart failure, and indeed renal failure or a history of stroke, are screened for sleep symptoms as a high proportion will have some form of sleep disordered breathing. Patients may complain of symptoms of poor sleep including sleepiness, sleep fragmentation, and insomnia. These symptoms however, can also be due to the underlying condition and may or may not resolve with treatment of sleep disordered breathing.

Other CSA syndromes include:

- Neurological disorders, in particular brainstem lesions.

- Use of certain medications, especially opioids and other respiratory depressants, which can also be associated with hypoventilation and hypercapnia.
- High altitude periodic breathing, whereby the normal response to altitude of periodic breathing is accompanied by symptoms of poor sleep, frequent awakenings, and dyspnoea.
- Treatment emergent CSA, also known as complex sleep apnoea, which is relatively new terminology describing the occurrence of central events following instigation of continuous positive airways pressure (CPAP) for OSA.
- Primary CSA- where all other causes of CSA are excluded.

Sleep-related hypoventilation disorders

Hypoventilation is defined by hypercapnia, which initially may be just nocturnal and subsequently may be present during the daytime. It can arise due to:

- Increased ventilatory load, such as due to obesity, chronic obstructive pulmonary disease, or kyphoscoliosis.
- Neuromuscular weakness, such as due to myotonic dystrophy, motor neurone disease, or post-intensive care neuropathy.
- Reduced ventilatory drive, such as due to sedatives, a brainstem lesion, excess oxygen supplementation in a patient who has compensated hypercapnia and thus is dependent upon hypoxia to maintain ventilatory drive, or (rarely) congenital hypoventilation syndrome (Ondine's curse). Vulnerability to ventilatory failure is increased during an intercurrent illness such as a chest infection or heart failure.

Typical symptoms of nocturnal hypoventilation are poor quality sleep, waking at night, day time somnolence and morning headaches due to hypercapnic cerebral vasodilatation, which are frontal, throbbing, and worse on bending forwards or straining. Symptoms of respiratory muscle weakness are breathlessness on exertion, lying flat, or when getting into water (these latter two situations displace the diaphragm upwards). Patients who are unable to generate sufficient expiratory flow are vulnerable to recurrent lower respiratory tract infections, and have difficulty with expectoration.

Objective assessment may include:

- *Morning blood gas*: isolated nocturnal hypoventilation is reflected by a normal daytime pCO_2 (<6 kPa), but raised bicarbonate and base-excess, whereas day- and nocturnal-hypoventilation by a raised pCO_2, bicarbonate, and base-excess.
- *Overnight pulse oximetry ± transcutaneous pCO_2*: where this is entirely normal throughout the night, hypoventilation is excluded. Early ventilatory failure is associated with oxygen desaturation/a rise in pCO_2 in rapid eye movement sleep alone, and subsequently desaturation throughout sleep. A more detailed respiratory sleep study may be required if co-existent OSA is suspected.
- *Spirometry*: a lung vital capacity (VC) of <60% predicted is associated with rapid eye movement (REM) sleep hypoventilation, <40% with nocturnal hypoventilation and <25% with awake hypoventilation (17). A fall in VC of >15% on lying flat suggests significant respiratory muscle weakness.
- *Peak cough flow*: a peak cough flow of <270 L/min may be associated with difficulty clearing secretions.

Narcolepsy syndrome type 1 and type 2

Narcolepsy syndrome is a disorder with debilitating excessive daytime sleepiness (EDS). The ICSD 3rd revision reclassified narcolepsy into type 1 when associated with cataplexy, and positive sleep study or near absence of CSF hypocretin, and type 2 without cataplexy, when sleep studies are supportive, but CSF hypocretin is not diagnostic or not done.

Classical type 1 narcolepsy has a prevalence of 0.02–0.18% in the USA and Western countries, higher in Japan. Onset is often in childhood and adolescence with half of patients reporting symptoms, prior to age 15 (18). A small percentage of children have onset before age 5, with other peaks between 35–45 years and around menopause in women (1). Cataplexy usually occurs within a year of onset of sleepiness, but can rarely precede sleepiness or can be delayed for up to 40 years. Cataplexy is reported to be the more prominent symptom in patients presenting after age 60 (19).

Typically, the naps in patients with narcolepsy are short, the patient waking refreshed though may sleep again in 2–3 h. However, patients especially children, may fight sleep, leading to long naps when they eventually fall asleep, with irritability on forced waking. Sleepiness is more likely to occur in sedentary situations, but can occur in unusual situations such as eating or driving and lead to automatic behaviour.

Cataplexy is characterized by sudden loss of muscle tone provoked by strong emotions, including laughter, anger, or surprise. Cataplexy can be partial, localized to the mouth, face, or neck muscles or a limb, but can also include all the skeletal muscle groups leading to falls and injury. Respiratory muscles are spared. Cataplexy episodes can last for a few seconds to several minutes; abrupt drug withdrawal can induce status cataplecticus. Cataplectic facies (20) has been described in children, with semi-permanent jaw and eyelid weakness, the mouth dropping open and tongue lolling, especially during periods of increased sleepiness, but not linked to strong emotions.

Other associate features of narcolepsy syndrome include hypnagogic hallucinations (vivid dream-like experiences occurring in transition from wake to sleep) and sleep paralysis, which with sleepiness and cataplexy form the classical tetrad of narcolepsy. Sleep disturbance with frequent waking is frequently reported at the onset of symptoms and weight gain is common.

The tight association with the human leucocyte antigen (HLA) subtypes DQB1 *06:02 point to an autoimmune cause for narcolepsy type 1. Almost all patients with cataplexy are HLA DQB1 *06:02 positive (12–38% in general population), with DQB1 *05:01 and DQB1 *06:01 protective in the presence of HLA DQB1 *06:02 (1). The recognition of near absence of CSF hypocretin-1 in patients with narcolepsy with cataplexy syndrome has led to the hypothesis that the condition is secondary to autoimmune destruction of approximately 70,000 hypothalamic neurons producing the neuropeptide hypocretin (18).

For a diagnosis of type 1 narcolepsy, daily periods of sleep should occur either with cataplexy and a mean sleep onset latency on the multiple sleep latency test (MSLT) of 8 min or less with 2 or more sleep onset REM (SOREMP = REM sleep within 15 min of sleep onset), though SOREMP on the preceding nocturnal PSG can replace one of the SOREMP in the MSLT, or the CSF hypocretin -1 concentration is 110 pg/ml or less or 1/3 of mean values obtained in normal subjects with the same standardized assay (1). In the absence of cataplexy, if the CSF hypocretin is more than 110 pg/ml or CSF hypocretin level not available, narcolepsy type 2 should be diagnosed, even if the sleep studies meet the diagnostic criteria for narcolepsy and there is no other explanation for the patients sleepiness, with the option to move the diagnosis to type 1, if cataplexy develops or CSF hypocretin level becomes available.

Sleep studies for the diagnosis of narcolepsy should include an MSLT preceded by nocturnal polysomnography (PSG), and at least 7–14 days sleep log or actigraphy, prior to the MSLT to confirm sustained sufficient nightly sleep. Sleep needs may vary between individuals and patients should be advised to increase night time sleep as much as possible leading up to the MSLT, with a minimum 7 h in bed/night. This is particularly important when considering the diagnosis of narcolepsy in shift workers and young patients who may be long sleepers and for whom 7 h of sleep may be insufficient. The patient must be free of drugs that can influence sleep for at least 14 days, or five times the half-life of the drug for longer acting metabolites.

Most patients with narcolepsy will have a mean sleep onset latency of 5 min or less with 2 or more SOREMP, but the 8 min cut off will include 90% of patients with narcolepsy (21). While a mean latency of 5 min or less definitely supports sleepiness and 10 min or more suggests normal alertness, the test has to be interpreted in the clinical context, as it is sensitive to sleep deprivation and circadian effect. Both false positive and false negative results can be obtained, especially if the protocol for the test is not strictly followed.

In clinical practice insufficient night sleep due to irregular bed times and disturbed sleep due to poor sleep hygiene and or psychiatric disorders are common causes for excessive daytime sleepiness, especially in adolescence and shift workers, including early and late starts. It is not unusual for patients to have sufficient sleep on the PSG preceding the MSLT, but 1–2 weeks actigraphy recording shows an irregular sleep–wake schedule; the accumulative sleep debt can give a low mean sleep onset latency on the MSLT (22) and even two or more SOREMP. Two or more SOREMP in the MSLT is reported in shift workers, occasionally in patients with severe SDB (23) and patients on antidepressants. Lumbar puncture for CSF hypocretin level should therefore be considered in older adults on medication for comorbid psychiatric/medical disorders, for a confident diagnosis of narcolepsy even with a positive MSLT, in the absence of definite cataplexy attacks.

If anticipated that the patient may not sleep well or cope with the PSG, arranging the MSLT after an actigraphy night, with the PSG arranged on another night can be an option, though there are no published data on this protocol. Our protocol for patients without cataplexy is to arrange the MSLT after the second PSG night, preceded by 10 days to 2 weeks of actigraphy including the 2 PSG nights. The recording is continued during the day and the study analysed for daytime naps during the day, in between the 2 PSG nights. This overcomes the problems of first night effect, and allows review of daytime naps, which can be informative, especially in young children with narcolepsy who may resist the MSLT protocol or are too young for the MSLT; in these children SOREMP in daytime naps in a 24-h recording, combined with SOREMP on the PSG nights can support the diagnosis of narcolepsy. Analysing daytime naps can also be useful in situations when patients for one reason or another struggle to fall asleep quickly during the MSLT, but the preceding 24-h PSG, records several naps, supporting their complaint of sleepiness due to narcolepsy or other sleep disorders.

The nocturnal PSG prior to the MSLT is recommended to exclude other sleep disorders, especially obstructive sleep apnoea. In narcolepsy, unless the patient has had a late evening/pre-bedtime nap, the sleep onset latency in the PSG is often short, less than 10 min; SOREMP is seen in 40–50% of patients with narcolepsy with cataplexy (24). Increase in wake periods with periodic and non-periodic leg movements is often present, and interrupted REM sleeps with increase in phasic and tonic EMG activity, with or without complex hallucinations and behaviours of dream enactment as in REM sleep Behaviour Disorder (RBD).

Other pitfalls in neurophysiological tests for diagnosing narcolepsy

There are reports of children with narcolepsy with cataplexy syndrome being misdiagnosed as epilepsy (25), by paediatricians more familiar with epilepsy. The situation can be compounded by incidental epileptiform discharge in the EEG, including Rolandic discharge or photosensitivity, both age related EEG abnormalities, or intrusion of 14 and 6/s positive spikes, a pseudo-epileptiform variant pattern in drowsiness. Of course, epilepsy is common and may co-occur with narcolepsy. Tiredness and sleepiness in children with epilepsy can be due to frequent clinical and subclinical seizures, especially from sleep, or due to medication. However, excessive daytime sleepiness of the severity seen in narcolepsy is uncommon in children with epilepsy, except if the child is in non-convulsive status state. Very rarely the initial presentation of a child with epilepsy can be with periods of long sleeps, coinciding with non-convulsive status state, seen in our centre in a boy with ring chromosome 20 abnormality. A routine EEG is not necessary when narcolepsy is suspected, but if sufficient EEG channels are included during the PSG and MSLT, interictal epileptiform abnormalities and the rare patient in non-convulsive status will not be missed, provided the sleep technologist is familiar with interpreting EEG.

Idiopathic hypersomnia

Idiopathic CNS hypersomnia is a less well understood sleep disorder associated with excessive daytime sleepiness without cataplexy (26). However, unlike narcolepsy type 2 these patients have undisturbed night sleep, but wake unrefreshed, and report sleep drunkenness, a term used to describe severe sleep inertia, difficulty waking in spite of setting several alarms, repeatedly going back to sleep, and irritability and confused behaviours on waking, lasting minutes to hours; sleep inertia also occurs on waking from daytime naps. While patients with narcolepsy find short (less than 30 min) naps refreshing, patients with idiopathic hypersomnia report more than 1-h unrefreshing naps during the day. Idiopathic hypersomnia can be associated with symptoms of autonomic system disturbance, including headaches, orthostatic hypotension, syncope, and Raynaud's type symptoms. The mean age of onset is 16.6–21.2 years. There is no association with a specific HLA type, and CSF hypocretin-1 concentration is normal (1). Co-morbid mild depression may be present in 15–25% of patients with idiopathic hypersomnia.

The main differential diagnosis in these patients is between sleepiness due to chronic insufficient night sleep and sleepiness associated with major psychiatric disorders, including atypical depression, dysthymia and bipolar depression, and occasionally chronic fatigue syndrome (27,28). Some patients with psychiatric disorders can also have mean sleep onset latency in MSLT in the sleepiness range, with prescribed medication compounding the difficulty in interpreting sleep studies.

For the diagnosis of idiopathic hypersomnia, ICSD-3 requires a mean sleep onset latency on the MSLT of 8 min or less and fewer than 2 SOREMP and no SOREMP in the previous night PSG. The PSG usually shows high sleep efficiency (>90%) after chronic insufficient sleep is excluded, in contrast to narcolepsy when disturbed sleep with reduced sleep efficiency is a PSG feature. As some patients with idiopathic hypersomnia may have mean sleep onset latency longer than 8 min (29) or the MSLT protocol cannot be followed as they struggle to wake in the mornings, the ICSD -3 diagnostic criteria allows the diagnosis based on demonstrating a total sleep time of 660 min or more (typically 12–14 h) with a 24-h PSG, after correction for chronic sleep deprivation. The long sleeps over 24 h can also be confirmed by wrist actigraphy or a sleep log over 7 days of unrestricted sleep, though nocturnal PSG is required for the other features including short sleep onset latency, high sleep efficiency, and often increase in slow wave sleep. The PSG of patients with sleepiness due to psychiatric disorders may show disturbed sleep with increase in light NREM stage 1 sleep and less slow wave sleep. PSG is also required to exclude other sleep disorders including OSA and increase in periodic limb movements (PLMs), though increase in PLMs in the absence of OSA or RLS does not preclude the diagnosis of idiopathic hypersomnia in patients who otherwise meet the criteria for this sleep disorder.

When suspecting idiopathic hypersomnia, as we routinely record over 2 days, we advise patients to wake naturally in the morning and have naps as needed during the first 24 h of the PSG recording, allowing calculation of total sleep need over 24 h. Following the second PSG night the patient is instructed to be woken by about 08.00 hours, to be followed by the MSLT, starting the test later if necessary. This is combined with 2 weeks of actigraphy. This protocol allows confirmation of increase in total sleep time and sleepiness in spite of sufficient night sleep. However, in spite of these extended studies, if there is any uncertainty about the diagnosis, it is important to keep the patient under review, and the sleep studies repeated, especially when associated with comorbid depression and as there are reports of remission rates of up to 14–25% (1,28) in patients with idiopathic hypersomnia.

Recurrent hypersomnia

The best-characterized sleep disorder associated with episodic hypersomnia is Kleine–Levin syndrome, with a subtype related to menstrual cycle in women (30). Episodes of sleepiness last for a few days to several weeks and recur 1–10 times a year. Patients may sleep from 16–18 h/day and, even when apparently awake, describe a feeling of unreality and confusion with only patchy memory for the duration of the episode. The sleep periods may be associated with irritability, binge eating, and disinhibited behaviours including hypersexuality. Transient dysphoria or elation with insomnia may signal the termination of the episode. These patients have normal sleep and behaviour in between episodes. Onset is usually in adolescence though onset in younger children and after age 20 has been reported. The episodes tend to lessen in severity and frequency over several years (median course 14 years) (1). In spite of extensive investigations the pathophysiology of the disorder is not yet understood.

The diagnosis of Kleine Levin syndrome is made clinically, though the initial episode may prompt investigations for encephalitis, structural brain pathology including tumours and hydrocephalus and epilepsy. A 24-h PSG during the sleepy phase will show prolonged total sleep time with reduced sleep efficiency and increase in wake periods after sleep onset during the night. An extended actigraphy may record the evolution of the sleep disorder, with long periods of sleep/inactivity during the sleepy phase with progressive recovery, leading to a couple of nights of insomnia before the baseline sleep–wake pattern/schedule returns.

Restless leg syndrome and periodic limb movements during sleep

RLS, also called Willis–Ekbom's syndrome, is a neurological sensorimotor disorder, characterized by a complaint of strong, irresistible urge to move the legs, usually accompanied or caused by an unpleasant sensation in the legs, the symptoms begin or worsen during periods of rest and inactivity, and are partially or totally relieved by movement, at least as long as the activity continues (31). The symptoms can also involve the arms and other body parts. There is a circadian pattern to the symptoms, being worse in the evenings or night, with relative relief in the mornings. Patients complain of poor sleep; RLS symptoms cause sleep onset and maintenance insomnia, with daytime fatigue and sometimes sleepiness, though in spite of the severe sleep disturbance the sleepiness is not of the severity seen in patients with narcolepsy, suggesting a degree of hyperaroused state (1).

The motor expression of the disorder is periodic limb movements (PLMs), during sleep (PLMS) and resting wakefulness (PLMW), the movements usually involving the legs, but can involve the arms. PLMs are increased in 80–90% of patients with RLS, but are not specific for RLS. PLMS leading to arousals can contribute to the sleep disturbance in patients with RLS.

The prevalence of RLS is between 5–10% in the North European population, more common in females, though a rigorous study suggested that prevalence with moderate negative impact is 1.9% and high negative impact 0.8% (31). The prevalence increases with age up to 60–70 yrs. More than 50% of patients with RLS report a familial pattern; autosomal dominant inheritance with high penetrance and anticipatory age of onset is observed. Although linkage analyses has identified several gene loci, no single gene loci has yet been found. The early age of onset (before 34–45 years) increases the risk of RLS occurrence in the family. Though considered to be a disorder of adulthood, there is increasing recognition that RLS can occur in children, and some children with presumed 'growing pains', may have early onset RLS (32). There may also be a link between RLS, PLMs, and attention deficit hyperactivity disorder (ADHD) (33).

Secondary RLS is associated with iron deficiency anaemia, not uncommon during pregnancy, and in patients with chronic renal failure. However, there is evidence that iron deficiency in the CNS (reflected by the serum ferritin level) is involved in primary RLS, even without anaemia. Several substances can induce or worsen RLS, including tricyclics and other antidepressants (except bupropion), lithium carbonate, D2 receptor blocking agents, such as neuroleptics, antihistamines, and alcohol with caffeine and smoking also incriminated. Medical disorders with possibly more than chance association of RLS include Parkinson's disease, peripheral neuropathy, multiple sclerosis, diabetes, rheumatoid arthritis, and obstructive sleep apnoea (1).

RLS is a clinical diagnosis and in most patients eliciting the 4 diagnostic criteria: the urge to move the legs due to the uncomfortable sensation, sensation worsening during periods of rest, relieved by movement, and occurring mainly in the evening or night, is sufficient for the diagnosis. PSG recording leg movements from the right and left tibialis anterior muscles may aid diagnosis when symptoms are equivocal, often due to co-morbid other medical/neurological disorders, or the sleep disturbance appears out of proportion to the mild to moderate RLS symptoms. A PLM index of more than 15/h including PLMW, in the overnight PSG, supports the diagnosis of RLS. The distribution of the PLMs, when more frequent during the early period of night sleep, favours RLS. Diagnosis of RLS in children between 2–12 years may need PSG, if the child meets the essential 4 adult criteria, but cannot describe the symptoms of discomfort in his or her own words. In this situation the diagnosis can be considered if two of the three other findings are present including sleep disturbance for age, if there is a biological parent or sibling with definite RLS, or the child has PSG documented PLMs of five or more per hour. In children increase in PLMs may precede the appearance of the diagnostic sensory symptoms (1).

Periodic limb movement disorder

PLMD is characterized by periodic episodes of repetitive, highly stereotyped movements that occur during sleep, disturbing sleep, when no other cause for sleep disturbance can be identified (1). It is essential to exclude disorders known to be associated with PLMs, especially RLS, sleep-related breathing disorder, REM sleep behaviour disorder (RBD) and narcolepsy. However, many other conditions have PSG showing PLMs, including neurological disorders such as Parkinson's disease, multi system atrophy, spinal cord injury, and multiple sclerosis. PLMs have also been reported in some patients with parasomnias, ADHD, Asperger's syndrome, post-traumatic stress disorder, low brain iron as measured by serum ferritin level and with use of medications that worsen RLS. In clinical practice it is not unusual to see PLMs in patients with chronic sleep disturbance due to any cause including some patients with insomnia or sleep disturbance due to chronic pain or skin disorders producing discomfort/itching in sleep. The diagnosis of PLMD as a standalone primary sleep disorder has been questioned. Nevertheless, it can be a useful working diagnosis in the small group of patients in whom the only finding contributing to the disturbed sleep is the marked increase in high amplitude PLMs with arousals, the movements often also involving the arms. The patient may report unrefreshing sleep, sleep onset or maintenance insomnia, daytime tiredness/sleepiness, or be unaware of the movements, their partner complaining of sleep disturbance and even injury from the movements.

The movements in PLMs are stereotypical and commonly involve extension of the big toe only or combined with partial extension of the ankle, flexion at the knee, and sometimes the hip. The movements can be associated with an autonomic arousal, cortical arousal, or awakening. As an arousal may precede, coincide with or follow the movement, it has been suggested that there may be a common central generator for the movement and arousal, the limb movements a visible marker of unstable sleep-related to dopaminergic impairment in conditions associated with increase in PLMs (1).

There can be night-to-night variability in frequency of PLMs, with up to 5 nights of recording recommended when increase in PLMs is clinically suspected (recorded with actigraphy). The criteria for identifying PLMs are based on Coleman's seminal publication (34) with some modification. The PSG records repetitive highly stereotyped limb movements with surface EMG electrodes placed longitudinally and symmetrically along the middle of the right and left tibialis anterior muscles, 2–3 cm apart or one third of the length of the muscle, whichever is shorter. Each movement should be 0.5–10 s in duration, the amplitude 8 μV or more than the resting EMG (or amplitude greater than or equal to 25% of toe dorsiflexion during calibration). A PLM series is defined as a sequence of 4 or more movements, separated by an interval of more than 5 s and less than 90 s, with movements on 2 different legs separated by less than 5 s between movements onset, counted as single leg movement (35). While a PLM index (series/h) of more than 5/h in children and more than 15/h in adults is considered significant in symptomatic patients, PLMs increase with age and normative data in asymptomatic older adults are not available, and there is overlap between symptomatic and asymptomatic PLM index values. Hence the PLM index should always be interpreted in the clinical context. PLMs also need to be differentiated from other movements in sleep including excessive fragmentary myoclonus, hypnagogic foot tremor, and alternating leg movement activation (ALMA), movements currently considered as of uncertain clinical significance, and classified under isolated symptoms or normal variants (1).

Circadian rhythm sleep disorders

The circadian rhythm is an endogenously generated, biological rhythm that occurs even in the absence of changes in the environment. The circadian rhythm for the sleep-wake cycle is *about* 24 h, with polymorphism in circadian clock genes determining the individual's endogenous cycle length to shorter or longer than 24 h, producing a population distribution between owls (strongly evening) and larks (strongly morning) (36,37). The biological master clock for all circadian rhythms is in the suprachiasmatic nucleus of the hypothalamus. An individual's endogenous rhythm has to be entrained to a 24-h day and the most important Zeitgeber (time giver) for this is light, although other non-photic Zeitgebers such as work schedule and meal times, also help. Light via the retinal ganglion cells influences the suprachiasmatic nucleus, which in turn influences the melatonin secreted by the pineal gland. Melatonin is suppressed by light; as daylight diminishes the melatonin levels begin to rise, followed by a drop in body temperature and sleep onset.

A circadian rhythm sleep disorder occurs when there is persistent or recurrent pattern of sleep disturbance due to either alteration of the circadian time keeping system or misalignment between the endogenous circadian rhythm and exogenous factors that affect the timing and duration of sleep (1). Patients with circadian rhythm disturbances may present with insomnia, excessive daytime sleepiness, or both with significant negative effect on educational, occupational, and social areas of functioning.

The ICSD 3 recognizes seven circadian rhythm sleep disorders (See Box 35.3).

Box 35.3 Circadian rhythm sleep disorders

- Delayed sleep phase disorder.
- Advanced sleep phase disorder.
- Irregular sleep-wake rhythm disorder.
- Non-entrained, non-24-h sleep-wake rhythm disorder.
- Shift work disorder.
- Jet lag disorder.
- Circadian sleep–wake disorder not otherwise specified, including alteration in sleep wake pattern secondary to medical, neurological and psychiatric disorders.

Source data from *Seizure*, **9**(2), Appleton R, Beirne M, Acomb B, Photosensitivity in juvenile myoclonic epilepsy, pp. 108–11, copyright (2000), Elsevier.

The circadian rhythm sleep disorder most frequently presenting in clinics is *delayed sleep phase syndrome* (DSPS), a chronic disorder characterized by habitual sleep times delayed usually by more than 2 h relative to conventional or socially acceptable times for age (Fig. 35.2). The disorder is common in adolescence and young adults, but also seen in children especially with ADHD. Polymorphism in the circadian clock genes and environmental factors contribute to the development of this sleep disorder (1). Patients with DSPS complain of difficulty falling asleep, with sleep onset often delayed to between 01.00 and 06.00 hours. The sleep is usually undisturbed once they fall asleep, though morning wake up time is delayed to late morning or afternoon. The sleep–wake pattern will be stable if allowed to sleep and wake at preferred times, but attempt to advance sleep onset and wake up times for school or work, with catch up sleep on days off, can lead to a chaotic sleep/wake pattern, bordering on a non 24-h cycle pattern. Forced morning waking can lead to confusion on waking. Also in an attempt to fall asleep earlier patients may use alcohol or become reliant on hypnotics or recreational substances, while others may use stimulants to overcome daytime sleepiness. Sleep hygiene (general advice that helps to promote a good sleep pattern), alone is insufficient to advance sleep onset times, leading to frustration and mood disorder.

In *advanced sleep phase syndrome* (ASPS) there is stable advancement of the major sleep period, so that sleep onset and wake up times are several hours earlier than conventional or desired wake up times. Patients complain of sleepiness in the late evening, fall asleep early and wake early, often in the early hours of the morning and may be mistakenly diagnosed with depression, a recognized cause for early morning waking. This disorder is often seen in middle age and older adults, usually associated with morning types, but also in pre-school children. Their preferred sleep onset is advanced to between 6 pm and 9 pm and wake up time between 2 am and 5 am; though social demands may force a later sleep onset time, wake up times may remain early, leading to reduced actual sleep times.

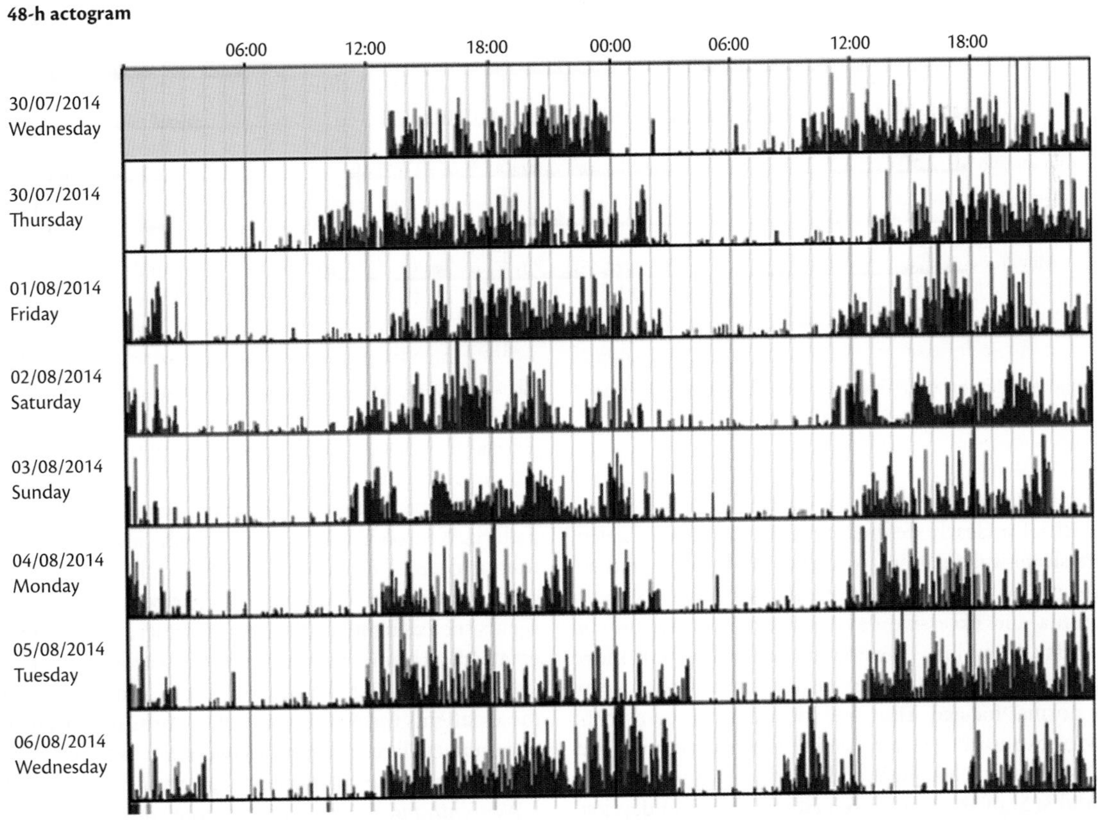

Fig. 35.1 Actigraphy recording of patient with delayed sleep phase syndrome: note sleep onset between 02.00 and 04.00 hours, with wake up times around midday; early waking on last night followed by a long nap after midday.

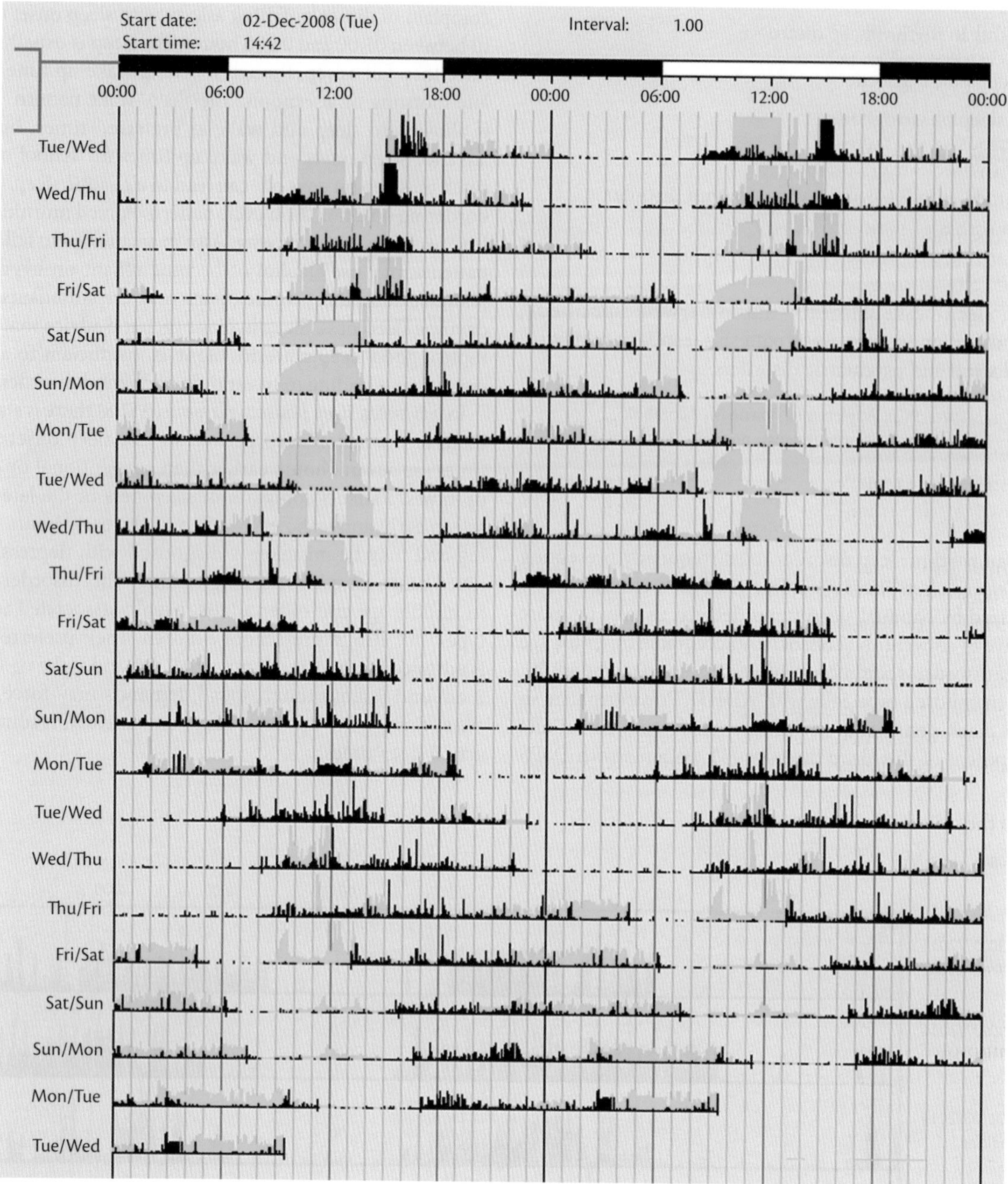

Fig. 35.2 Actigraphy recording of non-24-h sleep–wake cycle pattern with recording of environmental light: sleep onset times are progressively delayed leading to periods of sleep during daytime, with daylight during sleep periods.

Irregular sleep–wake rhythm circadian sleep disorder is characterized by lack of a clearly defined circadian rhythm of sleep and wake. The sleep–wake pattern is temporally disorganized so the sleep–wake periods are variable over the 24 h. Patients complain of insomnia and excessive daytime sleepiness depending on the time of day. Sleep logs or actigraphy recording will show multiple sleep periods over 24 h (at least three), though total sleep time remains normal for age. This sleep disorder can be associated with lack of exposure to external time givers entraining the circadian sleep rhythm such as light and activity, as in institutionalized older adults. Similar sleep–wake patterns may also develop in patients with medical and especially neurological disorders associated with neurocognitive impairment, though these patients may also have delayed or advanced or non-24-h sleep–wake cycle pattern. Neurological disorders may be associated with degeneration or decreased neuronal activity of the suprachiasmatic nuclei or decreased responsiveness of the circadian clock to entraining agents such as light (1). Patients on prescribed hypnotics, especially opiates for chronic pain, can present with variable periods of irresistible sleep during the day, which can lead to falls and injuries. In these patients the MSLT may not confirm sleepiness, but extended actigraphy will record the irregular sleep–wake cycle pattern, and a 24-h PSG will show multiple sleep periods.

Non-entrained, non-24-h sleep–wake disorder occurs when there is loss of entrainment of the endogenous rhythm that in these patients is usually slightly longer than 24 h. The sleep pattern can be variable, but if not forcibly adjusted, will show sleep and wake times drifting each day with a period longer than 24 h. As a result these patients will cyclically have periods of most of their sleep during night time, and other periods of sleep only during the day, with in between periods as sleep onset progressively delays (Fig. 35.2). This disorder can develop in children and adults who are blind with no light perception and who have not been able to entrain their circadian rhythm with non-photic time givers. It is also seen in young people with learning difficulties, or with psychiatric or personality disorders, especially when associated with social communication difficulties, who spend a lot of time indoors and have limited social interaction. Medical and neurological disorders can also lead to a non-entrained sleep–wake pattern.

In *jet lag sleep disorder* there is temporary mismatch between the sleep–wake times of the endogenous clock and that required by change in at least two time zones, associated with impairment in daytime functioning, malaise, and somatic symptoms including gastrointestinal disturbance within one or two days after travel. Jet lag is transient and does not require investigations. In contrast sustained disturbance of circadian rhythm and sleep time misalignment occurs in *shift work disorder*, with impaired cognitive and physical functioning, and can lead to chronic sleep disturbance and exacerbation of gastrointestinal and cardiovascular symptoms. This occurs when work hours are scheduled during usual sleep periods, and can include night shifts, early morning shifts, and rotating shifts. Patients complain of insomnia or excessive daytime sleepiness depending on the work schedule and the overlap with usual sleep times and often report poor quality sleep. Depending on the shift type, circadian preference of 'morningness' or 'eveningness' may affect the individual's ability to adjust to shift work (1).

The diagnosis of circadian rhythm disorders relies on a detailed clinical history, supported by sleep logs and actigraphy for at least 7 days (preferably 2 weeks or longer) to include appropriate periods such as weekends, long holidays, and various shift schedules (Fig. 35.1). The facility to monitor environmental light in many actiwatches can provide useful information regarding inappropriate light exposure during the night or inadequate light exposure during the day. PSG is usually not indicated for diagnosis. However, constant attempts to realign sleep to socially acceptable times or major sleep periods occurring during daylight hours, can lead to disturbed sleep and mood disorder, which may need treatment. If actigraphy shows significant sleep disturbance after sleep onset, a PSG may be indicated to exclude other sleep disorders including OSA and PLMD.

Insomnia

The majority of patients with insomnia do not need sleep studies. In most patients psychiatric/psychological factors will be maintaining the abnormal sleep pattern. In patients with psychophysiological insomnia, a life event may trigger the sleep disturbance, but insomnia persists after resolution of the trigger, personality traits, and excess focus and effort to sleep, maintaining the insomnia (38). These patients are often hyperaroused and in spite of poor sleep at night cannot sleep during the day. Sleep studies including actigraphy and PSG may be indicated in a small number of patients with insomnia, if the history suggests another sleep disorder, especially RLS or Circadian rhythm disorder or OSA. The threshold for PSG should be low in patients in whom insomnia is associated with excessive daytime sleepiness (not tiredness).

Parasomnias

Parasomnias are undesirable physical events or experiences that occur during entry into sleep, within sleep, or during arousals from sleep (1). These include abnormal sleep-related movements, behaviours, emotions, perceptions, dreams, with or without symptoms related to autonomic nervous system. Though many parasomnias are age-related and self-limiting, some persist with negative psychosocial effect, injury to self or others, and disturbed sleep with daytime tiredness. Table 35.1 lists the common parasomnias.

The ICSD separates out sleep-related movement disorders, though some of these conditions have features overlapping with the definition of a parasomnia.

Sleep-related movement disorders

- Restless leg syndrome.
- Periodic limb movement disorder.
- Sleep-related leg cramps.
- Sleep-related bruxism.
- Sleep-related rhythmic movement disorder.
- Benign sleep myoclonus in infancy.
- Propriospinal (truncal)myoclonus at sleep onset.
- Sleep-related movement disorder due to medical disorder.
- Sleep-related movement disorder due to drug or substance abuse.

There are sleep-related movements that are borderline pathological to normal variants, associated with sleep complaints or reflect sleep disturbance when exaggerated such as hypnic jerks (startle jerks at sleep onset), excessive fragmentary myoclonus (low amplitude asynchronous limb twitches in NREM sleep), hypnagogic foot tremor

Table 35.1 Parasomnias: modified from ICSD 3

Disorders of arousal from NREM sleep	Other parasomnias
◆ Confusional arousals ◆ Night terrors ◆ Sleep walking including sleep sex ◆ Sleep-related eating disorder	◆ Exploding head syndrome ◆ Sleep-related hallucinations ◆ Sleep enuresis ◆ Parasomnia due to medical disorder or substance use ◆ Sleep-related groaning (catathrenia) ◆ Sleep-related abnormal choking/panic-like attacks/laryngeal spasm ◆ Sleep-related dissociate behaviours in sleep/wake transition state
Parasomnias associated with REM sleep	**Isolated symptoms and normal variants**
◆ REM sleep behaviour disorder ◆ Recurrent isolated sleep paralysis ◆ Nightmares	◆ Sleep talking ◆ Be aware of malingering/pseudo parasomnia behaviour

Adapted from *International Classification of Sleep Disorders*, Third Edition (ICSD-3), copyright (2014), American Academy of Sleep Medicine.

and alternating leg muscle activation. These movements have potential to be misinterpreted as epileptic seizures and should be distinguished from PLMs. Familiarity with these movements often seen during PSG review is important. The AASM manual for scoring sleep includes criteria for recognizing these movements (35).

Most parasomnias are diagnosed clinically and do not need PSG recordings. However, sleep studies may be necessary if the behaviours are persistent, frequent, associated with high risk, potentially forensic behaviours, and when the diagnosis is clinically uncertain and sleep-related epilepsy is a possibility, or if the parasomnia has implication for future development of other neurological disorders as in REM Sleep Behaviour Disorder (RBD). Sleep studies are essential if there is any clinical suspicion that the sleep behaviours/movements are pretend or pseudo parasomnias, a situation clinicians are increasingly faced with as familiarity with sleep disorders increases; the pathophysiology and management for these is similar to non-epileptic attack disorders (NEAD) seen in epilepsy clinics. More often PSG studies are indicated to exclude other sleep disorders lowering the arousal threshold for the parasomnia, especially OSA or PLMs, including in young patients with persistent rhythmic movement disorders.

NREM arousal parasomnias may not occur in the laboratory setting or only minor confused arousals are recorded. The PSG may record unstable slow wave sleep cycles with abrupt arousals with or without clinical episodes in patients with frequent NREM arousal parasomnias, with sleep appearing more stable later in the night. Increase in cyclical alternating pattern (CAP) reflecting the high arousal oscillations in NREM sleep (39) may be seen. While these observations are of interest, their diagnostic value has not yet been evaluated. Similarly suggested protocols to induce episodes in forensic cases need more reliability data (40,41).

The two parasomnias commonly referred for assessments to sleep clinics are NREM arousal parasomnias and RBD. As both these parasomnias can have features overlapping with nocturnal epilepsy, especially nocturnal frontal lobe epilepsy (NFLE), the PSG should include video-telemetry with sufficient EEG channels to manipulate the data with seizure montages.

NREM arousal parasomnias occur when there is incomplete awakening usually from slow wave sleep and range from mild confusional behaviours to agitated waking of night terrors or complex behaviours associated with sleep walking. There is usually amnesia of the behaviour or there is limited dream memory. The episodes usually occur during the first third of the major sleep episode, with generally one episode during the night, but if there is increase in slow wave sleep cycles or lowering of arousal threshold secondary to another sleep disorder or environmental factors such as touch or noise, more than one episode can occur. Derry et al. (42,43) have attempted to characterize clinical features to aid differentiation between NREM arousal parasomnias and NFLE, but if clinically there are atypical features, or the episodes are frequent through the night, video-telemetry with PSG to record episodes and exclude other sleep disorders is essential. Interestingly an overlapping pathophysiology for the two disorders has been suggested (44) with co-occurrence of the two disorders reported (45).

Video-telemetry/PSG is indicated in patients who present with frequent night time waking with panic like episodes or with gasping, choking, or laryngeal spasm, usually from NREM sleep. In the absence of daytime panic attacks, other sleep disorders especially OSA, PLMs causing arousals, sleep-related gastro-oesophageal reflux, or cardiac arrhythmia have to be excluded. Occasionally NFLE or seizures from the insular cortex can produce choking episodes from sleep. Only if all these conditions are excluded should a parasomnia be diagnosed; the recorded episodes are usually from NREM sleep.

Sleep-related hallucinations include hypnagogic (at sleep onset) or hypnopompic (on waking) can occur in isolation, but are often seen in patients with narcolepsy. However, complex visual hallucinations represent a distinct sleep-related hallucination that may occur from NREM sleep. Patients report sudden waking without preceding dream memory and describe seeing complex scenes or vivid immobile images of people or animals (1), often jumping out of bed convinced there are bugs in the bed for instance. The hallucinations tend to disappear if a bedroom light is put on. These patients may also have other NREM arousal parasomnias. Sleep studies may confirm that the hallucinations occur from NREM sleep. The ability of the patient to remember the hallucination and react to the scene should not be mistaken for RBD, though when onset is in older adults, neurological disorders including Parkinson's disease, dementia with Lewy bodies, and mid brain and diencephalic pathology should be excluded.

RBD (46) is characterized by loss of normal REM sleep atonia with increase in phasic and tonic EMG activity in REM sleep, associated with increase in motor activity and dream enactment behaviour. Patients often report change in quality of their dreams, the content often involving aggression, leading to abusive vocalization, and punching, kicking movements leading to injury, including falling out of bed. These patients, if they manage to get out of bed, cannot get far, as the fragmented atonia, makes them unsteady and prone to falls, unlike NREM arousal sleep walking behaviours, when patients can bolt out of the room or walk long distance and negotiate objects. RBD has a male preponderance, emerging after aged 50, and is associated with neurodegenerative disorders, notably the synucleinopathies including Parkinson's disease, multisystem atrophy and dementia with Lewy body disease. Conversely, most patients with RBD if followed long enough, develop these diseases. RBD can be associated with other neurologic disorders, associated with interruption of REM atonia pathway (pedunculo-pontine nucleus) and/or disinhibition of brainstem motor pattern generation; in the young patient it is important to exclude narcolepsy. The RBD pattern in PSG is often related to prescribed medication, including antidepressants, especially venlafaxine, serotonin specific reuptake inhibitors (SSRIs), mirtazapine, beta blockers, and anticholinesterase inhibitors. There is currently insufficient data on whether drug associated RBD pattern in PSG also predates development of a synucleinopathy.

Parasomnia overlap disorder occurs in young patients with both NREM arousal parasomnias and RBD, and can be idiopathic or symptomatic. Status dissociatus or agrypnia excitata is a subtype of RBD (extreme form), with no slow wave sleep and absence of identifiable discrete sleep stages on PSG, replaced by fragmentary intrusions of NREM/REM and wake states. The patient is clinically either awake or asleep with increase in muscle twitching and vocalization, reporting dream like mentation on waking. The condition is seen with delirium tremens, Morvan's fibrillary chorea, and fatal familial insomnia, all associated with marked autonomic sympathetic activation, but can also occur after protracted withdrawal from alcohol abuse and other drugs (1).

PSG in patients with RBD shows an excess of sustained or intermittent loss of REM atonia or excess of phasic muscle activity of the submental and/or limb EMG in REM sleep. EMG from both upper and lower limbs monitoring is recommended, with a study showing that simultaneous recording of the mentalis and flexor digitorium superficialis, and extensor digitorium brevis, provides the highest rate of REM sleep phasic EMG activity in subjects with RBD (47). The AASM scoring manual (35) defines increase in tonic muscle activity in REM sleep as at least 50% of REM sleep epoch having a chin EMG amplitude greater than the minimal amplitude in NREM sleep. To quantitate increase in phasic EMG activity, the 30 s REM sleep epoch is divided into 3 s mini-epochs, and at least 5 (50%) of the mini epochs should contain bursts of transient muscle activity of 0.1–5 s duration and at least four times higher than background EMG activity. The PSG in RBD will also show increase in PLMs though less with arousals, the PLMs continuing in REM sleep. Paradoxical increase in slow wave sleep is often seen, but does not have diagnostic value.

Conclusions

Studies to aid diagnosis and monitor treatment of sleep disorders require familiarity with the range of sleep disorders and the pitfalls in interpreting neurophysiological sleep studies. If appropriately selected and standardized protocols followed and interpreted in the clinical context, these studies, though resource heavy can be a valuable tool in the diagnosis and management of sleep disorders.

References

1. Medicine AAoS. (2014). *International classification of sleep disorders*. Darien, IL: AAoS.
2. Aldrich, M.S. (1998). Diagnostic aspects of narcolepsy. *Neurology*, **50**, S2–7.
3. Johns, M.W. (1991). A new method for measuring daytime sleepiness: the Epworth sleepiness scale. *Sleep*, **14**, 540–5.
4. Greenstone, M. and Hack, M. (2014). Obstructive sleep apnoea. *British Medical Journal*, **348**, g3745.
5. Johns, M. and Hocking, B. (1997). Daytime sleepiness and sleep habits of Australian workers. *Sleep*, **20**, 844–9.
6. Hirshkowitz, M., Sarwar, A., and Sharafkhaneh, A. (2011). Evaluating sleepiness. In: M. Kryger, T. Roth, and W. Dement (Eds) *Principles and practice of sleep medicine*, 5th edn, pp. 1624–31. London: Elsevier Saunders.
7. Chung, F., Yegneswaran, B., Liao, P., et al. (2008). STOP questionnaire: a tool to screen patients for obstructive sleep apnea. *Anesthesiology*, **108**, 812–21.
8. Young, T., Palta, M., Dempsey, J., Skatrud, J., Weber, S., and Badr, S. (1993). The occurrence of sleep-disordered breathing among middle-aged adults. *New England Journal of Medicine*, **328**, 1230–5.
9. Young, T., Peppard, P.E., and Taheri, S. (2005). Excess weight and sleep-disordered breathing. *Journal of Applied Physiology*, **99**, 1592–9.
10. Eckert, D.J., White, D.P., Jordan, A.S., Malhotra, A., and Wellman, A. (2013). Defining phenotypic causes of obstructive sleep apnea. Identification of novel therapeutic targets. *American Journal of Respiratory & Critical Care Medicine*, **188**, 996–1004.
11. Kohler, M., Blair, E., Risby, P., et al. (2009). The prevalence of obstructive sleep apnoea and its association with aortic dilatation in Marfan's syndrome. *Thorax*, **64**, 162–6.
12. Qin, B., Sun, Z., Liang, Y., Yang, Z., and Zhong, R. (2014). The association of 5-HT2A, 5-HTT, and LEPR polymorphisms with obstructive sleep apnea syndrome: a systematic review and meta-analysis. *PLoS One*, **9**, e95856.
13. Young, T., Finn, L., Peppard, P.E., et al. (2008). Sleep disordered breathing and mortality: eighteen-year follow-up of the Wisconsin sleep cohort. *Sleep*, **31**, 1071–8.
14. Marshall, N.S., Wong, K.K.H., Liu, P.Y., Cullen, S., Knuiman, M.K., and Grunstein, R.R. (2008). Sleep apnea as an independent risk factor for all-cause mortality: the Busselton Health Study. *Sleep*, **31**, 1079–85.
15. Marin, J.M., Carrizo, S.J., Vicente, E., and Agusti, A.G. (2005). Long-term cardiovascular outcomes in men with obstructive sleep apnoea-hypopnoea with or without treatment with continuous positive airway pressure: an observational study. *Lancet*, **365**, 1046–53.
16. Ward, K.L., Hillman, D.R., James, A., et al. (2013). Excessive daytime sleepiness increases the risk of motor vehicle crash in obstructive sleep apnea. *Journal of Clinical Sleep Medicine*, **9**, 1013–21.
17. Ragette, R., Mellies, U., Schwake, C., Voit, T., and Teschler, H. (2002). Patterns and predictors of sleep disordered breathing in primary myopathies. *Thorax*, **57**, 724–8.
18. Aran, A., Einen, M., Lin, L., Plazzi, G., Nishino, S., and Mignot. E. (2010). Clinical and therapeutic aspects of childhood narcolepsy-cataplexy: a retrospective study of 51 children. *Sleep*, **33**, 1457–64.
19. Ohayon, M.M., Ferini-Strambi, L., Plazzi, G., Smirne, S., and Castronovo, V. (2005). How age influences the expression of narcolepsy. *Journal of Psychosomnography Results*, **59**, 399–405.
20. Serra, L., Montagna, P., Mignot, E., Lugaresi, E., and Plazzi, G. (2008). Cataplexy features in childhood narcolepsy. *Movement Disorder*, **23**, 858–65.
21. Morgenthaler, T.I., Kapur, V.K., Brown, T., et al. (2007). Practice parameters for the treatment of narcolepsy and other hypersomnias of central origin. *Sleep*, **30**, 1705–11.
22. Roehrs, T., Zorick, F., Sicklesteel, J., Wittig, R., and Roth, T. (1983). Excessive daytime sleepiness associated with insufficient sleep. *Sleep*, **6**, 319–25.
23. Singh, M., Drake, C.L., and Roth, T. (2006). The prevalence of multiple sleep-onset REM periods in a population-based sample. *Sleep*, **29**, 890–5.
24. Thorpy, M. (2001). Current concepts in the etiology, diagnosis and treatment of narcolepsy. *Sleep Medicine*, **2**, 5–17.
25. Stores, G. (1991). Confusions concerning sleep disorders and the epilepsies in children and adolescents. *British Journal of Psychiatry*, **158**, 1–7.
26. Bassetti, C. and Aldrich, M.S. (1997). Idiopathic hypersomnia. A series of 42 patients. *Brain*, **120**(Pt 8), 1423–35.
27. Abad, V.C. and Guilleminault, C. (2005). Sleep and psychiatry. *Dialogues in Clinical Neuroscience*, **7**, 291–303.
28. Bassetti, C. and Dauvilliers, Y. (2011). Idiopathic hypersomnolence. In: M. Kryger, T. Roth, and W. Dement (Eds) *Principles and practice of sleep medicine*. 5th edn, pp. 969–79. London: Elsevier Saunders.
29. Anderson, K.N., Pilsworth, S., Sharples, L.D., Smith, I.E., and Shneerson, J.M. (2007). Idiopathic hypersomnia: a study of 77 cases. *Sleep*, **30**, 1274–81.
30. Arnulf, I., Zeitzer, J.M., File, J., Farber, N., and Mignot, E. (2005). Kleine–Levin syndrome: a systematic review of 186 cases in the literature. *Brain*, **128**, 2763–76.
31. Allen, R.P., Stillman, P., and Myers, A.J. (2010). Physician-diagnosed restless legs syndrome in a large sample of primary medical care patients in western Europe: prevalence and characteristics. *Sleep Medicine*, **11**, 31–7.
32. Picchietti, D.L. and Walters, A.S. (1999). Moderate to severe periodic limb movement disorder in childhood and adolescence. *Sleep*, **22**, 297–300.
33. Cortese, S., Konofal, E., Lecendreux, M., et al. (2005). Restless legs syndrome and attention-deficit/hyperactivity disorder: a review of the literature. *Sleep*, **28**, 1007–13.
34. Coleman, R. (1982). Periodic movements in sleep (nocturnal myoclonus) and restless legs syndrome. In: C. Guilleminault (Ed.) *Sleeping and waking disorders: indication and techniques*, pp. 265–95. Menlo Park: Addison Wesley.
35. Iber, C., Ancoli-Israel, S., Chesson, A.L., Jr, and Quan, S.F. (2007). *The AASM manual for the scoring of sleep and associated events: rules,*

terminology and technical specifications. Westchester, IL: American Academy of Sleep Medicine.

36. Horne, J.A. and Ostberg, O. (1976). A self-assessment questionnaire to determine morningness-eveningness in human circadian rhythms. *International Journal of Chronobiology*, **4**, 97–110.
37. Chelminki, I., Ferraro, F.R., Petros, T.V., and Plaud, J.J. (1997). Horne and Ostberg questionnaire: a score distribution in a large sample of young adults. *Personality and Individual Differences*, **23**, 647–52.
38. Ebben, M.R. and Spielman, A.J. (2009). Non-pharmacological treatments for insomnia. *Journal of Behaviour Medicine*, **32**, 244–54.
39. Benbir, G., Kutlu, A., Gozubatik-Celik, G., and Karadeniz D. (2013). CAP characteristics differ in patients with arousal parasomnias and frontal and temporal epilepsies. *Journal of Clinical Neurophysiology*, **30**, 396–402.
40. Cartwright, R. (2000). Sleep-related violence: does the polysomnogram help establish the diagnosis? *Sleep Medicine*, **1**, 331–5.
41. Pilon, M., Montplaisir, J., and Zadra, A. (2008). Precipitating factors of somnambulism: impact of sleep deprivation and forced arousals. *Neurology*, **70**, 2284–90.
42. Derry, C.P., Davey, M., Johns, M., et al. (2006). Distinguishing sleep disorders from seizures: diagnosing bumps in the night. *Archives of Neurology*, **63**, 705–9.
43. Derry, C.P., Harvey, A.S., Walker, M.C., Duncan, J.S., and Berkovic, S.F. (2009). NREM arousal parasomnias and their distinction from nocturnal frontal lobe epilepsy: a video EEG analysis. *Sleep*, **32**, 1637–44.
44. Tassinari, C.A., Rubboli, G., Gardella, E., et al. (2005). Central pattern generators for a common semiology in fronto-limbic seizures and in parasomnias. A neuroethologic approach. *Neurology Science*, **26**(3), s225–32.
45. Tinuper, P., Provini, F., Bisulli, F., et al. (2007). Movement disorders in sleep: guidelines for differentiating epileptic from non-epileptic motor phenomena arising from sleep. *Sleep Medicine Reviews*, **11**, 255–67.
46. Schenck, C.H. and Mahowald, M.W. (2002). REM sleep behavior disorder: clinical, developmental, and neuroscience perspectives 16 years after its formal identification in SLEEP. *Sleep*, **25**, 120–38.
47. Frauscher, B., Iranzo, A., Högl, B., et al. (2008). Quantification of Electromyographic activity during REM sleep in multiple muscles in REM sleep behaviour disorder. *Sleep*, **31** (5), 724–31.

CHAPTER 36

Intraoperative monitoring

Marc R. Nuwer

Introduction

Intraoperative neurophysiological monitoring (IONM) aims to test any of several nervous system pathways to detect potential injury during surgery. By monitoring nervous system pathways during surgery, the neurophysiology team can alert the surgeon to changes that forewarn of imminent complications and does so in time to allow for interventions to prevent those complications from becoming worse or permanent. Testing during surgery also can identify structures so the surgeon can choose what to resect and what to leave intact.

Monitoring and testing in the operating room involve many types of neurophysiology procedures. Many are familiar to the outpatient clinical neurophysiology laboratory, such as somatosensory evoked potentials (SEPs) and brainstem auditory evoked potentials (BAEPs). Others are specific to the operating room, such as direct cortical stimulation techniques for localization of eloquent cortical functions.

IONM can be separated into two general categories of procedures: monitoring and testing. In monitoring, the neurophysiology team establishes baseline values early in a procedure. During the remainder of the procedure the baseline is compared to current status of amplitude, latencies, frequency content, or presence of discharges. Changes from baseline are the basis for alerts. The thresholds for alerts are set at typical predetermined levels, such as a 50% amplitude drop in the SEP cortical peak amplitude. Testing in the operating room seeks to identify nervous system structures techniques known to localize normal function or pathology, e.g. to find motor cortex.

Staffing IONM services involves three kinds of skills, knowledge, ability, training, and experience:

- A person is required in the operating room to apply and remove electrodes, and run the machine. If a monitoring professional were not in the room, this person would arrange for internet connectivity and assist with communication. A trained, certified EEG technologist usually fills this role.
- A professional with substantial knowledge, training, and experience in intraoperative monitoring supervises the technologist, assists in problem solving, and contributes to determining, which modalities to monitor in which patients.
- A licensed physician is a final team member of the team. This physician is knowledgeable about neurophysiology so as to integrate the IONM findings with the patient's medical history. This may influence the choice of modalities used and the responses when adverse changes occur during surgery.

The physician is in the best position to recommend surgical or medical interventions. The physician provides medical quality assurance and communicates on the medical issues with the surgeons and anaesthesiologist. Sometimes a person fills more than one of these three roles. For example, some physicians are sufficiently expert in IONM to serve both the second and third roles. In other institutions, a specially trained non-physician (e.g. PhD) neurophysiologist fills the second role while a physician fills the third role in the collaborative IONM team.

The preoperative assessment can help determining the types of monitoring needed. At the very least, a review of the patient's medical records can determine the monitoring and testing tactics. Based upon pre-existing conditions some changes to monitoring techniques may be indicated. Pre-existing conditions also will influence expectations about neurophysiological intraoperative findings.

Multiple modalities can be monitored simultaneously. The monitoring team can display the ongoing current status for these on a single screen. Fig. 36.1 shows a simultaneous display of somatosensory, motor, EMG, and EEG channels. Other screens are available for viewing individual trends over hours on particular channels. Fig. 36.2 shows a trend display over 4 h of upper extremity somatosensory potentials, two channels per limb. The effect of loose electrodes, a technical problem, is seen twice, which required the technologist to reapply the loose electrode each time. The second time it was reapplied incorrectly.

In a straightforward case, the monitoring neurophysiologist may be outside the operating room supervising remotely from a screen (1). In that case, excellent two-way communication with the operating room is needed. The remote monitoring screen should mimic what could be seen in the operating room, and allow the neurophysiologist to manipulate the data and change among various screens. Some neurophysiologists use this tactic to supervise several simultaneous cases when they are simple and no critical changes occur.

Somatosensory evoked potentials

SEPs typically are recorded from the median or ulnar nerve of the upper extremities and peroneal or posterior tibial nerves in the lower extremities. They are used for a variety of IONM purposes, most often for spinal cord monitoring.

Stimulation

Median and ulnar nerve stimulation is applied at the wrist. Stimulus intensity is above motor threshold. Motor threshold should be checked before any neuromuscular blockade. Intensity of

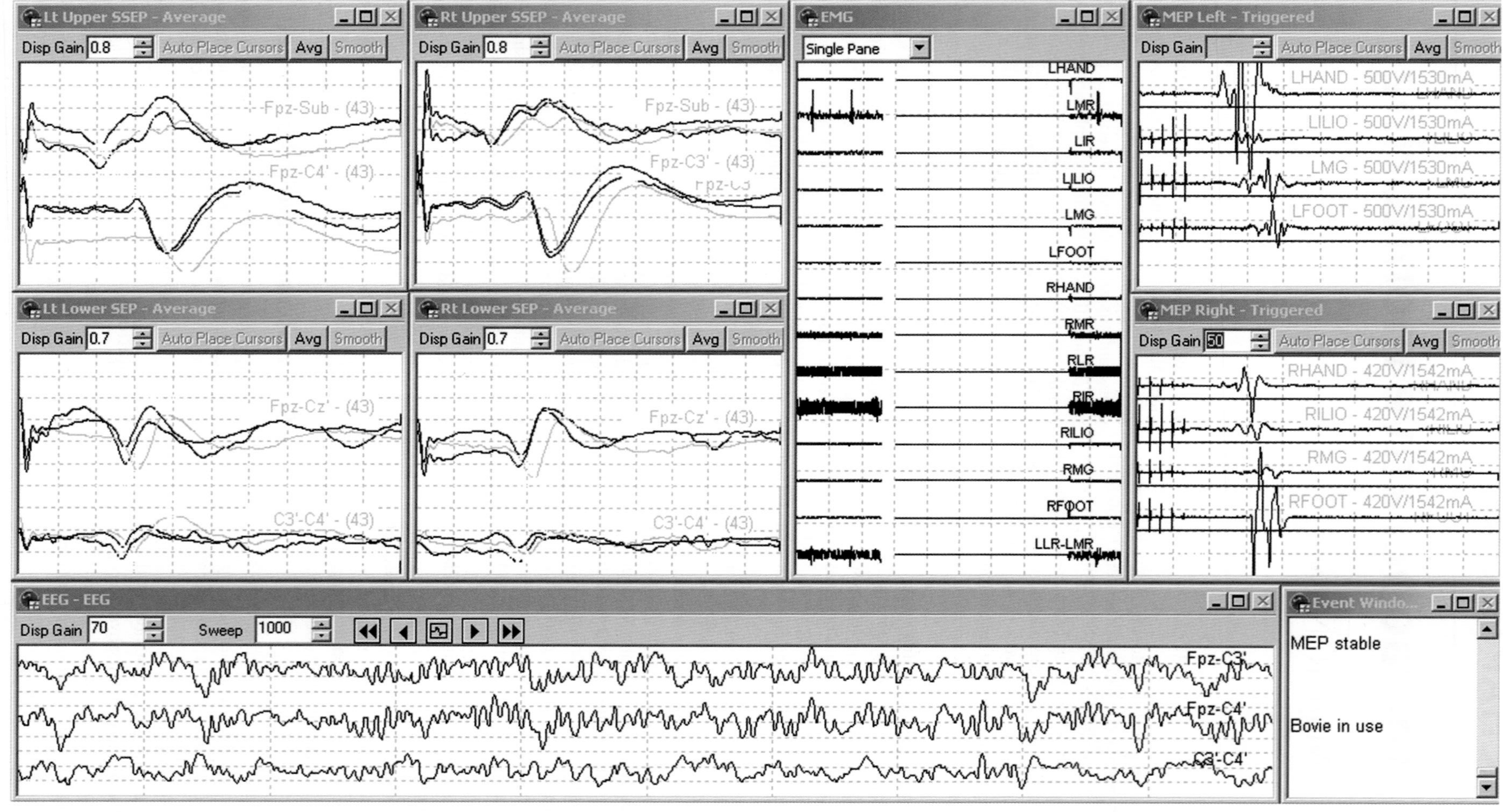

Fig. 36.1 Multimodality monitoring is performed assessing simultaneously four-limb somatosensory evoked potentials, four-limb motor evoked potentials, the multi-channel ongoing electromyograph, and a few channels of the ongoing electroencephalogram. A chat dialog box is used to communicate notes to a remote supervising neurophysiologist. This typical spinal surgery page allows the monitoring team an overview of many modalities. Other screen pages focus on specific modalities in greater detail.

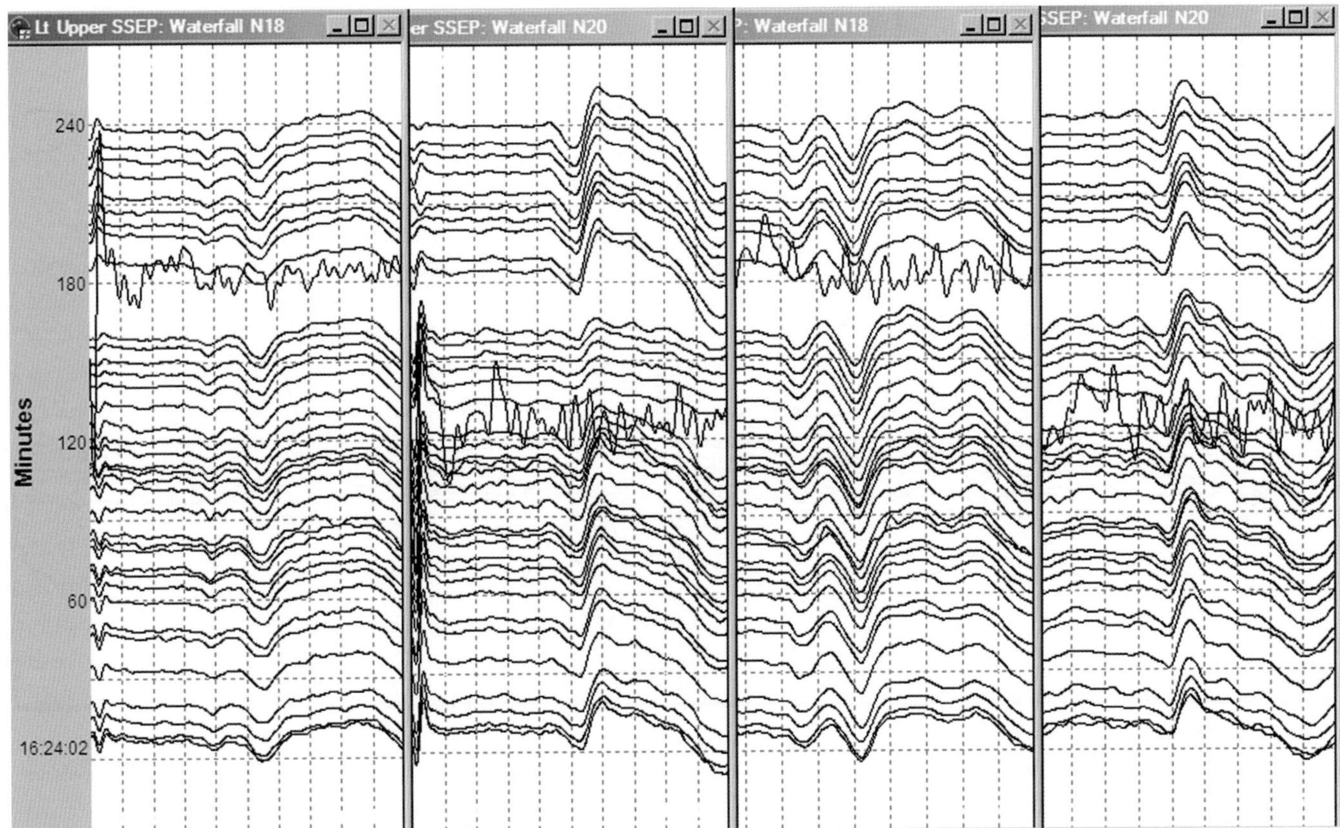

Fig. 36.2 Technical problems. Stable upper extremity somatosensory evoked potentials. This trend displays 4 h of two limbs, two channels per limb. Time is minutes-ago from the present (16:24 hours). Twice, a loose electrode occurred requiring the technologist to reapply. Note the relative reproducibility over 4 h, except for a greater than 50% drop in the second channel. That was technical—when the loose electrode was reapplied at 120 min previously, it was applied at an incorrect location. Problem solving determined that solution. Correcting the technical issue returned the tracings to baseline.

20–50 mA often is used. The median nerve is more often chosen for intracranial procedures because the cortical peaks are larger. The ulnar nerve is chosen for cervical procedures because the pathway enters the spinal cord several levels lower, giving more complete cervical pathway coverage. The ulnar often is used in lumbar cases because ulnar palsy due to the risk of brachial plexus impairment due to arm positioning during surgery.

Posterior tibial nerve stimulation is used most often for lower extremities, with stimulation just posterior to the medial malleolus. The peroneal nerve also is set up for monitoring patients with peripheral neuropathy, diabetes, or greater than 65 years of age. The peroneal nerve is stimulated as it crosses the fibular head just below the knee. By setting up both nerves for possible use, the team can choose whichever gives the better recording.

Stimulation rates are several per second. Stimulation can be as fast as 5/s (2). There is a trade-off between the stimulus rates and cortical peaks amplitude. Faster stimulation rates results in smaller amplitude peaks. Faster stimulation rates are desirable to produce evoked potential (EP) trials more quickly, i.e. reaching the desired 300–500 repetitions faster. In individual patients, the monitoring team adjusts the rate until the best rate-amplitude trade-off is found. When the baseline potentials are low amplitude, a typical strategy is to slow the stimulation rate so as to increase the peak amplitude.

Median nerve stimulation should produce a 1–2-cm thumb twitch. Ulnar nerve stimulation should produce a hypothenar movement. Posterior tibial nerve stimulation produces plantar bending of the foot. Peroneal nerve stimulation produces the dorsiflexion at the ankle. Identifying these movements helps set motor threshold and demonstrates that stimulus is working. Needle electrodes maintain contact through prolonged surgery without producing pressure sores. Bar or disc electrodes may require greater stimulation intensity through dry skin.

Commonly 300 repetitions are needed to obtain averaged SEPs. In some cases background noise and low amplitude peaks require larger sample sizes, e.g. 500–1000 repetitions, to produce each EP trial.

Recordings

These usually are made from the scalp and neck. Several channels are set up early in the procedure. Some users also employ a peripheral channel, e.g. over the shoulder for upper extremity SEPs or over the lumbar spine for lower extremity SEPs. Scalp sites use a modified 10% extension of the 10-20 system (3). In the 10% the system, the rows between F, C, P, O are referred to by letter combinations, e.g. CP is the row halfway between C and P. Numbered sites are interposed at halfway points, for example C1 is halfway between Cz and C3. For SEPs, site letters followed by a prime is located 2 cm posterior to the named site, for example C3' is 2 cm posterior to C3. The name Ci' refers to whichever of C3' or C4' is ipsilateral to the leg stimulated, and Cc' is contralateral. Channel Cc'-Ci' is commonly used to monitor cortical SEPs peaks. CPz or

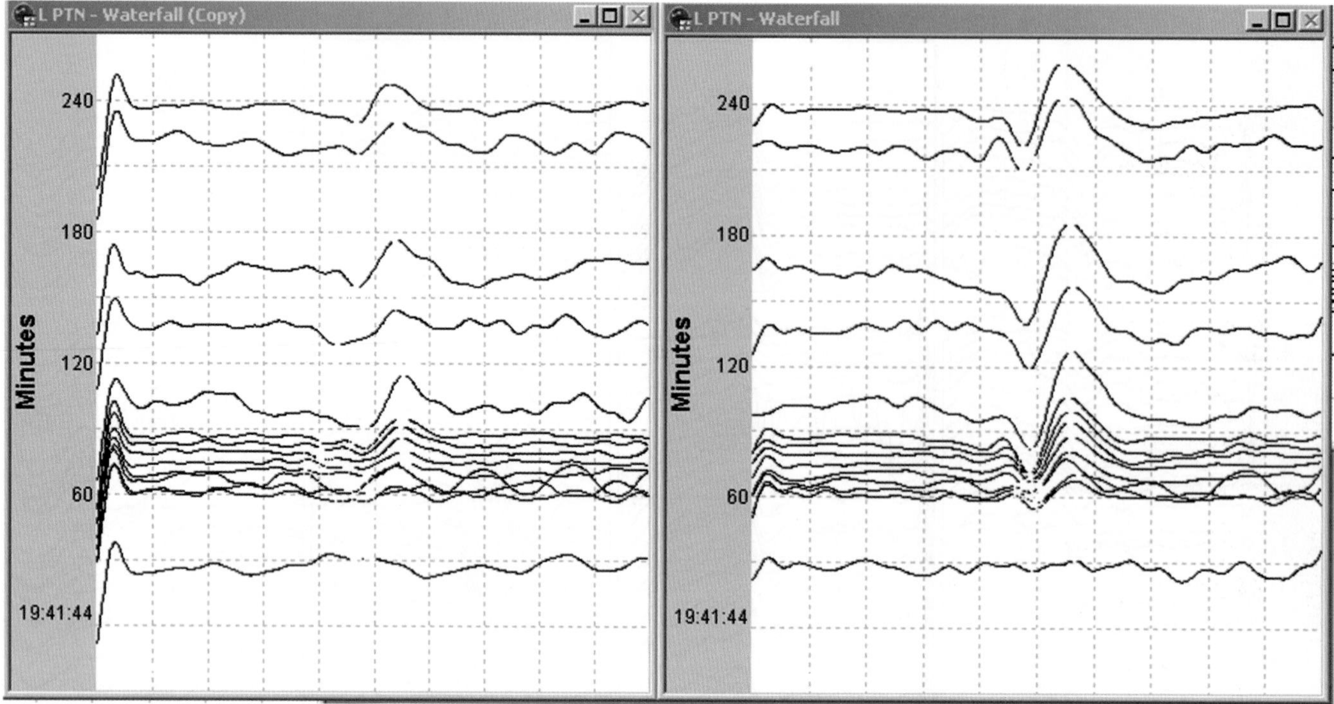

Fig. 36.3 Anaesthesia effect. Inhalation anaesthetic agent sevoflurane reduced then abolished the cortical somatosensory evoked potentials (SEP) at the end of this case. Two channels are showed for the left lower extremity SEP, one for a vertex-forehead and one for a transverse recording channel. Time is in minutes-ago from current time (19:41 hours). Anaesthesia switched from intravenous propofol to an inhalation agent during closing in preparation for post-operative awakening and recovery.

Cz' is also commonly used, because the cortical peak may have maximal amplitude at the midline (Cz') or off the mid-line (Ci'). If baseline potentials are difficult to find early in the procedure, the additional channels near the usual scalp recording channels may help to find the best amplitude cortical peaks. A cervical spine electrode is placed over the fifth cervical spine. During cervical surgery, a substitute site is mastoid or ear. Individual patients have peaks at different maximal locations on the scalp. Exploration early in the procedure will determine the best or find adequate locations for monitoring in individual patients. The monitoring team should not feel constrained to use a simple cookbook formula always to monitor every patient using the identical simple techniques.

The optimal low filter setting is usually 30 Hz. The users chose high filters from 500 to 3000 Hz. A higher setting of the high filter will record more background noise. A lower setting will attenuate the cortical peak amplitude. A notch filter removes the 50 or 60 Hz background noise in the electricity challenged operating room. Unfortunately, SEPs themselves have a basic frequency in the range of 50–60 Hz, so a notch filter can substantially reduce the desired peak itself. The notch filer also can produce an unwanted ringing artefact caused by the SEP stimulus artefact. The best tactic for electrical background noise is to identify and turning off culprit equipment rather than turning on the notch filter.

Interpreting changes

For upper extremity testing, the typical peaks sought are the cortical N20 and the cervical N13 peaks. The subcortical P14–N18 is also sometimes used especially for lumbar or thoracic spine surgery. These peaks are measured both for their latency and amplitude. For lower extremity SEPs, the principal peaks measured are the P37–N45 interpeak amplitude and the P37 latency. In many cases a subcortical P30 peak also can be found using cortical non-cephalic channel, i.e. a vertex scalp active electrode with the ear or neck as a reference. Classically, a 50% drop in amplitude is considered the criterion for raising an alarm. A 10% increase in latency is a secondary criterion for alarm.

Amplitude can be affected by anaesthesia. Inhalation anaesthetic agents can substantially diminish cortical peak amplitudes (Fig. 36.3). The subcortical peaks are much less affected by gas. Boluses of other medications also can affect the cortical peak amplitudes transiently, i.e. for 10–20 min. The gradual cumulative effect of anaesthesia over time is referred to as anaesthetic fade. That fade is more noticeable in the first 40 min when anaesthesia is taking effect after induction, and after that anaesthetic fade is present to a lesser extent over hours.

When changes occur, the monitoring team quickly must decide whether the change is due to technical problems, anaesthesia or systemic issues, or surgical problems. A variety of technical problems can occur when electrodes become disconnected or equipment malfunctions. Fig. 36.2 shows a technical problem from loose electrodes. Presence of a stimulus artefact demonstrates that stimuli still are being delivered. The raw data channel can show that real physiological data still is being collected. Systemic factors considered include hypothermia, hypotension, and hypoxia. Hypothermia typically causes latency increase according to the well-known basic neurophysiology effects of temperature on nerve conduction (Fig. 36.4). Anaesthesia levels and any recent bolus of medication can be quickly reviewed. If no obvious non-clinical causes are determined, the surgeon and anaesthesiologist are alerted to an SEP change.

Typical responses are to evaluate what the surgeon has done in the past 20 min that may have led to the observed change. Some

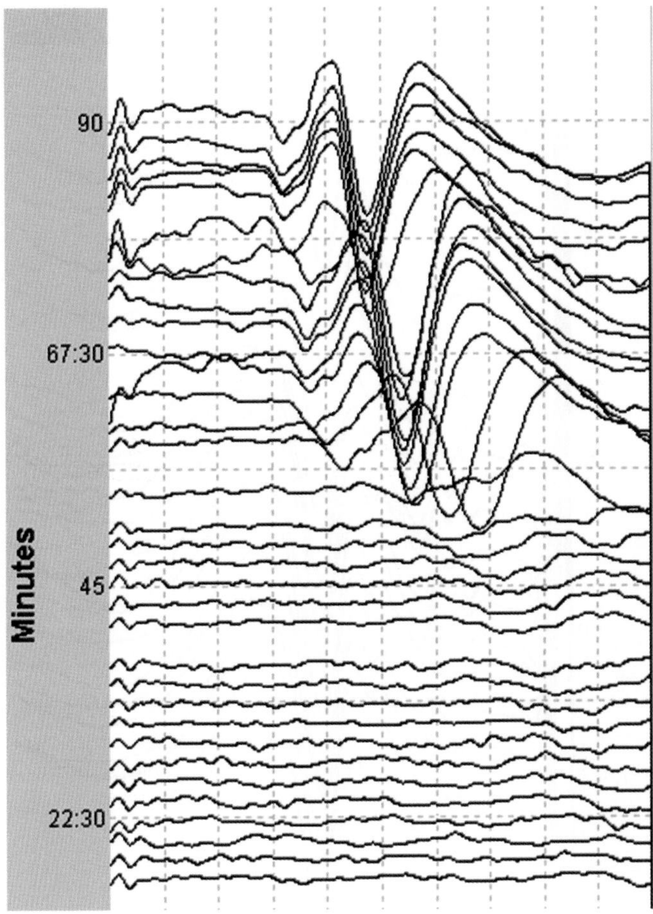

Fig. 36.4 Hypothermia effect. Median nerve somatosensory cortical evoked potential as the patient's temperature dropped from 37°C to 20°C. Time is minutes-ago from the current time. Note how the peak increases in latency then decreases in amplitude until it is lost at deep hypothermia. The hypothermia was used as cerebral protection during cardiac surgery. EEG (not shown) was used to monitor for the isoelectric state and cerebral recovery, and SEPs from four limbs were used to monitor the spinal cord during the systemic ischaemia of circulatory arrest. At the end of this case, the EEG and SEPs recovered uneventfully. Disappearance of EEG and SEPs reassured the surgeon that sufficient hypothermia was in place before circulatory arrest was used.

surgical manoeuvres or actions may not cause evoked potential changes for 20 min, perhaps due to the gradual accumulating effect of compression or stretching or a more marginal degrees of ischaemia or secondary autonomic vasospasm. Many times, the easiest clinical intervention to undertake is to raise the blood pressure. Sometimes pausing for a short while in surgery will allow the nervous system to recover from some abrupt change.

Clinical risk of change

Not all amplitude decreases predict an adverse neurological outcome. A 50–80% decrease in amplitude for several minutes poses only a small to moderate risks of post-operative deficits, especially if the SEP subsequently returns to baseline. The highest risk of post-operative occurs when the SEPs are abruptly completely lost and remain absent for the remainder of the operation. Even in such a grave circumstance, the risk of a new postoperative neurological impairment may be 50–75%. Fig. 36.5 shows loss of unilateral SEPs during an AVM resection, prompting responses (not shown) to increase blood pressure and postpone temporarily further resection at that location.

Motor evoked potentials

Motor evoked potentials (MEPs) monitor the corticospinal tracts. This is a particularly important monitoring modality because preserving motor function is a very important role for monitoring. In the operating room MEPs use transcranial electrical (tce) stimulation. Magnetic stimulation for MEPs can be used in awake outpatients, but unfortunately anaesthesia abolishes the magnetic MEP.

Stimulation

Stimulating electrodes are applied at the scalp. Corkscrew needle electrodes often are used to secure these electrodes. Electrodes are located near motor cortex on both hemispheres. A stimulating anode electrode is located about 2 cm anterior to C3 or C4 electrode sites and a cathode electrode is at Cz or CPz. In some patients, alternate scalp sites are used to obtain better responses.

Stimulus intensity is usually 250–400 mA. Sometimes stronger stimulation is needed up to 500mA, which may correspond to 1000–1200mV. A stimulus pulse width of 0.05 ms often is used, although a longer pulse width may be used when responses are difficulty to obtain.

Single pulses usually fail to produce good muscle responses. A brief stimulus train is used instead of a single pulse. Typical briefs trains are 5–7 individual pulses separated by a 2.0–4.0 ms interpulse interval. Such a brief stimulus train provokes a build-up of excitatory postsynaptic potentials at spinal cord anterior horn cells, resulting in the cell firing an action volley.

Double pulse trains are another tactic to excite anterior horn cells when single pulse trains of 5–7 pulses fail. In this, two trains of 3–4 pulses are separated by a several millisecond pause. The first brief train primes the anterior horn cell so that the second brief pulse train more effectively discharges the anterior horn cell. Each pulse discharges the corticospinal tracts within the cortical hemispheres. Stronger intensities discharge axons deeper in the hemispheric white matter. This is important to understand when using MEP for intracranial procedures. If MEPs are performed during a tumour resection, very strong stimulus intensity may actually discharge the corticospinal tract below the level of the hemispheric tumour. That may give a false sense of an intact motor pathway. Very strong stimuli, by discharging the corticospinal tract at deeper anatomical levels, shorten the latencies to muscle responses.

Recording

Typical recordings are made from electrodes in the arms and legs muscles. Both proximal and distal sites often are used. Having several sites in each limb is useful since the tceMEPs may produce good results in one muscle group, but only marginally recordable or absent results in another group. At baseline, the transcranial stimulation is gradually increased until adequate recordings are obtained.

Typical responses are polyphasic complex compound muscle action potentials (CMAPs) at each muscle tested. These recordings often are stable over the course of surgery. The shape varies greatly among patients and muscles. Measurements are the CMAP onset latency and overall amplitude.

An alternate method to record MEPs is from electrodes in the epidural space. These record axon volleys from the spinal cord.

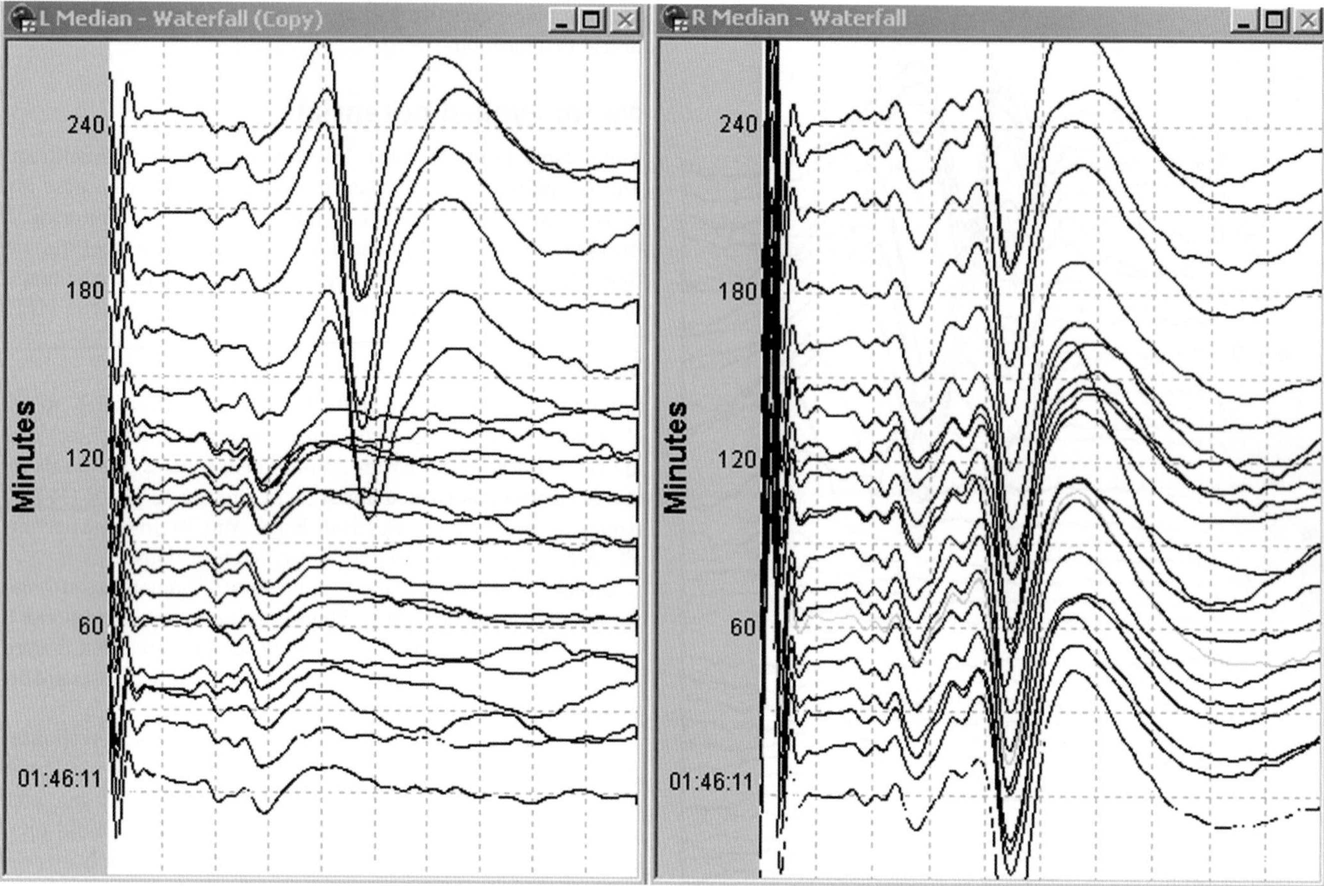

Fig. 36.5 Unilateral loss. During this arterial-venous malformation resection, the left median nerve somatosensory evoked potential cortical peak was lost abruptly. This lost was discussed with the surgeons, who then altered their surgical approach to minimize additional cortical ischaemia. In the end the patient awoke with only transient new weakness and sensory loss. The other hemisphere was unaffected. Time is in minutes-ago before the current time (1:46 hours).

This technique records the D wave, referring to the direct discharge of the corticospinal tract from electrical stimulation. D waves are recorded from these epidural electrodes from two closely spaced contacts or from one epidural contact compared to a nearby reference in soft tissue at the similar anatomical level. D waves are very small and are more easily obtained at a cervical and upper thoracic level. They can be difficult or impossible to obtain in a lower thoracic or lumbar level.

Safety

MacDonald's (4) review of safety found few adverse events caused by the stimulus itself. The most common is a tongue or lip laceration. The stimulus triggers directly a brisk jaw muscle contraction. As a precaution a mouth guard is placed prior to surgery. The mouth guard needs to be checked again after turning the patient prone for spine surgery. Seizures are rare and cardiac arrhythmia is not likely due to the tce stimulation. Minor scalp burns are rare. No spinal epidural recording electrode complications were found for the D wave technique. Relative contraindications include epilepsy, cortical lesions, convexity skull defects, raised intracranial pressure, cardiac disease, proconvulsant medications or anaesthetics, and cardiac pacemakers. More absolute contraindications include intracranial electrodes (e.g. Parkinsons deep brain stimulators), vascular aneurysm clips, and cochlear implant devices. Unexplained intraoperative seizures and cardiac arrhythmias are relative indications to avoid MEP stimulation. With appropriate precautions the benefits of MEP monitoring outweigh the associated risks.

Interpretation

Many monitoring teams use an all or none criterion for MEP alerts. This is supplemented by knowledge of the baseline amplitude from each muscle and the responses from other muscles in the same limb. If a limb has two muscles with good well-defined potentials, but one muscle's recording that is very small early in the procedure, the loss of the small potential may be of no clinical significance. The same anaesthesia fade seen in SEPs also applies in MEPs. Some variable amplitude changes are common. A small potential may disappear upon anaesthesia fade or for no particular reason. In the all or none method of interpretation, a sudden loss of all MEPs from a limb is very concerning. Sudden loss of all MEPs from both lower extremities is cause for great concern, especially if there is no anaesthetic reason to explain the loss. Other monitoring teams apply a more graded way of assessing concern. An 80% loss of amplitude may be considered sufficient to raise an alarm (Fig. 36.6). Some IOM groups track how polyphasic is a response. In that latter tactic, a very polyphasic response can prompt an alert if it suddenly becomes simplified to just two or three phases, especially when coupled with a modest loss of amplitude.

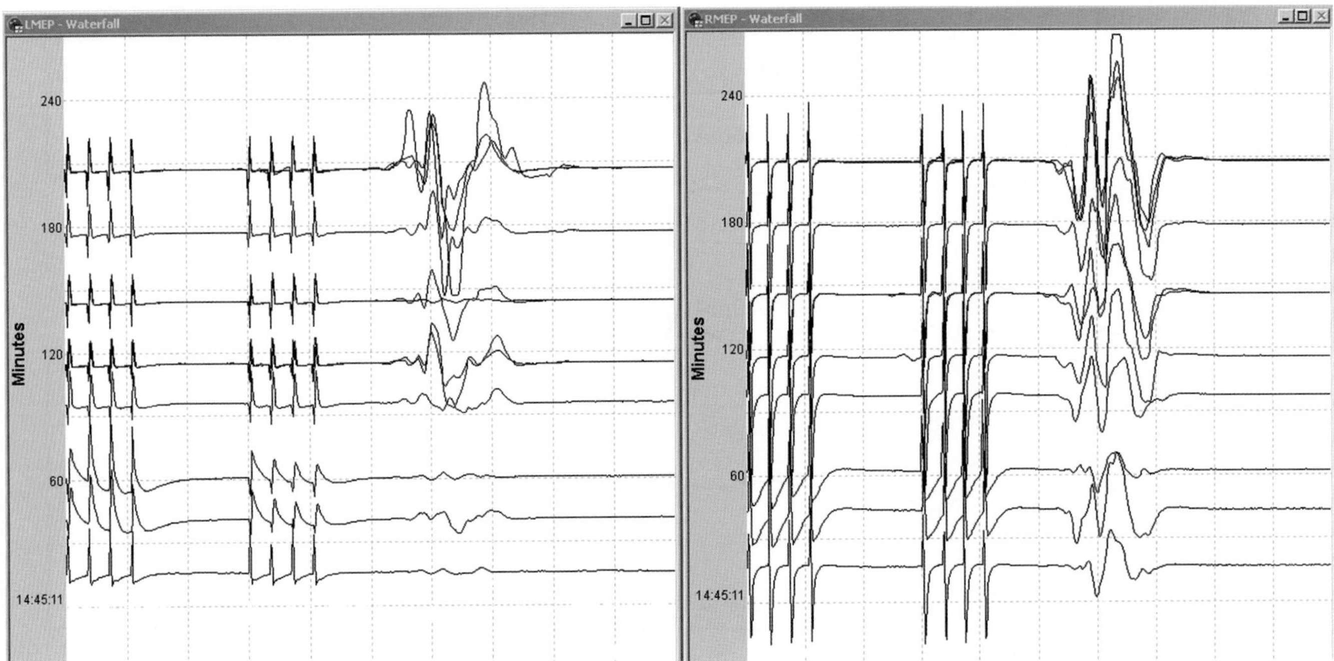

Fig. 36.6 Motor change. The motor evoked potentials (MEPs) decreased especially on the left side. Time is in minutes-ago from the current time (14:45 hours). Note the use of the double train stimulation technique, as indicated by the double set of stimulation artefacts. In this example the left side MEPs decreased more than 80% but did not entirely disappear.

The clinical decisions to raise a tceMEP alert can be complex, since one must integrate, which muscles changed, how many muscles changed, the change seen loss of phases or degree of amplitude change (80% loss versus total loss), and how robust were baseline recordings in those muscles (Fig. 36.7).

Anaesthesia plays a significant role in ability to find MEPs. Classically, total intravenous anaesthesia (TIVA) was required to find tceMEPs. Neuromuscular blockade must be avoided. Those two ideal requirements are not completely true. In many cases a small amount of inhalation anaesthesia can be tolerated, especially when the baseline MEPs are robust. This holds true more so among younger and middle aged patients with no pre-existing neurological condition. Inhalation anaesthesia may be poorly tolerated among patients who are older and have pre-existing neurological conditions, e.g. elderly patients with cervical myelopathy. Some groups use a continuous drip of small amounts of a neuromuscular blocker that tempers the degree of body movement with MEP stimulation. This blockade's balancing act is tricky, checking the train of four (TOF) test repeatedly to assure at least 3 twitches in the TOF response. Under those conditions a tceMEP often can be obtained and monitored.

Contraindications to tceMEP stimulation are cochlear implants or a metallic implant into the brain such as the deep brain stimulators used to treat Parkinsons. Concern has been raised about a number of other clinical settings. Plastic ventricular shunts, history of epileptic seizures, burr holes or craniotomy defects in the skull, cardiac arrhythmias, and vagus nerve stimulators do not exclude tceMEP stimulation. Usually MEP monitoring can be conducted despite those relative circumstances. One should consider monitoring EEG in patients with epilepsy. The electrical field during stimulation fills a small area in the head. It does not spread significantly to the thorax so that metal placed in the neck or thorax is generally considered relatively safe. That includes a cardiac pacemaker, although in that case anaesthesia should monitor the ECG when MEP stimulation is used.

Spinal cord monitoring

Spinal cord monitoring often uses both SEP and MEP monitoring. Some surgeons prefer to avoid MEPs because of the movements that occur from stimulation. Some anaesthesiologists prefer to avoid MEPs because it restricts the amount of inhalation agents. Otherwise both are used together. SEP can monitor continuously, and MEP intermittently as needed.

False positive monitoring events are when alerts occur without post-operative neurological deficits. Many of these false positive events are actually true detections of neurological risk, but the surgeon or anaesthesiologist intervened to avert a post-operative deficit. Those might be more properly referred to as 'true save' events. There is no practical way to differentiate between the two. Animal literature shows that failing to respond to SEP alerts carries a high risk of adverse post-operative events (5–10). Those studies also show that responding to alerts prevents post-operative deficits. That experimental evidence strongly supports the conclusion that SEP alerts are an effective way to reduce post-operative neurological deficits. So it is clinically recommended to respond clinically to SEP alarms.

False negative monitoring occurs when the patient awakens with new neurological deficits despite normal stable intraoperative SEPs. Some are due to delayed onset deficits that occur after monitoring was discontinued, i.e. during the several hours post-operative period. False negative events with SEP spinal cord monitoring are very rare (Table 36.1).

Clinical reports show that SEPs and MEPs detect spinal cord impairment relatively early, enough so to warn the surgeon in time to

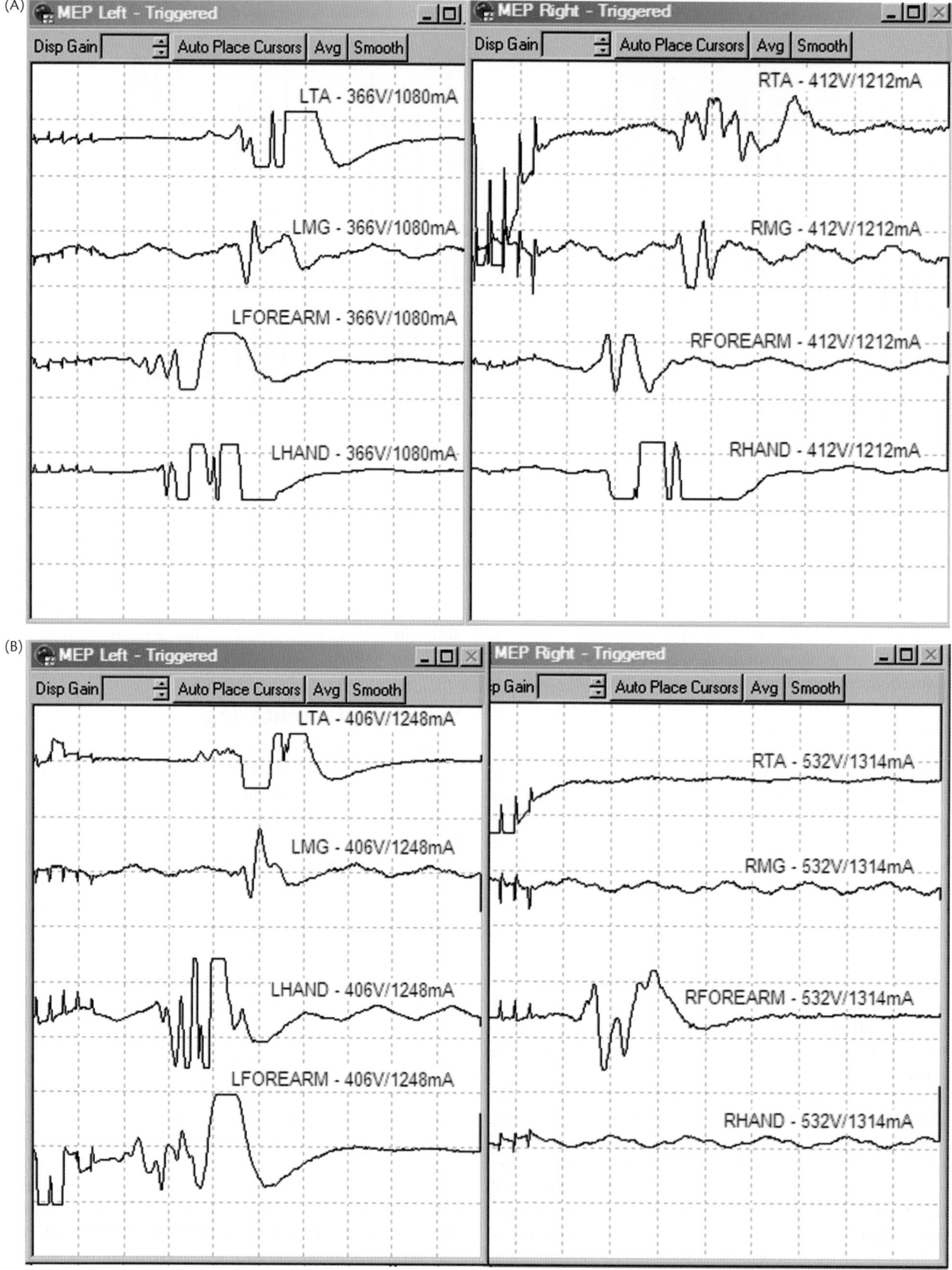

Fig. 36.7 The surgeon reported that he bumped the cervical spinal cord. (A) Baseline four-channel motor evoked potentials (MEPs) earlier in the case. (B) MEPs after bumping the spinal cord. Three of the four right MEPs were lost for a while after this mild cord contusion. Left MEPs were not affected. MEPs eventually returned, and the patient had no post-operative sequallae.

TA = tibialis anterior; MG = medial gastroc; FOREARM = flexor carpi ulnaris to flexor carpi radialis; HAND = adductor policis brevis to abductor digiti minimi.

Table 36.1 Neurological outcome prediction rates for SEP monitoring in spinal surgery

Total procedures monitored	**51,263**	**(100%)**
False-negative (FN) rate: neurological post-operative deficits despite stable SEPs		
Definite	34	(0.063%)
Equivocal	13	(0.025%)
Delayed onset	18	(0.035%)
Total:	65	(0.127%)
False-positive rate: no neurologic deficits despite SEP changes		
Definite	504	(0.983%)
Equivocal	270	(0.527%)
Total:	774	(1.510%)
True-positive (TP) rate: neurological deficits predicted by SEP changes		
Definite	150	(0.293%)
Equivocal	67	(0.131%)
Total:	217	(0.423%)
Neurologic deficits (FN plus TP)		
Definite	184	(0.356%)
Equivocal	80	(0.156%)
Delayed Onset	18	(0.035%)
Total:	282	(0.550%)
True negative rate: no neurological deficit and stable SEPs		
Total:	50,207	(97.94%)

These data are from the multicentre outcome study of SEP spinal cord monitoring in scoliosis organized through the Scoliosis Research Society (11). They were obtained from 153 US surgeon respondents. Note the very low rate of definite false negative cases (0.063%). Equivocal cases were transient or minor degrees of impairment. Delayed onset cases awoke from surgery intact, but developed impairment within the first day post-operatively.

Source data from *Electroencephalog Clin Neurophysiol.*, **96**(1), Nuwer MR, Dawson EG, Carlson LG, Kanim LEA, Sherman JE, Somatosensory evoked potential spinal cord monitoring reduces neurologic deficits after scoliosis surgery: results of a large multicenter survey, pp. 6–11, copyright (1995) Elsevier.

avoid deficits. The American Academy of Neurology and American Clinical Neurophysiology Society's spinal cord monitoring assessment (12) included several studies that assessed MEP and SEP alerts (13–24). That assessment concluded that IOM is established as effective to predict an increased risk of the adverse outcomes of paraparesis, paraplegia, and quadriplegia in spinal surgery. In the clinical setting the operating team should intervene to attempt to reduce the risk of adverse neurological outcomes when alerts occur.

As with all IOM it is important to have a knowledgeable, experienced team conducting both technical and clinical interpretation. An unfortunate example of this point occurred with a published case report claiming to show a failure of MEP spinal cord monitoring (25). Those authors reported that the tceMEP from the arms disappeared during the spinal procedure, but MEPs from the legs were preserved throughout. The patient awoke with new neurological deficits, which the authors claimed was a false negative. The authors published figures of their tracings at baseline and late in the procedure. The figures demonstrated that the authors confused the technical set up of their monitoring, and were unaware that they had mixed up the arms and the legs. In reality their example showed a true positive loss of MEPs from the legs during their spinal cord monitoring case—not a false negative lack of change. The lesson is that monitoring teams need to have substantial skills, knowledge, ability, and training experience so as to set up, recognize, and correctly interpret the tracings.

A large Scoliosis Research Society multi-centre study (11) evaluated more than 100,000 surgical cases. A US cohort of 184 surgeons tracked surgical outcomes over seven years. Approximately half of the cases were monitored with SEPs, the other half were not. The outcomes could be compared with and without monitoring, and compared against historical controls. Intraoperative neurophysiological monitoring was associated with a 60% reduction in paraparesis and paraplegia. The false negative rate was very low, especially after excluding cases with symptoms that were delayed in onset after surgery ended or symptoms that were only transient.

Sala et al. (26) reported a different kind of study. Historical controls were used to evaluate spinal cord IOM. They assessed motor status post-operatively in a group of neurosurgical patients. The McCormick grade of weakness was used for measuring motor outcome compared with pre-operative status. The aggregate pre-operative to post-operative McCormick grade improved + 0.28 in monitored patients, whereas without monitoring it deteriorated – 0.16 ($p < 0.002$ difference).

Studies have assessed both SEP and MEP monitoring. At the present time, there has been no definitive method to differentiate the relative utility of those two techniques. They monitor different spinal pathways, since SEP monitors dorsal columns whereas MEP monitors corticospinal tracts. SEPs can be conducted more or less continuously during surgery, which is a distinct advantage. MEPs are performed intermittently, which means they may be unmonitored during certain parts of the procedure. MEP testing also produces jerking movements when monitoring from extremities' muscles. Some surgeons dislike patient movements during the operation, so they use MEPs sparingly. Some published cases have shown MEP changes a few minutes earlier than SEPs during spinal cord alerts. This can lead to interventions being made such as raising the blood pressure when MEPs change. In fact, the blood pressure changes may be undertaken quickly enough so as to intervene even before the SEPs change.

Brainstem auditory evoked potentials

The brainstem auditory evoked potential (BAEP) can monitor the eighth cranial nerve and brainstem during posterior fossa procedures. Typically this is carried out along with other IOM modalities such as EMG of cranial nerves innervated muscles and SEP monitoring of the medial lemniscus.

Stimulation

Stimuli of 0.10 ms rarefaction square wave clicks are delivered at 11–21/s through a plastic tube inserted into the external ear canal. The tube resembles intravenous line tubing and is curved to fit into the ear. One end is connected to a hearing aid transducer that is secured to the side of the patient's neck. The tubing is taped securely into the external ear cannel. Otherwise it may fall out during surgery, or irrigation fluid may get into the ear and cause difficulty

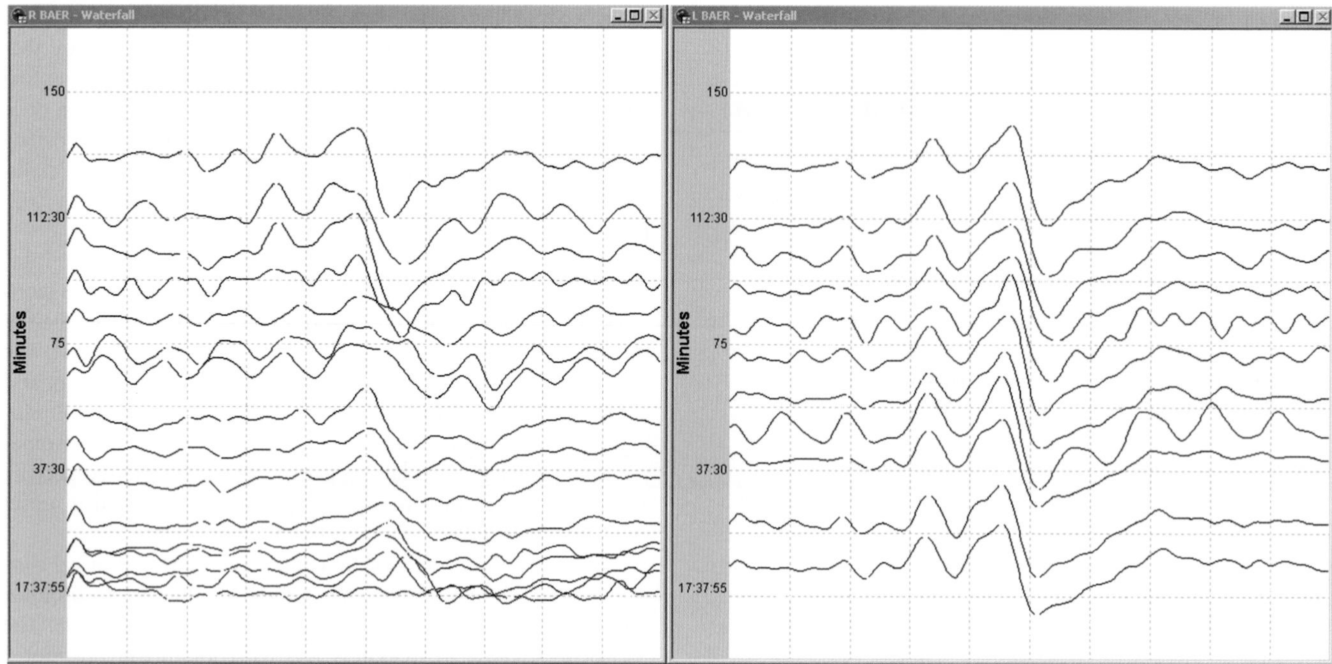

Fig. 36.8 The right brainstem auditory evoked potential (BAEP) deteriorated during this right acoustic neuroma resection. The left (BAEP) remained stable. Note that the wave V drifted out in latency, lost amplitude, and eventually disappeared entirely. Time is in minutes-ago before the current time (17:37 hours).

with air conduction. The tube tends to kink so it needs to be secured to minimize the likelihood of this. Some padding where the tube bends into the ear can help prevent kinking. Bone wax around the tube and tape over the ear can help keep water out and can help keep the tube in place throughout a long procedure. The ear in the head down position contralateral to the operating side is at a greater risk for that tube falling out by gravity.

Recording

BAEP recording channels are the same as for outpatient testing. Recording electrodes are located at the mastoid or earlobe, with a reference electrode at the scalp vertex. Both ipsilateral and contralateral channels are recorded. Filters of 100 and 3000 Hz are used. The average BAEP is calculated from 1000–2000 repetitions.

Interpretation

Typically attention is paid to waves I and V. Adverse changes are indicated by an increase the I–V interpeak interval or the wave I latency. The absolute latency of wave 1 is usually 0.8–1.0 ms longer during surgery than the latency recorded as the outpatient because of the distance of the stimulation tubing. A 10% latency increase is an alert criterion, which amounts to I–V interpeak interval increase of more than 0.4 ms. A 50% decrease in the wave I or V amplitude is another cause for alarm. The usual rules apply for interpreting the anatomic site of injury. A well-preserved wave I with absent wave V suggests a site of injury after the eighth nerve, and so forth. Loss of wave I suggests impairment at the eighth nerve or ear. Fig. 36.8 shows a gradual loss of the right wave V during a cerebellopontine angle tumour resection.

Anaesthesia does not significantly affect BAEP potentials. Certain other surgical factors can cause change. Irrigation with fluid at room temperature can decrease the temperature in these pathways and thus slow conduction. The use of body temperature irrigation fluid is recommended so as to avoid these confounding effects. Drilling can cause considerable loss of the BAEP for several reasons. Drilling the temporal bone near the cochlea or eighth nerve causes heat, which can adversely affect conduction. Drilling anywhere on the skull causes acoustic bone conducted noise and a masking effect on the click stimulus.

Clinical use

Clinical settings for BAEPs typically are posterior fossa procedures such as surgery around the cerebellopontine angle and posterior fossa. BAEP can be used to help as a predictor for post-operative impairment. In microvascular decompression, BAEP monitoring can warn of excess retraction or incidental impending injury to the brainstem or eighth nerve (27,28). Patients whose BAEPs show changes without return to baseline often have new post-operative deficits (29). A complete persistent disappearance of wave V is a risk for post-operative hearing loss (30–32). When change does occur, the usual BAEP interpretation rules apply to determining the likely site of impairment.

Electroencephalography

EEG recording has been performed in the operating room for many years. Recording from the scalp has been popular for carotid endarterectomy since the early 1970s. It is used for cardiothoracic surgery and aneurism surgery as well as other procedures that risk cerebral cortical damage from ischaemia. For patients with epilepsy, EEG can check for seizures under anaesthesia with neuromuscular blockade.

Recording

EEG is easily recorded with the standard electrode sites of the 10–20 electrode placement system. Sometimes the midline electrodes are deleted, and only the 16 temporal and parasagittal channels are used in a bipolar chain montage. Filters are typically set at 0.3–70 Hz. Using such a low filter helps to identify the slow activity present in the ischaemic injury. The notch 50–60 Hz filter can be used for EEG.

When monitoring for ischaemia, the page speed, i.e. paper speed, can be slowed below the usual 10 s/page. Slow EEG activity can be seen well with a horizontally compressed display running at slow page speed.

Modern digital equipment also allows for simultaneous display of two EEG segments. A typical tactic is to set the left window display with a baseline EEG segment, whereas the right window displays current EEG in real time. This allows the reader to use the left screen as a reminder of baseline EEG frequency content.

Interpreting change

EEG under anaesthesia typically consists of fast activity over the fronto-central regions bilaterally, along with smaller amounts of mixed background slow waves. Inhalation anaesthetic, especially at higher doses, produces frontopolar triangular waves. EEG is sensitive to depth of anaesthesia. EEG frequencies vary among patients. Classically changes in the type and depth of intravenous or inhalation anaesthetic agents regulate the characteristics of EEG under anaesthesia. The side-by-side two-window display method can help judge whether an EEG has changed from baseline. When changes are seen, one must assess whether the EEG change is due to changes in the depth of anaesthesia or a bolus of medication. Even a bolus of relatively innocuous medication, such as narcotics, can affect the EEG. A particular critical time during surgery is vascular clipping or clamping, such as clamping the carotid artery during carotid endarterectomy. Avoiding changes in anaesthesia during the 10 min prior to such a planned surgical event is desirable.

The first, lowest level of alarm for EEG monitoring is a 50% decrease in fast activity (33). Such a change is a sign of brain ischaemia. The second degree of EEG change is a 50% increase in slow activity. The third, worst degree of EEG change is a 50% decrease in EEG amplitude. Its worst extent is an EEG that becomes isoelectric. The latter is a sign of acute cortical ischaemia.

Studies of ischaemia have compared SEP and EEG. Both tend to decrease as the brain perfusion drops below 20 ml/100 g/min. The EEG becomes isoelectric around 7–15 ml/100g/min of blood flow, and the SEP cortical peaks become absent around 12–15 ml/100g/min (34). Fig. 36.9 compares the two techniques.

Clinical application

Carotid endarterectomy is the most common reason for intraoperative EEG monitoring. A vascular shunt is used in some patients to avoid risk of ischaemic injury (35). The selection of which patients to shunt is made in large part on whether a significant EEG change is seen upon cross-clamping the carotid. The monitoring team and anaesthetist need close communication to avoid any changes during the 10mins prior to clamping. The team will report any level of change, e.g. a 50% decrease in fast activity, as a reason to consider vascular shunting or other intervention.

Even though one carotid is clamped, both hemispheres may change. The circle of Willis can share equally the ischaemia between the two hemispheres. Therefore, monitoring teams avoid just relying on unilateral changes or hemispheric asymmetry as a means to detect ischaemia.

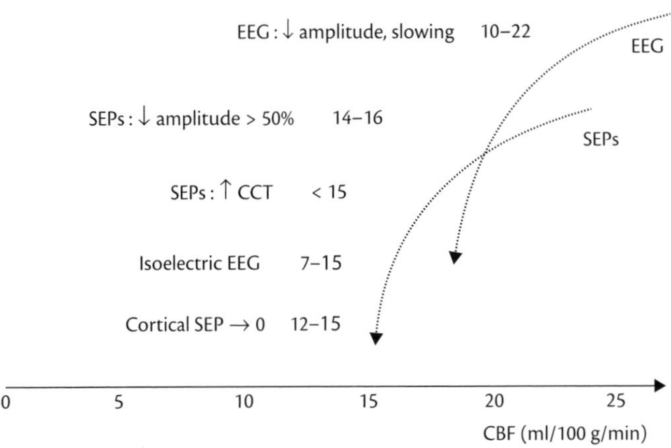

Fig. 36.9 Effects of ischaemia. Comparison of EEG and SEP changes during ischaemia, such as that caused in carotid endarterectomy. The arrows show the relative loss of amplitude of the two techniques as the blood flow decreases (CBF mg/100 g/min).

CCT = central conduction time.

Reproduced from *Neurophysiol Clin*, **34**(1), Florence G, Guérit JM, Gueguen B, Electroencephalography (EEG) and somatosensory evoked potentials (SEP) to prevent cerebral ischaemia in the operating room, pp. 17–32, copyright (2004), with permission from Elsevier.

During cardiac and aneurysm surgery, EEG is used both to evaluate for ischaemia and sometimes to determine if the patient has been adequately protected by hypothermia or barbiturates, which are used to protect against the damaging effects of ischaemia. Hypothermia and barbiturates are most effective when the EEG has reached an isoelectric state (36). Variable amounts of barbiturate or degrees of hypothermia are needed to achieve the isoelectric state in individual patients and the EEG can determine when an isoelectric state has been achieved.

Electromyography

The electromyograph (EMG) can be monitored from the face, limbs, or trunk during surgery. EMG monitoring can check for neurotonic discharges, or A-trains, which are signs of nerve injury (37). EMG also is used to assess proper placement of pedicle screws.

Recordings

Electrodes are placed in muscles at risk during surgery. Electrodes are uninsulated over an extended portion of the needle shaft, rather than using the traditional EMG electrodes typical in outpatient EMG testing. Recordings are monitored in real time without any averaging. Filters are set with high frequency of 3 k–10 kHz. Many channels are used, one for each muscle.

Cranial nerve surgery may risk injury to the seventh cranial nerve. In that case several facial muscles can be monitored such as orbicularis oris and orbicularis oculi. The fifth cranial nerve has a motor component, which can be monitored with electrodes placed in the masseter and temporalis muscles. Lower motor cranial nerves include the tongue, which can be monitored through electrodes placed from below the chin, and the accessory nerve, which can be monitored from the sternocleidomastoid or upper trapezius muscles. The vagus nerve is monitored typically with electrodes pasted to the outside the endotracheal tube so that they lie against the vocal cords. Some endotracheal tubes are manufactured with electrodes embedded in the tube. Monitoring the extraocular muscles is through needle electrodes placed in those muscles carefully by physicians with special training.

Needle electrodes in extremity muscles monitor risk to the peripheral nerves and plexus. Sites depend upon the nerves at risk. Appropriate muscles must be tested. This requires good medical knowledge to monitor a set of muscles appropriate for the level of surgery. Trunk muscles can be monitored, such as the abdominal rectus muscles or intercostal muscles. During cauda equina surgery, electrodes can be placed into the anal sphincter.

Interpretation

Recordings are monitored in real time continuously. During the baseline portion of the procedure any ongoing irregular background activity is observed, which may result from the pathophysiology that is the reason for the surgery. Often channels are all silent during the baseline. When a surgeon causes mechanical compression or stretches a nerve, such as placing a retractor too close to a nerve, a series of compound motor unit potentials may be seen. A greater degree can result in a continuous pattern. A classical sign of acute irritation or injury is the neurotonic discharge or A-train, which is a high frequency discharge often lasting 30–45 s (37). EMG monitoring is conducted in the absence of neuromuscular blockade, or with minimal continuous drip blockade with at least a 3 twitch in TOF testing.

Pedicle screw testing is performed to determine whether screw placement is correct. If placed improperly, the screw can encroach on the nerve or spinal cord. Placement of the pedicle screw would be easier if anatomy were normal and straightforward. However, many patients have pre-existing abnormalities that result in misaligned pedicles, for which the direction of screw placement can be difficult to determine. The surgeons have the option to use fluoroscopy to determine whether they are placing screw in the proper directions. The second technique available is EMG. A guide hole is drilled and the walls can be checked with electrical stimulation. The screw itself can be used to electrically stimulate as it is advanced or once it is in place. If the screw breaches the medial pedicle boney wall, low intensity electrical screw stimulation will activate a nearby nerve. Nerve stimulation results in EMG discharges from muscles in that dermatome. Pedicle screws are stimulated at constant current up to 20 mA. If a muscle response is obtained, stimulus intensity threshold is checked. In lumbar and lower thoracic spinal levels, a threshold 10 mA or higher is generally considered adequate. A threshold of 5 mA or lower is considered a sign of a wall breach. Values for thresholds are lower for higher spinal levels, i.e. cervical spine. Osteoporosis also produces lower thresholds through poorly mineralized bone. If testing suggests that a screw is malpositioned, then the surgeon can reposition it.

EMG monitoring is imperfect in several ways. First, the signs of nerve injury can appear transiently. If not watched carefully, a neurotonic discharge can pass by within a minute and be overlooked by monitoring technologist and physician. Not all nerve injuries produce EMG discharges. A nerve can be cut cleanly and produce no discharges (38). Chronically compressed nerves may be less sensitive to producing discharges. Since compressed nerves are the ones for, which surgery often is undertaken, EMG monitoring fails to detect some compressive, mechanical, or ischaemic nerve injuries.

EMG monitoring can identify nerves when they are buried in the resection field. This can be useful when a tumour surrounds or envelops a nerve. It also can be useful in the cauda equina to identify the filum terminale during a tethered cord sectioning. An electrical probe can survey a region to identify a nerve located within several millimeters of the probe tip. This can be a monopolar electrical stimulator with a distant ground or a bipolar stimulator. As long as the appropriate muscles are being monitored, this can determine whether a particular nerve is near the tip of the probe. Stimuli are delivered at several per second. In a posterior fossa procedure this technique can help identify, which nerve is which. When a surgeon has very limited view of the structure of the posterior fossa, stimulation can identify, which is the fifth and, which is the seventh cranial nerve. Typically, the fifth cranial nerve will produce EMG at a shorter latency and be maximal at masseter and temporalis. The seventh nerve, because of a longer anatomical pathway, has a longer latency and maximally affects facial muscles. Movement spread may cause non-target muscles to move, mimicking an EMG effect at other muscles. One tactic to differentiate nerve is to use smaller stimulus intensities, such as 0.1–0.2 mA for stimulation to identify the identity of a nerve. Higher intensities are used when scouting for any nerve located inside a region of pathology, for which 1–2 mA stimulation intensity is used.

Conclusions

Intraoperative neurophysiological monitoring involves many types of technique, the major ones of which are reviewed here. Many are familiar to traditional neurophysiologists from use in the outpatient laboratory. Some are unique to the operating room. The process can be supervised remotely, and more than one operation can be simultaneously monitored as long as too much difficulty is not encountered in any one case. A variety of tactics are now well known for obtaining good recordings in a surgical setting. Limits of normal variability have been established as alarm levels for alerting the surgeon. The use of intraoperative neurophysiological monitoring in experienced hands can reduce post-operative adverse neurological outcome, e.g. reduce by 60% the risk of paraplegia and paraparesis after spinal surgery. The monitoring team requires a technologist on site in the room, a knowledgeable expert in intraoperative monitoring and physician supervision. Many times a clinical neurophysiology physician will fill the letter two roles. This is an important role for health care prevention of neurological injury in surgical settings.

References

1. Nuwer, M.R., Cohen, B.H., and Shepard, K.M. (2013). Practice patterns for intraoperative neurophysiologic monitoring. *Neurology*, **80**, 1156–60.
2. Nuwer, M.R. and Dawson, E.G. (1984). Intraoperative evoked potential monitoring of the spinal cord: enhanced stability of cortical recordings. *Electroencephalography Clinical Neurophysiology*, **59**, 318–27.
3. Nuwer, M.R., Comi, Emerson, R.G., et al. (1998). Standards for digital recording of clinical EEG. *Electroencephalography & Clinical Neurophysiology*, **106**, 259–61.
4. MacDonald, D.B. (2002). Safety of intraoperative transcranial electrical stimulation motor evoked potential monitoring. *Journal Clinical Neurophysiology*, **19**, 416–29.
5. Coles, J.G., Wilson, G.J., Sima, A.F., Klement, P., and Tait, G.A. (1982). Intraoperative detection of spinal cord ischemia using somatosensory cortical evoked potentials during thoracic aortic occlusion. *Annals of Thoracic Surgery*, **34**, 299–306.
6. Kojima, Y., Yamamoto, T., Ogino, H., Okada, K., and Ono, K. (1979). Evoked spinal potentials as a monitor of spinal cord viability. *Spine*, **4**, 471–7.

7. Laschinger, J.C., Cunningham, J.N. Jr., Catinella, F.P., Nathan, I.M., Knopp, E.A., and Spencer, F.C. (1982). Detection and prevention of intraoperative spinal cord ischemia after cross-clamping of the thoracic aorta: use of somatosensory evoked potentials. *Surgery*, **92**, 1109–17.
8. Cheng, M.K., Robertson, C., Grossman, R.G., Foltz, R., and Williams, V. (1984). Neurological outcome correlated with spinal evoked potentials in a spinal cord ischemia model. *Journal of Neurosurgery*, **60**, 786–95.
9. Nordwall, A., Axelgaard, J., Harada, Y., Valencia, P., McNeal, D.R., and Brown, J.C. (1979). Spinal cord monitoring using evoked potentials recorded from feline vertebral bone. *Spine*, **4**, 486–94.
10. Bennett, M.H. (1983). Effects of compression and ischemia on spinal cord evoked potentials. *Experimental Neurology*, **80**, 508–19.
11. Nuwer, M.R., Dawson, E.G., Carlson, L.G., Kanim, L.E.A., and Sherman, J.E. (1995). Somatosensory evoked potential spinal cord monitoring reduces neurologic deficits after scoliosis surgery: results of a large multicenter survey. *Electroencephalography & Clinical Neurophysiology*, **96**, 6–11.
12. Nuwer, M.R., Emerson, R.G., Galloway, G., et al. (2012). Evidence-based guideline update: intraoperative spinal monitoring with somatosensory and transcranial electrical motor evoked potentials. *Journal of Clinical Neurophysiology*, **29**, 101–8.
13. Cunningham, J.N., Jr., Laschinger, J.C., and Spencer, F.C. (1987). Monitoring of somatosensory evoked potentials during surgical procedures on the thoracoabdominal aorta. IV: clinical observations and results. *Journal of Thoracic Cardiovascular Surgery*, **94**, 275–85.
14. Sutter, M., Eggspuehler, A., Grob, D., et al. (2007). The validity of multimodal intraoperative monitoring (MIOM) in surgery of 109 spine and spinal cord tumors. *European Spine Journal*, **16**, S197–S208.
15. Costa, P., Bruno, A., Bonzanino, M., et al. (2007). Somatosensory- and motor-evoked potential monitoring during spine and spinal cord surgery. *Spinal Cord*, **45**, 86–91.
16. Weinzierl, M.R., Reinacher, P., Gilsbach, J.M., and Rohde, V. (2007). Combined motor and somatosensory evoked potentials for intraoperative monitoring: intra- and postoperative data in a series of 69 operations. *Neurosurgery Reviews*, **30**, 109–16.
17. Etz, C.D., Halstead, J.C., Spielvogel, D., et al. (2006). Thoracic and thoracoabdominal aneurysm repair: is reimplantation of spinal cord arteries a waste of time? *Annals of Thoracic Surgery*, **82**, 1670–8.
18. May, D.M., Jones, S.J., and Crockard, H.A. (1996). Somatosensory evoked potential monitoring in cervical surgery: identification of pre- and post-operative risk factors associated with neurological deterioration. *Journal of Neurosurgery*, **85**, 566–73.
19. Lee, J.Y., Hilibrand, A.S., Lim, M.R., et al. (2006). Characterization of neurophysiologic alerts during anterior cervical spine surgery. *Spine*, **31**, 1916–22.
20. Pelosi, L., Lamb, J., Grevitt, M., Mehdian, S.M.H., Webb, J.K., and Blumhardt, L.D. (2002). Combined monitoring of motor and somatosensory evoked potentials in orthopaedic spinal surgery. *Clinical Neurophysiology*, **2113**, 1082–91.
21. Hilibrand, A.S., Schwartz, D.M., Sethuraman, V., Vaccaro, A.R., and Albert, T.J. (2004). Comparison of transcranial electrical motor and somatosensory evoked potential monitoring during cervical spine surgery. *Journal of Bone Joint Surgery*, **86A**, 1248–53.
22. Jacobs, M.J., Elenbass, T.W., Schurink, G.W.H., Mess, W.H., and Mochtar, B. (2000). Assessment of spinal cord integrity during thoracoabdominal aortic aneurysm repair. *Annals of Thoracic Surgery*, **74**, S1864–6.
23. Langeloo, D.D., Lelivelt, A., Journee, L., Slappendel, R., and de Kleuver, M. (2003). Transcranial electrical motor-evoked potential monitoring during surgery for spinal deformity: a study of 145 patients. *Spine*, **28**, 1043–50.
24. Khan, M.H., Smith, P.N., Balzer, J.B., et al. (2006). Intraoperative somatosensory evoked potential monitoring during cervical spine corpectomy surgery: experience with 508 cases. *Spine*, **31**, E105–13.
25. Modi, H.N., Suh, S.W., Yang, J.H., and Yoon, J.Y. (2009). False-negative transcranial motor evoked potentials during scoliosis surgery causing paralysis. *Spine*, **34**, E896–900.
26. Sala, F., Palandri, G., Basso, E., et al. (2006). Motor evoked potential monitoring improves outcome after surgery for intramedullary spinal cord tumors: a historical control study. *Neurosurgery*, **58**, 1129–43.
27. Radtke, R.A., Erwin, C.W., and Wilkins, R.H. (1989). Intraoperative brainstem auditory evoked potentials: significant decrease in postoperative morbidity. *Neurology*, **39**, 187–91.
28. Acevedo, J.C., Sindou, M., Fischer, C., and Vial, C. (1997). Microvascular decompression for the treatment of hemifacial spasm: retrospective study of a consecutive series of 75 operated patients—electrophysiologic and anatomical surgical analysis. *Stereotactic Function Neurosurgery*, **68**, 260–5.
29. Manninen, P.H., Patterson, S., Lam, A.M., Gelb, A.W., and Nantau, W.E. (1994). Evoked-potential monitoring during posterior-fossa aneurysm surgery—a comparison of 2 modalities. *Canadian Journal of Anaesthesia*, **41**, 92–7.
30. Schlake, H.P., Milewski, C., Goldbrunner, R.H., et al. (2001). Combined intra-operative monitoring of hearing by means of auditory brainstem responses (ABR) and transtympanic electrocochleography (ECochG) during surgery of intra- and extrameatal acoustic neurinomas. *Acta Neurochirurgia*, **143**, 985–95.
31. Thirumala, P., Carnovale, G., Habeych, M.E., Crammond, D.J., and Balzer, JR. (2014). Diagnostic accuracy of brainstem auditory evoked potentials during microvascular decompression. *Neurology*, **83**, 1747–52.
32. James, M.L. and Husain, A.M. (2005). Brainstem auditory evoked potential monitoring—when is change in wave V significant? *Neurology*, **65**, 1551–5.
33. Nuwer, M.R. (1993). Intraoperative electroencephalography. *Journal of Clinical Neurophysiology*, **10**, 437–44.
34. Florence, G., Guérit, J.M., and Gueguen, B. (2004). Electroencephalography (EEG) and somatosensory evoked potentials (SEP) to prevent cerebral ischaemia in the operating room. *Neurophysiology Clinic*, **34**, 17–32.
35. Blume, W.T. and Sharbrough, F.W. (2005). EEG monitoring during carotid endarterectomy and open heart surgery. In: E. Niedermeier and F. Lopes da Silva (Eds) *Electroencephalography: basic principles, clinical applications, and related fields*, pp. 8/15–27. Philadelphia, PA: Lippincott Williams and Wilkins.
36. Mizrahi, E.M., Patel, V.M., Crawford, E.S., Coselli, J.S., and Hess, K.R. (1989). Hypothermic-induced electrocerebral silence, prolonged circulatory arrest, and cerebral protection during cardiovascular surgery. *Electroencephalography & Clinical Neurophysiology*, **72**, 81–5.
37. Daube, J.R. and Harper, C.M. (1989). Surgical monitoring of cranial and peripheral nerves. In: J. E. Desmedt (Ed.) *Neuromonitoring in surgery*, pp. 115–38. Amsterdam: Elsevier.
38. Crum, B.A. and Strommen, J.A. (2007). Peripheral nerve stimulation and monitoring during operative procedures. *Muscle & Nerve*, **35**, 159–70.

Index